Exploration and Innovation in Sustainable Rubber Performance

Exploration and Innovation in Sustainable Rubber Performance

Guest Editors

Pilar Bernal-Ortega
Anke Blume
Rafał Anyszka

MDPI

Basel • Beijing • Wuhan • Barcelona • Belgrade • Novi Sad • Cluj • Manchester

Guest Editors

Pilar Bernal-Ortega
Department of Mechanics of Solids, Surfaces & Systems (MS3)
University of Twente
Enschede
The Netherlands

Anke Blume
Department of Mechanics of Solids, Surfaces & Systems (MS3)
University of Twente
Enschede
The Netherlands

Rafał Anyszka
Department of Mechanics of Solids, Surfaces & Systems (MS3)
University of Twente
Enschede
The Netherlands

Editorial Office
MDPI AG
Grosspeteranlage 5
4052 Basel, Switzerland

This is a reprint of the Special Issue, published open access by the journal *Polymers* (ISSN 2073-4360), freely accessible at: https://www.mdpi.com/journal/polymers/special_issues/G98K4FU9G3.

For citation purposes, cite each article independently as indicated on the article page online and as indicated below:

Lastname, A.A.; Lastname, B.B. Article Title. *Journal Name* **Year**, *Volume Number*, Page Range.

ISBN 978-3-7258-7573-3 (Hbk)
ISBN 978-3-7258-7574-0 (PDF)
https://doi.org/10.3390/books978-3-7258-7574-0

Contents

About the Editors

Pilar Bernal-Ortega

Pilar Bernal-Ortega earned her degree in Chemical Engineering from the Polytechnic University of Valencia (UPV), Spain. She went on to complete a master's program in High Specialization in Plastics and Rubber at the International University Menéndez Pelayo (UIMP), also in Spain. In 2016, she was awarded a government grant to pursue a PhD in rubber technology with the Elastomers Group at the Institute of Polymer Science and Technology (ICTP), Spain. She successfully defended her PhD thesis, titled "Sulfur-Modified Carbon Nanotubes for the Development of Rubber Tire Tread Compounds," on March 13, 2020. In December 2019, Pilar joined the Elastomer Technology & Engineering Group as a postdoctoral researcher, contributing to a collaborative project with the tire manufacturer Bridgestone. Since December 2022, she has been serving as an Assistant Professor within the same group.

Anke Blume

Anke Blume was born on April 6, 1969, in Hanover, Germany. She studied chemistry at the Leibniz University Hanover from October 1988 to July 1993. She first worked in depth with rubber during her master's thesis. Consequently, she carried out her PhD thesis at the "German Institute for Rubber Technology" (DIK), which she completed in mid-November 1995. She continued her work at the DIK as a post-doc for nine months. In 1996, she joined the Degussa company in Cologne, Germany (now named "Evonik Industries"), where she was in charge of the development of silica and silanes for rubber applications, as well as the management of Evonik's intellectual property regarding the use of silica and silanes in rubber applications. She became the head of the Elastomer Technology and Engineering department of the University of Twente on October 1, 2013, in addition to her work at Evonik. In April 2021, she decided to fully dedicate her work to the university. This allows her to follow her passion for teaching and rubber, to further explore its mysteries, and to inspire young people to engage with this ascinating material.

Rafał Anyszka

Rafał Anyszka got his Ph.D. in 2014 at the Lodz University of Technology, designing ceramifiable elastomer composites for fire protection systems. Afterward, he was involved in the development of sulfur-derived polymers and sulfur concretes. In 2017, he started Postdoc projects at the University of Twente, designing molecular Velcro systems for silica-rubber coupling and surface-functionalized nano-silica for high-voltage DC polypropylene insulation composites. In 2025, he finished an EU-funded MSCA Global Fellowship project focused on designing rubber for the Mars environment. Recently, his focus shifted to more interdisciplinary areas between rubber technology and socioeconomic studies. Currently, he is coordinating the establishment of the first ECIU association – ECIU Rubber Network.

Preface

Elastomers represent a fundamental element of modern materials engineering and are indispensable across a wide spectrum of industrial and consumer applications. From tires and hoses to seals, vibration dampers, and flexible mechanical components, rubber-based materials provide a unique combination of elasticity, durability, resilience, and chemical resistance. These properties have made elastomers essential to transportation, infrastructure, manufacturing, and everyday products. However, the increasing global demand for rubber materials has also intensified concerns regarding their environmental impact, resource consumption, and end-of-life management.

In recent years, the elastomer and rubber industries have increasingly focused on improving the sustainability of rubber products while maintaining their performance characteristics. Achieving this balance represents a significant scientific and technological challenge. Traditional rubber systems rely heavily on petroleum-derived components, energy-intensive processing methods, and crosslinked networks that are difficult to recycle. Consequently, developing more sustainable elastomer technologies requires a comprehensive approach that addresses material sourcing, product durability, processing efficiency, and waste management.

Several complementary strategies have emerged to reduce the environmental impact of elastomeric materials. One important approach focuses on improving the intrinsic performance of rubber compounds. By enhancing mechanical strength, wear resistance, fatigue behavior, and aging stability, elastomer products can achieve longer service lifetimes. Extended durability directly translates into reduced material consumption and lower waste generation over time, contributing significantly to sustainability goals.

Another promising direction involves the increasing use of naturally sourced or bio-based materials. Researchers and industry professionals are actively exploring renewable fillers, bio-derived polymers, and sustainable additives that can partially or fully replace conventional fossil-based ingredients. These developments aim to reduce reliance on non-renewable resources while maintaining compatibility with established processing technologies and performance requirements.

Equally important is the advancement of technologies that improve the recyclability and circularity of rubber products. Because traditional vulcanized elastomers form permanent crosslinked networks, recycling remains a complex challenge. Innovations in devulcanization, reprocessing methods, and dynamic crosslinking systems are creating new opportunities to recover value from rubber waste and reintegrate it into new products.

This Special Issue brings together contributions from researchers working at the forefront of elastomer science and technology. The collected papers present new systems and material concepts designed to enhance rubber performance, introduce advanced functionalities, improve recyclability, and expand the use of bio-based resources. Collectively, these studies provide valuable insights into emerging strategies that can guide the development of more sustainable elastomeric materials.

We hope that the research presented in this issue will stimulate further innovation and collaboration within the elastomer community and contribute to the broader objective of designing rubber materials that meet both performance and sustainability requirements for future applications.

Pilar Bernal-Ortega, Anke Blume, and Rafał Anyszka
Guest Editors

polymers

MDPI

Editorial

Closing Editorial to the Special Issue "Exploration and Innovation in Sustainable Rubber Performance"

Pilar Bernal-Ortega *, Anke Blume and Rafal Anyszka

Elastomer Technology & Engineering, Department of Mechanics of Solids, Surfaces & Systems (MS3), Faculty of Engineering Technology, University of Twente, Drienerlolaan 5, 7522 NB Enschede, The Netherlands; a.blume@utwente.nl (A.B.); r.p.anyszka@utwente.nl (R.A.)
* Correspondence: p.bernal@utwente.nl

It is with great pleasure that we present the collection of articles published in this Special Issue of MDPI's *Polymers*, dedicated to "Exploration and Innovation in Sustainable Rubber Performance". When we began this Special Issue, our goal was to gather innovative research addressing the ways in which rubber materials can be transformed to be more sustainable, whether that be through bio-based raw ingredients, improved recyclability, enhanced performance for extended durability, or novel strategies that reduce environmental footprints.

We are thrilled by the outcome: this issue features a diverse set of studies covering fundamental research, applied materials science, and engineering solutions. The contributions include investigations into recycled or bio-based fillers, novel processing methods, innovative elastomer formulations, studies on aging and durability, analyses of environmental impact, and explorations of material performance under real-world conditions. Together, they provide a complete overview of the current state of the art of the sustainability of rubber, reflecting both academic and industrial interests.

Through these works, this Special Issue addresses several key topics regarding sustainability within the rubber field and achieves the following results:

- Demonstrates viable pathways for integrating bio-based or recycled materials into elastomer matrices without compromising essential properties.
- Improves our understanding of structure–property relationships in sustainable rubber compounds, enabling the effective design of durable, high-performance materials.
- Highlights recycling, waste reduction, and life-cycle considerations for rubber products, contributing to circular economy approaches.
- Encourages innovation in filler, compounding, and processing techniques and collaborations across academia and industry.

We would like to extend our sincere gratitude to all authors, reviewers, and colleagues involved in the peer review and editorial process. Their dedication and scientific rigor have made this Special Issue possible. We also thank the Editorial Office of *Polymers* for their support and guidance throughout this process.

Finally, we hope that this collection will serve as a useful resource for researchers and industries aiming to push the boundaries of sustainable elastomer development. We encourage readers to explore the full set of papers and to build upon their findings, whether towards improved recycling processes, greener materials, or novel applications. We look forward to seeing how the insights gathered here will inspire future work and contribute to a more sustainable rubber industry.

Polymers **2026**, *18*, 576 https://doi.org/10.3390/polym18050576

Conflicts of Interest: The authors declare no conflict of interest.

https://doi.org/10.3390/polym18050576

polymers MDPI

Article

Synergistic Effects of Nitrosamine-Safe Accelerators for Enhanced Natural Rubber Latex Balloon in Sulfur Conventional Vulcanizing System

Thanakan Chang-In, Ekasit Anancharoenwong and Sunisa Suchat *

Faculty of Science and Industrial Technology, Prince of Songkla University, Surat Thani Campus, Surat Thani 84000, Thailand
* Correspondence: author: sunisa.su@psu.ac.th

Abstract

The vulcanization of natural rubber latex (NRL) relies on accelerators to achieve effective crosslinking by a sulfur conventional vulcanizing system and desirable properties such as tensile strength and elasticity. This study investigates the synergistic effects of carcinogenic nitrosamine-safe accelerators to develop a high-performance and safe NRL vulcanization system. A synergistic combination of 0.36 phr Zinc dibutyldithiocarbamate (ZBEC), 0.36 phr Tetrabenzylthiuram disulfide (TBzTD), and a trace amount of Zinc diethyldithiocarbamate (ZDEC), 0.03 phr, demonstrated optimal performance, yielding superior tensile strength (22.13 MPa), elongation at break (1153%), and thermal stability (Tmax 384.15 °C). Notably, this formulation exhibited below the detectable level limits of hazardous nitrosamines (*N*-nitrosodimethylamine (NDMA) and *N*-nitrosodibutylamine (NDBA)). The synergistic nitrosamine-safe accelerator system offers a promising strategy for producing environmentally responsible and consumer-safe NRL products with enhanced mechanical and thermal properties.

Keywords: natural rubber latex; nitrosamine-safe; synergistic accelerator system; mechanical properties

1. Introduction

The global balloon production volume is valued at around USD 1.85 billion in 2024, with growth expected to reach USD 3.38 billion by 2033, securing a significant market size and a consistent production volume increase across the world [1,2]. However, the production volume of natural rubber balloons is a small percentage of the total natural rubber production, as the global production of natural rubber in 2024 amounted to 28.8 million tons. Rubber balloons are thin gauge rubber products produced from natural rubber latex (NRL) compounds [3]. NRL balloons have superior elasticity, being capable of stretching to seven to eight times their original length while maintaining resilience: a quality not reached by synthetics. This exceptional performance has established rubber balloons as one of the most affordable and universally popular toys across all demographic segments [4]. Beyond children's entertainment, these versatile products serve multiple functions including celebratory decorations, retail environment enhancement, target practice activities, and festive displays, contributing to their sustained market growth.

However, this seemingly innocuous product harbors a concerning health risk: natural latex can break down to form nitrosamines, a group of known carcinogens detected in various latex products (rubber nipples and gloves) [5,6]. This discovery has prompted the development of methods for extraction and concentration of volatile nitrosamine from latex balloons, highlighting an urgent need to address safety concerns in this growing industry without compromising the product's widespread appeal and accessibility.

Most toxicological assessments of nitrosamines focus on their carcinogenic and mutagenic properties. Nitrosatable substances can be converted into carcinogenic nitrosamines through reactions with nitrosating agents such as nitrite and nitrogen oxides [7]. Endogenous formation of *N*-nitroso compounds occurs via nitrosation (by nitrite derived from nitrate) of ureas, guanidine, amides, amino acids, and (primary, secondary and aromatic) amines [8]. Of the over 300 different nitrosamines, RIVM (2003) [9] identified *N*-nitrosodimethylamine (NDMA) and *N*-nitrosodibutylamine (NDBA) as being the most prevalent ones, being detected in 96% and 61% of samples, respectively. Both are classified by IARC and the EU as category 2 carcinogens [10]. NDMA, with the generic chemical structure $R_2N\text{-}N{=}O$, is the most frequently detected carcinogenic nitrosamine, another form of nitrosamines, although it is of a less dangerous class (BAuA). The European Union has established specific limits for nitrosamines and nitrosatable substances in natural rubber balloons according to European standard EN71-12 [11,12]. The Federal Republic of Germany has proposed limits for nitrosamines and nitrosatable substances in balloons, which are 50 μg/kg and 1000 μg/kg, respectively. Traditionally, nitrosamines and nitrosatable substances are analyzed using thermal energy analyzers (TEA), but due to the limited availability of this expensive equipment type, an improved analysis method using the more widely available gas chromatography–mass spectrometry (GC-MS) technique has been developed based on the EN 12868:2017 standard [13–15].

Natural rubber latex balloons can release carcinogenic nitrosamines when certain vulcanization accelerators (carbamates such as zinc diethyldithiocarbamate; ZDEC) are used during balloon production [16]. These substances pose health risks when released during balloon inflation or mouthing, with children being particularly vulnerable. Research by Smith and Norris (2003) [17] identified hazardous exposure levels when children mouth balloons with high nitrosamine content. The primary problematic accelerators are ZDEC, which shows mutagenic activity in bacterial tests, and tetramethyl thiuram disulfide (TMTD), which is a suspected carcinogen: an avoidable threat [18]. Fortunately, safe alternatives exist, namely zinc n-butylbenzothiazole-2-sulfenamide (ZBEC) and tetrabenzylthiuram disulfide (TBzTD) [19]. These alternatives contain low nitrosatable substances, effectively eliminating nitrosamine formation while maintaining product quality. The rubber industry has the knowledge and the technology to produce safer balloon products by transitioning to these alternative accelerators, making this health risk entirely preventable.

This study aimed to comprehensively investigate the performance characteristics of a nitrosamine-safe binary accelerator system comprising TBzTD and ZBEC in the sulfur vulcanization of an NRL balloon. To evaluate the cure characteristics, the swelling index for dipping processability and chemical control were tested for NRL compounds containing various ratios of ZBEC and TBzTD accelerators. The mechanical properties (modulus, elongation at break and tensile strength) of vulcanizates before and after aging, FTIR of the functional groups and the crosslinks formed when prepared with different combinations of the TBzTD/ZBEC accelerator system were determined. The potential synergistic effects that occur when progressively replacing ZDEC with combinations of TBzTD and ZBEC accelerators were assessed. The synergistic nitrosamine-safe accelerator systems are essential to ensure that rubber products (such as balloons), achieve the required mechanical properties while fully complying with stringent safety standards concerning nitrosamine content.

https://doi.org/10.3390/polym18040438

2. Experimental

This research was divided into five steps, beginning with optimizing the formulation to be devoid of nitrosamine-producing substances, as follows.

2.1. Materials

(1) Materials in NRL films

The centrifuged concentrated latex (Hevea brasiliensis), as a high-ammonia latex (HA) with 60% dry rubber content (DRC), was supplied by Muang Mai Guthrie Public Company Limited, Thailand. The NRL compound formulations for balloon rubber products (Tables 1 and 2) were purchased from Thanodom Technology Co., Ltd., Bangkok, Thailand. Cellulose nanocrystals (CNC) were obtained as a white powder from rubber seed shells. The average molecular weight was 14,700 g/mol, with particle lengths ranging from 100 to 300 nm and diameters of 20 nm. A 5% CNC dispersion was prepared at a low filler loading of 2 phr.

Table 1. The compound formulations of NRL used in the present study.

Ingredient in Formulation	Content (phr [a])					Function
	1	2	3	4	5	
60% HA-NR latex	100	100	100	100	100	Backbone (main component)
10% Potassium hydroxide	1.0	1.0	1.0	1.0	1.0	Stabilizers 1
10% Potassium oleate	1.0	1.0	1.0	1.0	1.0	Stabilizers 2
50% Sulfur	1.5	1.5	1.5	1.5	1.5	Vulcanizing Agent
50% ZnO	1.5	1.5	1.5	1.5	1.5	Activator
5% CNC	2.0	2.0	2.0	2.0	2.0	Filler
50% Wingstay L	1.5	1.5	1.5	1.5	1.5	Polyphenol Antioxidant
50% Accelerators [b]	0.75	0.75	0.75	0.75	0.75	Accelerators

Note: [a] Parts per hundred rubber (phr) and [b] 50% accelerators in Table 2.

Table 2. The accelerator content in NRL compound formulations.

Formula No. #	Accelerator Type	Content (phr)			
		TBzTD	ZBEC	ZDEC	Total
1	ZDEC	0.75	-	-	0.75
2	ZBEC	-	0.75	-	0.75
3	TBzTD	-	-	0.75	0.75
4	ZBEC/TBzTL	0.375	0.375	-	0.75
5	ZBEC/TBzTL	0.36	0.36	0.03	0.75

2.2. Preparation of NRL Films

The NRL balloon compound formulation combinations of the TBzTD/ZBEC accelerator system were optimized to reduce the tested generation of nitrosamines. Specifically, concentrated latex was used with various types and volumes of accelerators, such as TBzTD, ZDEC, and ZBEC (singly), as well as combinations like TBzTD/ZBEC (binary) and TBzTD/ZBEC/ZDEC (ternary) in NRL. These choices were made to prevent carcinogenic nitrosamine formation. The NRL compound was prepared as a liquid mixture with additives, and mixed in a blender at 2 rpm for 20 min at room temperature. The final mixture was degassed to remove the entrapped air and then poured into glass molds, with eventual film thickness control by the total solid content. The films were cured at 80 °C

https://doi.org/10.3390/polym18040438

for 24 h to evaporate water and achieve uniform films. The dried films were peeled off the molds and stored for characterization. By adjusting the types and amounts of accelerators and controlling parameters such as 62% total solid content and a 103% swelling index, the synergistic activities in sulfur vulcanization of NRL films formulations were assessed, emphasizing safety and mechanical properties. A diagram of the experimental design is shown in Figure 1.

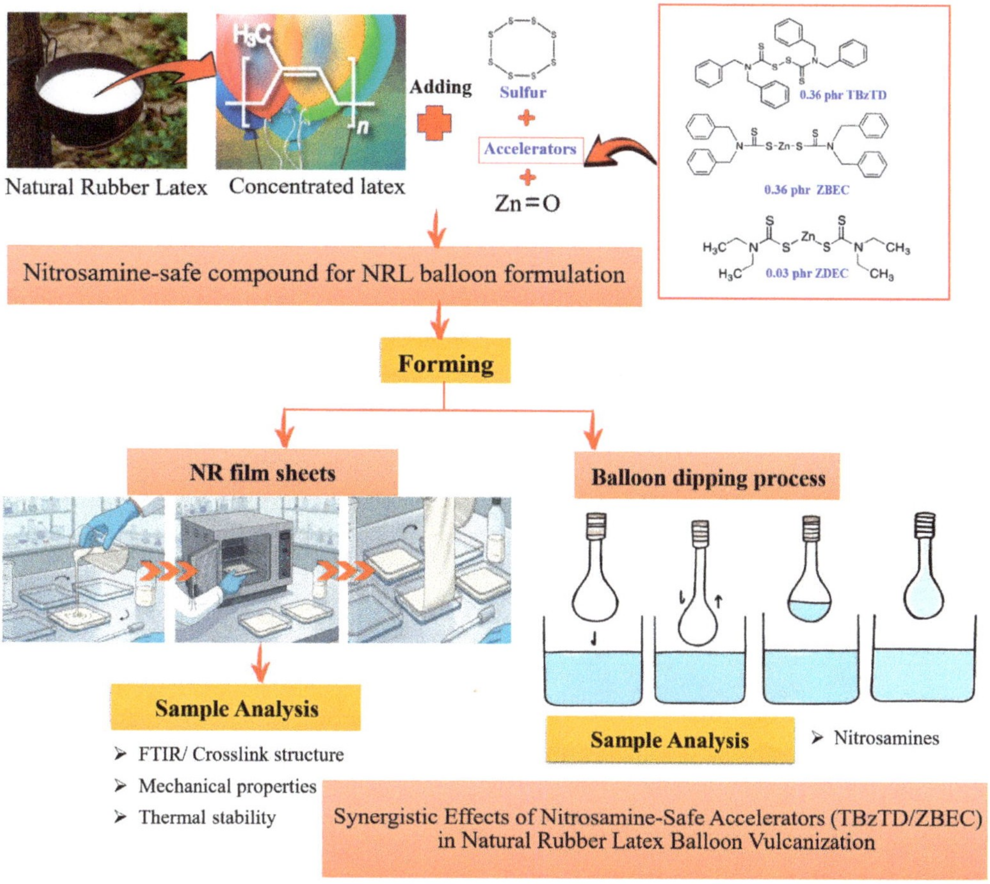

Figure 1. Experimental scheme to test synergistic activities in sulfur vulcanization of NRL balloon products.

2.3. NRL Balloon

The NRL balloon dipping process used a balloon mold, clean and dry, soaked in 10% calcium chloride coagulant and slightly dried at 45 °C, and then dipped in the NRL compound. The soaking time was 18 s, and to reuse the mold, it was washed with hot water at 70 °C for 10 min to remove coagulant or residues. The vulcanization dries the balloon in an oven at 100 °C for 1 h to complete the vulcanization process. Samples were removed from the oven and allowed to cool to room temperature. Talcum powder was applied to the balloon surface to prevent sticking.

The finished balloons were carefully removed from the mold, and samples without flaws or damage were kept for the nitrosamine test using GC-TEA, following the standard procedure based on EN 12868:2017 [20].

https://doi.org/10.3390/polym18040438

2.4. Characterization and Properties

2.4.1. The Swelling Test

The development of the NRL compound formula from concentrated latex involved varying the types and amounts of accelerator agents. In a sulfur vulcanized system, a stabilizer, activators, and an antioxidant were added to the NRL, and the mixture was matured until a swelling index of 103% was reached, indicating partial crosslinking via sulfur bridges. The compound was then poured onto a 2 mm thick layer on a glass plate, dried at 120 °C for 30 min, and finally cured by heating. The swelling test followed ISO 124, using 3.0 cm circular samples. The diameter was measured carefully at three points and the average was recorded, after which the specimens were immersed in toluene at 23 ± 2 °C for 3 h. The swelling (%) was calculated from the change in diameter caused by immersion, and the average is reported. The percentage swelling was computed as follows.

$$\%\ \text{Swelling} = [(\text{B} - \text{A})/\text{A}] \times 100 \quad (1)$$

where A is the initial diameter of the specimen before immersion in toluene, and B is the diameter of the specimen after immersion in toluene for 3 h.

For reliable results, this test was run in triplicates and the mean value was recorded, along with test conditions and sample identification.

2.4.2. FTIR Analysis

Fourier transform infrared spectroscopy (FTIR) was employed in transmission mode to analyze the structure, using a Spectrum Two FT-IR spectrometer equipped with a DTGS detector, manufactured by PerkinElmer, Inc., Waltham, MA, USA. Thin rubber sheets with smooth surfaces were analyzed in transmission mode, utilizing the attenuated total reflectance (ATR) accessory for solid-state sampling. FTIR spectra were recorded across the broad wavenumber range from 4000 to 400 cm^{-1} at a resolution of 4 cm^{-1}.

2.4.3. Mechanical Properties

The mechanical properties of the NRL film samples are reported as averages of five replicates. Tensile properties were measured, determining the modulus, strength and elongation at break, following the ASTM D 412 standard using a Tinius Olsen 10ST tensile testing machine (Honey Crock Lane, UK). Dumbbell Type C specimens were tested at a speed of 500 mm/min with a 5 kN load cell.

2.4.4. Thermal Stability

(1) Accelerated aging was conducted following ISO 188, where samples were subjected to thermal aging at 70 °C for 168 h. After aging, tensile strength, elongation at break, and modulus were tested.
(2) Thermogravimetric analysis (TGA) was performed using a TGA-SDTA 851 analyzer (Mettler Toledo, Zurich, Switzerland). Samples were heated from 30 °C to 600 °C under nitrogen atmosphere, then from 600 °C to 900 °C under oxygen atmosphere, both at a 10 °C/min heating rate.

2.4.5. Nitrosamines and Nitrosatable Substances

Nitrosamine testing of the balloon latex compound was done using a Gas Chromatography–Thermal Energy Analyzer (GC-TEA) at TÜV Rheinland Hong Kong Ltd., Tsuen Wan, Hong Kong. Nitrite migration from the samples was assessed following the European standard procedure, based on EN 12868:2017 [20]. This method determines nitrosamines by first extracting them into artificial saliva at 40 °C to simulate migration. Nitrosamines require an acidification step for conversion

https://doi.org/10.3390/polym18040438

prior to extraction and concentration. The prepared extracts are quantified using GC-TEA, due to its high specificity for nitroso compounds. The reported results are compared to standard limits (migratable nitrosamines, 0.05 mg/kg, and migratable nitrosatable substances, 1.0 mg/kg). The twelve target *N*-nitrosamines identified are *N*-nitrosodimethylamine (NDMA), *N*-nitrosodiethylamine (NDEA), *N*-nitrosodipropylamine (NDPA), *N*-nitrosodibutylamine (NDBA), *N*-nitrosodiisobutylamine (NDiBA), *N*-nitrosoethylphenylamine (NEPhA), *N*-nitrosopiperidine (NPIP), *N*-nitrosopyrrolidine (NPYR), *N*-nitrosomorpholine (NMOR), *N*-nitrosomethylphenylamine (NMPhA), *N*-nitrosodiisononylamine (NDiNA), and *N*-nitrosodibenzylamine (NDBzA).

3. Results and Discussion

3.1. Characterization and Properties

3.1.1. The Swelling Test

Figure 2 shows the results that are indicative of the crosslinking efficiency of individual accelerators (all tested) over time, with the swelling in all cases decreasing with the curing time, confirming continuous crosslink formation. The nitrosamine-safe alternatives, TBzTD/ZBEC/ZDEC, led to the fastest reduction in swelling, achieving the target crosslinking within 7 days. So, the curing time required to reach the 103% swelling index was reduced from 10 days (ZDEC and ZBEC) to 7 days. TBzTD and ZBEC binary and TBzTD alone were also effective, promoting sufficient crosslinking to reach the standard threshold (103% swelling) within 9 days.

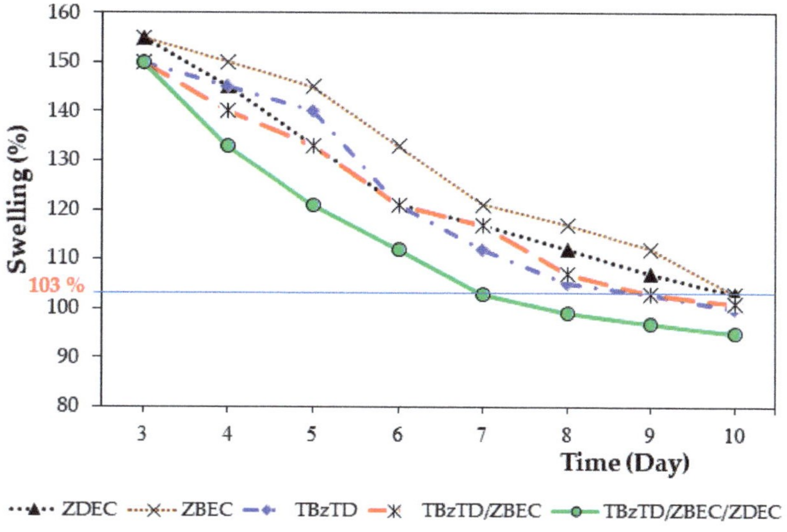

Figure 2. Swelling results are indicative of crosslink bond formation of compound rubbers with different accelerators.

Obtaining this 103% swelling index confirms the effectiveness of TBzTD/ZBEC/ZDEC as accelerators to reduce the curing time while maintaining the required crosslink density for product quality.

3.1.2. FTIR Analysis of Accelerator Types in Sulfur Vulcanization of NRL Films

The consulted expert sources focused on the formation and mitigation of nitrosamines in natural rubber latex balloons. One primary source details research into substituting problematic vulcanization accelerators like ZDEC with safer alternatives such as ZBEC

https://doi.org/10.3390/polym18040438

and TBzTD to eliminate the carcinogenic nitrosamine content while maintaining necessary mechanical properties like tensile strength and elasticity. The chemical and mechanical properties of the vulcanizate formed using a nitrosamine-free combination of 0.03 phr ZDEC with TBzTD/ZBEC 0.36/0.36 phr are compared to the properties of a vulcanizate with ZDEC in a neat recipe.

The experiment compared the use of two and three accelerators with the use of a single accelerator, specifically from ZBEC, TBzTD, and ZDEC. The NRL formulation uses a blend of two (TBzTD and ZBEC) or three types of accelerators with ZDEC added to enhance the accelerator system between TBzTD and ZBEC. The combinations of accelerators significantly improved the vulcanization efficiency, thermal stability, and mechanical properties, demonstrating the importance of synergistic accelerator selection for optimizing natural rubber balloons.

Characterization by Fourier transform infrared spectroscopy

The NRL formulation characterization in Figure 3 shows the Fourier transform infrared (FTIR) analysis that informs us of the functional groups and the crosslinks formed.

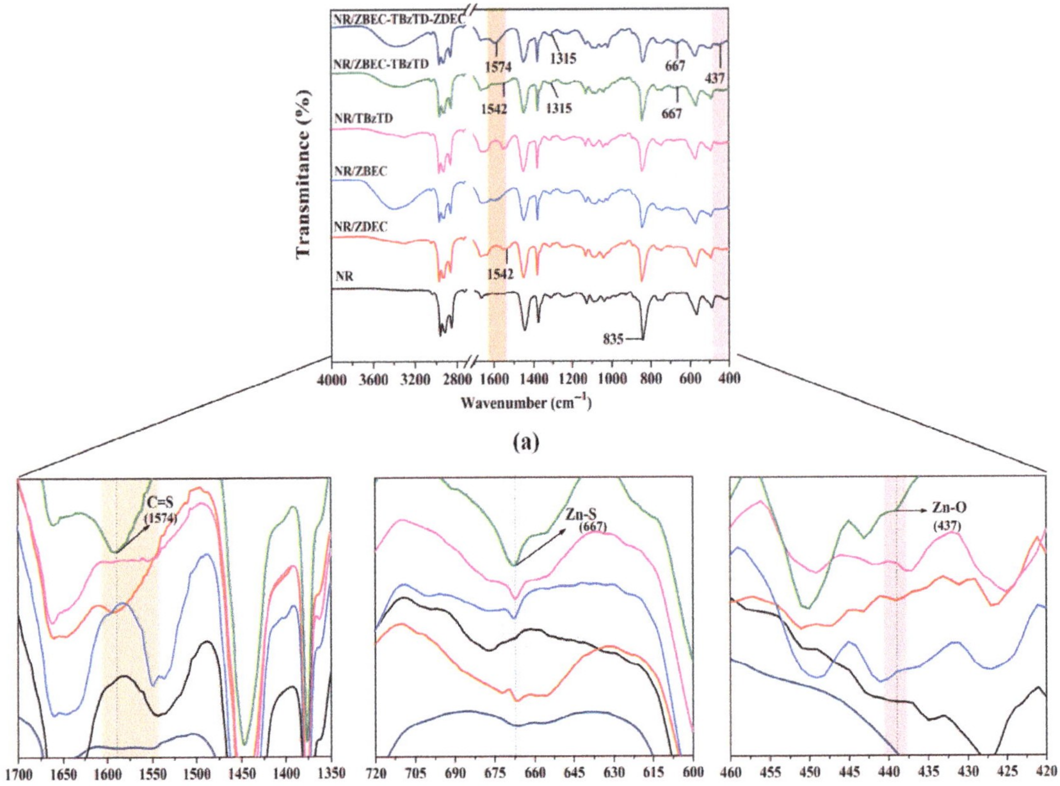

Figure 3. The FTIR spectra of NRL film formulations with different accelerators. (**a**) the FTIR spectra of the NRL; (**b**) the C=S band; (c) a cross-linking; (**d**) Zn–O bonding.

Figure 3a presents the FTIR spectra of the NRL formulations, demonstrating that the mixed-accelerator system (ZBEC/TBzTD/ZDEC) produces stronger C=S, C–S, Zn–S, and Zn–O peaks than single-accelerator systems. Around 1315 cm^{-1}, distinct C–S peaks indicate enhanced crosslink density, especially in formulations incorporating multiple

https://doi.org/10.3390/polym18040438

accelerators [21]. In Figure 3b, in the 1574 cm^{-1} region, the appearance and growth of the C=S band [22,23], together with its larger integrated area, confirms additional reactions of the accelerator with vulcanizing agents, leading to modified sulfur-containing structures. The distinct peak near 667 cm^{-1} confirms the formation of Zn–S bonds in Figure 3c, a key cross-linking of rubber chains in the sulfur vulcanization process [24]. In Figure 3d, the band at approximately 437 cm^{-1} evidences Zn–O bonding [25].

The spectral evidence confirms that the synergistic accelerator system (ZBEC/TBzTD/ZDEC) in NRL facilitates more efficient sulfur crosslinking and the formation of complex Zn–S, Zn–O, and C=S bonding networks. This enhanced crosslink density strengthens the mechanical performance and thermal resistance, outperforming single or dual accelerator systems (Table 3). Importantly, the formulation achieves these improvements without any or at a low generation of nitrosamine compounds.

Table 3. Mechanical properties of NRL formulation for balloons prepared with alternative accelerators.

Type of Accelerator	Mechanical Properties				
	Tensile Strength (MPa)	**Elongation at Break (%)**	**100% Modulus (MPa)**	**200% Modulus (MPa)**	**300% Modulus (MPa)**
Before aging					
ZBEC	18.74 ± 0.90	1079.31 ± 83	0.54 ± 0.01	0.81 ± 0.01	1.08 ± 0.03
TBzTD	19.57 ± 1.47	1136.74 ± 71	0.46 ± 0.01	0.65 ± 0.08	0.83 ± 0.03
ZDEC	21.36 ± 1.10	1040.86 ± 60	0.55 ± 0.07	0.88 ± 0.12	1.20 ± 0.14
ZBEC/TBzTD	17.91 ± 0.75	1115.82 ± 55	0.50 ± 0.13	0.67 ± 0.05	0.84 ± 0.10
ZBEC/TBzTD/ZDEC	22.13 ± 1.81	1152.90 ± 78	0.69 ± 0.02	0.95 ± 0.04	1.30 ± 0.07
After aging					
ZBEC	12.10 ± 2.11	626.50 ± 34	1.06 ± 0.01	1.81 ± 0.02	3.30 ± 0.02
TBzTD	13.25 ± 1.72	736.85 ± 26	1.24 ± 0.07	2.39 ± 0.12	3.21 ± 0.15
ZDEC	15.50 ± 0.89	851.9 ± 35	1.53 ± 0.02	2.00 ± 0.01	4.59 ± 0.10
ZBEC/TBzTD	13.85 ± 0.70	785.98 ± 15	0.95 ± 0.03	1.17 ± 0.03	3.50 ± 0.04
ZBEC/TBzTD/ZDEC	14.23 ± 1.81	892.91 ± 8	1.69 ± 0.02	2.48 ± 0.04	4.04 ± 0.07

3.1.3. Comparative Analysis of Mechanical Properties

The mechanical properties before and after accelerated aging in Table 3 indicate that changing the accelerator system greatly impacts the mechanical properties of natural rubber latex balloons. Formulations using single accelerators such as ZBEC, TBzTD or ZDEC provide moderate tensile strength and elongation at break, while binary combinations offer only limited improvements.

Notably, the ternary blend of 0.36 phr ZBEC/0.36 phr TBzTD/0.03 phr ZDEC, even with a minimal amount of ZDEC, achieved the highest tensile strength (22.13 ± 1.81 MPa) and elongation at break (~1153%) before aging, indicating a highly tough and well-developed crosslink network. The tensile strength of the optimized balloon compound remains within the internal specification range provided by the B.K. Latex Product Co., Ltd., with typical values around 25.792 MPa, elongation at break of approximately 1067%, and a 300% modulus of 1.569 MPa. In practice, the property that mainly influences whether a balloon is difficult to inflate is the 300% modulus, which should be relatively low to avoid excessive stiffness, while a sufficiently high tensile strength is still desirable to prevent rupture during inflation and use. This superior performance demonstrates a synergistic effect, where the combination of these three accelerators leads to more efficient vulcanization and greater reinforcement of the polymer structure. The ternary blend also resulted in the highest moduli at all elongation levels before aging, which is associated with increased initial stiffness and higher crosslink density in the rubber matrix. The thermal stability for natural rubber balloons, as described in standards such as TIS 685, BgVV, 2002 [26], and EN 71-12:2016 [12], is not defined by explicit numerical limits for thermal degradation,

https://doi.org/10.3390/polym18040438

but instead is evaluated indirectly through compliance with mechanical and physical tests after heat aging. Thermal stability is necessary so that the balloons continue to meet the required mechanical safety criteria (e.g., flexibility, absence of brittleness, and retention of tensile performance within acceptable range) following the specified heat-aging conditions, to ensure that the produce remains mechanically reliable throughout storage and use. After accelerated aging, the tensile strength and elongation when using the ternary system implies improved thermal resistance compared to single or binary accelerator systems. Each type of accelerator or combination thereof influences both the mechanical and thermal properties, with the three-accelerator system consistently yielding the best results. Overall, the synergistic accelerator system of ZBEC, TBzTD and ZDEC provides a high-performance, strong, and stable NRL balloon product, surpassing formulations with only one or two accelerators. The benefits impact both the mechanical properties and service life.

3.1.4. Thermal Stability of Sulfur Vulcanization

Figure 4 illustrates the thermal stability of NRL vulcanizates prepared with different accelerator systems, as evaluated by TGA and derivative thermogravimetry (DTG).

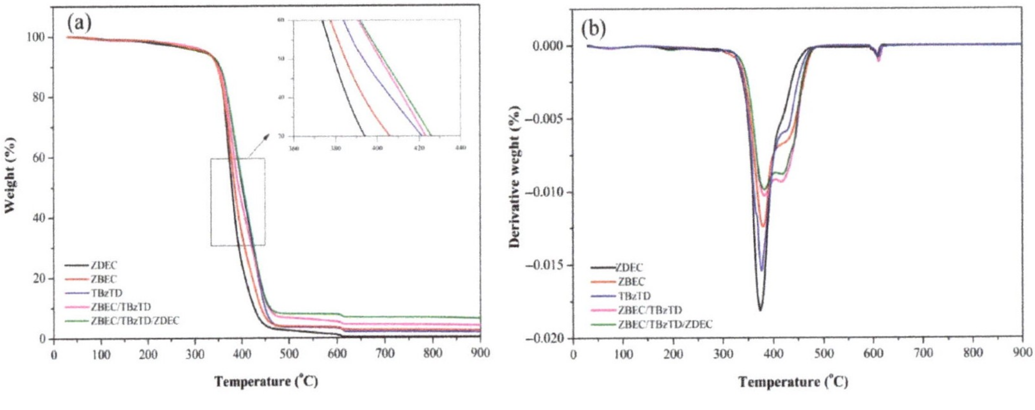

Figure 4. TGA (**a**) and DTG (**b**) graphs showing weight losses with the alternative types of accelerators in NRL formulation for balloons by sulfur vulcanization.

From Table 4, TGA curves in Figure 4a, and DTG in Figure 4b, it is found that all formulations exhibit a similar single-step degradation, with rapid weight loss between 180 and 320 °C being associated with the decomposition of the rubber backbone. The ZBEC/TBzTD/ZDEC formulation provides the greatest thermal stability, as evidenced by its highest T_{onset} (357.75 °C), T_{max} (384.15 °C), and final decomposition temperature (403.04 °C), together with the largest residual mass of 7.66%, indicating the presence of thermally stable species (e.g., ZnS, ZnO) within the measurement range. The DTG curves (Figure 4b) show the rate of that mass change and display a distinct shoulder at 400–500 °C arising from additional degradation of the crosslinked network and residue, confirming that the combined accelerator system enhances both the onset and completion temperatures of degradation compared with the other formulations.

https://doi.org/10.3390/polym18040438

Table 4. Thermal stability analysis of NRL film formulations with alternative accelerators for sulfur vulcanization.

Rubber Sample Type	T_{onset} * (°C)	T_{max} ** (°C)	Final Decomposition Temperature (FDT)	Final Residue (%)
ZBEC	356.01	374.78	379.79	0.38
TBzTD	356.13	378.01	391.14	4.47
ZDEC	358.05	376.47	384.47	2.63
ZBEC/TBzTD	342.16	382.62	400.56	2.05
ZBEC/TBzTD/ZDEC	357.75	384.15	403.04	7.66

Note: * T_{onset} is the decomposition onset temperature, T_{max} ** is the temperature at the maximum decomposition rate.

This balance between enhanced thermal stability and sulfur crosslinking makes the novel formulations suitable for applications requiring both mechanical performance and heat resistance [27–29]. These results confirm an accelerator synergistic effect in NRL compound sulfur vulcanizates, with increased thermal stability while maintaining adequate sulfur crosslinking for NRL balloon applications.

From the resulting increases in the mechanical strength, elasticity, and modulus of the NR compound, the synergistic accelerator system (ZBEC/TBzTD/ZDEC) in sulfur vulcanization is confirmed.

Figure 5 illustrates the interactions and cross-linking process between natural rubber latex (NRL), sulfur, zinc oxide, and cellulose nanocrystals, enhancing the entanglement and interfacial network within the rubber matrix. The CNC not only improves the dispersion of zinc oxide but also participates in the formation of ionic Zn–O–CNC crosslinks, which further connect to rubber chains via sulfur bridges. This leads to a robust NRL–S_8–Zn–O–CNC network, increasing the number of crosslink points and entanglement between rubber and filler.

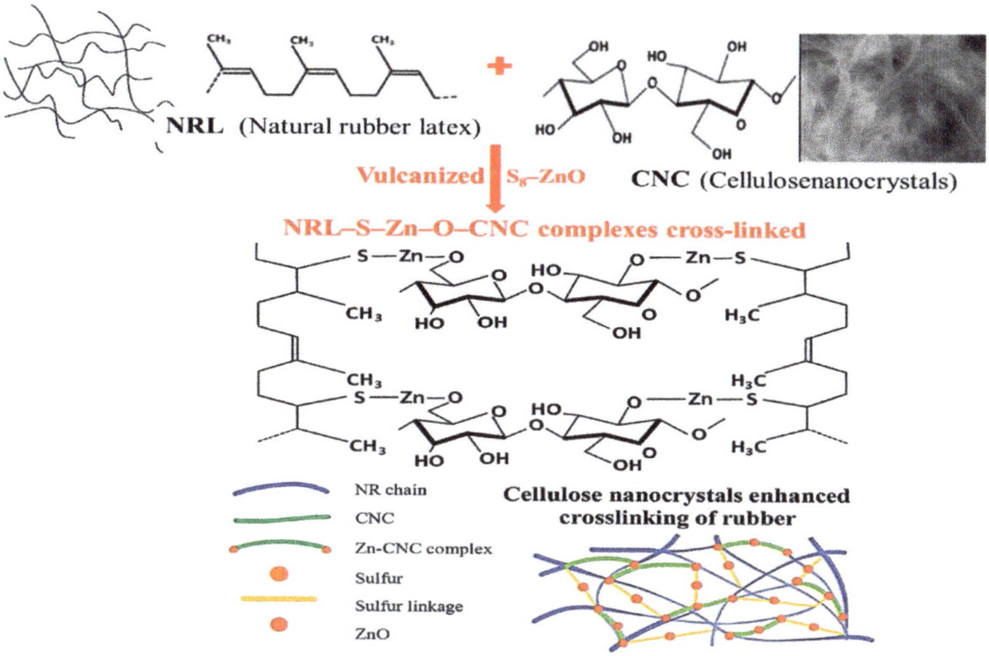

Figure 5. The formation of sulfur crosslinks in NRL–S_8–Zn–O–CNC complexes.

https://doi.org/10.3390/polym18040438

The curing mechanism indeed involves the CNC. The role of CNC is now exposed by reports that cellulose-based fillers can participate in zinc-containing curing domains and influence vulcanization behavior, which supports the plausibility of interactions between CNC surface groups, ZnO, and the sulfur–accelerator system. Regarding the proposed coupling of cellulose to the polymer via a Zn–S bridge, this structure is hypothetical and consistent with the generally accepted view that the active Zn complexes form at, or in association with, ZnO-containing domains. Additionally, we note that (i) Zn–accelerator–sulfur complexes generated at the ZnO surface can migrate into the rubber matrix, (ii) hydroxyl and modified functional groups on CNC can coordinate or interact with zinc species, and (iii) such interactions may contribute to a combined filler–curing network, rather than a single, unique structure. The proposed structures are one plausible realization among several possibilities, supported indirectly by the curing behavior and mechanical property trends, rather than by direct structural evidence. For related details and discussion, we refer to the works of Low et al. (2021) [30], Blanchard et al. (2020) [31], and Suchat et al. (2024, 2025) [13,32].

3.1.5. Nitrosamine Testing

The results of nitrosamine testing in the NRL formulation for balloons (Table 5), using GC-TEA according to EN 12868:2017, confirm that the synergistic activities of the novel combinations 0.03 phr ZDEC; 0.36 phr TBzTD/0.36 phr ZBEC and TBzTD, ZBEC, ZDEC, TBzTD/ZBEC, could be effective accelerator systems to replace the unsafe ZDEC neat for the sulfur vulcanization of NRL balloons.

Table 5. Test of migratable nitrosamines in NRL formulations for balloons.

Test Parameter	1-ZDEC		2-ZBEC		3-TBzTD		4-ZBEC + TBzTD		5-ZBEC + TBzTD + ZDEC	
	Nitrosamines (mg/kg)	Nitrosatable Substances (mg/kg)	Nitrosamines (mg/kg)	Nitrosatable Substances (mg/kg)	Nitrosamines (mg/kg)	Nitrosatable Substances (mg/kg)	Nitrosamines (mg/kg)	Nitrosatable Substances (mg/kg)	Nitrosamines (mg/kg)	Nitrosatable Substances (mg/kg)
1. NDMA	0.130	0.861	nd	nd	0.028	0.495	0.030	0.480	0.034	0.565
2. NDEA	0.100	0.405	nd	nd	nd	0.372	nd	0.250	0.006	0.329
3. NDPA	nd	nd	nd	nd	nd	nd	nd	nd	nd	nd
4. NDiBA	nd	nd	nd	nd	nd	nd	nd	nd	nd	nd
5. NDBA	nd	nd	0.045	0.300	nd	0.040	0.010	0.220	0.004	0.060
6. NPIP	nd	nd	nd	nd	nd	nd	nd	nd	nd	nd
7. NPYR	nd	nd	nd	nd	nd	nd	nd	nd	nd	nd
8. NMOR	nd	nd	nd	nd	nd	nd	nd	nd	nd	nd
9. NEPhA	nd	nd	nd	nd	nd	nd	nd	nd	nd	nd
10. NMPhA	nd	nd	nd	nd	nd	nd	nd	nd	nd	nd
11. NDiNA	nd	nd	nd	nd	nd	nd	nd	nd	nd	nd
12. NDBzA	nd	nd	nd	nd	0.015	0.076	0.004	0.035	0.004	0.040
Total	0.230	1.266	0.045	0.300	0.043	0.983	0.044	0.985	0.048	0.994
	Fail	Fail	pass *	pass **	pass *	pass **	pass *	pass **	pass *	pass **

Note: nd, not detected or lower than the reporting limit (* migratable nitrosamines, 0.05 mg/kg, and ** migratable nitrosatable substances 1.0 mg/kg). Requirements according to results of nitrosamine testing in the NRL formulation for balloons, using GC-TEA according to EN 12868:2017.

Table 5 indicates that migratable levels of nitrosamines were found: in particular, the detected levels of *N*-nitrosodimethylamine (NDMA) and *N*-nitrosodiethylamine (NDEA) were 0.13 and 0.10 mg/kg, respectively, in the NRL formulation for balloons using zinc diethyldithiocarbamate (ZDEC). The nitrosamine is a compound formed from the precursor diethylamine, which results from the decomposition of the accelerator ZDEC during the vulcanization. ZDEC is classified as a highly effective "super-accelerator" for NR latex but is also considered "unsafe" and is one of the "primary problematic accelerators" because it can form carcinogenic nitrosamines. The sources include diagrams showing the reaction mechanism in Figure 6. NDEA substance is classified by the International Agency for Research on Cancer (IARC) as a Group 2A carcinogen ("probably carcinogenic to humans") and is considered a very potent carcinogen [26].

https://doi.org/10.3390/polym18040438

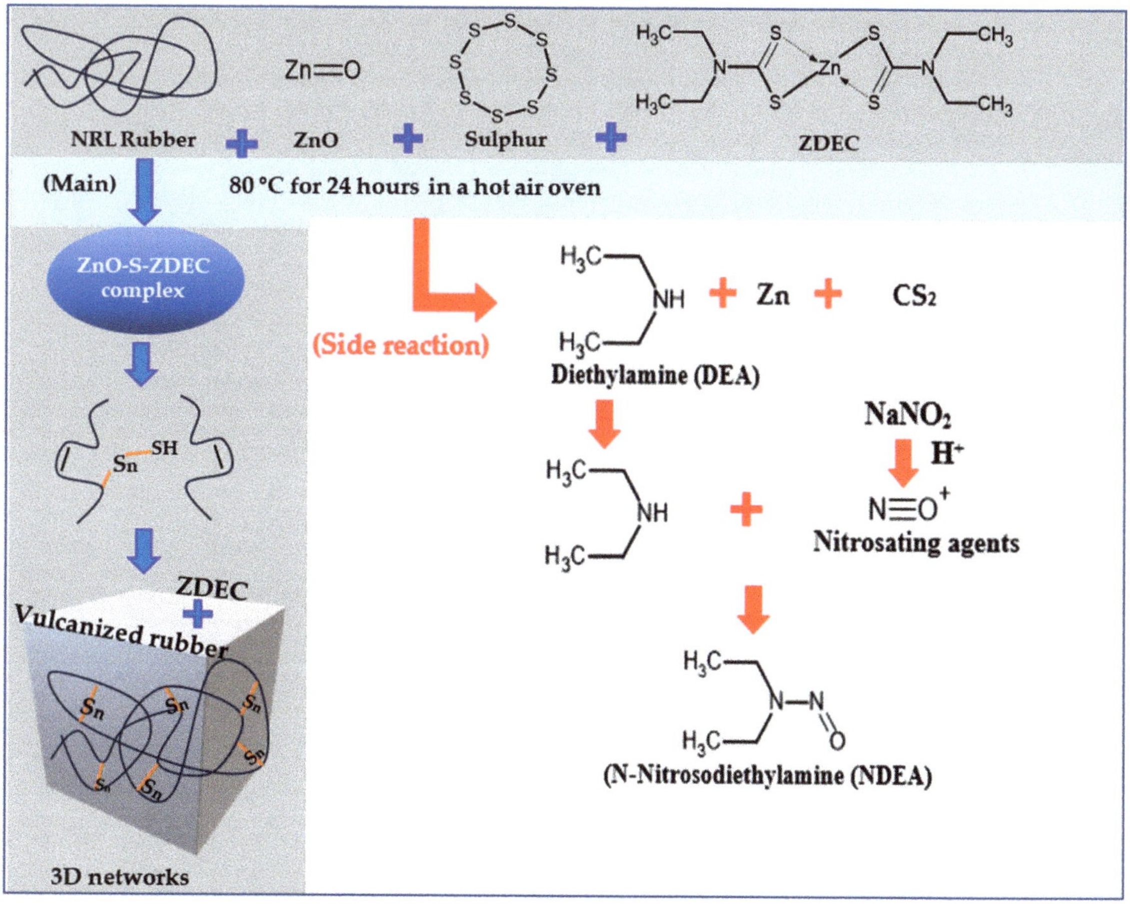

Figure 6. Diagram illustrating the reaction mechanisms that lead to the formation of nitrosodiethylamine in NRL formulation for balloons with ZDEC accelerator.

Figure 6 illustrates the mechanisms of NDEA formation from ZDEC that involve three steps: (1) decomposition of ZDEC during vulcanization, releasing diethylamine, a secondary amine; (2) nitrosation reaction where diethylamine reacts with nitrosating agents such as nitrites or nitrogen oxides, which may be present during the manufacturing process or in the environment; and (3) formation of NDEA, a nitrosamine classified as a carcinogen. The inductive and resonance effects of the amine group, along with steric factors, influence the formation of zinc–accelerator complexes.

The meaning of the red side-reaction arrow is to indicate that the chemical reaction leading to the release of *N*-nitrosodiethylamine (NDEA) does not occur in the developed formulation. The nitrosation pathway is suppressed by the novel accelerator system. The absence of NDEA formation has been verified through expert chemical review and confirmed by analytical testing, which showed non-detectable (nd) levels below the reporting limit. Regarding the mechanism, a conventional accelerator complex such as ZDEC can release NDEA because it decomposes during the heating for vulcanization, producing secondary amines (specifically diethylamine). These secondary amines then combine with nitrosating agents (such as nitrogen oxides (NOx) from the atmosphere or sodium nitrite ($NaNO_2$) used in processing) to form the carcinogenic *N*-Nitrosodiethylamine (NDEA).

https://doi.org/10.3390/polym18040438

The diagram aims to demonstrate that while conventional rubber compounds are prone to nitrosamine formation, the developed "nitrosamine-safe" formulation (using tetrabenzylthiuram disulfide (TBzTD) and zinc n-butylbenzothiazole-2-sulfenamide (ZBEC) as a big group) employs chemical structures that do not form secondary amines. This substitution effectively prevents the generation of regulated nitrosamines while maintaining the required mechanical properties for balloon production. While having chemical plausibility and consistency with established mechanistic understanding, the proposed reaction pathways remain indirectly supported by the works of Low et al. (2021) [30], Blanchard et al. (2020) [31] and Suchat et al. (2025) [32].

For safety reasons, we recommend choosing safer alternative accelerators, such as ZBEC and TBzTD, to replace ZDEC in the production of NRL products, especially those that come into contact with humans, such as balloons. The focus on the alternative accelerator compounds fulfilling the requirement of releasing no carcinogenic nitrosamines ensures that these alternative systems meet both the safety criterion of non-detectable carcinogenic nitrosamines and the required performance for balloon production, thereby strengthening the central message and focus of this study.

The use of multiple accelerants, whether two or three, compared to a single accelerant like TBzTD, still maintains the nitrosamines values below the latex balloon standard limits, as shown in Table 5 and Figures 7 and 8. The tables and figures corroborate that the novel formulations are safe for balloons made from natural rubber latex (NRL) [32,33].

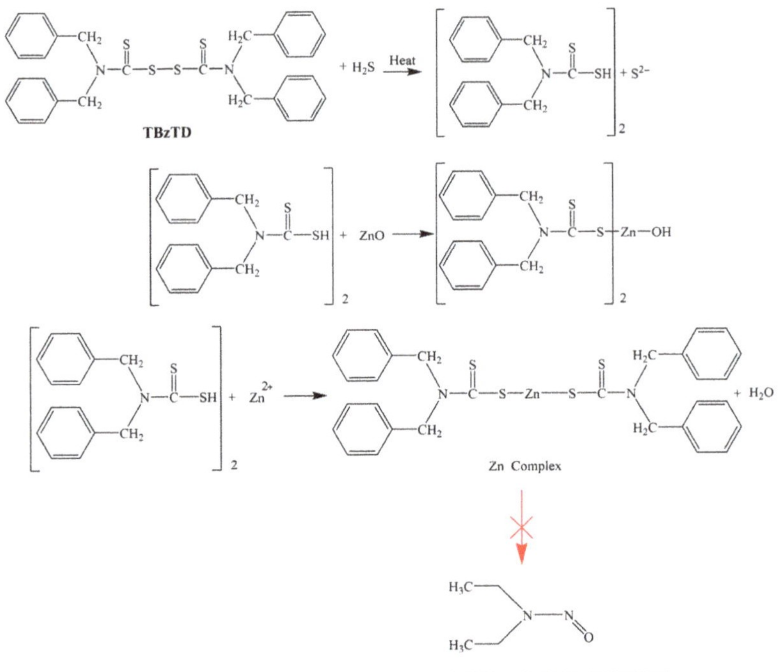

Figure 7. Diagram of the reaction mechanisms in the nitrosamine-free NRL formulation with TBzTD.

Tetrabenzyl thiuram disulfide (TBzTD), with the chemical formula C30H28S4N2 and a molecular weight of 542.78, undergoes a reaction, shown in Figure 7. The induction of the amine group, along with steric factors, influences the formation of zinc–accelerator complexes. Nonetheless, as shown in Table 5, *N*-nitrosodibenzylamine (NDBzA) is slightly produced, but this is safe and not a carcinogenic nitrosamine.

https://doi.org/10.3390/polym18040438

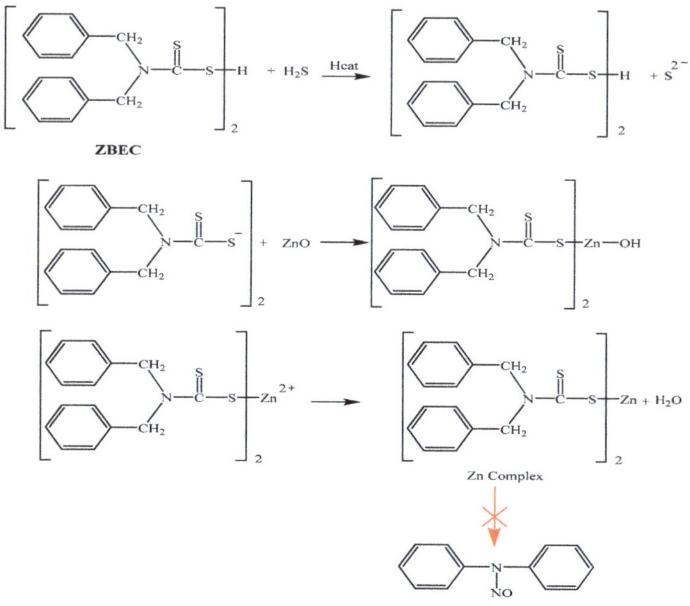

Figure 8. The reaction mechanisms in the nitrosamine-free NRL formulation for balloons with ZBEC.

Based on the data in Table 5 and Figure 8, when Zinc dibutyldithiocarbamate (ZBEC) is used in an NRL formulation for balloons, NDMA and NDEA are not detected (reporting limit 0.01 mg/kg). The inductive effect of the amine group, along with the steric factors, plays a significant role in influencing the formation of zinc complexes and accelerators during vulcanization. These factors affect how compounds interact and bond, thereby impacting the vulcanization. Comparative testing of NRL compounds showed clearly that the ZBEC accelerator contributed to sulfur vulcanization, and if a nitrosation reaction occurs, the resulting nitrosamine will be NDBA (nitrosodibutylamine) at a low concentration, which is safe and not a carcinogenic nitrosamine (Table 5).

It is also noteworthy that even though ZBEC could theoretically form NDBA, the test results shown in Table 4 have some NDBA, which is safe and not carcinogenic [19]. Using a ZBEC accelerator is an effective method for avoiding the formation of NDMA and NDEA in products, which is a key reason for its selection in formulations requiring safety by not carrying carcinogenic nitrosamines.

According to Table 5, the investigation of a novel synergistic accelerator system comprising a little bit of ZDEC (0.03 phr) with the combination of TBzTD (0.36 phr), and ZBEC (0.36 phr), suggests that this combination could be an effective and safer alternative to the unregulated ZDEC for the sulfur vulcanization of NRL balloons.

The formula of TBzTD/ZBEC and ZDEC in the NRL compound for balloons, incorporating the accelerators that complied with established nitrosamine and nitrosatable substances safety standards, appears to be suitable as a safer alternative to conventional accelerators. The bulky phenyl substituents exert steric hindrances that decrease the alkalinity of the amine. This reduction in alkalinity enhances the stability of the zinc–accelerator complex, which in turn suppresses the release of the reactive intermediates that are responsible for nitrosamine generation, thereby effectively reducing nitrosamine formation.

While TBzTD or ZBEC was specifically developed as a safer alternative to traditional, problematic accelerators like ZDEC, it cannot be considered completely safe. Here, the

https://doi.org/10.3390/polym18040438

balloon formulation was designed to have an effective accelerator system to replace the use of unsafe ZDEC neat.

Using a combination of TBzTD and ZBEC accelerators, supplemented with a very small amount (0.03 phr) of ZDEC, is a highly effective strategy for natural rubber latex balloon formulations. This approach achieves two critical goals.

(1) Enhanced product properties. Through a synergistic effect, it produces balloons with significantly higher tensile strength and thermal stability compared to other accelerator systems.

(2) High safety level. By relying on primarily safe accelerators and minimizing the precursor for harmful nitrosamines, it yields a final product with lower levels than the standard limits, thereby meeting the strict safety standards for rubber products that may be used by children.

The reaction mechanisms in nitrosamine-free NRL formulation for balloons with a synergistic accelerator combination of TBzTD and ZBEC are shown in Figure 9. The nitrosamine testing for the NRL balloon compounds that used the three alternative accelerator combinations, TBzTD/ZBEC/ZDEC, was done using GC-TEA according to EN 12868:2017 [15] and BgVV (2002) [26]. It confirmed that the total migration values of the nitrosamine and the nitrosatable substances were below the standard balloon limits (0.05 mg/kg for nitrosamines and 1.0 mg/kg for nitrosatable substances). Based on discussions with balloon manufacturers, exporters, and experts in this field, the proposed approach is considered acceptable, ensuring no exposure to carcinogenic nitrosamines for humans.

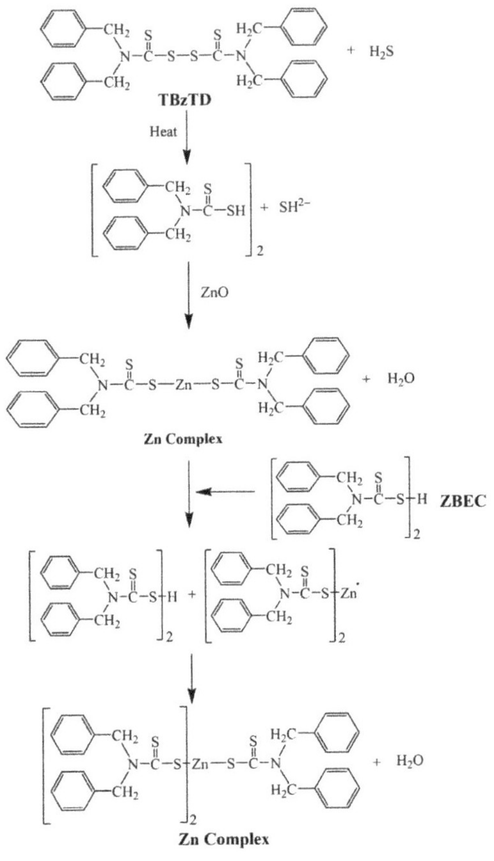

Figure 9. Reaction mechanisms in NRL formulation for balloons, based on a nitrosamine-free TBzTD/ZBEC combination.

https://doi.org/10.3390/polym18040438

4. Conclusions

This study developed a sulfur-vulcanized NRL compound formulation for balloon products, exceeding the required mechanical properties while controlling the carcinogenic nitrosamine content for compliance with the stringent safety standards. In experiments, the types and amounts of accelerators were manipulated, while controlling for the 103% swelling index and 62% total solid content targets. It was found that the synergistic accelerator system of TBzTD (0.36 phr) and ZBEC (0.36 phr), with a little bit of ZDEC (0.03 phr) gave significantly improved mechanical performance without detectable carcinogenic nitrosamines, outperforming formulations using a combination of TBzTD and ZBEC or each accelerator alone. Mechanical testing revealed the highest tensile strength (22.1 ± 1.8 MPa), elongation (1153 ± 78%), and 300% modulus (1.30 ± 0.07 MPa), complying with balloon product standards. The thermal stability was confirmed by its good mechanical properties after accelerated aging, and by thermogravimetric analysis, which showed a maximum decomposition temperature of 384.15 °C. Overall, the synergistic use accelerators enhanced the vulcanization efficiency, thermal stability, and mechanical properties, highlighting the critical role of accelerator synergy in optimizing natural rubber balloon performance, with safety standards concerning carcinogenic nitrosamine content.

Author Contributions: Conceptualization, S.S.; Methodology, T.C.-I. and E.A.; Validation, T.C.-I., E.A. and S.S.; Formal analysis, T.C.-I. and E.A.; Investigation, S.S.; Resources, T.C.-I.; Data curation, T.C.-I. and S.S.; Writing—original draft, S.S.; Writing—review & editing, S.S.; Visualization, E.A. and S.S.; Supervision, S.S.; Project administration, S.S.; Funding acquisition, S.S. All authors have read and agreed to the published version of the manuscript.

Funding: This research was supported by the National Science, Research and Innovation Fund (NSRF) and Prince of Songkla University, SIT6701127a, SIT6801101S and B.K. Latex Product Co., Ltd.

Institutional Review Board Statement: Not applicable.

Data Availability Statement: The original contributions presented in this study are included in the article. Further inquiries can be directed to the corresponding author.

Acknowledgments: We are grateful to Saranyoo Klaiklay and Seppo Karrila for their valuable assistance in improving the chemical reaction schemes and language of the draft manuscript.

Conflicts of Interest: The authors declare that this study received funding from B.K. Latex Product Co., Ltd. The funder was not involved in the study design, collection, analysis, interpretation of data, the writing of this article or the decision to submit it for publication.

References

1. Fortune Business Insights. *Xenon Market Size, Share & COVID-19 Impact Analysis, by Type (N3, N4. 5, and N5), by Application (Imaging & Lighting, Satellite, Electronics & Semiconductors, Medical and Others), and Regional Forcast, 2022–2029*; Fortune Business Insights: Pune, India, 2023.
2. Fortune Business Insights. Virtual Reality in Education Market Size, Share & Industry Analysis, by Component (Hardware, Software, and Content), by Application (K-12, Higher Education, and Vocational Training), and Regional Forecast, 2024–2032; Market Research Report. 2024. Available online: https://www.fortunebusinessinsights.com/industry (accessed on 29 January 2026).
3. Kędzia, J.R.; Sitko, A.M.; Haponiuk, J.T.; Lipka, J.K. Natural rubber latex-origin, specification and application. In *Application and Characterization of Rubber Materials*; IntechOpen: London, UK, 2022.
4. Stephens, H.L.; Bhowmick, A.K. *Handbook of Elastomers*; Dekker: New York, NY, USA, 2001.
5. Brunnemann, K.D.; Hoffmann, D. Analytical studies on tobacco-specific *N*-nitrosamines in tobacco and tobacco smoke. *Crit. Rev. Toxicol.* **1991**, *21*, 235–240. [CrossRef]
6. Biaudet, H.; Mouillet, L.; Debry, G. Migration of nitrosamines from condoms to physiological secretions. *Bull. Environ. Contam. Toxicol.* **1997**, *59*, 847–853. [CrossRef]
7. Farren, N.J.; Ramírez, N.; Lee, J.D.; Finessi, E.; Lewis, A.C.; Hamilton, J.F. Estimated exposure risks from carcinogenic nitrosamines in urban airborne particulate matter. *Environ. Sci. Technol.* **2015**, *49*, 9648–9656. [CrossRef]

https://doi.org/10.3390/polym18040438

8. Rao, T. *Genotoxicology of N-Nitroso Compounds*; Springer Science & Business Media: Berlin/Heidelberg, Germany, 2013.
9. van Bruggen, M.; van Putten, E.M.; Janssen, P.C.J.M. RIVM 2003. N-Nitrosamines in Balloons: Assessment of the Health Risk for Children. In *RIVM Report 609300002/2007 Nitrosamines Released from Rubber Crumb*; p. 17. Available online: https://www.rivm.nl/bibliotheek/rapporten/609300002.html (accessed on 29 January 2026).
10. World Health Organization. *Guidelines for Drinking-Water Quality: Incorporating the First and Second Addenda*; World Health Organization: Geneva, Switzerland, 2022.
11. Hwang, J.B.; Lee, J.E.; Kim, E.; Eom, K.Y.; Kim, H.-A.; Lee, S. Analysis of *N*-nitrosamines and *N*-nitrosatable substances from baby bottle rubber teats by liquid chromatography tandem mass spectrometry. *Food Addit. Contam. Part A* **2023**, *40*, 518–527. [CrossRef]
12. *EN 71-12:2016*; Safety of Toys—Part 12: N-Nitrosamines and N-Nitrosatable Substances. European Standard: Brussels, Belgium, 2016.
13. Suchat, S.; Boonrasri, S. Advancement of an environmentally friendly and innovative sustainable rubber wrap film with superior sealing properties. *Polymers* **2024**, *16*, 1499. [CrossRef]
14. Park, J.-e.; Seo, J.-e.; Lee, J.-y.; Kwon, H. Distribution of seven *N*-nitrosamines in food. *Toxicol. Res.* **2015**, *31*, 279–288. [CrossRef]
15. *EN 12868:2017*; Child Use and Care Articles—Method for Determining the Release of N-Nitrosamines and N-Nitrosatable Substances from Elastomer or Rubber Teats and Soothers. European Standard: Brussels, Belgium, 2017.
16. Fajen, J.; Carson, G.; Rounbehler, D.; Fan, T.; Vita, R.; Goff, U.; Wolf, M.; Edwards, G.; Fine, D.; Reinhold, V. *N*-nitrosamines in the rubber and tire industry. *Science* **1979**, *205*, 1262–1264. [CrossRef]
17. Smith, S.A.; Norris, B. Reducing the Risk of Choking Hazards: Mouthing Behaviour of Children Aged 1 Month to 5 Years. *Inj. Control. Saf. Promot.* **2003**, *10*, 145–154. [CrossRef] [PubMed]
18. Kędzia, J.; Haponiuk, J.; Formela, K. Natural Rubber Latex Wastes from Balloon Production as Valuable Source of Raw Material: Processing, Physico-Mechanical Properties, and Structure. *J. Compos. Sci.* **2024**, *8*, 365. [CrossRef]
19. Virdi, R.; Grover, B.; Ghuman, K. "Nitrosamine safe" thiuram disulfide. *Rubber Chem. Technol.* **2019**, *92*, 90–109. [CrossRef]
20. Park, S.J.; Jeong, M.J.; Park, S.R.; Choi, J.C.; Choi, H.; Kim, M. Release of *N*-nitrosamines and *N*-nitrosatable substances from baby bottle teats and rubber kitchen tools in Korea. *Food Sci. Biotechnol.* **2018**, *27*, 1519–1524. [CrossRef]
21. Murakami, L.M.S.; Azevedo, J.B.; Diniz, M.F.; Silva, L.M.; Dutra, R.d.C.L. Characterization of additives in NR formulations by TLC-IR (UATR). *Polímeros* **2018**, *28*, 205–214. [CrossRef]
22. Sujarit, J.; Phutdhawong, W. Synthesis of Zinc Diethyldithiocarbamate (ZDEC) and Structure Characterization using Decoupling ^{1}H NMR. In Proceedings of the Congress on Science and Technology of Thailand, Golden Jubilee Convention Hall, Khon Kean University (Thailand), Khon Kaen, Thailand, 20–22 October 2003.
23. Thajudin, N.N.; Shuib, R. Room temperature self-healing natural rubber based on ionic supramolecular network. *AIP Conf. Proc.* **2020**, *2267*, 020023. [CrossRef]
24. Liu, L.-N.; Dai, J.-G.; Zhao, T.-J.; Guo, S.-Y.; Hou, D.-S.; Zhang, P.; Shang, J.; Wang, S.; Han, S. A novel Zn (ii) dithiocarbamate/ZnS nanocomposite for highly efficient Cr^{6+} removal from aqueous solutions. *RSC Adv.* **2017**, *7*, 35075–35085. [CrossRef]
25. Sivakumar, P.; Thyagarajan, K.; Kumar, A. Investigations on physical properties of Zn ferrite nanoparticles using sol-gel auto combustion technique. *Dig. J. Nanomater. Bios.* **2018**, *13*, 1117–1122.
26. BgVV. Risk Assessment of N-Nitrosamines in Balloons. Federal Institute of Health Protection of Consumers and Veterinary Medicine, Expert Opinion. 11 April 2002. Available online: http://www.bgvv.de (accessed on 29 January 2026).
27. de Lima, D.R.; da Rocha, E.B.D.; de Sousa, A.M.F.; da Costa, A.C.A.; Furtado, C.R.G. Effect of vulcanization systems on the properties of natural rubber latex films. *Polym. Bull.* **2021**, *78*, 3943–3957. [CrossRef]
28. Larpkasemsek, A.; Raksaksri, L.; Chuayjuljit, S.; Chaiwutthinan, P.; Boonmahitthisud, A. Effects of sulfur vulcanization system on cure characteristics, physical properties and thermal aging of epoxidized natural rubber. *J. Met. Mater. Miner.* **2019**.
29. Boonkerd, K.; Deeprasertkul, C.; Boonsomwong, K. Effect of sulfur to accelerator ratio on crosslink structure, reversion, and strength in natural rubber. *Rubber Chem. Technol.* **2016**, *89*, 450–464. [CrossRef]
30. Low, D.Y.S.; Supramaniam, J.; Soottitantawat, A.; Charinpanitkul, T.; Tanthapanichakoon, W.; Tan, K.W.; Tang, S.Y. Recent developments in nanocellulose-reinforced rubber matrix composites: A review. *Polymers* **2021**, *13*, 550. [CrossRef]
31. Blanchard, R.; Ogunsona, E.O.; Hojabr, S.; Berry, R.; Mekonnen, T.H. Synergistic cross-linking and reinforcing enhancement of rubber latex with cellulose nanocrystals for glove applications. *ACS Appl. Polym. Mater.* **2020**, *2*, 887–898. [CrossRef]
32. Suchat, S.; Praksong, K.; Muangprathub, J.; Srisawat, T.; Abdullah, N.H.; Boonrasri, S. Agricultural performance of biomaterials mulch based on nitrosamine-free natural rubber. *Green Mater.* **2025**, *13*, 183–195. [CrossRef]
33. Suchat, S.; Boonrasri, S. Advancement of an Eco-Friendly and Innovative Sustainable Rubber Wrap Film Exhibiting Superior Sealing Properties. *Res. Sq.* **2024**. [CrossRef]

https://doi.org/10.3390/polym18040438

polymers

MDPI

Article

Alternative Natural Rubber Cross-Linking Utilizing a Disulfide-Containing Bismaleimide

Anureet Kaur [1], Maria Tucker [1], Keizo Akutagawa [1], Biqiong Chen [2] and James J. C. Busfield [1,*]

1 School of Engineering and Materials Science, Queen Mary University of London, London E1 4NS, UK; k.akutagawa@qmul.ac.uk (K.A.)
2 Department of Chemistry, University of Liverpool, Crown Street, Liverpool L69 7ZD, UK; biqiong.chen@liverpool.ac.uk
* Correspondence: j.busfield@qmul.ac.uk

Abstract

This study explores a disulfide-selective cross-linking strategy for natural rubber (NR) to formulate elastomeric materials with engineering-relevant mechanical properties. A disulfide-containing bismaleimide (BIS) cross-linker was synthesized from cystamine and maleic anhydride and compounded with NR. Three formulations were prepared: control (no inhibitor), $CuCl_2$-based, and copper(II) methacrylate (CuMA) based compounds, with BIS concentrations ranging from 3.55 to 8.88 phr. Rheological and mechanical testing revealed that $CuCl_2$ formulations suffered from molecular degradation, poor thermal stability, and mechanical brittleness due to oxidative reactions in the absence of antioxidants. In contrast, CuMA-based compounds exhibited intermediate molecular weights prior to curing, stable thermal behavior, and improved mechanical properties, including enhanced torque and tensile strength, indicating effective cross-linking and partial recyclability. The control formulations also performed reasonably well but did not match the mechanical strength of conventional sulfur-vulcanized NR. The results demonstrate that metal coordination, particularly with CuMA, can modulate disulfide metathesis kinetics and offer a pathway to thermally triggered network rearrangement. Overall, CuMA emerges as a promising candidate for developing high-performance, recyclable rubber materials, while $CuCl_2$ requires further stabilization strategies. This work establishes a baseline for future recyclability studies and advances the design of dynamic covalent networks in elastomers.

Keywords: bismaleimide; natural rubber; cross-linking; disulfide

1. Introduction

Conventional sulfur vulcanization (CV) yields robust three-dimensional polysulfide cross-linked networks that provide the mechanical performance required by engineering elastomers but render the resulting materials essentially unrecyclable [1]. Efforts to recover value from vulcanized rubber have therefore focused on devulcanization and other recycling routes, yet selective and efficient reclaiming of cross-linked networks remains an as yet unfulfilled challenge for the rubber industry [2,3]. One promising route to reconcile durable performance with recyclability is the introduction of dynamic covalent linkages into the cross-link network. Disulfide bonds (S–S) are known to behave as dynamic covalent bonds, which have been reported to undergo a free radical disulfide metathesis reaction following a radical-mediated [2 + 1] mechanism, consisting of cleavage of the disulfide bond and its subsequent reforming [4], that can occur spontaneously [5] or be thermally initiated [6]. The latter category is particularly useful for the development of recyclable elastomers, as the exchange reaction necessary to rearrange the cross-linked network can be triggered only under specific conditions, retaining elastomeric behavior in a wide range of service conditions. The lability of S–S bonds relative to typical carbon–carbon single or carbon–carbon double bonds underpins their utility for this purpose [7]. Controlling the temperature, and therefore the timing, of disulfide metathesis is critical: facile exchange at ambient temperature compromises dimensional and mechanical stability, whereas an exchange that is triggerable only at processing temperatures enables reprocessing without loss of in-service performance. Recent work demonstrates that metal coordination, for example via copper(II) methacrylate (CuMA) or $CuCl_2$, can be used to modulate disulfide metathesis kinetics and thus tune the rearrangement temperature window in sulfur-based vulcanizates, as the activation temperature is reached, the copper(II)-based complex catalysis can enable cross-over reactions between disulfide and polysulfides [8,9]. Notably, we have previously reported that optimizing a sulfur-based cure package with CuMA promotes a reversible disulfide/polysulfide network that can be recovered by thermal processing, thereby enabling recyclable sulfur-cured natural rubber formulations [1]. Motivated by these developments, the present study aimed to implement a different cross-linking architecture in an elastomer matrix by selectively introducing disulfide-only cross-links and evaluating the mechanical performance of the resulting pristine materials. The introduction of disulfide cross-links can only be achieved by either adjusting the sulfur/accelerator (S/A) ratio to >1 in a sulfur-based vulcanization—hence using the CV system, favoring the formation of polysulfide cross-links over monosulfides—or by introducing small organic molecules that contain disulfide functional groups. However, in CV-cured compounds that are characterized by a large number of polysulfides, once an S-S bond is cleaved and thiyl radicals are formed, they would statistically be keener to react with the NR chains, slowly shifting the cross-linked network to a semi-efficient system (SEV, where $S/A \approx 1$ and 50% of the cross-links are polysulfides) or efficient system (EV, where $S/A \ll 1$ and 80% of the cross-links are monosulfides) [10,11]. An example of disulfide-containing small organic molecules that can be used to cross-link natural rubber is disulfide-containing bismaleimides. These have been reported in the literature as capable of introducing self-healable and reprocessable properties in elastomers. Kitagawa et al. synthesized Jaffamine-derived polymers using bis(2-maleimidoethyl) disulfide and 4-aminophenyl disulfide as cross-linkers, achieving self-healing efficiencies higher than 90% after reprocessing at 180 °C [12]. Sim et al. cross-linked poly(furfuryl methacrylate) (PFMA) brushes with a disulfide-containing bismaleimide via the Diels–Alder (DA) reaction at 70 °C, where these could be decross-linked at 110 °C via the retro-Diels–Alder (rDA) reaction, and the S–S bond could be cleaved upon thermal or photo stimulus and regenerated through oxidative stimulus, offering another reversible decross-linking/cross-linking pathway for cross-linked PFMA brushes [13]. Usually, the rubber backbone is first grafted with maleic anhydride, which is then reacted with furfurylamine, and then, through DA, the bismaleimide molecule is introduced as a cross-linker [14]. More generally, bismaleimides to be introduced in polymers as cross-linkers require a furan functional group to react through the DA pathway [14–16]. However, grafting of maleic anhydride has been reported in the literature via application of shearing action in internal mixers and heat [17,18]. Following a similar approach, this work reports the cross-linking of NR with disulfide-containing bismaleimide via mixing in a micro compounder, followed by compression molding.

This work reports the pre-synthesis of a disulfide-containing bismaleimide (BIS) cross-linker, starting from desalination of cystamine dihydrochloride with subsequent reaction of the resulting cystamine with maleic anhydride, followed by compounding with NR, as shown in Figure 1. Three formulations were prepared: (i) control compounds without inhibitors, (ii) $CuCl_2$ compounds, and (iii) CuMA compounds. Copper(II) complexes act as catalysts; however, rather than lowering the activation energy, they increase the activation energy threshold to enable disulfide metathesis. Therefore, in this work, copper(II) compounds are referred to as inhibitors. In all three formulations, BIS concentration varied from 3.55 to 8.88 phr. Rheological cure behavior and standard mechanical tests were used to define cure parameters and to establish baseline tensile/elastic properties of the pristine compounds. This paper, therefore, (i) describes the design and processing of a disulfide-selective cross-linking strategy to form the intended dynamic and controlled cross-linked network, and (ii) reports the mechanical performance of the pristine materials as a baseline for any future recyclability studies. The results establish whether the selective introduction of disulfide cross-links and the incorporation of metal coordination agents can produce materials with engineering-relevant mechanical properties while offering a route to thermally trigger network rearrangement under controlled conditions. No additional accelerators or antioxidants were included in the formulations, as the primary objective of this study is to isolate and evaluate the specific influence of copper(II) compounds on disulfide-based cross-linking. Introducing accelerators or antioxidants could have introduced competing effects, such as altering cure rates or suppressing oxidative pathways, which would confound interpretation of the copper-mediated inhibition mechanism. By excluding these additives, we ensured that the observed changes in rheological and mechanical behavior could be attributed solely to the presence and coordination chemistry of the copper(II) species.

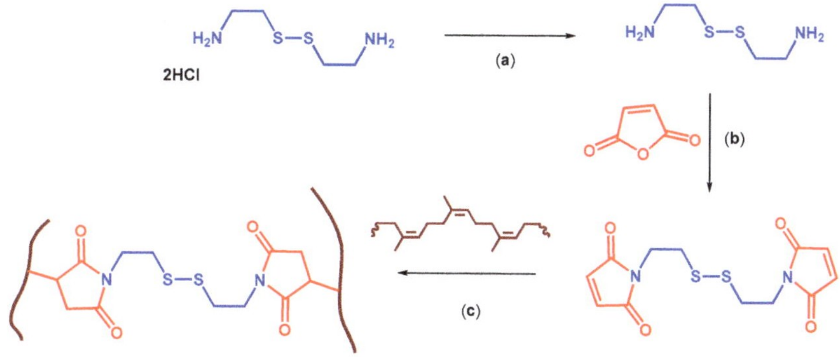

Figure 1. Schematic representation of rubber preparation. Conditions: (**a**) cystamine dihydrochloride, deionized water, NaOH, 12 h, room temperature; (**b**) 1. maleic anhydride, cystamine, toluene, 1 h, 30 °C, 2. hexamethyldisilazane, $ZnBr_2$, 12 h, 85 °C; (**c**) 1,1′-(disulfanediylbis(ethane-2,1-diyl))bis(1H-pyrrole-2,5-dione), NR, mixing in the micro compounder, followed by curing with hot press for the time required to reach 90% of maximum torque, t_{90}, at 180 °C.

2. Materials and Methods

2.1. Materials

Natural rubber (NR, SMR CV60 grade) was purchased from the Tun Abdul Razak Research Centre (Hertford, UK). Maleic anhydride (Man, purity ≥ 98%, technical grade) was purchased from VWR International (Leicestershire, UK). Copper(II) chloride ($CuCl_2$; purity ≥ 98%), hydrochloric acid (HCl; 37%, ACS reagent), hexamethyldisilazane (purity 99%), Chloroform-d ($CDCl_3$; 99.8 atom % D), methanol (purity ≥ 99.8%, ACS reagent), and dichloromethane (purity ≥ 99.9%, ACS reagent) were purchased from Sigma-Aldrich Co Ltd. (Dorset, UK). Cystamine dihydrochloride (Cys; purity 97%), sodium hydroxide (purity 98%), copper(II) methacrylate (CuMA, technical grade), zinc bromide (purity 98%, anhydrous), sodium hydrogen carbonate (purity 99%), and anhydrous sodium sulfate (purity 99%) were purchased from Fisher Scientific (Loughborough, UK). Toluene (purity $\geq$ 99.9%, HPLC grade) and tetrahydrofuran (THF; purity $\geq$ 99.9%, inhibitor free, HPLC grade) were purchased from Honeywell International Inc. (Bracknell, UK).

In the present work, different rubber formulations with and without Cu(II) inhibitors were tested in different conditions, before and after curing and recycling. The terminology adopted to refer to the formulations and blends in specific conditions is as follows:

- NR/BIS-X: formulations without inhibitors, acting as control formulations, with X referring to phr of Bismaleimide introduced; generally called Control formulations;
- NR/BIS-X/CuMA: formulations with CuMA as inhibitor, with X referring to phr of Bismaleimide introduced; CuMA is equimolar to Bismaleimide;
- NR/BIS-X/$CuCl_2$: formulations with $CuCl_2$ as inhibitor, with X referring to phr of Bismaleimide introduced; $CuCl_2$ is equimolar to Bismaleimide;
- Compound: something that has been mixed, but not cured;
- Cured compound: new cured compound;
- Recycled compound: recycled cured compound.

2.2. Preparation of Cystamine

Sodium hydroxide (10.65 g, 266.430 mmol) was dissolved in 25 mL of deionized water and was added dropwise to an aqueous solution of cystamine dihydrochloride (30 g, 133.215 mmol, in 25 mL of deionized water), in an ice bath. The resulting solution was stirred for 1 h at room temperature. The obtained solution was transferred into a separatory funnel, and dichloromethane was added (20 mL × 4). The organic phase was dried over anhydrous sodium sulfate and filtered before drying under vacuum in a rotatory evaporator to obtain cystamine as a pale-yellow oily liquid (yield 29.9%). The resulting product was characterized by ^1H and ^{13}C NMR spectroscopy with Bruker Advance III 400 MHz. ^1H NMR ($CDCl_3$, 400 MHz): chemical shift δ = 1.21, 2.61, 2.85 ppm. ^{13}C DEPTQ135 NMR ($CDCl_3$, 100 MHz): chemical shift δ = 40.56, 42.51 ppm.

2.3. Synthesis of Bismaleimide Cross-Linker

A suspension of maleic anhydride (0.6641 g, 6.77 mmol) and cystamine (0.6846 g, 3.39 mmol) was prepared in 80 mL of toluene and stirred at 30 °C for 1 h. Subsequently, zinc bromide (1.524 g, 6.77 mmol) and hexamethyldisilazane (1.093 g, 6.77 mmol), dissolved in 15 mL of toluene, were added to the reaction mixture. The resulting suspension was heated to 85 °C and refluxed for 2 h. Reaction progress was monitored by thin-layer chromatography (TLC) using silica gel F254 plates on aluminum sheets and a dichloromethane:methanol (9:1) eluent system (product retention factor = 0.0678), and the reaction was considered complete once the reactant stains disappeared. Upon completion, the mixture was cooled to room temperature and quenched with 50 mL of 1 M hydrochloric acid. The organic layer was separated, and the aqueous layer was extracted twice with 50 mL portions of ethyl acetate. The combined organic extracts were washed with a saturated aqueous solution of sodium hydrogen carbonate and brine, then dried over anhydrous sodium sulfate. The solution was filtered and concentrated under reduced pressure using a rotary evaporator. The crude product was purified by crystallization, and the concentrated solution was placed in a refrigerator overnight. The resulting crystals were collected by filtration, washed with cold methanol, and dried under reduced pressure for 2 h. The resulting Bismaleimide (BIS) product, 1,1′-(disulfanediylbis(ethane-2,1-diyl))bis(1H-pyrrole-2,5-dione) (yield 48%), was characterized by ^1H and ^{13}C NMR spectroscopy. ^1H NMR ($CDCl_3$, 400 MHz): δ = 2.95, 3.88, 6.74 ppm. ^{13}C DEPTQ135 NMR ($CDCl_3$, 100 MHz): δ = 35.92, 36.81, 134.05 ppm.

2.4. Compound Preparation and Curing

NR/BIS, NR/BIS/CuMA, and NR/BIS/$CuCl_2$ formulations are reported in Table 1. Compounds were prepared in an Xplore MC15 HT Micro Compounder. Mixing was performed at 100 °C, 100 rpm speed, and a maximum torque of 35 Nm. NR, the previously synthesized BIS cross-linker, and the copper inhibitor were added together and mixed for 3 min. The compounds were then molded in a manual hot press at 180 °C, applying a constant pressure of about 12 MPa, for a curing time equal to t_{90} determined using Monsanto Moving Die Rheometer (MDR) 2000 with the lower die moving at 1.66 Hz, which was different for each formulation, as reported in Table 1.

Table 1. NR/BIS, NR/BIS/CuMA, and NR/BIS/$CuCl_2$ formulations and curing time.

Formulation	NR/phr	BIS/phr	CuMA/phr	$CuCl_2$/phr	t_{90}/min
NR/BIS-3.55	100	3.55			42
NR/BIS-3.55/CuMA	100	3.55	2.38		42
NR/BIS-3.55/$CuCl_2$	100	3.55		1.37	49
NR/BIS-5.33	100	5.33			43
NR/BIS-5.33/CuMA	100	5.33	3.58		44
NR/BIS-5.33/$CuCl_2$	100	5.33		2.06	45
NR/BIS-7.10	100	7.10			46
NR/BIS-7.10/CuMA	100	7.10	4.77		45
NR/BIS-7.10/$CuCl_2$	100	7.10		2.74	43
NR/BIS-8.88	100	8.88			46
NR/BIS-8.88/CuMA	100	8.88	5.96		45
NR/BIS-8.88/$CuCl_2$	100	8.88		3.43	44

2.5. Gel Permeation Chromatography

The number average molecular weight ($\overline{M}_n$), the weight average molecular weight ($\overline{M}_w$), and the dispersity ($Đ$) of NR and each compound prior to vulcanization were determined by performing Gel Permeation Chromatography (GPC) using Agilent Technologies 1260 Infinity GPC/SEC system equipped with a refractive index (RI) detector. Calibration was carried out using polystyrene standards ($\overline{M}_w$ range from 162 g/mol to 6,570,000 g/mol) from Agilent Technologies, Inc. (Santa Clara, CA, USA). Samples were prepared at least 48 h prior to the analysis by dissolving approximately 10 mg of sample in 5 mL of THF. The solutions were then filtered through a 0.2 µm PTFE syringe filter and transferred into GPC vials. Each run was performed by injecting 100 µL into the columns, kept at a constant temperature of 25 °C.

2.6. Dynamic Differential Scanning Calorimetry (DSC)

Dynamic DSC was conducted to determine the glass transition temperature, T_g, of all uncured compounds and cured compounds. Both analyses were carried out using a TA Instruments DSC25. All samples were placed in Tzero Aluminium Hermetic pans, and the normalized heat flow was measured. The thermal behavior of the uncured samples was investigated by conducting heat–cool–heat experiments. This is shown as follows: (1) first heating ramp from 20 °C to 200 °C at a rate of 5 °C/min; (2) second cool ramp from 200 °C to 20 °C at a rate of 5 °C/min; (3) third heat ramp from 20 °C to 200 °C at a rate of 5 °C/min. For the cured samples the heat–cool–heat experiments were carried as follows: (1) first heating ramp from −90 °C to 250 °C at a rate of 5 °C/min; (2) second cool ramp from 250 °C to −90 °C at a rate of 5 °C/min; (3) third heat ramp from −90 °C to 250 °C at a rate of 5 °C/min. The T_g of the cured samples was calculated by evaluating the minimum point of the first derivative of the normalized heat flow plot.

2.7. Mechanical Characterization

Tensile testing until failure was carried out on cured compounds. Dumbbell shape samples were cut out using an ISO-37-4 [19] die cutter 24 h after curing was completed. Testing was carried out using an Instron 5967 machine equipped with a 1 kN load cell and pneumatic side action grips, using a rate of 1 mm/s, which corresponds to an approximate strain rate of 0.1/s. The width and the thickness were measured prior to the start of the test, while the initial length, L_0, was measured after gripping the sample, ensuring the width of the sample remained constant in the gauge length. Uniaxial stress and strain were calculated using the following Equations (1) and (2):

$$\sigma = \frac{F}{A_0} \tag{1}$$

$$\varepsilon = \frac{\Delta L}{L_0} \tag{2}$$

where σ is the engineering stress, F is the measured force, A_0 is the initial unstrained cross-sectional area, ε is the strain, and ΔL is the measured elongation.

2.8. Density

Density measurements were carried out using a Micromeritics AccuPyc II 1345 gas-displacement pycnometer (Micromeritics Instrument Corporation, Norcross, GA, USA) using the following settings: number of purges 10, purge fill pressure 134.45 kPa, number of cycles 10, cycle fill pressure 134.45 kPa, equilibration rate 3.45 kPa min^{-1}, and chamber size 1 cm^3.

3. Results and Discussion

During the mixing stage, rubber compounds may undergo molecular degradation due to mechanical and thermal stresses. Specifically, excessive mastication in the micro compounder can reduce $\overline{M}_n$ and $\overline{M}_w$, while elevated temperatures may promote premature chemical interactions among formulation components. Disulfide metathesis reactions are particularly favored under these conditions, potentially leading to exchange reactions prior to curing. Figure 2a,b presents the $\overline{M}_n$ and $\overline{M}_w$ values respectively following a 3 min mixing period. The control formulation exhibits $\overline{M}_n$ and $\overline{M}_w$ values comparable to raw NR, with NR/BIS-8.88 showing lower $\overline{M}_w$ (500,000 g/mol) than NR (700,000 g/mol), while $\overline{M}_n$ remains in the range of 180,000–240,000 g/mol. In contrast, the $CuCl_2$ formulation displays significantly reduced $\overline{M}_n$ and $\overline{M}_w$ (10,000–20,000 g/mol and 100,000–150,000 g/mol, respectively), indicating poor compatibility with NR and suggesting substantial chain scission during processing. CuMA samples show intermediate $\overline{M}_n$ and $\overline{M}_w$ values (100,000–110,000 g/mol and 300,000–310,000 g/mol), which remain suitable for further processing. Additionally reported in Figure 2c, $Đ$ values for NR, Control, and CuMA formulations range between 3 and 5, indicating relatively uniform chain lengths.

However, $CuCl_2$ formulations exhibit much broader $Đ$ values between 11 and 23, further supporting the hypothesis of polymer degradation and incompatibility in the absence of antioxidants. Overall, the molecular weights of CuMA and $CuCl_2$ compounds are lower than those of NR and Control compounds, primarily due to the mixing conditions of 100 °C at 100 rpm, and an additional 3 min mixing step required for incorporating the copper(II) inhibitors, which was not performed for the control formulation. DSC analysis was performed to determine the thermal behavior and curing characteristics of the uncured and cured rubber formulations. Dynamic DSC thermograms of uncured Control samples are reported in Figure 3a. Two distinct peaks are revealed: peak A between 115 °C and 120 °C, corresponding to the recrystallization of bismaleimide, and peak B (140–150 °C), associated with the reaction between NR and the cross-linker. As the cross-linker concentration increased from 3.55 to 8.88 phr, peak B shifted to higher temperatures, indicating delayed reaction onset. A curing temperature range of 170–190 °C was selected to accommodate all concentrations. DSC thermograms of CuMA samples reported in Figure 3b exhibited similar dual-peak behavior, with peak A representing bismaleimide crystallization, and peak B indicating NR-cross-linker interaction. The optimal curing temperature was again determined to be 170–190 °C. In contrast, $CuCl_2$ samples reported in Figure 3c showed irregular thermograms with no distinct peaks, suggesting degradation and incompatibility with NR. The absence of antioxidants likely led to oxidative degradation, producing hydrochloric acid and chlorine gas through complex redox reactions. Cured samples were also analyzed via DSC. Control formulations reported in Figure 4a displayed a single peak in the first heating cycle, possibly due to residual cross-linker reactions or bismaleimide copolymerization, as in the second heating cycle, no peak appears in the same region. CuMA samples in Figure 4b showed minor peaks, indicating further cross-linker activity. $CuCl_2$ samples in Figure 4c lacked distinct thermal transitions, reinforcing the conclusion of degradation. Glass transition temperatures, T_g, for all formulations were consistent with unfilled NR (~−65 °C), as reported in Figure 4d. Rheological measurements were employed to assess and confirm the curing behavior and network formation of the rubber compounds at 180 °C. As reported in Figure 5, NR, lacking curatives, exhibited a baseline torque of 1.5 dNm due to physical entanglements rather than chemical cross-links. As shown in Figure 5a, increasing the concentration of cross-linkers in the Control samples led to higher torque values from 3 to 6 dNm. NR/BIS-7.10 demonstrated optimal performance, with torque values nearly matching those of NR/BIS-8.88, suggesting efficient cross-link formation. CuMA formulations reported in Figure 5b exhibited a consistent increase in torque with cross-linker concentration going from 4 to 6.5 dNm. NR/BIS-8.88/CuMA showed the highest torque values, indicating robust and stable cross-link networks. $CuCl_2$ formulations reported in Figure 5c also showed increasing torque with higher cross-linker concentrations, going from 2 to 6 dNm. However, inconsistencies were observed, such as lower torque in NR/BIS-5.33/$CuCl_2$ compared with NR/BIS-3.55/$CuCl_2$, suggesting degradation. These results suggest that CuMA formulations are promising candidates for achieving high mechanical performance, while $CuCl_2$ formulations suffer from degradation and poor network stability. All formulations exhibit short or negligible scorch times, which could potentially hinder mold flow. However, full mold filling was achieved because the compounds were processed under controlled conditions that maintained sufficient mobility before significant network formation occurred. The marching cure behavior of these systems ensures that cross-linking progresses gradually rather than instantaneously; initial viscosity remains low enough for the material to flow and completely fill the mold before torque rises substantially. This dynamic curing characteristic is typical of disulfide-based systems, where bond exchange occurs progressively rather than locking the network immediately. Density measurements, presented in Figure 5d, were conducted to assess the structural integrity and packing efficiency within the rubber matrix. Among the formulations, $CuCl_2$ samples exhibited the highest density values between 0.98 and 1.00 g/mol, which suggests significant degradation. In contrast, CuMA formulations reported the lowest density values between 0.87 and 0.92 g/mol. This reduction may be attributed to the bulkier nature of the CuMA inhibitor, which could hinder tight packing and reduce the overall material density. Control samples reported higher density values compared with unfilled NR ($0.91 \pm 9.1 \times 10^{-4}$ g/mL), suggesting minimal structural disruption and effective cross-linking. Tensile testing was conducted on the cured rubber compounds to evaluate their mechanical performance, as shown in Figure 6. Due to extensive degradation during hot pressing, the $CuCl_2$ samples produced were too brittle to grip for tensile testing, highlighting their poor thermal and chemical stability in the absence of antioxidants. In contrast, the Control formulations reported in Figure 6a demonstrated a clear trend of increasing tensile strength with higher cross-linker concentrations. The maximum tensile stress recorded for Control NR/BIS-8.88 was 7.5 MPa, with a corresponding tensile strain of 11.5. Similarly, CuMA samples in Figure 6b exhibited comparable mechanical behavior, reaching a maximum tensile stress for NR/BIS-8.88/CuMA of 8.4 MPa and tensile strain of 11.1. The tensile strength is directly correlated with the density of cross-links formed during curing. Both Control and CuMA samples showed promising results, although they remain significantly below the tensile strength of conventionally sulfur-vulcanized unfilled NR, which typically reaches ~16 MPa [1]. This reduction in strength suggests that while bismaleimide-based cross-linking is effective, further optimization is required to match commercial compounds. To explore the recyclability of the materials, recycled samples were evaluated using rheology as a preliminary indicator of network retention, as torque values can provide insight into the extent of cross-link rearrangement and degradation upon reprocessing. As reported in Figure 7a, the recycled torque values of the Control samples ranged from 2 to 6 dNm. All samples showed reduced torque values compared with the respective pristine samples, suggesting partial retention of the cross-linked network and potential for limited recyclability. CuMA samples reported in Figure 7b showed better performance upon reprocessing, with torque values ranging from 2.5 to 4.5 dNm. NR/BIS-8.88/CuMA formulation retained the highest torque of 4.6 dNm, suggesting that the disulfide-based cross-links formed in this system are more resistant to breakdown and may allow for limited reprocessability. $CuCl_2$ samples reported in Figure 7c exhibited similar behavior, with reprocessed torque values ranging from 1 to 5.5 dNm.

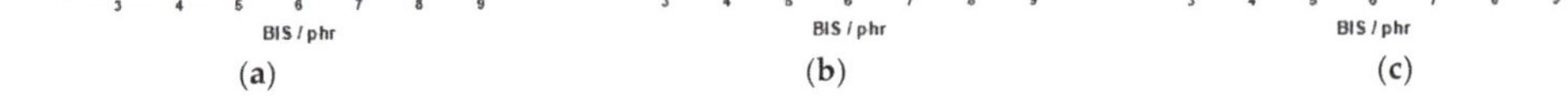

(a) (b) (c)

Figure 2. GPC analysis comparison between NR, Control, CuMA, and $CuCl_2$ formulations: (**a**) $\overline{M}_n$ results, (**b**) $\overline{M}_w$ results, (**c**) $Đ$ results.

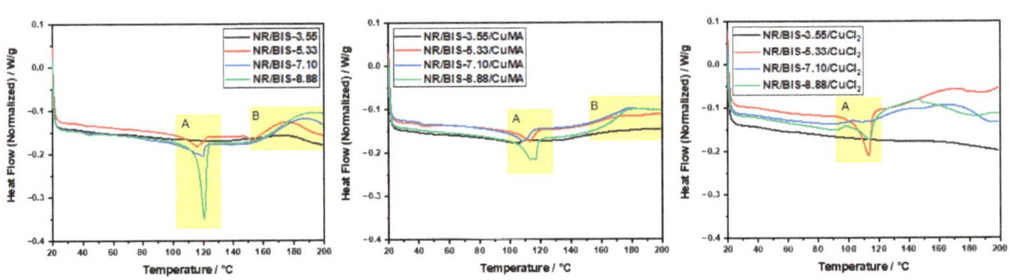

(a) (b) (c)

Figure 3. DSC analysis of uncured samples with peak A representing bismaleimide crystallization, and peak B indicating NR-cross-linker interaction: (**a**) Control, (**b**) CuMA, and (**c**) $CuCl_2$ formulations.

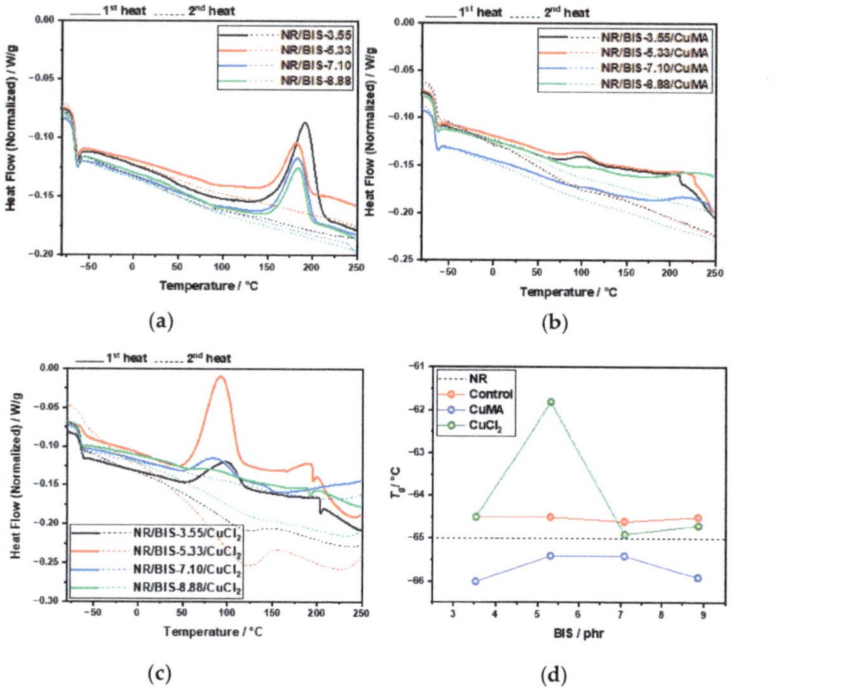

Figure 4. DSC analysis of cured samples: (**a**) Control, (**b**) CuMA, and (**c**) $CuCl_2$ formulations. (**d**) T_g of all samples.

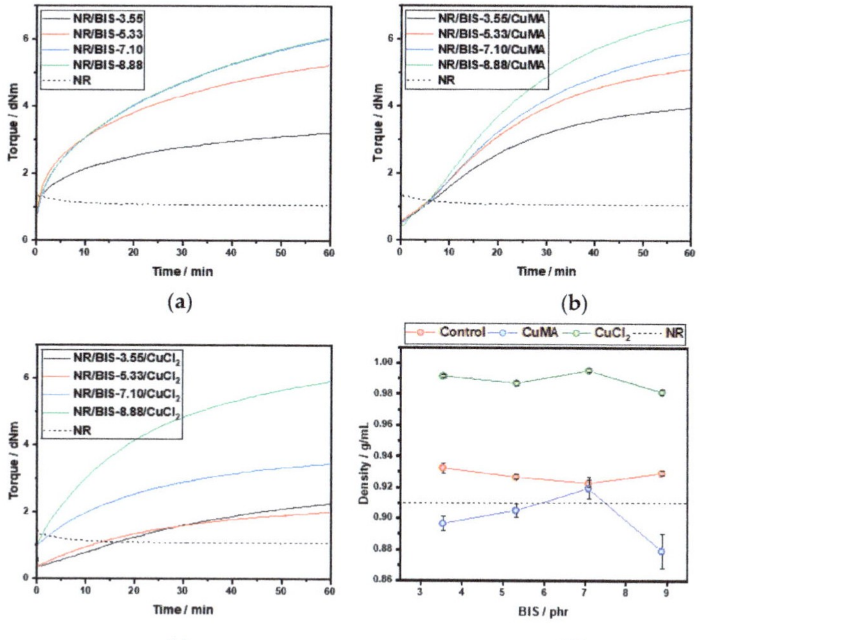

Figure 5. Rheology and density measurements. MDR analysis at 180 °C of (**a**) Control, (**b**) CuMA, and (**c**) $CuCl_2$ formulations. (**d**) Density vs. BIS concentration on the *x*-axis.

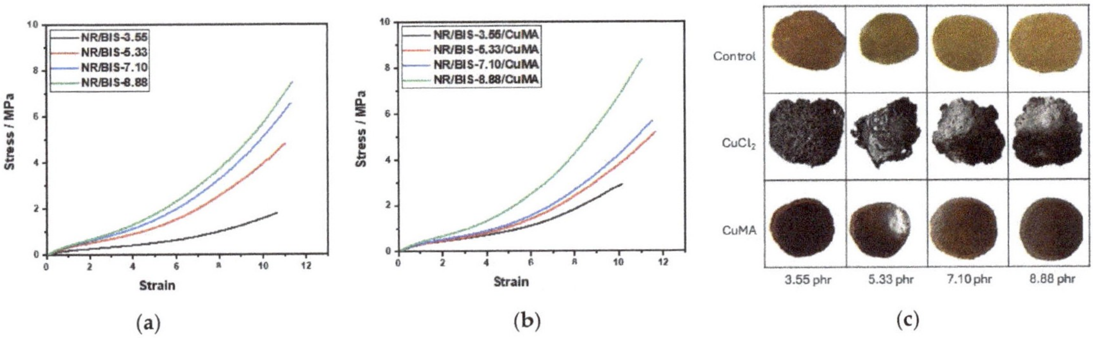

Figure 6. Tensile test analysis of (**a**) Control and (**b**) CuMA formulations. (**c**) Control, $CuCl_2$, and CuMA samples after vulcanization.

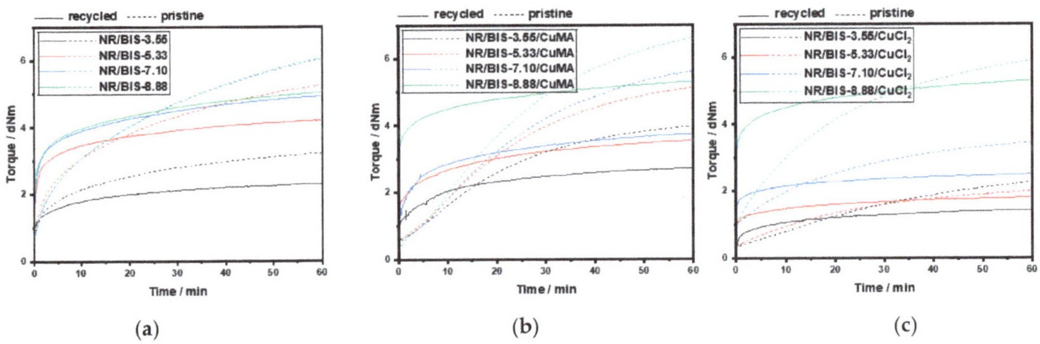

Figure 7. Rheology comparison between pristine and recycled samples: (**a**) Control, (**b**) CuMA, and (**c**) $CuCl_2$.

4. Conclusions

The study demonstrates that rubber formulations are highly sensitive to processing conditions and component compatibility. $CuCl_2$-based compounds exhibit significant main chain scission and degradation, poor thermal stability, and mechanical brittleness, largely due to oxidative reactions in the absence of antioxidants. In contrast, CuMA outperformed $CuCl_2$, likely because the methacrylate ligand provides a more stable coordination environment for copper(II) ions, thereby reducing their propensity to induce uncontrolled oxidation. As a result, CuMA formulations maintain intermediate molecular weights, stable thermal behavior, and promising mechanical properties, including enhanced torque and tensile strength, suggesting effective cross-linking and partial recyclability. Control formulations also perform well, though both CuMA and Control systems fall short of the mechanical strength observed in conventional sulfur-vulcanized NR. While recyclability testing was planned as a subsequent phase of the project, experimental progress ceased before those assessments could be completed. As such, recyclability data are not reported here. However, the observed molecular characteristics and mechanical resilience of CuMA formulations strongly support further investigation into their recyclability. Future studies could explore thermal reprocessing, chemical depolymerization, or mechanical recovery strategies to validate and optimize the recyclability of these materials. Overall, CuMA emerges as a viable alternative for high-performance, recyclable rubber materials, while $CuCl_2$ is unsuitable without further stabilization strategies.

Author Contributions: Conceptualization, A.K. and J.J.C.B.; methodology, A.K.; validation, A.K. and K.A.; formal analysis, A.K. and M.T.; investigation, A.K. and M.T.; resources, A.K. and J.J.C.B.; data curation, A.K. and M.T.; writing—original draft preparation, A.K. and M.T.; writing—review and editing, K.A., B.C., and J.J.C.B.; visualization, A.K. and M.T.; supervision, K.A., B.C., and J.J.C.B.; project administration, B.C. and J.J.C.B.; funding acquisition, B.C., and J.J.C.B. All authors have read and agreed to the published version of the manuscript.

Funding: This work was supported by the Engineering and Physical Sciences Research Council (EPSRC) grant EP/W018977/1.

Data Availability Statement: The original data presented in the study are openly available in the Queen Mary Research Online repository at https://qmro.qmul.ac.uk/xmlui/handle/123456789/114491 (accessed on 5 December 2025).

Conflicts of Interest: The authors declare no conflicts of interest.

References

1. Kaur, A.; Fefar, M.M.; Griggs, T.; Akutagawa, K.; Chen, B.; Busfield, J.J.C. Recyclable sulfur cured natural rubber with controlled disulfide metathesis. *Commun. Mater.* **2024**, *5*, 212. [CrossRef]
2. Markl, E.; Lackner, M. Devulcanization Technologies for Recycling of Tire-Derived Rubber: A Review. *Materials* **2020**, *13*, 1246. [CrossRef] [PubMed]
3. Coran, A.Y. Chemistry of the vulcanization and protection of elastomers: A review of the achievements. *J. Appl. Polym. Sci.* **2003**, *87*, 24–30. [CrossRef]
4. Nevejans, S.; Ballard, N.; Miranda, J.I.; Reck, B.; Asua, J.M. The underlying mechanisms for self-healing of poly(disulfide)s. *Phys. Chem. Chem. Phys.* **2016**, *18*, 27577–27583. [CrossRef] [PubMed]
5. Rekondo, A.; Martin, R.; Ruiz de Luzuriaga, A.; Cabañero, G.; Grande, H.J.; Odriozola, I. Catalyst-free room-temperature self-healing elastomers based on aromatic disulfide metathesis. *Mater. Horiz.* **2014**, *1*, 237–240. [CrossRef]
6. Kaur, A.; Gautrot, J.E.; Cavalli, G.; Watson, D.; Bickley, A.; Akutagawa, K.; Busfield, J.J.C. Novel Crosslinking System for Poly-Chloroprene Rubber to Enable Recyclability and Introduce Self-Healing. *Polymers* **2021**, *13*, 3347. [CrossRef] [PubMed]
7. Imbernon, L.; Oikonomou, E.K.; Norvez, S.; Leibler, L. Chemically crosslinked yet reprocessable epoxidized natural rubber via thermo-activated disulfide rearrangements. *Polym. Chem.* **2015**, *6*, 4271–4278. [CrossRef]
8. Xiang, H.P.; Qian, H.J.; Lu, Z.Y.; Rong, M.Z.; Zhang, M.Q. Crack healing and reclaiming of vulcanized rubber by triggering the rearrangement of inherent sulfur crosslinked networks. *Green Chem.* **2015**, *17*, 4315–4325. [CrossRef]
9. Xiang, H.P.; Rong, M.Z.; Zhang, M.Q. Self-healing, Reshaping, and Recycling of Vulcanized Chloroprene Rubber: A Case Study of Multitask Cyclic Utilization of Cross-linked Polymer. *ACS Sustain. Chem. Eng.* **2016**, *4*, 2715–2724. [CrossRef]
10. Kong, Y.; Chen, X.; Li, Z.; Li, G.; Huang, Y. Evolution of crosslinking structure in vulcanized natural rubber during thermal aging in the presence of a constant compressive stress. *Polym. Degrad. Stab.* **2023**, *217*, 110513. [CrossRef]
11. Wang, M.; Wang, R.; Chen, X.; Kong, Y.; Huang, Y.; Lv, Y.; Li, G. Effect of non-rubber components on the crosslinking structure and thermo-oxidative degradation of natural rubber. *Polym. Degrad. Stab.* **2022**, *196*, 109845. [CrossRef]
12. Kitagawa, S.; Ozawa, M.; Shibata, M. Self-healing and reprocessable bismaleimide-diamine thermosets containing disulfide linkages. *Polymer* **2023**, *278*, 126008. [CrossRef]
13. Sim, X.M.; Wang, C.-G.; Liu, X.; Goto, A. Multistimuli Responsive Reversible Cross-Linking–Decross-Linking of Concentrated Polymer Brushes. *ACS Appl. Mater. Interfaces* **2020**, *12*, 28711–28719. [CrossRef] [PubMed]
14. Polgar, L.M.; Cerpentier, R.R.J.; Vermeij, G.H.; Picchioni, F.; Duin, M.v. Influence of the chemical structure of cross-linking agents on properties of thermally reversible networks. *Pure Appl. Chem.* **2016**, *88*, 1103–1116. [CrossRef]
15. Gevrek, T.N.; Sanyal, A. Furan-containing polymeric Materials: Harnessing the Diels-Alder chemistry for biomedical applications. *Eur. Polym. J.* **2021**, *153*, 110514. [CrossRef]
16. Post, C.; van den Tempel, P.; Herrera Sánchez, P.; Maniar, D.; Bose, R.K.; Voet, V.S.D.; Folkersma, R.; Picchioni, F.; Loos, K. Thermoreversible Diels–Alder Cross-Linking of BHMF-Based Polyesters: Synthesis, Characterization and Rheology. *ACS Sustain. Chem. Eng.* **2025**, *13*, 3543–3553. [CrossRef] [PubMed]
17. Hayeemasae, N.; Sensem, Z.; Sahakaro, K.; Ismail, H. Maleated Natural Rubber/Halloysite Nanotubes Composites. *Processes* **2020**, *8*, 286. [CrossRef]

18. Ujianto, O.; Noviyanti, R.; Wijaya, R.; Ramadhoni, B. Effect of maleated natural rubber on tensile strength and compatibility of natural rubber/coconut coir composite. *IOP Conf. Ser. Mater. Sci. Eng.* **2017**, *223*, 012014. [CrossRef]
19. *ISO 37:2024(en)*; Rubber, Vulcanized or Thermoplastic—Determination of Tensile Stress-strain Properties. International Organization for Standardization: Geneva, Switzerland, 2024.

polymers

MDPI

Article

Analysis of Tyre Pyrolysis Oil as Potential Diesel Fuel Blend with Focus on Swelling Behaviour of Nitrile-Butadiene Rubber

Steffen Seitz [1,2,*], Tobias Förster [2] and Sebastian Eibl [2]

[1] Institute of Mechanics, Faculty of Aerospace Engineering, University of the Bundeswehr Munich, Werner-Heisenberg-Weg 39, 85579 Neubiberg, Germany

[2] Bundeswehr Research Institute for Materials, Fuels and Lubricants (WIWeB), Institutsweg 1, 85435 Erding, Germany; tobias1foerster@bundeswehr.org (T.F.); sebastianeibl@bundeswehr.org (S.E.)

* Correspondence: steffen.seitz@unibw.de

Abstract

This study examines the swelling behaviour of nitrile-butadiene rubber (NBR) when interacting with tyre pyrolysis oils (TPO), with a focus on the chemical composition of TPO and their interaction with rubber matrices. Initially, a comparative analysis with conventional diesel fuel (DF) was performed using advanced analytical techniques, including two-dimensional gas chromatography coupled to mass spectrometry (2D-GC/MS), infrared (IR) spectroscopy, and nuclear magnetic resonance (^{1}H-NMR) spectroscopy. The analysis revealed that TPO contains a significantly higher proportion of aromatic hydrocarbons than DF, along with unsaturated and oxygen-containing compounds not present in DF. Based on these compositional differences, blends of TPO and DF were formulated and evaluated for their suitability as liquid energy carriers according to the specifications of DF. In principle, blends with an addition of up to 5 vol% TPO in DF are technically suitable for use as fuel. Subsequently, the sorption behaviour of TPO, DF, and their blends in NBR was investigated. The swelling potential was determined based on mass, density, and volume, and the changes in the hardness and tensile strength of NBR were recorded. The results demonstrate that TPO induces pronounced swelling in NBR, as evidenced by a marked increase in mass uptake and volume expansion. A linear increase was observed between the degree of swelling and the increasing TPO content in the blends. Mechanical property assessments revealed a corresponding decrease in the hardness and tensile strength of NBR upon exposure to TPO, with the most severe effects associated with neat TPO. This work provides a comprehensive assessment of TPO as a potential blend component for DF. It highlights the need for careful consideration of material compatibility in practical applications.

Keywords: nitrile-butadiene rubber; tyre pyrolysis oil; chemical analysis; swelling behaviour; mechanical testing

1. Introduction

Around 3.4 million tons of tyre waste are generated across Europe every year. Processes for recycling tyres include combustion, which is predominantly used in the cement industry and material reuse, for which the plastic components have to be separated. Due to the complex composition of tyres, the challenge is to achieve complete separation of the components. The composition of tyres varies depending on their application. According to manufacturers, tyres typically consist of 41% rubbers, 30% fillers, 15% steel and textiles,

6% plasticizers, 6% vulcanization reagents, and 2% antioxidants [1]. The pyrolysis process is making the complete recycling of used tyres increasingly important. Textiles and steel can be separated from the rubber and recycled to produce new tyres. During pyrolysis, polymer chains in the tyre granulate are depolymerized under inert conditions to produce gas, oil, and coke residues [2–4]. The quantity of pyrolysis products depends on the type of tyres used and the process parameters [5–7]. T. Williams has evaluated various reactor types that were tested to preferably gain tyre pyrolysis oil (TPO) [8]. Pyrolysis gas is mostly used as an energy source and the solid residue can be reprocessed as recycled carbon black in tyres [3,9].

The individual pyrolysis mechanisms have been described by several authors for the elastomer types of natural rubber (NR), styrene-butadiene rubber (SBR) and butadiene rubber (BR) [10–15]. Due to the complexity of the tyre composition, further studies were carried out for corresponding elastomer mixtures [16–20]. In general, the polymer chains undergo depolymerization, disproportionation reactions or β-cleavages and are pyrolyzed into smaller molecules such as monomers, dimers, and oligomers. For example, NR predominantly forms isoprene and dipentene and BR produces butadiene, vinyl cyclohexene, and dipentene [10,15,21]. Diels–Alder reactions, addition reactions, cyclisation, aromatisation, and intramolecular hydrogen transfer convert the primary products into benzene, indene, and naphthalene with alkyl and alkenyl substituents [11,15,19]. SBR pyrolyzes mainly to styrene and methyl styrene and reacts via Diels–Alder reaction to alkylbenzenes or diaromatics like biphenyl, and with the BR and NR products to alkyl naphthalenes [12–14,16]. The oil component can be used as a fuel [22,23]. TPO is regarded as a possible blend for diesel fuel (DF) due to its similar composition [24]. Pyrolysis thus contributes to more sustainable tyre recycling and the conservation of fossil resources.

When TPO is used as a liquid energy carrier in civil or military transport vehicles, it is necessary to ensure operational availability and that TPO is compatible with the system. Compatibility with the elastomers must hence be analyzed. The resistance of sealing materials depends on the chemical composition of the liquid medium [25,26]. It is, therefore, crucial to determine the chemical composition of TPO when estimating its diffusion behaviour. This can be observed through swelling, extraction of additives, and changes in the mechanical properties of the elastomer [27–29]. Nitrile-butadiene rubber (NBR) is an elastomer widely utilized in the automotive and aerospace industries for applications such as seals, hoses, and tank linings. Its utilization is primarily attributed to its excellent resistance to non-polar liquids, including fuels, as well as its favourable mechanical and ageing properties [30]. With an increasing acrylonitrile content in the polymer, the polarity of the NBR increases, leading to better resistance to non-polar liquids [31,32].

This study aims to investigate the potential use of TPO as a replacement or blend for DF and to understand TPO's diffusion behaviour in the established sealing material NBR. On the one hand, the focus is on comparing the chemical composition of DF and TPO and determining the characteristic properties of TPO-DF blends according to DIN EN590 [33]. On the other hand, the work presents compatibility tests of the liquids with NBR with different acrylonitrile content to estimate its swelling potential and resulting influence on mechanical properties. In addition, the temperature dependence of the diffusion process of neat TPO and DF is investigated. The results contribute to an understanding of the sorption behaviour of TPO in contact with NBR and promote its potential use as a liquid energy source.

2. Materials

Polymer Service Merseburg (Merseburg, Germany) produced sulphur cross-linked nitrile-butadiene rubber (NBR) samples with an acrylonitrile content of 18 wt%, 28 wt%, and 39 wt% in the NB-copolymer. In this study, these are abbreviated as NBR18, NBR28, and NBR39. The exact formulations are shown in Table 1. The NBR samples contained additives, including DPHP and 6-PPD, which served as plasticizers and antioxidants. The cross-linking of the elastomer required zinc oxide, stearic acid, and sulphur, while N-Cyclohexyl-2-benzothiazolylsulfenamide (CBS) and Tetramethyl thiuram monosulfide (TMTM-80) acted as accelerators for the vulcanisation process. S2 Tension rods were punched out of the NBR sheets [34]. These were split to a thickness of 1 mm using a band knife splitting machine NAF470 from Fortuna GmbH (Weil der Stadt, Germany).

Table 1. Formulation of the three NBR elastomers with the specification in parts-per-hundred rubber (phr).

Ingredient	Concentration/phr		
Acrylonitrile content/wt%	18	28	39
Perbunan 1846	100		
Perbunan 2846		100	
Perbunan 3946			100
Carbon black (Typ N550)		60	
Di(2-propylheptyl) phthalate (DPHP)		20	
N-(4-Methylpentan-2-yl)-N-phenylbenzene-1,4-diamine (6-PPD)		2	
Zinc oxide		5	
Stearic acid		1	
Sulphur		2	
N-Cyclohexyl-2-benzothiazolylsulfenamide (CBS)		1.5	
Tetramethyl thiuram monosulfide (TMTM-80)		0.5	

To produce tyre pyrolysis oil (TPO), Pyrum Innovations AG from Dillingen/Saar (Germany) provided thermolysis oil from a mixture of truck, car, and bicycle tyres. The TPO was vacuum filtered for further use. Blends with the pyrolysis oil were created using a commercially available diesel fuel (DF) according to DIN EN 590. The mixing ratio was 1 vol%, 5 vol%, and 10 vol% of TPO in DF, abbreviated as DF99TPO1, DF95TPO5, and DF90TPO10.

3. Experiment

3.1. Analysis of Fuels

The liquids were tested using the A-test according to DIN EN 590 (2022-05) for DF [33]. The individual test methods and the standards used are shown in Table 2. In addition, the total acid number (TAN) was determined according to ASTM D664 [35]. A potentiometric titration with potassium hydroxide (KOH) was used for this determination. TAN represents the mass of KOH, in milligrams, required to neutralize one gram of the sample. Using the concentration (c) of the KOH solution, the sample mass (m_S), and the titration volumes of the blank (V_B) and the sample (V_S), the TAN can be calculated according to Equation (1).

$$\text{TAN} = \frac{56.1 \times c(V_S - V_B)}{m_S} \tag{1}$$

The iodine value is calculated in accordance with DIN EN ISO 3961 [36]. In this method, iodine chloride reacts with carbon–carbon double bonds, followed by the addition of potassium iodide to release iodine, which is subsequently titrated with sodium thiosulfate. The resulting iodine number is obtained from Equation (2), which relates the concentration of the sodium thiosulfate solution (c), the titration volume of the blank (V_B) and for the samples (V_S), and the mass of fuel sample (m_s).

$$\text{Iodine value} = \frac{12.69 \times c(V_B - V_S)}{m_S} \quad (2)$$

Table 2. Selected parameters from DIN EN 590 with standards used.

Parameter	Norm
Viscosity Density	ASTM D7042-21a [37]
Polyaromatic content	DIN EN 12916 [38]
Flash point	DIN EN ISO 2719 [39]
HFRR [1]-test	DIN EN ISO 12156-1 [40]
CFPP [2]	DIN EN 116 [41]
Cupper corrosion	DIN EN ISO 2160 [42]
Oxidation stability	DIN EN 16091 [43]
Water content	DIN 51777-2 [44]
Carbon residue	DIN EN ISO 10370 [45]
Ash	DIN EN ISO 6245 [46]
Sulphur content	DIN EN ISO 20884 [47]
Distillation process	DIN EN ISO 3405 [48]

[1] HFRR (high frequency reciprocating rig); [2] CFPP (cold filter plugging point).

Two-dimensional gas chromatography coupled to mass spectrometry (2D-GC/MS) analyses were performed with a Shimadzu GC-2010 Plus gas chromatograph equipped with a ZOEX ZX1 modulator and a Shimadzu GCMS-QP2010 Ultra mass spectrometer operated in electron ionization mode. The column combination used was a 30 m Restek Txi-1ms (0.25 mm; 0.25 μm film) with 3 m SGE Analytical Science BPX 50 (0.15 mm; 0.15 μm film). Measurements were performed in 1:10 split mode with helium as a carrier gas and a modulation time of 8 s. The oven temperature was maintained at 50 °C for 3 min, before heating to 320 °C at a constant heating rate of 5 °C min^{-1}, where it was held for 2 min.

In addition, 10 mL TPO was added to a HyperSep silica stationary solid phase extraction (SPE) column from Thermo Scientific and washed with 20 mL heptane and 10 mL acetone. The polar acetone fraction was analyzed by GC/MS using an HP5 column at 50 °C for 2 min and a 10 °C min^{-1} heated to 300 °C and held for 2 min.

^{1}H-Nuclear magnetic resonance (^{1}H-NMR) analyses were performed with a Spinsolve 80 Carbon from Magritek in Powerscan mode at 80 MHz and analyzed with MestReNova software. Octamethyl silane was added to TPO as reference.

Fourier transform infrared (FTIR) spectra were recorded with a Bruker Tensor 27 spectrometer, which was equipped with a 50 μm zinc selenid cuvette at a spectral resolution of 4 cm^{-1}. The spectra were averaged over 32 scans in a wavenumber range between 4000 cm^{-1} and 500 cm^{-1}. The spectra were baseline corrected and normalized to the band intensity at 2725 cm^{-1}.

3.2. Sorption Experiments

For preparation, the volatile components such as plasticizer DPHP and antioxidant 6-PPD were extracted from the NBR samples in a Soxhlet apparatus for 16 h with acetone. For the sorption tests, the NBR samples were stored in the liquids for 0.5 h, 1 h, 7 h, 24 h, 72 h, and 168 h. The NBR28 and NBR39 samples were additionally stored for 672 h and 2016 h and NBR39 for 4032 h. Additionally NBR18 were kept in TPO and DF at 50 °C and

80 °C. The relative mass change Δm_{grav} was calculated using Equation (3), where m_t is the mass after a certain storage time, and m_0 is the initial mass.

$$\Delta m_{grav} = \frac{(m_t - m_o)}{m_0} * 100\,\text{wt\%} \quad (3)$$

Using Fick's second diffusion law and the solution for one-dimensional diffusion, Equation (4) can be derived [49]. This Equation uses a sorption curve to determine the specific relative mass changes. The mean diffusion coefficient D can be determined by calculating the mass after specific time intervals m_t, the thickness of the NBR samples h (1 mm), the equilibrium mass m_∞ of the samples and the number of iteration steps n.

$$\frac{\Delta m_t}{\Delta m_\infty} = 1 - \frac{8}{\pi^2}\sum_{n=0}^{\infty}\left(\frac{1}{(2n+1)^2}exp\left(\frac{-(2n+1)^2\pi^2 Dt}{h^2}\right)\right) \quad (4)$$

After storage, the samples were immersed in 40/60 gasoline to remove any fuel adhering to the outside and blotted with a paper towel. To determine the density, the samples were weighed in water using the VF 4601 kit from Satorius (Göttingen, Germany) and calculated according to Archimedes' principle using Equation (5). The volume change in the NBR sample can be determined using Equation (6) with mass and density.

$$\rho_{NBR} = \frac{m_{NBR,Air} * \rho_{Water}}{m_{NBR,Air} - m_{NBR,Water}} \quad (5)$$

$$\Delta V = \frac{(V_t - V_o)}{V_0} * 100\,\text{vol\%} \quad (6)$$

A Bareiss® digi test II testing machine (Oberdischingen, Germany) was used to determine the micro-Shore-A hardness of the NBR according to DIN ISO 48-4:2021-02 [50]. Tensile tests according to DIN 53504 were carried out to determine the elongation at break and tensile strength using a Zwick A universal testing machine from ZwickRoell GmbH (Ulm, Germany) with a 500 N load cell and optical extensometer with an elongation rate of 0.167 s^{-1} [34].

4. Results

4.1. Comparison of the Chemical Composition of DF and TPO

4.1.1. GC/MS Analysis

Figure 1 presents the two-dimensional gas chromatogram (2D-GC/MS) of DF (Figure 1A) and TPO (Figure 1B). For a semi-quantitative analysis, the areas of the individual peaks from the 2D GC-MS chromatogram were integrated and assigned to a substance class based on the mass spectra. The results of this classification are summarized in Table 3. In region 1 TPO shows a homologous series of alkanes ranging from C_8 to C_{34}, along with the corresponding mono- and diunsaturated alkenes. In contrast, DF contains a broader range of alkanes (C_6 to C_{36}) with higher degree of structural variability in the branches and higher relative signal intensity (74 area%) compared to TPO (8 area%). Region 2 of the chromatogram reveals cyclic alkanes with methyl and ethyl substituents in DF (9 area%), whereas TPO (7 area%) is characterized by an intense peak of substituted cyclohexene (1-methyl-4-(1-methylethyl)-cyclohexene). Toluene is detectable in TPO at low retention time. Monoaromatic hydrocarbons are detected in region 3 for both DF and TPO. They consist of mono-, di-, tri-, or tetra-substituted methyl- or ethylbenzenes. Prominent examples include 1,3-diemehtylbenzene, 1-ethyl-2-methylbenzene, mesitylene, 1-methyl-4-(1-methylethyl)-benzene as well as with linear and/or branched C_3- to C_8-alkylbenzenes such as pentylbenzene. Similar substituents as in the aromatics are identified in region 4 for the various indenes (e.g., 1-methyl-1H-inden; 1,2,3-trimethylindene).

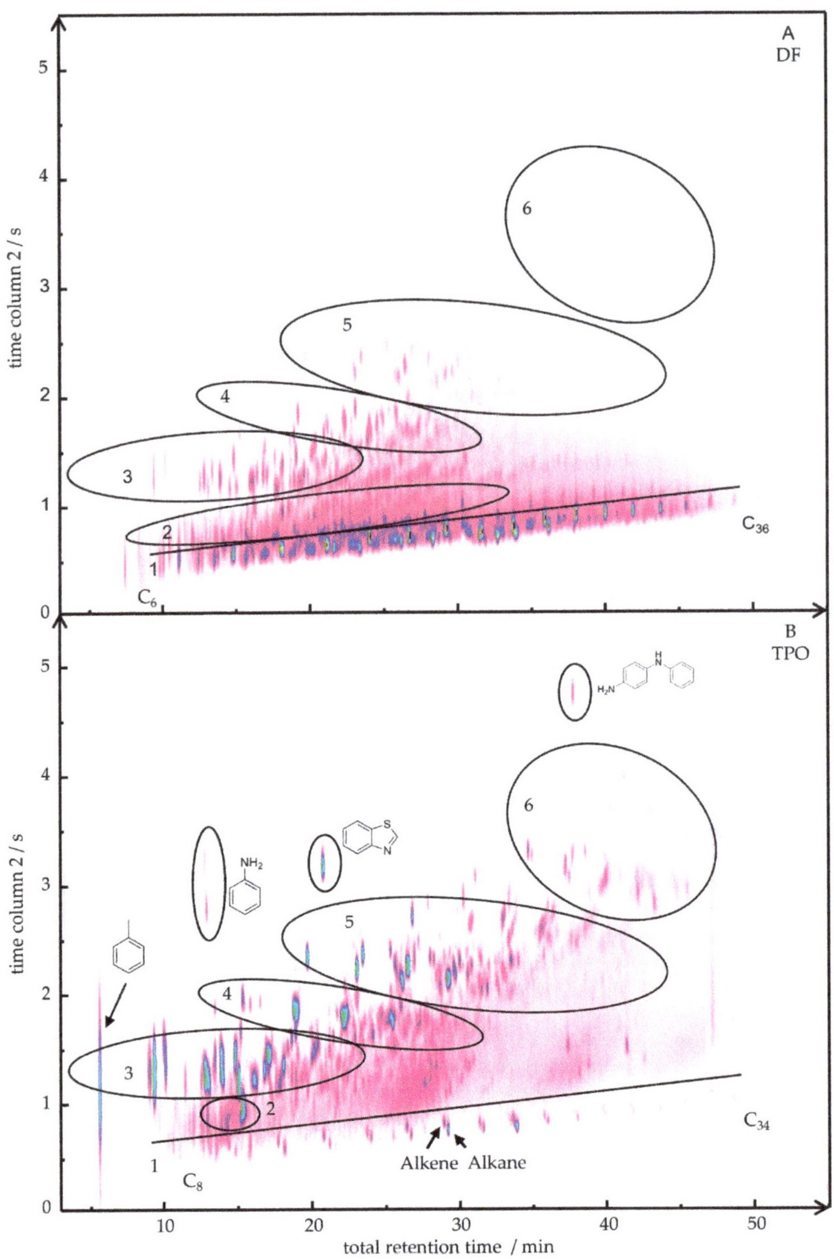

Figure 1. 2D-chromatograms of DF (**A**) and TPO (**B**) with marked areas of substance classes: 1 Alkanes and Alkenes; 2 Cyclic alkanes and alkenes, 3 Monoaromatics, 4 Indenes, 5 Diaromatics, 6 Polyaromatics.

Table 3. Intensity percentage in area% of the substance classes for DF and TPO.

Range	Substance Class	Intensity Percentage Area%	
		DF	TPO
1	Alkanes/Alkenes	74	8
2	Cycloalkanes	9	7
3	Aromatics	11	40
4	Indenes	4	12
5	Diaromatics	2	22
6	Polyaromatics	-	5
	Others *	≤1	6

* nitrogen-, oxygen-, and sulphur-containing compounds.

Naphthalenes substituted with mono-, di-, or trimethyl groups (e.g., 1-methylnaphthalene, 1,6-dimethylnaphthalene, 2,3,6-trimethylnaphthalene) and dihydronaphthalene (e.g., 1,2-dihydro-6-methylnaphthalene) can be seen in region 5. In addition, other diaromatic hydrocarbons including, e.g., biphenyl, fluorene, and chamazulene are detected in TPO. The most notable contrast in region 3 to 5 lies in the markedly higher peak intensity of the aromatic compounds in TPO. Polyaromatic hydrocarbons (PAHs) such as anthracene, pyrene, and retene, which are only present in TPO (5 area%), can be seen in range 6. Furthermore, TPO contains nitrogen-, oxygen-, and sulphur-containing compounds (6 area%) with the most intense occurrences of aniline, benzothiazole, and N-phenyl-1,4-benzenediamine, as illustrated in Figure 1.

The polar substances in the TPO were enriched by SPE and analyzed by GC/MS. Nitrogen-containing substances were detected, such as amines, with cyclic and linear hydrocarbons, and anilines, such as diphenyl diamines. Various oxygen-containing substances, such as phenol or stearic acid, palmitic acid, and myristic acid, could be detected in the TPO. The presence of heteroatoms can result from the additives used during tyre manufacturing. These include, for example, vulcanization agents (such as thiazoles, organic peroxides, nitro compounds), antioxidants (amines, phenols), plasticizers (aromatic esters), or processing aids (such as mineral oils and peptizers) [51]. In addition, during tyre usage, oxygen can diffuse into the material, initiating ageing processes such as oxidation. As a result, oxygen-containing compounds may form during pyrolysis [52].

In summary, significant differences in the chemical composition of TPO and DF were detected. DF showed a higher content and variability of linear and branched alkanes and cycloalkanes. TPO indicated linear alkanes and branched alkenes, and the corresponding content of monoaromatic and diaromatic compounds was much higher. Additionally, TPO showed PAHs and nitrogen-, oxygen- and sulphur-containing compounds, which were not detected in DF.

4.1.2. ^{1}H-NMR-Spetroscopy

Figure 2 presents the ^{1}H-NMR spectra of TPO and DF. The application of this technique facilitates the comparison of functional group distributions in complex liquid mixtures through the quantification of hydrogen environments. The majority of hydrogen atoms in DF and TPO are associated with aliphatic hydrocarbon chains with 91% for DF and 56% for TPO (0.5–3.3 ppm) [53–56]. These signals can be further separated and attributed to CH_3– (0.5–1 ppm), CH_2– (1–1.5 ppm) and CH–groups (1.5–2.0 ppm). Hydrogen in carbon chains adjacent to aromatic rings, like α-methyl carbons, appears between 2.0 ppm and 3.3 ppm and comprises 6% in DF and 21% in TPO [53,57]. A minor signal in TPO between 3.3 ppm and 4.5 ppm indicates a small amount (1%) of hydroxyl groups [54]. TPO also exhibits a signal (4.5 ppm and 6.5 ppm), representing hydrogen atoms bound to unsaturated hydrocarbons or phenolic structures, with a relative intensity of 6% [54,57]. Hydrogen atoms directly attached to monoaromatic rings can be located in the range of 6.5–7.25 ppm, with 3% in DF and more prominently with 13% in TPO. Signals between 7.25 ppm and 9.0 ppm, corresponding to polyaromatic structures, are observed in TPO (3%), as described by Rodríguez et al. [58,59]. The results differ from the 2D-GC/MS analysis, which detected polyaromatic compounds for TPO and Diaromatics for DF.

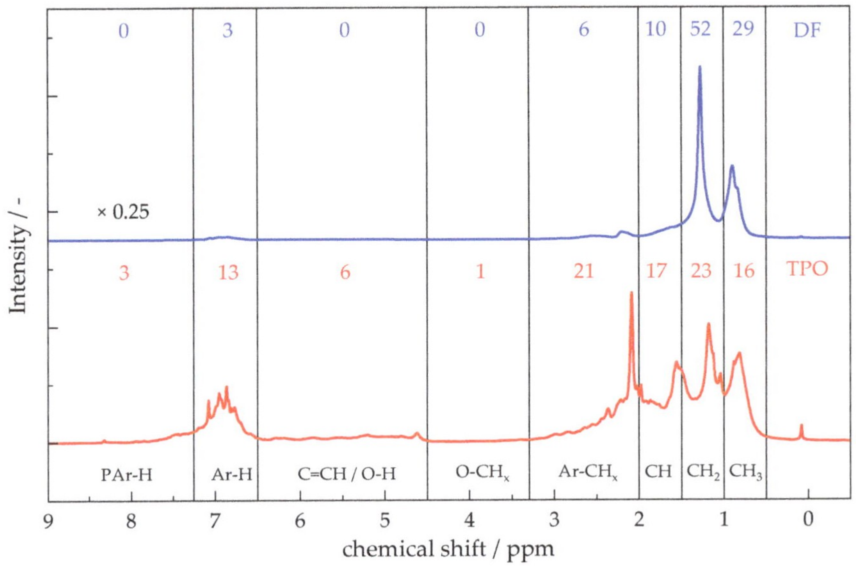

Figure 2. ^{1}H-NMR spectra of TPO an DF with the corresponding percent of hydrogen atoms in the defined areas. 0.5–1 ppm "CH_3 aliphatic"; 1–1.5 ppm "CH_2 aliphatic; 1.5–2 ppm, "CH aliphatic"; 2.0–3.3 ppm "aliphatic bonded to aromatic; 3.3–4.5 ppm "aliphatic bonded to oxygen"; 4.5–6.5 ppm "olefins and hydroxyl"; 6.5–7.25 ppm "monoaromatic"; 7.25–9 ppm "polyaromatic".

TPO exhibits a lower proportion of hydrogen in aliphatic carbon chains compared to DF, but a higher proportion of hydrogen atoms directly bound to aromatic rings. The proportion of carbon chains bound to an aromatic ring is also significantly higher in TPO. The presence of polyaromatic and unsaturated structures in TPO, which are absent in DF, highlights the considerable differences in molecular composition between the two substances. In summary, the results correspond to the 2D-GC/MS results.

4.1.3. ATR-FTIR-Spectroscopy

Figure 3 presents the ATR-FTIR spectra of TPO and DF. For both fluids the C-H valence vibrations of alkanes are observed between 3000 cm^{-1} and approx. 2700 cm^{-1}, and the corresponding deformation vibrations for CH_2 and CH_3 are at 1460 cm^{-1} and 1385 cm^{-1}, respectively [60,61]. A mixture of iso-alkanes can be assigned in TPO and DF at 1169 cm^{-1}, 1154 cm^{-1} and the in-plane bending at 960 cm^{-1} and 906 cm^{-1} [60–63]. DF shows n-alkanes with a band at 722 cm^{-1} [60–63]. TPO indicates a high intensity of tert-butyl groups with absorption bands at 1244 cm^{-1} and 1206 cm^{-1}, which are weakly expressed in DF [61,62]. A high intensity of cyclohexene in TPO can be characterized by a band at 890 cm^{-1} and aligns with the presence of 1-methyl-4-(1-methylethyl)-cyclohexene detected in 2D-GC/MS analysis. Aromatic compounds can be identified in TPO and DF with valence and deformation vibrations at 3020 cm^{-1} and 1605 cm^{-1} [60,61]. Monoaromatics such as alkyl benzenes can be assigned to the bands at 700 cm^{-1} and 740 cm^{-1}.

Di- and tri-substituted aromatics can be assigned to 770 cm^{-1}, 806 cm^{-1}, and 815 cm^{-1} [63]. Di-aromatics can be identified by the intermediate bands at 780 cm^{-1}. The ring vibration of the aromatics in TPO can be associated with the two split bands at 1515 cm^{-1} and 1494 cm^{-1} [61,62]. The intensity of the aromatics bands, especially of di- and triaromatics, is considerably higher for TPO than for DF, indicating a higher concentration of the aromatic compounds, agreeing with the 2D-GC/MS and ^{1}H-NMR results. Alkenes are also detected in the TPO spectrum, as indicated by C–H stretching at 3095 cm^{-1} and C=C stretching at 1640 cm^{-1} [61,62], whereas these features are absent in DF. These suggest that there are considerably less unsaturated compounds in DF than in TPO. These results confirm the findings from the 2D-GC/MS chromatograms and ^{1}H-NMR-spectra. In addition, various oxygen-containing functional groups are evident in the TPO spectra. Carboxylic acids are indicated by a broad O-H band at 3395 cm^{-1} and a raised background down to 2800 cm^{-1}. The presence of carboxylic acids can be confirmed by the C-O region between 990 cm^{-1} and 1120 cm^{-1} and the characteristic C=O band (1710 cm^{-1}). However, these bands can also originate from other compounds like ethers, alcohols, carboxylic acids, carboxylate, and ketones [61,62,64]. In summary, the ATR-FTIR spectra clearly demonstrate that TPO contains significantly higher concentrations of aromatic and unsaturated compounds, as well as oxygenated species, which are largely absent in DF. These spectral differences underscore the distinct chemical compositions of the two substances.

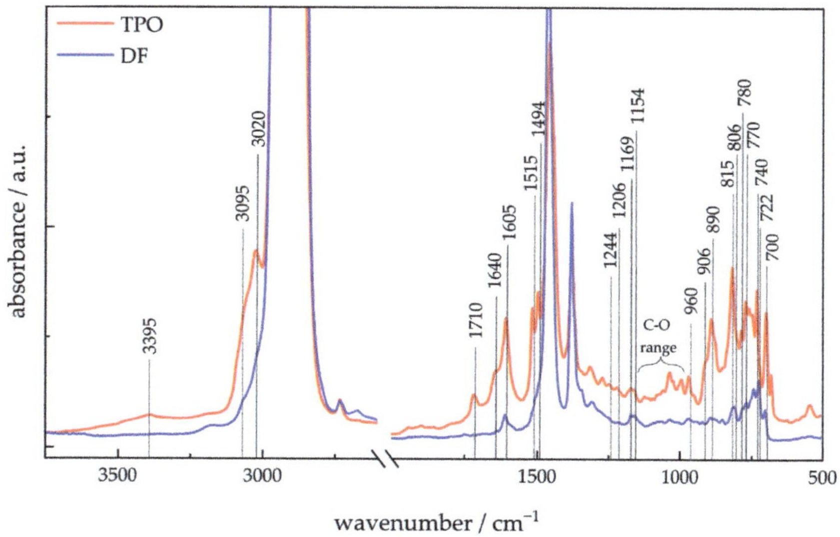

Figure 3. FTIR spectra of TPO and DF with indication of important bands.

4.1.4. Further Chemical Analysis

The total acid number (TAN) (see Table 4) corresponds to the amount of potassium hydroxide (KOH) required to neutralize the acids present in one gram of fuel. DF exhibits an acid number of 0 mg g^{-1}, indicating the absence of detectable acid compounds. With addition of TPO in DF, the TAN rises to 0.05 mg g^{-1} for DF99TPO1, 0.14 mg g^{-1} for DF95TPO5, 0.23 mg g^{-1} for DF90TPO10, and reaches a maximum of 2.08 mg g^{-1} for neat TPO. These findings corroborate the assumption derived especially from FTIR-spectroscopy, that acidic compounds are present in TPO. The iodine value serves as an indicator of the concentration of unsaturated compounds in olefines in a sample. The iodine value is calculated in accordance with DIN EN ISO 3961. The results show that the highest iodine value of 93 g 100 g^{-1} is observed for neat TPO, indicating a significant presence of unsaturated hydrocarbons. DF contains only a minimal fraction of unsaturated compounds, with an iodine value of 2.9 g 100 g^{-1}. This value increases slightly with the addition of TPO: 3.1 g 100 g^{-1} for DF99TPO1, 3.7 g 100 g^{-1} for DF95TPO5, and 14 g 100 g^{-1} for DF90TPO10. These findings are consistent with the results obtained from ^{1}H-NMR and FTIR-spectroscopy, confirming the presence of olefinic compounds in TPO. The aromatic content of TPO was quantified using High-Performance Liquid Chromatography (HPLC), according to DIN EN 12916. For this analysis, blends of DF and TPO were prepared. Based on linear regression (R^2 = 0.9989) of the data presented for DF, DF99TPO1, DF95TPO5, and DF90TPO10 in Table 4, the total aromatic content of TPO was determined to be ~60 vol%. This result corroborates the findings of the current analyses and agrees with previously published studies reporting a high aromatic content in TPO [7,16,17,65,66].

Table 4. Total acid number, iodine value, and aromatic content of DF, DF99TPO1, DF95TPO5, DF90TPO10, and TPO.

Characteristic	Units	DF	DF99TPO1	DF95TPO5	DF90TPO10	TPO
Total acid number	mg g^{-1}	0	0.05	0.14	0.23	2.08
Iodine value	g 100 g^{-1}	2.9	3.1	3.7	14	93
Aromatic content	wt%	24.1	24.4	26.0	27.7	~60 *

* with linear extrapolation (R^2 = 0.9989) of the data of DF and DF-TPO blends.

In summary, the chemical composition of TPO differs from that of DF. TPO has a significantly higher proportion of aromatic and polyaromatic hydrocarbons. In addition, olefinic compounds as well as oxygen- and nitrogen-containing compounds, were identified in TPO compared to DF.

4.2. Characteristics of TPO and DF-TPO-Blends

To assess the suitability of TPO as a blend component in DF, the key values according to DIN EN 590-2022 for DF, TPO, and the blends are determined. The results of the analysis and the corresponding specific target ranges are summarized in Table 5. With a distillation start at 36 °C, TPO shows a remarkably low initial boiling temperature for fuels fractions. This is plausible, as the pyrolysis process can result in molecules with low boiling points and vapour pressure. In comparison, usually DF has a distillation range from 150 °C to 390 °C, with the DF used in this study showing a distillation start at 171 C.

On this basis, TPO is expected to also have a low flash point. The addition of 10 vol% TPO to DF results in a linear drop of the flash point to 53.1 °C below the specified limit of 55 C. As the TPO content increases, the start of distillation shifts to lower temperatures, reaching 157 °C at 10 vol% TPO. However, over the course of distillation, a divergence in temperature progression becomes apparent. 50 vol% distillate volume is reached at 273 °C for DF and at 275 °C for DF90TPO10. This trend, which other authors have also described, is that DF-TPO blends require a higher temperature than neat DF to distil the same volume percent, and this trend persists until the end of the distillation [67]. The end of the distillation is recorded at 390 °C at a distillate volume of 92 vol% for TPO exceeding the result for DF at 358 °C. This indicates that TPO contains higher-boiling components, compared to DF. DF-TPO blends adhere to the target values for DF at 90 vol% and at 95 vol%.

Table 5. Characteristics of DF, DF99TPO1, DF95TPO5, DF90TPO10, and TPO according to DIN EN 590-2022.

Characteristic	Unit	Target	DF	DF99TPO1	DF95TPO5	DF90TPO10	TPO
Density	kg m^{-3}	822.0–845.0	833.3	834.2	838.1	843.1	917.1
Viscosity	mm^2 s^{-1}	2.000–4.500	2.840	2.836	2.830	2.826	1.985
Polyaromatic content	wt%	≤8.0	2.9	3.0	3.8	4.7	21.8 *
Flash point	°C	>55	65.5	64.0	58.0	53.1	-
HFRR [1]-Test	µm	≤460	420.5	359.5	222.5	209.5	175.5
CFPP [2]	°C	≤−20	−25	−25	−13	−14	-
Cu corrosion	extent	1	1a	1a	1a	1a	1a
Oxidation stability	min	≥60	85	196	183	93	22
Water content	mg kg^{-1}	≤200	37	30	60	40	2054
Coke residue	wt%	≤0.30	0.02	0.14	0.75	1.39	2.53
Ash	wt%	≤0.010	<0.001	0.001	0.001	0.001	-
Sulphur content	mg kg^{-1}	≤10.0	7.3	134	648	1292	11,600 (1.16%)
Distillation curve							
Start	vol%		171.3	177.9	165.3	157.0	36.0
10% v/v			210.7	208.0	203.8	200.4	124.0
50% v/v			273.2	275.1	274.3	274.8	226.0
90% v/v		≤360	334.5	336.2	338.1	342.2	387.0
95% v/v		≤360	349.4	349.9	353.1	358.8	-
End			358.6	356.9	357.9	363.4	390.1

[1] HFRR (high frequency reciprocating rig); [2] CFPP (cold filter plugging point), * with linear Extrapolation (R^2 = 0.9989) of the data of DF and DF-TPO blends.

A solid, slightly porous residue remains in the distillation flask, which is also reflected in the carbon residue measurements. At 2.53 wt%, TPO has a higher value of coke residues compared with 0.02 wt% in DF, a consequence attributed to the polyaromatic content of 21.8 wt% in TPO. Whereas DF show a significantly lower content of 2.9 wt%, the blends show a linear increase to 4.7 wt% at 10 vol% TPO. The CFPP, which evaluates fuel filterability at low temperatures weather conditions, is also affected by the polyaromatic content. At low temperatures, polyaromatics tend to precipitate, causing an increase in the CFPP with higher TPO content. The density and viscosity of the fuels are strongly affected by their aromatic and polyaromatic composition. Since polyaromatic compounds possess higher densities than aliphatic species, an increased aromatic content results in a higher overall fuel density. TPO, characterized by a higher polyaromatic fraction, therefore exhibits a higher density and a lower viscosity compared to DF. Upon blending TPO with DF, the density of the blends increases proportionally with the TPO content, following an approximately linear correlation with TPO content. Overall, these results reveal the interdependence of characteristic values on the aromatic content. An increase in the polyaromatic content might not only affect the emission of hazardous substances and particles, posing challenges for modern engine and exhaust technologies, but also promote the formation of residues due to agglomeration during thermal processing [68,69].

Oxidation stability was evaluated by measuring the time required for the consumption of 1 bar of hydrogen after applying an initial pressure of 10 bar H_2 to the fuel sample. Hydrogen reacts with unsaturated compounds present in the fuel, and the rate of consumption is proportional to their concentration. Thus, the higher the content of unsaturated compounds, the faster the pressure drops by 1 bar [70]. Due to the elevated concentration of unsaturated compounds formed during the pyrolysis process, TPO exhibits a low oxidation stability of 22 min, which is considerably lower than that of DF and the DF-TPO blends. The water content in TPO was determined to be 2053 mg kg^{-1}, which significantly exceeds the values measured in DF and the DF-TPO blends. This elevated level is likely influenced by the presence of unsaturated compounds and free alkalis in TPO, which may interfere with standard water quantification methods [71].

According to current EU regulations, DF may contain no more than 10 ppm sulphur to limit the formation of sulphur oxides and sulfuric acid, which contribute to environmental pollution, catalyst degradation, and engine corrosion [72]. No visible corrosion was observed in the copper strip test for any fuel blend. The DF used in this study complied with specifications at 7.3 mg kg^{-1} sulphur. In contrast, TPO contained 1.16 wt% sulphur, far exceeding regulatory limits. Consequently, sulphur levels in DF–TPO blends increase linearly with TPO content. The high sulphur content in TPO is attributed to vulcanized rubber in tyres, which contains sulphur-based compounds [73,74]. Fuel lubricity was assessed using the High-Frequency Reciprocating Rig (HFRR), where lower wear scar diameters indicate better lubrication. DF and its blends showed relatively high wear values, while TPO exhibited superior lubricity, likely due to its elevated sulphur content, as sulphur compounds enhance lubricating properties.

In summary, the applicability of TPO and its blends with DF according to DIN EN 590 can be considered limited. The addition of 5 vol% TPO to DF results in parameters such as sulphur content, carbon residue, and CFPP exceeding the specified limits. At 10 vol% TPO, the flash point also falls outside the acceptable range, while oxidation stability and density approach the threshold values. Due to the significant changes observed with increasing TPO content, the quality of the base DF is of particular importance and should be determined in advance to assess the feasible blending ratio of TPO. Several parameters exhibit mathematical dependencies, allowing for the determination of the precise dosage that can be applied.

4.3. Swelling Behaviour of NBR and Its Influence on Mechanical Properties

4.3.1. Mass and Volume Change in NBR After the Sorption

Figure 4 presents the relative mass and volume changes in NBR18 (Figure 4A,D), NBR28 (Figure 4B,E), and NBR39 (Figure 4C,F), along with the corresponding sorption curves based on Equation (4), following storage in DF, TPO, and various DF-TPO blends. The mass changes are plotted as a function of the square root of time. The equilibrium mass uptake and the diffusion coefficients derived from these experiments are summarized in Table 6. For NBR39, equilibrium was not reached in DF or the DF-TPO blends within the experimental timeframe. As a result, no sorption curves and diffusion coefficient according to Equation (4) were calculated for these. All calculated sorption curves show an initial linear increase, fulfilling the requirement for applying Fick's second diffusion law. The equilibrium mass uptake is lowest for DF across all elastomers, with values of 26 wt% for NBR18, 14 wt% for NBR28, and 6 wt% for NBR39. In contrast, TPO exposure results in the highest mass uptakes: 153 wt% (NBR18), 106 wt% (NBR28), and 74 wt% (NBR39), respectively. Identical trends can be observed for the relative volume changes in the NBR samples compared to the relative mass changes. With a volume increase of 199 vol% due to TPO, NBR18 exhibits an almost threefold expansion of the sample. NBR28 and NBR39 also exhibit a significant increase in volume with 134 vol% and 91 vol%, respectively, which underlines the pronounced influence of TPO on material behaviour.

To investigate the relationship between mass and volume of the NBR sample stored in TPO-DF blends, the relative mass change was plotted as a function of the relative volume change (see Figure 5). The data for pure TPO were excluded from the analysis, as their values are significantly higher than those of the blends, which would obscure a clear representation of the results. For NBR18, the blends exhibit a linear correlation between mass and volume changes. In contrast, NBR28 (for DF) and NBR39 (for DF and the blends) show a delayed volume expansion at the beginning of sorption, which is evident from the stage observed in the mass–volume relationship. With increasing TPO content, this delayed expansion diminishes. This behaviour can be attributed to the initially small mass uptake, during which the free volume of the NBR is filled before the material begins to swell. The addition of TPO leads to a more pronounced mass increase, thereby reducing this effect.

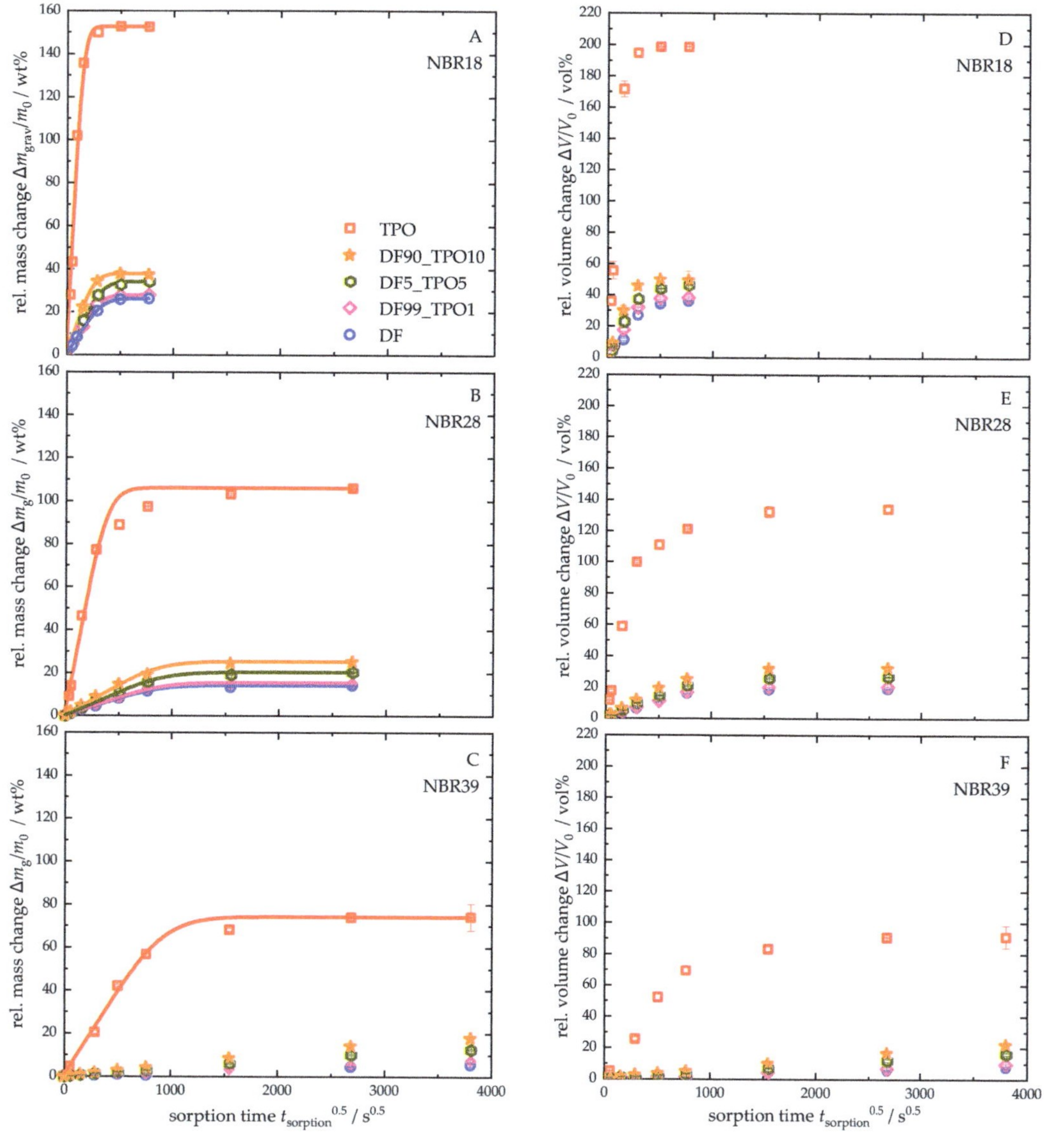

Figure 4. Relative mass and volume changes in NBR18 (**A**,**D**), NBR28 (**B**,**E**), and NBR39 (**C**,**F**) in DF, TPO and the mixtures with the ratios DF99TPO1, DF95TPO5, and DF90TPO10. The continuous lines show the calculated fit according to Equation (4). For NBR39 (**C**), the sorption curve could only be calculated for TPO, as these were in mass equilibrium.

Table 6. Determined diffusion coefficients D and the relative equilibrium mass increases m_∞ of NBR18, NBR28, and NBR39 after storage in DF, TPO, and the mixtures.

Fuel	NBR18		NBR28		NBR39	
	m_∞ wt%	$D \times 10^{-6}$ mm^2 s^{-1}	m_∞ wt%	$D \times 10^{-6}$ mm^2 s^{-1}	m_∞ wt%	$D \times 10^{-6}$ mm^2 s^{-1}
DF	26	1.6	14	0.21	-	-
DF99TPO1	28	1.8	16	0.23	-	-
DF95TPO5	34	2.8	20	0.23	-	-
DF90TPO10	38	4.3	25	0.24	-	-
TPO	153	11.8	106	1.2	74	0.21

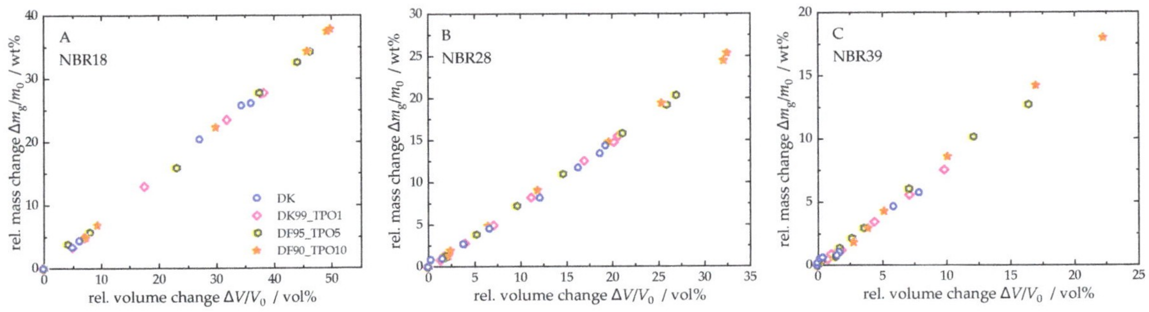

Figure 5. Relative mass change versus relative volume change in NBR18 (**A**), NBR28 (**B**), and NBR39 (**C**) for DF and TPO-DF blends.

The diffusion coefficients follow the same trend as the equilibrium mass uptakes. For DF, the values are 1.6 × 10^{-6} mm^2 s^{-1} (NBR18) and 0.21 × 10^{-6} mm^2 s^{-1} (NBR28). For TPO results the diffusion coefficients of 11.8 × 10^{-6} mm^2 s^{-1} (NBR18), 1.2 × 10^{-6} mm^2 s^{-1} (NBR28), and 0.21 × 10^{-6} mm^2 s^{-1} (NBR39). For the DF-TPO blends, the equilibrium mass uptake and the diffusion coefficients for the elastomers increase with rising TPO content. This relationship is clearly visible when plotting the equilibrium mass uptake and the mean diffusion coefficients against the TPO volume fraction in DF, as shown in Figure 6, where a linear correlation is observed.

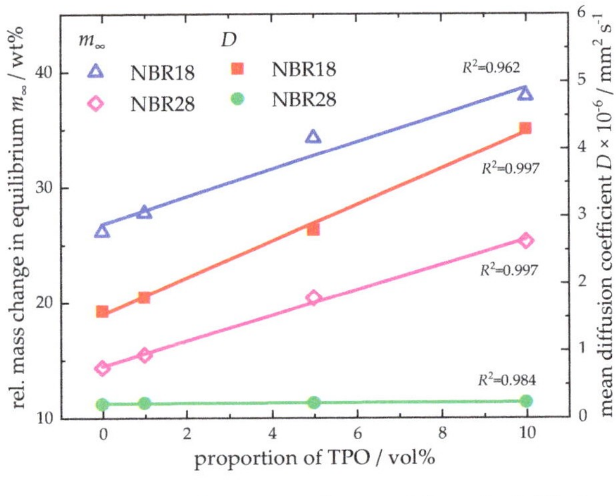

Figure 6. Relative equilibrium mass change and determined diffusion coefficients as a function of the volume fraction of TPO in DF. Linear fits with coefficient of determination (R^2) are also given.

The unusually high mass uptake observed for TPO can be attributed to its elevated aromatic content [25,75]. In addition to dispersion interactions, aromatic compounds in TPO also undergo π–π interactions with the acrylonitrile groups of the polymer matrix, which favours the sorption of these molecular classes [26]. Due to its higher acrylonitrile content, NBR39 exhibits stronger intramolecular interactions within the polymer chains, resulting in reduced free volume in the matrix. Consequently, NBR39 absorbs less fuel compared to NBR18 and NBR28. However, its greater affinity for the aromatic and polar compounds present in TPO results in a steeper initial slope in the sorption curves for NBR39, which means that equilibrium is reached faster when comparing TPO with DF. Additionally, low-molecular-weight aromatic compounds exhibit planar structures and less steric hindrance, which facilitates their diffusion into the polymer matrix more readily than larger, branched aliphatic molecules [25]. Since TPO contains polar compounds such as acids, enhanced swelling is expected due to the behaviour observed for fatty acid esters and alcohols in biodiesel [76–78]. These interactions reduce the intermolecular forces between the polymer chains, thereby increasing the matrix's free volume and enlarging the effective surface area in contact with the liquid medium [79]. Consequently, the material's surface tension can be affected, which reduces the resistance to sorption and facilitates further diffusion of TPO into the elastomeric network.

In summary, the pronounced and rapid diffusion of TPO into NBR, regardless of its high acrylonitrile content, typically selected for resistance to non-polar fuels, indicates that its use as a fuel in neat form is not recommended. However, depending on the blend ratio and the specific NBR formulation, the DF-TPO blends show mass uptakes and diffusion coefficients comparable to neat DF, suggesting that application in the form of blends appears possible.

4.3.2. Temperature Dependence of the Diffusion Process of DF and TPO

In Figure 7, the sorption curves of neat TPO and DF at 20 °C, 50 °C, and 80 °C are plotted against the square root of time. For NBR18, the equilibrium mass uptake increases with temperature in DF, while it remains nearly constant for TPO. The diffusion coefficients, calculated using Equation (4), show a temperature-dependent increase for both fluids. In the case of DF, the diffusion coefficient rises by a factor of approximately 21 from 0.73×10^{-6} mm^2 s^{-1} at 20 °C to 15.3×10^{-6} mm^2 s^{-1} at 80 °C. For TPO, the diffusion coefficient increases by a factor of about six, from 3.43×10^{-6} mm^2 s^{-1} to 21.8×10^{-6} mm^2 s^{-1}. The results are summarized in Table 7. It is well established that the diffusion of liquid media into elastomers increases with temperature. In a study by Blivernitz [26], the temperature dependence of model substances in NBR with 18 wt% acrylonitrile revealed that the equilibrium mass uptake of aliphatic compounds increases significantly with temperature, whereas aromatic compounds show either no change or only a slight increase. Consequently, the higher aliphatic content in DF results in a more pronounced temperature dependence. For DF-TPO blends, this implies that increasing the TPO content, thus raising the aromatic proportion, leads to a reduced temperature dependence of sorption behaviour.

The temperature-dependent diffusion coefficients can be used to determine the activation energy for diffusion according to Arrhenius (Equation (7)) [80]. In the Arrhenius plot (see Figure 8) the slopes (E_A/R) of the linear regressions result in activation energies of 48 kJ mol^{-1} for DF and 29 kJ mol^{-1} for TPO.

$$D = D_0 \exp\left(-\frac{E_A}{RT}\right) \rightarrow \ln(D) = \ln(D_0) - \frac{E_A}{R}\frac{1}{T} \quad (7)$$

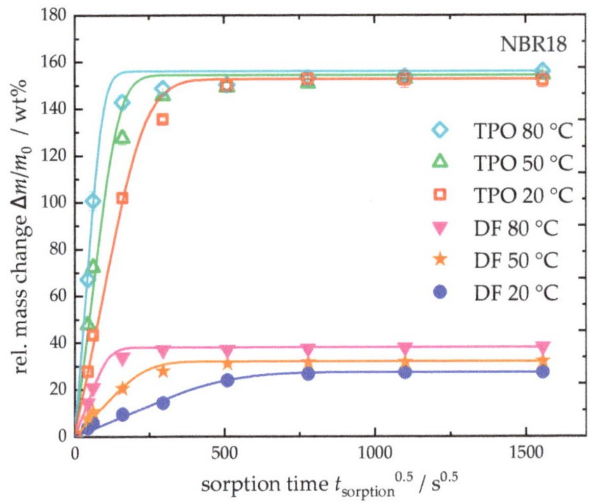

Figure 7. Relative mass change and sorption curves (Equation (4)) of NBR18 in neat DF and TPO at 20 °C, 50 °C and 80 °C.

Table 7. Mass change at equilibrium and determined diffusion coefficients of NBR18 in DF and TPO at 20 °C, 50 °C and 80 °C.

Sorption Temperature	DF		TPO	
	m_∞	$D \times 10^{-6}$	m_∞	$D \times 10^{-6}$
°C	wt%	$mm^2\ s^{-1}$	wt%	$mm^2\ s^{-1}$
20	27	0.73	153	3.43
50	32	3.36	154	8.83
80	38	15.3	156	21.8

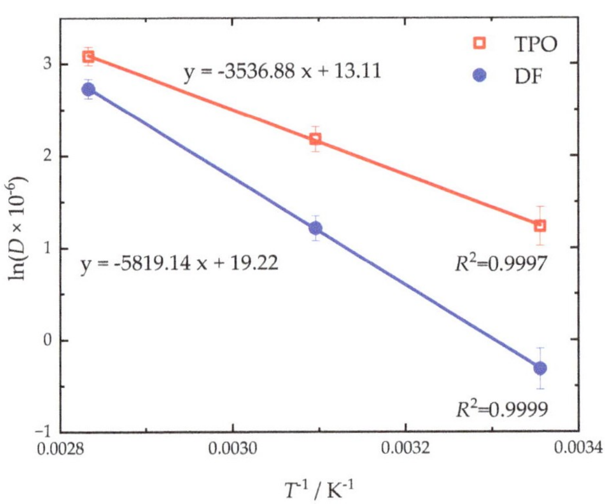

Figure 8. Arrhenius plot for the sorption of neat DF and TPO in NBR18. The coefficients of determination (R^2) for the linear regression are also given.

These results confirm the assumption that DF's diffusion rate is more temperature-dependent than TPO. However, no statement can be made about the equilibrium absorption. It can be assumed that the diffusion properties of blends of TPO and DF are less dependent on temperature than neat DF. Applications of blends at elevated temperatures should therefore not show any deterioration of swelling behaviour with NBR.

In summary, a rise in temperature results in a faster rate of mass increase and higher total mass uptake. This observation is more evident in DF, demonstrating its stronger temperature dependence relative to TPO.

4.3.3. Change in the Mechanical and Physical Properties of the NBR

Figure 9 illustrates the density and Shore A hardness of NBR18 (Figure 9A,D), NBR28 (Figure 9B,E), and NBR39 (Figure 9C,F) as a function of the square root of time. Both parameters decrease with increasing sorption time. The trend in density and hardness for NBR results from the mass uptake of TPO, as shown in Figure 4. During diffusion, both DF and TPO reduce the density of NBR, since the pristine material (ρ_{NBR18} = 1.19 g cm^{-3}; ρ_{NBR28} = 1.21 g cm^{-3}; ρ_{NBR39} = 1.23 g cm^{-3}) has a higher density than TPO, ρ_{TPO} = 0.917 g cm^{-3}, and DF, ρ_{DF} = 0.833 g cm^{-3}. The decrease in hardness is attributable to the plasticizing effect of the absorbed fluids, which are stored between the polymer chains, weakening intermolecular interactions and thereby reducing the counterforce of the material. Blends containing up to 10 vol% TPO exhibit slightly lower density and hardness values compared to neat DF, indicating a measurable influence of TPO. As the acrylonitrile content increases, the reduction in equilibrium density and hardness upon exposure to DF, TPO, and their blends becomes less pronounced, following the order: NBR18 > NBR28 > NBR39. Figure 10 shows the equilibrium density and hardness values plotted against the TPO volume fraction in DF. The results indicate that, with increasing acrylonitrile content, the influence of the material to changes in density and hardness due to TPO in DF becomes more pronounced. The initial increase in density and hardness observed for NBR39 in Figure 9 is attributed to the slower diffusion of DF and the blends. In this case, the free volume within the polymer is filled before swelling occurs, which temporarily restricts chain mobility and leads to an increase in the mechanical counterforce of the material. Furthermore, it is observed that NBR28 and NBR39 reach equilibrium in hardness after approximately three days, which is earlier than the mass uptake shown in Figure 4. This discrepancy is explained by the fact that hardness measurements primarily reflect changes at the surface, where the medium penetrates first, while full diffusion into the sample core occurs later in accordance with Fick's second law of diffusion.

Figure 11 presents the development of tensile strength at break (Figure 11A–C) and elongation at break (Figure 11D–F) for NBR18 (Figure 11A,D), NBR28 (Figure 11B,E), and NBR39 (Figure 11C,F) as a function of the square root of time. For all elastomers, a decrease in both mechanical parameters is observed with increasing sorption time. The influence of TPO on the sorption behaviour is particularly pronounced, as its higher and more rapid mass uptake accelerates and intensifies the decrease in tensile strength and elongation at break compared to DF. NBR18 in TPO exhibits the strongest decline in mechanical properties, with tensile strength decreasing from 13.5 MPa to 4 MPa and elongation at break from 413% to 137%. In contrast, storage in DF and the blends results in significantly lower reductions. As the acrylonitrile content increases, the extent of mechanical characteristics decreases. NBR28 shows a decline from 15.4 MPa and 345% to 6 MPa and 161% while NBR39 decreases from 15.9 MPa and 390% to 8 MPa and 210% over three months ($t^{0.5}$ = 2016 s$^{0.5}$). For NBR28, a smaller reduction is observed, though the measurement scatter increases. In NBR39, a more pronounced reduction in tensile strength is evident even with 5 vol% and 10 vol% TPO in DF, with values decreasing to 15.3 MPa

(DF) and 12.6 MPa (DF90TPO10). After six months of storage ($t^{0.5}$ = 3810 $s^{0.5}$), elongation at break for DF and its blends shows significant variation, ranging from 320% to 350%. The observed decrease in mechanical properties is directly related to the uptake of the liquid media. As previously described, sorption weakens the intermolecular interactions within the elastomer, facilitating chain mobility and thereby reducing strength under mechanical stress. The higher the mass uptake, the more severe the deterioration of mechanical integrity [81]. The particularly strong decrease caused by TPO can be predominantly attributed to its high aromatic content. This is explained by the specific interaction of aromatic compounds with the polar nitrile groups in the polymer, which disrupts the internal structure of the material.

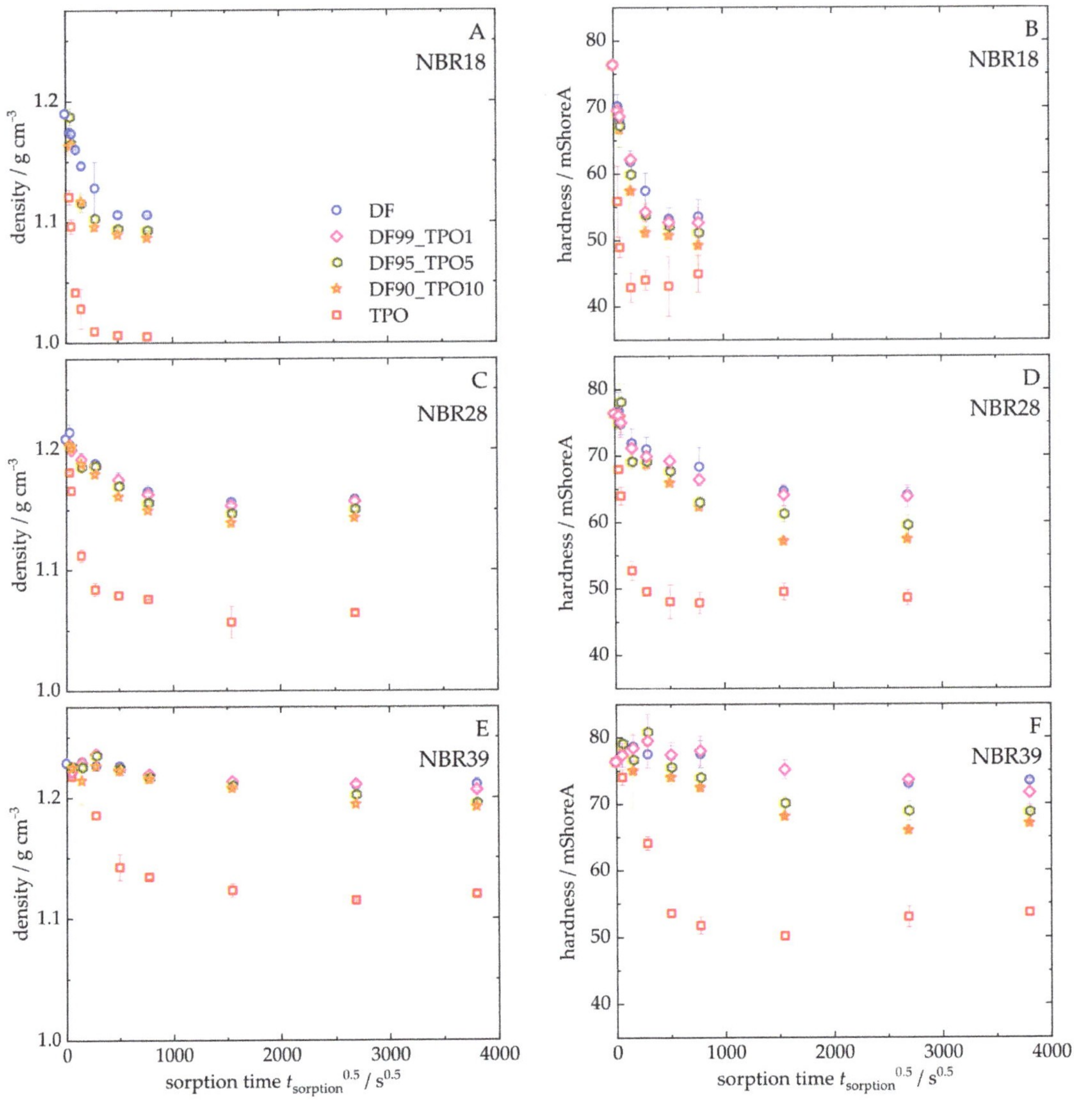

Figure 9. Density (**A–C**) and hardness (**D–F**) over sorption time of NBR18 (**A**,**D**), NBR28 (**B**,**E**), and NBR39 (**C**,**F**) after storage in DF, TPO, and the DF-TPO mixtures.

In summary, none of the examined NBR types is suitable for long-term contact with neat TPO due to the pronounced loss in mechanical performance. However, blends with low TPO content show a decrease in mechanical characteristics comparable to neat DF, suggesting their potential suitability for application as a fuel.

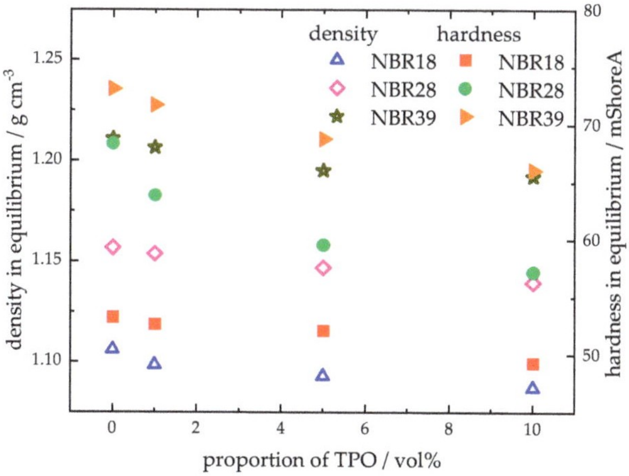

Figure 10. Density and hardness in equilibrium as a function of the volume fraction of TPO in DF.

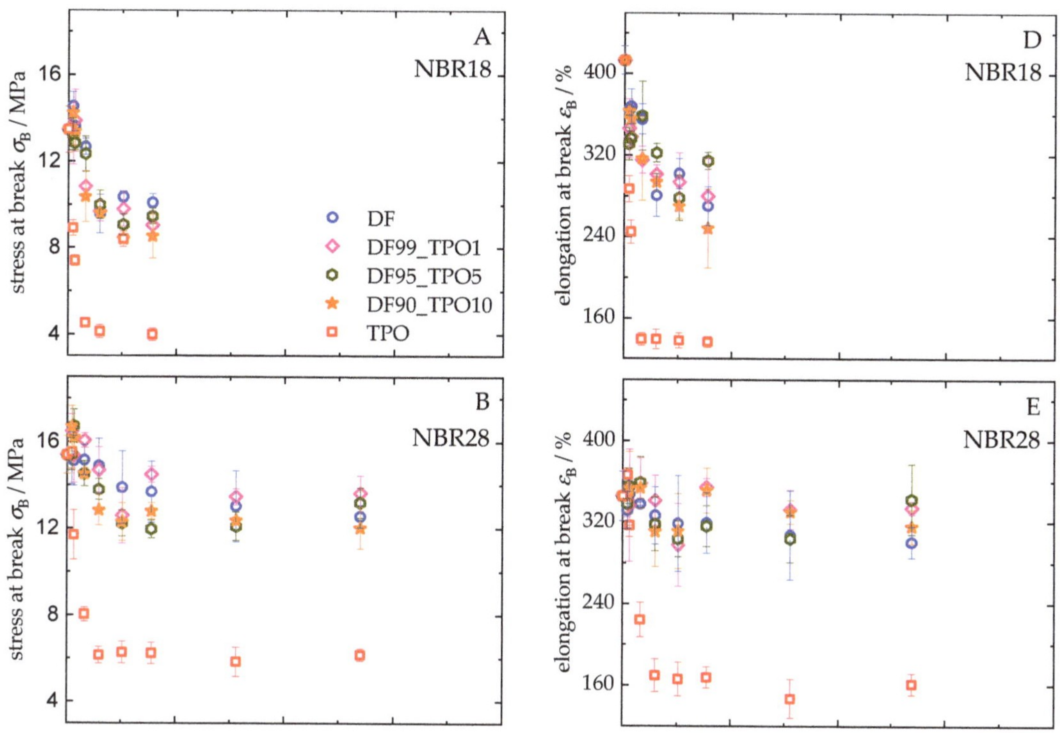

Figure 11. *Cont.*

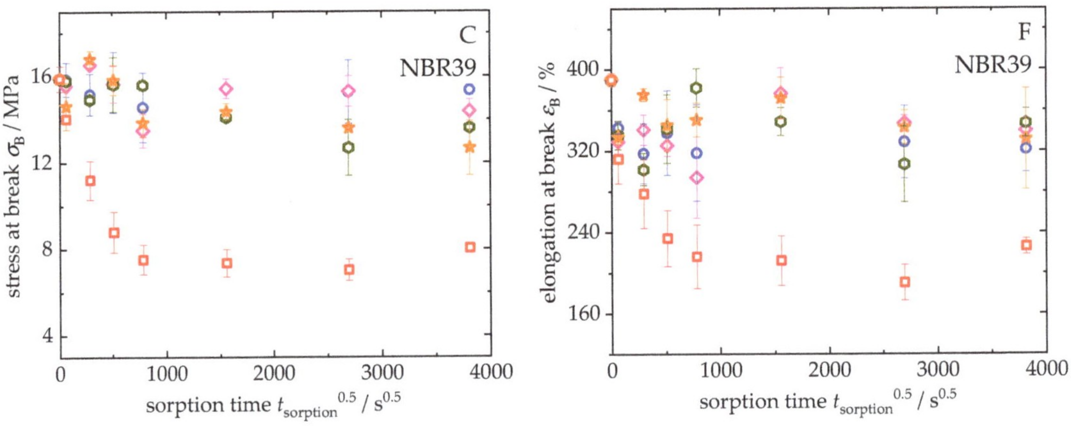

Figure 11. Stress (**A–C**) and elongation at break (**D–F**) over sorption time of NBR18 (**A**,**D**), NBR28 (**B**,**E**), and NBR39 (**C**,**F**) after storage in DF, TPO, and DF-TPO blends.

5. Conclusions

In this study, tyre pyrolysis oil (TPO) was analyzed, and the characteristic properties of its blends with diesel fuel (DF) were determined. The sorption behaviour of these blends and the resulting changes in the mechanical properties of nitrile-butadiene rubber (NBR) were investigated. The objective of the study was to understand the TPO diffusion behaviour in elastomeric materials and to evaluate the feasibility of TPO as a fuel component.

Through detailed analysis using 2D-GC/MS, complementary analytical techniques such as infrared and NMR spectroscopy become more informative, allowing for the acquisition of more precise data. This combination of analytical methods enables a comprehensive and detailed understanding of the chemical composition. Chemical analysis revealed several notable differences between the composition of TPO and DF. Compared to DF, TPO contains the following:

- Higher monoaromatic and polyaromatic contents. This property makes TPO a potential blending component for aromatic-free synthetic fuels.
- Higher sulphur content. Desulfurization is required to reduce the sulphur concentration to within the target range. Alternatively, due to its high sulphur content and correspondingly low value in the high frequency reciprocating rig test, TPO could be considered as a blend component for lubricating oils.
- Higher proportions of olefinic compounds. Hydrogenation is necessary to saturate olefins and improve oxidation stability, thereby expanding the possible applications of both neat TPO and its blends with other fuels and oils.
- Larger amounts of low- and high-boiling components. Distillation of TPO to separate these fractions could produce a cut similar to DF, potentially lowering the polyaromatic content, as these are typically found in the high-boiling fraction, and raising the flash point by removing low-boiling compounds.

Without further chemical processing, the use of TPO in its raw form may present challenges. When blended at 1 vol% and 5 vol% with DF, most characteristic parameters remain within the target ranges specified by DIN EN 590, with the exception of sulphur content. However, under specific circumstances, such as emergencies or fuel shortages, use could be permitted according to StVZO § 70(4).

TPO exhibits strong affinity towards NBR, leading to rapid and substantial mass uptake and increasing volume during sorption. This pronounced swelling significantly reduces mechanical stability properties such as hardness, tensile strength, and elongation at break, thereby permanently affecting the viscoelastic behaviour of the elastomer. With increasing nitrile content, TPO uptake decreases, but equilibrium mass uptake is reached markedly faster for TPO than for DF. Increasing TPO content in DF results in proportionally higher absorption and diffusion coefficients, with a linear relationship observed between these parameters and the TPO volume fraction. Blends containing 1 vol% and 5 vol% TPO show no critical deviations in mass uptake, volume expansion, or mechanical properties compared to neat DF.

Further research is required to elucidate the interactions between individual TPO components and NBR. Analytical methods could be used to monitor the sorption of individual compounds, as unsaturated components or acids in TPO may react chemically with NBR, potentially altering its molecular structure and thus permanently impairing its mechanical performance.

Author Contributions: Conceptualization, S.S.; methodology, S.S.; validation, S.S.; writing—original draft preparation, S.S.; writing—review and editing, T.F. and S.E.; visualization, S.S.; supervision, S.S.; funding acquisition, T.F. and S.E. All authors have read and agreed to the published version of the manuscript.

Funding: We acknowledge financial support by Universität der Bundeswehr München.

Institutional Review Board Statement: Not applicable.

Data Availability Statement: The original contributions presented in this study are included in the article. Further inquiries can be directed to the corresponding author(s).

Acknowledgments: The authors want to thank Kathrin Hufschmid for her experimental support. The authors would also like to thank Pyrum Innovations AG for providing the tyre pyrolysis oil.

Conflicts of Interest: The authors declare no conflicts of interest.

Abbreviations

The following abbreviations are used in this manuscript:

DF	Disel fuel
TPO	Tyre pyrolysis oil
NBR	Nitrile-butadiene rubber
NR	natural rubber
SBR	Styrene-butadiene rubber
BR	Butadiene rubber
phr	Parts-per-hundred rubber
DPHP	Di(2-propylheptyl)phthalate
6-PPD	N-(4-Methylpentan-2-yl)-N-phenylbenzene-1,4-diamine
TAN	Total acid number
HFRR	High frequency reciprocating rig
CFPP	cold filter plugging point
2D-GC/MS	Two-dimensional gas chromatography coupled to mass spectrometry
SPE	Solid phase extraction
GC/MS	Gas chromatography coupled to mass spectrometry
^{1}H-NMR	1H-Nuclear magnetic resonance
FTIR	Fourier transform infrared
HPLC	High-Performance Liquid Chromatography
PAHs	Polyaromatic hydrocarbons
KOH	Potassium hydroxide

References

1. Continental Reifen Deutschland GmbH. Reifengrundlagen. Available online: https://www.continental-reifen.de/tire-knowledge/tire-mixture/ (accessed on 19 August 2025).
2. Conesa, J.A.; Martín-Gullón, I.; Font, R.; Jauhiainen, J. Complete Study of the Pyrolysis and Gasification of Scrap Tires in a Pilot Plant Reactor. *Environ. Sci. Technol.* **2004**, *38*, 3189–3194. [CrossRef] [PubMed]
3. Roy, C.; Chaala, A.; Darmstadt, H. The Vacuum Pyrolysis of Used Tires End-Uses for Oil and Carbon Black Products. *J. Anal. Appl. Pyrolysis* **1999**, *51*, 201–221. [CrossRef]
4. Dai, X.; Yin, X.; Wu, C.; Zhang, W.; Chen, Y. Pyrolysis of Waste Tires in a Circulating Fluidized-Bed Reactor. *Energy* **2001**, *26*, 385–399. [CrossRef]
5. Uçar, S.; Karagöz, S.; Yanik, J.; Saglam, M.; Yuksel, M. Copyrolysis of Scrap Tires with Waste Lubricant Oil. *Fuel Process. Technol.* **2005**, *87*, 53–58. [CrossRef]
6. González, J.F.; Encinar, J.M.; Canito, J.L.; Rodríguez, J.J. Pyrolysis of Automobile Tyre Waste. Influence of Operating Variables and Kinetics Study. *J. Anal. Appl. Pyrolysis* **2001**, *58–59*, 667–683. [CrossRef]
7. De, I.; Rodriguez, M.; Laresgoiti, M.F.; Cabrero, M.A.; Torres, A.; Chomon, M.J.; Caballeró, B. Pyrolysis of Scrap Tyres. *Fuel Process. Technol.* **2001**, *72*, 9–22. [CrossRef]
8. Williams, P.T. Pyrolysis of Waste Tyres: A Review. *Waste Manag.* **2013**, *33*, 1714–1728. [CrossRef]
9. Okoye, C.O.; Jones, I.; Zhu, M.; Zhang, Z.; Zhang, D. Manufacturing of Carbon Black from Spent Tyre Pyrolysis Oil–A Literature Review. *J. Clean. Prod.* **2021**, *279*, 123336. [CrossRef]
10. Groves, S.A.; Lehrle, R.S.; Blazso, M.; Szckely, T. Natural Rubber Pyrolysis: Study of Temperature and Thickness-Dependence Indicates Dimer Formation Mechanism. *J. Anal. Appl. Pyrolysis* **1991**, *19*, 301–309. [CrossRef]
11. Chen, F.; Qian, J. Studies on the Thermal Degradation of Polybutadiene. *Fuel Process. Technol.* **2000**, *67*, 53–60. [CrossRef]
12. Huang, J.; Li, X.; Meng, H.; Tong, H.; Cai, X.; Liu, J. Studies on Pyrolysis Mechanisms of Syndiotactic Polystyrene Using DFT Method. *Chem. Phys. Lett.* **2020**, *747*, 137334. [CrossRef]
13. Gardner, P.; Lehrle, R. Polystyrene Pyrolysis Mechanisms-as Deduced from the Dependence of Product Yields in Film Thickness. *Eur. Polym. J.* **1993**, *29*, 425–435. [CrossRef]
14. Liqiu, F.; Jin, J.; Zang, Z.; Zhang, Q.; Sun, X.; Deng, L.; Zhang, J. Mechanism of Remarkable Interference of Styrene-Butadiene to Gasoline Identification for Fire Debris Analysis Based on TG-IR-GC/MS. *J. Anal. Appl. Pyrolysis* **2025**, *186*, 106958. [CrossRef]
15. Chen, F.; Qian, J. Studies on the Thermal Degradation of Cis-1,4-Polyisoprene. *Fuel* **2002**, *81*, 2071–2077. [CrossRef]
16. Williams, P.; Taylor, D. Aromatization of Tyre Pyrolysis Oil to Yield Polycyclic Aromatic Hydrocarbons. *Fuel* **1993**, *72*, 1469–1474. [CrossRef]
17. Williams, P.T.; Besler, S. Pyrolysis-Thermogravimetric Analysis of Tyres and Tyre Components. *Fuel* **1995**, *14*, 1277–1283. [CrossRef]
18. Sainz-Diaz, C.I.; Kelly, D.R.; Avenell, C.S.; Griffiths, A.G. Pyrolysis of Furniture and Tire Wastes in a Flaming Pyrolyzer Minimizes Discharges to the Environment. *Energy Fuels* **1997**, *11*, 1061–1072. [CrossRef]
19. Williams, P.T.; Brindle, A.J. Aromatic Chemicals from the Catalytic Pyrolysis of Scrap Tyres. *J. Anal. Appl. Pyrolysis* **2003**, *67*, 143–164. [CrossRef]
20. Fairburn, J.A.; Behie, L.A.; Svrcek, W.Y. Ultrapyrolysis of N-Hexadecane in a Novel Micro-Reactor. *Fuel* **1990**, *69*, 1537–1545. [CrossRef]
21. Brazier, D.W.; Schwartz, N. The Effect of Heating Rate on the Thermal Degradation of Polybutadiene. *J. Appl. Polym. Sci.* **1978**, *22*, 113–124. [CrossRef]
22. Frigo, S.; Seggiani, M.; Puccini, M.; Vitolo, S. Liquid Fuel Production from Waste Tyre Pyrolysis and Its Utilisation in a Diesel Engine. *Fuel* **2014**, *116*, 399–408. [CrossRef]
23. Arya, S.; Sharma, A.; Rawat, M.; Agrawal, A. Tyre Pyrolysis Oil as an Alternative Fuel: A Review. *Mater. Today Proc.* **2020**, *28*, 2481–2484. [CrossRef]
24. Goyal, A.; Kumar, N. Some Experimental Studies on the Use of Tyre Pyrolysis Oil (TPO) in an Agricultural Diesel Engine. *SAE Tech. Pap.* **2019**, *1*, 796. [CrossRef]
25. Graham, J.L.; Striebich, R.C.; Myers, K.J.; Minus, D.K.; Harrison, W.E. Swelling of Nitrile Rubber by Selected Aromatics Blended in a Synthetic Jet Fuel. *Energy Fuels* **2006**, *20*, 759–765. [CrossRef]
26. Blivernitz, A. Untersuchung Der Verträglichkeit von Elastomeren. Ph.D. Thesis, Universität der Bundeswehr, Neubibierg, Germany, 2019.
27. Musil, B.; Demmel, B.; Lion, A.; Johlitz, M. A Contribution to the Chemomechanics of Elastomers Surrounded by Liquid Media: Continuum Mechanical Approach for Parameter Identification Using the Example of Sorption Experiments. *J. Rubber Res.* **2021**, *24*, 271–279. [CrossRef]
28. Gormley, R.J.; Link, D.D.; Baltrus, J.P.; Zandhuis, P.H. Interactions of Jet Fuels with Nitrile O-Rings: Petroleum-Derived versus Synthetic Fuels. *Energy Fuels* **2009**, *23*, 857–861. [CrossRef]

29. Lou, W.; Zhang, W.; Wang, H.; Jin, T.; Liu, X. Influence of Hydraulic Oil on Degradation Behavior of Nitrile Rubber O-Rings at Elevated Temperature. *Eng. Fail. Anal.* **2018**, *92*, 1–11. [CrossRef]
30. Röthemeyer, F. *Kautschuk Technologie*, 3rd ed.; Hanser: München, Germany, 2013; Volume 3.
31. Starmer, P. Swelling of Nitrile Rubber Vulcanizates-Part 3: Factors Affecting Maximum Swelling. *Elastomers Plast.* **1993**, *25*, 188–215. [CrossRef]
32. Lachat, V. Understanding Oil Resistance of Nitrile Rubber Group Interactions at Interfaces. Ph.D. Thesis, University of Akron, Akron, OH, USA, 2008.
33. *DIN EN 590*; Kraftstoffe–Dieselkraftstoff–Anforderungen Und Prüfverfahren. DIN Deutsches Institut für Normung e. V: Berlin, Germany, 2022.
34. *DIN 53504:2017-03*; Testing of Rubber-Determination of Tensile Strength at Break, Tensile Stress at Yield, Elongation at Break and Stress Values in a Tensile Test. DIN Deutsches Institut für Normung e. V: Berlin, Germany, 2017.
35. *D7042-21a*; ASTM International Standard Test Method for Dynamic Viscosity and Density of Liquids by Stabinger Viscometer (and the Calculation of Kinematic Viscosity). ASTM International: West Conshohocken, PA, USA, 2022.
36. *DIN EN 12916*; Mineralölerzeugnisse–Bestimmung von Aromatischen Kohlenwasserstoffgruppen in Mitteldestillaten–Hochleistungsflüssigkeitschromatographie-Verfahren Mit-Detektion. DIN Deutsches Institut für Normung e. V: Berlin, Germany, 2022.
37. *DIN EN ISO 2719*; Bestimmung Des Flammpunktes–Verfahren Nach Pensky-Martens Mit Geschlossenem Tiegel. DIN Deutsches Institut für Normung e. V: Berlin, Germany, 2021.
38. *DIN EN ISO 12156-1*; Dieselkraftstoff –Bestimmung Der Schmierfähigkeit Unter Verwendung Eines-Prüfgerätes (HFRR). DIN Deutsches Institut für Normung e. V: Berlin, Germany, 2019.
39. *DIN EN 116*; Dieselkraftstoffe Und Haushaltsheizöle–Bestimmung Des Temperaturgrenzwertes Der Filtrierbarkeit–Verfahren Mit Einem Stufenweise Arbeitenden Kühlbad. DIN Deutsches Institut für Normung e. V: Berlin, Germany, 2018.
40. *DIN EN ISO 2160*; Mineralölerzeugnisse Auf Kupfer-Kupferstreifenprüfung. DIN Deutsches Institut für Normung e.V: Berlin, Germany, 1999.
41. *DIN EN 16091*; Bestimmung Der Oxidationsstabilität Mit Beschleunigtem Und Kleiner Probenmenge. DIN Deutsches Institut für Normung e. V: Berlin, Germany, 2022.
42. *DIN 51777*; Mineralölerzeugnisse–Bestimmung Des Wassergehaltes Durch Titration Nach Karl Fischer. DIN Deutsches Institut für Normung e. V: Berlin, Germany, 2020.
43. *DIN EN ISO 10370*; Mineralölerzeugnisse–Bestimmung Des Koksrückstandes–Mikroverfahren. DIN Deutsches Institut für Normung e.V: Berlin, Germany, 2015.
44. *DIN EN ISO 6245*; Mineralölerzeugnisse Bestimmung Der Asche. DIN Deutsches Institut für Normung e. V: Berlin, Germany, 2003.
45. *DIN EN ISO 20884*; Mineralölerzeugnisse–Bestimmung Des Schwefelgehaltes in Kraftstoffen–Wellenlängendispersive Röntgenfluoreszenz-Spektrometrie. DIN Deutsches Institut für Normung e. V: Berlin, Germany, 2022.
46. *DIN EN ISO 3405*; Mineralölerzeugnisse Und Verwandte Produkte Mit Natürlichem Oder Ursprung–Bestimmung Des Destillationsverlaufes Bei Atmosphärendruck. DIN Deutsches Institut für Normung e. V: Berlin, Germany, 2019.
47. *D664−24*; ASTM International Test Method for Acid Number of Petroleum Products by Potentiometric Titration. ASTM International: West Conshohocken, PA, USA, 2024.
48. *DIN EN ISO 3961*; Tierische Und Pflanzliche Fette Und Öle–Bestimmung Der Iodzahl. DIN Deutsches Institut für Normung e. V: Berlin, Germany, 2025.
49. John, C. *The Mathematics of Diffusion*, 2nd ed.; Clarendon Press: Oxford, UK, 1975; ISBN 0198533446.
50. *DIN ISO 48-4*; Elastomere Oder Thermoplastische Elastomere-Bestimmung Der Härte-Teil 4: Eindringhärte Durch Durometer-Verfahren (Shore-Härte). DIN Deutsches Institut für Normung e. V: Berlin, Germany, 2021.
51. Wagner, S.; Hüffer, T.; Klöckner, P.; Wehrhahn, M.; Hofmann, T.; Reemtsma, T. Tire Wear Particles in the Aquatic Environment-A Review on Generation, Analysis, Occurrence, Fate and Effects. *Water Res.* **2018**, *139*, 83–100. [CrossRef]
52. Di Salvi, W.; de Santo, C.E.; Matsumoto, A.; Calhabeu, A.M.; Lopes, É.S.N.; Gabriel, L.P. Characterization of Thermal-Oxidative Aging Mechanism of Commercial Tires. *Eng. Fail. Anal.* **2023**, *154*, 107631. [CrossRef]
53. Uebe, J.; Kryzevicius, Z.; Majauskiene, R.; Dulevicius, M.; Kosychova, L.; Zukauskaite, A. Use of Polypropylene Pyrolysis Oil in Alternative Fuel Production. *Waste Manag. Res.* **2022**, *40*, 1220–1230. [CrossRef]
54. Islam, M.R.; Tushar, M.S.H.K.; Haniu, H. Production of Liquid Fuels and Chemicals from Pyrolysis of Bangladeshi Bicycle/Rickshaw Tire Wastes. *J. Anal. Appl. Pyrolysis* **2008**, *82*, 96–109. [CrossRef]
55. Gupta, P.L.; Dogra, P.V.; Kuchhal, R.K.; Kumar, P. Estimation of Average Structural Parameters of Petroleum Crudes and Coal-Derived Liquids by 13C and 1H n.m.r. *Fuel* **1985**, *65*, 515–519. [CrossRef]
56. Banar, M.; Akyildiz, V.; Özkan, A.; Çokaygil, Z.; Onay, Ö. Characterization of Pyrolytic Oil Obtained from Pyrolysis of TDF (Tire Derived Fuel). *Energy Convers. Manag.* **2012**, *62*, 22–30. [CrossRef]
57. Elbaz, A.M.; Gani, A.; Hourani, N.; Emwas, A.H.; Sarathy, S.M.; Roberts, W.L. TG/DTG, FT-ICR Mass Spectrometry, and NMR Spectroscopy Study of Heavy Fuel Oil. *Energy Fuels* **2015**, *29*, 7825–7835. [CrossRef]

58. Rodriguez, J.; Tierney, J.W.; Wender, I. Evaluation of a Delayed Coking Process by 1H and 13C n.m.r. Spectroscopy: 1. Material Balances. *Fuel* **1994**, *73*, 1863–1869. [CrossRef]
59. Rodriguez, J.; Tierney, J.W.; Wender, I. Evaluation of a Delayed Coking Process by 1H and 13C n.m.r. Spectroscopy: 2. Detailed Interpretation of Liquid n.m.r. Spectra. *Fuel* **1994**, *73*, 1870–1875. [CrossRef]
60. Hummel, O.; Scholl, F. *Atlas of Polymer and Plastics Analysis*; Carl Hanser: Munich, Germany, 1988; Volume 2.
61. Socrates, G. *Infrared and Raman Characteristic Group Frequencies*, 2nd ed.; John Wiley & Sons: Chichester, UK, 1994.
62. Lin-Vien, D.; Colthup, N.B.; Fatley, W.G.; Grasselli, J.G. *The Handbook of Infrared and Raman Caracteristic Frequencies of Organic Molecules*; Academic Press: Cambridge, MA, USA, 1991.
63. Scheuermann, S.S.; Forster, S.; Eibl, S. In-Depth Interpretation of Mid-Infrared Spectra of Various Synthetic Fuels for the Chemometric Prediction of Aviation Fuel Blend Properties. *Energy Fuels* **2017**, *31*, 2934–2943. [CrossRef]
64. Smith, B. The C=O Bond, Part III Carboxylic Acids. *Spectroscopy* **2018**, *33*, 14–20.
65. Cunliffe, A.M.; Williams, P.T. Composition of Oils Derived from the Batch Pyrolysis of Tyres. *J. Anal. Appl. Pyrolysis* **1998**, *44*, 131–152. [CrossRef]
66. Mastral, A.M.; Murillo, R.; Callén, M.S.; García, T.; Snape, C.E. Influence of Process Variables on Oils from Tire Pyrolysis and Hydropyrolysis in a Swept Fixed Bed Reactor. *Energy Fuels* **2000**, *14*, 739–744. [CrossRef]
67. Day, M.; Cooney, J.D.; Shen, Z. Pyrolysis of Automobile Shredder Residue: An Analysis of the Products of a Commercial Screw Kiln Process. *J. Anal. Appl. Pyrolysis* **1996**, *37*, 49–61. [CrossRef]
68. Totton, T.S.; Misquitta, A.J.; Kraft, M. A Quantitative Study of the Clustering of Polycyclic Aromatic Hydrocarbons at High Temperatures. *Phys. Chem. Chem. Phys.* **2012**, *14*, 4081–4094. [CrossRef]
69. Elvati, P.; Violi, A. Thermodynamics of Poly-Aromatic Hydrocarbon Clustering and the Effects of Substituted Aliphatic Chains. *Proc. Combust. Inst.* **2013**, *34*, 1837–1843. [CrossRef]
70. Pullen, J.; Saeed, K. An Overview of Biodiesel Oxidation Stability. *Renew. Sustain. Energy Rev.* **2012**, *16*, 5924–5950. [CrossRef]
71. Lopez, G.; Alvarez, J.; Amutio, M.; Mkhize, N.M.; Danon, B.; van der Gryp, P.; Görgens, J.F.; Bilbao, J.; Olazar, M. Waste Truck-Tyre Processing by Flash Pyrolysis in a Conical Spouted Bed Reactor. *Energy Convers. Manag.* **2017**, *142*, 523–532. [CrossRef]
72. Stockwell, W.R.; Calvert, J.G. The Mechanism of the HO-SO2 Reaction. *Atmos. Environ.* **1983**, *17*, 2231–2235. [CrossRef]
73. Serefentse, R.; Ruwona, W.; Danha, G.; Muzenda, E. A Review of the Desulphurization Methods Used for Pyrolysis Oil. *Procedia Manuf.* **2019**, *35*, 762–768. [CrossRef]
74. Williams, P.T.; Bottrill, R.P. Sulfur-Polycyclic Aromatic Hydrocarbons in Tyre Pyrolysis Oil. *Fuel* **1995**, *74*, 736–742. [CrossRef]
75. Blivernitz, A.; Förster, T.; Eibl, S. Simultaneous and Time Resolved Investigation of Diffusion Processes of Individual Model Fuel Components in Acrylonitrile-Butadiene-Rubber in the Light of Swelling Phenomena. *Polym. Test.* **2018**, *70*, 47–56. [CrossRef]
76. Haseeb, A.S.M.A.; Masjuki, H.H.; Siang, C.T.; Fazal, M.A. Compatibility of Elastomers in Palm Biodiesel. *Renew. Energy* **2010**, *35*, 2356–2361. [CrossRef]
77. Kass, M.; Theiss, T.; Janke, C.; Pawel, S.; Technology, J.C.-S. Compatibility of Elastomers with Test Fuels of Gasoline Blended with Ethanol. *Seal. Technol.* **2012**, *2012*, 7–12. [CrossRef]
78. Zhu, L.; Cheung, C.S.; Zhang, W.G.; Huang, Z. Compatibility of Different Biodiesel Composition with Acrylonitrile Butadiene Rubber (NBR). *Fuel* **2015**, *158*, 288–292. [CrossRef]
79. Lachat, V.; Varshney, V.; Dhinojwala, A.; Yeganeh, M.S. Molecular Origin of Solvent Resistance of Polyacrylonitrile. *Macromolecules* **2009**, *42*, 7103–7107. [CrossRef]
80. George, S.; Varughese, K.T.; Thomas, S. Molecular Transport of Aromatic Solvents in Isotactic Polypropylene/ Acrylonitrile-Co-Butadiene Rubber Blends. *Polymer* **2000**, *41*, 579–594. [CrossRef]
81. Chai, A.B.; Andriyana, A.; Verron, E.; Johan, M.R. Mechanical Characteristics of Swollen Elastomers under Cyclic Loading. *Mater. Des.* **2013**, *44*, 566–572. [CrossRef]

polymers

MDPI

Article

Application of Sawdust-Derived Activated Carbon as a Bio-Based Filler in Vulcanized Rubber Bushings

Enasty Pratiwi Wulandari [1], Popy Marlina [2], Nasruddin [3,*], Lanjar [3], Heryoki Yohanes [3], Wahju Eko Widodo [3], S. Joni Munarso [3], Astuti [3], Eko Bhakti Susetyo [3], Yenni Bakhtiar [3], Haixin Guo [4] and Wahyu Bahari Setianto [3]

1 Mechanical Engineering Department, Tamansiswa University, Palembang 30126, Indonesia; enastypratiwi@gmail.com
2 The Palembang Center for Standardization and Industrial Services (BSPJI), Palembang 30128, Indonesia; popymarlina16@gmail.com
3 National Research and Innovation Agency (BRIN), KST B.J. Habibie, South Tangerang 15314, Indonesia; lanj001@brin.go.id (L.); hery009@brin.go.id (H.Y.); wahj003@brin.go.id (W.E.W.); sjon001@brin.go.id (S.J.M.); astu001@brin.go.id (A.); ekob002@brin.go.id (E.B.S.); yenn001@brin.go.id (Y.B.); wahy013@brin.go.id (W.B.S.)
4 Agro-Environmental Protection Institute, Ministry of Agriculture and Rural Affairs, Tianjin 300191, China; haixin_g@126.com
* Correspondence: nasr006@brin.go.id

Abstract

This study investigated sawdust-derived activated carbon (SAC) as a sustainable reinforcing filler for vulcanized rubber bushings (VRBs). Two types SAC200 (75 μm, carbonized at 200 °C) and SAC400 (38 μm, carbonized at 400 °C) were chemically activated and incorporated into natural rubber (NR) at 25–55 phr loadings, while SAC free VRBs served as controls. Fourier transform infrared (FTIR) analysis revealed that SAC400 exhibited stronger hydroxyl and carbonyl functional groups, indicating higher surface reactivity compared with SAC200. The incorporation of SAC increased cross-linking density, thereby enhancing both curing behavior and mechanical performance. VRBs reinforced with SAC400 demonstrated higher maximum torque (up to 38.07 kg·cm), shorter scorch time (5 min 58 s), and reduced cure time (11 min 05 s) relative to SAC200 and the control. Mechanical properties improved markedly, with hardness and tensile strength rising from 45 Shore A and 5.52 MPa in the control to 70 Shore A and 13.40 MPa in SAC400. Although elongation at break decreased slightly, it remained within the acceptable range for dynamic applications. Swelling resistance also increased, reaching 101.76% at 25 °C and 106.61% at 100 °C. Overall, SAC400 consistently outperformed SAC200 and the control, highlighting its potential as a renewable, biomass-derived filler for high-performance rubber bushings and promising a sustainable alternative to conventional fillers in industrial applications.

Keywords: activated carbon; bio-based filler; mechanical properties; sawdust-derived activated carbon; vulcanized rubber bushings

1. Introduction

Natural rubber (NR) is an elastomeric biopolymer composed of cis-1,4-polyisoprene [1,2]. It is one of the most widely used elastomers, with global production reaching 13.77 million tons in 2021 [3]. NR is extensively applied across diverse sectors, including automotive components, medical devices, petrochemicals, mining, and other industrial applications [4]. The growing diversity of NR-based applications has stimulated the development of advanced compounding systems, which incorporate various additives such as accelerators, activators, plasticizers [5], antioxidants [6], vulcanizing agents, and reinforcing fillers. Among these, reinforcing fillers play a pivotal role in enhancing the mechanical and dynamic performance of vulcanizates, including tensile strength, modulus, elongation at break, abrasion resistance, compression set, swelling resistance, and dimensional stability [7–9]. Nearly all commercial rubber vulcanizates utilize reinforcing fillers derived from petroleum-based carbon black (CB) [10,11]. Petroleum-derived CB is produced via incomplete combustion of hydrocarbons, where approximately two metric tons of petroleum feedstock are required to produce one metric ton of CB [12]. However, this reliance on petroleum-based CB imposes significant environmental burdens.

Among these, reinforcing fillers play a pivotal role in enhancing the mechanical and dynamic performance of vulcanizates, including tensile strength, modulus, elongation at break, abrasion resistance, compression set, swelling resistance, and dimensional stability [12–15]. In automotive applications, rubber bushings are essential for vibration isolation, enabling controlled movement between components while reducing noise and improving both comfort and safety [13–15]. To optimize these properties, conventional fillers such as CB are widely employed. Nevertheless, concerns regarding environmental impact and the depletion of fossil resources have driven interest in sustainable biomass-derived alternatives [16,17]. Bio-based fillers (bio-fillers) offer several advantages, including biodegradability, a lower environmental footprint, reduced density, and cost-effectiveness [18]. Their compatibility with rubber matrices further strengthens their potential for developing green composites [16].

The mechanical behavior of rubber bushings is strongly influenced by polymer type, filler loading, and compounding formulation [19]. Hence, the integration of bio-fillers into NR not only aligns with sustainable materials engineering but also provides promising mechanical reinforcement. Several bio-derived fillers have been investigated in NR systems. Cifriadi et al. [12] reported the utilization of cassava starch and tamarind wood, while Haslilywaty et al. [20] employed rice husk ash and tacca starch. Joshi et al. [21] demonstrated the reinforcing effects of banana fibers. These bio-fillers improved mechanical performance while also offering biodegradability [22]. Thomas et al. [23] highlighted the advantageous properties of cellulose-, chitin-, and starch-based fillers, which are cost-effective and environmentally friendly. Furthermore, Sowińska et al. and Suhu [24] reported improved thermal oxidation resistance and reinforcement using walnut shell powder and corn starch, respectively.

Activated carbon derived from lignocellulosic biomass, particularly sawdust, has recently emerged as a promising bio-filler due to its high surface area, porous structure, and abundant surface functionalities [25,26]. According to Negara et al., [27] sawdust-derived activated carbon exhibits mesoporous structures with an average pore diameter of 2.43 nm, a pore volume of 0.369 cm^3 g^{-1}, and a specific surface area 409.7 m^2 g^{-1}, enabling broad applicability. These properties facilitate both mechanical interlocking and physicochemical interactions with rubber matrices. Although its reinforcing potential is recognized, systematic studies remain limited on how processing parameters, particularly carbonization temperature and particle size affect its performance.

Rubber bushings operating under dynamic and cyclic loading conditions require high compressive strength, elasticity, and fatigue resistance [28]. While conventional fillers such as CB and silica provide excellent reinforcement [29,30], they often increase stiffness at the expense of elasticity. Consequently, bio-based fillers are sought as alternative reinforcements that can maintain mechanical performance while offering environmental benefits. Activated carbon from renewable precursors is an inexpensive and abundant alternative with the potential to generate added value [31].

Sawdust-derived activated carbon (SAC) typically exhibits high surface area, porosity, and functional group density, which promote strong interactions with polymer matrices.

For example, SAC has been reported to possess a hierarchical porous structure with micropores (0.55 nm) and mesopores (2.58 nm), resulting in a specific surface area of 621 m^2 g^{-1} and a pore volume of 0.35 cm^3 g^{-1} [32]. With such characteristics, SAC enhances interfacial interactions within rubber matrices. Previous studies synthesized SAC via chemical activation followed by carbonization 43. Chemically activated SAC exhibited favorable properties, including a BET surface area of ~1250 m^2 g^{-1}, a pore volume of 0.78 cm^3 g^{-1}, carbon content of ~93%, low ash content (7.36%), and an iodine number of 1161.3 mg g^{-1} [33]. Incorporation of SAC into rubber compounds at loadings up to 100 phr improved torque response and hardness.

Despite these advantages, most prior studies employed only a single SAC variant, limiting insights into how carbonization temperature and particle size govern reinforcement efficiency. Understanding these parameters is critical for tailoring filler morphology and functionality toward optimal performance in NR vulcanizates, particularly for high-demand automotive components such as bushings. This study aims to investigate the influence of carbonization temperature (200 °C and 400 °C) and particle size (75 μm and 38 μm) of SAC on the curing characteristics and mechanical properties of NR composites for automotive bushing applications. SAC200 and SAC400 were incorporated at various loadings and compared with an unfilled control. Fourier transform infrared spectroscopy (FTIR) was employed to explore surface chemistry variations, while vulcanization behavior (maximum torque, scorch time, and optimum cure time) and mechanical properties (tensile strength, elongation at break, hardness, and compression set) were systematically evaluated.

This work contributes to the development of environmentally friendly, high-performance rubber composites while supporting biomass waste valorization in advanced materials engineering. The incorporation of SAC into NR-based bushing compounds enhances interfacial polar interactions absent in conventional CB, resulting in superior reinforcement efficiency. Moreover, the findings emphasize that SAC surface chemistry plays a decisive role in determining the mechanical properties of rubber bushings.

Unlike previous studies, this study uses a single variant of activated carbon derived from sawdust. This study systematically explores the combined effects of carbonization temperature and particle size on the performance of reinforcing fillers in rubber vulcanizates. The novelty of this study lies in demonstrating how these processing parameters govern the drying behavior and mechanical response of rubber bushings under dynamic conditions. The results provide new insights into the tailoring of biomass-derived fillers for high-performance elastomer applications, while also promoting sustainable materials engineering through the utilization of sawdust waste.

2. Methods

2.1. Instrumentation

The preparation of the rubber compounds and vulcanized samples was conducted using the equipment listed in Table 1. Key instruments included a muffle furnace (Nabertherm GmbH) for sawdust carboniation, an XK-160 open two-roll mill for mastication and compounding, and a hydraulic hot press for vulcanization. Additional equipment included a digital balance for accurate weighing of materials and a Monsanto Moving Die Rheometer (MDR 2000, Alpha Technologies, Bellingham, WA, USA) for measuring vulcanization characteristics.

Table 1. Equipment specifications for the preparation of rubber compounds.

No	Equipment	Specifications
1	Muffle Furnace (Nabertherm GmbH, Lilienthal, Germany)	Max Temp: 1700 °C; Volume: 0.15 ft^3; Heating Elements: $MoSi_2$; Controller: Shimaden FP93
2	XK-160 Open Mill (Qingdao Ouli Machine Co., Ltd., Qingdao, China)	Roll Ø: 160 mm; Working length: 320 mm; Motor: 7.5 kW
3	Hydraulic Press (HTM10042T, Toyo Seiki Seisaku-sho, Ltd., Tokyo, Japan)	Pressure: 100 Ton; Platens: 2 levels; Temp: up to 300 °C
4	Digital Scale (Ohaus Corporation, Parsippany, NJ, USA)	Capacity: 5 kg; Resolution: 0.1 g; LCD display

2.2. Materials and Specifications

The materials utilized in this experiment are listed in Table 2.

Table 2. Materials used in the formulation of rubber composites.

No	Chemical Name	Source and Specifications
1	Natural Rubber (SIR-20)	PT. Prasidha Aneka Niaga Tbk, Indonesia; Compliant with SNI 1903:2017 (Ash: 0.25%, Dirt: 0.042%, Nitrogen: 0.24%, Volatile matter: 0.2%, PRI: 65%, Mooney viscosity: 70)
2	SAC200 (75 µm), SAC400 (38 µm)	Local supplier, Palembang, Indonesia
3	Zinc oxide (ZnO)	Shanghai, China; CAS No.: 1314-13-2; Purity: 99.7%
4	Stearic acid	Guangdong, China; CAS No.: 57-11-4; Purity: 99%
5	TMTD	Jiangsu, China; CAS No.: 137-26-8; Purity: 98%
6	CBS	Fujian, China; CAS No.: 95-33-2; Purity: 96%
7	TMQ	Hunan, China; CAS No.: 793-47-8; Purity: 99.9%
8	Paraffin oil (PO-60)	PT. Pertamina, Indonesia
9	6PPD	China; CAS No.: 793-24-8; Purity: 96%
10	PVI	Henan, China; CAS No.: 17796-82-6; Purity: 99%
11	Maleated Natural Rubber (MNR)	Local supplier
12	Phosphoric Acid (H_3PO_4)	Zhejiang, China; Purity: 85%
13	Sulfur	Tamil Nadu, India; CAS No.: 7704-34-9; Purity: 99.5%
14	Engine Oil 10W/40 4T	Local supplier, Indonesia

Note: All materials were procured from local Indonesian vendors.

2.3. Experimental Design

2.3.1. Preparation of Activated Charcoal

Sawdust was first washed with distilled water and oven-dried at 110 °C for 24 h. The carbonization process was conducted in a muffle furnace at 900 °C for 30 min to obtain raw charcoal, which was then ground and sieved to 200 mesh (75 µm) and 400 mesh (38 µm). The chemical activation process was initiated by soaking 5000 g of sieved charcoal in a 10% phosphoric acid (H_3PO_4) solution for 24 h at ambient temperature. Following impregnation, the samples were thoroughly rinsed with distilled water to remove residual acid. Subsequent carbonization was performed in a muffle furnace equipped with a gas circulation system utilizing hydrogen (H_2) and nitrogen (N_2), along with a hydrogen combustion safety unit. Two types of sawdust-derived activated charcoal were produced: SAC200—particle size of 75 µm, carbonized at 200 °C for 120 min; SAC400—particle size of 38 µm, carbonized at 400 °C for 120 min.

Post-carbonization, the activated charcoal was allowed to cool within the furnace for 24 h to prevent oxidation and then stored in airtight containers to maintain its integrity. The equipment utilized in this process is shown in Figure 1.

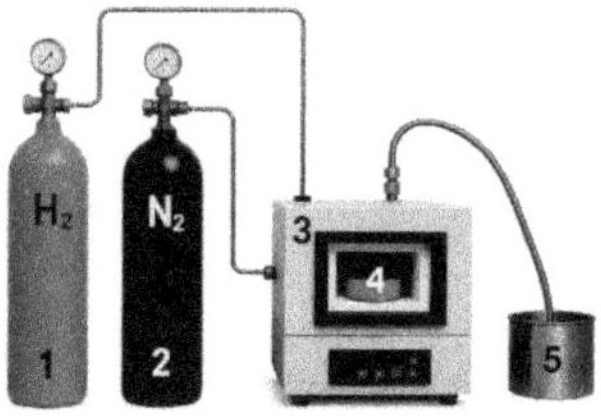

Figure 1. Schematic diagram of the carbonization equipment setup: (1) hydrogen gas cylinder, (2) nitrogen gas cylinder, (3) furnace, (4) carbonization reactor, and (5) nitrogen gas trap.

The quality parameters of the resulting activated charcoal were evaluated in accordance with SNI 06-3730-1995 standards (refer to Table 3) [34].

Table 3. Physicochemical characteristics of sawdust-based activated charcoal.

Parameter	Content
Water content	6.24%
Ash content	7.36%
Volatile matter	11.34%
Carbon	87.85%
Absorption of iodine	1161.34 mg/g

2.3.2. Rubber Compounding Procedure

Rubber compounding was conducted based on the formulation presented in Table 4 using a two-roll open mill. Natural rubber was first masticated for approximately 8 min until a plasticized state was achieved. The activator (ZnO) and co-activator (stearic acid) were then added and mixed for 3 min to ensure uniform dispersion. SAC was gradually incorporated and milled for 4 min until evenly distributed within the rubber matrix. Subsequently, the plasticizer, antioxidant, accelerator, scorch retarder, and coupling agent were sequentially introduced and mixed for 5 min to achieve a fully homogeneous compound. Finally, the vulcanizing agent (sulfur) was added and mixed for an additional 5 min until the compound reached complete uniformity.

Table 4. Formulations of natural rubber compounds with varying SAC particle sizes and loadings (phr).

Material	Formula (Per Hundred Rubber, phr)								
	1	2	3	4	5	6	7	8	9
SIR 20	100	100	100	100	100	100	100	100	100
ZnO	5	5	5	5	5	5	5	5	5
Stearic acid	1	1	1	1	1	1	1	1	1
6PPD	2	2	2	2	2	2	2	2	2
TMQ	1	1	1	1	1	1	1	1	1
SAC200	-	25	35	45	55	-	-	-	-
SAC400	-	-	-	-	-	25	35	45	55

Table 4. *Cont.*

Material	Formula (Per Hundred Rubber, phr)								
	1	2	3	4	5	6	7	8	9
Minarex oil	1	1	1	1	1	1	1	1	1
CBS	3	3	3	3	3	3	3	3	3
TMTD	2	2	2	2	2	2	2	2	2
MNR	1	1	1	1	1	1	1	1	1
Sulfur	2	2	2	2	2	2	2	2	2
PVI	0.5	0.5	0.5	0.5	0.5	0.5	0.5	0.5	0.5

2.3.3. Vulcanization

The rubber compounds were molded and vulcanized using a hydraulic hot press at 150 °C for 10 min under a pressure of 170 kgf/cm^2 to produce vulcanized rubber bushing (VRB) specimens.

2.4. Test Method

2.4.1. FTIR Spectroscopy

Fourier transform infrared (FTIR) spectroscopy was performed in this study using a Shimadzu IRTracer-100 spectrometer equipped with an Attenuated Total Reflectance (ATR) accessory to identify the surface functional groups present on the sawdust-derived activated carbon (SAC). Special attention was given to the detection of hydroxyl (–OH) and carbonyl (C=O) groups, which are known to promote interfacial interactions with elastomeric matrices and enhance filler–rubber compatibility. The presence of these oxygen-containing groups plays a crucial role in improving filler dispersion, interfacial adhesion, and overall reinforcement efficiency in rubber composites [35].

2.4.2. Curing Characteristics

Curing behavior was evaluated using a Moving Die Rheometer (MDR 2000) at 150 °C, following ISO 3417:2008 [36]. The parameters measured included minimum torque (Smin), maximum torque (Smax), scorch time (ts_2), and optimum cure time (t_{90}) [37,38]. The cure rate index (CRI) [39] was calculated using the following equation:

$$CRI = \frac{100}{tc_{90} - t_2} \times 100\% \quad (1)$$

2.4.3. Mechanical Testing Procedures

Mechanical properties were evaluated through standard test methods. Hardness was measured using a durometer in accordance with ASTM D2240-15 [40]. Tensile strength according to ASTM D395-18 (22 h at 100 °C) [40]. Swelling resistance was assessed following ASTM D471 [41] by immersing samples in engine oil at 100 °C for 72 h. The percentage swelling [42] was calculated using the following equation:

$$\%\ Swelling = \frac{W_2 - W_1}{W_1} \times 100\% \quad (2)$$

where W_1 is the mass of the specimen prior to immersion, and W_2 is the mass following immersion.

All mechanical tests were conducted in triplicate, and the results are presented as the mean ± standard deviation.

3. Results and Discussion

3.1. FTIR Analysis of Sawdust-Based Activated Carbon

The FTIR spectrum of sawdust-derived activated carbon (SAC) (Figure 2) revealed the presence of characteristic functional groups that contribute to the reinforcing capability of SAC in the natural rubber matrix. A broad absorption band centered at 3639–3138 cm^{-1} corresponds to O–H stretching vibrations from hydroxyl groups in phenolic, alcoholic, or adsorbed moisture species.

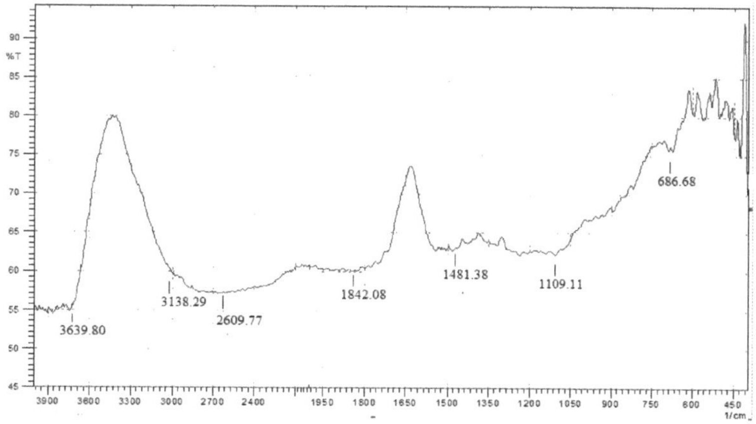

Figure 2. The FTIR spectrum of carbon black derived from sawdust indicates the presence of major functional groups.

These polar functionalities increase surface polarity, enabling hydrogen bonding and improved filler–rubber interactions. The bands observed at 2609.77 cm^{-1} and 1720–1842 cm^{-1} are assigned to C=O stretching vibrations, suggesting the presence of carbonyl groups from carboxylic acids, aldehydes, or ketones generated during oxidative activation. Such oxygenated groups are known to enhance interfacial reactivity and promote higher crosslinking density during vulcanization [36]. In addition, weak peaks near 2920 cm^{-1} (C–H stretching of $-CH_2/-CH_3$), 1481.38 cm^{-1} (C–H bending), and 1109.11 cm^{-1} (C–O stretching of secondary alcohols or ethers) further confirm the partial retention of lignocellulosic structures and the introduction of oxygen-based moieties. The fingerprint region (500–1100 cm^{-1}) also exhibits bands characteristic of alcohols, ethers, and esters, while the peak at 686.68 cm^{-1} corresponds to aromatic C=C stretching, indicating the presence of condensed aromatic structures derived from lignin carbonization [43]. While FTIR provides direct evidence of surface functional groups, the confirmation of elemental composition and chemical states requires complementary analysis. X-ray Photoelectron Spectroscopy (XPS) has been widely recommended in recent studies to quantify oxygen-containing groups (C–O, C=O, O–C=O) and to verify their chemical environment. For instance, XPS deconvolution of the C 1s region typically reveals sp^2-hybridized carbon (C=C), sp^3-hybridized carbon (C–C), and oxygenated species (C–O, C=O), in agreement with the oxygen functionalities detected by FTIR. The O 1s spectrum further distinguishes between hydroxyl, carbonyl, and carboxyl contributions. Therefore, the combination of FTIR and XPS provides a robust verification: FTIR identifies vibrational modes of functional groups, while XPS quantifies the atomic-level chemical states and confirms the abundance of oxygen functionalities that facilitate strong interfacial adhesion in SAC-filled rubber composites. The FTIR–XPS results in this study are in line with the report from Rosan et al. [44], Yang et al. [45], Marija Ilić et al. [46], where FTIR detects the characteristic vibrations of oxygenated groups (C–O, C=O), while XPS quantifies the contribution of oxygen species through the deconvolution of C1s and O1s peaks. This consistency strengthens the evidence that the presence of oxygen groups plays an important role in enhancing interfacial interactions in activated carbon-based materials.

This dual characterization not only validates the surface chemistry of SAC but also underpins its reinforcing role in natural rubber compounds formulated in Table 4. The simultaneous presence of hydroxyl, carbonyl, ether, and aromatic structures contributes to improved filler dispersion, increased cross-link density during vulcanization, and enhanced mechanical performance of vulcanized rubber bushings compared to conventional carbon black. Such complementary evidence highlights SAC as a sustainable and multifunctional bio-based filler with clear advantages in polymer reinforcement applications [47,48].

3.2. Curing Characteristics of Natural Rubber Compounds

The curing characteristics of natural rubber (NR) compounds filled with SAC200 and SAC400 at various loadings (0–55 phr) were evaluated using rheometric parameters, including maximum torque (S_{max}), minimum torque (S_{min}), torque difference ($\Delta_S = S_{max} - S_{min}$), scorch time ($ts_2$), optimum cure time ($t_{90}$), and cure rate index (CRI) [49]. The results, summarized in Table 5, reveal significant variations in vulcanization behavior across formulations. The chemical functionality of SAC, particularly the presence of oxygen-containing groups like hydroxyl and carbonyl, which improve interactions with curing agents and the rubber matrix, is primarily responsible for these variations [50]. FTIR test results indicate the presence of dominant hydroxyl and carbonyl groups in SAC400. The presence of these groups affects the hardening characteristics of the rubber compound. With the presence of this group, the maximum torque (Smax) of SAC400 vulcanizate increased significantly from 3.451 N·m to 3.733 N·m when compared with SAC200 and the control.

Table 5. Curing characteristics of natural rubber composites filled with SAC200 and SAC400.

Sample	Loading (phr)	Curing Characteristic					
		S_{max} (N·m)	S_{min} (N·m)	$S_{max} - S_{min}$ (N·m)	Opt Cure Time (tc_{90}) (min; s)	Scorch Time (ts_2) (min; s)	CRI (s^{-1})
NR/CB Without filler	0	3.683	0.874	2.810	12:35	6:79	0.316
Without filler	0	3.451	0.796	2.655	15:59	8:45	0.230
SAC200 (75 μm)	25	3.037	0.610	2.427	15:29	8:02	0.223
	35	3.065	0.584	2.480	14:45	7:56	0.244
	45	3.143	0.512	2.631	13:57	7:25	0.255
	55	3.247	0.708	2.539	13:07	7:10	0.273
Without filler	0	3.451	0.796	2.655	15:59	8:45	0.230
SAC400 (38 μm)	25	3.241	0.775	2.466	14:57	7:47	0.233
	35	3.526	0.834	2.693	13:15	7:00	0.267
	45	3.722	0.847	2.874	12:28	6:46	0.292
	55	3.733	0.884	2.850	11:05	5:58	0.326

Note: Torque values originally expressed in kg·cm were converted to N·m using the conversion factor 1 kg·cm = 0.0980665 N·m. NR: natural rubber; phr: per hundred rubber; CB: carbon black; SAC: sawdust-activated carbon.

The difference in filler loading has a significant effect on cross-link formation. At a filler loading of 30 phr, the cross-link density increases and the torque delta (Δ_S) changes. This implies that the network structure and stiffness of the rubber have improved [51]. However, at filler loadings above 30 phr, Δ_S decreases and the ts_2 and t_{90} values increase. This phenomenon implies that excessive amounts of filler can cause the rubber chain to become unstable and the resulting cross-link system to become less uniform.

Among the measured parameters, maximum torque (S_{max}) rose significantly with increasing SAC content, particularly for SAC400. A S_{max} of 38.07 kg·cm was recorded by SAC400 at 55 phr, indicating increased stiffness and cross-link density. This was ascribed to enhanced filler–matrix adhesion, which was facilitated by the polar surface and numerous active sites of SAC400 [52–55]. Conversely, minimum torque (S_{min}), which reflects compound viscosity before curing, showed only minor variation with SAC loading. Yet, SAC400 showed a marked increase in Δ_S at higher loadings, peaking at 29.31 kg·cm at 45 phr, likely due to its finer particle size (38 μm), which enhances surface area and dispersion efficiency [56]. SAC400, with a surface area of 38 μm, has a higher density of polar functional groups (–OH, –COOH, –C=O). These groups act as catalytic sites that can interact with sulfur during vulcanization. These groups accelerate the cross-linking reaction. Furthermore, vulcanizates loaded with SAC400 exhibited a faster ts_2 time than SAC200, resulting in a faster cross-linking reaction.

Similar trends have been reported in previous studies, where activated carbon with abundant surface functionalities accelerated the drying kinetics and enhanced the rubber–filler interactions. Li et al. [57] showed that activated carbon with very high surface area had abundant oxygen functionalities, while Ma et al. [58] reported that Sakura-derived activated carbon exhibited significant –OH, –C=O, and –COOH groups. The results indicate that the bio-based activated carbon filler accelerated the drying time and improved the mechanical performance of natural rubber composites, consistent with the faster ts_2 observed in the SAC400-filled vulcanizate.

The SAC material used in this study possessed high porosity and carbon content (72.32%), contributing to better surface reactivity and adsorption capacity [59]. These properties enabled more efficient heat transfer and molecular mobility during vulcanization, leading to denser crosslinking and improved mechanical response. Consequently, both scorch time (ts_2) and optimum cure time (t_{90}) decreased with SAC400 loading, reaching their lowest values at 5 min 58 sec and 11 min 05 sec, respectively. Acceleration of the curing process, in line with previous findings that finer particles increase thermal conductivity and internal friction, thereby shortening the overall curing time [60]. A shorter t_{90} increases energy efficiency in processing [61,62], while a lower ts_2 improves the processability of the compound.

The cure rate index (CRI), representing vulcanization kinetics, was highest for SAC400 at 55 phr, confirming its superior curing performance [63]. SAC400's enhanced surface properties high surface area, porosity, and polarity are key contributors to this outcome, promoting faster heat distribution and chain mobility. As observed in Table 5, the trend of increasing Smax with SAC loading suggests a corresponding increase in cross-link density, likely driven by interactions between SAC hydroxyl groups and the NR matrix [64]. Meanwhile, S_{min} values fluctuated moderately, likely influenced by the balance between filler dispersion and matrix viscosity.

At equivalent loadings, SAC400 consistently outperformed SAC200 in terms of ts_2, t_{90}, and CRI, confirming its superior curing efficiency. Variations in cross-link density and internal heat generation largely governed the differences in curing behavior from filler–rubber friction [65]. Overall, NR compounds filled with 45–55 phr SAC400 exhibited the most favorable curing characteristics, high Smax, low ts_2, t_{90}, and elevated CRI values. These findings position SAC400 not only as a reinforcing filler but also as an active contributor to vulcanization kinetics. Its dual role underscores its potential as an efficient, bio-based alternative to conventional fillers, particularly for demanding applications such as automotive rubber bushings [66].

3.3. Mechanical Properties and Performance Analysis

3.3.1. Hardness

Hardness is a key mechanical parameter reflecting the resistance of a vulcanizate to surface deformation and is directly related to cross-linking density and filler–matrix interactions. In automotive rubber bushings (VRBs), which must maintain dimensional stability under repeated mechanical stress, hardness serves as a crucial performance metric. As shown in Figure 3 and Table 4, the addition of SAC to natural rubber compounds can significantly increase hardness.

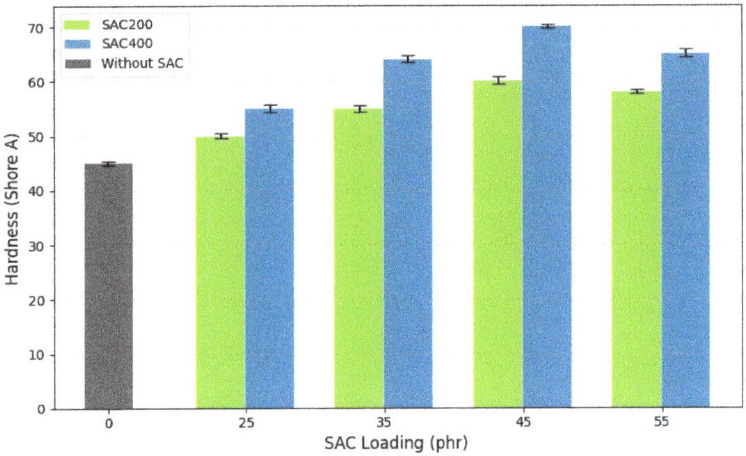

Figure 3. The effect of SAC type and loading on the Shore A hardness of vulcanized natural rubber bushings.

The unfilled control sample exhibited a baseline hardness of 45 Shore A. With increasing SAC loading, a progressive rise in hardness was observed. SAC400, carbonized at 400 °C and possessing a finer particle size (38 µm), consistently yielded higher hardness values than SAC200 (75 µm, 200 °C). At 55 phr, SAC400 reached 70 Shore A, fulfilling the industrial standard for VRB (70 ± 5 Shore A). Because SAC200 could only support 60 Shore A at the same loading, the reinforcement was less effective. Activated carbon with smaller particle size and richer surface functionality provides better reinforcement in the rubber matrix, resulting in higher hardness values compared to larger particle fillers.

The superior performance of SAC400 is attributed to its higher surface area and carbon purity, which promote effective filler dispersion and interfacial bonding. FTIR analysis (Figure 2) confirmed the presence of functional groups such as C=C and C=O, along with reduced –OH intensity in SAC400, indicating greater graphitization and hydrophobicity. The detection of sulfur-containing bonds (C–S, S–S) further indicates that SAC400 can participate in the vulcanization reaction, contribute to the formation of three-dimensional networks, and influence the VRB hardness value. The enhancement in hardness is closely related to increased crosslink density. The observed increase in hardness is primarily due to the presence of functional groups in SAC, as identified through FTIR analysis. These functional groups enhance interfacial interactions with the rubber matrix. Furthermore, cross-linking formed by sulfur, accelerated by accelerators and activators, increases the overall cross-linking density. This contributes to the increase in VRB hardness. These findings are consistent with previous findings on bio-based fillers [67]. The use of the ZnO stearic acid system, in combination with CBS and TMTD accelerators, supported efficient vulcanization and uniform crosslink distribution. However, hardness plateaued at loadings above 45 phr, likely due to filler agglomeration, which may hinder matrix uniformity and restrict further reinforcement.

The presence of coupling agents during mixing reduces interfacial tension and improves filler–rubber adhesion [68]. In modified natural rubber (MNR), polar functional groups can interact favorably with SAC, while non-polar domains align with the rubber matrix. Larger SAC particles may limit chain mobility during deformation, resulting in greater localized indentation and decreased hardness [69]. Overall, SAC400-filled NR compounds at 45 and 55 phr met commercial hardness requirements, demonstrating the material's capability for application in high-performance automotive bushings (VRB). The combination of optimized morphology and surface chemistry supports SAC400 as a viable bio-derived alternative to conventional fillers in elastomeric systems.

3.3.2. Tensile Strength

Tensile strength is a fundamental parameter for assessing the mechanical performance of vulcanized rubber (VRB), especially in reinforced systems. It represents the material's ability to withstand stress under tensile loading before failure. Figure 4 presents the tensile strength of VRB filled with SAC200 and SAC400 across various loadings.

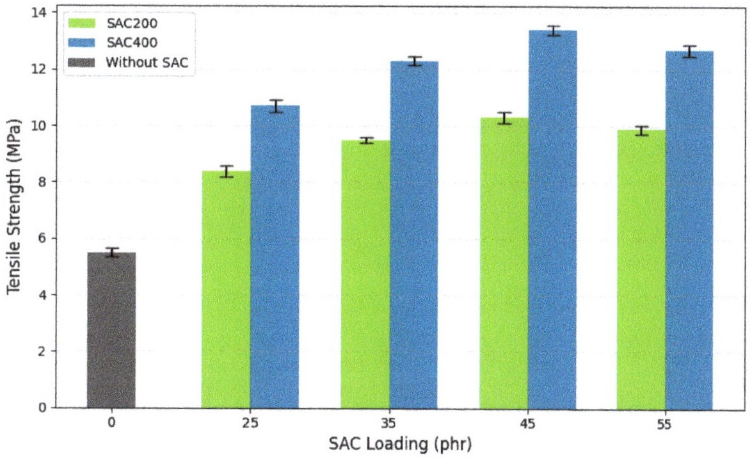

Figure 4. The effect of SAC type and loading on the tensile strength of vulcanized rubber bushings.

The unfilled compound exhibited a tensile strength of 5.52 MPa, establishing the baseline for comparison. As SAC content increased, tensile strength improved progressively, confirming SAC's reinforcing capability. The increased tensile strength is due to improved filler–matrix adhesion, interfacial stress transfer, and physical interlock, all of which are facilitated by SAC's porous structure and large surface area. These results are consistent with earlier studies on the enhancement of elastomeric properties by carbon fillers derived from biomass [70]. The existence of polar functional groups on the filler surface explains the higher tensile strength of VRB, particularly those reinforced with SAC400.

The polar groups actively encourage greater interfacial adhesion by forming hydrogen bonds and dipole–dipole interactions with the non-polar polyisoprene chains. Interfacial adhesion improves stress transfer effectiveness in the composite. Furthermore, the polar surface properties of SAC400 promote better dispersion into the NR matrix. The resulting interfacial adhesion interacts with ZnO, stearic acid, sulfur, and other additives to form tighter cross-links. This results in a significant increase in tensile strength.

Across all loadings, SAC400 outperformed SAC200. At 45 phr, SAC400 reached a peak tensile strength of 13.4 MPa, significantly higher than SAC200's 9.56 MPa at the same loading, an increase of approximately 40%. This value exceeds the tensile strength range typically reported for carbon black-filled NR (11–13 MPa at 30 phr) [71,72], highlighting SAC400's competitive performance. The superior performance of SAC400 is attributed to its finer particle size (38 μm), higher carbonization temperature (400 °C), and higher surface area, which collectively enhance dispersion and interfacial bonding. At 55 phr, SAC400 maintained a high tensile strength of 12.51 MPa, surpassing the minimum industry requirement of 11 MPa for automotive VRB. In contrast, SAC200 exhibited a marked reduction at this loading, likely due to agglomeration and poor dispersion, common challenges in particulate-filled elastomer systems.

The reinforcing mechanism involves both physical and chemical interactions. Physically, SAC particles interact with rubber chains via van der Waals forces. Chemically, functional groups on the SAC surface (–OH, C=O) form bonds with the rubber matrix during vulcanization [72]. The presence of maleated natural rubber (MNR) as a coupling agent further enhances compatibility and stress distribution. SAC400's optimized morphology also contributes to a more uniform crosslink network, resulting in higher tensile strength [73]. These results support earlier findings on the importance of filler particle size, surface area, and dispersion in achieving mechanical reinforcement [74]. Overall, SAC400 demonstrated superior performance compared to SAC200, positioning it as a highly effective, sustainable filler for high-performance vulcanized rubber bushings.

3.3.3. Elongation at Break

Elongation at break is an important parameter that reflects the flexibility and toughness of VRB under tensile stress. This parameter indicates the extent to which a material can stretch before breaking. As shown in Figure 5, the unfilled control sample (0 phr SAC) exhibited an elongation at break of 530%, demonstrating unrestricted chain mobility in the absence of fillers. The SAC200 and SAC400 loadings on each VRB significantly affected these properties. The loading of SAC200 and SAC400 on each VRB significantly affected these properties, consistent with previous findings that the addition of bio- or carbon-based fillers limits chain mobility and reduces elongation at break due to stronger filler–rubber interactions. According to Azura et al. [71], elongation at break is influenced by filler load, where increasing filler load will reduce elongation at break.

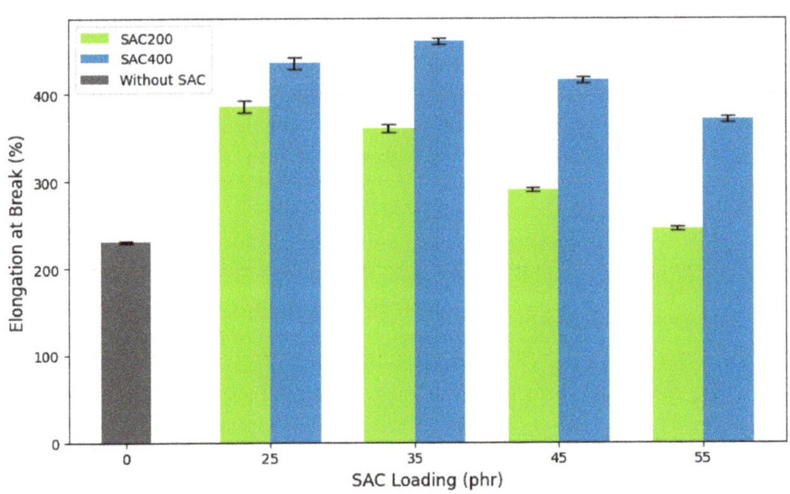

Figure 5. The effect of SAC type and loading on the elongation at break of vulcanized rubber bushings.

The gradual addition of SAC fillers reduced elongation due to restricted polymer chain mobility, as the rigid particles acted as physical crosslinking points. In SAC200-filled compounds, elongation values declined notably, especially at 35–55 phr, where values fell below the minimum industrial standard of 400%. This was attributed to poor dispersion, weak filler–rubber interfacial bonding, and filler agglomeration, which disrupted matrix homogeneity and increased stiffness.

At higher SAC content, the reduction in free volume further hindered chain flexibility. Excess filler led to saturation, preventing complete interaction between the SAC and the VRB. This condition compromises elasticity and promotes premature failure under tensile loading. In contrast, SAC400 demonstrated superior elongation-at-break performance. At 25 phr and 35 phr, elongation at break reached 435% and 460%, respectively, within the optimal range (400–500%) for VRB applications. However, further increasing the loading to 45 phr and 55 phr reduced elongation to 415% and 370%, respectively. This reduction was due to filler aggregation, which restricted chain extension and introduced stress concentration points.

The improved performance of SAC400 is largely due to its smaller particle size (38 μm) and higher carbonization temperature (400 °C), which results in greater surface area and porosity. These characteristics enhance stress distribution across the VRB, interfacial adhesion, and filler dispersion. SAC400's higher porosity resulting from thermal activation also allowed for tighter rubber filler contact and better stress transfer, as shown in similar studies using biomass-based carbon [54]. By comparison, SAC200 with a larger particle size (75 μm) led to uneven stress distribution, localized microvoids, and poor elongation results at higher loadings. These phenomena underline the importance of filler structure and dispersion quality in mechanical reinforcement [75].

The 35 phr SAC400 formulation represented the optimal filler level, where elongation at break peaked at 460%. Above this threshold, the performance declined below the 400% standard, reflecting the onset of filler overloading. Similar behavior was observed in carbon-filled composites post-thermal treatment, as reported [76]. Notably, SAC-filled VRB outperformed conventional carbon black systems, which typically exhibit maximum elongation between 350% and 400% at 20–30 phr [77]. In comparison, SAC400 reached 460% at 35 phr, demonstrating its superior reinforcement efficiency. This also highlights SAC400's viability as a sustainable alternative to petroleum-based fillers, combining mechanical reliability with environmental advantages, wide availability, and renewability. When considered together, the data confirm that SAC400 enhances both elasticity and toughness of VRB more effectively than SAC200, particularly at optimal loading. Its favorable elongation performance, alongside its bio-based origin, supports its use in the development of environmentally sustainable, high-performance rubber composites.

3.3.4. Compression Set

Compression set is an important parameter for evaluating the long-term dimensional stability and elastic recovery of rubber materials, especially in VRB applications. Compression set measures the ability of a rubber vulcanizate to return to its original shape after prolonged compression at elevated temperatures [78]. Lower VRB compression set values indicate superior elasticity, structural resistance, and the material's capacity to maintain mechanical integrity under thermal and compression loads [79]. Figure 6 presents the compression set values of VRB reinforced with various loadings of SAC200 and SAC400. All samples were aged at 100 °C for 72 h to simulate thermal degradation conditions. The control (unfilled) sample exhibited a compression set of 29.5%, typical of non-reinforced vulcanizates. In contrast, the addition of SAC fillers significantly reduced compression set values across all formulations, suggesting enhanced elastic recovery due to increased crosslink density and stronger filler–matrix interactions. Figure 6 presents the improved VRB compression set values with various SAC200 and SAC400 loadings.

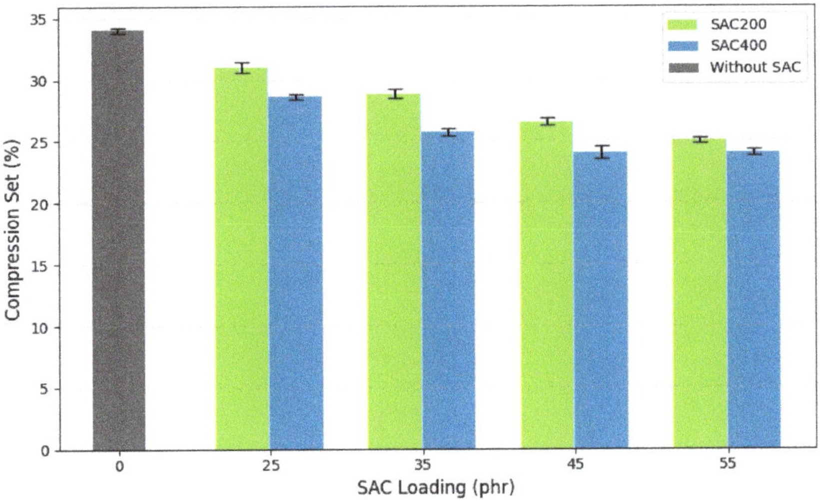

Figure 6. The effect of SAC type and loading on the compression set values of vulcanized rubber bushings.

For samples loaded with SAC200, the VRB compression set decreased from 24.2% at 10 phr to 20.3% at 30 phr, indicating improved dimensional stability. SAC400 consistently outperformed SAC200 at equivalent loads. At 30 phr, SAC400 achieved the lowest compression set value of 18.7%, representing a 36.6% reduction compared to the control. At 45 phr, SAC400 maintained a value around 24%, still lower than the SAC200 counterpart.

These improvements are attributed to SAC400's smaller particle size (38 μm), higher surface area, and more porous structure—resulting from carbonization at 400 °C, which facilitated better dispersion and stronger interfacial bonding with the rubber matrix [66,71]. The superior performance of SAC400 is further supported by its higher surface polarity and structural stability, which helped distribute applied pressure more evenly and restricted polymer chain relaxation [80]. This led to reduced permanent deformation and improved dimensional resilience. The enhanced filler–rubber bonding in SAC400 composites is due to greater contact area and a more favorable distribution of functional groups, consequences of its higher carbonization temperature and optimized microstructure.

A sharp decline in compression set was observed when filler loading exceeded 25 phr, suggesting the formation of a percolation network. This interconnected filler structure enhanced matrix reinforcement and minimized voids through increased physical contact and interfacial bonding [81]. Similar trends have been reported in elastomers filled with high-surface-area carbon materials, where enhanced stress transfer and limited chain mobility reduced permanent deformation [29]. These observations align with the study by Yan et al. [55], which demonstrated that finer carbon fillers significantly improve stress distribution and limit molecular relaxation. The 25 phr threshold appeared critical, beyond which compression set performance improved markedly [82]. Notably, SAC400-filled VRB in the 35–55 phr range showed compression set values between 24% and 30%, meeting the industrial requirement ($\leq$30%) for automotive and general-purpose bushing applications.

The excellent compression set resistance of SAC400 is also linked to its high porosity and functional surface chemistry, which promote physical entrapment of polymer chains and the formation of van der Waals and hydrogen bonds [83]. These interactions increase network compactness, reduce voids, and enhance the rubber's ability to recover from compressive strain. The synergy between small particle size, high carbonization temperature, and broad surface functionality highlights the potential of SAC400 as a sustainable active filler alternative derived from biomass.

3.3.5. Swelling Behavior

Swelling behavior in VRB is primarily governed by free volume, crosslink density, and interfacial interactions between the rubber matrix and the swelling medium. The swelling rate, defined as the percentage of absorbed oil, reflects the extent of solvent diffusion within the polymer network [42]. In this study, swelling tests were conducted on VRB using oil immersion at two temperatures (25 °C and 100 °C), as shown in Figure 7.

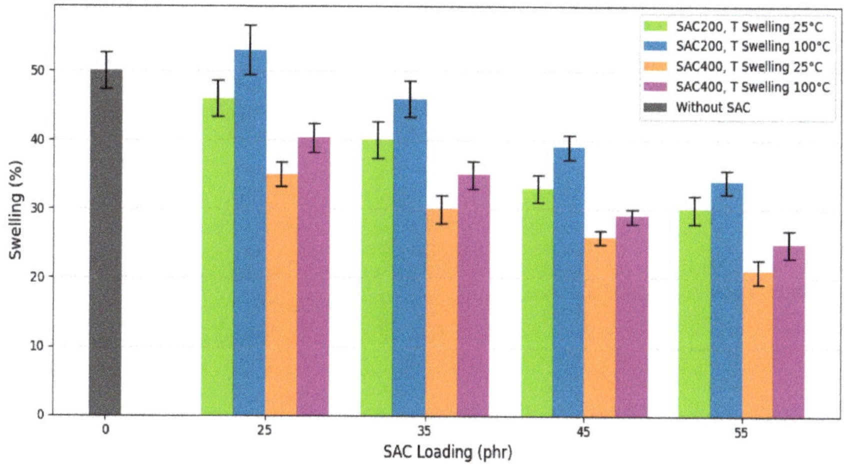

Figure 7. The effect of SAC type and loading on the oil swelling behavior of vulcanized rubber bushings.

The formulations are detailed in Table 4. At 25 phr filler loading, SAC200 and SAC400 exhibited the highest swelling ratios, indicating lower crosslink densities. SAC200 swelled up to 50% at 25 °C, while the unfilled compound showed ~55% swelling at 100 °C. This high swelling [84], is associated with insufficient crosslink formation, which permits solvent penetration and polymer chain expansion. Additionally, polar groups result in less swelling in non-polar solvents, which suggests improved network integrity and solvent resistance.

When the filler content increased to 55 phr, the swelling percentage of VRB decreased significantly. This phenomenon indicates increased cross-linking and improved compatibility of the filler with the rubber matrix due to the addition of SAC. VRB filled with SAC400 consistently showed lower swelling compared to SAC200. Furthermore, the denser three-dimensional network formed during vulcanization reduces solvent uptake by physically restricting chain mobility and void space. This effect is supported by FTIR results and aligns with previous findings that higher cross-linking density reduces matrix permeability and swelling [62]. Furthermore, the results of FTIR analysis showed that the presence of polar groups increased tissue integrity and solvent resistance by reducing the amount of swelling in non-polar solvents.

Swelling resistance was consistently lower at higher SAC loadings, particularly in the range of 35–55 phr for SAC400. These results emphasize the role of filler morphology and surface chemistry in limiting chain relaxation and solvent transport. This behavior is more pronounced at high temperatures (100 °C), where solvent diffusion is accelerated and mechanical integrity is reduced. Nevertheless, SAC-filled samples maintained better swelling resistance compared to unfilled VRB. Furthermore, strong interfacial bonding through hydrogen bonding, van der Waals forces, and chemical linkages contributed to improved matrix cohesion and solvent exclusion, This was especially evident in SAC400 systems, which offered superior dimensional and chemical stability under thermal and chemical stress. In summary, VRB filled with SAC400 exhibited superior swelling resistance compared to SAC200. This was attributed to the finer structure of SAC400, higher surface reactivity, and better integration into the rubber matrix. These findings support the use of SAC400 as a reinforcing filler to enhance cross-linking and reduce solvent-induced degradation of VRB in vulcanized rubber systems.

4. Conclusions

(1) Sawdust-derived activated carbon (SAC) successfully incorporates oxygenated and aromatic functional groups, as confirmed by FTIR and XPS.
(2) The addition of SAC improves the hardening properties, tensile strength, modulus, hardness, and tear resistance of NR vulcanizates.
(3) SAC200 retains more oxygenated groups, enhancing interfacial interactions, while SAC400 increases stiffness due to the increase in aromatic domains.
(4) Compared with conventional carbon black and other biomass-derived fillers, SAC exhibits superior sustainability and reinforcement potential.

Author Contributions: E.P.W.: Conceptualization, Data curation, Methodology, Writing—original draft, Writing—review & editing. P.M.: Conceptualization, Funding acquisition, Project administration, Writing—original draft. N.: Conceptualization, Funding acquisition, Methodology, Writing—original draft, Writing—review & editing. L.: Resources, Writing—review & editing. H.Y.: Data curation, Validation, Writing—review & editing. W.E.W.: Formal analysis, Investigation, Visualization. S.J.M.: Formal analysis, Investigation, Supervision. A.: Data curation, Project administration, Validation. E.B.S.: Data curation, Formal analysis, Resources, Investigation. Y.B.: Formal analysis, Investigation, Visualization. H.G.: Formal analysis, Investigation, Writing—review & editing. W.B.S.: Conceptualization, Supervision, Writing—original draft, Writing—review & editing. All authors have read and agreed to the published version of the manuscript.

Funding: This research received funding and research facilitation from the Palembang Industrial Research and Standardization Center, Industrial Research and Development Agency, Ministry of Industry of the Republic of Indonesia.

Institutional Review Board Statement: This research does not involve humans and animals as research objects.

Data Availability Statement: The original contributions presented in this study are included in the article material. Further inquiries can be directed to the corresponding author.

Acknowledgments: The authors are deeply grateful to the Research Organization for Agriculture and Food, a part of the National Research and Innovation Agency of the Republic of Indonesia, for the invaluable support for this work. They also express sincere appreciation for the funding and facilitation of this research by the Palembang Center for Standardization and Industrial Services (BSPJI). Furthermore, the authors would like to acknowledge the technical assistance provided by the technicians at the Rubber Laboratory of the Palembang Industrial Research and Standardization Center.

Conflicts of Interest: The authors declare that they have no known competing financial 566 interests or personal relationships that could have appeared to influence the work re- 567 ported in this paper.

Abbreviations

The following abbreviations are used in this manuscript:

NR	Natural rubber
SAC	Sawdust-derived activated carbon
VRB	vulcanized rubber bushings
SAC200	Activated carbon derived from sawdust with a particle size of 75 μm, carbonized at a temperature of 200 °C
SAC400	Activated carbon derived from sawdust with a particle size of 38 μm, carbonized at a temperature of 400 °C
ZnO	Zinc Oxide
CBS	N-Cyclohexyl2-benzothiazole sulfenamide
TMTD	Tetramethyl thiuram disulfide
MNR	Maleated natural rubber
SIR-20	Standar Indonesia Rubber
TMQ	Polymerized 2,2,4-trimethyl-1,2-dihydroquinoline
6PPD	N-(1,3-dimethylbutyl)-N′-phenyl-p-phenylenediamine
PVI	Pre-Vulcanization Inhibitor
H_3PO_4	Phosphoric Acid
CB	Carbon black
H_2	Hydrogen
N_2	Nitrogen
FTIR	Fourier transform infrared
ATR	Attenuated Total Reflectance
MDR	Moving Die Rheometer
CRI	Cure rate index
ASTM	American Society for Testing and Material
W	Weight
Mpa	Megapascal
°C	Degrees Celsius
%	Percent
Mikrometer	μm
kgf	Kilogram-force
cm	Centimeters
μm	micrometer
phr	parts per hundred of rubber
ts_2	Scorch time
t_{90}	Cure time
S_{max}	Maximum torque
S_{min}	Minimum torque
Δ_S	Torque difference

References

1. Andler, R. Bacterial and enzymatic degradation of poly(cis-1,4-isoprene) rubber: Novel biotechnological applications. *Biotechnol. Adv.* **2020**, *44*, 107606. [CrossRef]
2. Whba, R.; Su'ait, M.S.; Whba, F.; Sahinbay, S.; Altin, S.; Ahmad, A. Intrinsic challenges and strategic approaches for enhancing the potential of natural rubber and its derivatives: A review. *Int. J. Biol. Macromol.* **2024**, *276*, 133796. [CrossRef]
3. Marrero Nunes, F.; Emmel Silva, A.L.; May, J.; da Silva Szarblewski, M.; Flemming, L.; Assmann, E.E.; Ribas Moraes, J.A.; Machado, Ê.L. Environmental impacts associated with the life cycle of natural rubbers: A review and scientometric analysis. *Ind. Crops Prod.* **2025**, *224*, 120350. [CrossRef]
4. Dunuwila, P.; Rodrigo, V.H.L.; Goto, N. Sustainability of natural rubber processing can be improved: A case study with crepe rubber manufacturing in Sri Lanka. *Resour. Conserv. Recycl.* **2018**, *133*, 417–427. [CrossRef]
5. Rahman, M.M.; Oßwald, K.; Reincke, K.; Langer, B. Influence of bio-based plasticizers on the properties of NBR materials. *Materials* **2020**, *13*, 2095. [CrossRef]

6. Zhao, W.; He, J.; Yu, P.; Jiang, X.; Zhang, L. Recent progress in the rubber antioxidants: A review. *Polym. Degrad. Stab.* **2023**, *207*, 110223. [CrossRef]
7. Yang, J.; Wang, F.; Liang, C.; Zhou, S.; Huang, J.; Zhao, G.; Liu, Y. Trans-1,4-poly(isoprene-co-butadiene) rubber enhances abrasion resistance in natural rubber and polybutadiene composites. *Polymer* **2025**, *316*, 127855. [CrossRef]
8. Sethulekshmi, A.S.; Saritha, A.; Joseph, K. A comprehensive review on the recent advancements in natural rubber nanocomposites. *Int. J. Biol. Macromol.* **2022**, *194*, 819–842. [CrossRef]
9. Lee, S.H.; Park, G.W.; Kim, H.J.; Chung, K.; Jang, K.S. Effects of Filler Functionalization on Filler-Embedded Natural Rubber/Ethylene-Propylene-Diene Monomer Composites. *Polymers* **2022**, *14*, 3502. [CrossRef]
10. Farida, E.; Bukit, N.; Ginting, E.M.; Bukit, B.F. The effect of carbon black composition in natural rubber compound. *Case Stud. Therm. Eng.* **2019**, *16*, 100566. [CrossRef]
11. Mekbuntoon, P.; Kongpet, S.; Kaeochana, W.; Luechar, P.; Thongbai, P.; Chingsungnoen, A.; Chinnarat, K.; Kaewnisai, S.; Harnchana, V. The Modification of Activated Carbon for the Performance Enhancement of a Natural-Rubber-Based Triboelectric Nanogenerator. *Polymers* **2023**, *15*, 4562. [CrossRef]
12. Cifriadi, A.; Sugita, P.; Wismogroho, A.S.; Kemala, T.; Nikmatin, S.; Widayatno, W.B.; Amal, M.I.; Jayadi; Firdharini, C.; Kinasih, N.A. Development and Characterization of Biomass-Based Biocarbon Fillers for Natural Rubber Composites. *Int. J. Technol.* **2025**, *16*, 982–992. [CrossRef]
13. Adduci, R.; Vermaut, M.; Perrelli, M.; Cosco, F.; Vanpaemel, S.; Naets, F.; Mundo, D. A review of bushing modelling approaches for MultiBody simulations. *Mech. Mach. Theory* **2024**, *191*, 105496. [CrossRef]
14. Guo, P.; Li, J.; Lin, Z.; Lin, J. Analysis and Optimization of Driveline Bushing for Lateral Ride Vibration under Shock Excitation. *Appl. Sci.* **2021**, *11*, 2647. [CrossRef]
15. Ambarev, K.; Taneva, S. A Comparative Study of Vibrations in Front Suspension Components Using Bushings Made from Different Materials. *Eng. Proc.* **2025**, *100*, 42. [CrossRef]
16. Dadkhah, M.; Messori, M. A comprehensive overview of conventional and bio-based fillers for rubber formulations sustainability. *Mater. Today Sustain.* **2024**, *27*, 100886. [CrossRef]
17. Mokhothu, T.H.; John, M.J. Bio-based fillers for environmentally friendly composites. *Handb. Compos. Renew. Mater.* **2017**, *1–8*, 243–270. [CrossRef]
18. Wang, L.; Li, Q.; Jin, H.; Wang, J.; Wang, N.; Zhou, X. Research progress of bio-filler and its application in asphalt materials: Review and discussion. *Clean Technol. Environ. Policy* **2025**. [CrossRef]
19. Kiran, M.D.; Govindaraju, H.K.; Jayaraju, T.; Kumar, N. Review-Effect of Fillers on Mechanical Properties of Polymer Matrix Composites. *Mater. Today Proc.* **2018**, *5*, 22421–22424. [CrossRef]
20. Haslilywaty, G.; Siti Fatma, A.K.; Roslim, R.; Mohibbah, M.; Jefri, J. Effect of bio fillers on mechanical properties of natural rubber latex films. *Key Eng. Mater.* **2019**, *797*, 249–254. [CrossRef]
21. Joshi, M.; Dwivedi, C.; Manjare, S. A renewable cellulose-rich biofiller material extracted from waste banana stem fibers for reinforcing natural rubber composites. *J. Mater. Sci.* **2024**, *59*, 519–534. [CrossRef]
22. Tong, H.S.; Kabeb, S.M.; Abd Hamid, H.; Zulkifli, F.H. A review of biodegradability of natural rubber products: Physicochemical, thermal and mechanical properties. *Int. J. Biol. Macromol.* **2025**, *318*, 144973. [CrossRef] [PubMed]
23. Thomas, S.K.; Parameswaranpillai, J.; Krishnasamy, S.; Begum, P.M.S.; Nandi, D.; Siengchin, S.; George, J.J.; Hameed, N.; Salim, N.V.; Sienkiewicz, N. A comprehensive review on cellulose, chitin, and starch as fillers in natural rubber biocomposites. *Carbohydr. Polym. Technol. Appl.* **2021**, *2*, 100095. [CrossRef]
24. Sahu, N. Walnut Shells ' Potential Use as Biofillers in Natural Rubber Composites. *Int. J. Polym. Sci. Eng.* **2024**, *10*, 1–5.
25. Kumar Mishra, R.; Singh, B.; Acharya, B. A comprehensive review on activated carbon from pyrolysis of lignocellulosic biomass: An application for energy and the environment. *Carbon Resour. Convers.* **2024**, *7*, 100228. [CrossRef]
26. Eguare, K.O.; Ekabafe, L.O.; Ayo, M.D. Comparative Study of Acetylated Sawdust Powder as Filler in Natural Rubber Compounding. *NIPES—J. Sci. Technol. Res.* **2020**, *2*, 78–84.
27. Negara, D.N.K.P.; Nindhia, T.G.T.; Kencanawati, C.I.P.K.; Suriadi, I.G.A.K.; Widiyarta, I.M.; Lokantara, I.P.; Budiarsa, I.N. The Effect of Carbonisation Heating Rates on the Properties of N-Doped Teak Sawdust Waste Activated Carbon. *J. Phys. Sci.* **2023**, *34*, 1–20. [CrossRef]
28. Wang, A.; Wang, H.; Ling, C.; Wang, T.; Li, F.; Yang, S.; Zhao, M. Fatigue life prediction of rubber suspension bushings based on virtual road load spectrum pulses. *Results Eng.* **2024**, *23*, 102693. [CrossRef]
29. Fan, Y.; Fowler, G.D.; Zhao, M. The past, present and future of carbon black as a rubber reinforcing filler—A review. *J. Clean. Prod.* **2020**, *247*, 119115. [CrossRef]
30. Calaf-Chica, J.; Cea-González, V.; García-Tárrago, M.J.; Gómez-Gil, F.J. Fractional viscoelastic models for the estimation of the frequency response of rubber bushings based on relaxation tests. *Results Eng.* **2023**, *20*, 101465. [CrossRef]
31. Ying, D.; Addimulam, S. Innovative Additives for Rubber: Improving Performance and Reducing Carbon Footprint. *Asia Pac. J. Energy Environ.* **2022**, *9*, 81–88. [CrossRef]

32. Mohamad Aini, N.A.; Othman, N.; Hussin, M.H.; Sahakaro, K.; Hayeemasae, N. Lignin as Alternative Reinforcing Filler in the Rubber Industry: A Review. *Front. Mater.* **2020**, *6*, 329. [CrossRef]
33. Oladimeji, T.E.; Odunoye, B.O.; Elehinafe, F.B.; Obanla, O.R.; Odunlami, O.A. Production of activated carbon from sawdust and its efficiency in the treatment of sewage water. *Heliyon* **2021**, *7*, e05960. [CrossRef]
34. Gong, Y.; Chen, X.; Wu, W. Application of fourier transform infrared (FTIR) spectroscopy in sample preparation: Material characterization and mechanism investigation. *Adv. Sample Prep.* **2024**, *11*, 100122. [CrossRef]
35. Nguyen, V.H.; Nguyen, D.T.; Nguyen, T.T.; Nguyen, H.P.T.; Khuat, H.B.; Nguyen, T.H.; Tran, V.K.; Woong Chang, S.; Nguyen-Tri, P.; Nguyen, D.D.; et al. Activated carbon with ultrahigh surface area derived from sawdust biowaste for the removal of rhodamine B in water. *Environ. Technol. Innov.* **2021**, *24*, 101811. [CrossRef]
36. Candau, N.; Fernández Navarrete, A.; Lara Casanova, G.; Utrera-Barrios, S.; Hernandez Santana, M.; Stoclet, G.; Maspoch, M.L. Control by the curing time of the strain induced crystallization and elastocaloric properties in natural rubber and natural/waste rubber blends. *Polymers* **2024**, *312*, 127628. [CrossRef]
37. Poyraz, B. Poly(ethylene-co-(acrylic acid)) and maleic anhydride effect on micro crystalline cellulose and lignin-filled EPDM automotive sealing profiles. *Int. J. Biol. Macromol.* **2025**, *297*, 139602. [CrossRef]
38. Liu, X.; Qi, R.; Liao, W.; Dai, B.; Liu, D.; Wang, Q. Shear-sensing characteristics and theoretical model analysis of multiwalled carbon nanotube/natural rubber composite. *Compos. Commun.* **2025**, *53*, 102215. [CrossRef]
39. Methods, S.T. D395-14: Standard Test Methods for Rubber Property—Compression Set 1. *Current* **2003**, *9*, 1–6. [CrossRef]
40. *ASTM D471-16a*; Standard Test Method for Rubber Property—Effect of Liquids. ASTM—Amereican Society for Testing and Materials: West Conshohocken, PA, USA, 2021; pp. 1–16.
41. Darko, C. The link between swelling ratios and physical properties of EPDM rubber compound having different oil amounts. *J. Polym. Res.* **2022**, *29*, 325. [CrossRef]
42. Joselin, J.; Benila, B.S.; Brintha, T.S.S.; Jeeva, S. Phytochemical profiling, FT-IR spectroscopy, and antioxidant evaluation of select Lamiaceae species. *Intell. Pharm.* **2025**, *3*, 111–117. [CrossRef]
43. Rosson, E.; Sgarbossa, P.; Mozzon, M.; Venturino, F.; Bogialli, S.; Glisenti, A.; Talon, A.; Moretti, E.; Carturan, S.M.; Tamburini, S.; et al. Novel correlations between spectroscopic and morphological properties of activated carbons from waste coffee grounds. *Processes* **2021**, *9*, 1637. [CrossRef]
44. Yang, S.Y.; Bai, B.C.; Kim, Y.R. Effective Surface Structure Changes and Characteristics of Activated Carbon with the Simple Introduction of Oxygen Functional Groups by Using Radiation Energy. *Surfaces* **2024**, *7*, 12–25. [CrossRef]
45. Ilić, M.; Haegel, F.H.; Lolić, A.; Nedić, Z.; Tosti, T.; Ignjatović, I.S.; Linden, A.; Jablonowski, N.D.; Hartmann, H. Surface Functional Groups and Degree of Carbonization of Selected Chars from Different Processes and Feedstock. *PLoS ONE* **2022**, *17*, e0277365. [CrossRef] [PubMed]
46. Nakamura, S.; Tsuji, Y.; Yoshizawa, K. Role of Hydrogen-Bonding and OH-πInteractions in the Adhesion of Epoxy Resin on Hydrophilic Surfaces. *ACS Omega* **2020**, *5*, 26211–26219. [CrossRef] [PubMed]
47. Saito, K.; Xu, T.; Ishikita, H. Correlation between C•O Stretching Vibrational Frequency and p KaShift of Carboxylic Acids. *J. Phys. Chem. B* **2022**, *126*, 4999–5006. [CrossRef]
48. Kingkohyao, N.; Boonsiri, T.; Johns, J.; Nip, R.L.; Nakaramontri, Y. Synergistic effect of zinc ions from tire waste and modified zinc oxide on mechanical, thermo-mechanical and antibacterial properties in epoxidized natural rubber composites. *Ind. Crops Prod.* **2025**, *225*, 120577. [CrossRef]
49. Qiu, C.; Jiang, L.; Gao, Y.; Sheng, L. Effects of oxygen-containing functional groups on carbon materials in supercapacitors: A review. *Mater. Des.* **2023**, *230*, 111952. [CrossRef]
50. Hiranobe, C.T.; Ribeiro, G.D.; Torres, G.B.; Dos Reis, E.A.P.; Cabrera, F.C.; Job, A.E.; Paim, L.L.; Dos Santos, R.J. Cross-linked density determination of natural rubber compounds by different analytical techniques. *Mater. Res.* **2021**, *24* (Suppl. S1), e20210041. [CrossRef]
51. Wang, Y.; Su, S.; Liu, H.; Wang, R.; Liao, L.; Peng, Z.; Li, J.; Wu, H.; He, D. Effect of Proteins on the Vulcanized Natural Rubber Crosslinking Network Structure and Mechanical Properties. *Polymer* **2024**, *16*, 2957. [CrossRef]
52. Yan, Z.; Zaoui, A.; Zaïri, F. Physical and mechanical properties of vulcanized and filled rubber at high strain rate. *Chin. J. Phys.* **2023**, *86*, 12–23. [CrossRef]
53. Naik, P.; Pradhan, S.; Sahoo, P.; Acharya, S.K. Effect of filler loading on mechanical properties of natural carbon black reinforced polymer composites. *Mater. Today Proc.* **2019**, *26*, 1892–1896. [CrossRef]
54. Aboughaly, M.; Babaei-Ghazvini, A.; Dhar, P.; Patel, R.; Acharya, B. Enhancing the Potential of Polymer Composites Using Biochar as a Filler: A Review. *Polymer* **2023**, *15*, 3981. [CrossRef] [PubMed]
55. Sivaselvi, K.; Varma, V.S.; Harikumar, A.; Jayaprakash, A.; Sankar, S.; Krishna, C.Y.; Gopal, K. Improving the mechanical properties of natural rubber composite with carbon black (N220) as filler. *Mater. Today Proc.* **2020**, *42*, 921–925. [CrossRef]
56. Li, G.; Iakunkov, A.; Boulanger, N.; Lazar, O.A.; Enachescu, M.; Grimm, A.; Talyzin, A.V. Activated carbons with extremely high surface area produced from cones, bark and wood using the same procedure. *RSC Adv.* **2023**, *13*, 14543–14553. [CrossRef]

57. Ma, F.; Ding, S.; Ren, H.; Liu, Y. Sakura-based activated carbon preparation and its performance in supercapacitor applications. *RSC Adv.* **2019**, *9*, 2474–2483. [CrossRef]
58. Alfatah, T.; Mistar, E.M.; Supardan, M.D. Porous structure and adsorptive properties of activated carbon derived from Bambusa vulgaris striata by two-stage KOH/NaOH mixture activation for Hg^{2+} removal. *J. Water Process Eng.* **2021**, *43*, 102294. [CrossRef]
59. Mumtaz, N.; Li, Y.; Artiaga, R.; Farooq, Z.; Mumtaz, A.; Guo, Q.; Nisa, F.U. Fillers and methods to improve the effective (out-plane) thermal conductivity of polymeric thermal interface materials—A review. *Heliyon* **2024**, *10*, e25381. [CrossRef]
60. Ren, X.; Barrera, C.S.; Tardiff, J.L.; Gil, A.; Cornish, K. Liquid guayule natural rubber, a renewable and crosslinkable processing aid in natural and synthetic rubber compounds. *J. Clean. Prod.* **2020**, *276*, 122933. [CrossRef]
61. Gobetti, A.; Cornacchia, G.; La Monica, M.; Zacco, A.; Depero, L.E.; Ramorino, G. Assessment of the influence of electric arc furnace slag as a non-conventional filler for Nitrile Butadiene Rubber. *Results Eng.* **2023**, *17*, 100987. [CrossRef]
62. Hu, Y.; Cai, Y.; Wei, J.; Yin, Z.; Fan, H. Design and preparation of novel silicon rubber with homogeneous crosslinking network and its application in high-performance SR/EPDM blend. *Eur. Polym. J.* **2023**, *198*, 112381. [CrossRef]
63. Kratina, O.; Pöschl, M.; Stoček, R. The effect of apparent density of sulfidic crosslink and their chemical nature on self-heat build-up in carbon black filled Natural Rubber under cyclic mechanical loading. *Polym. Degrad. Stab.* **2024**, *227*, 110871. [CrossRef]
64. Wu, J.; Chen, L.; Su, B.L.; Wang, Y.S. Evolution of Payne effect of silica-filled natural rubber in curing process. *J. Rubber Res.* **2019**, *22*, 127–132. [CrossRef]
65. Lubura, J.; Kočková, O.; Strachota, B.; Bera, O.; Pavlova, E.; Pavličević, J.; Ikonić, B.; Kojić, P.; Strachota, A. Natural Rubber Composites Using Hydrothermally Carbonized Hardwood Waste Biomass as a Partial Reinforcing Filler—Part II: Mechanical, Thermal and Ageing (Chemical) Properties. *Polymer* **2023**, *15*, 2397. [CrossRef]
66. Ajay, C.; Das Gupta, S.; Mukhopadhyay, R.; Chattopadhyay, D.; Das, M. Exploring crosslink density in rubber vulcanisates—A comprehensive analysis using a dynamic mechanical analyser and an insight into mechanical properties. *J. Rubber Res.* **2025**, *28*, 305–323. [CrossRef]
67. Wang, W.; Yu, X.; Lai, C.; Li, G.; Zhu, P.; Sun, R. Systematic study of the effect of silane coupling agent on the hydrothermal aging resistance of the underfill epoxy resin and silica interface via molecular dynamics simulation. *Appl. Surf. Sci.* **2025**, *688*, 162313. [CrossRef]
68. Wei, Y.; Jia, X.; Chen, Y.; Ji, J. Single step carbonating and activating fir sawdust to activated carbon by recyclable molten carbonates and steam. *Sci. Total Environ.* **2022**, *818*, 151778. [CrossRef]
69. Xiao, J.; Zhang, Z.; Luan, Y.; Wu, S.; Wu, Y. Influence of carbon black surface characteristics on CB-NR interfacial interaction: Molecular simulation and experimental study. *Compos. Part A Appl. Sci. Manuf.* **2024**, *182*, 108198. [CrossRef]
70. Azura, A.R.; Leow, S.L. Effect of carbon black loading on mechanical, conductivity and ageing properties of Natural Rubber composites. *Mater. Today Proc.* **2019**, *17*, 1056–1063. [CrossRef]
71. Głowacka, K.; Klemenc, J.; Nagode, M.; Łagoda, T. Fatigue lifetime of rubber composites—State-of-the-art. *Polym. Test.* **2025**, *143*, 108713. [CrossRef]
72. Sultana, S.; Munna, N.; Sakib, T.U.; Ahmed, N.; Akanda, M.R.; Saha, M.S.; Zaman, M.N. High performance activated carbon derived from sawdust: Preparation, characterizations, methyl orange removal and kinetics investigation. *Biomass Convers. Biorefinery* **2025**, *15*, 17295–17307. [CrossRef]
73. Samal, S. Effect of shape and size of filler particle on the aggregation and sedimentation behavior of the polymer composite. *Powder Technol.* **2020**, *366*, 43–51. [CrossRef]
74. Sittitanadol, I.O.; Srakeaw, N.L.O.; Somdee, P.; Chumsamrong, P.; Noyming, S.; Singsang, W.; Prasoetsopha, N. Utilizing Coconut Biochar as a Bio-reinforcing Agent in Natural Rubber Composites. *Starch/Staerke* **2024**, *77*, 2400168. [CrossRef]
75. Chen, Q.; Yang, K.; Feng, Y.; Liang, L.; Chi, M.; Zhang, Z.; Chen, X. Recent advances in thermal-conductive insulating polymer composites with various fillers. *Compos. Part A Appl. Sci. Manuf.* **2024**, *178*, 107998. [CrossRef]
76. Kartal, İ.; Karagöz, İ. Enhancing natural rubber properties: A comprehensive study on the synergistic effects of wood sawdust and carbon black as fillers in rubber composites. *Polym. Bull.* **2025**, *82*, 2091–2109. [CrossRef]
77. Amoke, A.; Tenebe, O.G.; Ayo, M.D. Comparison of Mechanical Properties of Natural Rubber Vulcanizates Filled with Hybrid Fillers (Carbon Black/Palm Kernel Shell and Palm Kernel Shell/Sandbox Seed Shell). *Int. J. Res. Innov. Appl. Sci.* **2021**, *6*, 2454–6194.
78. He, S.; Zhang, F.; Liu, S.; Cui, H.; Chen, S.; Peng, W.; Chen, G.; Liao, X.; Liao, L. Influence of sizes of rubber particles in latex on mechanical properties of natural rubber filled with carbon black. *Polymer* **2022**, *261*, 125393. [CrossRef]
79. Li, X.; Ding, Y.; Zhang, H.; He, T.; Hao, J.; Wu, J.; Wu, Y.; Bai, H. Pine sawdust derived ultra-high specific surface area activated carbon: Towards high-performance hydrogen storage and supercapacitors. *Int. J. Hydrogen Energy* **2024**, *84*, 623–633. [CrossRef]
80. Bernal-Ortega, P.; Bernal, M.M.; González-Jiménez, A.; Posadas, P.; Navarro, R.; Valentín, J.L. New insight into structure-property relationships of natural rubber and styrene-butadiene rubber nanocomposites filled with MWCNT. *Polymer* **2020**, *201*, 122604. [CrossRef]

81. Wang, Z.; Wang, S.; Yu, X.; Zhang, H.; Yan, S. Study on the Use of CTAB-Treated Illite as an Alternative Filler for Natural Rubber. *ACS Omega* **2021**, *6*, 19017–19025. [CrossRef]
82. Buaksuntear, K.; Panmanee, K.; Wongphul, K.; Lim-arun, P.; Jansinak, S.; Shah, D.U.; Smitthipong, W. Enhancing mechanical properties and stabilising the structure of epoxide natural rubber using non-covalent interactions: Metal–ligand coordination and hydrogen bonding. *Polymer* **2024**, *291*, 126626. [CrossRef]
83. Mensah, B.; Onwona-Agyeman, B.; Nsaful, F.; Aboagye, I.A.; Sowah, N.L.; Angnunavuri, P.N.; Apalangya, V.A. Vulcanization kinetics and reinforcement behaviour of natural rubber-carbon black composites: Addition of Shea-butter versus aromatic oil as plasticizers. *Heliyon* **2024**, *10*, e25592. [CrossRef]
84. Zedler, L.; Colom, X.; Cañavate, J.; Saeb, M.R.; Haponiuk, J.T.; Formela, K. Investigating the impact of curing system on structure-property relationship of natural rubber modified with brewery by-product and ground tire rubber. *Polymer* **2020**, *12*, 545. [CrossRef]

polymers

MDPI

Article

Reduction in Sulfur Diffusion in Recycled Ground Rubber-Containing Compounds to Improve Tensile Strength †

Stefan Frosch [1,*], Volker Herrmann [1], Fabian Grunert [2] and Anke Blume [2]

[1] Faculty of Plastics Engineering and Surveying, Technical University of Applied Sciences Würzburg-Schweinfurt, 97070 Würzburg, Germany
[2] Elastomer Technology and Engineering (ETE), University of Twente, 7522 NB Enschede, The Netherlands
* Correspondence: stefan.frosch@thws.de; Tel.: +49-931-3511-8522
† In memory of Irene Dürr.

Abstract

Recycling end-of-life rubber to compound components for new formulations is one of the most promising ways to reach the sustainability goals of the rubber industry. Today, devulcanization and pyrolysis are both methods to reuse crosslinked elastomers. A third recycling approach is to process end-of-life rubber into ground rubber (GR), which is then added to green compounds. However, free sulfur diffuses during mixing, storage and vulcanization from the matrix material into the GR particles. As a result, the crosslink density in the matrix is reduced, which deteriorates the in-rubber properties of GR-containing vulcanizates compared to those that do not contain GR. Therefore, GR particles are mainly used today for rubber parts with less demanding dynamic-mechanical requirements, which limits the use of the particles. This study presents an approach for reducing the sulfur diffusion from the matrix into the GR particles by prevulcanizing the green matrix material. This leads to GR-containing vulcanizates with significantly improved mechanical properties. This new approach shows that the quality of the recycled rubber product can be significantly increased by blocking the sulfur diffusion. Even though such prevulcanization is currently only feasible under laboratory conditions, it might also pave the way for finding solutions in a production scale for an effective incorporation of GR into new rubber compounds.

Keywords: rubber; ground rubber; rubber recycling; sulfur diffusion; prevulcanization; crosslink density

1. Introduction

In terms of sustainability and the circular economy, the rubber industry is seeking ways to reduce energy consumption and CO_2 emissions in the manufacturing of rubber products. Recycling plays a key role in achieving such a reduction. Efforts are being made to further process polymer material from end-of-life tires (ELTs) by pyrolysis [1,2], devulcanization [3,4] or refining into ground rubber (GR) [5,6] as an additive for new products. This recycled additive reduces the absolute need for virgin material of a new vulcanizate and thus contributes to the overall goal of the circular economy. In this study, the recycling method of GR is examined in more detail, and a solution is presented to improve the tensile properties of GR-containing compounds.

To date, shredding and grinding ELTs into small particles is common practice. The size of these particles ranges roughly from diameters from 20 to 40 mm ("chips"), to 1–20 mm ("granulate"), and down to 0–0.4 mm ("powder") [7]. In this study, GR is referred to as particles with diameters between 0.2 and 2 mm, as previous measurements [8] have shown.

The incorporation of GR particles into green rubber during compounding is an approach for recycling the polymer fraction of ELTs. However, the properties of GR-containing vulcanizates are deteriorated in comparison to those that do not contain the recycled material [9–16]. The main reason for this effect is the diffusion of sulfur from the surrounding green matrix into the GR particles [9,14,15,17–19]. This diffusion out of the uncured compound leads to a reduction of the sulfur concentration in the matrix and consequently to a reduction in crosslink density (CLD) within the matrix [19,20]. At the same time, the sulfur concentration increases in the GR particles, which leads to particles with a higher CLD [19,20]. This diffusion effect causes an inhomogeneous vulcanizate in terms of different local sulfur concentrations and thus different local viscoelastic properties [19]. The combination of stiff GR particles in a flexible matrix is therefore responsible for the deteriorated in-rubber behavior of the vulcanizate.

It could be shown in [19] that the concentration difference of free, i.e., not covalently bound soluble sulfur (cyclo-S_8) between the matrix and GR particles is the driving force of the curative diffusion, as the concentration of free cyclo-S_8 is high in the green matrix. In contrast, the concentration of free cyclo-S_8 is low in the GR particles because these are vulcanized, and thus the sulfur is covalently bound. When GR is incorporated into the green matrix compound, free cyclo-S_8 diffuses from the matrix into the GR particles to reestablish a concentration equilibrium of free cyclo-S_8 in both phases. The diffusion ends when both phases contain the same amount of free cyclo-S_8. However, the diffusion also ends with the start of the vulcanization process in both phases. This also explains why the sulfur diffusion from the matrix into the GR particles is even more pronounced when the scorch time of the GR is shorter than that of the surrounding matrix: with a short scorch time of the GR, the sulfur diffuses from the surrounding matrix into the GR until final vulcanization, as the sulfur is only bound in the matrix at a later stage. Thus, the concentration difference of free cyclo-S_8 between matrix and GR is maintained until the vulcanization also starts in the matrix, which significantly reduces sulfur diffusion.

Phadke et al. [15,18] developed an approach for counteracting the effects of sulfur diffusion. They achieved higher sulfur concentrations in the vulcanized matrix by increasing the amount of both the sulfur and the also mobile accelerator in the green matrix compound before crosslinking. Sulfur diffusion continues to occur, as described above. However, the increased sulfur content in the matrix led to a higher sulfur equilibrium between both the matrix and GR particles. As the matrix became supersaturated with sulfur, it continued to release it through diffusion. However, due to the overall high sulfur concentration, the matrix retained a sufficient amount of sulfur. A satisfactory CLD in the matrix thus was formed and the deterioration of in-rubber properties of the entire vulcanizate is less severe. Until today, this method is a good way to counteract the negative influence due to the incorporation of GR particles [6,19,21].

Therefore, this technique is a first step in reducing the detrimental effects of sulfur diffusion. Nevertheless, sulfur enrichment in the GR continues to reduce the homogeneity between matrix and GR particles with respect to e.g., CLD. Furthermore, the use of a sulfur supersaturated matrix has the disadvantage of allowing diffusion to proceed unimpeded. For example, GR particle solubility and crosslinking kinetics will alter if its compound formulation changes. Thus, for every variation in GR type and quantity, the sulfur concentration may need to be adjusted. Determining the ideal ratio between GR quantity and sulfur supersaturated matrix requires long and laborious preliminary testing [19].

To minimize the deterioration of the properties of GR-containing vulcanizates, it therefore seems more appropriate to prevent sulfur diffusion. Three approaches were investigated in [19]. It could be shown that a sulfur donor system does not lead to reduced sulfur diffusion during vulcanization conditions. In contrast, the use of polymeric sulfur could significantly reduce the diffusion of sulfur during mixing and storage conditions. However, during the vulcanization process, the polymeric sulfur converts back into free cyclo-S_8 and thus leads to a diffusion behavior similar to that described above.

The third approach, however, showed promising results in reducing sulfur diffusion and thus increasing the in-rubber properties of GR-containing vulcanizates. It was shown that sulfur diffusion mainly occurs during mixing, storage and the scorch time of the curing process due to the elevated temperatures. At the same time, it was demonstrated that sulfur diffusion is mainly limited to the pure form of free cyclo-S_8. In contrast, sulfur species in combination with ZnO, stearic acid or accelerators are significantly less mobile in the elastomeric matrix. These species can be, e.g., ZnS or the active sulfurating agent complex that are formed during mixing and the scorch time of the crosslinking process [19].

For these reasons, the idea of prevulcanization of the matrix before the incorporation of GR particles was introduced. The current process thus enables the pure cyclo-S_8 to diffuse from the matrix into the GR particles during mixing, storage and vulcanization processes. In contrast, large proportions of the added free cyclo-S_8 are first converted into a less mobile sulfur species in the matrix during prevulcanization, as explained above. The reduced mobility of the sulfur-containing species is expected to lead to less sulfur diffusion from the matrix into the subsequently incorporated GR particles. This hypothesis of prevulcanization will be investigated in the following sections.

2. Materials and Methods

For the following investigations, the same experimental approach was used as already described in previous publications [19,20,22]. In total, the same compounds were produced five times together with the previous work cited and all vulcanization curves were compared with each other, so that in each case a very good match could be established and reproducibility was ensured.

2.1. Process Chain Overview

The entire process chain, from compounding to the finished test specimens, is illustrated in Figure 1. First, the natural rubber (NR, red) and styrene–butadiene rubber (SBR, blue) compounds are produced separately in mixing step 1. The vulcanization properties at 155 °C of both compounds are then determined in the rubber process analyzer (RPA). With this vulcanization information, the NR compound (red) is vulcanized to t_{95} and formed into 6 mm thick sheets (130 mm × 117 mm). The NR sheets are then converted into GR in a mill, as described in [8]. This material is thus defined as "NR-GR". Alternatively, the SBR compound (blue) is prevulcanized to the vulcanization time t_x with $x < 10$ in the available heating press to 6 mm thick sheets. This prevulcanization of the matrix compound is the only difference to the current state of the art in processing GR. To date, the GR is incorporated into a non-prevulcanized matrix. The material pairing of NR-GR in an SBR matrix was chosen because [19] showed that sulfur diffusion is pronounced in this material combination, since t_{95} of the NR-GR material is 4 times lower than that of the surrounding SBR matrix material.

In mixing step 2, the NR-GR is incorporated into the prevulcanized SBR matrix on a two-roll mill. The crosslinking characteristics of the new compound (red-and-blue-colored arrows) are then determined again in the RPA. With this information, rubber sheets with a thickness of 2 and 6 mm are vulcanized to t_{95}. These sheets are converted into test specimens for the following examinations.

The following sections describe the separate production steps and the equipment used in more detail.

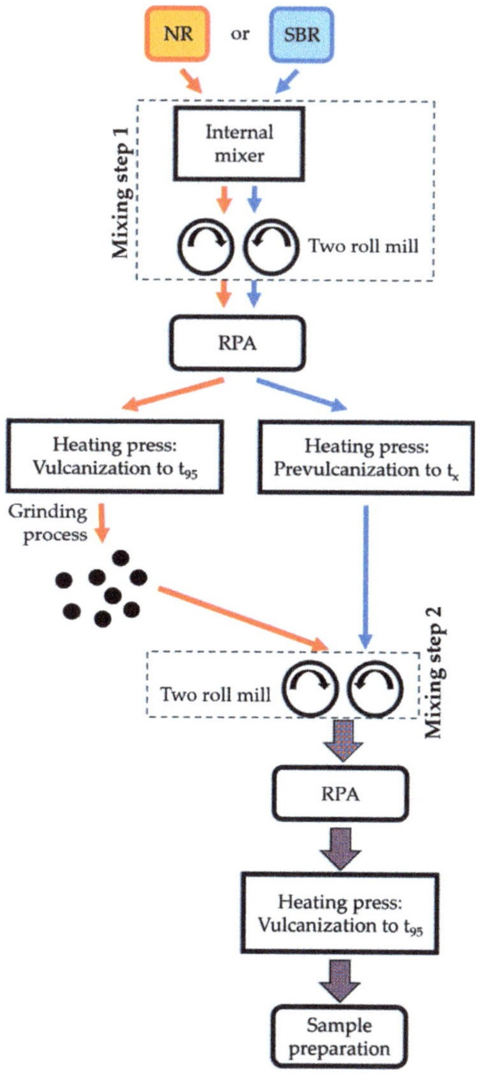

Figure 1. Process chain overview.

2.2. Compound Formulations and Mixing Conditions

This study investigates the diffusion behavior of soluble sulfur between the SBR matrix and the NR-GR material. The formulations, as well as the names and manufacturing companies of the individual compound components, are depicted in Table 1. The NR vulcanizate was manufactured to be subsequently converted into GR and then to be incorporated into compounds of SBR. In total, three SBR compounds were manufactured, each prevulcanized to different t_x. Details are explained in the following table.

Table 1. Compound formulations; all values are given in parts per hundred rubber (phr), adapted from [20].

	NR	SBR	Name	Company
NR	100	-	SIR 20	Weber & Schaer GmbH & Co. KG, Hamburg, Germany
E-SBR	-	100	Buna SE 1502 L	Arlanxeo Deutschland GmbH, Cologne, Germany
N330	50	50	Corax N330	Orion Engineered Carbons S.A., Eschborn, Germany
ZnO	5	5	Zinkoxid Rotsiegel	L. Brüggemann GmbH & Co. KG, Heilbronn, Germany
Stearic acid	3	3	Palmera B1804	KLK Emmerich GmbH, Emmerich am Rhein, Germany
Soluble sulfur	1.2	1.2	K46859483 542	Merck Chemicals GmbH, Darmstadt, Germany
CBS	1.2	1.2	Vulkacit CZ/EG-C	Lanxess Deutschland GmbH, Krefeld, Germany

The compound components described in Table 1 were mixed in mixing step 1 using an internal mixer (Werner & Pfleiderer, 1.5 l, PES3) and a laboratory mill (Schwabenthan 200 × 450). Mixing step 1 is a two-stage mixing process, divided into the preparation of the base and the final compounds. An overview of the mixing sequence and the key parameters of the base compounds can be found in Table 2. Prior to the compounding of the base compounds, 1 kg of the NR is masticated in the internal mixer for 20 min at 20 rpm and 25 °C.

Table 2. Production procedure of base compounds of mixing step 1 [19].

Time/Min	Processing Step of Base Compounds
0	Polymer
1	ZnO, stearic acid, carbon black
2.5	Cleaning step
4.5	Dump
	Weight check
	Sheeted off on laboratory mill (both rollers 20 rpm, gap 2.5 mm, 40 °C) for 1 min. Relaxation for 1 h.

Table 3 depicts manufacturing details of the final compounds of mixing step 1.

Table 3. Production procedure of final compounds of mixing step 1 [19].

Time/Min	Processing Step of Final Compounds
0	Base compound, sulfur, accelerator
3	Dump
	Weight check
	Sheeted off on laboratory mill (front roller 16 rpm, back roller 20 rpm gap 2.0 mm, 40 °C)
0	Final compound on laboratory mill
1	Cutting three times left and right, roll up and blend back at a gap of 1 mm
4	Gap: 2.5 mm, both rollers at 20 rpm, dump

The NR vulcanizate was then ground in a centrifugal mill (see Section 2.4). In contrast, the SBR compounds were prevulcanized to t_x and formed into 6 mm thick sheets with dimensions of 130 mm × 117 mm. Each sheet was then cut with common rubber scissors into four stripes to increase the processability on the two-roll mill in mixing step 2.

In the latter compounding process, 30 phr of NR-GR was incorporated on the two-roll mill into the prevulcanized SBR matrix (see mixing step 2 in Figure 1). This compound is referred to as "SBR&NR-GR" in the remainder of this article. Table 4 describes the compound constituents of SBR&NR-GR in detail.

Table 4. Compound formulation for mixing step 2: addition of 30 phr NR-GR into a prevulcanized SBR-matrix; all values are given in phr.

	SBR&NR-GR	
E-SBR	100	Prevulcanized to t_x with x < 10; cut into 6 mm thick stripes
N330	50	
ZnO	5	
Stearic acid	3	
Soluble sulfur	1.2	
CBS	1.2	
NR-GR	30	

The procedure of incorporating 30 phr NR-GR into a prevulcanized SBR-matrix in mixing step 2 is displayed in Table 5.

Table 5. Production procedure for mixing step 2.

Time/Min	Processing Step of SBR&NR-GR
	Laboratory mill (both rollers 20 rpm, gap 6 mm, 40 °C)
0	Start feeding stripes of the prevulcanized SBR compound onto both rollers
5	All stripes are incorporated; reduction of gap to 2.5 mm
6	Friction: front roller at 16 rpm, back roller at 20 rpm
7	Reduction of gap to 2.0 mm
8	Friction: front roller to 20 rpm, back roller to 16 rpm
9	Cutting two times left and right
10	Adding 30 phr ground rubber
14	3 x roll up and blend back at gap of 1.0 mm
15	Both rollers at 20 rpm, gap of 2.0 mm, dump

2.3. Vulcanization Behavior

Subsequent to mixing steps 1 and 2, specimens were cut from the compound sheets with the SIS-VS punch press from TA Instruments (New Caste, Delaware, USA, former Scarabaeus) for the following investigations with the RPA. The latter is a SIS V50 from TA Instruments (former Scarabaeus) and was used to determine the curing behavior of all samples at 155 °C, with a frequency of 1.667 Hz and an oscillation angle amplitude of 0.50° within a measurement time of 60 min. Every compound was measured three times. The rheometer was equipped with the software SQS—RPA Monitor Version 6.15.

2.4. Mill

A centrifugal mill, ZM200, from the company Retsch (Haan, Germany), with a 12 tooth rotor and a sieve with a mesh size of 2 mm was used to grind the NR at a frequency of 18.000 rpm. To prepare the samples for this grinding process, the 6 mm thick sheets of NR were cut with common rubber scissors into approximately 6 × 6 × 6 mm^3 cubes. Subsequently, these cubes were cooled below their glass transition temperature (Tg) using liquid nitrogen. Then, they were placed in portions of several 10 g into the mill. In the centrifugal mill, the vulcanizate cubes were further reduced to GR. The SBR material was not ground.

2.5. Heating Press

The heating press Polystat 200 T from Servitec (Wustermark, Germany) was used to prevulcanize the SBR matrix to t_x and to vulcanize both the NR and SBR&NR-GR to t_{95}. The heating press was equipped with two molds to produce both 2 and 6 mm thick sheets with dimensions of 130 mm × 117 mm.

2.6. Micro X-Ray Fluorescence Analysis

A micro X-ray fluorescence analysis (μ-XRF) was employed to investigate areas of 10 mm × 10 mm of the vulcanized 6 mm thick sheets. This technique is used to qualify and quantify chemical elements on (rubber) surfaces in an imaging manner [22–25]. For the investigations, an M4 Tornado from Bruker Nano Analytics (Berlin, Germany) was equipped with a rhodium tube that was set to 200 μA and 50 kV. The target material was rhodium, no filter was chosen. The diameter of the X-ray beam was set to 25 μm, identical to the resolution (distance between two measurement spots). Each spot was measured three times for 40 ms. The fluorescence radiation was detected with two silicon drift detectors. To enhance the sulfur signal by eliminating the argon present in the ambient air, the sample chamber was evacuated to an absolute pressure of 20 mbar for the analysis. This was necessary, since the fluorescence energies of both elements are similar [26]. The software Esprit M4 (version: 1.6.0.286) was used to qualify and quantify the local sulfur concentrations [19].

2.7. Tensile Tests

S2 tensile test specimens were cut out of the 2 mm thick vulcanized rubber sheets using a Zwick punch press 7103 (Ulm, Germany). These specimens were then investigated using a Zwick 1474 tensile testing machine (Ulm, Germany) based on DIN 53,504 with a crosshead speed of 50 mm min^{-1}. The software Doli Test & Motion, version 4.6.0.6, was used to process the data.

2.8. Measurement Errors

For all results, mean values were calculated, which are listed in the following chapter. These mean values are presented with their measurement error, calculated using the Student's t-test with a 95% confidence interval. This means that the mean values reported in this article, including their measurement errors, are valid with a 95% probability [27,28].

3. Results

The results of the different test methods are described in the following section.

3.1. Vulcanization Behavior

The crosslinking behavior is depicted in Figure 2. The elastic torque S′ in dNm is plotted over the measurement time in min. The median curves for S'_{max} from three measurements of the NR and SBR1 (one of three investigated SBR compounds) are illustrated. The NR shows a relatively short scorch time, after which S′ rises to a maximum value of 16.6 ± 0.2 dNm. In the following, S′ decreases again, indicating that the material shows reversion. Compound SBR1, however, behaves differently: the material reaches a plateau with a value of 16.5 ± 0.4 dNm for S′; therefore, the maximum S′ values of NR and SBR1 are at an identical level, taking the measurement error into account. Furthermore, the scorch time of SBR1 is significantly longer than that of NR, although identical concentrations of sulfur and accelerator were used in both compounds. This increased scorch time could be due to the retarding effect of the added ZnO in the SBR, as described in [29]. In this article, it was found that the postulated vulcanization process described by Morrison and

Porter [30] applies to NR. In contrast, it was found that for other polymers, such as high vinyl SBR, other ring opening and vulcanization reactions occur, and that ZnO can exhibit a retarding effect in these processes. For example, sulfur can accumulate in the ZnO crystal and can therefore not participate in vulcanization reactions [31].

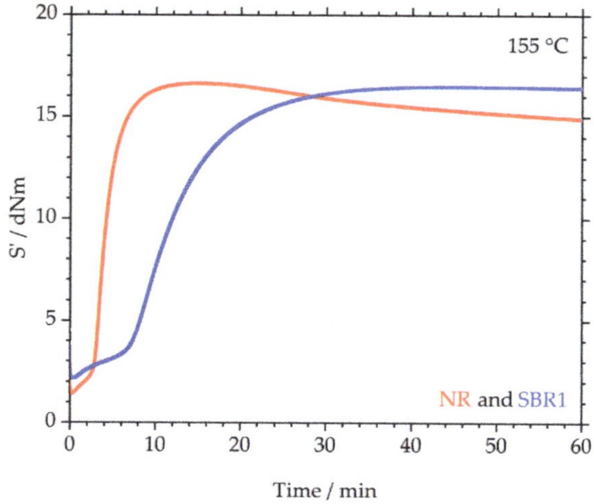

Figure 2. Crosslinking behavior of NR and SBR1: median curves (maximum of S') of three measurements each.

In Table 6, characteristic values of the RPA measurements of all analyzed samples are listed. Each compound was measured three times, with the mean values including errors for t_{10}, t_{50}, t_{95} and S'_{max} given in each case. It is evident that the NR compound cures significantly faster than the three compounds made of SBR (SBR1, SBR2 and SBR3).

Table 6. Crosslink characteristics of the investigated compounds.

Compound	t_{10}/min	t_{50}/min	t_{95}/min	S'_{max}/dNm
NR	3.0 ± 0.0	4.2 ± 0.2	8.5 ± 0.2	16.6 ± 0.2
SBR1	6.9 ± 0.6	11.7 ± 0.9	26.9 ± 3.0	16.5 ± 0.4
SBR2	7.2 ± 0.3	12.1 ± 0.5	29.2 ± 2.1	16.9 ± 0.3
SBR3	6.8 ± 0.3	11.4 ± 0.8	26.6 ± 2.9	17.1 ± 0.6
SBR_0min&NR-GR *	3.1 ± 0.1	8.6 ± 0.1	28.8 ± 1.3	11.3 ± 0.8
SBR1_2min&NR-GR	3.9 ± 0.1	8.7 ± 0.8	24.6 ± 3.7	12.0 ± 0.3
SBR2_4min&NR-GR	2.5 ± 0.1	6.3 ± 0.2	19.9 ± 0.6	12.0 ± 0.3
SBR3_6min&NR-GR	1.2 ± 0.0	4.0 ± 0.2	15.2 ± 1.6	11.6 ± 0.2

* Reference sample without prevulcanization but identical compound formulation; adapted from [12,32].

The three SBR compounds each have a t_{10} of approximately 7 min. Preliminary tests were carried out, which showed that GR could not be incorporated into t_{10} prevulcanized matrix compounds. This can be explained by the lack of viscous processing behavior of the compound because first crosslinking reactions initiated the transformation of the matrix into a vulcanizate. Accordingly, the prevulcanization time was shortened to enable the incorporation of the recycled material. Consequently, SBR1, SBR2 and SBR3 were prevulcanized to different t_x with $x < 10$. Prevulcanization times of 2, 4 and 6 min (t_2, t_5 and t_7 accordingly) were selected for SBR1, SBR2 and SBR3, before 30 phr of NR-GR was incorporated into each. The GR-containing and prevulcanized compounds are referred to as "SBR1_2min&NR-GR", "SBR2_4min&NR-GR" and "SBR3_6min&NR-GR".

Figure 3 depicts the median (S'_{max}) vulcanization curves of these three compounds. The S′-values of all three GR-containing samples settle on a plateau of approximately 12 dNm. These plateaus are approx. 4 dNm below that of SBR1, which was not prevulcanized, does not contain GR and serves as a reference here. Because the three prevulcanized specimens reach the same plateau, it appears that the degree of prevulcanization has no influence on the resulting S'_{max}. Thus, the 4 dNm difference between SBR1 and the three prevulcanized samples corresponds to the general influence of the GR. This influence is underlined by "SBR_0min&NR-GR", a sample with an identical formulation to the first three specimens. The difference from the other three samples is that the matrix was not prevulcanized before the incorporation of 30 phr of NR-GR. It is therefore a sample that represents the current manufacturing process in the rubber industry. The sample has already been examined in more detail in [12,19]; however, the results are illustrated here again because, together with those of the three prevulcanized specimens, they demonstrate the influence of prevulcanization. It is therefore evident that the reduction of S'_{max} is an effect of the addition of GR and not of prevulcanization (see also Table 6).

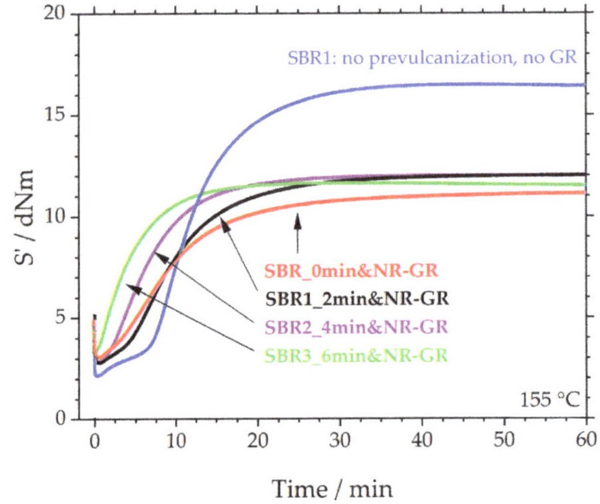

Figure 3. Crosslinking behavior of the prevulcanized and the 30 phr NR-GR-containing compounds SBR1_2min&NR-GR, SBR2_4min&NR-GR and SBR3_6min&NR-GR with the reference compounds SBR1 (compound that was not prevulcanized and does not contain GR) and SBR_0min&NR-GR (not prevulcanized but contains 30 phr NR-GR). Median curves (maximum of S′) were derived from three measurements each.

In comparison, however, the scorch time decreases significantly with the degree of prevulcanization. The reactions of all three prevulcanized samples start earlier than the SBR1 reference. The scorch time is reduced with increasing prevulcanization: t_{10} decreases from 6.9 $\pm$ 0.6 min for SBR1, to 3.9 $\pm$ 0.1 min for SBR1_2min&NR-GR, 2.5 $\pm$ 0.1 min for SBR2_4min&NR-GR and 1.2 $\pm$ 0.0 min for SBR3_6min&NR-GR. This reduction in scorch time can be explained by the matrix material having already been prevulcanized, thereby initiating and interrupting scorch reactions. This scorch status is maintained during the incorporation of NR-GR and then continues at crosslinking temperatures.

The non-prevulcanized reference SBR_0min&NR-GR also exhibits a shorter scorch time compared to SBR1. This is consistent with the findings of Gibala and Hamed [14] and can be explained by the diffusion processes of accelerator fragments from the GR particles to the matrix material, speeding up vulcanization.

The comparison with SBR1 (blue) reveals that S'_{min} is higher for the other four samples. This is because each of them contains 30 phr of NR-GR, i.e., already crosslinked material, which increases the initial torque. It becomes apparent that the 6 min prevulcanized specimen (green) exhibits the highest initial torque. The possible reason for this could be that parts of the matrix had already crosslinked slightly, increasing S'_{min}.

The exact t_{10}, t_{50} and t_{95} can be found in Table 6 and in Figure 4. The latter shows t_{10}, t_{50} and t_{95} of the investigated samples from Figure 3 during the prevulcanization of the matrix material. It is evident that all three characteristic times decrease linearly as prevulcanization progresses. This reduction can be explained by the matrix material having already been prevulcanized, thereby both initiating and interrupting scorch reactions. Longer prevulcanization times of the matrix compound thus lead to shorter characteristic vulcanization times of the NR-GR-containing compounds.

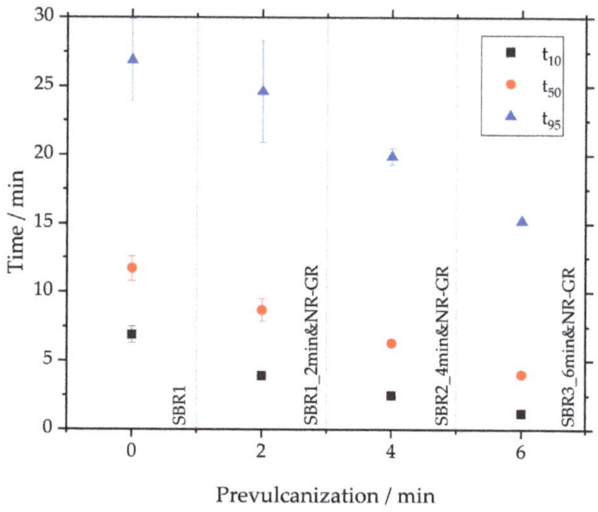

Figure 4. The influence of prevulcanization on t_{10}, t_{50} and t_{95}.

3.2. Micro X-Ray Fluorescence Analysis

Using μ-XRF, the local sulfur concentrations of the three samples with different degrees of prevulcanization SBR1_2min&NR-GR, SBR2_4min&NR-GR and SBR3_6min&NR-GR, as well as those of the non-prevulcanized reference SBR_0min&NR-GR, were determined, investigating areas of 10 mm × 10 mm. The results can be found in Figure 5.

The non-prevulcanized sample SBR_0min&NR-GR—top left in Figure 5—shows an inhomogeneous sulfur distribution because of the diffusion processes during the mixing, storage and crosslinking steps [19]. The GR particles can be identified by their green-to-red coloration. The local sulfur concentration is therefore approx. 1.8–3.5 wt%. Considering that the initial concentration according to the formulation was only 0.93 wt%, it can be concluded that the sulfur concentration in the GR particles is locally doubled to quadrupled. The reason for the sulfur enrichment is that the latter diffuses from the matrix into the GR particles due to the concentration differences in free sulfur between the matrix and GR [19]. This depletes the matrix of sulfur, which can be distinguished from the GR particles by its dark blue coloration. A sulfur concentration of approx. 0.4 wt% is achieved within the matrix, i.e., approximately halving the sulfur concentration according to the initial formulation (0.93 wt%). In [19], it is shown that the local viscoelastic properties also change with the local inhomogeneous sulfur distribution, whereby the in-rubber properties of the entire vulcanizate are deteriorated.

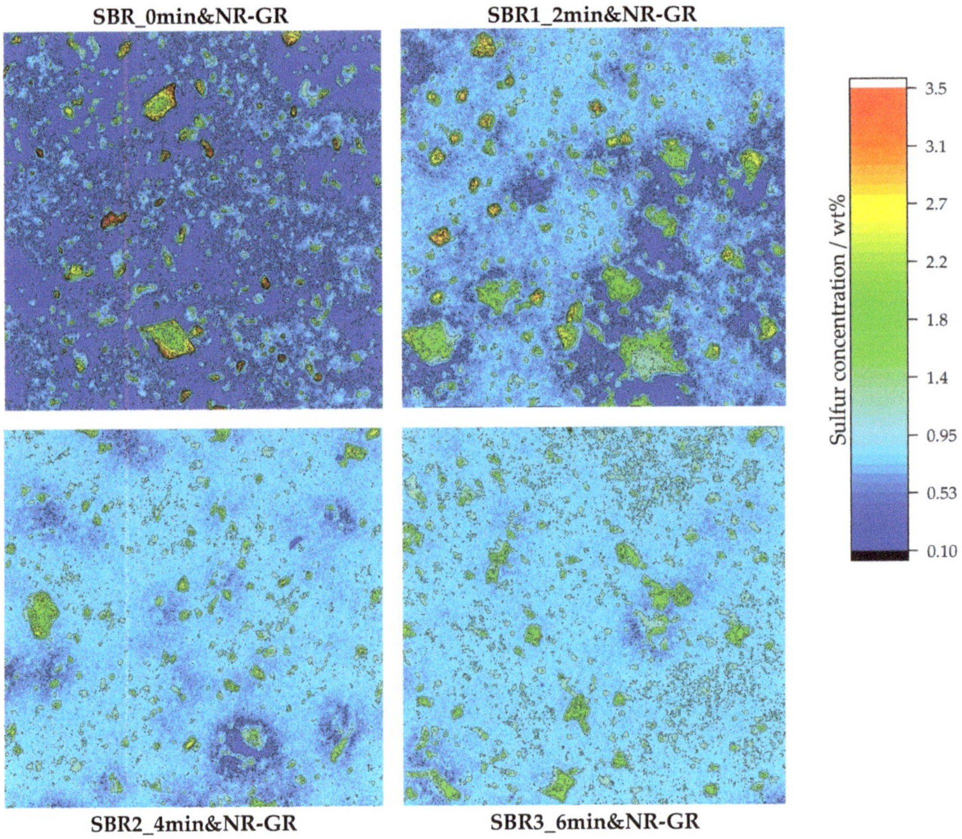

Figure 5. Local sulfur concentrations detected via μ-XRF on different vulcanizate surfaces with measurement areas of 10 mm × 10 mm for every specimen. The GR particles are identifiable by their higher sulfur concentration compared to the surrounding matrix. With increasing prevulcanization of the matrix material, the homogeneity of the sulfur distribution increases. Levels of matrix prevulcanization: 0, 2, 4 and 6 min (from top left to bottom right).

Sample SBR1_2min&NR-GR (top right in Figure 5) also shows an inhomogeneous local sulfur distribution. Higher sulfur concentrations are found in the GR particles compared to in the surrounding matrix. Compared to sample SBR_0min&NR-GR, however, it is evident that there are fewer red zones with sulfur concentrations of 3.5 wt% in the GR particles. At the same time, more GR particles are colored green, which indicates a local sulfur concentration of approx. 1.8 wt%. It is therefore clear that less sulfur accumulates in the GR particles if the matrix was prevulcanized for 2 min before incorporating the NR-GR. There are still regions in the matrix with a concentration of approximately 0.4 wt% because 2 min of prevulcanization could not fully stop the sulfur diffusion from the matrix into the GR. However, there are also areas in the matrix which, at around 0.9 wt%, are within the target range according to the compound formulation. It is therefore evident that a matrix prevulcanization of 2 min already restricts sulfur diffusion.

If the prevulcanization is extended to 4 min—bottom left in Figure 5—it reveals that sulfur diffusion is further restricted. Most GR particles are displayed in green, which corresponds to a local sulfur concentration of approx. 1.8 wt%, and thus a doubling of the initial concentration (0.93 wt%) and at the same time a halving in comparison to the concentrations of the non-prevulcanized sample (3.5 wt%). The extremely low-sulfur zones in the matrix become steadily smaller. Most of the matrix has a sulfur content of approx. 0.75 wt%, which is closer to the target range (0.93 wt%) according to the compound formulation. Accordingly, sulfur diffusion seems to be further restricted with prolonged prevulcanization.

This trend continues with 6 min of prevulcanization (see bottom right in Figure 5). This specimen exhibits the most uniform sulfur homogeneity of all four analyzed samples. At the same time, an increased sulfur concentration is still determined in the GR particles. However, the difference from the concentration in the matrix is the lowest in this compound in comparison to all other compounds. An extension of the prevulcanization time therefore leads to a more homogeneous distribution of sulfur within the sample.

A summary of the μ-XRF investigations of the local sulfur concentration is depicted in Figure 6 in the form of distribution curves. The locally measured sulfur concentrations are plotted on the abscissa. According to the formulation, both compounds—i.e., NR and SBR—contain 0.93 wt% sulfur. This concentration is plotted on the abscissa as a dashed line. The distribution curve of the sulfur concentration was determined using the matrix material SBR, which has not yet been prevulcanized and does not yet contain NR-GR, shown here as an example (blue curve). The data originates from [19]. The most frequently detected sulfur concentration in SBR is 0.9 wt%, but there are also zones with lower and higher local concentrations of sulfur. The curve of sample SBR_0min&NR-GR (red curve, also originating from [19]) shifts towards lower sulfur concentrations. This means that the matrix has a significantly lower sulfur concentration due to diffusion processes.

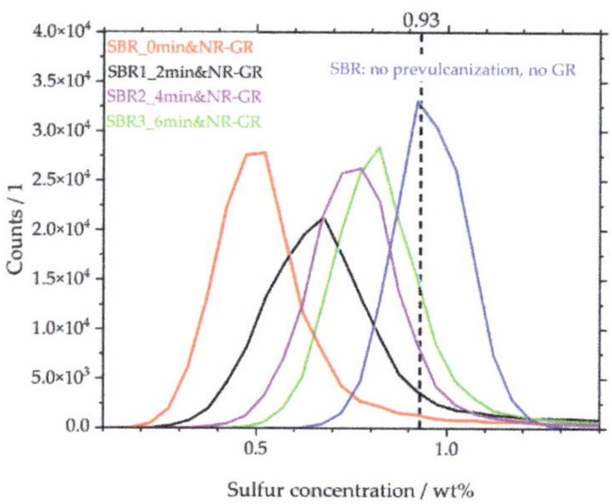

Figure 6. Distribution curves of local sulfur concentrations in the matrices of different vulcanizates.

However, if the matrix is prevulcanized for 2 min (black curve) before NR-GR is incorporated, the local sulfur concentration changes. The distribution curve shifts to higher concentrations compared to that for the non-prevulcanized sample (red). More sulfur remains in the matrix because of the reduced diffusion behavior of the sulfur species formed during prevulcanization. This behavior increases with longer prevulcanization times of 4 min (purple) and 6 min (green). Compared to the non-prevulcanized sample (red), with a maximum of approx. 0.45 wt%, the vulcanizate with a matrix prevulcanized for 6 min reaches a peak at 0.8 wt%. This behavior supports the hypothesis proposed by [19], that species formed during crosslinking—e.g., the active sulfurating agent postulated by Morrison and Porter [30]—are less mobile than the free sulfur initially present. It is thus evident that with increasing prevulcanization of the matrix, less free sulfur is able to diffuse into the GR particles. As a result, more sulfur remains in the matrix, and the distribution curves shift with increasing prevulcanization to the SBR-reference (blue). Even during the relatively short prevulcanization of 2 min, the content of the less mobile species formed is so high that the distribution curve (black) differs significantly from that of the non-prevulcanized curve (red).

The distribution curves indicate that, apart from the matrix peak, there is no other peak representing the sulfur concentration in the GR particles. Instead, there are higher sulfur concentrations at moderate counts to the right of the matrix peaks. It therefore indicates that the sulfur concentration in the GR particles is less uniform and instead extends over a wide concentration range, e.g., from concentrations of approx. ≥1 wt% in the 2 min prevulcanized sample (black).

Compared to the SBR reference (blue), the matrices of the prevulcanized compounds are comparatively less sulfur-depleted, which leads to the expectation of increased mechanical properties compared to the non-prevulcanized specimen (red), for example. This is investigated in the following section.

3.3. Tensile Tests

To assess the influence of prevulcanization on mechanical properties, tensile tests were carried out.

In Figure 7 and Table 7, the sample SBR1 (blue) reaches a tensile strength of approx. 22.5 MPa at an elongation at break of 435%. For the same matrix material to which 30 phr NR-GR was added, and for which the matrix was not prevulcanized (SBR_0min&NR-GR, red), the tensile strength and elongation at break are significantly reduced. SBR_0min&NR-GR is again derived from [12,19] and serves as a reference to assess the influence of the prevulcanization.

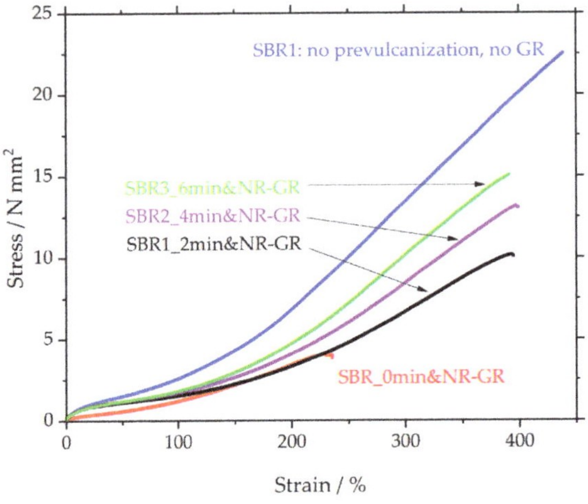

Figure 7. Stress strain curves of SBR1 and vulcanizates containing ground rubber with different degrees of prevulcanization.

Table 7. Summary of tensile properties of the investigated vulcanizates.

Specimen	Number of Valid Tests	Mean Stress /N mm^2 Incl. 95% Error	Mean Strain /% Incl. 95% Error
SBR1	12	21.8 ± 1.0	421 ± 13
SBR_0min&NR-GR	10	4.0 ± 0.3	232 ± 12
SBR1_2min&NR-GR	12	10.1 ± 0.8	379 ± 19
SBR2_4min&NR-GR	11	12.9 ± 0.8	395 ± 17
SBR3_6min&NR-GR	11	15.0 ± 0.5	372 ± 11

This influence of prevulcanization is demonstrated by the curves of SBR1_2min&NR-GR, SBR2_4min&NR-GR and SBR3_6min&NR-GR. The samples differ from SBR_0min&NR-GR (red) only in terms of the prevulcanization of the matrix for 2, 4 or 6 min, respectively. A 2 min prevulcanization already leads to a more than doubled tensile strength of 10 MPa compared to the red reference curve, exhibiting only 4.8 MPa. The elongation at break increases from 240% to almost 400%. This significant effect is the result of the prevulcanization step prior the incorporation of GR particles. Extending the prevulcanization time leads to a further increase in tensile strength with the same elongation at break. The significant increase in tensile strength and elongation at break can be explained by the higher sulfur concentration in the matrix and therefore an assumed higher CLD. The fact that the elongation at break for all three prevulcanized specimens remains identical at just under 400% could be due to the GR particles increasing the intrinsic elongation of the surrounding matrix. This leads to a local rise in strain in the vicinity of the particles, leading to failure of the entire specimen when this strain exceeds the elongation at break of the matrix material. The tensile tester, however, detected only one non-local strain value of the entire specimen, which is listed in Table 7 to be between 372 and 395%. This value might not be sensitive enough to distinguish differences between the different prevulcanized samples.

4. Discussion

The rubber industry is making efforts to drive towards a more circular economy to increase its sustainability. The recycling of rubber components such as end-of-life tires (ELTs) is an essential part of this procedure. Recycling rubber is challenging due to the covalent crosslinks between the polymer chains formed during the vulcanization process. These bonds define the viscoelastic properties of the vulcanizate. One recycling method involves grinding the ELTs into ground rubber (GR) particles. These particles can then be incorporated into green compounds to reduce the need for raw material. However, GR-containing compounds exhibit significantly deteriorated properties compared to those without GR. This is firstly because the GR particles perform as an inactive filler, increasing the intrinsic elongation of the matrix material in their vicinity. Secondly, free sulfur diffuses from the matrix into the GR particles during compounding, storage and vulcanization, resulting in matrix systems with lower and GR particles with higher crosslink density (CLD).

To improve the properties of GR-containing vulcanizates, the influence of matrix prevulcanization upon the diffusion behavior of sulfur was investigated in this article. It was found that the mobility of sulfur is significantly reduced when it forms different intermediates or species during the vulcanization process. Therefore, it was investigated whether the incorporation of GR particles into a prevulcanized matrix leads to less sulfur diffusion and thus increased tensile properties compared to a compound with the particles being incorporated into a non-prevulcanized matrix.

RPA measurements showed that prevulcanization reduces the scorch time of GR-containing compounds, because the heating phase continues the already started but interrupted vulcanization process during prevulcanization. An influence of the prevulcanization upon the maximum torque could not be detected.

It was demonstrated, by using micro X-ray fluorescence analysis (μ-XRF), that sulfur diffusion significantly decreases with the degree of prevulcanization. This indicates that the mobility of sulfur is reduced in prevulcanized matrix compounds in comparison to those that have not been prevulcanized. In contrast to the conventional production process without prevulcanization, the sulfur content in the matrix could almost be doubled, because less sulfur diffused out of the matrix into the GR particles.

The increased sulfur concentration in the matrix results in the formation of a higher CLD, which has a positive effect both on tensile strength and elongation at break. It was thus possible to increase tensile strength from about 4 MPa to 15 MPa, improving the elongation at break at the same time from about 230% to almost 400%. Besides that, the elongation at break reaches a maximum, indicating the influence of the GR particles in form of an inactive filler.

Prevulcanization is a comparatively simple method on a laboratory scale because it does not need any additional machinery or changes of compound formulations besides the prevulcanization step itself. However, the prevulcanization will face challenges bringing it into industrial scale production because of higher resulting viscosities (compounding, energy consumption) and vulcanization processes (scorching). If the scorching behavior could be tailored with respect to torque rise over a relatively long time, the prevulcanization could be installed into the rubber production procedure. Therefore, prevulcanization might pave the way for effective incorporation of GR into new rubber compounds, even at production scale.

Author Contributions: Conceptualization, S.F., V.H., F.G. and A.B.; methodology, S.F.; software, S.F.; validation, S.F., V.H., F.G. and A.B.; formal analysis, S.F., V.H., F.G. and A.B.; investigation, S.F.; resources, S.F. and V.H.; data curation, S.F.; writing—original draft preparation, S.F.; writing—review and editing, S.F., V.H., F.G. and A.B.; visualization, S.F.; supervision, S.F.; project administration, S.F.; funding acquisition, S.F. and V.H. All authors have read and agreed to the published version of the manuscript.

Funding: This research was financially supported by DKG (German Rubber Society). In addition, it was supported by the publication fund of the Technical University of Applied Sciences Würzburg-Schweinfurt.

Data Availability Statement: Dataset available on request from the authors.

Acknowledgments: The authors would like to thank the companies L. Brüggemann GmbH & Co. KG, KLK Emmerich GmbH and Lanxess Deutschland GmbH for their material donations of ZnO, stearic acid and accelerator, respectively. The authors would also like to express their gratitude to Stefan Schneider for his help mixing the compounds.

Conflicts of Interest: The authors declare no conflicts of interest. The funders had no role in the design of this study; in the collection, analyses, or interpretation of data; in the writing of the manuscript; or in the decision to publish the results.

Abbreviations

The following abbreviations are used in this manuscript:

μ-XRF	Micro X-ray fluorescence analysis
CLD	Crosslink density
cyclo-S_8	Soluble sulfur
ELTs	End-of-life tires
GR	Ground rubber
RPA	Rubber process analyzer
S′	Elastic torque
S'_{max}	Maximum elastic torque
S'_{min}	Minimum elastic torque
t_{10}	Time to reach 10% of full vulcanization
t_{50}	Time to reach 50% of full vulcanization
t_{95}	Time to reach 95% of full vulcanization
Tg	Glass transition temperature
t_x	Time to reach x % of full vulcanization

References

1. Norris, C.J.; Cerdán, A.L.; Ter Haar, P. Understanding recovered Carbon Black. *Rubber Chem. Technol.* **2023**, *96*, 196–213. [CrossRef]
2. Anjum, A. Recovered Carbon Black from Waste Tire Pyrolysis: Characteristics, Performance, and Valorisation. Ph.D. Thesis, University of Twente, Enschede, The Netherlands, 2021.
3. van Hoek, J.W. Closing the Loop: Reuse of Devulcanized Rubber in New Tires. Ph.D. Thesis, University of Twente, Enschede, The Netherlands, 2022.
4. Dorigato, A.; Rigotti, D.; Fredi, G. Recent advances in the devulcanization technologies of industrially relevant sulfur-vulcanized elastomers. *Adv. Ind. Eng. Polym. Res.* **2022**, *6*, 288–309. [CrossRef]
5. Karger-Kocsis, J.; Mészáros, L.; Bárány, T. Ground tyre rubber (GTR) in thermoplastics, thermosets, and rubbers. *J. Mater. Sci.* **2013**, *48*, 1–38. [CrossRef]
6. Schüwer, N.; Clayfield, T.; Martineau, P.; Estudante Ventura, M.J.; Marshall, S.M.; Hofmann, S.; Ilisch, S.; Stratton, T. Eigenschaften S-SBR-basierter Vulkanisate mit recyceltem Gummimehl. *Gummi Fasern Kunststoffe* **2021**, *74*, 100–109.
7. KURZ Karkassenhandel. (18 August 2020). Available online: https://kurz-karkassenhandel.de/gummi-produkte/gummiprodukte-anfragen/ (accessed on 30 January 2024).
8. Herrmann, V.; Speyer, Y.; Leiber, J. Herstellung und Verarbeitung von sortenreinem Gummimehl im Labormaßstab. *Gummi Fasern Kunststoffe* **2017**, *70*, 240–247.
9. Kim, S.W.; Park, H.Y.; Lim, J.C.; Jeon, I.R.; Seo, K.H. Cure characteristics and physical properties of ground-rubber-filled natural rubber vulcanizates: Effects of the curing systems of the ground rubber and rubber matrix. *J. Appl. Polym. Sci.* **2007**, *105*, 2396–2406. [CrossRef]
10. Hrdlička, Z.; Brejcha, J.; Šubrt, J.; Vrtiška, D.; Malinová, L.; Čadek, D.; Kadeřábková, A. Ground tyre rubber produced via ambient, cryogenic, and waterjet milling: The influence of milling method and particle size on the properties of SBR/NR/BR compounds for agricultural tyre treads. *Plast. Rubber Compos.* **2022**, *51*, 497–506. [CrossRef]
11. Rodgers, B.; D'Cruz, B. Recycling of Rubber. In *Rubber Compounding—Chemistry and Applications*; Taylor & Francis Group: Boca Raton, FL, USA; London, UK; New York, NY, USA, 2016.
12. Herrmann, V.; Frosch, S.; Maslowski, T.; Werner, D. Einsatz von definiertem Gummimehl aus NR und SBR in unterschiedlichen Matrixwerkstoffen. *Gummi Fasern Kunststoffe* **2021**, *74*, 200–207.
13. Han, S.-C.; Han, M.-H. Fracture behavior of NR and SBR vulcanizates filled with ground rubber having uniform particle size. *J. Appl. Polym. Sci.* **2002**, *85*, 2491–2500. [CrossRef]
14. Gibala, D.; Hamed, G.R. Cure and Mechanical Behavior of Rubber Compounds Containing Ground Vulcanizates. Part I—Cure Behavior. *Rubber Chem. Technol.* **1994**, *67*, 636–648. [CrossRef]
15. Phadke, A.A.; Bhowmick, A.K.; De, S.K. Effect of cryoground rubber on properties of NR. *J. Appl. Polym. Sci.* **1986**, *32*, 4063–4074. [CrossRef]
16. Herrmann, V.; Schulz, A. Der Einfluss von Lkw-Reifenmehl als Additiv auf die Eigenschaften einer Lkw-Laufflächenmischung. *Gummi Fasern Kunststoffe GAK* **2016**, *69*, 698–705.
17. Kim, S.-W.; Hong, K.-H.; Seo, K.-H. Effects of ground rubber having different curing systems on the crosslink structures and physical properties of NR vulcanizates. *Mater. Res. Innov.* **2003**, *7*, 149–154. [CrossRef]
18. Phadke, A.A.; Chakraborty, S.K.; De, S.K. Cryoground Rubber-Natural Rubber Blends. *Rubber Chem. Technol.* **1984**, *57*, 19–33. [CrossRef]
19. Frosch, S. Diffusion Processes in Ground Rubber Containing Compounds. Ph.D. Thesis, University of Twente, Enschede, The Netherlands, 29 October 2024.
20. Frosch, S.; Herrmann, V.; Schülein, T.; Grunert, F.; Blume, A. Sulfur Diffusion Studies Imitating Recycled Ground-Rubber-Containing Compounds. *Polymers* **2024**, *16*, 3112. [CrossRef]
21. Herrmann, V.; Hanning, S.; Kreyenschmidt, M.; Wolff, A.; Ludwig, D.; Ludwig, J. Untersuchungen zur Diffusion des Schwefels in Rezyklat-Kautschukmischungen: Teil 2: Schwefeldiffusion in Mischungen mit Gummimehl. *Gummi Fasern Kunststoffe* **2018**, *71*, 340–349.
22. Frosch, S.; Herrmann, V.; Grunert, F.; Blume, A. Quantification of sulfur distribution on rubber surfaces by means of µ-X-ray fluorescence analysis. *Polym. Test.* **2023**, *128*, 108237. [CrossRef]
23. Haschke, M. *Laboratory Micro-X-Ray Fluorescence Spectroscopy: Instrumentation and Applications*; Springer: Cham, Switzerland, 2014; Volume 55.
24. Mellawati, J.; Sumarti, M.; Menry, Y.; Surtipanti, S.; Kump, P. Application of X-ray fluorescence spectrometry in multielement analysis of rubber samples. *Appl. Radiat. Isot.* **2001**, *54*, 881–885. [CrossRef] [PubMed]
25. Brouwer, P. *Theory of XRF: Getting Acquainted with the Principles*, 5th ed.; Malvern Panalytical: Almelo, The Netherlands, 2018.
26. Markowicz, A.A. X-Ray Physics. In *Handbook of X-Ray Spectrometry*, 2nd ed.; van Grieken, R.E., Markowicz, A.A., Eds.; Marcel Dekker: New York, NY, USA, 2002; Volume 29, pp. 1–94.

27. Ahsanullah, M.; Kibria, B.M.G.; Shakil, M. *Normal and Students t Distributions and Their Applications*; Atlantis Press: Paris, France, 2014.
28. Brell, C.; Brell, J.; Kirsch, S. (Eds.) Verteilungen. In *Statistik von Null auf Hundert: Mit Kochrezepten Schnell Zum Statistik-Grundwissen*; Springer: Berlin/Heidelberg, Germany, 2017; pp. 153–157. [CrossRef]
29. Blume, A.; van Elburg, F.; Grunert, F.; Talma, A. "Re-Think" Sulfur Curing. *Molecules* **2024**, *29*, 5198. [CrossRef]
30. Morrison, N.J.; Porter, M. Temperature Effects on the Stability of Intermediates and Crosslinks in Sulfur Vulcanization. *Rubber Chem. Technol.* **1984**, *57*, 63–85. [CrossRef]
31. Banerjee, P.; Jain, P.K. Mechanism of sulfidation of small zinc oxide nanoparticles. *RSC Adv.* **2018**, *8*, 34476–34482. [CrossRef] [PubMed]
32. Maslowski, T. Untersuchungen zum Einsatz von Nicht Sortenreinem Gummimehl in Kautschukmischungen: NR-Mehl in Einer SBR-Matrix. Bachelor's Thesis, University of Applied Sciences Würzburg-Schweinfurt, Würzburg, Germany, 2020.

polymers

MDPI

Article

The Importance of Feedstock and Process Control on the Composition of Recovered Carbon Black

Christopher Norris [1,*], Antonio Lopez-Cerdan [1], Peter Eaton [2], Richard Moon [3] and Mark Murfitt [1]

1 Murfitts Industries, Station Road, Lakenheath, Suffolk IP27 9AD, UK; antonio.lopez-cerdan@murfittsindustries.com (A.L.-C.); mark@murfittsindustries.com (M.M.)
2 The Bridge, Joseph Ruston Building, University of Lincoln, Edgewest Road, Lincoln, Lincolnshire LN6 7EL, UK; peaton@lincoln.ac.uk
3 Materials Innovation, Avon Protection, Hampton Park West, Semington Road, Melksham, Wiltshire SN12 6NB, UK; richard.moon@avon-protection.com
* Correspondence: chris.norris@murfittsindustries.com; Tel.: +44-1842-860220

Abstract

Pyrolysis has emerged as a commercially viable material recovery process that supports circularity in the tyre industry. Here, it is demonstrated that a high degree of control can be imparted over the UK tyre waste stream and that statistically different feedstocks can be used to produce different grades of rCB based on their ash contents. The lower ash content rCB produced from truck tyres had superior in-rubber properties, closely matching those of the N550 reference. Silica, when not paired with a coupling agent, is known to be less reinforcing than CB, lowering the reinforcing behaviour of the high ash content rCB variant produced from car tyres. This justifiably places ash content within the classification and specification development discussion. However, a proximate analysis of UK waste tyres suggests that the typical rCB ash specifications of <20 wt% are unrealistic. Such limits would force producers to consider modifying process conditions to allow the deposition of carbonaceous residues to artificially dilute the ash content. This study investigates this process philosophy but conclusively demonstrates that carbonaceous residue is more detrimental to rCB performance than ash content. As such, carbonaceous residue content demands far more attention from the industry than it is currently afforded.

Keywords: rCB; pyrolysis; waste tyres; recycling; rubber; reinforcement; sustainability

1. Introduction

The UK generates approximately 50 million waste tyres, equivalent to 700,000 tonnes, each year. Since the EU ban on landfilling came into effect in 2003 (EC Directive 1999/31), most of these waste tyres have been utilised in civil engineering applications, sports and recreational surfaces and for energy recovery as fuel in cement kilns [1]. The emergence of policy drivers to combat climate change and to promote material circularity, such as the European Commission Green Deal [2], has increased pressure on the industry to recover materials from waste tyres. Now, all the major tyre brands have imposed stringent sustainability goals, with most targeting the construction of tyres from 100% sustainable materials by 2050. This necessitates a movement away from traditional mechanical recycling technologies to those that can chemically deconstruct waste tyres into reusable raw materials. Pyrolysis has emerged as the frontrunning commercially capable material recovery process to support circularity in the tyre industry. As well as offering recovered products with low global warming potential and avoiding the consumption of fossil resources [3,4], properly controlled tyre pyrolysis comes with minimal emissions or residual waste [5].

Pyrolysis is defined as the thermal decomposition of organic materials in the absence of oxygen [6]. In the case of tyres, this involves cracking the long organic polymer chains into a hot gas phase, which consists of low-molecular-weight non-condensable gas and condensable tyre pyrolysis oil (TPO) fractions. The removal of organic matter leaves behind a solid mixture consisting of carbon black (CB), inorganic compounding ingredients and other pyrolysis residues, defined as 'raw recovered carbon black (rCB)' by ASTM committee D36 [7]. Assuming the raw rCB is of sufficient quality, it undergoes further milling and pelletising processes to alter its physical properties to meet the specification requirements of the end user [8]. This paper focusses on demonstrating how the properties of rCB are controlled by both feedstock selection and pyrolysis conditions.

The key to any material recovery process, including pyrolysis, is to produce consistent products that always meet the specification requirements of the end user. This poses a significant challenge to rCB producers, given that tyres are structurally complex articles comprising several different rubber components, including the tread, carcass and sidewall, alongside steel and textile reinforcements [9]. Each of the rubber components will typically contain one or more elastomers, reinforcing fillers such as CB and silica, plasticising agents, antidegradants, processing aides, curatives and other additives. It has been reported that more than one hundred ingredients can be added to the tyre, the precise nature of which depends on its specific use requirements [10,11]. To tackle the challenges associated with product consistency, it seems prudent to narrow the compositional variance of pyrolysis feedstocks by segregating the waste stream into structurally and compositionally similar tyres. However, it is acknowledged that manufacturer variability will persist even when segregating the waste stream by type. In the UK and the EU, the two main classes of tyres are passenger car and truck, accounting for approximately 70 wt% and 20 wt% of the total tyre waste stream, respectively. Passenger car tyres contain a higher synthetic rubber content and a greater utilisation of silica filler and textile reinforcements [12]. In contrast, truck tyres contain higher levels of steel, natural rubber and carbon black [13]. An additional benefit of feedstock selection is to narrow the aggregate size distribution of the resultant rCB, as passenger car and truck tyres contain different grades of CB. A broader aggregate size distribution has been linked to a reduced reinforcing potential for rCB [8,14].

The first aim of this study was to assess the ability to produce compositionally different pyrolysis feedstocks by segregating car and truck tyres from the UK waste stream, and, further, to assess the impact on the composition and reinforcing ability of the resultant rCB products. The major difference between the feedstocks is the higher silica content of the passenger car tyre, arising from the drive to enhance fuel efficiency following the introduction of labelling rules by the European Commission in 2012 [15]. Assuming a complete pyrolysis of the tyre rubber, the resultant solid rCB product is a mixture of CB, silica and other inorganic compounding ingredients. rCB produced from passenger car tyres is, therefore, expected to contain a higher silica (ash) content. The ash content of an rCB has been reported as an influential factor on rubber reinforcing potential [16–18], the dominant factor being that silica, when not paired with a coupling agent, is a less reinforcing filler than CB, with laboratory produced rCB samples demonstrated to decrease in reinforcing potential with increasing silica content [8]. It is also documented that silica interacts with zinc during compounding, leading to the removal of soluble zinc from its normal accelerator activating function, subsequently reducing crosslink density [19]. Based on these observations, it is understandable that tyre manufacturers have developed rCB specifications that stipulate a requirement for ash content to be $\leq$20 wt% [20]. However, these observations are only relevant to 'clean', low-residue-content rCB produced from the complete pyrolysis of the rubber feedstock.

The goal of tyre pyrolysis is the complete conversion of all organic matter into TPO and non-condensable gas products, yielding an rCB free of pyrolysis residues. Process optimisation with this in mind has two major benefits:

1. By preventing the formation of pyrolysis residues, the surface of the recovered fillers is accessible in the elastomer phase, thus promoting reinforcing behaviour.
2. The biogenic carbon contained within the feedstock is recovered in the TPO and gas phases, where it adds value by reducing scope 1 emissions when using these materials directly, or after refinement as fuels.

rCB pyrolysis residues arise from two primary routes. Firstly, the incomplete removal of organic matter will leave oily residues on the surface of the rCB which, at high levels, have been conclusively linked to poor reinforcing properties [21]. Such residues pose little challenge to the industry, as they are easily identifiable by both the producer and end user using standard ASTM techniques, such as toluene discolouration (D1618) and thermogravimetric analysis (D8474) [22,23]. The second type of residue arises from polymer decomposition reactions that lead to the formation of additional fixed carbon, often referred to as carbonaceous residue [21,24,25]. Carbonaceous residue contents of >20 wt% have been reported and linked to a significantly reduced surface activity of rCB [8,26–30]. This residue is much harder to differentiate from the CB content of rCB, and so it often goes unnoticed. The net result of these pyrolysis residues is a reduction in ash content by the dilution effect, and this may be seen as an opportunity to meet the stringent ash content specifications applied in the rubber industry. The second aim of this study was to intentionally produce a high-carbonaceous-residue-content rCB from car tyres to explore an alternative strategy to meet ash specifications, and to assess the impact on morphological and in-rubber properties.

2. Materials and Methods

2.1. Tyre Granulate Preparation and Characterisation

Through careful selection of waste tyres, pyrolysis feedstocks were prepared from 100% car tyres or 100% truck tyres using Murfitts Industries commercial processing line at Lakenheath in the UK. The multi-stage process comprises shredding, granulation and separation steps to produce a $\leq$8 mm granulate with at least 99% steel and 98% fibre removal.

Proximate and ultimate analyses of twenty batches each of the car and truck feedstocks were completed. Proximate analysis followed the procedure outlined in ASTM D8474 [23], using a Perkin Elmer TGA4000 (Waltham, MA, USA) to yield volatile matter (VM), fixed carbon (FC) and ash contents.

2.2. rCB Materials Preparation

Pyrolysis trials were conducted using a two-stage continuous process comprising two in-sequence ETIA Technologies Spirajoule® (Lacroix-Saint-Ouen, France) reactors with an overall throughput capacity of 90 kg/h. The operating conditions of the first stage were designed to complete $\geq$90% of the pyrolysis, with the second stage operating at a higher temperature to remove the residual organic matter from the raw rCB. Although the full process conditions cannot be disclosed for proprietary reasons, a brief description of how the process was optimised to produce the different rCB materials for this study is provided below.

Two grades of rCB were produced from the different feedstocks using process conditions optimised to minimise the rCB yield. This means that the target yield was based on the FC + ash content from the feedstock proximate analysis, producing material that is as free as possible from pyrolysis residues. The car tyre-derived rCB is referred to as Mi360+ and the truck tyre-derived grade as Mi360HP.

A further rCB material was produced from the car tyre feedstock but using very different process conditions to minimise the ash content. The different process conditions, including higher temperatures, were selected to promote the formation of carbonaceous residue on the rCB, thus diluting the inorganic components. This material is referred to as HCR (high carbonaceous residue).

All three materials were finalised in the same manner. The particle size was reduced to a target D97 of 10 ± 0.5 μm using a Hosokawa jet mill (Augsburg, Germany) before wet pelletising with a MARS Minerals (Mars, PA, USA) pin mixer and subsequent drying to produce pellets of <80 gf maximum hardness.

2.3. rCB Material Characterisation

A series of ASTM standardised techniques developed for carbon black and rCB were used to characterise the samples: this included BET surface area (D6556) [31], toluene discolouration (D1618) [22], bulk composition by TGA (D8474) [23], milled particle size distribution (WK87480) [32] and pellet hardness (D5230) [33].

To determine zinc and silicon contents, 250 mg of sample was digested on a hotplate with sulphuric and perchloric acids. When cold, they were transferred to plastic flasks where hydrofluoric acid was added to digest the silicon. Quantitative elemental analysis was performed using an Agilent 5800 (Agilent, CA, USA) inductively coupled plasma optical emission spectrometer (ICP-OES).

rCB morphology and inorganic distribution assessments were made using a Thermo Scientific F200i scanning (Waltham, MA, USA)/transmission electron microscope (S/TEM) fitted with a Bruker SDD XFlash 6.100 Energy Dispersive X-ray (EDX) detector (Berlin, Germany). The TEM was operated at 200 kV, and Velox software v3.19 used to acquire and process the elemental maps. Sample preparation involved a modified version of ASTM D3849 [34], as described by Grulke et al. [35].

2.4. In-Rubber Characterisation Protocol

The ASTM D3191 styrene butadiene rubber (SBR) formulation was used to evaluate the in-rubber behaviour of the rCB samples and an N550 reference [36]. A filler loading of 70 parts per hundred rubber (pphr) was used, as recommended in ASTM D8491 for rCB materials [37]. Compounds were produced using a 78 cc Haake Rheomix OS/3000 (Waltham, MA, USA) with Banbury style rotors set at 40 °C, with a rotor speed of 60 rpm.

Moving die rheometer (Alpha Technologies MDR2000, Akron, OH, USA) testing (to ASTM D5289) at 160 °C was used to assess the cure characteristics of each compound and to allow preparation of cured test sheets using a cure time to T90 + 5 min [38]. The cured rubber sheets were tested as follows:

- Tensile properties were determined using a Llyod LR5K (AMETEK, Bognor Regis, UK) following ASTM D412 [39].
- Shore A Hardness was determined using a Wallace H17A (Walace Instruments, Dorking, UK) in accordance with ASTM D2240 [40].
- Filler dispersion was assessed by examining surfaces cut with fresh razor blades at 250× magnification using a Hitachi TM3030 Scanning Electron Microscope (SEM) (Tokyo, Japan). Surface roughness plots and average surface roughness (Ra) values were generated using 3D-Image Viewer software v2.0 (Denshi Kougaku Kenkyusyo Co., Ltd., Tokyo, Japan).

- Strain sweeps were conducted using a Perkin-Elmer DM8000 (Waltham, MA, USA) configured in tension mode. During this step, 2 × 2 × 10 mm specimens were tested at 40 °C, 10 Hz and a double strain amplitude (DSA) range of ~0.04 to 4%.

3. Results and Discussion

3.1. Feedstock Characterisation

Estimates suggest that four million tons of waste tyres are generated globally every year [41]. Such statistics give the impression that this waste stream represents an almost infinite resource that can be misleading in relation to securing a consistent pyrolysis feedstock locally. Without doubt, compositionally consistent pyrolysis outputs necessitate a compositionally consistent feedstock. To emphasise this point, twenty batches of two compositionally different feedstocks, from 100% car or 100% truck tyres, were prepared, and their bulk compositional properties were determined through a proximate analysis. An example TGA weight loss profile is provided in Figure 1, showing the determination of volatile matter (VM), fixed carbon (FC) and ash contents, as adopted by Bowles and Fowler in their review of published rubber pyrolysis feedstocks [13]. Assuming pyrolysis is fully complete, the VM value derived from the TGA test should correlate with the hot gas yield of a pyrolysis process. Likewise, the sum of the FC and Ash values should offer a theoretical rCB yield, assuming that pyrolysis is complete and no process losses occur.

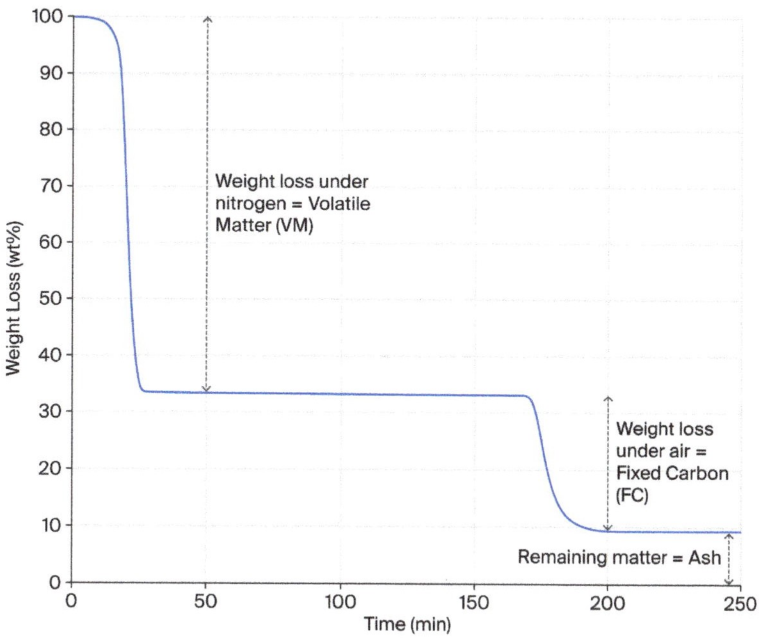

Figure 1. Example of rubber granulate TGA weight loss profile.

A summary of the bulk compositional properties of the two feedstocks is provided in Table 1, showing that the truck granulate contains a higher VM, higher FC and lower ash content. Two sample T-tests for each of the parameters yielded *p*-values of 0, confirming that the differences between the feedstocks are statistically significant. The narrow band of the 95% confidence intervals also demonstrates that it is possible to impart a high level of feedstock compositional control. Based on this data, the predicted pyrolysis rCB yields (FC + ash) are 35 wt% and 38 wt% when using truck and car granulate feedstocks, respectively. This data provides a valuable reference point when optimising the process to limit rCB contamination from pyrolysis residues. The data can also be used to predict the ash content of the resultant rCB. It is worthy of note that, even with the lower-ash-content truck feedstock, the theoretical ash content of the rCB is marginally above the industry standard of <20 wt%. This alone questions the practicality of the current specifications, especially when considering that the higher-ash-content car tyre rubber represents the bulk of the waste stream.

Table 1. Mean proximate analysis values with 95% confidence intervals in brackets.

Parameter	Unit	Truck Granulate	Car Granulate
VM Content	wt%	64.7 (64.0→65.3)	62.5 (61.8→63.0)
FC Content	wt%	28.2 (27.6→28.8)	24.5 (23.4→25.7)
Ash Content	wt%	7.1 (6.5→7.8)	13.1 (12.1→14.0)
Theoretical rCB Yield (FC + Ash)	wt%	35.3	37.6
Theoretical rCB Ash Content	wt%	20.1	34.8

3.2. rCB Characterisation

The properties of the three different rCB materials are summarised in Table 2. The rCB yields of samples optimised for low residue content, Mi360HP (truck) and Mi360+ (car), were in good agreement with the theoretical yields derived from the proximate analysis of the feedstock. Compared to the Mi360+, the high carbonaceous residue (HCR) sample produced from the same feedstock had an elevated FC content, confirming that the changes in the pyrolysis process conditions had the intended effect of increasing the carbonaceous residue content. The difference in yield between the Mi360+ and HCR suggests that approximately 17 wt% of the feedstock was converted into carbonaceous residue for the latter. All samples had comparably low organic residue contents, as demonstrated by both the toluene extract transmittance and VM content values. This parameter can therefore be disregarded when considering the differences in rubber reinforcing behaviour between the samples. The VM fraction will contain a contribution from moisture and a small amount of residual organic matter.

Table 2. Summary of rCB sample properties.

Parameter	Unit	Truck rCB (Mi360HP)	Car rCB (Mi360+)	Car rCB HCR
Measured rCB Yield	wt%	36	38	55
Toluene Transmission	%	98	99	100
VM–TGA	wt%	1.3	1.7	1.8
FC–TGA	wt%	79.0	67.6	79.4
Ash–TGA	wt%	18.9	30.8	18.7
Si–ICP	wt%	3.7	9.1	7.0
SiO_2–Calculated *	wt%	7.9	19.5	15.0
Zn–ICP	wt%	5.0	3.5	2.5
ZnS–Calculated *	wt%	7.5	5.2	3.7
BET Surface Area–Raw rCB	m^2/g	79.3	82.7	44.7
BET Surface Area–rCB	m^2/g	89.9	85.8	75.0
Milled Particle Size, d97	µm	10.3	9.5	10.3
Average Pellet Hardness	gF	33	35	39

* ICP derived atomic mass data converted to molecular mass of associated chemical.

The measured ash contents of the Mi360 grades were slightly below that predicted from the feedstock analysis. However, a good agreement between theoretical and measured rCB yields would suggest that the residue content is very low. The discrepancy may arise from natural variability (i.e., only one sample measured compared to the average of twenty feedstock samples) and may also result from the different temperature profiles used for the TGA test and pyrolysis process. The key point is that the deposition of carbonaceous residues within the car tyre-derived HCR sample reduced the ash content in line with the truck tyre-derived Mi360HP. This provides the basis for assessing the relative effects of ash and carbonaceous residues on the reinforcing behaviour of the rCB materials.

One of the strategies adopted by rCB producers to meet ash content targets is to dilute the higher-ash-content car tyre feedstock with lower-ash-content truck tyres. Using the ash contents of the low-residue Mi360HP and Mi360+, predictions can be made on the necessary blend ratio to meet the <20 wt% ash content, as seen in Figure 2. The prediction assumes that, by blending the feedstock, the residue content of the rCB does not differ to those produced from the segregated waste streams. Based on this data, a feedstock blend containing no less than 91 wt% truck tyres would be necessary to meet the target and is by no means reflective of the relative abundance of this waste tyre category. Such ash content targets almost necessitate the need to form carbonaceous residue during pyrolysis.

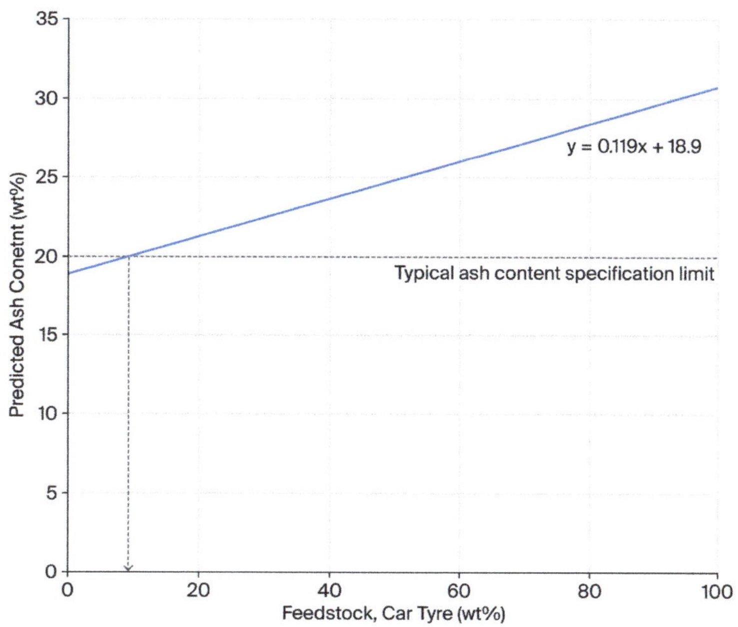

Figure 2. Relationship between rCB ash content and the percentage of car tyres used in a feedstock blend with truck tyres.

Of course, ash is a poor descriptor of the inorganic components contained within an rCB product. Most of the inorganic matter arises from the use of silica filler and zinc oxide additives [8,13,16–18,21], as confirmed by an ICP-OES analysis of acid digests. Car tyres contain a much higher loading of silica, but it is also apparent that silica has found use in truck tyres. The data also confirm that truck tyres contain a higher zinc loading compared to car tyres. The rCB materials will also contain low levels of other inorganic matter, such as iron and other trace metals originating from the feedstock, but these were not quantified as part of this study.

STEM/EDX mapping of the rCB samples was performed to better understand the distribution of these inorganic components. An example of an Mi360+ fused agglomerate is displayed in Figure 3, showing that the silica clusters are intimately entwined amongst the CB aggregates. Zinc was found to be in the form of zinc sulphide, appearing as discrete particles at the surface of the fillers, in agreement with other microscopy assessments of rCB [16,18]. As such, no evidence was identified to suggest that the zinc sulphide particles influenced agglomeration. The pyrolysis gas phase contains an abundance of hydrogen sulphide, with levels of up to 5.1 vol% reported [42,43]. It is highly likely that the simple reaction between zinc oxide and hydrogen sulphide yields the observed particles of zinc sulphide:

$$ZnO + H_2S \rightarrow ZnS + H_2O$$

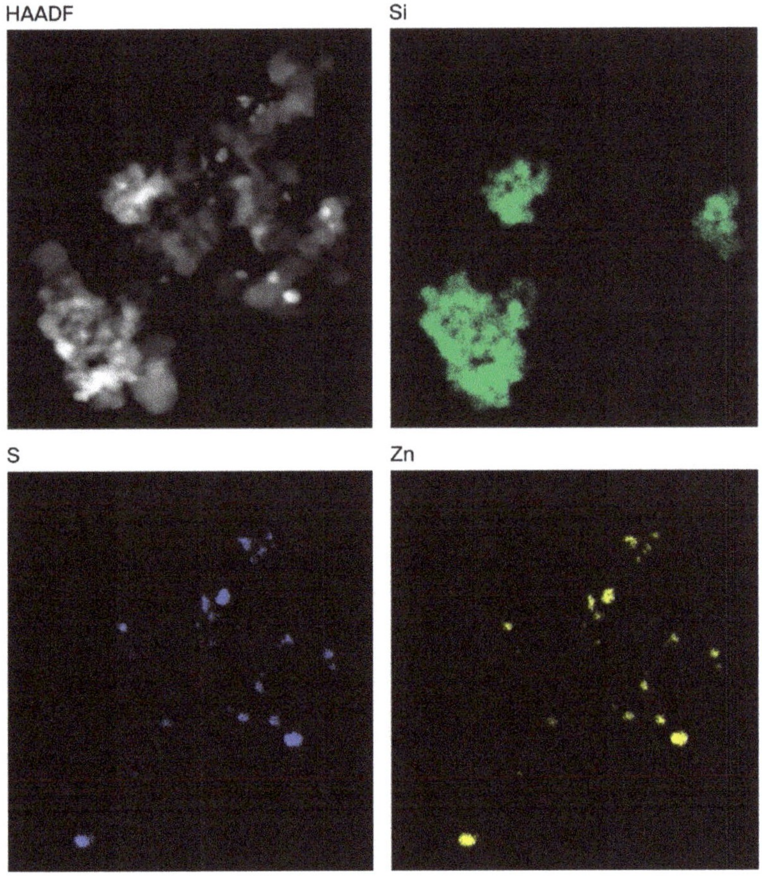

Figure 3. STEM HAADF (electron) image of rCB fused agglomerate alongside silicon, sulphur and zinc distribution maps.

The conversion of ZnO to ZnS may not be true for all pyrolysis processes, depending on the process conditions utilised.

The same observations were made for the Mi360HP, but with lower amounts of silica present.

The presence of carbonaceous residue within the HCR sample had a dramatic effect on morphology, with the aggregate nature much less apparent in many of the particles studied, as seen in Figure 4. This observation aligns with the BET measurements of the raw rCBs (prior to milling and pelletising), with the HCR exhibiting 46% less surface area than the Mi360+ produced from the equivalent feedstock. As structure level and surface area are regarded as fundamental properties of rubber reinforcement [44–46], the HCR would be expected to exhibit an inferior reinforcing behaviour.

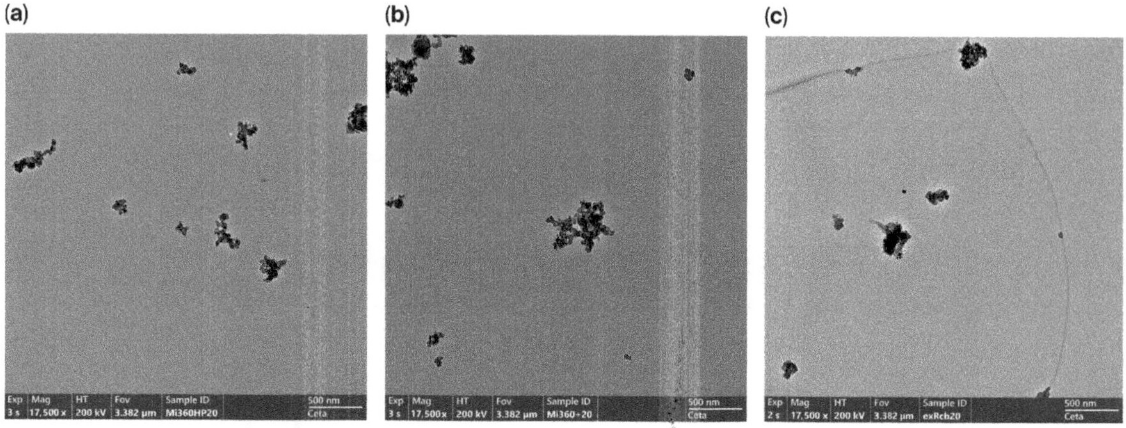

Figure 4. TEM images showing rCB morphology; (**a**) Mi360HP, (**b**) Mi360+ and (**c**) HCR.

The milled particle size target of d97 10 ± 0.5 µm was achieved for all samples. This parameter is directly related to dispersion [47], and so it is important that all were size-reduced to the same level to allow for a fair comparison of in-rubber properties. All samples were noted to exhibit an increased surface area post milling: Mi360HP by 13%, Mi360+ by 4% and HCR by 68%. These data suggest that the presence of carbonaceous residue is linked to a significant increase in milled surface area, closing the gap between the HCR and the Mi360 grades. Traditional carbon black colloidal tests have long been established as poor predictors of the in-rubber properties of rCB [8,21]. These observations further challenge the value of surface area in rCB specifications, especially when considering that the end-user cannot test the material in the pre-milled state.

Finally, all samples were successfully pelletised to industry standard levels to facilitate proper handling and rubber compounding behaviour.

3.3. In-Rubber Characterisation

The rCB materials were compounded into the SBR-based ASTM D3191 formulation to investigate the effects of the compositional differences on their in-rubber properties, as seen in Table 3. Comparing the properties of Mi360HP and Mi360+ offers insight into the effect of ash content, whilst the differences between the Mi360+ and HCR are wholly related to the carbonaceous residue content.

Surface roughness maps indicating the level of filler dispersion are shown in Figure 5. The dispersion of the Mi360 grades almost matched that of the N550. It should be noted that N550 is one of the easiest grades to disperse, with its low surface area and high structure level. Although milled to the same particle size, the HCR was found to be harder to disperse compared to the Mi360 grades. This would suggest that the Mi360 grades underwent a level of agglomerate breakdown during the mixing process. It is likely that the presence of carbonaceous residue on the HCR prevents effective de-agglomeration, leading to an inferior dispersion. The presence of larger agglomerates has been linked to poorer mechanical properties [14], as discussed below.

Table 3. Summary of in-rubber properties.

Parameter	Unit	N550	Truck rCB (Mi360HP)	Car rCB (Mi360+)	Car rCB HCR
Dispersion–Ra	μm	0.34	0.43	0.46	0.75
Shore A Hardness	°	80	78	76	70
M100%	MPa	8.38	8.51	5.08	4.74
M200%	MPa	17.92	17.39	11.72	10.72
Break Stress	MPa	21.6	21.3	17.1	13.6
Break Strain	%	252	247	296	303
E'_0	MPa	24.1	23.0	19.4	15.0
E'_∞	MPa	14.6	14.6	13.0	11.3
$\Delta E'$	MPa	9.5	8.4	6.4	3.8
Tan δ_{max}	-	0.19	0.18	0.17	0.16

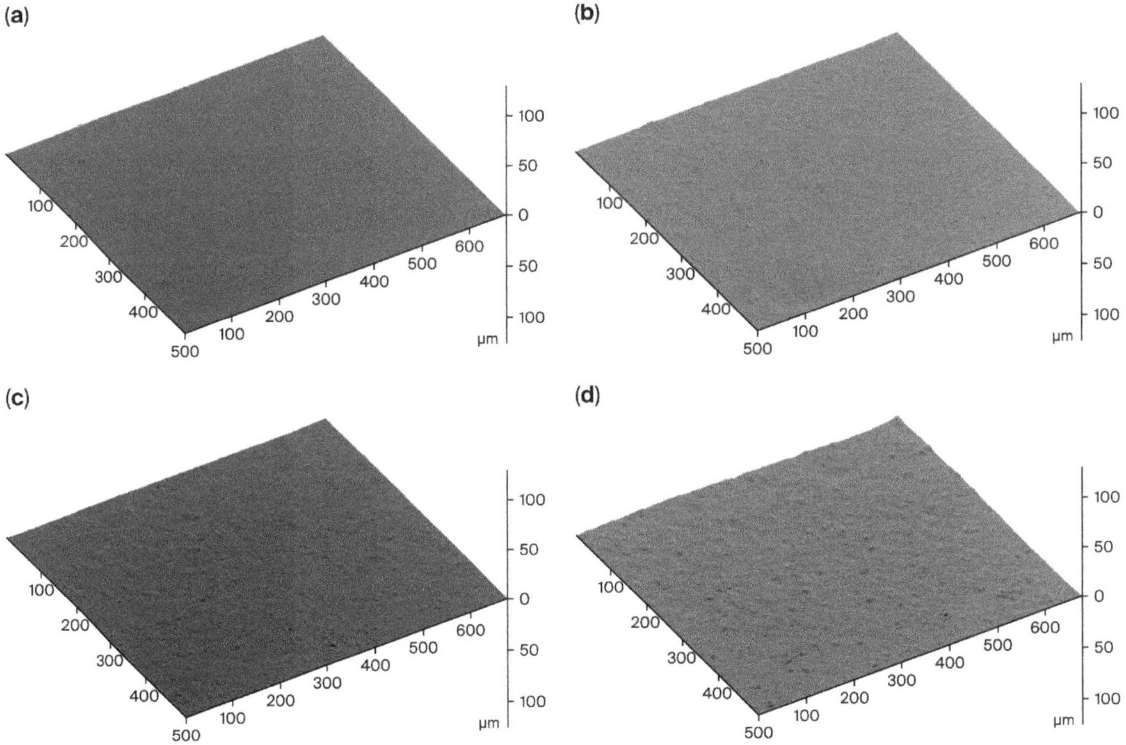

Figure 5. SEM Surface roughness maps of sectioned surfaces of model SBR compounds containing (**a**) N550, (**b**) Mi360HP, (**c**) Mi360+ and (**d**) HCR.

The mechanical properties of the Mi360HP were a good match to the N550 reference. The higher ash content Mi360+ had significantly reduced stiffness values. Most of the difference between the Mi360HP and the Mi360+ is attributable to the higher silica content of the latter. Silica, without a coupling agent, is known to be less reinforcing than carbon black [19]. In addition, silica can remove soluble zinc from its normal accelerator activating function, resulting in a reduced crosslink density [18]. Another factor is likely to be the different grades of CB recovered from the truck and car tyre feedstocks. Truck tyres utilise more N100 series in the tread sections and N300 series in the sidewalls to provide the necessary abrasion and stiffness characteristics. Car tyres tend to use N300 series in the treads and N500 to N600 in the sidewalls. In short, the CB content of Mi360HP is expected to be more reinforcing than the CB content of Mi360+.

Although the ash content matched that of the Mi360HP, the deposition of carbonaceous residues in the HCR reduced the hardness and modulus values further compared to the Mi360+. This conclusively demonstrates that carbonaceous residues are more detrimental to reinforcing behaviour than ash. In particular, there is an apparent loss of filler–polymer interaction at higher tensile strains above approximately 200%, as seen in Figure 6. This aligns with the loss of structure identified during the TEM assessment and reports that the carbonaceous residue masks the surface activity of the original carbon black.

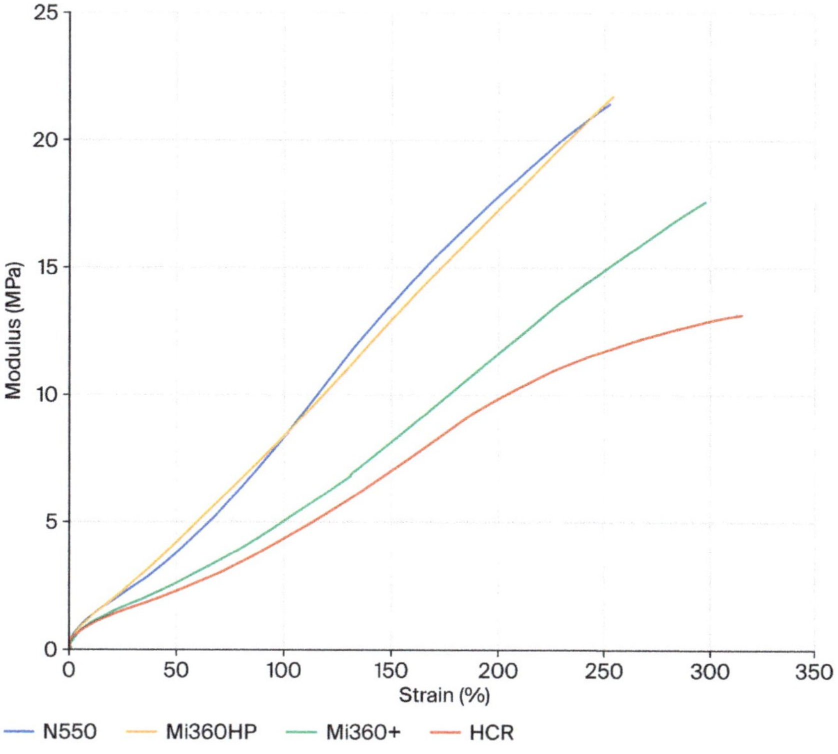

Figure 6. Tensile curves of the different rCB variants compared to N550 in the model SBR formulation.

Given that the ultimate goal of tyre pyrolysis is to provide raw materials back into the tyre industry, it was prudent to evaluate the dynamic properties of the rCB materials. Elastic modulus plots showing the non-linearity of compounds filled with N550 and different rCB materials are provided in Figure 7. At a low strain (E'_0), the dynamic stiffness contains the full contribution of the filler network. As the dynamic strain increases, the filler–filler interactions are gradually disrupted, leading to a plateau at E'_∞. This phenomenon is known as the Payne effect [48]. The difference between the low and high strain modulus ($\Delta E' = E'_0 - E'_\infty$) can therefore be used as a measure of the filler network. Assuming equal filler loadings and dispersion, the two key parameters that influence the level of filler–filler interaction are the surface area and surface chemistry. For standard grades of CB, $\Delta E'$ scales directly with surface area, as all exhibit the same surface chemistry. Given the higher surface area of the rCB materials, they would be expected to exhibit higher $\Delta E'$ values than N550 if their surface chemistry was the same. This is clearly not the case for the rCBs, with all exhibiting a lower filler networking efficiency, in line with previous studies that attributed the networking reduction to the presence of carbonaceous residue masking the active sites of CB [8,21]. Considering the effect of ash, there is a general trend for reducing filler–filler interactions from N550→Mi360HP→Mi360+ as ash content increases. In some respects, the rCBs behave like a dual-filler system, with silica and zinc sulphide particles acting to disrupt the CB network. The clear effect of carbonaceous residue is seen when comparing the HCR to the Mi360+, with further reduced filler–filler interactions for the former. This will be a function of both the reduced surface area and surface activity of the HCR.

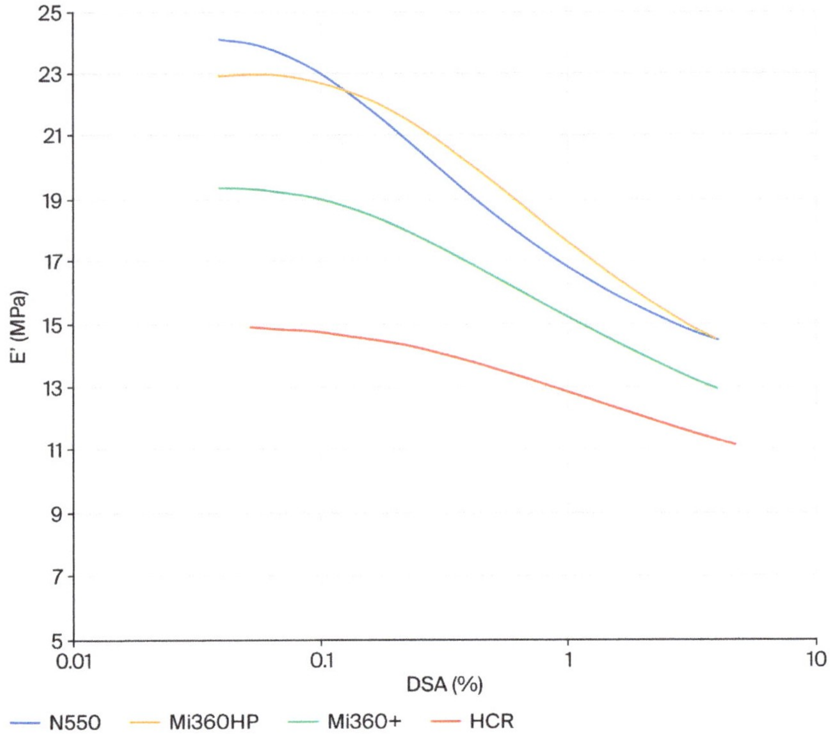

Figure 7. Strain dependency plots of the different rCB variants compared to N550 in the model SBR formulation.

At a high strain (E'_∞), the dynamic stiffness is a combination of gum modulus, hydrodynamic effects and filler–polymer interactions [48]. Considering the formulation and filler loadings were constant, E'_∞ can be used as a measure of the filler–polymer interaction. In this case, the Mi360HP exhibits a similar level of filler–polymer interaction as the N550, in good agreement with the physical properties data. As the silica level increases for the Mi360+, the filler–polymer interaction lessens. A previous study has shown that the silica content of an rCB is amenable to silane coupling, offering a route to enhancing the polymer–filler interaction of residue-free rCB [49]. As for the other parameters, the effects of carbonaceous residue on reducing the polymer–filler interaction were pronounced.

4. Conclusions

This study demonstrates that it is possible to impart a high degree of control to prepare statistically different feedstocks to produce different grades of rCB. Furthermore, a detailed understanding of the feedstock composition can be used to predict both rCB yields and rCB ash contents, offering useful tools in the optimisation of a pyrolysis process. The fixed carbon and ash content of a pyrolysis feedstock should correspond to the rCB yield, if free from organic and carbonaceous residues. This approach was used in the production of two grades of rCB, Mi360HP and Mi360+, produced from truck and car tyres, respectively. The primary difference between the two is their ash content, at 19% for the Mi360HP and 31% for the Mi360+. The term ash is a poor descriptor of the inorganic components of rCB, with TEM/EDX and ICP-OES analyses confirming the bulk to be silica filler clusters entangled with the CB aggregates, being more prevalent in the car tyre derived Mi360+. The other key inorganic component was zinc sulphide, found to reside as discrete particulates on the surface of the filler aggregates.

It has long been established that, for equivalent colloidal properties, rCB does not exhibit the same in-rubber behaviour as CB. Clearly, ash content, and to be more precise silica content, is a contributing factor. Silica, without a coupling agent, exhibits a lower polymer interaction compared with CB, manifesting as lower quai-static and dynamic modulus values for the Mi360+ when compared to the lower-silica-content Mi306HP. Ash content, therefore, performs an important role in the development of rCB specifications. However, the typical rCB ash content specifications of <20 wt% are unrealistic from UK waste tyres, especially when considering that the theoretical rCB ash contents from truck and car tyres were determined as 20 and 35 wt%, respectively. Furthermore, the Mi360HP was close to this limit (19 wt%), but still closely matched the tensile properties and filler–polymer interaction of the N550.

To demonstrate that ash is far from the only consideration regarding rCB reinforcing behaviour, a high carbonaceous residue (HCR) sample was prepared from car tyres to dilute the ash content to 19 wt%, i.e., within typical ash content specifications. This material ranked worse than the high-silica-content Mi360+ for all of the in-rubber properties, including dispersion, tensile and dynamic behaviour. TEM and colloidal assessments identified the carbonaceous residues as significantly reducing the surface area and structure of the rCB, both of which are fundamental to rubber reinforcement. Although ash content attracts a lot of attention, this study conclusively shows that carbonaceous residues are more detrimental to rCB performance.

Many have studied rCB ash reduction [16,17,50–52], but little attention has been given to controlling and limiting carbonaceous residues. This is likely linked to the availability of well-established test methods for the easy determination of ash content [23,53]; however, the same is not true for carbonaceous residues, which are difficult to differentiate from the CB content of an rCB. A fast and reliable method for carbonaceous residue determination would be extremely beneficial to the industry, offering the missing piece of the jigsaw in the classification and specification debate. Furthermore, the ability to silane-couple the silica component of a residue-free rCB provides an opportunity to benefit from the growing silica content of road tyres.

Author Contributions: Conceptualization, C.N. and M.M.; data curation, C.N., R.M. and P.E.; formal analysis, C.N., P.E. and R.M.; methodology, A.L.-C., P.E. and R.M.; project administration, C.N.; resources, C.N.; supervision, M.M.; writing—original draft, C.N.; writing—review and editing, C.N. All authors have read and agreed to the published version of the manuscript.

Funding: This research received no external funding.

Institutional Review Board Statement: Not applicable.

Data Availability Statement: The original contributions presented in this study are included in the article. Further inquiries can be directed to the corresponding author.

Acknowledgments: The authors would like to thank the Itochu Corporation of Japan for supporting this R&D activity.

Conflicts of Interest: Authors Christopher Norris, Antonio Lopez-Cerdan and Mark Murfitt were employed by the company Murfitts Industries. Author Richard Moon was employed by the company Avon Protection. The remaining author declares that the research was conducted in the absence of any commercial or financial relationships that could be construed as a potential conflict of interest.

Abbreviations

The following abbreviations are used in this manuscript:

rCB	Recovered carbon black
TPO	Tyre pyrolysis oil
CB	Carbon black
VM	Volatile matter
FC	Fixed carbon
HCR	High carbonaceous residue
BET	Brunauer–Emmett–Teller
TGA	Thermogravimetric analysis
ICP-OES	Inductively coupled plasma optical emission spectrometer
EDX	Energy dispersive X-ray
S/TEM	Scanning/transmission electron microscope
MDR	Moving die rheometer
SEM	Scanning Electron Microscope

References

1. Zerin, N.; Rasul, M.; Jahirul, M.; Sayem, A. End-of-life tyre conversion to energy: A review on pyrolysis and activated carbon production processes and their challenges. *Sci. Total Environ.* **2023**, *905*, 166981. [CrossRef]
2. European Commission. *The European Green Deal, COM(2019) 640 Final*; European Commission: Brussels, Belgium, 2019.
3. Banar, M. Life cycle assessment of waste tire pyrolysis. *Fresenius Environ. Bull.* **2015**, *24*, 1215.
4. Briones-Hidrovo, A.; Costa, S.M.; Maganinho, C.; Silva, C.M.; Rocha, J.; Dias, A.C.; Portugal, I.; Silva, C.M. Recovering carbon black from end-of-life tires: A consequential life cycle assessment. *Environ. Impact Assess. Rev.* **2025**, *115*, 108044. [CrossRef]
5. Zhang, Y.; Hwang, J.Y.; Peng, Z.; Andriese, M.; Li, B.; Huang, X.; Wang, X. Microwave Absorption Characteristics of Tire. In *Characterization of Minerals, Metals, and Materials*; Carpenter, J.S., Bai, C., Escobedo, J.P., Hwang, J.-Y., Ikhmayies, S., Li, B., Li, J., Monteiro, S.N., Peng, Z., Zhang, M., Eds.; Springer: Cham, Switzerland; Hoboken, NJ, USA, 2015; pp. 235–243.
6. Inamuddin; Asiri, A.M.; Kanchi, S. (Eds.) Green rubber composites. In *Green Sustainable Process for Chemical and Environmental Engineering and Science*; Elsevier: Amsterdam, The Netherlands, 2022; Chapter 14; pp. 273–312.
7. *ASTM D8178-20*; Standard Terminology Relating to Recovered Carbon Black (rCB). ASTM International: West Conshohocken, PA, USA, 2020.
8. Norris, C.J.; Lopez-Cerdan, A.; ter Haar, P. Understanding Recovered Carbon Black. *Rubber Chem. Technol.* **2023**, *96*, 196. [CrossRef]
9. Xiao, Z.; Pramanik, A.; Basak, A.; Prakash, C.; Shankar, S. Material recovery and recycling of waste tyres—A review. *Clean. Mater.* **2022**, *5*, 100115. [CrossRef]
10. Martínez, J.D.; Puy, N.; Murillo, R.; García, T.; Navarro, M.V.; Mastral, A.M. Waste tyre pyrolysis—A Review. *Renew. Sustain. Energy Rev.* **2013**, *23*, 179. [CrossRef]
11. Hoyer, S.; Kroll, L.; Sykutera, D. Technology comparison for the production of fine rubber powder from end of life tyres. *Procedia Manuf.* **2020**, *43*, 193. [CrossRef]
12. Valentini, F.; Pegoretti, A. End-of-life options of tyres. A review. *Adv. Ind. Eng. Polym. Res.* **2022**, *5*, 203. [CrossRef]
13. Bowles, A.J.; Fowler, G.D. Assessing the impacts of feedstock and process control on pyrolysis outputs for tyre recycling. *Resour. Conserv. Recycl.* **2022**, *182*, 106277. [CrossRef]
14. Paul, S.; Rahaman, M.; Ghosh, S.K.; Das, P.; Katheria, A.; Ghosh, T.; Das, N.C. Effects of tire-derived pyrolytic carbon black and pyrolytic heavy oil on the curing and mechanical properties of styrene-butadiene rubber composites. *Polym. Eng. Sci.* **2023**, *63*, 2942. [CrossRef]

15. European Union. Regulation (EU) 2020/740 of the European Parliament and of the Council of 25 May 2020 on the labelling of tyres with respect to fuel efficiency and other parameters, amending Regulation (EU) 2017/1369 and repealing Regulation (EC) No 1222/2009. *Off. J. Eur. Union* **2020**, *177*, 1–31.
16. Bowles, A.J.; Wilson, A.L.; Fowler, G.D. Synergistic benefits of recovered carbon black demineralisation for tyre recycling. *Resour. Conserv. Recycl.* **2023**, *198*, 107124. [CrossRef]
17. Silva, C.M.; Maganinho, C.; Mendes, A.; Rocha, J.; Portugal, I.; Silva, C.M. Recovered carbon black: A comprehensive review of activation, demineralization, and incorporation in rubber matrices. *Carbon Resour. Convers.* **2025**, 100334. [CrossRef]
18. Moulin, L.; Da Silva, S.; Bounaceur, A.; Herblot, M.; Soudais, Y. Assessment of Recovered Carbon Black Obtained by Waste Tires Steam Water Thermolysis: An Industrial Application. *Waste Biomass Valor.* **2017**, *8*, 2757. [CrossRef]
19. Hewitt, N. *Compounding Precipitated Silica in Elastomers*; William Andrew Publishing: Norwich, NY, USA, 2007.
20. Bridgestone & Michelin, Recovered Carbon Black Guidelines. 2023. Available online: https://business.michelinman.com/blog/articles/michelin-and-bridgestone-tackle-environmental-initiative (accessed on 26 August 2025).
21. Norris, C.J.; Hale, M.; Bennett, M. Pyrolytic carbon: Factors controlling in-rubber performance. *Plast. Rubber Compos.* **2014**, *43*, 245. [CrossRef]
22. *ASTM D1618-18*; Standard Test Method for Carbon Black Extractables—Transmittance of Toluene Extract. ASTM International: West Conshohocken, PA, USA, 2018.
23. *ASTM D8474-23*; Standard Test Method for Recovered Carbon Black (rCB)—Compositional Analysis by Thermogravimetry (TGA). ASTM International: West Conshohocken, PA, USA, 2023.
24. Darmstadt, H.; Roy, C.; Kaliaguine, S.; Sahouli, B.; Blacher, S.; Pirard, R.; Brouers, F. Fractal Analysis of Commercial and Pyrolytic Carbon Blacks Using Nitrogen Adsorption Data. *Rubber Chem. Technol.* **1995**, *68*, 330. [CrossRef]
25. Karthikeyan, S.; Sathiskumar, C.; Moorthy, R. Effect of process parameters on tire pyrolysis: A review. *J. Sci. Ind. Res. India* **2012**, *71*, 309.
26. Darmstadt, H.; Roy, C.; Kaliaguine, S. ESCA characterisation of commercial carbon blacks and of carbon blacks from vacuum pyrolysis of used tires. *Carbon* **1994**, *32*, 1399.
27. Yang, Y.; Wang, J.; Yan, L. Surface character of pyrolytic carbon black. *J. Chem. Ind. Eng.* **2005**, *56*, 720.
28. Pantea, D.; Darmstadt, H.; Kaliaguine, S.; Roy, C. Heat-treatment of carbon blacks obtained by pyrolysis of used tires. Effect on the surface chemistry, porosity and electrical conductivity. *J. Anal. Appl. Pyrol.* **2003**, *67*, 55. [CrossRef]
29. Darmstadt, H.; Roy, C.; Kaliaguine, S.; Cormier, H. Surface energy of commercial and pyrolytic carbon blacks by inverse gas chromatography. *Rubber Chem. Technol.* **1997**, *70*, 759. [CrossRef]
30. Chen, J.; Hu, M.; Li, Y.; Li, R.; Qing, L. Significant Influence of Bound Rubber Thickness on the Rubber Reinforcement Effect. *Polymers* **2023**, *15*, 2051. [CrossRef]
31. *ASTM D6556-21*; Standard Test Method for Carbon Black—Total and External Surface Area by Nitrogen Adsorption. ASTM International: West Conshohocken, PA, USA, 2021.
32. *ASTM WK87480*; New Test Method for Dry Powder rCB Particle Size Analysis via Laser Diffraction. ASTM International: West Conshohocken, PA, USA, 2023. Available online: https://www.astm.org/membership-participation/technical-committees/workitems/workitem-wk87480 (accessed on 26 August 2025).
33. *ASTM D5230-21*; Standard Test Method for Carbon Black—Automated Individual Pellet Hardness. ASTM International: West Conshohocken, PA, USA, 2021.
34. *ASTM D3849-22*; Standard Test Method for Carbon Black-Morphological Characterization of Carbon Black Using Electron Microscopy. ASTM International: West Conshohocken, PA, USA, 2022.
35. Grulke, E.A.; Rice, S.B.; Xiong, J.; Yamamoto, K.; Yoon, T.H.; Thomson, K.; Saffaripour, M.; Smallwood, G.J.; Lambert, J.W.; Stromberg, A.J.; et al. Size and Shape distributions of carbon lack aggregates by transmission electron microscopy. *Carbon* **2018**, *130*, 822. [CrossRef]
36. *ASTM D3191-10*; Standard Test Methods for Carbon Black in SBR (Styrene-Butadiene Rubber)—Recipe and Evaluation Procedures. ASTM International: West Conshohocken, PA, USA, 2010.
37. *ASTM D8491-23*; Standard Test Method for Recovered Carbon Black—Rheological Non-Linearity of a Rubber Compound by Fourier Transform Rheology. ASTM International: West Conshohocken, PA, USA, 2023.
38. *ASTM D5289-19a*; Standard Test Method for Rubber Property—Vulcanization Using Rotorless Cure Meters. ASTM International: West Conshohocken, PA, USA, 2019.
39. *ASTM D412-16*; Standard Test Methods for Vulcanized Rubber and Thermoplastic Elastomers—Tension. ASTM International: West Conshohocken, PA, USA, 2016.
40. *ASTM D2240-15*; Standard Test Method for Rubber Property—Durometer Hardness. ASTM International: West Conshohocken, PA, USA, 2015.
41. Hu, Y.; Yu, X.; Ren, J.; Zeng, Z.; Qian, Q. Waste tire valorization: Advanced technologies, process simulation, system optimization, and sustainability. *Sci. Total Environ.* **2024**, *942*, 173561. [CrossRef]

42. Czajczyńska, D.; Czajka, K.; Krzyżyńska, R.; Jouhara, H. Waste tyre pyrolysis—Impact of the process on its products on the environment. *TSEP* **2020**, *20*, 100690. [CrossRef]
43. Czajczyńska, D.; Krzyżyńska, R.; Ghazal, H.; Jouhara, H. Experimental investigation of waste tyres pyrolysis gas desulfurization through absorption in alkanolamines solutions. *Int. J. Hydrogen Energy* **2022**, *52*, 1006. [CrossRef]
44. Tricás, N.; Vidal-Escales, E.; Borrós, S.; Gerspacher, M. Influence of carbon black amorphous phase content on rubber filled compounds. *Compos. Sci. Technol.* **2003**, *63*, 1155. [CrossRef]
45. Medalia, A.I. Effective degree of immobilization of rubber occluded within carbon black aggregates. *Rubber Chem. Technol.* **1972**, *45*, 1171. [CrossRef]
46. Dannenberg, E.M. Bound rubber and carbon black reinforcement. *Rubber Chem. Technol.* **1986**, *59*, 512. [CrossRef]
47. Hallett, J. Dispersion kinetics of carbonaceous materials during elastomer mixing. In Proceedings of the Tire Tech, Hannover, Germany, 4–6 March 2025.
48. Payne, A.R. Dynamic mechanical properties of filler loaded vulcanisates. *Rubber Plast. Age* **1961**, *42*, 963.
49. Norris, C.J.; Lopez-Cerdan, A.; Murfitt, M. Ash. In Proceedings of the Recovered Carbon Black, Amsterdam, The Netherlands, 5–6 November 2024.
50. Costa, S.M.R.; Fowler, D.; Carreira, G.A.; Portugal, I.; Silva, C.M. Production and Upgrading of Recovered Carbon Black from the Pyrolysis of End-of-Life Tires. *Materials* **2022**, *15*, 2030. [CrossRef] [PubMed]
51. Zainal, N.; Halim, S. A review on demineralization, activation, and potential applications of the solid fraction obtained from the pyrolyzed waste tires. *Clean. Waste Syst.* **2025**, *11*, 100285. [CrossRef]
52. Jiang, G.; Pan, P.; Sun, Y. Recovery of high pure pyrolytic carbon black from waste tires by dual acid treatment. *J. Clean. Prod.* **2022**, *374*, 133893. [CrossRef]
53. *ASTM D1506-15*; Standard Test Methods for Carbon Black—Ash Content. ASTM International: West Conshohocken, PA, USA, 2015.

polymers

MDPI

Article

The Influence of Zinc Stearate Complexes on the Sulfur Vulcanization of Ethylene–Propylene–Diene Monomer

Krzysztof Kaczewiak [1,2,*], Piotr Głąb [2] and Magdalena Maciejewska [3]

[1] Interdisciplinary Doctoral School, Lodz University of Technology, 116 Żeromskiego Street, 90543 Lodz, Poland
[2] Hutchinson Poland Sp z o.o., 130 Kurczaki Street, 90543 Lodz, Poland
[3] Institute of Polymer and Dye Technology, Lodz University of Technology, 16 Stefanowskiego Street, 90543 Lodz, Poland
* Correspondence: krzysztof.kaczewiak@dokt.p.lodz.pl

Abstract

This article explores the influence of synthesized zinc stearate (ZnSt) complexes with different molar ratios of zinc oxide (ZnO) and stearic acid (StA) on the sulfur vulcanization and properties of the ethylene–propylene–diene monomer (EPDM). The aim of study was to reduce the amount of ZnO in elastomer composites without affecting their compression set. The research involves synthesizing zinc complexes, including the addition of sulfur, and using them as activators in EPDM compounds. The synthesized products were characterized using DSC and FTIR to identify their composition and thermal properties. The results of the compression set and equilibrium swelling tests indicated that ZnO was used in excess as an activator. Zinc complexes allowed for a significant reduction in zinc content compared to ZnO. The stoichiometric salt of ZnO and StA was the most effective in terms of vulcanization parameters and the compression set. Non-stoichiometric complexes were less effective, proportionally to their content. SEM images showed that the dispersion of components was relatively homogeneous in vulcanizates with active ZnO and commercial ZnSt. However, the blooming of ZnSt from vulcanizates was observed. Thus, using synthesized materials could reduce the zinc content in EPDM composites while maintaining good properties, but a further reduction is needed to avoid blooming.

Keywords: EPDM; cross-linking density; compression set; elastomers; carbon footprint

1. Introduction

Cross-linked rubber is a material with exceptional elastic properties, which plays a key role in many areas of our lives. Its ability to stretch and return to its original shape makes it irreplaceable in the production of tires, seals, medical gloves, and many other rubber products. Thanks to its unique properties, rubber is a material that has revolutionized many industries and continues to play a key role in the modern world.

The production of rubber composites is a complex process that requires precision and advanced technology to obtain products of high quality and durability. Rubber composites consist of both rubber and a number of other functional additives, e.g., fillers, plasticizers, activators, accelerators, or antioxidants, which improve their processing and facilitate properties of the vulcanizates. One of the substances widely used in the vulcanization process is zinc oxide (ZnO) [1,2]. It acts as an activator, speeding up the vulcanization reaction by activating the vulcanization accelerators [3]. When combined with stearic acid (StA), it increases the effectiveness of sulfur in the vulcanization process. Moreover, adding

 https://doi.org/10.3390/polym17212875

ZnO to a rubber compound improves the tensile strength, abrasion resistance, and elasticity of vulcanizates. Due to the key role of ZnO in elastomer technology, it is estimated that over 50% of ZnO production is consumed by the rubber industry [4,5]. Other industries consuming ZnO include agriculture, ceramics, and concrete. ZnO is also used on a smaller scale in the electronics industry and cosmetics production. ZnO used in the rubber industry is most often obtained in a metallurgical process by roasting zinc ore in the presence of air [6,7]. Alternatively, ZnO can be obtained by using a number of chemical methods, not always applicable to the industrial-scale production of ZnO due to the low efficiency of synthesis and economics [8]. ZnO production also generates carbon dioxide (CO_2) and other greenhouse gas emissions, mainly due to the energy processes involved in roasting ores or drying and roasting the products of chemical synthesis methods. Depending on the method used, the carbon footprint of ZnO production can vary [9,10]. Metallurgical processes are usually more energy-intensive and generate larger CO_2 emissions. On the other hand, chemical methods can be less energy-intensive, but require the use of other reagents, which also has an impact on the environment. The production of ZnO, regardless of the method, is therefore a harmful process for the natural environment; hence, it is justified to look for possibilities of reducing its consumption in various industries, especially in the case of the rubber industry, which is the main consumer of this material.

Regarding the rubber industry, ZnO is considered as the best activator for sulfur cross-linking due to its higher reaction kinetics and promotion of short sulfur cross-links that increase the overall cross-linking density of vulcanized rubber [11,12]. It also acts as a filler and scavenger of harmful reaction products, a cure agent for halogenated or carboxylated rubbers, and even provides protection from molded product shrinkage or mold fouling. In addition, ZnO has a significant effect on some of the physical properties of vulcanizates. Owing to its poor dispersibility in the rubber matrix, ZnO demands a high dosage to ensure efficient vulcanization. From an ecological point of view, a high amount of ZnO raises a concern about its release due to the exploitation of rubber products and its toxicity for the environment [1]. Generally, ZnO is considered as a substance safe for humans, but based on the international standard, i.e., Globally Harmonized System of Classification and Labelling of Chemicals (GHS), it is classified as H400: very toxic to aquatic life. Hence, for the sake of the environment, it is recommended to reduce the amount of ZnO in rubber products. It should be noticed that, when ZnO is used in vulcanization in the form of microsized ZnO crystal particles, only a reduced portion is able to react with the curing compounds, leading to a residual amount of unreacted ZnO in the rubber matrix. This unreacted ZnO can penetrate into groundwater while rubber products are stored in landfills, or in the form of dust, which is generated as a result of rubbing tires on the road surface. In the case of tires, a consequence of the wear of tire treads is zinc leaching, which may generate an excess of zinc concentration in the environment, potentially increasing in the future due to the widespread and expanding use of ZnO. This makes the topic of ZnO reduction very important for the rubber industry, as one of the largest users of ZnO.

The efficiency of ZnO in vulcanization and its influence on the properties of rubber composites strongly depend on the dispersion degree of ZnO particles in the elastomer matrix. Obtaining a good dispersion of ZnO in the elastomer matrix, and consequently an availability of ZnO particles for reaction with other components of the cross-linking system, can significantly improve its activity in the cure process and the final mechanical properties of the vulcanizates. Sreethu and Naskar [11] compared the activation effectiveness and impact on mechanical properties and the cure characteristics of different types of ZnO, i.e., conventional, active, and nanopowder, in the vulcanization of natural rubber filled with carbon black. Their investigation showed that a more homogeneous dispersion and higher surface area of ZnO can help to reduce its content without decreasing the mechanical properties of the rubber composites. The addition of fatty acids, such as StA, also plays a significant role in vulcanization [1,13]. They improve the dispersion of ZnO particles in the rubber matrix by boosting the contact surface of zinc ions with vulcanization accelerators. ZnO reacts with StA, forming zinc stearate (ZnSt), which is assumed as an essential cure activator. Junkong et al. [14] performed in situ research to study the activity of different fatty acids in comparison with StA and ZnSt in the vulcanization of isoprene rubber. It was shown that ZnO creates 2:2 bidentate zinc/stearate complexes, with StA also having some hydroxyl groups. These complexes act as catalysts to accelerate the sulfur cross-linking reactions in the rubber matrix. Heideman et al. [15] studied various metal oxides and a clay loaded with Zn^{2+} ions as a solution that can help reduce the zinc content in rubber composites. Hence, zinc stearate, zinc 2-ethylhexanoate, and zinc m-glycerolate were employed instead of ZnO to activate the sulfur vulcanization of the ethylene–propylene–diene monomer (EPDM) and styrene–butadiene (SBR) rubber. Zinc m-glycerolate, with the highest amount of zinc per 1 g, showed the highest activity in the vulcanization. However, no reduction in the zinc amount was provided compared to rubber compounds cured with ZnO, since the authors of the publication used zinc complexes in an amount equivalent to the content of zinc ions in the amount of ZnO used. Therefore, the possibility of reducing the amount of zinc in rubber products was not investigated, but only the possibility of replacing ZnO with other zinc compounds. Przybyszewska et al. [16] reported that the use of organic zinc activators in the form of zinc salts or zinc complexes with 1,3-diketones, depending on their structure, allowed for a reduction in the amount of zinc in rubber compounds by 70–90 wt.% compared to those containing ZnO and stearic acid. However, vulcanizates with a lower cross-link density compared to those containing ZnO were obtained, which can have a significant impact on their mechanical properties, e.g., the compression set, which is crucial for rubber seals. Thus, the results reported in the literature are quite optimistic, but they need to be investigated further due to a lack of clear knowledge of all variables that affect the vulcanization process.

Considering the vulcanization, there is a lot of information in the literature about ZnO and StA complex formation. Scotti [12] prepared an extensive review, which indicated that stoichiometric ZnSt is not an efficient activator of sulfur vulcanization. However, Junkong et al. [14] observed that non-stoichiometric salts of ZnO and fatty acids can effectively activate vulcanization. It was found that, at high temperatures, a dominant structure of the product formed in the ZnO and StA reaction was a bridging bidentate zinc/stearate complex with a molar ratio of 2:2. It involved two zinc ions bridged by two stearate (StA) ligands, where each stearate molecule coordinated to both zinc ions with its carboxylate group. This structure was represented as $(Zn_2(\mu\text{-}O_2CC_{17}H_{35})_2)^{2+}(OH^-)_2 \cdot R_1R_2$, where R_1 and R_2 can be water molecules or rubber segments, respectively. The most likely structure of the binuclear type bridging bidentate zinc/stearate complex is presented in Figure 1 [17].

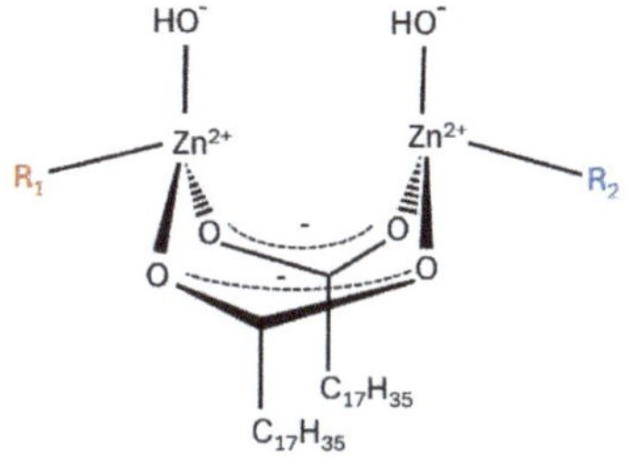

Figure 1. Binuclear type bridging bidentate zinc/stearate complex, where R_1 and R_2 are water and/or a rubber segment, based on [17].

A good example of a non-stoichiometric complex of ZnO and StA was also described by Setianto et al. [18]. The aim of this study was to produce ZnSt; however, ZnO and StA were mixed in non-stoichiometric proportions. Instead of 2 moles of StA per 1 mole of ZnO, the synthesis involved 3 moles of ZnO per 1 mole of StA, which resulted in a 6-fold excess of ZnO. Surprisingly, the tensile strength and hardness of the vulcanizates obtained by replacing ZnO and StA with a synthesis product showed significant improvement. This suggests that the non-stoichiometric complex of ZnO with StA can be potentially as active as the high amount of active ZnO, which is typically used in rubber compounds in the amount of 3 phr. Hence, the use of non-stoichiometric zinc complexes with StA can potentially contribute to the reduction in the amount of zinc in elastomer composites compared to those containing active ZnO, or even to the elimination of active ZnO. On the other hand, there is no information in the literature about sulfur-ZnO complex formation during the vulcanization of rubber compounds. The past research showed that there is a reaction between ZnO and sulfur, with the creation of zinc sulfide (ZnS), which is deposited on the surface of the vulcanization molds [19]. This proves that, during the vulcanization of rubber, like EPDM, some amount of ZnO is transformed into ZnS, which can influence the rubber cross-linking density and consequently the vulcanizate properties [3]. Thus, in this work, we synthesized zinc complexes using different molar ratios of ZnO:StA, while additionally introducing sulfur into some of these complexes. Standard zinc stearate (ZnSt) contains two moles of monovalent StA anions for one mole of divalent cation of zinc. Taking into consideration that, in the literature, there are plenty of studies confirming the formation of non-stoichiometric complexes between stearic acid and ZnO, we decided to check how they affect the curing characteristics of rubber composites and properties of the vulcanizates. During previously conducted research, we have also observed that heating sulfur with ZnO in a masterbatch before adding an accelerator pack improved the functional parameters of vulcanizates, i.e., the compression set, despite the cross-linking density reduction [20]. Therefore, in this research, samples with sulfur were also prepared. Moreover, syntheses in the presence of sulfur were performed to verify the potential formation of complexes between carboxylic acid, sulfur, and ZnO. Then, we used the obtained synthesis products as activators of sulfur vulcanization as an alternative to systems containing ZnO, StA, and sulfur separately. We believed that the use of these complexes would reduce the amount of ZnO in elastomer composites, and thus the harmfulness of ZnO to the environment and the carbon footprint associated with its application.

2. Materials and Methods

2.1. Materials

The masterbatch of the ethylene-propylene-diene monomer compound with the hardness of 74 Shore A (74ShA EPDM) used for the study was a serial product used on extrusion. This compound was vulcanized with a sulfur cure system. Its composition is not revealed due to the Hutchinson confidentiality policy. In case of cure system adjustments, the 74ShA compound was mixed in the form of a masterbatch that did not contain any components of the curing system, i.e., sulfur, ZnO, or StA. The accelerators were commercially available predispersions purchased from Konimpex Chemicals Sp. Z o.o. (Konin, Poland), with the following concentration of active ingredient: caprolactam disulfide (CLD)—80%, 2,2′-benzothiazyl disulfide (MBTS)—70%, N-cyclohexyl-2-benzothiazole sulfenamide (CBS)—80%, and zinc dialkyldithiophosphate (ZDTP)—50%. Zinc oxide with a specific surface area of 10 m^2/g (Huta Będzin, Będzin, Poland) and active zinc oxide with a specific surface area of 45 m^2/g (Silox India, Vadodara, India), together with stearic acid (Uniqema, Goole, UK), were used to synthesize zinc complexes. Zinc stearate with a purity > 98.5%

was purchased from Warchem (Trakt Brzeski, Poland), while sulfur was manufactured by Siarkopol (Tarnobrzeg, Poland).

2.2. Synthesis of Zinc Complexes and Salts

The sets of synthesis products were prepared in a flask heated with methyl silicone oil to 160 °C. Mixing was ensured by a magnetic stirrer. The synthesis station is presented in Supplementary Materials (Figure S1). The mixtures were prepared according to the recipes expressed in moles in Table 1.

Table 1. Composition of the mixtures of ZnO with StA and/or S in moles of each ingredient.

No.	Sample Symbol	Stearic Acid (mole)	Zinc Oxide [1] (mole)	Sulfur [2] (mole)
1	StZn_s_1	2	1	-
2	StZn_s_2	2	2	-
3	StZnS_s_1	2	1	8
4	StZnS_s_2	2	2	8
5	StZn_a_1	2	1 A	-
6	StZn_a_2	2	2 A	-
7	StZnS_a_1_7	2	1 A	8
8	StZnS_a_2_8	2	2 A	8
9	StZnS_a_1_9	2	1 A	8
10	StZnS_a_2_10	2	2 A	8

[1] The ZnO used for the synthesis was in two variants: standard ZnO and active ZnO with increased specific surface area, marked with the letter A in the table. [2] Sulfur in samples No. 3, 4, 7, and 8 was added after reaction between ZnO and StA. Sulfur in samples No. 9 and 10 was added to StA before ZnO.

For samples 1 to 8 given in Table 1, ZnO was added to the melted technical StA and stirred at 160 °C. After a turbulent synthesis reaction with the release of water vapor, sulfur was added according to the recipes presented in Table 1, and all samples were further heated and stirred for 45 min. Samples 9 and 10 were prepared according to the same recipe as samples 7 and 8 but differed in the order of dosing the individual components. Sulfur was added to the melted stearic acid, and, after it had melted, active ZnO was added. After the synthesis reaction had taken place, the mixture was stirred and heated for 45 min at 160 °C.

Stearic acid used for the synthesis was a standard material used in rubber industry, containing 57% of stearic acid ($C_{17}H_{35}COOH$), 35% of palmitic acid ($C_{15}H_{31}COOH$), and other fatty acids from C14 to C20. To calculate the amount of ZnO to technical StA in complex preparations, the assumption was made as for StA of 100% purity. Considering the difference in molecular weight between palmitic and stearic acid, a 3% excess of carboxylic acid vs. calculated molar ratio for 100% purity of StA should be expected. However, we decided to accept the possible slight excess of carboxylic acids in stoichiometric reactions.

The TD-GC-MS chromatogram presented in Figure 2 confirmed the compliance of the material with the manufacturer's declaration. The first peak (~21 min) corresponds to palmitic acids and the second (~23 min) to stearic acid. There are visible traces of shorter and longer carboxylic acids on the chromatogram as well.

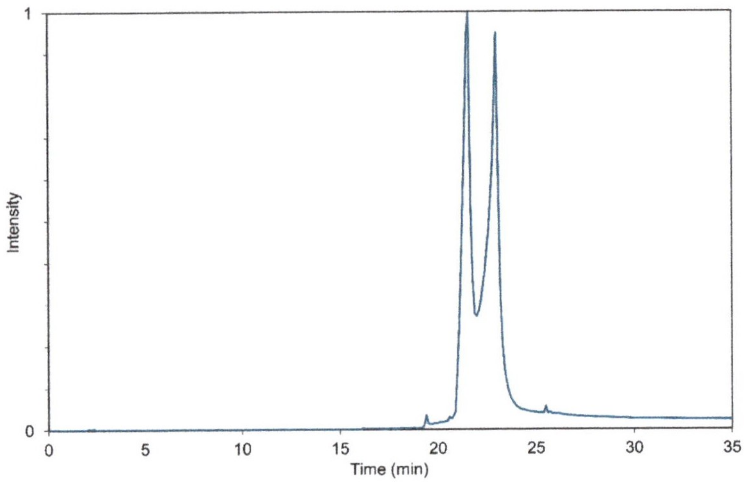

Figure 2. TD-GC-MS chromatogram of technical stearic acid used for synthesis reactions.

2.3. Preparation of Rubber Compounds

Three series of rubber compounds, with compositions given in Tables 2 and 3, were prepared and studied.

Table 2. Compositions of rubber compounds containing synthesized products described in Table 1.

Rubber Compound	Description	ZnO phr Equivalent	StA/Zn [mole/mole]
Ref0	Reference compound with standard ZnO.	3	0.15
Ref1	Reference compound with active ZnO.	3	0.15
Ref2	Reference compound with active ZnO and sulfur thermally transformed (200 °C/5 min) in masterbatch before adding accelerator set—experiment following the previous study [13].	3	0.15
Test1	Rubber compound with StZn_a_1. Targeted to stoichiometric ZnSt. Amount of StZn_a_1 equivalent to StA concentration in reference. A total of 13.6 times less Zn than the reference.	0.22	2
Test2	Rubber compound with StZn_a_1. Targeted to stoichiometric ZnSt. Amount of StZn_a_1 to obtain Zn equivalent 3 times less than the reference.	1	2
Test3	Rubber compound with StZn_a_2. Targeted to ZnSt complex. Amount of StZn_a_2 equivalent to stearic acid concentration in reference. A total of 6.7 times less Zn than the reference.	0.44	1
Test4	Rubber compound with StZn_a_2. Targeted to ZnSt complex. Amount of StZn_a_2 to obtain Zn equivalent 3 times less than the reference.	1	1
Test5	Rubber compound with StZn_s_1. Targeted to stoichiometric ZnSt. Amount of StZn_s_1 equivalent to StA concentration in reference. A total of 13.6 times less Zn than the reference.	0.22	2

Table 2. *Cont.*

Rubber Compound	Description	ZnO phr Equivalent	StA/Zn [mole/mole]
Test6	Rubber compound with StZn_s_1. Targeted to stoichiometric ZnSt. Amount of StZn_s_1 to obtain Zn equivalent 3 times less than the reference.	1	2
Test7	Rubber compound with StZn_s_2. Targeted to ZnSt complex. Amount of StZn_s_2 equivalent to StA concentration in reference. A total of 6.7 times less Zn than the reference.	0.44	1
Test8	Rubber compound with StZn_s_2. Targeted to ZnSt complex. Amount of StZn_s_2 to obtain Zn equivalent 3 times less than in reference.	1	1
Test9	Rubber compound with StZnS_a_1_7. Targeted to ZnSt/sulfur complex. Sulfur in amount equivalent to S_8 per one Zn, added at the end. Amount of StZnS_a_1_7 equivalent to StA concentration in reference. A total of 13.6 times less Zn than the reference. [1]	0.22	2
Test10	Rubber compound with StZnS_a_1_9. Targeted to ZnSt/sulfur complex. Sulfur in amount equivalent to S_8 per one Zn, added before ZnO. Amount of StZnS_a_1_9 equivalent to StA concentration in reference. A total of 13.6 times less Zn than the reference. [1]	0.22	2
Test11	Rubber compound with StZnS_a_1_7. Targeted to ZnSt/sulfur complex. Sulfur in amount equivalent to S_8 per one Zn, added at the end. Amount of StZnS_a_1_7 equivalent to double StA concentration in reference. A total of 6.7 times less Zn than the reference. [1]	0.44	2
Test12	Rubber compound with StZnS_a_2_8. Targeted to ZnSt/sulfur complex. Sulfur in amount equivalent to S_8 per one Zn, added at the end. Amount of StZnS_a_2_8 equivalent to StA concentration in reference. A total of 6.7 times less Zn than the reference. [1]	0.44	1
Test13	Rubber compound with StZnS_a_2_10. Targeted to ZnSt/sulfur complex. Sulfur in amount equivalent to S_8 per one Zn, added before ZnO. Amount of StZnS_a_2_10 equivalent to StA concentration in reference. A total of 6.7 times less Zn than the reference. [1]	0.44	1
Test14	Rubber compound with StZnS_s_1. Targeted to ZnSt/sulfur complex. Sulfur in amount equivalent to S_8 per one Zn, added at the end. Amount of StZnS_s_1 equivalent to StA concentration in reference. A total of 13.6 times less Zn than the reference. [1]	0.22	2
Test15	Rubber compound with StZnS_s_2. Targeted to ZnSt/sulfur complex. Sulfur in amount equivalent to S_8 per one Zn, added at the end. Amount of StZnS_s_2 equivalent to stearic acid concentration in reference. A total of 6.7 times less Zn than the reference. [1]	0.44	1

[1] NOTE: sulfur did not completely react with ZnSt during mixing in the reactive flask—the fraction of sulfur in the complex was lower than assumed; that is why a regular amount of sulfur was added. This can generate some excess of sulfur in the rubber compound.

Table 3. Composition of rubber compounds containing commercial ZnSt and synthesized products described in Table 1.

Rubber Compound	Description	ZnO phr Equivalent	StA/Zn [mole/mole]
Ref3	Reference compound with commercial ZnSt and 1 phr of StA. Amount of ZnSt to obtain Zn equivalent 3 times less than Ref1 (Table 2).	1	2.28
Test16	Rubber compound with commercial ZnSt and 1 phr of StA premixed in Brabender with masterbatch. Amount of ZnSt to obtain Zn equivalent like in Ref3.	1	2.28
Test17	Rubber compound with sulfur, ZnSt, and 1 phr of StA premixed in Brabender with masterbatch. Amount of ZnSt to obtain Zn equivalent like in Ref3.	1	2.28
Test18	Rubber compound with StZn_a_1. Targeted to stoichiometric ZnSt. Amount of StZn_a_1 to obtain Zn equivalent 3 times less than the reference. Composition like Test2 (Table 2) but added 0.5 phr of stearic acid. Similar StA/Zn ratio like in Ref3.	1	2.13
Test19	Rubber compound with StZn_a_2. Targeted to ZnSt complex. Amount of StZn_a_2 to obtain Zn equivalent 3 times less than the reference. Composition like Test4 (Table 2) but added 1 phr of stearic acid. Reduced StA/Zn ratio like in Ref3.	1	1.29

The first series, which consisted of the reference rubber compounds (Ref0, Ref1, and Ref2, respectively), was manufactured in a standard way using a two-roll mill (David Bridge & Co., Rochdale, UK). The reference rubber compounds contained standard ZnO or active ZnO and other components of the cross-linking system, i.e., sulfur, accelerators, and StA, added separately. During the preparation of the reference rubber compounds, the masterbatch was first plasticized for 3 min, and then ZnO, StA, and sulfur were added successively and mixed using the two-roll mill. The total time for preparing the rubber compound was 10 min. The mixing process was carried out at temperatures up to 50 °C, since in such conditions no chemical reaction between ZnO and stearic acid was expected before vulcanization.

The other two series of rubber compounds, which contained the previously prepared synthesis products obtained in the reaction of stearic acid, zinc oxide, and sulfur (S_8), were prepared using a Brabender mixer. These materials contained synthesis products together with a rubber masterbatch without the ingredients used to prepare the synthesis products and accelerators included in the cross-linking system. The aim of this task was to homogeneously mix the melted ingredients together with the masterbatch before mixing them with the rest of the masterbatch and vulcanization accelerators, which would not be possible at the later stage without starting the vulcanization process. Initial Brabender chamber temperature was set at 115 °C, with a mixing speed of 45 revolutions per minute (rpm). After masterbatch dosing, the speed was kept at 45 rpm for 1 min before other ingredients were dosed. After performing complete dosing, the speed was changed to 70 rpm for 5 min and then to 90 rpm. The discharge of premixing was when the temperature of mixing rose to a minimum of 130 °C. After introducing the synthesis products into the masterbatch in the molten state using a Brabender mixer, vulcanization accelerators and sulfur were added successively using a two-roll mill. and the sheets of rubber compounds were prepared after 10 min of mixing.

To check the differences between synthesized materials and commercial ZnSt, additional rubber compounds described in Table 3 were prepared. Two additional rubber composites (Test18 and Test19, respectively) were also manufactured to check if a higher amount of StA is necessary for accelerators and can have some positive impact on vulcanizate properties.

All rubber compounds and test conceptions are summarized in Figure 3 below.

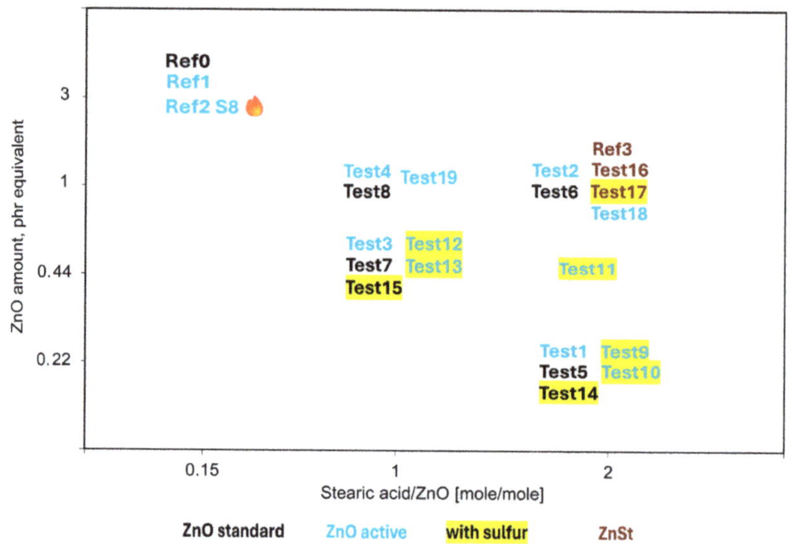

Figure 3. Drawing of the test concept for rubber compounds described in Tables 2 and 3.

2.4. Research Methods

Rheometric tests for rubber compounds were performed using Production Moving Die Rheometer MDR-P (Alpha Technologies, Hudson, OH, USA). The cure characteristics of EPDM compounds were studied at a temperature of 180 °C for the test time of 7 min.

Viscosity tests for rubber compounds were made using the MV 2000 Mooney viscometer (Alpha Technologies, Hudson, OH, USA) at a temperature of 125 °C according to the scorch test.

Samples of vulcanizates for compression set (CS) measurements and swelling tests were prepared using the Argenta AW 03 vulcanization press (Argenta, Kielce, Poland) with bottom and top heating plates and a temperature control system. Plates of vulcanizates with a thickness of approximately 2 mm were obtained at 180 °C and time of 10 min. There were five samples for each test prepared to minimize faults of measurements.

The equilibrium swelling tests were conducted for EPDM vulcanizates. Samples of rubber cut from 2 mm plates of vulcanizates were weighed on a laboratory scale and put in steel cans with xylene (mixture of isomers, purity 98%, Avantor Performance Materials S.A., Gliwice, Poland) for 48 h at a room temperature of approximately 23 °C. Then, the swollen samples were dried using a paper towel and secondly weighed. The equilibrium swelling (Q_w) of the vulcanizates was calculated using Equation (1):

$$Q_w = \frac{m_1 - m_0}{m_0} \cdot 100\%, \quad (1)$$

where

Q_w—equilibrium swelling (%),

m_1—weight of the swollen sample (mg),

m_0—weight of the sample before swelling (mg).

Compression set tests were performed using the standard procedure described in ISO 815-1:2019 method B [21]. The samples with a thickness of 6 mm were prepared from three layers of 2 mm plates. The height of the samples was measured, and then they were compressed by 25% for 22 h at 70 °C. After 2 h of conditioning at room temperature in a compressed state and 30 min of relaxation after removing the external stress, the height of the samples was secondly measured, and CS was calculated using Equation (2):

$$CS = \frac{h_0 - h_2}{h_1}, \tag{2}$$

where

h_0—initial thickness of the sample before test (mm),

h_1—height of compression (mm),

h_2—final thickness of the sample after test (mm).

The mechanical properties of vulcanizates were explored using the Zwick Roell 1435 (ZwickRoell, Ulm, Germany) universal testing machine. Tensile tests were carried out for four paddle-shaped samples with a thickness of approximately 1 mm and the width of the measuring section of 4 mm. All the samples were cut on press from vulcanizates cured at the temperature of 180 °C for 10 min. The crosshead speed during tensile tests was 500 mm/min.

Differential Scanning Calorimetry (DSC) tests were performed for the synthesized products employing the DSC1 analyzer (Mettler Toledo, Greifensee, Switzerland). Measurements were carried out at a temperature range from 25 °C to 200 °C, with a heating rate of 10 °C/min. The flow rate of nitrogen during the measurement was 80 mL/min. Based on DSC data, the expected heat of zinc stearate fusion was determined for synthesized products using Equations (3)–(8):

$$mStA = \frac{m_s \cdot LStAs}{LStA}, \tag{3}$$

where

$mStA$—mass of stearic acid in DSC sample (mg),

m_s—mass of DSC sample (mg),

$LStAs$—DSC measured sample heat of melting about StA phase transition temperature (J/g),

$LStA$—DSC measured heat of melting for pure StA (J/g).

$$m_0StA = \frac{m_s \cdot 2 \cdot MStA}{2 \cdot MStA + a \cdot MZnO}, \tag{4}$$

where

m_0StA—initial mass of stearic acid in sample (mg),

$MStA$—molar mass of StA (g/mol),

$MZnO$—molar mass of ZnO (g/mol),

a—molar proportion of ZnO in recipe (Table 1).

$$mRStA = m_0StA - mStA, \tag{5}$$

where

mRStA—reacted stearic acid during synthesis (mg).

$$m_yZnSt = \frac{mRStA \cdot MZnSt}{2 \cdot MStA}, \quad (6)$$

where

m_yZnSt—zinc stearate yielded (mg),
MZnSt—molar mass of ZnSt (g/mol).

$$LeZnSt = \frac{m_yZnSt \cdot LZnSt}{m_s}, \quad (7)$$

where

LeZnSt—expected heat of melting of zinc stearate yielded (J/g),
LZnSt—expected heat of melting of commercial zinc stearate (J/g).

$$\%y = \frac{mRStA}{m_0StA} \cdot 100\% \quad (8)$$

where

%y—relative effectiveness of stearic acid conversion to carboxylic acid salt for stoichiometry products (%).

Fourier Transform Infrared Spectroscopy (FTIR) measurements for synthesized products were carried out using a ThermoScientific Nicolet 6700 FT-IR spectrometer (Thermo Fisher Scientific, Waltham, MA, USA) equipped with OMNIC 3.2 software. The attenuated total reflectance (ATR) technique with a single diamond ATR crystal was used for the tests. FTIR absorbance spectra were recorded in the wavenumber range of 4000–400 cm^{-1} using 128 scans.

Scanning Electron Microscopy (SEM) was used to investigate the distribution of solid particles in the rubber matrix. SEM images of the vulcanizates were taken using an SEM LEO 1450 microscope (Carl Zeiss AG, Oberkochen, Germany). Before measurement, the vulcanizates were fractured after cooling in liquid nitrogen, and their surfaces were covered with a thick layer of carbon. SEM with Energy-Dispersive X-ray Spectroscopy (EDS) was used to study the composition of agglomerates. SEM/EDS studies were performed using a HITACHI S-4700 (Hitachi, Mannheim, Germany) SEM microscope.

Thermal Desorption–Gas Chromatography–Mass Spectrometry (TD-GC-MS) was used for stearic acid composition verification. The apparatus used for analysis was from Perkin-Elmer (Shelton, CT, USA). Tests were performed by desorption at 200 °C on TurboMatrix 650. The column used for chromatography was Clarus 680, with He gas flow 2 mL/min. For identification of materials, mass spectroscopy on Clarus 600 C was used in EI+ ionization mode.

3. Results and Discussion

3.1. Characterization of the Zinc Salts and Complexes

While heating the components of the mixtures presented in Table 1, the synthesis reaction of ZnSt was observed, along with the release of water. In the case of active ZnO, this process was much more turbulent than for standard ZnO. In the case of samples with added sulfur, it was observed that the mixture delaminated due to the separation of melted sulfur in the lower part of the reaction flask. The prepared synthesis products were poured into aluminum containers and left to cool. The obtained products are presented in Figure 4.

Figure 4. Images of synthesized products: (**a**) StZn_s_1; (**b**) StZn_a_1; (**c**) StZn_a_2; (**d**) StZnS_a_1_7.

The obtained synthesis products were subjected to DSC analysis in order to observe the differences in thermal transitions occurring during their heating and identify the synthesis products, as well as the presence of unreacted substrates. DSC curves for the synthesis products are presented in Figure 5a,b. Additionally, DSC measurements were conducted for technical stearic acid and pure sulfur (S8) (Figure 5c).

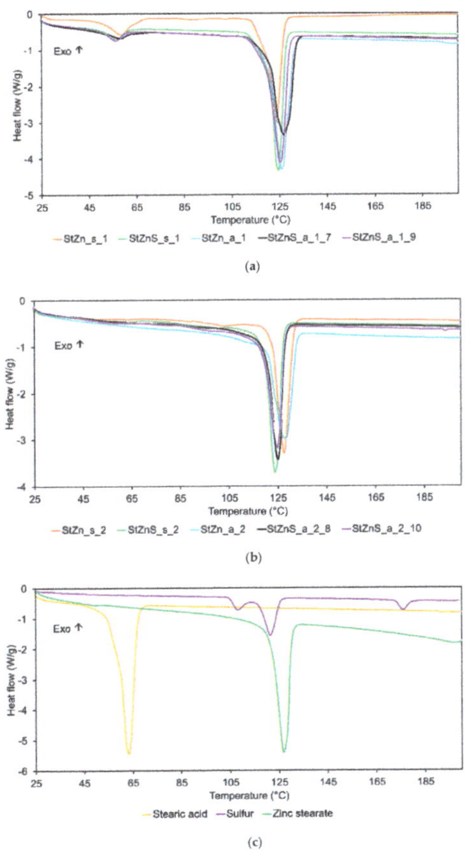

Figure 5. DSC results for synthesis products described in Table 1: (**a**) stoichiometric; (**b**) non-stoichiometric; (**c**) raw materials, i.e., commercial StA, S, and ZnSt.

Analyzing the DSC curves presented in Figure 5a, it was observed that the products obtained from the mixture containing 2 moles of StA per 1 mole of ZnO showed a melting of residual carboxylic acid in the temperature range of 50–70 °C. The residual carboxylic acid content resulted from the composition of technical stearic acid, which was a blend of palmitic and stearic acid. As mentioned earlier, the average molecular weight of acid blends is lower than that of the pure stearic acid, which resulted in the excess of carboxylic acid in the prepared mixture. The second intensive endothermic peak on the DSC curve in the temperature range of 110–130 °C was attributed to the melting of the products resulting from the reaction between StA and ZnO. However, for non-stoichiometric products, no melting peak for the residual carboxylic acid was observed, but only the sharp endothermic peak for melting of the synthesis products (Figure 5b).

On the other hand, for stoichiometric products, the DSC curves did not change significantly regardless of whether a standard or active ZnO was used. The addition of sulfur had no significant effect on the melting peak of the main synthesis product. Some residual endothermic peaks were observed on DSC curves at a temperature of 105 °C for synthesis products containing sulfur, which resulted from the allotropic transformation of sulfur.

From DSC measurements, the heat of the melting and peak temperatures for synthesized products and raw materials were determined and summarized in Table 4.

Table 4. Heat of melting and peak temperatures of synthesized products and raw materials determined by DSC.

Sample	Phase Change 1		Phase Change 2	
	Heat of Melting [J/g]	Peak [°C]	Heat of Melting [J/g]	Peak [°C]
StZn_s_1	−25.0	57.8	−102.6	121
StZn_s_2	-	-	−95.9	126
StZnS_s_1	−12.7	57.8	−134.2	123
StZnS_s_2	-	-	−108.5	122
StZn_a_1	−11.5	56.1	−162.4	124
StZn_a_2	-	-	−121.9	127
StZnS_a_1_7	−6.3	57.7	−142.3	125
StZnS_a_2_8	-	-	−104.9	123
StZnS_a_1_9	−8.9	56.6	−149.6	123
StZnS_a_2_10	-	-	−104.9	123
StA	−194.5	61.5	-	-
S	−9.1	108.0	−33.5	121
ZnSt	-	-	−153.3	126

The heat of melting determined for the zinc stearate sample was 153.3 J/g, lower than reported in the literature. Konkoly-Thege et al. [22] reported it as 103 kJ/mol, which gives 162.9 J/g. These researchers also determined the heat of melting for zinc palmitate as 93 kJ/mol, which gives 161.4 J/g. The melting temperatures were reported as 130 °C for zinc stearate and 134 °C for zinc palmitate, also a bit higher than the peak temperatures determined from DSC data in our study. The significantly higher heat of melting for residual carboxylic acid in the case of standard ZnO (sample StZn_s_1) compared to active ZnO (sample StZn_a_1) revealed that the standard ZnO was not fully converted during the reaction. It is also visible that the heat of melting of the carboxylic acid salt for the standard ZnO (sample StZn_s_1) was significantly lower than that for the active ZnO (sample StZn_a_1). A similar dependence was observed in the case of ZnO excess; the heat of melting for zinc salt produced from standard ZnO (sample StZn_s_2) was significantly lower than that for active ZnO (sample StZn_a_2).

Based on the DSC data, the expected heat of zinc stearate fusion was determined for synthesized products without sulfur and listed in Table 5. Mixtures with sulfur were excluded from estimations because part of the sulfur from those mixtures was not incorporated into the synthesized products and stayed on the bottom of the aluminum containers. On the other hand, the part of sulfur which was incorporated into the synthesis product had a similar melting point to ZnSt.

Table 5. Expected heat of melting calculated from DSC results for selected synthesis products described in Table 1.

Sample	Mass of DSC Sample [mg]	Mass of StA in DSC Sample [mg]	Initial Mass of StA in Sample [mg]	StA Reacted [mg]	ZnSt Yielded [mg]	Expected Heat of Melting [J/g]	Reacted StA [%]
StZn_s_1	11.8	1.52	10.3	8.8	9.8	−127.6	85
StZn_s_2	16.5	0.00	12.8	12.8	14.3	−132.9	100
StZn_a_1	13.6	0.80	11.9	11.1	12.3	−139.4	93
StZn_a_2	14.3	0.00	11.1	11.1	12.4	−132.9	100

The relative effectiveness of stearic acid conversion to carboxylic acid salt for stoichiometry products was 85% for standard ZnO and 93% for active ZnO (excess of acid), respectively. The higher effectiveness of the reaction for excess of ZnO (samples StZn_s_2 and StZn_a_2, respectively) can suggest that the reaction efficiency was linked to the size of the ZnO particles. The presence of a carboxylic acid melting peak suggests that there is no formation of non-stoichiometric ZnSt containing more carboxylic acid. However, this observation does not eliminate the possibility that a non-stoichiometric complex with an excess of Zn has been created. The heat of the melting of StA is higher than would be expected from a 3% excess of StA, which means that not all the ZnO has reacted. ZnO was dissolved by the carboxylic acid gradually. Thus, there is a residue of both carboxylic acid and ZnO or its impurities in the preparation. Based on the calculated results of reacted StA from DSC measurements and its 3% excess, non-reacted or complexed ZnO is about 12% for StZn_s_1 (standard ZnO) and 4% for StZn_a_1 (active ZnO).

The heats of melting for synthesized products differed from those determined theoretically based on DSC-measured stearic acid remains. For standard ZnO, they were lower than expected, but for active ZnO the results were opposite. The heats of melting determined through DSC for products from non-stoichiometric mixtures were significantly lower than for the stoichiometric one. On the other hand, the counted expected heats of melting were quite similar. All those differences could result from some non-stoichiometric zinc stearate particles formed as a binuclear bidentate or crystal structure of products yielded during synthesis. Zinc stearate can exhibit different crystal structures depending on the synthesis conditions, with two main types of geometry observed in zinc soaps, i.e., a highly symmetrical packing for long-chain saturated soaps and an alternating packing for zinc soaps with shorter, unsaturated, or dicarboxylic chains [23]. The method used to synthesize ZnSt can influence its crystal structure. For example, precipitation and fusion processes can yield ZnSt with slightly different melting points and crystal morphologies. The phase transition in ZnSt is not just a simple melting of the alkyl chains [24]. It also involves changes in the coordination of the stearate ions to the zinc ions, potentially forming ionic clusters in the liquid phase. In ZnSt, each zinc ion is coordinated to two stearate anions. This bidentate coordination is believed to contribute to the relatively high melting point and enthalpy of fusion compared to simple alkanes. At the temperature of the industrial vulcanization process, i.e., around 180 °C, the crystal structure of the yielded ZnSt should not have an impact on the vulcanization course and efficiency because those products should be in a liquid state.

The transitions of substrates into zinc salts in the reactive mixtures were checked using FTIR Spectroscopy. The FTIR spectra of StA used for the synthesis and commercial ZnSt were collected to compare them with those obtained for the reaction products prepared in stoichiometric (StZn_s_1) and non-stoichiometric (StZn_s_2) conditions (Figure 6). Regarding the FTIR spectrum for StA, the absorption peaks at about 2915 and 2849 cm^{-1} in the high-frequency region were attributed to $-CH_2-$ asymmetric and symmetric stretching vibrations, respectively. They were accompanied by stretching vibrations of the $-CH_2$- band at a wave number of approximately 1472 cm^{-1} [25]. Furthermore, the strong and sharp peak at approximately 1700–1710 cm^{-1}, characteristic of the C=O stretching vibrations in the carboxylic acid group, and peaks at 1300–1420 cm^{-1}, characteristic of the O-H bending vibrations of the hydroxyl group in carboxylic acids, were observed. In addition, peaks corresponding to −C–O stretching vibrations in −COOH were identified in the wave number range of 930–1100 cm^{-1} [26]. It should be noticed that, in both spectra for synthesized products, the band at a wave number of approximately 1700 cm^{-1} characteristic for the double bond between carbon and oxygen in carboxylic acids, i.e., C=O band, changed to a sharp peak at approximately 1535 cm^{-1}, which is typical for asymmetric stretching vibrations of $-COO^-$ in carboxylate salts, accompanied by sharp peaks corresponding to symmetric stretching vibrations of $-COO^-$ at approximately 1400–1450 cm^{-1} [26]. This confirmed that a reaction took place between ZnO and stearic acid, and a carboxylic acid salt was formed. On the spectrum of StZn_s_1, it is also visible that a part of StA remained unreacted, since the small peak at approximately 1700–1710 cm^{-1}, characteristic of the C=O stretching vibrations in the carboxylic acid group, is still present. On the other hand, in the FTIR spectrum of StZn_s_2, the presence of non-reacted ZnO can be concluded, since the intensity of the peak at a wave number of approximately 400–600 cm^{-1}, which is typical for Zn–O stretching vibrations [27], is higher compared to the FTIR spectrum of the product synthesized in stoichiometric conditions, i.e., StZn_s_1 or pure ZnSt. The intensity of the bands in this wave number range is higher than for the product synthesized in stoichiometric conditions, even though this is already the limit of the measuring range of the device. Due to fact that, in this range of wave numbers, the absorbance is imposed with a ZnSt answer, i.e., Zn-O stretching, which confirms metal–oxygen bonding, it is difficult to determine the presence of unreacted ZnO for ZnSt_s_1. There is no difference in the spectra of ZnSt_s_1 and ZnSt_s_2 other than the lack of bands at 1700 cm^{-1} (typical for C=O group in carboxylic acids) and approximately 420 cm^{-1} (typical for Zn-O vibrations) in the spectrum of ZnSt_s_2. This observation supports the findings concluded based on DSC results that only carboxylic acids salts were formed without non-stoichiometric complexes.

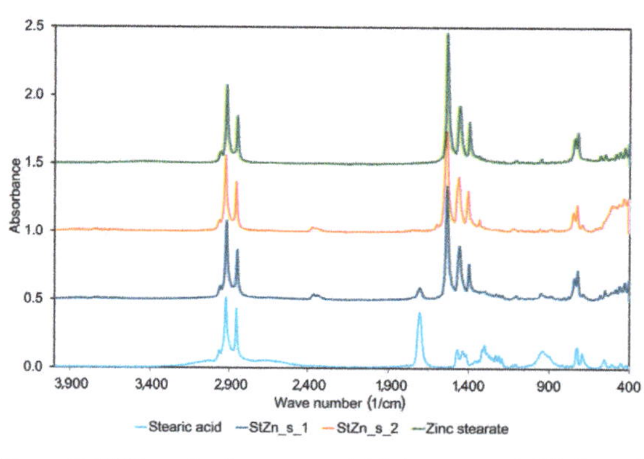

Figure 6. FTIR spectra for commercial StA, commercial ZnSt, and product of synthesis, i.e., StZn_s_1 and StZn_s_2, respectively.

3.2. Characterization of the EPDM Composites Containing the Synthesized Zinc Salts and Complexes

Having characterized the products of the performed synthesis, we used them to prepare rubber compounds. Then, the effects of standard ZnO, active ZnO, and the synthesized zinc salts and complexes on the rheometric properties of EPDM compounds were explored. The results are presented in Table 6.

Table 6. MDR and Mooney test results for compounds described in Tables 2 and 3 (MDR results: ML—minimum torque; MH—maximum torque; t_{s2}—time to 2% rise in torque; MH–ML—torque increment; t_{90}—time to 90% rise in torque; Mooney results: ML—minimum viscosity; t_5—time to increase the viscosity by 5 Mooney units (MU) from the ML value).

Compound	MDR Results					Mooney Results	
	ML [dNm]	MH [dNm]	MH–ML [dNm]	t_{s2} [min]	t_{90} [min]	ML [MU]	t_5 [min]
Ref0	1.54	7.79	6.25	0.97	1.79	32.8	14.0
Ref1	0.96	9.05	8.09	0.93	1.94	26.1	12.4
Ref2	1.05	9.53	8.48	0.93	2.01	28.3	11.8
Test1	1.17	8.15	6.98	1.00	1.78	31.8	12.7
Test2	1.05	9.21	8.16	1.21	2.88	26.1	18.6
Test3	1.23	8.82	7.59	0.95	1.78	32.0	12.0
Test4	1.10	8.95	7.85	1.10	2.26	28.4	17.2
Test5	1.06	8.02	6.96	0.98	1.74	27.9	12.1
Test6	0.93	8.11	7.18	1.13	2.52	23.4	17.9
Test7	1.18	8.43	7.25	0.96	1.70	29.0	13.0
Test8	1.06	7.47	6.41	1.10	1.88	26.4	16.7
Test9	1.05	8.07	7.02	0.93	1.93	27.8	12.6
Test10	0.99	7.99	7.00	0.88	1.62	27.0	11.4
Test11	1.03	8.32	7.29	0.95	2.27	25.0	16.3
Test12	1.11	8.43	7.32	0.90	1.73	27.9	13.4
Test13	1.04	8.65	7.61	0.86	1.88	27.6	12.8
Test14	1.08	8.11	7.03	0.95	1.80	28.3	12.5
Test15	1.10	8.45	7.35	0.91	1.66	28.8	12.3
Ref3	0.82	6.90	6.08	1.22	3.46	20.9	25.1
Test16	0.98	7.60	6.62	1.32	2.66	22.9	24.1
Test17	1.01	8.24	7.23	1.34	2.87	23.7	24.1
Test18	0.83	6.71	5.88	0.89	2.20	21.9	14.7
Test19	0.94	6.59	5.65	0.75	1.90	24.8	11.6

Analysis of the measurement results performed using the MDR and Mooney viscometer (Table 6) showed that, for the reference rubber compounds, the application of active ZnO (Ref1) increased the torque increment during the rheometric test (MH-ML), which is an indirect measure of the cross-linking degree of the rubber compound, whereas a slightly longer optimal vulcanization time t_{90} (time to reach 90% of maximum torque during rheometric test) compared to standard ZnO (Ref0) was achieved. This confirmed the higher activity of active ZnO compared to standard ZnO, which was due to the larger specific surface area of active ZnO, resulting in a larger contact area between the components of the cross-linking system during vulcanization. Applying materials synthesized using active ZnO (Test1 to 4) allowed us to achieve the same trend compared with the materials obtained based on standard ZnO (Test5 to 8), but the differences were not so significant as in the case of the rubber compounds containing raw zinc oxides. Rubber compounds with a 3-times reduced quantity of ZnO showed both a higher MH–ML and longer t_{90} for active ZnO vs. standard ZnO (Test2 vs. 6 and Test4 vs. 8, respectively), which additionally confirmed the greater activity in vulcanization of both active ZnO and the synthesis product obtained using it. For rubber compounds containing a zinc amount equivalent to stearic acid in the reference rubber compound, the differences in MDR and Mooney test results compared to the standard and active ZnO were not so significant (Test1 vs. 5 and Test3 vs. 7). The rubber compounds containing materials synthesized with sulfur exhibited MDR and Mooney values similar to rubber compounds containing the synthesis products obtained using only StA and ZnO with the same ZnO equivalent (Test1, 9, 10 vs. 3, 12, 13). Adding sulfur before ZnO during the synthesis resulted in the decrease in t_{90} only for rubber compounds containing materials obtained in stoichiometric conditions (Test9 and 10, respectively). This indicated the lack of formation of sulfur complexes with ZnO and StA during the reaction—only sulfur melting and changes in its crystalline form took place during the mixing of the ingredients in the reaction flask. Applying commercial laboratory grade ZnSt (Ref3, Test16 and 17) resulted in torque values on similar levels as rubber compounds containing materials synthesized from standard ZnO, but with a significantly longer t_{90}. The differences in the rheometric data obtained for Ref3 and premixed in the Brabender mixer rubber compounds (Test16 and Test17) revealed that mixing in the molted state, aiming to obtain a perfect dispersion of the curatives, improved the torque increment MH-ML and decreased the cure time, thus having a beneficial effect on the effectiveness of vulcanization. Additional amounts of stearic acid, which can influence the activity of accelerators, significantly reduced the torque increment MH-ML, deteriorating the cross-linking degree of the elastomer and slightly decreasing the t_{90}, thus improving the rate of vulcanization (Test2 vs. 18 and Test4 vs. 19).

Having investigated the influence of standard and active ZnO, as well as their salts or complexes, with stearic acid on the vulcanization parameters, we then examined their impact on the functional properties of the EPDM vulcanizates, i.e., the compression set (CS), which is crucial for sealing applications. The results are presented in Table 7.

Table 7. Equilibrium swelling and compression set test results for rubber compounds described in Tables 2 and 3.

Compound	Compression Set [%]			Equilibrium Swelling [%]		
Ref0	72.5	+/−	1.7	96.1	+/−	2.0
Ref1	60.9	+/−	2.3	85.6	+/−	0.4
Ref2	57.9	+/−	1.4	82.6	+/−	1.1
Test1	73.7	+/−	1.1	103.9	+/−	1.9
Test2	52.9	+/−	2.1	74.7	+/−	0.4
Test3	66.8	+/−	1.7	95.2	+/−	0.5
Test4	56.5	+/−	0.9	81.6	+/−	0.5
Test5	74.9	+/−	1.6	105.9	+/−	1.4

Table 7. *Cont.*

Compound	Compression Set [%]			Equilibrium Swelling [%]		
Test6	56.3	+/−	1.2	81.3	+/−	1.1
Test7	69.3	+/−	1.1	100.2	+/−	2.0
Test8	61.9	+/−	1.5	93.9	+/−	1.2
Test9	70.4	+/−	0.8	102.1	+/−	2.5
Test10	71.4	+/−	1.6	103.7	+/−	1.4
Test11	68.4	+/−	0.8	92.0	+/−	1.1
Test12	69.2	+/−	1.5	98.9	+/−	1.3
Test13	69.2	+/−	0.2	97.8	+/−	1.3
Test14	70.3	+/−	1.7	102.3	+/−	1.4
Test15	67.9	+/−	1.6	97.1	+/−	0.5
Ref3	44.5	+/−	0.7	82.6	+/−	1.7
Test16	42.5	+/−	2.3	78.3	+/−	1.6
Test17	41.5	+/−	0.8	77.3	+/−	1.4
Test18	51.1	+/−	1.5	85.1	+/−	1.2
Test19	52.7	+/−	0.8	97.1	+/−	1.6

As it is widely known, the kind of ZnO and its quantity have a significant impact on vulcanizate properties. Analyzing the data collected in Table 7, rubber composites with active ZnO exhibited better properties, i.e., a lower CS and equilibrium swelling, compared to standard ZnO. However, when the synthesized materials based on these types of ZnO were applied, the differences between the obtained results were not so significant as in the case of reference composites with both types of ZnO. The CS and equilibrium swelling values decreased for EPDM vulcanizates with active ZnO (Ref1 and Ref2) compared to standard ZnO (Ref0). The lower equilibrium swelling allowed us to conclude that the active ZnO ensured that the vulcanizates had a higher cross-linking density compared to the standard ZnO. This correlates well with the results of rheometric tests. Ref3 with zinc stearate exhibited the best functional properties, i.e., the lowest CS, among all references, despite the fact that its zinc equivalent was three times lower than for other reference rubber compounds. A better dispersion of zinc and sulfur for rubber compounds premixed in the Brabender probably led to an improvement in the vulcanizate properties despite the zinc content reduction. This trend was observed for both kinds of ZnO and for zinc stearate.

The addition of stoichiometric zinc salt in an amount that was 3 times lower than in the reference (Test2) resulted in a significant improvement in CS (52.9% vs. 60.9%) obtained for Ref1, with a reduction in equilibrium swelling (74.7% vs. 85.6%, respectively), and, consequently, an increase in the cross-linking density of the vulcanizate. A further decrease in the zinc salt amount to a 13.6-times (Test1) lower equivalent of ZnO resulted in a CS on a level similar to vulcanization with standard ZnO, i.e., 73.7% vs. 72.5% (Ref0), whereas the equilibrium swelling increased significantly to 103.9% due to the decrease in the cross-linking density of the vulcanizate. Stoichiometric zinc salts produced from standard ZnO resulted in only a slightly higher CS of vulcanizates compared to zinc salts produced from active ZnO (Test5 vs. Test1 and Test6 vs. Test2, respectively), which was also in line with the increase in equilibrium swelling. Slight differences (on the level of measurement uncertainty) between the results obtained for vulcanizates containing zinc salts synthesized from active ZnO and standard ZnO suggest that only the salt amount is important for the properties of vulcanizates, not the kind of ZnO used for the synthesis.

Non-stoichiometric products of synthesis, which contained almost half of the unreacted ZnO and half of the ZnSt, resulted in a slightly lower CS of vulcanizates compared to stoichiometric ZnSt only for Test3 vs. Test1, but for Test4 vs. Test2 the relationship was opposite. Those changes are in line with the equilibrium swelling results (lower equilibrium swelling and consequently higher cross-linking density accompanied by lower CS of vulcanizates). Hence, in order to obtain the proper CS of vulcanizates, the more important factor is not the type of vulcanization activator, but the cross-linking density of the vulcanizate, which can be obtained by using this particular activator. For vulcanizates containing the products synthesized from standard ZnO, the effects of additional amounts of ZnO on the CS and equilibrium swelling of vulcanizates have the same trend. Applying the mixtures containing ZnO, StA, and sulfur resulted in vulcanizates with CS and swelling results slightly lower for a very low ZnO equivalent (Test9 and 10 vs. Test5, 13.6-times lower zinc amount than in Ref1) and slightly higher for a low ZnO equivalent (Test12 and 13 vs. Test3, 6.7-times lower zinc amount than in Ref1). This suggests that sulfur did not form any complex with ZnO and StA during mixing; the obtained differences in CS of vulcanizates can be only due to the impact of ZnSt and sulfur content. A comparison of the CS and swelling results of vulcanizates for Test13 and 12, which contained the products obtained when sulfur was added before or after ZnO addition during the synthesis, showed no significant differences considering the measurement uncertainty. This supports the earlier statement that there was no formation of sulfur complex during the reaction of sulfur, ZnO, and StA. A comparison of the results obtained for vulcanizates containing synthesized products using standard ZnO vs. active ZnO for stoichiometric salts allowed us to observe that there were no significant differences between CS and swelling values for Test14 (standard ZnO) and Test9 (active ZnO). CS values for both vulcanizates were approximately 70.4%, and the equilibrium swelling of these vulcanizates was also similar, which indicates a similar cross-linking density. This further confirmed that only the amount of salt is important. A similar relationship was observed for Test15 vs. 12, i.e., zinc salts obtained from standard ZnO vs. active ZnO in non-stoichiometric conditions. An analysis of the obtained results allowed us to conclude that the excess of StA combined with ZnSt had a very important effect on the CS of the vulcanizates. For vulcanizates with ZnSt, i.e., Ref3, Test16, and Test17, CS was reduced to 41.5–44.5% from approximately 60.9% (Ref1—with active ZnO). This is an even better result than for Test2 (stoichiometric synthesis product, 1 phr of active ZnO), which exhibited a CS of approximately 52.9%. It proved that raw ZnSt ensured better curing conditions than a mix of other zinc salts, with ZnSt received in the product synthesized from commercial StA. The equilibrium swelling also decreased significantly, indicating a marked increase in the cross-linking density of the vulcanizates, which probably contributed to the decrease in CS. This confirmed that a better mixing of ZnSt can improve the availability of zinc ions during cross-linking reactions. A comparison of the results obtained for Test18 (1 phr of stoichiometric ZnSt from active ZnO) vs. Test2 (with addition of StA) showed no improvement in the CS; however, the equilibrium swelling increased, indicating a deterioration of the cross-linking density of the vulcanizate. This suggests that a small excess of StA can be beneficial for the activation of accelerators. On the contrary, a comparison of the results for Test19 (1 phr of non-stoichiometric ZnSt from active ZnO) vs. Test4 (with addition of StA) revealed an improvement in the CS, while Test4 did not contain any excess of StA. Thus, it was concluded that StA acted not only as a ZnSt precursor, but also as an activator of accelerators.

Most importantly, the performed studies proved that adding materials synthesized in the reaction of ZnO, StA, and sulfur could help to reduce the content of zinc in rubber composites, while keeping a good level of both important manufacturing parameters, like the viscosity and cure time of rubber compounds, and also functional properties, i.e., the CS or cross-linking density, of the vulcanizates.

In the next step of the study, the tensile properties of the vulcanizates were examined. The results are presented in Table 8.

Table 8. Tensile test results of the vulcanizates described in Tables 2 and 3 (SE_{100}—stress at relative elongation of 100%, TS—tensile strength, E_b—elongation at break).

Compound	SE_{100} [MPa]			TS [MPa]			E_b [%]		
Ref0	2.6	+/−	0.2	3.0	+/−	0.6	123	+/−	17
Ref1	2.1	+/−	0.1	6.3	+/−	0.3	499	+/−	40
Test2	2.2	+/−	0.1	7.9	+/−	0.7	421	+/−	55
Test4	2.2	+/−	0.1	8.7	+/−	0.1	412	+/−	10
Test6	2.6	+/−	0.1	9.0	+/−	0.1	415	+/−	12
Test8	3.5	+/−	0.2	10.6	+/−	0.4	335	+/−	15
Test9	2.4	+/−	0.2	7.6	+/−	0.1	445	+/−	69
Test14	2.3	+/−	0.1	7.7	+/−	0.1	506	+/−	25
Ref3	2.3	+/−	0.1	8.5	+/−	0.2	435	+/−	8
Test16	2.7	+/−	0.1	8.5	+/−	0.2	352	+/−	23

The reference vulcanizate Ref0 with standard ZnO exhibited the lowest tensile strength (TS) and elongation at break (E_b). It should be noted that the E_b of this vulcanizate was approximately 120%, while for most other vulcanizates the E_b was above 400%. This indicates a very low elasticity of this vulcanizate compared to the others, which may be caused by its significantly higher degree of cross-linking. The vulcanizate with active ZnO (Ref1) showed twice the TS compared to that with standard ZnO, and a significantly better elasticity (E_b of approximately 500%). The vulcanizate with ZnSt (Ref3) also showed better tensile parameters, i.e., TS and E_b, compared to that with standard ZnO, despite having a 3-fold lower zinc equivalent. Most importantly, vulcanizates containing the synthesized zinc complexes exhibited a significantly higher TS and E_b than vulcanizates with standard (Ref0) or active ZnO (Ref1). Surprisingly, vulcanizates with a 13.6-times lower zinc amount (Test9 and 14) compared to the reference samples showed much better tensile properties despite a significantly higher equilibrium swelling and, consequently, lower cross-linking degree. This allowed us to assume that the ZnO-containing vulcanizates were slightly over-cross-linked compared to those containing synthesis products, which caused a deterioration of their TS and an increase in their brittleness. On the other hand, a high amount of StA in rubber compounds creates a huge issue to solve, i.e., the whitening of vulcanizates (blooming) [28]. With time, additives like the zinc salts of fatty acids migrate on the surface of vulcanizates, causing visual aspects and influencing the adhesion [29]. A higher content of StA in the recipe of rubber compounds results in a higher amount of ZnSt in the vulcanizates. In Figure 7 the problem is presented for the vulcanizates containing active ZnO with the standard amount of StA (Ref1), the reference with ZnSt (Ref3), Test16 with an StA equivalent about 5-times higher than in Ref1, Test2 with StA equivalent 6.7-times higher than in Ref1, and Test11 with an StA equivalent 2-times higher than in Ref1. Vulcanizates with a 2-times higher content of StA compared to Ref1, i.e., Test11, did not show any visual blooming aspects. The amount of created ZnSt in the reaction of StA with ZnO during the preparation of the rubber compound and the curing process should also be twice as high accordingly. The increasing of the ZnSt amount in the vulcanizate by a higher amount of StA led to an unacceptable migration on the vulcanizate surface (Ref3, Test2, and Test16). For rubber composites containing synthesized materials, the reduction of zinc in the amount that did not impact their functional properties was not enough. In turn, a further reduction in the zinc amount to keep the equivalent StA as in the standard rubber compound (Ref1) or not creating blooming issues (Test11) did not guarantee keeping the demanded properties of the vulcanizates.

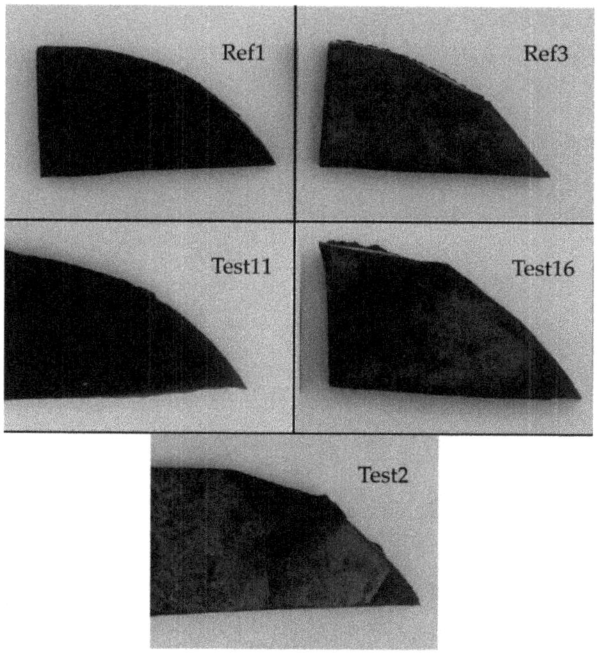

Figure 7. Blooming of vulcanizates described in Tables 2 and 3.

Parts of vulcanizates were examined using SEM to check whether their composition or the method of introducing the crucial cross-linking components (ZnO, StA, sulfur, ZnSt) separately or as products of the reaction influences their dispersion in the elastomer matrix. SEM images of the vulcanizate fractures are presented in Figure 8.

In the case of the reference vulcanizate with active ZnO, the overall dispersion of the components was relatively homogeneous (Figure 8a), although some agglomerates with a complex morphology and a size of several micrometers, representing clusters of primary particles less than 1 micrometer in size, also occurred (Figure 8b). These agglomerates were quite well wetted by the elastomer; most of the agglomerate visible in the photo was surrounded by an elastomer film. In the case of the vulcanizate containing stoichiometric ZnSt synthesized from active ZnO, the SEM image (Figure 8c) shows areas of both uniform particle dispersion and micrometric agglomerates. These agglomerates resulted from the agglomeration of particles smaller than 1 micrometer and demonstrated a better wettability by the elastomer compared to the agglomerates present in the SEM image of the vulcanizate containing active ZnO (Figure 8d). The elastomer film is clearly visible, penetrating between the agglomerate particles and surrounding them on all sides. The use of the product obtained by adding sulfur in the synthesis process of zinc salts/complexes did not have a significant effect on the overall dispersion of the mixture components (Figure 8e), although it seems that the wettability of the agglomerate visible in the SEM image (Figure 8f) is slightly worse in relation to the vulcanizate containing the synthesis product without sulfur. In the case of the vulcanizate containing commercial ZnSt, the dispersion of the components can be considered the most homogeneous (Figure 8g). The wettability of the particles and their aggregates by the elastomer is very good, as all of them are coated with an elastomer film and embedded in it (Figure 8h). Overall, this vulcanizate exhibited one of the lowest CSs and highest cross-linking densities. In the case of the vulcanizate containing stoichiometric ZnSt synthesized from active ZnO with an additional amount of StA, the overall dispersion of the components in the elastomer matrix is quite good (Figure 8i), but several micrometer-sized agglomerates of varying morphology are also visible, representing clusters of relatively large primary particles with a fairly regular platelet or block shape (Figure 8j). These agglomerates demonstrate a low wettability by the elastomer.

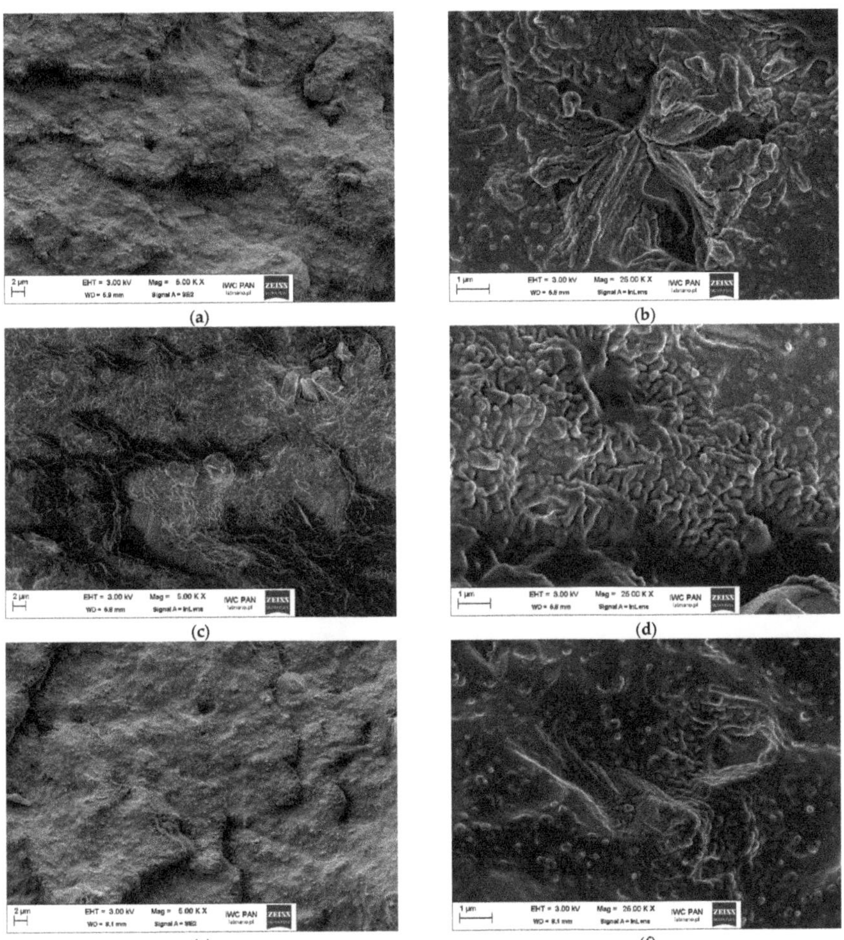

Figure 8. *Cont.*

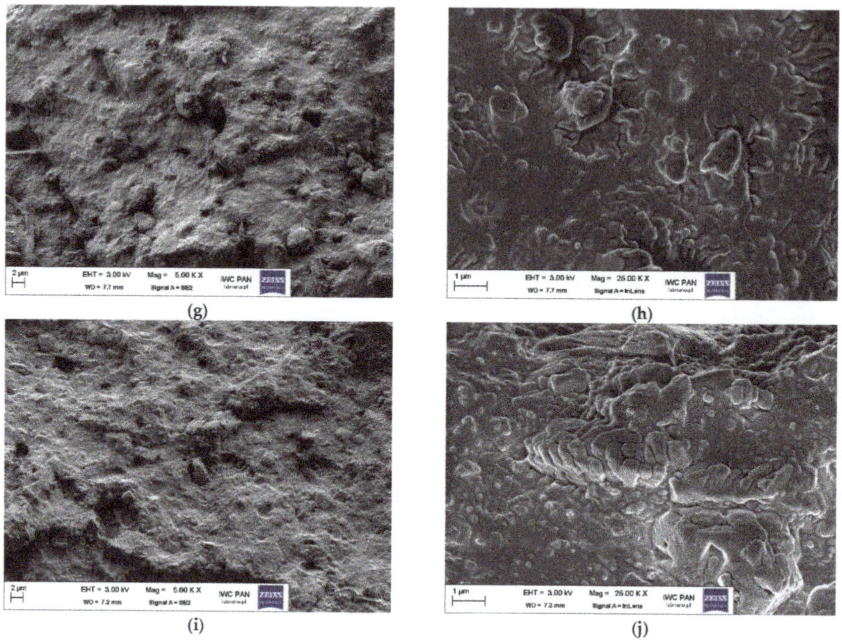

Figure 8. SEM images of vulcanizate fractures: (**a**,**b**) Ref1; (**c**,**d**) Test2; (**e**,**f**) Test11; (**g**,**h**) Test16; (**i**,**j**) Test18.

To identify the composition of the agglomerates visible in the above SEM images (Figure 8), SEM with EDS analysis was employed. The results in the form of EDS maps are presented in Figure 9.

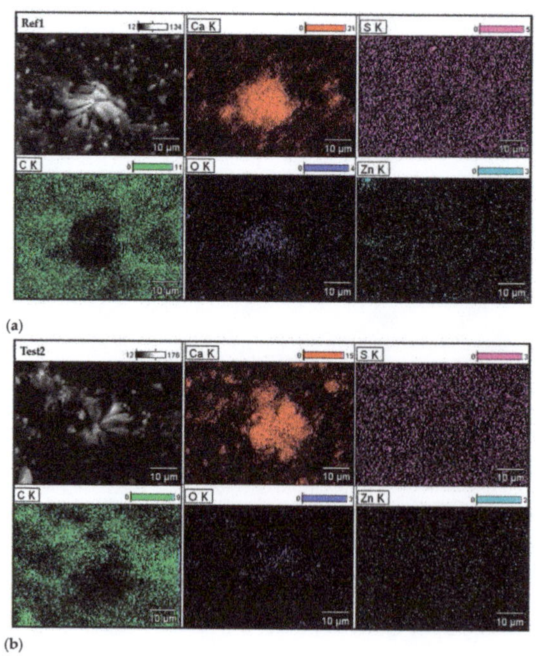

Figure 9. *Cont.*

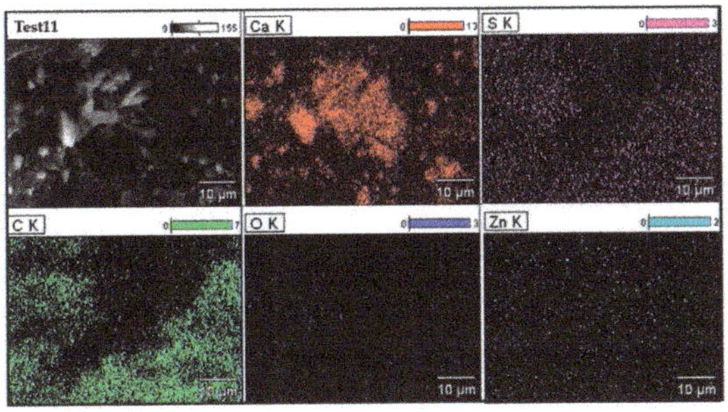

(c)

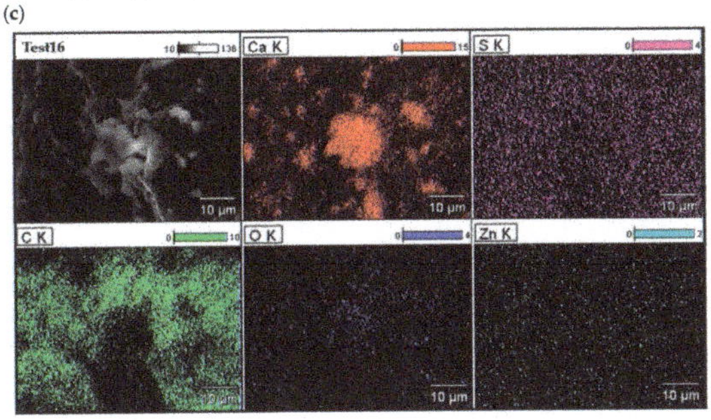

(d)

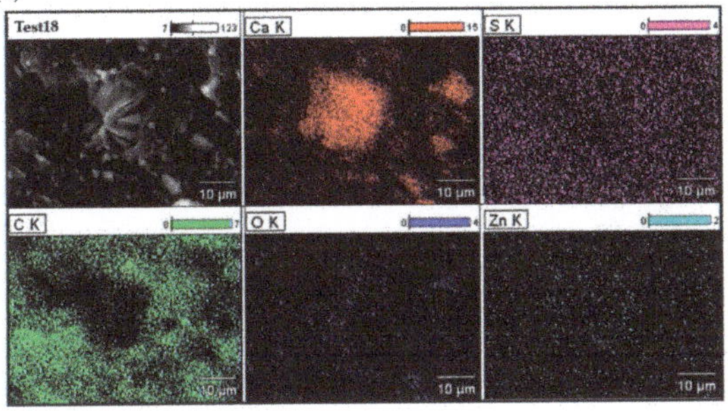

(e)

Figure 9. SEM images of vulcanizate fractures with EDS mapping for Ca, S, C, O, and Zn elements: (**a**) Ref1; (**b**) Test2; (**c**) Test11; (**d**) Test16; (**e**) Test18. Analysis of ESD mapping revealed that the agglomerates present in SEM images of the vulcanizates resulted from the agglomeration of calcium compounds present in the masterbatch composition, e.g., $CaCO_3$. On the other hand, EDS analysis showed that components of the curing system, i.e., sulfur and zinc compounds, were homogeneously dispersed in the elastomer matrix, regardless of whether ZnO or the synthesized products obtained in the reaction of ZnO and StA under stoichiometric or non-stoichiometric conditions were used as the vulcanization activator. Hence, it can be concluded that the composition of the elastomer composite, i.e., the type and amount of the activator, did not significantly influence the general dispersion of the rubber composite components.

4. Conclusions

In this research, zinc complexes with different molar ratios of ZnO and StA with or without the presence of sulfur were synthesized and studied as potential vulcanization activators for EPDM compounds. The vulcanization parameters and properties of the rubber composites containing synthesized materials were compared with those containing commercial ZnSt to verify the hypothesis on the key role in sulfur vulcanization played by non-stoichiometric zinc complexes.

DSC results confirmed that the reaction between ZnO and StA resulted in the formation of ZnSt. Complexes with different molar ratios of ZnO and StA contained the amount of ZnSt proportional to amount of StA used in the synthesis. In addition, an FTIR study of reaction products did not reveal any absorption bands different from those characteristic for StA and ZnSt. Active ZnO exhibited a higher reactivity in the formation of ZnSt than the standard ZnO, due to the higher specific surface area and lower size of particles that are consumed during the reaction. The heats of melting determined through DSC for the products from non-stoichiometric mixtures were significantly lower than for the stoichiometric ones. These differences could result from the difference in the crystallization process depending on the reaction conditions, since ZnSt particles can form binuclear bidentate or a crystal structure that differs in the heat of melting.

The stoichiometric salt of ZnO and StA was the most effective in terms of vulcanization parameters and the compression set of rubber composites. Non-stoichiometric complexes were less effective, proportionally to their content. However, they can also be a potential alternative to traditional ZnO and StA systems in the vulcanization of EPDM composites. The comparison with stoichiometric ZnSt revealed that the effect on the vulcanization state and compression set of rubber composites is mostly driven by stoichiometric salt. In case of non-stoichiometric complexes, the influence on the vulcanization parameters, cross-linking density, and compression set was proportional to the content of stoichiometric salt. Moreover, the amount of salt was more important than the type of ZnO used during the synthesis. Th excess of StA can improve the availability of zinc during cross-linking reactions, leading to improved properties of the vulcanizates. The addition of sulfur to zinc complexes did not significantly affect the vulcanization and properties of the EPDM composites, which disproved the hypothesis that sulfur formed complexes with StA and ZnO.

Supplementary Materials: The following supporting information can be downloaded at https://www.mdpi.com/article/10.3390/polym17212875/s1. Figure S1: Research station used for the synthesis of StA with ZnO and/or S reaction products.

Author Contributions: Conceptualization, K.K. and P.G.; methodology, K.K. and P.G.; validation, P.G. and M.M.; formal analysis, K.K.; investigation, K.K.; resources, K.K.; data curation, K.K., P.G. and M.M.; writing—original draft preparation, K.K.; writing—review and editing, P.G. and M.M.; visualization, K.K.; supervision, P.G.; project administration, K.K.; funding acquisition, M.M. All authors have read and agreed to the published version of the manuscript.

Funding: This work was supported by Ministerstwo Edukacji i Nauki (DWD/6/0559/2022).

Institutional Review Board Statement: Not applicable.

Data Availability Statement: The data presented in this study are available on request from the corresponding author due to the confidentiality policy of Hutchinson.

Acknowledgments: This work was completed while Krzysztof Kaczewiak was the Implementation Doctoral Candidate in the Interdisciplinary Doctoral School at the Lodz University of Technology, Poland.

Conflicts of Interest: Authors Krzysztof Kaczewiak and Piotr Głąb are employed by Hutchinson Poland Sp z o.o. company. The remaining authors declare that the research was conducted in the absence of any commercial or financial relationships that could be construed as a potential conflict of interest.

Abbreviations

The following abbreviations are used in this manuscript:

ZnSt	Zinc stearate
EPDM	Ethylene–propylene–diene monomer
ZnO	Zinc oxide
StA	Stearic acid
LCA	Life Cycle Analysis
GHS	Globally Harmonized System of Classification and Labelling of Chemicals
CLD	Caprolactam disulfide
MBTS	2,2′-benzothiazyl disulfide
CBS	N-cyclohexyl-2-benzothiazole sulfenamide
ZDTP	Zinc dialkyldithiophosphate
MDR	Moving Dye Rheometer
CS	Compression set
FTIR	Fourier Transform Infrared Spectroscopy
DSC	Differential Scanning Calorimetry
SEM	Scanning Electron Microscopy
TD-GC-MS	Thermal Desorption–Gas Chromatography–Mass Spectrometry
EDS	Energy-Dispersive X-ray Spectroscopy

References

1. Moezzi, A.; McDonagh, A.M.; Cortie, M.B. Zinc Oxide Particles: Synthesis, Properties and Applications. *Chem. Eng. J.* **2012**, *185–186*, 1–22. [CrossRef]
2. Yan, M.F. Zinc Oxide. In *Concise Encyclopedia of Advanced Ceramic Materials*; Elsevier: Amsterdam, The Netherlands, 1991; pp. 523–525. [CrossRef]
3. Heideman, G.; Datta, R.N.; Noordermeer, J.W.M.; van Baarle, B. Influence of Zinc Oxide during Different Stages of Sulfur Vulcanization. Elucidated by Model Compound Studies. *J. Appl. Polym. Sci.* **2005**, *95*, 1388–1404. [CrossRef]
4. Borysiewicz, M.A. ZnO as a Functional Material, a Review. *Crystals* **2019**, *9*, 505. [CrossRef]
5. Wilaiwong, W.; Smitthipong, W. Study of Non-Negligible Chemical Reduction of ZnO in the Rubber Industry. *IOP Conf. Ser. Mater. Sci. Eng.* **2022**, *1234*, 012014. [CrossRef]
6. Kołodziejczak-Radzimska, A.; Jesionowski, T. Zinc Oxide—From Synthesis to Application: A Review. *Materials* **2014**, *7*, 2833–2881. [CrossRef]
7. Macarrão, I.M.; Dutra, J.C. Sustainable Development Assessment of Zinc Oxide Production Focused on Agribusiness. *J. Eng. Exact. Sci.* **2020**, *6*, 577–584. [CrossRef]
8. Wojnarowicz, J.; Chudoba, T.; Lojkowski, W. A Review of Microwave Synthesis of Zinc Oxide Nanomaterials: Reactants, Process Parameters and Morphologies. *Nanomaterials* **2020**, *10*, 1086. [CrossRef]
9. Grund, S.C.; van Genderen, E. Life Cycle Assessment and Product Carbon Footprint for SHG Zinc Production. *J. Phys. Conf. Ser.* **2024**, *2738*, 012038. [CrossRef]
10. Buendia, E.; Tanabe, K.; Kranjc, A.; Jamsranjav, B.; Fukuda, M.; Ngarize, S.; Osako, A.; Pyrozhenko, Y.; Shermanau, P.; Federici, S. *2019 Refinement to the 2006 IPCC Guidelines for National Greenhouse Gas Inventories*; IPCC: Geneva, Switzerland, 2019.
11. Sreethu, T.K.; Naskar, K. Zinc Oxide with Various Surface Characteristics and Its Role on Mechanical Properties, Cure-Characteristics, and Morphological Analysis of Natural Rubber/Carbon Black Composites. *J. Polym. Res.* **2021**, *28*, 183. [CrossRef]
12. Mostoni, S.; Milana, P.; Di Credico, B.; D'Arienzo, M.; Scotti, R. Zinc-Based Curing Activators: New Trends for Reducing Zinc Content in Rubber Vulcanization Process. *Catalysts* **2019**, *9*, 664. [CrossRef]
13. Heideman, G.; Datta, R.N.; Noordermeer, J.W.M.; van Baarle, B. Activators in Accelerated Sulfur Vulcanization. *Rubber Chem. Technol.* **2004**, *77*, 512–541. [CrossRef]
14. Junkong, P.; Morimoto, R.; Miyaji, K.; Tohsan, A.; Sakaki, Y.; Ikeda, Y. Effect of Fatty Acids on the Accelerated Sulfur Vulcanization of Rubber by Active Zinc/Carboxylate Complexes. *RSC Adv.* **2020**, *10*, 4772–4785. [CrossRef]
15. Heideman, G.; Noordermeer, J.W.M.; Datta, R.N.; Van Baarle, B. Various Ways to Reduce Zinc Oxide Levels in S-SBR Rubber Compounds. *Macromol. Symp.* **2006**, *245*, 657–667. [CrossRef]
16. Przybyszewska, M.; Zaborski, M.; Jakubowski, B.; Zawadiak, J. Zinc Chelates as New Activators for Sulphur Vulcanization of Acrylonitrile-Butadiene Elastomer. *Express Polym. Lett.* **2009**, *3*, 256–266. [CrossRef]

17. Ikeda, Y.; Yasuda, Y.; Ohashi, T.; Yokohama, H.; Minoda, S.; Kobayashi, H.; Honma, T. Dinuclear Bridging Bidentate Zinc/Stearate Complex in Sulfur Cross-Linking of Rubber. *Macromolecules* **2015**, *48*, 462–475. [CrossRef]
18. Setianto, W.B.; Yohanes, H.; Astuti; Maisaroh; Atmaji, G. Synthesis of Palm Oil-Base Zinc Stearate and Its Application on Manufacture of Rubber Component. *IOP Conf. Ser. Earth Environ. Sci.* **2022**, *963*, 012028. [CrossRef]
19. Bukhina, M.F.; Morozov, Y.L.; Ven, P.M.; Noordermeer, J.W.M. Mold Fouling of EPDM Rubber Compounds. *Kautsch. Und Gummi Kunststoffe* **2003**, *56*, 172–183.
20. Kaczewiak, K.; Głąb, P.; Maciejewska, M. Vulcanization Temperature Impact on Cross-Linking Density and Compression Set of Sulfur-Cured Ethylene–Propylene–Diene Monomer. *J. Appl. Polym. Sci.* **2025**, *142*, e57188. [CrossRef]
21. *ISO 815-1:2019*; Rubber, Vulcanized or Thermoplastic—Determination of Compression set, Part 1: At Ambient or Elevated Temperatures. International Organization for Standardization: Geneva, Switzerland, 2019.
22. Konkoly-Thege, I.; Ruff, I.; Adeosun, S.O.; Sime, S.J. Properties of Molten Carboxylates Part 6. A Quantiative Differential Thermal Analysis Study of Phase Transitions in Some Zinc and Cadmium Carboxylates. *Thermochim. Acta* **1978**, *24*, 89–96. [CrossRef]
23. Zhang, X.; Wei, F.; Wang, Z.; Li, G.; Yang, S.; Feng, J. Comparative Investigation of the Structural Evolution of Zinc Stearate and Calcium Stearate in a Polypropylene Random Copolymer upon Heating and Cooling. *Polymer* **2023**, *267*, 125646. [CrossRef]
24. Ishioka, T.; Maeda, K.; Watanabe, I.; Kawauchi, S.; Harada, M. Infrared and XAFS Study on Structure and Transition Behavior of Zinc Stearate. *Spectrochim. Acta A Mol. Biomol. Spectrosc.* **2000**, *56*, 1731–1737. [CrossRef]
25. Zhu, J.; Liu, B.; Li, L.; Zeng, Z.; Zhao, W.; Wang, G.; Guan, X. Simple and Green Fabrication of Super-Hydrophobic Surface by One-Step Immersion for Continuous Oil/Water Separation. *J. Phys. Chem. A* **2016**, *120*, 5617–5623. [CrossRef]
26. Larkin, P.J.; Jackson, A. Interpretation of the Infrared Spectra of Metal-Stearate Salts. *Appl. Spectrosc. Pract.* **2024**, *2*, 27551857241253834. [CrossRef]
27. Albarakaty, F.M.; Alzaban, M.I.; Alharbi, N.K.; Bagrwan, F.S.; Abd El-Aziz, A.R.M.; Mahmoud, M.A. Zinc oxide Nanoparticles, Biosynthesis, Characterization and Their Potent Photocatalytic Degradation, and Antioxidant Activities. *J. King Saud. Univ. Sci.* **2023**, *35*, 102434. [CrossRef]
28. Choi, S.; Chung, H.; Joo, Y.; Yang, K.; Lee, S. Analysis of Whitening Phenomenon of EPDM Article by Humid Aging. *J. Appl. Polym. Sci.* **2012**, *123*, 2451–2457. [CrossRef]
29. Dimopoulos, M.; Roy Choudhury, N.; Ginic-Markovic, M.; Matisons, J.; Williams, D.R.G. Surface Studies on the Additive Migration and Diffusion in the Windowseal Rubber Component Influencing Adhesion to Coating. *J. Adhes. Sci. Technol.* **1998**, *12*, 1377–1390. [CrossRef]

polymers

MDPI

Article

Sustainable Rubber Solutions: A Study on Bio-Based Oil and Resin Blends

Frances van Elburg [1,*], Fabian Grunert [1], Claudia Aurisicchio [2], Micol di Consiglio [2], Auke Talma [1], Pilar Bernal-Ortega [1] and Anke Blume [1,*]

[1] Elastomer Technology & Engineering, Department of Mechanics of Solids, Surfaces & Systems (MS3), Faculty of Engineering Technology, University of Twente, Drienerlolaan 5, 7522 NB Enschede, The Netherlands; f.grunert@utwente.nl (F.G.); a.g.talma-1@utwente.nl (A.T.); m.d.p.bernalortega@utwente.nl (P.B.-O.)

[2] Bridgestone EU NV/SA, Italian Branch–Technical Center, Via del Fosso del Salceto 13/15, 00128 Rome, Italy

[*] Correspondence: f.a.vanelburg@utwente.nl (F.v.E.); a.blume@utwente.nl (A.B.)

Abstract

One of the most important challenges the tire industry faces is becoming carbon-neutral and using 100% sustainable materials by 2050. Utilizing materials from renewable sources and recycled substances is a key aspect of achieving this goal. Petroleum-based oils, such as Treated Distillate Aromatic Extract (TDAE), are frequently used in rubber compounds, and a promising strategy to enhance sustainability is to use bio-based plasticizer alternatives. However, research has shown that the replacement of TDAE oil with bio-based oils or resins can significantly alter the glass transition temperature (T_g) of the final compound, influencing the tire properties. In this study, the theory was proposed that using a plasticizer blend, comprising oil and resin, in a rubber compound would result in similar T_g values as the reference compound containing TDAE. To test this, the cycloaliphatic di-ester oil Hexamoll DINCH, which can be made out of bio-based feedstock by the BioMass Balance approach, was selected and blended with the cycloaliphatic hydrocarbon resin Escorez 5300. Various oil-to-resin ratios were investigated, and a linear increase in the T_g of the vulcanizate was obtained when increasing the resin content and decreasing the oil content. Additionally, a 50/50 blend, consisting of 18.75 phr Hexamoll DINCH and 18.75 phr Escorez 5300, resulted in the same T_g of −19 °C as a compound containing 37.5 phr TDAE. Furthermore, this blend resulted in similar curing characteristics and cured Payne effect as the reference with TDAE. Moreover, a similar rolling resistance indicator (tan δ at 60 °C = 0.115), a slight deterioration in wear resistance (ARI = 83%), but an improvement in the stress–strain behavior (M300 = 9.18 ± 0.20 MPa and Ts = 16.3 ± 0.6 MPa) and wet grip indicator (tan δ at 0 °C = 0.427) were observed. The results in this work show the potential of finding a balance between optimal performance and sustainability by using plasticizer blends.

Keywords: plasticizers; oils; resins; tire tread; sustainability

1. Introduction

Tire manufacturers aim to reduce their carbon footprint and even become carbon-neutral by 2050 [1]. One of the approaches to achieve this is the transition to using 100% sustainably produced raw materials. These sustainable materials are obtained from renewable sources or recycled substances, which reduces the CO_2 footprint compared to fossil-based materials [2,3]. Due to the increasing interest in using sustainable materials in

 https://doi.org/10.3390/polym17152111

tire applications, the availability of scientific papers about the influence of these ingredients in rubber materials is also progressively increasing. Various review articles have been published in the last decade, which together provide an overview of the many sustainable polymers, fillers, plasticizers, and other additives investigated in rubber compounds [4–10].

One of the ingredient groups under investigation is the plasticizer. Together with the rubber polymer and reinforcing filler, plasticizers take up a large part of the rubber compound. In conventional rubber formulations, petroleum-based plasticizers are added to improve processing and increase the flexibility of the rubber compound [11]. The petroleum-based oil Treated Distillate Aromatic Extract (TDAE) is commonly used in tire tread compounds [12]. A previous study [13] showed that replacing TDAE with various vegetable oils shows potential, but certain material properties exhibit degradation. An important factor contributing to this deterioration in properties is the decrease of the glass transition temperature (T_g) to lower temperatures. This decrease in T_g was widely described in the literature when plant-based oils were used in rubber compounds [8].

Another material that has gained increasing interest in the tire industry is resin [14]. A resin is a low molecular weight (800–4000 g/mol) polymeric material, either in liquid or solid state at room temperature, depending on the softening point [15]. Various resins are available, with differences in chemical structure, molecular weight, and softening point. Resins can be derived from fossil feedstock (e.g., coumarone-indene and hydrocarbon resins) or natural sources (e.g., rosin and polyterpene resins) [15]. These materials act not only as tackifiers but also as softeners and plasticizers. Bernal-Ortega et al. investigated the use of hydrocarbon resins [16] and bio-based resins [17] as replacements for TDAE in simplified tire tread compounds. Based on these studies, it can be concluded that resins with a high softening point and glass transition temperature of the pure resin elevate the T_g of the rubber compound, in contrast to most vegetable oils that shift the T_g to lower temperatures. This phenomenon occurs when the resin and polymer used are compatible [15].

To achieve optimum performance in the rubber compound, a sustainable alternative to the commonly used petroleum-based plasticizer that results in a similar glass transition temperature is needed. A possible solution for this might be using oil and resin plasticizer blends. The practice of blending oils with resins has been addressed in the literature. Lopitaux [18] described a rubber compound containing 1,2-cyclohexane dicarboxylate diester and a resin. Although similar in-rubber properties were observed, the glass transition temperatures of the vulcanized materials were not analyzed. In another patent, the combination of vegetable oil with resin in rubber is claimed [19]. The experimental example of this patent shows silica-filled SBR/BR compounds with various combinations of petroleum-based oil, soybean oil, and traction resin. Compared to the compound with only petroleum-based oils, the plasticizer blends result in improved wet traction, a similar or worse abrasion resistance, and a slight deterioration in rolling resistance, but a clear improvement in tear strength. However, the T_g values of the various compounds were not reported as well.

In this study, the use of plasticizer blends comprising a di-ester oil and a cycloaliphatic hydrocarbon resin as alternatives to TDAE is proposed. It is hypothesized that blending the bio-based oil with resin will shift the glass transition temperature to values more comparable to compounds with TDAE, thereby positively influencing the final material properties of the vulcanized compound. In the first section, the cycloaliphatic di-ester oil Hexamoll DINCH is blended with the Escorez 5300 resin in various ratios to identify the optimal oil-to-resin blend ratio. A detailed evaluation of various plasticizer blend ratios is expected to reveal new trends and correlations in rubber performance properties. The second section explores the use of different di-ester oils (Hexamoll DINCH, Palatinol 10P, and Plastomoll DOA) combined with Escorez 5300 in 50/50 blends, to assess the effects of various chemical structures present in the oils. High concentrations (80 phr) of the plasticizer blends are evaluated to gain a deeper understanding of the effects of replacing TDAE with plasticizer blends on various properties. Because the three selected di-ester oils can be created with the BioMass Balance approach, making these oils up to 100% bio-based [20–22] and reducing the CO_2 footprint by approx. 60% [23]. Therefore, replacement of the petroleum-based TDAE with one of these di-ester oils will improve the sustainability of the final rubber compound.

2. Materials and Methods

2.1. Materials

The polymers used in this study were Styrene Butadiene Rubber (S-SBR) (SPRINTAN® SLR 4601, styrene content 21%, molecular weight = 450 kg/mol) (Synthos Schkopau GMBH, Schkopau, Germany) and Butadiene Rubber (BR) (Buna® CB24, cis-1,4 content > 96%, molecular weight = 620 kg/mol) (Arlanxeo, Dormagen, Germany). ULTRASIL® 7000 GR (specific surface area (CTAB) 160 m2/g) and bis(triethoxysilylpropyl) disulfide (TESPD) were supplied by Evonik Industries (Wesseling, Germany), and used as highly reinforcing filler and coupling agent, respectively. The curing package consisted of the curing activators zinc oxide (ZnO) (Umicore Zinc Chemicals, Angleur, Belgium) and stearic acid (Emery Oleochemicals GmbH, Düsseldorf, Germany), sulfur (Zolfindustria, Trecate, Italy), the primary accelerator N-tert-butyl-benzothiazole sulfonamide (TBBS) (General Química S.A., Lantarón, Spain) and the secondary accelerator N,N′-diphenylguanidine (DPG) (MLPC International (Arkema Group), Rion-des-Landes, France).

Treated Distillate Aromatic Extract (TDAE) (Hansen & Rosenthal, Hamburg, Germany) was used as a plasticizer in the reference compound. Additionally, the di-ester oils Hexamoll® DINCH, Palatinol® 10P, and Plastomoll® DOA (BASF, Ludwigshafen am Rhein, Germany) were evaluated, as well as the resin Escorez™ 5300 (ExxonMobil, Houston, TX, USA). The di-ester oils were selected based on their availability, potential of being fully sustainable, and saturated nature. Because these oils contain no double bonds in their tail structures, less interference with other rubber ingredients is expected [13]. The Escorez 5300 resin was evaluated in a previous study [16] and showed promising results on the tire performance. Moreover, this resin is widely available and is expected to be compatible with the selected di-ester oils and the used polymers, due to the similarities in Hansen Solubility Parameter (HSP). The HSP values, as well as the molecular weight (Mw), density (ρ), glass transition temperature (T_g) and chemical structure of the plasticizers are summarized in Table 1. The rubber formulations of the various rubber compounds are shown in the introductions of the separate chapters.

Table 1. Main characteristics according to the technical data sheets of the studied plasticizers.

Ingredient	Mw in g/mol	ρ in g/cm³	T_g * in °C	HSP ** in MPa$^{1/2}$	Chemical Structure
TDAE	x	0.95	−52	$\delta_D = 15.2$ $\delta_P = 11.2$ $\delta_H = 1.0$ $\delta_{tot} = 18.9$	Carbon distribution [12]: $C_{aromatic}$ = 25 wt% $C_{napthene}$ = 30 wt% $C_{paraffin}$ = 45 wt%
Hexamoll® DINCH	425	0.95	−90	$\delta_D = 16.1$ $\delta_P = 1.9$ $\delta_H = 2.6$ $\delta_{tot} = 16.4$	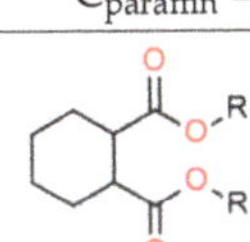 where R = C_9H_{19}

Table 1. *Cont.*

Ingredient	Mw in g/mol	ρ in g/cm³	T_g * in °C	HSP ** in MPa$^{1/2}$	Chemical Structure
Palatinol® 10P	447	0.96	−79	$\delta_D = 16.8$ $\delta_P = 4.2$ $\delta_H = 2.3$ $\delta_{tot} = 17.4$	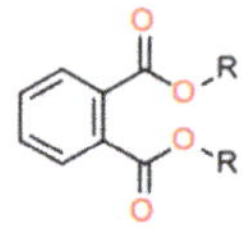 where R = $C_{10}H_{21}$
Plastomoll® DOA	371	0.92	−106	$\delta_D = 16.2$ $\delta_P = 2.7$ $\delta_H = 3.4$ $\delta_{tot} = 16.7$	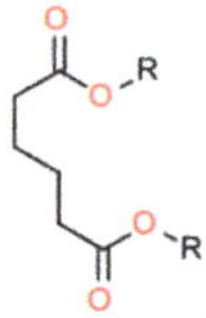 where R = C_8H_{17}
Escorez™ 5300	670	1.01	45	$\delta_D = 17.9$ $\delta_P = 0.1$ $\delta_H = 0.1$ $\delta_{tot} = 17.9$	Cycloaliphatic hydrocarbon resin

* Obtained with Differential Scanning Calorimetry measurements using the DSC 214 Polyma with a heat rate of 10 °C/min. ** Calculated with the Hansen Solubility Parameter in Practice (HSPiP) software (5th edition, version 5.4.04), except for the values of TDAE, which were obtained from [24]. In addition, the HSP values for the S-SBR/BR blend were calculated (δ_D = 18.5 MPa$^{1/2}$, δ_P = 0.9 MPa$^{1/2}$, δ_H = 1.7 MPa$^{1/2}$, δ_{tot} = 18.9 MPa$^{1/2}$).

2.2. Mixing Procedure

The Brabender Plasticorder 350S (Duisburg, Germany), an internal mixer with a chamber volume of 390 cm^3, was used to mix the rubber compounds, according to the mixing procedure displayed in Table 2. After each mixing stage, the compounds were sheeted out on a Polymix 80 T two-roll mill (Schwabenthan, Berlin, Germany).

Table 2. Mixing procedure.

Time	Action
m:s	**Stage 1:** *pre-heating 80 °C, 70 rpm, fill factor 72%*
0:00	Addition of rubber
1:00	Addition of 2/3 silica, 2/3 silane
2:30	Addition of 1/3 silica, 1/3 silane, ZnO, Stearic acid, and plasticizer
4:00	15-s ramp sweep
4:15	Increase of torque (increase temperature to 130 °C)
7:00	Stop mixing (reaching 140 °C)
m:s	**Stage 2:** *pre-heating 80 °C, 80 rpm, fill factor 69%*
0:00	Addition of elastomer masterbatch
0:50	Addition of DPG
1:00	Increase of torque (increase temperature to 130 °C)
5:00	Stop mixing (reaching 140 °C)
m:s	**Stage 3:** *pre-heating 50 °C, 50 rpm, fill factor 66%*
0:00	Addition of elastomer masterbatch, curatives (sulfur, and TBBS)
3:00	Stop mixing

2.3. Analytical Testing

2.3.1. Vulcanization

To determine the optimal vulcanization time, the Rubber Process Analyzer TA Elite (TA Instruments, New Castle, DE, USA) was employed. A frequency of 1.667 Hz and a deformation of 6.98% were applied, with the temperature maintained at 160 °C. The curing behavior was monitored for one hour. The time to achieve 90% conversion (t90) was subsequently utilized to cure 2 mm thick test samples for stress–strain and dynamic mechanical analysis measurements. For the 6 mm thick abrasion, and the 12 mm thick hardness specimens, the samples were cured for an additional 3 and 5 min respectively, to ensure that the compounds were cured entirely. The rubber samples were vulcanized at 160 °C under a pressure of 100 bar using a Wickert WLP 1600 hydraulic press (Wickert, Landau in der Pfalz, Germany).

2.3.2. Cured Payne Effect

The Rubber Process Analyzer TA Elite (TA Instruments, New Castle, DE, USA) was utilized to determine the cured Payne effect. Samples were initially cured at 160 °C within the device, according to their t90 values. Immediately following this curing process, the storage modulus (G′) was measured at 100 °C. Strain sweeps ranging from 0.1% to 100% were applied to the cured samples at a frequency of 1.667 Hz. The difference between 0.56% and 100% strain was calculated and used as the cured Payne effect.

2.3.3. Hardness

The hardness of the compounds was evaluated using a Zwick 3150 hardness tester (Zwick, Ulm, Germany) according to ISO 48-4 [25]. Measurements were performed on the Shore A scale.

2.3.4. Stress–Strain Behavior

The stress–strain behavior of the compounds was measured using the Universal Mechanical Tester Zwick Z01 (Zwick, Ulm, Germany) in accordance with ISO 37 [26] (die type 2). A crosshead speed of 500 mm/min was employed, and the tensile tests were conducted at room temperature. Five samples of each compound were tested, and the average values of these measurements were analyzed. Additionally, the median values for each compound are presented in stress–strain graphs.

2.3.5. Dynamical Properties

Dynamic mechanical analysis was conducted using the Gabo-Netzsch Eplexor tester (Netzsch, Germany). Rectangular samples with a thickness of 2 mm and a width of 5.5 mm were measured in tension mode. The analysis utilized a static strain of 0.3% and a dynamic strain of 0.1%. The test was performed over a temperature range from −80 to 80 °C with a heating rate of 2 °C/min and a dynamic load frequency of 10 Hz.

2.3.6. DIN Abrasion

The DIN abrader Montech ABR 3000 Rubber Abrasion Tester (MonTech USA GmbH, Columbia, IN, USA) was selected to analyze the abrasion resistance of the vulcanized in accordance with ISO 4649 [27] method A (non-rotating test piece configuration). The test specimens were cylindrical, with a diameter of 16 mm and a height of 6 mm. The percentage ratio of the volume loss of the test rubber to that of a standard rubber was calculated to obtain the abrasion resistance index (ARI, %). Three samples were tested for each compound, and the average volume loss from these samples was used to determine the ARI.

3. Results and Discussion

3.1. Various Oil and Resin Ratios

Compounds with different ratios between the oil and resin were evaluated to investigate how the oil and resin influence the properties of the material. As a reference, a compound with 37.5 phr TDAE oil was evaluated. The selected oil and resin were Hexamoll DINCH and Escorez 5300. Due to the cycloaliphatic nature of both substances, it was expected that they would be compatible with each other, enhancing the influence on the final properties of the rubber compound. Additional characteristics of the two substances are displayed in Table 1. Three different ratios between the Hexamoll DINCH and Escorez 5300 were evaluated, namely 20:80, 50:50, and 80:20. Table 3 shows the rubber formulations of these compounds.

Table 3. Rubber formulations for compounds with various ratios between oil and resin.

Compounds	S-SBR in phr	BR in phr	Silica in phr	TDAE in phr	Hexamoll DINCH in phr	Escorez 5300 in phr
TDAE	80	20	80	37.5	X	X
E20HE80	80	20	80	X	30	7.5
E50HE50	80	20	80	X	18.75	18.75
E80HE20	80	20	80	X	7.5	30

Furthermore, the compounds consisted of the same quantities of the remaining ingredients: 6.2 phr TESPD, 2.5 phr stearic acid, 2.5 phr zinc oxide, 1.4 phr sulfur, 2.0 phr TBBS, and 1.5 phr DPG.

In addition to the plasticizer blends, a compound with 37.5 phr of Hexamoll DINCH (H100) and a compound with 37.5 phr Escorez 5300 (E100) were mixed and evaluated. These results are shown and discussed in the evaluation of the DMA measurements.

3.1.1. Rheometer Curves and Payne Effect

Figure 1 depicts the rheometer curves of the compounds with various oil and resin ratios, and Table 4 contains additional curing characteristics like the scorch time (ts2), the optimum cure time (t90), maximum torque (MH), minimum torque (ML), and difference in torque (MH-ML).

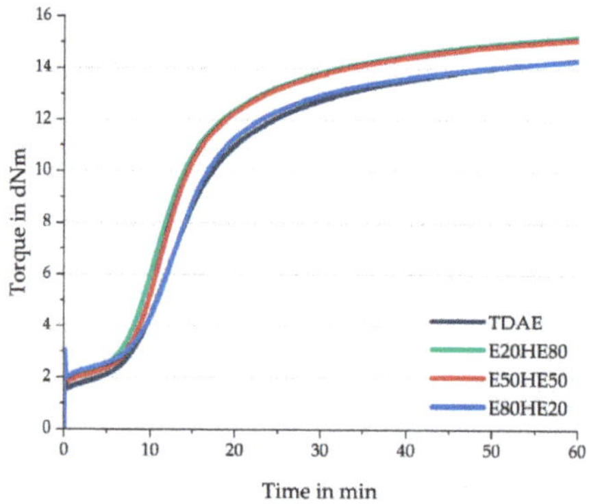

Figure 1. Rheometer curves of the compounds with various ratios between oil and resin.

Table 4. Curing characteristics and Payne effect data of the compounds with various ratios between oil and resin.

Compounds	ts2 in min	t90 in min	MH in dNm	ML in dNm	MH-ML in dNm	$G'_{0.56–100\%}$ in kPa	$G'_{100\%}$ in kPa
TDAE	8.9	33.1	14.3	1.6	12.7	719	525
E20HE80	7.9	30.6	15.2	1.9	13.3	724	566
E50HE50	8.6	30.3	15.1	1.9	13.2	738	564
E80HE20	9.5	31.5	14.3	2.0	12.3	640	610

Figure 1 and Table 4 reveal that the ML value is lowest for TDAE, while it is slightly higher for E20HE80 and E50HE50, and the highest value is obtained for E80HE20. The ML is related to the viscosity of the compound, where a higher ML indicates a higher compound viscosity [28]. Plasticizers are added to the compound to lower the viscosity of the compound and improve processing and dispersion [11]. Therefore, the high value obtained for the E80HE20 compound can be explained by the relatively high resin content, which has a higher molecular weight compared to the used oil (Table 1). In general, an increase in molecular weight is related to an increase in viscosity. Although the resin will change from a solid to a low viscous liquid at processing and curing temperatures [11], the viscosity is likely higher than the used di-ester oil at these temperatures.

A clear trend is visible between the oil and resin ratio and the scorch time. A higher concentration of oil, and a lower concentration of resin, results in a shorter scorch time. This can be explained by the concentration of ester-based oil in the compound. The ester groups might interact with the silica surface [29]. Due to these interactions, the silica surface is unable to absorb curing activators and accelerators, which become available for the vulcanization process. This could explain the faster scorch time, but also the higher maximum torque values for E20HE80 and E50HE50 because these are the two compounds with the highest concentration of ester-based oil. Usually, the maximum torque in the cure curve is correlated to polymer–polymer, polymer–filler, and filler–filler interactions [28]. As a result of more available curatives for the actual curing reaction in these compounds, more polymer–polymer interactions can be formed, causing an increase in the maximum torque.

Figure 2 shows the storage modulus of the cured samples over a range of low to high strain values. The difference between the storage modulus at low and high strains is the Payne effect, in this study indicated with $G'_{0.56–100\%}$. The Payne effect is associated with the breakdown of filler–filler interactions and a lower Payne effect is usually related to an improved micro-dispersion of the filler in the compound [30]. The $G'_{0.56–100\%}$ and the storage modulus at 100% strain ($G'_{100\%}$) are displayed in Table 4.

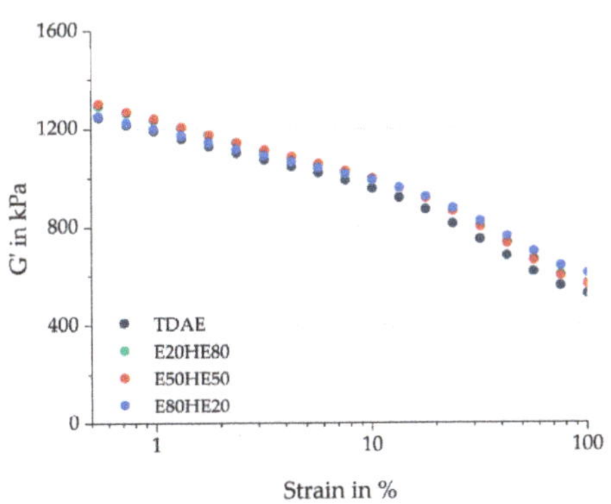

Figure 2. Cured Payne curves of the compounds with various ratios between oil and resin.

Although the figure shows similar Payne effect curves for all four compounds, when considering the values in the table, slight differences are visible. Especially E80HE20 results in different values, with a lower cured Payne effect, but a higher storage modulus at 100% strain. The lower cured Payne effect can be caused by the higher molecular weight of the resin, leading to higher shear forces during mixing, improving the silica dispersion. A similar effect was observed by Bernal-Ortega et al. [16], who used the Escorez 5300 resin in combination with TDAE oil in a silica-filled S-SBR/BR compound. The slightly higher Payne effect values for E20HE80 and E50HE50 indicate slightly more filler–filler interactions, which might also contribute to the higher maximum torque values obtained for these two compounds in the cure curves (Figure 1) [28]. However, the differences are small, and therefore it is more likely that the higher maximum torque is caused by an increase in crosslink density. Equilibrium swelling is commonly used to evaluate the crosslink density of rubber samples. The widely used procedure, as described by Bernal-Ortega et al. [31], proved unsuitable for this study. It was observed that the resin did not dissolve during the initial extraction step using acetone, while the oil was effectively removed. As a result, the measurements were significantly affected by the presence of the undissolved resin, leading to invalid outcomes. Additional trials without the extraction step were performed, but these trials also resulted in non-reproducible and invalid values.

The storage modulus at 100% strain is influenced by the polymer network, the hydrodynamic effects, and in-rubber structures [30]. In this study, the higher $G'_{100\%}$ is most likely influenced by an increase in in-rubber structures, for example, due to more polymer–filler interactions. Bernal-Ortega et al. [17] studied the use of bio-based resins in silica-filled S-SBR/BR compounds and concluded that the pure resins could act as "glue" between the fillers, making the filler–filler interactions stronger. A similar phenomenon could occur when using the E80HE20 blend. This plasticizer blend might act as "glue" between the polymer and the filler, making these interactions stronger, and increasing the $G'_{100\%}$ slightly. This effect is less pronounced for the other plasticizer blends, due to the higher oil content.

Based on all characteristics obtained from the cure curves and the cured Payne measurements (Table 4), it can be concluded that the values of the E20HE80 and E50HE50 compounds are similar to each other, while the E80HE20 blend shows deviating behavior. This could indicate that the influence of the oil is more dominant when a similar or higher amount of oil is used compared to the resin. Because Hexamoll DINCH is more polar than Escorez 5300, as can be concluded from the δ_p values in Table 1, more interactions between the oil and the other rubber ingredients are expected. When the concentration of resin becomes higher, and the concentration of oil lower, the influence of the polar oil on the material properties becomes less substantial. Nonetheless, the three plasticizer blends lead to relatively similar curing behavior and cured Payne effect as the reference compound with TDAE.

3.1.2. Stress–Strain Behavior and Tire Performance

Figure 3 shows the stress–strain behavior of the compounds. In Table 5 the stress–strain properties like the modulus at 100% strain (M100) and 300% strain (M300), the tensile strength (Ts), and the elongation at break (E_{ab}) are presented. This table also contains the hardness values of the compounds.

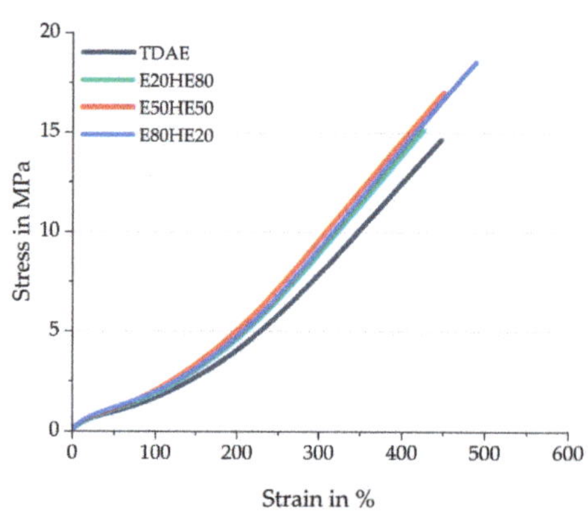

Figure 3. Stress–strain behavior of the compounds with various ratios between oil and resin.

Table 5. Stress–strain data and hardness values of the compounds with various ratios between oil and resin.

Compounds	Hardness in Shore A	M100 in MPa	M300 in MPa	E_{ab} in %	Ts in MPa
TDAE	52 ± 0	1.71 ± 0.04	7.95 ± 0.15	460 ± 30	15.2 ± 1.4
E20HE80	51 ± 0	1.82 ± 0.09	8.52 ± 0.27	440 ± 20	15.0 ± 1.0
E50HE50	54 ± 0	1.94 ± 0.05	9.18 ± 0.20	450 ± 10	16.3 ± 0.6
E80HE20	54 ± 0	1.94 ± 0.05	8.82 ± 0.18	490 ± 30	17.9 ± 1.5

An increasing value of tensile strength and elongation at break is obtained when increasing the resin content and decreasing the oil content in the compound, as shown in Figure 3. The stress–strain behavior was measured at room temperature. When mixing the pure Escorez 5300 in various ratios with the pure Hexamoll DINCH above the softening point of the resin, and cooling it down to room temperature, the mixture stays in a liquid form and no phase separation was observed. However, the ratio between oil and resin did influence the viscosity of the mixture, where a higher resin content resulted in a more viscous liquid than the mixture with a higher oil content. The higher viscosity of the E80HE20 plasticizer blend likely results in higher shear forces during mixing, enhancing the filler dispersion. Because of this, more filler–polymer interactions might be formed, increasing the strength of the compound.

All three plasticizer blends result in higher M100 and M300 values, compared to the reference compound with TDAE oil. This is presumably caused by the increased polymer–polymer and polymer–filler interactions in these three plasticizer blend compounds, which is in line with the higher $G'_{100\%}$ observed in the cured Payne effect measurements.

The hardness of the compounds as displayed in Table 5 is similar for all compounds, because the values lie within the measurement error of the device (±2 Shore A). Therefore, it is not possible to conclude that there is a significant hardness increase when using more resin in the plasticizer blend.

Based on the stress–strain behavior and hardness, the E50HE50 and E80HE20 show more similar values to each other, compared to the E20HE80 blend. This is in contrast with the cure characteristics and cured Payne effect, where the E50HE50 blend showed more similarities with E20HE80. A possible explanation is that at temperatures significantly higher than the glass transition temperature of the plasticizers, both plasticizers are liquid and behave similarly. This is evident in measurements of the cure characteristics at 160 °C and the Payne effect at 100 °C. At these high temperatures, the chemical differences between the plasticizers, such as the polarity of the di-ester oil, play a larger role than the physical state. At room temperature, the oil is still above the T_g, while the resin is below the T_g (Table 1). Therefore, the physical state of the resin differs and has a dominant influence on the final properties of the rubber compound.

Table 6 shows the normalized ARI values of the three plasticizer blends, compared to the reference with TDAE. All three plasticizer blends result in a lower resistance against wear, which has a negative influence on the durability of the tread compound. No clear trend between the oil and resin ratio and the wear resistance can be observed.

Table 6. Normalized abrasion resistance index values of the compounds with various ratios between oil and resin.

Compounds	ARI in %
TDAE	100
E20HE80	92
E50HE50	83
E80HE20	86

However, an inverse linear correlation between the ARI and the M300 is obtained for the compounds evaluated in this study, as shown in Figure 4. Although typically an increase in stiffness leads to improved wear resistance, excessive stiffness can lead to the opposite effect [32]. The material might be slightly more brittle, which makes it more prone to the formation of cracks.

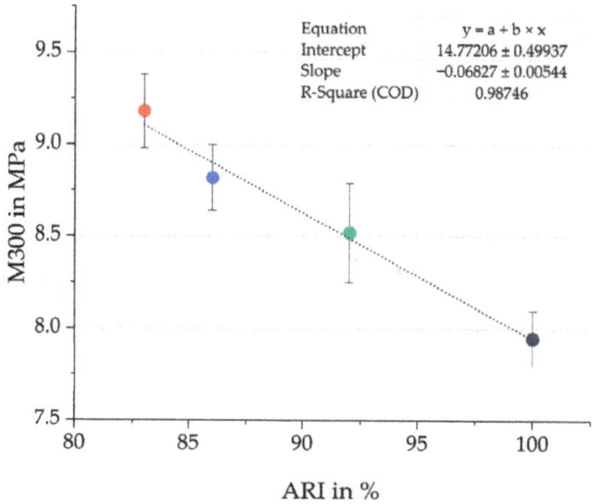

Figure 4. Correlation between wear resistance (ARI) and modulus at 300% strain (M300) of all four compounds. The colors correspond to the evaluated compounds (gray = TDAE, green = E20HE80, red = E50HE50 and blue = E80HE20).

Figure 5 depicts the tan δ curves of the compounds with various plasticizer blends. The peak position can be related to the glass transition temperature of the compound [33]. To understand the influence of the T_g of the pure oil and resin on the compound, the compounds with 37.5 phr of Hexamoll DINCH (H100), and the compound with 37.5 phr of Escorez 5300 (E100) are also included in Figure 5. Moreover, the tire tread indicators like the wet grip (tan δ at 0 °C) and rolling resistance (tan δ at 60 °C), as well as the glass transition temperature (T_g), are displayed in Table 7.

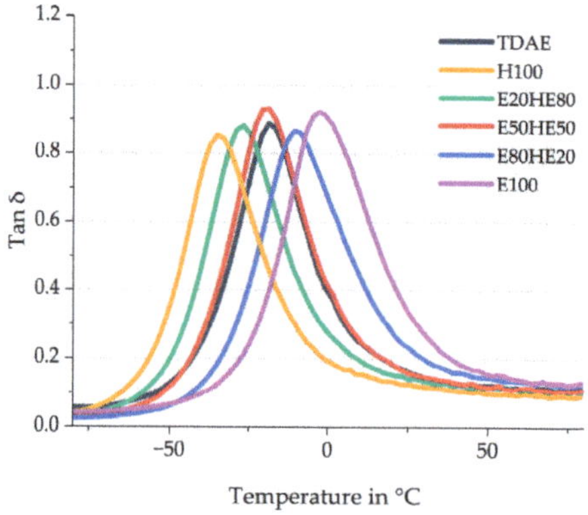

Figure 5. Tan δ curves of the compounds with various ratios between oil and resin.

Table 7. Tire properties of the compounds with various ratios between oil and resin.

Compounds	T_g in °C	tan δ at 60 °C	tan δ at 0 °C
TDAE	−19	0.117	0.406
E20HE80	−27	0.114	0.287
E50HE50	−19	0.115	0.427
E80HE20	−11	0.136	0.692

A higher content of oil results in a lower T_g value, while a higher resin concentration leads to a higher T_g. These shifts in T_g are even more pronounced for the H100 and E100 compounds. Table 1 displays the T_g of the pure plasticizers, and it can be observed that the T_g of the di-ester oil is much lower than that of the resin. Because a relatively large part of the compound consists of a plasticizer, the T_g of the pure plasticizer strongly influences the T_g of the final compound. This explains why the T_g of the compound with a high oil content is much lower than the T_g of the compounds with high resin content. Figure 6 shows the correlation graph of the T_g obtained with DMA measurements against the Escorez 5300 content.

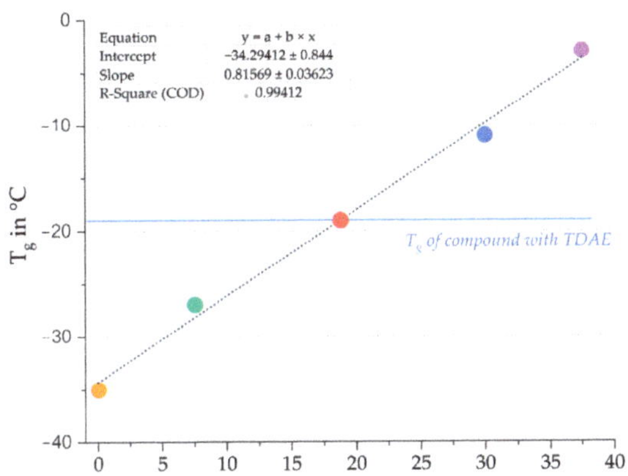

Figure 6. Glass transition temperature obtained with DMA plotted against the Escorez 5300 content. All compounds contain 37.5 phr of plasticizer (Escorez 5300 and Hexamoll DINCH) as shown in Table 3. The colors of the data points correspond to the evaluated compounds and match those used in Figure 5.

Based on Figure 6, it becomes clear that there is a linear correlation between the T_g of the compounds and the plasticizer blend ratios. Using various oil and resin blends makes it possible to modify the T_g of the vulcanizate, which is an efficient approach for tailoring the glass transition temperature according to specific customer requirements and the desired properties of the final product. Furthermore, the 50/50 blend of Hexamoll DINCH and Escorez 5300 results in a similar tan δ peak position as the reference compound with TDAE.

The wet grip and rolling resistance indicators can be obtained from the tan δ curves: the tan δ at low temperatures (0 to 20 °C) indicates the wet grip, and the tan δ at higher temperatures (40 to 80 °C) indicates the rolling resistance [34]. In this study, the tan δ at 0 °C and 60 °C are used as wet grip and rolling resistance indicators, respectively, and displayed in Table 7. A high tan δ at 0 °C and low tan δ at 60 °C are preferred. It needs to be considered that the wet grip indicator is less reliable than the rolling resistance indicator. The wet grip evaluated with Dynamic Mechanical Analysis (DMA) measurements does not always correlate with the wet grip obtained during actual breaking tests, especially when resins are introduced [35]. Due to the influence of the resin on the T_g, shifting it to temperatures closer to 0 °C, the wet grip rises as well. Because the rolling resistance indicator is measured at temperatures further away from the T_g, this indicator is observed to be less influenced by the shift in T_g and is therefore more reliable. Additional measurements are needed to get better insight into the tire properties of the compounds. However, in this study, the DMA indicators are shown to give a first indication. Of the three plasticizer blends, the E80HE20 results in the highest wet grip indicator, due to the high resin content. However, this compound also results in the highest rolling resistance indicator. The best balance in tire performance properties obtained from the tan δ curves is obtained for the E50HE50 compound.

To summarize the most important tire properties, the magic triangle of tire performance is displayed in Figure 7. In this figure, the values are normalized, and the results obtained for the reference compound with TDAE are set to 100. For all three properties, a higher value is preferred.

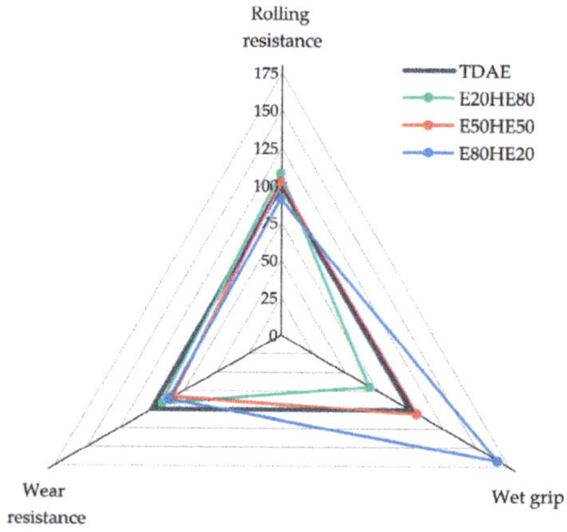

Figure 7. Magic triangle of the compounds with various oil and resin ratios.

Based on the results in Figure 7 it can be concluded that the various oil and resin blend ratios influence the wet grip to the greatest extent. Using a blend with a higher resin content, in this case 80%, results in a severe increase in the wet grip indicator. However, decreased values for the rolling resistance and wear resistance are obtained for this blend. The best balance can be observed for the 50/50 blend, which results in a slightly higher wet grip, a similar rolling resistance, and only slightly decreased wear resistance, compared to the reference compound. The blend ratio with only 20% of resin shows relatively good results for the wear resistance and rolling resistance, but a much lower value for the wet grip indicator. Based on these results and the T_g values, it can be concluded that the replacement of TDAE with 18.75 phr of Hexamoll DINCH and 18.75 phr of Escorez 5300 is the most feasible.

3.2. Plasticizer Blends with Resin and Different Di-Ester Oils

To further evaluate the plasticizer blends, three different di-ester oils were selected to blend with Escorez 5300. All oils are di-ester oils, with different chemical structures between the two ester groups. These structures and additional characteristics are displayed in Table 1. The oils were blended with the resin in a 50:50 ratio, because this was the most successful plasticizer blend ratio based on the tire properties, and this ratio resulted in a similar glass transition temperature as the reference compound with TDAE. In addition to the more conventional concentration of 37.5 phr plasticizer, the compounds were mixed with high plasticizer concentrations of 80 phr (40 phr of oil, and 40 phr of resin). This way, the influence of the used plasticizer blends on the compound properties was assumed to become more visible. The rubber formulations of both the conventional concentration plasticizer and the high concentration are presented in Table 8. In the figures and tables presented in this chapter, the high-plasticizer compounds (80 phr) are indicated with an (H).

Table 8. Rubber formulations of the compounds with resin and different di-ester oils.

Compounds	S-SBR in phr	BR in phr	Silica in phr	TDAE in phr	Hexamoll DINCH in phr	Palatinol 10P in phr	Plastomoll DOA in phr	Escorez 5300 in phr
TDAE	80	20	80	37.5	X	X	X	X
E50HE50	80	20	80	X	18.75	X	X	18.75
E50PA50	80	20	80	X	X	18.75	X	18.75
E50PL50	80	20	80	X	X	X	18.75	18.75
TDAE (H)	80	20	80	80	X	X	X	X
E50HE50 (H)	80	20	80	X	40	X	X	40
E50PA50 (H)	80	20	80	X	X	40	X	40
E50PL50 (H)	80	20	80	X	X	X	40	40

Furthermore, the compounds consisted of the same quantities of the remaining ingredients: 6.2 phr TESPD, 2.5 phr stearic acid, 2.5 phr zinc oxide, 1.4 phr sulfur, 2.0 phr TBBS, and 1.5 phr DPG.

For the high-concentration plasticizer compounds, the resin was added in three separate portions to the compound. The portions were added in the first stage after 1 min, 2.5 min, and 4 min of mixing. Furthermore, the mixing procedure was kept the same as shown in Table 2.

3.2.1. Rheometer Curves and Payne Effect

In Figure 8 the rheometer curves of the compounds are shown. The solid lines indicate the conventional plasticizer concentrations, while the high-concentration plasticizer compounds are indicated with the dotted lines. Table 9 displays the additional curing characteristics.

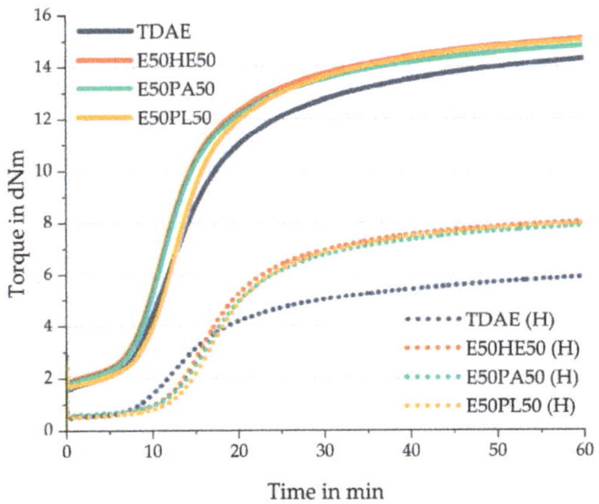

Figure 8. Rheometer curves of the compounds with resin and different di-ester oils. Compounds with high concentrations of plasticizer are indicated with (H).

Table 9. Curing characteristics and Payne effect data of the compounds with resins and different di-ester oils. Compounds with high concentrations of plasticizer are indicated with (H).

Compounds	ts2 in min	t90 in min	MH in dNm	ML in dNm	MH-ML in dNm	$G'_{0.56–100\%}$ in kPa	$G'_{100\%}$ in kPa
TDAE	8.9	33.1	14.3	1.6	12.7	719	525
E50HE50	8.6	30.3	15.1	1.9	13.2	738	564
E50PA50	8.6	29.6	14.8	1.8	13.0	723	579
E50PL50	9.7	30.8	15.0	1.7	13.4	766	567
TDAE (H)	13.0	37.6	5.9	0.5	5.4	440	201
E50HE50 (H)	14.7	35.1	8.1	0.5	7.5	526	306
E50PA50 (H)	15.2	35.6	7.9	0.6	7.3	522	282
E50PL50 (H)	15.7	35.2	8.0	0.5	7.5	489	280

All plasticizer blends show similar cure curves and curing characteristics compared to each other. When focusing on the compounds with 37.5 phr, the scorch time of TDAE is similar to the scorch times observed for the compounds with plasticizer blends. When comparing this characteristic in the high-concentration plasticizer compounds, a shorter scorch time is obtained for TDAE (H) than for the blends. According to the characteristics of TDAE as described by Rathi [12], this oil contains a small content of 0.8 wt% of sulfur. At lower concentrations, this sulfur in TDAE does not have a significant effect on the curing reaction. However, when higher concentrations of TDAE oil are used, relatively more sulfur is added to the compound, increasing the scorch time slightly.

Compared to the reference compound with TDAE, the plasticizer blends all show higher maximum torque values. This could be related to a higher crosslink density for these compounds. When evaluating the 80 phr plasticizer compounds, the difference between the maximum torque of the 50/50 blends and the reference is even more severe. This could be due to two influences. On the one hand, it might be that the plasticizer blends enhance the curing reaction. The ester groups in the plasticizer blends could interact with the silica surface, shielding this filler, which might result in more curing ingredients available in the compound. However, it would be expected that this would also result in a shorter scorch time, which is not observed for the high-concentration plasticizer blend compounds. On the other hand, TDAE could interfere with the curing process, for example by reacting with the sulfur and therefore hindering the vulcanization process. A reason for this can be that TDAE might contain some unsaturation, which can interact with the sulfur, leading to a lower crosslink density [13].

In general, much lower torque values are obtained for the high plasticizer compounds than for the compound with 37.5 phr plasticizer. When increasing the plasticizer concentration without changing the quantities of the other ingredients there are relatively fewer polymer, filler, and curing chemicals available in the compound. This influences the rubber network structure of the material, for example, fewer crosslinks can be formed.

The cured Payne effect curves are depicted in Figure 9. In this figure, the darker-colored dots are related to the compounds with conventional concentrations of plasticizer. The hollow hexagon-shaped markings indicate the results of the high-plasticizer compounds. The cured Payne effect values and storage moduli at 100% strain can be seen in Table 9.

The cured Payne effect of the compounds with 37.5 phr of plasticizers are all similar. A slight shift to higher G′ values compared to TDAE can be observed which is likely due to a slightly higher crosslink density. These results are in agreement with the maximum torque values obtained from the rheometer curves (Figure 8).

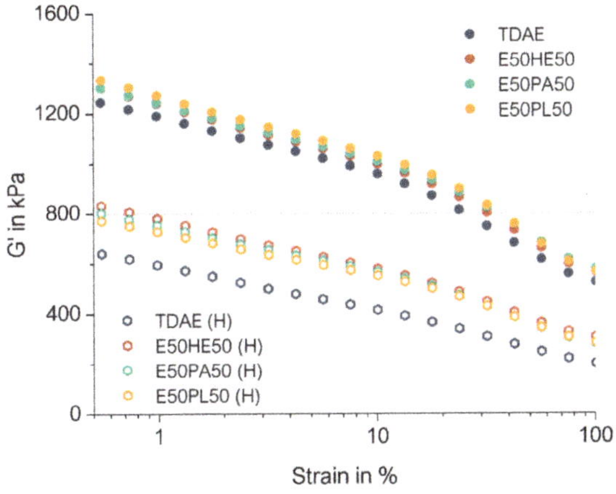

Figure 9. Cured Payne curves of the compounds with resin and different di-ester oils. Compounds with high concentrations of plasticizer are indicated with (H).

Comparing the cured Payne effect of the 37.5 phr plasticizer compounds with the 80 phr plasticizer compound, it becomes clear that much lower values are obtained for the high-concentration plasticizer compounds. This can be explained by the fact that the compounds contain relatively much more plasticizer, and therefore less filler and polymer. This results in less filler–filler, filler–polymer, and polymer–polymer interactions, resulting in lower storage modulus values, as well as lower cured Payne effect.

The compounds with a high content of plasticizer blends show a clear shift in the G′ throughout the measurement, compared to TDAE (H). As explained previously, TDAE likely interferes with the curing reaction by interacting with the curing ingredients. Therefore, a lower crosslink density is obtained, leading to lower G′ values during the applied strain sweep. This effect is more pronounced when high concentrations of the TDAE are used.

3.2.2. Stress–Strain Behavior and Tire Performance

The stress–strain behavior of the compounds is shown in Figure 10. The solid lines indicate the compounds with 37.5 phr of plasticizer, while the dotted lines indicate the high-plasticizer compounds. Table 10 contains additional stress–strain properties and the hardness of the compounds.

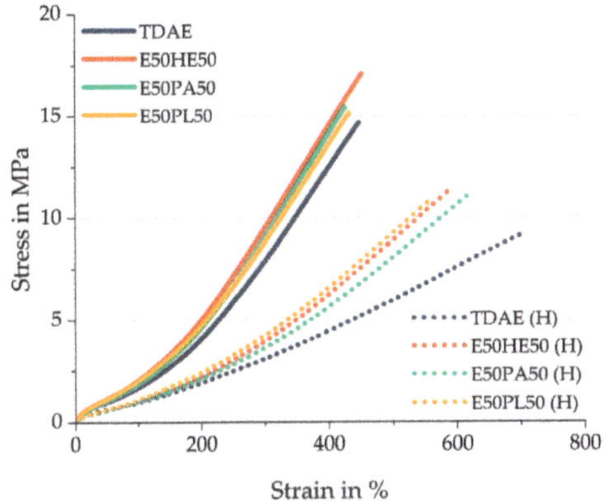

Figure 10. Stress–strain behavior of the compounds with resin and different di-ester oils. Compounds with high concentrations of plasticizer are indicated with (H).

Table 10. Stress–strain data and hardness values of the compounds with resin and different di-ester oils Compounds with high concentrations of plasticizer are indicated with (H).

Compounds	Hardness in Shore A	M100 in MPa	M300 in MPa	E_{ab} in %	Ts in MPa
TDAE	52 ± 0	1.71 ± 0.04	7.95 ± 0.15	460 ± 30	15.2 ± 1.4
E50HE50	54 ± 0	1.94 ± 0.05	9.18 ± 0.20	450 ± 10	16.3 ± 0.6
E50PA50	54 ± 0	1.94 ± 0.07	9.20 ± 0.09	430 ± 20	15.4 ± 0.9
E50PL50	54 ± 0	1.93 ± 0.05	8.85 ± 0.22	430 ± 10	14.9 ± 0.5
TDAE (H)	33 ± 0	0.93 ± 0.05	3.00 ± 0.12	700 ± 20	9.0 ± 0.3
E50HE50 (H)	39 ± 1	1.04 ± 0.05	4.05 ± 0.10	580 ± 20	11.1 ± 0.4
E50PA50 (H)	40 ± 0	1.03 ± 0.05	3.75 ± 0.19	620 ± 30	10.9 ± 0.7
E50PL50 (H)	39 ± 0	1.05 ± 0.04	4.15 ± 0.09	550 ± 30	10.6 ± 0.7

The 37.5 phr plasticizer compounds with 50/50 oil and resin blends show all three slightly higher modulus values compared to the reference compound with TDAE oil. These results are caused by the possible higher crosslink density and higher polymer–filler interactions, which is consistent with the observed findings from the curing behavior and cured Payne effect.

The E50HE50 shows the most distinct behavior compared to the two other blends, with a higher tensile strength and elongation at break. Because the results of the rheometer curves and the cured Payne effect are similar, it is not likely that the E50HE50 results in significant differences in network structure. The higher tensile strength and elongation at break might be due to an improved compatibility between the oil and the resin. Comparing the Hansen Solubility Parameter values in Table 1 of the resin with the three di-ester oils, it can be concluded that the Hexamoll DINCH results in a more similar δd and δh in comparison to Escorez 5300.

Comparing the M100 and M300 values of the high-concentration plasticizer compounds, the difference between the plasticizer blends and TDAE becomes more pronounced. This further illustrates that the plasticizer blends enhance the reinforcement of the vulcanizates. As previously mentioned, this can be explained by the possible higher crosslink density and the higher polymer–filler interactions.

The low tensile strength and high elongation at break obtained for TDAE (H) can be explained by the lower crosslink density of this sample, as explained in the previous subchapter. These stress–strain results are in line with the rheometer curves and cured Payne effect obtained for this compound.

Figure 11 shows the tan δ curves of the analyzed 50/50 blends, compared to the reference with TDAE. The dotted lines show the high-concentration plasticizer compounds. The values of the tire performance properties as well as the T_g are shown in Table 11.

Table 11. Tire properties of the compounds with resin and different di-ester oils. Compounds with high concentrations of plasticizer are indicated with (H).

Compounds	T_g in °C	tan δ at 60 °C	tan δ at 0 °C
TDAE	−19	0.117	0.406
E50HE50	−19	0.115	0.427
E50PA50	−19	0.120	0.453
E50PL50	−25	0.111	0.309
TDAE (H)	−21	0.151	0.377
E50HE50 (H)	−21	0.138	0.463
E50PA50 (H)	−19	0.139	0.571
E50PL50 (H)	−31	0.130	0.285

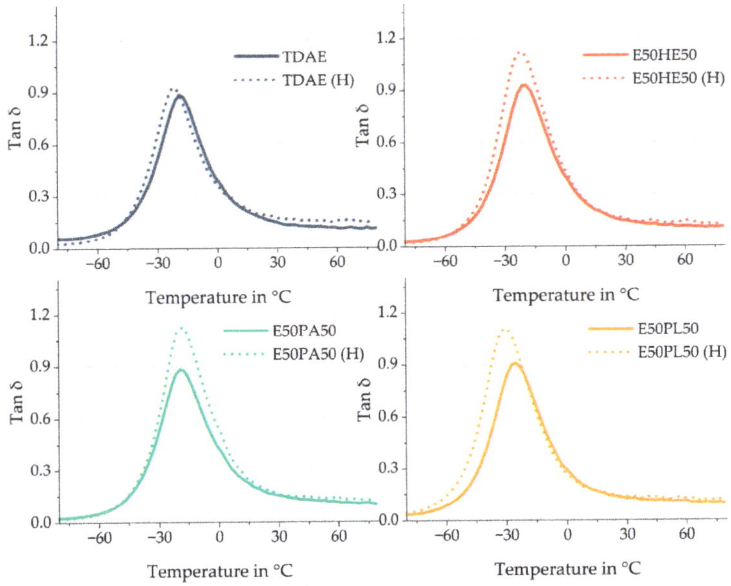

Figure 11. Tan δ curves of the compounds with resin and different di-ester oils. Compounds with high concentrations of plasticizer are indicated with (H).

One of the main differences that can be observed in Figure 11 is the height of the tan δ. For the high-concentration plasticizer blends, the peak is higher with values of approximately 1.1, compared to the compounds with the conventional amount of plasticizer which show tan δ values of approx. 0.9. The tan δ is the ratio between the loss modulus E'' and the storage modulus E'. A higher tan δ often indicates higher dissipation behavior, while a lower tan δ is related to a higher elastic behavior [33]. Increasing the plasticizer results in a larger viscous and dissipating part, increasing the loss modulus.

The T_g is also influenced by the higher concentration of plasticizer in the compounds with the E50PA50 and E50PL50 blends. This effect is most pronounced for E50PL50, which can be explained by the lower T_g of the pure Plastomoll DOA di-ester oil (Table 1). Because this oil inhibits a lower T_g, and the amount of oil is increased, the influence of the oil on the compound T_g becomes larger. For the E50PA50 a similar effect is obtained. However, the higher T_g of the pure Palationol 10P results in a slightly higher T_g of the compound compared to the TDAE (H) compound. E50HE50 shows the same T_g as TDAE when using 37.5 phr of plasticizer and when using 80 phr. This indicates that the combination of 50% Hexamoll DINCH and 50% Escorez 5300 results in the most similar T_g as TDAE.

Table 12 includes the normalized ARI values of the compounds with 37.5 phr of plasticizer. Based on these results, it becomes clear that the best wear resistance is obtained for the compound with the blend between Escorez 5300 and Plastomoll DOA. Veith [36] observed a linear correlation between the glass transition temperature and the wear resistance of SBR/BR rubber blends. A lower T_g of the compound resulted in better wear resistance. Therefore, it is assumed that the improved wear resistance of the E50PL50 compound compared to the other two plasticizer blends is due to the lower T_g of this rubber material. The plasticizer blends with a more similar T_g to the reference compound, namely the E50HE50 and E50PA50, show worse wear resistance. Full replacement of the TDAE oil with Escorez 5300 resin in the study of Bernal-Ortega et al. [16] showed an improved abrasion resistance index. This is explained by the possible better filler dispersion due to the higher shear forces obtained when mixing with resin. In this study, the cured Payne effect of all

50/50 plasticizer blends is slightly higher, which might indicate a slight deterioration in the filler dispersion compared to TDAE. The worse filler dispersion could be a reason for the lower wear resistance. Moreover, the 50/50 plasticizer blends show higher moduli in the stress–strain curves and higher hardness values compared to TDAE. This slight increase in stiffness and hardness could be another cause for the decrease in wear resistance.

Table 12. Normalized abrasion resistance index of the compounds with resin and different di-ester oils.

Compounds	ARI in %
TDAE	100
E50HE50	83
E50PA50	83
E50PL50	95

The magic triangle of tire performance is displayed in Figure 12. The rolling resistance, wet grip, and wear resistance are normalized based on the reference compound with TDAE. In the figure, a higher value indicates an improved performance. Only the compounds with 37.5 phr of plasticizer are displayed in this figure because the high-concentration plasticizer compounds are not feasible for tire tread applications due to their poor mechanical properties.

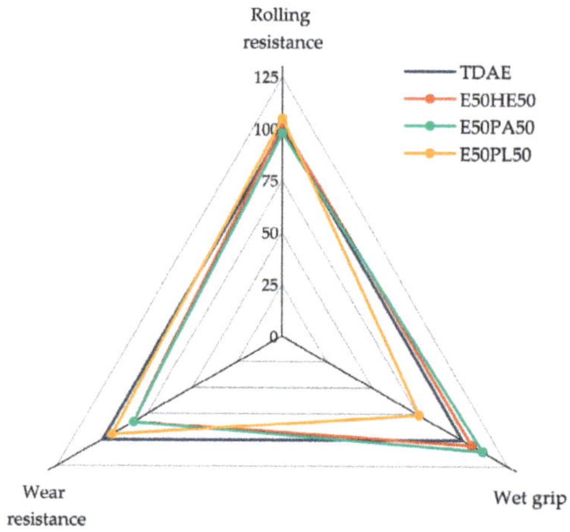

Figure 12. Magic triangle of compounds with resin and different di-ester oils.

Based on the magic triangle, it can be concluded that the E50PL50 compound shows the best rolling resistance and wear resistance, but the worst wet grip. The other two blends (E50HE50 and E50PA50) result in similar rolling resistance values as TDAE, slightly increased wet grip indicators, but lower wear resistance values. However, the differences are relatively small, and in general, the 50/50 blends result in similar values as the reference with TDAE, which is promising.

The differences between the chemical structures in the selected di-ester oils do not result in a severe influence on the final properties of the rubber compounds. It is more likely that the physical characteristics of the oils, like molecular weight and glass transition temperature, influence the slight differences obtained in this study. High-concentration plasticizer compounds show that the plasticizer blends possibly enhance the crosslink formation of the compounds, and result in improved reinforcement, compared to the compound with TDAE.

4. Conclusions

When using bio-based oils in rubber compounds, e.g., vegetable oils, the glass transition temperature is observed to shift towards lower temperatures. The opposite occurs when using resins. These shifts in T_g can influence the performance of the final rubber vulcanizate, which is not always advantageous. To find a plasticizer that is more sustainable, and results in a similar T_g to a compound with commonly used petroleum-based plasticizers, it was proposed to use an oil and resin plasticizer blend.

In the first part of this study, three different oil and resin blend ratios were used in a silica-filled S-SBR/BR compound, and compared to a reference compound with TDAE. The di-ester oil Hexamoll DINCH and resin Escorez 5300 were selected due to their cycloaliphatic nature. Various properties were evaluated, among which the curing characteristics, cured Payne effect, stress–strain behavior, hardness, viscoelastic behavior, and wear resistance. Most measurements showed relatively similar results for the compounds with plasticizer blends based on Hexamoll DINCH and Escorez 5300 compared to the compound with TDAE. However, a clear influence of the oil and resin blend ratio on the viscoelastic behavior of the material was observed. A correlation plot between the T_g of the compound and the Escorez 5300 concentration revealed a linear correlation. This suggests that the T_g of the compound can be easily modified using various plasticizer blend ratios, without influencing the curing behavior or strength of the material. This approach can be used to tailor the T_g of the rubber compound based on the application of the material. However, the various plasticizer blends do influence the laboratory predictors for the wet grip and rolling resistance. The best balance in properties was obtained for the 50/50 blend.

In the second part of this work, various di-ester oils were selected and used in 50/50 plasticizer blends with Escorez 5300. Compounds with more conventional concentrations of plasticizer (37.5 phr) as well as compounds with high plasticizer content (80 phr) were mixed. All three plasticizer blends showed similar results, except for some slight differences in the E50PL50 compounds. This could be explained by the lower molecular weight and lower T_g of the pure di-ester oil. The differences between the plasticizer blends and TDAE in the high-concentration plasticizer compounds were more severe. These compounds showed even more clearly that all three blends enhance the crosslinking formation and result in an improved reinforcement compared to the compound with TDAE. Among the three blends, the E50HE50 blend showed the most similar T_g in both plasticizer concentrations.

Because Hexamoll DINCH can be created via the BioMass Balance approach, the E50HE50 blend is 50% bio-based. This study demonstrates that using the E50HE50 blend as a replacement for TDAE results in similar performance while increasing the sustainability of the compound.

This study primarily focused on evaluating the performance characteristics of the final rubber compound. However, it is recognized that replacement of the plasticizer might influence the processability of the compound. Therefore, it is recommended that future research contains a more detailed analysis of processing properties, for example, the Mooney viscosity.

Author Contributions: Conceptualization, F.v.E., F.G., C.A., M.d.C., A.T. and A.B.; Methodology, F.v.E., F.G., C.A., M.d.C., A.T. and A.B.; Validation, F.v.E.; Formal analysis, F.v.E.; Investigation, F.v.E.; Resources, F.v.E., F.G., C.A., M.d.C. and A.B.; Data curation, F.v.E., F.G. and A.B.; Writing—original draft, F.v.E.; Writing—review & editing, F.G., C.A., M.d.C. and P.B.-O.; Visualization, F.v.E.; Supervision, F.G., C.A., M.d.C., A.T. and A.B.; Project administration, F.v.E., F.G. and A.B. All authors have read and agreed to the published version of the manuscript.

Funding: This research received no external funding.

Institutional Review Board Statement: Not applicable.

Data Availability Statement: The raw data supporting the conclusions of this article will be made available by the authors on request.

Acknowledgments: The authors express their gratitude to Bridgestone EU NV/SA for both scientific and financial support. They also thank BASF SE for supplying the di-ester oils and ExxonMobil Petroleum & Chemical BV for providing the resin used in this study.

Conflicts of Interest: Authors Claudia Aurisicchio and Micol di Consiglio were employed by the company Bridgestone EU NV/SA. The remaining authors declare that the research was conducted in the absence of any commercial or financial relationships that could be construed as a potential conflict of interest.

References

1. Top 10 Companies in Sustainable Tire Market in 2024 Shaping Global Industry Trends. Available online: https://www.emergenresearch.com/blog/top-10-companies-in-sustainable-tire-market?srsltid=AfmBOoqLPuRRTjrjbIVCTvBHVXWQfEaw6UA07K39yMqVMwREIQj3ztL_ (accessed on 25 June 2025).
2. Bavkar, V. Sustainable Tire Materials Market: Opportunities, Trends, and Industry Growth. Available online: https://www.marketsandmarkets.com/blog/CM/Sustainable-Tire-Materials-Market-Opportunities-Trends-Industry-Growth (accessed on 5 June 2025).
3. Dhamija, D. The Road to Sustainability: Exploring the Green Tires Market. Available online: https://blog.bccresearch.com/the-road-to-sustainability-exploring-the-green-tires-market (accessed on 5 June 2025).
4. Sarkar, P.; Bhowmick, A.K. Sustainable Rubbers and Rubber Additives. *J. Appl. Polym. Sci.* **2018**, *135*, 45701. [CrossRef]
5. Bardha, A.; Prasher, S.; Dumont, M.J. Waste Biomass-Derived Rubber Composite Additives: Review of Current Research and Future Investigations Into Biowaste Tire Formulation. *Biomass Bioenergy* **2024**, *183*, 107149. [CrossRef]
6. Chang, B.P.; Gupta, A.; Muthuraj, R.; Mekonnen, T.H. Bioresourced Fillers for Rubber Composite Sustainability: Current Development and Future Opportunities. *Green Chem.* **2021**, *23*, 5337–5378. [CrossRef]
7. Boonmahitthisud, A.; Boonkerd, K. Sustainable Development of Natural Rubber and Its Environmentally Friendly Composites. *Curr. Opin. Green Sustain.* **2021**, *28*, 100446. [CrossRef]
8. Mohamed, N.R.; Othman, N.; Shuib, R.K.; Hayeemasae, N. Perspective on Opportunities of Bio-Based Processing Oil to Rubber Industry: A Short Review. *Iran. Polym. J.* **2023**, *32*, 1455–1475. [CrossRef]
9. Tang, S.; Li, J.; Wang, R.; Zhang, J.; Lu, Y.; Hu, G.; Wang, Z.; Zhang, L. Current Trends in Bio-Based Elastomer Materials. *SusMat* **2022**, *2*, 2–33. [CrossRef]
10. Dadkhah, M.; Messori, M. A Comprehensive Overview of Conventional and Bio-Based Fillers for Rubber Formulations Sustainability. *Mater. Today Sustain.* **2024**, *27*, 100886. [CrossRef]
11. Moneypenny, H.G.; Menting, K.-H.; Gragg, F.M. General Compounding. In *Rubber Compounding: Chemistry and Applications*; Rogers, B., Ed.; CRC Press: Boca Raton, FL, USA, 2016; pp. 333–378, ISBN 9781482235500.
12. Rathi, A. Investigating Safe Mineral-Based and Bio-Based Process Oils for Tire Tread Application. Ph.D. Thesis, University of Twente, Enschede, The Netherlands, 2019.
13. van Elburg, F.; Grunert, F.; Aurisicchio, C.; di Consiglio, M.; di Ronza, R.; Talma, A.; Bernal-Ortega, P.; Blume, A. Exploring the Impact of Bio-Based Plasticizers on the Curing Behavior and Material Properties of a Simplified Tire-Tread Compound. *Polymers* **2024**, *16*, 1880. [CrossRef] [PubMed]
14. IndustryARC. Plasticizers & Tackifiers in Tyres Market by Type 2027. Available online: https://www.industryarc.com/Report/19142/plasticizers-tackifiers-in-tyres-market (accessed on 4 September 2024).
15. Duddey, J.E. Resins. In *Rubber Compounding: Chemistry and Applications*; Rodgers, B., Ed.; CRC Press: Boca Raton, FL, USA, 2016; pp. 379–418, ISBN 9781482235500.
16. Bernal-Ortega, P.; Gaillard, E.; van Elburg, F.; Blume, A. Use of Hydrocarbon Resins as an Alternative to TDAE Oil in Tire Tread Compounds. *Polym. Test.* **2023**, *126*, 108168. [CrossRef]
17. Bernal-Ortega, P.; van Elburg, F.; Araujo-Morera, J.; Gojzewski, H.; Blume, A. Heading Towards a Fully Sustainable Tire Tread Compound: Use of Bio-Based Resins. *Polym. Test.* **2024**, *133*, 108406. [CrossRef]
18. Lopitaux, G. Rubber Tyre Composition Containing Diester Plasticizer. U.S. Patent US8637597 B2, 28 January 2014.
19. Sandstrom, P.H.; Rodewald, S.; Ramanathan, A. Tire with Rubber Tread Containing Combination of Resin and Vegetable Oil, Particularly Soybean Oil. U.S. Patent US2014/0135437 A1, 15 May 2014.
20. BASF. Plastomoll® DOA BMB. Available online: https://products.basf.com/global/en/cp/plastomoll-doa-bmb (accessed on 19 September 2024).

21. BASF. Palatinol® 10P BMB. Available online: https://products.basf.com/global/en/cp/palatinol-10-p-bmb (accessed on 19 September 2024).
22. BASF. Hexamoll® DINCH BMB. Available online: https://products.basf.com/global/en/cp/hexamoll-dinch-bmb (accessed on 19 September 2024).
23. BASF's New Plastic Additives Reduces CO_2 Emissions by 60%. Available online: https://carboncredits.com/basfs-new-plastic-additives-reduces-co2-footprint-by-60/ (accessed on 17 March 2025).
24. Farshchi, N.; Abbasian, A. Inverse Gas Chromatography Study of Hansen Solubility Parameters of Rubber Process Oils (DAE, TDAE, MES, and NAP). *Rubber Chem. Technol.* **2020**, *93*, 297. [CrossRef]
25. *ISO 48-4:2018*; Rubber, Vulcanized or Thermoplastic—Determination of Hardness. ISO: Geneva, Switzerland, 2018.
26. *ISO 37:2017*; Rubber, Vulcanized or Thermoplastic—Determination of Tensile Stress-Strain Properties. ISO: Geneva, Switzerland, 2017.
27. *ISO 4649:2024*; Rubber, Vulcanized or Thermoplastic—Determination of Abrasion Resistance Using a Rotating Cylindrical Drum Device. ISO: Geneva, Switzerland, 2024.
28. Aggarwal, A.; Grunert, F.; Ilisch, S.; Stratton, T.; Blume, A. Investigation of Different Ingredients Contributing to the Cure Torque in Silica-Filled SSBR Compounds. In Proceedings of the IRC RubberCon, Edinburgh, UK, 9 May 2023.
29. Siwarote, B.; Sae-Oui, P.; Wirasate, S.; Suchiva, K. Effects of Bio-Based Oils on Processing Properties and Cure Characteristics of Silica-Filled Natural Rubber Compounds. *J. Rubber Res.* **2017**, *20*, 1–19. [CrossRef]
30. Fröhlich, J.; Niedermeier, W.; Luginsland, H.D. The Effect of Filler-Filler and Filler-Elastomer Interaction on Rubber Reinforcement. *Compos. Part A Appl. Sci. Manuf.* **2005**, *36*, 449. [CrossRef]
31. Bernal-Ortega, P.; Anyszka, R.; Morishita, Y.; di Ronza, R.; Blume, A. Determination of the Crosslink Density of Silica-Filled Styrene Butadiene Rubber Compounds by Different Analytical Methods. *Polym. Bull.* **2024**, *81*, 995–1018. [CrossRef]
32. MonTech USA. Key Factors That Affect DIN Abrasion Testing. Available online: https://www.montechusa.com/key-factors-din-abrasion (accessed on 28 October 2024).
33. Whitcomb, K. Measurement of Glass Transition Temperatures by Dynamic Mechanical Analysis and Rheology. Available online: https://www.tainstruments.com/pdf/literature/RH100.pdf (accessed on 17 March 2025).
34. Colvin, H. General-Purpose Elastomers. In *Rubber Compounding: Chemistry and Applications*; Rodgers, B., Ed.; CRC Press: Boca Raton, FL, USA, 2016; pp. 34–76, ISBN 9781482235500.
35. Babazadeh, K.; Blume, A.; Zeeman, R.; Bandzierz, K. Optimized Laboratory Prediction of Wet Grip Performance of Tires. In Proceedings of the Tire Technology Expo, Hannover, Germany, 23 March 2023.
36. Veith, A.G. Tire Treadwear: The Joint Influence of Tg, Tread Composition and Environmental Factors. A Proposed "Two-Mechanism" Theory of Treadwear. *Polym. Test.* **1987**, *7*, 177. [CrossRef]

polymers

MDPI

Article

Bouncy Idea or Solid Practice? Exploring Industry Barriers in the Incorporation of Devulcanized Rubber into Compounds for Rubber Products

Eric Roetman [1,2,*], Jelle Joustra [1], Geert Heideman [2] and Ruud Balkenende [1]

[1] Faculty of Industrial Design Engineering, Delft University of Technology, 2628 CE Delft, The Netherlands; j.j.joustra@tudelft.nl (J.J.); a.r.balkenende@tudelft.nl (R.B.)

[2] Polymer Engineering, University of Applied Sciences Windesheim, 8017 CA Zwolle, The Netherlands; g.heideman@windesheim.nl

* Correspondence: e.c.roetman@tudelft.nl

Abstract

Devulcanization has the potential to help meet circular economy goals by recovering end-of-life rubber. However, the adoption of devulcanized rubber by manufacturers remains low at the industry level. Devulcanization value chains are complex and involve multiple stakeholders, including waste collectors, sorters, recyclers, compounders, manufacturers and regulatory bodies. This study investigated the barriers compounders and manufacturers face when incorporating devulcanized rubber into new compounds and identified primary underlying causes. The research was conducted through in-depth interviews with compounders and manufacturers of tires and general rubber goods, focusing on the technical, market, institutional, and cultural factors related to incorporating recycled materials, specifically devulcanized rubber. From the results, we identified a number of barriers faced by the industry. A key barrier was the heterogeneity of devulcanized rubber, which made it more difficult to add to new rubber compounds with consistent quality. Other barriers included a lack of standardization and coordination, along with misaligned regulations that hamper the market adoption of devulcanized rubber. This implies that increasing the uptake of devulcanized rubber at the industry level will not be achieved through technological advancements alone or isolated market interventions; instead, it requires comprehensive, systemic solutions.

Keywords: circular economy; rubber; recycling; devulcanization; barriers

1. Introduction

Although devulcanization techniques for rubber have been claimed for decades [1], recent years have seen significant technological advancements, making it increasingly suitable for rubber recycling and raising the potential to meet the goals of the circular economy [2,3]. Despite these technological advancements and considerable investments, the market uptake of devulcanized rubber remains low compared to other rubber recycling routes [2,4].

Devulcanization is an interesting technology that can be used to meet circular economy goals. During rubber goods production, rubber is made firm, elastic, and durable by vulcanization. In that process, sulphur cross-links are created on a molecular level between the rubber polymer main chains. Vulcanization enhances rubber's durability and heat resistance but creates significant recycling challenges. Devulcanization is a group of recycling technologies with the objective to reverse vulcanization by selectively breaking the sulphur cross-links. Devulcanization processes aim to recycle rubber while retaining most of its original properties [3,5,6]. Devulcanized rubber can be added to new rubber during the compounding stage when raw rubbers are compounded with various additives and fillers to achieve specific properties [6–8]. A rough estimation of the total current European production capacity for devulcanized rubber is 100 kton/year (In Roetman et al. [4], 17 companies were identified in Europe that devulcanized rubber. The interviewed companies had an average capacity of 1.45 ton/h. The assumption is that these companies can run 4000 h on a yearly basis, leading to a total annual capacity of 100 kton of devulcanized rubber) [4].

Devulcanization is a unique approach to rubber recycling because it specifically aims to preserve the original properties of the recycled rubber [2,9]. For comparison, an overview of the main recycling methods is presented in Appendix A. Devulcanization is unique because of its aim to preserve the material properties, as opposed to other methods that often lead to degradation or deterioration of the rubber. Devulcanization is important because of the significance of rubber for the circular economy. Synthetic rubber is primarily made from fossil fuels through energy-intensive methods [10,11]. Natural rubber is a critical and potentially scarce raw material and its regeneration presents challenges [12]. Thus, devulcanization contributes to a circular economy by offering a solution by preserving both material groups. Further, it fits the circular economy goal of maintaining performance and value as much as possible [13].

Even though devulcanized rubber can be used to replace primary, virgin rubber, its composition is different. Devulcanized rubber contains raw, natural or synthetic rubber as well as the additives of the original compounds [2]. These additives include processing aids, vulcanization agents and reinforcing fillers [14,15]. Additionally, the performance and quality of devulcanized rubber depends on the devulcanization technology, as well as the properties, quality, and consistency of the rubber waste collected [5,6,16].

Despite its potential to at least partially replace virgin rubber compounds and thus reduce the carbon footprint [2], there is a low uptake of this material by manufacturers and compounders who could potentially add devulcanized rubber to their compounds. Their perspective is important to investigate, as a study on companies that devulcanize rubber in Europe identified the low uptake of devulcanized rubber as a significant barrier for their business case [4].

Rubber compounders and manufacturers perform a crucial role in the rubber value chain by developing formulations and blending rubbers with additives to meet specific requirements [15]. Rubber compounders are specialized companies that develop rubber formulations and blend rubbers with additives to achieve the desired properties. While some rubber manufacturers have their own compounding divisions, their primary goal is production of goods containing rubber. We distinguish two types of manufacturers: tire manufacturers and manufacturers of rubber goods. Tire manufacturers consume the majority of the global rubber supply. Table 1 shows the consumption of different types of rubber on a global level and Table 2 compares the consumption of general rubber goods and tire production in Europe. Table 3 presents import and export data for rubber products within Europe, highlighting the significance of rubber production in the region.

General rubber goods manufacturers produce finished or semi-finished rubber products, such as conveyor belts, shoe soles, waterproof materials, electrical insulation, adhesives, O-rings, seals, hoses and belts, and industrial gloves for various industries [15,17]. In these goods general-purpose rubbers, such as SBR and natural rubber, as well as specialty rubbers are used that meet specific industry requirements (e.g., fluorocarbon (FPM), which is resistant chemical and solvents) [18]. The rubber compounders and manufacturers are in the position to add devulcanized rubber to a wide range of compounds.

Table 1. Global rubber consumption in 2020 [19,20].

	Natural Rubber		Synthetic Rubber	
Region	%	**Million Metric Tons**	%	**Million Metric Tons**
Europe	8%	1.02	14%	1.99
China	43%	5.46	31%	4.40
India	8%	1.02	4%	0.57
Japan	5%	0.64	5%	0.71
USA	6%	0.76	11%	1.56
Rest of the world	30%	3.81	35%	4.97
Total	100%	12.7	100%	14.2

Table 2. Rubber consumption by industry in the EU and UK in 2020 [19,20].

	Natural Rubber		Synthetic Rubber	
	%	**Million Metric Tons**	%	**Million Metric Tons**
General rubber goods	27%	0.25	55%	1.09
Tires	73%	0.77	45%	0.90
Total	100%	1.02	100%	1.99

Table 3. Economic value of rubber products imported and exported from the EU in 2020 (excl. UK) [21].

	Imported	**Exported**
	€ Billion	**€ Billion**
General rubber goods	5.6	5.2
Tires	5.8	5.6
Total	11.4	10.8

In the academic literature, the main barriers identified for manufacturing to come to a circular economy are high startup costs, complex supply chains, challenging collaboration amongst companies, compromises on quality, and the high costs of disassembling products [22]. However, the specific reasons for the low uptake of devulcanized rubber have not been extensively investigated. Therefore, the main research question that will be addressed in this article is: What barriers do compounders and manufacturers encounter when incorporating devulcanized rubber into new rubber compounds, and what are the major underlying factors that contribute to these barriers? The studied factors include technical, market, institutional, and cultural factors, which are recognized in the academic literature as key to the circular economy transition [23].

The methods section outlines the selection of respondents, the type of interviews, and interpretation of these interviews. The results section presents key findings from the interviews. This section addresses the main subjects for the technical, economic, institutional, and cultural factors, while the discussion section interprets these findings by reflecting on potential interventions. Finally, the conclusion section summarizes the overall insights regarding the opportunities and barriers for devulcanization as a means to transition to a circular economy for the rubber industry.

2. Methods

For this study, a qualitative research method was used to identify the barriers for the uptake of devulcanized rubber and their underlying factors. We identified these barriers by interviewing representatives of compounding and manufacturing industries, the potential users of devulcanized rubber. We used purposive sampling [24], deliberately selecting organizations rather than randomly choosing them, and conducted in-depth interviews to identify key topics and gain an understanding of the underlying complexity [25,26].

Representatives for the interviews were selected from companies with in-depth knowledge and ample experience related to the addition of recycled materials to new compounds. The interviews were conducted in-person or via video conferencing, recorded, and transcribed for an accurate representation. The interviews were then coded and interpreted according to the coding scheme in Appendix C.

2.1. Selecting Respondents

This study focused on both compounders and manufacturers with experience of incorporating devulcanized rubber. We identified and approached potential participants through organizations with extensive networks in the rubber industry and through national sector specific associations for the rubber and tire industry in Europe. Companies based in Europe or with European subsidiaries were prioritized, although many were part of international corporations or holding companies. This generated an international view on the topic with an emphasis on Europe. We approached both tire manufacturers as well as general rubber goods manufacturers. Amongst the tire manufacturers, respondents were specifically selected from the 15 companies with the highest global turnover in 2023 [27].

Eight respondents were interviewed. Seven of these respondents had experience working with reclaimed and/or devulcanized rubber. Three of these respondents did not add devulcanized rubber to compounds for commercial products at the time of the interview but continued to do tests with new types of devulcanized rubber. One respondent, a compounder, had stopped adding devulcanized material because it could not be used as a drop-in, a direct replacement of virgin rubber. One respondent did not have any experience of working with devulcanized rubber but did have experience of working with non-vulcanized production waste. This allowed him to draw comparisons between the two materials, providing valuable insights and context for positioning devulcanized rubber in the analysis.

All interviewed companies were operating internationally, and they provided information not only about the European context but frequently also compared this context to the situation in other continents. Table 4 provides general information about the respondents and their companies.

Table 4. Information about the respondents.

#	Type of Company	Role Within Company	Reference Code
1	Tire manufacturer	Research and Development Manager	T1
2	Tire manufacturer	Research and Development Manager	T2
3	Tire manufacturer	Corporate Affairs Manager	T3
4	Compounder	Director	C1
5	Compounder	Research and Development Manager	C2
6	Compounder	Research and Development Manager	C3
7	Producer of general rubber goods	Director	G1
8	Producer of general rubber goods	Production Manager	G2

2.2. Interview Topics and Coding Scheme

The interviews lasted about one hour. Prior to the interview, the respondents received an opening statement, including the interview topics (Appendix B). The interview topics were derived from the literature on barriers to the circular economy transition. A common approach in these research papers is grouping the identified topics into sets of factors: technological, market, institutional, and cultural. This emphasizes the need for a multidisciplinary framework to identify the barriers [23,28–30]. This framework was also applied to plastics recycling in supply chains in other studies [31–33]. Table 5 provides an overview of the factors and research themes identified from the literature.

Table 5. Interview topics.

Technical factors are the material-related aspects and the industrial activities to incorporate recycled ingredients into compounds for new rubber.		
Identified research themes	**Sources**	**Key words**
Overall quality of supplied recycled materials.	[31,33–36]	Quality
Required changes in compound formulations and product designs to use recycled materials.	[31,36,37]	Compound formulation, product design
Consistency in both the quality and quantity of the supply of recycled materials.	[33,38]	Consistency, quality, supply
Market Factors are the aspects related to the cost, pricing and market acceptance.		
Identified research themes:	Sources	
Investment costs for compounding with recycled materials.	[31,33]	Investments, costs
Costs and pricing of recycled materials versus virgin materials.	[31–33,35]	Cost, pricing, recycled, virgin
Market dynamics for products with recycled materials.	[32,33,39]	Market dynamics
Cultural factors are related to how a culture is collaborative and focused on engaging in the circular economy, as well as aligning value to facilitate the use of recycled materials.		
Identified research themes:	Sources	
Position of sustainability in company culture.	[31,33,39]	Sustainability, company/organisation, culture
Coordinated efforts to align and increase the use of recycled materials in new products.	[33,35,40]	Coordination, alignment
End user appreciation of increasing the uptake of recycled materials.	[32,33]	End user, customer, appreciation
Institutional factors are the policies, rules, regulations, taxes and infrastructure that directly or indirectly influence the decision-making of companies.		
Identified research themes:	Sources	
Efforts towards industry alignment and standardization to increase the uptake of recycled materials.	[31,33]	Alignment, standards
Accountability mechanisms of manufacturers and compounders that impact the use of recycled materials.	[33,35,39]	Accountability/ responsibility, mechanisms
Alignment of regulations across sectors and geographical regions that impact the use of recycled materials.	[31,35]	Alignment, regulations, sectors, regions

2.3. Data Analysis

The interviews were coded using structural coding, an approach for analysing large segments of qualitative data along broad categories. This creates an overall framework that can be applied for in-depth analysis of these topics [41]. After the interviews, coding was done using QDA Minor. Supplemental codes were added to capture additional topics that emerged during the interviews. The interviews were interpreted and described along these codes. The coding scheme was adapted after analysing the interviews, leading to the grouping of some codes and the addition of others. The final coding scheme was based on the topics in Table 5 and can be found in Appendix C.

2.4. Comparison of Interview Data with the Literature

The literature on the use of devulcanized rubber by the industry is limited. Therefore, two recycled materials were selected from the academic literature for comparison, to gain a more in-depth understanding of the identified barriers and the extent to which these barriers are uniquely related to this type of recycling. This approach was used to determine if the identified barriers are found more often and to identify interventions that could facilitate the uptake of devulcanized rubber. Two materials were selected: recovered carbon black and mechanically recycled plastic.

Recovered carbon black was chosen as it is a recycled material with the potential to be added to rubber compounds in large quantities. It is retrieved from end-of-life rubber through pyrolysis and can be blended into new rubber compounds as a replacement for carbon black [42]. However, recovered carbon black has lower quality than the original material. One of the reasons is that its surface is often contaminated [43,44]. In a consultancy report, 12 companies were listed that in total recovered 20 kton/year of recovered carbon black in Europe, mainly in pilot factories. These companies announced plans to increase their production to 232 kton/year in the coming years [45]. Although the production capacity is currently relatively small, similar to devulcanization, it is expected to grow significantly in the coming years.

The second material selected was recycled granulate from the mechanical recycling of end-of-life thermoplastic waste [46]. This represents the bulk source of recycled polymer, accounting for more than 400 million metric tons in 2023 worldwide [47]. It is produced from a mixed range of input materials and is downgraded compared to the virgin raw materials.

By entering combinations of the main keywords of the identified research themes of Table 5 into Google Scholar and ScienceDirect and using terms such as 'recovered carbon black', the literature on these two groups of materials was identified. Only articles published after 2020 were included to obtain recent research findings on encountered barriers and interventions. The first 40 results were reviewed for each combination of keywords. If no academic literature was identified about recovered carbon black or thermoplastic recyclate, information was sought from white papers and consultancy reports. If information from such non-academic sources was used, this is explicitly stated in the text.

3. Results

The findings are presented according to the four categories outlined in the methods section: technical, market, institutional, and cultural factors. In each of these categories, the factors are described along with the specific results that emerged in the interviews. The results are provided in Tables 6–9, which also show which respondents referred to particular topics by differentiating between compounders (C), manufacturers of general rubber goods (G), and tire manufacturers (T). In the next column, the identified references to comparable barriers found in the literature on recycled carbon black (rCB) and thermoplastic recycled granulate (TRG) are listed.

3.1. Technical Factors

Table 6 lists the barriers that were found related to the technical factors. The introduction of a devulcanized rubber as a new ingredient impacts the overall composition of rubber compounds. All six respondents that used devulcanized rubber first did additional research and development and then reformulated their rubber compounds. If the quality of devulcanized rubber was not fully consistent, even more research was needed.

Table 6. Barriers related to the technical factors.

Barriers identified in interviews with compounders (C), general rubber goods (G), and tire manufacturers (T)	***Respondents*** C	G	T	*Similar barriers identified in the literature about recovered carbon black (rCB) and thermoplastic recycled granulate (TRG)*
Research theme: overall quality of supplied recycled ingredients.				
Not a drop-in replacement: devulcanized rubber is not considered to be a drop-in replacement for virgin rubber, as the properties are not similar to virgin materials.	2	2	2	**rCB:** the scientific literature only mentions different properties for rCB [44,48]. A white paper: [49]. **TRG:** [31,50–52].
Research theme: consistency in both the quality and quantity of the supply of recycled materials.				
Heterogeneity: respondents developed their own, exclusive compounds for unique products, which causes heterogeneous waste streams.	2	2	2	Causes of heterogeneity: **rCB:** [43,53]. **TRG:** [54].
Reduced material properties: devulcanized rubber was derived from mixed sources, which resulted in a heterogeneous mixture with inconsistent quality.	3	1	1	**rCB:** [44,55]. **TRG:** [32,34,46,51,56,57].
Unintended side effects: the addition of recycled material could lead to unexpected side effects during processing or use.	2	1	1	**rCB:** [44,58]. **TRG:** [57,59,60].
Research theme: Required changes in compound formulations and product designs to use recycled ingredients.				
Need to reformulate compounds: compounds needed to be reformulated to compensate for the composition of devulcanized rubber.	2	1	2	**rCB:** [44,58]. **TRG:** [61,62].
Numerous formulation variations: general rubber goods manufacturers and compounders often had dozens of different compound formulations for specific products, all of which needed to be adapted.	3	1	-	**rCB:** No literature identified. However, the literature does mention the limitations of rCB for specific purposes [44,48] **TRG:** No literature identified but limitations of TR were mentioned for specific purposes, such as applications in the automotive industry [56,63].

In recent years, more types of devulcanized rubber have entered the market, mainly sourced from different types of vulcanized production waste and rubber recovered from end-of-life tires. Respondents who tested these new types of devulcanized rubber mentioned that the technical properties were better than those of materials previously available. Despite these improvements, the devulcanized rubber was only added in small quantities for critical applications; respondent T2 mentioned that only 1–2 percent was added to the most critical compounds. Adding a larger amount to these compounds was deemed too risky.

The barriers, as experienced by the respondents, primarily arose from issues related to the inconsistent quality of devulcanized rubber, which was partially caused by the variability in feedstock from rubber waste streams. The negative effect of heterogeneously sourced material was also a challenge for recovered carbon black [44,48] and recycled granulate from thermoplastic waste [31,50,51]. Until recently, recovered carbon black was rarely used in the rubber industry due to inconsistent properties [64]. Recent studies have shown improved quality consistency by reducing the ash content [42,43,65]. As these methods are still under development, the recovered carbon black industry also focusses on negating any negative effects [66], as is currently the case for devulcanized rubber.

In the case of the mechanical recycling of thermoplastic waste, the composition was made more homogeneous by automated separation and sorting [67] and with detection of the types of plastics by applying state-of-the-art vision technologies [68,69]. However, the quality was still inconsistent due to, e.g., layered or non-separable plastics and contaminations on the plastic [34,46,51,56]. An underlying reason was the municipal systems of plastic waste collection and separation before recycling, leading to additional contamination and mixing of materials [70]. This mirrors the challenges associated with car tires as a source of devulcanized rubber, as they consist of multiple layers of different rubber compounds and are gathered through diverse collection systems across Europe, each with its own processes and efficiencies [4].

3.2. Market Factors

Table 7 lists the barriers observed related to the market factors. Although devulcanized rubber had a lower purchasing price than virgin rubber, respondents did have to incur additional costs. These costs primarily arose from the effect of this new ingredient. Respondents noted that they had additional costs to compensate for the lower consistency in the quality of the material. More research and development were also needed to make the devulcanized rubber compatible with the rubber compound. Respondents C2 and G2 also included costs for troubleshooting for recycled materials in their calculations if unexpected effects occurred during processing or to address possible customer complaints. To account for these costs, respondent C2 mentioned that when new compounds needed to be developed for a customer, they required the customer to share in the associated development costs.

Table 7. Barriers related to the market factors.

Barriers identified in interviews with compounders (C), general rubber goods (G), and tire manufacturers (T).	***Devulcanization respondents***			*Similar barriers identified in the literature about recovered carbon black (rCB) and thermoplastic recycled granulate (TRG).*
	C	*G*	*T*	
Research theme: investment costs for compounding with recycled ingredients.				
Additional expenditures: additional expenditures to compensate for a less consistent quality of the material.	2	2	2	**rCB:** no literature identified. **TRG:** [31,33,35,51,56].
Research theme: costs of recycled ingredients versus virgin ingredients.				
Low purchasing cost: devulcanized rubber was cheaper than virgin rubber and less valuable on the market.	1	1	2	**rCB:** [71,72]. **TRG:** [31,73].
Research theme: market dynamics for products with recycled ingredients.				
More customer requests, greater diversity of supply: in business-to-business sales, more customers were requesting rubber compounds that included biobased or recycled materials, and an increasing number of suppliers were offering recycled materials.	2	2	2	**rCB:** [72]. In additional to this source, the topic was raised by tire manufacturers in their white paper [74]. **TRG:** the literature describes an imbalance between supply and demand in terms of availability, quality, and pricing [31,75].
Increased obligations: more requirements for recycled materials than for virgin materials; companies had to undergo additional audits and environmental impact assessment from their customers.	3	1	3	**rCB:** no literature identified. **TRG:** [29,33,76].

The market dynamics were also different from virgin materials. A reason was that respondents had to engage with new potential suppliers of recycled materials. For devulcanization, many of these suppliers were startups that needed launching customers.

Another reason was that respondents T2, T3, C1, C2, and C3 noted more requests for rubber products that contained recycled or biobased materials. The sustainability goals set by the automotive industry were an important driver. Respondent T2 cited an example of a car manufacturer that requested a tire that contained at least 40% sustainable materials.

Respondents C2, C3, and G1 explicitly mentioned that when they supplied products with recycled materials to such customers, they were expected to collaborate on the sustainability reporting of their customers. They needed to provide additional data to their customers related to the customer sustainability goals, such as a reduced CO_2 footprint. This included information about the recycled materials and environmental assessments.

According to the literature, market demand of recycled materials is vulnerable to fluctuations in the prices of the virgin materials, which is described for carbon black [71,77], plastics [31,73], and rubber [78]. In Europe, plastic recyclers went out of business due to the low prices of virgin plastics in 2024 [79]. In contrast, the business case for recovered carbon black through tire pyrolysis is not only dependent on carbon black but also on pyrolysis oil (naphtha), which can be blended into marine fuel and allows shipping operators to meet their obligations under the FuelEU Maritime Regulation [80]. Thus, although the market dynamics for these recycled materials differ, their prices are still influenced by factors such as supply and demand, production costs, and the prices of competing virgin materials.

3.3. Cultural Factors

Table 8 lists the barriers observed related to cultural factors. Respondents highlighted many of their efforts to increase the use of recycled materials in rubber compounds. However, respondents T1, T2, T3, C1, C2, C3, and G2 emphasized that product quality and overall costs were not allowed to be compromised. In contrast, G1 actively sought markets that accepted products made with recycled materials.

Table 8. Barriers related to the cultural factors.

Barriers identified in interviews with compounders (C), general rubber goods (G), and tire manufacturers (T)	*Devulcanization respondents*			*Similar barriers identified in the literature about recovered carbon black (rCB) and thermoplastic recycled granulate (TRG)*
	C	*G*	*T*	
Research theme: position of sustainability in company culture.				
Limited room for compromises on quality and costs: companies set clear sustainability goals, but have no willingness to compromise on either cost or quality.	2	2	1	**rCB:** no literature identified. **TRG:** [31,51,56,81].
Research theme: collaborative efforts to align and increase the use of recycled materials in new products.				
No coordination: coordinated industry wide efforts were identified to increase recovered carbon black but not devulcanized rubber.	-	-	3	**rCB:** no academic literature identified but other sources mention that tire manufacturers collaborated with rCB suppliers and other stakeholders [74]. **TRG:** Although no academic literature was identified, several inter-company collaborations were recognized: [82,83].
Research theme: end user appreciation for increasing the uptake of recycled ingredients.				
Low consumer demand for recycled materials: no demand was identified for tires made from sustainable materials.	2	-	2	**rCB:** no literature identified. **TRG:** [31,33,84].

The demand for more recycled materials in rubber came from customers from the industry, not from consumers. According to respondents T2 and T3 the demand from end users was low and they did not spot any willingness among consumers to pay a premium for rubber products containing recycled material, such as tires. Respondent T2 mentioned that his company wanted to contribute to a campaign aimed at raising awareness among consumers to increase the acceptance of recycled and biobased materials in tires.

No industry-wide collaboration or coordination was mentioned by any of the respondents for an industry-wide commitment to the uptake of devulcanization. In contrast, respondents did mention such collaboration for recovered carbon black, which was also mentioned in consultancy reports and a white paper by the rubber industry [45,74]. For recycling of thermoplastic waste, several industry wide collaborations were identified that aimed to develop more recyclable products, coordinate waste collection systems and innovate recycling techniques [82,83,85].

3.4. Institutional Factors

Table 9 lists the barriers observed related to the institutional factors. Respondents T1, T2, and T3 considered quality standards to be important for the uptake of recycled materials. However, they identified such standards for recovered carbon black but not for devulcanized rubber. This is in line with the literature, in which standards were identified for recovered carbon black [55,86] and thermoplastic recycled granulate [87–89] but not for devulcanized rubber [3,4].

Table 9. Institutional factors.

Barriers identified in interviews with compounders (C), general rubber goods (G), and tire manufacturers (T).	*Devulcanization respondents*			*Similar barriers identified in the literature about recovered carbon black (rCB) and thermoplastic recycled granulate (TRG).*
	C	*G*	*T*	
Research theme: coordinated efforts for alignment and standardization to increase the uptake of recycled materials.				
No specific standards: respondents mentioned standards for other recycled materials but were not aware of such standards for devulcanization.	-	-	3	**rCB:** a specific ASTM standard, including a test method were developed for rCB [90,91]. **TRG:** Standards were developed by industries for thermoplastic recyclate [76].
Research theme: accountability mechanisms of manufacturers that impact the use of recycled ingredients.				
Liability concerns: Companies were accountable for their products, including liability for any damages caused by those products or their components. This made the companies more risk-averse and reluctant to increase the amount of recycled materials.	2	1	1	**rCB:** no literature identified. **TRG:** no literature identified.
Research theme: alignment of regulations across sectors and geographical regions that impact the use of recycled ingredients.				
National regulatory differences: differences in interpretations of EU waste laws made it increasingly difficult to ship rubber across borders.	1	1	1	**rCB:** no literature identified. **TP:** [28,35].
Regulations from other domains: a key concern was meeting multiple regulations concerning environmental and safety goals.	1	1	3	**rCB:** no literature was identified **TRG:** [28,31,35,92,93].
Need for alignment of regulations: respondents indicated a need for better governmental alignment of regulations that directly or indirectly impact the adoption of recycled materials.	-	1	3	**rCB:** no literature was identified. **TRG:** [28,31,35,92].

Respondents T1, T2, and T3 noted that legislation could lead to an increase in the uptake of recycled materials in rubber but T1, T2, T3, C2, and G1 also expressed concerns about the implementation of upcoming legislation, such as a blending obligation of non-fossil ingredients in rubber. This was considered to have negative side effects by respondents T3, C1, and G1. The reasons were: effect on product quality, increased regulatory pressure, and its impact on pricing compared to international competition.

The current legislative framework in Europe was identified as a barrier to increase the uptake of recycled materials. A specific concern expressed by respondents T1, T2, and T3 was that they already had to meet multiple environmental and safety regulations. These could conflict with the aim to increase recycling rates. Respondents T1, T2, T3, and G3 also mentioned uncertainties related to the effect of upcoming and future considered regulations that could make it more difficult to rely on recycled materials. For this reason, respondent T1 mentioned that they not only conducted a technical check but also a legal check to guarantee that such materials could be used in the coming 10 years. A first example mentioned by T2, T3 and G1 was the EU Deforestation Regulations [94], which could imply that the origin of recycled natural rubber had to be demonstrated, which is challenging in an open-loop recycling system. A second example mentioned by respondents T2, T3, and G1 was that recycled rubber is now considered exempt from Registration, Evaluation, Authorisation, and Restriction of Chemicals (REACH) because the material has already been registered during the initial production. However, the European Chemicals Agency is considering that recycled materials, such as rubber, need to be demonstrably free of substances of concern [95]. This uncertainty of such upcoming regulation is also described in the literature about thermoplastic recycling [92,96].

Respondents T3 and G2 noted that European member states are allowed their own interpretation and implementation of EU legislation for waste and trade, which made it more difficult for their companies to work and ship across borders [97,98], directly leading to higher costs. For this reason, respondents T1, T2, T3, C2, C3, and G1 considered better alignment and more coherent implementation of regulations an important condition to increase recycling rates. As became evident from both the interviews and the literature in Table 9, the regulatory issues are a widely shared concern within the plastics and rubber industry.

4. Discussion

This study identified barriers across the technical, market, institutional, and cultural factors faced by compounders and manufacturers when incorporating devulcanized rubber into new rubber compounds, along with their underlying causes. Respondents from tire manufacturers, general rubber goods manufacturers, and compounders were interviewed as experts on processing recycled materials, with a specific focus on devulcanized rubber. The respondents reflected on their experiences with the addition of devulcanized rubber and other types of recycled materials to their compounds. In this discussion we reflect on these observations by focusing on how devulcanized rubber can potentially be incorporated in larger quantities into new compounds and identifying system interventions to achieve this goal.

A recurring topic among the technical factors is that devulcanized rubber has a heterogeneous composition, which is a clear barrier for the uptake of rubber. This heterogeneity is partially inherent to devulcanization, as a devulcanized rubber contains both the original rubber polymers as well as the additives that were used to create the new rubber compound [6]. This also arises from the heterogeneity of rubber waste sources. If such a large variation in rubber products enters waste systems, the devulcanization material becomes inconsistent, thus reducing quality [6,16]. Therefore certain companies devulcanize rubber from specific sources, such as truck tire tread rubber, to obtain devulcanized rubber with a more consistent quality [4,5,16]. This is an example of a closed-loop/closed-source system, implying that end-of-life materials from a well-controlled source are recycled at relatively high quality and value [99]. Other research showed that the closed-loop/closed-source concept is based on the premise that the original equipment manufacturers take responsibility for collection to arrive at a relatively homogeneous input stream from selected sources [100].

An important barrier among the market factors is that recycled materials are valued lower by compounders and manufacturers than new, virgin rubber. A reason is the required additional research and development costs, as well as potential risks related to product liability. An underlying reason is a lack of standardization for devulcanization, which could be improved by setting standards, as were identified for recovered carbon black [90] and thermoplastic recycled granulate [76]. Additionally, compounders could specialize in blending devulcanized rubber with virgin rubber compounds to create materials with devulcanized content and a more consistent quality for manufacturers. Companies that specialize in this type of blending were identified for recovered carbon black, such as [66], and thermoplastic recycled granulate, such as [101]. Thus, by standardizing and consolidating the devulcanization value chain, opportunities could arise to enhance quality and potentially increase the value of devulcanized rubber.

A key barrier among the cultural factors is the lack of industry-wide engagement in preserving the value and usability of rubber. Devulcanization research and development strongly focuses on technology development, hardly addressing other factors that limit implementation [4]. In addition to technological solutions for improved sorting and separating, companies could be creating distinct material recycling loops within their respective value chains, as identified in plastics recycling [102]. This requires an involvement from compounders, manufacturers, distributors, and recyclers and necessitates infrastructure and appropriate regulations to facilitate these closed-loop recycling systems [99]. Given the diverse range of rubber products and markets, multiple recycling chains can be established to create effective closed-loop/closed-source systems, each tailored to specific material streams where applicable.

A recurring institutional barrier is the absence of appropriate regulations. When looking at improvements, it is important to also take plans by the European Union into consideration. In the Circular Economy Action Plan for 2030 [103] a European blending obligation was announced, which mandates the inclusion of non-fossil ingredients (either biobased or recycled) in plastics and rubber. This obligation could boost the uptake of recycled materials in general, but the results of this study showed that compounders and manufacturers have a strong preference for materials that can be blended into compounds without having to make any modifications to their recipes. However, modifications are needed in compounds to compensate for the additives in devulcanized rubber [6]. For this reason, such an obligation may increase the uptake of other types of biobased and recycled materials, but not necessarily devulcanized rubber.

Another potential legislative intervention is extended producer responsibility (EPR), which can be effective for collecting waste and creating a secondary raw material market [104,105]. These systems are a collective responsibility of manufacturers but also importers, organized by producer responsibility organizations. Such organizations exist in several European countries and are aimed at the collecting of end-of-life tires [106]. This has not yet resulted in the most optimal material reprocessing from a circular economy perspective [106,107]. Due to the collective responsibility of all manufacturers and importers, ERP systems are geared towards open-loop recycling. This poses a particular obstacle for devulcanization, as a closed-loop recycling system is desirable to reduce the heterogeneity of devulcanized rubber. Thus, an ERP system, in which manufacturers carry a collective responsibility does not provide the incentive needed to stimulate rubber recycling by devulcanization.

The drawbacks associated with the blending obligations for non-fossil ingredients and EPR systems highlight the need for a more comprehensive approach to address the complexities, as outlined in this study. A system would be needed that focusses on creating both the infrastructure as well as industry collaboration to achieve the highest possible closed-loop recycling. To achieve this, policymakers should reconsider which stakeholders need to be involved to achieve the most optimal use and recycling of rubber. It is recommended as a topic for future research to study how an integrated approach could increase the circular economy for the rubber industry, specifically for devulcanization. This study demonstrates that achieving this goal requires more than just technological advancements; to reach higher quality standards on an industrial scale, collaboration across the supply chain is essential to enable closed-loop recycling via devulcanization.

The analytical framework depends on choices made during the analysis, which could restrict the scope of interpretations. Since most respondents were from Europe, the geographical validity of the results may be limited.

5. Conclusions

This study focused on identifying the barriers compounders and manufacturers face when using devulcanized rubber in new rubber compounds and finding the underlying causes to these barriers. It is demonstrated that significant changes are needed to increase the uptake of devulcanized rubber in the transition to a circular economy. It provides greater insight into the barriers, interconnections, and underlying causes, while also identifying opportunities for improved policy and collaboration, ultimately supporting the transition to the circular economy.

Most academic research on the topic of devulcanization has focused strongly on the technical aspects [2,3,108,109], but this study explored other factors influencing the uptake of devulcanized rubber that are essential to progress devulcanization beyond its niche in the industry. We investigated the technical, economic, institutional, and cultural factors that affect the uptake of devulcanized rubber as an ingredient in new rubber compounds.

Reflecting on the current situation, respondents identified barriers stemming from a variety of interrelated factors. Similar challenges have also been documented for recovered carbon black and thermoplastic recycled granulate, such as the lower homogeneity of recycled materials compared to new, virgin materials, leading to inconsistencies in quality. Barriers that were more specific to devulcanized rubber include limited industry collaboration and the need for greater standardization, both of which are important to improve efficiency and ensure consistency across the industry.

To produce high-quality devulcanized rubber, multiple interconnected barriers must be addressed. Overcoming these challenges on the industry level calls for systemic changes and effective collaboration among key stakeholders. Several approaches developed for other recycled materials could be worthwhile to increase the uptake of devulcanized rubber: industry standards, harmonized regulations, and closed-loop/closed-source recycling strategies.

Author Contributions: Conceptualization, E.R., J.J., G.H. and R.B.; formal analysis, E.R.; funding acquisition, G.H. and E.R.; investigation, E.R.; methodology, E.R., J.J. and R.B.; supervision, J.J. and R.B.; visualization, E.R.; data curation, E.R.; writing—original draft, E.R.; and writing—review and editing, J.J., G.H. and R.B. All authors have read and agreed to the published version of the manuscript.

Funding: This research was conducted as a part of the 'Hoogwaardige recycling door devulkanisatie' project (high-quality recycling by devulcanization). This research was funded by Regieorgaan SIA, which is a part of the NWO (Nederlandse Organistatie voor Wetenschappelijk Onderzoek). This research was part of the RAAK PRO subsidy framework, grant number RAAK.PRO04.027.

Institutional Review Board Statement: The study was conducted in accordance with the Declaration of Helsinki, and the protocol was approved by the Human Research Ethics Committee of the Delft University of Technology of 4351 on 7 July 2024.

Data Availability Statement: The data utilized in this study cannot be made publicly available as it contains proprietary company-specific information.

Acknowledgments: We would like to thank the respondents for the interviews and for sharing their knowledge and insights regarding the topic of devulcanization.

Conflicts of Interest: The authors declare no conflicts of interest. The funders had no role in the design of the study; in the collection, analyses, or interpretation of data; in the writing of the manuscript; or in the decision to publish the results.

Appendix A. End-of-Life Methods for Rubber

The following methods were identified in the literature [2,110].

	Feedstock	Working Principle	Applications
Devulcanization	Rubber crumbs or powder	Devulcanized rubber compounds are aimed at a higher quality than reclaimed rubber. Devulcanization is a process that aims to selectively break the cross-links while keeping the rubber polymers intact. This is achieved through various methods, such as exposing the rubber to heat, chemicals, and shear strength. The result is a rubber compound that can be re-vulcanized [3,9,110].	Devulcanized rubber can be used to replace virgin rubber and additives. In most cases, this material is blended into new rubber compounds to compromise for any loss of quality [5,6]. Additionally, devulcanized rubber is added to bitumen to partially replace the plasticizer SBS [4,111].
Reclaimed rubber	Rubber crumbs or powder	On the molecular scale, in a reclaim process, both the rubber polymers and cross-links and are randomly broken. The input of an industrial reclaim process consists of rubber crumbs or powder, which are converted into a rubber compound that can be re-vulcanized [3,9].	Reclaimed rubber is mostly added to rubber compounds and then re-vulcanized. It is mostly used for less critical applications or added in small quantities to new rubber compounds [9].
Binder systems	Rubber crumbs or powder	For certain applications, the granulated rubber is bound with a polyurethane or a thermoplast [112,113]. Additionally, the surface of the rubber powder can be treated for better bonding in a rubber compound [114].	This material can be used for various products, such as rubber tiles and mats and underlayment for floors [113].
Pyrolysis	Rubber crumbs or powder	Pyrolysis is a thermal process in which rubber compounds are decomposed by heat (400–800 degrees Celsius) in the absence of oxygen into smaller molecules, which leads to oil, carbon black, and gas [110].	Pyrolysis oil can be used as a fuel and in theory for producing new monomers to create new plastics and rubber [115]. The carbon black can replace certain grades of virgin carbon black [116]. Lastly, the gas can be a source for energy [117].

	Feedstock	Working Principle	Applications
Retreading	Passenger car tires, tuck tires, airplane tires, and off-the-road tires	In retreading the tread of a worn-out tire is removed. A new rubber tread compound is vulcanized onto the tire. Retreading is different from the recycling process, because the carcass of a tire can be reused [118]. The removed rubber from the tread can be recycled by another method, such as devulcanization [4]	Retreading is primarily used for tires from planes, truck, agricultural tires and sometimes passenger car tires [119].
Mechanical grinding	Different types of tires and rubber products, such as conveyor belts	Rubber is sorted from steel, canvas and other contaminants and is then mechanically processed into crumbs or powder of various sizes [113]. An alternative to mechanical processing is cryogenic shattering [120].	Rubber crumb or powder is used as a feedstock for various recycling methods, such as reclaim and devulcanization [9]. Additionally, rubber crumbs can be utilized as artificial turf on sports fields or recreational surfaces [120]. Rubber powder is also used as a simple filler material that can be added in small quantities to new rubber compounds [114]. Rubber crumbs are used as an additive for tarmac/asphalt [113].
Incineration	Tires and other end-of-life rubber products	Tires are disposed of by incineration, but this process produces toxic emissions [110]. Rubber is incinerated in cement plants [2] as an alternative source of heat, which produces less toxic emissions than coal [121]	Incineration can be used to recover 32.6 MJ/kg of embedded energy in tires [110].
Landfill	Tires and other end-of-life rubber products	Globally, 55 percent of all tires are dumped in landfills [122]	Large quantities of landfilled tires are known to pose a significant fire hazard. Additionally, they are toxic to soils and groundwater, and when burned, they cause air pollution [123]

Appendix B. Opening Statement and Topic List Used for Interviews

Appendix B.1. Opening Statement

My name is Eric Roetman, and I am a researcher at Windesheim University of Applied Sciences and a PhD candidate at TU Delft. I am conducting interviews with representatives from European companies that have devulcanized tires. I would like to interview you as a representative of one of these companies. During the interview, I aim to gather information about the key opportunities and barriers of adding recycled rubber to new rubber compounds.

During the interview, I will ask a set of questions on this topic. By posing these questions to you and representatives of other companies, we hope to gain an overview of the opportunities and barriers. The information will be recorded using a voice recorder and then transcribed. You will receive a copy of the main conclusions and have the opportunity to respond, suggest changes, or address any misinterpretations. If we wish to use non-anonymized quotes, or quotes traceable back to you, from the interview, we will contact you and only proceed with your explicit approval.

The aim of these interviews is to collect data for publication in an academic article and a PhD thesis. The information collected will be handled with great care, and the outcomes will be published in a manner that prevents identification of you or your company. Information traceable back to you will not be shared beyond the research team. The information from the interviews will be stored in a data repository by Hogeschool Windesheim (University of Applied Sciences).

Your answers in this study will be kept confidential to the best of our ability. Your participation in this study is entirely voluntary, and you can withdraw at any time.

If you have any questions or concerns about the interview or handling of the data, you can contact me directly by sending an email to e.c.roetman@tudelft.nl or calling me at [phone number].

Appendix B.2. Interview Topics

General

1. Why did you start to incorporate recycled materials in rubber products/compounds?
2. What type of recycled materials?
3. Since when do you incorporate recycled rubber into your products/compounds? Why at that moment?
4. Have you applied recycled material to a new type of product or to existing products?
5. What new skills, facilities and resources did you need in your company to add recycled materials to rubber?
6. Up to what degree do you experience an advantages or disadvantages of being one of the first companies that incorporates recycled material in rubber?

Technical factors

1. What types of recycled source material (devulcanized, reclaimed, micronized rubber material, etc.) are most and least suitable for you to add to your rubber compounds? How does devulcanization fit in this list?
2. What product categories are most suitable products to incorporate such recycled materials?
3. How do you experience the performance and processability of recycled source material compared to virgin rubber material?
4. To what degree do the properties of your recycled source material pose challenges, and how can you mitigate these challenges?
5. How would you rate the availability of the recycled source materials for production?
6. How do you assess the quality of the supplied rubber material, specifically regarding contamination such as the presence of PAHs?
7. Did you have to change your production process to process recycled rubber, and why?
8. Up to what extent did you have to change your research and development process for processing recycled rubber?

Market factors

1. How would you evaluate the overall reliability of suppliers of recycled source material?
2. Up to what extent is it an added value for your customers that you include recycled content in products, or do they have reservations?
3. Up to what extent did encounter additional costs when incorporating recycled materials into your products? Includes upfront costs?

Institutional factors

1. Did upcoming regulations positively or negatively influence your decision to incorporate recycled rubber into your products?

2. Up to what extent have you benefitted from financial incentives (e.g., subsidies) that encouraged you to use recycled materials in manufacturing processes?
3. (How) did industry standards positively or negatively influence your decision to incorporate recycled rubber into your products?

Cultural factors

1. Up to what degree do you perceive a willingness amongst customers to add recycled materials?
2. Up to what extent is your engagement with your suppliers of recycled material different from suppliers of virgin materials?
3. How does the addition of recycled material to products affect your relationship with your compounder(s)?

Closing the interview

1. Is there any topic that is not addressed during this interview and you would like to address?

Appendix C. Interview Coding

Technical factors are the material-related aspects and the industrial activities to process recycled ingredients into compounds for new rubber.
Identified research themes
Overall quality of supplied recycled materials • Performance devulcanized material • Performances (other) recycled materials • Technical challenges recycled materials • Suitability recycled materials for rubber products
Required changes in compound formulations and product designs to use recycled materials • Required changes to rubber compounds • Additional research and development • Required changes to production processes
Consistency in both the quality and quantity of the supply of recycled materials • Availability of recycled materials • Quality of recycled materials
Market factors are the aspects related to the cost, pricing, and market acceptance.
Identified research themes:
Investment costs for compounding with recycled materials • Required skills, facilities, resources • Additional costs to include recycled materials
Costs and benefits of recycled materials versus virgin materials • Additional costs of adding recycled materials • Financial benefits to include recycled materials
Market dynamics for products with recycled materials • Availability suppliers and supplies • Reliability of suppliers • Effect of adding recycled materials on trade

Cultural factors are related to the extent to which a culture is collaborative and focused on engaging in the circular economy, as well as aligning value to facilitate the use of recycled materials.
Identified research themes:
Position of sustainability in company culture • Willingness to include sustainability standards • Communication of sustainability goals
Coordinated efforts to align and increase the use of recycled materials in new products • Industry-wide collaborations • Engagement of suppliers • Engagement of business-to-business customers
Consumer appreciation and involvement aimed at increasing the uptake of recycled materials • Engagement of consumers
Institutional factors are the policies, rules, regulations, taxes, and infrastructure that directly or indirectly influence the decision-making of companies
Identified research themes:
Efforts for alignment and standardization to increase the uptake of recycled materials • Industry standards for devulcanization • Industry standards (general)
Accountability mechanisms of manufacturers that impact the use of recycled materials • Identified accountability mechanisms • Identified collaborations in industry
Alignment of regulations across sectors and regions that impact the use of recycled materials • Regulations (general) • Regulations across sectors • Regulations across national borders

References

1. Winkeman, H.A.; Busenburg, E.B. Effect of Stearic Acid on Reclaimed Rubber. *Ind. Eng. Chem. Ind. Ed.* **1929**, *21*, 730–732. [CrossRef]
2. Markl, E.; Lackner, M. Devulcanization technologies for recycling of tire-derived rubber: A review. *Materials* **2020**, *13*, 1246. [CrossRef] [PubMed]
3. Wiśniewska, P.; Wang, S.; Formela, K. Waste tire rubber devulcanization technologies: State-of-the-art, limitations and future perspectives. *Waste Manag.* **2022**, *150*, 174–184. [CrossRef] [PubMed]
4. Roetman, E.; Joustra, J.; Heideman, G.; Balkenende, R. Does the Rubber Meet the Road? Assessing the Potential of Devulcanization Technologies for the Innovation of Tire Rubber Recycling. *Sustainability* **2024**, *16*, 2900. [CrossRef]
5. Ghosh, R. *Transforming Silica-Silane Reinforced Rubber into a High Quality Devulcanizate*; University of Twente: Enschede, The Netherlands, 2024. [CrossRef]
6. van Hoek, H. Closing the Loop. Reuse of Devulcanized Rubber in New Tires. Ph.D. Thesis, University of Twente, Enschede, The Netherlands, 2022.
7. Bandyopadhyay, S. Chapter 1 Fundamentals of rubber compounding. In *Rubber Products*; Bireswar, B., Ed.; De Gruyter: Berlin, Germany; Boston, MA, USA, 2024; pp. 1–52.
8. De, S.K.; White, J.R. *Rubber Technologist's Handbook*; iSmithers Rapra Publishing: Shawbury, UK, 2001; Volume 1.
9. Saiwari, S.; Dierkes, W. 7—Regeneration and devulcanization. In *Tire Waste and Recycling*; Letcher, T.M., Shulman, V.L., Amirkhanian, S., Eds.; Academic Press: Cambridge, MA, USA, 2021; pp. 97–144.
10. Shanbag, A.; Manjare, S. Life cycle assessment of tyre manufacturing process. *J. Sustain. Dev. Energy Water Environ. Syst.* **2020**, *8*, 22–34. [CrossRef]
11. Wiśniewska, P.; Haponiuk, J.T.; Colom, X.; Saeb, M.R. Green Approaches in Rubber Recycling Technologies: Present Status and Future Perspective. *ACS Sustain. Chem. Eng.* **2023**, *11*, 8706–8726. [CrossRef]

12. European Commission. *Study on the Critical Raw Materials for the EU 2023—Final Report*; Publications Office of the European Union: Luxembourg, 2023.
13. Ellen MacArtur Foundation. What Is a Circular Economy? Available online: https://www.ellenmacarthurfoundation.org/topics/circular-economy-introduction/overview (accessed on 10 March 2025).
14. Rodgers, B.; Waddell, W. Chapter 9—The Science of Rubber Compounding. In *The Science and Technology of Rubber*, 4th ed.; Mark, J.E., Erman, B., Roland, C.M., Eds.; Academic Press: Boston, MA, USA, 2013; pp. 417–471.
15. Sisanth, K.S.; Thomas, M.G.; Abraham, J.; Thomas, S. 1—General introduction to rubber compounding. In *Progress in Rubber Nanocomposites*; Thomas, S., Maria, H.J., Eds.; Woodhead Publishing: Cambridge, UK, 2017; pp. 1–39.
16. Lacroix, F.; Seghar, S. *In-Product-Sorting End-of-Life Tires for Effective Recycling by Devulcanization*; Academic Press: Cambridge, MA, USA, 2021.
17. Boon, Z.H.; Teo, Y.Y.; Ang, D.T.-C. Recent development of biodegradable synthetic rubbers and bio-based rubbers using sustainable materials from biological sources. *RSC Adv.* **2022**, *12*, 34028–34052. [CrossRef]
18. Samsuri, A. Degradation of Natural Rubber and Synthetic Elastomers. *Shreir's Corros.* **2010**, *3*, 2407–2438.
19. Malaysian Rubber Board. Consumption of Natural and Synthetic Rubber Worldwide from 1990 to H1 2024 (in Million Metric Tons) [Graph]. In Statista. 2024. Available online: https://www.statista.com/statistics/275399/world-consumption-of-natural-and-synthetic-caoutchouc/ (accessed on 18 November 2024).
20. ETRMA. The ETRMA Statistics Report. Available online: https://www.etrma.org/wp-content/uploads/2021/12/20211030-Statistics-booklet-2021VF.pdf (accessed on 1 May 2025).
21. ETRMA. *European Tyre & Rubber Industry Statistics Report*; ETRMA: Brussels, Belgium, 2011.
22. Jaeger, B.; Upadhyay, A. Understanding barriers to circular economy: Cases from the manufacturing industry. *J. Enterp. Inf. Manag.* **2020**, *33*, 729–745. [CrossRef]
23. Kirchherr, J.; Piscicelli, L.; Bour, R.; Kostense-Smit, E.; Muller, J.; Huibrechtse-Truijens, A.; Hekkert, M. Barriers to the Circular Economy: Evidence From the European Union (EU). *Ecol. Econ.* **2018**, *150*, 264–272. [CrossRef]
24. Campbell, S.; Greenwood, M.; Prior, S.; Shearer, T.; Walkem, K.; Young, S.; Bywaters, D.; Walker, K. Purposive sampling: Complex or simple? Research case examples. *J. Res. Nurs.* **2020**, *25*, 652–661. [CrossRef] [PubMed]
25. Chaminade, C.; Edquist, C. From theory to practice: The use of the systems of innovation approach in innovation policy. In *Innovation, Science, and Institutional Change: A Research Handbook*; Oxford University Press: Oxford, MS, USA, 2006; pp. 141–163. Available online: https://www.sciencedirect.com/book/9780128206850/tire-waste-and-recycling (accessed on 11 March 2025).
26. Knott, E.; Rao, A.H.; Summers, K.; Teeger, C. Interviews in the social sciences. *Nat. Rev. Methods Primers* **2022**, *2*, 73. [CrossRef]
27. Tyrepress. The World's Leading Tyre Makers of 2023. Available online: https://www.tyrepress.com/wp-content/uploads/2023/06/Leading-Companies-2023.pdf (accessed on 18 October 2024).
28. Grafström, J.; Aasma, S. Breaking circular economy barriers. *J. Clean. Prod.* **2021**, *292*, 126002. [CrossRef]
29. Paletta, A.; Leal Filho, W.; Balogun, A.-L.; Foschi, E.; Bonoli, A. Barriers and challenges to plastics valorisation in the context of a circular economy: Case studies from Italy. *J. Clean. Prod.* **2019**, *241*, 118149. [CrossRef]
30. de Jesus, A.; Mendonça, S. Lost in Transition? Drivers and Barriers in the Eco-innovation Road to the Circular Economy. *Ecol. Econ.* **2018**, *145*, 75–89. [CrossRef]
31. van der Vegt, M.; Velzing, E.-J.; Rietbergen, M.; Hunt, R. Understanding Business Requirements for Increasing the Uptake of Recycled Plastic: A Value Chain Perspective. *Recycling* **2022**, *7*, 42. [CrossRef]
32. Baldassarre, B.; Maury, T.; Mathieux, F.; Garbarino, E.; Antonopoulos, I.; Sala, S. Drivers and Barriers to the Circular Economy Transition: The Case of Recycled Plastics in the Automotive Sector in the European Union. *Procedia CIRP* **2022**, *105*, 37–42. [CrossRef]
33. Pinheiro, M.F.; Ferreira, L.M.D.F.; Azevedo, S.G.; Magalhães, V.S.M. Barriers to the Circular Economy in the Plastics Industry: A Systematic Literature Review. In Proceedings of the Flexible Automation and Intelligent Manufacturing: Establishing Bridges for More Sustainable Manufacturing Systems, Porto, Portugal, 18–22 June 2023; Springer: Cham, Switzerland, 2024; pp. 850–858.
34. Shen, L.; Worrell, E. Chapter 31—Plastic recycling. In *Handbook of Recycling*, 2nd ed.; Meskers, C., Worrell, E., Reuter, M.A., Eds.; Elsevier: Amsterdam, The Netherlands, 2024; pp. 497–510.
35. Bening, C.R.; Pruess, J.T.; Blum, N.U. Towards a circular plastics economy: Interacting barriers and contested solutions for flexible packaging recycling. *J. Clean. Prod.* **2021**, *302*, 126966. [CrossRef]
36. Alassali, A.; Picuno, C.; Chong, Z.K.; Guo, J.; Maletz, R.; Kuchta, K. Towards Higher Quality of Recycled Plastics: Limitations from the Material's Perspective. *Sustainability* **2021**, *13*, 13266. [CrossRef]
37. Milios, L.; Holm Christensen, L.; McKinnon, D.; Christensen, C.; Rasch, M.K.; Hallstrøm Eriksen, M. Plastic recycling in the Nordics: A value chain market analysis. *Waste Manag.* **2018**, *76*, 180–189. [CrossRef]
38. McKinnon, D.; Bakas, I.; Herczeg, M.; Vea, E.B.; Busch, N.; Christensen, L.H.; Christensen, C.; Damgaard, C.K.; Milios, L.; Punkkinen, H. *Plastic Waste Markets: Overcoming Barriers to Better Resource Utilisation*; Nordic Council of Ministers: Copenhagen, Denmark, 2018.

39. Agarwal, S.; Tyagi, M.; Garg, R.K. Conception of circular economy obstacles in context of supply chain: A case of rubber industry. *Int. J. Product. Perform. Manag.* **2023**, *72*, 1111–1153. [CrossRef]
40. Schultz, F.C.; Valentinov, V.; Kirchherr, J.; Reinhardt, R.J.; Pies, I. Stakeholder governance to facilitate collaboration for a systemic circular economy transition: A qualitative study in the European chemicals and plastics industry. *Bus. Strategy Environ.* **2024**, *33*, 2173–2192. [CrossRef]
41. Saldana, J. *The Coding Manual for Qualitative Researchers*; SAGE Publications: Thousand Oaks, CA, USA, 2015.
42. Cardona-Uribe, N.; Betancur, M.; Martínez, J.D. Towards the chemical upgrading of the recovered carbon black derived from pyrolysis of end-of-life tires. *Sustain. Mater. Technol.* **2021**, *28*, e00287. [CrossRef]
43. Bowles, A.J.; Fowler, G.D. Assessing the impacts of feedstock and process control on pyrolysis outputs for tyre recycling. *Resour. Conserv. Recycl.* **2022**, *182*, 106277. [CrossRef]
44. Costa, S.M.R.; Fowler, D.; Carreira, G.A.; Portugal, I.; Silva, C.M. Production and Upgrading of Recovered Carbon Black from the Pyrolysis of End-of-Life Tires. *Materials* **2022**, *15*, 2030. [CrossRef]
45. Simon, P.; Plessis, S.; Ghanimi, A. Evaluating the Path to a Sustainable Tire Industry: Unlocking the Potential of Recovered Carbon Black. 2024. Available online: https://www.emerton.co/news/evaluating-the-path-to-a-sustainable-tire-industry-unlocking-the-potential-of-recovered-carbon-black (accessed on 24 November 2024).
46. Schyns, Z.O.G.; Shaver, M.P. Mechanical Recycling of Packaging Plastics: A Review. *Macromol. Rapid Commun.* **2021**, *42*, 2000415. [CrossRef]
47. Plastics Europe. Annual Production of Plastics Worldwide from 1950 to 2022 (in Million Metric Tons) [Graph] in Statista. Available online: https://www.statista.com/statistics/282732/global-production-of-plastics-since-1950/ (accessed on 18 November 2024).
48. Dwivedi, C.; Manjare, S.; Rajan, S.K. Recycling of waste tire by pyrolysis to recover carbon black: Alternative & environment-friendly reinforcing filler for natural rubber compounds. *Compos. Part B Eng.* **2020**, *200*, 108346. [CrossRef]
49. Semperit. *Sustainability Report 2023*; Semperit: Vienna, Austria, 2024.
50. Demets, R.; Van Kets, K.; Huysveld, S.; Dewulf, J.; De Meester, S.; Ragaert, K. Addressing the complex challenge of understanding and quantifying substitutability for recycled plastics. *Resour. Conserv. Recycl.* **2021**, *174*, 105826. [CrossRef]
51. Auer, M.; Schmidt, J.; Diemert, J.; Gerhardt, G.; Renz, M.; Galler, V.; Woidasky, J. Quality Aspects in the Compounding of Plastic Recyclate. *Recycling* **2023**, *8*, 18. [CrossRef]
52. Jager, R.; Vlugter, J.; Gort, I. *Gevolgen National Norm Circulaire Plastics. Verkenning naar de gevolgen van de Verplichting Voor een Minimumaandeel Recyclaat en/of Biogebaseerd Kunststof en Rubber Voor Nederlandse Converters*; Research was Commissioned by Rijksdienst voor Ondernemend Nederland; Rijksoverheid: The Hague, The Nederlands, 2023.
53. Kratina, O.; Stoček, R.; Voldánová, J. How the rubber compounds of different tire's components heat up? In *Constitutive Models for Rubber XII*; CRC Press: Boca Raton, FL, USA, 2022; pp. 199–204.
54. Wiesinger, H.; Wang, Z.; Hellweg, S. Deep Dive into Plastic Monomers, Additives, and Processing Aids. *Environ. Sci. Technol.* **2021**, *55*, 9339–9351. [CrossRef]
55. Anjum, A. Recovered Carbon Black from Waste Tire Pyrolysis: Characteristics, Performance, and Valorisation. University of Twente: Enschede, The Netherlands, 2021.
56. Gall, M.; Freudenthaler, P.J.; Fischer, J.; Lang, R.W. Characterization of Composition and Structure–Property Relationships of Commercial Post-Consumer Polyethylene and Polypropylene Recyclates. *Polymers* **2021**, *13*, 1574. [CrossRef] [PubMed]
57. Hahladakis, J.N.; Velis, C.A.; Weber, R.; Iacovidou, E.; Purnell, P. An overview of chemical additives present in plastics: Migration, release, fate and environmental impact during their use, disposal and recycling. *J. Hazard. Mater.* **2018**, *344*, 179–199. [CrossRef] [PubMed]
58. Norris, C.J.; Cerdán, A.L.; ter Haar, P. Understanding Recovered Carbon Black. *Rubber Chem. Technol.* **2023**, *96*, 196–213. [CrossRef]
59. Schlossnikl, J.; Pinter, E.; Jones, M.P.; Koch, T.; Archodoulaki, V.-M. Unexpected obstacles in mechanical recycling of polypropylene labels: Are ambitious recycling targets achievable? *Resour. Conserv. Recycl.* **2024**, *200*, 107299. [CrossRef]
60. Chea, J.D.; Yenkie, K.M.; Stanzione, J.F.; Ruiz-Mercado, G.J. A generic scenario analysis of end-of-life plastic management: Chemical additives. *J. Hazard. Mater.* **2023**, *441*, 129902. [CrossRef]
61. Ghosh, A. Performance modifying techniques for recycled thermoplastics. *Resour. Conserv. Recycl.* **2021**, *175*, 105887. [CrossRef]
62. Dorigato, A. Recycling of polymer blends. *Adv. Ind. Eng. Polym. Res.* **2021**, *4*, 53–69. [CrossRef]
63. Orzan, E.; Janewithayapun, R.; Gutkin, R.; Lo Re, G.; Kallio, K. Thermo-mechanical variability of post-industrial and post-consumer recyclate PC-ABS. *Polym. Test.* **2021**, *99*, 107216. [CrossRef]
64. Moulin, L.; Da Silva, S.; Bounaceur, A.; Herblot, M.; Soudais, Y. Assessment of recovered carbon black obtained by waste tires steam water thermolysis: An industrial application. *Waste Biomass Valorization* **2017**, *8*, 2757–2770. [CrossRef]
65. Kong, D.; Wang, S.; Shan, R.; Gu, J.; Yuan, H.; Chen, Y. Characteristics and chemical treatment of carbon black from waste tires pyrolysis. *J. Anal. Appl. Pyrolysis* **2024**, *178*, 106419. [CrossRef]
66. Cabot Corporation. *Industry User Guide: Reclaimed Carbon Blending Process*; Cabot Corporation: Boston, MA, USA, 2024.

67. Lubongo, C.; Bin Daej, M.A.A.; Alexandridis, P. Recent Developments in Technology for Sorting Plastic for Recycling: The Emergence of Artificial Intelligence and the Rise of the Robots. *Recycling* **2024**, *9*, 59. [CrossRef]
68. Yang, J.; Xu, Y.-P.; Chen, P.; Li, J.-Y.; Liu, D.; Chu, X.-L. Combining spectroscopy and machine learning for rapid identification of plastic waste: Recent developments and future prospects. *J. Clean. Prod.* **2023**, *431*, 139771. [CrossRef]
69. Neo, E.R.K.; Yeo, Z.; Low, J.S.C.; Goodship, V.; Debattista, K. A review on chemometric techniques with infrared, Raman and laser-induced breakdown spectroscopy for sorting plastic waste in the recycling industry. *Resour. Conserv. Recycl.* **2022**, *180*, 106217. [CrossRef]
70. Nyuk Khui, P.L.; Rahman, M.R.; Matin, M.M.; Bin Bakri, M.K. 7—Recycled rubber waste plastic and its composites. In *Recycled Plastic Biocomposites*; Rahman, M.R., Bin Bakri, M.K., Eds.; Woodhead Publishing: Cambridge, UK, 2022; pp. 147–163.
71. Okoye, C.O.; Jones, I.; Zhu, M.; Zhang, Z.; Zhang, D. Manufacturing of carbon black from spent tyre pyrolysis oil—A literature review. *J. Clean. Prod.* **2021**, *279*, 123336. [CrossRef]
72. Martınez, J.D.; Gisele Jung, C.; Bouysset, J.P. Recovered carbon black: Potential markets. *Tire Waste Recycl.* **2021**, 165–224. [CrossRef]
73. Faraca, G.; Martinez-Sanchez, V.; Astrup, T.F. Environmental life cycle cost assessment: Recycling of hard plastic waste collected at Danish recycling centres. *Resour. Conserv. Recycl.* **2019**, *143*, 299–309. [CrossRef]
74. Bridgestone, Michelin. Whitepaper Recovered Carbon Black Guidelines. 2023. Available online: https://cxf-prod.azureedge.net/b2b-experience-production/attachments/clp2vx72q0l6y01odzd738y3d-rcb-white-paper.pdf (accessed on 19 November 2024).
75. Kahlert, S.; Bening, C.R. Why pledges alone will not get plastics recycled: Comparing recyclate production and anticipated demand. *Resour. Conserv. Recycl.* **2022**, *181*, 106279. [CrossRef]
76. Calzolari, T.; Genovese, A.; Brint, A. The adoption of circular economy practices in supply chains—An assessment of European Multi-National Enterprises. *J. Clean. Prod.* **2021**, *312*, 127616. [CrossRef]
77. Donnet, J.-B. *Carbon Black: Science and Technology*; Routledge: London, UK, 2018.
78. Kumar, A.; Pinto, P.; Hawaldar, I.T.; Spulbar, C.M.; Birau, F.R. Crude oil futures to manage the price risk of natural rubber: Empirical evidence from India. *Agric. Econ.* **2021**, *67*, 423–434.
79. Taylor, B. BIR Autumn 2024: A Year of Setbacks. Profits Have Been Hard to Come by for Plastics Recyclers and Reprocessors in 2024, with Europe's Dismal Landscape a Particular Disappointment. Available online: https://www.recyclingtoday.com/news/plastic-bir-recycling-meeting-challenges-pricing-mandates-unep-singapore-2024/ (accessed on 19 November 2024).
80. European Union. *Regulation (EU) 2023/1805 of the European Parliament and of the Council of 13 September 2023 on the Use of Renewable and Low-Carbon Fuels in Maritime Transport, and Amending Directive 2009/16/EC*; European Union: Brussels, Belgium, 2023.
81. Chapman, A.; Sen, K.K.; Mochida, T.; Yoshimoto, Y.; Kishimoto, K. Overcoming barriers to proactive plastic recycling toward a sustainable future. *Environ. Chall.* **2024**, *17*, 101040. [CrossRef]
82. European Plastics Converters. Sustainability/Circular Economy. Available online: https://www.plasticsconverters.eu/ (accessed on 19 November 2024).
83. Plastics Europe. Sustainability. Available online: https://plasticseurope.org/sustainability/ (accessed on 19 November 2024).
84. Tan, J.; Tan, F.J.; Ramakrishna, S. Transitioning to a circular economy: A systematic review of its drivers and barriers. *Sustainability* **2022**, *14*, 1757. [CrossRef]
85. Alliance to End All Plastic Waste. Advancing a Circular Economy for Plastic. Available online: https://www.endplasticwaste.org/ (accessed on 5 March 2025).
86. Fan, Y.; Fowler, G.D.; Zhao, M. The past, present and future of carbon black as a rubber reinforcing filler—A review. *J. Clean. Prod.* **2020**, *247*, 119115. [CrossRef]
87. Cecon, V.S.; Da Silva, P.F.; Curtzwiler, G.W.; Vorst, K.L. The challenges in recycling post-consumer polyolefins for food contact applications: A review. *Resour. Conserv. Recycl.* **2021**, *167*, 105422. [CrossRef]
88. Vilaplana, F.; Karlsson, S. Quality Concepts for the Improved Use of Recycled Polymeric Materials: A Review. *Macromol. Mater. Eng.* **2008**, *293*, 274–297. [CrossRef]
89. Pfaendner, R. Restabilization—30 years of research for quality improvement of recycled plastics review. *Polym. Degrad. Stab.* **2022**, *203*, 110082. [CrossRef]
90. Marcinowski, M. Committee D36 on Recovered Carbon Black (rCB). Available online: https://www.astm.org/membership-participation/technical-committees/committee-d36 (accessed on 18 November 2024).
91. ASTM. *Resolve 2023 Annual Report*; ASTM: West Conshohocken, PA, USA, 2024.
92. De Tandt, E.; Demuytere, C.; Van Asbroeck, E.; Moerman, H.; Mys, N.; Vyncke, G.; Delva, L.; Vermeulen, A.; Ragaert, P.; De Meester, S.; et al. A recycler's perspective on the implications of REACH and food contact material (FCM) regulations for the mechanical recycling of FCM plastics. *Waste Manag.* **2021**, *119*, 315–329. [CrossRef]
93. Waszczyłko-Miłkowska, B.; Bernat, K.; Szczepański, K. Assessment of the Quantities of Non-Targeted Materials (Impurities) in Recycled Plastic Packaging Waste to Comply with EU Regulations and Sustainable Waste Management. *Sustainability* **2024**, *16*, 6226. [CrossRef]

94. European Union. *Commission Notice. Guidance Document for Regulation (EU) 2023/1115 on Deforestation-Free Products (C/2024/6789)*; European Union: Brussels, Belgium, 2023.
95. Manžuch, Z.; Akelytė, R.; Camboni, M.; Carlander, D. Chemical Recycling of Polymeric Materials from Waste in the Circular Economy. 2021. Available online: https://www.actu-environnement.com/media/pdf/news-38548-rapport-echa-recyclage-chimique.pdf (accessed on 18 November 2024).
96. de Römph, T.J.; Van Calster, G. REACH in a circular economy: The obstacles for plastics recyclers and regulators. *Rev. Eur. Comp. Int. Environ. Law* **2018**, *27*, 267–277. [CrossRef]
97. European Union. *Regulation (EU) 2024/1157 of the European Parliament and of the Council of 11 April 2024 on Shipments of Waste, Amending Regulations (EU) No 1257/2013 and (EU) 2020/1056 and Repealing Regulation (EC) No 1013/2006*; European Union: Brussels, Belgium, 2024.
98. European Commission. Guidance on the Interpretation of Key Provisions of Directive 2008/98/EC on Waste. Available online: https://ec.europa.eu/environment/pdf/waste/framework/guidance_doc.pdf (accessed on 18 November 2024).
99. Kara, S.; Hauschild, M.; Sutherland, J.; McAloone, T. Closed-loop systems to circular economy: A pathway to environmental sustainability? *CIRP Ann.* **2022**, *71*, 505–528. [CrossRef]
100. Bakker, C.; Balkenende, R.; Poppelaars, F. Design for product integrity in a circular economy. In *Designing for the Circular Economy*; Routledge: London, UK, 2018; pp. 148–156.
101. PureCycle. Multiple Compounds Successfully Created Using PureCycle Resin. Available online: https://www.purecycle.com/blog/multiple-compounds-successfully-created-using-purecycle-resin?utm_source=chatgpt.com (accessed on 14 February 2025).
102. Uekert, T.; Singh, A.; DesVeaux, J.S.; Ghosh, T.; Bhatt, A.; Yadav, G.; Afzal, S.; Walzberg, J.; Knauer, K.M.; Nicholson, S.R.; et al. Technical, Economic, and Environmental Comparison of Closed-Loop Recycling Technologies for Common Plastics. *ACS Sustain. Chem. Eng.* **2023**, *11*, 965–978. [CrossRef]
103. European Commission. *A New Circular Economy Action Plan for a Cleaner and More Competitive Europe*; European Commission: Brussels, Belgium, 2020.
104. Maitre-Ekern, E. Re-thinking producer responsibility for a sustainable circular economy from extended producer responsibility to pre-market producer responsibility. *J. Clean. Prod.* **2021**, *286*, 125454. [CrossRef]
105. Tumu, K.; Vorst, K.; Curtzwiler, G. Global plastic waste recycling and extended producer responsibility laws. *J. Environ. Manag.* **2023**, *348*, 119242. [CrossRef] [PubMed]
106. Winternitz, K.; Heggie, M.; Baird, J. Extended producer responsibility for waste tyres in the EU: Lessons learnt from three case studies—Belgium, Italy and the Netherlands. *Waste Manag.* **2019**, *89*, 386–396. [CrossRef] [PubMed]
107. Campbell-Johnston, K.; Calisto Friant, M.; Thapa, K.; Lakerveld, D.; Vermeulen, W.J.V. How circular is your tyre: Experiences with extended producer responsibility from a circular economy perspective. *J. Clean. Prod.* **2020**, *270*, 122042. [CrossRef]
108. Joseph, A.M.; George, B.; Madhusoodanan, K.; Alex, R. The current status of sulphur vulcanization and devulcanization chemistry: Devulcanization. *Rubber Sci.* **2016**, *29*, 62–100.
109. Bockstal, L.; Berchem, T.; Schmetz, Q.; Richel, A. Devulcanisation and reclaiming of tires and rubber by physical and chemical processes: A review. *J. Clean. Prod.* **2019**, *236*, 117574. [CrossRef]
110. Saputra, R.; Walvekar, R.; Khalid, M.; Mubarak, N.M.; Sillanpää, M. Current progress in waste tire rubber devulcanization. *Chemosphere* **2021**, *265*, 129033. [CrossRef]
111. Zhang, L.; Wang, H.; Zhang, C.; Wang, S. Laboratory testing and field application of devulcanized rubber/SBS composite modified asphalt. *Case Stud. Constr. Mater.* **2023**, *19*, e02285. [CrossRef]
112. Akhtar, A.Y.; Tsang, H.-H. Dynamic leaching assessment of recycled polyurethane-coated tire rubber for sustainable engineering applications. *Chem. Eng. J.* **2024**, *495*, 153351. [CrossRef]
113. Lapkovskis, V.; Mironovs, V.; Kasperovich, A.; Myadelets, V.; Goljandin, D. Crumb Rubber as a Secondary Raw Material from Waste Rubber: A Short Review of End-Of-Life Mechanical Processing Methods. *Recycling* **2020**, *5*, 32. [CrossRef]
114. Lazorenko, G.; Kasprzhitskii, A.; Mischinenko, V. Rubberized geopolymer composites: Effect of filler surface treatment. *J. Environ. Chem. Eng.* **2021**, *9*, 105601. [CrossRef]
115. Faussone, G.C.; Cecchi, T. Chemical Recycling of Plastic Marine Litter: First Analytical Characterization of The Pyrolysis Oil and of Its Fractions and Comparison with a Commercial Marine Gasoil. *Sustainability* **2022**, *14*, 1235. [CrossRef]
116. Zaino, M.M.; Mustaffa, N.; Hassan, A.H.; Mustapa, A.N.; Mamat, M.H.; Syed-Hassan, S.S.A.; Rahman, S.N.F.S.A.; Azli, A.A.M.; Hamdan, K. A Mini Review on Carbon Black Production as Rubber Reinforcement in Tyre Application. *Chem. Eng. Trans.* **2024**, *113*, 235–240.
117. Czajczyńska, D.; Krzyżyńska, R.; Jouhara, H.; Spencer, N. Use of pyrolytic gas from waste tire as a fuel: A review. *Energy* **2017**, *134*, 1121–1131. [CrossRef]
118. Gaidhane, J.; Ullah, I.; Khalatkar, A. Tyre remanufacturing: A brief review. *Mater. Today Proc.* **2022**, *60*, 2257–2261. [CrossRef]
119. Dabić-Ostojić, S.; Miljuš, M.; Bojović, N.; Glišović, N.; Milenković, M. Applying a mathematical approach to improve the tire retreading process. *Resour. Conserv. Recycl.* **2014**, *86*, 107–117. [CrossRef]

120. Myhre, M.; MacKillop, D.A. Rubber recycling. *Rubber Chem. Technol.* **2002**, *75*, 429–474. [CrossRef]
121. Vasiliu, L.; Gencel, O.; Damian, I.; Harja, M. Capitalization of tires waste as derived fuel for sustainable cement production. *Sustain. Energy Technol. Assess.* **2023**, *56*, 103104. [CrossRef]
122. Ferdous, W.; Manalo, A.; Siddique, R.; Mendis, P.; Zhuge, Y.; Wong, H.S.; Lokuge, W.; Aravinthan, T.; Schubel, P. Recycling of landfill wastes (tyres, plastics and glass) in construction—A review on global waste generation, performance, application and future opportunities. *Resour. Conserv. Recycl.* **2021**, *173*, 105745. [CrossRef]
123. Dabrowska, D.; Rykala, W.; Nourani, V. Causes, types and consequences of municipal waste landfill fires—Literature review. *Sustainability* **2023**, *15*, 5713. [CrossRef]

polymers

MDPI

Article

A Study on the Aging Mechanism and Anti-Aging Properties of Nitrile Butadiene Rubber: Experimental Characterization and Molecular Simulation

Min Zhu [1], Hanyuan Huang [1], Haiyan Li [1,*], Gui Huang [1,*], Jingjing Lan [1], Jing Fu [1], Juqin Fan [1], Yujun Liu [1], Zhiwu Ke [2], Xiaojie Guo [2], Hongkuan Zhou [2] and Yan Li [1]

[1] Naval University of Engineering, Wuhan 430033, China; min0zhu@163.com (M.Z.); h20209999@163.com (H.H.); tangerine1@163.com (J.L.); fujing_2005@sina.com (J.F.); 15172515814@163.com (J.F.); lyj2424@163.com (Y.L.); yacareft@163.com (Y.L.)

[2] China State Shipbuilding Corporation, Limited (CSSC) 719th Research Institute, Wuhan 430205, China; keylab_rndl@163.com (Z.K.); lucky_gxj_0372@163.com (X.G.); hongkuanzhou@hust.edu.cn (H.Z.)

* Correspondence: haiyanli1818@163.com (H.L.); julianxiaoban@163.com (G.H.)

Abstract

To tackle the degradation of sealing performance in nitrile butadiene rubber (NBR) seals due to material aging during long-term service, this study integrates experimental and molecular simulation methods to elucidate the aging mechanism. Experimental results reveal that the contents of C=C and C=O functional groups significantly decrease during aging, accompanied by enhanced hydrophobicity and increased crosslink density of NBR, indicating that crosslinking reactions dominate the aging process with the participation of C=C and C=O groups. Quantum mechanics (QM) and molecular dynamics (MD) simulations further demonstrate that α-H, C=C, and C≡N groups are preferentially oxidized due to their low bond energies. The oxidation of NBR generates unstable epoxy intermediates, which undergo chain scission to form ketones, aldehydes, and ultimately crosslinked structures. Using a multi-dimensional evaluation system based on bond dissociation energy (BDE), solubility parameter (Δδ), and migration coefficient (MSD), four antioxidants (4010NA, 4010, MC, and BHT) were screened. BHT emerges as the optimal choice, exhibiting superior free radical scavenging ability (BDE = 346.3 kJ/mol), good matrix compatibility (Δδ = 2.95), and anti-migration properties. The MD-based screening method established herein provides a theoretical basis for designing antioxidant systems in high-performance rubber materials, facilitating the development of advanced rubber products.

Keywords: rubber seal; aging; molecular model

1. Introduction

Nitrile butadiene rubber is widely used as a sealing ring in mechanical equipment due to its excellent wear resistance, airtightness, and waterproof performance. During long-term operation under humid, high-temperature, and compressive conditions, seals experience stress relaxation and compression set, leading to reduced contact stress with clamping rings and gradual loss of sealing efficiency. Therefore, investigating the aging mechanism of NBR seals is critical for ensuring equipment safety.

Analyzing failure mechanisms serves as the foundation for describing and confirming material degradation. Current research has established a multi-scale characterization system to dissect this process: aging-induced mechanical property degradation is accelerated under humid–heat conditions, with compression set rates significantly higher than those in dry environments [1]. NBR aging damage is severe and non-uniform, with the most pronounced degradation occurring at edge regions. Crosslinking dominates the aging process [2,3], during which tensile strength initially increases with crosslink density but declines after reaching a peak [3]. FTIR and X-ray photoelectron spectroscopy (XPS) reveal the hydrolysis of nitrile groups during aging, accompanied by the formation of hydroxyl and carbonyl groups [4–6]. FTIR results also indicate plasticizer migration to the rubber surface during thermo-oxidative aging, with a competitive interplay between plasticizer volatilization and medium diffusion in oil environments—an effect more pronounced than in hot air [7].

While experimental studies have significantly advanced aging mechanism exploration, challenges remain: aging tests are time-consuming, and complex degradation pathways are difficult to characterize [8]. With breakthroughs in computational chemistry, multi-scale simulation techniques have emerged as a new paradigm for unraveling rubber aging mechanisms. At the microscale, quantum mechanics (QM) simulations are widely used to quantitatively analyze reaction energetics, such as transition state barriers and reaction free energies, and to derive thermodynamic parameters for hydrogen bonding interactions [9,10].

To enhance the antioxidant capacity of rubber composites, antioxidants are typically incorporated. Selection criteria involve analyzing their interactions with rubber matrices, with priority given to matrix compatibility—a parameter directly impacting composite mechanical properties—and key factors such as radical scavenging activity and migration resistance [11]. Traditional antioxidant screening for rubber composites requires extensive experimental work, whereas molecular dynamics (MD) simulations enable rapid prescreening by calculating solubility parameters (δ) to assess additive-matrix compatibility [12] and evaluating migration resistance through metrics like Mean square displacement (MSD) and free volume fraction [13].

In this study, we comprehensively evaluated the antioxidant capabilities and compatibility with nitrile–butadiene rubber (NBR) of four commercially available antioxidants, amine-type antioxidants 4010NA (N-isopropyl-N′-phenyl-p-phenylenediamine) and 4010 (N-cyclohexyl-N′-phenyl-p-phenylenediamine), phenolic antioxidant BHT (2,6-di-tert-butyl-p-cresol), and reactive antioxidant MC (N-(4-anilinophenyl)maleimide), whose molecular structures are presented in Figure 1. The amine-type antioxidants demonstrated strong free-radical scavenging abilities, making them particularly suitable for high-temperature environments; however, their discoloration and contamination tendencies limit their application mainly to dark-colored or black rubber and plastic products [14]. In contrast, the phenolic antioxidant BHT relied on its phenolic hydroxyl group to provide effective antioxidant protection at medium and low temperatures but suffered from limitations such as relatively low polarity matching degree, limited compatibility, and insufficient high-temperature stability [15]. Notably, the reactive antioxidant MC exhibited a unique dual functionality by covalently bonding to the rubber matrix or filler surface through chemical grafting of its double bond and aniline group, thereby achieving both "permanent" antioxidant properties and improved filler–rubber compatibility, which significantly inhibited antioxidant migration and enhanced interfacial bonding strength [16]. These works collectively provide valuable insights and guidance for the development of high-performance nitrile–butadiene rubber formulations.

In this study, we simulated the service environment of NBR and conducted thermo-mechanical accelerated aging tests to investigate surface property and crosslink density changes using FTIR, crosslink density analysis, and contact angle measurements. MD simulations were employed to analyze bond energies in NBR's typical chemical structures, identify degradation-prone sites during aging, and elucidate the underlying aging mechanisms. Four antioxidants—4010NA, 4010, MC, and BHT—were systematically evaluated for their compatibility, antioxidant efficiency, and migration resistance in NBR matrices. This work provides theoretical guidance for developing polymer composites with optimized anti-aging and mechanical properties through multi-scale analysis of antioxidant-NBR interactions.

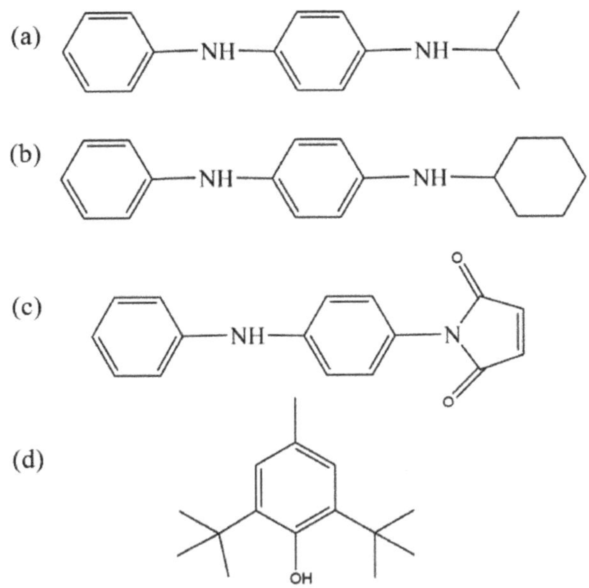

Figure 1. Chemical structures of antioxidants: (**a**) 4010NA; (**b**) 4010; (**c**) MC; and (**d**) BHT.

2. Methods

2.1. Aging of the Sample

The NBR used in this study was provided by Qingdao Yike Rubber & Plastics Co., Ltd. in Qingdao, China (specific formulation details were withheld by the company for commercial confidentiality). Test specimens were prepared according to GB/T 1683 standards [17] by cutting the NBR into small cylinders with dimensions of φ 10 mm × 10 mm (±0.2 mm height). Accelerated aging tests were conducted under a 30% compression ratio to simulate the actual service conditions of NBR seals. The aging procedure was conducted in a blast drying oven (HW-490 high-temperature aging chamber, manufactured by Aerospace Reliance Corporation in Tianjing, China). The aging temperature was set at 80 °C, which, based on our prior experimental results, exhibits a failure mechanism consistent with that at ambient temperature while enabling a moderate aging rate [18]. Specimens were extracted for characterization after aging at 80 °C under a 30% compression ratio for durations of 19 h, 76 h, 152 h, and 228 h, respectively.

2.2. Performance Test

Attenuated Total Reflection–Fourier Transform Infrared Spectroscopy (ATR-FTIR): Surface functional group changes during NBR aging were monitored via ATR-FTIR. A Nicolet 80 spectrometer (Thermo Fisher Scientific, Waltham, MA, USA) was used with a wavenumber range of 4000–500 cm^{-1}, a resolution of 4 cm^{-1}, and 32 co-added scans to reduce noise.

The sessile drop method using an SL200 contact angle analyzer (KINO, New York, NY, USA) was executed under controlled conditions maintained by HVAC systems: droplet volume = 2.0 ± 0.1 μL (0.5 mm needle), T = 25 ± 0.5 °C, and RH = 50–60%. Three independent measurements were taken at distinct regions (edge, center, and transition zone) per sample, with outliers excluded prior to mean calculation.

Crosslink Density Test: Crosslink density evolution during aging was measured using a modified equilibrium swelling method. Sulfured rubber samples (0.12–0.13 g, m_0) were swollen in a solution of 20 mL toluene and 5 mL ammonia (28 wt%) for 48 h at room temperature. Here, the role of ammonia water is to break the hydrogen bond interaction between the macromolecular chains and the fillers, ensuring that the obtained crosslinking density is derived more from the chemical crosslinking of the molecular chains. Swollen samples were dried to constant mass (m_2) in a vacuum oven at 70 °C. Crosslink density was calculated using the Flory–Rehner equation [19,20]. Three replicates per condition were averaged.

$$v_e = -\frac{\ln(1 - v_f) + v_f + \chi v_f^2}{V_S(v_f^{1/3} - v_f/2)} \quad (1)$$

where V_S represents the molar volume of toluene (106.5 mL/mol), χ denotes the Flory–Huggins interaction parameter between the polymer and toluene (0.435) [16], and v_f is the volume fraction of the polymer at equilibrium swelling, which is calculated using Equation (2):

$$v_f = \frac{m_2/\rho_p}{m_2/\rho_p + (m_1 - m_2)/\rho_s} \quad (2)$$

where ρ_p is the density of the NBR (1.02 g/cm^3) and ρ_s is the density of toluene (0.86 g/cm^3).

2.3. *Molecular Dynamics Simulation*

Bond Dissociation Energy (BDE) Calculation: The Materials Studio (MS) 2020 software (20.1 X64) was used to construct NBR models, and bond dissociation energies of NBR's typical chemical structures were calculated using the Dmol^3 module. The BDE calculation formula is as follows:

$$\Delta G = E(A\bullet) + E(B\bullet) - E(AB) \quad (3)$$

where *E*(*A*), *E*(*B*), and *E*(*AB*) denote the energies of cleavage products *A*, *B*, and the reactant molecule *AB*, respectively.

MD simulations were performed using MS to construct unit cells for pure NBR, 4010NA, 4010, MC, and BHT. Solubility parameters of these systems were analyzed, followed by the creation of NBR/4010NA, NBR/4010, NBR/MC, and NBR/BHT composite cells to evaluate the MSD of antioxidants (denoted as AO) in NBR matrices. Each NBR molecular chain contained 100 repeating units with an acrylonitrile-to-butadiene ratio of 1:4. The pure NBR unit cell consisted of 4 NBR chains, while AO cells contained 30 antioxidant molecules. Composite NBR/AO cells included 4 NBR chains and 10 AO molecules.

The QM simulations in this study were implemented using the DMol^3 module of Materials Studio 2020. The specific settings are as follows: The electronic structure calculation is based on the Perdew–Burke–Ernzerhof (PBE) functional of generalized gradient approximation (GGA), combined with the dual-numerical polarization basis set (DNP+) to describe the system wave function. The characteristic of this basis set is that it contains the polarization *p* functions of all hydrogen atoms and has high accuracy. The Semi-core Pseudopotentials method was selected for external processing. The spin multiplicity is set according to the electronic characteristics of the system: For a system containing single-electron free radicals, the spin multiplicity = the number of lone pairs of electrons +1. To improve the convergence of SCF, the Fermi–Amaldi hybrid algorithm (with an initial hybrid coefficient of 0.2) was adopted, and the Fermi level smoothing parameter (smearing width = 0.005 eV) was set. The convergence criteria for geometric optimization were set as energy change <1 × 10^{-5} Hartree and maximum atomic displacement <0.001 A, and the structure was optimized through the BFGS algorithm. The frequency analysis was carried out using the Vibrational Analysis command in the Tools tool to ensure that only one virtual frequency occurs in the transition state structure and the vibration mode is correct.

For MD simulations, the COMPASSII all-atom force field [21], validated for polymer systems, was employed with 50 annealing cycles (200 K → 500 K) to eliminate initial conformation bias. The equilibration protocol consisted of two phases: 200 ps pre-equilibration under the NVT ensemble (298 K, Nosé-Hoover thermostat, 1 ps relaxation time), followed by 500 ps equilibration under the NPT ensemble (298 K, 0.1 MPa, Berendsen barostat, compressibility 4.5 × 10^{-5} bar^{-1}) to achieve energy convergence (fluctuation < 0.5 kcal/mol/atom). The final 480 ps trajectory from the production phase was used to calculate solubility parameters via the Hildebrand equation and to analyze molecular mobility through mean squared displacement (MSD) [9,13,16,22,23].

3. Results and Discussion

3.1. FTIR Analysis

FTIR is a powerful analytical tool that can be used to evaluate the changes in the chemical structure of NBR during the aging process. By comparing the FTIR spectra at different aging stages, characteristic changes related to oxidative aging can be identified. Figure 2 shows the Fourier–transform infrared spectra of NBR before and after thermo-oxidative aging. In Figure 2, the peak at 3365 cm^{-1} corresponds to O-H stretching vibration, while those at 2916 cm^{-1}, 2850 cm^{-1}, and 1418 cm^{-1} are caused by the asymmetric stretching vibrations, symmetric stretching, and bending vibration of C–H in -CH_2 [24,25]. The peaks at 1731 cm^{-1}, 1540 cm^{-1}, and 1162 cm^{-1} arise from C=O stretching and C-O-C stretching vibrations [24,26], and those at 958 cm^{-1} and 903 cm^{-1} are assigned to the vibrations of C-H adjacent to unsaturated bonds in trans-1,4-polybutadiene and vinyl-1,2-polybutadiene units [26–28].

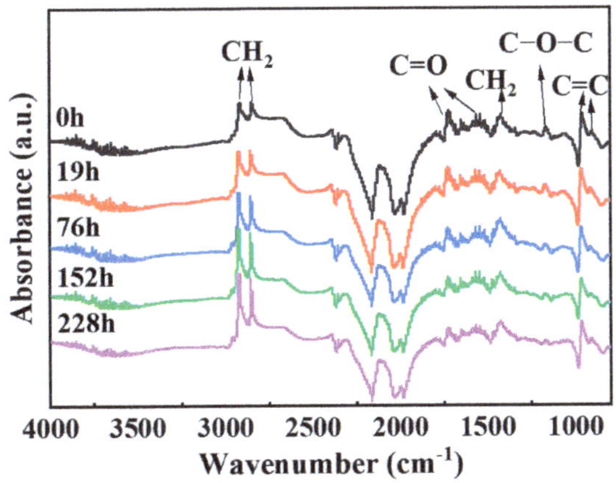

Figure 2. Infrared spectrum changes of nitrile butadiene rubber during aging.

Analysis of the intensity changes of each functional group was conducted by examining the peak areas of their respective vibration bands: -CH_2 (2945–2805 cm^{-1}), -C=O (1745–1675 cm^{-1}), C=C (976–847 cm^{-1}), and C-O-C (1190–1130 cm^{-1}). The peak areas are listed in Table 1. There are a large number of methylene groups (-CH_2) in NBR. Theoretically, the change of methylene groups during the aging process is relatively small [29]. However, due to the crosslinking and degradation of NBR caused by aging, which leads to changes in the surface chemical environment, the relative intensity of the $-CH_2$ absorption peak is enhanced. In this paper, taking the intensity of the $-CH_2$ absorption peak as a standard, the changes in the intensities of each absorption peak relative to that of the $-CH_2$ absorption peak are shown in Table 2. The results show that the intensities of the absorption peaks of C=C, -C=O, and C-O-C all decrease during the aging process [3,30]. This is because during the aging process of NBR materials, the C=C and C=O groups participate in crosslinking reactions, leading to a reduction in their associated infrared absorption peaks as the degree of aging increases [26].

Table 1. Changes in characteristic peaks during the aging process of nitrile butadiene rubber.

Aging Time	CH_2	C=O (1731)	C=C (958)	C-O-C (1162)
0 h	3.41	2.40	7.19	0.82
19 h	3.95	2.20	7.46	0.77
76 h	4.81	1.63	5.59	0.61
152 h	5.50	1.25	5.30	0.50
228 h	5.52	1.37	6.03	0.27

Table 2. Change in characteristic peak strength relative to methylene peak during the aging of NBR.

Aging Time	CH_2	C=O	C=C	C-O-C
0 h	1.00	0.70	2.11	0.24
19 h	1.00	0.56	1.89	0.19
76 h	1.00	0.34	1.16	0.13
152 h	1.00	0.23	0.96	0.09
228 h	1.00	0.25	1.09	0.05

3.2. Crosslinking Density Change

The crosslinking density of nitrile rubber during the aging process was measured using the equilibrium swelling method. The evolution of NBR's crosslink density during aging is shown in Figure 3. As observed from the figure, crosslinking dominates the aging process, with crosslink density increasing progressively with aging time—a trend consistent with the previous literature [2,3].

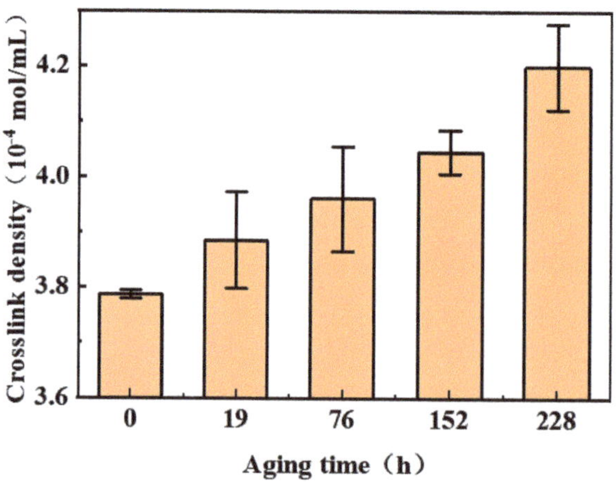

Figure 3. Variation in NBR crosslinking density with aging time.

3.3. Contact Angle Analysis

Contact angle variations reflect changes in NBR's surface polarity: more polar functional groups correspond to smaller water contact angles, while fewer polar groups result in larger angles. In NBR, $-CH_2$ groups exhibit strong hydrophobicity, whereas oxygen-containing groups like -C=O and C-O-C are highly polar.

The contact angle evolution of NBR during aging is shown in Figure 4. The results show that the contact angle of NBR first increases and then decreases with the extension of time. Moreover, the increase in the contact angle (enhanced hydrophobicity) was significantly correlated with the decrease in the relative content of C=O, indicating that the change in surface chemical composition dominated the wettability evolution.

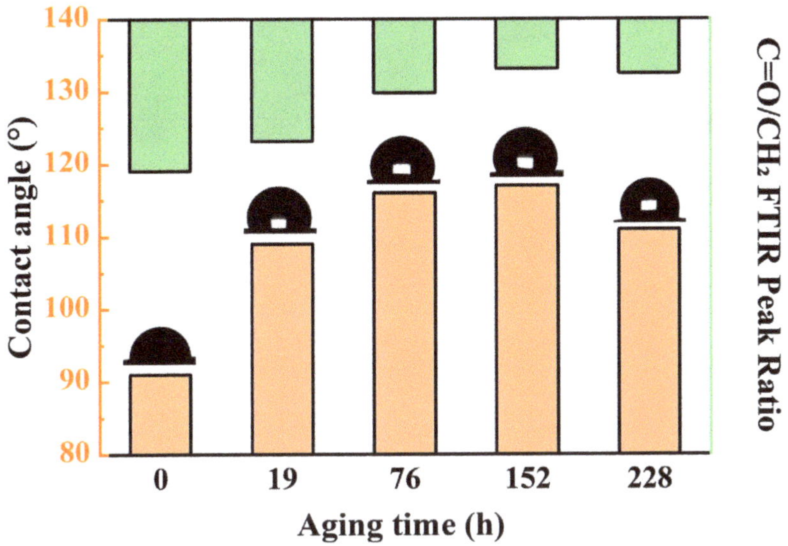

Figure 4. Change in the surface contact angle of NBR during the aging process.

3.4. Molecular Dynamics Analysis of the Aging Mechanism

In previous studies, changes in surface chemical groups and polarity of NBR during aging were characterized using FTIR and contact angle measurements. This section investigates the underlying mechanisms of these chemical changes through MD simulations, which reveal reaction sites and pathways in NBR molecular chains during thermo-oxidative aging.

NBR is synthesized from butadiene and acrylonitrile monomers with added fillers and antioxidants. This analysis focuses solely on the NBR matrix, excluding additive effects on aging. First, a typical structural unit of NBR was constructed using Materials Studio. QM simulations were then performed to calculate the BDE of NBR's representative molecular structures, followed by transition state barrier analysis for further decomposition and crosslinking of oxidation products. These computations provide mechanistic insights into NBR aging processes.

3.4.1. Bond Dissociation Energy

The typical structures of butadiene and acrylonitrile monomers in NBR were analyzed, and the BDE of these typical structures was calculated. The positions of the bonds for which the dissociation energies were calculated are shown in Figure 5, and the calculated dissociation energies are summarized in Table 3.

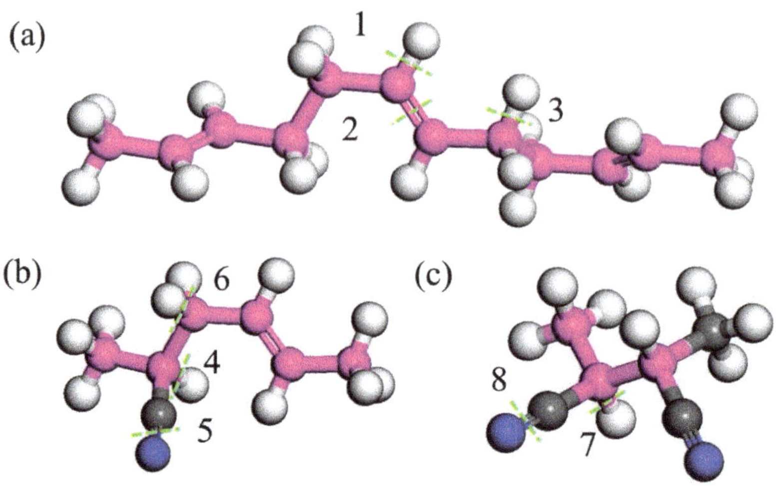

Figure 5. The solution of the typical structure of butylenitrile rubber: (**a**) the butadiene representative unit, (**b**) the acrylonitrile representative unit, and (**c**) acrylonitrile representative unit 2. Where pink beads represent carbon atoms on the carbon skeleton, gray beads represent carbon atoms on the branches, white beads represent hydrogen atoms, and blue beads represent nitrogen atoms. The green dashed line represents the location of key breakage, and its key energy corresponds one-to-one with positions 1–8 in Table 3.

Table 3. Energy and corresponding bond dissociation energies for different structures.

Structural Units	Dissociation Energy (kJ/mol)
H·	-
Butadiene Representative Unit	-
Acrylonitrile Representative Unit	-
Acrylonitrile Representative Unit 2	-
Position 1	467.1
Position 2	269.2
Position 3	371.4
Position 4	377.1
Position 5 (generates C≡N)	266.6
Position 5-2 (generates C−N)	438.9
Position 6	370.4
Position 7	390.7
Position 8 (generates C≡N)	314.0
Position 8-2 (generates C−N)	276.7
Position 9/9′	353.2/380.7
Position 10	351.4
Position 11	372.6
Position 12	377.3
Position 13	346.3

From the table, it can be seen that the C=C double bond (Structure 2 in Figure 5) and C≡N triple bonds (Structures 5 and 8 in Figure 5) exhibit low bond dissociation energies (BDEs). Additionally, the C≡N and C=N bonds in Structure 4c also have low cleavage energies, indicating that these structures are preferentially attacked by O_2 during aging—a finding consistent with the reduced peak intensities of C=C and C≡N bonds observed in FTIR spectra. We also observed that α-hydrogens adjacent to functional groups (Structures 3, 4, and 6) possess low BDEs. During aging, allylic structures like Structure 3 retain high stability and persist for extended durations [31], making α-hydrogens highly reactive and prone to oxidation. As shown in Figure 6, during oxidation, α-hydrogens or C=C double bonds are activated by environmental factors to form radicals, which react with oxygen to generate new radicals and eventually epoxy structures [9,23].

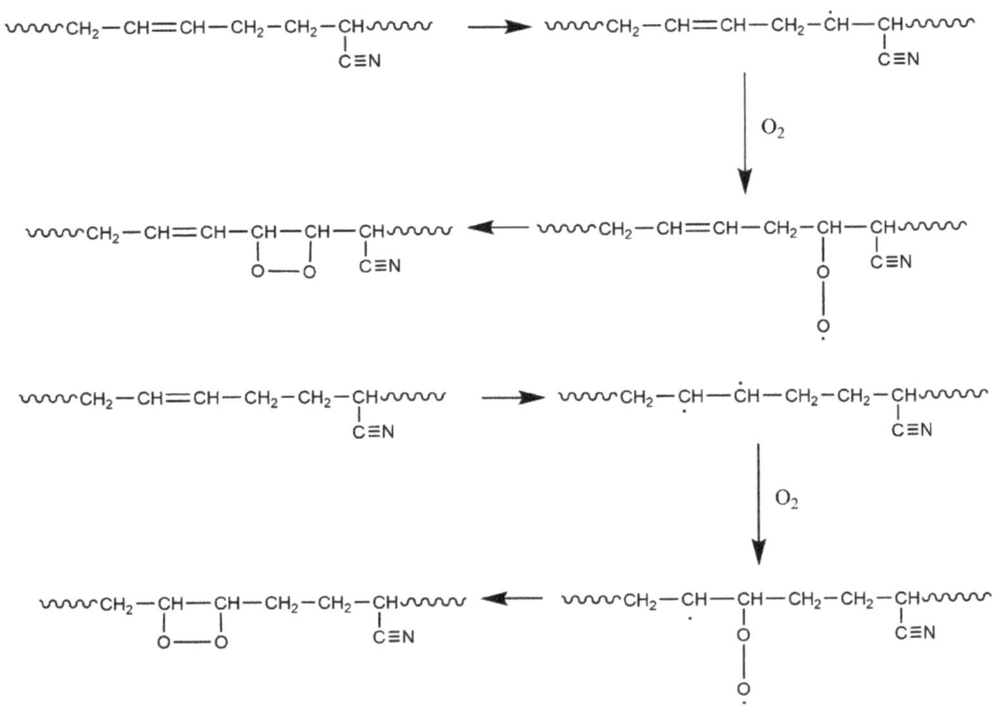

Figure 6. Reaction mechanism of α-H and double bonds with O_2.

3.4.2. Reaction Energy Barriers

The epoxy species formed during NBR oxidation are unstable and further react to generate C=O or -OH groups. This section reports transition state barrier calculations for epoxy decomposition into C=O and -OH via reaction pathway searches.

In physical chemistry, the occurrence of chemical reactions should not only consider thermodynamics but also dynamics. The reaction energy barrier reflects the reaction rate. Generally speaking, the lower the reaction energy barrier, the faster the chemical reaction rate. In order to further analyze the oxidation process of NBR, we simulated the $E_{barrier}$ of the NBR reaction with oxygen by QM. In the calculation process, we use the structure in Figure 5b as the representative unit, which contains both C≡N, C=C, and α-H, and the corresponding bond energy is low, which can affect the reaction process of NBR and oxygen.

Reaction pathways for epoxy decomposition into C=O and -OH (Figure 6) were analyzed. The calculation results (Figure 7a,b) reveal that epoxy groups do not directly form -OH but proceed through a C=O intermediate. The formation of C=O is accompanied by main chain scission, reducing crosslink density. This conclusion is consistent with the degradation pathway of the anti-degradation agent 6PPD reported by Rossomme, where an epoxy intermediate converts into aromatic compounds (aniline and catechol) and carbonyl products (aldehydes and ketones) [23]. Zheng et al. [32] and Ozonation et al. [33] also reported similar behavior, wherein the O-O bond in the epoxy intermediate state is unstable under thermodynamic or ultraviolet conditions, breaks and decomposes into carbonyl products, and leads to chain opening. These findings support the plausibility of our proposed reaction pathway.

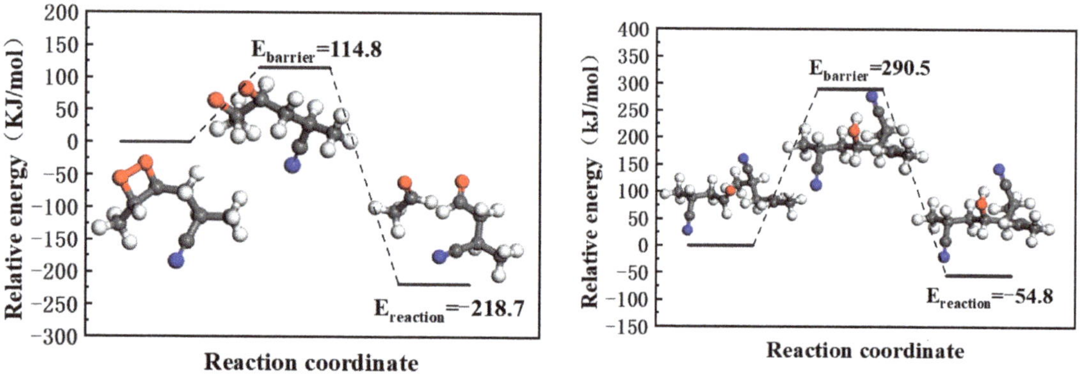

(**a**) Decomposition of epoxy structure into C=O analysis (**b**) Analysis of C=O Decomposition Path

Figure 7. Relationship between energy and chemical structure.

Our calculations also show (Figure 7b) that the oxidized C=O group can react with the unstable hydrogen on the NBR chain to form -OH, triggering crosslinking and spreading aging (with an activation energy of −54.8 kJ/mol). The crosslinking reaction pathway of C=O was confirmed by our experimental evidence and relevant literature. Our infrared spectroscopy results show a significant decrease in the C=O peak intensity during aging (Figure 2), accompanied by an increase in contact angle (Figure 4), indicating reduced surface polarity. This trend corroborates Luo et al.'s [26] conclusion that C=O groups participate in crosslinking reactions during aging. Together, these experimental observations and theoretical calculations provide multi-faceted support for our conclusions.

3.5. Antioxidant Screening

This section evaluates the BDEs, solubility characteristics, and migration/diffusion behaviors of four common AOs—4010NA, 4010, MC, and BHT—using MD simulations. Specifically, BDEs serve as critical indicators of radical-trapping efficiency, directly influencing antioxidant performance; solubility parameters reflect the uniformity of AO dispersion in rubber matrices; and diffusion coefficients correlate with AO durability and extraction resistance.

3.5.1. The Free Energy of Dissociation

As analyzed in Section 3.4, isolated double bonds, nitrile groups, and α-H in rubber macromolecules exhibit low bond dissociation energies (BDEs), making them prone to radical generation under thermo-oxidative conditions. Antioxidants (AOs) suppress oxidation by scavenging radicals, thereby delaying aging. The BDE of labile hydrogens in AOs is a critical metric for evaluating antioxidant efficacy. MD calculations of the labile hydrogen BDEs in Figure 8 (summarized in Table 3) show that all AO BDEs exceed those of double bonds and nitrile groups, indicating that AOs cannot protect these structures. AO BDEs (346.3–377.3 kJ/mol) closely match α-H BDE (371.4 kJ/mol), enabling partial protection against α-H decomposition. Among the tested AOs, BHT has the lowest BDE (346.3 kJ/mol) and the best protective effect.

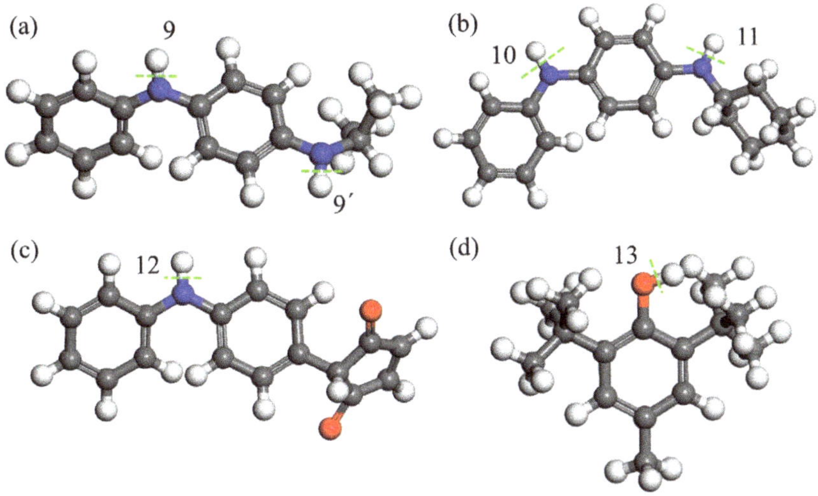

Figure 8. Bond dissociation energies of labile hydrogens: (**a**) 4010NA, (**b**) 4010, (**c**) MC, and (**d**) BHT molecular structure diagram, in which the underlined part is the active hydrogen site. Where red beads represent carbon atoms on the carbon skeleton, gray beads represent carbon atoms on the branches, white beads represent hydrogen atoms, and blue beads represent nitrogen atoms. The green dashed line represents the location of key breakage, and its key energy corresponds one-to-one with positions 9–13 in Table 3.

3.5.2. Solubility Parameters

The solubility parameters of each structure were obtained by statistically analyzing the equilibrium structure according to the MD calculation process in Section 2.3 (taking the average of the three segments). Solubility parameters of AOs and their solubility differences (Δδ) with NBR are listed in Table 4. Solubility parameters evaluate the compatibility of blended materials. According to semi-empirical methods, materials are fully miscible when Δδ < 2.05 $MPa^{1}/^{2}$ and partially miscible when Δδ ranges from 2.05 $MPa^{1}/^{2}$ to 10.02 $MPa^{1}/^{2}$ [9]. In Table 4, the solubility order of AOs in NBR is as follows: BHT > 4010 > 4010 Na > MC. However, all AOs can partially dissolve in NBR and serve as antioxidants for NBR.

Table 4. Solubility parameters of AOs and their solubility differences (Δδ) with NBR.

Structure	NBR	4010NA	4010	MC	BHT
Solubility parameter δ	31.60 ± 0.07	38.03 ± 0.12	35.01 ± 0.13	39.09 ± 0.10	28.65 ± 0.11
△δ	-	6.43	3.41	7.49	2.95

3.5.3. Mean Square Displacement

Based on the AOs/NBR composite cell model constructed in Section 2.3, the composite cells were annealed and pre-balanced according to MD, and the balanced cells were statistically analyzed. And the MSD and diffusion coefficient are used to characterize the anti-extraction property of AO in the rubber matrix, which is expressed as follows [26,34]:

$$MSD(t) = \frac{1}{N}\sum_{i=1}^{N}\left\langle |r_i(t+t_0) - r_i(t_0)|^2 \right\rangle \tag{4}$$

$$D = \frac{1}{6N}\lim_{x\to\infty}\frac{d}{dt}\sum_{i=1}^{N}\left\langle |r_i(t+t_0) - r_i(t_0)|^2 \right\rangle = \frac{1}{6}\lim_{x\to\infty}\frac{d}{dt}MSD(t) \tag{5}$$

where N is the number of molecules, D is the diffusion coefficient, $r_i(t_0)$ is the displacement at t_0, and $r_i(t + t_0)$ is the displacement at $t + t_0$. The value of the diffusion coefficient in the text is equal to $1/6$ of the slope of the MSD linear region.

The MSD curves and diffusion coefficients of four antioxidants in NBR are shown in Figure 9. Analysis shows that MC has the lowest migration rate in NBR, followed by BHT, and 4010 and 4010NA have poor anti-migration ability in NBR.

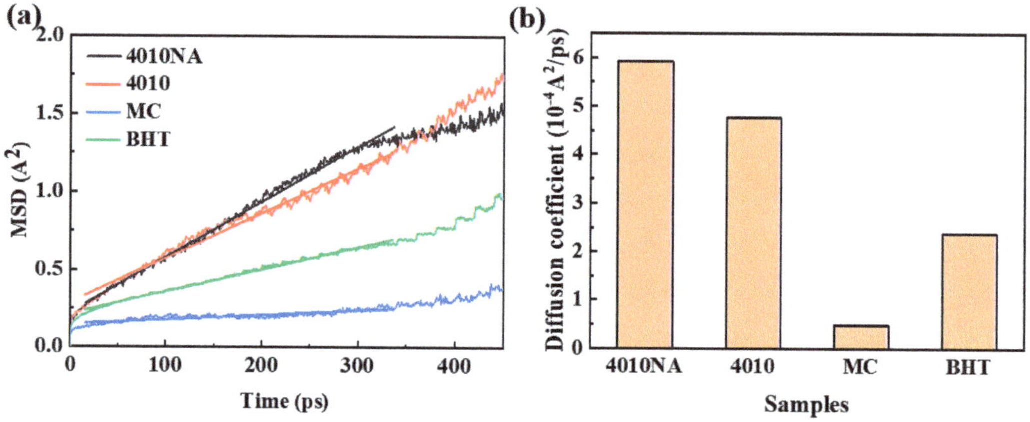

Figure 9. (**a**) MSD curves and (**b**) diffusion coefficient of 4 selected antioxidants in the NBR matrix.

3.5.4. Summary

Table 5 comprehensively summarizes the performance evaluations of four antioxidants, namely 4010NA, 4010, MC, and BHT, in terms of free radical scavenging ability, solubility, and diffusion coefficient within nitrile–butadiene rubber (NBR). The results indicate that MC exhibits outstanding anti-migration properties, maintaining a relatively stable distribution within the NBR matrix. However, it shows poor free radical scavenging efficiency and limited compatibility with NBR, potentially compromising its antioxidant effectiveness in practical applications. Due to their similar molecular structures, 4010NA and 4010 demonstrate comparable performance in free radical scavenging, solubility, and diffusion coefficient in NBR. Notably, 4010 outperforms 4010NA across all indicators, presenting a more balanced and efficient antioxidant profile. Among these antioxidants, BHT stands out with its superior free radical scavenging ability, excellent solubility, and good anti-migration properties. Its comprehensive performance makes it the most optimal antioxidant choice for NBR systems.

Table 5. Summary of free radical trapping ability, compatibility, and migration of four AOs in nitrile rubber.

Structure	4010NA	4010	MC	BHT
Free radical scavenging capacity	3 (353.2)	2 (351.4)	4 (377.3)	1 (346.3)
$\triangle\delta$	3 (6.43)	2 (3.41)	4 (7.49)	1 (2.95)
Diffusion capacity	4	3	1	2

4. Conclusions

This study systematically reveals the thermo-oxidative aging mechanism of NBR under thermo-mechanical coupling environments through a multi-scale research approach combining experimental characterization and MD simulations. A multi-dimensional evaluation framework for antioxidants was established, providing an important theoretical basis for the development of high-performance rubber materials.

Our results showed that absorption peaks of C=C and C=O groups decreased during aging, corroborated by quantum mechanics (QM) simulations confirming preferential oxidation of α-H, C=C, and C≡N groups in NBR. Crosslink density measurements revealed increased crosslinking after aging, indicating dominant chain reorganization during oxidation.

Multidimensional evaluation results from molecular dynamics simulations of antioxidants demonstrate that BHT exhibits the most outstanding comprehensive performance among the four candidate antioxidants. In terms of free radical scavenging ability, BHT has an extremely low OH bond dissociation energy (BDE) of 346.3 kJ/mol, indicating that it can efficiently quench free radicals by releasing hydrogen atoms with minimal energy consumption. Regarding compatibility with the nitrile–butadiene rubber (NBR) matrix, BHT shows a solubility parameter difference ($\Delta\delta$) of only 2.95 with NBR, significantly lower than that of other antioxidants. This suggests strong intermolecular forces between BHT and NBR, enabling uniform dispersion within the rubber matrix. Additionally, the simulation results confirm that BHT has a moderate diffusion rate within NBR, endowing it with excellent anti-migration properties and effectively preventing performance degradation caused by antioxidant loss due to migration. Through quantitative analysis of free radical scavenging efficiency, compatibility, and anti-migration characteristics, BHT is identified as the optimal choice among the four antioxidants, providing an ideal solution for enhancing the antioxidant performance of NBR materials.

This research provides a multi-faceted analysis of NBR's thermo-mechanical aging mechanism, deepening understanding of rubber degradation under coupled environments. Additionally, it establishes a multi-dimensional evaluation method for antioxidants in NBR, offering theoretical guidance for preparing high-performance antioxidant NBR materials.

Author Contributions: Conceptualization, M.Z.; Methodology, M.Z. and Y.L. (Yan Li); Software, H.H., G.H., J.F. (Jing Fu) and J.F. (Juqin Fan); Validation, H.L., G.H., Y.L. (Yujun Liu), Z.K. and H.Z.; Formal analysis, J.L. and J.F. (Jing Fu); Investigation, H.H. and H.L.; Resources, M.Z., J.F. (Jing Fu), Z.K. and X.G.; Data curation, J.L. and J.F. (Juqin Fan); Writing—original draft, H.L.; Writing—review & editing, G.H. and J.F. (Juqin Fan); Visualization, Y.L. (Yujun Liu); Supervision, Y.L. (Yujun Liu), Z.K., X.G. and H.Z.; Project administration, J.L., Z.K., X.G. and H.Z.; Funding acquisition, Y.L. (Yan Li). All authors have read and agreed to the published version of the manuscript.

Funding: This research received no external funding.

Institutional Review Board Statement: Not applicable.

Data Availability Statement: The original contributions presented in this study are included in the article. Further inquiries can be directed to the corresponding authors.

Conflicts of Interest: The authors declare that they have no conflicts of interest.

References

1. Gao, W.; Du, G.; Fan, W.N.; Wang, Y.; Wang, X.L.; Jia, Z.D. Hygrothermal aging characteristics and mechanism of nitrile rubber for transformer sealing. *High Volt. Eng.* **2023**, *49*, 608–617. (In Chinese)
2. Kömmling, A.; Jaunich, M.; Wolff, D. Effects of heterogeneous aging in compressed HNBR and EPDM O-ring seals. *Polym. Degrad. Stab.* **2016**, *134*, 133–141. [CrossRef]
3. Liu, J.; Li, X.; Xu, L.; Zhang, P. Investigation of aging behavior and mechanism of nitrile-butadiene rubber (NBR) in the accelerated thermal aging environment. *Polym. Test.* **2016**, *54*, 59–66. [CrossRef]
4. Wei, X.Q.; Wu, H.L.; Zhang, L.W.; Zhang, S.Y.; Xiao, Y.; Luo, T.Y. Failure Analysis of Nitrile Rubber O-Rings Static Sealing for Packaging Barrel. *J. Fail. Anal. Prev.* **2018**, *18*, 628–634. [CrossRef]

5. Lou, W.; Zhang, W.; Jin, T.; Liu, X.; Wang, H. Stress–thermal oxidative aging behavior of hydrogenated nitrile rubber seals. *J. Appl. Polym. Sci.* **2018**, *136*, 47156. [CrossRef]
6. Zheng, J.; Xiang, K.W.; Huang, G.S. Infrared spectroscopy study on aging mechanism and life prediction of butyl rubber. *Aerosp. Mater. Technol.* **2013**, *43*, 89–92. (In Chinese)
7. Li, G.; Zhuo, W.Y.; Yang, G.; Niu, Y.H.; Li, G.X. Aging behavior and mechanism of nitrile rubber under multi-factor coupling. *Acta Polym. Sin.* **2021**, *52*, 762–774. (In Chinese)
8. Zhang, F.; Yang, R.; Lu, D. Investigation of polymer aging mechanisms using molecular simulations: A review. *Polymers* **2023**, *15*, 1928. [CrossRef]
9. Lu, L.; Luo, K.; Yang, W.; Zhang, S.; Wang, W.; Xu, H.; Wu, S. Insight into the anti-aging mechanisms of natural phenolic antioxidants in natural rubber composites using a screening strategy based on molecular simulation. *RSC Adv.* **2020**, *10*, 21318–21327. [CrossRef]
10. Wang, X.; Huang, C.; Wang, X.; Luo, Y.; Wang, X. Multiscale simulation study on radiation aging of EPDM and preparation of radiation-resistant materials. *Compos. Sci. Technol.* **2024**, *252*, 110595. [CrossRef]
11. Zhu, M.; Shao, Z.Q.; Zhu, C.H.; Yang, Y. Research progress on additives for radiation-resistant rubber aging. *J. Nav. Univ. Eng.* **2024**, *36*, 57–63. (In Chinese)
12. Zhou, J.; Ranjith, P.G. Insights into interfacial behaviours of surfactant and polymer: A molecular dynamics simulation. *J. Mol. Liq.* **2022**, *346*, 117865. [CrossRef]
13. Yang, J.; Lou, W. Molecule dynamics simulation of the effect of oxidative aging on properties of nitrile rubber. *Polymers* **2022**, *14*, 226. [CrossRef] [PubMed]
14. Puskarova, I.; Breza, M. DFT studies of the effectiveness of p-phenylenediamine antioxidants through their Cu(II) coordination ability. *Polym. Degrad. Stab.* **2016**, *128*, 15–21. [CrossRef]
15. Zhao, W.; He, J.; Yu, P.; Jiang, X.; Zhang, L. Recent progress in the rubber antioxidants: A review. *Polym. Degrad. Stab.* **2023**, *207*, 110223. [CrossRef]
16. Song, M.; Yue, X.; Chang, C.; Cao, F.; Yu, G.; Wang, X. Investigation of the compatibility and damping performance of graphene oxide grafted antioxidant/nitrile-butadiene rubber composite: Insights from experiment and molecular simulation. *Polymers* **2022**, *14*, 736. [CrossRef]
17. *GB/T 1683-2018*; National Technical Committee for Standardization of Rubber and Rubber Products (SAC/TC 35). Determination Method for Constant Compression Permanent Deformation of Vulcanized Rubber. China Standard Press: Beijing, China, 2018.
18. Zhu, M.; Ma, D.; Zhou, Y.; Huang, H.; Shao, Z.; Wu, F.; Li, B. The change of sealing property in the aging process of NBR sealing equipment based on finite element analysis. *Coatings* **2024**, *14*, 1178. [CrossRef]
19. Sun, Y.; Cheng, Z.; Zhang, L.; Jiang, H.; Li, C. Promoting the dispersibility of silica and interfacial strength of rubber/silica composites prepared by latex compounding. *J. Appl. Polym. Sci.* **2020**, *137*, 49526. [CrossRef]
20. Liu, Q.; Huang, W.; Liu, B.; Wang, P.C.; Chen, H.B. Gamma radiation chemistry of polydimethylsiloxane foam in radiation-thermal environments: Experiments and simulations. *ACS Appl. Mater. Interfaces* **2021**, *13*, 41287–41302. [CrossRef]
21. Wan, S.; Li, T.; Chen, S.; Huang, X.; Cai, S.; He, X. Effect of multi-modified layered double hydroxide on aging resistance of nitrile-butadiene rubber. *Compos. Sci. Technol.* **2020**, *195*, 108193. [CrossRef]
22. Yang, B.; Li, Y.; Wang, S.; Nie, R.; Wang, Q. Aminosilane modified graphene oxide for reinforcing nitrile butadiene rubber: Experiments and molecular dynamic simulations. *Compos. Sci. Technol.* **2023**, *235*, 109956. [CrossRef]
23. Rossomme, E.; Hart-Cooper, W.M.; Orts, W.J.; McMahan, C.M.; Head-Gordon, M. Computational studies of rubber ozonation explain the effectiveness of 6PPD as an antidegradant and the mechanism of its quinone formation. *Environ. Sci. Technol.* **2023**, *57*, 5216–5230. [CrossRef]
24. Lou, W.; Zhang, W.; Liu, X.; Dai, W.; Xu, D. Degradation of hydrogenated nitrile rubber (HNBR) O-rings exposed to simulated servo system conditions. *Polym. Degrad. Stab.* **2017**, *140*, 199–208. [CrossRef]
25. Wang, H.; Liu, S.; Liu, G. Investigation on the Thermo—Oxidative Aging Resistance of Nitrile—Butadiene Rubber/Polyamide Elastomer Blend and the Swelling Behaviors in Fuels Predicted via Hansen Solubility Parameter Method. *Polym. Degrad. Stab.* **2023**, *217*, 110512. [CrossRef]
26. Luo, R.; Kang, D.; Huang, C.; Yan, T.; Li, P.; Ren, H.; Zhang, Z. Mechanical Properties, Radiation Resistance Performances, and Mechanism Insights of Nitrile Butadiene Rubber Irradiated with High—Dose Gamma Rays. *Polymers* **2023**, *15*, 3723. [CrossRef]
27. Redon, A.; Le Cam, J.-B.; Robin, E.; Miroir, M.; Fralin, J.-C. Aging Characterization of Different Nitrile Butadiene Rubbers for Sealing in a Pneumatic System: Linking the Change of the Physicochemical State to the Mechanical Properties. *J. Appl. Polym. Sci.* **2023**, *140*, e54068. [CrossRef]
28. Nguyen, D.M.; Tran, D.T.; Nguyen, T.T.; Grillet, A.C.; Kim, N.H.; Lee, J.H. Enhanced Mechanical and Thermal Properties of Recycled ABS/Nitrile Rubber/Nanofil N15 Nanocomposites. *Compos. B Eng.* **2016**, *93*, 280–288.
29. Ammineni, S.P.; Nagaraju, C.; Raju, D.L. Modal performance degradation of naturally aged NBR. *Polym. Test.* **2022**, *114*, 107683. [CrossRef]

30. Li, B.; Li, S.X.; Shen, M.X.; Xiao, Y.L.; Zhang, J.; Xiong, G.Y.; Zhang, Z.N. Tribological behaviour of acrylonitrile-butadiene rubber under thermal oxidation ageing. *Polym. Test.* **2021**, *93*, 106954. [CrossRef]
31. Martínez-Morlanes, M.J.; Castell, P.; Alonso, P.J.; Martinez, M.T.; Puértolas, J.A. Multi-walled carbon nanotubes acting as free radical scavengers in gamma-irradiated ultrahigh molecular weight polyethylene composites. *Carbon* **2012**, *50*, 2442–2452. [CrossRef]
32. Zheng, T.; Zheng, X.; Zhan, S.; Zhou, J.; Liao, S. Study on the ozone aging mechanism of Natural Rubber. *Polym. Degrad. Stab.* **2021**, *186*, 109514. [CrossRef]
33. Ozonation of anilines: Kinetics, stoichiometry, product identification and elucidation of pathways. *Water Res.* **2016**, *98*, 147–159. [CrossRef] [PubMed]
34. Zhou, M.; Liu, J.; Hou, G.; Yang, H.; Zhang, L. Study on structures, dynamics and mechanical properties of styrene butadiene rubber (SBR)/silica interfaces: A fully atomistic molecular dynamics. *Polymer* **2021**, *218*, 123523. [CrossRef]

polymers

MDPI

Article

Improved Rubber Performance Through Phenolic Resin-Modified Silica: A Novel Coupling Mechanism for Enhanced Recyclability

Pilar Bernal-Ortega [1,*], Rafal Anyszka [1], Raffaele di Ronza [2], Claudia Aurisicchio [2] and Anke Blume [1]

[1] Department of Mechanics of Solids, Surfaces & Systems (MS3), Faculty of Engineering Technology, University of Twente, 7522 NB Enschede, The Netherlands; r.p.anyszka@utwente.nl (R.A.); a.blume@utwente.nl (A.B.)
[2] Bridgestone EU NV/SA, Italian Branch—Technical Center, Via del Fosso del Salceto, 00128 Rome, Italy; claudia.aurisicchio@bridgestone.eu (C.A.)
[*] Correspondence: m.d.p.bernalortega@utwente.nl

Abstract

Passenger car tires (PCTs) usually consist of a silica/silane-filled Butadiene Rubber (BR) or Solution Styrene Butadiene (SSBR) tread compound. This system is widely used due to improvements observed in rolling resistance (RR) as well as wet grip compared to carbon black-filled compounds. However, the covalent bond that couples silica via silane with the rubber increases the challenge of recycling these products. Furthermore, this strong covalent bond is unable to reform once it is broken, leading to a deterioration in tire properties. This work aims to improve these negative aspects of silica-filled compounds by developing a novel coupling system based on non-covalent interactions, which exhibit a reversible feature. The formation of this new coupling was accomplished by reacting silica with silane and a phenolic resin in order to obtain simultaneous π–π interactions and hydrogen bonding. The reaction was performed using two different silanes (amino and epoxy silane) and an alkyl phenol–formaldehyde resin. The implementation of the new coupling resulted in improved crosslink density, better mechanical performance, superior fatigue behavior, and a similar rolling resistance indicator.

Keywords: silica; tires; silane; sustainability; coupling; recyclability; phenolic resin

1. Introduction

One of the main challenges to achieve sustainability in the rubber industry is reducing waste generation and improving the existing recycling methods. Managing and treating rubber waste, particularly from End-of-Life Tires (ELTs), is a major challenge [1–3]. Tires have a complex composition and structure. For example, the tread of a Passenger Car Tire (PCT) consists of a mixture of natural and synthetic rubbers, silica, silane, carbon black, oils and chemical additives. Additionally, the chemical crosslinks between the polymer chain as well as between silica and the polymer via silane, which are formed during the vulcanization process, make rubber recycling even more difficult [2,4,5].

Nowadays, a large number of tires is recycled, especially in Europe [2]. The main methodologies used for recycling of tires are mechanical recycling, pyrolysis, and devulcanization [3,6–8]. The ground rubber obtained after the mechanical recycling process can be used in low-value applications such as asphalt or playground surfaces [2]. In the case of pyrolysis, useful products like recovered carbon black or oil are obtained that can be used again in the production of tires. However, the pyrolysis process consumes high amounts of energy and the recovered carbon blacks obtained have a low reinforcing potential [9]. Lastly, the devulcanization of rubber aims to break down the chemical sulfur crosslinks formed during the vulcanization process and go back to the original state of the material to be able to use it again in the production of new tires [6,10]. Still, devulcanization is a complex process, in which only di- and polysulfidic bonds can be broken. The obtained materials show a high deterioration of the in-rubber properties compared to the virgin rubber, which limits its application.

Today, modern passenger car tire tread compounds consist of a silica–SBR-reinforced system. This system is widely used due to the improvements observed in the rolling resistance compared to carbon black-filled compounds [11–13]. Nevertheless, the polar silica particles do not have a good interaction with the non-polar rubber. To improve the compatibility of silica with rubber, silane coupling agents are needed. Silanes are bi-functional structures that act as "bridges" between the silica surface and the rubber [14–16]. They create a chemical bond with the filler and the elastomer, enhancing in this way the in-rubber properties of silica-filled compounds. However, the presence of these covalent bonds is an additional challenge when recycling tire waste. Another drawback of the silica–rubber coupling is that it is short and stiff, and once it is broken under strain, it is not able to re-connect [17,18]. An additional disadvantage in the devulcanization of silica-filled systems is the nature of the silica−silane−polymer bonds. These systems are primarily composed of C−S−C monosulfidic bonds, which are highly stable and resistant to cleavage. As a result, selective breaking the polymer−sulfur−polymer bonds becomes nearly impossible. Consequently, the devulcanization process yields low-quality outcomes.

In recent years, one of the more studied approaches is the implementation of reversible bonds on rubber materials instead of strong covalent bonds to facilitate the devulcanization process [18–24]. In this regard, non-covalent interactions have been widely studied. Reversible non-covalent interactions are weaker than covalent bonds, but they show the ability to be (re-)formed and broken under certain conditions or stimuli, such as pressure, temperature, or the application of a magnetic or electric field [21,25]. This characteristic makes these bonds extremely attractive for polymeric materials in order to achieve products with higher durability or self-healing properties. Some examples of the most studied non-covalent interactions in rubber compounds are hydrogen bonding, π–π stacking, or cation–π interactions [19,20,26–28].

This study presents the development of a novel silica–rubber coupling based on non-covalent interactions with the aim of providing self-healing properties and facilitate the recycling of silica-filled compounds. This new coupling, based on π–π stacking interactions, was achieved through the use of a phenolic resin and silane. Phenolic resins are synthetic polymers derived from the reaction between phenol and formaldehyde [29–31]. They are known for their high thermal stability, mechanical strength, and chemical resistance [29,30,32]. In rubber compounds, they are mainly used as reinforcing, tackifying, or bonding agents. The addition of phenolic resins in rubber compounds leads to improvement in properties such as heat resistance, mechanical strength, and adhesion with other materials [33–36]. The aromatic rings present in the chemical structure of the phenolic resin might interact with the styrene units of the styrene–butadiene rubber through π–π stacking. Furthermore, if the resin has good compatibility with the SBR, physical interactions between the resin and the rubber might occur.

In this work, the formation of this new coupling was accomplished by the reaction of silica with silane and afterwards with a phenolic resin in order to obtain simultaneous π–π interactions and hydrogen bonding. The reaction was performed using two different silanes, 3-Aminopropyltriethoxysilane and 3-Glycidyloxypropyltrimethoxysilane. Additionally, an alkyl phenol–formaldehyde resin was used. The in-rubber results obtained for this novel coupling were compared with a traditional silica–silane-rubber coupling using bis(triethoxysilylpropyl) disulfide as the coupling agent.

2. Materials and Methods

2.1. Materials

The rubber compounds analyzed in this work were prepared using Solution Styrene Butadiene Rubber (SSBR) and Butadiene Rubber (BR) as the polymers and silica as the filler. The polymers used in this work were Sprintan 4601 (SSBR, Synthos, Schkopau, Germany) and Buna CB24 (BR, Arlanxeo, Dormagen, Germany). Silica employed is a highly dispersible and reinforcing precipitated silica, ULTRASIL 7000 GR (Evonik Industries, Wesseling, Germany) with a specific BET surface area of approximately 175 m^2/g. For the preparation of the reference samples, Bis(triethoxysilylpropyl) disulfide silane (TESPD) (Evonik Industries, Wesseling, Germany) was used as a coupling agent. For the in situ modification of silica, 3-Aminopropyltriethoxysilane (Sigma Aldrich, Zwijndrecht, The Netherlands) and 3-Glycidyloxypropyltrimethoxysilane (Sigma Aldrich, Zwijndrecht, The Netherlands) were used as silanes and an alkyl phenol–formaldehyde resin (DUREZ 19900, Sumitomo Bakelite Europe, Ghent, Belgium) as the resin. The rest of ingredients used in the rubber formulations were as follows: zinc oxide (ZnO) and stearic acid were used as activators (Millipore Sigma, Hamburg, Germany); sulfur and N-tert-butyl-benzothiazole sulfenamide (TBBS) (Caldic B.V., Rotterdam, The Netherlands) as curatives, and Treated Distillate Aromatic Extracted (TDAE) (Hansen and Rosenthal, Hamburg, Germany) as oil.

2.2. Compounding and Mixing

The samples studied in this work were prepared according to the formulation shown in Table 1. The compounds studied were as follows:

- Reference: SSBR/silica compound using TESPD as a silane coupling agent;
- AMR5: SSBR/silica compound using 3-Aminopropyltriethoxysilane as silane and 5% of the phenolic resin;
- AMR10: SSBR/silica compound using 3-Aminopropyltriethoxysilane as silane and 10% of the phenolic resin;
- EPR5: SSBR/silica compound using 3-Glycidyloxypropyltrimethoxysilane as silane and 5% of the phenolic resin;
- EPR10: SSBR/silica compound using 3-Glycidyloxypropyltrimethoxysilane as silane and 10% of the phenolic resin.

Table 1. Formulation of the rubber compounds.

Compounds	SBR (phr)	BR (phr)	Silica (phr)	TESPD (phr)	Amino Silane (phr)	Epoxy Silane (phr)	Resin (phr)
Reference	80	20	80	6.2	-	-	-
AMR5	80	20	80	-	5.78	-	4
AMR10	80	20	80	-	5.78	-	8
EPR5	80	20	80	-	-	6.2	4
EPR10	80	20	80	-	-	6.2	8

The amount of resin added to the samples was 5% or 10% of the amount of silica in the compounds. The amount of silane was adjusted equimolarly for each compound based on the amount of TESPD in the reference sample.

The quantities of used ingredients were the same for all compounds: TDAE—37.5 phr, ZnO—2.5 phr, stearic acid—2.5 phr, sulfur—1.4 phr, and TBBS—2 phr.

The rubber compounds were mixed in an internal mixer (Brabender Plasticorder 350S, Duisburg, Germany) with a fill factor of 0.7, an initial temperature of 100 °C for steps 1 and

2, and an initial temperature of 50 °C for step 3, at an initial rotor speed of 50 rpm. The complete mixing procedure is shown in Table 2.

Table 2. Mixing procedure of rubber compounds.

Time (min)	Action
Stage 1. Pre-heating 100 °C—rotor speed 50 rpm	
0.00	Addition of rubber
1.20	Addition of 1/3 filler, 1/2 silane
2.40	Addition of 1/3 filler, 1/2 silane, TDAE
4.00	Addition of 1/3 filler and resin
5.00	Increase in temperature
10.00	End of mixing (reaching 140 °C)
Stage 2. Pre-heating 100 °C—rotor speed 50 rpm	
0.00	Addition of batch from stage 1
1.00	Addition of ZnO and Stearic Acid
1.20	Increase in temperature
5.00	End of mixing (reaching 140 °C)
Stage 3. Pre-heating 50 °C—rotor speed 50 rpm	
0.00	Addition of batch from stage 2
1.30	Addition of curatives (Sulfur, TBBS)
3.00	End of mixing

2.3. *In-Rubber Tests*

The following in-rubber characterizations were performed for all rubber compounds.

2.3.1. Cure Behavior

The vulcanization properties of rubber compounds were analyzed by using a Rubber Process Analyzer, RPA elite from TA instruments (New Castle, DE, USA), at 160 °C, by applying a deformation of 6.98% at a frequency of 1.667 Hz. The t_{90}, the time to reach 90% conversion of each compound was used as a molding time for vulcanizing the rubber samples in a hydraulic press (Wickert Maschinenbau GmbH, Landau, Germany).

2.3.2. Payne Effect

The micro-dispersion of the filler and the degree of filler–rubber interactions were analyzed by the study of the Payne effect. The Payne effect test was performed on a Rubber Process Analyzer, RPA elite from TA instruments (New Castle, DE, USA), with strain sweeps from 0.1% to 100% for cured samples at a frequency of 1.6 Hz and a temperature of 60 °C. The cured samples were vulcanized beforehand inside the equipment chamber according to the vulcanization conditions. The Payne effect is the difference between the storage moduli at 0.56% and 100% of strain.

2.3.3. Macro-Dispersion of the Filler

The macro-distribution of the silica particles in the rubber compounds was studied by the analysis of the macro-dispersion of the filler. The macro-dispersion was studied by means of a Dispergrader, Dispersion Tester Alpha View (Alpha Technologies, Hudson, OH, USA). The macro-dispersion of a filler describes the degree of a filler distribution at a scale of 2 µm up to 100 µm. The rubber samples were investigated by optical light microscopy with a 30° irradiation angle and at 100× magnification.

2.3.4. Crosslink Density

The crosslink density (CLD) of the compounds was studied using two different methods: equilibrium swelling experiments and stress–strain experiments using the Mooney–Rivlin approach.

- The crosslink density by equilibrium swelling experiments was obtained using the Flory Rehner equation [37]. Five vulcanized samples (~0.25 g) of each compound were swollen in toluene at room temperature for a period of 7 days. The samples were extracted before with acetone for 24 h in order to remove the oil and unreacted curing agents.
- The determination of the crosslink density by the Mooney–Rivlin approach was carried out by stress–strain assays performed in a universal testing machine Zwick Z010 (Zwick, Ulm, Germany). Test specimens were 2 mm thick, with a test length of 20 ± 0.5 mm and a width of the narrow portion of 4 ± 0.1 mm, according to the ISO 37 (dye type 2) standard [38]. The stress–strain assays consisted of two different steps:
  (i) In the first step, the samples were pre-cycled 10 times until 200% strain at a crosshead speed of 500 mm/min. This first step was performed in order to destroy the filler–filler network.
  (ii) In the second step, the specimens were stretched until a strain of 200% at a crosshead speed of 10 mm/min.

2.3.5. Stress–Strain Behavior

The mechanical properties of the compounds at room temperature were measured by a universal testing machine Zwick Z05 (Zwick, Ulm, Germany) operating with a crosshead speed of 500 mm/min according to the ASTM D412 standard [39]. For each compound, five samples were tested, and the average result of these samples was determined.

2.3.6. Dynamical Properties

Dynamic mechanical measurements of the vulcanized samples were carried out on a Gabo-Netzsch Eplexor tester (Netzsch, Selb, Germany). The measurements were performed with a frequency of 10 Hz, with a dynamic strain of 1% below 0 °C and 3% above 0 °C. The change in the strain was chosen because the rubber becomes softer at higher temperatures, which can generate noise in the measurements. The temperature range of the test was from −80 to 80 °C. The loss factor (tan δ) of each compound was evaluated at low (0 °C) and at high temperatures (60 °C).

2.3.7. Re-Connectivity of Bonds

After the study of how the new coupling systems affect the in-rubber properties, the best compound was selected and compared with the reference. These two compounds were tested to quantify the degree of the re-connectivity of the non-covalent interactions.

The re-connectivity of the new bonds created with the silica modification was analyzed by:

- The study of the mechanical response of the compounds at high temperatures

The mechanical properties of the compounds were measured by a universal testing machine Zwick Z010 (Zwick, Ulm, Germany) operating with a crosshead speed of 500 mm/min and with a limit strain of 330%, which is the thermal chamber limit. The tests were performed in a temperature chamber at 100 °C.

- Analysis of the changes in the dynamical properties after submitting the compounds to cycling (fatigue) test

The dynamic mechanical properties of the samples were compared before and after the cyclic strain at 100 °C. The measurements of the dynamic properties of the vulcanized compounds before and after tensile cycling were performed in the same way as the dynamic mechanical measurements explained above. The tensile cycling of the samples was performed in a universal testing machine Zwick Z010 (Zwick, Ulm, Germany) operating with a crosshead speed of 500 mm/min. All the samples were cyclically strained from 0% to 100% five times at 100 °C. Immediately after performing the cycling, the dynamical mechanical properties were tested.

- Creep experiments

The creep experiments were measured by a universal testing machine Zwick Z010 (Zwick, Ulm, Germany). A force of 15 N was applied to the samples. This force was maintained for 1 h. After this time, the force was released and the changes in the strain of the samples were measured for another hour. The temperature of the experiments was 100 °C.

- Recycling test

The recyclability of the materials was tested by the study of the re-use of ground material in new compounds. The samples were cryo-ground, added to new compounds, and characterized. The samples were cryogenically ground in a grinder from Fritsch GmbH, Idar-Oberstein, Germany. The granules were frozen with liquid nitrogen and ground with a 0.6 mm sieve. This step was performed in order to facilitate the incorporation into the rubber. The properties of the compounds containing ground materials were compared to the ones of the original samples. For the preparation of the new compounds, the formulation shown in Table 1 was used, substituting 15 phr of the filler with the ground material. The ground material obtained from the selected best-performing sample was introduced to the same virgin compound, and the ground material from the reference was added to the virgin reference compound.

3. Results

The samples with the new coupling were characterized by different methodologies and compared to a reference compound with a traditional silica/silane coupling using TESPD as the coupling agent. The in situ modification of silica was performed in all cases during the mixing process. The new coupling system was developed by reacting silica with silane, followed by the reaction with a phenolic resin, to establish π–π interactions between the resin and the styrene–butadiene rubber. Two different silanes were studied, 3-Aminopropyltriethoxysilane and 3-Glycidyloxypropyltrimethoxysilane. A schematic representation of the possible reactions that can occur in the system is shown in Figure 1. In Figure 1a, the reaction using 3-Aminopropyltriethoxysilane is shown. First, the amino silane reacts covalently with the hydroxyl groups of the silica surface, and subsequently, the amino group (-NH_2) of silane can form hydrogen bonds with the hydroxyl groups of the phenolic resin. However, in extreme conditions (e.g., extremely high temperatures, catalysis), covalent bonding may also occur; however, under the conditions used in this study, hydrogen bonding is assumed to be the dominant interaction in this system. 3-Glycidyloxypropyltrimethoxysilane (Figure 1b), as amino silane, first reacts covalently with the silica surface. Afterwards, silane can form a covalent bond with the phenolic resin via the epoxide ring-opening reaction caused by the hydroxyl groups of the resin. In addition to covalent bonding, hydrogen bonding could occur between the hydroxyl groups on the phenolic resin and the newly formed secondary hydroxyl groups on the silane after the opening of the epoxide ring (Figure 1c). During the mixing process, an additional side reaction could occur. While mixing, silane coupling agents are exposed to atmospheric moisture, leading to their hydrolysis and subsequent condensation to form siloxane networks within the rubber matrix. This reaction pathway is particularly relevant in the presence of amino silanes, which can catalyze these processes. Under such conditions, mechanisms similar to the Stöber process [40]—originally described for the controlled synthesis of monodisperse silica particles from alkoxysilanes in alcoholic media—may occur. In the compounding process, the presence of moisture and amine groups can enhance the in situ formation of silica-like domains via hydrolysis and polycondensation of silanes. This phenomenon could influence the filler dispersion and therefore the mechanical and dynamic properties of the final compound.

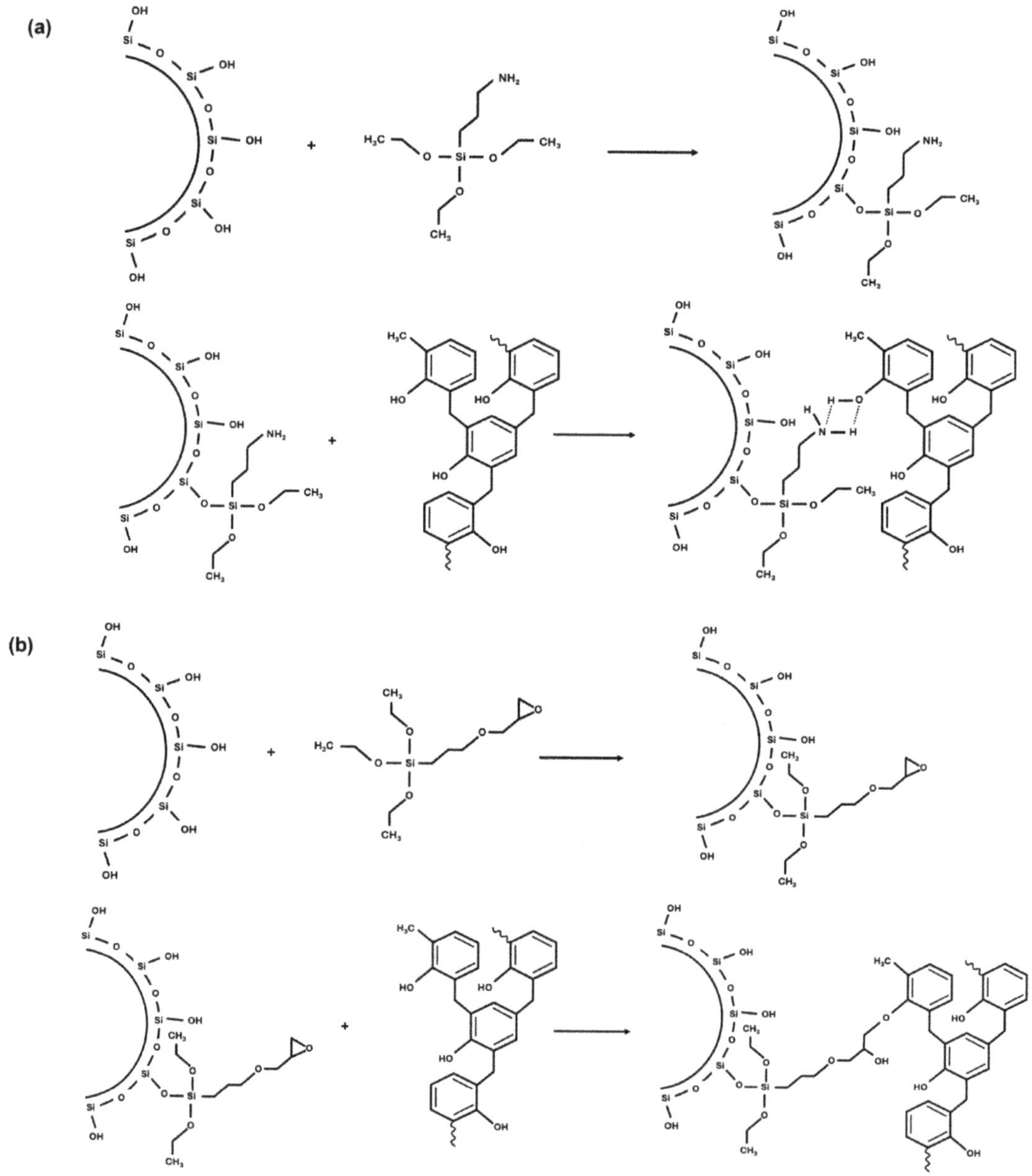

Figure 1. *Cont.*

(c)

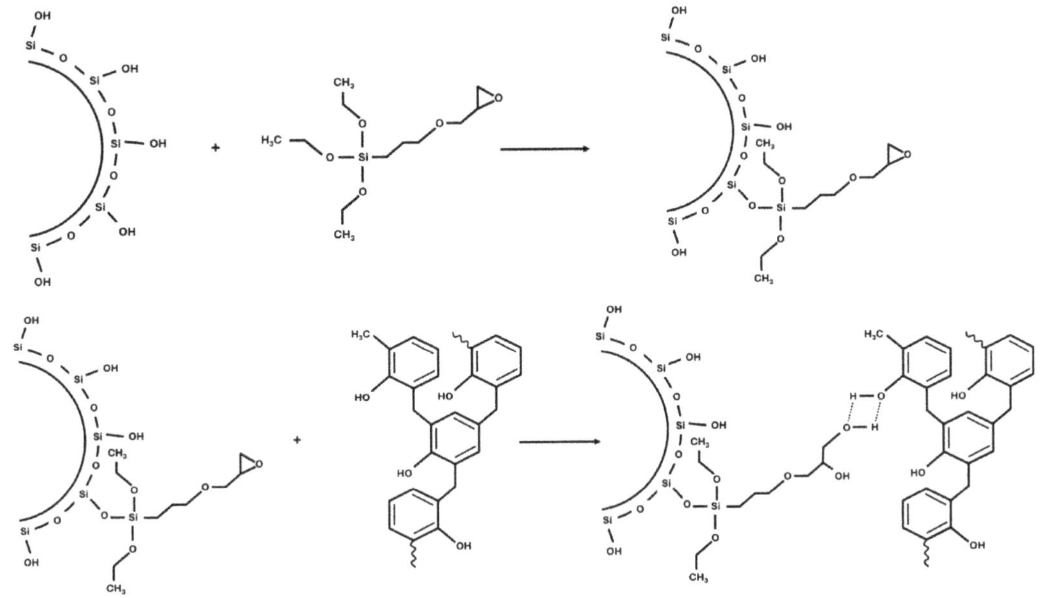

Figure 1. Schematic representation of the silica surface modification mechanism with the phenol resin using (**a**) amino silane, (**b**), and (**c**) epoxy silane.

3.1. Vulcanization

Figure 2 shows the cure behavior of the studied samples. In the graph, it can be observed that samples containing amino silane (AMR5 and AMR10) show a faster vulcanization compared to the samples with epoxy silane and the reference. This is caused by the presence of the amino group (-NH_2). It has been demonstrated that the amino group accelerates vulcanization because it is a base and promotes the opening of the sulfur ring and its conversion to more reactive forms [41,42]. The slow curing behavior of the reference compound is caused by the presence of only one accelerator in this sample. The compound was designed in this way to reduce the complexity of the system to avoid interference with the newly developed coupling systems. Comparing AMR5 and AMR10, the compound with only 5% resin has a faster vulcanization rate. This might be caused by a dilution effect—the higher the content of the resin, the lower the amount of elastomer per volume. Additionally, the phenolic resin can interact with the accelerator, deactivating it for the vulcanization process. The same occurs in the compounds with epoxy silane, where EPR5 undergoes faster vulcanization than EPR10. In the samples with amino silane, two opposite reactions occur in the system: the fastening of the crosslinking reaction due to the presence of the amino group and the deceleration of it due to the phenolic resin. In sample AM5, the effect of the amino silane is more predominant, but in sample AM10, with the increase in the phenolic resin content, the second reaction becomes more significant.

Figure 2 also shows that the samples with 5% of resin (AMR5 and EPR5), independently of silane, show the higher values of the maximum torque (M_H), with AMR5 reaching the highest value in all samples. The addition of 10% of resin in both systems leads to lower M_H. Compared to the reference, all samples with the new couplings show higher values of M_H. Commonly a higher value of maximum torque is an indication of higher crosslink density (CLD). Therefore, these results might suggest that the samples containing resins show a higher CLD than the reference. However, a higher maximum torque can also be caused by other factors. It can also be associated with higher filler–filler interactions and thus worse dispersion in these compounds. This increase in the filler–filler interactions might be a consequence of the presence of many different interactions in these systems: filler–filler interactions (e.g., via silica–resin–silica) or polymer–filler interactions (via silica–resin–polymer).

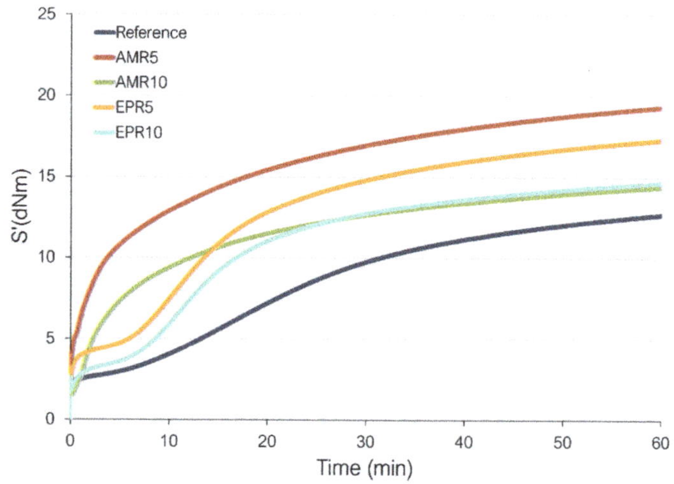

Figure 2. Vulcanization curves of the studied compounds.

3.2. Payne Effect

Figure 3 shows the reduction in the storage modulus (G′) with an increasing strain. This phenomenon is associated with the filler network breakdown, described in the literature as the Payne effect [43–45]. The Payne effect is an indirect method of analyzing the micro-dispersion of the filler in the rubber matrix [43–45]. A higher Payne effect, calculated as the difference between G′ at high and low strains, generally suggests a worse micro-dispersion. The Payne effect also contributes to the energy dissipation in rubber compounds, which is a key parameter in the performance of tires (wet grip and rolling resistance).

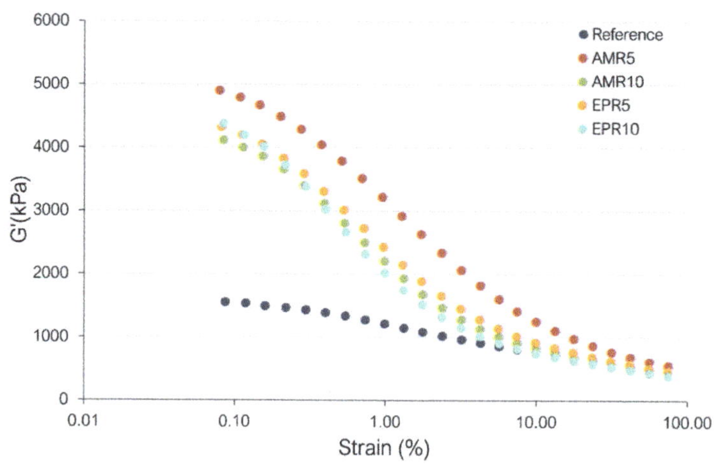

Figure 3. Payne effect curves of the studied compounds.

Figure 3 and Table 3 display that the reference sample has the lowest Payne effect of all samples. The compounds with the grafted resin, either with amino or epoxy silane, present significantly higher values. The sample AMR5 shows the highest Payne effect of all compounds. These results are in agreement with the higher maximum torques observed for these samples in the analysis of the cure behavior. Indicating that the higher torques observed are most likely to be caused by higher filler–filler interactions in these compounds instead of a higher CLD. Figure 3 also shows that in the case of the samples containing resin, the decay of G′ occurs faster than in the reference compound. Due to the increased polarity of the system in the compounds with the new coupling, the interaction possibilities between silica particles have significantly increased. It is possible that silanes (both amino and epoxy silane) can connect different silica particles. Both silanes can chemically couple to silica via the ethoxy group but can also form a hydrogen bond to another Si-OH from another silica particle. The distance between these silica particles might be bigger and more flexible than the one of a direct particle–particle interaction. A schematic representation of these interactions is shown in Figure 4.

Table 3. Payne effect ($\Delta G'$) and G′ at 100% strain of the studied compounds.

Compound	$\Delta G'$ (kPa)	G′ at 100% Strain (kPa)
Reference	1150	400
AMR5	4340	530
AMR10	3700	420
EPR5	3880	440
EPR10	4010	370

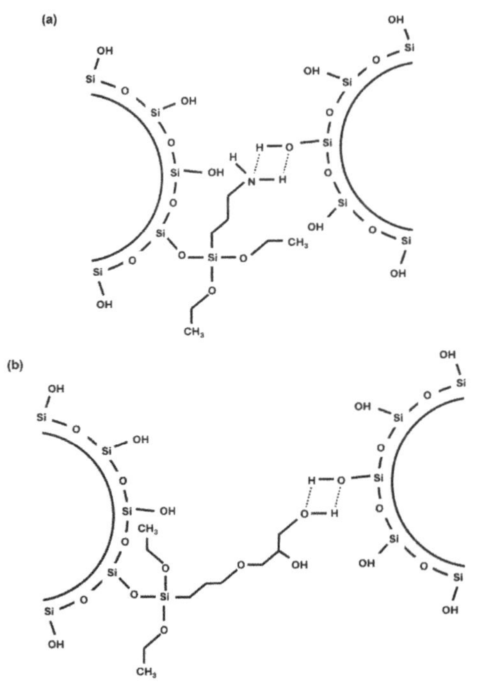

Figure 4. Schematic representation of the interactions between silica particles connected by (**a**) amino and (**b**) epoxy silane.

Another possible explanation for the higher Payne effect in the samples with the new coupling is that not only filler–filler interactions are breaking with the applied strain. It is possible that the non-covalent interactions present in these systems, such as hydrogen bonding or π–π stacking break when strain is applied due to its weaker nature compared to covalent bonding. Therefore, in these cases, the measurement of the Payne effect is the combination of two different effects, the filler–filler and the filler–rubber interactions [18]. This effect is also known from previous studies of functionalized polymers [46], where the interaction of the functionalization is based on hydrogen bonding and the Payne effect curves show similar behavior. At low strains, the non-covalent links remain intact, contributing to a higher value of G' as can be seen in Figure 3. As the strain increases, these weaker interactions start to break down, causing a sharp drop in G'.

Comparing the effect of the different silanes (amino and epoxy silane) it is observed that with 5% of resin, the sample with amino silane shows a substantially higher Payne effect than the one with epoxy silane. In the case of the epoxy silane, it has more direct reactivity towards the phenolic resin through the ring opening. This reaction can lead to the formation of some covalent bonds, stronger and more stable. The amino silane primarily interacts with the resin via hydrogen bonding, and therefore the interactions are weaker and more prone to be broken when deformed.

As can be seen in Table 3 and Figure 3, all samples show similar values of G' at high strains. This parameter is associated with higher rubber–filler interactions. Therefore, this result indicates that the new compounds present similar polymer–filler interactions than the reference that could result in a comparable mechanical performance. This also indicates that not all non-covalent interactions are broken down when the strain increases. It also could be an indication that some of the non-covalent interactions are reformed.

3.3. Macro-Dispersion of the Filler

The macro-dispersion of the filler in the analyzed compounds was evaluated using the optical Dispergrader (Alpha Technologies, Hudson, OH, USA) device. The resulting images are presented in Figure 5. All samples, except for AMR5, show a similar macro-dispersion of the filler. Sample AMR5 presents slightly poorer macro-dispersion, with the presence of larger clusters being noticeable. The implementation of the new coupling system and all the new possible interactions in the system (Figure 4) does not seem to have a significant effect on the macro-dispersion of the silica particles, contrary to what was observed in the Payne effect study which delivers hints on the micro-dispersion. However, both measurements refer to two different length scales and are two independent processes [43–45].

3.4. Crosslink Density

The crosslink density of the studied compounds is shown In Figure 6. The results obtained with equilibrium swelling correspond to the solid color bars and the ones obtained with the Mooney–Rivlin approach to the bars with stripes.

As can be seen in Figure 6, the results of both techniques show some differences. This might be caused by the differences between the two methods such as the use of a solvent in the equilibrium swelling experiments or the pre-treatment to destroy the filler network of the samples in the Mooney–Rivlin tests: In Figure 6, it is shown that the samples containing 5% resin, with both silanes, have a higher CLD than the reference compound. This is more evident in the equilibrium swelling results. In samples containing amino and epoxy silanes, the crosslinks are formed through the vulcanization process via the sulfur addition, but all the additional interactions facilitated by the new coupling system (hydrogen bonds, π–π stacking, or covalent bonding) are also present. These new interactions could contribute to the calculated CLD by swelling, enhancing the overall network structure.

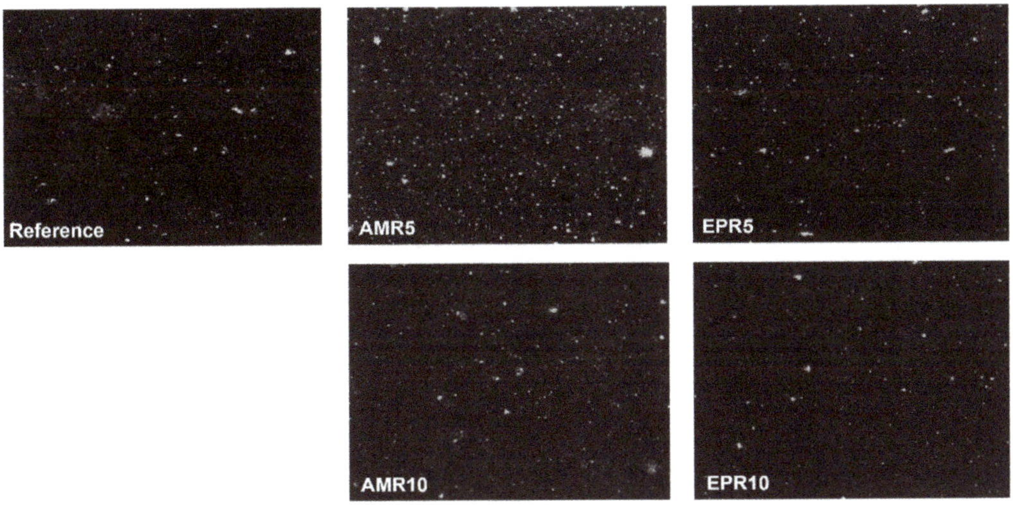

Figure 5. Dispergrader images of the analyzed rubber compounds (150 × 250 μm area).

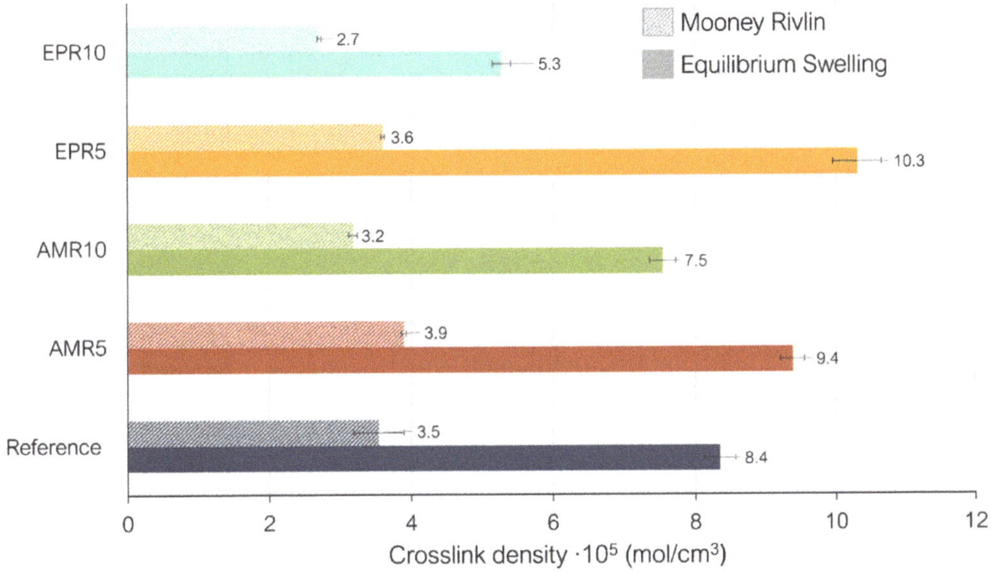

Figure 6. Crosslink density measured by equilibrium swelling and the Mooney–Rivlin approach applied to the studied compounds.

Furthermore, if the resin has good compatibility with the SBR, some physical interactions can be formed. The resin can be entangled in between the polymer chains of the SBR. All these additional interactions can contribute to a higher value of CLD measured by swelling in these compounds compared to the reference sample with TESPD. However, in the case of the CLD results obtained using the Mooney–Rivlin method, samples containing 5% resin exhibit a CLD comparable to that of the reference compound. This discrepancy with the swelling test results may arise from the fact that the Mooney–Rivlin method distinguishes between chemical crosslinks and physical entanglements. As such, the weaker additional interactions in the modified compounds are more likely to contribute to entanglement density rather than to measurable crosslink density. This is supported by the results in Figure 7, where all modified samples show higher entanglement densities than the reference, with AMR5 displaying the highest value.

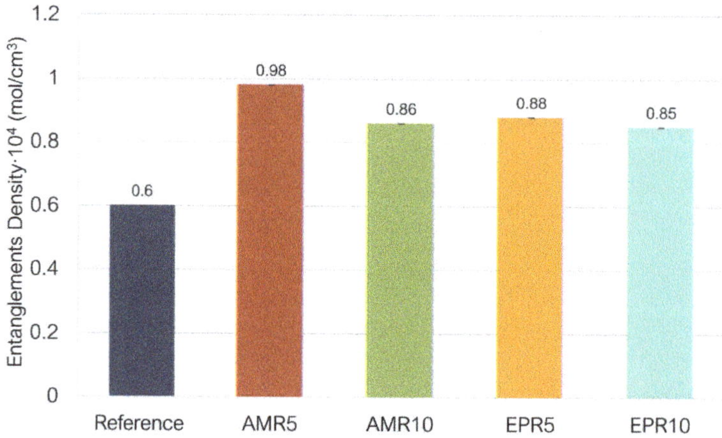

Figure 7. Entanglements density measured by the Mooney–Rivlin method applied to the studied compounds.

When 10% of the resin is added, a slight decrease in the CLD is observed for the compounds containing both silanes. As mentioned in the vulcanization analysis, it could be that 5% of resin is an optimum amount of the material, and 10% is an excessive quantity. The increasing resin content can elevate the possibility that the phenolic resin interacts with the accelerator, deactivating it during the vulcanization process and leading to lower values of CLD. These results, combined with the ones obtained for the Payne effect, seem to indicate that the higher values of the Payne effect and the maximum torques observed in the curing curves are a combination of the CLD, filler–filler, and polymer–filler interactions.

3.5. Stress–Strain Behavior

The mechanical properties of the compounds studied in this work are shown in Figure 8 and Table 4. As can be seen, all samples including resin have a higher tensile strength (T_s) and elongation at break (ϵ_b) than the reference. The notable increase in elongation at break compared to the reference can be attributed to the slippage of entangled polymer chains connected to silica, facilitated by the additional interactions in the new system among silica, silanes, resin, and elastomer [18]. These increased interactions in the newly developed systems contribute to the higher tensile strength observed in these samples.

Table 4. Mechanical properties of rubber compounds at RT.

Compound	Reinforcement Index ($M_{300\%}/M_{100\%}$)	T_s (MPa)	ε_b (%)
Reference	3.6 ± 0.01	10.7 ± 1.8	470 ± 60
AMR5	3.6 ± 0.1	17.5 ± 0.7	600 ± 80
AMR10	3.9 ± 0.1	19.0 ± 1.2	680 ± 70
EPR5	2.8 ± 0.5	15.7 ± 1.9	780 ± 90
EPR10	3.0 ± 0.2	16.6 ± 1.1	880 ± 70

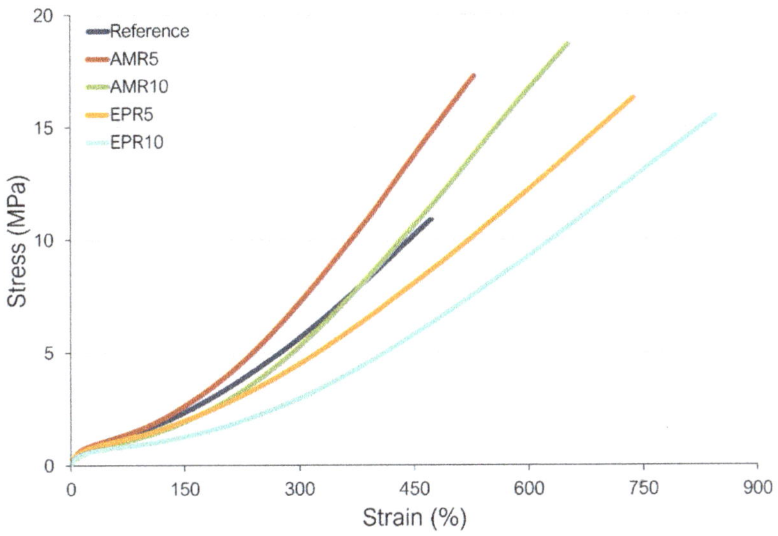

Figure 8. Stress–strain curves at RT of the rubber compounds.

The samples containing amino silane also present a higher reinforcement index (ratio between the modulus at 300% strain and the modulus at 100% strain). Both samples with amino silane exhibit an improved mechanical performance compared to the sample with TESPD and to the samples with epoxy silane. The sample with amino silane and 5% of resin (AMR5) shows the best performance of all compounds. These results are in agreement with the ones obtained for the Payne effect, in which this sample had the highest G′ at high strains, indicating higher filler–rubber interactions in this compound. As observed already in the cure behavior, the Payne effect, and CLD, an increase in the amount of resin to 10% seems to have a negative effect on some in-rubber properties. Independently of silane used (amino or epoxy silane), the addition of 10% of resin caused a deterioration of the moduli compared to the samples with 5%. This result is another hint that 10% of resin is above the optimum amount that should be added into the rubber compounds. As previously mentioned, at this higher material concentration, the resin appears to deactivate the accelerator, resulting in a significantly lower CLD compared to the reference as well as the compounds containing 5% resin.

3.6. Dynamic Properties

The impact of the implementation of the new coupling on the dynamic properties of the rubber compounds was studied through the analysis of the loss factor (tan δ) as a function of temperature. The loss factor is defined as the ratio between the loss and the storage moduli, and it is employed as an indicator of tire performance. Usually, the values of tan δ at low (0 °C) and high temperatures (60 °C) are used as indicators of the wet grip (WG) and the rolling resistance (RR), respectively [47]. These two properties are key in tire performance and form—together with the abrasion resistance—the Magic Triangle of tires. However, the prediction for the RR is more accurate than the one of the WG. The reason for this is that the wet grip is a complex phenomenon that takes place within a high-frequency range of 10^4 to 10^7 Hz, and these frequencies are too high to be measured using DMA [47].

The variation in the loss factor as a function of temperature is shown in Figure 9.

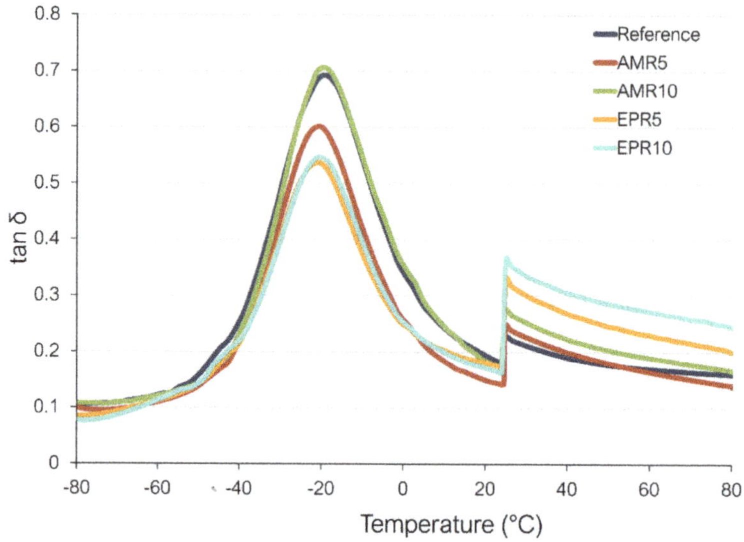

Figure 9. Variation in the loss factor (tan δ) as a function of temperature of the studied compounds (the sharp drop in the lost factor at 25 °C is caused by the increase in applied strain).

As can be seen in Figure 9, the application of the new coupling system has a significant impact on the dynamical properties of the compounds. Substantial differences can be observed at both low and high temperatures. Observing the values of the loss factor at 0 degrees, commonly used as the wet grip indicator, it is shown that samples AMR5, EPR5 and EPR10, present a lower value than the reference. While sample AMR10, shows a slightly higher loss factor at 0 degrees. The lower loss factor for the samples AMR5, EPR5, and EPR10 could be caused by a stronger network in these compounds. As discussed before, it is assumed that these samples contain more interactions in the system than the reference, leading to a more elastic material and therefore to lower hysteresis. These samples contain not only the crosslinks formed during the vulcanization but also additional non-covalent interactions between silica, the resin, and silanes. Regarding the sample AMR10, with 10% resin, it could be, as previously discussed, that the optimum amount of resin is exceeded and that the resin is also acting as a plasticizer, showing a result more similar to the reference. This behavior is not observed in the case of the sample with 10% of resin and epoxy silane, because as discussed beforehand, the epoxy silane has a higher reactivity towards the resin due to the ring opening. This could lead to the presence of stronger (covalent) interactions compared to the sample with amino silane. However, it is known [48] that the prediction of the wet grip using this method is not as reliable as for the rolling resistance when two different systems are compared. Therefore, only a real tire test can show the real grip performance.

Regarding the rolling resistance indicator (tan δ) at 60 °C, some differences to the reference compound are observed. In this case, the behavior of the new coupling is extremely influenced by the type of silane used. The samples AMR5 and AMR10 with amino silane show a significantly lower RR indicator than the samples with epoxy silane. Samples containing amino silane exhibit a behavior which is more similar to the reference, with AMR5 displaying the lowest RR indicator among all compounds. This sample shows a CLD similar to the reference and exhibits stronger rubber–filler interactions, as evidenced by the highest G′ value at high strains in the Payne effect measurement. These interactions contribute to greater elasticity and, consequently, reduced energy dissipation. In the case of samples with epoxy silane, it could be that the higher energy dissipation at higher energy is caused by the bonding and debonding of the different interactions present in the system. This result could be a hint that the amount of these non-covalent interactions is higher in the compounds with epoxy silane.

Figure 9 also shows that the resin has good compatibility with the rubber matrix. When a non-compatible resin is used, a second peak in the tan δ curve is typically observed [49–52]. Therefore, in this case, without having such a second peak, the resin presents a good compatibility with the SBR.

3.7. Re-Connectivity of Bonds

After the analysis of the effects of the new coupling system on the in-rubber properties, the best-performing compound was selected for further analysis of its reversible nature and a recyclability study. The chosen compound was AMR5, in which 3-Aminopropyltriethoxysilane was used as silane together with 5% resin. The selection of this compound was performed because it showed the best overall in-rubber performance from all the newly designed compounds tested.

- Study of the mechanical response of the compounds at high temperatures

The behavior of the new coupling at high temperatures was analyzed by testing the mechanical response of the compounds at 100 °C. The mechanical properties of the reference and the sample AMR5 are shown in Figure 10.

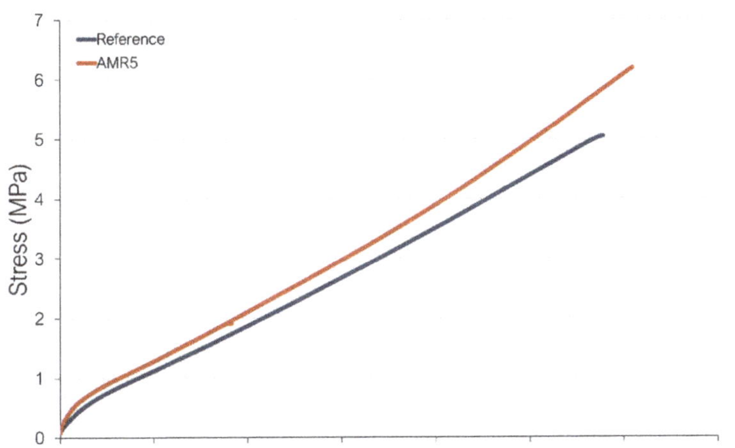

Figure 10. Stress–strain curves at 100 °C of the reference and AMR5.

As observed in Figure 10, the sample AMR5 shows higher moduli at 100 and 300% strain, tensile strength, and elongation at break at 100 °C than the reference. This improved mechanical performance at elevated temperatures could be caused by the reversible nature of the non-covalent interactions present in this system. When a connection is broken in the compound AMR5 due to the applied strain, it is possible to re-connect. This leads to observed higher resistance to applied strain at elevated temperatures than in the reference compound with TESPD.

- Creep experiments

In Figure 11, the results of the creep experiment at 100 °C are depicted.

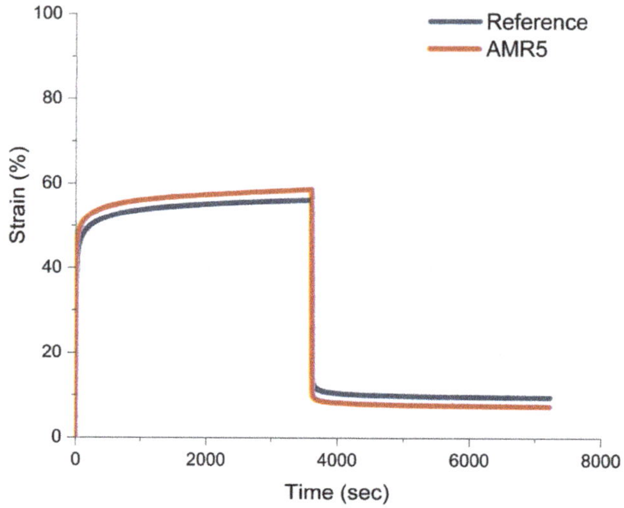

Figure 11. Variation in the strain during the creep experiments for the reference and AMR5.

When the initial force of 15 N is applied, the sample AMR5 reaches a slightly higher initial strain than the reference compound, approximately 50% for AMR5 and 47% for the reference. The stress is then maintained for one hour, in which the strain increases slowly until the stress is released. At this point, the sample AMR5 reaches a maximum strain of ~59% and the reference ~55%. Subsequently, when the stress is released, the compounds go back to their original shape. Both compounds almost regained their previous form. However, it can be observed that sample AMR5 shows a slightly higher recovery than the reference. This could indicate that the non-covalent interactions present in the new system provide the compound with a slightly higher thermo-mechanical fatigue resistance.

- Analysis of the changes in the dynamical properties after subjecting the vulcanizates to a cycling (fatigue) test

The dynamic properties of the selected compounds were analyzed before and after a fatigue test. Compound AMR5 and the reference were subjected to five tensile cycles, each stretched to 100% strain at 100 °C. Following this procedure, the dynamic properties were measured and compared with the values recorded before the cycling test. Figure 12 shows the evolution of the loss factor with temperature. Table 5 depicts the values of tan δ at 0 and 60 degrees.

Table 5. Values of tan δ at 60 °C and 0 °C of the studied compounds before and after the thermo-mechanical treatment.

Compound	tan δ at 60 °C	tan δ at 60 °C After Cycling 5 Times at 100 °C	tan δ at 0 °C	tan δ at 0 °C After Cycling 5 Times at 100 °C
Reference	0.169	0.203	0.352	0.244
AMR5	0.164	0.192	0.266	0.299

As depicted in Figure 12, the variations between the curves before and after fatigue testing are less pronounced for the AMR5 sample compared to the reference compound. For the reference compound, the rolling resistance (RR) indicator rose by approximately 20%, while the wet grip (WG) indicator dropped by around 30%. In comparison, the AMR5 sample showed a 17% increase in the RR indicator and a 12% rise in the WG indicator. The compound with the new coupling shows better wet grip and rolling resistance indicators after the cycling test than the reference with TESPD. This suggests that the new coupling offers better resistance to higher temperatures and fatigue than the conventional silica/silane coupling.

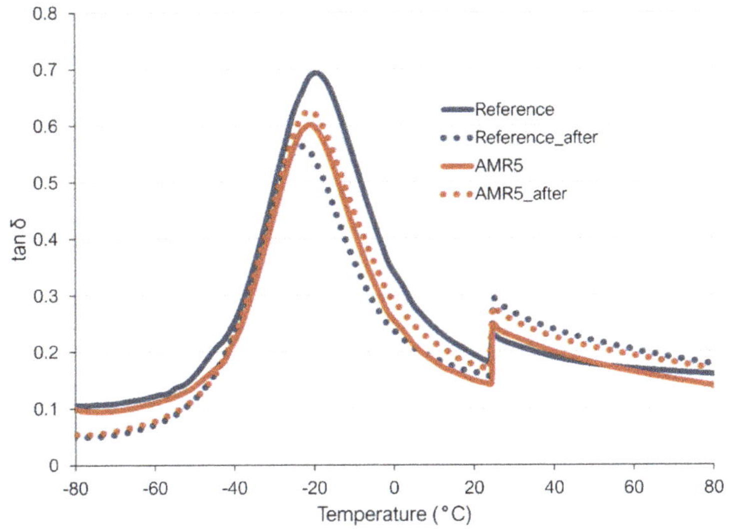

Figure 12. Variation in the loss factor as a function of temperature before and after the thermo-mechanical treatment (the sharp drop in the loss factor at 25 °C is caused by the increase in applied strain).

3.8. Recycling Study

One of the main goals of the development of a new coupling based on non-covalent interactions is to facilitate the recyclability of silica-filled rubber compounds. In order to explore the advantages of the new coupling regarding recyclability, a recycling study was conducted with samples AMR5 and the reference. The primary objective of this test is to determine whether the introduction of a reversible coupling positively influences the recyclability of the rubber compounds. The samples were subjected to a mechanical treatment in which they were ground to smooth its incorporation into virgin rubber compounds as explained in the experimental section of this manuscript. Figure 13 depicts the curing behavior, the Payne effect, stress/strain curves, and the loss factor as a function of temperature of the compounds containing ground rubber material in comparison with the original compounds. In both cases, the ground particles are included in a compound of the same composition as the ground rubber, e.g., AMR5 ground particles were included in the virgin AMR5 compound. As can be seen, all the compounds containing ground rubber show a deterioration in the in-rubber properties. Such behavior was reported in previous studies [6,7,10].

Figure 13a presents the cure curves for the compounds. AMR5_ground exhibits a higher maximum torque than AMR5, while the opposite is true for the reference compound, where the compound with ground rubber shows a decrease in this parameter. As discussed before, the maximum torque can be related to the CLD, filler–filler and polymer–filler interactions. In the case of AMR5_ground, the increase in M_H could be caused by a combination of both, a higher CLD and an increase in all the interactions present in this system. For the reference compound, most likely the decrease in the maximum torque is caused by a decrease in the CLD. In a previous study [52], it was reported that the presence of ground rubber particles causes diffusion of sulfur towards these ground rubber particles. This phenomenon leads to an inhomogeneous distribution of the sulfur in the rubber samples and therefore an inhomogeneous CLD. This behavior could be less pronounced in the samples with the new coupling because the penetration of the sulfur is more difficult than in the reference compound. This can be explained by a higher overall network (crosslinks, filler–filler, and polymer filler interactions) in these samples, due to the possible reformation of the non-covalent interactions after breaking them during the grinding process.

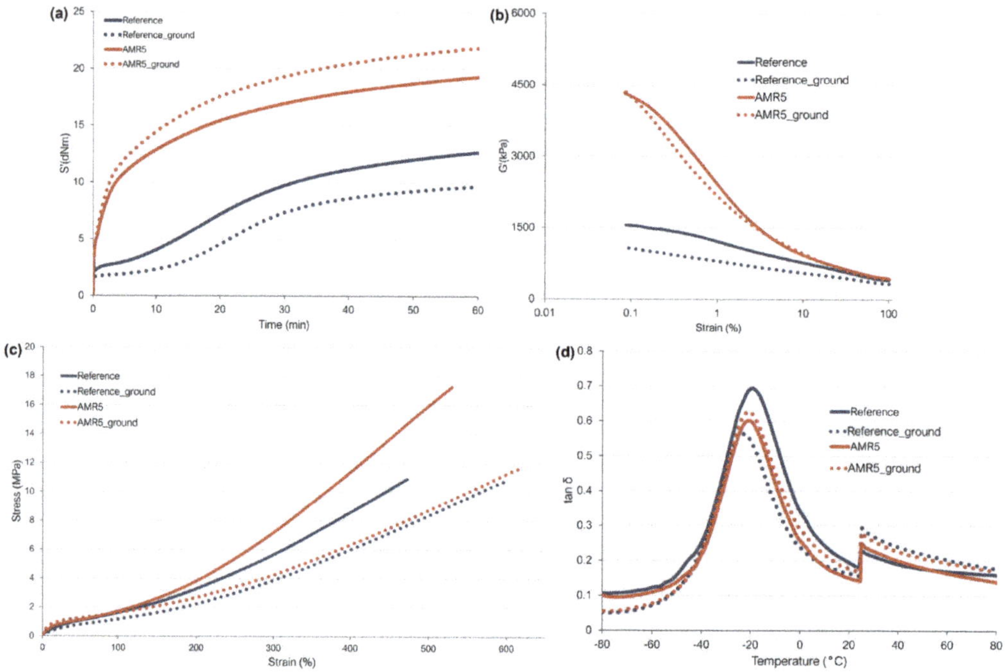

Figure 13. (**a**) Vulcanization curves, (**b**) Payne effect plots, (**c**) stress–strain curves and (**d**) variation in the loss factor as a function of temperature of the compounds analyzed in the recyclability study.

In Figure 13b, the Payne effect curves are depicted. Sample AMR5 with ground material shows a similar Payne effect compared to its original counterpart. However, as previously considered, it is also possible that the Payne effect in these samples is a combination of both the breakdown of the filler network and the break of the non-covalent bonds present in the system. It might be possible that in sample AMR5_ground also the non-covalent interactions present are breaking with the applied strain. In the case of the reference, a significant decrease in the Payne effect is observed. This phenomenon can be attributed to the reduced quantity of fillers in these compounds. In the sample containing ground material, 15 phr of the silica filler was replaced by ground rubber. Since this material consists of rubber, fillers, curatives, and other components, the overall filler content in the compounds that include ground rubber is lower than in the original compound. This reduction results in decreased particle–particle interactions, which in turn leads to a reduced Payne effect.

The mechanical performance of the samples is shown in Figure 13c and Table 6. It can be seen that the compounds with ground material show a significant worsening of the mechanical properties. As explained before, this can also be caused by the inhomogeneity in

CLD caused by the sulfur diffusion to the ground particles [53]. This behavior occurs in both the reference and the sample with the new coupling system. However, the deterioration of the mechanical performance is more pronounced in the sample with the new coupling system. The higher degree of deterioration of the properties for the sample AMR5_ground could be explained in addition by a worse micro-dispersion in this sample as observed by the Payne effect results.

Table 6. Mechanical properties of the compounds analyzed in the devulcanization study.

Compound	Reinforcement Index ($M_{300\%}/M_{100\%}$)	T_s (MPa)	ε_b (%)
Reference	3.6 ± 0.12	11.6 ± 1.8	470 ± 60
Reference_ground	3.2 ± 0.12	10.7 ± 1.0	630 ± 30
AMR5	3.6 ± 0.1	17.5 ± 0.7	600 ± 80
AMR5_ground	3.1 ± 0.2	9.5 ± 0.3	620 ± 40

Lastly, the dynamical mechanical properties of the compounds were analyzed. The loss factor as a function of temperature of the studied samples is shown in Figure 13d. As with the other properties, significant changes can be observed in the samples with ground material. However, in the case of the sample AMR5, the changes in the tan δ curve are smaller than in the reference compound. As mentioned before, the changes in the dynamical properties can be explained by the diffusion of sulfur to the ground particles [53] that is less prominent for the samples with the new coupling.

4. Conclusions

This study explores the development of a new silica/rubber coupling mechanism based on non-covalent interactions. The silica surface was modified with phenolic resin, which was grafted onto silica by reacting with silane that had previously interacted with the silica surface. Two different silanes were analyzed, 3-Aminopropyltriethoxysilane and 3-Glycidyloxypropyltrimethoxysilane. The modification of silica was performed during the mixing of the rubber compounds. The in-rubber results of the compounds containing the new coupling were compared to a conventional silica/silane system with TESPD.

This work revealed that the implementation of this novel coupling based on non-covalent interactions using amino silane and 5% of resin (sample AMR5) leads to an improvement in crosslink density, mechanical performance, and rolling resistance indicator. Tensile test results showed that the samples containing amino silane exhibit superior modulus, tensile strength, and elongation at break compared to the reference, indicating more efficient stress transfer between the filler and rubber matrix, likely due to enhanced filler dispersion and stronger interfacial interactions. Furthermore, these samples also show a promising dynamic behavior, with similar energy dissipation at 60 degrees (rolling resistance indicator) than the reference. However, the use of the new coupling has a negative effect on the Payne effect. Nevertheless, it is possible that the breaking and re-connection of the non-covalent interactions present in these systems is contributing to the higher values of the Payne effect. The improvement in the properties of the new compounds can be explained by the presence of a high number of additional interactions and a good compatibility of the phenolic resin with the rubber matrix. The new coupling also exhibits a reversible nature, which results in better performance when tested at high temperatures and after being subjected to mechanical fatigue. The creep testing supported this theory, showing thar under prolonged loading the new compound had reduced permanent deformation. Furthermore, the new coupling also shows encouraging results in the recyclability study, with a smaller deterioration of the properties compared to the virgin material.

This work shows the capability of the use of non-covalent interactions to develop a new silica/rubber coupling. This system has the potential to be optimized to obtain a higher improvement in rubber performance and recyclability.

Author Contributions: Conceptualization, P.B.-O., R.A. and A.B.; Methodology, P.B.-O., R.A., R.d.R., C.A. and A.B.; Validation, R.A. and A.B.; Formal analysis, P.B.-O.; Investigation, P.B.-O., R.A., R.d.R., C.A. and A.B.; Resources, R.d.R. and A.B.; Data curation, P.B.-O.; Writing—original draft, P.B.-O.; Writing—review & editing, R.A., C.A. and A.B.; Visualization, P.B.-O. and A.B.; Supervision, R.A., R.d.R., C.A. and A.B.; Project administration, R.d.R., C.A. and A.B.; Funding acquisition, R.d.R. and A.B. All authors have read and agreed to the published version of the manuscript.

Funding: This research was funded by Bridgestone EU NV/SA, Italian Branch—Technical Center.

Institutional Review Board Statement: Not applicable.

Data Availability Statement: Data are contained within the article.

Conflicts of Interest: Author Claudia Aurisicchio was employed by the company Bridgestone EU NV/SA. The remaining authors declare that the research was conducted in the absence of any commercial or financial relationships that could be construed as a potential conflict of interest.

References

1. Ramarad, S.; Khalid, M.; Ratnam, C.T.; Chuah, A.L.; Rashmi, W. Waste Tire Rubber in Polymer Blends: A Review on the Evolution, Properties and Future. *Prog. Mater. Sci.* **2015**, *72*, 100–140. [CrossRef]
2. Araujo-Morera, J.; Verdejo, R.; López-Manchado, M.A.; Hernández Santana, M. Sustainable Mobility: The Route of Tires through the Circular Economy Model. *Waste Manag.* **2021**, *126*, 309–322. [CrossRef] [PubMed]
3. Grammelis, P.; Margaritis, N.; Dallas, P.; Rakopoulos, D.; Mavrias, G. A Review on Management of End of Life Tires (Elts) and Alternative Uses of Textile Fibers. *Energies* **2021**, *14*, 571. [CrossRef]
4. Shulman, V.L. Tire Recycling. In *Waste*; Elsevier: Amsterdam, The Netherlands, 2019; pp. 489–515.
5. De, S.K.; Isayev, A.I.; Khait, K. Rubber Recycling. *Choice Rev. Online* **2006**, *43*, 43–2826. [CrossRef]
6. Kumar, A.; Dhanorkar, R.J.; Mohanty, S.; Gupta, V.K. Advances in Recycling of Waste Vulcanized Rubber Products via Different Sustainable Approaches. *Mater. Adv.* **2024**, *5*, 7584–7600. [CrossRef]
7. van Hoek, H. *Closing the Loop: Reuse of Devulcanized Rubber in New Tires*; University of Twente: Enschede, The Netherlands, 2022.
8. Bockstal, L.; Berchem, T.; Schmetz, Q.; Richel, A. Devulcanisation and Reclaiming of Tires and Rubber by Physical and Chemical Processes: A Review. *J. Clean. Prod.* **2019**, *236*, 117574. [CrossRef]
9. Anjum, A.; Ramani, B.; Bramer, E.; Brem, G.; Dierkes, W.; Blume, A. Role of Recovered Carbon Black Ash Content Composition on In-Rubber Performance. In Proceedings of the International Rubber Conference, IRC 2019, London, UK, 3–5 September 2019.
10. Wiśniewska, P.; Haponiuk, J.T.; Colom, X.; Saeb, M.R. Green Approaches in Rubber Recycling Technologies: Present Status and Future Perspective. *ACS Sustain. Chem. Eng.* **2023**, *11*, 8706–8726. [CrossRef]
11. Dannenberg, E.M. The Effects of Surface Chemical Interactions on the Properties of Filler-Reinforced Rubbers. *Rubber Chem. Technol.* **1975**, *48*, 410–444. [CrossRef]
12. Collins, J.M.; Jackson, W.L.; Oubridge, P.S. Relevance of Elastic and Loss Moduli of Tire Components to Tire Energy Losses. *Rubber Chem. Technol.* **1965**, *38*, 400–414. [CrossRef]
13. Waddell, W.H.; Evans, L.R. Use of Nonblack Fillers in Tire Compounds. *Rubber Chem. Technol.* **1996**, *69*, 377–423. [CrossRef]
14. Mittal, K.L. *Silanes and Other Coupling Agents*; Brill Academic Pub: Leiden, The Netherlands, 2007; Volume 4, ISBN 9789047414025.
15. Goerl, U.; Hunsche, A.; Mueller, A.; Koban, H.G. Investigations into the Silica/Silane Reaction System. *Rubber Chem. Technol.* **1997**, *70*, 608–623. [CrossRef]
16. Blume, A. Kinetics of the Silica-Silane Reaction. *KGK Kautsch. Gummi Kunststoffe* **2011**, *64*, 38–43.
17. Anyszka, R.; Beton, K.; Szczechowicz, M.; Bielinski, D.M.; Blume, A. Velcro-Inspired Supramolecular System for Silica-Rubber Coupling. *Rubber Chem. Technol.* **2020**, *93*, 672–682. [CrossRef]
18. Bernal-Ortega, P.; Anyszka, R.; Gojzewski, H.; di Ronza, R.; Aurisicchio, C.; L Valentín, J.; M Salamanca, F.; Blume, A. New Silica/Polymer Coupling by the Use of Cation-π Interactions. *Ind. Eng. Chem. Res.* **2024**, *63*, 3525–3538. [CrossRef]
19. Li, L.; Kim, J.K. Thermoreversible Cross-Linking Maleic Anhydride Grafted Chlorobutyl Rubber with Hydrogen Bonds (Combined with Ionic Interactions). *Rubber Chem. Technol.* **2014**, *87*, 459–470. [CrossRef]

20. Jiang, Z.; Bhaskaran, A.; Aitken, H.M.; Shackleford, I.C.G.; Connal, L.A. Using Synergistic Multiple Dynamic Bonds to Construct Polymers with Engineered Properties. *Macromol. Rapid Commun.* **2019**, *40*, e1900038. [CrossRef]
21. Wojtecki, R.J.; Meador, M.A.; Rowan, S.J. Using the Dynamic Bond to Access Macroscopically Responsive Structurally Dynamic Polymers. *Nat. Mater.* **2011**, *10*, 14–27. [CrossRef]
22. Wang, D.; Zhang, H.; Cheng, B.; Qian, Z.; Liu, W.; Zhao, N.; Xu, J. Dynamic Cross-Links to Facilitate Recyclable Polybutadiene Elastomer with Excellent Toughness and Stretchability. *J. Polym. Sci. A Polym. Chem.* **2016**, *54*, 1357–1366. [CrossRef]
23. Wemyss, A.M.; Bowen, C.; Plesse, C.; Vancaeyzeele, C.; Nguyen, G.T.M.; Vidal, F.; Wan, C. Dynamic Crosslinked Rubbers for a Green Future: A Material Perspective. *Mater. Sci. Eng. R Rep.* **2020**, *141*, 100561. [CrossRef]
24. Bernal-Ortega, P.; Anyszka, R.; di Ronza, R.; Aurisicchio, C.; Blume, A. Dynamic Imine Bonds in Tire Tread Compounds: A Pathway to a Circular Economy and Reduced Waste. *ACS Sustain. Chem. Eng.* **2025**, *13*, 3209–3221. [CrossRef]
25. Zou, W.; Dong, J.; Luo, Y.; Zhao, Q.; Xie, T. Dynamic Covalent Polymer Networks: From Old Chemistry to Modern Day Innovations. *Adv. Mater.* **2017**, *29*, 1606100. [CrossRef]
26. Sattar, M.A.; Gangadharan, S.; Patnaik, A. Design of Dual Hybrid Network Natural Rubber-SiO2 Elastomers with Tailored Mechanical and Self-Healing Properties. *ACS Omega* **2019**, *4*, 10939–10949. [CrossRef]
27. Deng, Y.; Zhang, Q.; Feringa, B.L. Dynamic Chemistry Toolbox for Advanced Sustainable Materials. *Adv. Sci.* **2024**, *11*, e2308666. [CrossRef]
28. Berki, P.; Khang, D.Q.; Minh, D.Q.; Hai, L.N.; Tung, N.T.; Karger-Kocsis, J. Interphase Tailoring via π-Cation Interaction in Graphene and Graphene Oxide Containing NR Nanocomposites Prepared by Latex Compounding. *Polym. Test.* **2018**, *67*, 46–54. [CrossRef]
29. Ibeh, C.C. Phenol-Formaldehyde Resins. *Handb. Thermoset Plast.* **1998**, *2*, 23–71.
30. Santos, C.; Santos, T.; Fonseca, R.; Melo, K.; Aquino, M. Phenolic Resin and Its Derivatives. In *Phenolic Polymers Based Composite Materials*; Jawaid, M., Asim, M., Eds.; Springer: Singapore, 2021; pp. 1–11. ISBN 978-981-15-8932-4.
31. Gardziella, A.; Pilato, L.A.; Knop, A. *Phenolic Resins: Chemistry, Applications, Standardization, Safety and Ecology*; Springer Science & Business Media: Cham, Switzerland, 2013; ISBN 3662041014.
32. Pilato, L. *Phenolic Resins: A Century of Progress*; SpringerLink: Springer e-Books; Springer: Berlin/Heidelberg, Germany, 2010; ISBN 9783642047145.
33. Shojaei, A.; Faghihi, M. Physico-Mechanical Properties and Thermal Stability of Thermoset Nanocomposites Based on Styrene-Butadiene Rubber/Phenolic Resin Blend. *Mater. Sci. Eng. A* **2010**, *527*, 917–926. [CrossRef]
34. Kaynak, C.; Cagatay, O. Rubber Toughening of Phenolic Resin by Using Nitrile Rubber and Amino Silane. *Polym. Test.* **2006**, *25*, 296–305. [CrossRef]
35. Guo, Y.; Hu, L.; Jia, P.; Zhang, B.; Zhou, Y. Enhancement of Thermal Stability and Chemical Reactivity of Phenolic Resin Ameliorated by NanoSiO2. *Korean J. Chem. Eng.* **2017**, *35*, 298–302. [CrossRef]
36. Ming, L.Y.; Feng, C.P.; Siddig, E.A.A. Effect of Phenolic Resin on the Performance of the Styrene-Butadiene Rubber Modified Asphalt. *Constr. Build. Mater.* **2018**, *181*, 465–473. [CrossRef]
37. Flory, P.J.; Rehner, J. Statistical Mechanics of Cross-Linked Polymer Networks II. Swelling. *J. Chem. Phys.* **1943**, *11*, 521–526. [CrossRef]
38. *ISO 37:2024*; Rubber, Vulcanized or Thermoplastic—Determination of Tensile Stress-Strain Properties. ISO: Geneva, Switzerland, 2024.
39. *ASTM D412-16(2021)*; Standard Test Methods for Vulcanized Rubber and Thermoplastic Elastomers—Tension. ASTM: West Conshohocken, PA, USA, 2021.
40. Sivolapov, P.; Myronyuk, O.; Baklan, D. Synthesis of Stober silica nanoparticles in solvent environments with different Hansen solubility parameters. *Inorg. Chem. Commun.* **2022**, *143*, 109769. [CrossRef]
41. Hayichelaeh, C.; Reuvekamp, L.A.E.M.; Dierkes, W.K.; Blume, A.; Noordermeer, J.W.M.; Sahakaro, K. Reinforcementof Natural Rubber by Silica/Silane in Dependence of Different Amine Types. *Rubber Chem. Technol.* **2017**, *90*, 651–666. [CrossRef]
42. Krebs, H. The Mechanism of Accelerator Action. *Rubber Chem. Technol.* **1957**, *30*, 962–971. [CrossRef]
43. Grunert, F. *Analytical Method Development to Predict the In-Rubber Dispersibility of Silica*; University of Twente: Enschede, The Netherlands, 2018; ISBN 9789036546553.
44. Jin, J.; Kaewsakul, W.; Noordermeer, J.W.M.; Dierkes, W.K.; Blume, A. The Relation between Macro- and Micro-Dispersion of Silica Filler in Rubber Matrix. In Proceedings of the International Rubber Conference, London, UK, 3–5 September 2019; IRC: London, UK, 2019; pp. 1–8.
45. Jin, J.; Kaewsakul, W.; Noordermeer, J.W.M.; Dierkes, W.K.; Blume, A. Macro- and Micro-Dispersion of Silica in Tire Tread Compounds: Are They Related? *Rubber Chem. Technol.* **2021**, *94*, 355–375. [CrossRef]
46. Yamada, C. *Influence of Functionalized S–SBR on Silica–Filled Rubber Compound Properties*; University of Twente: Enschede, The Netherlands, 2023.
47. Conant, F.S. Tire Temperatures. *Rubber Chem. Technol.* **1971**, *44*, 397–439. [CrossRef]

48. Heinz, M. Customer Seminar—Dynamic Testing of Rubber—An Introduction. 2018.
49. Indriasari; Noordermeer, J.; Dierkes, W. Incorporation of Oligomeric Hydrocarbon Resins for Improving the Properties of Aircraft Tire Retreads. *Appl. Sci.* **2021**, *11*, 9834. [CrossRef]
50. Vleugels, N.; Pille-Wolf, W.; Dierkes, W.K.; Noordermeer, J.W.M. Understanding the Influence of Oligomeric Resins on Traction and Rolling Resistance of Silica-Reinforced Tire Treads. *Rubber Chem. Technol.* **2015**, *88*, 65–79. [CrossRef]
51. Bernal-Ortega, P.; van Elburg, F.; Araujo-Morera, J.; Gojzewski, H.; Blume, A. Heading towards a Fully Sustainable Tire Tread Compound: Use of Bio-Based Resins. *Polym. Test.* **2024**, *133*, 108406. [CrossRef]
52. Bernal-Ortega, P.; Gaillard, E.; van Elburg, F.; Blume, A. Use of Hydrocarbon Resins as an Alternative to TDAE Oil in Tire Tread Compounds. *Polym. Test.* **2023**, *126*, 108168. [CrossRef]
53. Frosch, S.; Herrmann, V.; Grunert, F.; Blume, A. Quantification of Sulfur Distribution on Rubber Surfaces by Means of μ-X-Ray Fluorescence Analysis. *Polym. Test.* **2023**, *128*, 108237. [CrossRef]

polymers

MDPI

Article

Effect of Proteins on the Network Formation and Degradation of Peroxide Cross-Linked Natural Rubber Elucidated by Time-Domain NMR

Adun Nimpaiboon [1,2], Antonio González-Jiménez [3,4], Roberto Pérez-Aparicio [5], Fernando Martín-Salamanca [5], Zenen Zepeda-Rodríguez [5], Juan López-Valentín [5,*] and Jitladda Sakdapipanich [6,*]

1 Rubber Technology Research Centre (RTEC), Faculty of Science, Mahidol University, Nakhon Pathom 73170, Thailand; adun.nim@mahidol.ac.th
2 Department of Chemistry and Center of Excellence for Innovation in Chemistry, Faculty of Science, Mahidol University, Bangkok 10400, Thailand
3 Plastic Research Center (AIMPLAS), València Parc Tecnològic, C/Gustave Eiffel 4, 46980 Paterna, Spain; angonzalez@aimplas.es
4 Escuela Superior de Ingeniería y Tecnología (ESIT), Universidad Internacional de La Rioja (UNIR), Av. de la Paz 137, 26006 Logroño, Spain
5 Institute of Polymer Science and Technology (CSIC), C/Juan de la Cierva 3, 2800 Madrid, Spain; rperezaparicio@gmail.com (R.P.-A.); fms@ictp.csic.es (F.M.-S.); zenen@ictp.csic.es (Z.Z.-R.)
6 Department of Chemistry and Center of Excellence for Innovation in Chemistry, Faculty of Science, Mahidol University, Nakhon Pathom 73170, Thailand
* Correspondence: jlvalentin@ictp.csic.es (J.L.-V.); jitladda.sak@mahidol.ac.th (J.S.)

Abstract

The importance of sustainable polymers has increased greatly in the last years since most polymers are derived from non-renewable sources. Sustainable polymers (i.e., biopolymers) such as natural rubber (NR) are proposed as a solution for this concern. A comparative study between NR and deproteinized NR (DPNR) was carried out to elucidate the role of proteins on the network formation and degradation of peroxide cross-linked NR using time-domain NMR experiments. The ^{1}H multiple-quantum (MQ) NMR experiments provided information on the cross-link density and its spatial distribution, while the actual fraction of non-coupled network defects was obtained by exploiting the Hahn echo approach measured on swollen samples. The results showed that proteins influenced the network formation during the vulcanization process of NR, leading to a higher number of non-elastic network defects and promoting the creation of additional cross-links with a broader spatial distribution. The formation of network heterogeneities in different length scales deeply influences the mechanical properties of NR samples. On the other hand, the proteins showed a pro-oxidant activity on the degradation behavior by accelerating the degradation process of peroxide cross-linked NR.

Keywords: biopolymers; natural rubber; proteins; sustainability; multiple-quantum NMR

1. Introduction

Sustainable rubbers are natural materials or materials derived from renewable, recycled, and waste carbon resources and their combinations, which at the end of life can be recycled, biodegraded, or composted [1–3]. Sustainable rubbers also exhibit a reduced environmental impact throughout their life cycle. Biopolymers can be derived from renewable resources, or they can be partially made from renewables and synthesized like traditional polymers, as in the case of bio-based polymers [4]. Rubber from *Hevea brasiliensis*, known as natural rubber (NR), has been recognized as the most important biopolymer in the rubber industry, although more than 2000 species of plants and fungi have also been found to produce rubber hydrocarbon. NR is a very high-molecular-weight and long-chain branched biopolymer, composed primarily of *cis*-1,4-polyisoprene containing about 6% of non-rubber constituents such as proteins, lipids, carbohydrates, and other ingredients [5]. NR is a unique material for many industrial as well as medical applications, such as aircraft and truck tires, surgical gloves, and condoms, owing to its superior crack growth resistance and excellent tensile properties in comparison with other naturally occurring and synthetic polyisoprenes. The structural characterization of NR using high-resolution ^{13}C and ^{1}H NMR revealed that the fundamental structure of NR is long-chain isoprene units with end-groups, i.e., with initiating ω-terminal groups and terminating α-terminal groups. The ω-terminal group is not revealed yet but has been presumed to interact with proteins [6,7]. The α-terminal group is proposed as mono- or di-phosphate linked with phospholipids [8,9]. These findings suggest that the interaction of proteins and lipids at the molecular chain ends of NR possibly forms a three-dimensional pseudo-end-linked network, contributing to the numerous characteristics of un-vulcanized NR, such as gel formation [7,10] or strain-induced crystallization [11]. Therefore, proteins are one of the essential components believed to bring about these marvelous properties of NR. However, some proteins exuding from NR products are found to be associated with the severe allergic reactions termed a *Type I* allergy, which is potentially serious or even life-threatening. The U. S. Food and Drug Administration (FDA) issued a medical alert to rubber device producers in 1991 after a number of patients died of this protein allergy [12]. Proteins in latex can usually be removed by the enzymatic decomposition of proteins in the presence of surfactant, followed by centrifugation [11,13,14].

The cross-linking process of rubber is generally known as vulcanization, where three-dimensional networks are formed by chemical or physical methods [15,16]. The most useful vulcanizing systems in rubber technology are sulfur/accelerator systems and organic peroxides [17–21]. The role of proteins on cross-linked NR was studied based on accelerated sulfur systems. It was proven that the proteins and nitrogenous compounds in NR act as natural accelerators for sulfur vulcanization [22], as evidenced by the deproteinized NR (DPNR) that demonstrates a slower cure as well as a lower cross-link density compared to raw NR. In addition, the proteins play a strong role in the network structure and distribution of the three types of sulfidic cross-linking [22]. Peroxide vulcanization is preferred in some applications due to the good heat resistance, low compression set, and high transparency of NR products. Moreover, the absence of some additives offers a solution for solving the nitrosamine contamination as a carcinogen and *Type IV* allergy (allergic contact dermatitis) caused by the amine-based accelerators in sulfur vulcanization [23]. As the peroxide vulcanization process takes place via a radical mechanism, the proteins in NR that might associate with radicals need to be clarified.

Several techniques such as inverse gas chromatography [24], osmometry [25], mechanical analysis [26], equilibrium swelling experiments [27], dielectric measurements [28], high-resolution solid-state NMR [29,30], or scattering techniques [31,32] have been used to partially elucidate the network structure of elastomers. Recently, time-domain solid-state NMR has been shown to be a robust and versatile technique to obtain direct and local evidence about the different parameters that define the polymer network structure [33–39]. This is based on the direct measurement of residual dipolar couplings that persist because of the presence of cross-links and other topological constraints, such as entanglements, that restrict the number of accessible segmental conformations, which leads to local chain ordering caused by the non-isotropic fast segmental motions. Many NMR experiments, e.g., ^{1}H transverse relaxation [40], two-dimensional (2D) magnetization exchange spectroscopy [41], and Hahn, together with solid echoes [42], were used to determine various parameters controlled by residual dipolar couplings. Nevertheless, it is already well known that ^{1}H double-quantum (DQ) (or more generally multiple-quantum, MQ) NMR can be considered the most quantitative and reliable method for the measurement of weak residual dipolar couplings [37,43]. By suitable data analysis, the coherent dipolar effect, related to the number of constraints and their spatial distribution (inhomogeneities), could be independently separated from the temperature-dependent segmental dynamics for rubber networks [37].

The aim of this work is to study the role of proteins in the network formation and degradation process of NR vulcanized with an organic peroxide system by means of different time-domain NMR experiments performed in low-field spectrometers. In this sense, the ^{1}H MQ-NMR experiment was used to evaluate the evolution of cross-link density and the spatial distribution of cross-links for NR and DPNR samples during the peroxide vulcanization as well as during the degradation process. Additionally, the actual fraction of non-coupled network defects, e.g., free chain ends, dangling chain ends, or loops, was obtained by applying Hahn echo experiments on swollen samples with deuterated solvent. The contributions of those proteins to the network structure and its influence on the network formation in NR is crucial to determine quantitative knowledge about the most important biopolymer in the rubber industry.

2. Experimentation

2.1. Sample Preparation

NR was prepared by diluting fresh NR (FNR) latex to 30% dry rubber content (DRC) and adding 1% (w/v) formic acid to coagulate the latex. The coagulum was rinsed with water several times and then dried in an oven at 70 °C for 24 h.

DPNR was carried out by incubating 30% DRC of the FNR latex with 0.04% (w/v) proteolytic enzyme (KP-3939, Kao Co., Tokyo, Japan) and 1% (w/v) sodium dodecyl sulfate (SDS) at 37 °C for 12 h, followed by centrifugation twice. The cream phase was cast on a glass plate and dried under the same method as the NR.

The NR and DPNR were compounded with 1 part per hundred rubber (phr) of dicumyl peroxide (DCP) in an open two-roll mill using the standard procedure and then vulcanized by compression molding at 160 °C by varying times.

2.2. Characterization

The protein content was measured using nitrogen analysis using a LECO Nitrogen Analyzer (model FP 528) (LECO, St. Joseph, MO, USA) and generally calculated using a universal conversion factor of 6.25.

The content of the long-chain fatty acid ester group was determined by FTIR measurements based upon a calibration using a series of mixtures of methyl stearate and synthetic *cis*-1,4-polyisoprene (Kuraprene IR10). The quantity of fatty acid ester groups per weight of rubber was determined by the intensity ratio of the bands at 1739 cm^{-1} (C=O) to 1664 cm^{-1} (C=C).

The quantitative analysis of the metal content was determined by inductively coupled plasma atomic emission spectroscopy (ICP-AES).

The cure behavior and the optimum cure time (t_{95}) were monitored by the Rubber Process Analyzer (RPA) (Alpha Technologies® RPA 2000™, Heilbronn, Germany), where t_{95} is the time required for the torque of the rheometer to increase up to 95% of the total torque at 160 °C.

The time-domain proton NMR experiments were performed on a Bruker minispec mq20 spectrometer (Bruker, Billerica, MA, USA) operating at a resonance frequency of 20 MHz, with a 90° pulse length of 3 μs and a dead time of 12 μs. The sample temperature was controlled with an air-operated BVT3000 heater (Bruker, Billerica, MA, USA).

The tensile tests were performed on type 2 dumbbell specimens according to ISO 37 in an Instron model 3365 dynamometer (Instron, Norwood, AR, USA) equipped with a video camera to measure deformations at a strain rate of 500 mm/min.

2.2.1. ^{1}H DQ-NMR Experiment

Time-domain ^{1}H DQ-NMR spectroscopy is based on the measurement of the residual dipolar couplings (NMR observable) between protons originating from the orientational anisotropy due to constraints, such as cross-links and entanglements. Such local anisotropy can be described by a non-zero dynamical orientation of the polymer backbone $\langle P_2(\cos\theta)\rangle$ (second-order Legendre polynomial). The NMR experiments are in the timescale related to the length R and orientation α of the end-to-end vector. Assuming a freely jointed rigid rod chain model, the average segment orientation is as follows:

$$\langle P_2(\cos\theta)\rangle \approx \frac{3}{5}\frac{R^2}{N^2b^2}\frac{3\cos^2\alpha - 1}{2} \approx S_b\frac{3\cos^2\alpha - 1}{2} \quad (1)$$

where N is the number of statistical segments between constraints and b is the statistical segment length. The overall NMR signal is the contribution of the individual monomer or segment signals, which are at least partially time-averaged over the distribution of the end-to-end vectors between constraints. From this, an average value of the local backbone orientation, S_b, with respect to the end-to-end vector, arises (Equation (1)). For ideal chains $R^2 \approx Nb^2$, it may be written as follows:

$$S_b \approx \frac{3}{5}\frac{R^2}{N^2b^2} \propto \frac{1}{N} \approx v \quad (2)$$

being proportional to the constraint density v. Since the proton dipolar coupling depends on molecular orientation, S_b is detected in NMR because it gives a non-zero residual dipolar coupling and, then, is calculated from the experimental average residual dipolar coupling constant D_{res}, by comparison with its static counterpart, D_{stat}, as follows (k is a correction factor <1 accounting for the spin arrangement and motions within a statistical segment):

$$S_b = k\frac{D_{res}}{D_{stat}} \quad (3)$$

According to Equations (2) and (3), D_{res} is inversely proportional to the average molecular weight of the rubber chains between constraints M_x. There are different models [44] that describe the contribution of cross-links (M_c) and entanglements (M_e) to the NMR observable (D_{res}) and the measured order parameter (S_b). Assuming the additivity of entanglement and cross-link constraints [45,46], we may write $D_{res} \propto 1/M_c + 1/M_e$.

The ^{1}H DQ-NMR experiment technique is based on enhanced Baum–Pines pulse sequences [37,47]. The raw experimental data from such experiments need to be normalized in such a way that the temperature-independent network structure effect (at temperatures far above the T_g) can be separated from the temperature-dependent segmental dynamics without invoking any specific model, as was widely explained elsewhere [37,43,47]. A proper analysis of the NMR signals by means of a numerical inversion procedure based on the fast Tikhonov regularization analysis [48] provides the actual distribution of dipolar couplings that can be defined by the average apparent coupling constant D_{res}, and the standard deviation, σ, characterizing the distribution width of the residual dipolar couplings.

From published spin dynamic simulations and assuming a reasonable model for the intra-segmental motions, the relationship between the experimental residual dipolar coupling constant, D_{res}, and the molecular weight between constraints, M_x, was obtained in the NR samples as the following equation [37]:

$$M_x^{(NR)} = \frac{617\ \text{Hz}}{D_{res}/2\pi}[\text{kg/mol}] \tag{4}$$

assuming a constant entanglement density (then, $M_x^{(NR)} \propto M_c^{(NR)}$), tetra-functional cross-links, and a phantom network model (to take into account the non-affinity of real systems at the scale of the cross-links) [49], the cross-link density, υ_{NMR}, can be determined by $\upsilon_{NMR} = 1/M_c$.

2.2.2. Hahn Echo Experiments Measured on Swollen Samples

The fraction of non-coupled network defects has been investigated by ^{1}H DQ-NMR and Hahn echo experiments [37,49]. In the dry state, the fraction of non-coupled defects depends on temperature, and it is always underestimated because chain packing or topological constraints promote some restrictions in the fast-segmental motions of dangling chain segments and loops [43]. Thus, some fractions of network defects are, in fact, observed as dipolar coupled and elastically active segments because the slow arm retraction process is not able to relax the dangling chains in the experimental NMR timescale [50,51]. The topological restrictions are released, and the chain dynamics are speeded up by swelling the rubber networks with a suitable solvent, leading to fast isotropic motions. Consequently, the NMR experiments on the swollen samples provide a reliable measure of the actual fraction of non-coupled and elastically inactive network defects [49]. In the present work, the rubber samples were swollen in deuterated toluene at room temperature for 24 h in sealed dark vials in order to prevent them from degradation [27].

The complex thermodynamics and heterogeneous nature of the swelling process [27,52] hide substantial structural information from the stretched network, making nonsensical the use of complex DQ-NMR experiments to perform a deeper analysis of a direct signal that codifies the dipolar coupling information. For that reason, simpler Hahn echo experiments (which also have a dipolar origin for the transverse relaxation) were performed to identify and quantify the actual fraction of non-coupled polymer segments in the swollen samples.

3. Results and Discussion

3.1. *Characterization of Rubber Samples*

The characteristics of the rubber samples are given in Table 1. The NR samples contained 2.62% (w/w) of proteins that could be selectively removed by deproteinization, as evidenced by the very low protein content of DPNR. The ester content of the NR and DPNR, which represented the lipid part, showed similar values. The metal content (Cu^{2+}, Fe^{2+}, Mn^{2+}) of both the NR and DPNR, which strongly influences the radical mechanism [53], was negligible.

Table 1. The characteristics of the rubber samples.

Samples	Protein Content (% w/w)	Ester Content (mmol/kg-Rubber)	Metal Content (% w/w)		
			Cu^{2+}	Fe^{2+}	Mn^{2+}
NR	2.62	14.28	ND	0.0016	ND
DPNR	0.06	14.23	ND	0.0014	ND

3.2. Pseudo-End-Linked Network

Although the main goal of this work is to investigate the network structure formed by peroxide vulcanization, un-vulcanized NR also contains a pseudo-end-linked network that has been believed to bring about superior elastic properties rather than synthetic polyisoprene (IR) [54]. The cross-linking point of the end-linked network is classified into two types, i.e., (i) the intermolecular interaction of the proteins via hydrogen bonding at the initiating ω-terminal group [6,7] and (ii) the micelle formation or association of the polar head-groups of phospholipid molecules at the terminating α-terminal group [8,9]. Thus, the deproteinization process decomposes the network formed by the proteins in the DPNR, wherein the network derived from the phospholipids still remains. Additional trans-esterification treatment (explained in the Supplementary Information) removes the phospholipids and avoids the formation of the end-linked network. The proposed pseudo-end-linked network structure of NR, including the structural change after deproteinization and trans-esterification, can be seen in the Supplementary Information, Figure S1 [10].

In the un-vulcanized NR samples, the existence of the pseudo-end-linked network has been proved to cause an apparent increment in molecular weight and the related development of entanglements and their dependent properties [55]. In this sense, it is well known that D_{res} is sensitive and directly proportional to any constraints (independently of its nature) that restrict the segmental motion of polymer chains [56–58]. Therefore, ^{1}H DQ-NMR experiments have been successfully used to analyze the segmental autocorrelation function of un-vulcanized (linear melt) samples [59,60], which is dominated by the restrictions imposed by entanglements and topological constraints (caused by the polymer molecular weight and chain architecture). Note that D_{res} for a linear melt is an apparent quantity that depends on temperature and time. According to these statements, the variation in the average D_{res} for the different un-vulcanized samples studied in this work (see Figure 1) must be related to changes in the topological constraints and chain mobility.

Figure 1 shows the temperature dependence of the squared local backbone orientation parameter, S_b^2, which corresponds to the orientation autocorrelation function at the timescale of the DQ-NMR experiments for the un-vulcanized samples: u-NR, u-DPNR, trans esterified DPNR (u-TE-DPNR), and synthetic polyisoprene (u-IR). S_b^2 decreases in the chosen temperature range and the actual NMR experimental time window is related to changes in segmental conformations inside the constrained space caused by faster dynamics in the reputation regime [61], where the dimensions of the apparent tube are determined by the real molecular weight between entanglements, M_e, for each sample. In the case of u-NR and u-DPNR, they present higher S_b^2 values than the trans-esterified counterpart (u-TE-DPNR) at any temperature, which might reflect the presence of the pseudo-end-linked network that can promote higher anisotropy in the rubber motions. In this sense, the importance of phospholipids (absent in the u-TE-DPNR sample) on the formation of thermo-sensitive cross-link points should be more evident in comparison with the almost negligible effect that the proteins (u-DPNR) imposed on the chain motions of the u-NR sample. A synthetic u-IR sample was also investigated in order to compare it with the u-TE-DPNR sample. The differences between u-IR and u-TE-DPNR should be related to changes in the molecular weight. These results are completely aligned with the elastic properties of these samples: u-NR and u-DPNR showed almost the same properties, while u-TE-DPNR was quite similar to u-IR [54]. It can be confirmed that the topological constraints in NR should be dominated by the entanglements and the pseudo-end-linked network derived from phospholipids (with an almost negligible contribution from proteins). In this sense, the actual molecular weight between junction points was not properly estimated in the un-vulcanized samples because (i) this pseudo-end-linked network was unstable at temperatures above 100 °C [62] and (ii) at temperatures below 100 °C, the segmental dynamics in NR and DPNR were not fast enough to average out all the available conformations according to the restrictions caused by them.

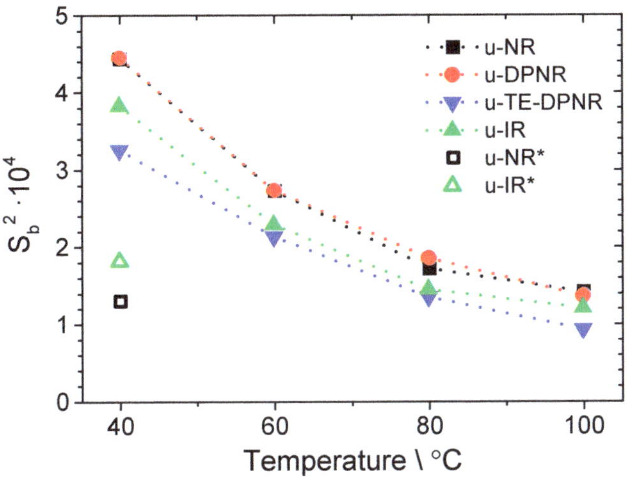

Figure 1. Temperature dependence of the squared local backbone orientation parameter (S_b^2) for un-vulcanized linear melt samples: natural rubber (u-NR), deproteinized natural rubber (u-DPNR), trans-esterified natural rubber DPNR (u-TE-DPNR), and synthetic polyisoprene (u-IR), respectively. On the other hand, u-NR* and u-IR* are samples subjected to mastication treatment on a two-roll mill for 15 min. The preparation details of the u-TE-DPNR and u-IR samples are explained in the Supplementary Information.

It is worth pointing out that those samples were not subjected to any mechanical treatment, unlike the vulcanized samples, which were masticated during their compounding. In order to consider these effects on the rubber samples, Figure 1 also shows the S_b^2 value for the natural rubber (u-NR*) and synthetic polyisoprene (u-IR*) samples (measured at 40 °C) after a mastication treatment on a two-roll mill for 15 min (equivalent to the compounding step on the vulcanized samples). The results indicate a stronger effect of the mechanical treatment on the rubber chain molecular weight and, hence, on the entanglement effect in the case of the u-NR* compared to the synthetic counterpart. These results could be related to the effect of mastication on the pseudo-end-linked network in natural rubber.

The temperature dependence of D_{res} for the un-vulcanized rubber samples (see Figure 1) was completely suppressed in the vulcanized samples (3 min at 160 °C) in the temperature range from 40 °C to 120 °C for both the NR and DPNR samples, as shown in Figure 2A. This means that the measuring temperature is far above the rubber T_g, and the segmental dynamics are fast enough to average out all the accessible conformations in the time window imposed by the NMR measurement [43]. Consequently, in those samples, D_{res} reflects the constraints that define the rubber network structure without any interference from the polymer dynamics (temperature-dependent effects).

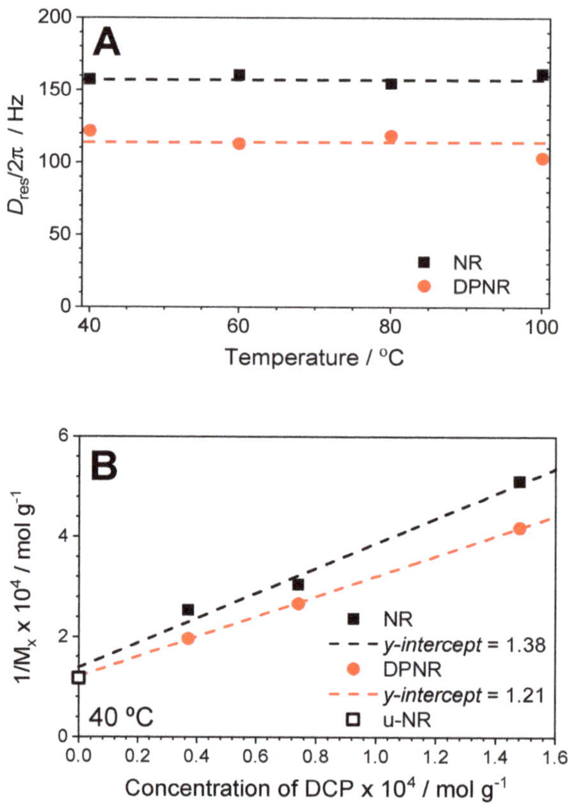

Figure 2. (**A**) Residual dipolar couplings, D_{res}, for vulcanized NR and DPNR with 1 phr of DCP (3 min at 160 °C) as a function of the NMR measuring temperature. (**B**) Relationship between the density of constraints (given as $1/M_x$ after 3 min of vulcanization time at 160 °C) and the concentration of DCP for the NR and DPNR samples measured at 40 °C. The y-intercept value represents the contribution of the entanglements ($1/M_e$) to the total amount of constraints assuming additive contributions ($1/M_c + 1/M_e$). The result shown in (B) for the un-vulcanized sample subjected to mastication treatment on a two-roll mill for 15 min (u-NR*) reflects the apparent (time and temperature-dependent) restrictions obtained from the measured D_{res} for that linear rubber melt (see also Figure 1).

In order to demonstrate the linear relationship between the cross-link density measured by NMR and the concentration of peroxide for the NR and DPNR samples (see Figure 2B), additional samples containing 1, 2, and 4 phr of DCP were prepared under the same conditions, and vulcanized at 160 °C for 3 min, as expected. According to the previous explanation, the molecular weight between constraints, M_x, measured by NMR, in fact, codifies the actual number of constraints imposed by (i) cross-links, (ii) entanglements (including those that are trapped), and (iii) the pseudo-end-linked network. Therefore, the extrapolated y-intercept value in Figure 2B represents the constraints in terms of $1/M_e$ for the NR and DPNR after the vulcanization process. The NR and DPNR show quite similar y-intercept values, which match with the D_{res} value obtained in the un-vulcanized NR (u-NR*) sample after the mastication process (see Figure 1). These results point out the strong effect of mastication and vulcanization dominating over the pseudo-end-linked network structure formed by the proteins and phospholipids. The latter governs the outstanding properties of un-vulcanized NR (in terms of green strength) but has a negligible effect in the vulcanized samples, whose elastic properties mainly depends on the network structure formed by C-C cross-links (in the case of the peroxide-cured samples). Although the presence of proteins has a minimal effect on the topological constraints for the vulcanized samples (y-intercept), the slope in Figure 2B likely indicates a strong influence of these non-rubber substances on the vulcanization process, which undoubtedly enhanced the cross-linking efficiency in the NR, compared to the DPNR.

At this point, it is important to remark on some limitations by other experimental approaches to quantitatively analyze the effect of proteins on the actual network structure of NR in comparison with the ^{1}H DQ-NMR experiments. On the one hand, the widely used swelling experiments provide only partial and qualitative information about the cross-link density because of the uncertainties associated with the complex thermodynamics of the swelling process and the elastic model that defines the behavior of swollen networks [27]. On the other hand, the most complete information about the NR network structure obtained by the small-angle neutron scattering (SANS) technique was intensely disturbed by the strong scattering of the protein aggregates [63].

Because of these statements, the evolution of the main parameters that define the network structure, e.g., the cross-link density and its spatial distribution, as well as the non-coupled network defects, during the vulcanization process have been deeply studied in the following sections by using ^{1}H DQ-NMR experiments. These measurements were carried out at 40 °C because, at this temperature, the chain dynamics of the studied samples are fast enough to reach the plateau value of D_{res} (see Figure 2A). Additionally, this low temperature prevents the subsequent reactions of unreacted peroxides (especially at the initial steps of the vulcanization process) and NR degradation.

3.3. Evolution of Cross-Link Density During the Vulcanization Process

The formation of cross-links between polymer chains during the vulcanization process transforms rubber samples into an elastic material. It strongly influences their mechanical properties [64]. Figure 3 shows the evolution of cross-link formation for the NR and DPNR with the vulcanization time determined by NMR and rheometer measurements. A good correlation was clearly found between both techniques showing a characteristic curve for peroxide vulcanization. Peroxide vulcanization has a radical pathway that initiates the cross-linking reaction immediately after the homolytic scission of peroxide, as evidenced by the lack of scorch time in the vulcanization curve. The increased cross-link density (as measured by NMR) with the vulcanization time was reflected in the torque measured by the rheometer until a plateau was reached at the end of the process.

By comparing the NR and DPNR vulcanization curves, it is obvious that the proteins in NR increased the cross-link density without any influence on the vulcanization rate, since the optimum vulcanization time, defined as t_{95}, is quite similar for both samples. In that sense, the proteins might act as a coagent, improving the efficiency of peroxide as a cross-linking agent. An enhancement in the efficiency of vulcanization via radical reaction could be associated with different mechanisms, such as the suppression of the non-network forming the side reactions, i.e., chain scission and disproportionation [65,66], as well as the formation of bridges between polymer chains contributing to additional cross-links [67]. The coagent bridges which are covalently bonded to the polymer chains are possibly a grafting of coagents [67,68], an interpenetrating network of homo-polymerized coagents [69], and an aggregation of coagents like filler particles [70,71], depending on several factors, such as the readiness of the coagent to homo-polymerize, the concentration polarity, and the solubility of the coagent [17,68]. The role of the existing protein in NR as a coagent was supported by the presence of radicals on the molecules of protein extracted from the NR. The electron paramagnetic resonance (ESR) technique (see Supplementary

Information, Figure S2) can detect the presence of radicals on the proteins of NR, for which the g value at 2.004 corresponds to the carbon radicals on the backbone and the side chain of proteins similar to the results from the studies of amino acids and proteins examined elsewhere [72–74].

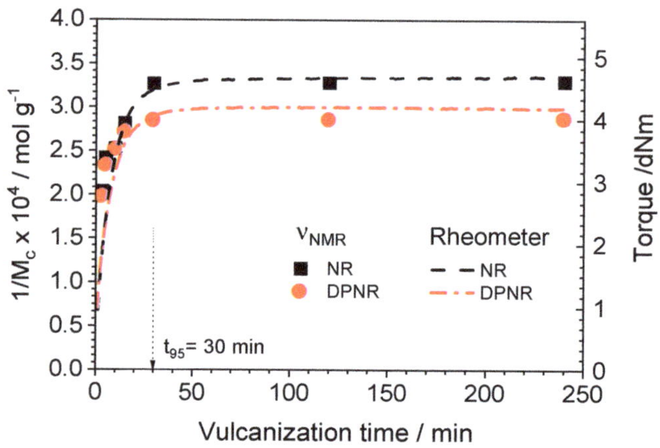

Figure 3. Evolution of cross-link formation for the NR and DPNR samples with 1 phr of DCP at 160 °C as a function of vulcanization time determined by NMR (symbols) and rheometer (dashed lines), respectively.

The proteins in NR are rather polar molecules, and therefore, their miscibility with the non-polar NR chains is quite poor. Such proteins tend to separate and then form aggregates of cross-linked proteins caused by the recombination of the protein radicals and homo-polymerized proteins at the unsaturated bond of the side chain, e.g., phenylalanine and tryptophan. The possible mechanism reaction of proteins as a coagent in peroxide cross-linked NR adapted from ref. [75] is given in Figure 4. When the radical is formed on the NR chain, the protein radical (a single protein molecule and/or an aggregate of cross-linked proteins) with more than one radical on that molecule is attached to the NR chain. Finally, the termination reaction can occur by the combination of radical intermediates, i.e., radicals on the NR and NR-proteins, leading to extra cross-links.

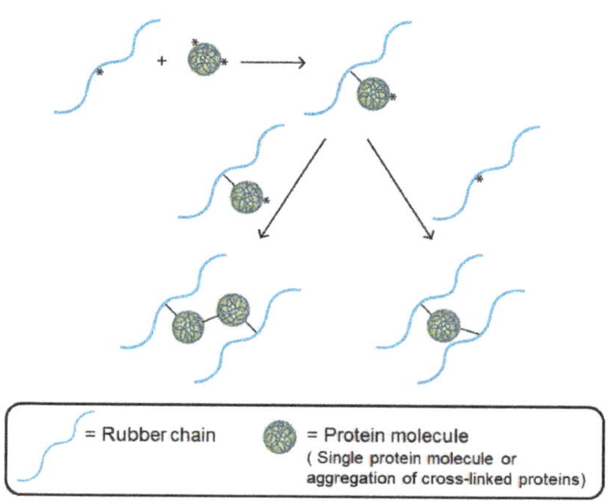

Figure 4. Schematic representation of the possible mechanism reaction of proteins as a coagent in peroxide cross-linked NR, where radicals have been represented by the symbol *. Adapted from ref. [75].

In addition, proteins can be oxidized by the presence of radicals and O_2 to generate protein hydroperoxides [76–79] (more details and evidence of protein degradation will be discussed in the following sections). Although the concentration of O_2 during the vulcanization process is quite low, some proteins might be oxidized by peroxide along with trace amounts of O_2 to generate additional radicals in the vulcanization process.

NR samples always show higher cross-link density than DPNR because of the presence of proteins independently of the involved mechanism, which seems to be related to the higher concentration of radicals during the vulcanization process.

3.4. Evolution of Non-Elastic Network Defects During the Vulcanization Process

Even though most of the literature in the rubber field supposes perfect networks obtained from the cross-linking reaction to explain the elastic properties of those materials, non-elastic network defects, mainly dangling chain ends and loops, have recently been reported as an important factor to consider when describing the structure of vulcanized rubber samples [43,49] and understanding their stress–relaxation processes [80–82].

The non-elastic network defects for the NR and DPNR samples during peroxide vulcanization are shown in Figure 5A. Similar tendencies on the evolution of non-elastic network defects for the NR and DPNR up to the optimum vulcanization time (t_{95}) were observed: the network defects decreased with the vulcanization time because of the formation of cross-links. At longer times, a slight increase in the fraction of network defects could be measured for the NR samples, whereas the DPNR seems to be a more stable sample with a constant number of defects above the optimum vulcanization times.

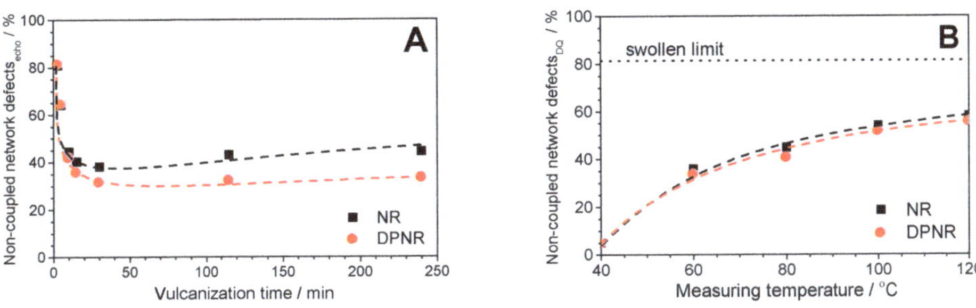

Figure 5. Evolution of non-coupled network defects for NR and DPNR as (**A**) a function of vulcanization time (measured at 40 °C in the swollen state by using Hahn echo experiments) and (**B**) as a function of the measuring temperature for samples vulcanized for 3 min (measured in the dry state by using DQ-NMR experiments). The swollen limit in (**B**) reflects the actual fraction of non-coupled defects for samples vulcanized for 3 min as measured by NMR in a swollen state at 40 °C from (**A**).

The first phenomenon is due to the progressive cross-linking reaction that reduces the fraction of un-cross-linked chains and shortens the molecular weight of the chain ends. Nevertheless, in any case, the actual fraction of non-coupled network defects for the optimum cross-linked NR samples reaches values of 30–40%. This means that peroxide vulcanization seems to be a quite ineffective process to form elastic cross-linked network structures in NR, whereas the side reactions that provide dangling chains (e.g., chain scission processes) seem to be favored in comparison with other elastomers [43].

Obviously, this scenario depends on temperature (see Figure 5B) and the observed timescale, minimizing the effect of network defects on the mechanical properties. The actual fraction of non-coupled defects (measurable only in a swollen state) is always underestimated when it is measured in dry samples (see Figure 5B) because chain packing or topological constraints promote some restrictions on the chain ends that provide some anisotropy in the movement to those entangled segments. Therefore, the fraction of non-elastic network defects measured by NMR decreases drastically at lower temperatures. The clearest example is given by the samples vulcanized for 3 min, whose actual fraction of defects is around 80% (as measured in the swollen state), but only 5% of those polymer segments behave as non-elastic defects at 40 °C in the dry state in the NMR timescale.

These statements support the assumption that cross-linking and chain scission reactions are competing and sequential reactions during the vulcanization process of NR with peroxides [83]. Before the optimum vulcanization time, t_{95}, the overall process seems to be dominated by the cross-linking reaction: an increase in cross-link density and elastic torque and a decrease in the non-coupled network defects. However, the chain scission reaction seems to govern this process after achieving the network formation at t_{95} for the NR samples. It is important to remark that the increase in non-elastic network defects observed above the optimum vulcanization time for the NR sample (Figure 5A) has neither a visible effect on the average molecular weight between constraints nor on the elastic torque measured in the rheometer as demonstrated in Figure 3. This could be explained by the combined (and opposite) changes in the network structure caused by the presence of competing reactions, the cross-linking and the chain scission that modify the spatial distribution of cross-links, as will be discussed in the following sections.

By comparing the NR and DPNR samples, it is possible to establish that the proteins have a negative effect in this complex process since the NR showed a higher fraction of non-elastic network defects than the DPNR. In this sense, the higher concentration of radicals due to the presence of proteins increases the cross-link density. It enhances the chain scission reactions that occur during the NR vulcanization with peroxides, contributing to the higher fraction of defects of the NR. In addition, the proteins can form ineffective cross-links, i.e., proteins bound to a single rubber chain without forming any bridge with other chains that create dangling and pendant proteins grafted to NR chains. Proteins can raise the fraction of network defects in NR vulcanized with peroxides independently of the assumed mechanism involved in the process.

3.5. Evolution of Spatial Distribution of Cross-Links During the Vulcanization Process

The ^{1}H DQ-NMR experiments not only allow the calculation of an average cross-link density, but also provide information on the spatial distribution of cross-links. Previous works have demonstrated that NR samples vulcanized with peroxide exhibited a broader distribution of cross-links than sulfur-based networks [43,57], and, in some cases (according to the peroxide concentration, vulcanization temperature, and type of elastomer), they contained highly cross-linked areas (inhomogeneities) termed as clusters [84], which were formed by the subsequent radical addition mechanism (similar to a polymerization reaction) to the double bond on the chain backbone [85–87].

The ratio between the standard deviation and the average value of the dipolar coupling distribution measured by NMR [47] is a robust parameter to describe a variation in the spatial distribution of the cross-links with the vulcanization time (see Figure 6). The quite broad cross-link distribution in the initial steps of vulcanization for both the NR and DPNR progressively decreased with the vulcanization time up to the optimum vulcanization time when it reached the narrowest distribution. At longer times (above the optimum vulcanization time), the scenario turned, and the distribution became broader again. This trend can be explained by the coexistence of two competitive reactions, i.e., the formation of cross-links and the chain scission reactions during the vulcanization process.

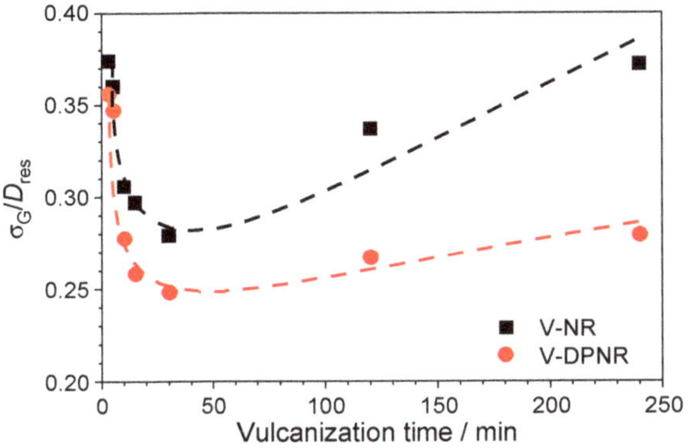

Figure 6. Variation in the relative width of the distribution of cross-link density as a function of vulcanization time for NR and DPNR, where σ represents the standard deviation and D_{res} the average value of the dipolar coupling distribution.

The spatial distribution of cross-links became narrower with the progress of vulcanization because the formation of additional cross-links randomly distributed (according to the radical pathway that governs the vulcanization based on peroxides) was dominant over other side reactions. Consequently, the cross-link density and spatial homogeneity of the formed network increased during this process.

To understand the network evolution over the optimum vulcanization time, it is mandatory to analyze in detail the actual spatial distribution of cross-links (see Figure 7). In the case of the NR, the distribution of cross-links became broader when the vulcanization time increased over t_{95} (at 120 and 240 min, respectively) because both extremes in the distribution shifted to lower and higher cross-link densities, respectively. This is molecular evidence about the opposite effects of the two main reactions during the vulcanization process: (i) chain scission reactions that increase the fraction of non-elastic network defects (see Figure 5A) and the molecular weight between cross-links (M_c), and (ii) the cross-linking reactions that promote the formation of areas with higher cross-link densities. Both opposite processes seem to compensate for their effects, and, consequently, the average cross-link density (measured as the average value of D_{res} distribution) and the elastic torque measured in the rheometer remain constant, as demonstrated in Figure 3.

The deproteinized sample, DPNR, did not show any effect derived from chain scission reactions over the optimum vulcanization time (completely aligned with the almost constant value in the non-coupled network defects, as shown in Figure 5A). In that case, the broadening of the spatial cross-link distribution was caused by the formation of additional cross-links thanks to the persistence of radicals in the sample.

Since the proteins in the NR increased the concentration of radicals in the system and enhanced both the cross-linking and the chain scission reactions, the NR sample showed a higher fraction of network defects, higher cross-link density, and broader spatial distribution of cross-links than the DPNR sample. Chain scission reactions were minimized, and their negative effect on the network structure was almost negligible. These effects were amplified at vulcanization times over t_{95}, where the proteins were active in this complex process.

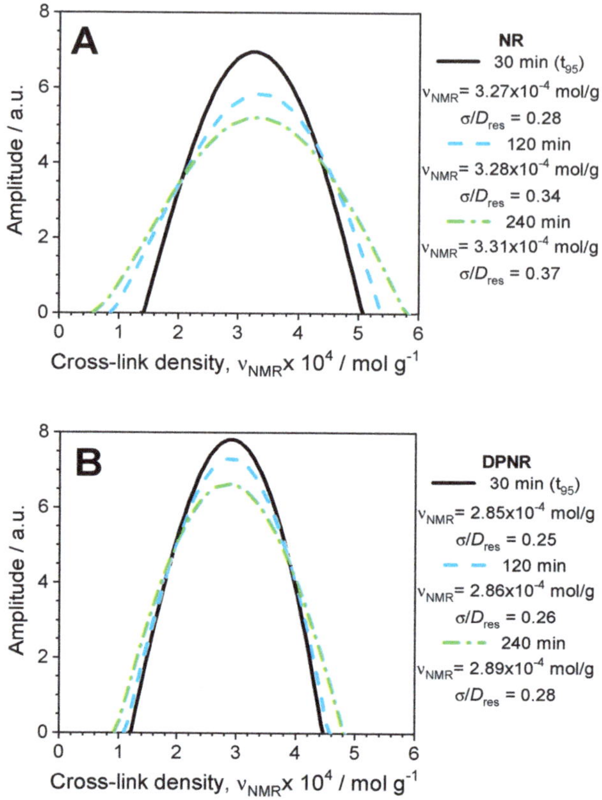

Figure 7. Spatial distribution of cross-links at various vulcanization times for (**A**) NR and (**B**) DPNR according to the regularization analysis of NMR data. Average cross-link density (υ_{NMR}) and the relative width of the distributions (σ/D_{res}) are indicated.

Nevertheless, it is important to point out that in both cases, the low concentration of added peroxide (only 1 phr) cannot allow for the formation of highly cross-linked areas or clusters as was previously demonstrated in NR for samples with higher concentrations of peroxide [43]. This type of local inhomogeneity in the network structure for peroxide vulcanized samples should not be confused with the larger length-scale inhomogeneities caused by protein aggregates in both the un-cross-linked [87] and peroxide cross-linked [63] NR samples as was observed by atomic force microscopy (AFM) and SANS techniques, respectively. The latter is a larger-scale inhomogeneity that is undetectable by the more local NMR technique, and it has no influence on the actual structural information analyzed from the NMR experiments.

Consequently, the protein aggregation in the peroxide cross-linked NR needed to be confirmed, and this could be clarified by AFM and nitrogen analysis. The protein aggregations in the peroxide cross-linked NR were found even after leaching with hot water, which is an effective method for removing proteins from sulfur cross-linked NR products [88–92]. Furthermore, the nitrogen content of the peroxide cross-linked NR insignificantly changed after the leaching process (see Supplementary Information, Figure S3). These results signify that protein aggregates can be covalently bonded to the rubber backbone during the vulcanization of NR with peroxide, unlike the sulfur cross-linked NR.

It should be noted that NMR experiments are sensitive to very local information on the length scale of 10 nm, whereas AFM is measured in the micrometer scale. These techniques provide complementary information on different length scales, allowing us to conclude that the NR samples vulcanized with peroxide show inhomogeneities on different length scales caused by the consecutive addition reaction to the double bond (the formation of clusters of cross-links) and by the grafting of protein aggregates to the rubber chains.

Based on the cross-link density, the fraction of non-elastic network defects, and the spatial cross-link distribution results obtained by NMR measurements and the AFM results in this work, the proposed network structures derived from the presence of proteins in peroxide cross-linked NR are schematically illustrated in Figure 8, where some additional cross-links are derived from protein hydroperoxides, the bridge of proteins, and the aggregation of cross-linked proteins according to AFM results, while the dangling and pendant proteins grafted to the NR chain do not form any cross-links.

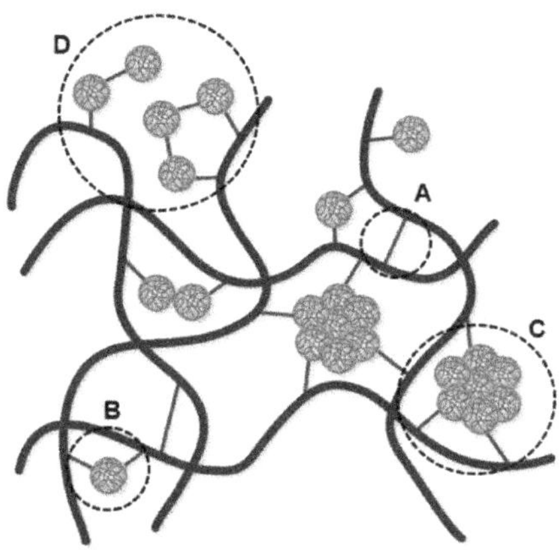

Figure 8. Schematic representation of the network structures associated with proteins in peroxide cross-linked NR. Cross-links are derived from (A) protein hydroperoxides, (B) the bridge of proteins, and (C) the aggregation of cross-linked proteins. No cross-link is derived from (D) dangling and pendant proteins grafted to the NR chain.

3.6. Effect of Network Heterogeneities on the Mechanical Properties

The mechanical properties of vulcanized rubber depend on the different local parameters that define the network structure: the fraction of dangling chain ends; the cross-link density; the nature, functionality, and spatial distribution of cross-links; the effect of entanglements; and the junction mobility [82,93–96]. Several of these structural parameters are drastically modified during the vulcanization process, making it quite difficult to study the influence of each individual factor on the mechanical properties of rubber, which is usually dominated by the effect of cross-link density. Nevertheless, above the optimum vulcanization time, the variation in the cross-link density is almost neglectable for both the NR and DPNR samples, which provides an opportunity to evaluate the effect of other structural parameters in the tensile properties of those networks.

Figure 9 shows that tensile strength decreases with the vulcanization time when the optimum time has been exceeded (t_{95} = 30 min) for the vulcanized NR and DPNR samples with 1 phr of DCP at 160 °C. The reported tensile strength values are the median of five test specimens, and the error bars represent their standard deviation. As discussed above, at those long vulcanization times, the coexistence of cross-linking reactions and chain scission processes in NR drives the formation of a higher fraction of non-elastic network defects and a broader spatial distribution of cross-links, but maintains the almost constant cross-link density in the sample. The absence of proteins reduces these negative reactions and the subsequent effects on the network structure of DPNR, showing higher tensile performance with lower cross-link density compared with NR.

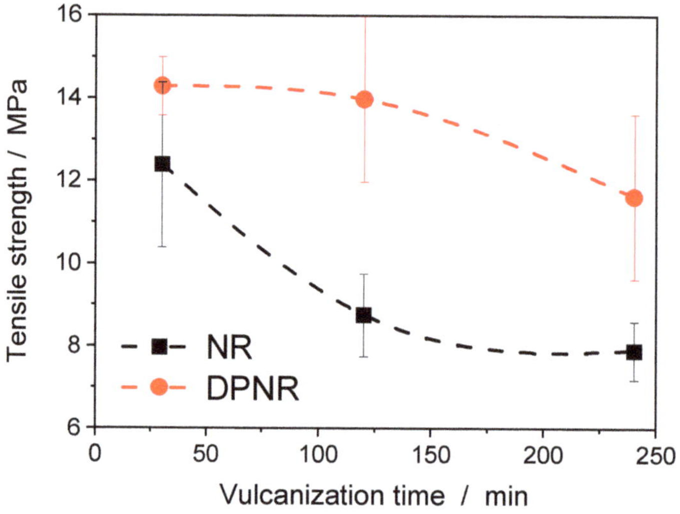

Figure 9. Variation in tensile strength with the vulcanization time for the NR and DPNR samples vulcanized with 1 phr of DCP at 160 °C.

On the one hand, Figure 10 shows a clear relationship between the decrease in the tensile strength and the relative width of the spatial distribution of cross-links, demonstrating that network inhomogeneities strongly influence the tensile strength of these rubber samples. On the other hand, the effect of non-elastic network defects on the tensile strength is difficult to evaluate. Although the presence of a high fraction of dangling chain ends (30–40%) was demonstrated in those samples, their elastic response depends on temperature and the observed timescale, minimizing the effect of these network defects on the mechanical properties.

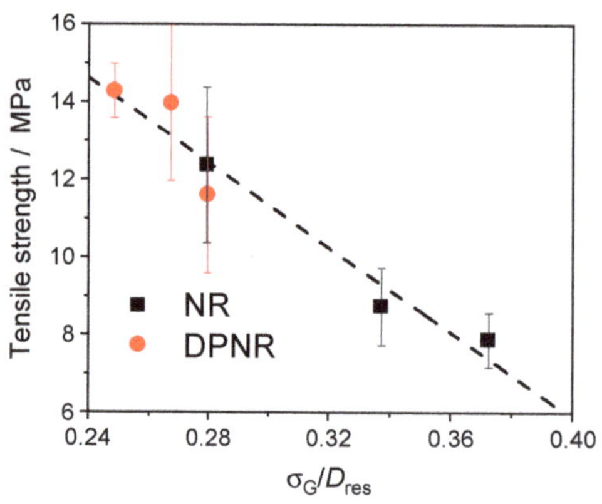

Figure 10. Relationship between tensile strength and the relative width of the spatial distribution of cross-links for the vulcanized rubber samples (NR and DPNR) with 1 phr of DCP at 160 °C.

These results suggest that rubber materials are not ideal networks because they contain non-elastic network defects (which should be considered) and a heterogeneous spatial distribution of cross-links. The latter seems to be a critical structural factor in understanding the ultimate tensile properties of rubber samples.

3.7. Thermal–Oxidative Degradation of Peroxide Cross-Linked NR

The effects of some non-rubber constituents on the degradation behavior of NR have been reported. For example, tocotrienols exhibited antioxidant activity [97], whereas copper accelerated the degradation as a pro-oxidant [98]. The role of proteins has been also investigated on the degradation of NR, concluding that proteins showed either insignificant protection [99] or antioxidant activity [100] against thermal–oxidative aging. This might be due to a variety of testing conditions and analyzing techniques used for that purpose. To clarify this issue, the rubber samples were aged at 100 °C under air atmosphere for 6, 24, and 48 h in order to monitor the progressive degradation and evolution of the network structure by using NMR experiments. Figure 11 shows the changes in the cross-link density for NR and DPNR (vulcanized at their optimum vulcanization times, t_{95}) at various aging times.

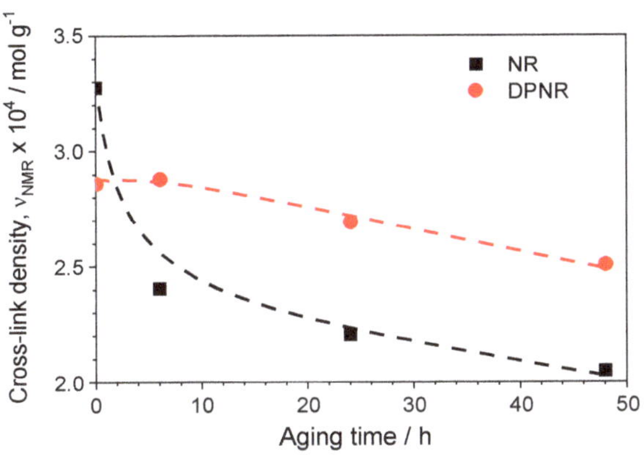

Figure 11. Evolution of cross-link density for the NR and DPNR samples (vulcanized at t_{95}) as a function of aging time. Lines are a guide for the eyes.

It was found that after 6 h of aging, the cross-link density for the NR remarkably decreased, while the DPNR sample showed no significant change with respect to the original structure. After that time, the cross-link density for both samples decreased with aging times up to 48 h in a similar way. These results evidence the active role of proteins in this process since they affect the degradation behavior of NR by increasing both the degradation rate and the final state of degradation. Nevertheless, it is important to remark that the effect of proteins seems to be limited to the initial stages of this thermo-oxidative process.

Additional details about how networks are degraded were provided by the analysis of the actual spatial distribution of cross-links at various aging times, as shown in Figure 12. The degradation behavior for the DPNR and NR samples is, in both cases, complex. However, it is different for each sample: the spatial distribution of cross-links for the aged NR was only shifted towards lower cross-link densities by the preferential effect of the chain scission reaction (see Figure 12A), whereas the distribution for the DPNR sample was widened by both extremes, at lower and higher cross-link densities, respectively, due to the coexistence of the chain scission and cross-linking reactions during the degradation process (see Figure 12B).

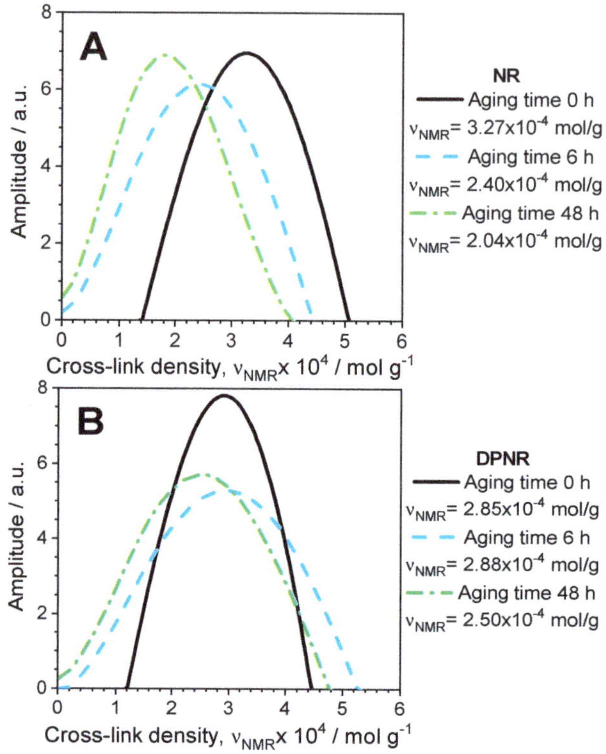

Figure 12. Spatial distribution of cross-link density at various aging times for (**A**) NR and (**B**) DPNR according to the regularization analysis.

Reactive oxygen species damage proteins at the protein backbones and the 20 common amino acid side chains, contributing to numerous protein degradation products such as carbonyl compounds, alcohols, and protein hydroperoxides [76–79]. The protein hydroperoxides can generate secondary radicals that induce damage to other proteins and other molecules as a chain reaction [78,101,102]. This is consistent with the assumption that proteins raise the rate and state of degradation in NR.

The analysis of the protein hydroperoxides of NR was carried out using extracted proteins as the model compound to confirm the formation of hydroperoxides caused by protein degradation (see Supplementary Information, Figure S4).

This confirmatory evidence indicates that the proteins in NR create hydroperoxides during the aging process, accelerating the degradation of other protein aggregates (further hydroperoxides can be generated as a chain reaction) and rubber chains. Consequently, the proteins in NR can prove to be pro-oxidant molecules that actively affect the degradation behavior of peroxide cross-linked NR and minimize the cross-linking reactions via radical mechanisms in the presence of oxygen.

4. Conclusions

Biopolymers are materials that can be derived from renewable resources or synthesized using a combination of renewable feedstocks and traditional polymerization methods, as observed in bio-based polymers. These biopolymers and bio-based polymers present a sustainable alternative to conventional petroleum-based polymers, being sourced from diverse feedstocks such as agricultural products (e.g., corn or soybeans) and non-traditional sources like algae or food waste. Additionally, these polymers are engineered with properties including biodegradability and composability.

The role of proteins on the pseudo-end-linked network, which characterizes the un-vulcanized NR, and their effects on the network formation and degradation of peroxide cross-linked NR was elucidated by time-domain ^{1}H-NMR experiments. This experimental procedure has proven to be a successful approach to obtain quantitative and complete local information about the actual fraction of non-elastic network defects, the cross-link density, and its spatial distribution in NR without any interference from proteins and their aggregates. This is a superior advantage shown by NMR approaches with respect to qualitative swelling experiments and the small-angle neutron scattering (SANS) technique that is intensely disturbed by the strong scattering phenomenon of protein aggregates.

In un-vulcanized NR samples, the accessible conformational space for chain segments is always restricted with respect to the synthetic polyisoprene (IR) counterpart because of the presence of a pseudo-end-linked network as was demonstrated by ^{1}H DQ-NMR experiments. The successive subtraction of proteins and phospholipids from the NR samples has demonstrated the main role of micelle formation or the association of the polar head-groups of phospholipid molecules at the α-terminal of the rubber chain as the cross-linking points of the naturally occurring networks, whereas the effect of proteins as cross-linkers at the NR chain ends seems to be minimal.

Although proteins have a negligible effect on the pseudo-end-linked network formed in un-vulcanized NR samples, they played an active role as coagents during the peroxide vulcanization of this material. It has been demonstrated that this type of vulcanization is a highly ineffective complex process for the NR matrix because at least two main competing reactions take place: cross-linking and chain scission phenomena. The existence of proteins in NR increases the concentration of radicals and provokes the effect on both opposite reactions. Consequently, the NR always showed a higher fraction of network defects, higher cross-link density, and broader (more heterogeneous) cross-link distribution than its deproteinized counterpart (DPNR). It is important to note that the extraction of proteins minimizes the importance of chain scission reactions, making their effects on the network structure negligible during the vulcanization process (in the absence of oxygen).

In addition to the local spatial inhomogeneities observed by NMR, NR samples vulcanized with peroxides also contain aggregates of proteins covalently grafted to the rubber backbone, as demonstrated by AFM. Those aggregates form larger length-scale inhomogeneities (in the range of microns) compared to the size of the highly cross-linked areas (in the range of tens of nm) in the rubber network structure. The latter has a deep impact on the tensile strength properties of these rubber samples.

Finally, proteins have a detrimental effect on the thermo-oxidative degradation of peroxide cross-linked NR because they act as pro-oxidant agents generating hydroperoxides that accelerate the degradation process. As a consequence, chain scission reactions are enlarged during this radical process to increase both the degradation rate and the final state of degradation. Contrary to this behavior, both opposite processes take place in deproteinized NR, and the spatial distribution of cross-links expands to both extremes, the higher and lower cross-link densities, although the chain scission reactions are predominant.

Supplementary Materials: The following supporting information can be downloaded at: https://www.mdpi.com/article/10.3390/polym17081063/s1, Figure S1. Scheme of a proposed pseudo-end-linked network structure of NR (proteins at the ω-terminal and phospholipid molecules at the α-terminal). Figure S2. ESR spectrum and g value of the extracted proteins. Figure S3. AFM micrographs and nitrogen content for vulcanized NR, vulcanized DPNR and leached NR samples. The phase images were obtained by AFM tapping mode. Figure S4. Hydroperoxides content generated from extracted proteins at various aging times. Reference [103] is cited in the supplementary materials.

Author Contributions: A.N., J.L.-V. and J.S.: writing—original draft, investigation, data acquisition, formal analysis, methodology, and conceptualization; A.G.-J. and R.P.-A.: investigation and methodology.; F.M.-S. and Z.Z.-R.: supervision, writing—review and editing; J.L.-V. and J.S.: funding acquisition and project administration. All authors have read and agreed to the published version of the manuscript.

Funding: This research was partially funded by the Spanish Ministry of Science and Innovation (MINECO) grant number PID2020-119047RB-I00, PLEC2021-00779 and PID2023-147542OB-I00 and Mahidol University and the National Research Council of Thailand grant number NRCT5-TRG63009-09.

Institutional Review Board Statement: Not applicable.

Data Availability Statement: The original contributions presented in this study are included in the article/Supplementary Material. Further inquiries can be directed to the corresponding authors.

Acknowledgments: The authors would like to acknowledge the Faculty of Science, Mahidol University, the Center of Excellence for Innovation in Chemistry (PERCH-CIC), Ministry of Higher Education, Science, Research, and Innovation. Sincere appreciation is extended to the Thai Rubber Latex Corporation (Thailand) Public Co., Ltd., for supplying the NR samples. The authors also acknowledge the Spanish Ministry of Science and Innovation (MINECO) (PID2020-119047RB-I00, PLEC2021-00779 and PID2023-147542OB-I00) and the Mahidol University and the National Research Council of Thailand: (NRCT5-TRG63009-09) for the partial financial support of this work.

Conflicts of Interest: The authors declare no conflicts of interest.

References

1. Saleh, T.A. *Polymer Hybrid Materials and Nanocomposites: Fundamentals and Applications*; William Andrew: Jeddah, Saudi Arabia, 2021.
2. Blume, A.; van Elburg, F.; Grunert, F.; Talma, A. "Re-Think" Sulfur Curing. *Molecules* **2024**, *29*, 5198. [CrossRef] [PubMed]
3. Zhu, Y.; Romain, C.; Williams, C. Sustainable polymers from renewable resources. *Nature* **2016**, *540*, 354–362. [CrossRef]
4. Mohanty, A.K.; Wu, F.; Mincheva, R.; Hakkarainen, M.; Raquez, J.M.; Mielewski, D.F.; Narayan, R.; Netravali, A.N.; Misra, M. Sustainable polymers. *Nat. Rev. Methods Primers* **2022**, *2*, 46. [CrossRef]
5. Allen, P.W.; Bloomfield, G.F. Natural rubber hydrocarbon. In *Chemistry and Physics of Rubber-like Substance*; Bateman, L., Ed.; MacLaren and Sons: London, UK, 1963; Chapter 1, pp. 1–17.
6. Eng, A.H.; Tanaka, Y.; Gan, S.N. FTIR studies on amino groups in purified Hevea rubber. *J. Nat. Rubber Res.* **1992**, *7*, 152–155.
7. Tangpakdee, J.; Tanaka, Y. Characterization of sol and gel in Hevea natural rubber. *Rubber Chem. Technol.* **1997**, *70*, 707–713. [CrossRef]
8. Tarachiwin, L.; Sakdapipanich, J.T.; Ute, K.; Kitayama, T.; Bamba, T.; Fukusaki, E.; Kobayashi, A.; Tanaka, Y. Structural characterization of α-terminal group of natural rubber 1. Decomposition of branch-points by lipase and phosphatase treatments. *Biomacromolecules* **2005**, *6*, 1851–1857. [CrossRef] [PubMed]
9. Tarachiwin, L.; Sakdapipanich, J.T.; Ute k Kitayama, T.; Tanaka, Y. Structural characterization of α-terminal group of natural rubber 2. Decomposition of branch-points by phospholipase and chemical treatments. *Biomacromolecules* **2005**, *6*, 1858–1863. [CrossRef]
10. Tarachiwin, L.; Sakdapipanich, J.; Tanaka, Y. Structure and origin of long-chain branching and gel in natural rubber. *Kautsch. Gummi Kunstst.* **2005**, *58*, 115–122.
11. Amnuaypornsri, S.; Sakdapipanich, J.; Toki, S.; Hsiao, B.; Ichikawa, N.; Tanaka, Y. Strain-induced crystallization of natural rubber: Effect of proteins and phospholipids. *Rubber Chem. Technol.* **2008**, *81*, 753–766. [CrossRef]
12. Food and Drug Administration. *Allergic Reaction to Latex-Containing Medical Devices*; FDA Medical Alert 1991; Food and Drug Administration: Silver Spring, MD, USA, 1991; MDA91-1.
13. Ichikawa, N.; Aik-Hwee, E.; Tanaka, Y. Properties of deproteinised natural rubber. In *Natural Rubber: Current Development in Product Manufacture and Application, Proceedings of the International Rubber Technology Conference 1993, Kuala Lumpur, Malaysia, 14–16 June 1993*; Kadir, A.A.S.A., Ed.; Rubber Research Institute of Malaysia: Kuala Lumpur, Malaysia, 1993; pp. 101–110.
14. Nakade, S.; Kuga, A.; Hayashi, M.; Tanaka, Y. Highly purified natural rubber. IV. Preparation and characteristics of gloves and condoms. *J. Nat. Rubb. Res.* **1997**, *12*, 33–42.
15. Coran, A.Y. Vulcanization: Conventional and dynamic. *Rubber Chem. Technol.* **1995**, *68*, 351–375. [CrossRef]
16. Aprem, A.S.; Joseph, K.; Thomas, S. Recent developments in crosslinking of elastomers. *Rubber Chem. Technol.* **2005**, *78*, 458–488. [CrossRef]

17. Dluzneski, P.R. Peroxide vulcanization of elastomers. *Rubber Chem. Technol.* **2001**, *74*, 451–492. [CrossRef]
18. Coleman, M.M.; Shelton, J.R.; Koenig, J.L. Sulfur vulcanization of hydrocarbon diene elastomers. *Ind. Eng. Chem. Prod. Res. Dev.* **1974**, *13*, 154–165. [CrossRef]
19. Krejsa, M.R.; Koenig, J.L. A review of sulfur crosslinking fundamentals for accelerated and unaccelerated vulcanization. *Rubber Chem. Technol.* **1993**, *66*, 376–410. [CrossRef]
20. Akiba, M.; Hashim, A.S. Vulcanization and crosslinking in elastomers. *Prog. Polym. Sci.* **1997**, *22*, 475–521. [CrossRef]
21. Heideman, G.; Datta, R.N.; Noordermeer, J.W.M. Activators in accelerated sulfur vulcanization. *Rubber Chem. Technol.* **2004**, *77*, 512–541. [CrossRef]
22. Suchiva, K.; Kowitteerawut, T.; Srichantamit, L. Structure properties of purified natural rubber. *J. Appl. Polym. Sci.* **2000**, *78*, 1495–1504. [CrossRef]
23. Conde-Salazar, L.; del-Río, E.; Guimaraens, D.; González Domingo, A. Type IV allergy to rubber additives: A 10-year study of 686 cases. *J. Am. Acad. Dermatol.* **1993**, *29*, 176–180. [CrossRef]
24. Tan, Z.; Jaeger, R.; Vancso, G.J. Crosslinking studies of poly(dimethylsiloxane) networks: A comparison of inverse gas chromatography, swelling experiments and mechanical analysis. *Polymer* **1994**, *35*, 3230–3236. [CrossRef]
25. Arndt, K.F.; Schreck, J. Netzwerkcharakterisierung durch dampfdruckosmotische Messungen. *Acta Polym.* **1985**, *36*, 56–57. [CrossRef]
26. Patel, S.K.; Malone, S.; Cohen, C.; Gillmor, J.R.; Colby, R.H. Elastic modulus and equilibrium swelling of poly(dimethylsiloxane) networks. *Macromolecules* **1992**, *25*, 5241–5251. [CrossRef]
27. Valentín, J.L.; Carretero-González, J.; Mora-Barrantes, I.; Chassé, W.; Saalwächter, K. Uncertainties in the determination of cross-link density by equilibrium swelling experiments in natural rubber. *Macromolecules* **2008**, *41*, 4717–4729. [CrossRef]
28. Imanishi, Y.; Adachi, K.; Kotaka, T.J. Further investigation of the dielectric normal mode process in undiluted cis-polyisoprene with narrow distribution of molecular weight. *Chem. Phys.* **1988**, *89*, 7585–7592.
29. Zaper, A.M.; Koenig, J.L. Solid State Carbon-13 NMR Studies of Vulcanized Elastomers. II, Sulfur Vulcanization of Natural Rubber. *Rubber Chem. Technol.* **1987**, *60*, 252–277. [CrossRef]
30. Orza, R.A.; Magusin, P.C.M.M.; Litvinov, V.M.; van Duin, M.; Michels, M.A.J. Mechanism for peroxide cross-linking of EPDM rubber from MAS 13C NMR Spectroscopy. *Macromolecules* **2009**, *42*, 8914–8924. [CrossRef]
31. Pyckout-Hintzen, W.; Springer, T.; Forster, F.; Gronski, W. Small-angle neutron scattering investigation of a multilinked polybutadiene network crosslinked in solution. *Macromolecules* **1991**, *24*, 1269–1274. [CrossRef]
32. Ikeda, Y.; Higashitani, N.; Hijikata, K.; Kokubo, Y.; Morita, Y.; Shibayama, M.; Osaka, N.; Suzuki, T.; Endo, H.; Kohjiya, S. Vulcanization: New focus on a traditional technology by small-angle neutron scattering. *Macromolecules* **2009**, *42*, 2741–2748. [CrossRef]
33. Litvinov, V.M.; De, P.P. (Eds.) *Spectroscopy of Rubbers and Rubbery Materials*; Rapra Technology Ltd.: Shawbury, UK, 2002.
34. Cohen-Addad, J.P. Effect of the anisotropic chain motion in molten polymers: The solidlike contribution of the nonzero average dipolar coupling to NMR signals theoretical description. *J. Chem. Phys.* **1973**, *60*, 2440–2453. [CrossRef]
35. Brereton, M.G. NMR transverse relaxation function calculated for constrained polymer chains: Application to entanglements and networks. *Macromolecules* **1990**, *23*, 1119–1131. [CrossRef]
36. Graf, R.; Heuer, A.; Spiess, H.W. Chain-order effects in polymer melts probed by 1H double-quantum NMR spectrocopy. *Phys. Rev. Lett.* **1998**, *80*, 5738–5741. [CrossRef]
37. Saalwächter, K. Proton multiple-quantum NMR for the study of chain dynamics and structural constraints in polymeric soft materials. *Prog. Nucl. Magn. Reason. Spectrosc.* **2007**, *51*, 1–35. [CrossRef]
38. Vieyres, A.; Pérez-Aparicio, R.; Albouy, P.-A.; Sanseau, O.; Saalwächter, K.; Long, D.R.; Sotta, P. Sulfur-Cured Natural Rubber Elastomer Networks: Correlating Cross-Link Density, Chain Orientation, and Mechanical Response by Combined Techniques. *Macromolecules* **2013**, *46*, 889–899. [CrossRef]
39. Basterra-Beroiz, B.; Rommel, R.; Kayser, F.; Westermann, S.; Valentín, J.L.; Heinrich, G. New Insights into Rubber Network Structure by a Combination of Experimental Techniques. *Rubber Chem. Technol.* **2017**, *90*, 347–366. [CrossRef]
40. Litvinov, V.M. EPDM/PP thermoplastic vulcanizates as studied by proton NMR relaxation: Phase composition, molecular mobility, network structure in the rubbery phase, and network heterogeneity. *Macromolecules* **2006**, *39*, 8727–8741. [CrossRef]
41. Demco, D.E.; Hafner, S.; Fülber, C.; Graf, R.; Spiess, H.W. Two-dimensional proton magnetization-exchange NMR spectroscopy in cross-linked elastomers. *J. Chem. Phys.* **1996**, *105*, 11285–11296. [CrossRef]
42. Callaghan, P.T.; Samulski, E.T. Molecular ordering and the direct measurement of weak proton-proton dipolar interactions in a rubber network. *Macromolecules* **1997**, *30*, 113–122. [CrossRef]
43. Valentín, J.; Posadas, P.; Fernández-Torres, A.; Malmierca, M.A.; González, L.; Chassé, W.; Saalwächter, K. Inhomogeneities and chain dynamics in diene rubbers vulcanized with different cure systems. *Macromolecules* **2010**, *43*, 4210–4222. [CrossRef]

44. Basterra-Beroiz, B.; Rommel, R.; Kayser, F.; Valentín, J.L.; Westermann, S.; Heinrich, G. Revisiting Segmental Order: A Simplified Approach for Sulfur-Cured Rubbers Considering Junction Fluctuations and Entanglements. *Macromolecules* **2018**, *51*, 2076–2088. [CrossRef]
45. Saalwächter, K.; Herrero, B.; López-Manchado, M.A. Chain Order and Cross-Link Density of Elastomers As Investigated by Proton Multiple-Quantum NMR. *Macromolecules* **2005**, *38*, 9650–9660. [CrossRef]
46. Saleesung, T.; Reichert, D.; Saalwächter, K.; Sirisinha, C. Correlation of crosslink densities using solid state NMR and conventional techniques in peroxide-crosslinked EPDM rubber. *Polymer* **2015**, *56*, 309–317. [CrossRef]
47. Saalwächter, K.; Ziegler, P.; Spyckerelle, O.; Haidar, B.; Vidal, A.; Sommer, J.U. 1H multiple-quantum nuclear magnetic resonance investigations of molecular order distributions in poly(dimethylsiloxane) networks: Evidence for a linear mixing law in bimodal systems. *J. Chem. Phys.* **2003**, *119*, 3468–3482. [CrossRef]
48. Chassé, W.; Valentín, J.L.; Genesky, G.D.; Cohen, C.; Saalwächter, K. Precise dipolar coupling constant distribution analysis in proton multiple-quantum NMR of elastomers. *J. Chem. Phys.* **2011**, *134*, 044907. [CrossRef] [PubMed]
49. Chassé, W.; Lang, M.; Sommer, J.-U.; Saalwächter, K. Cross-link density estimation of PDMS networks with precise consideration of networks defects. *Macromolecules* **2012**, *45*, 899–912. [CrossRef]
50. Acosta, R.H.; Vega, D.A.; Villar, M.A.; Monti, G.A.; Vallés, E.M. Double quantum NMR applied to polymer networks with low concentration of pendant chains. *Macromolecules* **2006**, *39*, 4788–4792. [CrossRef]
51. Acosta, R.H.; Monti, G.A.; Villar, M.A.; Vallés, E.M.; Vega, D.A. Transiently trapped entanglements in model polymer networks. *Macromolecules* **2009**, *42*, 4674–4680. [CrossRef]
52. Sommer, J.-U.; Chassé, W.; Valentín, J.L.; Saalwächter, K. Effect of excluded volume on segmental orientation correlations in polymer chains. *Phys. Rev. E* **2008**, *78*, 051803. [CrossRef]
53. Iqbal, J.; Bhatia, B.; Nayyar, N.K. Transition metal-promoted free-radical reactions in organic synthesis: The formation of carbon-carbon bonds. *Chem. Rev.* **1994**, *94*, 519–564. [CrossRef]
54. Tanaka, Y. Structural characterization of natural polyisoprenes: Solve the mystery of natural rubber based on structural study. *Rubber Chem. Technol.* **2001**, *74*, 355–375. [CrossRef]
55. Toki, S.; Che, J.; Rong, L.; Hsiao, B.S.; Amnuaypornsri, S.; Nimpaiboon, A.; Sakdapipanich, J. Entanglements and networks to strain-induced crystallization and stress—strain relations in natural rubber and synthetic polyisoprene at various temperatures. *Macromolecules* **2013**, *46*, 5238–5248. [CrossRef]
56. Valentin, J.L.; Lopez, D.; Hernandez, R.; Mijangos, C.; Saalwachter, K. Structure of Poly(vinyl alcohol) Cryo-Hydrogels as Studied by Proton Low-Field NMR Spectroscopy. *Macromolecules* **2009**, *42*, 263–272. [CrossRef] [PubMed]
57. Che, J.; Toki, S.; Valentin, J.L.; Brasero, J.; Nimpaiboon, A.; Rong, L.; Hsiao, B.S. Chain Dynamics and Strain-Induced Crystallization of Pre- and Postvulcanized Natural Rubber Latex Using Proton Multiple Quantum NMR and Uniaxial Deformation by in Situ Synchrotron X-ray Diffraction. *Macromolecules* **2012**, *45*, 6491–6503. [CrossRef]
58. Malmierca, M.A.; Mora-Barrantes, I.; Posadas, P.; Rodríguez, A.; Ibarra, L.; Gonzalez-Jiménez, A.; Nogales, A.; Saalwachter, K.; Valentín, J.L. Characterization of Network Structure and Chain Dynamics of Elastomeric Ionomers by Means of 1H Low-Field NMR. *Macromolecules* **2014**, *47*, 5655–5667. [CrossRef]
59. Vaca Chávez, F.; Saalwächter, K. Time-domain NMR observation of entangled polymer dynamics: Analytical theory of signal functions. *Macromolecules* **2011**, *44*, 1560–1569. [CrossRef]
60. Vaca Chávez, F.; Saalwächter, K. Time-domain NMR observation of entangled polymer dynamics: Universal behavior of flexible homopolymers and applicability of the tube model. *Macromolecules* **2011**, *44*, 1549–1559. [CrossRef]
61. Doi, M.; Edwards, S.F. *The Theory of Polymer Dynamics*; Claredon Press: Oxford, UK, 1986.
62. Amnuaypornsri, S.; Sakdapipanich, J.; Tanaka, Y. Green strength of natural rubber: The origin of the stress–strain behavior of natural rubber. *J. Appl. Polym. Sci.* **2009**, *111*, 2127–2133. [CrossRef]
63. Suzuki, T.; Osaka, N.; Endo, H.; Shibayama, M.; Ikeda, Y.; Asai, H.; Higashitani, N.; Kokubo, Y.; Kohjiya, S. Nonuniformity in cross-linked natural rubber as revealed by contrast-variation small-angle neutron scattering. *Macromolecules* **2010**, *43*, 1556–1563. [CrossRef]
64. Whelan, A.; Lee, K.S. (Eds.) *Development in Rubber Technology*; Applied Science: London, UK, 1979; Volume 1.
65. García-Quesada, J.C.; Gilbert, M. Peroxide crosslinking of unplasticized poly(vinyl chloride). *J. Appl. Polym. Sci.* **2000**, *77*, 2657–2666. [CrossRef]
66. Busci, A.; Szocs, F. Kinetics of radical generation in PVC with dibenzoyl peroxide utilizing high-pressure technique. *Macromol. Chem. Phys.* **2000**, *201*, 435–438.
67. Keller, R.C. Peroxide curing of ethylene-propylene elastomers. *Rubber Chem. Technol.* **1988**, *61*, 238–254. [CrossRef]
68. Murgic, Z.H.; Jelencic, J.; Murgic, L. The mechanism of triallylcyanurate as a coagent in EPDM peroxide vulcanization. *Polym. Eng. Sci.* **1998**, *38*, 689–692. [CrossRef]
69. Class, J. A review of the fundamentals of crosslinking with peroxides. *Rubber World* **1999**, *220*, 35–39.

70. Oh, S.J.; Koenig, J.L. Studies of peroxide curing of cis-1,4-Polyisoprene/Diallyl phthalate blends by spectroscopic techniques. *Rubber Chem. Technol.* **1999**, *72*, 334–342. [CrossRef]
71. Oh, S.J.; Koenig, J.L. Studies of peroxide curing of polybutadiene/zinc diacrylate blends by fast FT-IR imaging. *Rubber Chem. Technol.* **2000**, *73*, 74–79. [CrossRef]
72. Schaich, K.M.; Karel, M. Free radical reactions of peroxidizing lipids with amino acids and proteins: An ESR study. *Lipids* **1976**, *11*, 392–400. [CrossRef]
73. Gardner, H.W. Lipid hydroperoxide reactivity with proteins and amino acids: A review. *J. Agric. Food Chem.* **1979**, *27*, 220–229. [CrossRef]
74. Saeed, S.; Fawthrop, S.A.; Howell, N.K. Electron spin resonance (ESR) study on free radical transfer in fish lipid–protein interaction. *J. Sci. Food Agric.* **1999**, *79*, 1809–1816. [CrossRef]
75. Endstra, W.C. Application of coagents for peroxide crosslinking. *Kautsch. Gummi Kunstst.* **1990**, *43*, 790–793.
76. Simpson, J.A.; Narita, S.; Gieseg, S.; Gebicki SGebick, J.M.; Dean, R.T. Long-lived reactive species on free-radical-damaged proteins. *Biochem. J.* **1992**, *282*, 621–624. [CrossRef]
77. Gebicki, S.; Gebicki, J.M. Formation of peroxides in amino acids and proteins exposed to oxygen free radicals. *Biochem. J.* **1993**, *289*, 743–749. [CrossRef]
78. Dean, R.T.; Fu, S.; Stocker, R.; Davies, M.J. Biochemistry and pathology of radical-mediated protein oxidation. *Biochem. J.* **1997**, *324*, 1–18. [CrossRef] [PubMed]
79. Hawkins, C.L.; Davies, M.J. Generation and propagation of radical reactions on proteins. *Biochim. Biophys. Acta* **2001**, *1504*, 196–219. [CrossRef]
80. Curro, J.G.; Pincus, P. A theoretical basis for viscoelastic relaxation of elastomers the long-time limit. *Macromolecules* **1983**, *16*, 559–562. [CrossRef]
81. Roth, L.E.; Vega, D.A.; Valles, E.M.; Villar, M.A. Viscoelastic properties of networks with low concentration of pendant chains. *Polymer* **2004**, *45*, 5923–5931. [CrossRef]
82. Batra, A.; Cohen, C.; Archer, L. Stress relaxation of end-linked Polydimethylsiloxane elastomers with long pendent chains. *Macromolecules* **2005**, *38*, 7174–7180. [CrossRef]
83. Moore, C.G.; Scanlan, J.J. Determination of degree of crosslinking in natural rubber vulcanizates. Part VI. Evidence for chain scission during the crosslinking of natural rubber with organic peroxides. *Polym. Sci.* **1960**, *43*, 23–33. [CrossRef]
84. Mark, H.F. Elastomers-Past, Present, and Future. *Rubber Chem. Technol.* **1988**, *61*, G73–G96. [CrossRef]
85. González, L.; Rodríguez, A.; Marcos, A.; Chamorro, C. Crosslink reaction mechanisms of diene rubber with dicumyl peroxide. *Rubber Chem. Technol.* **1996**, *69*, 203–214. [CrossRef]
86. Valentín, J.L.; Fernández-Torres, A.; Posadas, P.; Marcos-Fernandez, A.; Rodríguez, A.; González, L. Measurements of freezing-point depression to evaluate rubber network structure. Crosslinking of natural rubber with dicumyl peroxide. *J. Polym. Sci. Part B Polym. Phys.* **2007**, *45*, 544–556. [CrossRef]
87. Karino, T.; Ikeda, Y.; Yasuda, Y.; Kohjiya, S.; Shibayama, M. Nonuniformity in natural rubber as revealed by small-angle neutron scattering, small-angle x-ray scattering, and atomic force microscopy. *Biomacromolecules* **2007**, *8*, 693–699. [CrossRef]
88. Pailhories, G. Reducing proteins in latex gloves: The industrial approach. *Clin. Rev. Allergy* **1993**, *11*, 391–402. [CrossRef]
89. Ng, K.P.; Yip, E.; Mok, K.L. Production of natural rubber latex gloves with low extractable protein content: Some practical recommendations. *J. Nat. Rubber Res.* **1994**, *9*, 87–95.
90. Ghazaly, H.M. Factory production of examination gloves from low protein latex. *J. Nat. Rubber Res.* **1994**, *9*, 96–108.
91. Dalrymple, S.J.; Audley, B.G. Allergenic proteins in dipped products: Factors influencing extractable protein levels. *Rubber Dev.* **1992**, *45*, 51–60.
92. Leynadier, F.; Xuan, T.T.; Dry, J. Allergenicity suppression in natural latex surgical gloves. *Allergy* **1991**, *46*, 619–625. [CrossRef]
93. Klüppel, M.; Heinrich, G. Network Structure and Mechanical Properties of Sulfur Cured Rubbers. *Macromolecules* **1994**, *27*, 3595–3603. [CrossRef]
94. Wu, B.; Chassé, W.; Peters, R.; Brooijmans, T.; Dias, A.A.; Heise, A.; Duxbury, C.J.; Kentgens, A.P.M.; Brougham, D.F.; Litvinov, V.M. Network Structure in Acrylate Systems: Effect of Junction Topology on Cross-Link Density and Macroscopic Gel Properties. *Macromolecules* **2016**, *49*, 6531–6540. [CrossRef]
95. Lang, M. Relation between Cross-Link Fluctuations and Elasticity in Entangled Polymer Networks. *Macromolecules* **2017**, *50*, 2547–2555. [CrossRef]
96. Vilgis, T.A.; Heinrich, G. Statics and dynamics of heterogeneous polymer networks. *Macromol. Theory Simul.* **1971**, *3*, 271–293. [CrossRef]
97. Nadarajah, M.; Tirimanne, A.S.L.; Coomarasamy, A.; Kasinathan, S. Some naturally occurring antioxidants in Hevea brasiliensis latex. *J. Rubb. Res. Inst. Sri Lanka* **1971**, *48*, 202–211.
98. Tuampeomsab, S.; Sakdapipanich, J.; Tanaka, Y. Influence of some non-rubber components on aging behavior of purified natural rubber. *Rubber Chem. Technol.* **2007**, *80*, 159–169. [CrossRef]

99. Hasma, H. Role of some non-rubber constituents on thermal oxidative aging of natural rubber. *J. Nat. Rubb. Res.* **1990**, *5*, 1–18.
100. Tuampeomsab, S.; Sakdapipanich, J. Role of naturally occurring lipids and proteins on thermal aging behaviour of purified natural rubber. *Kautsch. Gummi Kunstst.* **2007**, *12*, 678–684.
101. Halliwell, B.; Gutteridge, J.M.C. *Free Radicals in Biology and Medicine*, 3rd ed.; Clarendon Press: Oxford, UK, 1999.
102. Gebicki, J.M. Protein hydroperoxides as new reactive oxygen species. *Redox Rep.* **1997**, *3*, 99–110. [CrossRef]
103. Tanaka, Y.; Tarachiwin, L. Recent advances in structural characterization of natural rubber. *Rubber Chem. Technol.* **2009**, *82*, 283–314. [CrossRef]

polymers

MDPI

Article

Sulfur and Peroxide Cross-Linking of Lignosulfonate-Filled Compounds Based on Acrylonitrile–Butadiene Rubber and Styrene–Butadiene Rubber

Ján Kruželák [1,*], Michaela Džuganová [1], Andrea Kvasničáková [1], Ján Hronkovič [2], Jozef Preťo [2], Ivan Chodák [3] and Ivan Hudec [1]

1 Department of Plastics, Rubber and Fibres, Faculty of Chemical and Food Technology, Slovak University of Technology in Bratislava, Radlinského 9, 812 37 Bratislava, Slovakia; michaela.dzuganova@stuba.sk (M.D.); andrea.kvasnicakova@stuba.sk (A.K.); ivan.hudec@stuba.sk (I.H.)

2 VIPO a.s., Gen. Svobodu 1069/4, 958 01 Partizánske, Slovakia; jan.hronkovic@vipo.sk (J.H.); jpreto@vipo.sk (J.P.)

3 Slovak Academy of Sciences, Polymer Institute, Dúbravská Cesta 9, 845 41 Bratislava, Slovakia; ivan.chodak@savba.sk

* Correspondence: jan.kruzelak@stuba.sk

Abstract

Calcium lignosulfonate was incorporated into rubber compounds based on styrene–butadiene rubber (SBR) and acrylonitrile–butadiene rubber (NBR) in amounts ranging from 10 to 60 phr. A sulfur-based curing system and a peroxide curing system consisting of dicumyl peroxide in combination with methacrylic acid zinc salt were used for cross-linking of the compounds. The aim of the work was to investigate the influence of lignosulfonate and curing system composition of the cross-linking process, morphology, physical–mechanical and dynamic–mechanical characteristics of the composites. The achieved results showed that peroxide cured composites demonstrated higher cross-link density, which was found not to be influenced by the content of lignosulfonate. The cross-link density of sulfur-cured composites was lower and showed a decreasing tendency with increasing amounts of the biopolymer. A lower cross-linking degree was reflected in a higher elongation at break and higher increase in the elongation at break of the corresponding composites. On the other hand, peroxide-cured composites exhibited a higher modulus M100 and higher hardness. The microscopic analysis revealed that co-agent in peroxide vulcanization contributed to the improvement of adhesion between the biopolymer and the rubber resulting in higher tensile strength of the equivalent composites. The higher cross-link density of peroxide-cured composites caused higher restriction of the chain segments' mobility, due to which these composites exhibited a higher glass transition temperature.

Keywords: rubber; biopolymer; sulfur curing system; peroxide curing system; cross-linking; properties

1. Introduction

The depletion of sources, global warning, and negative effects on the environment have shifted the awareness towards bio-economy and green technologies. Lignocellulosic raw materials represent suitable alternatives to petroleum-based products, as they are highly available and sustainable and have a positive effect on global greenhouse gas reduction. Lignin is the second most widespread biopolymeric material in the world and thus bears a huge application potential. The annual production of technical lignin is about 50–70 million tons, of which only 1–2% is used to produce the value-added goods. The rest is landfilled or burned to generate energy or to recover chemicals [1,2]. Lignin exhibits highly branched amorphous aromatic structure with diversity of functional groups as carboxyls, carbonyls, hydroxyls, or methoxy groups. Due to these groups, lignin exhibits attractive properties including antimicrobial, antioxidant, UV absorption, adhesive properties, hydrophobicity, etc. [3–7].

There are several types of lignin depending on the processing method and extraction procedures [7–11]. Sulfur and sulfur-free extraction processes are the most common procedures to obtain lignin from lignocellulosic parts of wood and plants. Sulfur-based lignins include Kraft lignin and lignosulfonates. Kraft lignin is derived by Kraft pulping using sodium sulfide and sodium hydroxide at pH between 13 and 14. The sulfite pulping process is acid-catalyzed and basic (pH between 1 and 2) using an aqueous sulfur dioxide and a sodium, calcium, ammonium, or magnesium salt-based acid. When compared to Kraft lignin, lignosulfonates contain less free carboxyl and phenolic hydroxyl groups and higher amount of sulfur, mainly in the form of sulfonate groups [12,13]. Due to the presence of sulfonate groups, they are water-soluble. Lignosulfonates have been widely used as dispersant agents, surfactants, binders, compatibilizers, additives to concretes, and composites [14–17]. In addition, due to their aromatic structure, high amount of carbon, good mechanical stability, and viscoelastic properties, they are suitable materials as additives and fillers in rubber-based compounds and composites [18–22]. To fabricate polymer composite materials with good applicable and utility properties, good adhesion and interactions between the filler and the rubber matrix on their interfacial region must be formed. Lignins and lignosulfonates are polar materials and thus their compatibility with non-polar matrices is usually poor, often resulting in weak physical–mechanical properties of the final composites. On the other hand, polar functional moieties provide space for surface modification and thus increasing interactions with rubber matrices. Good adhesion and formation of interactions between the filler and the rubber on the filler–rubber interface is the basic principle for rubber composites' reinforcement.

The final utility properties of rubber goods are achieved during the vulcanization or curing. Vulcanization is a physical–chemical process during which a plastic rubber compound is transformed into cross-linked highly elastic products—vulcanized by chemical reactions between functional groups of rubber and curing additives. The cross-linked network improves the dimensional stability of rubber materials; increases elasticity, mechanical, and dynamical properties; and reduces hysteresis or plasticity. A variety of curing systems have been applied for the cross-linking of rubber compounds, such as sulfur-based systems, organic peroxides, phenol–formaldehyde resins, quinones, metal oxides, amines, ureas, etc. Among them, sulfur curing systems are the most commonly used, accounting for about 90% or rubber-based articles, followed by organic peroxides.

Sulfur vulcanization is the oldest method used for cross-linking unsaturated rubbers. Sulfur is always used in combination with activators and accelerators. Sometimes, retarders of vulcanization or prevulcanization inhibitors can be part of sulfur curing systems, too. The mechanism and chemistry of sulfur vulcanization is very complex and still not comprehensibly elucidated. During the process, rubber chains segments are linked with sulfur-based cross-links having different lengths and amounts of sulfur in sulfur bridges (monosulfidic, disufidic, and polysulfidic cross-links) [23,24]. In general, sulfur-cured vulcanizates demonstrate good tensile behavior, good elastic and dynamic properties, good abrasion resistance, or good resistance to dynamic fatigue. On the other hand, due to the low bonding energy of sulfidic cross-links, their thermo-oxidative stability is usually weak [25–27].

Organic peroxides can be applied to vulcanize both saturated and unsaturated rubbers. By the application of organic peroxides, carbon–carbon bonds are formed between the chains [28]. C-C bonds have higher dissociation energy when compared to sulfidic cross-links, and thus the main feature of peroxide-cured rubbers is higher thermal resistance and higher resistance to degradation factors, like oxygen and heat [29,30]. On the other side, when compared to sulfur-cured vulcanizates, they usually have worse tear and tensile strength and weaker dynamic properties.

To boost the cross-linking process with organic peroxides, double bonds containing organic compounds, termed as co-agents, are often used [31,32]. The co-agents help to increase the efficiency of peroxide curing process by enhancing the cross-link density and by modifying the cross-links' structure. Couple of reaction mechanisms have been proposed for particular types of co-agents. In general, network enhancement through the homopolymerization and grafting of co-agents between the chain segments are the main co-agents' reaction mechanisms [33–36]. As a lot of co-agents are polar materials, they also increase the polarity of rubber compounds and enhance adhesion towards polar fillers and additives in the rubber compounding.

In this study, calcium lignosulfonate in various amounts was incorporated into NBR- and SBR-based formulations. A sulfur curing system and organic peroxide in combination with a co-agent were applied for cross-linking. The main aim was to compare the effect of vulcanization systems on the cross-linking process, physical–mechanical, and dynamic–mechanical properties of the composites.

2. Experimental

2.1. Materials

Styrene–butadiene rubber (SBR, Kralex 1502, styrene content—23.5 wt.%, Mooney viscosity ML 1 + 4 (100 °C)—52) prepared by cold emulsion polymerization was supplied from Synthos Kralupy, a.s., Kralupy nad Vltavou, Czech Republic. Acrylonitrile–butadiene rubber (NBR, SKN 3345, acrylonitrile content—31–35%, Mooney viscosity ML 1 + 4 (100 °C)—45) was supplied from Sibur International, Moscow, Russia. Calcium lignosulfonate having trade name Borrement CA120 was provided by Borregaard Deutschland GmbH, Karlsruhe, Germany. The molecular weight of lignosulfonate was 24,000 $g \cdot mol^{-1}$ and the specific surface area was 3.9 $m^2 \cdot g^{-1}$. In addition to carbon (46.63 wt.%), the biopolymer contained hydrogen (5.35 wt.%), nitrogen (0.14 wt.%), sulfur (5.62 wt.%), and hydroxyl groups (1.56 wt.%). Lignosulfonate was applied into rubber formulations in a concentration scale ranging from 10 to 60 phr. For the cross-linking of rubber compounds, sulfur and peroxide curing systems were used. The sulfur-based curing system consisted of zinc oxide (3 phr) and stearic acid (2 phr) as activators, accelerator *N*-cyclohexyl-2-benzothiazole sulfenamide CBS (1.5 phr), and sulfur as curing agent (1.5 phr). The additives of the sulfur curing system were supplied from Vegum a.s., Dolné Vestenice, Slovak Republic. The peroxide curing system consisted of dicumyl peroxide DCP as curing agent (1 phr) and methacrylic acid zinc salt ZDMA (10 phr) as co-agent. DCP and ZDMA were supplied from Sigma-Aldrich, St. Louis, MO, USA.

2.2. Methods

2.2.1. Fabrication and Curing

The fabrication of the composites was performed in a laboratory kneader Brabender (Brabender GmbH & Co. KG, Duisburg, Germany) at 90 °C and 55 rpm. The mixing process was carried out in two steps.

First, rubber was plasticated for 1 min, then the activators of the sulfur curing system were added. Calcium lignosulfonate was introduced after 2 min, and the rubber mixture was compounded for the next 4 min. After the first step of mixing, the rubber compounds were additionally homogenized and sheeted by using a two-roll mill. Sulfur and the accelerator were introduced in the second step, and the mixing process continued in the kneader for 4 min at 90 °C with rotor speed set up to 55 rpm. Finally, the rubber compounds were sheeted in a two-roll mill.

The compounding procedure of rubber formulations with the peroxide curing system proceeded following the same conditions (90 °C, 55 rpm, overall time of mixing 10 min). The rubber and the filler were compounded in the first step, which took 6 min, and additives of curing systems were applied in the second step. Final homogenization and sheeting were accomplished in a two-roll mill.

The curing process was carried out using a Fontijne hydraulic press (Fontijne, Vlaardingen, The Netherlands) following the optimum cure time of each rubber compound. The temperature of curing was 170 °C and the press was 15 MPa. After curing, thin sheets with dimensions of 15 × 15 cm and a thickness of 2 mm were obtained.

2.2.2. Determination of Curing Characteristics

An MDR 2000 oscillatory rheometer (Alpha Technologies, Akron, OH, USA) was used to determine curing isotherms and curing characteristics.

The investigated curing parameters were as follows:

M_L (dN·m)—minimum torque;
M_H (dN·m)—maximum torque;
ΔM (dN·m)—torque difference, $\Delta M = M_H - M_L$;
t_{c90} (min)—optimum curing time;
t_{s1} (min)—scorch time.

2.2.3. Determination of Cross-Link Density

To calculate the cross-link density ν, dried and weighted samples were immersed in xylene, in which they swelled. The samples were taken out from the solvent every hour, wiped out of solvent, and weighted. When the weight of the samples was constant, equilibrium swelling was reached and used for the determination of cross-link density by the Krause-modified Flory–Rehner equation [37].

2.2.4. Investigation of Physical–Mechanical Characteristics

The tensile tests were performed according to valid technical standards using Zwick Roell/Z 2.5 tearing equipment (Zwick GmbH & Co. KG, Ulm, Germany). The cross-head speed was set up to 500 mm·min^{-1} with a gauge length of 25 mm. Dumbbell-shaped specimens (width 6.4 mm, length 80 mm) were cut with a special knife from a 2 mm thick cured rubber plate. The hardness in Shore A was measured by using a durometer.

2.2.5. Microscopic Analysis

The surface morphology was evaluated by a JEOL JSM-7500F scanning electron microscope (Jeol Ltd., Tokyo, Japan). The samples were first cooled down in liquid nitrogen under glass transition temperature and then fractured into small fragments with surface area of 3 × 2 mm. The fractured surface was covered with a thin layer of gold and put into the microscope. The source of electrons was a cold cathode UHV field emission gun, the accelerate voltage ranged from 0.1 kV to 30 kV, and the resolution was 1.0 nm at 15 kV and 1.4 nm at 1 kV. SEM images were captured by a CCD-Camera EDS (Oxford INCA X-ACT, Jeol Ltd., Tokyo, Japan).

3. Results and Discussion

3.1. Curing Process and Cross-Link Density

The corresponding vulcanization isotherms of rubber compounds cured with the sulfur and peroxide systems are illustrated in Figure 1. It becomes apparent from them that the course of curing isotherms was influenced by the type of rubber as well as by the curing system composition. From Figure 2, it becomes apparent that the minimum torque M_L of both SBR- as well as NBR-based compounds show an increasing trend with increasing content of the biopolymer. The minimum torque, to a certain extent, relates to the viscosity of the compounds before the curing process started, which suggests that the incorporation of the biopolymer resulted in an increase in viscosity. The highest minimum torque was exhibited by rubber compounds based on SBR cured with the peroxide system, followed by the equivalent compounds cured with the sulfur system. The compounds based on NBR cured with both vulcanization systems demonstrated a lower M_L. This points to the higher viscosity of SBR-based rubber formulations, as SBR exhibited higher Mooney viscosity. When comparing curing systems, higher minimum torque and thus higher viscosity were manifested by rubber compounds with the peroxide system applied. The maximum torque showed an increasing trend with increasing amounts of the biopolymer, too. As shown in Figure 1A,B, the M_H of the compounds based on SBR and NBR cured with the sulfur system was very similar. When comparing both types of rubber formulations cured with the peroxide system (Figure 1C,D), one can see that higher maximum torque was exhibited by those based on SBR.

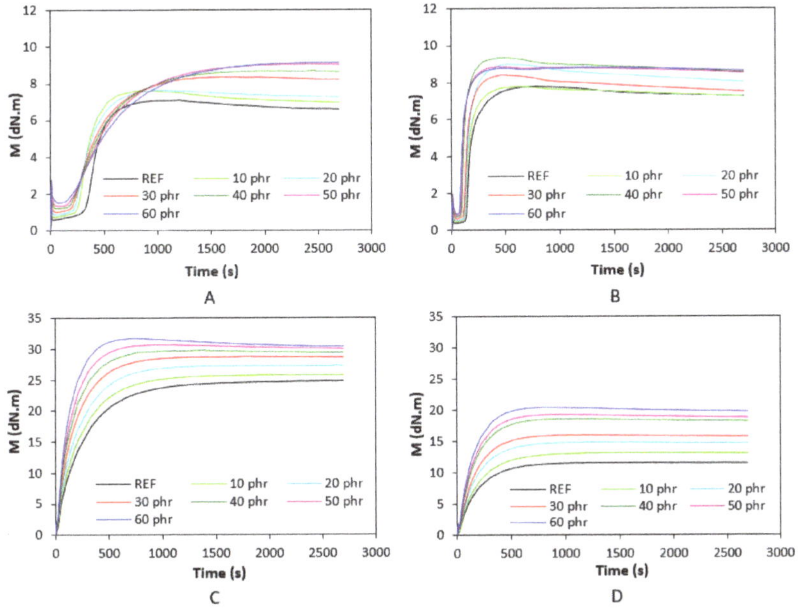

Figure 1. Curing isotherms of SBR-based composites cured with sulfur system (**A**). NBR-based composites cured with sulfur system (**B**). SBR-based composites cured with peroxide system (**C**). NBR-based composites cured with peroxide system (**D**).

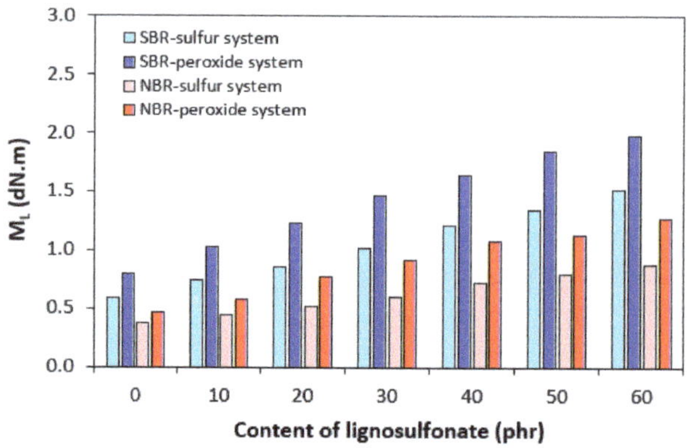

Figure 2. Influence of lignosulfonate content and curing system composition on minimum torque M_L of composites.

The highest difference between the maximum and minimum torque ΔM was demonstrated by SBR-based formulations cured with peroxide system, followed by the equivalent compounds based on NBR (Figure 3). The SBR- and NBR-based compounds cured with sulfur system exhibited lower ΔM. The difference between maximum and minimum torque ΔM is to a large extent proportional to the cross-link density, and when comparing Figures 3 and 4, one can see a very close correlation between both characteristics. The highest cross-link density was exhibited by peroxide-cured composites based on SBR having the highest ΔM. On the other hand, SBR-based composites cured with the sulfur system with the lowest torque difference demonstrated the lowest degree of cross-linking (Figure 4). The incorporation of lignosulfonate resulted in a slight decrease in cross-link density for both SBR- and NBR-based compounds cured with the sulfur system. On the other hand, almost no influence of the biopolymer on cross-link density was recorded for rubber formulations cured with the peroxide system. The corresponding swelling indexes are presented in Figure 5. The higher the degree of cross-linking, the shorter the chain segments between the cross-links and the lower the free volume in the matrix. With a lower free volume, a lower amount of solvent can diffuse into the compounds. As seen in Figure 5, the highest amount of xylene was diffused into the samples in the initial stages of swelling, up to 5–6 h. Equilibrium swelling was achieved after 24 h.

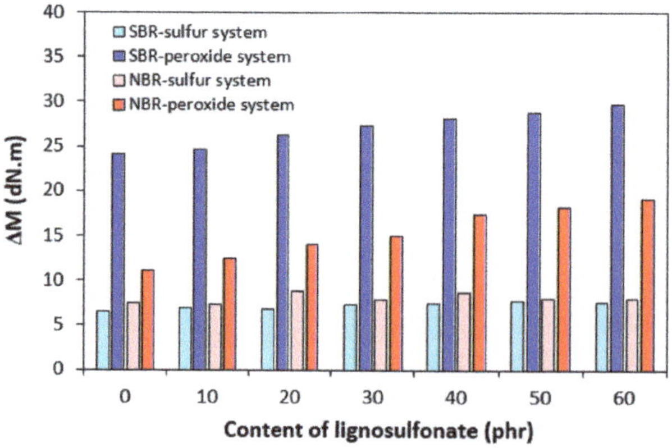

Figure 3. Influence of lignosulfonate content and curing system composition on torque difference ΔM of composites.

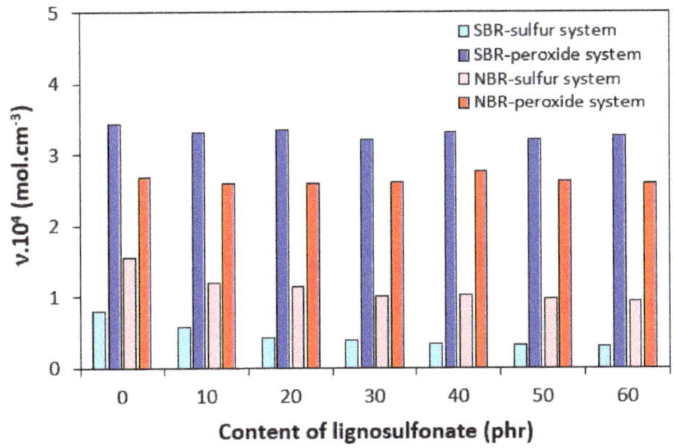

Figure 4. Influence of lignosulfonate content and curing system composition on cross-link density v of composites.

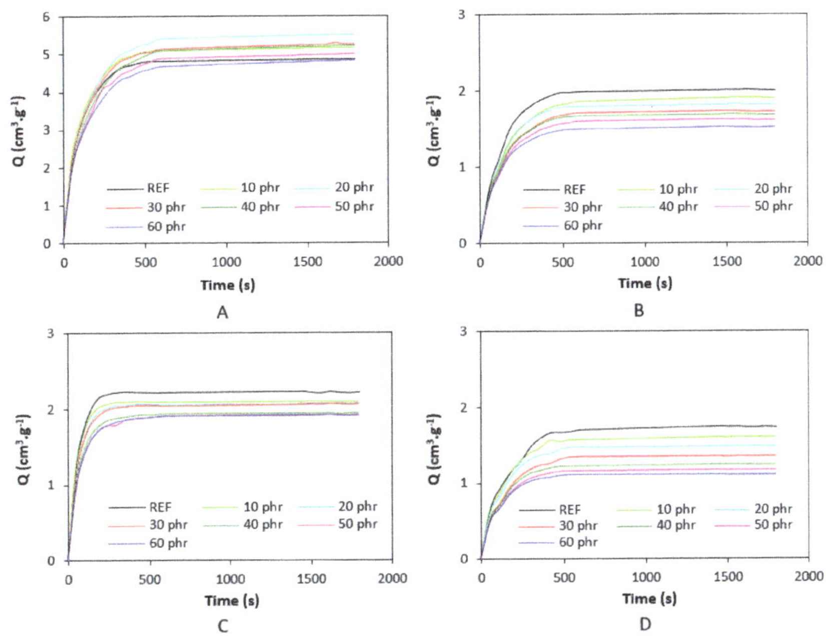

Figure 5. Swelling indexes of SBR-based composites cured with sulfur system (**A**). NBR-based composites cured with sulfur system (**B**). SBR-based composites cured with peroxide system (**C**). NBR-based composites cured with peroxide system (**D**).

The application of lignosulfonate caused a decrease in scorch time t_{s1} for rubber formulations cured with the sulfur system (Figure 6). The scorch time of the compounds cured with the peroxide system was much lower and was found to be independent on the content of the biopolymer. With the exception of the compounds based on SBR cured with the sulfur system, the optimum cure time t_{c90} showed a decreasing trend with increasing content of lignosulfonate (Figure 7). On the other hand, a significant prolongation of the optimum cure time was recorded for sulfur-cured SBR-based formulations with high biopolymer content.

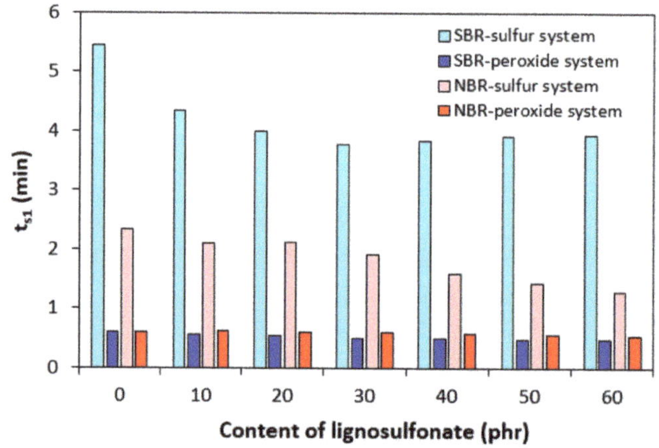

Figure 6. Influence of lignosulfonate content and curing system composition on scorch time t_{s1} of composites.

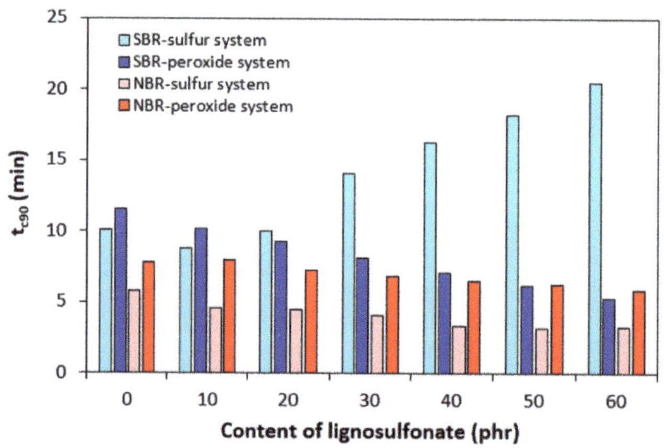

Figure 7. Influence of lignosulfonate content and curing system composition on optimum cure time t_{c90} of composites.

Peroxide vulcanization is a radical process, during which organic peroxide first undergoes homolytic cleavage via breaking labile oxygen–oxygen bonds at a curing temperature [38]. The formed radical species promptly react with rubber chain segments to form cross-links between them. To that corresponds a very low scorch time (Figure 6). The regulation of scorch time during peroxide vulcanization is very complicated as it is achieved only by the type of peroxide and its dissociation rate at a curing temperature [39–41]. It becomes apparent from Figure 6 that scorch time for both types of rubber formulations cured with the peroxide system was very similar and fluctuated by only around half a minute.

Both NBR- and SBR-based compounds cured with the peroxide system exhibited very similar curing kinetics, i.e., scorch time and optimum cure time, but differed in torque difference and cross-link density. Thus, the insight into the mechanism of vulcanization should be outlined. SBR and NBR are copolymers of styrene or acrylonitrile, respectively. Styrene and acrylonitrile structural units do not provide active sites for cross-linking. Cross-linking of both rubbers is performed exclusively via butadiene structural units [27,36,42–44]. The active radical species formed from peroxide decomposition can react with rubber chains in two ways. The first mechanism involves the abstraction of reactive allylic hydrogens from rubber chains to form macromolecular radicals. During the second mechanism, the peroxide-derived radicals attack the double bonds in the rubber structure, again followed by the formation of macromolecular radicals. The formed macroradicals can either mutually recombine or participate in addition reactions with the free double bonds in rubber chains, mainly those situated in 1,2-butadiene structural units. Pendant vinyl units are very prone to radical addiction mechanisms. Moreover, they are less sterically hindered and thus more accessible for radical species when compared to double bonds in the main chains (cis/trans) [45–47]. The chain character of addition reactions results in high cross-linking effectiveness. The highest torque difference and cross-link density were exhibited by peroxide-cured composites based on SBR. The peroxide cross-linking efficiency of NBR is lower, as NBR contains lower amount of butadiene structural units. Also, electron-withdrawing acrylonitrile groups make double bonds in NBR less reactive, so addition reactions have been reported to be of lower importance than in the case of SBR [36,48]. Thus, the equivalent NBR-based composites exhibited lower torque difference and lower cross-link density. The co-agent used in combination with dicumyl peroxide contributed to the increase in the cross-link density of both composite types, too. The specific reaction mechanism of methacrylic acid zinc salt in peroxide vulcanization is discussed in following section.

On the other hand, sulfur vulcanization of rubber compounds is a very complex process. Although it is still not clearly understood, it has been proposed that it proceeds in three general stages. In the first stage, the accelerator reacts with the activators to form a transition complex, which then reacts with sulfur to generate an active sulfurating agent. This stage is called the induction period, and the length of induction period is dependent on the composition of the sulfur curing system, mainly on the type of accelerator. The used accelerator CBS is the class of delayed-action accelerators, which is characterized by a prolonged induction period [49,50]. To that corresponds the longer scorch time of sulfur-cured formulations when compared to those cured with the peroxide system (Figure 6). In the second stage, the active sulfurating agent substitutes the allylic hydrogens in rubber chains via a sulfur bridge, forming cross-link precursors. A primary vulcanizate network with a prevalence of polysulfidic cross-links is formed. In the third phase, the restructuration of the formed polysulfidic cross-links and modification of the rubber chains occur, and the final three-dimensional vulcanizate network is generated [27,51].

The experimental results showed that sulfur-cured rubber compounds based on NBR demonstrated a lower scorch time and lower optimum cure time when compared to corresponding compounds based on SBR. Simultaneously, both characteristics showed a decreasing trend with increasing content of the biopolymer, demonstrating its accelerating effect on the sulfur vulcanization of NBR-based formulations. However, the prolongation of the optimum cure time of SBR-based formulations with increasing content of lignosulfonate suggests that it has a retarding effect in the sulfur vulcanization of SBR.

Sulfur-cured SBR-based formulations demonstrated lower cross-link density, which seems to be a bit surprising as SBR contains a higher amount of butadiene structural units and thus a higher amount of active cross-linking sites. The higher cross-link density of composites based on NBR might be attributed to higher branching and entanglements of the rubber chains and polar structural units of NBR. Upon generation of chemical cross-links between adjacent entangled and branched rubber chains, the physical entanglements can also act as cross-linking points. Due to the presence of polar acrylonitrile units, stronger intermolecular interactions are formed between the chain segments, which could contribute to the higher cross-linking degree of NBR-based formulations, too. As also shown in Figure 4, the cross-link density of sulfur-cured composites based on NBR as well as SBR showed a slight decreasing trend with increasing content of calcium lignosulfonate. A possible explanation for this is that lignosulfonate could sterically hinder the formation of the cross-links between the chain segments. On the other hand, no negative influence of the biopolymer on the cross-link density of peroxide-cured composites suggest that organic peroxide in combination with a co-agent could cause the cross-linking of lignosulfonate or contribute to the formation of linkages between the biopolymer and the rubber matrix.

3.2. Physical–Mechanical Properties and Morphology

The incorporation of lignosulfonate resulted in an increase of the composites' hardness, suggesting that the hardness of the biopolymer was higher than that of the rubbers (Figure 8). The dependence of hardness on the type of rubber matrix and curing system composition was directly proportional to cross-link density (Figure 4). The highest cross-link density of composites based on SBR cured with the peroxide system was responsible for the highest hardness. On the other hand, the equivalent composites cured with the sulfur system with the lowest cross-linking degree exhibited the lowest hardness. The hardness of composites based on NBR ranged between the sulfur- and peroxide-cured composites based on SBR and was higher for peroxide-cured counterparts, again following the trend of cross-link density.

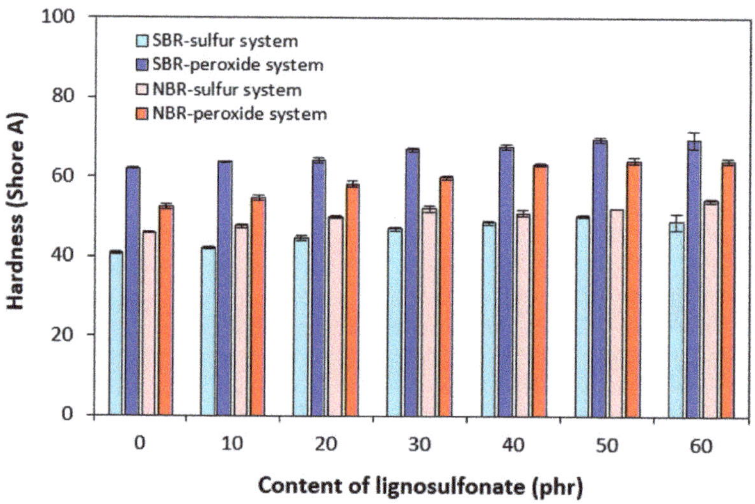

Figure 8. Influence of lignosulfonate content and curing system composition on hardness of composites.

The higher cross-linking degree of both composite types cured with peroxide system was reflected in the higher modulus M100 (Figure 9). The M100 of NBR-based composites cured with the peroxide system was found to increase with increasing content of calcium lignosulfonate. The M100 of the peroxide-cured reference based on SBR was not possible to detect as this sample ruptured before reaching 100% elongation. The composites cured with the sulfur system demonstrated almost the same modulus with no influence on the biopolymer content. As also shown, the M100 of sulfur-cured composites was much lower than that of peroxide-cured equivalents due to their much lower cross-link density.

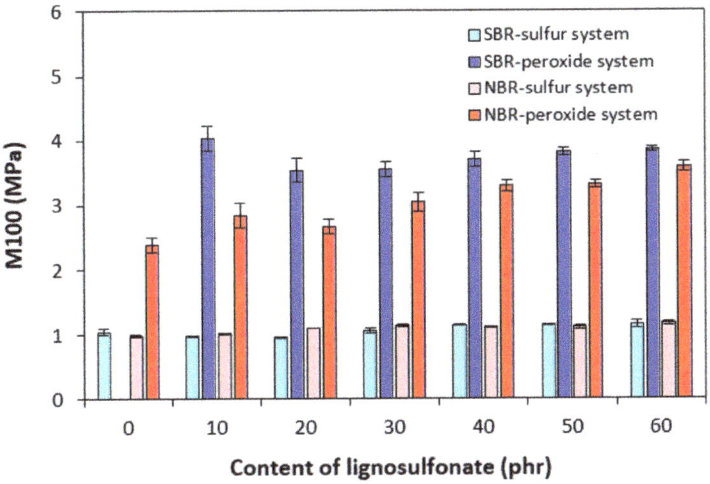

Figure 9. Influence of lignosulfonate content and curing system composition on modulus M100 of composites.

On the other hand, the lower cross-link density of composites cured with the sulfur system resulted in their higher elongation at break (Figure 10). The elongation at break of the sulfur-cured composites based on NBR and SBR was very similar and was significantly dependent on the amount of the biopolymer. At the maximum lignosulfonate content, the elongation at break increased by less than 500% for SBR-based composites (from about 340% for the reference up to 830% for the composite with 60 phr of the biopolymer). In the case of sulfur-cured NBR-based composites, the elongation at break increased from 460% up to 790% by an increase in biopolymer content from 0 up to its maximum content. The higher the cross-link density, the higher the restriction of rubber chain segments' elasticity and mobility, resulting in a decrease in elongation at break. The lowest elongation at break was exhibited by the SBR-based composites cured with the peroxide system with the highest cross-linking degree. As the cross-link density of peroxide-cured composites almost did not change with the change in lignosulfonate content, the elongation at break of the equivalent composites was much less dependent on lignosulfonate content. When compared to the references, the elongation at break increased by about 90% and 140% for the maximally filled composites based on NBR or SBR, respectively. Another factor contributing to the different elongation at break of sulfur- and peroxide-cured composites is the structure of the formed cross-links. Longer and more flexible sulfidic cross-links facilitate micro-Brownian motion of the chain segments between the cross-links, which results in better elastic properties in the sulfur cured rubber materials. Higher elasticity of the chain segments causes higher redistribution of the deformation strains within the rubber matrix, resulting in higher tensile characteristics, too. On the other hand, shorter and rigid carbon–carbon bonds restrict the orientation and mobility of the chain segments, which results in formation of local stress concentrations. To that correspond the worse dynamic and tensile behavior of peroxide-cured vulcanizates in general [27,29,52].

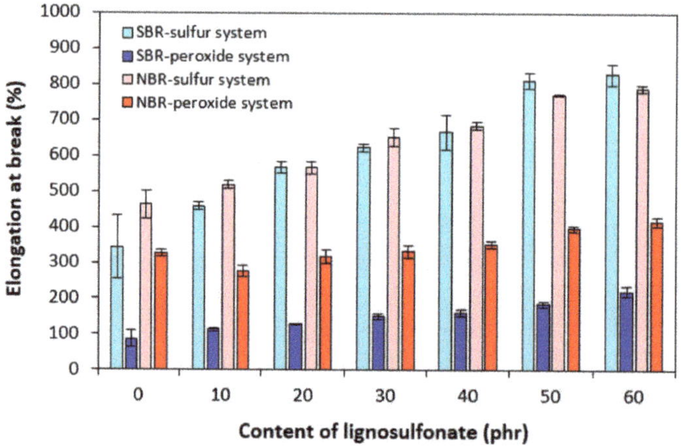

Figure 10. Influence of lignosulfonate content and curing system composition on elongation at break of composites.

From Figure 11, it becomes apparent that with the exception of the peroxide-cured reference sample based on NBR, the other reference samples exhibited roughly the same tensile strength. The tensile strength of sulfur-cured composites based on SBR was almost not influenced by the amount of the biopolymer, and it fluctuated only in a very low range of experimental values. The influence of lignosulfonate on the tensile strength of sulfur-cured NBR-based composites was also very low, though highly filled NBR-based composites exhibited slightly higher tensile strength when compared to equivalent sulfur-cured composites based on SBR. The application of the peroxide curing system resulted in the enhanced tensile behavior of composites, and the highest tensile strength was manifested by peroxide-cured composites based on NBR. By the incorporation of 10 phr of the biopolymer, an improvement in the tensile strength of both peroxide-cured composites was recorded when compared to the references. Then, no significant influence of lignosulfonate content on tensile strength was observed. The higher tensile characteristics of peroxide-cured vulcanizates can be attributed to the presence of a co-agent in peroxide vulcanization.

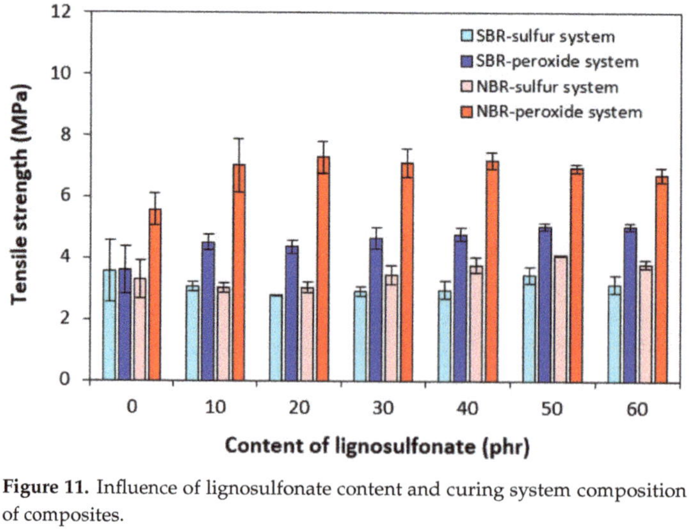

Figure 11. Influence of lignosulfonate content and curing system composition on tensile strength of composites.

It has been reported that methacrylic acid zinc in the presence of organic peroxide can undergo in situ polymerization within the rubber matrix. Polymerized molecules of ZDMA can be physically adsorbed or chemically grafted onto rubber chains [53–55]. Thus, they form physical and chemical linkages in the rubber matrix. In addition, having polar-character ZDMA exhibits strong adhesion to polar materials [56,57]. Due to the presence of zinc ions, ZDMA can form ionic cross-links or ionic clusters, which can interact with polar functional groups of lignosulfonate. This leads to improvement in the adhesion and compatibility between the rubber and the biopolymer on the filler–rubber interface. Subsequent improvement in tensile strength was achieved. It becomes apparent from Figure 11 that higher tensile strength was demonstrated by composites based on NBR. As NBR, calcium lignosulfonate, and ZDMA are polar materials, their adhesion, compatibility, and formation of physical interactions between the components is higher when compared to non-polar SBR.

The surface morphology of composites was studied by performing scanning electron microscopy (SEM). SEM images of the SBR-based composites cured with the sulfur system are presented in Figure 12 on the left side, while SEM images of the equivalent composites based on NBR are depicted in Figure 12 on the right side. Similarly, SEM images of peroxide-cured composites based on SBR and NBR are presented in Figure 13 on the left or right side, respectively. It is shown in Figure 12 that lignosulfonate tended to agglomerate in sulfur-cured composites. The agglomeration and worse dispersion of lignosulfonate within the rubber matrices cured with the sulfur system are the main reasons why lignosulfonate does not behave as a reinforcing filler. However, it should be remarked that although the application of lignosulfonate did not result in improvement in the tensile strength, no negative effect on the characteristic was recorded even at high lignosulfonate contents. Looking at Figure 13, one can see that the surface fractures of composites cured with peroxide system seem to be more compact, and better adhesion between the rubber and lignosulfonate on the interfacial region was achieved due to the presence of ZDMA. Some surface cracks visible on SEM images can be attributed to lignosulfonate, which was first melted during vulcanization at high temperatures. Then, the samples were frozen in liquid nitrogen, which might result in the brittleness of lignosulfonate. The cracks could then be caused by the fracturing of frozen samples for SEM analysis.

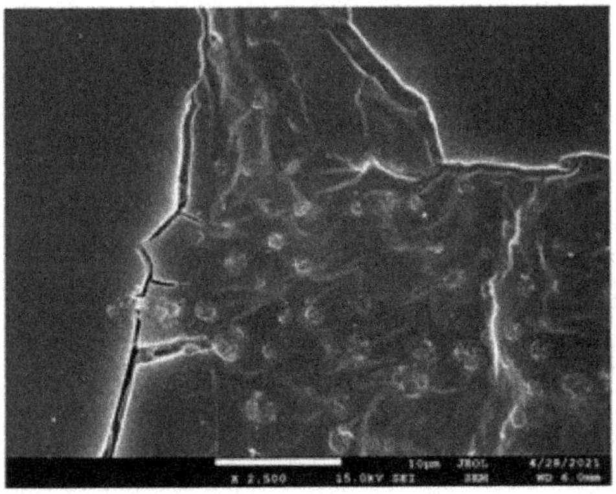

A

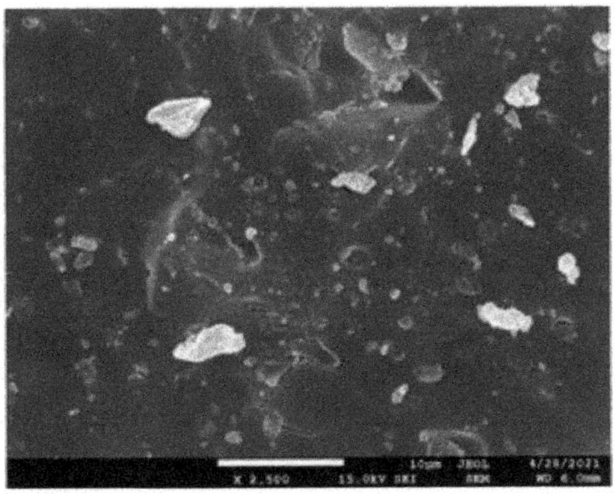

B

Figure 12. *Cont.*

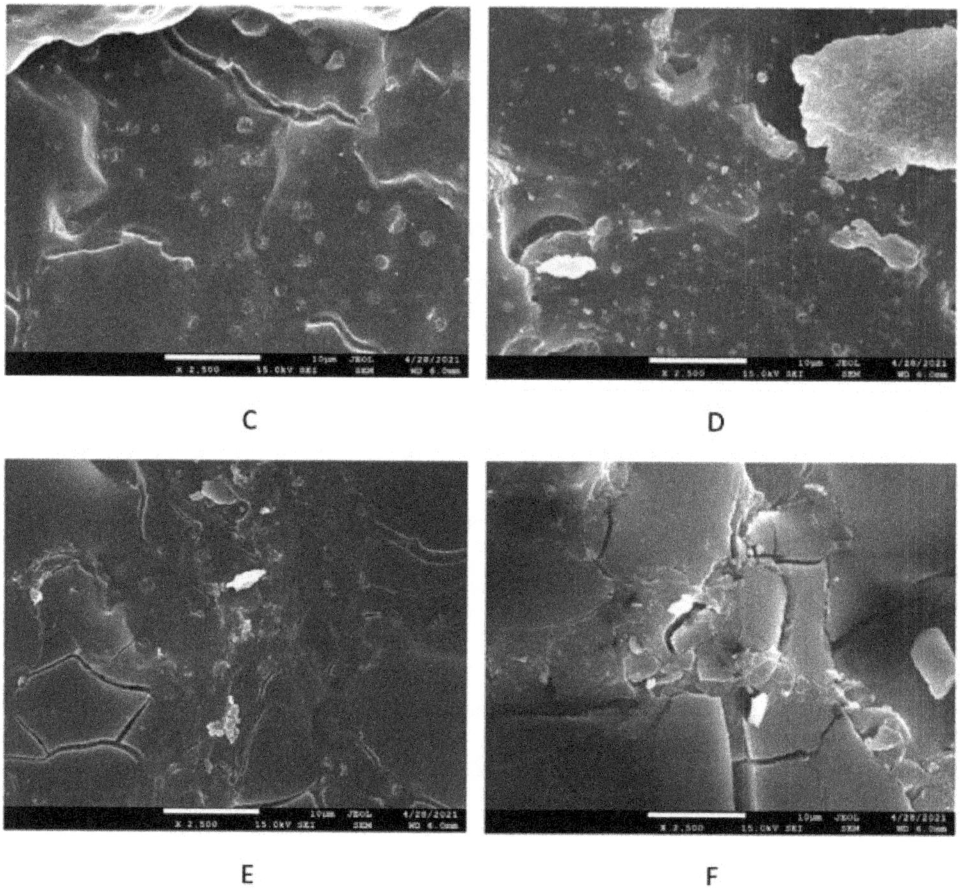

Figure 12. SEM images of sulfur-cured SBR-based composites filled with 10 phr of lignosulfonate (**A**), 30 phr of lignosulfnate (**C**), and 60 phr of lignosulfonate (**E**); sulfur-cured NBR-based composites filled with 10 phr of lignosulfnate (**B**), 30 phr of lignosulfnate (**D**), and 60 phr of lignosulfonate (**F**).

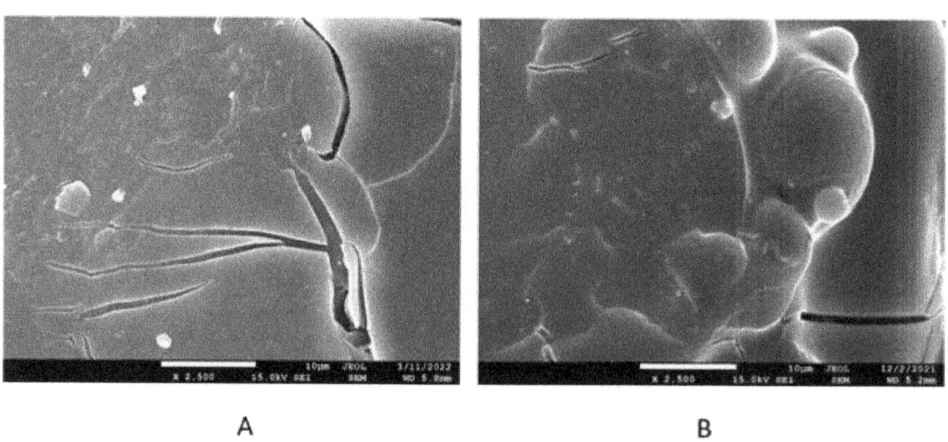

Figure 13. *Cont.*

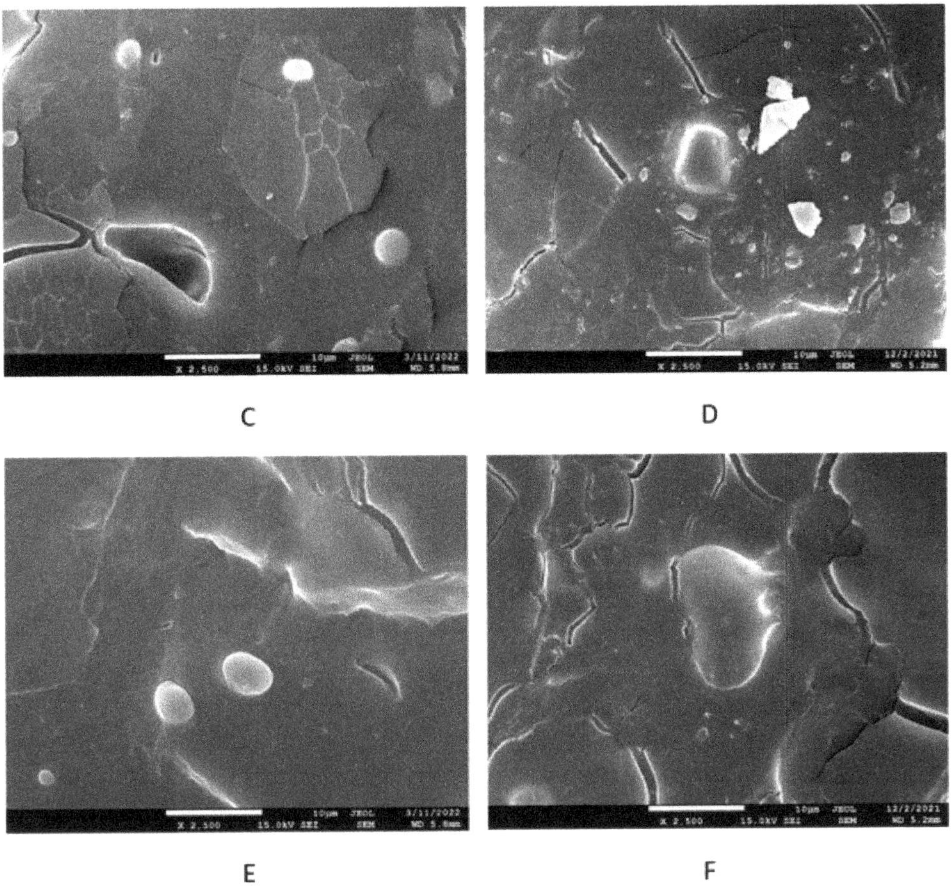

Figure 13. SEM images of peroxide-cured SBR-based composites filled with 10 phr of lignosulfonate (**A**), 30 phr of lignosulfnate (**C**), and 60 phr of lignosulfonate (**E**); peroxide cured NBR-based composites filled with 10 phr of lignosulfnate (**B**), 30 phr of lignosulfnate (**D**), and 60 phr of lignosulfonate (**F**).

3.3. Dynamic–Mechanical Analysis

The temperature dependences of loss factor tan δ for tested materials are illustrated in Figure 14, while the values of tan δ at chosen temperatures are summarized in Tables 1–4. Tan δ is a ratio of loss and storage modulus and corresponds to the energy lost to energy absorbed and returned by the system pre unit cycle. With the rise in temperature, tan δ reaches the maximum in the transition region, followed by a subsequent decrease in the rubbery region. The peak maximum corresponds to the glass transition temperature Tg. In glassy region below Tg, the damping is low, as the chain segments are immobilized and deformations are mostly elastic due to limited propensity of the chain segments to viscous flow. In the rubbery region, damping is low, because the rotational movement of the chain segments enables free movement with minimum resistance to flow. In the transition region, damping and energy dissipation is high due to the inception of micro-Brownian movement in the chains and chain stress relaxation, although not all the segments are able to participate in such relaxations together. As shown in Figure 14, the application of lignosulfonate resulted in the lowering of the tan δ peak and a reduction in area in the transition region, resulting in lower energy dissipation, which can be attributed to a higher storage modulus in this region. In all cases, the lowest area in a transition region was demonstrated by composites filled with maximum lignosulfonate content.

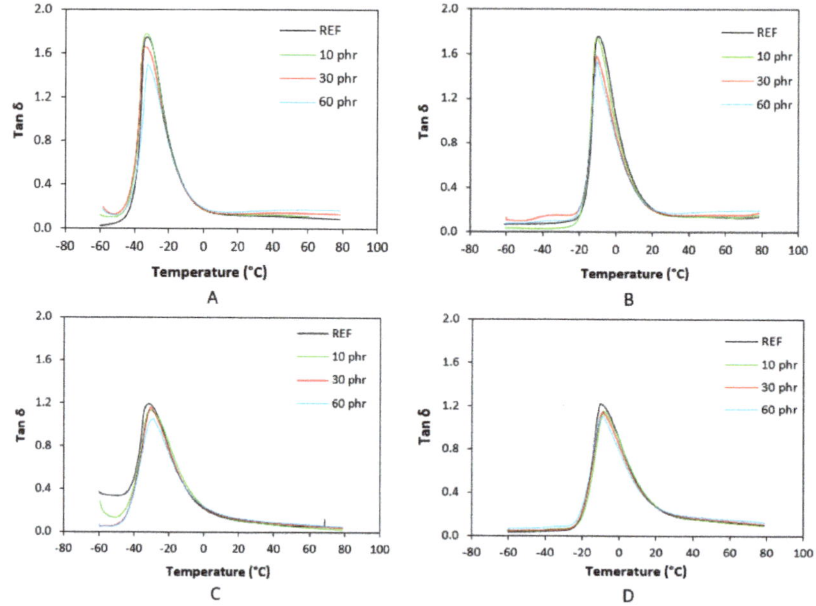

Figure 14. Temperature dependences of loss factor tan δ for SBR-based composites cured with sulfur system (**A**), NBR-based composites cured with sulfur system (**B**), SBR-based composites cured with peroxide system (**C**), and NBR-based composites cured with peroxide system (**D**).

Table 1. Glass transition temperature Tg and loss factor tan δ of SBR-based composites cured with sulfur system.

Lignosulfonate (phr)	Tg (°C)	tan δ at Tg	tan δ (−50 °C)	tan δ (−20 °C)	tan δ (0 °C)	tan δ (20 °C)	tan δ (50 °C)
0	−32.4	1.74	0.06	0.83	0.18	0.12	0.11
10	−33.0	1.77	0.11	0.81	0.17	0.13	0.12
30	−33.6	1.66	0.14	0.79	0.18	0.14	0.15
60	−32.4	1.50	0.12	0.77	0.18	0.15	0.17

Table 2. Glass transition temperature Tg and loss factor tan δ of SBR-based composites cured with peroxide system.

Lignosulfonate (phr)	Tg (°C)	tan δ at Tg	tan δ (−50 °C)	tan δ (−20 °C)	tan δ (0 °C)	tan δ (20 °C)	tan δ (50 °C)
0	−31.3	1.20	0.33	0.78	0.22	0.10	0.06
10	−30.7	1.13	0.14	0.82	0.24	0.11	0.05
30	−30.4	1.16	0.06	0.77	0.23	0.12	0.07
60	−28.9	1.05	0.06	0.73	0.23	0.13	0.08

Table 3. Glass transition temperature *Tg* and loss factor tan δ of NBR-based composites cured with sulfur system.

Lignosulfonate (phr)	*Tg* (°C)	tan δ at *Tg*	tan δ (−50 °C)	tan δ (−20 °C)	tan δ (0 °C)	tan δ (20 °C)	tan δ (50 °C)
0	−9.8	1.76	0.07	0.15	1.03	0.21	0.13
10	−10.9	1.74	0.04	0.14	0.96	0.20	0.14
30	−10.7	1.58	0.10	0.21	0.88	0.21	0.16
60	−10.7	1.53	0.08	0.21	0.82	0.21	0.19

Table 4. Glass transition temperature *Tg* and loss factor tan δ of NBR-based composites cured with peroxide system.

Lignosulfonate (phr)	*Tg* (°C)	tan δ at *Tg*	tan δ (−50 °C)	tan δ (−20 °C)	tan δ (0 °C)	tan δ (20 °C)	tan δ (50 °C)
0	−10.1	1.22	0.04	0.22	0.89	0.27	0.14
10	−8.2	1.15	0.05	0.18	0.88	0.28	0.14
30	−8.9	1.15	0.05	0.23	0.83	0.27	0.15
60	−8.9	1.10	0.07	0.28	0.77	0.27	0.16

From Figure 14 and Tables 1–4, it becomes apparent that sulfur- as well as peroxide-cured composites based on SBR exhibited a lower *Tg* when compared to the corresponding composites based on NBR. It is a logical reflection of the higher amount of rubbery butadiene units in SBR's structure. The *Tg* of sulfur-cured composites based on SBR moved around −33 °C, while the *Tg* of peroxide cured composites was slightly higher (around −31 °C). This can be attributed to the higher cross-linking degree of peroxide-cured composites, suggesting that increased cross-link density restricts the chain segments' elasticity and mobility, thus elevating *Tg*. The same can be applied to composites based on NBR. The higher cross-link density of peroxide cured composites resulted in an increase in *Tg* by about 2 °C (from about −11 °C for sulfur-cured composites to around −9 °C for peroxide-cured equivalents).

It can also be seen from Tables 1–4 that the *Tg* in some samples increases, while in others, it slightly decreases with increasing filler content. However, there was no recorded direct influence of the filler on *Tg* regarding the type of rubber matrix or curing system composition. It can also be stated that the values tan δ in the glassy and rubbery region were significantly influenced neither by the amount of lignosulfonate nor by the type of curing system applied. This may suggest a lack of strong filler–rubber interactions, implying no molecular-level changes.

4. Conclusions

Calcium lignosulfonate-filled formulations based on SBR and NBR were cured with sulfur as well as peroxide curing systems. The results revealed that peroxide-cured composites exhibited a lower scorch time, higher torque difference, and higher cross-link density. Higher cross-link density resulted in a higher hardness and modulus and a lower elongation at break in the corresponding composites. Peroxide-cured composites demonstrated higher tensile strength, which can be attributed to higher compatibility and adhesion and between the filler and the rubber on their interface due to the presence of methacrylic acid zinc salt in peroxide vulcanization. The highest tensile strength of composites based on NBR can be attributed to the polar nature of the rubber, the biopolymer, and the co-agent resulting in the best adhesion and physical interactions between the components. The application of lignosulfonate in all rubber systems resulted in an increase in elongation at break and hardness, while a more significant influence on elongation at break was recorded for sulfur-cured counterparts. The presence of the biopolymer in the compounds did not significantly influence the Tg of the composites and tan δ in the glassy and rubbery region. The decrease in the tan δ peak and lowering of the area in the transition region with the increase in lignosulfonate content point to reduced damping in that region.

Author Contributions: Methodology, J.K. and J.H.; Validation, J.P.; Formal analysis, I.C.; Investigation, M.D. and A.K.; Writing—original draft, J.K.; Supervision, I.H. All authors have read and agreed to the published version of the manuscript.

Funding: This work was funded by the Slovak Research and Development Agency under the contract No. APVV-19-0091, APVV-22-0011 and by the Slovak Scientific Grant Agency VEGA under the contract No. 1/0056/24.

Institutional Review Board Statement: Not applicable.

Data Availability Statement: The original contributions presented in this study are included in the article. Further inquiries can be directed to the corresponding author.

Conflicts of Interest: Authors Ján Hronkovič and Jozef Preťo were employed by VIPO a.s. The remaining authors declare that the research was conducted in the absence of any commercial or financial relationships that could be construed as a potential conflict of interest.

References

1. Ruwoldt, J. A critical review of the physicochemical properties of lignosulfonates: Chemical structure and behavior in aqueous solution, at surfaces and interfaces. *Surfaces* **2020**, *3*, 622–648. [CrossRef]
2. Hemmilä, V.; Hosseinpourpia, R.; Adamopoulos, S.; Eceiza, A. Characterization of wood-based industrial biorefinery lignosulfonates and supercritical water hydrolysis lignin. *Waste Biomass Valori* **2020**, *11*, 5835–5845.
3. Lu, X.; Gu, X.; Shi, Y. A review on lignin antioxidants: Their sources, isolations, antioxidant activities and various applications. *Int. J. Biol. Macromol.* **2022**, *210*, 716–741. [PubMed]
4. Morena, A.G.; Bassegoda, A.; Natan, M.; Jacobi, G.; Banin, E.; Tzanov, T. Antibacterial properties and mechanisms of action of sonoenzymatically synthesized lignin-based nanoparticles. *ACS Appl. Mater. Interfaces* **2022**, *14*, 37270–37279.
5. Wang, H.; Qiu, X.; Liu, W.; Fu, F.; Yang, D. A novel lignin/ZnO hybrid nanocomposite with excellent UV absorption ability and its application in transparent polyurethane coating. *Ind. Eng. Chem. Res.* **2017**, *56*, 11133–11141.
6. Younesi-Kordkheili, H.; Pizzi, A. A comparison among lignin modification methods on the properties of lignin–phenol–formaldehyde resin as wood adhesive. *Polymers* **2021**, *13*, 3502. [CrossRef]
7. Lisý, A.; Ház, A.; Nadányi, R.; Jablonský, M.; Šurina, I. About hydrophobicity of lignin: A review of selected chemical methods for lignin valorisation in biopolymer production. *Energies* **2022**, *15*, 6213. [CrossRef]
8. Sugiarto, S.; Leow, Y.; Li Tan, C.; Wang, G.; Kai, D. How far is Lignin from being a biomedical material? *Bioact. Mater.* **2022**, *8*, 71–94.
9. Fabbri, F.; Bischof, S.; Mayr, S.; Gritsch, S.; Bartolome, M.J.; Schwaiger, N.; Guebitz, G.M.; Weiss, R. The biomodified lignin patform: A review. *Polymers* **2023**, *15*, 1694.
10. Saadan, R.; Alaoui, C.H.; Ihammi, A.; Chigr, M.; Fatimi, A. A brief overview of lignin extraction and isolation processes: From lignocellulosic biomass to added-value biomaterials. *Environ. Earth Sci. Proc.* **2024**, *31*, 3. [CrossRef]
11. Alam, M.M.; Greco, A.; Rajabimashhadi, Z.; Corcione, C.E. Efficient and environmentally friendly techniques for extracting lignin from lignocellulose biomass and subsequent uses: A review. *Clean. Mater.* **2024**, *13*, 100253. [CrossRef]
12. Shorey, R.; Salaghi, A.; Fatehi, P.; Mekonnen, T.H. Valorization of lignin for advanced material applications: A review. *RSC Sustain.* **2024**, *2*, 804. [CrossRef]
13. Gonçalves, S.; Ferra, J.; Paiva, N.; Martins, J.; Carvalho, L.H.; Magalhães, F.D. Lignosulphonates as an alternative to non-renewable binders in wood-based materials. *Polymers* **2021**, *13*, 4196. [CrossRef]
14. Madyaratri, E.W.; Ridho, M.R.; Aristri, M.A.; Lubis, M.A.R.; Iswanto, A.H.; Nawawi, D.S.; Antov, P.; Kristak, L.; Majlingová, A.; Fatriasari, W. Recent advances in the development of fire-resistant biocomposites—A review. *Polymers* **2022**, *14*, 362. [CrossRef]
15. Guterman, R.; Molinari, V.; Josef, E. Ionic liquid lignosulfonate: Dispersant and binder for preparation of biocomposite materials. *Angew. Chem. Int. Ed.* **2019**, *58*, 13044–13050. [CrossRef]

16. Breilly, D.; Fadlallah, S.; Froidevaux, V.; Colas, A.; Allais, F. Origin and industrial applications of lignosulfonates with a focus on their use as superplasticizers in concrete. *Constr. Build. Mater.* **2021**, *301*, 124065. [CrossRef]
17. Antov, P.; Mantanis, G.I.; Savov, V. Development of wood composites from recycled fibres bonded with magnesium lignosulfonate. *Forests* **2020**, *11*, 613. [CrossRef]
18. Mohamad Aini, N.A.; Othman, N.; Hussin, M.H.; Sahakaro, K.; Hayeemasae, N. Lignin as alternative reinforcing filler in the rubber industry: A review. *Front. Mater.* **2020**, *6*, 329. [CrossRef]
19. Thungphotrakul, N.; Dittanet, P.; Loykulnunt, S.; Tanpichai, S.; Parpainainar, P. Synthesis of sodium lignosulfonate from lignin extracted from oil palm empty fruit bunches by acid/ alkaline treatment for reinforcement in natural rubber composites. *IOP Conf. Ser. Mater. Sci. Eng.* **2019**, *526*, 012022. [CrossRef]
20. Kruželák, J.; Džuganová, M.; Kvasničáková, A.; Preťo, J.; Hronkovič, J.; Hudec, I. Influence of plasticizers on cross-linking process, morphology, and properties of lignosulfonate-filled rubber compounds. *Polymers* **2025**, *17*, 393. [CrossRef]
21. An, D.; Cheng, S.; Jiang, C.; Duan, X.; Yang, B.; Zhang, Z.; Li, J.; Liu, Y.; Wong, C.P. A novel environmentally friendly boron nitride/lignosulfonate/natural rubber composite with improved thermal conductivity. *J. Mater. Chem. C* **2020**, *8*, 4801–4809.
22. Džuganová, M.; Stoček, R.; Pöschl, M.; Kruželák, J.; Kvasničáková, A.; Hronkovič, J.; Preťo, J. Strategy for reducing rubber wear emissions: The prospect of using calcium lignosulfonate. *Express Polym. Lett.* **2024**, *18*, 1277–1290.
23. Nardelli, F.; Calucci, L.; Carignani, E.; Borsacchi, S.; Cettolin, M.; Arimondi, M.; Giannini, L.; Geppi, M.; Martini, F. Influence of sulfur-curing conditions on the dynamics and crosslinking of rubber networks: A time-domain NMR study. *Polymers* **2022**, *14*, 767. [CrossRef] [PubMed]
24. Kaur, A.; Fefar, M.M.; Griggs, T.; Akutagawa, K.; Chen, B.; Busfield, J.J.C. Recyclable sulfur cured natural rubber with controlled disulfide metathesis. *Commun. Mater.* **2024**, *5*, 212.
25. Naebpetch, W.; Junhasavasdikul, B.; Saetung, A.; Tulyapitak, T.; Nithi-Uthai, N. Influence of accelerator/sulphur and co-agent/peroxide ratios in mixed vulcanisation systems on cure characteristics, mechanical properties and heat aging resistance of vulcanised SBR. *Plast. Rubber Compos.* **2016**, *45*, 436–444.
26. Shahrampour, H.; Motavalizadehkakhky, A. The Effects of sulfur curing systems (insoluble-rhombic) on physical and thermal properties of the matrix polymeric of styrene butadiene rubber. *Pet. Chem.* **2017**, *57*, 700–704.
27. Kruželák, J.; Sýkora, R.; Hudec, I. Sulphur and peroxide vulcanisation of rubber compounds—Overview. *Chem. Pap.* **2016**, *70*, 1533–1555. [CrossRef]
28. Rodríguez Garraza, A.L.; Mansilla, M.A.; Depaoli, E.L.; Macchi, C.; Cerveny, S.; Marzocca, A.J.; Somoza, A. Comparative study of thermal, mechanical and structural properties of polybutadiene rubber isomers vulcanized using peroxide. *Polym. Test.* **2016**, *52*, 117–123.
29. Peidayesh, H.; Nógellová, Z.; Chodák, I. Effects of peroxide and sulfur curing systems on physical and mechanical properties of nitrile rubber composites: A comparative study. *Materials* **2024**, *17*, 71.
30. Wei, B.X.; Yi, X.T.; Xiong, Y.J.; Wei, X.J.; Wu, Y.D.; Huang, Y.D.; He, J.M.; Bai, Y.P. The preparation and characterization of a carbon fiber-reinforced epoxy resin and EPDM composite using the co-curing method. *RSC Adv.* **2020**, *10*, 20588.
31. Bhattacharya, A.B.; Gopalan, A.M.; Chatterjee, T.; Vennemann, N.; Naskar, K. Exploring the thermomechanical properties of peroxide/co-agent assisted thermoplastic vulcanizates through temperature scanning stress relaxation measurements. *Polym. Eng. Sci.* **2021**, *61*, 2466–2476.
32. Laing, B.; De Keyzer, J.; Seveno, D.; Van Bael, A. Effect of co-agents on adhesion between peroxide cured ethylene–propylene–diene monomer and thermoplastics in two-component injection molding. *J. Appl. Polym. Sci.* **2020**, *137*, 48414.
33. Hayichelaeh, C.; Boonkerd, K. Enhancement of the properties of carbon-black-filled natural rubber compounds containing soybean oil cured with peroxide through the addition of coagents. *Ind. Crop Prod.* **2022**, *187*, 115306.
34. Kruželák, J.; Kvasničáková, A.; Hložeková, K.; Hudec, I. Influence of dicumyl peroxide and Type I and II co-agents on cross-linking and physical–mechanical properties of rubber compounds based on NBR. *Plast. Rubber Compos.* **2020**, *49*, 307–320.
35. Zhao, X.; Cornish, K.; Vodovotz, Y. Synergistic mechanisms underlie the peroxide and coagent improvement of natural-rubber-toughened poly(3-hydroxybutyrate-co-3-hydroxyvalerate) mechanical performance. *Polymers* **2019**, *11*, 565. [CrossRef]
36. Kruželák, J.; Sýkora, R.; Hudec, I. Vulcanization of rubber compounds with peroxide curing systems. *Rubber Chem. Technol.* **2017**, *90*, 60–88.
37. Kraus, G. Swelling of filler-reinforced vulcanizates. *J. Appl. Polym. Sci.* **1963**, *7*, 861–871.
38. Hosseini, S.M.; Razzaghi-Kashani, M. On the role of nano-silica in the kinetics of peroxide vulcanization of ethylene propylene diene rubber. *Polymer* **2017**, *133*, 8–19.
39. Nikolova, S.; Mihaylov, M.; Dishovsky, N. Mixed peroxide/sulfur vulcanization of ethylene-propylene terpolymer based on composites. Curing characteristics, curing kinetics and mechanical properties. *J. Chem. Technol. Metall.* **2022**, *57*, 881–894.
40. Wang, H.; Zhuang, T.; Shi, X.; Van Duin, M.; Zhao, S. Peroxide cross-linking of EPDM using moving die rheometer measurements. II. Effects of the process oils. *Rubber Chem. Technol.* **2018**, *91*, 561–576.
41. George, B.; Alex, R. Stable free radical assisted scorch control in peroxide vulcanization of EPDM. *Rubber Sci.* **2013**, *27*, 135–145.

42. Choi, S.S.; Kim, J.C. Lifetime prediction and thermal aging behaviors of SBR and NBR composites using crosslink density changes. *J. Ind. Eng. Chem.* **2012**, *18*, 1166–1170. [CrossRef]
43. Valentín, J.L.; Posadas, P.; Fernández-Torres, A.; Malmierca, M.A.; González, L.; Chassé, W.; Saalwächter, K. Inhomogeneities and chain dynamics in diene rubbers vulcanized with different cure systems. *Macromolecules* **2010**, *43*, 4210–4222. [CrossRef]
44. González, L.; Rodríguez, A.; Marcos-Fernández, A.; Valentín, J.L.; Fernández-Torres, A. Effect of network heterogeneities on the physical properties of nitrile rubbers cured with dicumyl peroxide. *J. Appl. Polym. Sci.* **2007**, *103*, 3377–3382. [CrossRef]
45. Kruželák, J.; Kvasničáková, A.; Hudec, I. Peroxide curing systems applied for cross-linking of rubber compounds based on SBR. *Adv. Ind. Eng. Polym. Res.* **2020**, *3*, 120–128. [CrossRef]
46. Liu, X.; Zhou, T.; Liu, Y.; Zhang, A.; Yuan, C.; Zhang, W. Cross-linking process of cis-polybutadiene rubber with peroxides studied by two-dimensional infrared correlation spectroscopy: A detailed tracking. *RSC Adv.* **2015**, *5*, 10231–10242. [CrossRef]
47. González, L.; Rodríguez, A.; Del Campo, A.; Marcos-Fernández, A. Effect of heterogeneities on the physical properties of elastomers derived from butadiene cured with dicumyl peroxide. *Polym. Int.* **2004**, *53*, 1426–1430. [CrossRef]
48. Valentín, J.L.; Rodríguez, A.; Marcos-Fernández, A.; Gonzáles, L. Dicumyl peroxide cross-linking of nitrile rubbers with different content in acrylonitrile. *J. Appl. Polym. Sci.* **2005**, *96*, 1–5. [CrossRef]
49. Charoeythornkhajhornchai, P.; Samthong Ch Somwangthanaroj, A. Influence of sulfenamide accelerators on cure kinetics and properties of natural rubber foam. *J. Appl. Polym. Sci.* **2017**, *134*, 44822. [CrossRef]
50. Ghosh, J.; Ghorai, S.; Jalan, A.K.; Roy, M.; De, D. Manifestation of accelerator type and vulcanization system on the properties of silica-reinforced SBR/devulcanize SBR blend vulcanizates. *Adv. Polym. Technol.* **2018**, *37*, 2636–2650. [CrossRef]
51. Lian, Q.; Li, Y.; Li, K.; Cheng, J.; Zhang, J. Insights into the vulcanization mechanism through a simple and facile approach to the sulfur cleavage behavior. *Macromolecules* **2017**, *50*, 803–810. [CrossRef]
52. Sheng, C.; Hu, Z.; Martin, H.; Duan, Y.; Zhang, J. Effect of a small amount of sulfur on the physical and mechanical properties of peroxide-cured fully saturated HNBR compounds. *J. Appl. Polym. Sci.* **2015**, *132*, 41612. [CrossRef]
53. Strohmeier, L.; Balasooriya, W.; Schrittesser, B.; van Duin, M.; Schlögl, S. Hybrid in situ reinforcement of EPDM rubber compounds based on phenolic novolac resin and ionic coagent. *Appl. Sci.* **2022**, *12*, 2432. [CrossRef]
54. Siaw, C.; Baharulrazi, N.; Che Man, S.H.; Othman, N. Effect of zinc dimethacrylate concentrations on properties of emulsion styrene butadiene rubber/butadiene rubber blends. *Plast. Rubber Compos.* **2023**, *52*, 315–329. [CrossRef]
55. Li, C.; Yuan, Z.; Ye, L. Facile construction of enhanced multiple interfacial interactions in EPDM/zinc dimethacrylate (ZDMA) rubber composites: Highly reinforcing effect and improvement mechanism of sealing resilience. *Compos. A Appl. Sci.* **2019**, *126*, 105580. [CrossRef]
56. Cao, Y.; Gu, N.; Hu, S.; Jin, R.; Zhang, J. Preparation and properties of zinc-diacrylate-modified montmorillonite/rubber nanocomposite. *Appl. Mech. Mater.* **2012**, *182–183*, 47–51. [CrossRef]
57. Henning, S.K.; Costin, R. Fundamentals of curing elastomers with peroxides and coagents. *Rubber World* **2006**, *233*, 28–35.

polymers

MDPI

Article

Concentrated Pre-Vulcanized Natural Rubber Latex Without Additives for Fabricating High Mechanical Performance Rubber Specimens via Direct Ink Write 3D Printing

Lin Liu [1,2], Jizhen Zhang [2], Zirong Luo [2], Na Kong [2], Xu Zhao [2], Xu Ji [1], Jihua Li [3], Shenbo Huang [2], Pengfei Zhao [2], Shuang Li [2], Yanqiu Shao [1,*] and Jinlong Tao [2,3,*]

1 Heilongjiang Key Laboratory of Photoelectric Functional Materials, College of Chemistry and Chemical Engineering, Mudanjiang Normal University, Mudanjiang 157011, China; liulin990705@163.com (L.L.); 18345260314@163.com (X.J.)

2 Hainan Provincial Key Laboratory of Natural Rubber Processing, Agricultural Products Processing Research Institute, Chinese Academy of Tropical Agricultural Sciences, Zhanjiang 524001, China; quickly3000@126.com (J.Z.); luozirong3@163.com (Z.L.); kongna109101@126.com (N.K.); zx19960211@163.com (X.Z.); a641167221@163.com (S.H.); pengfeizhao85ac@163.com (P.Z.); shuangl163163@163.com (S.L.)

3 Rubber Research Institute, Chinese Academy of Tropical Agricultural Sciences, Haikou 571101, China; foodpaper@126.com

* Correspondence: 1001002@mdjnu.edu.cn (Y.S.); j.l.tao@catas.cn (J.T.)

Abstract

Direct ink writing (DIW) is an economical, straightforward, and relatively energy-efficient 3D printing technique that has been used in various domains. However, the utilization of rubber latex for DIW remains limited due to its high fluidity and inadequate support, which makes it challenging to meet the required ink rheological characteristics for DIW. In this study, a concentrated pre-vulcanized natural rubber latex (CPNRL) ink with a high solid content of 73% without additives is developed for DIW 3D printing. The CPNRL ink is concentrated using superabsorbent polymer (SAP) beads, which demonstrates good colloidal stability, favorable rheological properties, and superior printability. The impact of printing angles on the mechanical properties of the rubber specimens based on the CPNRL-73 ink is explored in detail, wherein the tensile strength of the specimen printed at a 90° angle reaches an impressive 26 MPa and a strain of approximately 800%, which surpasses the majority of 3D-printed rubber latex specimens. Additionally, the CPNRL ink can be used to print a wide range of intricate shapes, demonstrating its advantages in excellent formability. The preparation of 3D printable ink using the absorption method will expand the application of elastomers in fields such as customized flexible sensing and personalized rubber products.

Keywords: concentrated pre-vulcanized natural rubber latex; superabsorbent polymer beads; additive-free; high mechanical performance; direct ink writing

1. Introduction

In contrast to conventional manufacturing processes, 3D printing enables swift and precise model creation [1–4] as well as facilitates a reduction in material waste, the optimization of shape, and an improvement in production efficiency [5–7]. The 3D printing of elastomers has garnered significant attention in recent years, particularly within the domains of bionic organs, soft robotics, and intelligent healthcare [8,9]. Natural rubber, as a bio-based elastomer, exhibits exceptional elasticity, high tensile strength, and superior tensile recovery [10,11]. These properties render it indispensable across various domains [12]. The swift advancement of 3D printing technology paves a novel pathway for the development and advancement of natural rubber [13–20].

Currently, rubber is mainly printed by fused deposition modeling (FDM) [21–24], by vat photo-polymerization (VPP) [1,25–27] using photo-curable precursors, and by direct ink writing (DIW) [28–30]. They have been used to successfully fabricate various complicated elastic rubber structures. However, they also demonstrate some limitations. For example, the 3D printing of FDM needs a much higher temperature, preventing its wide application in rubber and restraining the use of additives that are not resistant to high temperatures. For the VPP technique, toxic photo-initiators are required. Regarding the DIW technique, it is only necessary to regulate the rheological properties of the ink, for instance, by introducing rheological modifiers to satisfy specific requirements. In comparison with the two aforementioned methods, DIW exhibits lower costs and simpler operational procedures [28]. In summary, compared to FDM and VPP, the DIW 3D printing of rubber elastomers is regarded as one of the most promising technologies. It eliminates the need for high temperatures and photosensitizers while enabling the incorporation of a wider range of additives in 3D printing inks. However, the addition of substances such as rheological regulators may result in increased costs and volume shrinkage of printed samples. Moreover, these additives may compromise the mechanical properties of the printed objects. Consequently, there is an urgent need for an additive-free and scalable method to manufacture rubber samples with superior mechanical properties through direct ink writing 3D printing.

In this paper, a new type of CPNRL ink with a high solid content and without the addition of any thickening agents was fabricated through the absorption concentration method based on SAP beads, which is specifically suitable for DIW 3D printing (Figure 1). The colloidal and rheological properties of the CPNRL ink were systematically analyzed. The optimal printing conditions were identified by evaluating a range of extrusion pressures and linear velocities, and the printing accuracy was also explored. Furthermore, the mechanical properties of the printed specimens can be adjusted by raster orientation, and the tensile strength reached an impressive 26.4 MPa when printed in the 90°direction, accompanied by an elongation at break close to 800%. Its mechanical properties surpass the film prepared using the casting method and the majority of 3D-printed specimens based on rubber latex. Additionally, the CPNRL ink can be used to print a wide variety of complex structures and shapes with high precision. The exceptional mechanical properties of natural rubber are fully retained, providing new insights into the processing of complex natural rubber structures with high mechanical performance.

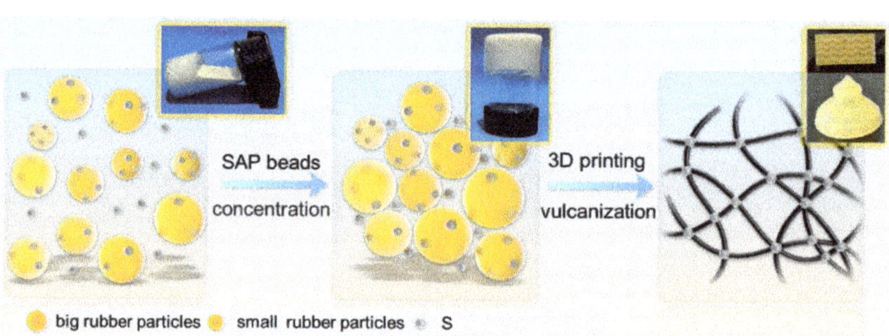

Figure 1. Schematic illustration of the preparation of CPNRL ink using SAP beads and a vulcanization diagram.

2. Materials and Methods

2.1. Materials

Concentrated natural rubber latex (60% in dry rubber content) was obtained from GuangKen Group Company in Guangdong (Yangjiang, China). Zinc oxide (ZnO), sulfur, zinc diethyl dithiocarbamat (ZDC), potassium oleate, sodium fluorosilicate, formaldehyde, antiaging agent-264, diffusible agent-N, casein, and 2-mercaptobenzene were purchased from Shanghai Aladdin Biochemical Technology Co., Ltd. (Shanghai, China). The commercial SAP beads were procured from Dinghao Corporation (Yiwu, China). The extrusion 3D bio-printer EFL-BP-6601 was purchased from Suzhou Yong qinquan Intelligent Equipment Co., Ltd. (Suzhou, China), and all the printed specimens in this article were completed by this equipment.

2.2. Preparation of PNRL and CPNRL

Pre-vulcanized natural rubber latex (PNRL) is usually fabricated using concentrated natural rubber latex and a curing system that includes sulfur, an accelerator, and an activator. According to the formulation in Table S1, the mixture was poured into a ball milling tank and milled at 40 rpm$\cdot$min^{-1} for 72 h to obtain a 50% mass fraction of the vulcanized hybrid dispersion. It is important to note that the dispersion should be filtered through a sieve to remove any bulk material and foam before use. The stabilizer and vulcanizing dispersant were slowly added to the concentrated natural rubber latex while stirring, and the mixing ratios are shown in Table S2. After the addition, the mixed system was heated and stirred at 52~60 °C for 3 h. At the end of the reaction, the rubber temperature was cooled to room temperature, and PNRL with 58% solids was obtained.

Concentrated pre-vulcanized natural rubber latex (CPNRL) was prepared as follows: 3 wt% SAP beads were slowly added into PNRL with low-speed stirring at 180 rpm$\cdot$min^{-1}. By regulating the duration of the concentration process, CPNRL with a solid content ranging from 62% to 73% can be achieved. CPNRL-62, CPNRL-65, CPNRL-69, CPNRL-71, and CPNRL-73 represent the solid contents of CPNRL of 62%, 65%, 69%, 71%, and 73%, respectively.

2.3. Scale Accuracy of 3D-Printed Rubber Specimens

All 3D-printed models were processed using Slic3r software (V1.2.3 (230313)). The design of the rectangular tracing model comprised a length of 20 cm and a width of 1 cm (Figure S3). The height of the gap between the tip of the nozzle and the substrate was set to 90% of the diameter of the nozzle, which was 21 G (inner diameter 0.51 mm). The linear speed was fixed at 1000 mm/min, and the pressure was controlled from 35 to 55 kPa. Additionally, the extrusion pressure was fixed at 45 kPa, and the linear speed was from 100 to 2000 mm/min.

For accuracy testing, the cuboid structures were 3D-printed using a 1 × 2 × 3 cuboid model integrated into the Slic3r software (V1.2.3 (230313)). These structures were printed with a 21 G nozzle head at 100% infill density. Five cube samples were printed for testing accuracy, and the average values were used for analysis.

2.4. Preparation of the 3D-Printed Tensile Model and Film Using the Casting Method Based on CPNRL-73

The 3D-printed specimen based on CPNRL-73 for studying tensile properties was designed to be a standardized dog bone tensile model with 100% packing density. Afterward, the printed specimens were vulcanized at 90 °C in an oven for 4 h. For comparison, the CPNRL-73 film was prepared using the casting method. In accordance with the GB/T 18011-2008 (ISO 498:1992, IDT) standard [31,32], CPNRL-73, which initially had a solid content of 73%, was diluted to achieve a solid content of 62% to prepare the casting film. It was filtered using a 200-mesh screen and then casted into the glass plate to keep static conditions for 48 h at normal temperature. Finally, it was heated in an oven at 35 °C for 48 h to complete the vulcanization.

2.5. Printing of Complex Rubber Structure Based on CPNRL-73 Ink

The models of letters, simplified tire treads, associated rotating components, and a hollow air rubber bag were created using Slic3r software. Subsequently, these models were fabricated via the CPNRL ink. Finally, the printed complex structures underwent a four-hour heat treatment at 90 °C to complete the vulcanization.

2.6. Characterization

The particle size and Zeta potential of PNRL and CPNRL were determined by using Malvern Zetasizer Nano ZS90 (Malvern Instruments Ltd., Worcestershire, UK). Scanning electron microscopy (HITACHI S4800, Tokyo, Japan) was used to study the morphology of the rubber particles present in PNRL and CPNRL and the surface and cross-section morphology of the 3D-printed specimens. TPNRL and CPNRL were diluted 50 times and stained with a 2 wt% osmium tetroxide aqueous solution before testing. The viscoelastic behavior of CPNRL with various solid contents was measured using a 20 mm plate device on a Rheometer (Thermo Scientific™ HAAKE™ MARS, Waltham, MA, USA) at room temperature. The steady-state mode was employed, with shear rates ranging from 0.01 to 1000 s^{-1}. The storage modulus and loss modulus were determined in the vibration mode at an angular frequency of 10 rad·s^{-1}. The range of shear stress applied was from 0 to 1000 Pa. The tensile tests of the casting rubber film and 3D-printed specimens were conducted utilizing an electronic universal testing machine (ETM103C). Each sample underwent five sets of parallel experiments, and the average values were used for analysis. According to GB/T 528-2009 [33], a dog-bone-shaped specimen was selected and tested at a speed of 500 mm/min to obtain the stress–strain curve.

The equilibrium swelling method was employed to determine the cross-linking density of the vulcanized sample. After precisely weighing 0.2 g of the vulcanized sample, we added 100 mL of toluene solution and stored it in darkness for 7 days. After removing the swollen sample's surface-attached toluene solution, its weight was measured. The Flory–Rehner Formula (1a)–(1c) was utilized to calculate the cross-link density of several groups of samples.

$$\Phi_r = \frac{m_1/\rho_p + m_2/\rho_s}{m_1/\rho_p} \tag{1a}$$

$$-\ln(1-\Phi_r) - \Phi_r - X_r\Phi_r^2 = nV_0(\Phi_r^{1/3} - \frac{1}{2}\Phi_r) \tag{1b}$$

$$M_c = \frac{\rho}{n} \tag{1c}$$

The formula includes the swelling ratio (Φ_r), density of natural rubber (ρ_p) (0.930 g/cm^3), density of toluene (ρ_s) (0.886 g/cm^3), molar volume of toluene (V_0 = 106.2 mL/mol), interaction parameter between natural rubber and toluene (X_r = 0.393), and average molecular mass between cross-linking points (M_c). The molar volume of toluene and the interaction parameters between natural rubber and toluene are considered constants, while the remaining values are experimentally determined.

3. Results and Discussion

3.1. Colloid Properties of PNRL and CPNRL

The initial solid content of pre-vulcanized natural rubber latex (PNRL) before concentration is 58%, which is insufficient for direct application in DIW 3D printing due to its high fluidity and inadequate support. This study employed the absorption concentration method based on SAP beads to increase the solid content of PNRL and regulate the rheological properties of CPNRL. CPNRL with various solid contents of 62%, 65%, 69%, 71%, and 73% can be prepared by precisely controlling the adsorption concentration time. As shown in Figure S1, CPNRL-62, CPNRL-65, and CPNRL-69 all exhibit excellent mobility. However, the mobility of CPNRL starts to reduce when the solid content reaches 71% and continues to reduce as the solid content reaches 73%, indicating that CPNRL-73 should be suitable for DIW 3D printing. The stability of PNRL, as assessed by the Zeta potential, was examined both before and after water removal using SAP (Figure 2a). The Zeta potential of PNRL was approximately −37.8 mV, while it increased to around −38.3 mV for CPNRL-73 after concentration by removing the water present in PNRL using SAP beads. The Zeta potential value serves as an indicator of the stability of PNRL [34,35]. The similar potential value of PNRL and CPNRL-73 demonstrates that concentration does not reduce the stability of CPNRL-73.

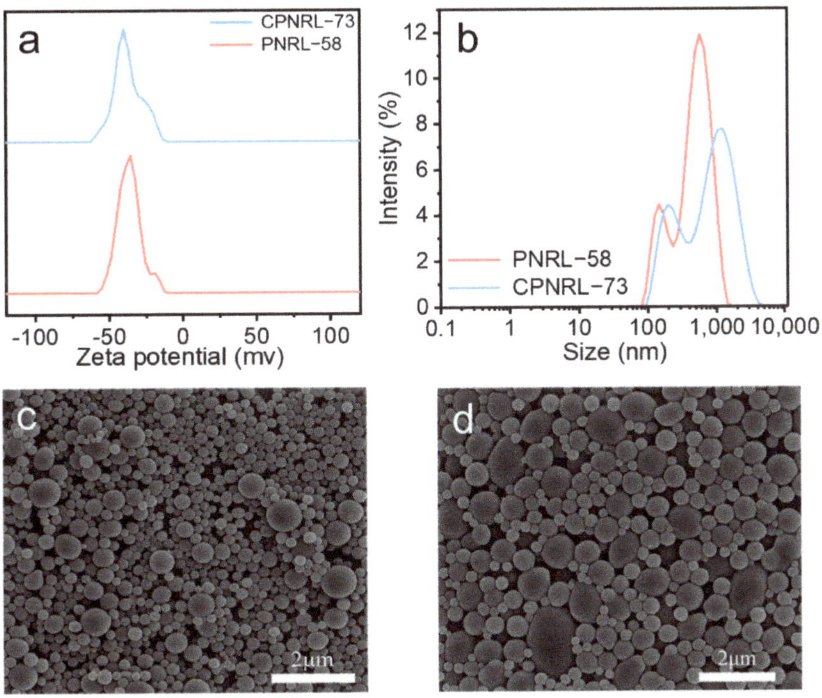

Figure 2. Characterization of the colloidal properties and morphology of PNRL-58% and CPNRL-73%: (**a**) Zeta potential, (**b**) particle size, (**c**,**d**) SEM images.

The particle size of CPNRL-73 exhibited a significant increase (Figure 2b), with the average diameter expanding from 354.6 nm to 505.3 nm. The figure illustrates that both smaller and larger rubber particles underwent an increase in size, likely attributed to the fusion and cross-linking of these particles as a result of water reduction during the concentration process. Furthermore, the SEM images revealed that the rubber particles were predominantly spherical, with smaller particles being more abundant prior to concentration, interspersed with some larger particles (Figure 2c). However, following the concentration treatment, the number of smaller particles decreased, while pear-shaped and larger particles became more prevalent (Figure 2d).

3.2. Rheological Properties of CPNRL

Three-dimensional printed inks, which are typically utilized in DIW printing, should demonstrate suitable rheological properties, including apparent viscosity, yield stress under both shear and compressive forces, and viscoelastic characteristics (i.e., loss and elastic moduli) [36–38]. The results of the rheological test of CPNRL are presented in Figure 3a,b. As depicted in the figure, the viscosity of CPNRL is influenced by shear strain, indicating the significance of enhancing the rheological properties of printing ink for improved quality. High viscosity hinders the proper extrusion and shape retention of printed samples after extrusion from the nozzle under external stress. This phenomenon demonstrates the shear thinning behavior characteristic of non-Newtonian fluids, where fluid viscosity decreases with increasing shear stress [39]. Additionally, it is evident that when the solid content reaches 71%, the CPNRL ink exhibits excellent shear thinning ability, initially increasing and then declining. In the shear modulus diagram (Figure 3b), it can also be observed that $G' > G''$ for CPNRL with solid contents of 71% and 73% under low shear stress conditions. However, under higher shear forces, $G' < G''$ demonstrates remarkable performance in terms of shear thinning behavior. Consequently, we conclude that CPNRL ink with a solid content of 73% or higher possesses favorable characteristics for 3D printing applications, aligning well with the rheological properties of inks required by DIW 3D printing.

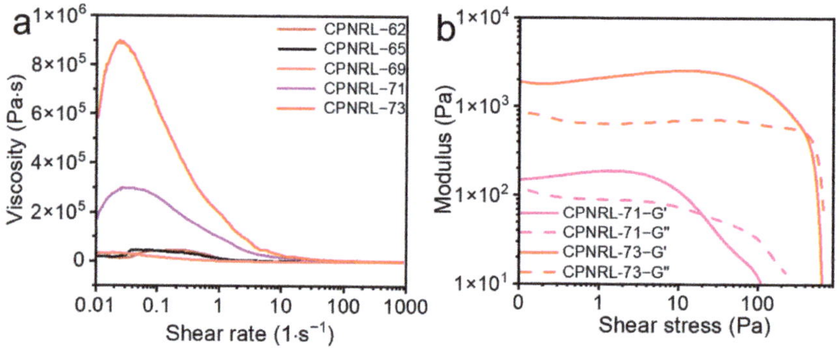

Figure 3. (**a**) Viscosity as a function of the shear rate of different CPNRL samples. (**b**) The correlation between modulus and shear stress of CPNRL suitable for 3D printing.

3.3. The Stability of Line Printing, the Stacking of CPNRL Ink, and the Analysis of Scale Accuracy

The fundamental principle of DIW 3D printing to create a self-supporting extrusion layer is directly dependent on the printability and shape fidelity of the ink [38,40]. Machine parameters, such as nozzle diameter and printing velocity, significantly influence the accuracy and resolution of the printed output [41]. Typically, a nozzle with a smaller diameter can enhance printing resolution. However, this necessitates higher extrusion pressure and extended printing time to prevent nozzle clogging. Likewise, reduced print speeds generally lead to improved shape tolerances and fidelity, although this also prolongs the printing process [38].

In order to determine optimal printing parameters, the CPNRL-73 ink was subjected to testing under different printing pressures and speeds. The printing results under various combinations of pressure and speed are shown in Figure 4a,b. To ascertain the optimal printing parameters, the CPNRL-73 ink was evaluated across a range of extrusion pressures and printing velocities. Figure 4a illustrates the line width at varying speeds from 100 to 2000 mm·min^{-1}, with a constant extrusion pressure of 45 kPa. The average line width of the CPNRL-73 ink is approximately double the nozzle diameter at a speed of 100 mm·min^{-1}. The line width aligns closely with the nozzle diameter within the range of 500 to 1000 mm·min^{-1}. Figure 4b illustrates the line width of the ink within the pressure range of 35 to 55 kPa, with increments of 5 kPa and a constant speed of 1000 mm·min^{-1}. The CPNRL-73 ink can be smoothly extruded from the nozzle under a pressure of 35 kPa. In contrast, when the pressure reaches 50 kPa, the average printing line width exceeds twice the nozzle diameter. A slow line speed or excessive pressure results in excessive ink extrusion, which hinders the formation of a stable support structure, making it challenging to meet the requirements for multi-layer printing. At a printing speed of 1500 mm·min^{-1}, the lines exhibit excellent form, with their width being precisely three-quarters (75%) of the nozzle diameter. Consequently, after thorough consideration, we determined that a pressure range of 45 kPa, in conjunction with a constant linear velocity of 1500 mm·min^{-1}, constituted the optimal printing conditions.

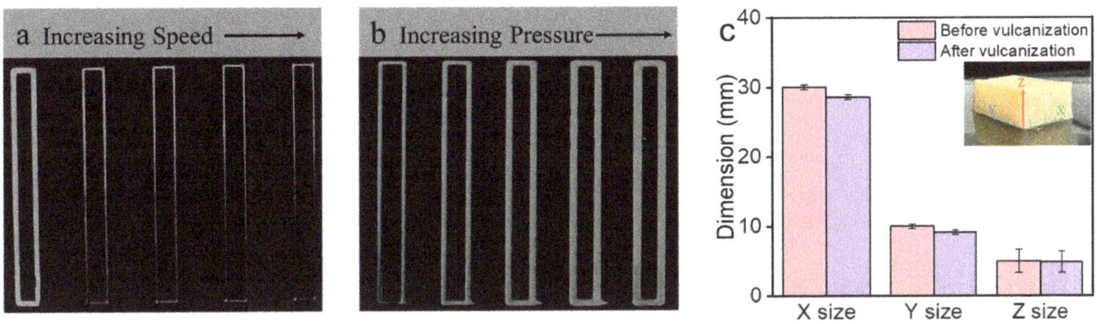

Figure 4. Characterization of the printing line drawings and accuracy of the CPNRL-73 ink. (**a**) At a constant pressure of 45 kPa across various speeds. (**b**) At a constant speed of 1000 mm/s under different pressures. (**c**) The dimension of the accuracy of the cubic sample in the XYZ direction after vulcanization.

The shrinkage of a specimen is inevitable when using hydrogel or rubber latex ink for 3D printing [28]. The shrinkage phenomenon of 3D-printed specimens based on the CPNRL-73 ink is also observed in this study. To evaluate the accuracy of the 3D-printed samples, we selected a solid rubber cube as a model. The inserted image in Figure 4c illustrates the XYZ orientation of the solid rubber cube after vulcanization. As depicted in Figure 4c, the dimensional fidelity of the vulcanized samples was 95.3%, 91.6%, and 96.6% in the XYZ direction, respectively. There was a decrease in volume for the printed rubber cube after vulcanization. Prior to vulcanization, the CPNRL-73 ink still contains some water, which evaporates during the vulcanization process due to heating. This results in the overall shrinkage of the structure and subsequently reduces the volume of the 3D-printed specimens.

3.4. Mechanical Properties

To compare the mechanical properties of the 3D-printed rubber specimens with different orientations and those prepared using the traditional casting method, a standardized dog bone tensile model was created through CAD modeling and printed by using CPNRL-73. The results indicate that different printing orientations have certain effects on the mechanical properties of the 3D-printed rubber specimens. For instance, when printed at

0°, only a strain of 720% and a stress of up to 22 MPa can be achieved. However, as the printing angle changes, there is a significant increase in stress for the samples printed at 45° and 90°. Simultaneously, the tensile strength at 90° and the elongation at break can achieve 26 MPa and approximately 800%, respectively. This phenomenon may be attributed to the anisotropic nature of the material. An increase in the angle relative to the direction of the applied force results in a reduction in the sample's strength, which is particularly evident in polymeric materials [29]. When printed at a 90° raster angle, the contact area between the lines is maximized, and the lines align with the direction of the applied force during stretching. Conversely, the mechanical properties at 0° and 45° grating angles tend to diminish due to the misalignment between the angle and the direction of the applied force [29]. Additionally, compared to traditional casting films, the 3D-printed models exhibit higher cross-linking density across different printing angles. This discrepancy may be attributed to the lower vulcanization temperature used in traditional casting methods, leading to a reduced cross-linking density post-vulcanization compared to 3D-printed models.

Owing to the excellent interfacial compatibility between the printing layers (as shown in Figure 5), the mechanical properties of the 3D-printed rubber specimens are markedly superior to those reported in other studies (Figure 6f) [1,24–28,42]. The surface and cross-sectional morphology of the 3D-printed rubber specimen were examined using SEM. Distinct linear structures are evident on the surface of the rubber specimen, suggesting it was printed and fabricated layer by layer. However, no distinct interlayer interface was observed in the cross-section of the rubber specimen, indicating that excellent cross-linking between layers was formed during the vulcanization process and demonstrating a high degree of uniformity and consistency. Furthermore, the mechanical properties of the 3D-printed rubber specimen at various angles were evaluated. The device printed at 90° exhibited the highest tensile strength at 26 MPa, whereas the device printed at 0° demonstrated a strength of 22 MPa, indicating only a slight reduction. This suggests that the cross-linking between layers remained high during the vulcanization process, preventing significant delamination and preserving the superior mechanical properties of the 3D-printed rubber specimen.

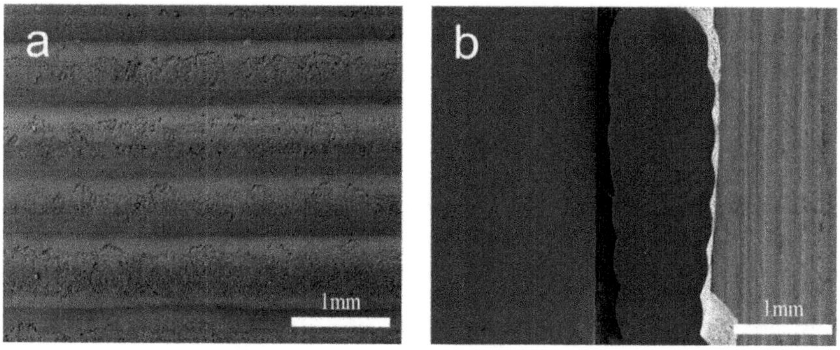

Figure 5. SEM images of the cross-section and surface of the 3D-printed rubber specimen. (**a**) Surface topography. (**b**) Cross-section morphology.

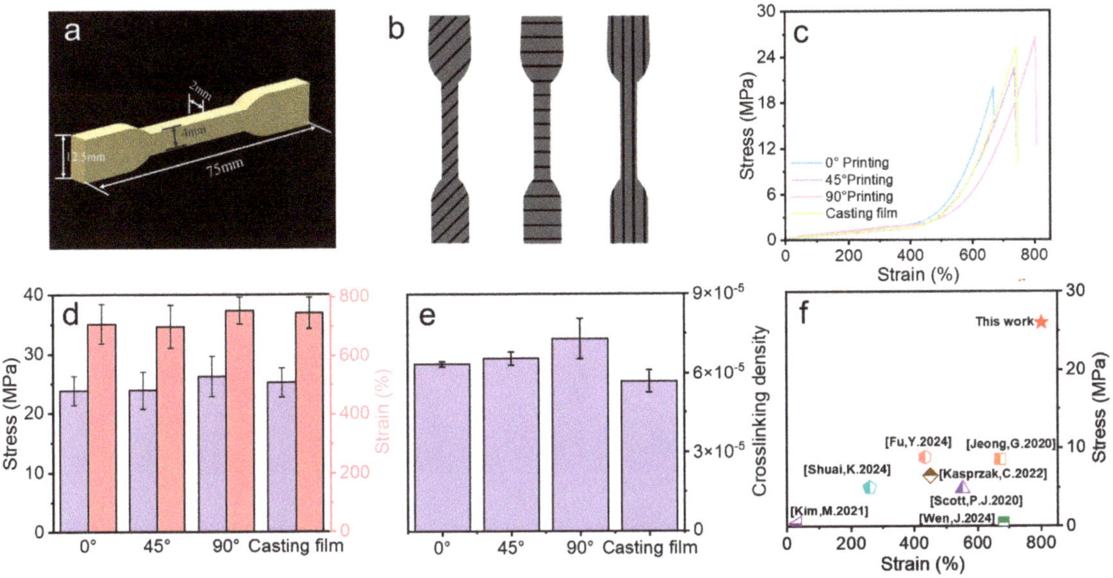

Figure 6. Mechanical testing and characterization. (**a**) Drawing and dimensions of CAD-generated dog bone. (**b**) Diagrams of the models for the tensile test prepared in various printing orientations. (**c**) Stress–strain curves for specimens printed in different directions. (**d**) Comparison diagram of stress–strain. (**e**) Cross-linking density of specimens printed in different directions and prepared using the casting method. (**f**) A comparative analysis of the mechanical properties of various 3D-printed elastomers [1,24–28,42].

3.5. The 3D Structures Fabricated Using CPNRL-73 Ink

Four distinct structural models were designed using CAD software and printed with the CPNRL-73 ink: a letter model, a tire tread model, rotary-shaped components, and an air rubber bag. In Figure 7, the computer-generated model, the un-vulcanized rubber specimens, and the vulcanized rubber specimens are displayed from left to right. The un-vulcanized printed specimens appear white in color because of the presence of water in the CPNRL-73 ink, and they become light yellow after vulcanization and drying by heating. The surface of the 3D-printed tire tread model was printed according to a 100% infill density, indicating that the ink was smoothly squeezed out during accumulation without any residual ink around the nozzle. The remaining 3D-printed models were stacked by rubber lines, and no collapse or inclination was observed, proving that the CPNRL-73 ink has good support and stability. Following vulcanization, the product exhibits enhanced precision with a reduced rate of shrinkage and deformation. The printed rubber product retains a shape nearly identical to the original design model.

Figure 7. Demonstration of different 3D-printed shapes before and after vulcanization. (**a**) Alphabetic model. (**b**) Tread model. (**c**) Special-shaped part model. (**d**) Air rubber bag.

4. Conclusions

In this study, we developed a new type of PNRL ink using the absorption concentration method based on SPA beads for DIW 3D printing without using additives. The total solid content of the CPNRL ink is up to 73%, and the resulting ink demonstrates good colloidal stability and 3D printability, proving that the absorption concentration method is an effective approach to fabricating rubber latex ink for DIW 3D printing. The optimal extrusion pressure was found to be approximately 45 kPa, while the optimal extrusion speed was determined to be 1500 mm/min. Furthermore, the tensile strength of the printed specimen reached an impressive 26.4 MPa when printed in the 90°direction, accompanied by a strain exceeding approximately 800%, which surpasses the majority of rubber latex specimens produced by 3D printing. The superior mechanical properties may be derived from excellent interface compatibility between the printing layers. Additionally, a wide variety of complex elastic structures with high precision can be 3D-printed by using CPNRL ink, demonstrating excellent printing adaptability and support. This study not only develops a scalable method to fabricate rubber latex ink for DIW 3D printing but also offers fresh insights into processing and structural design possibilities for complex and multifunctional rubber latex products.

Supplementary Materials: The following supporting information can be downloaded at https://www.mdpi.com/article/10.3390/polym17030351/s1: Table S1: The vulcanization system formula of pre-vulcanized natural rubber latex; Table S2: The formula of pre-vulcanized natural rubber latex; Table S3: Comparative analysis of the mechanical properties of various 3D-printed elastomers. Figure S1. Photographs of CPNRL with different solid contents. Figure S2. Scale precision model of 3D-printed rubber sample.

Author Contributions: L.L.: Writing—original draft, review, and editing, visualization, software, formal analysis, and data curation. J.Z.: Writing—review and editing, resources, methodology, funding acquisition, conceptualization, and investigation. Z.L.: Writing—review and editing and supervision. N.K.: Software, methodology, supervision, and investigation. X.Z.: Methodology. X.J.: Formal analysis. J.L.: Methodology. S.H.: Methodology. P.Z.: Methodology. S.L.: Methodology. Y.S.: Formal analysis. J.T.: Writing—review and editing, project administration, and funding acquisition. All authors have read and agreed to the published version of the manuscript.

Funding: This work was supported by the Foundation of National Key R&D Program of China (No. 2024YFD2300904), the Science and Technology Program of Hainan Province (ZDYF2020230, ZDYF2022GXJS226), the Hainan Provincial Natural Science Foundation of China (521QN0931), the Guangdong Basic and Applied Basic Research Foundation (2024A1515011788), the Central Public-interest Scientific Institution Basal Research Fund for Chinese Academy of Tropical Agricultural Sciences (1630122022007), and the Chinese Academy of Tropical Agricultural Sciences for Science and Technology Innovation Team of National Tropical Agricultural Science Center (CATASCXTD202306).

Data Availability Statement: The original contributions presented in this study are included in the article/Supplementary Materials. Further inquiries can be directed to the corresponding author.

Conflicts of Interest: The authors declare that they have no known competing financial interests or personal relationships that could have appeared to influence the work reported in this paper.

References

1. Jeong, G.; Park, C.H.; Kim, B.-Y.; Kim, J.; Park, S.-D.; Yang, H.; Lee, W.S. Photocurable Elastomer Composites with SiO_2-Mediated Cross-Links for Mechanically Durable 3D Printing Materials. *ACS Appl. Polym. Mater.* **2020**, *2*, 5228–5237. [CrossRef]
2. Ligon, S.C.; Liska, R.; Stampfl, J.; Gurr, M.; Mülhaupt, R. Polymers for 3D printing and customized additive manufacturing. *Chem. Rev.* **2017**, *117*, 10212–10290. [CrossRef] [PubMed]
3. Nadgorny, M.; Ameli, A. Functional polymers and nanocomposites for 3D printing of smart structures and devices. *ACS Appl. Mater. Interfaces* **2018**, *10*, 17489–17507. [CrossRef]
4. Farahani, R.D.; Chizari, K.; Therriault, D. Three-dimensional printing of freeform helical microstructures: A review. *Nanoscale* **2014**, *6*, 10470–10485. [CrossRef]
5. Bean, R.H.; Rau, D.A.; Williams, C.B.; Long, T.E. Rheology guiding the design and printability of aqueous colloidal composites for additive manufacturing. *J. Vinyl Addit. Technol.* **2023**, *29*, 607–616. [CrossRef]
6. Chaturvedi, I.; Jandyal, A.; Wazir, I.; Raina, A.; Ul Haq, M.I. Biomimetics and 3D printing—Opportunities for design applications. *Sens. Int.* **2022**, *3*, 100191. [CrossRef]
7. Hisham, M.; Dileep, D.; Jacob, L.; Butt, H. Additive manufacturing of carbon nanocomposites for structural applications. *J. Mater. Res. Technol.* **2024**, *28*, 4674–4693. [CrossRef]
8. Wallin, T.J.; Pikul, J.; Shepherd, R.F. 3D printing of soft robotic systems. *Nat. Rev. Mater.* **2018**, *3*, 84–100. [CrossRef]
9. Zhou, L.-Y.; Fu, J.; He, Y. A Review of 3D Printing Technologies for Soft Polymer Materials. *Adv. Funct. Mater.* **2020**, *30*, 2000187. [CrossRef]
10. Dasgupta, A.; Dutta, P. Printability of elastomer as a 3D printing material for additive manufacturing. *J. Rubber Res.* **2024**, *27*, 137–157. [CrossRef]
11. Li Na, S.; Abdul Kariem, A. Elastomers and Their Potential as Matrices in Polymer Electrolytes. In *Elastomers*; Nevin, C., Ed.; IntechOpen: Rijeka, Croatia, 2017.
12. Whba, R.; Su'ait, M.S.; Whba, F.; Sahinbay, S.; Altin, S.; Ahmad, A. Intrinsic challenges and strategic approaches for enhancing the potential of natural rubber and its derivatives: A review. *Int. J. Biol. Macromol.* **2024**, *276*, 133796. [CrossRef]
13. Alkadi, F.; Lee, J.; Yeo, J.-S.; Hwang, S.-H.; Choi, J.-W. 3D Printing of Ground Tire Rubber Composites. *Int. J. Precis. Eng. Manuf. Green Technol.* **2019**, *6*, 211–222. [CrossRef]
14. Alkadi, F.; Lee, K.-C.; Bashiri, A.H.; Choi, J.-W. Conformal additive manufacturing using a direct-print process. *Addit. Manuf.* **2020**, *32*, 100975. [CrossRef]
15. Adhikari, B.; De, D.; Maiti, S. Reclamation and recycling of waste rubber. *Prog. Polym. Sci.* **2000**, *25*, 909–948. [CrossRef]
16. Fang, Y.; Zhan, M.; Wang, Y. The status of recycling of waste rubber. *Mater. Des.* **2001**, *22*, 123–128. [CrossRef]
17. De, D.; Das, A.; De, D.; Dey, B.; Debnath, S.C.; Roy, B.C. Reclaiming of ground rubber tire (GRT) by a novel reclaiming agent. *Eur. Polym. J.* **2006**, *42*, 917–927. [CrossRef]
18. Wu, B.; Zhou, M.H. Recycling of waste tyre rubber into oil absorbent. *Waste Manag.* **2009**, *29*, 355–359. [CrossRef] [PubMed]

19. Sunthonpagasit, N.; Duffey, M.R. Scrap tires to crumb rubber: Feasibility analysis for processing facilities. *Resour. Conserv. Recycl.* **2004**, *40*, 281–299. [CrossRef]
20. Fiksel, J.; Bakshi, B.R.; Baral, A.; Guerra, E.; DeQuervain, B. Comparative life cycle assessment of beneficial applications for scrap tires. *Clean Technol. Environ. Policy* **2011**, *13*, 19–35. [CrossRef]
21. Gao, Y.; Li, Y.; Hu, X.; Wu, W.; Wang, Z.; Wang, R.; Zhang, L. Preparation and Properties of Novel Thermoplastic Vulcanizate Based on Bio-Based Polyester/Polylactic Acid, and Its Application in 3D Printing. *Polymers* **2017**, *9*, 694. [CrossRef]
22. Hu, X.; Kang, H.; Li, Y.; Geng, Y.; Wang, R.; Zhang, L. Preparation, morphology and superior performances of biobased thermoplastic elastomer by in situ dynamical vulcanization for 3D-printed materials. *Polymer* **2017**, *108*, 11–20. [CrossRef]
23. Wissamitanan, T.; Dechwayukul, C.; Kalkornsurapranee, E.; Thongruang, W. Proper Blends of Biodegradable Polycaprolactone and Natural Rubber for 3D Printing. *Polymers* **2020**, *12*, 2416. [CrossRef]
24. Fu, Y.; Shi, Z.; Feller, K.D.; Schulz, M.D.; Long, T.E.; Williams, C.B. Vat photopolymerization of silica reinforced styrene-butadiene rubber elastomeric nanocomposites. *Appl. Mater. Today* **2024**, *40*, 102370. [CrossRef]
25. Wen, J.; Bean, R.H.; Nayyar, G.; Feller, K.; Scott, P.J.; Williams, C.B.; Long, T.E. Vat Photopolymerization of Synthetic Isoprene Rubber Latexes. *ACS Appl. Polym. Mater.* **2024**, *6*, 2169–2176. [CrossRef]
26. Scott, P.J.; Meenakshisundaram, V.; Hegde, M.; Kasprzak, C.R.; Winkler, C.R.; Feller, K.D.; Williams, C.B.; Long, T.E. 3D Printing Latex: A Route to Complex Geometries of High Molecular Weight Polymers. *ACS Appl. Mater. Interfaces* **2020**, *12*, 10918–10928. [CrossRef]
27. Kasprzak, C.; Brown, J.R.; Feller, K.; Scott, P.J.; Meenakshisundaram, V.; Williams, C.; Long, T. Vat Photopolymerization of Reinforced Styrene–Butadiene Elastomers: A Degradable Scaffold Approach. *ACS Appl. Mater. Interfaces* **2022**, *14*, 18965–18973. [CrossRef] [PubMed]
28. Kim, M.; Choi, J.-W. Rubber ink formulations with high solid content for direct-ink write process. *Addit. Manuf.* **2021**, *44*, 102023. [CrossRef]
29. Chansoda, K.; Suvanjumrat, C.; Wiroonpochit, P.; Kaewprakob, T.; Chookaew, W. Support medium development for 3D printing natural rubber latex via direct ink writing in the support bath technique. *Clean. Mater.* **2024**, *13*, 100257. [CrossRef]
30. Banks, J.D.; Emami, A. Influence of filler concentration and type on rheological and mechanical properties of liquid isoprene rubber for high-strength parts via material extrusion. *Addit. Manuf.* **2023**, *78*, 103851. [CrossRef]
31. *GB/T 18011-2008*; Natural Rubber Latex Concentrate—Preparation of Dry Films. China Standards Press: Beijing, China, 2008.
32. *ISO 498:1992*; Natural Rubber Latex Concentrate—Preparation of Dry Films. ISO: Geneva, Switzerland, 1992.
33. *GB/T 528-2009*; Rubber, Vulcanized or Thermoplastic—Determination of Tensile Stress—Strain Properties. China Standards Press: Beijing, China, 2009.
34. Yusof, N.H.; Singh, M.; Mohd Rasdi, F.R.; Tan, K.S. Properties of concentrated skim rubber latex using membrane separation process and its comparison with other natural rubber latexes. *J. Rubber Res.* **2023**, *26*, 169–177. [CrossRef]
35. Huang, S.; Luo, Z.; Zhang, J.; Kong, N.; Li, J.; Zhao, P.; Wu, J.; Tan, S.; Tao, J. An eco-friendly, high-yield and scalable method for processing concentrated natural rubber latex with superabsorbent polymer beads. *Chem. Eng. J.* **2024**, *502*, 158013. [CrossRef]
36. Neumann, T.V.; Dickey, M.D. Liquid Metal Direct Write and 3D Printing: A Review. *Adv. Mater. Technol.* **2020**, *5*, 2000070. [CrossRef]
37. Smay, J.E.; Cesarano, J.; Lewis, J.A. Colloidal Inks for Directed Assembly of 3-D Periodic Structures. *Langmuir* **2002**, *18*, 5429–5437. [CrossRef]
38. Saadi, M.A.S.R.; Maguire, A.; Pottackal, N.T.; Thakur, M.S.H.; Ikram, M.M.; Hart, A.J.; Ajayan, P.M.; Rahman, M.M. Direct Ink Writing: A 3D Printing Technology for Diverse Materials. *Adv. Mater.* **2022**, *34*, 2108855. [CrossRef] [PubMed]
39. Kavalli, T.; Wolf, R.; Lalevée, J. Ultrafast Epoxy-Anhydride Photopolyaddition Reaction. *Macromol. Chem. Phys.* **2020**, *221*, 2000236. [CrossRef]
40. M'Barki, A.; Bocquet, L.; Stevenson, A. Linking Rheology and Printability for Dense and Strong Ceramics by Direct Ink Writing. *Sci. Rep.* **2017**, *7*, 6017. [CrossRef]
41. Maguire, A.; Pottackal, N.; Saadi, M.A.S.R.; Rahman, M.M.; Ajayan, P.M. Additive manufacturing of polymer-based structures by extrusion technologies. *Oxf. Open Mater. Sci.* **2021**, *1*, itaa004. [CrossRef]
42. Shuai, K.; Zhang, K.; Yao, S.; Ni, Z.; Shi, D.; Miao, J.-T.; Chen, M. 3D Printing of Branched Polyurethane with Tough Mechanical Properties for Stretchable Sensors. *Adv. Eng. Mater.* **2024**, *26*, 2301103. [CrossRef]

polymers

MDPI

Article

Evaluating Oil Palm Trunk Biochar and Palm Oil as Environmentally Friendly Sustainable Additives in Green Natural Rubber Composites

Narong Chueangchayaphan, Manop Tarasin, Wimonwan Phonjon and Wannarat Chueangchayaphan *

Faculty of Science and Industrial Technology, Prince of Songkla University, Surat Thani Campus, Surat Thani 84000, Thailand; narong.c@psu.ac.th (N.C.); manop.ta@psu.ac.th (M.T.); phonjonwimonwan@gmail.com (W.P.)
* Correspondence: wannarat.p@psu.ac.th

Abstract

This research examines the possibility of palm oil and oil palm trunk biochar (OPTB) from pyrolysis effectively serving as alternative processing oils and fillers, substituting petroleum-based counterparts in natural rubber (NR) composites. Chemical, elemental, surface and morphological analyses were used to characterize both carbon black (CB) and OPTB, by using Fourier transform infrared spectroscopy (FTIR), X-ray diffraction (XRD), Brunauer–Emmett–Teller (BET) gas porosimetry, and scanning electron microscopy (SEM). The influences of OPTB contents from 0 to 100 parts per hundred rubber (phr) on thermal, dielectric, dynamic mechanical, and cure characteristics, and the key mechanical properties of particulate NR-composites were investigated. OPTB enhanced the characteristics of the composites, as demonstrated by a rise in dielectric constant, thermal stability, storage modulus, glass transition temperature (T_g), hardness and modulus at 300% elongation, along with a decrease in the loss tangent (tan δ). Tear strength exhibited an increase with OPTB content up to a specific threshold, whereas tensile strength and elongation at break declined. This implies a compromise between the various mechanical properties when incorporating OPTB as a filler. This work supports the potential application of OPTB as a renewable substitute for CB in the rubber industry, particularly in tire production and other industrial rubber applications, which would also bring environmental, sustainability, and economic benefits for the palm oil-related industry.

Keywords: oil palm trunk; biochar; palm oil; sustainable additive; natural rubber; green composites

1. Introduction

Biopolymers have gained a lot of attention in recent years due to their potential to lessen the need for fossil fuel resources and nonbiodegradable polymers, both of which contribute significantly to environmental pollution [1]. Natural rubber (NR), derived from the *Hevea brasiliensis* plant, is a naturally occurring polymer consisting of a combination of poly(*cis*-1,4-isoprene) and biological components. This unique composition imparts distinct characteristics to the material [2]. Thus, there has been a consistent rise in the significance of NR for human beings due to its inherent qualities of elasticity, low hysteresis, high strength, exceptional toughness, and versatile production capabilities [3]. However, it should be noted that raw NR possesses a soft texture and exhibits relatively low mechanical properties. Consequently, the incorporation of additives and the process of vulcanization

 https://doi.org/10.3390/polym17020223

have become imperative in order to enhance its overall performance [4]. Processing oil is an additive needed in NR compounds because it acts as an internal lubricant to achieve acceptable processability, improves flex life at a low temperature, filler dispersion, end-use properties, and gives a cost benefit [5,6]. Vegetable oil has replaced petroleum-based oil for decades because of its broad availability, non-toxicity, affordability, environmental sustainability, and being a renewable resource [7]. Vegetable oils have been extensively studied as alternative processing aids in rubber compounds due to their environmental benefits and ability to replace petroleum-based oils. Castor oil has been investigated as a plasticizer in natural rubber (NR), showing improved processability and mechanical properties [8]. Similarly, the natural oils soybean, palm, and sunflower oil have been explored for their effectiveness as processing aids and activators in carbon black-filled NR [9]. Soybean oil and its modified forms have demonstrated potential in replacing petroleum oils across various elastomers, including NR, styrene-butadiene rubber (SBR), and ethylene-propylene-diene monomer rubber (EPDM), contributing to enhanced filler dispersion, processing characteristics, and mechanical performance [10–15]. Palm oil, a readily available and renewable resource, has also been shown to improve the processing and mechanical properties of EPDM and silica-filled NR composites [16,17]. In nitrile butadiene rubber (NBR), palm-based processing aids have optimized curing systems and physical properties [18]. These findings highlight the versatility and potential of vegetable oils to advance sustainable practices in rubber manufacturing. In this study, palm oil was selected as the processing oil for the fabrication of green composite natural rubber. This selection was based on a review of the relevant literature, highlighting its status as the most extensively manufactured vegetable oil globally, accounting for a substantial share of the vegetable oil industry and offering a reasonable price for large-scale industrial applications [19,20]. Palm oil is derived from renewable resources and contains key fatty acids, including palmitic, oleic, and linoleic acids [21], which contribute to its excellent lubricating properties, high heat resistance [9] and compatibility with rubber matrices. Additionally, its renewable and environmentally friendly nature aligns with the objectives of developing sustainable and eco-friendly rubber composites. These factors collectively support the suitability of palm oil as an effective alternative to petroleum-based processing oils in rubber compounding.

Filler is one of the compounding ingredients required to enhance the properties of rubber, and common ones are carbon black, clay, calcium carbonate, silica, etc. [22]. The type of filler most frequently utilized in the rubber industry as a reinforcing substance is carbon black (CB) [23]. Despite the advantageous physical, mechanical, thermal, and barrier qualities that CB imparts to rubber composites, it is not without its drawbacks. CB is produced from the incomplete combustion of heavy petroleum products. The incomplete combustion of petroleum products can result in the emission of poisonous fumes and the generation of hazardous waste [24]. As a result, the production and utilization of CB give rise to numerous risks of health complications and environmental damage. Thus, it is imperative that the rubber industry adopts alternative, renewable, and environmentally favorable filler materials in place of CB. Biochar is a solid carbon-rich substance like CB; it is obtained by subjecting biomass to elevated temperatures in a sealed chamber with varying degrees of oxygen lack. In this controlled environment, biomass undergoes a thermochemical transformation, producing biochar. Because of its affordability and environmentally friendly nature, biochar has garnered significant interest for diverse uses such as enhancing soil quality, managing waste, mitigating greenhouse gas emissions, and generating energy [25,26]. Biomass derived from oil palm plantations presents itself as a highly promising and viable option as fuel and feedstock of carbon black, serving as a commendable alternative to conventional renewable and sustainable resources.

Palm oil is the predominant oil produced on a global scale, playing a substantial role in the vegetable oil market [20]. The increasing worldwide need for palm oil has resulted in intensified production in the oil palm estates. This heightened cultivation has resulted in the significant production of substantial amounts of biomass from milling operations and plantations. Among the various categories of residual oil palm biomass are empty fruit bunches (EFBs), palm kernel shells (PKSs), mesocarp fiber (MF), oil palm fronds (OPFs), and oil palm trunks (OPTs). OPTs hold significant appeal due to their consistent year-round availability following the felling of oil palm trees [27,28]. The substantial volume of OPTs generated poses inevitable challenges in waste management. Instead of resorting to landfill disposal or open burning, which result in significant air pollution, there is an opportunity to transform this waste into value-added materials. To achieve a significant reduction in the organic compound content, OPT is subjected to pyrolysis to generate biochar. In contrast to unprocessed biomass, biochar is more advantageous owing to its superior qualities, which include reduced moisture content, elevated calorific value, decreased density for storage purposes, and long-lasting durability [29]. This approach can effectively address and mitigate waste disposal concerns. Other than its application, as previously mentioned, the biochar produced possesses the ability to serve as an additive in rubber composites. This would provide an alternative to carbon black, which is derived from petroleum. In the prior literature, there are reports on the utilization of biochar from oil palm biomass as bio-filler in rubber composites, for example the incorporation of palm kernel shell biochar (PKSB) in NR composites [22,30–32] and in carboxylated nitrile butadiene rubber (XNBR) [22]. The viability of using OPTB as a flame-retardant component has been examined in two varieties of specialty natural rubber (SpNR) latex foam, epoxidized and deproteinized, providing a limited study of the application of OPTB as a filler in vulcanized natural rubber [33].

The concept of a sustainable circular economy has gained significant attention for its economic, social, and environmental benefits. By utilizing green and renewable materials, it supports industrial applications that benefit both current and future generations [19]. This study is innovative in simultaneously utilizing two renewable materials—OPTB as a filler and palm oil as a processing oil—to replace carbon black- and petroleum-based processing oils in NR composites. To our knowledge, no previous research has combined these materials in a single formulation. The functional groups, chemical composition, and morphology of OPTB were characterized using FTIR, XRD, and SEM. The effects of OPTB content (0–100 phr) on the cure characteristics and thermal, dynamic mechanical, dielectric, and mechanical properties of NR composites were analyzed. Comparative samples with CB as a filler were prepared using the same methodology. This approach offers a significant advance in reducing fossil fuel consumption and developing eco-friendly additives by transforming agricultural waste into high-value materials, aligning with sustainability and circular economy principles.

2. Materials and Methods

2.1. Materials

Palm oil was obtained from Suksomboon Vegetable Oil Co., Ltd. (Chonburi, Thailand). Standard Thai Rubber (STR20) was used in the rubber formulation and the other chemicals were N-phenyl-p-phenylenediamine (6PPD), zinc oxide (ZnO), stearic acid (SA), diphenyl guanidine (DPG), 2-Mercaptobenzothiazyl disulfide (MBTS), carbon black (N550), and sulfur (S), supplied by Bossoftical Public Co., Ltd. (Songkhla, Thailand). All the chemicals were utilized as received, without further purification. OPTs were gathered from a nearby palm plantation located in Surat Thani, Thailand. OPTs used in this study were from the Surat Thani 2 cultivar, harvested from a 22-year-old plantation located in Mueang District, Surat Thani, Thailand. The palms had an average trunk height of 9.5 m. For the experiment, the trunks were cut into 10 cross-sectional discs (thickness: 5 cm each) at intervals of 1 m along the length of the trunk to ensure a representative sampling of the material. These discs were pre-dried at 100 ± 5 °C in a controlled oven until their weights stabilized, which is essential for achieving efficient pyrolysis and minimizing energy consumption during the process. The pre-dried samples were then arranged evenly in the pyrolysis chamber to optimize heat distribution and ensure uniform carbonization. The OPTs were heated gradually from ambient temperature and held at 600 °C for 6 h. The resulting pyrolytic OPTB was allowed to cool for 24 h before being ground using a grinder. Following this, the ground material was sifted through a 250-mesh sieve and dried in an oven at 105 °C until no further changes in quality were observed.

2.2. Characterization of CB and OPTB Powder

FTIR spectra were obtained using a Vertex70 FTIR spectrometer (Thermo Scientific, Bruker, Ettlingen, Germany) with the KBr pellet technique. XRD analysis was conducted with Cu-Kα radiation, operating at 40 kV and 30 mA, scanning from 10° to 90° (2θ), and equipped with a 0.154 monochromator. Specific surface area was determined using Brunauer–Emmett–Teller (BET) gas porosimetry (ASAP2060, Micromeritics, Norcross, GA, USA) employing the BET N2 method. Particle size and morphology of CB and OPTB powders were analyzed using a particle size analyzer (Beckman Coulter LS 230, Vernon Hills, IL, USA) and SEM (QUANTA400, Thermo Fisher Scientific, Brno, Czech Republic), respectively. Prior to SEM imaging, a thin layer of gold was sputter-coated onto the surfaces of CB and OPTB powders under a vacuum.

2.3. Preparation and Characterizations of NR Composites

The compositions of the compounds, expressed in parts per hundred of rubber (phr), are detailed in Table 1.

Table 1. Formulations used for NR composites with, alternatively, OPTB or CB filler.

Component	Case of OPTB (phr)				
	OPTB0	**OPTB10**	**OPTB30**	**OPTB50**	**OPTB100**
STR20	100	100	100	100	100
ZnO	5	5	5	5	5
SA	1	1	1	1	1
6PPD	1	1	1	1	1
Palm oil	10	10	10	10	10
MBTS	1	1	1	1	1
DPG	0.5	0.5	0.5	0.5	0.5
C	2.5	2.5	2.5	2.5	2.5
OPTB or CB	0	10	30	50	100

NR composites were prepared by blending in a two-roll mill for 30 min. Following sheeting out, the compound was left to rest for 16–24 h before undergoing vulcanization at 160 °C utilizing an electrically heated plate press (Charoen Tut Co., LTD., Bangkok, Thailand) operating at a pressure of 1500 psi for the optimal curing time determined using a moving die rheometer (MDR) (Alpha-Technologies Co., Ltd., Hudson, OH, USA) in accordance with ASTM D5289 standards, resulting in 2 mm thick sheets.

2.4. Characterization of NR Composites

2.4.1. Rheological Property Measurement

The Mooney viscosity (ML(1 + 4), 100 °C) was determined at 100 °C using a large rotor for a duration of 4 min following 1 min of preheating. Results were recorded as ML(1 + 4) and obtained using a MonTech Mooney viscometer (MV2020) in compliance with ASTM D1646 standards. The Payne effect assessment was conducted utilizing a rubber process analyzer (RPA2000, Alpha Technologies, Hudson, OH, USA) operating in the strain sweep mode. This involved evaluating the storage modulus of uncured NR composites at various strain amplitudes. Strain sweep tests were employed to measure the storage shear moduli (G_0) of samples across different shear strain amplitudes, ranging from 0.56 to 1000%, with a fixed oscillation frequency of 1 Hz at a temperature of 100 °C.

2.4.2. Thermal and Dynamic Mechanical Properties

Thermogravimetric analysis (TGA/DSC 3+ model, Mettler Toledo, Greifensee, Switzerland) was conducted using a simultaneous thermal analyzer, employing a heating rate of 10 °C/min within the temperature range from 25 to 1000 °C under a nitrogen atmosphere. Dynamic mechanical thermal analysis was performed using a PerkinElmer DMTA 8000. Samples were subjected to tension mode testing from −100 to 150 °C at a heating rate of 3 °C/min, with fixed frequency and strain set at 10 Hz and 0.1%, respectively.

2.4.3. Dielectric and Mechanical Properties

The dielectric constant (ε') was determined using an impedance analyzer across the frequency range of 10^3–10^6 Hz at an AC potential of 1 V. Durometer (Instron, Norwood, MA, USA) measurements were employed to assess the hardness of 6 mm thick samples in accordance with ASTM D2240 standards. Tear and tensile tests were conducted using a universal tensile testing machine (Tinius Olsen, model 10ST, Salfords, UK) on dumbbell-shaped test specimens prepared according to ASTM D624 and ASTM D412, respectively, at a crosshead speed of 500 mm/min. For each test, five duplicate samples were used to ensure the accuracy and reproducibility of the results.

3. Results and Discussion

3.1. Characterization of CB and OPTB

Pyrolytic chars are the most economically viable substitutes for carbon black in the rubber industry, used as carbon fillers. To assess feasibility, a pyrolytic OPTB was analyzed and compared to a commercial CB, N550. The FTIR analysis was performed in order to ascertain the chemical functional groups present in both CB and OPTB, as shown in Figure 1a. It is evident that the weaker peaks in both cases, located between 2800 and 2975 cm^{-1}, which are attributed to the stretching vibrations of -CH bonds [34], could be the consequence of pyrolysis-induced alkane breakdown resulting in the significantly reduced hydrogen content of the carbons [35]. In addition, the prominent absorption peak at approximately 3400 cm^{-1}, corresponding to the stretching vibrations of -OH of hydroxyl, indicates carboxylic and phenolic functional groups present in both the materials [36]. For the case of CB, the peak at around 1625 cm^{-1} was exclusively seen, and it is attributed to C=C stretching vibrations [37,38]. Meanwhile, Zang et al. (2023) ascribed the observed peaks to the stretching of C=O bonds, indicating the presence of carboxyl, ketenes, lactones [39], and pyrones [40]. In addition, the CH stretching of the aromatic molecule was shown by the weak intensity absorption at 1380 cm^{-1} [38]. The vinyl ether was detected at a wavenumber of 1023 cm^{-1}, corresponding to the stretching vibrations of the C-O bond. The spectrum for OPTB exhibited clearly visible bands at 1580 cm^{-1}, attributed to an aromatic skeleton vibration of the C–C bond [41]. This signifies the aromatic ring of lignin [42,43]. According to Zappielo et al. (2016), a noticeable peak around 1580 cm^{-1} indicates the presence of carbonyl or hydroxyl groups. The absorption bands at 1424 cm^{-1} represent C–H deformation by the bending of alkanes or alkyl groups, while 1258 cm^{-1} is for the C–O–C stretching of ester groups [37]. The peaks detected at 874, 814, and 755 cm^{-1} likely originated from the bending of C–H bonds in aromatic rings [44,45]. The aforementioned results show that, while the FTIR peaks of the two raw materials, OPTB and CB, were slightly different, the chemical structures of CB and OPTB had certain identical functional groups, such as C=C stretching in aromatic and oxygen-containing groups, including carboxylic, hydroxyl, phenolic (hydroxyl), carbonyl, lactone, carboxylic acid anhydride, and cyclic peroxide groups, as reported by [46–48]. This was most likely caused by the varied sources and methods used to prepare OPTB and CB.

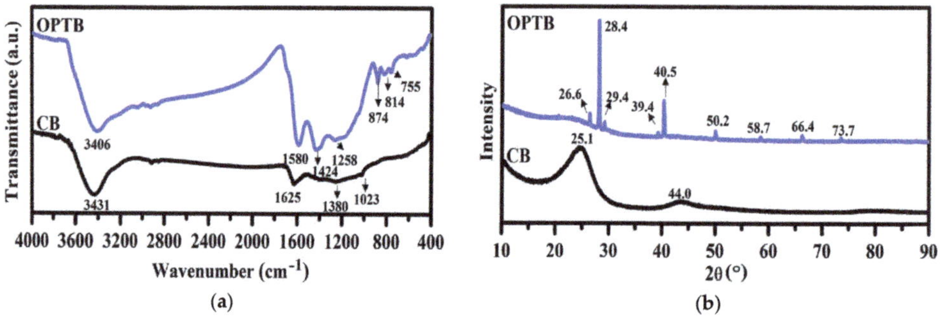

Figure 1. (**a**) FTIR spectra and (**b**) XRD analyses of CB and OPTB.

XRD was used to characterize the CB and OPTB fillers, as shown in Figure 1b. It was found that CB demonstrated typical amorphous carbon in the diffractograms. The carbons show no signs of prominent Bragg peaks caused by mineral impurities. Nevertheless, a broad diffraction peak at 2θ values ranging from 20 to 30° (002) and 40 to 50° (100) indicates the existence of graphitic carbon alongside amorphous carbon [34–36,49,50]. In contrast, the XRD patterns of OPTB exhibited several sharp diffraction peaks reflecting different mineralogical phases, indicating miscellaneous mineral crystals and other inorganic components. It was noted that several peaks overlapped and provided multiple-peak combinations. The major crystalline peaks of the OPTB were for potassium chloride (KCl), silicon oxide (SiO_2), and calcium carbonate ($CaCO_3$). These mineral elements (silicate, potassium salt, and calcium carbonate) made up the majority of the ash that was released as the cellulose and lignin components of OPT were progressively broken down [51].

The BET surface area of the CB that was determined in this investigation (37 m^2/g) was comparable to that which previous studies have found [40]. The BET specific surface area of OPTB was higher than that of CB, as shown in Table 2. Despite this, the CB particles demonstrated smaller diameters, not only in the mean value (D_{50}), but also for 10% and 90% thresholds in sample volume (D_{10} and D_{90}, respectively) when compared to OPTB. Apparently, the abundant inorganic materials and metal elements in OPTB (shown in XRD analysis) generated pores or catalytically accelerated the creation of pores in the carbonaceous material [36].

Table 2. Specific surface areas (from BET) and particle size distribution statistics for CB and OPTB.

Parameter	CB	OPTB
S_{BET} (m^2/g)	37.06	68.09
D_{10} (μm)	0.5	6.4
D_{50} (μm)	2.2	17.9
D_{90} (μm)	7.2	35.7

Figure 2 provides SEM images of CB and OPTB, demonstrating significant differences between the samples. It was found that CB showed different shapes of carbon black aggregates, often spheroidal or ellipsoidal with a rough surface. OPTB had larger sized particles with pores on their surfaces. The images distinctly depict the porous structure of OPTB, with pores evident on the surface and displaying discernible sizes and shapes of both cylindrical and polygonal cross-sections. The formation of these porous channels resulted from the gradual breakdown of the lignocellulosic constituents, cellulose, hemicelluloses, and lignin. The residual non-volatile components were transformed into biochar [52]. It is the porosity of the larger particles in OPTB that explains its larger specific surface.

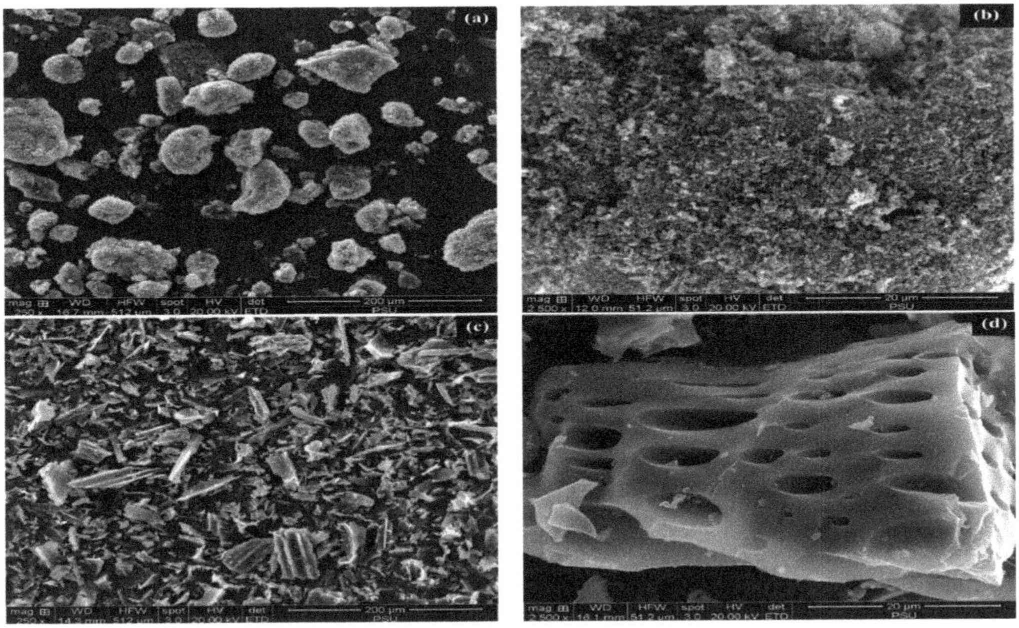

Figure 2. SEM images of CB and OPTB showing the morphologies of (**a**) CB at ×250 magnification, (**b**) CB at ×2500 magnification, (**c**) OPTB at ×250 magnification, and (**d**) OPTB at ×2500 magnification.

3.2. *Thermal Stability Study*

The influence of OPTB content on NR composites is shown in Figure 3. The thermal behavior of all specimens showed mutual resemblance, characterized by a primary decomposition peak. It can be noted that the TGA curves gradually declined until reaching a stage where the NR composites started to decompose. Thermal degradation of pure NR occurred between 195 and 430 °C, with T_{max} around 373 °C. It was also observed that the T_{max} slightly shifted to higher temperatures with increasing OPTB content, as listed in Table 3. This suggests that adding OPTB to NR composites increased their thermal stability. Pure NR produced the least final residue at 600 °C among the composites examined in the study. The ultimate residual amount rose with increasing OPTB content, attributable to the buildup of a greater amount of char capable of enduring high temperatures. Moreover, OPTB exhibits a slower decomposition rate in comparison to NR. Consequently, with the rise in OPTB concentration, a greater residual weight was noted [33].

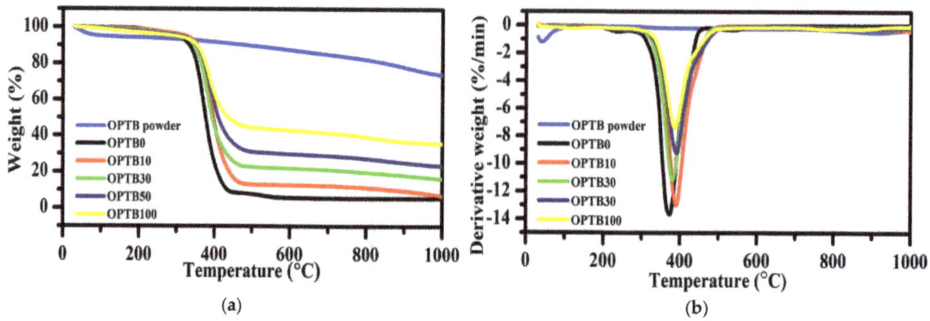

Figure 3. Thermal properties of OPTB powder and NR composites: (**a**) TGA curves and (**b**) DTG curves.

Table 3. Thermal properties and dynamic mechanical properties of NR composites according to DMA and TGA.

Case	DMA			TGA	
	tan δ (at 0 °C)	tan δ (at 60 °C)	T_g (°C)	T_{max} (°C)	Residue (%)
OPTB0	0.047	0.000	−50.20	373	5.46
OPTB10	0.076	0.004	−47.15	390	6.85
OPTB30	0.074	0.013	−44.75	381	16.31
OPTB50	0.082	0.023	−44.30	391	23.41
OPTB100	0.108	0.049	−44.50	386	35.70

3.3. Dielectric Properties

Figure 4 depicts the relative dielectric constant for NR composites with variable OPTB contents, at various frequencies and environmental conditions. It can be noted that as frequency increased, the dielectric constant declined across all NR composites. This is because the polarization of the specimen is unable to synchronize with a high-frequency electric field [53], as well as a gradual reduction in dipole movement or alteration in orientation. Pure NR exhibited the lowest dielectric constant (1.50 at 1 kHz) due to its hydrophobic (nonpolar) structure. The addition of OPTB raised the dielectric constant of NR composites over the whole frequency range. For example, 100 phr OPTB-NR composite demonstrated the highest dielectric constant (3.80 at 1 kHz), for an increase by 153%, suggesting that OPTB is more prone to polarization by an electric field than NR. The increased dielectric constant can be due to the significant interfacial polarization in OPTB/NR composites. Interfacial polarization is generated by charge accumulation at the phase interfaces of the filler (here biochar) and rubber matrix [54]. In addition, the presence of carbonaceous elements within biochar could be pivotal in fostering significant polarization within the material. The FTIR results demonstrated π electrons found in carbon materials, which have the freedom to move across wide areas, presenting a particularly intriguing phenomenon [55]. Moreover, the dielectric constant increased with OPTB loading due to the mineral phases identified by XRD. As OPTB content rises, the mineral concentration in the matrix increases, resulting in more charge carriers for polarization and a higher dielectric constant.

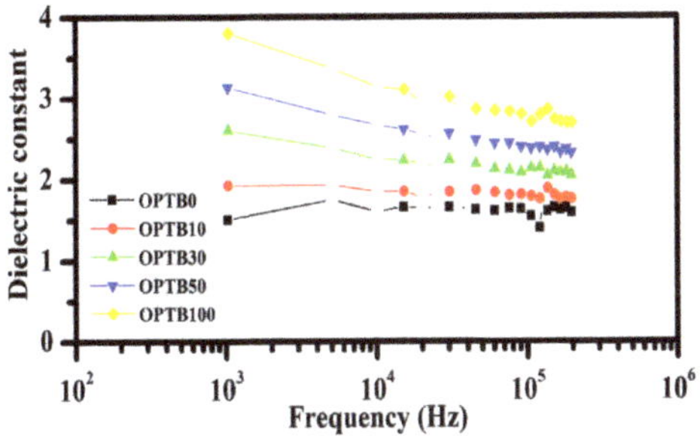

Figure 4. Dielectric constants of NR composites by frequency of excitation.

3.4. Dynamic Behaviors

It is commonly known that elastic modulus and stored energy are both quantified by the storage modulus. The temperature dependence of storage modulus is reported in Figure 5a. The results show that when temperature increased, the storage modulus generally decreased, because chain mobility increased and physical bonding and chain entanglements decreased [56]. The NR composite with 100 phr OPTB had the highest storage modulus in the glassy region. When OPTB was introduced to NR composites, the storage modulus rose in comparison to pure NR. This is because the NR chain motion is restricted by the high-stiffness OPTB filler.

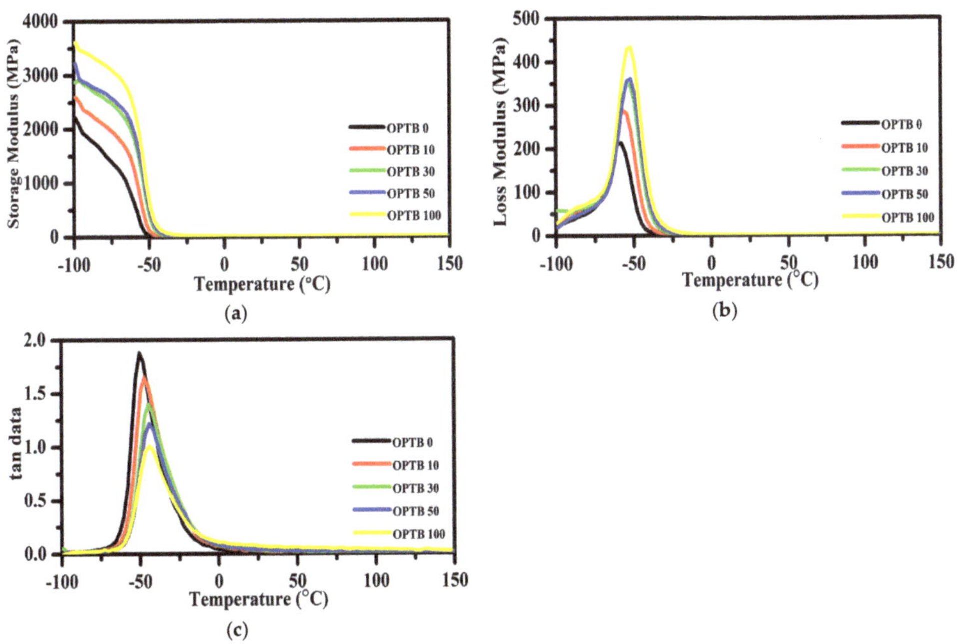

Figure 5. (**a**) Storage modulus, (**b**) loss modulus, and (**c**) tan δ for the experimental NR composites.

The loss modulus (E″) represents the viscous response and damping as a result of energy dissipation as heat. The addition of OPTB to the NR matrix broadened the loss modulus peak, as seen in Figure 5b. This could be explained by the fact that the addition of filler results in a larger number of chain segments inside the composites, which inhibits relaxation [57].

Figure 5c depicts the temperature-dependent evolution of the damping factor, also known as tan δ (loss to storage modulus ratio). The OPTB in NR composites resulted in a much lower height of the tan δ peak, indicating a lower lost-to-stored energy ratio in the T_g zone. The decrease in tan δ_{max} may be attributed to the dense and entangled OPTB filler network, which significantly hinders chain mobility during glass transition. It has been documented that natural rubber is among the polymer materials exhibiting the highest damping factor, attributed to the inherent movement of the polyisoprene backbone. The presence of additional materials in its matrix tends to limit damping capacity [56].

The transition to the maximum of a tan δ peak is closely related to the glass transition temperature T_g. The tan δ peak parameters are summarized in Table 3. The glass transition temperatures of NR composites were impacted by the addition of OPTB. When OPTB was added to NR composites, a shift in T_g to a higher temperature was noted. It resulted from filler–rubber interactions limiting the mobility of the rubber chains. The rolling resistance and wet traction properties of NR are extremely significant for car tire applications. It is common practice to use the dynamic mechanical loss angle tan δ at 60 °C of vulcanized compounds as a predictor of rolling resistance. As a result, a low tan δ at 60 °C is of importance, since it indicates that the vulcanized compound is performing better. When rolling resistance is reduced, less energy is needed to maintain the speed of the vehicle, which conserves fuel for the internal combustion engine and lowers the emissions. Results show that the rolling resistance of NR composites tended to decrease with OPTB content. The energy loss of the rubber vulcanizate, or tan δ at 0 °C, is used to assess the good wet-grip qualities. A high tan δ at 0 °C indicates good wet-grip. The highest tan δ at 0 °C was recorded for 100 phr OPTB filled NR composite, while tan δ at 0 °C was lower or the wet-grip characteristics were poorer for pure NR.

It is well known that the addition of fillers to a rubbery matrix significantly alters its viscoelastic behavior. The Payne effect characterizes alterations in the characteristics of an elastomeric component in relation to the applied strain. It can be explained in terms of a filler–filler network in the rubber matrix. The variation in storage modulus of NR filled with varying loadings of OPTB as a function of dynamic strain amplitude and a summary of findings are demonstrated in Figure 6 and Table 4. It was found that the storage modulus decreased with strain, which is attributed to the filler–filler network breaking down at larger dynamic strains. In addition, the Payne effect increased with OPTB content. The maximum storage modulus and Payne effect magnitude were observed for NR composite filled at 100 phr with OPTB, indicating the strongest effects of a filler–filler network in the elastomer matrix. This is, however, an excessive filler content causing a loss of key material properties.

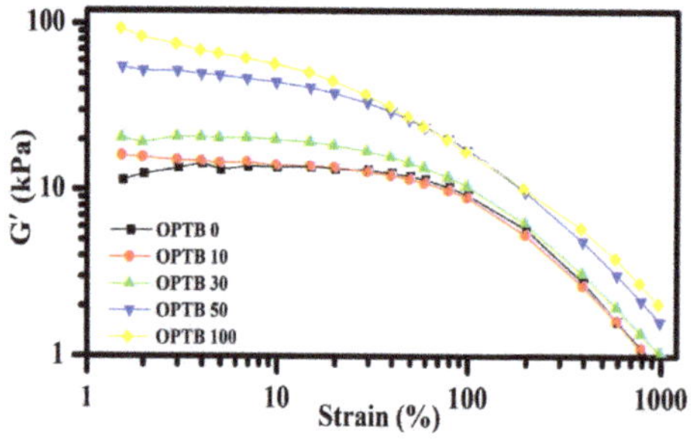

Figure 6. Storage modulus vs. strain for the NR composites.

Table 4. Payne effect for the NR composites.

Case	$G'_{1.56}$	G'_{100}	G'_{1000}	$\Delta G' = G'_{1.56} - G'_{100}$	$\Delta G' = G'_{1.56} - G'_{1000}$
OPTB0	11.49	9.31	0.79	2.18	10.7
OPTB10	16.27	9.13	0.83	7.14	15.44
OPTB30	20.94	10.71	1.06	10.23	19.88
OPTB50	55.51	17.65	1.64	37.86	53.87
OPTB100	94.32	17.25	2.11	77.07	92.21

3.5. Rheological Properties

The rheological responses of NR compounds, acquired at 160 °C using an MDR rheometer, are displayed in Figure 7, and the summarizing data for scorch time (t_{s1}), cure time (t_{c90}), cure rate index (CRI), minimum torque (M_L), maximum torque (M_H), and delta torque (M_H–M_L) are given in Table 5. t_{s1} for pure NR and NR composites were similar, with the exception of the 100 phr OPTB case, which had a 50% decrease in t_{s1} compared to pure NR. On the other hand, t_{c90} displayed a marginally rising trend with OPTB content. This may be because the surface activity of OPTB, which includes oxygen-containing compounds such as hydroxyl functional groups (see the previously discussed FTIR analysis of OPTB), retards t_{c90}. The M_L and M_H denote viscosity and rigidity, as well as the stiffness of the rubber compound, respectively. It was found that the characteristics for the vulcanization of all prepared rubber compounds had the same trend of increasing M_L and M_H with filler content, because the filler was limiting the molecular motions of NR chains. These results are supported by the Mooney viscosity of unvulcanized rubber, often influenced by the macrostructure and flow properties of the rubber composite. It revealed that a rise in Mooney viscosity in unvulcanized rubber composites was associated with an increase in OPTB content, attributed to the hydrodynamic impact of the filler within the rubber. The crosslink density from both chemical and physical crosslinks can be assessed from delta torque (M_H–M_L). Pure NR exhibited the least M_H–M_L, whereas OPTB-filled NR composites displayed a rising trend with OPTB loading. This could be attributed to crosslinking between filler and NR matrix phases.

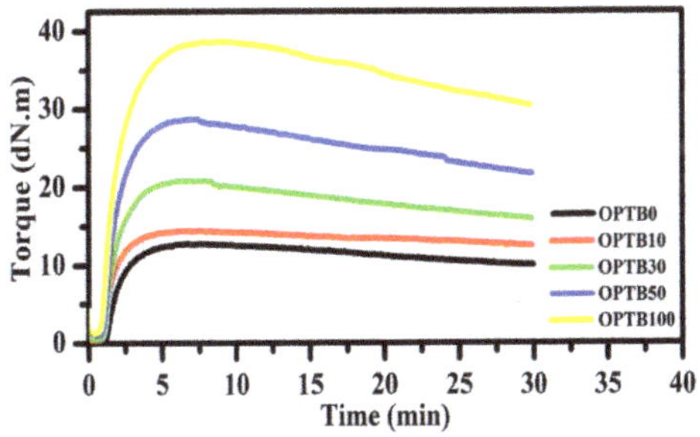

Figure 7. Time profiles of torque for the NR compounds.

Table 5. Cure characteristics and Mooney viscosities of the NR compounds.

Case	Mooney Viscosity (MU)	M_L (Ib/In)	M_H (Ib/In)	M_H–M_L (Ib-In)	t_{s1} (min)	t_{c90} (min)	CRI (min^{-1})
OPTB0	8.80	0.38	12.78	12.40	1.17	3.47	27.65
OPTB10	9.30	0.25	14.51	14.26	0.57	3.21	30.58
OPTB30	14.10	0.48	20.88	20.40	1.07	3.41	28.26
OPTB50	24.50	1.06	28.71	27.65	1.04	3.45	27.95
OPTB100	33.10	1.18	38.76	37.58	0.58	4.19	23.29

3.6. Mechanical Properties

The incorporation of OPTB resulted in increases in hardness (Figure 8a), modulus at 100% elongation (M100), and modulus at 300% elongation (M300) (Figure 8d) in the OPTB filled NR composites. This can be ascribed to the notably high hardness and modulus of OPTB, along with the strong interfacial adhesion between OPTB filler and NR matrix. Consequently, OPTB particles could proficiently resist the load, enhancing the indentation resistance of the composites [58]. The tear strength exhibited an increasing trend with OPTB loading up to 70 phr (Figure 8b). This enhancement can be attributed to improved interfacial adhesion between the OPTB filler and NR matrix. The adhesion mechanism involves hydrogen bonding between the carboxylic (-COOH) and phenolic (-OH) groups present on the OPTB surface and the proteins and phospholipids found in the non-rubber components of NR, as evidenced by the FTIR analysis discussed previously. However, the tear strength shows a significant decrease when OPTB loading exceeds 70 phr, reaching its lowest value at 100 phr. Obviously, the tensile strength decreased gradually with OPTB loading, from pure NR onwards (Figure 8e). Nevertheless, incorporating CB enhanced tensile strength, attributed to the interactions between CB filler and rubber matrix, alongside its heightened activity and smaller particle size compared to OPBT [4]. The sizable particles of OPTB induced localized stresses within the composites, which undermined the strength of the composite, in particular reducing the tensile strength [59]. Furthermore, the greater particle size of OPTB may prevent the strain-induced crystallization of NR, also reducing tensile strength. This is the opposite effect to that of adding biochar to synthetic rubber [4,22,60,61]. There are studies seeking to improve the tensile strength of biochar/polymer (or rubber) composites by reducing the particle size of biochar [62,63]. When OPTB was increased to 100 phr, there was a significant reduction in tensile strength and tear strength. This indicates an excessive filler content that caused the filler to aggregate, hindering appropriate filler–matrix interactions and forming stress-concentration points initiating fracture failures. A substantial incorporation of OPTB markedly reduced the elongation at break of the NR composites (Figure 8f), signifying diminished ductility and toughness. Typically, as filler content increases, ductility and toughness decrease due to the filler particles serving as hindrances, impeding the slippage and movement of rubber chains, and consequently reducing ductility [58].

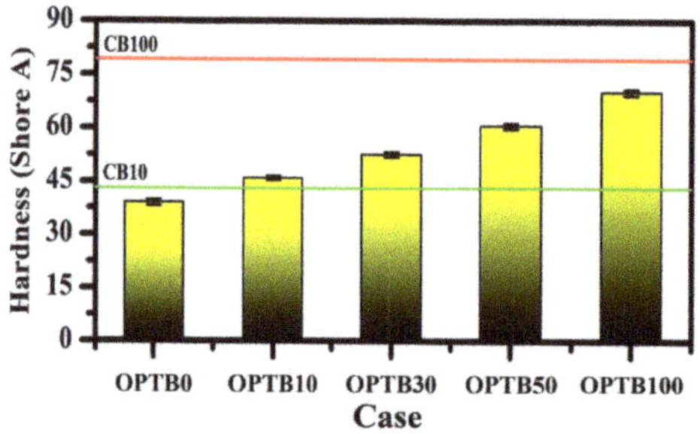

(a)

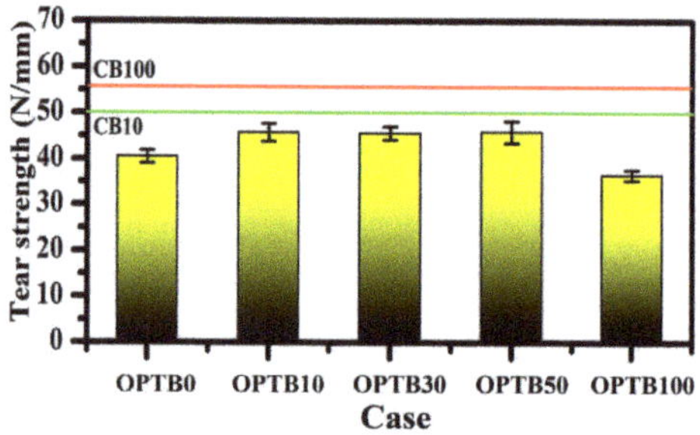

(b)

Figure 8. *Cont.*

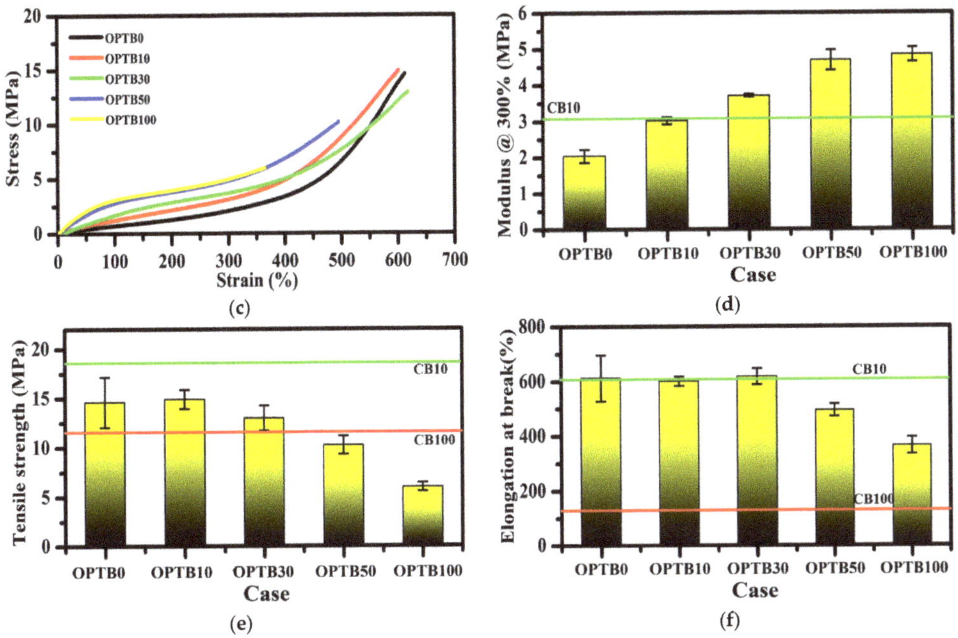

Figure 8. Physical and mechanical properties of NR composites: (**a**) hardness, (**b**) tear strength, (**c**) stress–strain curves, (**d**) modulus @300%, (**e**) tensile strength, and (**f**) elongation at break.

4. Conclusions

This study investigated the feasibility of using pyrolytic OPTB as a replacement for commercial CB in the rubber industry. FTIR analysis revealed the presence of similar functional groups in both CB and OPTB, including C=C and oxygen-containing groups. However, XRD analysis showed that CB possessed an amorphous structure, while OPTB contained various crystalline mineral phases. The addition of OPTB significantly enhanced key composite characteristics. The glass transition temperature (T_g) increased by approximately 5.7 °C, indicating reduced polymer chain mobility and greater rigidity due to the filler's reinforcing effect. Thermal stability improved by 13 °C, reflecting enhanced heat resistance, while the dielectric constant rose by about 153%, demonstrating improved electrical energy storage capabilities. However, the mechanical properties were not uniformly improved. Hardness, modulus, and tear strength increased with OPTB contents up to a certain point, while tensile strength and elongation at break decreased. This suggests a trade-off between the various mechanical properties when using OPTB as a filler. Overall, the results suggest that OPTB has the potential to partly replace CB in rubber composites. Nonetheless, challenges remain, including optimizing filler content and particle size to achieve a better balance of properties. Additionally, the large-scale adoption of OPTB faces hurdles such as feedstock variability, high production costs, and resistance from industries reliant on carbon black. Addressing these issues through precise process control and sustainable practices is essential for establishing OPTB as a competitive, eco-friendly filler.

Author Contributions: Conceptualization, N.C., M.T. and W.C.; methodology, N.C. and W.P.; validation, N.C., W.P. and W.C.; formal analysis, N.C.; investigation, W.P.; resources, M.T.; data curation, N.C. and W.C.; writing—original draft preparation, N.C. and W.C.; writing—review and editing, W.C.; visualization, W.P.; supervision, N.C.; project administration, N.C. and M.T.; funding acquisition, M.T. All authors have read and agreed to the published version of the manuscript.

Funding: This research was funded by the Research and Development Office (RDO), Prince of Songkla University, Grant No. SIT6601265S and Prince of Songkla University, Surat Thani Campus, 2024.

Institutional Review Board Statement: Not applicable.

Data Availability Statement: The original contributions presented in this study are included in the article. Further inquiries can be directed to the corresponding author.

Acknowledgments: We are grateful to Seppo Karrila for his assistance with manuscript preparation.

Conflicts of Interest: The authors declare no conflicts of interest.

References

1. Sintharm, P.; Phisalaphong, M. Green natural rubber composites reinforced with black/white rice husk ashes: Effects of reinforcing agent on film's mechanical and dielectric properties. *Polymers* **2021**, *13*, 882. [CrossRef] [PubMed]
2. Vaysse, L.; Bonfils, F.; Thaler, P.; Sainte-Beuve, J. Natural rubber. In *Sustainable Solutions for Modern Economies*; Höfer, R., Ed.; RSC Publishing: Cambridge, UK, 2009; pp. 339–367. [CrossRef]
3. Fan, Y.; Fowler, G.D.; Zhao, M. The past, present and future of carbon black as a rubber reinforcing filler—A review. *J. Clean. Prod.* **2020**, *247*, 119115. [CrossRef]
4. Lubura, J.; Kojić, P.; Ikonić, B.; Pavličević, J.; Govedarica, D.; Bera, O. Influence of biochar and carbon black on natural rubber mixture properties. *Polym. Int.* **2022**, *71*, 1347–1353. [CrossRef]
5. Pechurai, W.; Chiangta, W.; Tharuen, P. Effect of vegetable oils as processing aids in SBR compounds. *Macromol. Symp.* **2015**, *354*, 191–196. [CrossRef]
6. Hayichelaeh, C.; Nun-Anan, P.; Purbaya, M.; Boonkerd, K. Unfilled natural rubber compounds containing bio-oil cured with different curing systems: A comparative study. *Polymers* **2022**, *14*, 2479. [CrossRef]
7. Fu, Q.; Tan, J.; Wang, F.; Zhu, X. Study on the synthesis of castor oil-based plasticizer and the properties of plasticized nitrile rubber. *Polymers* **2020**, *12*, 2584. [CrossRef]
8. Raju, P.; Nandanan, V.; Kutty, S.K.N. A study on the use of castor oil as plasticizer in natural rubber compounds. *Prog. Rubber Plast. Recycl. Technol.* **2007**, *23*, 169–180. [CrossRef]
9. Jayewardhana, W.G.D.; Perera, G.M.; Edirisinghe, D.G.; Karunanayake, L. Study on natural oils as alternative processing aids and activators in carbon black filled natural rubber. *J. Natl. Sci. Found. Sri Lanka* **2009**, *37*, 187–193. [CrossRef]
10. Petrović, Z.S.; Ionescu, M.; Milić, J.; Halladay, J.R. Soybean oil plasticizers as replacement of petroleum oil in rubber. *Rubber Chem. Technol.* **2013**, *86*, 233–249. [CrossRef]
11. Wang, Z.; Han, Y.; Zhang, X.; Huang, Z.; Zhang, L. Plasticization effect of transgenic soybean oil. I. on ethylene propylene diene monomer (EPDM), as substitute for paraffin oil. *J. Appl. Polym. Sci.* **2013**, *130*, 4457–4463. [CrossRef]
12. Boontawee, H.; Nakason, C.; Kaesaman, A.; Thitithammawong, A.; Chewchanwuttiwong, S. Benzyl esters of vegetable oils as processing oil in carbon black-filled SBR compounding: Chemical modification, characterization, and performance. *Adv. Polym. Technol.* **2015**, *36*, 320–330. [CrossRef]
13. Li, J.; Isayev, A.I.; Ren, X.; Soucek, M.D. Toward replacement of petroleum oils by modified soybean oils in elastomers. *Rubber Chem. Technol.* **2016**, *89*, 608–630. [CrossRef]
14. Petrović, Z.S.; Milić, J.; Ionescu, M.; Halladay, J.R. EPDM rubber plasticized with polymeric soybean oil of different molecular weights. *Rubber Chem. Technol.* **2017**, *90*, 667–682. [CrossRef]
15. Siriwong, C.; Khansawai, P.; Boonchiangma, S.; Sirisinha, C.; Sae-Oui, P. The influence of modified soybean oil as processing aids in tire application. *Polym. Bull.* **2020**, *78*, 3589–3606. [CrossRef]
16. Wang, Z.; Peng, Y.; Zhang, L.; Zhao, Y.; Vyzhimov, R.; Tan, T.; Fong, H. Investigation of palm oil as green plasticizer on the processing and mechanical properties of ethylene propylene diene monomer rubber. *Ind. Eng. Chem. Res.* **2016**, *55*, 2784–2789. [CrossRef]
17. Siwarote, B.; Sae-oui, P.; Wirasate, S.; Suchiva, K. Effects of bio-based oils on processing properties and cure characteristics of silica-filled natural rubber compounds. *J. Rubber Res.* **2017**, *20*, 1–19. [CrossRef]
18. Hanif, H.M.; Yong, K.C.; Lee, S.Y. Effects of a newly developed palm-based processing aid and curing systems on the physical properties of nitrile rubber composites. *Sains Malays.* **2020**, *49*, 2187–2196. [CrossRef]
19. Roy, K.; Poompiew, N.; Pongwisuthiruchte, A.; Potiyaraj, P. Application of different vegetable oils as processing aids in industrial rubber composites: A sustainable approach. *ACS Omega* **2021**, *6*, 31384–31389. [CrossRef]
20. Aprianti, N.; Kismanto, A.; Supriatna, N.K.; Yarsono, S.; Nainggolan, L.M.T.; Purawiardi, R.I.; Fariza, O.; Ermada, F.J.; Zuldian, P.; Raksodewanto, A.A.; et al. Prospect and challenges of producing carbon black from oil palm biomass: A review. *Bioresour. Technol. Rep.* **2023**, *23*, 101587. [CrossRef]

21. Ribeiro, W.B.; Damo, T.; Zattera, A.J.; Brandalise, R.N. Green compounds of SBR with nanofibrilated cellulose and palm oil in replacement to traditionally compounds used. *J. Elastomers Plast.* **2022**, *54*, 635–655. [CrossRef]
22. Abidin, Z.Z.; Mamauod, S.N.L.; Romli, A.Z.; Sarkawi, S.S.; Zainal, N.H. Synergistic effect of partial replacement of carbon black by palm kernel shell biochar in carboxylated nitrile butadiene rubber composites. *Polymers* **2023**, *15*, 943. [CrossRef]
23. Farida, E.; Bukit, N.; Ginting, E.M.; Bukit, B.F. The effect of carbon black composition in natural rubber compound. *Case Stud. Therm. Eng.* **2019**, *16*, 100566. [CrossRef]
24. Dominic, M.; Joseph, R.; Begum, P.M.S.; Kanoth, B.P.; Chandra, J.; Thomas, S. Green tire technology: Effect of rice husk derived nanocellulose (RHNC) in replacing carbon black (CB) in natural rubber (NR) compounding. *Carbohydr. Polym.* **2020**, *230*, 115620. [CrossRef]
25. Panwar, N.L.; Pawar, A.; Salvi, B.L. Comprehensive review on production and utilization of biochar. *SN Appl. Sci.* **2019**, *1*, 168. [CrossRef]
26. Muda, N.A.; Yoshida, H.; Ishak, H.; Ismail, M.H.S.; Izhar, S. Production and properties of solid biochar from oil palm trunk waste using sub-critical water technology. *J. Oil Palm Res.* **2020**, *32*, 526–536. [CrossRef]
27. Razali, N.; Kamarulzaman, N.Z. Chemical characterizations of biochar from palm oil trunk for palm oil mill effluent (POME) treatment. *Mater. Today Proc.* **2020**, *31*, 191–197. [CrossRef]
28. Pulingam, T.; Lakshmanan, M.; Chuah, J.A.; Surendran, A.; Zainab-L, I.; Foroozandeh, P.; Uke, A.; Kosugi, A.; Sudesh, K. Oil palm trunk waste: Environmental impacts and management strategies. *Ind. Crops Prod.* **2022**, *189*, 115827. [CrossRef]
29. Nudri, N.A.; Ghani, W.A.W.A.K.; Bachmann, R.T.; Baharudin, B.T.H.T.; Ng, D.K.S.; Said, M.S.M. Co-combustion of oil palm trunk biocoal/sub-bituminous coal fuel blends. *Energy Conv. Manag X* **2021**, *10*, 100072. [CrossRef]
30. Abbas, K.; Ghazali, A.M.M.; Ong, S.K. The effect of particle size of palm kernel shell on the mechanical properties and physical properties of filled natural rubber vulcanizates. *Mater. Today Proc.* **2019**, *19*, 1599–1607. [CrossRef]
31. Malomo, D.; Olasupo, A.D.; Adesigbin, A.M.; Egharevba, O.; Adewuyi, S.O.; Odubunmi, J.O.; Idemudia, L.; Abimbade, S.F.; Ogunyemi, E.K. Comparative studies on the physico-mechanical properties of natural rubber filled with CB/CPKS and CB/APKS vulcanizates. *JCBPS* **2020**, *10*, 211–226. [CrossRef]
32. Arraudhoh, M.T.N.; Haziq, M.F.M.; Zafirah, Z.A.; Liyana, M.S.N.; Hayawin, Z.N. Comparative studies on cure characteristics and mechanical properties of oil palm biomass filled natural rubber composites. *J. Oil Palm Res.* **2022**, *36*, 128–139. [CrossRef]
33. Ramli, R.; Chai, A.B.; Kamaruddin, S.; Ho, J.H.; Rasdi, F.R.M.; De Focatiis, D.S.A.; Ong, S.K.; Bachmann, R.T. Effects of oil palm trunk biochar on the thermal stability and acoustic properties of specialty natural rubber latex foam. *J. Rubber. Res.* **2023**, *26*, 1–15. [CrossRef]
34. Paul, S.; Rahaman, M.; Ghosh, S.K.; Kateria, A.; Das, T.K.; Patel, S.; Das, N.C. Recycling of waste tire by pyrolysis to recover carbon black: An alternative reinforcing filler. *J. Mater. Cycles Waste Manag.* **2023**, *25*, 1470–1481. [CrossRef]
35. Borah, D.; Satokawa, S.; Kato, S.; Kojima, T. Characterization of chemically modified carbon black for sorption application. *Appl. Surf. Sci.* **2008**, *254*, 3049–3056. [CrossRef]
36. Wang, M.; Zhang, L.; Li, A.; Irfan, M.; Du, Y.; Di, W. Comparative pyrolysis behaviors of tire tread and side wall from waste tire and characterization of the resulting chars. J. *Environ. Manag.* **2019**, *232*, 364–371. [CrossRef]
37. Zappielo, C.D.; Nanicuacua, D.M.; dos Santos, W.N.L.; da Silva, D.L.F.; Dall'Antônia, L.H.; de Oliveira, F.M.; Clausen, D.N.; Tarley, C.R.T. Solid phase extraction to on-line preconcentrate trace cadmium using chemically modified nano-carbon black with 3-mercaptopropyltrimethoxysilane. *J. Braz. Chem. Soc.* **2016**, *27*, 1715–1726. [CrossRef]
38. Sugatri, R.I.; Wirasadewa, Y.C.; Saputro, K.E.; Muslih, E.Y.; Ikono, R.; Nasir, M. Recycled carbon black from waste of tire industry: Thermal study. *Microsyst. Technol.* **2018**, *24*, 749–755. [CrossRef]
39. Zang, G.; Jiang, Y.; Wang, S.; Zhang, Y. Influence of a novel coupling agent on the performance of recovered carbon black filled natural rubber. *Compos. B Eng.* **2023**, *255*, 110614. [CrossRef]
40. Yepes, W.U.; Uribe, N.C.; Isaza, C.A.V.; Martínez, J.D. Incorporating the recovered carbon black produced in an industrial-scale waste tire pyrolysis plant into a natural rubber formulation. *J. Environ. Manag.* **2021**, *287*, 112292. [CrossRef]
41. Nagalakshmi, T.V.; Emmanuel, K.A.; Babu, C.S.; Chakrapani, C.; Divakar, P.P. Preparation of mesoporous activated carbon from jackfruit PPI-1 waste and development of different surface functional groups. *Int. Lett. Chem. Phys. Astron.* **2015**, *54*, 189–200. [CrossRef]
42. Som, A.M.; Wang, Z.; Al-Tabbaa, A. Palm frond biochar production and characterization. *Earth Environ. Sci. Trans. R. Soc. Edinb.* **2012**, *103*, 39–50. [CrossRef]
43. Razali, N.; Kamarulzaman, N.Z. Chemical characterizations of biochar from oil palm frond for palm oil mill secondary effluent treatment. *Malays. J. Anal. Sci.* **2022**, *26*, 370–383.
44. Barhtiar, M.H.A.M.; Sari, N.B.A.; Yaacob, A.B.; Yunus, M.F.B.M.; Ismail, K.B. Characterization of oil palm empty fruit bunch (EFB) biochar activated with potassium hydroxide under different pyrolysis temperature. *JESTEC* **2019**, *14*, 2792–2807.
45. Hassan, N.; Abdullah, R.; Khadiran, T.; Elham, P.; Vejan, P. Biochar derived from oil palm trunk as a potential precursor in the production of high-performance activated carbon. *Biomass Convers. Biorefin.* **2023**, *13*, 15687–15703. [CrossRef]

46. Boehm, H.P. Some aspects of the surface chemistry of carbon blacks and other carbons. *Carbon* **1994**, *32*, 759–769. [CrossRef]
47. Leblanc, J.L. Rubber–filler interactions and rheological properties in filled compounds. *Prog. Polym. Sci.* **2002**, *27*, 627–687. [CrossRef]
48. Sajjadi, B.; Chen, W.Y.; Egiebor, N.O. A comprehensive review on physical activation of biochar for energy and environmental applications. *Rev. Chem. Eng.* **2019**, *35*, 735–776. [CrossRef]
49. Saleh, T.A.; Danmaliki, G.I. Adsorptive desulfurization of dibenzothiophene from fuels by rubber tyres-derived carbons: Kinetics and isotherms evaluation. *Process Saf. Environ. Prot.* **2016**, *102*, 9–19. [CrossRef]
50. Liang, Y.; Wang, Y.; Ding, N.; Liang, L.; Zhao, S.; Yin, D.; Cheng, Y.; Wang, C.; Wang, L. Preparation and hydrogen storage performance of poplar sawdust biochar with high specific surface area. *Ind. Crops Prod.* **2023**, *200*, 116788. [CrossRef]
51. Wang, T.; Liu, H.; Duan, C.; Xu, R.; Zhang, Z.; She, D.; Zheng, J. The eco-friendly biochar and valuable bio-oil from caragana korshinskii: Pyrolysis preparation, characterization, and adsorption applications. *Materials* **2020**, *13*, 3391. [CrossRef]
52. Sianipar, E.M.; Hutapea, S.; Panjaitan, E.; Mardiana, S. Applicability assessment of oil palm trunk biochar for use as soil amendment: Morphology, structure, and chemical properties. *Sci. Technol. Asia* **2024**, *29*, 256–270.
53. Ni, Y.; Yang, D.; Wei, Q.; Yu, L.; Ai, J.; Zhang, L. Plasticizer-induced enhanced electromechanical performance of natural rubber dielectric elastomer composites. *Compos. Sci. Technol.* **2020**, *195*, 10820. [CrossRef]
54. Qian, M.; Zou, B.; Shi, Y.; Zhang, Y.; Wang, X.; Huang, W.; Zhu, Y. Enhanced mechanical and dielectric properties of natural rubber using sustainable natural hybrid filler. *Appl. Surf. Sci. Adv.* **2021**, *6*, 10017. [CrossRef]
55. Salema, A.A.; Yeow, Y.K.; Ishaque, K.; Ani, F.N.; Afzal, M.T.; Hassan, A. Dielectric properties and microwave heating of oil palm biomass and biochar. *Ind. Crops Prod.* **2013**, *50*, 366–374. [CrossRef]
56. Oboh, J.O.; Okafor, J.O.; Kovo, A.S.; Abdulrahman, A.S. Dynamic mechanical properties of crosslinked natural rubber composites reinforced with cellulosic nanoparticles. *Niger. J. Technol.* **2018**, *37*, 668–673. [CrossRef]
57. Jyoti, J.; Arya, A.K. EMI shielding and dynamic mechanical analysis of graphene oxide-carbon nanotube-acrylonitrile butadiene styrene hybrid composites. *Polym. Test.* **2020**, *91*, 106839. [CrossRef]
58. Li, S.; Xu, Y.; Jing, X.; Yilmaz, G.; Li, D.; Turng, L.S. Effect of carbonization temperature on mechanical properties and biocompatibility of biochar/ultra-high molecular weight polyethylene composites. *Compos. B Eng.* **2020**, *196*, 10812. [CrossRef]
59. Peterson, S.C. Coppiced biochars as partial replacement of carbon black filler in polybutadiene/natural rubber composites. *J. Compos. Sci.* **2020**, *4*, 147. [CrossRef]
60. Jiang, C.; Bo, J.; Xiao, X.; Zhang, S.; Wang, Z.; Wang, Z.; Yan, G.; Wu, Y.; Wong, C.; He, H. Converting waste lignin into nano-biochar as a renewable substitute of carbon black for reinforcing styrene-butadiene rubber. *Waste Manag.* **2020**, *102*, 732–742. [CrossRef]
61. Silva, V.D.; Ferrari, M.D.; Brandalise, R.N.; Benvenutti, E.V.; Schrekker, H.S.; Amico, S.C. EPDM with biochar, carbon black, aramid pulp and ionic liquid-compatibilized aramid pulp. *Fibers Polym.* **2021**, *22*, 1180–1188. [CrossRef]
62. Richard, S.; Rajadurai, J.S.; Manikandan, V. Influence of particle size and particle loading on mechanical and dielectric properties of biochar particulate-reinforced polymer nanocomposites. *Int. J. Polym. Anal. Charact.* **2016**, *21*, 462–477. [CrossRef]
63. Peterson, S.; Kim, S. Reducing biochar particle size with nanosilica and its effect on rubber composite reinforcement. *J. Polym. Environ.* **2020**, *28*, 317–322. [CrossRef]

polymers

MDPI

Article

Enhancing Mechanical and Antibacterial Performance of Tire Waste/Epoxidized Natural Rubber Blends Using Modified Zinc Oxide–Silica

Napasorn Kingkohyao [1], Tanit Boonsiri [2], Jobish Johns [3], Raymond Lee Nip [4] and Yeampon Nakaramontri [1,*]

1 Sustainable Polymer & Innovative Composite Materials Research Group, Department of Chemistry, Faculty of Science, King Mongkut's University of Technology Thonburi, Bangkok 10140, Thailand; napasorn.king@mail.kmutt.ac.th
2 Department of Microbiology, Phramongkutklao College of Medicine, Bangkok 10140, Thailand; boonsiri-t@hotmail.com
3 Department of Physics, Rajarajeswari College of Engineering, Bangalore 560074, India; jobish_johns@rediffmail.com
4 Global Chemical Co., Ltd., Bangpoo Industrial Estate, Muang 10280, Samutprakarn, Thailand; rlnip@glochem.com
* Correspondence: yeampon.nak@kmutt.ac.th

Abstract

This study investigates the synergistic effects of incorporating modified zinc oxide–silica (ZnO-SiO_2) into tire waste (TW) and epoxidized natural rubber (ENR) blends, with a focus on crosslinking dynamics, mechanical reinforcement, and antibacterial activity. The addition of ZnO-SiO_2 significantly enhanced crosslink density, as evidenced by increased torque and accelerated cure rates. An optimal concentration of 10 phr was found to yield the highest performance. This optimal balance between chemical activation and mechanical reinforcement resulted in exceptional tensile properties, including notable improvements in Young's modulus, tensile strength, and strain-induced crystallization (SIC). These enhancements were attributed to the strong interactions between ENR molecular chains and SiO_2 surfaces. However, excessive ZnO-SiO_2 concentrations caused filler agglomeration, which reduced both mechanical and antibacterial performances. An antibacterial analysis revealed a remarkable 99.9% bacterial reduction at 10 phr ZnO-SiO_2, attributed to the Zn^{2+} ion release and reactive oxygen species (ROS) generation, with sustained activity even after thermal aging. This durability underscores the composites' potential for long-term applications. The findings establish ZnO-SiO_2 as a dual-functional filler that optimizes crosslinking, enhances mechanical properties, and provides durable antibacterial efficiency. These results highlight the potential of TW/ENR blends while offering critical insights into mitigating filler agglomeration to improve overall material performance.

Keywords: tire waste; zinc oxide; antibacterial efficiency; epoxidized natural rubber; silica

1. Introduction

The global demand for sustainable materials has intensified the focus on recycling and upcycling industrial waste, particularly tire waste (TW), which is a major contributor to environmental pollution. Each year, millions of tires are discarded, posing significant ecological and disposal challenges. However, TW, which is rich in rubber, carbon black, and other additives, holds immense potential as a resource for advanced composite materials [1,2]. When blended with polymers such as natural rubber (NR), synthetic rubbers, and

 https://doi.org/10.3390/polym17010109

modified NR, specifically, epoxidized natural rubber (ENR) [3], TW can be transformed into value-added materials. These composites benefit from the synergistic effects of the active chemicals in TW and the addition of new ingredients, improving the rate of crosslinking, mechanical and dynamic mechanical properties, and thermal stability [4,5]. Due to the unknown ingredients in TW, particularly the mix of hydrophobic and hydrophilic fillers along with various rubber types, epoxidized natural rubber (ENR), derived from renewable NR, offers unique properties, such as enhanced polarity, oil resistance, and compatibility with polar fillers. Additionally, the hydrocarbon chains in ENR can interact physically with the non-polar regions of TW through induced dipole–induced dipole intermolecular forces [6,7]. These attributes make ENR an excellent matrix for enhancing the utility of TW in composite formulations. However, the limited reinforcing ability of residual fillers in TW, such as carbon black (CB), poses a significant challenge. These fillers often exhibit poor dispersion and weak interactions with fresh ENR molecular chains [8,9] due to the presence of embedded rubber molecules from tire preparation. This limitation underscores the need for innovative strategies to enhance filler performance and optimize the properties of TW-based composites.

Recent studies have highlighted the potential of hybrid fillers, such as modified zinc oxide combined with particulate fillers like calcium carbonate (ZnO-$CaCO_3$), titanium dioxide (ZnO-TiO_2), and silica (ZnO-SiO_2), in addressing these challenges. Previously, ZnO-$CaCO_3$ and ZnO-TiO_2 were used in NR latexes for film and foam applications [10–13]. These hybrids demonstrated the synergistic effects of ZnO, a vulcanization accelerator with the reinforcing and interfacial adhesion properties of $CaCO_3$ and TiO_2 [14,15], widely used fillers in rubber materials. Modified ZnO hybrids have been shown to enhance crosslink density, mechanical strength, and functional properties, including the antibacterial activity of NR films and foams. The dual functionality of ZnO-$CaCO_3$ and ZnO-TiO_2 also allows them to act as reactive fillers, improving the interactions between rubber chains and promoting the absorption of volatile organic compounds (VOCs) [16]. However, excessive filler concentrations can lead to agglomeration, which diminishes both mechanical performance and antibacterial efficacy. In addition to mechanical reinforcement, antibacterial functionality has become increasingly important for advanced composite materials, especially in healthcare and hygiene applications. ZnO is well documented for its antibacterial properties, primarily attributed to the release of zinc ions (Zn^{2+}) and the generation of reactive oxygen species (ROS) [17,18]. A key challenge lies in maintaining antibacterial efficiency during prolonged use and under thermal aging conditions, which is critical for ensuring the reliability of these composites in real-world applications. However, the incorporation of ZnO-$CaCO_3$ and ZnO-TiO_2 into rubber composites using melt blending technology has not been previously examined, particularly regarding the elucidation of the ZnO effect in relation to residual chemicals in TW, such as ZnO, fillers, accelerators, and curing agents. As discussed, the compatibilization of TW with rubbers—especially polar rubbers like ENR—is promising and warrants further investigation, particularly with the use of modified ZnO combined with polar fillers like SiO_2. The ability of SiO_2 to form hydrogen bonds with ENR molecular chains enhances the dispersion and distribution of ZnO within the bulk matrix, thereby improving the overall performance of the composite [19].

This study aims to bridge the gap in understanding the multifunctional role of ZnO-SiO_2 in TW/ENR composites. By systematically evaluating the effects of ZnO-SiO_2 loading on chemical crosslinking behavior, tensile performance, and antibacterial activity, the study identifies the optimal ZnO-SiO_2 concentration to achieve a balance between mechanical and bioactive properties. Heat-aging tests further assess the long-term durability of these blends, focusing on the retained activity of ZnO within the TW. The insights gained from this research not only advance the design of high-performance, sustainable materials, but also address critical challenges in recycling TW for high-value applications. The findings hold broad implications for industries seeking to develop eco-friendly, multifunctional materials that combine superior mechanical properties with durable antibacterial functionality.

2. Materials and Methods

2.1. Materials

Epoxidized natural rubber containing 25 mol% epoxide (ENR25) was obtained from Muang Mai Guthrie Co., Ltd., located in Surat Thani, Thailand. Waste tire rubber (TW) was supplied by Thai Rubb Tech Co., Ltd., based in Bangkok, Thailand, specifically from truck tire waste. The TW was derived from the tire tread of a truck tire after undergoing reclamation processes, which involved the random cleavage of crosslinked rubber chains, both along the rubber main chains and at the -C-S- bonds. With a viscosity of ML (1+4@100 °C) of 52.17, TW may contain an unknown rubber matrix, along with activators such as ZnO and stearic acid, as well as fillers including carbon black (CB), silica (SiO_2), and/or calcium carbonate ($CaCO_3$). Silicon dioxide-modified zinc oxide nanoparticles (ZnO-SiO_2), with a ZnO:SiO_2 weight ratio of 90:10 through wet processes, were collaborated and received from Global Chemical Co., Ltd., located in Samut Prakan, Thailand. The ZnO-SiO_2 nanoparticles possess a surface area of approximately 100 m^2/g, with the aggregate sizes ranging from 100 to 300 nm and individual particle sizes below 10 nm. Stearic acid and sulfur were procured from Imperial Chemical Co., Ltd., in Pathum Thani, Thailand, while 2,2′-dithiobis(benzothiazole) (MBTS) was supplied by Flexsys Inc., located in Termoli, Italy.

2.2. Preparation of TW/ENR Blends Filled with Modified ZnO

The compounding process involved masticating the TW/ENR with modified ZnO in an internal mixer (MX500, Charoen Tut Co., Ltd., Samutprakarn, Thailand) using various ratios of ZnO-SiO_2 concentrations at 0–20 phr. The mixing conditions were standardized with a temperature of 80 °C, a fill factor of 75%, and a rotor speed of 60 rpm. Initially, the ENR was masticated until its viscosity matched that of TW, after which TW was introduced into the mixer. Subsequently, additional chemicals listed in Table 1, including the ZnO-SiO_2 with the different concentrations, were incorporated. The total mixing process lasted 8 min, followed by the rolling of the mixture through a two-roll mill (Charoen Tut Co., Ltd., Samutprakarn, Thailand) to enhance dispersion and distribution. Rheometer testing was conducted to determine the production time before the blends were shaped via compression molding at 160 °C. It is important to note that the present formulation was specifically designed to elucidate the synergistic effects between the residual ZnO in TW and the additional ZnO-SiO_2 in the blends. Consequently, only the key potential chemicals, namely, the co-activator, accelerator, and sulfur curing agent, were included in the formulation.

Table 1. Formulation compositions of TW/ENR blends filled with modified ZnO-SiO_2.

Ingredients	Content (phr *)
TW/ENR	50/50
Stearic acid	1
ZnO-SiO_2	0, 10, 15, and 20
MBTS	1
Sulfur	2.5

* phr refers to part per hundred rubber.

2.3. Cure Characterization

The crosslink propagation of the TW/ENR compounds, along with the details of their chemical crosslinking, was analyzed using a moving die rheometer (MDR3000M, Mon-Tech, Buchen, Germany) operated at 160 °C for 30 min, with a fixed oscillation frequency of 1.66 Hz and an amplitude of 0.5° arc. The key parameters, including scorch time (T_{s1}), cure time (T_{90}), minimum torque (M_L), maximum torque (M_H), and the torque difference (M_H–M_L), were measured to evaluate the compounds' curing characteristics and crosslinking efficiency.

2.4. Tensile Properties

The tensile properties of TW/ENR blends filled with different concentrations of ZnO-SiO_2 were evaluated by conducting tensile tests on specimens cut from crosslinked sheets. These sheets were prepared through compression molding into dumbbell-shaped specimens in accordance with ISO 37 Type 2 standards [ISO 37: Rubber, vulcanized or thermoplastic—Determination of tensile stress-strain properties]. The tests were performed using a universal testing machine (Zwick Z 1545, Zwick GmbH & Co. KG, Ulm, Germany) at a crosshead speed of 500 mm/min under room temperature conditions. The tensile strength, elongation at break, and moduli at 100% and 300% elongation were measured and analyzed to assess the mechanical performance of the blends.

2.5. Morphologies

The morphologies and elemental compositions of the fractured surfaces of TW/ENR blends were analyzed using scanning electron microscopy (SEM) coupled with energy-dispersive X-ray spectroscopy (EDS). High-resolution images were obtained with an FE-SEM (Nova NanoSEM 450, EDS X Flash 6 series, FEI, Australia) at an accelerating voltage of 10 kV. To prepare the cross-sectional surfaces for analysis, the specimens were cryogenically fractured in liquid nitrogen. The fractured surfaces were then coated with a thin layer of gold to prevent electrical charging during the measurements, ensuring accurate imaging and compositional analysis. In addition, in order to characterize the surface roughness of a wider area, the optical microscope (OM) (Carl Zeiss Microscope GmbH, Oberkochen, Germany) was applied. The blends were quickly cut with a razor blade to create a smooth surface before capturing microscopic photographs. Also, a transmission electron microscope (TEM) (Model JEOL, JEM-1400, Tokyo, Japan), operated at an accelerating voltage of 80 kV, was also used to provide a clearer visualization of the particles. The powder specimens were dispersed in ethanol and sonicated for 30 min, then deposited onto carbon-coated copper grids. The solvent was allowed to evaporate completely before analysis.

2.6. Antibacterial Efficiency

The antimicrobial efficiency of the tire waste (TW)/epoxidized natural rubber (ENR) blends was evaluated by quantitatively measuring their ability to inhibit the growth of *Staphylococcus aureus* (*S. aureus*, ATCC 25923) and *Escherichia coli* (*E. coli*, ATCC 25922). Bacterial suspensions were prepared in liquid culture media (Nutrient Broth, NB) with an initial concentration of approximately 10^6 colony-forming units per milliliter (CFU/mL). A volume of 0.4 mL of each bacterial suspension was placed onto sterilized 5×5 cm^2 samples of the TW/ENR blends, which were kept in a Petri dish. To maintain contact between the bacterial suspension and the surface of the samples while preventing drying, the droplet was covered with a sterilized 4×4 cm^2 polyethylene (PE) film. The samples were then incubated at 37 °C for 24 h. After incubation, both the sample and the PE film were washed with 10 mL of soybean-casein digest medium (SCDLP). Following the washing step, the bacterial suspensions underwent a 10-fold dilution and were plated onto a nutrient agar

(NA) medium. After incubation at 37 °C for 24 h, the number of viable bacteria was counted. The percentage reduction in bacterial count (%R) was calculated according to Equation (1):

$$\%R = [(A - B)/A] \times 100 \quad (1)$$

where A is the number of viable bacteria in the treated samples after incubation and B is the number of viable bacteria in the control samples after incubation.

3. Results and Discussion

3.1. Cure Characteristics

The cure characteristics of the compounds and vulcanizate composites after exposure to a constant temperature for various durations is illustrated in Figure 1, which depicts the changes in torque as a function of time. When considering the over-cure performance of the composites, reversion curves appear with the addition and increasing concentration of ZnO-SiO_2. However, the maximum torque tends to increase, which can be attributed to two factors: (i) The higher ZnO content enhances the crosslink density by increasing activator levels in the crosslinking system. (ii) The greater SiO_2 loading within the bulk TW/ENR matrix improves the reinforcement efficiency, particularly in the ENR phase. Nonetheless, as the degree of polysulfide crosslinking intensifies, thermo-oxidative degradation sets in, leading to a decline in torque after the optimal crosslinking levels are reached [20]. Examining the cure rate slope reveals that a ZnO-SiO_2 loading of 10 phr results in the steepest slope, indicating the fastest crosslinking rate due to a balanced chemical loading that fulfills multiple roles in activation and reinforcement. However, at concentrations exceeding 10 phr, SiO_2 agglomeration becomes evident, impacting the overall performance. In addition, the thermal stability of the blends filled with varying concentrations of ZnO-SiO_2 was evaluated by monitoring the torque as a function of time at a fixed temperature of 160 °C, revealing significant differences. It was observed that the pure TW exhibited a constant torque over time, presenting a characteristic plateau curve. However, when blended with ENR, reversion curves appeared, showing a delayed decrease in torque as the ZnO-SiO_2 concentration increased. This reversion in ENR is attributed to thermo-oxidative degradation, resulting from the high reactivity of ENR's epoxide groups with oxygen. Additionally, the increase in polysulfidic bonds leads to the decomposition of sulfur linkages between TW and ENR molecules. As shown in Figure 1, the initiation of reversion was slightly delayed with increasing ZnO-SiO_2 concentrations. This indicates that the enhanced crosslink density, driven by the higher ZnO-SiO_2 content, slows down the degradation of rubber chains.

Figure 1 also illustrates the crosslinking curves, while Table 2 summarizes the minimum torque (M_L), maximum torque (M_H), and the torque difference (M_H–M_L), which corresponds to the estimated crosslink density of the TW/ENR composites. The table additionally provides the scorch time (T_{s1}) and cure time (T_{90}) of the compounds, highlighting variations in the production time of the composites. The observed increase in M_H with rising ZnO-SiO_2 concentrations reflects a higher crosslink density, as crosslinking restricts chain mobility and forms chemical bonds between TW/ENR molecules, while also enables the absorption of rubber chains onto the surfaces of ZnO and SiO_2 particles [21]. Furthermore, Figure 2 shows the morphologies of ZnO-SiO_2 in comparison to the active ZnO and commercial ZnO white seal. Active ZnO particles are significantly smaller than commercial ZnO. For ZnO-SiO_2, the chemical absorption of ZnO onto SiO_2 surfaces results in improved ZnO dispersion due to controlled interparticle distances. The rough surface of SiO_2 enhances interactions with rubber molecular chains, particularly ENR, forming a bound rubber layer that contributes to reinforcement. However, at higher SiO_2 concentrations within the bulk TW/ENR matrix, the polar functional groups on the SiO_2 surface promote filler–filler interactions, leading to agglomeration and a subsequent reduction in reinforcement efficiency [22].

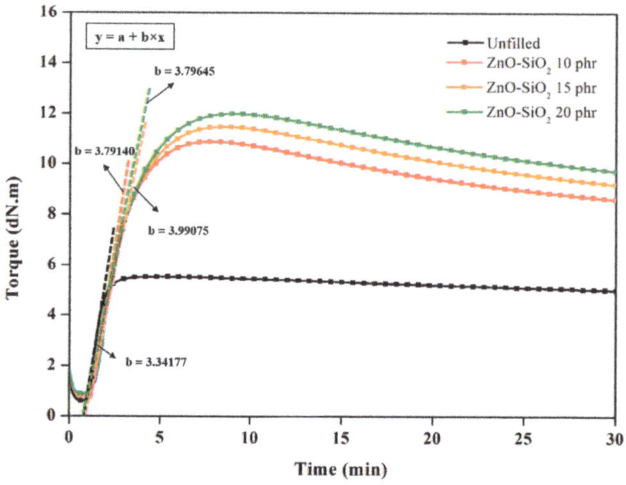

Figure 1. Cure characteristic curves of TW/ENR blend with different modified $ZnO-SiO_2$ loading. "Unfilled" refers to the TW:ENR ratio of 50:50 phr without $ZnO-SiO_2$.

Table 2. Crosslink properties of TW/ENR blends from compounds to vulcanizates with varying concentrations of the $ZnO-SiO_2$ activator.

Formulations	T_{s1} (min)	T_{90} (min)	M_L (d.Nm)	M_H (d.Nm)	M_H–M_L (d.Nm)
Unfilled	1.37	2.18	0.60	5.59	4.98
$ZnO-SiO_2$ 10 phr	1.72	4.65	0.77	10.86	10.09
$ZnO-SiO_2$ 15 phr	1.77	5.00	0.82	11.48	10.60
$ZnO-SiO_2$ 20 phr	1.81	5.28	0.89	12.00	11.11

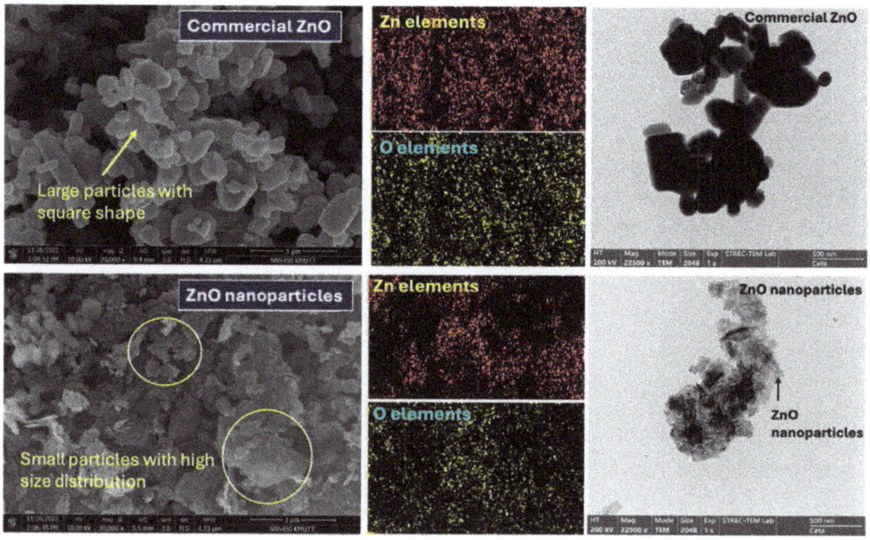

Figure 2. *Cont.*

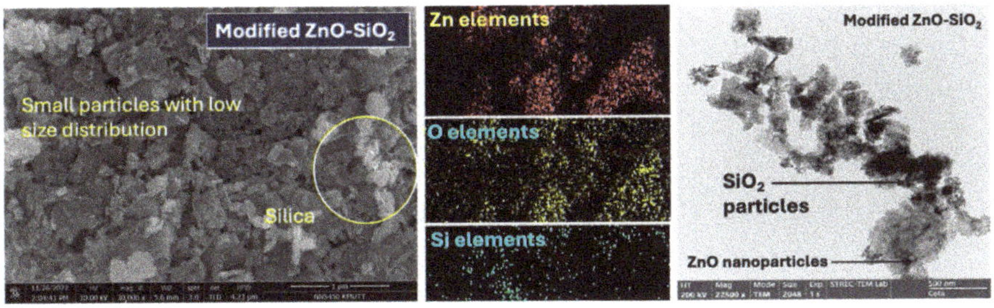

Figure 2. Morphologies of the solely $ZnO\text{-}SiO_2$, relative to the commercial ZnO.

3.2. Tensile Properties

Figure 3 shows the stress–strain plots, performed according to mechanical measurement standards. The formation of crosslinks that provide resistance to breakage, and three distinct regions of the tensile curves, are interpreted as follows:

(i) **Reinforcement region:** Without $ZnO\text{-}SiO_2$, the blends exhibit poor vulcanization, and their strength arises mainly from chain entanglement contributed by additional ENR. Young's modulus and the 100% modulus reflect these effects, with both increasing as $ZnO\text{-}SiO_2$ concentration rises. This behavior confirms an increase in chemical crosslink density, which enhances the blends' resistance to chain breakage. Notably, the modulus of blends without $ZnO\text{-}SiO_2$ is very low, indicating that the residual carbon black (CB) in TW does not provide significant reinforcement due to the insufficient diffusion of ENR molecules onto the particle surfaces. Conversely, the addition of $ZnO\text{-}SiO_2$ facilitates the activation of chemical crosslinking between the remaining TW and un-crosslinked ENR chains. The SiO_2 particles function as a reinforcing filler, interacting with rubber molecules to prevent blend breakage under applied mechanical deformation [23]. However, at concentrations exceeding 10 phr, no significant improvement is observed due to SiO_2 particle agglomeration.

(ii) **Rubber chain relaxation:** This region corresponds to the physical interactions between rubber chains and filler surfaces. An increased slope in this region indicates stronger rubber–rubber and rubber–filler interactions, resulting in greater friction among chains during extension. The slope of this region increases with $ZnO\text{-}SiO_2$ concentrations in the range of 10–20 phr, which can be attributed to the enhanced crosslink density and improved $ENR\text{-}SiO_2$ interactions.

(iii) **Strain-induced crystallization (SIC) and rubber failure:** SIC occurs as molecular chains align into a crystalline structure during elongation, typically noticeable at approximately 400% strain. Figure 3 shows that the increased crosslink density from ZnO activation enhances the degree of crystallization. Furthermore, hydrogen bonding between ENR molecules and SiO_2 particles accelerates the alignment of molecular chains along the direction of extension, as depicted in the proposed interaction mechanism in Figure 4 (Zone 3) [23,24]. However, differences in SIC improvement are reflected in the tensile strength and elongation-at-break data presented in Table 3. This difference highlights the need to consider the entire TW/ENR matrix, as shown in Figure 4 (Zones 1 and 2). In the TW proportion of the blends, filler particles with rubber molecular chain absorption are present. Without pyrolysis, rubber chains remain embedded on the filler surfaces as glassy or sticky bound rubber, with only the outer chains interacting with the SiO_2 surface and ENR molecules [25]. These interactions occur via induced dipole–dipole forces between the SiO_2 surface and polar groups on ENR molecules (Zone 2), as well as induced dipole–induced dipole interactions on the non-polar regions of ENR (Zone 1) [26]. However, at excessive

$ZnO-SiO_2$ concentrations, SiO_2 agglomeration inhibits interactions between the TW and ENR phases, leading to reduced tensile strength and limited elongation at break. It is important to note that ENR chains can also interact with ZnO particles due to the polarity of ZnO, particularly at higher concentrations. However, this interaction primarily occurs during the mixing and compounding stages. During vulcanization, the applied heat facilitates the release of Zn^{2+} ions and ROS from the ZnO surfaces, shifting ZnO's role to that of an activator for chemical crosslinking. Additionally, the residual ZnO in the rubber matrix continues to release ions, which predominantly contributes to properties such as antibacterial activity. This indicates that the reinforcement efficiency of the blends is primarily associated with the SiO_2 particles.

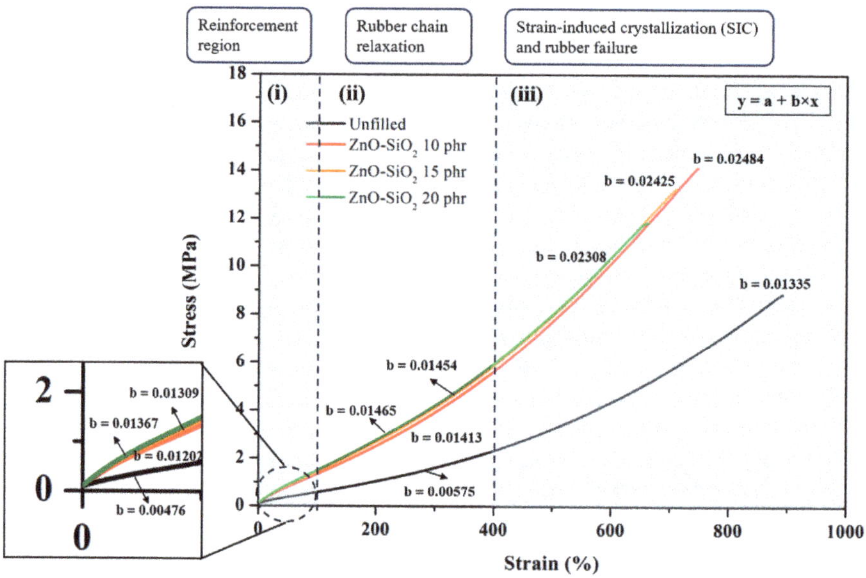

Figure 3. Stress–strain curves of TW/ENR filled with different concentrations of $ZnO-SiO_2$ to elucidate crosslinking behavior.

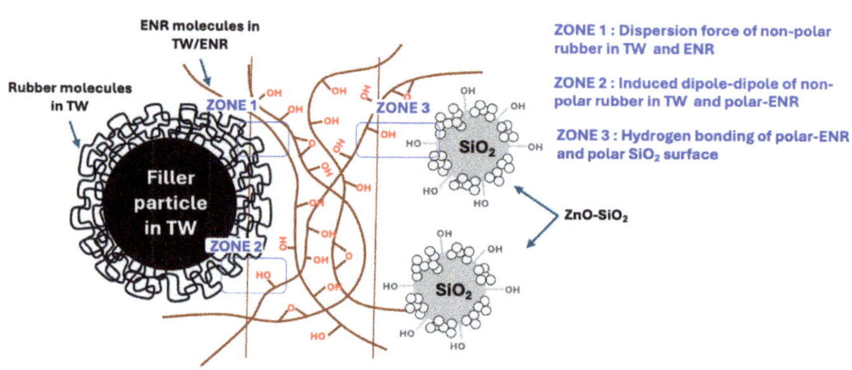

Figure 4. Proposed interaction of ENR with SiO_2 in $ZnO-SiO_2$, as well as the TW phases separation.

Table 3. Mechanical properties of TW/ENR blend with different modified $ZnO\text{-}SiO_2$ loading.

Formulations	100% Modulus (MPa)	300% Modulus (MPa)	Tensile Strength (MPa)	Elongation at Break (%)
Unfilled	0.60 ± 0.02	1.64 ± 0.04	8.79 ± 0.15	879.72 ± 22.09
$ZnO\text{-}SiO_2$ 10 phr	1.35 ± 0.02	3.91 ± 0.05	14.24 ± 0.16	749.77 ± 22.86
$ZnO\text{-}SiO_2$ 15 phr	1.45 ± 0.01	4.16 ± 0.03	13.59 ± 0.19	715.51 ± 25.22
$ZnO\text{-}SiO_2$ 20 phr	1.50 ± 0.04	4.25 ± 0.03	12.34 ± 0.38	668.86 ± 37.80

3.3. Antibacterial Properties and Morphologies

The antibacterial efficiency of the TW/ENR blends is attributed to the release of zinc ions (Zn^{2+}) onto the blend surfaces, facilitated by the thickness of the bound rubber layer surrounding the ZnO embedded in TW and the additional $ZnO\text{-}SiO_2$ at varying concentrations. This occurs because the excess ZnO in both TW and ENR continues to release Zn^{2+} and reactive oxygen species (ROS) onto the rubber surfaces. The ROS react with bacterial cell walls, while Zn^{2+} penetrates the bacterial cells, interacting with their DNA and cytoplasm [27–29]. This mechanism, previously described in the literature, highlights the antibacterial activity of ZnO. Based on the crosslinking curves in Figure 1, it is evident that ZnO remains active even though the sulfur has been fully consumed during tire manufacturing. This demonstrates that the antibacterial properties of the blend arise from the positive synergy between residual ZnO in TW and the newly introduced $ZnO\text{-}SiO_2$.

To investigate the ability of Zn^{2+} and ROS to migrate through the rubber layer covering the ZnO surfaces, a swelling test was conducted to measure the total bound rubber (β) and bound rubber layer thickness (δ') in the blends with increasing $ZnO\text{-}SiO_2$ concentrations. Blend compounds without accelerators or sulfur were pressed at 160 °C for 5 min, and approximately 2 g of the sample powders were placed in specific sieves and immersed in toluene at room temperature for 7 days. The weight of the blends was recorded, and the β value was calculated using Equation (2), as provided below:

$$\beta\ (\%) = (W_{fg} - W_f)/W_p \quad (2)$$

where W_f is the approximated weight of the filler in TW received from the remaining ash of the thermogravimetric analysis (TGA) and additional $ZnO\text{-}SiO_2$ concentrations. The factor W_p is the weight of the rubber, estimated from the TGA and ENR existing in the blends, and W_{fg} is the weight of the remaining filler with absorbed rubber after toluene extraction. Also, the δ' was evaluated through Equation (3) as follows:

$$\delta'\ (nm) = (m_2 - m_1(C_f))/\rho_r m_1(C_f)(S_f) \quad (3)$$

where m_1 is the mass of the rubber compound before extraction, and m_2 is the mass of the rubber–filler gel consisting of the non-dissolving bound rubber part and the filler. Parameters C_f and S_f are the mass concentration and the specific surface area of fillers in the blends regarding carbon black and SiO_2, respectively, whereas ρ_r refers to the density of solely ENR matrix.

The results of the β and δ' values are summarized in Table 4. It was found that the TW/ENR blend exhibited approximately 20% bound rubber, correlating with the rubber absorption on the filler surfaces in TW and indicating a δ' value of 14.12%. This is attributed to the absence of pyrolysis reactions during the reclamation processes. On the other hand, the addition of $ZnO\text{-}SiO_2$ to the blends increased the β value while reducing the δ' value. This behavior is due to the increase in SiO_2 filler content, which enhances rubber molecule absorption on the rough surface of the filler [30]. Consequently, the δ' value tends to decrease significantly. With an increasing $ZnO\text{-}SiO_2$ content, the highest β value was observed at 10 phr, accompanied by the lowest δ' value. This is likely due to the onset of ZnO-SiO_2 agglomeration within the TW/ENR matrix. This phenomenon highlights the enhanced opportunity for ion movement through the blend's surfaces, enabling further reactions with bacterial cell walls.

Table 4. Results of bound rubber and bound rubber layer thickness of the TW/ENR blends for elucidating the Zn^{2+} and ROS ions diffusion.

Formulations	β (%)	δ' (nm)
Unfilled	20.43 ± 0.13	14.12 ± 0.11
ZnO-SiO_2 10 phr	30.44 ± 1.05	2.99± 0.01
ZnO-SiO_2 15 phr	29.78 ± 1.12	4.34 ± 0.13
ZnO-SiO_2 20 phr	27.93 ± 1.78	4.95 ± 0.02

The antibacterial efficiency was initially evaluated using qualitative disk diffusion and quantitative cell counting methods under controlled conditions for 8 and 24 h. As shown in Figure 5, the qualitative test revealed no distinct inhibition zone. This suggests that the antibacterial activity, attributed to Zn^{2+} ions and reactive oxygen species (ROS), is likely confined to the surface of the blend, even at ZnO-SiO_2 concentrations of up to 20 phr, and does not extend to the surrounding areas. This behavior contrasts with the broader diffusion observed with silver (Ag) and copper (Cu) nanoparticles [31–33]. Thus, the use of ZnO-SiO_2 to release Zn^{2+} ions and ROS does not lead to the contamination of adjacent surfaces.

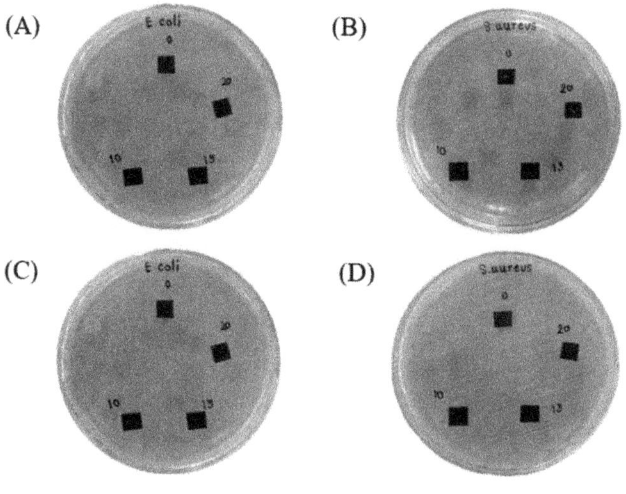

Figure 5. Disk diffusion qualitative measurement of TW/ENR filled with different concentrations of ZnO-SiO_2 against *Gram-negative E. coli* and *Gram-positive S. aureus* was performed at 8 h (**A**,**B**) and 24 h (**C**,**D**).

To evaluate the durability of the antibacterial performance after prolonged use, the blends were subjected to thermal aging at 70 °C for 72 h in a hot air oven, followed by quantitative antibacterial testing. As shown in Figure 6, the antibacterial efficiency remained largely unchanged, except for the TW/ENR blend without ZnO-SiO_2. This result indicates that the release of Zn^{2+} ions and ROS under photocatalytic conditions is limited in blends containing only ZnO. In contrast, blends with at least 10 phr of ZnO-SiO_2 maintained their antibacterial activity even under these harsh conditions. Figure 7 illustrates the surface morphologies and elemental composition analyzed using OM and SEM-EDX. As shown in Figure 7A, the TW/ENR blend without ZnO-SiO_2 exhibits a smooth surface without agglomeration, as ZnO-SiO_2 is absent from the bulk matrix. However, in the TW/ENR matrix with ZnO alone, the ZnO particles were dispersed unevenly, potentially limiting the release of Zn^{2+} ions and ROS to the surface.

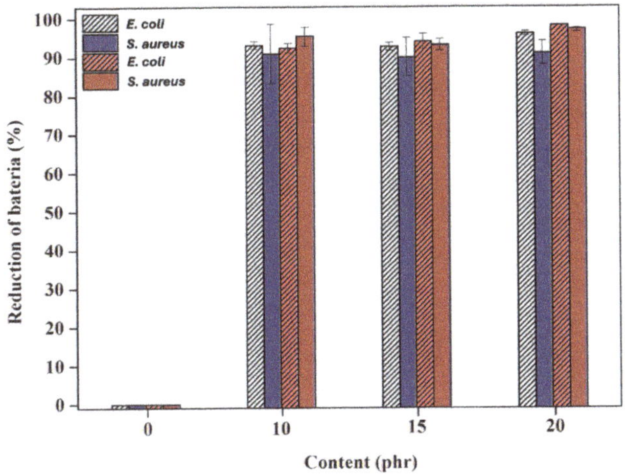

Figure 6. Antibacterial activities of the TW/ENR filled with different concentrations of ZnO-SiO_2 against *E. coli* and *S. aureus* within 8 h before and after aging at 70 °C for 72 h.

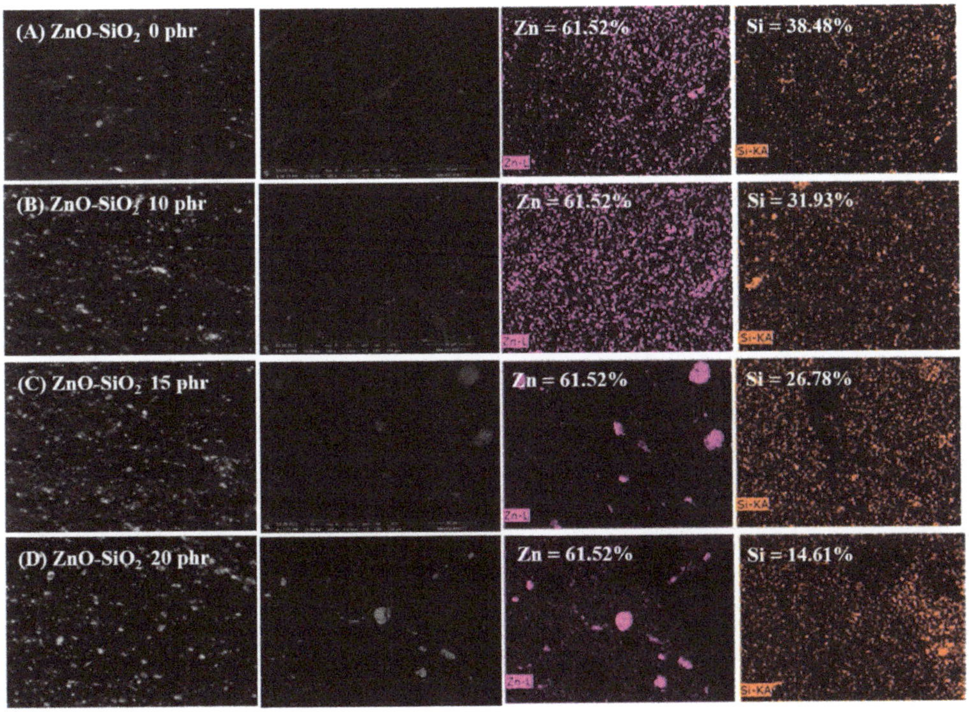

Figure 7. OM and SEM images, along with EDX elemental analysis, of TW/ENR blends filled with different concentrations of ZnO-SiO_2 after aging for 72 h at 70 °C: blend without ZnO-SiO_2 (**A**), blends with ZnO-SiO_2 at concentrations of 10 phr (**B**), 15 phr (**C**), and 20 phr (**D**) together with atom percentage % within TW/ENR matrix.

In blends containing 10 phr of ZnO-SiO_2, the optimal dispersion and distribution of ZnO throughout the TW/ENR matrix was achieved, as seen in Figure 7B. On the other hand, strong agglomeration was observed in blends with 15 or 20 phr of ZnO-SiO_2, indicating that the excessive SiO_2 content from ZnO modification led to particle clustering (Figure 7C,D) although a high concentration of Zn elements was found at the blend's surfaces. This agglomeration not only reduced antibacterial efficiency but also negatively affected the tensile properties, including tensile strength and elongation at break. Based on these findings, the optimal concentration of ZnO-SiO_2 was identified as 10 phr. At this concentration, ZnO acts effectively as an activator to enhance the antibacterial efficiency of the blends, while SiO_2 contributes to reinforcement. However, concentrations exceeding 10 phr result in pronounced agglomeration, compromising both mechanical and antibacterial properties.

4. Conclusions

The present study explored the effects of incorporating ZnO-SiO_2 into TW/ENR blends, at a ratio of 50:50 phr, on cure characteristics, mechanical properties, and antibacterial efficiency, with the key findings summarized below:

- The incorporation of ZnO-SiO_2 significantly enhanced the crosslink density, as indicated by the increased maximum torque and shorter cure times. At the optimal ZnO-SiO_2 loading of 10 phr, the composite exhibited the fastest crosslinking rate, attributed to the balanced effects of chemical activation and reinforcement. However, higher SiO_2 concentrations (>10 phr) resulted in filler agglomeration, which negatively impacted the crosslinking efficiency and mechanical properties.
- Tensile properties were enhanced at the optimal concentration of ZnO-SiO_2 at 10 phr. The enhancements in Young's modulus and tensile strength were attributed to the synergistic interactions between ENR molecules and SiO_2 surfaces, as well as strain-induced crystallization, which further reinforced the matrix during extension. However, at concentrations beyond 10 phr, particle agglomeration reduced the reinforcement efficiency, leading to reduced tensile strength and elongation at break.
- ZnO-SiO_2 contributed significantly to antibacterial activity, achieving up to 99.9% bacterial reduction at 10 phr. This efficiency was attributed to the release of Zn^{2+} ions and reactive oxygen species (ROS), which interact with bacterial cell walls and DNA. Notably, the antibacterial performance remained stable even after heat aging, confirming the durability of the ZnO-SiO_2-modified blends. However, higher ZnO-SiO_2 concentrations led to agglomeration, reducing the surface exposure and antibacterial effectiveness.
- The study identified 10 phr as the optimal ZnO-SiO_2 concentration, offering a balance between mechanical reinforcement, crosslinking efficiency, and antibacterial performance. While ZnO served as an activator and antibacterial agent, SiO_2 provided mechanical reinforcement through its rough surface interactions with ENR molecules.

These findings highlight the dual functionality of ZnO-SiO_2 in TW/ENR composites, showcasing its potential applications as a result of enhanced mechanical performance and antibacterial activity. This study lays the foundation for optimizing rubber-based materials for a wide range of applications, including healthcare and engineering. Further investigations into alternative filler designs and advanced processing methods could provide additional opportunities to enhance performance while minimizing agglomeration effects.

Author Contributions: N.K.: conceptualization, methodology, validation, formal analysis, writing—original draft preparation; T.B.: conceptualization, methodology, validation, formal analysis; J.J.: writing—review and editing; R.L.N.: writing—review and editing; Y.N.: visualization, supervision, formal analysis, writing—review and editing, funding acquisition. All authors have read and agreed to the published version of the manuscript.

Funding: This research was funded by The Asahi Glass Foundation and King Mongkut's University of Technology Thonburi under the KIRIM ID: 27480, as well as the Petchra Pra Jom Klao Master's Degree Research Scholarship from King Mongkut's University of Technology Thonburi.

Data Availability Statement: All the data are available from the authors and can be provided on request.

Acknowledgments: This work was supported by King Mongkut's University of Technology Thonburi under the supported scholarships. Also, the Phramongkutklao College of Medicine is also acknowledged.

Conflicts of Interest: Author Raymond Lee Nip was employed by the company Global Chemical Co., Ltd. The remaining authors declare that the research was conducted in the absence of any commercial or financial relationships that could be construed as a potential conflict of interest.

References

1. Bhalla, G.; Kumar, A.; Bansal, A. Performance of scrap tire shreds as a potential leachate collection medium. *Geotech. Geol. Eng.* **2010**, *28*, 661–669. [CrossRef]
2. Martínez, J.D.; Cardona-Uribe, N.; Murillo, R.; García, T.; López, J.M. Carbon black recovery from waste tire pyrolysis by demineralization: Production and application in rubber compounding. *Waste Manag. Res.* **2019**, *85*, 574–584. [CrossRef] [PubMed]
3. Baker, C.S.L.; Gelling, I.R.; Newell, R. Epoxidized natural rubber. *Rubber Chem. Technol.* **1985**, *58*, 67–85. [CrossRef]
4. Darestani Farahani, T.; Bakhshandeh, G.R.; Abtahi, M. Mechanical and viscoelastic properties of natural rubber/reclaimed rubber blends. *Polym. Bull.* **2006**, *56*, 495–505. [CrossRef]
5. Li, S.; Lamminmäki, J.; Hanhi, K. Improvement of mechanical properties of rubber compounds using waste rubber/virgin rubber. *Polym. Sci. Eng.* **2005**, *45*, 1239–1246. [CrossRef]
6. Kroshefsky, R.D.; Price, J.L.; Mangaraj, D. Role of compatibilization in polymer nanocomposites. *Rubber Chem. Technol.* **2009**, *82*, 340–368. [CrossRef]
7. Olejnik, O.; Masek, A. Recent progress in bio-based elastomers with intrinsic self-healing mechanisms-part I: Natural rubber modifications. *J. Saudi Chem. Soc.* **2023**, *27*, 101676. [CrossRef]
8. Mensah, B.; Kang, S.I.; Wang, W.; Nah, C. Effect of graphene on polar and nonpolar rubber matrices. *Mech. Adv. Mater. Mod. Process.* **2018**, *4*, 1. [CrossRef]
9. Krainoi, A.; Boonkerd, K. Mechanical/electrical properties and strain sensibility of epoxidized natural rubber nanocomposite filled with carbon nanotube: Effect of sodium alginate as a surfactant on latex technology process. *Express Polym. Lett.* **2023**, *17*, 850–866. [CrossRef]
10. Krainoi, A.; Prasert, W.; Kesakomol, P.; Thongdee, P.; Nitchaphanit, S.; Sungsirin, N.; Boonsiri, T.; Watanaveeradej, V.; Ounjai, K.; Nakaramontri, Y. Effect of modified zinc oxide nanoparticles on enhancement of mechanical, thermal and antibacterial properties of disinfectant natural rubber latex foams. *Mater. Today Commun.* **2023**, *35*, 105601. [CrossRef]
11. Krainoi, A.; Poomputsa, K.; Kalkornsurapranee, E.; Johns, J.; Songtipya, L.; Nip, R.L.; Nakaramontri, Y. Disinfectant natural rubber films filled with modified zinc oxide nanoparticles: Synergetic effect of mechanical and antibacterial properties. *Express Polym. Lett.* **2021**, *15*, 1081–1100. [CrossRef]
12. Toh-ae, P.; Saramolee, P.; Chiarakorn, S.; Waraho-Zhmayev, D.; Kamthong, A. Enhanced photocatalysis of natural rubber foams filters boosted by modified-titanium oxide hybrid fillers: Gaseous benzene removal, antibacterial properties and air permeability. *Express Polym. Lett.* **2022**, *16*, 1229–1252. [CrossRef]
13. Laurentowska, A.; Ciesielczyk, F.; Siwińska-Stefańska, K.; Jesionowski, T. Precipitation of ZnO-SiO_2 oxide composites in the presence of natural rubber latex and selected non-ionic surfactants. *Pigm. Res. Technol.* **2022**, *41*, 199–209. [CrossRef]
14. Norazlina, H.; Fahmi, A.R.M.; Hafizuddin, W.M. $CaCO_3$ from seashells as a reinforcing filler for natural rubber. *J. Mech. Eng. Sci.* **2015**, *8*, 1481–1488. [CrossRef]
15. Meera, A.P.; Said, S.; Grohens, Y.; Luyt, A.S.; Thomas, S. Tensile stress relaxation studies of TiO_2 and nanosilica filled natural rubber composites. *Ind. Eng. Chem. Res.* **2009**, *48*, 3410–3416. [CrossRef]
16. Shayegan, Z.; Lee, C.S.; Haghighat, F. TiO_2 photocatalyst for removal of volatile organic compounds in gas phase—A review. *J. Chem. Eng.* **2018**, *334*, 2408–2439. [CrossRef]

17. Mendes, C.R.; Dilarri, G.; Forsan, C.F.; Sapata, V.D.M.R.; Lopes, P.R.M.; de Moraes, P.B.; Montagnolli, R.N.; Ferreira, H.; Bidoia, E.D. Antibacterial action and target mechanisms of zinc oxide nanoparticles against bacterial pathogens. *Sci. Rep.* **2022**, *12*, 2658. [CrossRef] [PubMed]
18. Zhang, L.; Jiang, Y.; Ding, Y.; Daskalakis, N.; Jeuken, L.; Povey, M.; O'neill, A.J.; York, D.W. Mechanistic investigation into antibacterial behaviour of suspensions of ZnO nanoparticles against *E. coli*. *J. Nanopart. Res.* **2010**, *12*, 1625–1636. [CrossRef]
19. Raksaksri, L.; Chuayjuljit, S.; Chaiwutthinan, P.; Boonmahitthisud, A. Use of TBzTD as noncarcinogenic accelerator for ENR/SiO_2 nanocomposites: Cured characteristics, mechanical properties, thermal behaviors, and oil resistance. *Int. J. Polym. Sci.* **2017**, *2017*, 9721934. [CrossRef]
20. Wang, M.; Wang, R.; Chen, X.; Kong, Y.; Huang, Y.; Lv, Y.; Li, G. Effect of non-rubber components on the crosslinking structure and thermo-oxidative degradation of natural rubber. *Polym. Degrad. Stab.* **2022**, *196*, 109845. [CrossRef]
21. Peng, Z.; Kong, L.X.; Li, S.D.; Chen, Y.; Huang, M.F. Self-assembled natural rubber/silica nanocomposites: Its preparation and characterization. *Compos. Sci. Technol.* **2007**, *67*, 3130–3139. [CrossRef]
22. Li, G.; Zhao, T.; Zhu, P.; He, Y.; Sun, R.; Lu, D.; Wong, C.P. Structure-property relationships between microscopic filler surface chemistry and macroscopic rheological, thermo-mechanical, and adhesive performance of SiO_2 filled nanocomposite underfills. *Compos.—A Appl. Sci.* **2019**, *118*, 223–234. [CrossRef]
23. Chen, Y.; Peng, Z.; Kong, L.X.; Huang, M.F.; Li, P.W. Natural rubber nanocomposite reinforced with nano silica. *Polym. Sci. Eng.* **2008**, *48*, 1674–1677. [CrossRef]
24. Wang, Y.; Liao, L.; Zhong, J.; He, D.; Xu, K.; Yang, C.; Luo, Y.; Peng, Z. The behavior of natural rubber–epoxidized natural rubber–silica composites based on wet masterbatch technique. *J. Appl. Polym. Sci.* **2016**, *133*, 43571–43580. [CrossRef]
25. Sengloyluan, K.; Sahakaro, K.; Dierkes, W.K.; Noordermeer, J.W.M. Silica-reinforced tire tread compounds compatibilized by using epoxidized natural rubber. *Euro. Polym. J.* **2014**, *51*, 69–79. [CrossRef]
26. Tang, X.; Chen, Z.; Liu, J.; Chen, Z.; Xie, W.; Evrendilek, F.; Buyukada, M. Dynamic pyrolysis behaviors, products, and mechanisms of waste rubber and polyurethane bicycle tires. *J. Hazard. Mater.* **2021**, *402*, 123516. [CrossRef]
27. Susteric, Z.; Kos, T. Elastomer-Clay nanocomposites: Effect of Elastomer Polarity. *Macromol. Symp.* **2010**, *296*, 311–315. [CrossRef]
28. Wang, L.; Hu, C.; Shao, L. The antimicrobial activity of nanoparticles: Present situation and prospects for the future. *Int. J. Nanomed.* **2017**, *12*, 1227–1249. [CrossRef] [PubMed]
29. Sirelkhatim, A.; Mahmud, S.; Seeni, A.; Kaus, N.H.M.; Ann, L.C.; Bakhori, S.K.M.; Hasan, H.; Mohamad, D. Review on zinc oxide nanoparticles: Antibacterial activity and toxicity mechanism. *Nanomicro Lett.* **2015**, *7*, 219–242. [CrossRef] [PubMed]
30. Nordin, R.; Yusof, R.; Nawawi, W.I.; Salihin, M.Z.; Ishak, Z.A.M. Effect of several commercial rubbers as substrates for zinc oxide in the photocatalytic degradation of methylene blue under visible irradiation. *Express Polym. Lett.* **2020**, *14*, 838–847. [CrossRef]
31. Choi, S.S.; Ko, E. Novel test method to estimate bound rubber formation of silica-filled solution styrene-butadiene rubber compounds. *Polym. Test.* **2014**, *40*, 170–177. [CrossRef]
32. Amany, A.; El-Rab, S.F.G.; Gad, F. Effect of reducing and protecting agents on size of silver nanoparticles and their antibacterial activity. *Pharm. Chem.* **2012**, *4*, 53–65.
33. Ruparelia, J.P.; Chatterjee, A.K.; Duttagupta, S.P.; Mukherji, S. Strain specificity in antimicrobial activity of silver and copper nanoparticles. *Acta Biomater.* **2008**, *4*, 707–716. [CrossRef]

MDPI AG
Grosspeteranlage 5
4052 Basel
Switzerland
Tel.: +41 61 683 77 34

Polymers Editorial Office
E-mail: polymers@mdpi.com
www.mdpi.com/journal/polymers

www.ingramcontent.com/pod-product-compliance
Lightning Source LLC
Chambersburg PA
CBHW061758210926
51850CB00006B/586

PRINCIPLES OF MECHANICAL ENGINEERING

For the Students of B.E./B.Tech. of
Maharshi Dayanand University (MDU), Rohtak
and
Kurukshetra University, Kurukshetra

Latest Question Papers with Solutions and other Solved Questions from University Examinations

Dr. SADHU SINGH
Professor & Head, Mechanical Engineering Department (Former)
&
Dean, Faculty of Engineering & Technology, (Former)
Govind Ballabh Pant University of Agriculture & Technology,
PANTNAGAR – 263 145
Director (Colleges), Punjab Technical University, (Former)
JALANDHAR

S. CHAND & COMPANY LTD.
(An ISO 9001 : 2000 Company)
RAM NAGAR, NEW DELHI - 110 055

S. CHAND & COMPANY LTD.
(An ISO 9001 : 2000 Company)
Head Office: 7361, RAM NAGAR, NEW DELHI - 110 055
Phone: 23672080-81-82, 9899107446, 9911310888
Fax: 91-11-23677446
Shop at: **schandgroup.com**; e-mail: **info@schandgroup.com**

Branches:

AHMEDABAD	: 1st Floor, Heritage, Near Gujarat Vidhyapeeth, Ashram Road, **Ahmedabad** - 380 014, Ph: 27541965, 27542369, ahmedabad@schandgroup.com
BENGALURU	: No. 6, Ahuja Chambers, 1st Cross, Kumara Krupa Road, **Bengaluru** - 560 001, Ph: 22268048, 22354008, bangalore@schandgroup.com
BHOPAL	: Bajaj Tower, Plot No. 243, Lala Lajpat Rai Colony, Raisen Road, **Bhopal** - 462 011, Ph: 4274723. bhopal@schandgroup.com
CHANDIGARH	: S.C.O. 2419-20, First Floor, Sector - 22-C (Near Aroma Hotel), **Chandigarh** -160 022, Ph: 2725443, 2725446, chandigarh@schandgroup.com
CHENNAI	: 152, Anna Salai, **Chennai** - 600 002, Ph: 28460026, 28460027, chennai@schandgroup.com
COIMBATORE	: No. 5, 30 Feet Road, Krishnasamy Nagar, Ramanathapuram, **Coimbatore** -641045, Ph: 0422-2323620 coimbatore@schandgroup.com **(Marketing Office)**
CUTTACK	: 1st Floor, Bhartia Tower, Badambadi, **Cuttack** - 753 009, Ph: 2332580; 2332581, cuttack@schandgroup.com
DEHRADUN	: 1st Floor, 20, New Road, Near Dwarka Store, **Dehradun** - 248 001, Ph: 2711101, 2710861, dehradun@schandgroup.com
GUWAHATI	: Pan Bazar, **Guwahati** - 781 001, Ph: 2738811, 2735640 guwahati@schandgroup.com
HYDERABAD	: Padma Plaza, H.No. 3-4-630, Opp. Ratna College, Narayanaguda, **Hyderabad** - 500 029, Ph: 24651135, 24744815, hyderabad@schandgroup.com
JAIPUR	: A-14, Janta Store Shopping Complex, University Marg, Bapu Nagar, **Jaipur** - 302 015, Ph: 2719126, jaipur@schandgroup.com
JALANDHAR	: Mai Hiran Gate, **Jalandhar** - 144 008, Ph: 2401630, 5000630, jalandhar@schandgroup.com
JAMMU	: 67/B, B-Block, Gandhi Nagar, **Jammu** - 180 004, (M) 09878651464 **(Marketing Office)**
KOCHI	: Kachapilly Square, Mullassery Canal Road, Ernakulam, **Kochi** - 682 011, Ph: 2378207, cochin@schandgroup.com
KOLKATA	: 285/J, Bipin Bihari Ganguli Street, **Kolkata** - 700 012, Ph: 22367459, 22373914, kolkata@schandgroup.com
LUCKNOW	: Mahabeer Market, 25 Gwynne Road, Aminabad, **Lucknow** - 226 018, Ph: 2626801, 2284815, lucknow@schandgroup.com
MUMBAI	: Blackie House, 103/5, Walchand Hirachand Marg, Opp. G.P.O., **Mumbai** - 400 001, Ph: 22690881, 22610885, mumbai@schandgroup.com
NAGPUR	: Karnal Bag, Model Mill Chowk, Umrer Road, **Nagpur** - 440 032, Ph: 2723901, 2777666 nagpur@schandgroup.com
PATNA	: 104, Citicentre Ashok, Govind Mitra Road, **Patna** - 800 004, Ph: 2300489, 2302100, patna@schandgroup.com
PUNE	: 291/1, Ganesh Gayatri Complex, 1st Floor, Somwarpeth, Near Jain Mandir, **Pune** - 411 011, Ph: 64017298, pune@schandgroup.com **(Marketing Office)**
RAIPUR	: Kailash Residency, Plot No. 4B, Bottle House Road, Shankar Nagar, **Raipur** - 492 007, Ph: 09981200834, raipur@schandgroup.com **(Marketing Office)**
RANCHI	: Flat No. 104, Sri Draupadi Smriti Apartments, East of Jaipal Singh Stadium, Neel Ratan Street, Upper Bazar, **Ranchi** - 834 001, Ph: 2208761, ranchi@schandgroup.com **(Marketing Office)**
SILIGURI	: 122, Raja Ram Mohan Roy Road, East Vivekanandapally, P.O., **Siliguri**-734001, Dist., Jalpaiguri, (W.B.) Ph. 0353-2520750 **(Marketing Office)**
VISAKHAPATNAM	: Plot No. 7, 1st Floor, Allipuram Extension, Opp. Radhakrishna Towers, Seethammadhara North Extn., **Visakhapatnam** - 530 013, (M) 09347580841, visakhapatnam@schandgroup.com **(Marketing Office)**

First Edition 2010
Reprint with Additional Matter 2011

ISBN : 81-219-3526-1 **Code** : 10 443

PRINTED IN INDIA

By Rajendra Ravindra Printers Pvt. Ltd., 7361, Ram Nagar, New Delhi -110 055 and published by S. Chand & Company Ltd., 7361, Ram Nagar, New Delhi -110 055.

Dedicated to

my wife

Smt. Manjit Kaur

PREFACE

Principles of Mechanical Engineering is a comprehensive Textbook for first year B.Tech./B.E. students of all branches of M.D.U. (Rohtak) and Kurukshetra University, Kurukshetra as per the new syllabus. It is also useful to students of all other Indian Universities.

The book contains 17 chapters on Introduction to Mechine Tools; Basic Concepts of Thermodynamics; Properties of Steam; Steam Boilers; Refrigeration and Air – Conditioning; Hydraulic Turbines, Pumps and Devices; Internal Combustion Engines; Steam Turbines; Condensers; Gas Turbines; Simple Lifting Machines; Power Transmission Methods and Devices; Stresses and Strains; Principal Stresses and Strains; Shear Force and Bending Moment; Flexural and Torsional Loading; and Introduction to Manufacturing Systems.

The book contians a large number of solved and unsolved problems. This has been supplemented with Multichoice questions, review questions, true and false and fill in the blanks type of questions.

It is hoped that the book shall be quite useful to the students and teachers. Suggestions for the further improvement of the book are welcome and shall be duly acknowledged.

The moral support received from Mrs. Narinderpal Kaur, my daughter – in – law, and grand children Kanupreet and Amitoej is praise worthy.

The support and cooperation received from the staff of S. Chand & Company Ltd., and in particular from Navin Joshi, VP (Publishing), Mr. P.J. Singh, Regional Manager, Jalandhar office, Mr. R.S. Saxena, Advisor (Publishing) , is highly acknowledged.

AUTHOR

SYLLABUS

MAHARSHI DAYANAND UNIVERSITY (MDU), ROHTAK

ME – 101- F

Section A

Introduction to Machine Tools:

Lathe, Shaper, Planer, Milling, Drilling, Slotter, Introduction to Metal Cutting.

Basic concept of thernodynamics:

Introdution, States, Work, Heat, Temperature, Zeroth, 1st, 2nd and 3rd law of thermodynamics, concept of internal energy, enthalpy and entropy. Problems

Properties of Steam and Steam Generator:

Formation of Steam of constant pressure, Thermodynamic properties of Steam. Use of steam tables, Measurement of dryness fraction by throttling calorimeter.

Section B

Refrigeration and Airconditioning :

Introduction to refrigeration and air - conditioning, Rating of refrigeration machines, Coefficient of performance, Simple refrigeration vapour compression cycle, Psychrometric charts and its use, Human comforts.

Hydraulic Turbines and Pumps:

Introduction, Classification, Construction details and working of Pelton, Francis and Kaplan turbines, Specific speed and selection of turbines Classification of water pumps and their working.

Section C

Power Transmission Methods and Devices:

Introduction to power transmission, Belt, Rope, Chain and Gear drive. Types and functioning of clutches.

Stresses and Strains:

Introduction, Concepts and types of Stresses and strains, Poison's ratio, stresses and strains in simple and compound bars under axial, flexure and torsional loading, Stress - strains diagrams, Hooks law, Elastic constants and their relationships.

Section D

Introduction to Manufacturing Systems, Fundamentals of Numerical Control (NC), Advantage of NC systems, Classifications of NC, Comparison of NC and CNC.

SYLLABUS

KURUKSHETRA UNIVERSITY

First & Second Semester

ME-101E

UNIT-I

Properties of Steam and Boilers: Formation of steam at constant pressure, Thermodynamics properties of steam. Condition of steam, Steam tables, Measurement of dryness fraction by throttling calorimeter, Classification of boilers, Comparison of water and fire tube boiler mounting and accessories with their functions, Constructional and operational details of Cochran and Babcock and Wilcox boilers, Problems.

Steam Turbines and Condensers: Classification of turbines, Working principle of impulse and reaction turbine, Compounding of impulse turbine, Comparison of impulse and reaction turbines, Types of condensers. Cooling ponds and cooling towers, Condenser and vacuum efficiencies.

UNIT-II

I.C. Engines and Gas Turbines: Introduction, Classification, Constructional details and working of two-stroke and four-stroke diesel and petrol engines, Otto, Diesel and Dual cycles, Working principle of gas turbine, Constant pressure gas turbine cycle.

Water Turbines, Pumps and Hydraulic Devices: Introduction, Classification, Construction details and working of Pelton, Francis and Kaplan turbines, Specific speed and selection of turbines, Classification of water pumps and their working. Hydraulic jack and lift.

UNIT-III

Simple Lifting Machines: Definition of machine, Velocity retio, Mechanical advantage, Efficiency, Laws of machines, Reversibility of machine, Wheel and axle, Differential Pulley block, Single, double and triple start worm and worm wheel, Single and double purchase winch crabs, Simple and Compound screw jacks. Problems.

Power Transmission Methods and Devices: Introduction of Power transmission, Belt drive, Rope drive, Chain drive, Pulley. Gear drive, Types of gears, Gear train, Clutches, Types and function of clutches, Types and function of brakes, Power measurement by dynamometer, Types of dynamometers.

UNIT-IV

Stresses and Strains: Introduction, Concept and types of Stresses and strains, Poison's ratio, stresses and strains in simple and compound bars under axial loading, Stress-strain diagrams, Hooks law, Elastic constants and their relationships, Principles stresses and strains and principal- planes, Mohr's circle of stresses. Numerical problems.

Bending Moment and Shear Force: Definitions, SF and BM diagrams for cantilever and simply supported beam. Calculation of maximum SF, BM and point of contraflexure under the loads of (i) concentrated load (ii) uniformly distributed load (iii) combination of concentrated and uniformly distributed loads. Problems.

CONTENTS

1. INTRODUCTION TO MACHINE TOOLS 1 – 55

1 INTRODUCTION TO MACHINE TOOLS

1.1 INTRODUCTION

A machine tool is a power driven machine for making articles of a given shape, size and accuracy (according to the blueprints) by removing metal from workpieces in the form of chip. Machine tools are factory equipment for producing machines, instruments and tools of all kinds. So, it can be said that the machine tool is the mother of all machines.

Most machine tools perform the following four functions :

1. Hold the job.
2. Hold the cutting tools.
3. Move one or both of these (rotary motion or reciprocating motion).
4. Provide a feeding motion for one of these.

1.2 CLASSIFICATION OF MACHINE TOOLS

(*a*) **According to field of Application**

1. **General purpose machine tools :** General purpose or universal machine tools are used for performing a great variety of machining operations on a wide range of workpieces. These are employed chiefly in piece and small-lot production and for repair work. Machine tools used for a particularly wide range of work are known as multipurpose machine tools. Especially versatile machine tools are also called "Omniversal". General purpose machine tools include : plain turning lathes, turret lathes, milling machines, drilling machines, grinding machine etc.

2. **Single purpose machine tools :** These machine tools are designed to perform a single definite machining operation, *e.g.*, broaching, thread cutting, gear shaping and hobbing machines, machines for machining pistons, crank shafts, camshafts and for turning the can contours on camshafts etc.

3. **Limited purpose machine tools :** These machine tools are capable of a narrow range of operations on a wide variety of workpieces, *e.g.*, automatic cutting off machines.

4. **Production machine tools :** These are mainly used in batch and mass production and feature high power and rigidity. These machine tools include : multi-tool lathes, single and multi-spindle automatics, and semi-automatic lathes, plunge-cut cylindrical grinders, centreless, planer-type milling machines, thread rolling machines for tap production, numerically controlled machine tools etc.

5. **Specialized machine tools :** These are used for machining articles similar in shape but different in size. This group includes unit built machine tools. These machine tools allow the machining of several surfaces in different planes. Their advantage is that they are readily changed over from one job to another. This is done by mounting additional unit heads, positioning them

at an angle to a horizontal or vertical plane or otherwise. These machines are used mainly in large-lot production.

6. **Special machine tools :** These machine tools are designed and manufactured individually and are intended for performing a certain definite operation in machining a certain definite workpiece. These machine tools include : machines for sharpening round threading dies, for grinding relief surfaces at the chamfer of round threading dies, for marking round threading dies and shank-type tools, for threading by die taps, for grinding flutes on taps and reamers, tap chamfers, flutes on twist drills etc. These machines find applications in large-lot and mass production.

The general purpose machine tools have the following characteristics :

(*i*) Usually less initial investment in equipment.

(*ii*) Greater machine flexibility.

(*iii*) Fewer machines may be required.

(*iv*) Less maintenance cost.

(*v*) Less set up and debugging time.

(*vi*) Less danger of obsolescence.

The special purpose machines have the following characteristics :

(*i*) Uniform product flow.

(*ii*) Reduced in-process inventory.

(*iii*) Reduces manpower requirements.

(*iv*) Reduced factory floor space.

(*v*) Higher output.

(*vi*) Higher product quality.

(*vii*) Reduces inspection cost.

(*viii*) Reduced operator skill requirements.

(*b*) **According to Accuracy of Machine Tools**

1. **Normal accuracy :** Machine tools of normal or standard accuracy include the majority of the general purpose machine tools.

2. **Higher accuracy :** Machine tools of higher accuracy are manufactured to the same drawings as the normal accuracy models, except that higher requirements are made to the accuracy with which the critical parts are manufactured, as well as to the quality of assembly and adjustments.

3. **Precision** machine tools have certain parts that have been specially designed with the aim of maintaining the high accuracy standards. In addition, narrow tolerances are stipulated for the machining of all the parts, as well as for their assembly and adjustment as a whole.

4. **High precision** machine tools are manufactured according to even more rigorous accuracy requirements than "Precision" class machine tools.

5. **Supper-high precision** or Master machine tools are intended for machining the parts which determine the accuracy of machine tools belonging to "precision" and "high precision" class.

To ensure the required accuracy of their operation, the machine tools of the last three classes are to be installed in special constant-temperature rooms, with automatically controlled (Constant) temperature and humidity.

(*c*) **According to Weight**

Machine tools can be classified as : Light weight (upto 1 tonne), medium weight (upto 10 tonnes), and heavy-weight (over 10 tonnes). The last group can be further divided into subgroups as : Large size (10 to 30 tonnes), Heavy (30 to 100 tonnes) and Extra-heavy (over 100 tonnes).

(*d*) **According to the processing operations**

1. **Lathes :** Engine and facing lathes, cutting-off lathes, multiple-tool lathes, Turret lathes, Automatics and Semiautomatics (Single spindle, multiple spindle). Vertical turning and boring mills, and specialized machine tools.

Vertical turning and boring mills are used to machine blanks of large diameter and relatively small height. Most of the larger machines (some can accommodate work 25 m in diameter) are called vertical boring mills and the smaller models are usually known as vertical turret lathes.

2. **Drilling and Boring machines :** Upright drill presses, Semiautomatics single-spindle drilling machines, Semi-automatic multiple-spindle drilling machines Jig borers, Radial drills, Boring machines, Precision boring machines, and horizontal drilling machines.

3. **Planers, Shapers, Slotters and Broaching machines :** Open side Planers, Double-housing planers, Shapers, Slotters, Horizontal broaching machines, Vertical broaching machines.

4. **Milling machines :** Horizontal knee-type milling machines. Vertical knee-type milling machines, Tracer controlled milling and engraving machines, Coutinuous milling machines, Vertical spindle compound-table milling machine, Fixed-bed and planer-type milling machines, and Ram-head milling machines.

5. **Grinding and micro-finishing machines :** Cylindrical grinders, Internal grinders, Snagging grinders, Specialized grinders, Tool and cutter grinders, Surface grinders, and Micro-finishing machines.

6. **Gear and thread cutting machines :** Shapers and planers for Spur and Helical gears, Bevel gear generators, Hobbers for Spur and helical gears and splined shafts, Worm Wheel and Worm cutting machines, Gear tooth Chamfering machines Thread-milling machines, gear finishing machines, Gear and thread grinders.

7. **Combination machine tools :** General purpose machines, Semi-automatic machines and Automatic machines.

8. **Cutting-off machines :** Cutting off lathes, Abrasive cutting machines, Circular-saw friction cutting machines, Straightening and cutting off machines, Band saw cutting-off machines, Circular Cold sawing machines, and Power hacksawing machines.

9. **Miscellaneous machine tools :** Coupling and pipe thread cutting machines, Saw-cutting machines, Centreless bar turning and straightening machine, Tool Testing machines, Dividing machines, and Balancing machines.

1.3 LATHE

The lathe is one of the most important machines in any workshop. Its main objective is to remove material from outside by rotating the work against a cutting tool. Though a lathe is used to produce cylindrical work, yet it may also be used for many other purposes such as drilling, threading, grinding, milling etc.

1.3.1 Working Principle of a Lathe

The working principle of a lathe is shown in Fig 1.1. In a lathe, the workpiece is held in a chuck or between centres and rotated about its axis at a uniform speed. The cutting tool held in the tool post is fed into the woekpiece for a desired depth and in the desired direction (*i.e.*, in the linear, transverse or lateral direction). Since there exists a relative motion between the workpiece and the cutting tool, therefore the material is removed in the form of chips and the desired shape in obtained.

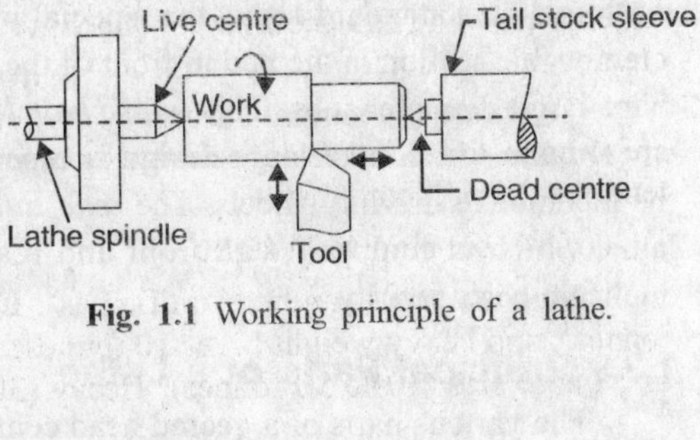

Fig. 1.1 Working principle of a lathe.

1.3.2 Types of Lathe

Though the fundamental principle of operation of all lathes is same and perform the same function, yet they are classified according to the design, type of drive, arrangement of gears, and purpose. The following are important types of lathes.

1. *Speed lathe.* It is one of the simplest type of all lathes. It is driven by power and consists of a bed, a headstock, a tail stock and an adjustable slide for supporting the tool. Since the tool is fed into the work by hand and cuts are very small, therefore this type of lathe is driven at high speeds usually from 1200 to 3600 r.p.m. The work may be held between centres or attached to a face plate on the headstock. The speed lathe is used mainly for wood working, centering, metal spinning, polishing etc.

2. *Engine or centre lathe.* It is a general purpose lathe and is widely used in workshops. It differs from a speed lathe that it has additional mechanism for controlling the spindle speed and for supporting and controlling the feed of the fixed cutting tool. The cutting tool may be fed both in cross and longitudinal direction with reference to the lathe axis with the help of a carriage.

The engine lathe, depending upon the design of headstock for receiving power, may be classified as belt driven lathe, motor driven lathe and geared head lathe.

3. *Bench lathe.* The bench lathe is so small that it can be mounted on a bench. All the types of operation can be performed on this lathe that may be done on an ordinary speed or engine lathe. This is used for small work usually requiring considerable accuracy such as in the production of gauges, punches and beds for press tools.

4. *Tool room lathe.* The tool room lathe is similar to an engine lathe and is equipped with all the accessories needed for accurate tool work. It has an individually driven geared headstock with a wide range of spindle speeds. Since this lathe is used for precision work on tools, gauges, dies, jigs and other small parts, therefore greater skill is needed to operate the lathe.

5. *Capstan and turret lathe.* The capstan and turret lathes are the modification of engine lathes and is particularly used for mass production of identical parts in the minimum time. These lathes are semi-automatic and are fitted with multi tool holding devices, called capstan and turret heads. The advantage of capstan and turret lathe is that several different types of operations can be performed on a workpiece without resetting of work or tools.

6. *Automatic lathes.* The automatic lathes are so designed that the tools are automatically fed to the work and withdrawn after all the operations are complete to finish the work. Since the entire operation is automatic, these lathes require little attention of the operator. These lathes are used for mass production of identical parts.

7. *Special purpose lathes.* The work which can not be conveniently accommodated or machined on a standard lathe, the special purpose lathes are used. The *gap bed lathe* which has a removable section in the bed in front of the headstock to provide a space or gap, is used to swing extra large diameter jobs. The *crankshaft lathe* is especially used for turning crankshafts. The wheel *lathe* which is of large design is especially used for finishing the journals and for turning the locomotive driving wheels. The *axle lathe* is used for turning car axles. The manufacturing lathe, which is equipped with front and rear tool block and other features, is used for turning duplicate parts rapidly.

1.3.3 Principal Parts of a Lathe

The various parts of a geared head centre lathe are shown in Fig 1.2. The principal parts are as follows :

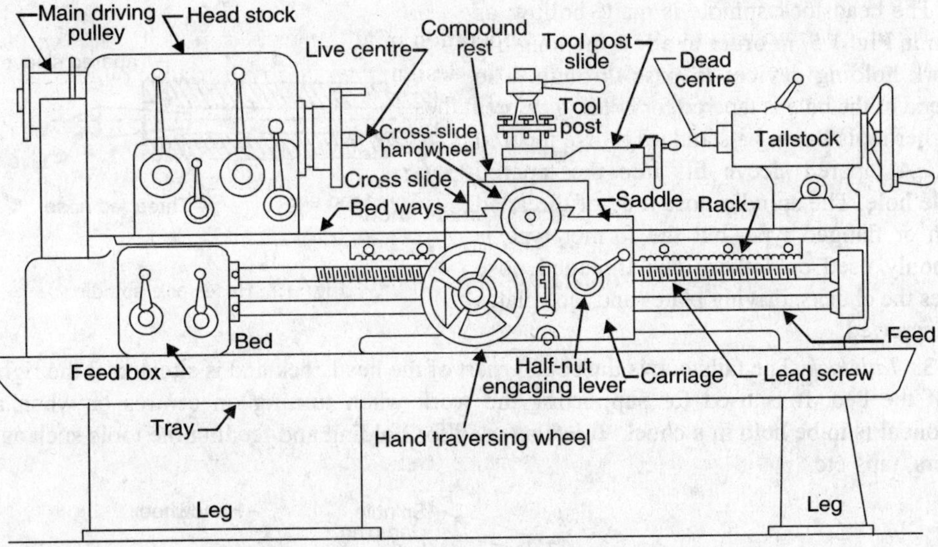

Fig. 1.2 Geared head centre lathe.

1. *Bed*. The bed is the base or foundation of the lathe. It is a heavy, rigid casting made in one piece. It holds or supports all other parts of the lathe. The top of the bed is planed to form guide ways for the carriage and tail stock. The guide ways are of two types, *i.e.*, wide flat guide ways as shown in Fig. 1.3 (*a*) or inverted Vee guide ways. Generally, the combination of both the flat and inverted Vee guide ways is used, as shown in Fig. 1.3 (b).

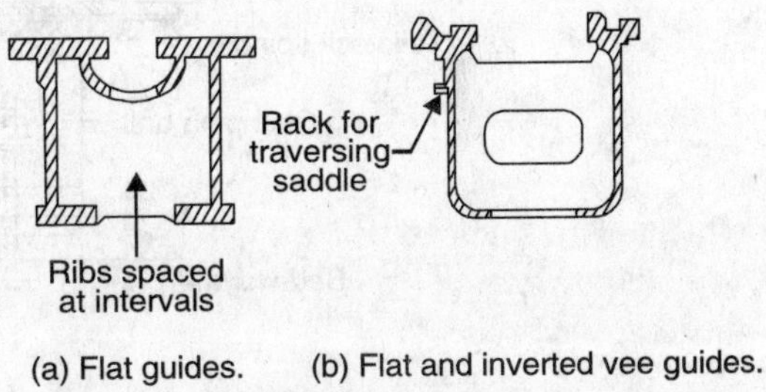

Fig. 1.3 Bed.

2. *Headstock*. The headstock is permanently fastened on the inner ways at the left hand end of the bed. It serves to support the spindle and driving arrangements. All lathes receive their power through the headstock, which may be equipped with a step-cone pulleys or a geared head drive. The modern lathes are provided with all-geared type headstock to get large variation of spindle speeds. Fig. 1.4 shows a 9-speed all-geared headstock.

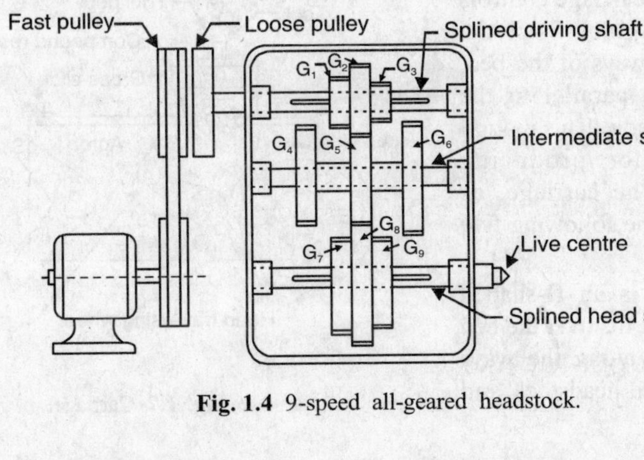

Fig. 1.4 9-speed all-geared headstock.

The headstock spindle is made hollow, as shown in Fig. 1.5, in order to allow the long bar or work holding devices to pass through. The front end of the hole is tapered for holding centres and other tools having a standard morse tapered shank. A tapered sleeve fits into the tapered spindle hole. The spindle nose may of threaded design or flanged type, but the former type is commonly used on lathes. The threaded nose enables the chucks, driving plates and face plates to be screwed.

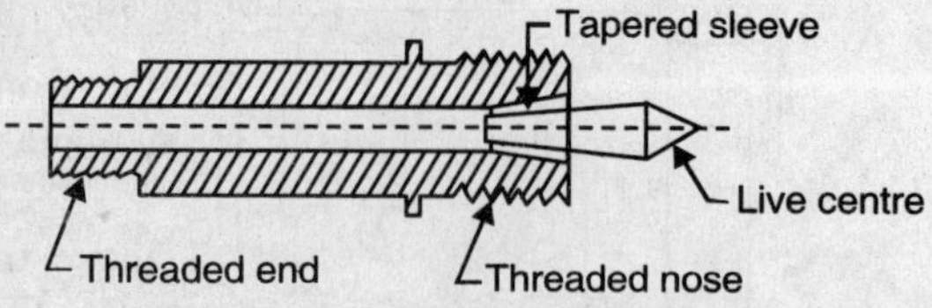

Fig. 1.5 Headstock spindle.

3. *Tailstock*. The tailstock is the counterpart of the headstock and is situated at the right hand end of the bed. It is used for supporting the work when turning on centres or when a long component is to be held in a chuck. It is also used for holding and feeding the tools such as drills, reamers, taps etc.

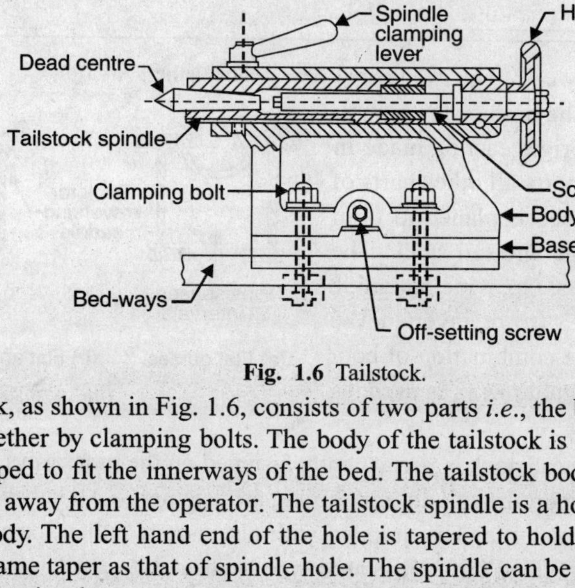

Fig. 1.6 Tailstock.

The tailstock, as shown in Fig. 1.6, consists of two parts *i.e.*, the base and the body. The two parts are held together by clamping bolts. The body of the tailstock is a casting and placed on its base which is shaped to fit the innerways of the bed. The tailstock body is adjustable and can be moved towards or away from the operator. The tailstock spindle is a hollow shaft and is mounted in the tailstock body. The left hand end of the hole is tapered to hold the dead centre and other tools having the same taper as that of spindle hole. The spindle can be firmly clamped in position by means of a clamping lever.

4. *Carriage*. The carriage controls and supports the cutting tool. Since the carriage moves on the ways of the bed, therefore its motion is parallel to the rotational axis of the spindle. This motion is commonly used for producing cylindrical surfaces. The carriage, as shown in Fig 1.7, has the following five major parts:

(*a*) *Saddle*. It is an H-shaped casting that fits over the bed and slides along the ways between the headstock and tailstock.

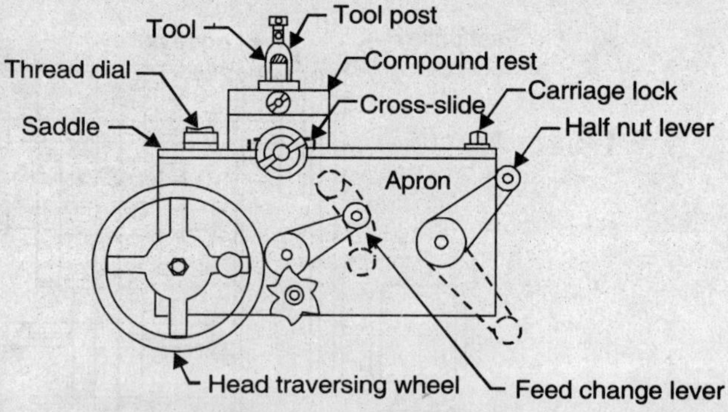

Fig. 1.7 Carriage.

(*b*) *Cross-slide.* It is mounted on the saddle. It supports the compound rest. The cross-slide also has a dovetail that fits over the saddle dovetail. The transverse movement of the cross-slide is obtained by turning the cross-slide hand wheel.

(*c*) *Compound rest.* It is mounted on top the cross-slide. It can be swivelled to any desired angle in a horizontal plane and its base is graduated in degrees to indicate its angular setting.

(*d*) *Tool post.* It is mounted above the compound rest. A T-slot is machined in the compound rest to accommodate the tool post. It serves to rigidly clamp the cutting tool or tool holder, in a desired position.

(*e*) *Apron.* It is fastened to the saddle and hangs over the front of the bed. It contains gears, clutches and levers for operating the carriage by hand and power feeds. The apron hand wheel can be turned to moved the carriage longitudinally back and forth by hand.

5. *Feed mechanism.* The feed mechanism of a lathe is employed for imparting various feeds to the cutting tool. The feed *i.e.*, the movement of the tool relative to the work may be longitudinal, cross and angular, as shown in Fig. 1.8

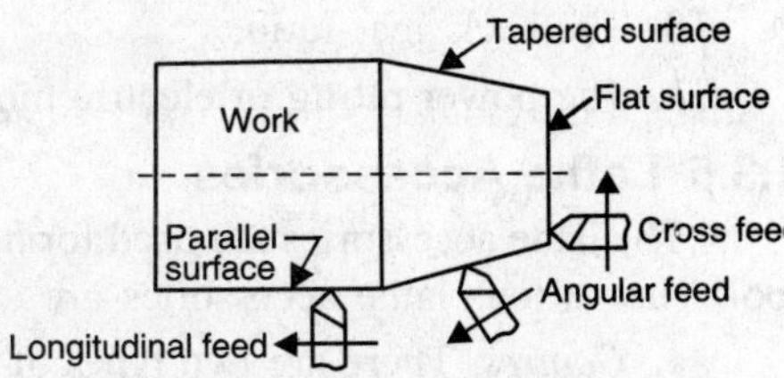

Fig. 1.8 Longitudinal, cross and angular feeds.

The feed is said to be longitudinal when the tool moves parallel to the work. It is obtained by the movement of the carriage which can be operated by hand or power.

The feed is said to be cross when the tool moves perpendicular to the work. It is obtained with the help of a cross-slide which can be operated by hand or power.

The feed is said to be angular when the tool moves at an angle to the work. It is obtained by swivelling the compound slide which can be operated by hand only.

The feed mechanism consists of feed reverse lever, tumbler reversing mechanism, change gears, feed gear box, quick change gear box lead screw, feed rod, apron mechanism and half nut mechanism.

1.3.4 Size and Specifications of a Lathe

The size of a lathe is determined by

1. The height of the centres above the top of the bed. This is shown by H in Fig. 1.9
2. The swing or the maximum diameter of the work that can be rotated over the ways of the bed. This is equal to twice the height of the centres. This is shown by D in Fig 1.9.
3. The maximum length of the work that can be accommodated between the lathe centres. This is shown by L in Fig 1.9.
4. The maximum diameter of the work that can be rotated over the lathe saddle. This is shown by D_1 in Fig 1.9.
5. The maximum diameter of the work that can be accommodated in the gap for a gap bed lathe. This is shown by D_2 in Fig. 1.9.

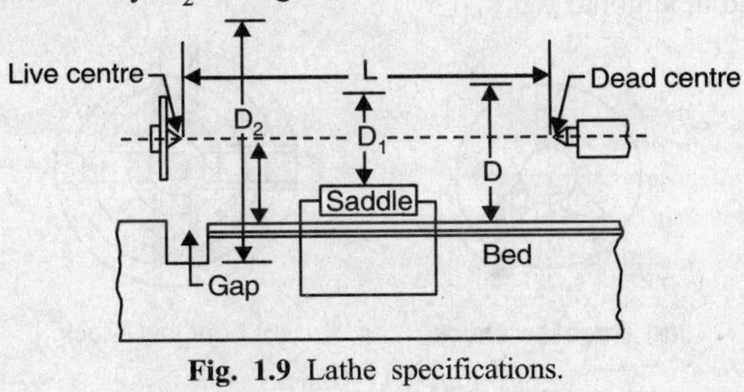

Fig. 1.9 Lathe specifications.

In addition to the above, the following specifications are necessary to provide while ordering a lathe :

1. The length, width and depth of bed.
2. The depth and width of the gap, if it is a gap lathe.
3. The swing over gap.
4. The number and range of spindle speeds.
5. The number of feeds.
6. The lead screw diameter.
7. The number and range of metric threads that can be cut.
8. The tailstock spindle travel.
9. The tailstock spindle set over.
10. The back gear-ratio.
11. The power rating of electric motor.

1.3.5 Lathe Accessories

The lathe accessories are used for holding and supporting the work or for holding the cutting tool. The various lathe accessories are discussed as follows :

1. *Centres.* There are two types of centres *i.e.*, live centre and dead centre. A centre which fits into the headstock spindle and revolves with the work is called live centre while the centre which is used in a tailstock spindle and does not revolve is called dead centre. These centres have a standard Morse taper shank at one end and a 60° point at the other end. The various types of centres, used for different works, are shown in Fig. 1.10.

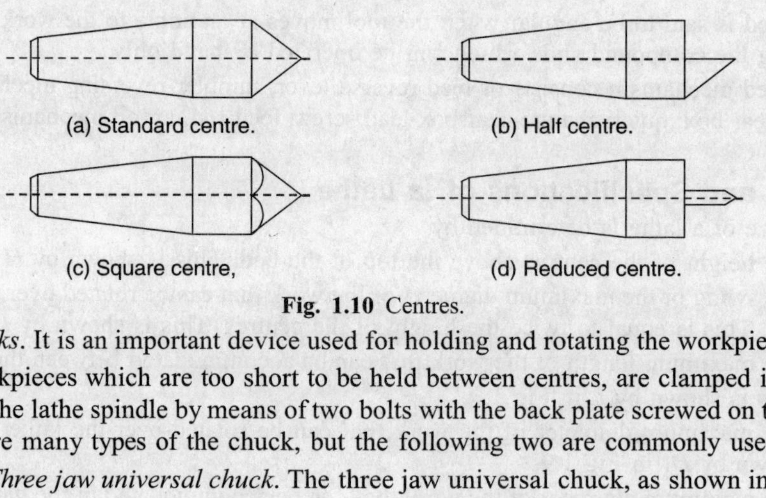

(a) Standard centre. (b) Half centre.

(c) Square centre, (d) Reduced centre.

Fig. 1.10 Centres.

2. *Chucks*. It is an important device used for holding and rotating the workpiece in a lathe.

The workpieces which are too short to be held between centres, are clamped in a chuck. It is attached to the lathe spindle by means of two bolts with the back plate screwed on to the spindle nose. There are many types of the chuck, but the following two are commonly used.

(*a*) *Three jaw universal chuck*. The three jaw universal chuck, as shown in Fig 1.11 (*a*) is also called self centering chuck or scroll chuck. Thus chuck is used for holding round and hexagonal work.

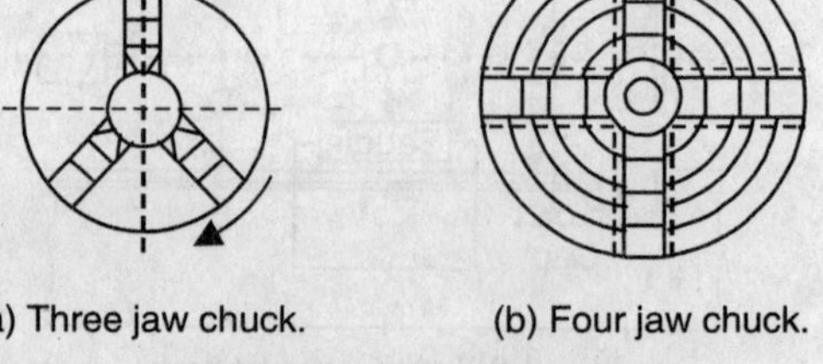

(a) Three jaw chuck. (b) Four jaw chuck.

Fig. 1.11 Chucks.

(*b*) *Four jaw independent chuck*. The four jaw independent chuck, as shown in Fig. 1.11 (*b*), has four reversible jaws, each of which may be independently adjusted to accommodate the work it supports This type of chuck can hold square, round and irregular shape of work in either a concentric or eccentric position.

Note : The other type of the chucks are combination chucks, magnetic chuck, collect chuck, drill chuck, and air or hydraulic chuck.

3. *Lathe dog or carrier*. The work placed on a mandrel or held between centres is rotated positively by clamping the dog or carrier to the end of the work. This is engaged with a pin attached to the drive plate or face plate. The lathe dog or carrier may be of straight type or bent type as shown in Fig. 1.12 (*a*) and (*b*) respectively.

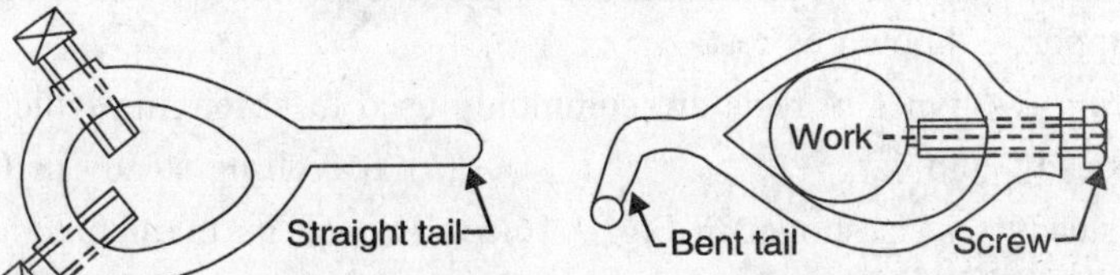

(a) Straight tail type. (b) Bent tail type.

Fig. 1.12 Lathe dog or carrier.

4. *Drive plate*. The drive plate, as shown in Fig. 1.13, is a circular plate which is bored out and threaded so that it can be attached to the spindle nose. It also carries a hole for the pin which is used only when work is held in a lathe dog having straight tail. When bent-tail dog is used, this pin is taken out and the bent portion of the tail is inserted into the hole.

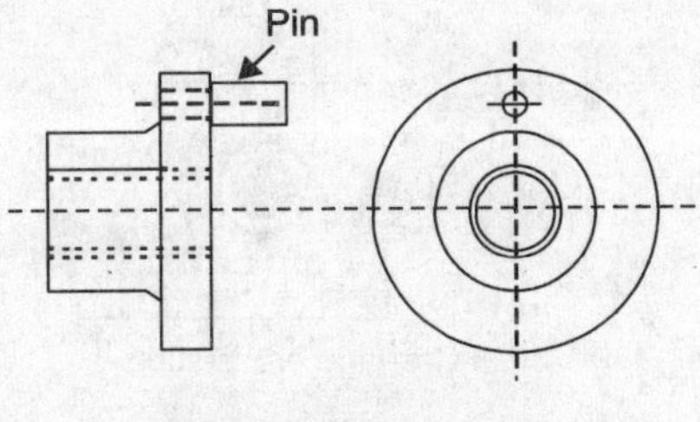

Fig. 1.13 Drive plate.

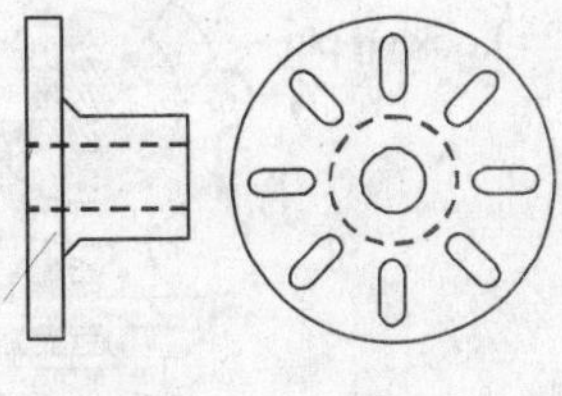

Fig. 1.14 Face plate

5. *Face plate*. The face plate, as shown in Fig. 1.14, is similar to drive plate except that it is larger in diameter. It contains more open slots or T-slots so that bolts may be used to clamp the workpiece to the face of the plate. The faceplate is used for holding workpieces which cannot be conveniently held in a chuck.

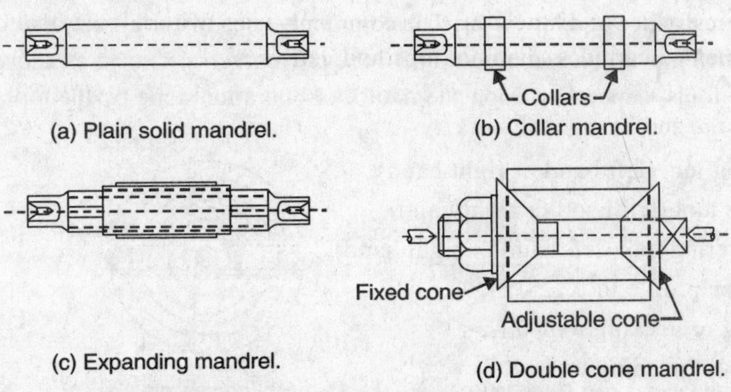

(a) Plain solid mandrel. (b) Collar mandrel.

(c) Expanding mandrel. (d) Double cone mandrel.

Fig. 1.15 Types of mandrels.

6. *Angle plate*. An angle plate is simply a cast iron plate with two faces planed at right angles to each other and having slots in various positions for the clamping bolts. It is always used with the plate for holding such parts which can not be clamped against the vertical surface of the face plate.

7. *Mandrels*. The lathe mandrel is a cylindrical bar with centre hole at each end. It is used to hold hollow workpieces to machine their external surface. The work revolves with the mandrel which is mounted between the centres of the lathe. The various types of mandrels used for different classes of work are shown in Fig. 1.15.

8. *Rests*. When a long slender workpieces are to be turned between the lathe centres, it must be supported at some intermediate point in order to prevent springing or bending of the workpiece. This auxiliary support is known as rest.

The following two types of rests are commonly used in an engine lathe.

(*a*) Fixed steady, and (*b*) Travelling steady or follower rest.

The fixed steady rest is shown in Fig. 1.16(*a*). Though its main function is to provide a support for long thin workpiece that is turned between centres, yet it can be used for supporting one end of a long workpiece for facing, drilling and boring operations. It may be noted that when steady rest is used, the carriage can not be fed to the full length of the work.

The travelling steady or follower rest, as shown in Fig 1.16(b), is used to support the work which is close to the cutting tool. It is bolted to the cross-slide on the saddle and travels with the tool as it feeds along the work.

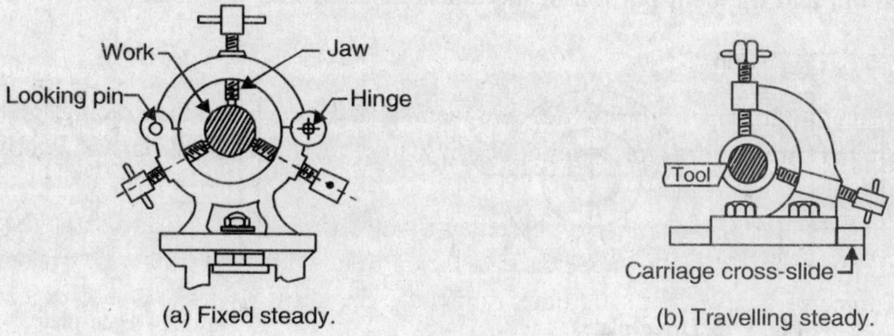

Fig. 1.16 Rests.

1.3.6 Lathe Tools

The tool used in a lathe, for general purpose work, is a single point tool, but for special operations multi-point tools may be used. The material used for lathe tools should have hardness, toughness, heat resistance and low wear. The commonly used materials are high carbon steel, high speed steel, cemented carbides, diamond tips and ceramics.

The lathe tools, depending upon the nature of operation done by the tool, are classified as follows :

1. Turning tool (left hand or right hand),
2. Facing tool (left hand or right hand),
3. Chamfering tool (left hand or right hand),
4. Form or profile tool,
5. Parting or necking tool,
6. External threading tool,
7. Internal threading tool,

8. Boring tool, and
9. Knurling tool.

These tools are shown in Fig. 1.17.

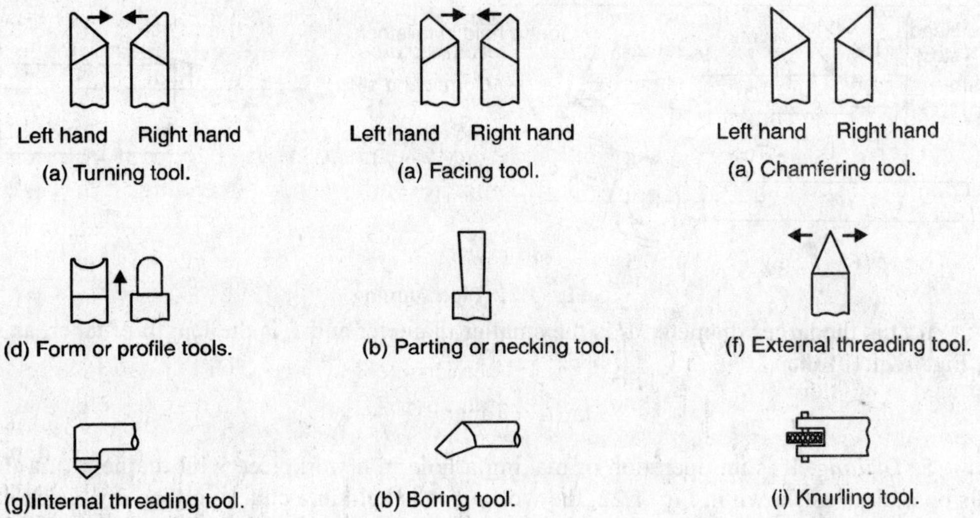

Left hand Right hand
(a) Turning tool.
Left hand Right hand
(a) Facing tool.
Left hand Right hand
(a) Chamfering tool.
(d) Form or profile tools.
(b) Parting or necking tool.
(f) External threading tool.
(g)Internal threading tool.
(b) Boring tool.
(i) Knurling tool.

Fig. 1.17 Lathe tools

1.3.7 Lathe Operations

The most common operations which can be carried out on a lathe are facing, plain turning, step turning, taper turning, drilling, reaming, boring, undercutting, threading, and knurling. These operations are discussed as follows :

1. *Facing.* This operation is almost essential for all works. In this operation, as shown in Fig. 1.18, the workpiece is held in the chuck and the facing tool is fed from the centre of the workpiece towards the outer surface or from the outer surface to the centre, with the help of a cross-slide.

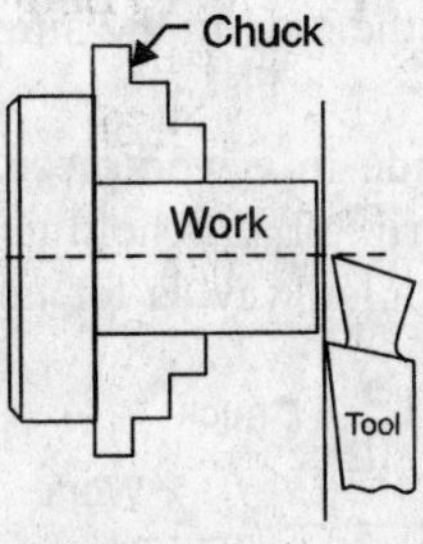

Fig. 1.18 Facing

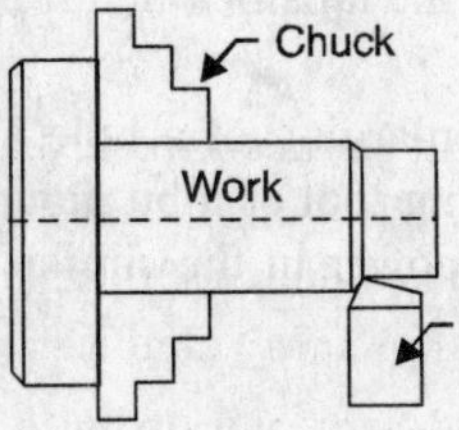

Fig. 1.19 Plain turning

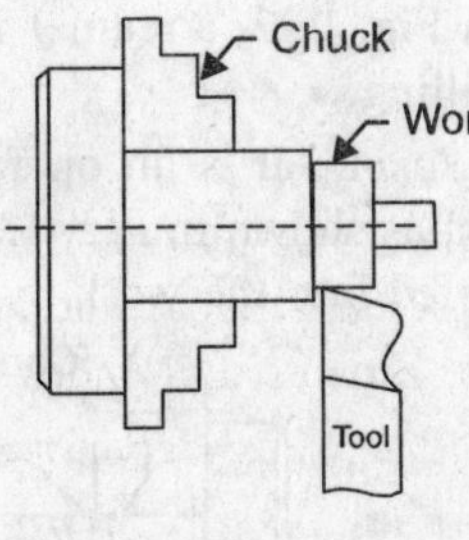

Fig. 1.20 Step turning

2. *Plain turning.* It is an operation of removing excess amount of material from the surface of the cylindrical workpiece. In this operation, as shown in Fig. 1.19, the work is held either in the chuck or between centres and the longitudinal feed is given to the tool either by hand or power.

3. *Step turning.* It is an operation of producing various steps of different diameters in the workpiece, as shown in Fig 1.20. This operation is carried out in the similar way as plain turning.

4. *Taper turning.* It is an operation of producing an external conical surface on a workpiece. A small taper may be produced with the help of a forming tool or chamfering tool, but the larger tapers are produced by swivelling the compound rest, as shown in Fig. 1.21 (a) at the required angle or by offsetting the tailstock or by taper turning attachment.

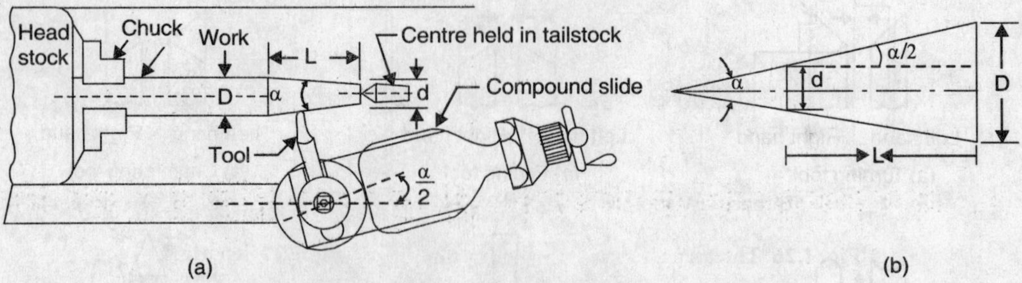

Fig. 1.21 Taper turning.

It D is the larger diameter, d is the smaller diameter and L is the length of taper, as shown in Fig. 1.21 (b) then

$$\tan\alpha = \text{Taper} = \frac{D-d}{L}$$

5. *Drilling.* It is an operation of making a hole in a workpiece with the help of a drill. In this operation, as shown in Fig. 1.22, the workpiece is held in a chuck and the drill is held in the tailstock. This drill is fed manually, into the rotating woekpiece, by rotating the tailstock hand wheel.

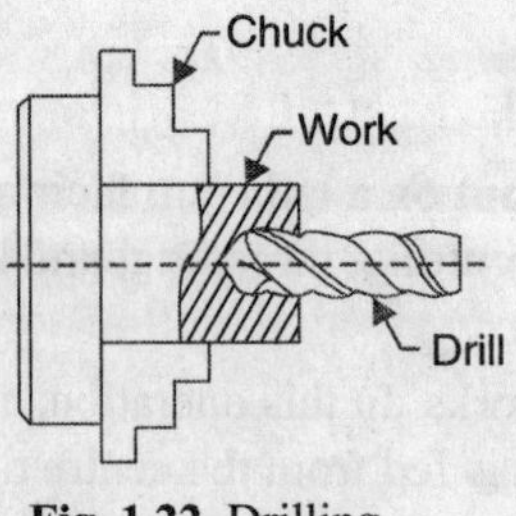

Fig. 1.22 Drilling.

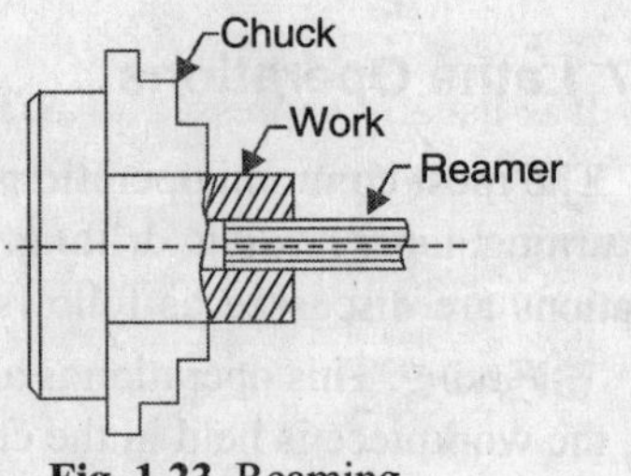

Fig. 1.23 Reaming.

6. *Reaming.* It is operation of finishing the previously drilled hole. In this operation, as shown in Fig. 1.23, a reamer is held in the tailstock and it is fed into the hole in the similar way as for drilling.

7. *Boring.* It is an operation of enlarging of a hole already made in a workpiece. In this operation, as shown in Fig. 1.24, a boring tool or a bit mounted on a rigid bar is held in the tool post and fed into the work by hand or power in the similar in the similar way as for turning.

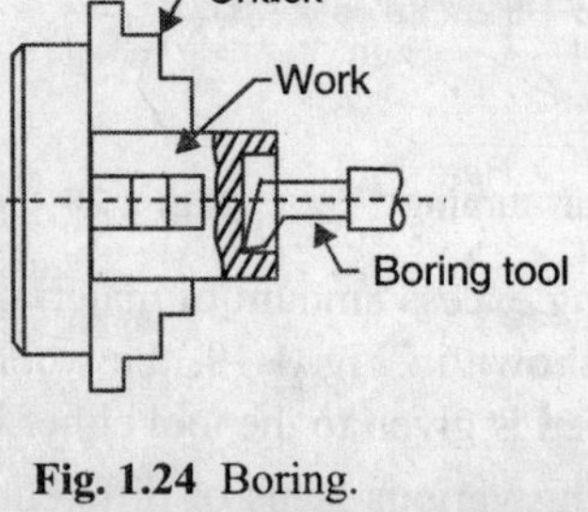

Fig. 1.24 Boring.

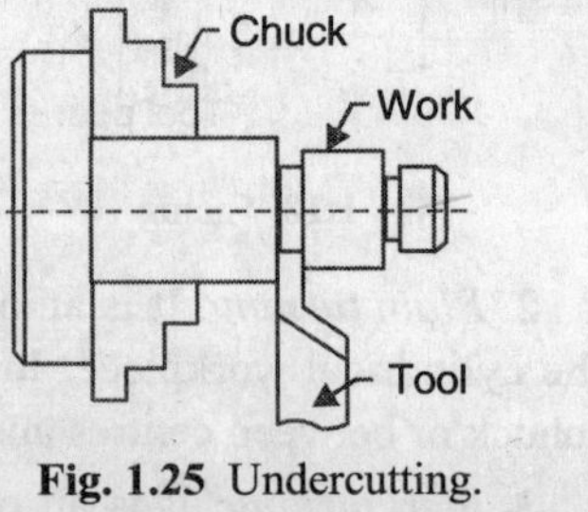

Fig. 1.25 Undercutting.

8. *Undercutting or Grooving.* It is an operation of reducing the diameter of a workpiece over a very narrow surface. In this operation, as shown in Fig. 1.25, a tool of appropriate shape is fed into the revolving work upto the desired depth at right angles to the centre line of the workpiece.

9. *Threading.* It is an operation of cutting helical grooves on the external cylindrical surface of workpiece. In this operation, as shown in Fig 1.26, the work is held in a chuck or between centres and the threading tool is fed longitudinally to the revolving work. The longitudinal feed is equal to the pitch of the thread to be cut.

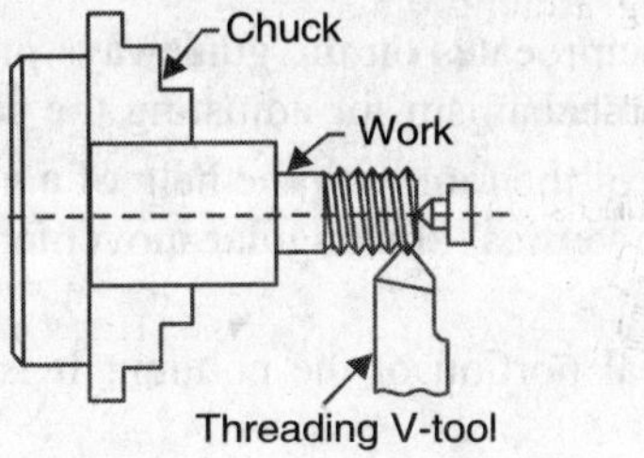

Fig. 1.26 Threading.

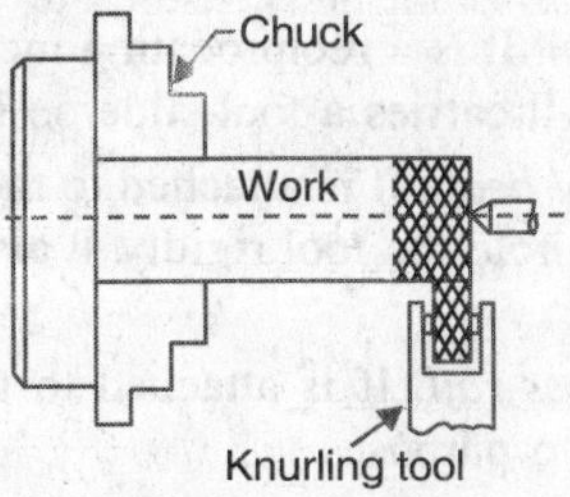

Fig.1.27 Knurling.

10. *Knurling.* It is an operation of providing knurled surface on the workpiece. In this operation, as shown in Fig. 1.27, a knurled tool is moved longitudinally to a revolving workpiece surface.

The projections on the knurled tool reproduce depressions on the work surface.

1.3.8 Terms Used in Lathe Machining

The following terms are commonly used while machining a workpiece on lathe.

1. *Cutting speed.* It is defined as the speed at which the metal is removed by the tool from the workpiece. In other words, it is the peripheral speed of the work past the cutting tool. It is usually expressed in metres per minute.

2. *Feed.* It is defined as the distance which the tool advances for each revolution of the work. It is usually expressed in millimetres.

3. *Depth of cut.* It is defined as the depth of penetration of the tool into the workpiece during machining. In other words, it is the perpendicular distance measured from the machined surface to the unmachined surface of the workpiece. It is usually expressed in millimetres.

1.4 SHAPER

The shaper (also called shaping machine) is a reciprocating type of machine tool used for producing small flat surfaces with the help of a single point tool reciprocating over the stationary workpiece. The flat surface may be horizontal, inclined or vertical. The reciprocating motion of the tool is obtained either by the crank and slotted lever quick return motion mechanism or Whitworth quick return motion mechanism.

1.4.1 Principal Parts of a Shaper

The principal parts of a shaper, as shown in Fig. 1.28, are as follows :

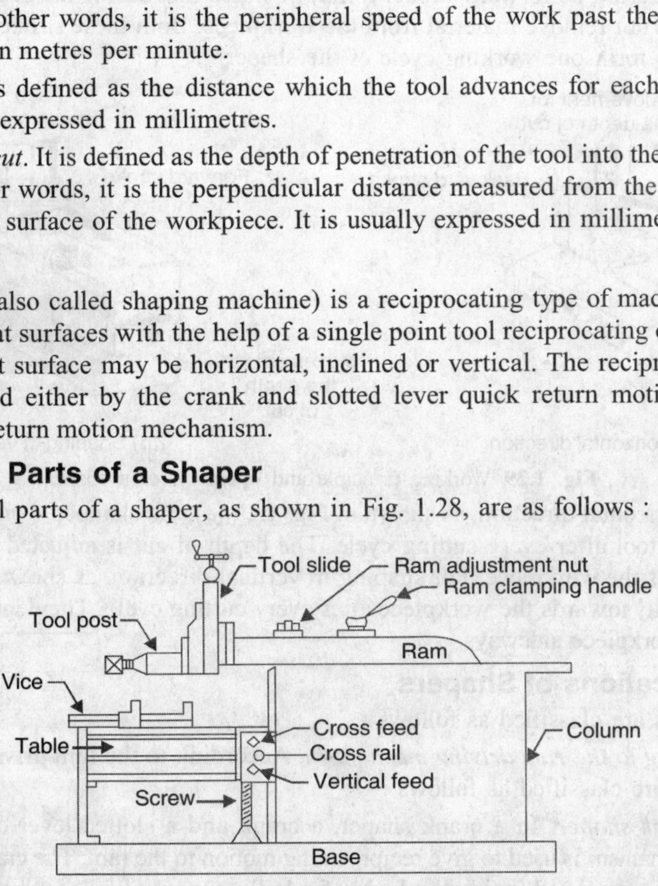

Fig. 1.28 Principal parts of a shaper.

1. *Base.* It is a heavy structure of cast iron which supports other parts of a shaper.

2. *Column.* It is a box-like structure made up of cast iron and mounted upon the base. It contains the driving mechanism and is provided with two machined guide ways on the top of it on which the ram reciprocates.

3. *Ram.* It is a reciprocating member which reciprocates on the guideways provided above the column. It carries a tool-slide on its head and a mechanism for adjusting the stroke length.

4. *Tool head.* It is attached to the front portion of the ram with the help of a nut and a bolt. It is used to hold the tool rigidly. It also provides the vertical and angular movements to the tool for cutting.

5. *Cross-rail.* It is attached to the front vertical portion of the column. It is used for the following two purposes

(*a*) It helps in elevating the table over the column in the upward direction, and

(*b*) The table can be moved in a direction perpendicular to the axis of the ram over this cross-rail.

6. *Table.* It is used for holding the work piece. It can be adjusted horizontally and vertically with the help of spindles.

1.4.2 Working Principle and Operation of a Shaper

We have already discussed that in a shaper, a single point cutting tool reciprocates over the stationary workpiece. The tool is held in the tool post of the reciprocating ram and performs the cutting operation during its forward stroke. It may be noted that during the backward stroke of the ram, the tool does not remove material from the workpiece. Both these strokes (*i.e.*, forward and backward strokes) form one working cycle of the shaper.

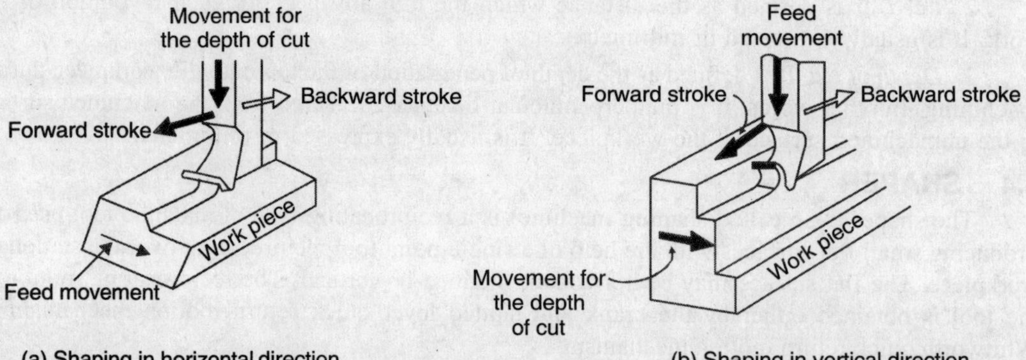

Fig. 1.29 Working principle and operation of a shaper.

For shaping in horizontal direction, as shown in Fig. 1.29 (*a*), the clamped workpiece is fed against the reciprocating tool after every cutting cycle. The depth of cut is adjusted by moving the tool downward towards the workpiece. For shaping in vertical direction, as shown in Fig. 1.29(*b*) the tool is fed vertically towards the workpiece after every cutting cycle. The depth of cut is adjusted by moving the workpiece sideways.

1.4.3 Classifications of Shapers

The shapers are classified as follows :

1. *According to the ram driving mechanism.* According to the ram driving mechanism, the shapers are classified as follows :

(*a*) *Crank shaper.* In a crank shaper, a crank and a slotted lever quick return motion mechanism is used to give reciprocating motion to the ram. The crank arm is adjustable and is arranged inside the body of a bull gear (also called crank gear).

(*b*) *Geared shaper*. In a geared shaper, the ram carries a rack below it, which is driven by a spur gear. This type of shaper is not widely used.

(*c*) *Hydraulic shaper*. In a hydraulic shaper, a hydraulic system is used to drive the ram. This type of shaper is more efficient than the crank and geared shaper.

2. *According to position and travel of ram*. According to the position and travel of ram, the shapers are classified as follows :

(*a*) *Horizontal shaper*. In a horizontal shaper, the ram moves or reciprocates in a horizontal direction. This type of shaper, is mainly used for producing flat surfaces.

(*b*) *Vertical shaper.* In a vetical sharper, the ram reciprocates vertically in the downward as well as in upward direction. This type of shaper is very convenient for machining internal surfaces, keyways, slots or grooves.

3. *According to the direction of cutting stroke*. According to the direction of cutting stroke, the shapers are classified as follows :

(*a*) *Push-cut shaper*. In a push-cut shaper, the ram pushes the tool across the work during cutting operation. In other words, forward stroke is the cutting stroke and the backward stroke is an idle stroke. This is the most general type of shaper used in common practice.

(*b*) *Draw - cut shaper*. In a draw-cut shaper, the ram draws or pulls the tool across the work during cutting operation. In other words, the backward stroke is the cutting stroke and forward stroke is an idle stroke.

4. *According to the design of the table*. According to the design of the table, the shapers are classified as :

(*a*) *Standard or plain shaper*. In a standard or plain shaper, the table has only two movements *i.e.*, horizontal and vertical, to give the feed. It can not be swivelled or tilted.

(*b*) *Universal shaper.* In a universal shaper, in addition to the above two movements, the table can be swivelled about a horizontal axis parallel to the ram and the upper portion of the table can be tilted about the other horizontal axis perpendicular to the first axis. This type of shaper is mostly used in tool room work.

1.4.4 Specifications of a Shaper

The shaper is specified as follows :

1. Maximum length of stroke in millimetres,
2. Size of the table, *i.e.*, length, width and depth of the table,
3. Maximum vertical and horizontal travel of the table,
4. Maximum number of strokes per minute,
5. Power of the drive motor,
6. Type of quick return mechanism.
7. Floor space required, and
8. Weight.

Horizontal shapers range in size from small bench models with stroke of 175 or 200 mm to heavy duty models with strokes as much as 900 mm. Shaping machines are commonly provided with power feeds ranging from 0.2 to 0.5 mm/stroke.

1.4.5 Quick Return Mechanism of a Shaper (Crank and Slotted Lever Type)

The crank and slotted lever quick return motion mechanism for a shaper is shown in Fig. 1.30 An electric motor drives the pinion at a uniform angular speed which, in turn, drives a bull gear. The crank CB is contained in the bull gear and may be adjusted in length by screw mechanism. The crank rotates at a uniform angular speed with the bull gear about the fixed centre C. A sliding block is attached to the crankpin at *B* which is fitted within the slotted lever. This sliding block slides in the guideways of the slotted lever, the bottom end of which is pivoted at A to the base of the shaper while the upper end P is forked and connected to the ram by a short link PR.

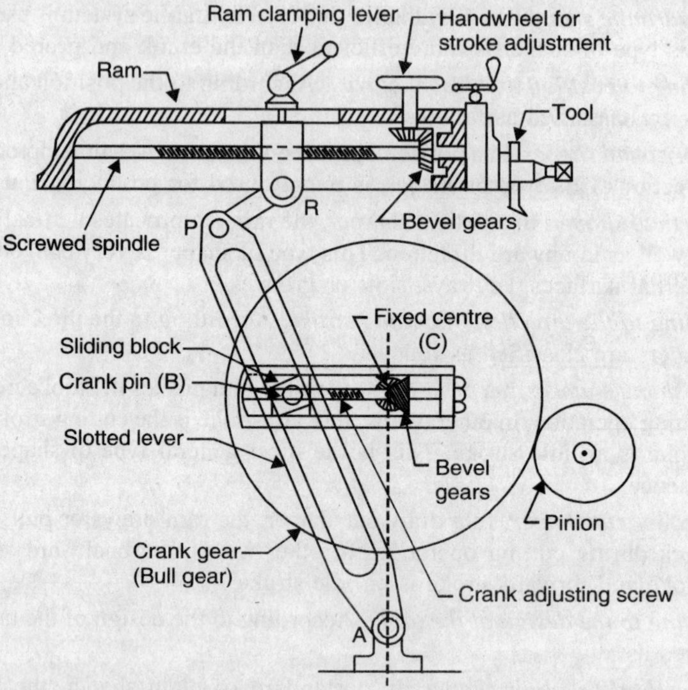

Fig. 1.30 Crank and slotted lever quick return motion mechanism.

When the bull gear rotates, the crank and the sliding block attached to the crank pin also rotates and at the same time slides up and down in the slotted lever causing it to oscillate about the pivoted point A. The short link PR transmits motion from AP to ram which carries the tool and reciprocate along the line of stroke R_1R_2 as shown in Fig. 1.31. Thus in a shaper, the rotary motion of the crank or bull gear is converted into reciprocating motion of the ram. The line of stroke (i.e., R_1R_2) is perpendicular to the line AQ produced.

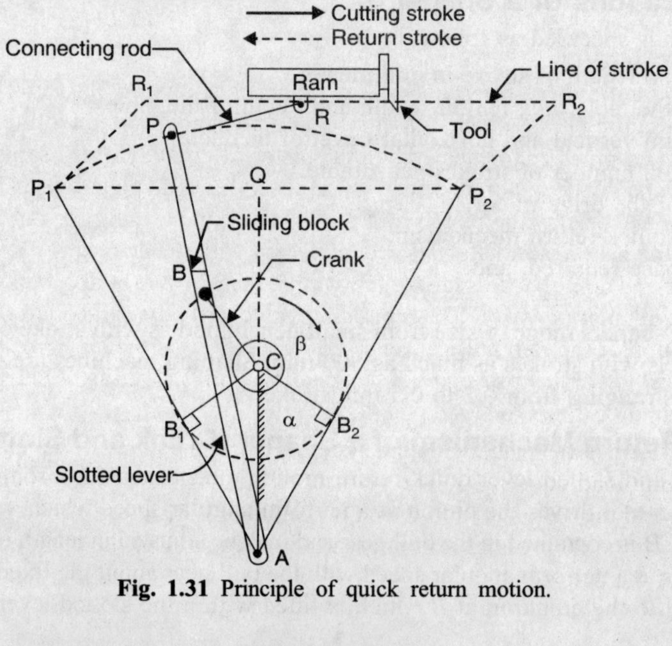

Fig. 1.31 Principle of quick return motion.

The principle of quick return motion can be easily understood from the configuration diagram of the mechanism as shown in Fig. 3.31. In the extreme, positions, the slotted lever *AP* occupies the positions AP_1 and AP_2 which are tangential to the crank-pin circle and the cutting tool is at the end of the stroke. The forward or cutting stroke occurs when the crank rotates from the positions CB_1 to CB_2 (or through an angle β) in the clockwise direction. The backward or return stroke occurs when the crank rotates from the position CB_2 to CB_1 (or through an angle α) in the clockwise direction. Since the crank rotates at a uniform speed, therefore,

$$\frac{\text{Time of cutting stroke}}{\text{Time of return stroke}} = \frac{\beta}{\alpha} = \frac{\beta}{360° - \beta}$$

$$\text{Length of stroke} = R_1R_2 = P_1P_2 = 2P_1Q = 2AP \times \frac{CB}{AC}$$

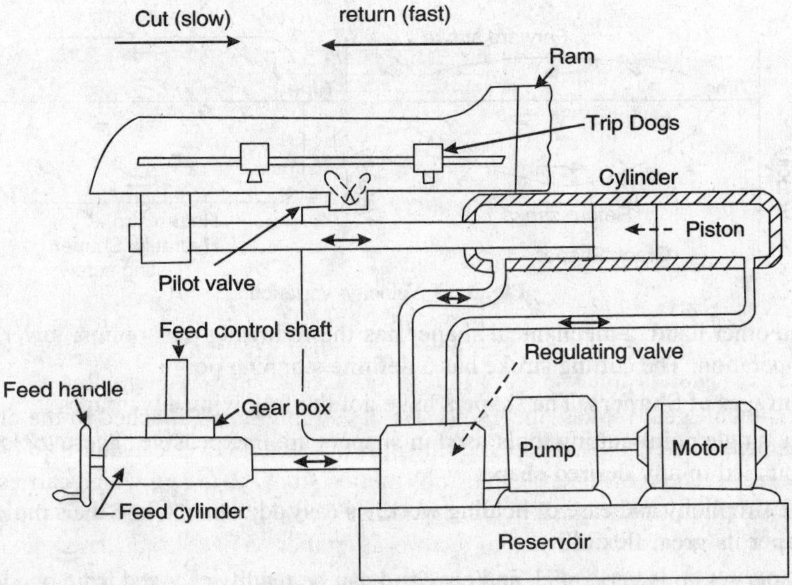

Fig. 1.32 Hydraulic Drive.

1.4.6 Hydraulic Drive

A typical hydraulic drive for a horizontal shaper is shown in Fig. 1.32. A constant speed motor drives a hydraulic pump which delivers oil at a constant pressure to the line. A regulating valve admits oil under pressure to each end on the piston alternately, at the same time allowing oil from the opposite end of the piston to return to the reservoir. The piston is pushed by the oil and, being connected to the ram by the piston rod, pushes the ram carrying the tool. The admission of oil to each end of the piston, alternately, is accomplished with the help of trip dogs and pilot valve. As the ram moves and completes its stroke (forward or return) a trip dog will trip the pilot valve which operates the regulating valve. The regulating valve will admit the oil to the other side of the piston and the motion of the ram will get reversed. It is clear that the length of the ram stroke will depend upon the position of the trip dogs. The length of the ram stroke can be changed by unclamping and moving the trip dogs to the desired positions.

The above system is a constant pressure system. Hence, the ram will travel to slightly different speeds under varying load conditions. The velocity of the ram travel will be directly proportional to the oil pressure and the piston area to which it is applied. The return stroke is quicker, since the piston area on which the oil pressure acts is greater as compared to the other end for which it gets reduced because of the piston rod.

Another oil line is connected to a smaller feed cylinder to change the hydraulic power to mechanical power for feeding the work past the tool.

Advantages of Hydraulic Drive

1. Offers greater flexibility of speed.
2. Smoother operation.
3. Ability to slip when the tool is over loaded.
4. Ability to stall against an obstruction without damage to the tool or the machine.
5. Ability to change length and position of stroke or speed while the machine is running.
6. The hydraulic drive shows a very nearly constant velocity as compared with a mechanical drive, which has a constantly changing velocity because the horizontal component of the crankpin moving about its circle is constantly changing, Fig. 1.33.

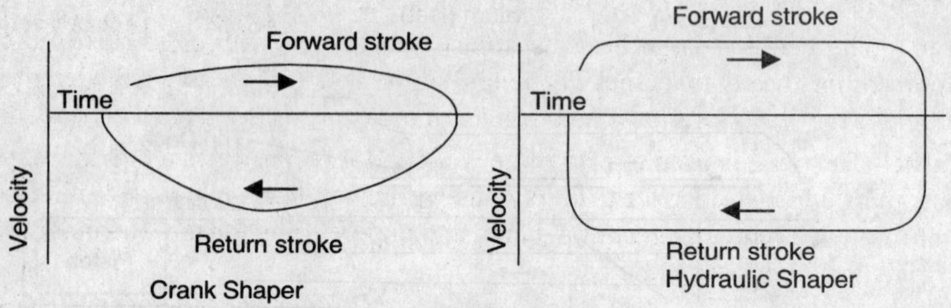

Fig. 1.33 Velocity Variation.

On the other hand, a mechanical shaper has the following plus points: lower first cost and simpler in operation. The cutting stroke has a definite stopping point.

Advantages of Shapers. The shapers have got the following advantages :

1. The single point cutting tools used in shapers are inexpensive. These tools can be easily grounded to any desired shape.
2. The simplicity and ease of holding work, its easy adjustment, and the simple tool give the shaper its great flexibility.
3. Shaper set up is very quick and easy and can be readily changed from one job to another.
4. Thin or fragile jobs can be conveniently machined on shapers because of lower cutting forces.

However, a shaper is by nature a slow machine, because of its straight line, forward and return (idle) stroke. The single point tool requires many strokes to complete a work. Also, cutting speeds in these types of machining operations (shaping, planing and slotting) are not usually very high since difficulties are encountered in designing machine tools with high speeds of reciprocating motion due to high inertia forces developed in the motion of the units and components of the machine. For all these reasons, the shaper does not find ready adaptability for assembly and production line. But because of its advantages discussed above, the shaper finds best use in toolroom, jobbing shops and die or jig making shops (for small or medium-sized work), where its flexibility is important and the relative slowness unimportant because so few identical pieces are made.

1.5 PLANER

In conventional planers, the work held on the table reciprocates past the tool/tools. 'However, in unconventional planers (pit planer and edge planer), the table is stationary and the tool head reciprocates. The planer is the largest of the reciprocating machines (shapers and millers). It has got the following advantages :

1. Larger work can be handled as compared to shapers and millers.

2. Capable of taking much heavier cuts as compared to shapers and millers.

3. There are no overhanging parts such as a ram. Therefore, there is no work or tool deflection or distortion.

4. The work is mounted on a table which is supported throughtout its entire movement to give maximum support.

The work customarily machined on a planer include the following : bases and tables of all kinds of machine tools, frames for printing processes, textile machines, locomotive frames and engines, forging hammer die blocks, large fixtures and moulds, rolling mill parts, parts for large hydraulic presses and altering rolled structural parts.

1.5.1 Principal parts of a Planer

The block diagram of a double housing (column) planer is shown in fig. 1.34. Its principal parts are : Beb, table, housings, cross-rail, and tool heads.

1. **Bed** - The bed is a large heavy casting strengthened with cross-ribs to provide a stiff, heavy support for this heavy machine. The length of the bed is about twice the length of the table. The bed provides the ways for the table to run on. It also houses the drive mechanism for the table

2. **Table** - The table is used to clamp the work and it reciprocates along the ways of the bed. It has T-slots in its upper surface for clamping the workpiece. The table is powered from a variable speed motor through a reducing gear and a rack and pinion arrangement. The other drives can be : hydraulic drive, fast and loose pulley drive.

3. **Housings** - At the sides of the bed, two vertical housings, columns or uprights are arranged. The front face of each column is machined as a set of ways on which travel the cross-rail which carries two tool heads. Most of the planers carry two side tool heads. The housings enclose the various mechanisms which transmit power to the upper parts of the machine from the main motor drive. This includes the vertical feed shaft, rail elevating screw, feed bar for cross feed etc.

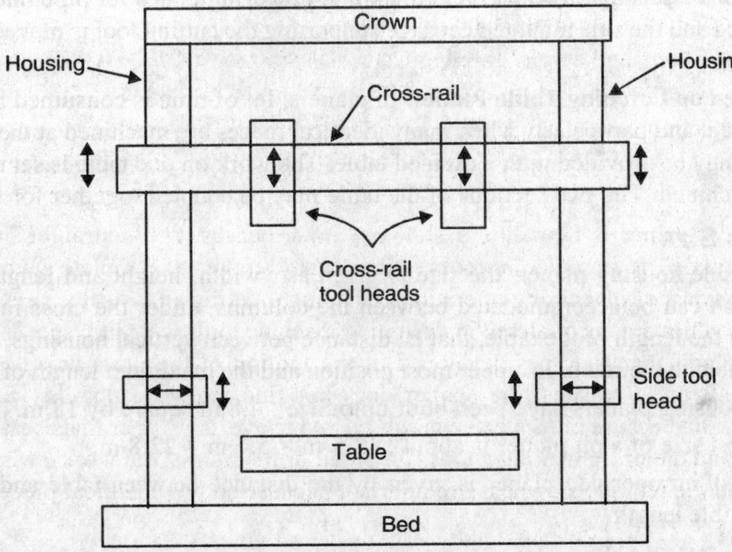

Fig. 1.34 Double housing planer

4. **Cross-rail.** The cross-rail is a horizontal member of heavy cross-section connecting the two housings to increase the rigidity of the machine. Its face is machined to provide ways for the two cross-rail tool heads. The housings are connected together at the top by a cross member called a crown. It provides additional strength and rigidity to the structure. Cross-rail can be traversed

vertically along ways on the housings by means of a drive powered from a separate motor.

5. **Tool heads.** There are a maximum of four tool heads on a double housing planer, two on the cross - rail and two on the uprights. They are mounted by means of saddles. The cross-rail tool heads are used for planing horizontal surfaces. The side tool heads are used for planning vertical surfaces.

1.5.2 Types of Planers

The planers may be classified as follows :

1. **Double housing planer.** It limits the width of the work that can be machined due to the two housings.

2. **Open side planer.** It has only one column. It can carry maximum of 3 tool heads. It can accommodate too wide jobs.

3. **Planer Miller or Grinder.** It is essentially the same as a double housing planer except that it has a milling cutter or a grinding head in place of one or more of the conventional tool heads on the cross rail.

4. **Pit Type Planer.** When the work becomes extremely large and heavy, its reciprocation becomes very difficult. In such cases, the table is kept stationary and the tool reciprocates. Massive rails are mounted on both sides of the table. Tool heads (usually two) are mounted on the cross rails for two way planning. This design (Pit planer) is used to plane the largest kind of work (bases and slabs) which no other machine could handle. A pit planer is one of the largest machine tools. Because of the size of the parts machined on these planers, it is convenient and economical to move the machine than the work.

5. **Plate Planer.** In plate planer, the job/table is stationary while the tool and operator more bark and forth along the work for feed and cut. It is a special purpose (single purpose) machine tool, designed specifically for squaring or levelling the edges of the heavy steel plates (armour plate, plate for ships and for vessels upto about 20 cm thick and preparing plates for pipe lines). The plate is clamped to a bed and the side mounted carriage supporting the cutting tool is moved back and forth along the edge.

6. **Divided or Latching Table Planer.** In planers, lot of time is consumed in setting up the work. To reduce it and particularly when many identical pieces are machined at the same time, any type of planer may be provided with a divided table. The work on one table is set up while that on the other is machined. The two sections of the table may be coupled together for long work.

1.5.3 Planer Size

For a double housing planer, the size is given as : width, height and length of the largest workpiece which can be accommodated between the columns, under the cross rail in its highest position and on the length of the table, that is, distance between vertical housings, the height from the top of the table to the rail in its upper most position and the maximum length of the table travel.

Double housing planers have been built upto Size : 4.8 m square by 18 m.

The largest size of a pit planer is about : 12.5 m × 5.4 m × 22.8 m.

The size of an openside planer is given by the distance between table and maximum rail height and by table length.

1.5.4 Planer Drive

The two popular drives for reciprocating the table are: with d.c. reversible motor and the hydraulic drive.

1. **Variable speed, Reversible d.c. Motor Drive.** This drive has the advantage that the speed of the motor can be varied according to the field current supplied. The full speed forward to full

speed return can be changed so quickly that it is hardly detectable. The common return spoke of a planer is three or four times as fast as the cutting stroke. The power from the motor goes through a clutch and a gear train to a rack mounted on the underside of the table. Adjustable trip dogs at the side of the table determines the length and the position of the stroke.

2. Hydraulic Drive. The hydraulic drive is quite similar to that used for a horizontal shaper. More than one hydraulic cylinder may be used to give a wide range of speeds. The chief drawback of the hydraulic drive on long planers is spasmodic movement of the table due to the compressibility of the hydraulic fluid.

The mechanical drive for cross feed movement of tool heads is similar to that used on shapers.

1.5.5 Planing and Shaping Tools

The tools used in planing and shaping are similar to single-point lathe tools, consisting of the shank and the point. Planing and shaping tools are made of H.S.S. or tipped with cemented carbides. In the latter case, the shanks are made of carbon tool steel. It is a good practice to use planing and

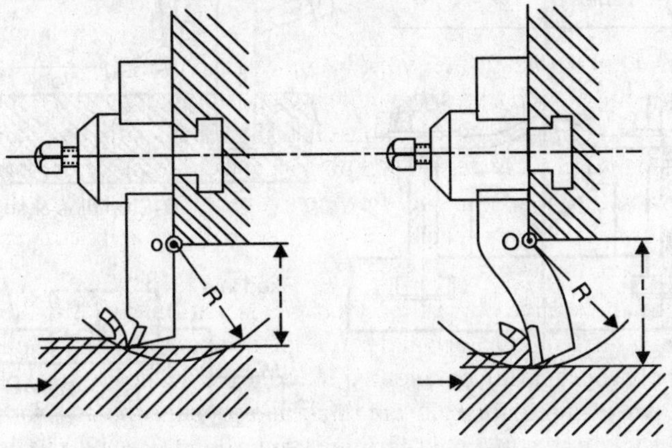

(*a*) (*b*)

Fig. 1.35 Shanks of Planing Tools

shaping tools with an offset, or gooseneck shank to reduce jamming of the tool during operation. When a straight shank tool is used, backward spring of the tool about point *O*, owing to the cutting forces acting on the tool, leads to gouging of the tool nose into the work, [Fig. 1.35(*b*)]. The tendency for the tool to dig into the work will be less for a gooseneck shank, and will be completely eliminated if the tool nose is lined up with the base of the tool shank [Fig. 1.35 (*b*)] since the radius *R* along with the tool nose springs is equal to the overhang *L*.

1.5.6 Work-holding Devices for Shapers, Slotters and Planers

The commonly used work holding devices on these machine tools are given below:

1. T-Bolts and Clamps. The tables of machine tools such as shapers, slotters, planers, drilling machines and milling machines are provided with T-slots. Most jobs can be held on the machine table by means of T-bolts, blocks and clamp plates, Fig. 1.36 (*a*, *b*, *c*). A simple block or a step block (to accommodate jobs of different heights) can be used as the fulcrum block. For round jobs, V-blocks are used for the proper location of the jobs, Fig. 1.36 (*b*, *c*). T-bolt should be nearer to the job than to the fulcrum.

2. Stop pins. Fig 1.36 (*d*, *e*).

3. Stop pins and Toe-dogs. Fig 1.36 (*f*). This arrangement is used for holding thin jobs (particularly on planers).

4. Vices. A vice is a common work holding device. It has two hardened jaws, one fixed and the other movable with the help of a handle. The job is held between the two jaws. The base of

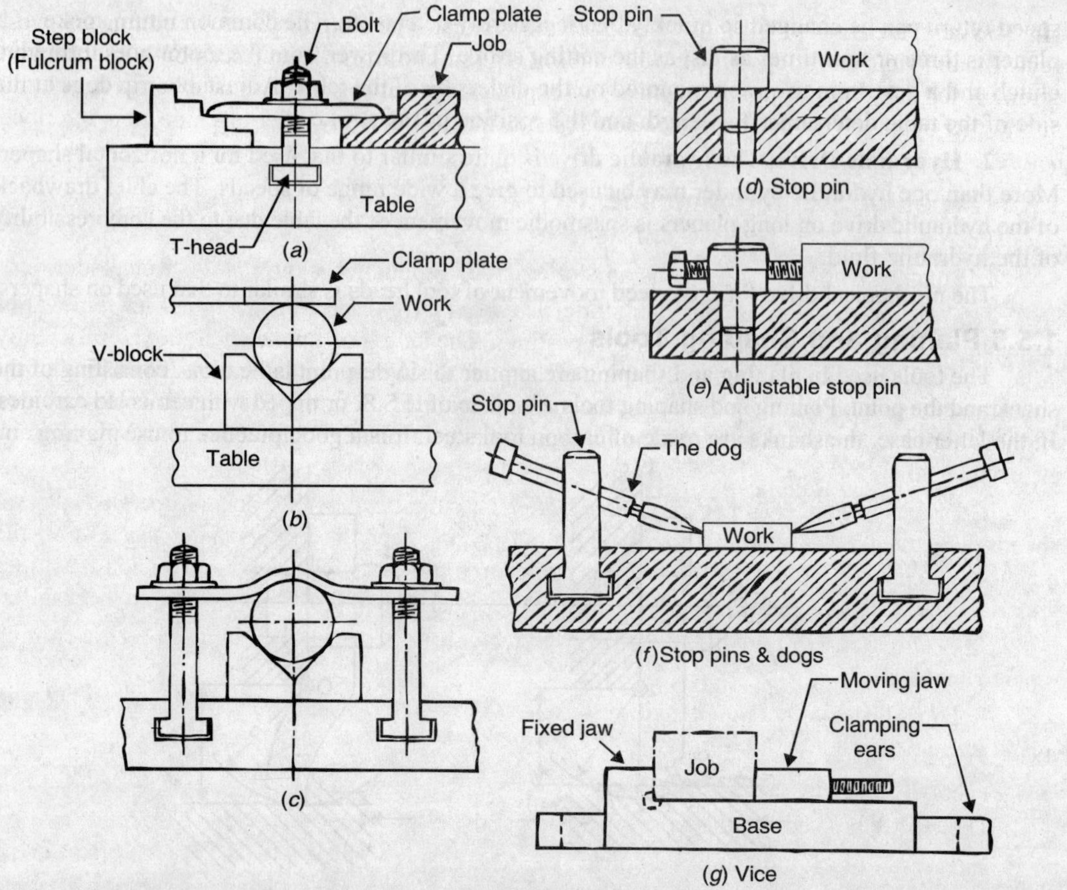

Fig. 1.36 Work-holding devices

the vice carries slots or clamping cars to fix the vise to the table of the machine tool, with the jaws either parallel or perpendicular to T-slots. Special jaws can be attached (with the help of set screws) to the plain jaws of the vise to suit the different contours of jobs. For round jobs, Vees may also be fitted instead of standard plain jaws.

The 'Swivel vice' is made in two parts. It consists of a round swivel base over which the vise body is mounted. The base is graduated into 360°. The body is clamped to the base by means of clamping bolts. The body can be adjusted at any angle relative to the base and so the jaws can be set at any angle. If the vise, apart being swivelled in the horizontal plane, can also be tilted in the vertical plane, it is called 'Universal Vice'.

5. Fixtures. If a job cannot be held in the general purpose work holding devices mentioned above, then a special purpose work holding device, known as fixture, will have to be designed and fabricated.

1.6 SLOTTING MACHINE

The slotting machine falls under the category of reciprocating type of machine tool similar to a shaper or a planer. It operates almost on the same principle as that of a shaper. The major difference between a slotter and shaper is that in a slotter the ram holding the tool reciprocates in a vertical axis, whereas in a shaper the ram holding the tool reciprocates in a horizontal axis. A vertical shaper and a slotter are almost similar to each other as regards to their construction, operation, and use. The only difference being, in the case of a vertical shaper, the ram holding the tool may also reciprocate at an angle to the horizontal table in addition to the vertical stroke.

The ram can be swivelled not more than 5° to the vertical. The slotter is used for cutting grooves, keyways and slots of various shapes, for making regular and irregular surfaces both internal and external, for handling large and awkward workpieces, for cutting internal or external gears and many other operations which cannot be conveniently machined in any other machine tool described before.

1.6.1 Types of Slotting Machine

There are mainly four classes of slotter:

1. Puncher slotter 2. Precision slotter 3. Production slotter 4. Key slotter

1. **Puncher slotter:** The puncher slotter is a heavy, rigid machine designed for removal of large amount of metal from large forging or castings. The length of a puncher slotter is sufficiently large. It may be as long as 1800 to 2000 mm. The puncher slotter ram is usually driven by a spiral pinion meshing with the rack teeth cut on the underside of the ram. The pinion is driven by a variable speed reversible electric motor similar to that of a planer. The feed is also controlled by electrical gears.

2. **Precision slotter :** The precision slotter is a lighter machine and is operated at high speeds. The machine is designed to take light cuts giving accurate finish. Using special jigs, the machine can handle a number of identical works on a production basis. The precision machines are also used for general purpose work and are usually fitted with Whitworth quick return mechanism.

3. **Production slotter** –This is of very heavy construction and can be used for machining tapered jobs.

4. **Key slotter** – It is a special purpose slotter which is designed for cutting keyways in gears, pulleys, cams and similar parts.

1.6.2 Applications of Slotter

The slotter is used for the following purposes:

1. Internal machining of blind holes.
2. Work requiring machining on internal sections such as splines, keyways, various slots and grooves, and teeth.
3. Cutting of teeth on ratchet or gear rings which require primarily rotary feed.
4. Machining of die, punches, straight and curved slots.

1.6.3 Slotter Size

The size of a slotter like that of a shaper is specified by the maximum length of stroke of the ram, expressed in mm. The size of a general purpose or precision slotter usually ranges from 80 to 900 mm. To specify a slotter correctly the diameter of the table in mm. amount of cross and longitudinal travel of the table expressed in mm, number of speeds and feeds available, power (kW) of the motor, floor space required, etc. should also be stated.

1.6.4 Slotting Machine Parts

Fig. 1.37 illustrates a slotting machine. The different parts of a slotting machine are:

1. Base.
2. Column.
3. Saddle.
4. Cross-slide.
5. Rotary table.
6. Ram and toolhead assembly.
7. Ram drive mechanism.
8. Feed mechanism.

1. **Base or bed :** The base is rigidly built to take up all the cutting forces and entire load of the machine. The top of the bed is accurately finished to provide guideways on which the saddle is mounted. The guideways are perpendicular to the column face.

2. **Column:** The column is the vertical member which is cast integral with base and houses driving mechanism of the ram and feeding mechanism. The front vertical face of the column is accurately finished for providing ways on which the ram reciprocates.

3. **Saddle:** The saddle is mounted upon the guideways and may be moved toward or away from the column either by power or manual control to supply longitudinal feed to the work. The top face of the saddle is accurately finished to provide guideways for the cross-slide. These guideways are perpendicular to the guideways on the base.

4. **Cross-slide :** The cross-slide is mounted upon the guideways of the saddle and may be moved parallel to the face of the column. The movement of the slide may be controlled either by hand or power to supply crossfeed.

5. **Rotary table :** The rotary table is a circular table which is mounted on the top of the cross-slide. The table may be rotated by rotating a worm which meshes with a worm gear connected to the underside of the table. The rotation of the table may be effected either by hand or power. In some machines the table is graduated in degrees that enables the table to be rotated for indexing or dividing the periphery of a job in equal number of parts. T-slots are cut on the top face of the table for holding the work by different clamping devices. The rotary table enables a circular or contoured surface to be generated on the workpiece.

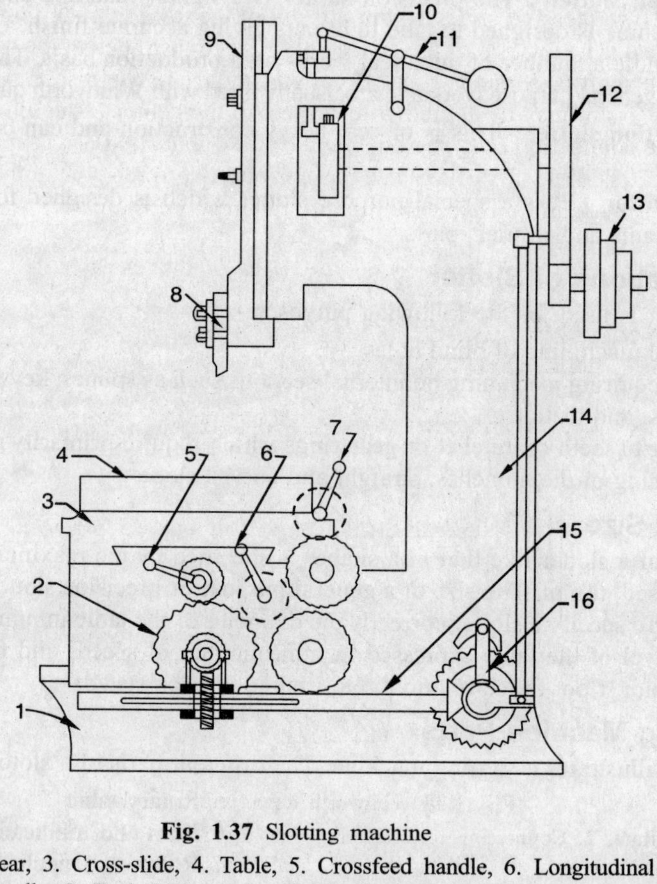

Fig. 1.37 Slotting machine

1. Base, 2. Feed gear, 3. Cross-slide, 4. Table, 5. Crossfeed handle, 6. Longitudinal feed handle, 7. Circular feed handle, 8. Tool, 9. Ram, 10. Crank disc, 11. Lever for counterbalance weight, 12. Bull gear, 13. Cone pulley, 14. Column, 15. Feed shaft, 16. Pawl actuating crank.

6. **Ram and toolhead assembly:** The ram is the reciprocating member of the machine mounted on the guideways of the column. It supports the tool at its bottom end on a toolhead. A slot is cut on the body of the ram for changing the position of stroke. In some machines, special type of toolholders are provided to relieve the tool during its return stroke.

7. **Ram drive mechanism :** A slotter removes metal during downward cutting stroke only whereas during upward return stroke no metal is removed. To reduce the idle return time, quick return mechanism is incorporated in the machine. The usual types of ram drive mechanisms are

(*i*) Whitworth quick return mechanism.

(*ii*) Variable speed reversible motor drive mechanism.

(*iii*) Hydraulic drive mechanism.

8. **Whitworth quick return mechanism:** The Whitworth quick return mechanism is most widely used in a medium sized slotting machine for driving the ram. As shown in Fig. 1.38, the bullgear 7 located at the back of the machine receives its motion from the pinion which is driven by an electric motor. The gear 7 is mounted on a fixed pin or hub 9 attached to the machine frame. The driving plate 8 is mounted on the shaft 6 which passes through the fixed hub 9. The shaft 6 is placed eccentrically with respect to the bullgear centre. A crank pin is mounted on the face of the bullgear which holds a slide block 10. The slide block 10 is fitted within a radial slot provided at the innerside of the driving plate 8. As the bullgear rotates, the crank pin and the slide block 10 rotate in a circular path, but owing to the eccentricity of the bullgear and the driving plate 8, the slide block 10 rotates and slides within the slot of the driving plate imparting it and the shaft 6 rotary movement. The rotation of the driving plate 8 is transmitted to the disc 5 which is attached to the end of the shaft 6. A radial T-slot is cut on the face of the disc 5. The position of the pin fitted within the T-slot may be altered with respect to the center of the disc 5 and then clamped at one end of the connecting rod. The other end of the connecting rod is attached to the ram 1 by a clamping bolt 2. The rotation of the disc 5 is converted into reciprocating movement of the ram by the connecting rod and the pin eccentrically mounted on the disc 5.

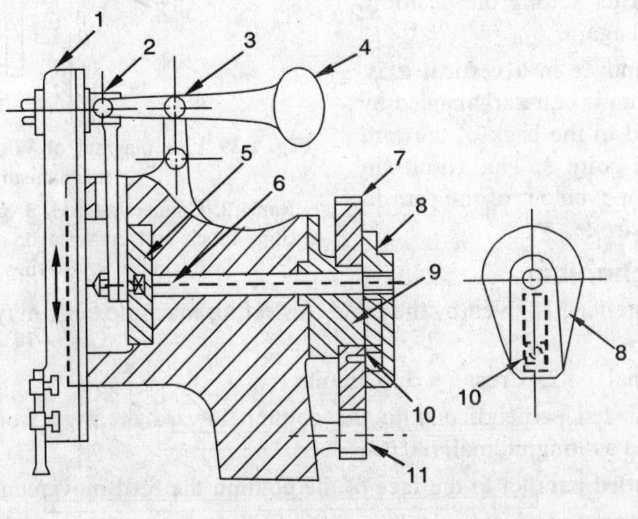

Fig. 1.38 Whitworth quick return mechanism

1. Ram, 2. Connecting rod clamping bolt, 3. Pivot, 4. Counter balance weight, 5. Crank disc, 6. Driving shaft, 7. Bull gear, 8. Driving plate, 9. Fixed hub, 10. Crankpin with slide block, 11. Driving pinion.

The principle of quick return mechanism can be understood from the line diagram shown in Fig. 1.39. *A* and *B* are the fixed centres of the bull gear 7 and the driving plate 8 as shown in the Fig. 1.38. The crank pin and the slide block 4 rotate in a circular path at a constant speed, rotating the plate 8 about *B*. This causes the driving disc 5 attached to the shaft 6 to rotate. The pin 3 on the disc rotates in a circular path about the fixed point *B*. The length of the ram is equal to twice the throw of eccentricity, which is equal to 2 × 3*B*. From the diagram 1.39 it is clear that when the block 4 is in the position *C*, the ram is at the maximum upward position of the stroke and when it is at *D*, the maximum downward position of the ram has been reached. Thus if the bullgear rotates in anti clockwise direction, when the block 4 rotates through an angle *CAD*, the downward cutting stroke is performed, whereas when the block rotates through an angle *DAC*, the return stroke completed. As the block 4 rotates at a constant speed the rotation of the crank pin through an angle *CAD* during the cutting stroke takes longer time than rotation through an angle *DAC* during the return stroke. Thus the quick return motion is obtained. The cutting time and return time is related by the formula :

$$\frac{\text{Cutting time}}{\text{Return time}} = \frac{\angle CAD}{\angle DAC}$$

The length of stroke of the ram can be varied by altering the position of pin 3 with respect to the centre *B*, i.e. the centre of the disc 5. Further the position of the pin 3 with respect to the disc centre, greater will be the throw of eccentricity and longer will be the stroke length.

The position of stroke of the ram can be adjusted by releasing the nut 2 and then by altering the position of the connecting rod clamping bolt within the slot provided on the body of the ram. After setting the position, the nut is tightened again.

As the ram moves in a vertical axis, the weight of the ram is counterbalanced by a weight 4 attached to the back of the ram and is pivoted at a point 3. This results in even and jerk free movement of the ram in cutting and return stroke.

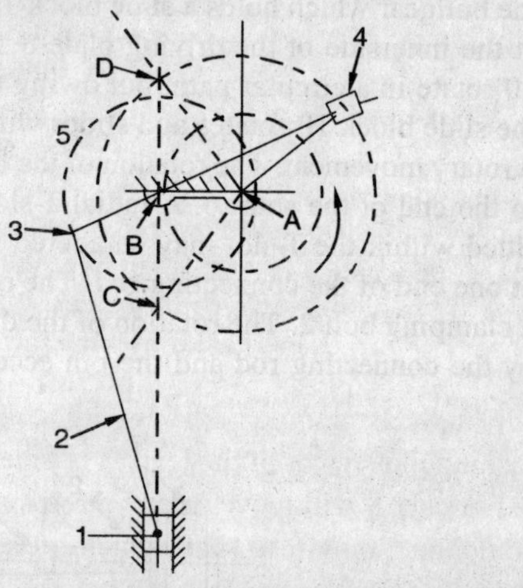

Fig. 1.39 Line diagram of Whitworth quick return mechanism

1. Ram, 2. Connecting rod, 3. Crankpia on disc "5". 4. Slide block and crankpia on driving plate. *A*. Bull gear centre *B*. Driving plate centre

1.6.5 Feed Mechanism

In a slotter, the feed is given by the table. A slotting machine table may have three types of feed movements :

1. Longitudinal 2. Cross 3. Circular

If the table is fed perpendicular to the column toward or away from its face, the feed movement is termed as longitudinal.

If the table is fed parallel to the face of the column the feed movement is termed as cross.

If the table is rotated on a vertical axis, the feed movement is termed as circular.

Like a shaper or a planer, the feed movement of a slotter is intermittent and supplied at the beginning of the cutting stroke. The feed movement may by supplied either by hand or power. The hand feed is supplied by rotating the individual feed screws.

The power feed mechanism is shown in Fig. 1.40 A cam groove 1 is cut on the face of the bull gear in which a roller 2 slides. As the bull gear rotates, the roller attached to a lever 3 follows the contour of the cam groove and moves up and down only during a very small part of revolution of the bull gear. The cam groove may be so cut that the movement of the lever 3 will take place

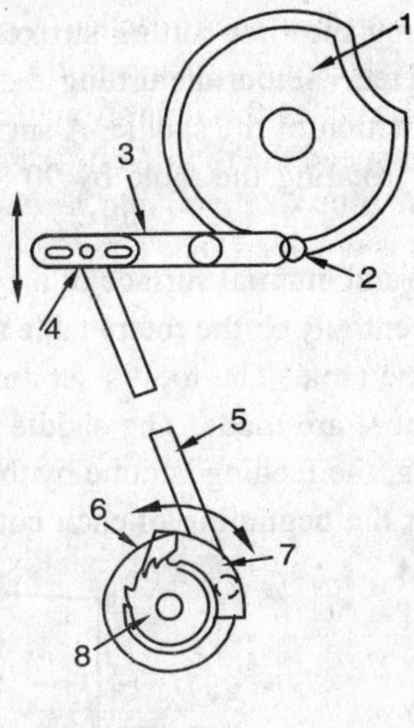

Fig 1.40 Power feed mechanism

1. Cam groove on bull gear. 2. Roller, 3. Lever, 4. Feed adjustment pin, 5. Connecting rod, 6. Pawl, 7. Pawl lever, 8. Ratchet wheel.

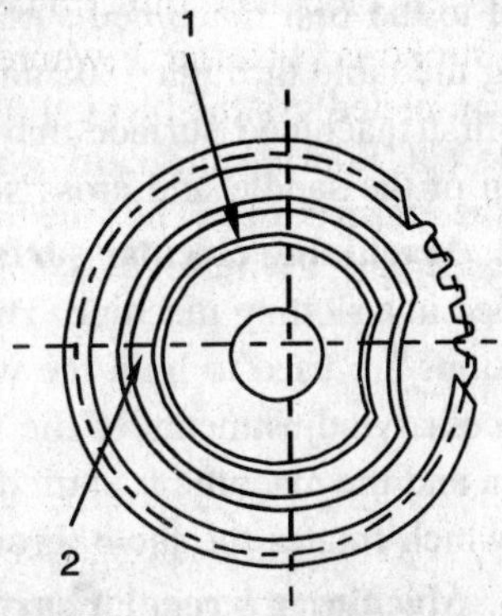

Fig. 1.41 Cam groove on bull gear

1. Cam groove, 2. Roller.

only at the beginning of the cutting stroke. Fig. 1.41 shows the cam groove cut on a bull gear. The rocking movement of the lever 3 is transmitted to the ratchet and pawl mechanism 6 and 8, so that the ratchet 8 will move in one direction only during this short period of time. The ratchet wheel is mounted on a feed shaft which may be engaged with cross, longitudinal or rotary feed screws individually or together to impart power feed movement to the table.

1.6.6 Work Holding Devices

The work is held on a slotter table by a vice, T-bolts and clamps or by special fixtures. T-bolts and clamps are used for holding most of the work on the table. Before clamping, pieces are placed below the work so as to allow the tool to complete the cut without touching the table. Fixtures are used for holding repetitive work.

1.6.7 Slotter Operations

The operations performed in a slotter are :

1. Machining flat surface.
2. Machining cylindrical surface.
3. Machining irregular surface and cam machining.
4. Machining slots, keyways and grooves.

1. **Machining flat surfaces:** The external and internal flat surfaces may be generated on a workpiece easily in a slotting machine. The work to be machined is supported on parallel strips so that the tool will have clearance with the table when it is at the extreme downward position of the stroke. The work is then clamped properly on the table and the position and the length of stroke is adjusted. A clearance of 20 to 25 mm is left before the beginning of cutting stroke, so that the feeding movement may take place during the ideal part of the stroke. The table is clamped to prevent any longitudinal or rotary travel and the cut is started from one end of the work. The

crossfeed is supplied at the beginning of each cutting stroke and the work is completed by using a roughing and a finishing tool. While machining an internal surface, a hole is drilled in the work piece through which the slotting tool may pass during the first cutting stroke. A second surface parallel to the first machined surface can be completed without disturbing the setting by simply rotating the table through 180° and adjusting the position of the saddle. A surface perpendicular to the first machined surface may be completed by rotating the table by 90° and adjusting the position of the saddle and cross slide.

2. **Machining circular surfaces:** The external and internal surface of a cylinder can also be machined in a slotting machine. The work is placed centrally on the rotary table and packing pieces and clamps are used to hold the work securely on the table. The tool is set radially on the work and necessary adjustments of the machine and the tool are made. The saddle is clamped in its position and the machine is started. While machining, the feeding is done by the rotary table feed screw which rotates the table through a small arc at the beginning of each cutting stroke.

3. **Machining irregular surfaces or cams:** The work is set on the table and necessary adjustments of the tool and the machine are made as detailed in other operations. By combining cross, longitudinal and rotary feed movements of the table any contoured surface can be machined on a workpiece.

4. **Machine grooves or keyways :** Internal and external grooves are cut very conveniently on a slotting machine. A slotter is specially intended for cutting internal grooves which are difficult to produce in other machines. External or internal gear teeth can also be machined in a slotter by cutting equally spaced grooves on the periphery of the work. The indexing or dividing the periphery of the work is done by the graduations on the rotary table.

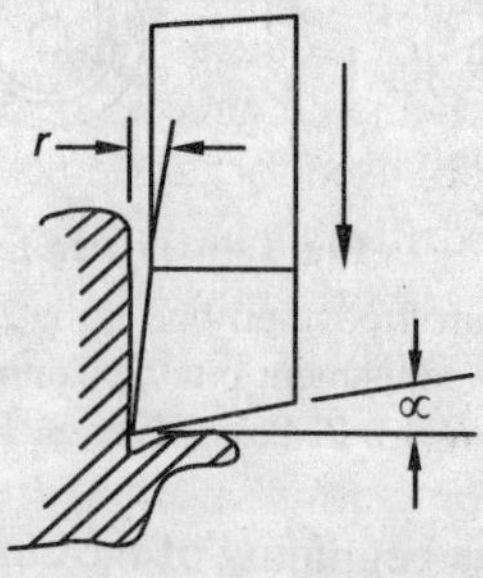

Fig. 1.42 Slotter tool angles

α Top rake angle, r Front clearance angle.

1.6.8 Slotter Tools

A slotting machine tool differs widely from a shaper or a planer tool as the tool in a slotter removes metal during its vertical cutting stroke. This changed cutting condition presents a lot of difference in the tool shape. In a lathe, shaper or a planer tool the cutting pressure acts perpendicular to the tool length, whereas in a slotter the pressure acts along the length of the tool. The rake and the clearance angle of a slotter tool apparently look different from a lathe or a shaper tool as these angles are determined with respect to a vertical plane rather than the horizontal. Fig. 1.42. illustrates a typical slotter tool with tool angles. Slotter tools are provided with top rake, front clearance and side clearance, but no side rake is given. The nose of the tool projects slightly beyond the shank to provide clearance. The amount of rake angle given is similar to that of a shaper tool.

The slotter tools are robust in cross-section and are usually of forged type; of course, bit type tools fitted in heavy duty tool holders are also used. Fig. 1.43 illustrates different slotter tools used in different operations. Keyway cutting tools are thinner at the cutting edges. Round nose tools are used for machining circular or contoured surfaces. Square nosed tools are used for machining flat surfaces.

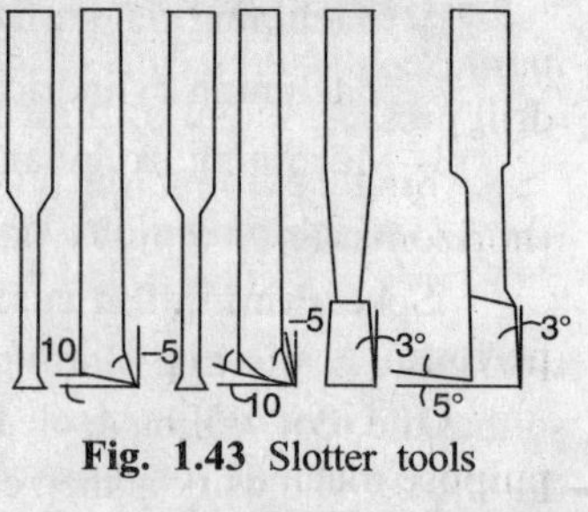

Fig. 1.43 Slotter tools

1.6.9 Cutting Speed, Feed and Depth of Cut

Similar to a shaper, the *cutting speed* of a slotter is defined by the rate with which the metal is removed during downward cutting stroke and is expressed in meters per minute.

Feed is the movement of the work per double stroke expressed in mm.

Depth of cut is the perpendicular distance measured between the machined surface and unmachined surface expressed in min.

1.7 DRILLING MACHINE

The basic purpose of a drilling machine is to drill cylindrical holes in workpieces (metallic and non-metallic materials). The holes are cut out of the material with a cutting tool, which is known as drill. The drill is fixed in rotating spindle and can be fed towards the workpiece which may be fixed to the table or to the base of the machine. The speed of the spindle and the feed can be adjusted according to the workpiece.

1.7.1 Working Principle of a Drilling Machine

We have already discussed that a metal cutting operation that produces cylindrical holes or enlarges existing holes with an end cutting tool is called as drilling. The end cutting tool used for drilling holes in the workpiece is called as twist drill and is provided with two cutting edges. In order to cut off chips, two motions are given to the drill simultaneously as shown in Fig. 1.44.

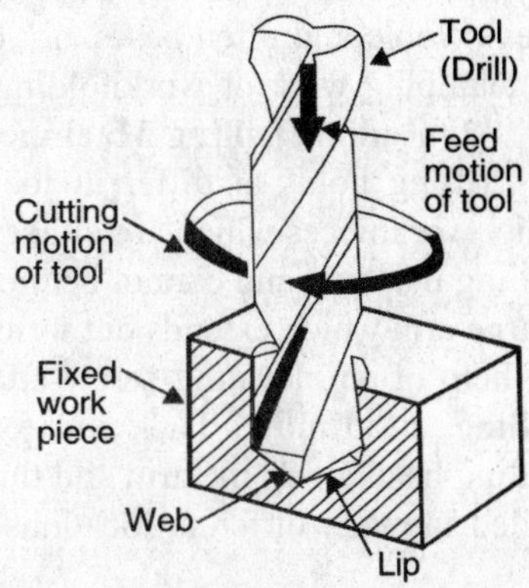

Fig. 1.44 Motions of a twist drill.

1. *Rotary motion*. The rotary motion is called as main motion or cutting motion. In some cases, this motion is given to workpieces also (e.g., while drilling on the lathe). The cutting motion is generally measured in m/min. The highest cutting speed is on the periphery of the drill and it decreases towards the centre of the drill.

2. *Liner motion towards the fixed workpiece*. The motion is called as feed and it controls the thickness of the chip. In small bench type drilling machines, instead of giving feed to the drill, the workpiece is moved towards the rotating drill by raising the table. The feed is generally measured in mm/rev. As the drill is provided with two cutting edges, the thickness of the chip is half the feed.

By the simultaneous double action of cutting (or main motion) and the feed, each cutting edge of the drill describes a spiral and thereby produces a constant flow of chips.

1.7.2 Classification of Drilling Machines

Drilling machines are classified into : general purpose, single purpose and specialised drilling machines.

General purpose Machines. Any hole making operation can be performed in the general purpose machines. This class includes : bench type drill press, Single- and multiple-spindle upright drill presses, radial drilling machines and others.

Single purpose Machines. This class includes : Centre-drilling and deep hole drilling machines (horizontal and vertical types) ; semi-automatic and automatic drilling machines.

Specialised Machines. Special unit-built multiple-spindle drilling machines, intended for performing a single definite operation, are used in mass production.

1. General purpose Drilling Machines. The block diagram of a single-spindle general purpose machine (drill press) is shown in Fig. 1.45. It main parts are : the base, column, table and head. The upright column, table and head. The upright column is supported by a heavy base. The table is suspended from the column and may be moved up and down and clamped at the desired height. The drill head is mounted on the top of the column and consists of the main operating parts.

The cutting tools (held in spindle) are power driven from the motor by a pair of cone pulleys-V belt or through a speed gear box to obtain various speeds. The axial feed of the tool is controlled by hand (on small machines) but larger machines provide power feed for the drill spindle. The spindle has a keyway or spline, so that it may be moved up and down along its axis for feeding, but still retains its drive at any point. The rack pinion is rotated through a worm gearing (in power feeding) and hand feed lever (in manual feeding. The pinion meshes with a rack secured to a non-revolving sleeve or quill in which the spindle freely rotates. The spindle is fed in the axial (vertical) direction together with the sleeve. Table and often bases, are provided with T slots for clamping work or workholding devices.

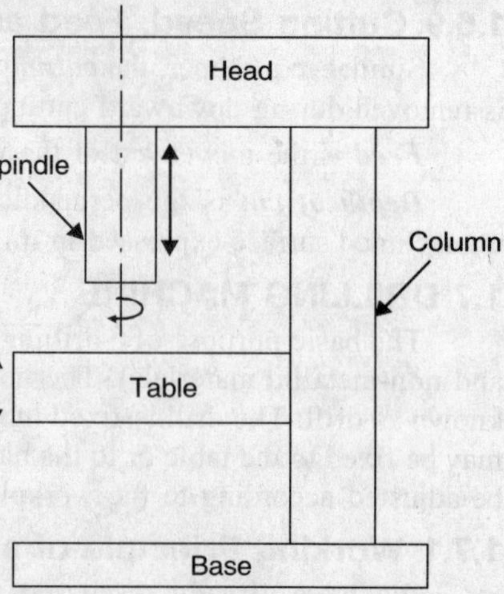

Fig. 1.45 Block Diagram of a Drill press.

2. Radial Drilling Machine. These machines are for drilling holds at different locations on heavy and bulky workpieces which are either inconvenient or impossible to mount on the table of an upright drilling machine and cannot be moved easily. The radial drilling machine shown in Fig. 1.46 has a large arm which extends out form the column. It can be raised or lowered along the column with the help of an elevating screw and it also swings in a complete circle around the column. The drilling head moves back and forth on this arm. Due to the back and forth movement of the drilling head along the arm and the swinging motion of the arm around the column, holes can be drilled in many different locations without moving the workpiece.

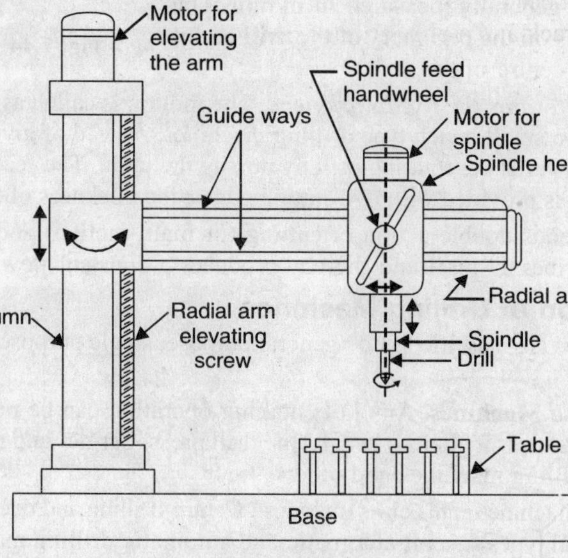

Fig. 1.46 Radial Drilling Machine.

3. Pillar Drilling machine. A pillar drilling machine is shown in Fig. 1.47. It consists of a (1) base plate, (2) column, (3) main drive, (4) drill spindle, (5) feed drive and (6) table. The spindle imparts rotating motion to the drilling tool.

Feed drive is a mechanism for feeding the tool into the work. The work rests on the table. It is a single purpose machine primarily used for drilling a hole.

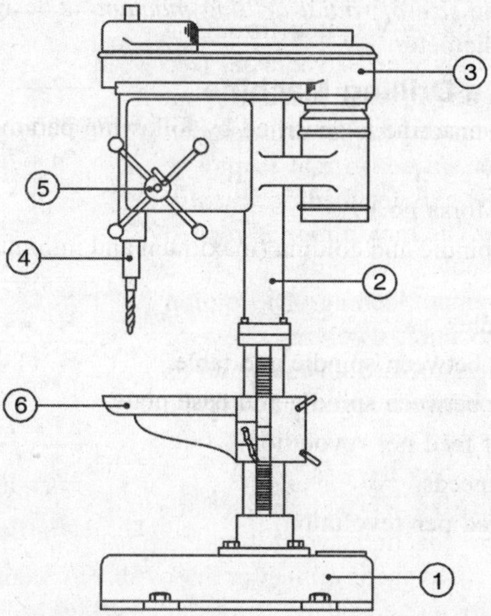

Fig. 1.47 Pillar drilling machine.

4. Bench Drilling Machine. It is the simplest drilling machine used for light duty working. The bench drilling machine shown in Fig. 1.48 consists of a cast iron base having a fixed table over it. The vertical column carries a swivelling table whose height can be adjusted. It can also be swung to any desired position. The drive is provided at the top of the column, which consists of cone pulleys with V-bells and electric motor. Vertical movement to the spindle is given by the feed handle through a rack and pinion arrangement. It can be used for drilling holes upto 2. mm diameter.

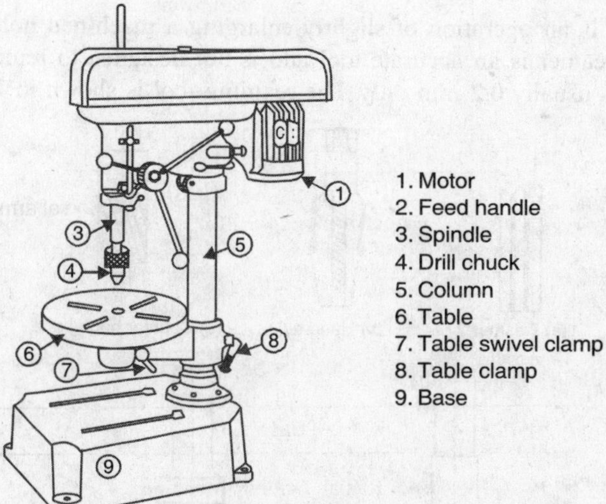

Fig. 1.48 Bench drilling machine.

1.7.3 Size of a Drilling Machine

Portable drilling machines are designated by the maximum diameter of drill they can hold. *Sensitive and up-right drilling machines* are specified by the diameter of the largest workpiece that can be accommodated which is twice the clearance between the centre line of the drill and the machine pillar of frame. The *size of radial drilling machine* is designated by the length of the radial arm and the column diameter.

1.7.4 Specification of a Drilling Machine

A heavy duty drilling machine is specified by following parameters.

1. Drilling capacity,
2. Taper in spindle (Morse no.),
3. Distance between spindle and column (maximum and minimum) in case of radial drilling machine,
4. Transverse of spindle,
5. Minimum distance between spindle and table,
6. Minimum distance between spindle and base plate,
7. Working surface of feed per revolution,
8. Range of spindle speeds,
9. Range of power feed per revolution,
10. Motor speed, and
11. Motor power.

1.7.5 Drilling Operations

The most common operations which can be carried on a drilling machine are drilling, boring, reaming, tapping, counter boring, spot facing and countersinking. These operations are discussed as follows:

1. *Drilling*. It is an operation of producing a circular hole in a workpiece by forcing a drill against it.

2. *Boring*. It is an operation of enlarging a hole that has already been drilled by a single point tool, so as to make it true to required size.

3. *Reaming*. It is an operation of slightly enlarging a machined hole to proper size with smooth finish. The reamer is an accurate tool and is not designed to remove much metal. The reaming allowance is usually 0.2 mm only. The reaming tool is shown in Fig 1.49 (a).

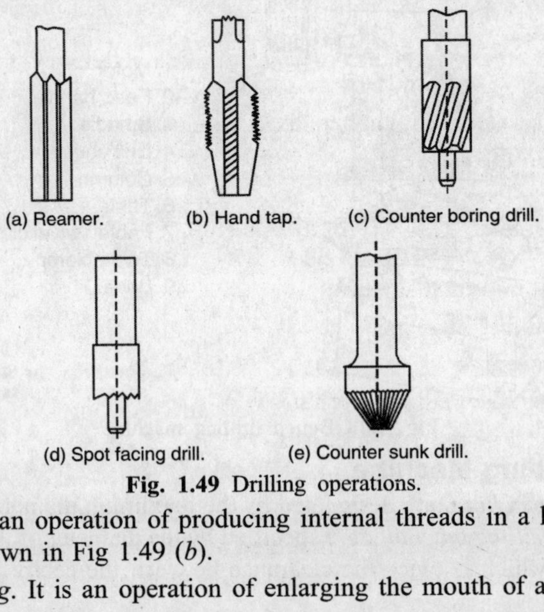

Fig. 1.49 Drilling operations.

4. *Tapping*. It is an operation of producing internal threads in a hole by means of a tool called tap. A tap is shown in Fig 1.49 (*b*).

5. *Counter boring*. It is an operation of enlarging the mouth of a drilled hole to set bolt

heads and nuts below the surface so that they may not project out from the surface level. It is done by a counter boring tool as shown in Fig. 1.49 (*c*) which is similar to an end mill or a single blade ground to give two cutting edges and is guided by a pilot pin which fits into the drilled hole.

6. *Spot facing*. It is an operation to finish off a small portion of rough surface around a drilled hole to provide smooth seat for bolt head. Spot facing tool is a single blade; as shown in Fig. 1.49 (*d*) (which is little larger than the bolt head diameter) ground to cutting edges and is guided by a stem which runs in a pilot hole.

7. *Counter-sinking*. It is an operation to bevel the top of a drilled hole for making a conical seat for a flat head screw. Ordinary flat drill ground to correct angle is used for countersinking holes. A counter sunk drill is shown in Fig. 1.49 (*e*).

1.7.6 Drilling Speed and Feed

The cutting speed in drilling is the peripheral speed of the drill relative to the work. Mathematically,

$$\text{Cutting speed,} \quad v_c = \frac{\pi DN}{1000} m/\min$$

where D = Diameter of drill in mm, and

N = Work speed in r.p.m.

The cutting speed depends upon the material being cut, the tool material, capacity of machine, quality of surface required and the coolant used.

Feed of drill is the distance it moves into the work each revolution of the spindle. It is generally measured in mm/revolution of the spindle.

$$\text{Machining time, } t_m = \frac{l + 0.3d}{\text{feed}(f) \times \text{rpm of drill}(N)} \text{ min}$$

where l = depth of hole to be drilled, mm

f = feed, mm/rev

1.8 MILLING MACHINE

Milling machines are used to hold and rotate milling cutter, hold the workpiece and feed the workpiece to the milling cutter in one of several directions. Milling machines can machine flat, contoured and helical surfaces, cut gears and do various other jobs.

1.8.1 Types of Milling Machines

Milling machines can be classified in different ways :

(*a*) Location of axis of spindle

(*i*) Horizontal milling machines

(*ii*) Vertical milling machines

(*b*) According to purpose

(*i*) General purpose milling machines

(*ii*) Special purpose milling machines.

1. General purpose milling machines

The most common kinds of milling machines is the column and knee type model shown in Fig. 1.50. The structure of the machine is mounted on the base or bed of the machine. It provides rigidity and strength to the machine. It may also act as reservoir for the cutting fluid. The column

is a heavy box like structure. Inside this is space fox motor and much of the driving mechanism. The "knee" slides vertically up and down on the column face. A feed or elevating screw extends from the base to the knee. On top of the knee is a "saddle". The saddle can move transversely (cross movement toward and away from the column). On top of the saddle is the "table". The table moves longitudinally
(back and forth) in front of the column. The surface of the table is cut with T-slots to which work holding devices such as the vise can be clamped.

The driving power for all the cutting tools comes from the rotation of the spindle. This is a hollow shaft that rotates in bearings.

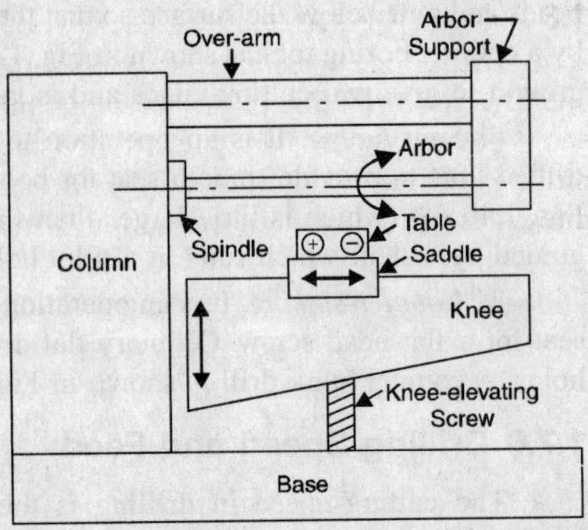

Fig. 1.50 Plain Column and Knee type Milling Machine.

In a Universal milling machine, the table can be swivelled to enable helical groves to be milled.

2. Vertical Milling Machine

A vertical milling machine is shown in Fig. 1.51. It consists of head, swivel plate, spindle nose, work table, saddle, knee, knee elevating screw, column and base. The spindle is held in a vertical position.

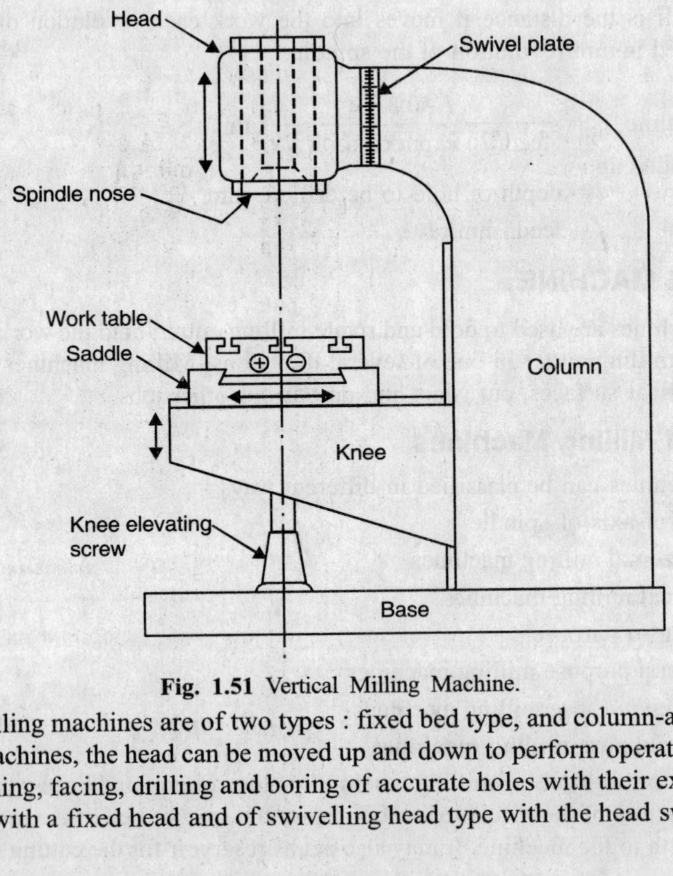

Fig. 1.51 Vertical Milling Machine.

Vertical milling machines are of two types : fixed bed type, and column-and-knee type (Fig. 3.39). In some machines, the head can be moved up and down to perform operations like grooving, slotting, die forming, facing, drilling and boring of accurate holes with their exact locations. The machine can be with a fixed head and of swivelling head type with the head swivelling around a horizontal axis.

1.8.2 Milling Operations

The milling operations can be classified into two major groups, namely, 1. horizontal milling, and 2. Vertical milling. In the horizontal milling operation, the cutter axis is horizontal. Fig. 1.52 shows some common horizontal million operations. These are : slab milling, slot milling and form milling. Horizontal milling can be further divided into two groups depending on the relative directions of cutting and feed motion. When the arrangement is like that as sown in Fig 1.53(*a*),

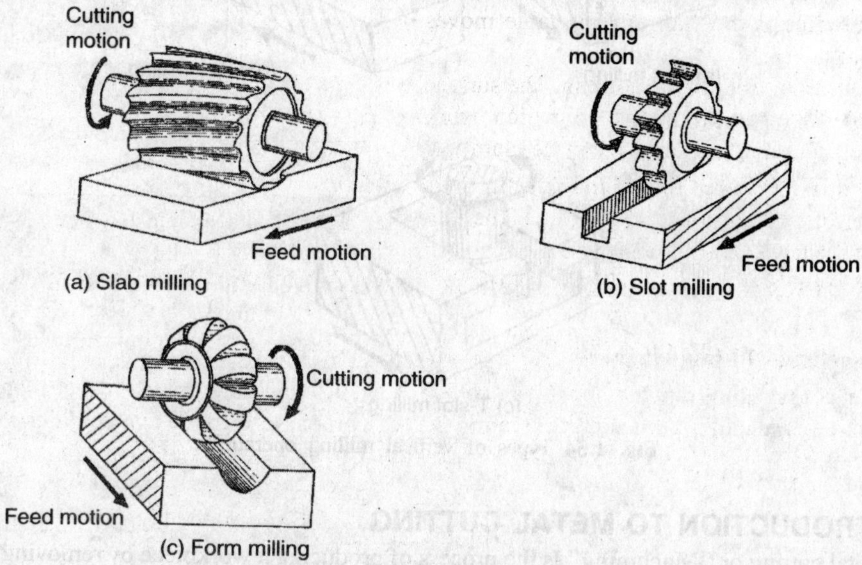

Fig. 1.52 Types of milling operations.

the operation is called up milling. When the cutting and feed motion are in the same direction as shown in Fig. 1.53 (*b*), the operation is called down milling. Since in down milling there is a tendency of the job being dragged into the cutter, up milling is safer and is commonly used. However, down milling gives better surface finish and longer tool life.

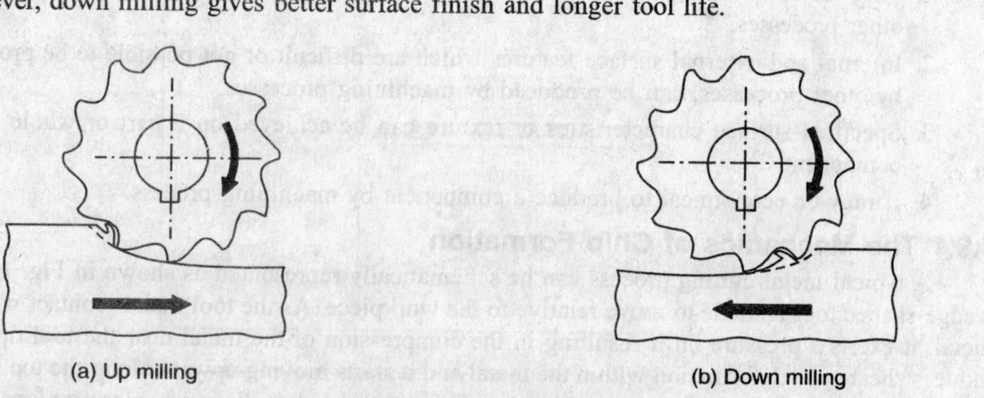

Fig. 1.53 Upmilling and down milling.

The cutter axis is vertical and perpendicular to the work surface in vertical milling. Some common vertical milling operations are shown in Fig. 1.54. Theses operations are slot milling, face milling and T-slot milling.

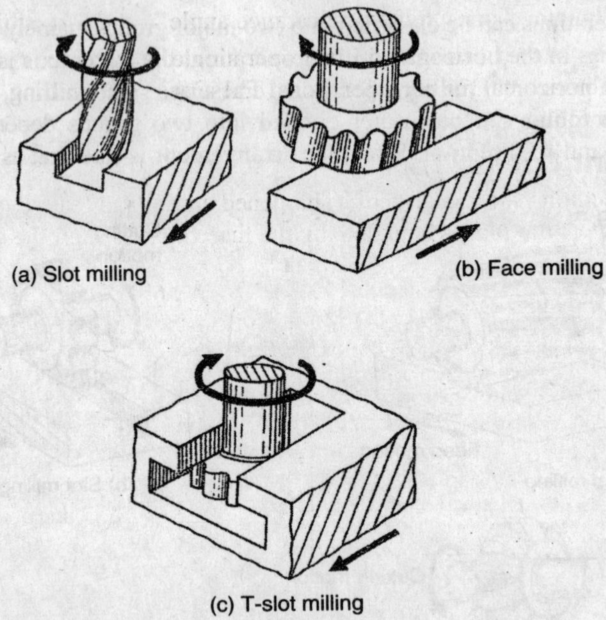

Fig. 1.54 Types of vertical milling operations.

1.9 INTRODUCTION TO METAL CUTTING

Metal cutting or "Machining" is the process of producing a workpiece by removing unwanted material from a block of metal, in the form of chips. This process is most important since almost all the products get their final shape and size by metal removal, either directly or indirectly. The major drawback of the process is loss of material in the form of chips. Inspite of these drawbacks, the machining process has the following characteristics :

1. They improve the dimensional accuracy and tolerances of the components produced by other processes.
2. Internal and external surface features which are difficult or not possible to be produced by other processes, can be produced by machining processes.
3. Specified surface characteristics or texture can be achieved on a part or whole of the component.
4. It may be economical to produce a component by machining process.

1.9.1 The Mechanics of Chip Formation

A typical metal cutting process can be schematically represented as shown in Fig. 1.55. A wedge-shaped tool is made to move relative to the workpiece. As the tool makes contact with the metal, it exerts a pressure on it resulting in the compression of the metal near the tool tip. This induces shear-type deformation within the metal and it starts moving upward along the top face of the tool. As the tool advances, the material ahead of it is sheared continuously along a plane called the "Shear plane". This shear plane is actually a narrow zone (of the order of about 0.025 mm) and extends from the cutting edge of the tool to the surface of the workpiece. The cutting edge of the tool is formed by two intersecting surfaces. The surface along which the chip moves upwards is called "Rake surface" and the other surface which is relieved to avoid rubbing with the machined surface, is called "Flank". The angle between the rake surface and the normal is known as "Rake angle", α (which may be positive or negative), and the angle between the flank and the horizontal machined surface is known as the "relief or clearance angle", γ. Most cutting processes have the same basic features as shown in Fig. 1.55, where a single point cutting tool is used (a milling cutter, a drill, and a broach can be regarded as several single-point tools joined together and are known as multi-point tools).

1.9.2 Single Point Cutting Tool

A single point cutting tool consists of a sharpened cutting part called its point and the shank, (Fig. 1.56). The point of the tool is bounded by the face (along which the chips slide as they are

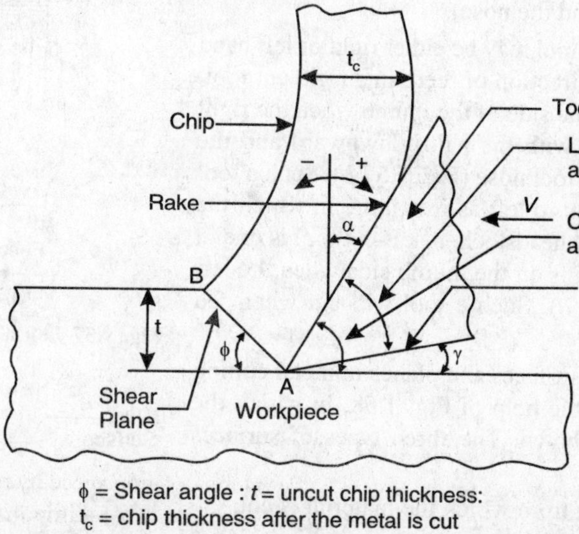

Fig. 1.55 Schematic Representation of Machining Process.

cut by the tool), the side flank or major flank the end flank, or minor flank and the base. The side cutting edge, *a-b*, is formed by the intersection of the face and side flank. The end cutting edge *a-c* is formed by the intersection of the face and the end flank. The chips are cut from the work piece by the side-cutting edge. The point '*a*' where the end and side-cutting edges meet is called the "nose" of the tool. Fig 1.56 is for a right hand tool. Below, we give the definitions of the various tool elements and tool angles :–

Shank. It is the main body of the tool.

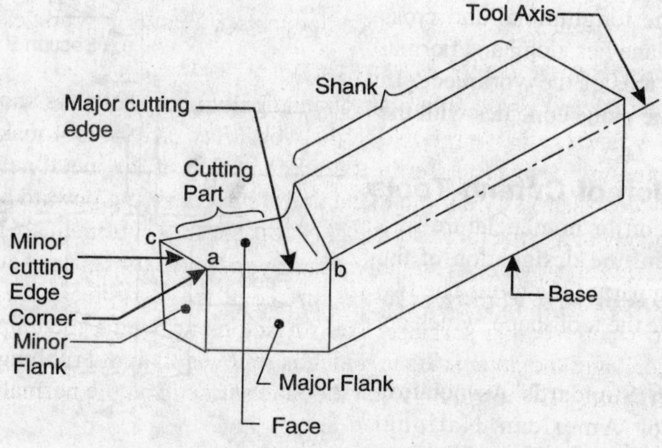

Fig. 1.56 A Single Point Cutting Tool.

Flank. The surface or surfaces below and adjacent to the cutting edge is called flank of the tool.

Face. The surface on which the chip slides is called the face of the tool.

Heel. It is the intersection of the flank and the base of the tool.

Nose. It is the point where the side cutting edge and end cutting edge intersect.

Cutting edge. It is the edge on the face of the tool which removes the material from the workpiece. The total cutting edge consists of side cutting edge (major cutting edge), end cutting edge (minor cutting edge and the nose).

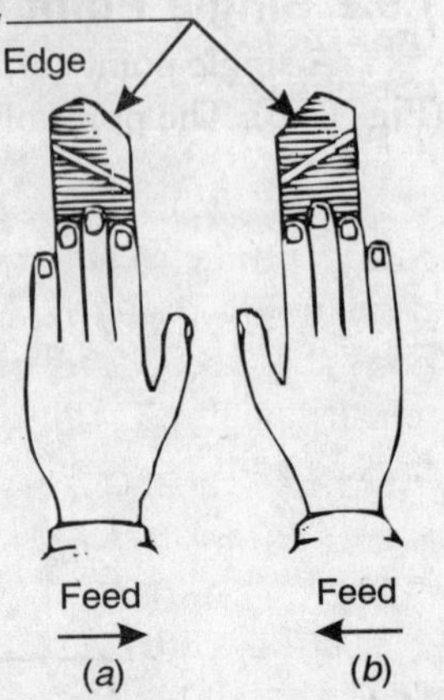

Fig. 1.57 Left and Right Cut Tools.

A single point cutting tool may be either right or left hand cut tool depending on the direction of feed. In a right cut tool, the side cutting edge is on the side of the thumb when the right hand is placed on the tool with the palm downward and the fingers pointed towards the tool nose (Fig. 1.57 *b*). Such a tool will cut when fed from right to left as in a lathe in which the tool moves from tailstock to headstock. A left-cut tool is one in which the side cutting edge is on the thumb side when the left hand is applied (Fig. 1.57 *a*). Such a tool will cut when fed from left to right.

The various types of surfaces and planes in metal cutting are explained below with the help of Fig. 1.58, in which the basic turning process is shown. The three types of surfaces are :–

(1) the work surface, from which the material is cut.

(2) the machined surface which is formed or generated after removing the chip.

(3) the cutting surface which is formed by the side cutting edge of the tool.

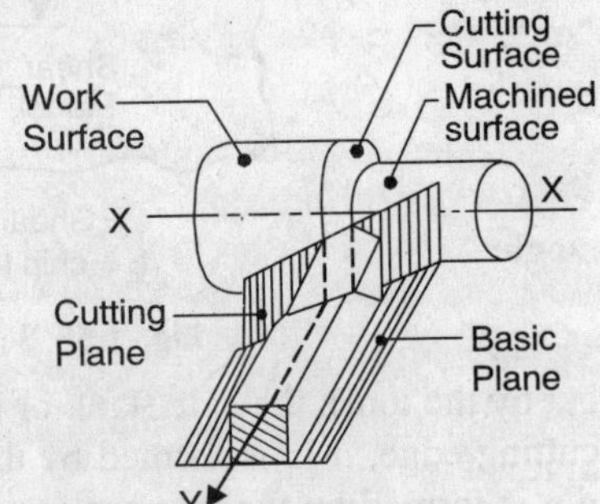

Fig. 1.58 Principal Surfaces and Planes in Metal Cutting.

The references from which the tool angles are specified are the 'cutting plane' and the 'basic plane' or the 'principal plane'. The cutting plane is the plane tangent to the cutting surface and passing through and containing the side cutting edge. The basic plane is the plane parallel to the longitudinal and cross feeds, that is, this plane lies along and normal to the longitudinal axis of the workpiece. In a lathe tool, the basic plane concides with the base of the tool.

1.9.3 Designation of Cutting Tools

By designation or nomenclature of a cutting tool is meant the designation of the shape of the cutting part of the tool., The two systems to designate the tool shape, which are widely used, are :–

1. American Standards Association System (ASA) or American National Standards Institute (ANSI).

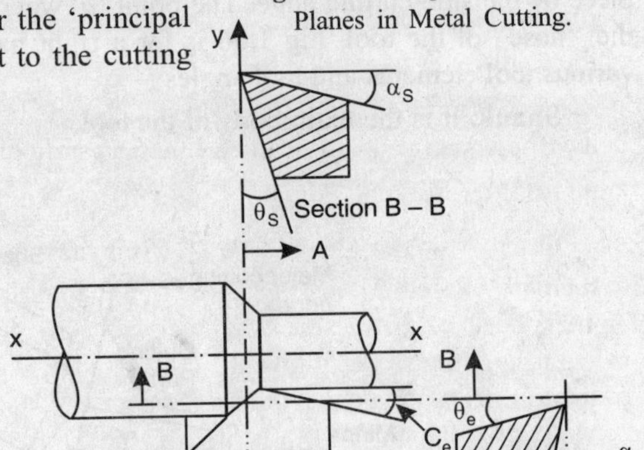

Fig. 1.59 ASA System.

2. Orthogonal Rake System (ORS).

ASA System. In the ASA system, the angles of tool face, that, is its slope, are defined in two orthogonal planes, one parallel to and the other perpendicular to, the axis of the cutting tool, both planes being perpendicular to the base of the tool. For simple turning operation, this system is illustrated in Fig. 1.59.

The typical right hand single point cutting tool terminology is given in Fig. 1.60 (*a*). Fig. 1.60 (*b*) gives the three views of the single point cutting tool, with all the details marked on it.

The various tool angles are defined and explained below :

Side Cutting Edge Angle (SCEA). Side cutting edge angle, C_s also known as lead angle, is the angle between the side cutting edge and the side of the tool shank.

The complimentary angle of SCEA is called the "Approach angle"

End Cutting-Edge Angle (ECEA). This is the angle between the end cutting edge and a line normal to the tool shank C_e.

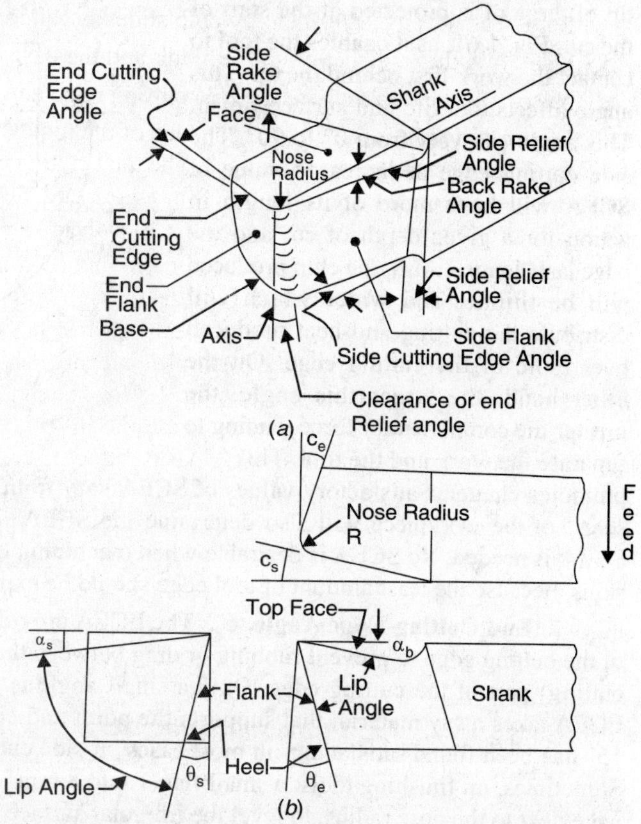

Fig. 1.60 Tool Terminology.

Side Relief Angle (SRA). It is the angle between the portion of the side flank immediately below the side cutting edge and a line perpendicular to the base of the tool, and measured at right angle to the side flank θ_s.

End Relief Angle (ERA). It is the angle between the portion of the end flank immediately below the end cutting edge and a line perpendicular to the base of the tool, and measured at right angle to the end flank θ_e.

Back-Rake Angle (BRA), α_b. It is the angle between the face of the tool and a line parallel to the base of the tool and measured in a plane (perpendicular) through the side cutting edge. This angle is positive, if the side cutting edge slopes downwards from the point towards the shank and is negative if the slope of the side cutting edge is reverse. So this angle gives the slope of the face of the tool from the nose towards the shank.

Side-Rake Angle (SR), α_s. It is the angle between the tool face and a line parallel to the base of the tool and measured in a plane perpendicular to the base and the side cutting edge. This angle gives the slope of the face of the tool from the cutting edge. The side rake is negative if the slope is towards the cutting edge and is positive if the slope is away from the cutting edge.

Importance of Tool Angles :

1. **Side Cutting-Edge Angle, C_s.**

It is the angle which prevents interference as the tool enters the work material. The tip of the tool is protected at the start of the cut, Fig. 1.61, as it enables the tool to contact the work first behind the tip. This angle affects tool life and surface finish. This angle can vary from 0° to 90°. The side cutting edge at increased value of SCEA will have more of its length in action for a given depth of cut and the edge lasts longer. Also, the chip produced will be thinner and wider which will distribute the cutting and heat produced over more of the cutting edge. On the other hand, the larger this angle, the greater the component of force tending to separate the work and the tool. This promotes chatter. Satisfactory values of SCEA vary from 15° to 30°, for general machining. The shape of the workpiece will also determine the SCEA. To produce a 90° shoulder, zero degree SCEA is needed. No SCEA is desirable when machining castings and forgings with hard and scaly skins, because the least amount of tool edge should be exposed to the destructive action of the skin.

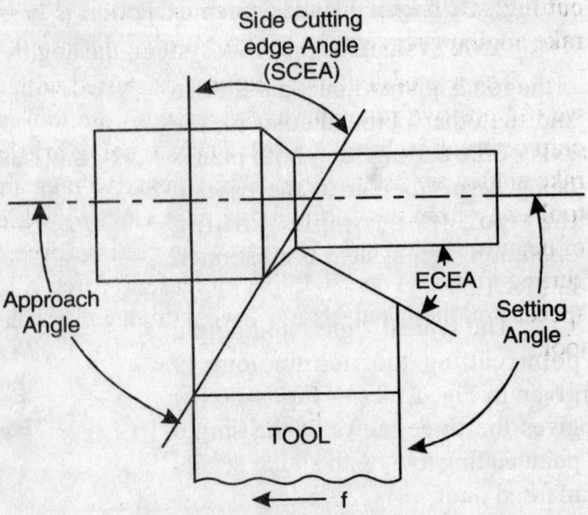

Fig. 1.61 SCEA and ECEA.

2. **End Cutting-Edge Angle, C_e.** The ECEA provides a clearance or relief to the trailing end of the cutting edge to prevent rubbing or drag between the machined surface and the trailing (non-cutting) part of the cutting edge. Only a small angle is sufficient for this purpose. Too large an ECEA takes away material that supports the point and conducts away the heat. An angle of 8° to 15° has been found satisfactory in most cases on side cutting tools, like boring and turning tools. Sometimes, on finishing tools, a small flat (1.6 to 8 mm long) is ground on the front portion of the edge next to the nose radius, to level the irregular surface produced by a roughing tool. End cutting tools, like cut off and necking tools often have no end cutting-edge angle.

3. **Side Relief Angel, (SRA) and End Relief Angle (ERA).** These angles (denoted as θ_s and θ_e in the figure) are provided so that the flank of the tool clears the workpiece surface and there is no rubbing action between the two. Relief angles range from 5° to 15° for general turning. Small relief angles are necessary to give strength to the cutting edge when machining hard and strong materials. Tools with increased values of relief angles penetrate and cut the workpiece material more efficiently and this reduces the cutting forces. Too large relief angles weaken the cutting edge and there is less mass to absorb and conduct the heat away from the cutting edge.

4. **Back and Side Rake Angle (α_b, α_s).** The top face of the tool over which the chip flows is known as the rake face. The angle which this face makes with the normal to the machined surface at the cutting edge is known as "Back-rake angle, α_b", and the angle between the face and a plane parallel to the tool base and measured in a plane perpendicular to both the base of the tool holder and the side cutting edge, is known as "Side-rake angle, α_s" The rake angles may be positive, zero, or negative. Cutting angle and the angle of shear are affected by the values for rake angles. Larger the rake angle, smaller the cutting angle (and larger the shear angle) and the lower the cutting force and power. However, since, increasing the rake angle decreases the cutting angle, this leaves less metal at the point of the tool to support the cutting edge and conduct away the heat. A practical rake angle represents a compromise between a large angle for easier cutting and a small angle for tool strength. In general, the rake angle is small for cutting hard materials and large for

cutting soft ductile materials. An exception is brass which is machined with a small or negative rake angle to prevent the tool form digging into the work.

The use of negative rake angles started with the employment of carbide cutting tools. When we use positive rake angle, the force on the tool is directed towards the cutting edge, tending to chip or break it, Fig. 1.62 (*a*). Carbide being brittle lacks shock resistance and will fail if positive rake angles are used with it. Using negative rake angles, directs the force back into the body of the tool away from the cutting edge, Fig. 1.62 (*b*), which gives protection to the cutting edge. The use of negative rake angle, increases the cutting force. But at higher cutting speeds, at which carbide cutting tools can be used, this increase in force is less than at normal cutting speeds. High cutting speeds are, therefore, always used with negative rakes, which requires ample power of the machine tool.

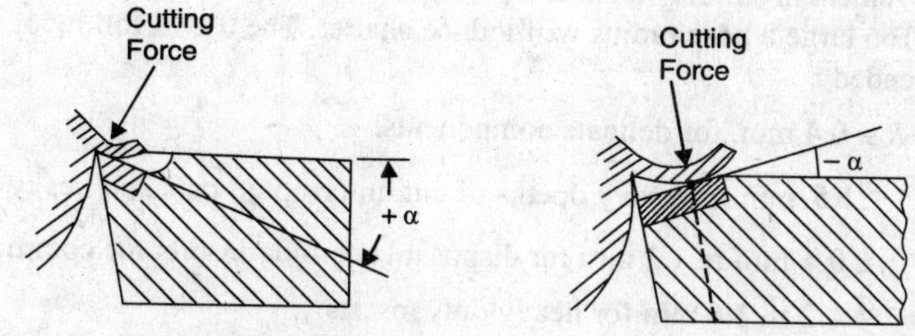

(*a*) With Positive Rake. (*b*) With Negative Rake.

Fig. 1.62 Cutting with Positive and Negative Rake Tools.

The use of indexable inserts has also promoted the use of negative rake angles. An insert with a negative rake angle has twice as many cutting edges as an equivalent positive rake angle insert (as will be discussed ahead). So, to machine a given number of components, smaller number of negative rake inserts are needed as compared to positive rake inserts.

The use of positive rake angles is recommended under the following conditions :

1. When machining low strength ferrous and non-ferrous materials and work-hardening materials.
2. When using low power machines.
3. When machining long shafts of small diameters.
4. When the set up lacks strength and rigidity.
5. When cutting at low cutting speeds.

The use of negative rake angles is recommended under the following conditions :

1. When machining high strength alloys.
2. When there are heavy impact loads such as in interrupted machining.
3. For rigid set ups and when cutting at high speeds.

Recommended rake angles are given in Table 1.1.

Table 1.1 Recommended Rake Angles

Work Material	*Tool Material*					
	H.S.S. and Cast Alloys		*Cemented Carbide Brazed*		*+ Throw away*	
	Back	*Side*	*Back*	*Side*	*Back*	*Side*
Free Machining Steels	10	12	0	6	– 5	– 5
Mild Steel	8	10	0	6	– 5	– 5
Med. Carbon Steels	0	10	0	6	– 5	– 5

(Contd.)

Alloy tool Steels	0	10	– 5	– 5	– 5	– 5
Stainless Steels	0	10	0	6	– 5	– 5
Cast Iron	5	5 to 10	0	6	– 5	– 5
Aluminium Alloys	20	15	3	15	0	5
Copper Alloys	5	10	0	8	0	5
Magnesium Alloys	20	15	3	15	0	5
Titanium Alloys	0	5	0	6	– 5	– 5

5. Nose Radius. Nose radius is favourable to long tool life and good surface finish. A sharp point on the end of a tool is highly stressed, short lived and leaves a groove in the path of cut. There is an improvement in surface finish and permissible cutting speed as nose radius is increased from zero value. Too large a nose radius will induce chatter. The use of following values for nose radius is recommended :

$R = 0.4$ mm, for delicate components.

≥ 1.5 mm for heavy dpeths of cut, interrupted cuts and heavy feeds.

$= 0.4$ mm to 1.2 mm for disposable carbide inserts for common use.

$= 1.2$ to 1.6 mm for heavy duty inserts.

The rules of thumb for selection of Nose radius are :-

(*i*) For a strong cutting edge, select the largest prossible Nose radius.

(*ii*) A large nose radius permits higher feeds.

(*iii*) Select a smaller nose radius if there is a tendency to vibrate.

For rough machining, the most commonly used nose radii are 1.2 to 1.6 mm

Tool Designation. The tool designation or tool signature, under ASA system is given in the order given next :

Back rake, Side rake, End relief, Side relief, End cutting edge angle, Side cutting edge angle, and nose radius that is,

$$\alpha_b - \alpha_s - \theta_e - \theta_s - C_e - C_s - R$$

If tool designation is :

8 – 14 – 6 – 6 – 6 – 15 – 3, it means that,

$\alpha_b = 8°$, $\alpha_s = 14°$

$\theta_e = 6°$, $\theta_s = 6°$

$C_e = 6°$, $C_s = 15°$

and $R = 3$ mm.

In ASA system of tool angles, the angles are specified independently of the position of the cutting edge. It, therefore, does not give any indication of the behaviour of the tool in practice. Therefore, in actual cutting operation, we should include the side cutting edge (principal cutting edge) in the scheme of reference planes. Such a system is known as Orthogonal Rake System (ORS).

Orthogonal Rake System (ORS). As mentioned above, in this system the planes for designating tools are the plane containing the principal or side cutting edge and the plane normal to it. In the plane NN which is normal to the principal cutting edge and is known as Orthogonal plane or the chief plane, we have the following angles : side relief angle γ, the side rake angle (known as Orthogonal rake angle) α wedge (lip angle) and the cutting angle (see Fig. 1.63).

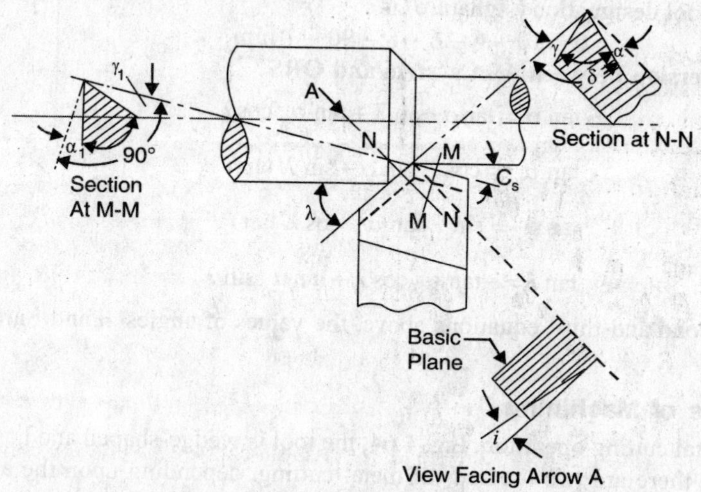

Fig. 1.63 ORS of Tool Angles.

The side relief angle is the angle between the side (main) flank and the cutting plane. The side rake angle, α, is the angle between the toolface and a plane normal to the cutting plane and passing through the main cutting edge. This angle is positive when the face slopes downward from the plane perpendicular to the cutting plane (as shown in Fig. 1.63), equal to zero when the face is perpendicular to the cutting plane and negative when the face slopes upwards. The "wedge angle, β " is the angle between the tool face and the main flank. The "cutting angle, δ " is the angle between the tool face and the cutting plane. When α is positive, we have,

$$\alpha + \gamma + \text{ wedge angle} = 90°$$

$$\gamma + \beta = \delta$$

The usual values of α and γ are :

$$\alpha = -10° \text{ to} + 15°, \ \gamma = 6° \text{ to } 12°$$

In the ORS, the back rake angle is the inclination angle (*i*) between the principal cutting edge and a line passing through the point of the tool parallel to the principal plane. This angle is measured in a plane passing through the main cutting edge and perpendicular to the basic plane. In Fig. 1.63, the angle i is negative with tool nose being the highest point of the cutting edge. It will be zero when the cutting edge is parallel to the basic plane and positive if the cutting edge is towards the right (Fig. 1.63) of the line passing through the point of the tool and parallel to the principal (basic) plane, that is, the tool nose is the lowest point of the cutting edge.

In addition to the angles discussed above, angles are also measured in the plane MM (known as Auxiliary reference plane) which is normal to the projection of the end cutting edge on the basic plane. These angles are the end relief angle γ_1, and the back rake angle α_1 (also called auxiliary rake angle). The plan angles are the Approach angle or entering angle λ which is equal to $(90° - C_s)$ and the end cutting edge angle, C_e.

$$\gamma_1 = 8° \text{ to } 10°, \ \lambda = 30° \text{ to } 70°, \ C_e = 10° \text{ to } 15°$$

The tool designation under ORS is :

$$i-\alpha-\gamma-\gamma_1-C_e-\lambda-R$$

A typical tool designation (signature) is :

$$0-10-6-6-8-90-1 \text{ mm}$$

Interconversion between ASA system and ORS

$$\tan\alpha = \tan\alpha_s \sin\lambda + \tan\alpha_b \cos\lambda$$

$$\tan\alpha_b = \cos\lambda \tan\alpha + \sin\lambda \tan i$$

$$\tan\alpha_s = \sin\lambda \tan\alpha - \cos\lambda \tan i$$

$$\tan i = -\tan\alpha_s \cos\lambda + \tan\alpha_b \sin\lambda$$

In the second and third equations above, the values of angles α and i are taken with their signs.

1.9.4 Methods of Machining

In the metal cutting operation, Fig. 1.64, the tool is wedge-shaped and has a straight cutting edge. Basically, there are two methods of metal cutting, depending upon the arrangement of the cutting edge with respect to the direction of relative work-tool motion :

1. Orthogonal cutting or two dimensional cutting.
2. Oblique cutting or three dimensioning cutting.

In orthogonal cutting, Fig. 1.64, the cutting edge of the tool is arranged perpendicular to the cutting velocity vector, V, whereas in oblique cutting, it is set at some angle other than 90° to the cutting velocity vector, which gives an "inclination angle *i*". The analysis of oblique cutting being very complex, the relatively simple arrangement of orthogonal cutting is, therefore, widely used in theoretical and experimental work.

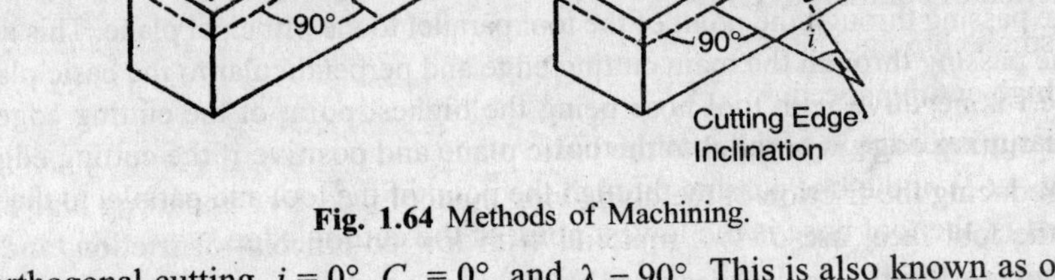

Fig. 1.64 Methods of Machining.

In pure orthogonal cutting, $i = 0°$, $C_e = 0°$, and $\lambda = 90°$. This is also known as orthogonal system of second kind. When $i = 0$, and $0 < \lambda < 90°.0$, it is called as orthogonal system of first kind. A common example of pure orthogonal cutting process is the turning of a thin pipe with a straight edged tool set normal to the longitudinal axes.

1.9.5 Types of Chips

· Whatever the cutting conditions can be, the chips produced may belong to one of the following three types, (Fig. 1.65) :

1. Discontinuous Chips
2. Continuous Chips
3. Continuous Chips with build-up-edge (BUE).

1. Discontinuous Chips. These types of chips are usually produced when cutting more brittle materials like grey cast iron, bronze and hard brass. These materials lack the ductility necessary for appreciable plastic chips formation. The material ahead of the tool edge fails in a brittle fracture manner along the shear zone. This produces small fragments of discontinuous chips. Since the chips break up into small segments, the friction between the tool and the chips reduces, resulting in better surface finish. These chips are convenient to collect, handle and dispose of. Discontinuous chips are also produced when cutting more ductile materials under the following conditions :

(*i*) large chip thickness.
(*ii*) low cutting speed.
(*iii*) small rake angle of the tool.
(*iv*) cutting with the use of a cutting fluid.

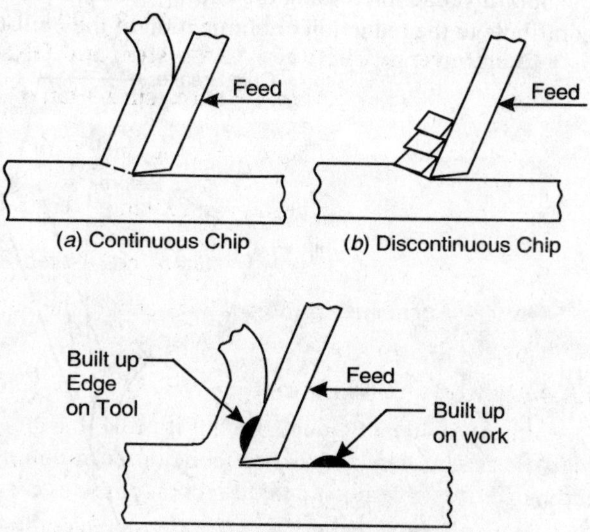

Fig. 1.65 Types of Chips

2. Continuous Chips. These types of chips are produced when, machining more ductile materials. Due to large plastic deformations possible with ductile materials, longer continuous chips are produced. This type of chip is the most desirable, since it is stable cutting, resulting in generally good surface finish. On the other hand, these chips are difficult to handle and dispose off. The chips coil in a helix (chip curl) and curl around the work and the tool and may injure the operator when break loose. Also, this type of chip remains in contact with the tool face for a longer period, resulting in more frictional heat. These difficulties are usually avoided by attaching to the tool face or machine on the tool face, a 'chip breaker', (Fig. 1.66). The function of chip breaker is to reduce the radius of curvature of the chip and thus break it. The following cutting conditions also help in the production of continuous chips :

(*i*) small chip thickness.
(*ii*) high cutting speed.
(*iii*) large rake angle of the cutting tool.
(*iv*) reducing the friction of the chip along the tool face, by : imparting high surface finish to the tool face, use of tool material with low co-efficient of friction, and use of a good cutting fluid.

3. Continuous Chips with Built-up-edge (BUE). When machining ductile materials, conditions of high local temperature and extreme pressure in the cutting zone and also high friction in the tool-chip interface, may cause the work material to adhere or weld to the cutting edge of the tool forming the built-up edge. Successive layers of work material are then added to the built-up edge. When this edge becomes larger and unstable, it breaks up and part of it is carried up the face of the tool alongwith the chip while the remaining is left over the surface being machined, which contributes to the roughness of the surface. The built-up edge changes its size during the cutting operation. It first increases, then decreases, then again increases etc. This cycle is a source of vibration and poor surface finish. Although, the built-up edge protects the cutting edge of the tool, it changes the geometry of the cutting tool. Low cutting speed also contributes to the formation of the built-up edge. Increasing the cutting speed, increasing the rake angle and using a cutting fluid contribute to the reduction or elimination of the built-up edge.

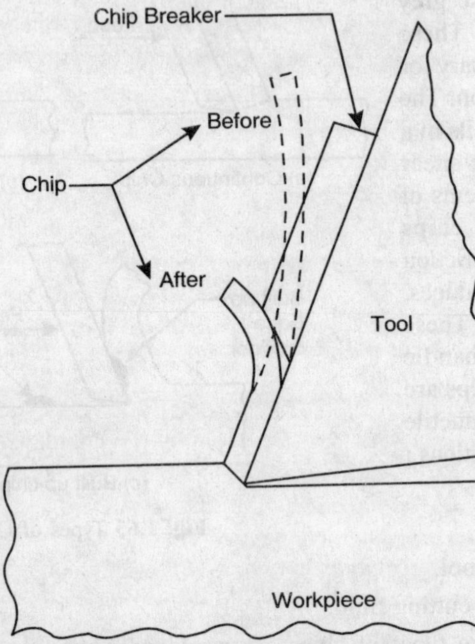

Fig. 1.66. Chip Breaker.

From the above discussion, we can summarize the factors (Table 1.2) that are likely to influence the formation of various types of chips.

Table 1.2 Factors affecting type of chips.

Factors	*Types of Chips*		
	Continuous	*Continuous with B.U.E.*	*Discontinuous*
Material	Ductile	Ductile	Brittle
Tool :-			
Rake angle	Large	Small	Small
Cutting edge	Sharp	Dull	–
Cutting conditions :			
Speed	High	Low	Low
Feed	Low	High	High
Friction	Low	High	–
Cutting Fluid	Efficient	Poor	–

Example 1.1 *Explain the construction of an engine lathe.*

Solution. Engine Lathe. It is so called because the first of this type of lathe was driven by a steam engine. It is also called. "Centre lathe" because, it has two centres between which the job can be held and rotated. A very high percentage of all lathe work is turned between centres.

The main parts of a centre lathe are : Bed, Head stock, Tail stock, Carriage and the Electric drive, (Fig 1.67)

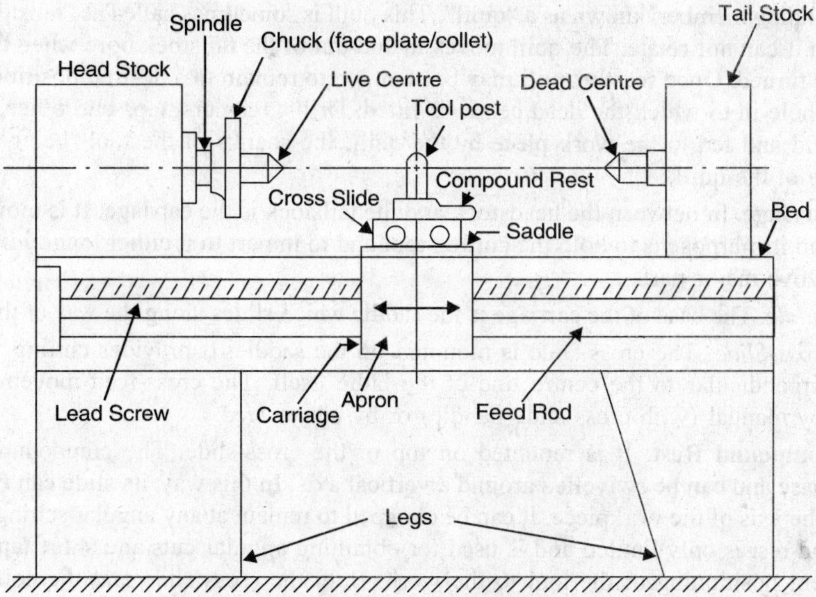

Fig. 1.67 Block Diagram of an Engine Lathe.

1. Bed. The bed is the base or foundation of the lathe. It is a massive (heavy) and rigid casting made in one piece to resist deflection and vibrations. It holds or supports all other parts, that is, head stock, tailstock and carriage etc. The top of the bed is planed to form "guides" or "ways". Ways are accurate rails which support carriage and the tailstock. More expensive lathes have a combination of V ways and flat ways. Less expensive lathes have flat ways. Directly under the front way on the bed is a rack. A pinion gear meshes with the rack for moving the carriage when the handwheel is turned. The bed is usually fastened to steel legs so the lathe can be bolted to the shop floor.

2. Head stock. The headstock assembly is permanently fastened to the left hand end of the lathe. It serves to support the first operative unit of the lathe, that is, the spindle. The spindle revolves in bearings, one at each end of headstock. The spindle is rotated by a combination of gears and cone pulleys or by gears alone. Present day lathes have individual motor drives and most of them have geared headstocks. The steel spindle is hollow to take long bar stock. The spindle has a definite taper at the front end for holding centres and other tools having a tapered shank. The hole through the spindle makes it possible to use a knockout bar to remove such tools. Work-holding attachments such as driving plate, face plate or various types of chucks may be mounted on the threaded spindle nose. Some type of work may also be held in a collet which is inserted into the hollow headstock spindle. A taper sleeve fits into taper spindle hole. The headstock or live centre fits into the sleeve. This centre is called live centre because it turns with the work. The centre is a tapered metal part with a pointed end. This supports the end of the workpiece as it turns between the centres. All centre points have a 60-degree included angle.

3. Tail stock. Tail stock is on the other end of the bed from the head stock. Its chief function is to hold the dead centre so that long workpieces ($L/D > 4$) can be supported between centres. It can be moved along the bed and clamped to the bed at the various desired locations to suit the length of the workpiece.

Tailstock consists of two main parts. The lower part rests directly on the bed ways, and the upper part rests on the lower part. Adjusting screws hold the two together. The upper casting can be moved toward or away from the operator to offset the tailstock for taper turning and to re-align the tailstock centre for straight turning. The body of the tailstock has a bore for the hollow cylindrical sliding member, known as a "quill". This quill is sometimes called as "tailstock spindle" even though it can not rotate. The quill moves in and out of the tailstock bore when the tailstock handweel is turned. Once set, the quill may be clamped to remain in a desired position. The quill has a taper hole in to which the dead centre is fitted. Drills, reamers, taps and other end cutting tools are held and fed to the work piece by the quill, the shanks of the tools being held in the tapered hole of the quill.

4. Carriage. In between the headstock and the tailstock is the carriage. It is movable on the bed ways and its purpose is to hold the cutting tool and to impart to it either longitudinal or cross feed. It has five major parts :

(*a*) *Saddle*. The base of the carriage is the saddle which slides along the way of the lathe bed.

(*b*) *Cross-Slide*. The cross-slide is mounted on the saddle. It provides cutting tool motion which is perpendicular to the centre line of the lathe itself. The cross-feed movement may be controlled by manual (with cross slide handle) or by power feed.

(*c*) **Compound Rest.** It is mounted on top of the cross-slide. The compound rest has a graduated base and can be swivelled around a vertical axis. In this way, its slide can be set at any angle with the axis of the workpiece. It can be clamped to remain at any angular setting. The range of compound rest is only limited and is used for obtaining angular cuts and short tapers, as well as convenient positioning of the tool to the work. Both the cross slide and the compound rest screws are equipped with micrometer collars. These are used in making accurate adjustments when turning workpieces to close measurements, and when cutting screw threads. There is no power feed for the compound rest.

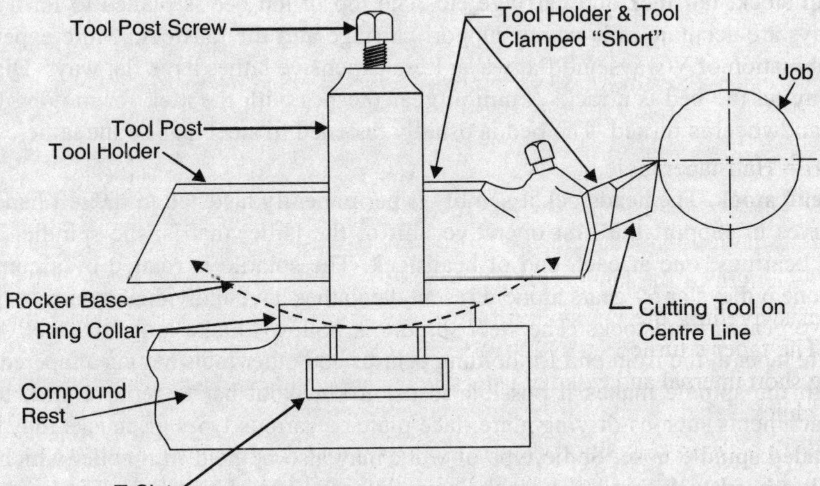

Fig. 1.68 Tool Post.

(*d*) **Tool post.** The tool post is mounted on the compound rest and slides in a T-slot. Cutting tool/tool holder is firmly held in it. The tool can be swivelled as well as tilted by means of a rocker and a concave ring collar, (Fig. 1.68)

(*e*) **Apron.** The apron is secured underneath the saddle and hangs over the front of the bed. It contains the gears, clutches, and levers for operating the carriage by hand and power feeds. The apron hand wheel can be turned to move the carriage longitudinally by hand. This hand wheel is attached to a pinion that meshes with the rack under the front of the bed. The apron also contains friction clutches for automatic feeds and a splitnut or halfnut. The split nut can be closed over the lead screw threads and is used only when cutting screw threads.

Example 1.2 *Explain the various taper turning methods on a lathe.*

Solution. Taper turning. Tapered surfaces can be turned by employing one of the following methods :

1. Compound rest Method. The compound rest is swivelled to the required angle, (Fig. 1.69). The angle is determined by the formula :

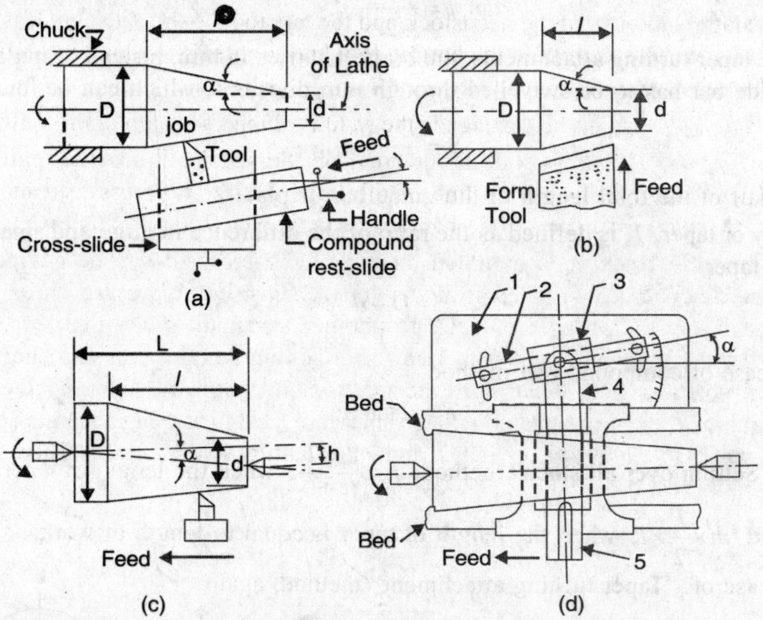

Fig. 1.69 Taper Turning Methods.

where α = Half taperangle.

D = larger diameter of the taper

d = smaller diameter

l = length of the taper

After swivelling the compound rest to this angle about the vertical axis, it is clamped in position. The taper s turned by hand wheel by rotating the handle. The method can be employed for turning short internal and external tapers with a large angle of taper. The workpiece is commonly held in a chuck.

2. With a Form tool. Short external tapers with various angles of taper can be turned with a form tool, Fig 1.69, using cross feed. The width of the from tool slightly exceeds that of the taper being turned. The work is held in a chuck or clamped on a faceplate.

3. Setting over the Tail-stock. Long workpieces with a small angle of taper (not exceeding 8°) are usually turned by setting over the tailstock centre, (Fig 1.69 *c*). Tailstock setting over or offset can be determined from the formula,

$$h = \frac{L(D-d)}{2l} \text{ mm}$$

where, L = full length of the workpiece

l = length to be taper turned

4. Taper turning attachment. Long tapered work is frequently turned with a taper turning attachment, Fig. 1.69(*d*). Bracket 1 is attached to the lathe bed. It carries guide bar 2 which can be turned to the required angle and clamped in place. Guide block or slide 3, linked to the lathe cross slide 5 by tie member 4, is free to slide along the guide bar. The cross slide is disengaged from the cross feed screw. Upon travel of the saddle along the bed ways, the guide block slides along the guide bar, and through the tie member, forces the cross slide with the cutting tool to travel parallel to the guide bar at the given angle and the work is turned to the specified taper. The angle of taper can be found out by the relation :

$$\tan\alpha = \frac{D-d}{2l}$$

In some taper turning attachments, the bracket carries in mm, instead of in degrees. In such cases, the guide bar has to be swivelled through mm divisions, which can be found out as :

$$S = \frac{D-d}{2l} \times lg$$

where lg = Half of the total length of the guide bar or plate.

Conicity or taper, T, is defined as the ratio of the difference in large and small diameters to the length of taper.

$$\therefore \quad T = \frac{D-d}{l}$$

In the case of compound rest method,

$$T = 2\tan\alpha$$

In the "Setting over all stock" method, $h = \dfrac{L \times T}{2}$, when the length of taper is less than of workpiece and $h \times \dfrac{T}{2} \times l$, when the length of taper is equal to length of work.

In the case of "Taper turning attachment" method, again

$$T = 2\tan\alpha.$$

Example 1.3 *Discuss the cutting speed, feed and depth of cut in shaping, planning and slotting operations.*

Solution.

The cutting speed, V, in planing, shaping and slotting operations is the speed of the tool (or the work) in the direction of the primary cutting motion during the working stroke when the metal is being cut. The cutting speed may be either constant or variable depending upon the design of the machine tool.

Shapers with crank and slotted arm mechanisms have a variable cutting speed and a variable return speed (Fig. 1.70) of the tool. The speed is minimum at the beginning and end of the stroke and maximum in the middle of the stroke. The cutting speed can be determined by the formula:

$$V = \frac{L\,N(1+K)}{1000}\ \text{m/min}.$$

where L = Length of ram stroke, mm

N = number of full strokes/min,

that is, a full stroke comprises working and return strokes.

K = Ratio of return time to cuttingtime.

Planers with a rack and pinion drive for the table have a constant cutting speed, which can be determined from the formula :

$$V = \frac{2N\,L}{1000},\ \text{m/min}.$$

The feed, *f*, in planing, shaping or slotting (Fig. 1.70) is the relative movement of the tool or work in a direction perpendicular to the primary cutting motion per full stroke and is expressed in mm/stroke.

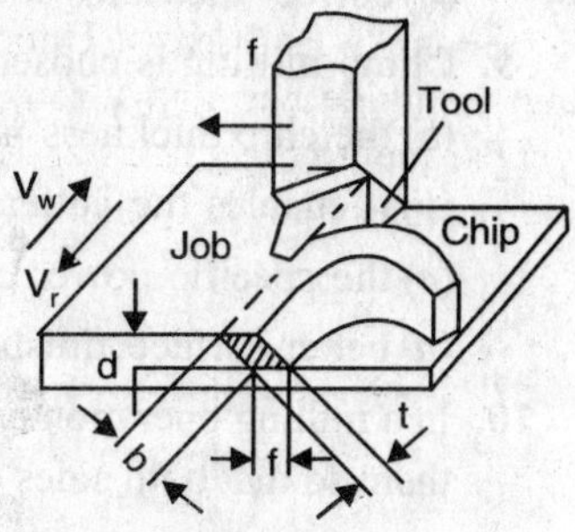

Fig. 1.70 *Cutting Action in a Planer.*

The depth of cut, *d*, is the thickness of layer of metal removed in one cut or pass. It is measured in a direction perpendicular to the machined surface.

The cross-sectional area of the undeformed chip is the product of the thickness, *t*, and the width, *b*, of the undeformed chip :

Area of undeformed chip = $t \times b = f \times d$ mm^2.

Cutting speeds in these kinds of machining operations are not usually very high, since difficulties are encountered in designing machine tools with high speed of reciprocating motion due to high inertia forces developed in the motion of the units and components of machine. Thus in planing cast iron, the recommended cutting speed is 15 to 20 m/min for roughing and 4 to 12 m/min for finishing. The corresponding depths of cut are 0.5 to 0.8 mm, and 0.08 mm respectively, with feed ranging from 1.50 to 2.00 mm/stroke.

MULTI-CHOICE QUESTIONS

1. The chuck used for machining a casting on lathe is
(*a*) three jaw chuck (*b*) four jaw chuck
(*c*) collet (*d*) magnetic chuck

2. The lathe post not provided with power feed is
(*a*) lead screw (*b*) carriage
(*c*) compound rest (*d*) cross slide

3. External taper can be obtained by
(*a*) taper attachment (*b*) form tool
(*c*) compound rest (*d*) tail stock offset.

4. A cone shaped recess at the top of drilled hole can be made by
(*a*) countersinking (*b*) counterboring
(*c*) taper drilling (*d*) grooing.

5. Power feed is not provided on the following drilling machine :
(*a*) bench drill press (*b*) gang drilling machine
(*c*) radial drilling machine (*d*) all of the above.

6. Reaming operation is performed to
(*a*) improve the finish of a hole (*b*) obtain correct diameter
(*c*) enlarge the hole (*d*) make a hole.

7. Which of the machine tools can be used for boring ?
1. Lathe 2. Drilling machine
3. Vertical milling machine 4. Horizontal milling machine.
(*a*) 1,2,3, (*b*) 1,3,4
(*c*) 2 and 4 (*d*) 1, 2, 3, 4

8. The main purpose of boring operation as compared to drilling is to :
(*a*) drill a hole (*b*) finish drilled hole

(*b*) correct the hole (*d*) enlarge the existing hole.

9. Climb milling is chosen while machining because

(a) the chip thickness increases gradually

(*b*) it enables the cutter to dig in and start the cut

(*c*) the specific power consumption is reduced

(*d*) better surface finish can be obtained.

10. In a milling operation two side milling cutters are mounted with a desired distance between them so that both sides of a work piece can be milled simultaneously. This set up is called

(*a*) gang milling (*b*) straddle milling

(*c*) string milling (*d*) side milling.

11. Match List-I (Type of drill) with List II (Application) and select the correct answer using the codes given below the lists :

List-I	*List-II*
A. Straight shank	1. Soft materials
B. Toper shank	2. Deep holes
C. Single flute	3. General purpose
D. High helix	4. Small hole diameter

Codes :

	A	*B*	*C*	*D*
(*a*)	3	4	1	2
(*b*)	3	4	2	1
(*c*)	4	3	2	1
(*d*)	4	3	1	2

12. Consider the following statements :

In Up milling process,

1. the cutter starts the cut from the machined surface and proceeds upwards.
2. the cutter starts the cut from the top surface and proceeds downwards.
3. the job is fed in a direction opposite to that of cutter rotation.
4. the job is fed in the same direction as that of cutter so rotation.

Of these statements :

(*a*) 1 and 3 are correct (*b*) 1 and 4 are correct

(*c*) 2 and 3 are correct (*d*) 2 and 4 are correct.

13. Match List-I with List-II and select the correct answer using the codes given below the lists :

List-I	*List-II*
A. Reaming	1. Smoothing and squaring surface around the hole for proper seating.
B. Counter-boring	2. Sizing and finishing the hole.
C. Counter-sinking	3. Enlarging the end of the hole.
D. Spot facing	4. Making a conical enlargement of the end of the hole.

Codes :

	A	*B*	*C*	*D*
(*a*)	3	2	4	1
(*b*)	2	3	1	4
(*c*)	3	2	1	4
(*d*)	2	3	4	1

14. Match List-I (drill bits) with List -II (Applications) and select the correct answer using the codes given below the Lists:

List-I	*List-II*
A. Core drill	1. To enlarge a hole to a certain depth so as to accommodate the bolt head of a screw.
B. Reamer	2. To drill and enlarge an already existing hole in a casting.
C. Counter-bore drill	3. To drill a hole making internal threads
D. Tap drill	4. To improve the surface finish and dimensional accuracy of the already drilled hole.

Codes :

	A	*B*	*C*	*D*
(*a*)	1	3	2	4
(*b*)	2	3	1	4
(*c*)	2	4	1	3
(*d*)	3	2	4	1

15. Match List-I (Machine tools) with List-II (Machine tool parts) and select the correct answer using the codes given below the lists :

List-I (*Machine tools*)	*List-II* (*Machine tool parts*)
A. Lathe	1. Lead screw
B. Milling machine	2. Rocker arm
C. Shaper	3. Universal indexing
D. Drilling machine	4. Flute

Codes :

	A	*B*	*C*	*D*
(*a*)	4	2	3	1
(*b*)	1	3	2	4
(*c*)	4	3	2	1
(*d*)	1	2	3	4

16. In milling machine, the cutting tool is held in position by

(*a*) chuck (*b*) spindle (*c*) arbor (*d*) tool holder.

17. Consider the following statements associated with the lathe accessories :

1. Steady rest is used for supporting a long job in between head stock and tail stock.
2. Mandrel is used for turning small cylindrical jobs.
3. Collects are used for turning disc-shaped job.

Of these statements :

(*a*) 1 and 2 are correct (*b*) 2 and 3 are correct

(*c*) 3 alone is correct (*d*) 1 alone is correct.

18. Match List-I (parts) with dist-II (Method of holding on a lathe) and select the correct answer using the codes given below the lists :

List-I (*parts*)	*List-II* (*Method of holding on a lathe*)
A. Cylindrical parts	1. Circular face plate
B. Non-cylindrical parts	2. Collet
C. Very complicated shapes	3. Three-jaw chuck
D. Material in bar form on small lathes.	4. Four-jaw chuck

Codes :

	A	*B*	*C*	*D*
(*a*)	3	1	4	2
(*b*)	2	4	1	3
(*c*)	3	4	1	2
(*d*)	2	1	4	3

ANSWERS

1. (*d*) **2.** (*c*) **3.** (*d*) **4.** (*a*) **5.** (*a*) **6.** (*a*) **7.** (*a*)
8. (*d*) **9.** (*a*) **10.** (*b*) **11.** (*a*) **12.** (*a*) **13.** (*d*) **14.** (*c*)
15. (*b*) **16.** (*c*) **17.** (*a*) **18.** (*c*)

REVIEW QUESTIONS

1. Define a machine tool.
2. What are the function performed by a machine tool?
3. How machine tools are classified according to the following criteria :
(*a*) Field of application.
(*b*) Accuracy
(*c*) Weight
(*d*) Processing operations.
4. What are the characteristics of general purpose machine tools ?
5. List the characteristics of special purpose machine tools.
6. Explain the working principle of a lathe with a neat sketch.
7. How do you classify a lathe? Explain briefly.
8. Describe the principle parts of a lathe with a neat sketch.
9. How a lathe is specified? Explain with a neat diagram.
10. List the various lathe accessories.
11. Differentiate between live and dead centres of a lathe.
12. List the various lathe tools.
13. What are the various lathe operation?
14. What is a shaper? What are its principal parts? Explain with a neat sketch.
15. Explain the working principal of a shaper.
16. How shapers are classified?
17. How a shaper is specified?

18. Describe the quick return motion mechanism for a shaper.
19. What is the purpose of a drilling machine ? Explain its working principle.
20. How drilling machines are classified?
21. Explain the working of following drilling machines with a neat diagram:
    (*a*) General purpose drilling machine.
    (*b*) Pillar drilling machine.
    (*c*) Bench drilling machine.
    (*d*) Redial drilling machine.
22. How do you specify a drilling machine?
23. List the various drilling operations.
24. What are the main functions of a milling machine?
25. List the various types of milling machines.
26. Describe the construction and working of the following :
    (*a*) General purpose milling machine.
    (*b*) Vertical milling machine.
27. How the size of milling machine is specified?
28. Describe the various milling operations.
29. Describe the construction and working of a centre lathe.
30. Describe the various taper turning methods on a lathe.
31. Describe various types of slotters in brief.
32. How a slotter is specified?
33. Describe the main parts of a slotting machine.
34. Describe the various ram drive mechanisms of a slotler.
35. Describe the various slotter operations.
36. Describe various feed movements in a slotting machine.
37. Expalin the difference in operation between a planer and a shaper.
38. Describe the principal parts of a double housing planer.
39. What are the advantages of a planer ?
40. Explain the machining process with a neat sketch.
41. Explain the various elements of a single point cutting tool with a neat sketch.
42. Explain the two systems of designating a cutting tool.
43. Describe the various types of chips produced during metal machining.
44. What is the use of chip breakers ?

2 BASIC CONCEPTS OF THERMODYNAMICS

2.1 INTRODUCTION

Thermodynamics may be defined as follows :

- The science that deals with energy interactions between material systems.
- The science of energy transfer and its effect on the physical properties of substances.
- The science which deals with the transformation of energy of all kinds from one form to another.

The scope of applied thermodynamics is restricted to the study of heat and work and the conversion of one into the other.

The applications of thermodynamics are :

- Design of energy converting devices, such as, steam engines, internal combustion engines, steam and gas turbines, fuel cells, thermoelectric and thermionic generators.
- Design of refrigerators, air-conditioners, heat transfer and reaction equilibrium, etc.

In this chapter, we shall study the basic concepts of thermodynamics.

2.2 THERMODYNAMIC SYSTEM

A thermodynamic system is defined as a prescribed region or space of finite quantity of matter surrounded by an envelope called the boundary. The boundary may be a real physical surface, such as the walls of a vessel, internal combustion engine cylinder, etc. or it may be an imaginary surface enclosing some matter, such as, steam, gas, vapour, etc. The boundary may be either fixed or moving.

In a broad sense, a thermodynamic system may be defined as a specified region wherein changes due to transfer of mass or energy or both are studied.

Surroundings (or Environment). The space and matter external to the thermodynamic system is called the surroundings. The system is separated from the surroundings by the system boundary. The thermodynamic system, boundary and surroundings are shown in Fig. 2.1.

Universe. The thermodynamic system and surroundings put together is called the universe.

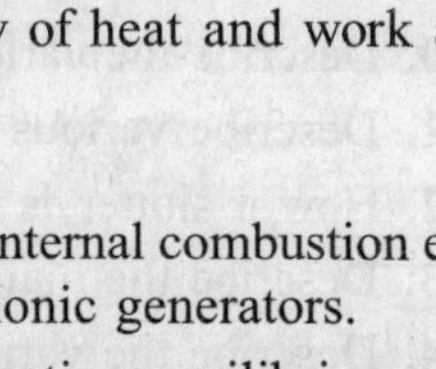

Fig. 2.1 Representation of system, boundary and surroundings.

2.2.1 Types of Systems

There are three types of systems : (1) Closed system, (2) open system, and (3) Isolated system.

1. Closed System : In a closed system, the mass within the boundary of the system remains constant and only the energy (heat and work) may transfer across its boundary. Such a closed system is shown in Fig. 2.2. Some examples of closed system are :

(*i*) The gas confined between a piston and cylinder.
(*ii*) A rubber balloon filled with air and tightly closed.
(*iii*) Pressure cooker.
(*iv*) Dash pot.

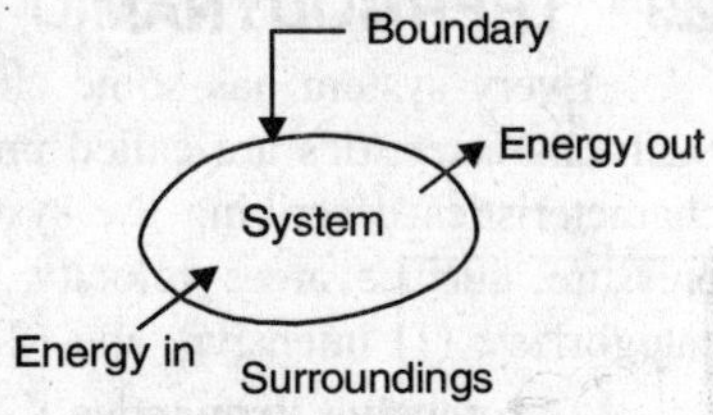

Fig. 2.2 A closed sytem

2. Open System : In an open system, the mass as well as the energy transfer across its boundaries may take place. Such an open system is shown in Fig. 2.3. Most of the thermodynamic systems are of the open type. Some examples of open systems are :

(*i*) Air compressors
(*ii*) Boilers
(*iii*) Turbines
(*iv*) Water pumps
(*v*) Nozzles and diffusers
(*vi*) Steam engines, and
(*vii*) Internal combustion engines, etc.

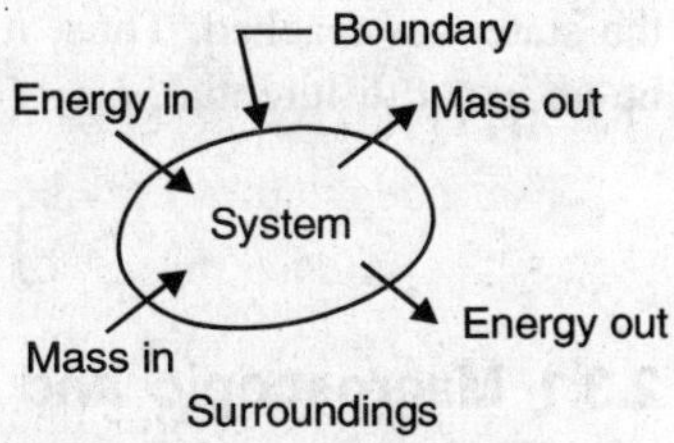

Fig. 2.3 An open system

3. Isolated System : In an isolated system, neither mass nor energy transfer takes place across its boundaries. Such an isolated system is shown in Fig. 2.4.

Some examples of isolated system are :

(*i*) Thermos flask
(*ii*) The universe.

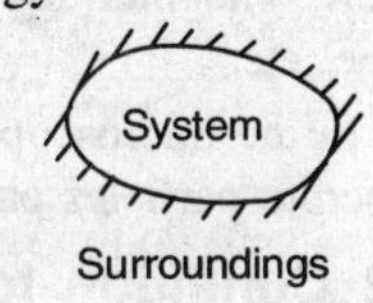

Fig. 2.4 An isolated system

2.2.2 Control Volume (CV)

It is a fixed region in space through which mass, momentum and energy may flow. The surface of the control volume is called the **control surface (CS)**. Matter as well as energy crosses the control surface. The control volume may be stationary or moving with a constant velocity. The concept of control volume and control surface is depicted with the help of an air compressor open system in Fig. 2.5.

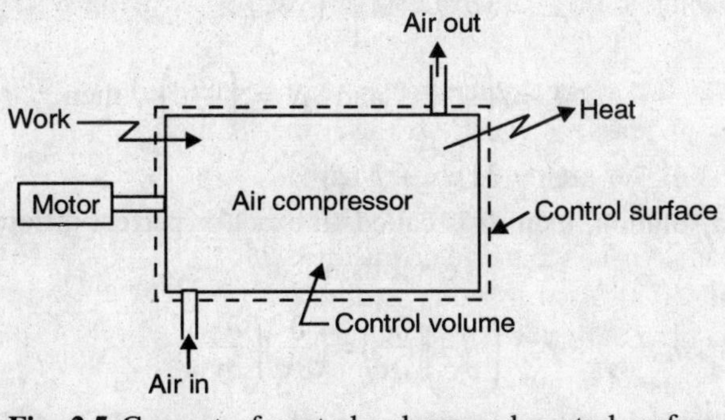

Fig. 2.5 Concept of control volume and control surface

2.2.3 Concept of Continuum

From the continuum point of view, the matter is seen as being distributed through space and treats the substance as being continuous disregarding the action of individual molecules. The properties are taken at a point.

2.3 THERMODYNAMIC PROPERTIES

Every system has some characteristics by which its physical condition may be described. Such characteristics are called properties of the system. A property of a system is its measurable characteristics describing the system, such as, mass, volume, physical composition, temperature, pressure, surface area, velocity, electrical potential, etc. Properties may be classified into two categories : (1) intensive, and (2) extensive.

1. **Intensive Properties :** These properties are independent of the mass of the system, *e.g.*, temperature, pressure, density, velocity, height, viscosity, etc.
2. **Extensive Properties :** These properties are dependent on the mass of the system, *e.g.*, volume, energy (internal, potential and kinetic), surface area, electric charge, magnetization, etc.

The property depends only on the state of the system and not upon the manner in which the state was reached. Thus, it is a point function. If property is represented by ϕ, then $d\phi$ should be an exact differential, *i.e.*,

$$\int_1^2 d\phi = \phi_2 - \phi_1$$

2.3.1 Macroscopic and Microscopic Approach

In the macroscopic approach, a certain quantity of matter is considered without taking into account the events occurring at molecular level. This is also called classical thermodynamics point of view. The thermodynamic system is explained by measurable properties, such as, pressure, volume, etc.

In the microscopic approach, the system is considered to be made up of very large number of discrete particles called molecules, This is called statistical thermodynamics point of view.

2.3.2 Condition for a Function to be a Property

Consider the function :

$$f(x, y, z) = 0$$

Let $x = x(y, z)$, $y = y(z, x)$, and $z = z(x, y)$

Now if we suppose that z is a dependent variable and x, y are independent variables, then

$$dz = \left(\frac{\partial z}{\partial x}\right)_y dx + \left(\frac{\partial z}{\partial y}\right)_x dy$$

If $M = \left(\frac{\partial z}{\partial x}\right)_y$ and $N = \left(\frac{\partial z}{\partial y}\right)_x$, then

$$dz = M\,dx + N\,dy \qquad \ldots(1)$$

If Eqn. (1) has a solution, then dz is called an exact or perfect differential, and is integrable.

If $z = z(x, y)$, then

$$\left(\frac{\partial}{\partial y}\right)\left(\frac{\partial z}{\partial x}\right) = \left(\frac{\partial}{\partial x}\right)\left(\frac{\partial z}{\partial y}\right)$$

or

$$\left(\frac{\partial M}{\partial y}\right)_x = \left(\frac{\partial N}{\partial x}\right)_y \qquad \ldots(2)$$

Eqn. (2) is a necessary and sufficient condition for the existence of a function of x and y. A function which satisfies Eqn. (2) is called a point function or the property of the system. Let us consider the following cases :

(*i*) $\int p\,dV + \int V.dp = \int (p.dV + V.dp) = \int d(pV) = pV$. Hence, it is an exact differential, and is a property of the system.

(*ii*) $\int p.dV$ is not an exact differential, hence it is not a property of the system because p and V are inter-related to each other and the integral can be evaluated only if the relationship between p and V is known.

(*iii*) $\int V.dp$ is also not an exact differential and so it is not a property.

2.4 PHASE

If a quantity of matter is homogeneous or uniform throughout in physical structure and chemical composition, it is termed as a phase. The phase may be single, double or multiple constituents. Some examples of a single phase are :

- A piece of iron.
- Gas contained in a vessel.
- A mixture of gases.
- Pure water.
- Liquid oxygen.
- Salt dissolved in water.
- Sugar dissolved in milk.

Some examples of two phases are :

- Ice in water.
- Alcohol in water.
- Steam and water.
- Oil in water.
- Mild steel.

Some examples of three phases are :

- Mixture of solid, liquid and gas.
- Steam, water and ice at the triple point.

On the basis of phases, the system may also be classified as :

(*a*) **Homogeneous System.** If a system consists of a single phase, it is called a homogeneous system, *e.g.*, mixture of air and water vapour, solution of ammonia in water (aqua ammonia), water and nitric acid, etc.

(*b*) **Heterogeneous System.** If the system consists of more than one phase, it is called a heterogeneous system, *e.g.*, water and steam, ice and water, water and oil, etc.

2.5 STATE AND PATH

State. The state is a unique condition of the system. It is the condition of a system at an instant of time described by its properties, such as, pressure, volume, temperature, density, etc.

Path. A thermodynamic system passing through a series of states constitutes a path.

In Fig. 2.6, point (p_1, v_1) and point (p_2, v_2) represent the states at 1 and 2. Along curve 12, the states of pressure and volume change from point to point.

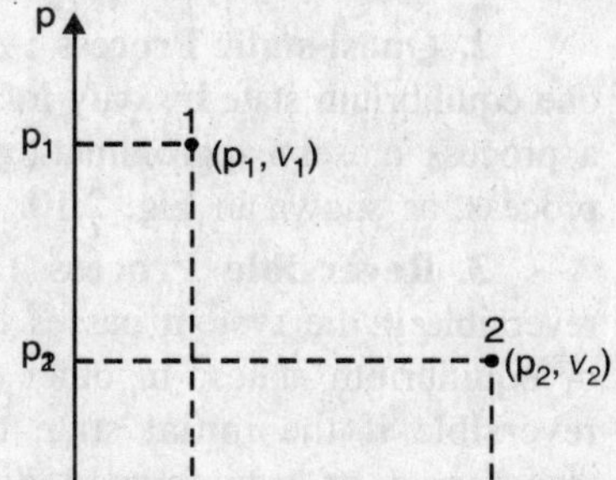

Fig. 2.6 Representation of state.

2.6 PROCESS AND CYCLE

Process. A process is defined as a transition in which a system changes from one initial state to a final state. In a process, the path is completely specified. In Fig. 2.7 the line 12 is the process that has taken place.

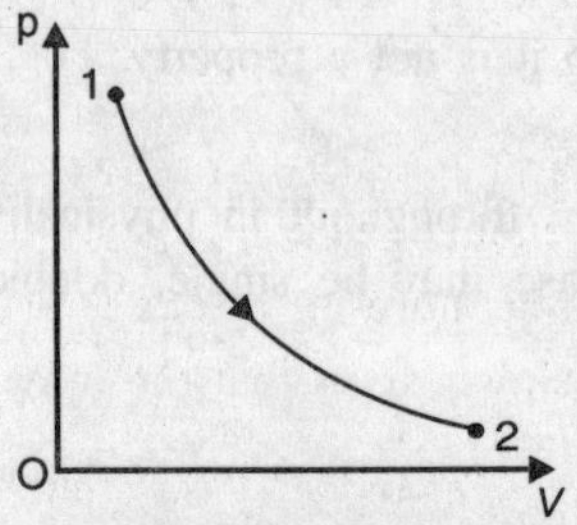

Fig. 2.7 Representaiton of a process.

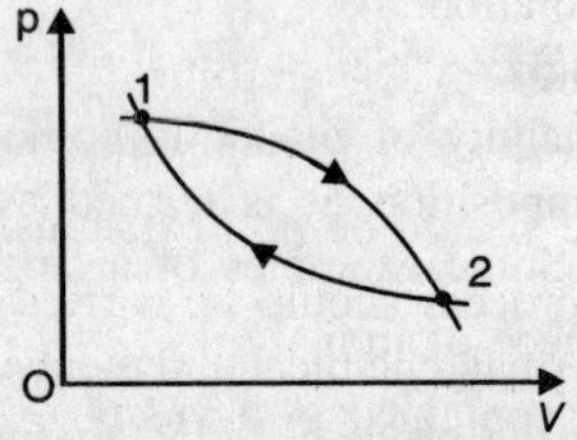

Fig. 2.8 Representation of a cycle.

Cycle. A thermodynamic cycle is defined as a series of state changes such that the final state is identical with the initial state. As shown in Fig. 5.8, 1 – 2 – 1 is the cycle.

2.6.1 Types of Processes

The various types of thermodynamic processes are :

1. Cyclic process.
2. Quasi – static process.
3. Reversible process.
4. Irreversible process.

1. Cyclic Process : In a cyclic process, the system undergoes through a series of processes from one state and ends with the same initial state by forming a complete cycle so that the properties of the system at the end of the cycle are the same as at the beginning. A cyclic process is shown in Fig. 2.9 on $p - v$ diagram.

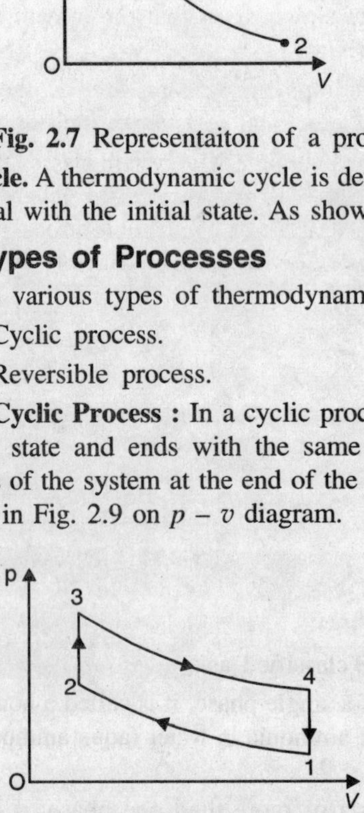

Fig. 2.9 Cyclic process.

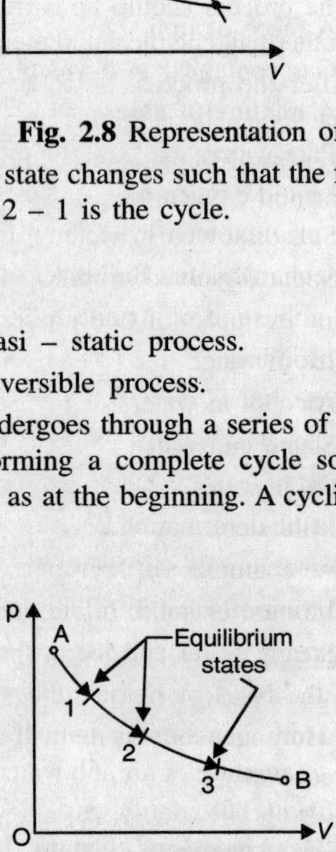

Fig. 2.10 A quasi-static process.

2. Quasi-static Process : A quasi – static process is one in which the system deviates from one equilibrium state by only infinitesimal amounts throughout the entire process. In other words, a process closely approximating to a succession of equilibrium states is called quasi – static process, as shown in Fig. 2.10. It is a nearly reversible process.

3. Reversible Process : A process is said to be reversible if the system passes through a continuous series of equilibrium states. In other words, a process is called reversible if the initial state together with all energies transferred or transformed during the process can be completely restored in both system and environment. A process will be reversible only if there is no friction in the system. A reversible process is shown in Fig. 2.11. Some examples of reversible process are:

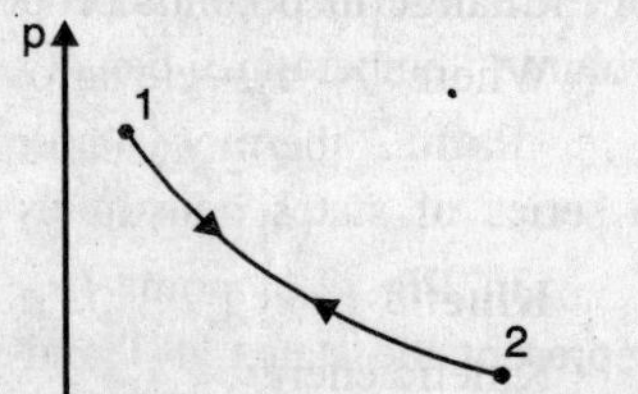

Fig. 2.11 Reversible process.

- Frictionless adiabatic expansion and compression of a fluid.
- Polytropic expansion and compression of a fluid.
- Isothermal expansion or compression of a gas.
- Extension and compression of a spring.
- Electrolysis.
- Evaporation.

Conditions for Reversibility

The following conditions must be satisfied by a reversible process :

- There should be no heat transfer across finite temperature difference.
- The process should be carried out with absolute slowness so that the system is always in thermal, chemical and mechanical equilibrium.
- After the process, both the system and surroundings must be restored to the original state, *i.e.*, there should be no friction, viscosity, electrical resistance, inelasticity, etc.

4. Irreversible Process : A process is said to be irreversible if a system passes through a sequence of non-equilibrium states. An irreversible process will not retrace the reverse path to restore the original state, as shown in Fig. 2.12. Some examples of irreversible process are :

- Mechanical and fluid friction.
- Combustion of air and fuel.
- Throttling.
- Free expansion.
- Adiabatic mixing.
- Heat transfer by convection.
- Plastic deformation.
- Dissolving of salt in water.
- Mixing of sugar in milk.

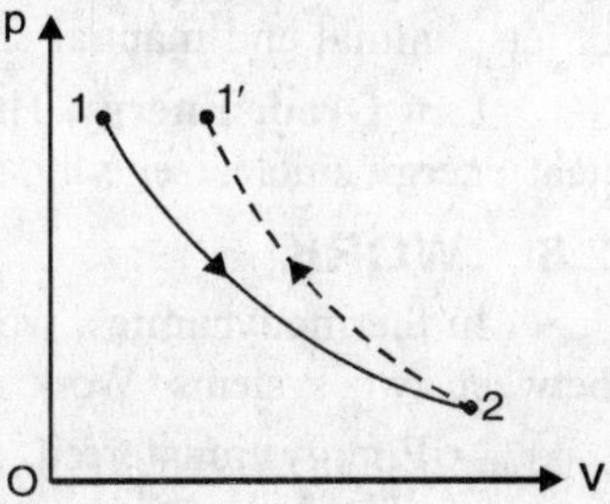

Fig. 2.12 Irreversible process.

2.7 ENERGY AND ITS FORMS

Energy is defined as the capacity to do work. The SI units of energy are Joule (J). Energy may be broadly classified as : stored energy, and energy in transition.

Stored Energy. It is contained within the system boundaries, *e.g.*, potential energy, kinetic energy, and internal energy, etc.

Energy in Transition. This type of energy crosses the system boundaries, *e.g.*, heat, work, electricity, etc.

Potential Energy. This energy is due to the position of mass of the body with respect to the Earth.

Potential energy, $PE = mgz$

Change in potential energy, $\Delta PE = mg\,(z_2 - z_1)$

Where z = elevation of mass with respect to the Earth.

g = acceleration due to gravity

m = mass of the body

Kinetic Energy. This energy is due to the velocity of the mass of the body.

Kinetic energy, $KE = \frac{1}{2}.mc^2$

Where c = velocity of mass

Change in kinetic energy, $\Delta KE = \frac{1}{2}.m\,(c_2^2 - c_1^2)$

Internal Energy. It is a property consisting of the combined molecular kinetic and potential energies. It is determined by the intensive properties, such as, pressure and temperature. The various energies which contribute towards internal energy are :

- Translational, rotational and vibrational energy of molecules.
- Spin energy of electrons.
- Potential energy of molecules.
- Molecular binding energy.
- Magnetic or electric dipole energy.

For gases, internal energy is defined as :

$$U = mC_v \, \Delta T$$

Where C_v = specific heat at constant volume.

ΔT = rise in temperature.

Total energy, $E = U + PE + KE$

$$= U + mgz + ½.mc^2$$

High Grade Energy. It is that energy which can be completely transformed into shaft work without any loss, *e.g.*, mechanical and electrical work; water, wind and tidal energy; kinetic energy of jets; animal and manual energy.

Low Grade Energy. This energy cannot be completely converted into mechanical work, *e.g.*, heat energy, nuclear energy, heat from combustion of fuels like coal, wood, oil, etc.

2.8 WORK

In thermodynamics, work is defined as energy transition, *i.e.* a form of energy interaction between two systems. Work may be defined as follows :

- Energy transferred, without transfer of mass across the boundary of a system because of an intensive property difference other than temperature that exists between system and surroundings.
- An interaction between a system and surroundings, and is done by a system if the sole effect external to the boundaries of the system could have been the raising of a weight.

The work done is represented by the area under the p –V diagram, as shown in Fig. 2.13. Thus,

Work, $$W_{1-2} = \int_{V_1}^{V_2} p.dV$$

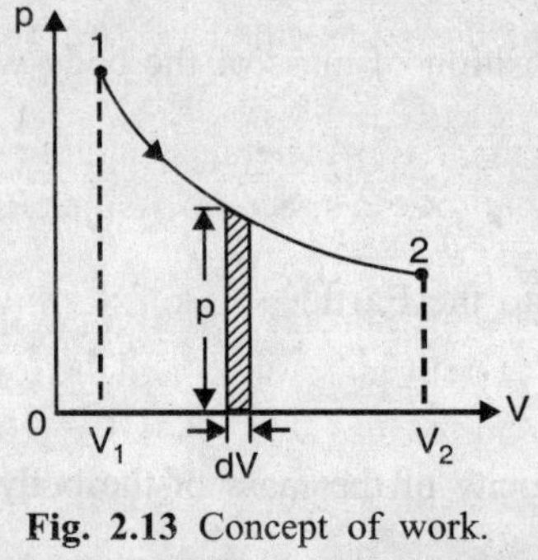

Fig. 2.13 Concept of work.

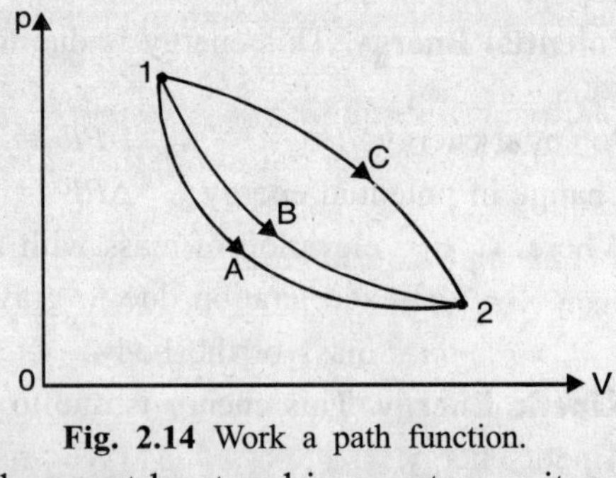

Fig. 2.14 Work a path function.

The SI units of work are N.m or Joule. Work cannot be stored in a system as it exists only during the transition. The work transferred into a system exists as stored energy. Work is a path function, as shown in Fig. 2.14.

$$W_A \neq W_B \neq W_C, \text{ as}$$

$$W_{1-2} = \int_1^2 \delta W$$

Sign Convention for Work. Work done by the system is considered positive and work done on the system as negative.

2.9 HEAT

Heat is an energy interaction between the system and the surroundings. Heat may be defined as the energy transferred without transfer of mass across the boundary of a system due to a temperature difference between the system and the surroundings. The SI units of heat are Joule.

A wall which does not allow the flow of heat is called an adiabatic wall where as a wall which permits the flow of heat is a **diathermal** wall. The similarities and dissimilarities between work and heat are :

Similarities :

- Energy interaction.
- Transient phenomenon.
- Boundary phenomenon.
- Not the property of the system.
- Path functions.

Dis-similarities :

- Heat transfer is the energy interactions due to temperature difference only. All other energy interactions are work interactions.
- Heat is a low grade energy while work is a high grade energy.

Sign Convention for Heat. Heat received by the system is considered positive and heat rejected by the system as negative.

2.10 THERMODYNAMIC EQUILIBRIUM

A system is said to be in a state of thermodynamic equilibrium if the value of the property is same at all points in the system. An isolated system always reaches, in due course of time, a state of thermodynamic equilibrium and can never depart from it spontaneously.

A system will be in a state of thermodynamic equilibrium, if the conditions for the following three types of equilibrium are satisfied :

Mechanical Equilibrium. A system is said to be in a state of mechanical equilibrium if there exists no unbalanced force either in the interior of the system or between the system and the surroundings.

Chemical Equilibrium. A system is said to be in a state of chemical equilibrium if there exists no chemical reaction or transfer of matter from one part of the system to another, such as, diffusion or solution.

Thermal Equilibrium. A system is said to be in thermal equilibrium if there exists an uniformity of temperature throughout the system or between system and surroundings.

A system is said to be in thermodynamic equilibrium only if the system is capable of undergoing a spontaneous change in any macroscopic property when it is isolated.

2.11 POINT FUNCTION AND PATH FUNCTION

When the thermodynamic property has a definite value for a given state, it is called a point function. For example, pressure, temperature, volume, etc. are point functions. These properties depend only on the initial and final states of the system. The differentials of point functions are exact, *e.g.*,

$$\int_{V_1}^{V_2} dV = V_2 - V_1$$

The quantities which are not thermodynamic properties are called path functions. For example, heat and work are path functions because they depend upon the path taken from one state to the other. The work and heat are inexact differentials and are written as dW and dQ.

$$\int_1^2 \delta W = W_{1\text{-}2}$$

and

$$\int_1^2 \delta Q = Q_{1\text{-}2}$$

2.12 CONCEPT OF AN IDEAL GAS

Many gases which do not change their phase during a thermodynamic process and obey a set of common rules governing change of their properties are called ideal (or perfect) gases. For an ideal gas, the following conditions are satisfied :

1. Gas law :

$$pv = RT$$

Where
p = pressure, kPa
v = specific volume, m^3/kg
R = 0.287 kJ/kg. K, characteristic gas constant
T = absolute temperature, K

2. The internal energy is a function of temperature alone, *i.e.*

$$U = f(T) \text{ only}$$

None of the gases existing in nature are ideal in real sense. However, most of the gases like nitrogen, oxygen, helium, hydrogen, carbon dioxide and air obey gas laws at normal temperature and pressure. These gases may be regarded as ideal or perfect gases for all practical purposes.

2.12.1 Characteristic Gas Equation

By combining Boyle's and Charles' laws a relation between pressure, temperature and volume may be found. The resulting equation is called the equation of state for a gas. Consider one kg mass of a perfect gas undergoing changes from p_1, v_1, T_1 to p_2, v_2, T_2 by the two processes 1 – A and A – 2, as shown in Fig. 2.15. 1 – A is a constant pressure process obeying Charles' law and A – 2 is an isothermal process obeying Boyle's law.

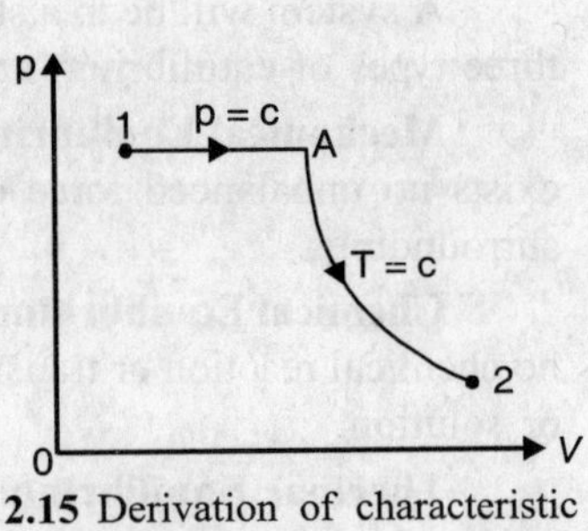

Fig. 2.15 Derivation of characteristic gas equation.

Process 1 – A :

$$\frac{v_1}{T_1} = \frac{v_A}{T_A}$$

or

$$v_A = \frac{v_1 T_A}{T_1} = \frac{v_1 T_2}{T_1} \qquad (\because T_2 = T_A) \qquad \ldots(1)$$

Process A – 2 :

$$p_2 v_2 = p_A v_A = p_1 v_A \qquad (\because p_A = p_1)$$

or $$v_A = \frac{v_2 p_2}{p_1} \quad \ldots(2)$$

From Eqs. (1) and (2), we get

$$\frac{v_1 T_2}{T_1} = \frac{v_2 p_2}{p_1}$$

or $$\frac{p_1 v_1}{T_1} = \frac{p_2 v_2}{T_2}$$

or $$\frac{pv}{T} = \text{Const.} = R$$

This is called the characteristic gas equation.

For a gas having m kg mass,

$$p_V = mRT$$

$$pv_{mole} = R_0 T$$

$$pV = nR_0 T$$

where V = total volume of gas, m^3

n = number of kg. moles of a gas = m/M

m = mass of gas, kg

M = molecular weight of a gas, kg/kg moles

v_{mole} = mole volume = $\frac{V}{n}$, m^3/kg mole

R_0 = universal gas constant

= 8.314 kJ/kg mole.K

2.13 ZEROTH LAW OF THERMODYNAMICS

This law states that if two systems are each in thermal equilibrium with a third system, they are also in thermal equilibrium with each other.

Let us consider three bodies A, B and C. as shown in Fig. 2.16. If bodies A and B are in thermal equilibrium with body C, then according to this law, bodies A and B will also be in thermal equilibrium with each other.

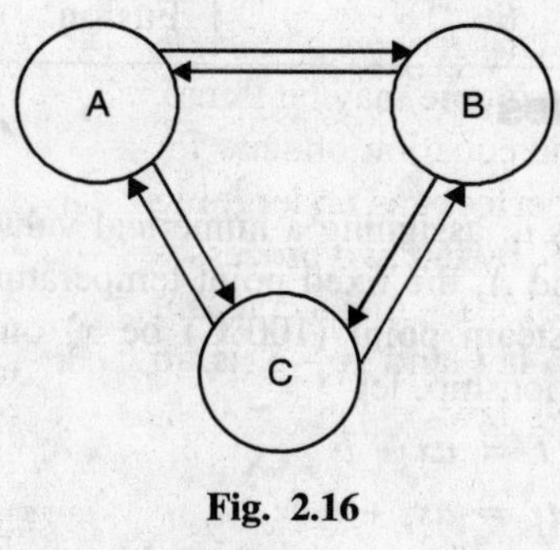

Fig. 2.16

2.14 CONCEPT OF TEMPERATURE

Temperature may be defined as follows :

- It is the thermal state of a body considered with reference to its ability to communicate heat to other bodies.

- It is that property of a system which determines whether or not it is in thermal equilibrium with the other systems.
- It is a manifestation of activity of the molecules.

Temperature is an intensive property of the system because it is independent of mass. It is not a function of path but a function of state of the system, *i.e.* it is a point function.

2.15 PRINCIPLE OF THERMOMETRY

Thermometry may be defined as the art or device of measuring temperature with scientific precision. The basic principle of thermometry is that the two bodies in contact and in thermal equilibrium have the same temperature. To measure temperature, heat is transferred to a measuring instrument and the change due to temperature in some property is recorded. The property which changes in value as a function of temperature is called thermometric property. The substance which has the thermometric property is called the thermometric substance. In practice, thermometry includes :

- Construction of an instrument called **thermometer**.
- Calibration of instrument.
- Sensitiveness of the instrument.

2.15.1 Types of Thermometers

The various types of thermometers with their thermometric properties are given in Table 2.1.

Table 2.1 Thermometers and thermometric properties.

Sl.No.	Thermometers	Thermometric property	Symbol
1.	Liquid – in – glass thermometer	Length	L
2.	Constant volume gas thermometer	Pressure	p
3.	Constant pressure gas thermometer	Volume	V
4.	Electrical resistance thermometer	Resistance	R
5.	Thermocouple	Thermal e.m.f.	ε
6.	Radiation pyrometer	Thermal radiation	E
7.	Optical pyrometer	Monochromatic radiation	E_λ
8.	Pyrometric cone	Fusion	–

2.15.2 Thermometric Scales

(*a*) Two Point Scale

A temperature scale amounts to assigning a numerical value to the thermometer. Suppose for a liquid thermometer with a liquid *A*, the fixed point temperatures be represented by t_i and t_s. Let the ice point (0°C) be x_i and steam point (100°C) be x_s on the temperature scale shown in Fig. 2.17 (*a*). For the linear relationship, let

$$t = ax + b$$

So that $$t_i = ax_i + b$$

and $$t_s = ax_s + b$$

This gives, $$a = \frac{t_s - t_i}{x_s - x_i}$$

$$b = t_i - \left[\frac{t_s - t_i}{x_s - x_i}\right] x_i$$

$$t = t_i + (t_s - t_i)\left[\frac{x - x_i}{x_s - x_i}\right]$$

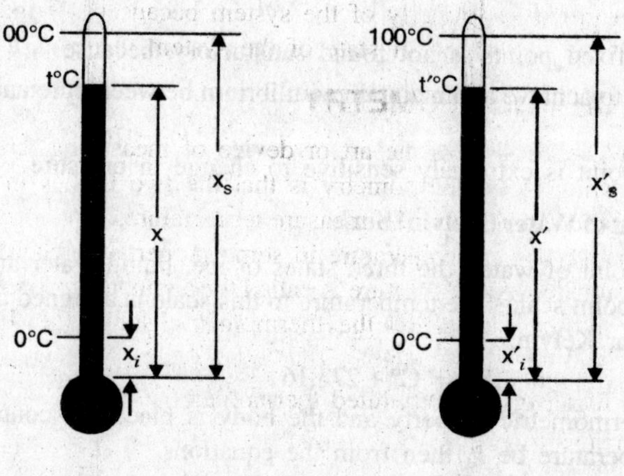

(*a*) Liquid A thermometer (*b*) Liquid B thermometer

Fig. 2.17

For the Celsius scale :

$$t_i = 0°C \text{ and } t_s = 100°C$$

$$\therefore \quad t°C = 100\left[\frac{x - x_i}{x_s - x_i}\right] \quad \ldots(1)$$

For the Fahrenheit scale :

$$t_i = 32°F \text{ and } t_s = 212°F$$

$$t°F = 32 + 180\left[\frac{x - x_i}{x_s - x_i}\right] \quad \ldots(2)$$

Comparing Eqs. (1) and (2), we get

$$\frac{C}{100} = \frac{F - 32}{180}$$

or

$$t°C = \frac{F - 32}{1 \cdot 8} \quad \ldots(3)$$

For Reaumer scale :

$$t_i = 0°R \text{ and } t_s = 80°R$$

$$t°R = 80\left[\frac{x - x_i}{x_s - x_i}\right] \quad \ldots(4)$$

From Eqs. (1) and (4), we get

$$\frac{C}{100} = \frac{R}{80}$$

or

$$t°C = 1.25R \quad \ldots(5)$$

Combining Eqs. (3) and (5), we get

$$\frac{C}{5} = \frac{F - 32}{9} = \frac{R}{4} \quad \ldots(6)$$

For liquid *B* (Fig. 1.16 (*b*)) :

$$t\,°C = 100\left[\frac{x' - x'_i}{x'_s - x'_i}\right]$$

The use of two fixed points is not found satisfactory, because

(*i*) It is difficult to achieve and maintain equilibrium between pure and saturated air with pure water.

(*ii*) The steam point is extremely sensitive to change in pressure.

(*b*) **Triple – point of Water (Kelvin) Scale**

At the triple – point of water, the three states of ice, liquid water and vapour coexist.. This gives only one fixed point scale. The temperature to this scale is assigned an arbitrary value 273.16 K, where K stands for Kelvin.

$$K = C° + 273.16$$

Let x_i be the thermometric property and the body is placed in contact with the triple point of water whose temperature be t_t, then from the equations,

$$t = ax \text{ and } t_t = ax_t, \text{ we have}$$

$$a = \frac{t_i}{x_t} = \frac{273.16}{x_t}$$

$$\therefore \quad t = 273.16\left(\frac{x}{x_t}\right)$$

According to above equation, temperature recorded by different thermometers will be as given below in Table 2.2.

Table 2.2 Temperature recorded by various thermometres

Sl.No.	Thermometer	Temperature
1.	Liquid thermometer	$t(L) = 273.16\left(\frac{L}{L_t}\right)$
2.	Constant volume gas thermometer	$t(p) = 273.16\left(\frac{p}{p_t}\right)$
3.	Constant pressure gas thermometer	$t(V) = 273.16\left(\frac{V}{V_t}\right)$
4.	Electrical resistance thermometer	$t(R) = 273.16\left(\frac{R}{R_t}\right)$
5.	Thermocouple	$t(\varepsilon) = 273.16\left(\frac{\varepsilon}{\varepsilon_i}\right)$

Absolute temperature, $T = [°C + 273.16]\ K$

$$= [°F + 459.67]\ K$$

2.15.3 Perfect Gas Temperature Scale

Consider a constant volume gas thermometer, which comprises of a sensing bulb containing a gas, connected to a U-tube manometer shown in Fig. 2.18. The U-bend of the manometer is flexible so that its arm B can be raised or lowered, thus maintaining the meniscus in the other arm A at the mark indicated in the figure.

For measuring a temperature, the gas bulb is first placed in a constant temperature bath at the triple point of water and the volume of the gas is brought upto the mark in arm A by moving the manometer arm B. Let the pressure reading taken at this temperature T_{tp} be p_{tp}. Then the bulb is placed in a location where the temperature T is to be measured. Again, in a similar manner, the meniscus is brought upto the mark in arm A. Let the new pressure reading be p. The required temperature is then given by:

$$T = 273.16\frac{p}{p_{tp}}$$

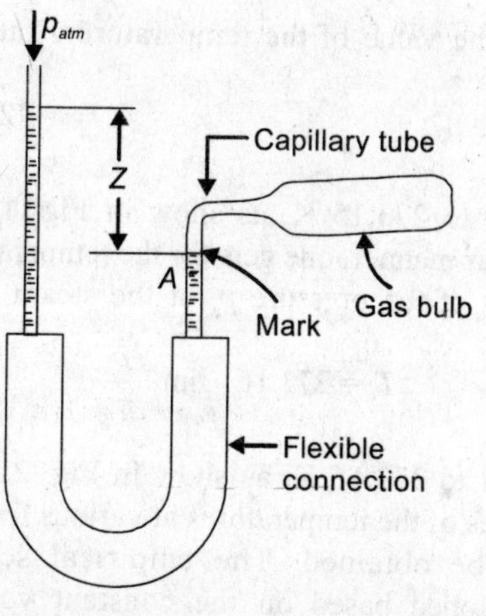

Fig. 2.18 Constant volume gas thermometer

Let us consider that gas A is filled in the bulb such that initially the pressure p_{tp} of the gas at the triple point of water is 1000 mm Hg, and the pressure at the measured temperature is p_1 mm Hg. Then the measured temperature is:

$$273.16\left(\frac{p_1}{1000}\right) = T_1 \text{ (say)}$$

Now let some gas be leaked out of the bulb so that the pressure initially measured at the triple point of water is 500 mm Hg, and the pressure at the measured temperature is p_2 mm Hg. Then, for the measured temperature, we get

$$273.16\left(\frac{p_2}{500}\right) = T_2 \text{ (say)}$$

Similarly, the initial pressure of the gas may be lowered further and every time we may get a slightly different value of temperature T, say, T_1, T_2, T_3, ... etc. Let the values of these temperatures be plotted against the pressure p_{tp} of the gas. A curve will result as shown in Fig 2.19. If this curve is extrapolated to zero pressure, it gives the required temperature in the limit as:

$$273.16 \lim_{p_{tp}\to 0} \frac{p}{p_{tp}} = T$$

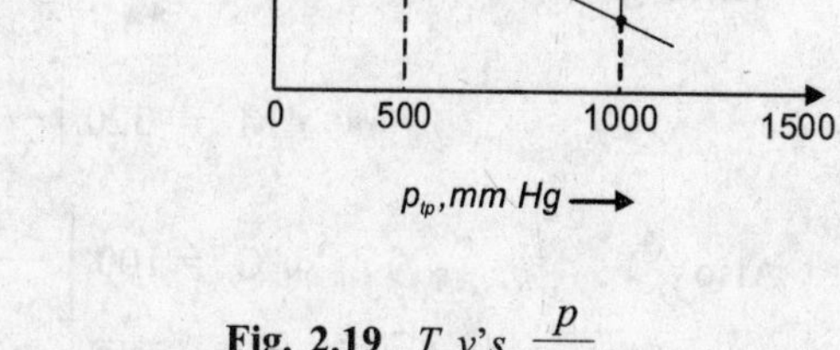

Fig. 2.19 T v's $\frac{p}{p_{tp}}$.

This is the true value of the temperature. The above experiment may be repeated for gas B and C in the bulb to obtain another set of values for the temperature T. When extrapolated to zero pressure, the T-graph will intersect the ordinate giving the same value of the true temperature T. Thus, as pressure tends to zero different gases indicate the same ratio $\frac{T}{T_{tp}} = \frac{p}{p_{tp}}$ and hence the same temperature T. This temperature is referred to as the ideal gas temperature.

By measuring the pressure p_{tp} and p_i respectively at the triple point and the ice point, one gets the value of the temperature T_i at the ice point as:

$$T_i = 273.16 \lim_{p_{tp}\to 0} \frac{p_i}{p_{tp}}$$

equal to 273.15 K, as show in Fig. 1.19. In a similar manner, one gets for the temperature T_s in terms of the pressure p_s at the steam point as:

$$T_s = 273.16 \lim_{p_{tp}\to 0} \frac{p_s}{p_{tp}}$$

equal to 373.15 K, as show in Fig. 2.20. Thus, values of the temperatures at various fixed points can be obtained. The empirical scale, thus developed based on the constant volume gas thermometer is commonly referred to as the ideal gas temperature scale.

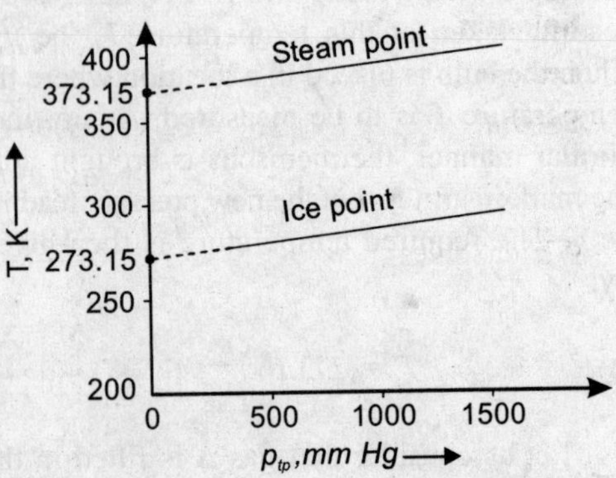

Fig. 2.20 Steam point and ice point.

Alternatively, if plots of p_s/p_i v's p_{tp} are made with different gases in the bulb we get different curves and when extrapolated to zero pressure all curves converge, and the ratio p_s/p_i tend to a constant value 1.366, so that

$$\frac{T_s}{T_i} = \lim_{p_{tp}\to 0} \frac{p_s}{p_i} = 1.366$$

This value may be considered as a universal constant of nature.

Example 2.1 *The ice and steam points on a temperature scale °M are 80°M and 400°M respectively. Correlate this with the Celsius scale. The °N reading on this scale is a certain number of degrees on a corresponding absolute temperature scale. Find this absolute temperature at °N.*

Solution.

Let $\quad t = ax + b$

So that $\quad 80 = ax_i + b$

$$400 = ax_s + b$$

Which gives, $\quad a = \dfrac{320}{x_s - x_i},\; b = 80 - 320\left[\dfrac{x_i}{x_s - x_i}\right]$

$$t°M = 320\left[\frac{x - x_i}{x_s - x_i}\right] + 80$$

Also $\quad t°C = 100\left[\dfrac{x - x_i}{x_s - x_i}\right]$

$\therefore \quad t°M = 3.2\, t°C + 80$

Now $\quad \dfrac{T_s}{T_i} = 1.366$

And $\quad T_s - T_i = 400 - 80 = 320$

$$T_i = \frac{320}{0.366} = 874.3°N$$

$$T_s = 1194.3°N$$

$$T°N\ abs = (t°N - 80) + 874.3 = t°N + 794.3$$

At t°N = 0°N, we get

$$T°\text{N } abs = 0°\text{N} + 794.3 = 794.3$$

Example 2.2 *A thermometer reads 32°F when placed in melting ice and 212°F when placed in boiling water. What will it read when the measured temperature is 290 K ?*

Solution.

Let $t = ax + b$

So that $32 = ax_i + b$

$$212 = ax_s + b$$

Which gives, $a = \dfrac{180}{x_s - x_i}$, $b = 32 - 180\left[\dfrac{x_i}{x_s - x_i}\right]$

$$t°\text{F} = 180\left[\frac{x - x_i}{x_s - x_i}\right] + 32$$

Also $t°\text{C} = 100\left[\dfrac{x - x_i}{x_s - x_i}\right]$

∴ $t°\text{F} = 1.8\, t°\text{C} + 32$

Now $t°\text{C} = 290 - 273.15 = 16.85$

∴ $t°\text{F} = 1.8 \times 16.85 + 32 = 62.33$

Example 2.3 *The temperature scale of a thermometer is given by : t = a ln p + b. If at the ice point and steam point the thermometric properties are found to be 2 and 8 respectively, what will be the temperature corresponding to the thermometric property of 4 on Celsius scale ?*

Solution.

$$t = a \ln p + b$$

At ice point, $0 = a \ln 2 + b$

At steam point, $100 = a \ln 8 + b$

Which gives, $a = \dfrac{100}{\ln 4} = 72.135$

$$b = -50$$

$$t = 72.135 \ln p - 50$$

At $p = 4$, $t = 72.135 \ln 4 - 50 = 50°\text{C}$

Example 2.4 *Define a new temperature scale, say °F, in which the steam point and ice point are 212°F and 32°F respectively. The °F reading on this scale is a number of degrees on a corresponding absolute temperature scale (also called Rankine scale). What is this temperature at 10°F ?*

Solution.

$$\frac{T_s}{T_i} = 1.366$$

$$T_s - T_i = 212 - 32 = 180$$

$$(1.366 - 1)\, T_i = 180$$

$$T_i = \frac{180}{0.366} = 491.7°\text{F abs or } 491.7°\text{R}$$

$$T_s = 180 + 491.7 = 671.7°\text{R}$$

$$T°\text{R} = (T°\text{F} - 32) + 491.7 = t°\text{F} + 459.7$$

At 10°F, $T°\text{R} = 10 + 459.7 = 469.7$

Example 2.5 *A cylinder contains compressed helium gas. It is used to inflate rubber balloon to a volume of $0.8m^3$. If the atmospheric pressure is 1 bar, find the work done upon the atmosphere by the balloon. Sketch the system before and after the process.*

Solution. The cylinder – balloon system is shown in Fig. 2.21.

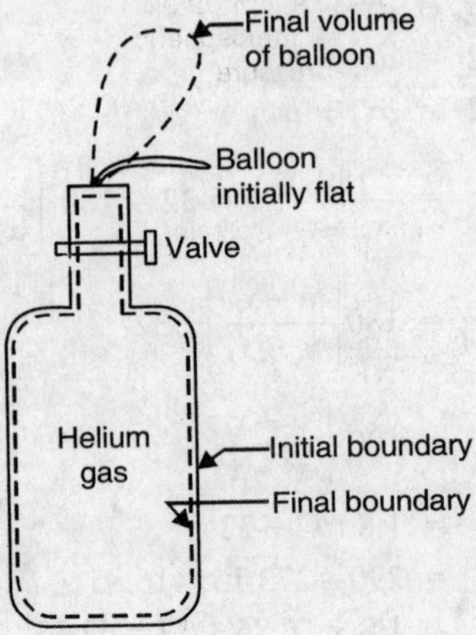

Fig. 2.21

Net displacement work done,

$$W_{net} = \int_{cylinder} p.dV + \int_{balloon} p.dV$$

$$= 0 + 1 \times 10^2 \times 0.8 = 80 \text{ kJ}$$

No work is done on the cylinder as its volume is constant.

Example 2.6 *Air enters into an evacuated cylinder from the atmosphere when the valve is opened. Calculate the work done if atmospheric pressure is 1 bar and 2 m^3 of air at atmospheric pressure enters the cylinder.*

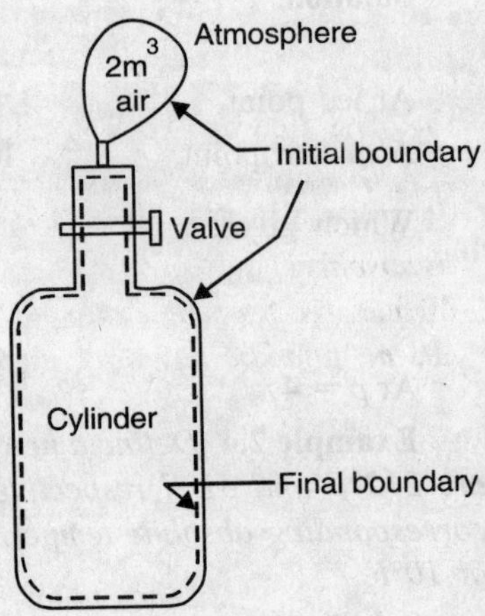

Fig. 2.22

Solution.

The cylinder – atmosphere system is shown in Fig. 2.22.

Net work done, $$W_{net} = \int_{cylinder} p.dV + \int_{free-air\ boundary} p.dV$$

$$= 0 + \int_2^0 1 \times 10^5 \, dV$$

$$= 1 \times 10^5 (0 - 2) = -200 \text{ kJ}$$

Example 2.7 *An engine piston has an area of 50 cm^2 and moves 6 cm during part of suction stroke. 350 cm^3 of fresh air is drawn in from the atmosphere. The pressure in the cylinder during suction stroke is 0.9 bar and atmospheric pressure is 1 bar. The flow resistance in the suction pipe and inlet valve is accounted for by the difference of atmospheric pressure and suction pressure. Calculate the net work done during the process.*

Solution.

The system conditions are shown in Fig. 2.23.

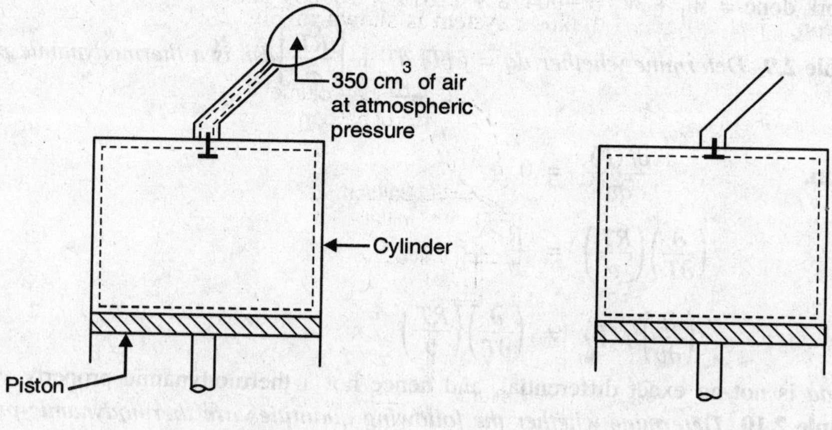

(*a*) Initial condition (*b*) Final condition

Fig. 2.23

Net work done,

$$W_{net} = \underset{\text{Piston}}{\int p.dV} + \underset{\text{free – air Boundary}}{\int p.dV}$$

$$= 0.9 \times 10^5 \times 50 \times 6 \times 10^{-6} - 1 \times 10^5 \times 350 \times 10^{-6}$$

$$= 27 - 35 = -8 \text{ J}$$

The free air work is negative because its boundary contracts.

Example 2.8 *A piston and cylinder machine containing a fluid system has a stirring device as shown in Fig. 2.24. The piston is frictionless and is held down against the fluid due to atmospheric pressure of 1 bar. The stirring device is turned 9600 revolutions with an average torque against the fluid of 1.5 N.m. In the mean time the piston of 0.6 m diameter moves out 0.8 m. find the net work done for the system.*

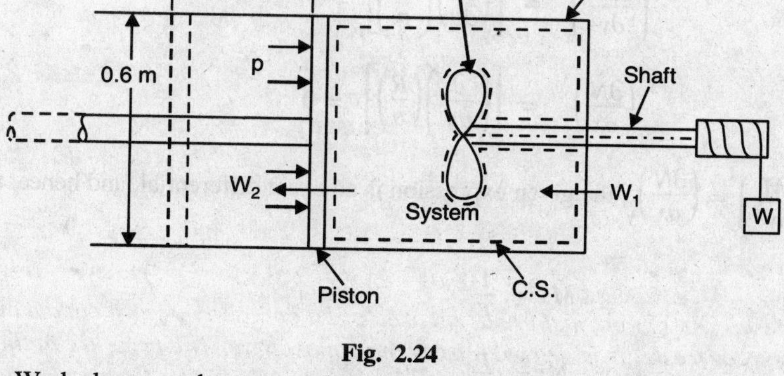

Fig. 2.24

Solution. Work done on the system by stirring device,

$$W_1 = -2\pi NT = -2\pi \times 9600 \times 1.5 = -90478 \text{ N.m}$$

Work done by the system, $W_2 = pAL = \dfrac{1\times10^5 \times\pi\times(0.6)^2\times0.8}{4} = 22619$ N.m

Net work done $= W_1 + W_2 = -90478 + 22619 = -67859$ N.m

Example 2.9 *Determine whether* $dq = f(T)\,dT + \left[\dfrac{RT}{v}\right]dv$ *is a thermodynamic property or not.*

Solution.
$$\frac{\partial f(T)}{\partial v} = 0$$

and
$$\left(\frac{\partial}{\partial T}\right)\left(\frac{RT}{v}\right) = \frac{R}{v}$$

Hence
$$\left(\frac{\partial}{\partial v}\right) f(T) \neq \left(\frac{\partial}{\partial T}\right)\left(\frac{RT}{v}\right)$$

Thus dq is not an exact differential, and hence not a thermodynamic property.

Example 2.10 *Determine whether the following quantities are thermodynamic properties or not :*

(*a*) $\left(\dfrac{v}{T}\right)dp + \left(\dfrac{P}{T}\right)dv$ (*b*) $\left(\dfrac{dT}{T}\right) + \left(\dfrac{p}{T}\right)dv$

(*c*) $\left(\dfrac{dT}{T}\right) + \left(\dfrac{p}{v}\right)dv$ (*d*) $\left(\dfrac{dT}{T}\right) - \left(\dfrac{v}{T}\right)dp$

Solution. A quantity $dz = Mdx + Ndy$ would be an exact differential, and hence a thermodynamic property, if

$$\left(\frac{\partial M}{\partial y}\right)_x = \left(\frac{\partial N}{\partial x}\right)_y$$

(*a*) Here
$$M = \frac{v}{T} = \frac{R}{p} \qquad (pv = RT)$$
$$N = \frac{p}{T} = \frac{R}{v}$$
$$x = p \text{ and } y = v$$
$$\left(\frac{\partial M}{\partial y}\right)_x = \left[\left(\frac{\partial}{\partial v}\right)\left(\frac{R}{p}\right)\right]_p = 0$$
$$\left(\frac{\partial N}{\partial x}\right)_y = \left[\left(\frac{\partial}{\partial p}\right)\left(\frac{R}{v}\right)\right]_v = 0$$

Since $\left(\dfrac{\partial M}{\partial y}\right)_x = \left(\dfrac{\partial N}{\partial x}\right)_y$, the given expression is an exact differential, and hence thermodynamic property.

(*b*) Here
$$M = \frac{1}{T}$$
$$N = \frac{p}{T}$$
$$x = T \text{ and } y = v$$

$$\left(\frac{\partial M}{\partial y}\right)_x = \left[\left(\frac{\partial}{\partial v}\right)\left(\frac{1}{T}\right)\right]_T = 0$$

$$\left(\frac{\partial N}{\partial x}\right)_y = \left[\left(\frac{\partial}{\partial T}\right)\left(\frac{R}{v}\right)\right]_v = 0 \qquad \left(\frac{p}{T} = \frac{R}{v}\right)$$

Since $\left(\frac{\partial M}{\partial y}\right)_x = \left(\frac{\partial N}{\partial x}\right)_y$, the given expression is an exact differential, and hence a thermodynamic property.

(*c*) Here

$$M = \frac{1}{T}$$

$$N = \frac{p}{v}$$

$$x = T \text{ and } y = v$$

$$\left(\frac{\partial M}{\partial y}\right)_x = \left[\left(\frac{\partial}{\partial v}\right)\left(\frac{1}{T}\right)\right]_T = 0$$

$$\left(\frac{\partial N}{\partial x}\right)_y = \left[\left(\frac{\partial}{\partial T}\right)\left(\frac{p}{v}\right)\right]_v \neq 0$$

Since $\left(\frac{\partial M}{\partial y}\right)_x \neq \left(\frac{\partial N}{\partial x}\right)_y$, the given expression is not an exact differential, and hence not a thermodynamic property.

(*d*) Here

$$M = \frac{1}{T}$$

$$N = \frac{v}{T}$$

$$x = T \text{ and } y = p$$

$$\left(\frac{\partial M}{\partial y}\right)_x = \left[\left(\frac{\partial}{\partial p}\right)\left(\frac{1}{T}\right)\right]_T = 0$$

$$\left(\frac{\partial N}{\partial x}\right)_y = \left[\left(\frac{\partial}{\partial T}\right)\left(\frac{R}{P}\right)\right]p = 0 \qquad \left(\because \frac{v}{T} = \frac{R}{p}\right)$$

Since $\left(\frac{\partial M}{\partial y}\right)_x = \left(\frac{\partial N}{\partial x}\right)_y$, the given expression is an exact differential, and hence a thermodynamic property.

Example 2.11 *Calculate the work done when the volume changes from 5 m^3 to 10 m^3 in a non – flow quasi – static process in which pressure is given by :*

$$p = (5V - 5) \text{ bar.}$$

Solution.

Work done,

$$W_{1\cdot 2} = \int p.dV = \int_5^{10} (5V - 5) \times 10^5 \, dV$$

$$= 10^5 \left| 2.5V^2 - 5V \right|_5^{10} = 16250 \text{ kJ}$$

Example 2.12 *A non – flow reversible process occurs for which $p = 3V^2 + \frac{1}{V}$, where p is in bar and V in m^3. What will be the work done when V changes from 0.5 m^3 to 1 m^3 ?*

Solution.

Work done, $$W_{1\text{-}2} = \int p.dV = \int_{0.5}^{1} \left(3V^2 + \frac{1}{V}\right) dV$$

$$= \left| V^3 + \ln V \right|_{0.5}^{1} = 69.3 \text{ kJ}$$

Example 2.13 *A fluid is contained in a cylinder behind a piston of specific volume 0.2 m^3/kg at 5 bar pressure. The fluid is compressed reversibly to a pressure of 0.5 bar according to the law : $pv^{1.5} = C$, where C is a constant. Calculate the work done by the fluid on the piston.*

Solution.

$$v_2 = v_1 \left(\frac{p_1}{p_2}\right)^{\frac{1}{1.5}} = 0.2\left(\frac{5}{0.5}\right)^{\frac{1}{1.5}} = 0.92832 \text{ m}^3/\text{kg}$$

Work done, $$w_{1\text{-}2} = \int_{v_1}^{v_2} p.dv = C\int_{0.2}^{0.92832} \frac{dv}{v^{1.5}} = C\left[\frac{-v^{-0.5}}{0.5}\right]_{0.2}^{0.92832}$$

$$= 2C\,[0.2^{-0.5} - 0.92832^{-0.5}] = 2.39635 \text{ C}$$

Now $$C = p_1 v_1^{1.5} = 5 \times 10^5 \times 0.2^{1.5} = 44.72 \times 10^3$$

$$w_{1\text{-}2} = 2.39635 \times 44.72 \times 10^3 = 107.168 \text{ kJ/kg}$$

Example 2.14 *A new scale N of temperature is devised in such a way that the freezing point of ice is 100°N and boiling point is 400°N. What is the temperature reading on this new scale when the temperature is 150°C? At what temperature both the Celsius and new temperature scale reading would be the same?*

Solution.

Let $$t°\text{N} = ax + b$$

So that $$100 = ax_i + b$$

$$400 = ax_s + b$$

$$a = \frac{300}{x_s - x_i}$$

$$b = 100 - \frac{300\,x_i}{x_s - x_i}$$

$$t°\text{N} = \frac{300\,x}{x_s - x_i} + 100 - \frac{300\,x_i}{x_s - x_i}$$

$$= \frac{300\,(x - x_i)}{x_s - x_i} + 100$$

Also $$t°\text{C} = 100\,\frac{100\,(x - x_i)}{x_s - x_i}$$

∴ $$t°\text{N} = 3\,t°\text{C} + 100$$

When $t°\text{C} = 150$, then

$$t°\text{N} = 3 \times 150 + 100 = 550$$

For $$t°\text{N} = t°\text{C}$$

$$t°\text{C} = 3t°\text{C} + 100$$

$$t°\text{C} = -50$$

Example 2.15 *Two Celsius thermometers A and B with temperature readings T_a and T_b agree at ice point (0°C) and steam point (100°C) but elsewhere they are related by:*

$$T_a = l + m\,T_b + n\,T_b^2$$

where l, m and n are constants. When they are immersed in oil bath, A shows 51°C while B shows 50°C. Determine T_b when T_a reads 25°C.

Solution.

Given $T_a = T_b = 0°C$, $T_a = T_b = 100°C$ and $T_a = 51°C$ when $T_b = 50°C$

So that
$$0 = l \quad ...(1)$$
$$100 = l + 100\,m + 100^2\,n = 100\,m + 100^2\,n \quad ...(2)$$
$$51 = l + 50\,m + 50^2\,n = 50\,m + 50^2\,n$$

or
$$102 = 100\,m + 2 \times 50^2\,n \quad ...(3)$$

Substracting Eqn. (2) from (3), we have
$$2 = (5000 - 100^2)n$$
$$n = -4 \times 10^{-4}$$
$$100 = 100\,m + 100^2 \times (-4 \times 10^{-4})$$
$$m = +1.04$$

$\therefore$
$$T_a = +1.04\,T_b - 4 \times 10^{-4}\,T_b^2$$
$$25 = +1.04\,T_b - 4 \times 10^{-4}\,T_b^2$$

or $4 \times 10^{-4}\,T_b^2 - 1.04\,T_b + 25 = 0$

$$T_b = \frac{+1.04 \pm \sqrt{(1.04)^2 - 4 \times 25 \times 4 \times 10^{-4}}}{2 \times 4 \times 10^{-4}}$$
$$= \frac{+1.04 \pm 1.02059}{8 \times 10^{-4}} = 24.265°C \text{ (taking –ve sign)}$$

Example 2.16 *Define a new temperature scale, say °F, in which the steam point and ice point are 212°F and 32°F. The °F reading on this scale is a number of degrees on a corresponding absolute temperature scale (also called Rankine scale). What is this absolute temperature at 10°F?*

Solution.

$$\frac{T_s}{T_i} = 1.366$$

$$T_s - T_i = 212 - 32 = 180$$

$\therefore$
$$(1.366 - 1)\,T_i = 180$$

$$T_i = \frac{180}{0.366} = 491.7°F \text{ abs or } 491.7°R$$
$$T_s = 180 + 491.7 = 671.7°R$$
$$T°R = (t°F - 32) + 491.7$$
$$= t°F + 459.7$$

At
$$10°F,\ T°R = 10 + 459.7 = 469.7°R$$

2.16 FIRST LAW OF THERMODYNAMICS

This law may be stated as follows :

- For a system undergoing a cyclic change the algebraic sum of the work delivered to the surroundings is equal to the algebraic sum of heat taken from the surroundings, *i.e.*

$$\oint \delta Q = \oint \delta W$$

where $\oint$ = integral over a cycle

δQ = infinitesimal heat transfer

δW = infinitesimal work transfer

- Heat and work are mutually convertible one into the other.
- The total energy of an isolated system, in all its forms, remains constant.
- It is impossible to construct a perpetual motion machine of the first kind (PMM – 1).
- PMM – 1 is capable of producing energy without corresponding expenditure of the energy.

2.16.1 Verification of the First Law

The first law of thermodynamics can be verified by the Joule's experiment. The experimental set up is shown in Fig. 2.15(*a*). Work is done on the fluid kept in an insulated vessel by stirring of the paddle wheel. This work input to the fluid causes a rise in the temperature of the fluid. The amount of work on the fluid is calculated by the product of the weight and the vertical height through which the weight descends. The system is then immersed in a water bath, as shown in Fig. 2.25(*b*), after removing the insulation. The heat is transferred from the fluid to the water bath till the original temperature is reached, which will be indicated by the thermometer. Thus the system undergoes a complete cycle. The amount of heat rejected by the fluid is equal to the increase of energy of the water bath. Many such experiments were performed by Joule. It was found that the net work input W was directly proportional to the net heat transferred from the system. Thus

$$\oint \delta Q \propto \oint \delta W$$

or $$\oint \delta Q = \oint \delta W, \text{ for the same units of } Q \text{ and } W.$$

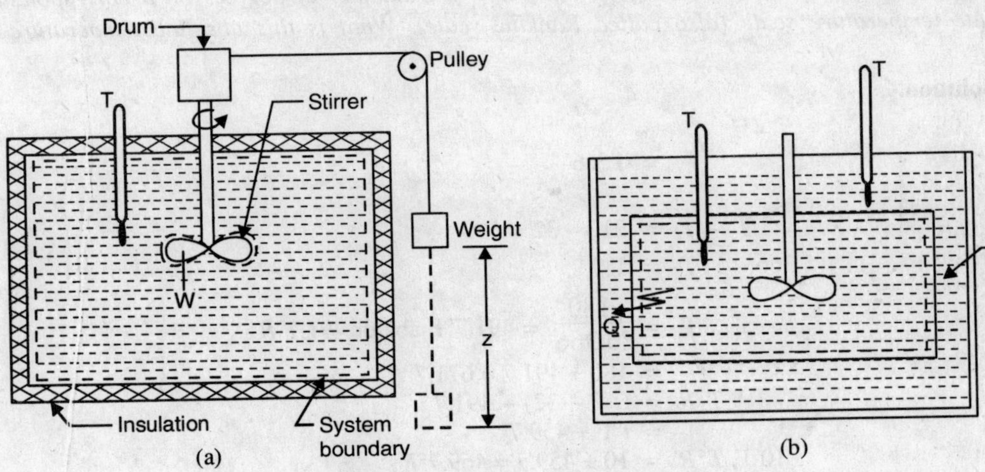

Fig. 2.25 Joule's experiment.

2.16.2 Corollaries of the First Law

Corollary 1 : System executing a process

For a system executing a process, change in stored energy of the system,

$$dE = dQ - dW$$

where E = energy stored in the system.

Corollary 2 : Isolated system

For an isolated system,

$$dQ = 0 \text{ and } dW = 0$$

Hence, $dE = 0$

or E = const.

Thus, the energy of an isolated system remains unchanged.

Corollary 3 : PMM – 1

It is impossible to construct a perpetual motion machine of the first kind. The PMM – 1 violates the first law.

As per the law of conservation of energy, no engine can produce mechanical work continuously without some other form of energy disappearing simultaneously.

2.17 APPLICATIONS OF FIRST LAW

2.17.1 System Undergoing a Process

One very important consequence of the first law of thermodynamic is that the energy of the system is a property. To prove this, consider a closed system undergoing a change of state from 1 to 2 as shown in Fig. 2.26. The system may proceed from state 1 to state 2 along the path A and then return to the original state 1 along two paths B and C (say).

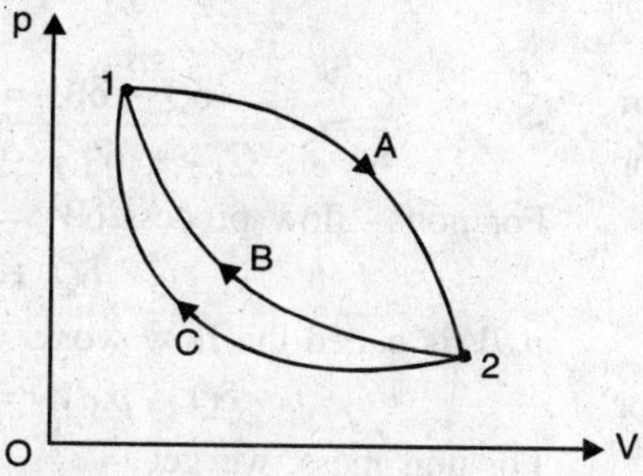

Fig. 2.26 Existance of internal energy.

For cycle 1 – A – 2 – B – 1, we have

$$\oint (\delta Q - \delta W) = 0$$

or

$$\int_1^2 (\delta Q - \delta W)_A + \int_2^1 (\delta Q - \delta W)_B = 0 \quad \ldots(1)$$

For cycle 1 – A – 2 – C – 1, we have

$$\int_1^2 (\delta Q - \delta W)_A + \int_2^1 (\delta Q - \delta W)_C = 0 \quad \ldots(2)$$

Comparing Eqs. (1) and (2), we get

$$\int_2^1 (\delta Q - \delta W)_B = \int_2^1 (\delta Q - \delta W)_C \quad \ldots (3)$$

The Eq. (3) suggests that when a system operates between states 1 and 2, the quantity $\int_2^1 (\delta Q - \delta W)$ is constant irrespective of the path along which the system proceeds. Thus the value of $\int_1^2 (\delta Q - \delta W)$ is solely fixed by the end states and is independent of the path of the process.

Therefore, the quantity $\int_1^2 (\delta Q - \delta W)$ is a property, as it is a point function. This is called the total energy E of the system and is an exact differential. Thus

$$\int_1^2 (\delta Q - \delta W) = \int_1^2 dE = E_2 - E_1$$

or $$\delta Q - \delta W = dE$$

Now $$E = U + KE + PE + \ldots\ldots\ldots\ldots$$

$$\int_1^2 (\delta Q - \delta W) = \int_1^2 [dU + d(KE) + d(PE) + \ldots\ldots]$$

or $$(\delta Q - \delta W) = \delta U + d(\tfrac{1}{2}.mc^2) + d(mgz)$$

or $$Q_{1\text{-}2} - W_{1\text{-}2} = (U_2 - U_1) + \tfrac{1}{2}.m\,(c_2^2 - c_1^2) + mg\,(z_2 - z_1)$$

Considering internal energy only, we have

$$\int_1^2 (\delta Q - \delta W) = \int_1^2 dU$$

or $$\delta Q - \delta W = dU$$

or $$Q_{1\text{-}2} - W_{1\text{-}2} = U_2 - U_1$$

For non – flow process, $\delta W = p.dV$

$$\therefore \quad \delta Q = dU + p.dV$$

$p.dV$ is called the flow work.

or $$\delta Q - p.dV = dU$$

For unit mass, we get

$$\delta q - \delta w = du$$

Therefore, the first law of thermodynamics enables us to define a property of the system called **internal energy.**

2.17.2 System Undergoing a Cycle

Consider a closed system undergoing a cyclic process as shown in Fig. 2.19. Applying the first law of thermodynamics to each process, we have

$$Q_{1\text{-}2} = E_2 - E_1 + W_{1\text{-}2}$$
$$Q_{2\text{-}3} = E_3 - E_2 + W_{2\text{-}3}$$
$$Q_{3\text{-}4} = E_4 - E_3 + W_{3\text{-}4}$$
$$Q_{4\text{-}1} = E_1 - E_4 + W_{4\text{-}1}$$

Adding, we get

$$Q_{1\text{-}2} + Q_{2\text{-}3} + Q_{3\text{-}4} + Q_{4\text{-}1} = W_{1\text{-}2} + W_{2\text{-}3} + W_{3\text{-}4} + W_{4\text{-}1}$$

$$\oint \delta Q = \oint \delta W$$

$$\oint (\delta Q - \delta W) = 0$$

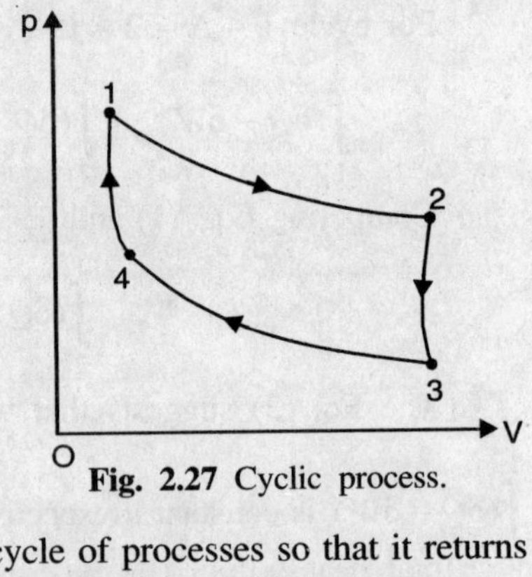

Fig. 2.27 Cyclic process.

This equation shows that if a system is taken through a cycle of processes so that it returns to the same state from which it started, the sum of heat and work transfers will be zero.

2.18 ENTHALPY

The summation of internal energy and flow energy is defined as enthalpy, and is given by:

Enthalpy, $$H = U + p.dV$$

Specific enthalpy, $$h = u + p.dv$$

Where u = specific internal energy, v = specific volume

Enthalpy is an extensive property, as it depends upon mass.

Now $$dq = du + p.dv$$
$$= du + d(pv) - v.dp$$
$$= d(u + pv) - v.dp$$
$$= dh - v.dp$$

For a constant pressure process, $dp = 0$.

$\therefore$ $$dq = dh$$

Thus, enthalpy change is the heat supplied at constant pressure.

Also for a perfect gas, $u = c_vT$ and $pv = RT$

$\therefore$ $$h = c_vT + RT$$

where C_v = specific heat at constant volume.

or $$h = T(c_v + R) = c_PT$$

where c_P = specific heat at constant pressure

or $$H = mc_PT$$

2.19 SPECIFIC HEATS

Specific heat of a substance may be defined as the amount of heat required to raise the temperature of a unit mass of the substance through one degree. The SI units of specific heat are J/kg.K. The solids and liquids have only one specific heat whereas gases have two specific heats, *i.e.*, at constant volume, and constant pressure.

2.19.1 Specific Heat at Constant Volume, c_v

The quantity of heat required to raise the temperature of unit mass of the gas through one degree at constant volume is called specific heat at constant volume, c_v.

Consider a quantity of gas contained in a fixed container shown in Fig. 2.28. When heat is supplied to the gas, the volume of the gas will not change because the container is fixed but there will be an increase in pressure and temperature of the gas.

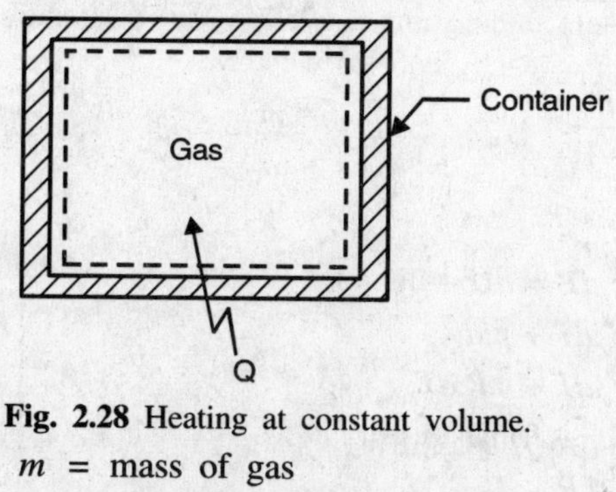

Fig. 2.28 Heating at constant volume.

Let m = mass of gas

T_1 = initial temperature of gas

T_2 = final temperature of gas

Total heat supplied to gas at constant volume,

$$Q_{1-2} = mc_v(T_2 - T_1) = mc_vdT$$

Since volume of gas remains constant, therefore, no work is done by the gas and the whole heat supplied remains within the mass of gas. This represents the increase in internal energy of the gas.

$$u = c_v.dT$$

or $$U = mc_v.dT$$

2.19.2 Specific Heat at Constant Pressure, c_p

Consider a quantity of gas contained in a cylinder tightly fitted with a frictionless piston carrying a load W, as shown in Fig. 2.29. When heat is supplied to the gas, there will be increase in temperature of the gas. But the tendency for the pressure to rise will be counter – balanced by the piston rising upward, thereby lifting the load and increasing the volume of gas. Hence the heat supplied increases the temperature and volume of gas and does work.

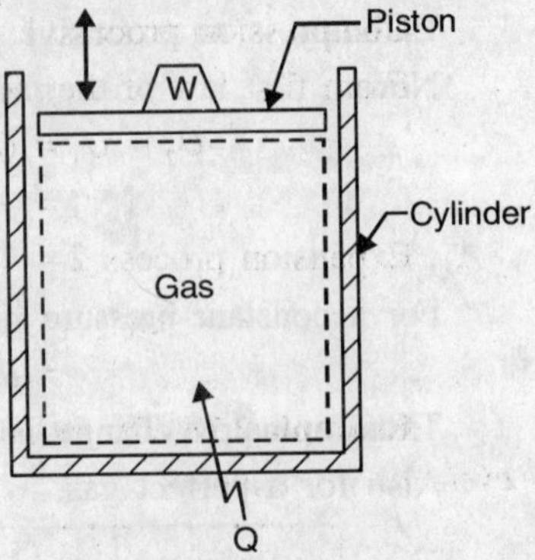

Fig. 2.29 Heating at constant pressure.

The specific heat at constant pressure, c_p, may be defined as the quantity of heat required to raise the temperature of unit mass of a gas through one degree.

Let m = mass of gas

T_1, T_2 = initial and final temperatures respectively

V_1, V_2 = initial and final volumes of gas respectively

$Q_{1\text{-}2} = mc_p\,(T_2 - T_1)$

The heat supplied to the gas at constant pressure is utilized for two purposes:

(*i*) To increase the internal energy of the gas.

(*ii*) For doing external work.

Thus, the heat energy supplied is utilized in increasing the enthalpy of gas.

Now $$dU = mc_v\,(T_2 - T_1)$$

And $$W_{1\text{-}2} = p(V_2 - V_1) = mR\,(T_2 - T_1)$$

2.19.3 Relation Between the Specific Heats

or $$c_p - c_v = R \quad \text{... (1)}$$

or $$c_v = \frac{R}{(\gamma - 1)} \quad \text{...(2)}$$

$$c_p = \frac{\gamma R}{(\gamma - 1)} \quad \text{...(3)}$$

At constant pressure,

$$Q_{1-2} = mc_p dT = dU + W_{1-2}$$
$$= mc_v.dT + p.dv$$
$$= mc_v.dT + mR.dT$$
$$= m(c_v + R)\,dT$$

$\therefore$ $$c_p = c_v + R$$

Example 2.17 *An air cycle is completed by two processes as follows :*

(*i*) *Compression process : 240 kJ of work is done on air while 100 kJ of heat is rejected to the surroundings.*

(*ii*) *Expansion process : air does 320 kJ of work.*

Calculate the quantity of heat added to the cycle.

Solution.

The cycle is shown in Fig. 2.30.

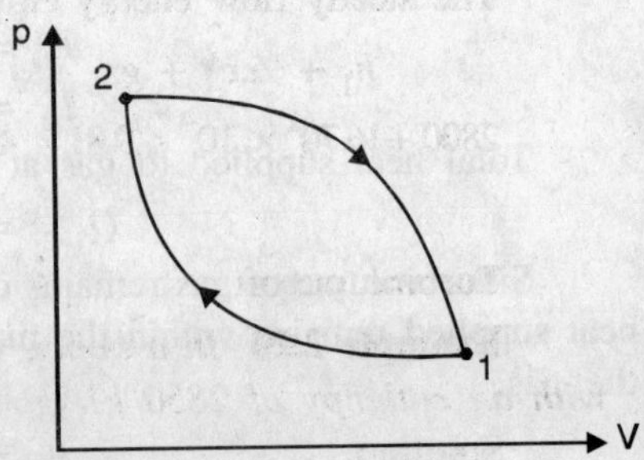

Fig. 2.30

Compression process 1 – 2 :

From first law of thermodynamics, we have

$$U_2 - U_1 = Q_{1-2} - W_{1-2}$$
$$= -100 - (-240) = 140 \text{ kJ}$$

Expansion process 2 – 1 :

$$Q_{1-2} = (U_1 - U_2) + W_{1-2}$$
$$= -140 + 320 = 180 \text{ kJ}$$

Example 2.18 *The following data refers to a closed system which undergoes a thermodynamic cycle as given below :*

Process	*Heat Transfer kJ/min*	*Work Transfer kJ/min*
1 – 2	Nil	– 1.5
2 – 3	+ 45	Nil
3 – 4	– 4.5	+ 25
4 – 1	– 12.5	– 1.5

Show whether the data is consistent with the first law of thermodynamics or not.

Solution.

$$\oint \delta Q = 0 + 45 - 4.5 - 12.5 = 28 \text{ kJ/min}$$

$$\oint \delta W = -1.5 + 0 + 25 - 1.5 = 22 \text{ kJ/min}$$

Since $\oint \delta Q \neq \oint \delta W$, therefore, first law of thermodynamics is violated.

Example 2.19 *A steam turbine operates under steady flow conditions. It receives 7500 kg/h of steam from the boiler. The steam enters the turbine at 2800 kJ/kg enthalpy, 70 m/s velocity, and an elevation of 4 m. The steam leaves the turbine at 2000 kJ/kg enthalpy, 140 m/s velocity, and an elevation of 1.5 m. Heat losses from the turbine to surroundings amount to 1600 kJ/h. Calculate the output of the turbine.*

Solution.

$$m = \frac{7500}{3600} = 2.083 \text{ kg/s}$$

$$q = \frac{-1600}{2.083 \times 3600} = -0.213 \text{ kJ/kg}$$

The steady flow energy equation per kg mass is :

$$h_1 + ½.c_1^2 + gz_1 + q = h_2 + ½.c_2^2 + gz_2 + w$$

$$2800 + ½.70^2 \times 10^{-3} + 9.81 \times 4 \times 10^{-3} - 0.213 = 2000 + ½.140^2 \times 10^{-3} + 9.81 \times 1.5 \times 10^{-3} + w$$

$$w = 791.46 \text{ kJ/kg}$$

Turbine output, $\dot{W} = \dot{m}w = 2.083 \times 791.46 = 1648.6 \text{ kW}$

Example 2.20 *In a boiler, water enters with an enthalpy of 167.5 kJ/kg and steam leaves with an enthalpy of 2850 kJ/kg. Find the heat transferred per kg of stream.*

Solution.

For a boiler, $PE = 0$, $KE = 0$ and $w = 0$. Hence, the steady energy equation becomes :

$$h_1 + q = h_2$$

Or $$q = h_2 - h_1 = 2850 - 167.5 = 2682.5 \text{ kJ/kg}$$

2.20 CLOSED SYSTEM : NON-FLOW PROCESSES

According to first law of thermodynamics, for a closed system, we have

$$Q_{1-2} - W_{1-2} = (\Delta E)_{system}$$

Where $$(\Delta E)_{system} = (\Delta U + \Delta KE + \Delta PE)_{system}$$

If ΔKE and ΔPE are negligible, then

$$Q_{1-2} - W_{1-2} = \Delta U$$

2.20.1 Constant volume (or Isochoric) Process

Consider a completely closed vessel filled with a perfect gas as shown in Fig. 2.31(*a*). Let Q units of heat be supplied to the system. This results in an increase of pressure and temperature of the system at constant volume as represented by process 1-2 on the *p-v* diagram shown in Fig. 2.31(*b*). Since there is no change in volume, therefore

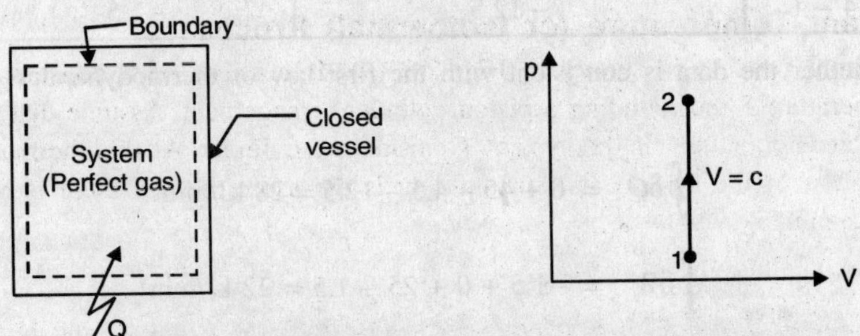

(a) Constant volume vessel (b) p-V diagram

Fig. 2.31 Isochoric process.

$$\Delta W = \int p.\, dV = 0$$

Hence

$$Q_{1-2} = dU = mc_v\,(T_2 - T_1)$$

2.20.2 Constant Pressure (or Isobaric) Process

Consider a cylinder with a piston carrying perfect gases shown in Fig. 2.32(*a*). When heat Q is supplied to the system, its temperature will rise and it will expand, forcing the piston to move upward. Thus a displacement work is done by the system against a constant force. The above process is shown in Fig. 2.32(*b*) on *p-V* diagram.

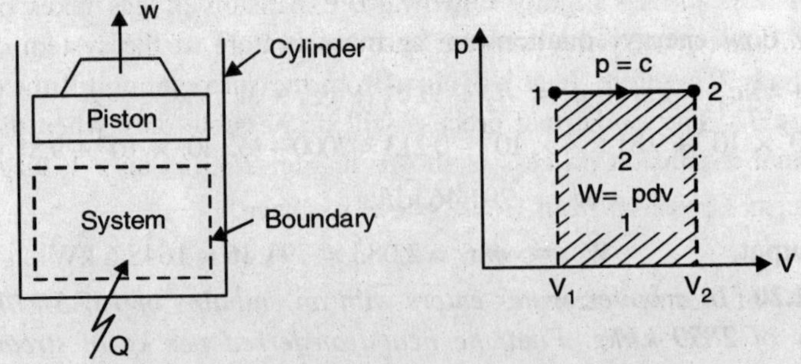

(a) Constant pressure heat addition (b) p-v diagram

Fig. 2.32 Isobaric process.

Workdone by the system, $\delta W = p.dV$

Or $$W_{1-2} = \int_1^2 p.dV = p(V_2 - V_1)$$

From first law of thermodynamics, we have

$$\begin{aligned}\delta Q &= \delta W + dU \\ &= \delta(p.dV) + dU \\ &= d(pV) + dU \\ &= d(U + pV) \\ &= dH\end{aligned}$$

Or $$Q_{1-2} = H_2 - H_1 = mc_p(T_2 - T_1)$$

Or $$q_{1-2} = c_p(T_2 - T_1)$$

It can be seen that during isobaric process the heat transfer is equal to the change in enthalpy.

2.20.3 Constant Temperature (or Isothermal) Process

An isothermal process is shown in Fig. 2.33(a). It consists of a constant temperature reservoir at temperature T_1 surrounding a piston cylinder arrangement. Assume that a perfect gas is at any instant at temperature of the system T_1 inside the cylinder. At this thermal equilibrium state the temperature of the system and surrounding is the same. Hence, there is no transfer of heat across the boundary.

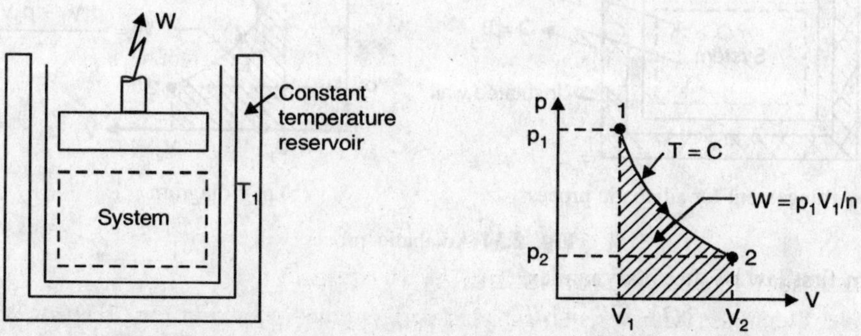

(a) Arrangement for isothermal process. (b) p-V diagram

Fig. 2.33 Isothermal process.

If the piston now moves slightly downward, expansion of gas takes place increasing its volume by dV and consequently the pressure and temperature of the system drop by an amount dp and dT respectively. Therefore, heat will flow from the surrounding till the system reaches the original temperature T_1. The isothermal process will be possible only when the process is quasi-static. The isothermal expansion process is shown in Fig. 2.33(b) on p-V diagram.

For an isothermal process, from Boyle's law, we have

$$pV = \text{const.} = C$$

From first law of thermodynamics,

$$\delta Q - \delta W = dU$$

But dU = 0 for constant temperature.

$$\therefore \quad \delta Q = \delta W$$

$$= p.dV$$

$$Q_{1-2} = W_{1-2} = \int_1^2 p.dV$$

Or $$W_{1-2} = C\int_1^2 \frac{dV}{V} = C \mid \ln V \mid_1^2 = C \ln\left(\frac{V_2}{V_1}\right)$$

But $$p_1V_1 = p_2V_2 = C$$

$$W_{1-2} = p_1V_1 \ln\left(\frac{V_2}{V_1}\right) = p_1V_1 \ln\left(\frac{p_1}{p_2}\right) = RT_1 \ln\left(\frac{V_2}{V_1}\right)$$

2.20.4 Reversible Adiabatic (or Isentropic) Process

An adiabatic process is one in which the system undergoes no heat transfer between the system and surrounding but the boundary of the system moves giving displacement work. The arrangement for the adiabatic process is shown in Fig. 2.26(*a*). It consists of a piston cylinder arrangement where the cylinder is insulated from all sides to prevent heat transfer. Since $\delta Q = 0$, therefore, δW is only due to dU. The p–V diagram is shown in Fig. 2.34(*b*).

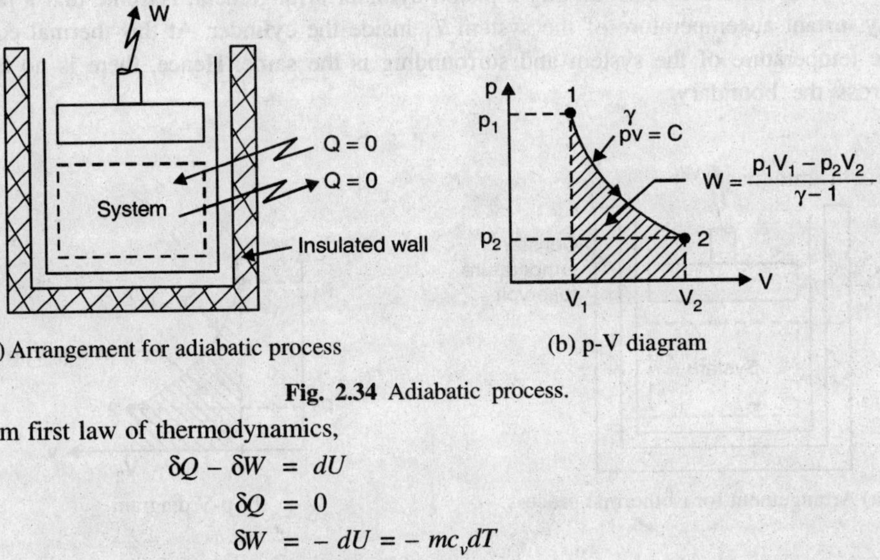

(a) Arrangement for adiabatic process (b) p-V diagram

Fig. 2.34 Adiabatic process.

From first law of thermodynamics,

$$\delta Q - \delta W = dU$$

As $$\delta Q = 0$$

$\therefore$ $$\delta W = -\,dU = -\,mc_v dT$$

For a quasi-static (friction less) process,

$$\delta W = p.dV$$

$\therefore$ $$p.dV = -\,dU$$

Or $$p.dV + dU = 0$$

For a perfect gas, $$dU = mc_v dT$$

$\therefore$ $$p.dV + mc_v dT = 0$$

Now $$pV = mRT$$

So that $$p.dV + V.dp = mRdT$$

$$dT = \frac{p.dV + V.dp}{mR}$$

$$\therefore \quad p.dV + mc_v \frac{p.dV + V.dp}{mR} = 0$$

Or
$$R\,p.dV + c_v\,(p.dV + V.dp) = 0$$

But
$$R = c_p - c_v$$

$$\therefore \quad (c_p - c_v)\,p.dV + c_v\,(p.dV + V.dp) = 0$$

Or
$$c_p\,p.dV + C_v.V.dp = 0$$

Dividing by $c_p pV$, we get

$$\frac{dV}{V} + \frac{c_v}{c_p}.\frac{dp}{p} = 0$$

Or
$$\frac{dp}{p} + \frac{c_p}{c_v}.\frac{dV}{V} = 0$$

Now
$$\gamma = \frac{c_p}{c_v}$$

$$\therefore \quad \frac{dp}{p} + \frac{\gamma dV}{V} = 0$$

Integrating, we get

$$\ln p + \gamma \ln V = \ln \text{Const.}$$

Or
$$\ln (pV^\gamma) = \ln \text{Const.}$$

Or
$$pV^\gamma = \text{Const.}$$

For a unit mass,
$$pv^\gamma = \text{Const.}$$

Now
$$p_1 V_1^\gamma = p_2 V_2^\gamma$$

Or
$$\frac{V_2}{V_1} = \left(\frac{p_1}{p_2}\right)^{\frac{1}{\gamma}}$$

For a perfect gas,

$$\frac{p_1 V_1}{T_1} = \frac{p_2 V_2}{T_2}$$

$$\frac{T_2}{T_1} = \frac{p_2}{p_1}.\frac{V_2}{V_1} = \left(\frac{V_1}{V_2}\right)^\gamma.\frac{V_2}{V_1} = \left(\frac{V_1}{V_2}\right)^{\gamma-1}$$

Or
$$\frac{T_2}{T_1} = \left(\frac{p_2}{p_1}\right)^{\frac{\gamma-1}{\gamma}}$$

Work done,
$$W_{1-2} = \int_1^2 p.dV$$

Now
$$pV^\gamma = C$$

Or
$$p = \frac{C}{V^\gamma}$$

$$W_{1-2} = C\int_1^2 \frac{dV}{V^\gamma} = C\int_1^2 V^{-\gamma} dV$$

$$= C\left|\frac{V^{-\gamma+1}}{-\gamma+1}\right|_1^2$$

$$= \left[\frac{C}{1-\gamma}\right][V_2^{1-\gamma} - V_1^{1-\gamma}]$$

But $$C = p_1V_1^{\gamma} = p_2V_2^{\gamma}$$

$$W_{1-2} = \frac{p_2V_2 - p_1V_1}{1-\gamma} = \frac{p_1V_1 - p_2V_2}{\gamma-1}$$

For a unit mass,

$$W_{1-2} = \frac{p_1V_1 - p_2V_2}{\gamma-1} = \frac{R(T_1-T_2)}{\gamma-1}$$

2.20.5 Polytropic Process

The polytropic law is expressed as,

$$pV^n = \text{Const.}$$

Process	Isochoric	Isobaric	Isothermal	Isentropic	Polytropic
n	∞	0	1	γ	n
c_n	c_v	c_p	∞	0	$\frac{c_p - nc_v}{(1-n)}$

Work done, $$W_{1-2} = \int_1^2 p.dV = C\int_1^2 V^{-n}dV = \frac{C[V_2^{1-n} - V_1^{1-n}]}{(1-n)}$$

But $$C = p_1V_1^n = p_2V_2^n$$

$$W_{1-2} = \frac{p_1V_1 - p_2V_2}{n-1} = \frac{mR(T_1-T_2)}{n-1}$$

For a unit mass, $$w_{1-2} = \frac{p_1V_1 - p_2V_2}{n-1} = \frac{R(T_1-T_2)}{n-1}$$

Change in internal energy,

$$dU = U_2 - U_1 = mc_v(T_2 - T_1)$$

Or $$du = c_v(T_2 - T_1)$$

Heat transfer, $$Q_{1-2} = dU + \int_1^2 p.dV$$

$$= mc_v\,(T_2 - T_1) + \frac{mR(T_2-T_1)}{1-n}$$

$$= m\,\frac{c_v + R}{1-n}\,(T_2 - T_1)$$

$$= m\,\frac{c_v + (c_p - c_v)}{1-n}\,(T_2 - T_1)$$

$$= m\,\frac{c_p - nc_v}{1-n}\,(T_2 - T_1)$$

$$= mc_v\left[\frac{\gamma-n}{1-n}\right](T_2-T_1)$$

$$= \left[\frac{mR}{\gamma-1}\right]\cdot\left[\frac{\gamma-n}{1-n}\right](T_2-T_1)$$

$$= mc_v(T_2-T_1)$$

Or $$q_{1-2} = c_v\left[\frac{\gamma-n}{1-n}\right](T_2-T_1)$$

$$= \left[\frac{R}{\gamma-1}\right]\left[\frac{\gamma-n}{1-n}\right](T_2-T_1)$$

The polytropic process is shown in Fig. 2.35 on p–V diagram.

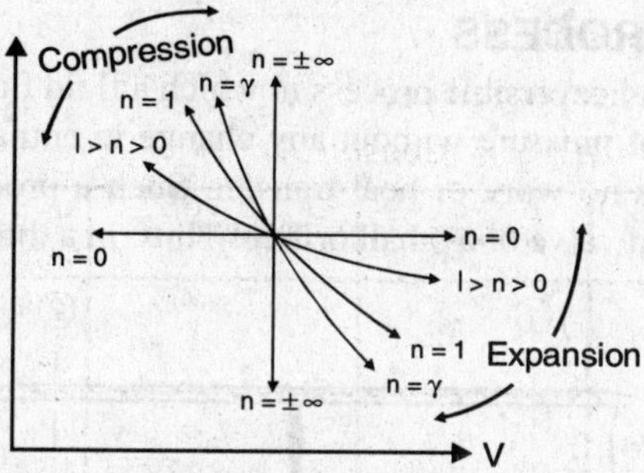

Fig. 2.35. Polytropic process on p-V diagram.

Now $$p_1V_1^{\,n} = p_2V_2^{\,n}$$

Or $$\frac{p_2}{p_1} = \left(\frac{V_1}{V_2}\right)^n$$

Also $$\frac{p_1V_1}{T_1} = \frac{p_2V_2}{T_2}$$

Or $$\frac{T_2}{T_1} = \left(\frac{p_2}{p_1}\right)\left(\frac{V_2}{V_1}\right)$$

$$= \left(\frac{V_1}{V_2}\right)^n \times \left(\frac{V_2}{V_1}\right) = \left(\frac{V_1}{V_2}\right)^{n-1} = \left(\frac{p_2}{p_1}\right)^{\frac{n-1}{n}}$$

2.20.6 Constant Internal Energy (or Free Expansion) Process

Consider a system consisting of an insulated vessel in which there are two compartments

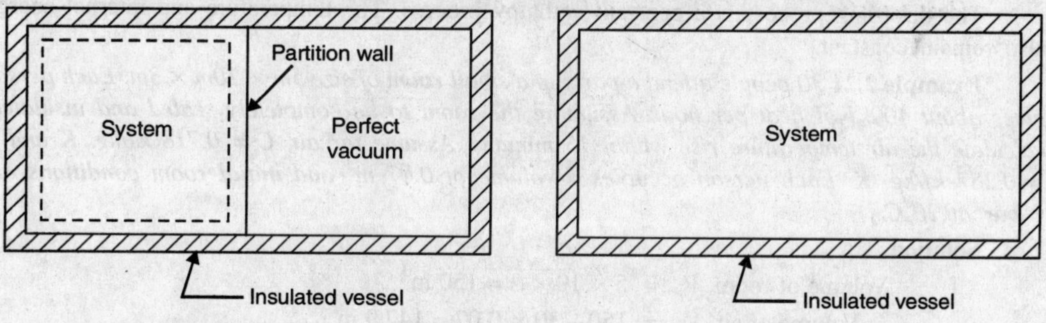

Fig. 2.36 Free expension process.

separated by a partition wall as shown in Fig. 2.36(*a*). There is a perfect gas in one compartment and the other compartment is in a state of perfect vacuum. Now if the partition is punctured the system will expand and fill up the whole space as shown in Fig. 2.36(*b*). The free expansion process is shown in Fig. 2.37 on *V–U* diagram. The process is adiabatic but irreversible and no work is done by the system. There is also no change in the temperature of the system. For this process,

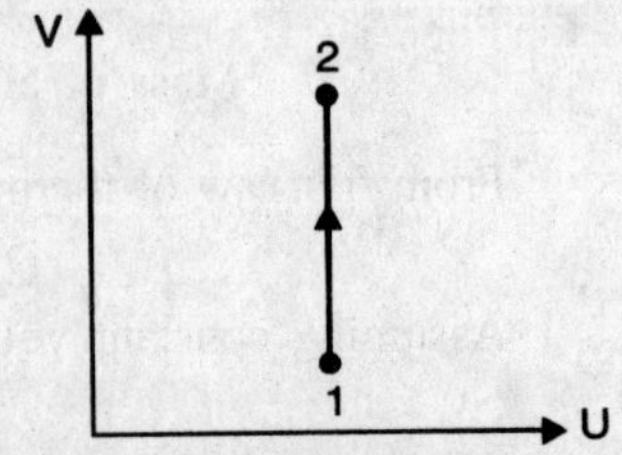

Fig. 2.37 *V-U* diagram for free expansion process.

$$dU = \delta Q - \delta W = 0$$

Or $$U_2 = U_1$$

$\therefore$ $$T_2 = T_1 \text{ and } \Delta h = 0$$

There is no change in enthalpy and the system is also called constant enthalpy process.

2.21 THROTTLING PROCESS

Throttling process is an irreversible process in which a fluid flows across a restriction in such a manner that there is a drop of pressure without any change in enthalpy of the fluid, kinetic energy, potential energy, and there is no work or heat transfer. Such a process occurs in the flow through porous plug, a partially closed valve or a small orifice. Thus, in a throttling process (Fig. 5.30).

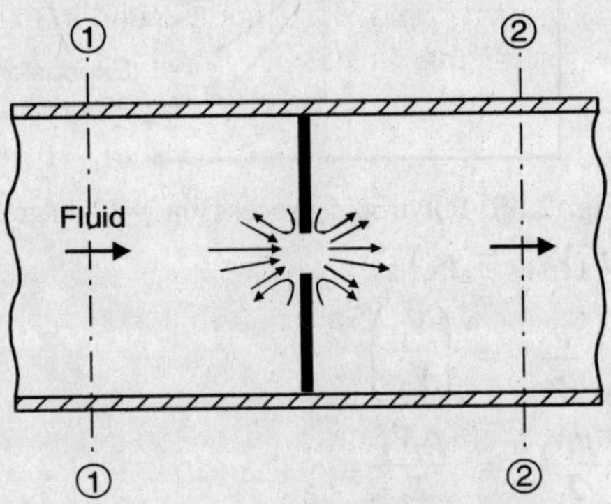

Fig. 2.38 Throttling process.

$$W_{1-2} = 0$$

$$q_{1-2} = 0$$

$$dPE = 0, \text{ i.e. } z_1 = z_2$$

$$dKE = 0, \text{ i.e. } c_1 = c_2$$

$$h_1 = h_2$$

$$u_1 = u_2 \text{ and } T_1 = T_2$$

The throttling process is a constant enthalpy process. The temperature and internal energy also remain constant.

Example 2.21 *30 people attend a party in a small room of size 5m × 10m × 3m. Each person gives about 400kJ of heat per hour. Assuming the room to be completely sealed and insulated, calculate the air temperature rise within 15 minutes. Assume for air, C_v= 0.718 kJ/kg. K and R = 0.287 kJ/kg. K. Each person occupies a volume of 0.07 m^3 and initial room conditions are 1 bar at 20^0C.*

Solution.

$$\text{Volume of room, } V_r = 5 \times 10 \times 3 = 150 \text{ m}^3$$

$$\text{Volume of air, } V_a = 150 - 30 \times 0.07 = 147.9 \text{ m}^3$$

$$\text{Mass of air, } m_a = \frac{pV_a}{RT} = \frac{1 \times 10^2 \times 147.9}{0.287 \times 293} = 175.88 \text{ kg}$$

From first law of thermodynamics,

$$Q - W = dU$$

Assuming constant volume process, $W = 0$

$$\therefore \quad Q = dU = 400 \times 30 = 12000 \text{ kJ/h}$$

For 15 minutes period,

$$dU = 12000 \times 15/60 = 3000 \text{ kJ}$$

$$dU = m_a c_v \, dT$$

$$dT = \frac{3000}{175.88 \times 0.718} = 23.76^0\text{C}$$

Example 2.22 *A perfect gas changes from equilibrium state 1 to state 2 by three quasi-static processes A, B, C as shown in Fig. 2.39. The internal energy at 1 is 800 J. The work transfer during paths A, B, and C are 120 J, 180J and 150 J respectively. Calculate the internal energy at state 2 and heat energy transferred during these processes.*

Solution.

The internal energy is a property and does not depend on the path. Therefore, internal energy at state 2 will be same for all three paths.

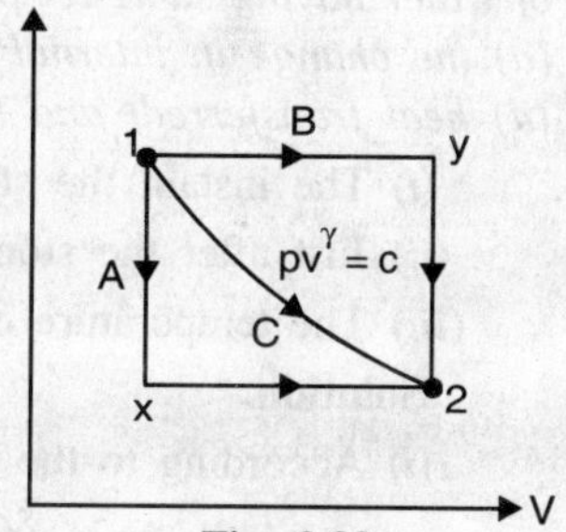

Fig. 2.39

Path C:

$$Q_{1-2} = 0$$

$$-dU = W_{1-2}$$

$$U_1 - U_2 = 150$$

$$U_2 = 800 - 150 = 650 \text{ J}$$

Path A:

$$Q_{1-2} = (U_2 - U_1) + W_{1-2}$$

$$= -150 + 120 = -30 \text{ J}$$

Path B:

$$Q_{1-2} = (U_2 - U_1) + W_{1-2}$$

$$= -150 + 180 = +30 \text{ J}$$

Example 2.23 *In a vessel 10 kg of air is heated in a reversible, non-flow, constant volume process so that pressure of air is increased two times that of the initial value. The initial temperature is 20^{0}C. Calculate (a) the final temperature, (b) change in internal energy, (c) change in enthalpy, and (d) heat transfer. Take R = 0.287 kJ/kg. K and C_v = 0.718 kJ/kg. K for air.*

Solution.

For constant volume process:

(*a*) $T_2 = (p_2/p_1)\, T_1 = 2 \times 293 = 586$ K or 313^{0}C

(*b*) $dU = mC_v\,(T_2 - T_1) = 10 \times 0.718\,(586 - 293) = 2103.74$ kJ

(*c*) $dH = mC_p(T_2 - T_1) = 10 \times (0.718 + 0.287)(586 - 293) = 2944.65$ kJ

From First law of thermodynamics,

$$Q_{1-2} = dU + W_{1-2}$$

For constant volume process, $W_{1-2} = 0$

$$\therefore \quad Q_{1-2} = dU = 2103.74 \text{ kJ}$$

Example 2.24 *A process receives 250 kJ of heat at constant volume process and rejects 270 kJ of heat at constant pressure and 50 kJ of work is done on the system. The system is* *brought* to the original state by an adiabatic process. Calculate the adiabatic work. If the initial internal energy is 240 kJ, then calculate the value of internal energy at all points.

Solution.

The $p-V$ diagram is shown in Fig. 2.40.

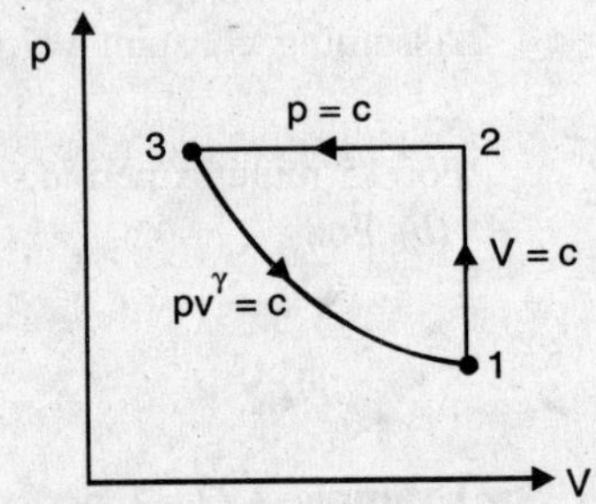

Fig. 2.40

Process 1–2 : $V = C, dW = 0$

$$Q_{1-2} = 250 = U_2 - U_1$$

$$U_2 = 250 + 240 = 490 \text{ kJ}$$

Process 2–3 : $p = C$

$$Q_{2-3} = (U_3 - U_2) + W_{2-3}$$

$$-270 = (U_3 - 490) - 50$$

$$U_3 = 270 \text{ kJ}$$

Process 3–1 : $pV^\gamma = C, Q_{3-1} = 0$

$$\oint \delta Q = 250 - 270 = -20 \text{ kJ}$$

According to First law of thermodynamics,

$$-20 = -50 + W_{3-1}$$

$$W_{3-1} = 30 \text{ kJ}$$

Example 2.25 *A system consists of a stone of mass 10 kg and a bucket containing 220 kg of water having same temperature. The stone falls into water from a height of 100 m. Calculate (a) the change in internal energy, (b) change in kinetic energy, (c) change in potential energy, (d) heat transferred, and (e) work transferred for the following cases:*

(*i*) The instant the stone is about to enter water.

(*ii*) Just after the stone comes to rest in the bucket.

(*iii*) The temperature of stone and water remains the same.

Solution.

(*i*) According to the first law of thermodynamics,

$$\delta Q - \delta W = dU + d(KE) + d(PE)$$

$$\text{If } \delta Q = 0 \text{ and } \delta W = 0, \text{ then}$$

$$dU + d(KE) + d(PE) = 0$$

For a non-flow process, $dU = 0$

$$\therefore \quad d(KE) = -d(PE)$$

Or $\Delta KE = -\Delta PE = -mg(z_2 - z_1)$

$$= -10 \times 9.81(0 - 100) = 9810 \text{ J or } 9.81 \text{ kJ}$$

$$\Delta PE = -9.81 \text{ kJ}$$

(*ii*) $\delta Q = 0, \delta W = 0 \text{ and } d(KE) = 0$

$$d(PE) = -dU$$

Or $\Delta PE = -\Delta U$

$$\Delta PE = mg(z_2 - z_1) = 10 \times 9.81(0 - 100) = -9.81 \text{ kJ}$$

(*iii*) $dU = 0, \ d(KE) = 0, \text{ and } \delta W = 0$

$$\therefore \quad Q_{1-2} = \Delta PE = mg\,(z_2 - z_1) = -9.81 \text{ kJ}$$

Example 2.26 *A fluid is compressed reversibly from a volume of 8 m^3 to 1m^3 by the law pV^n = const. The initial temperature and pressure of fluid is 25^{0}C and 1 bar. Calculate the final temperature, pressure, work done, and heat transfer, if n = 1, 1.3, and 1.4. Take c_v = 0.718 kJ/kg. K, c_p = 1.005 kJ/kg. K, and R = 0.287 kJ/kg.K.*

Solution.

(*a*) For $n = 1$

$$p_1V_1 = p_2V_2$$
$$p_2 = 1 \times 8 \therefore p_2 = 8 \text{ bar}$$
$$T_1 = T_2 = 25 + 273 = 298 \text{ K}$$
$$W_{1-2} = Q_{1-2} = p_1V_1 \ln \frac{V_2}{V_1} = 1 \times 10^2 \times 8 \ln \frac{1}{8} = -1663.6 \text{ kJ}$$

(*b*) For $n = 1.3$

$$p_1V_1^{1.3} = p_2V_2^{1.3}$$
$$p_2 = p_1\left(\frac{V_1}{V_2}\right)^{1.3} = 1\times 8^{1.3} = 14.928 \text{ bar}$$
$$\frac{T_2}{T_1} = \left(\frac{V_1}{V_2}\right)^{n-1}$$
$$T_2 = 298\,(8)^{0.3} = 556.1 \text{ K}$$
$$W_{1-2} = \frac{p_1V_1 - p_2V_2}{n-1}$$
$$= 10^2\left(\frac{1\times 8 - 14.928\times 1}{1.3-1}\right)$$
$$= -2309.3 \text{ kJ}$$
$$Q_{1-2} = mC_n(T_2 - T_1)$$
$$= m\,\frac{c_p - nc_v}{1-n} \times (T_2 - T_1)$$
$$m = \frac{p_1V_1}{RT_1} = \frac{1\times 10^2 \times 8}{0.287 \times 298} = 9.354 \text{ kg}$$
$$Q_{1-2} = 9.354\left[\frac{1.005 - 1.3\times 0.718}{1-1.3}\right] \times (556.1 - 298)$$
$$= -576.2 \text{ kJ}$$

(*c*) For $n = \gamma = 1.4$

$$p_2 = p_1\left(\frac{V_1}{V_2}\right)^{\gamma} = 1 \times 8^{1.4} = 18.379 \text{ bar}$$
$$T_2 = T_1\left(\frac{V_1}{V_2}\right)^{\gamma-1} = 298 \times 8^{0.4} = 684.6 \text{ K}$$
$$W_{1-2} = \frac{p_1V_1 - p_2V_2}{\gamma-1}$$
$$= 10^2\left(\frac{1\times 8 - 18.379\times 1}{1.4-1}\right)$$
$$= -2594.7 \text{ kJ}$$
$$Q_{1-2} = 0$$

Example 2.27 *A system containing 0.25 m^3 of air at a pressure of 4 bar and 150^{0}C expands isentropically to a pressure of 1bar and after this the gas is heated at constant pressure till the* *enthalpy increases by 60 kJ. Calculate the work done.*

If these processes are replaced by a single reversible polytropic process producing the same work between initial and final states, find the index of this expansion. Take c_p = 1.005 kJ/kg. K for air.

Solution.

The $p - V$ diagram is shown in Fig. 2.41.

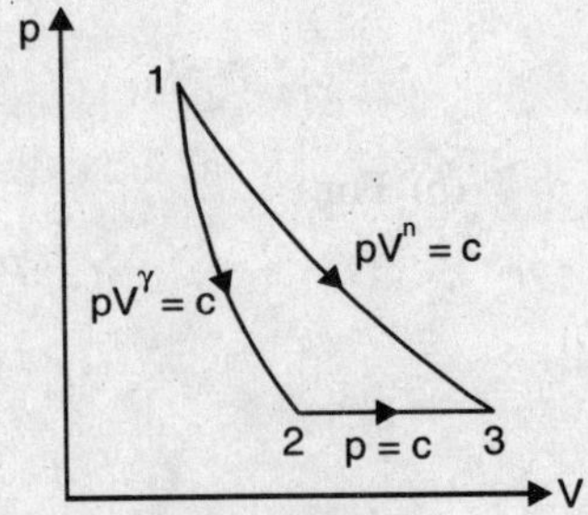

Fig. 2.41

For isentropic process 1-2,

$$V_2 = V_1\left(\frac{p_1}{p_2}\right)^{\frac{1}{\gamma}} = 0.25\left(\frac{4}{1}\right)^{\frac{1}{1.4}} = 0.673 \text{ m}^3$$

$$T_2 = T_1\left(\frac{p_2}{p_1}\right)^{\frac{\gamma-1}{\gamma}} = 423\left(\frac{1}{4}\right)^{\frac{1.4-1}{1.4}} = 284.66 \text{ K}$$

$$m = \frac{p_1V_1}{RT_1} = \frac{4\times10^2\times0.25}{0.287\times423} = 0.8237 \text{ kg}$$

$$Q_{2-3} = mc_p(T_3 - T_2)$$

$$60 = 0.8237\times1.005\,(T_3 - 284.66)$$

$$T_3 = 357.14 \text{ K}$$

$$V_3 = \frac{V_2T_3}{T_2} = \frac{0.673\times357.14}{284.66} = 0.844 \text{ m}^3$$

$$W_{1-2-3} = W_{1-2} + W_{2-3}$$

$$= \left(\frac{p_1V_1 - p_2V_2}{\gamma-1}\right) + p_2(V_3 - V_2)$$

$$= 10^2\left(\frac{4\times0.25-1\times0.673}{1.4-1}\right) + 1\times10^2(0.844-0.673)$$

$$= 81.75 + 17.1 = 98.85 \text{ kJ}$$

$$W_{1-3} = \frac{p_1V_1 - p_2V_3}{n-1}$$

$$98.85 = 10^2\left(\frac{4\times0.25-1\times0.844}{n-1}\right)$$

$$n = 1.158$$

Example 2.28 *3 m^3 of air at a pressure of 1bar and 25^0C is compressed isentropically to 5 bar. It is then expanded isothermally and restored to original volume by constant volume heat rejection process. Determine (a) pressure, volume and temperature at the end of each operation, (b) heat added during the isothermal process, (c) heat rejected during constant volume process, and (d) change in internal energy during each process.*

Assume R = 0.287 kJ/kg. K and c_p = 1.005 kJ/kg. K.

Solution.

The p–v diagram is shown in Fig. 2.42.

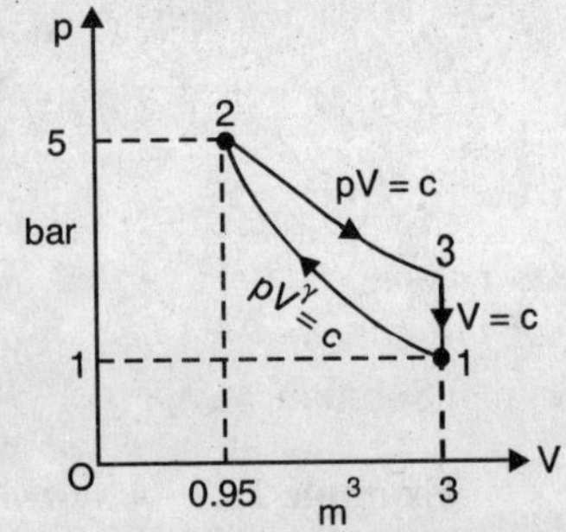

Fig. 2.42

$$c_v = c_p - R = 1.005 - 0.287$$

$$= 0.718 \text{ kJ/kg.K}$$

$$\gamma = \frac{c_p}{c_v} = \frac{1.005}{0.718} = 1.4$$

Process 1–2:

$$V_2 = V_1\left(\frac{p_1}{p_2}\right)^{\frac{1}{\gamma}}$$

$$= 3\left(\frac{1}{5}\right)^{\frac{1}{1.4}} = 0.95 \text{ m}^3$$

$$T_2 = T_1\left(\frac{p_2}{p_1}\right)^{\frac{\gamma-1}{\gamma}}$$

$$= 298\left(\frac{5}{1}\right)^{\frac{(1.4-1)}{1.4}} = 471.98 \text{ K}$$

$$W_{1-2} = \frac{p_1V_1 - p_2V_2}{\gamma-1}$$

$$= 10^2\left(\frac{1\times3-5\times0.95}{1.4-1}\right) = -437.5 \text{ kJ}$$

From first law of thermodynamics,

$$\delta Q - \delta W = dU$$

For isentropic process, $\delta Q = 0$

$$\therefore \quad -W_{1-2} = \Delta U = 437.5 \text{ kJ}$$

Process 2-3:

$$p_2V_2 = p_3V_3 = p_3V_1$$

$$p_3 = \frac{5\times0.95}{3} = 1.583 \text{ bar}$$

$$T_3 = T_2 = 471.98 \text{ K}$$

For isothermal process, $\Delta U = 0$

$$\therefore \quad Q_{2-3} = W_{2-3} = p_2V_2 \ln\frac{V_3}{V_2}$$

$$= 5\times10^2\times0.95 \ln\left(\frac{3}{0.95}\right) = 546.2 \text{ kJ}$$

Process 3 –1:

$$Q_{3-1} = mc_v(T_1 - T_3)$$

$$m = \frac{p_1V_1}{RT_1} = \frac{1\times10^2\times3}{0.287\times298} = 3.507 \text{ kg}$$

$$Q_{3-1} = 3.507\times0.718\,(298 - 471.98) = -438 \text{ kJ}$$

During constant volume process, $\delta W = 0$

$$\therefore \quad Q_{3-1} = \Delta U = -438 \text{ kJ}$$

Example 2.29 *A cylinder contains 0.2 m³ of a gas at 1bar and 100°C. This is compressed to volume 0.05 m³ so that the pressure becomes 6 bar. Calculate (a) the mass of gas, (b) energy of the gas, and (c) heat transfer during compression. Assume γ = 1.4 and R = 0.3 kJ/kg. K.*

Solution.

(*a*)
$$m = \frac{pV}{RT} = \frac{1\times10^2\times0.2}{0.3\times373} = 0.17873 \text{ kg}$$

(*b*)
$$p_1V_1^{\,n} = p_2V_2^{\,n}$$
$$1 \times (0.2)^n = 6 \times (0.05)^n$$
$$\left(\frac{0.2}{0.05}\right)^n = 6$$
$$n \ln 4 = \ln 6$$
$$n = 1.2925$$
$$T_2 = T_1\left(\frac{p_2}{p_1}\right)^{\frac{n-1}{n}} = 373\,(6)^{0.2263} = 559.5 \text{ K}$$

(*c*)
$$\Delta U = mc_v(T_2 - T_1) = \left[\frac{mR}{\gamma - 1}\right](T_2 - T_1)$$
$$= \left[\frac{0.17873 \times 0.3}{1.4 - 1}\right](559.5 - 373) = 25 \text{ kJ}$$

(*d*)
$$W_{1-2} = \frac{p_1V_1 - p_2V_2}{n-1} = \frac{mR(T_1 - T_2)}{n-1}$$
$$= \frac{0.17873 \times 0.3\,(373 - 559.5)}{1.2925 - 1} = -\,34.188 \text{ kJ}$$

Now
$$Q_{1-2} = \Delta U + W_{1-2}$$
$$= 25 - 34.188 = -\,9.188 \text{ kJ}$$

Example 2.30 *A reciprocating air compressor takes in air at 1 bar and 20^0C and delivers at 6 bar. Calculate the work done, heat transfer, and change in internal energy per kg of air compressed if the compression process is (a) isothermal, (b) isentropic and (c) polytropic: $pv^{1.35} = C$. Take $\gamma = 1.4$ and $R = 0.287$ kJ/kg.K. Neglect change in potential and kinetic energies.*

Solution.

(*a*) Isothermal process :

Work done,
$$w = p_1v_1 \ln \frac{v_2}{v_1} = RT_1 \ln\left(\frac{p_1}{p_2}\right)$$
$$= 0.287 \times 293 \ln \frac{1}{6} = -\,150.67 \text{ kJ/kg}$$

Change in internal energy, $dU = 0$

Heat transfer, $q = w = -\,150.67$ kJ/kg

Isentropic Process :
$$T_2 = T_1\left(\frac{p_2}{p_1}\right)^{\frac{\gamma-1}{\gamma}} = 293\left(\frac{6}{1}\right)^{\frac{1.4-1}{1.4}} = 488.87 \text{ K}$$
$$w = \left[\frac{\gamma}{\gamma-1}\right](p_1v_1 - p_2v_2) = \left[\frac{\gamma}{\gamma-1}\right]R(T_1 - T_2)$$
$$= \left[\frac{1.4}{1.4-1}\right] \times 0.287\,(293 - 488.87) = -\,196.75 \text{ kJ/kg}$$

For isentropic process, $q = 0$
$$du = c_v(T_2 - T_1) = \left[\frac{R}{\gamma-1}\right](T_2 - T_1)$$

$$= \left[\frac{0.287}{1.4-1}\right](488.87 - 293) = 140.536 \text{ kJ/kg}$$

Polytropic process :

$$T_2 = T_1\left(\frac{p_2}{p_1}\right)^{\frac{n-1}{n}} = 293 \times 6^{\frac{1.35-1}{1.35}} = 466.24 \text{ K}$$

$$w = \left[\frac{n}{n-1}\right](p_1v_1 - p_2v_2) = \left[\frac{n}{n-1}\right]R(T_1 - T_2)$$

$$= \left[\frac{1.35}{1.35-1}\right] \times 0.287\,(293 - 466.24) = -191.78 \text{ kJ/kg}$$

From steady flow energy equation,

$$h_1 + q = h_2 + w$$

$$q = w + (h_2 - h_1)$$

$$= w + c_p(T_2 - T_1)$$

$$= w + \left[\frac{\gamma}{\gamma-1}\right]R(T_2 - T_1)$$

$$= -191.78 + \left[\frac{1.4}{1.4-1}\right] \times 0.287\,(466.24 - 293)$$

$$= -17.76 \text{ kJ/kg}$$

$$du = c_v(T_2 - T_1) = \left[\frac{R}{\gamma-1}\right](T_2 - T_1)$$

$$= \left(\frac{0.287}{0.4}\right)(466.24 - 293) = 124.3 \text{ kJ/kg}$$

Example 2.31 *A cylinder closed at its both ends is divided into two compartments by a frictionless and freely moving piston. Both piston and cylinder are insulated. One compartment contains air and the second one contains nitrogen. Initially both air and nitrogen have pressure 1bar, volume 0.02 m^3 and temperature 298 K. Heat is supplied from the side of air compartment by an electric heater till the volume of nitrogen is reduced to 0.01 m^3. Calculate final temperature of air and heat supplied to it. Assume for air, c_v = 0.718 kJ/kg. K, R = 0.287 kJ/kg. K, and for Nitrogen, γ = 1.4.*

Solution.

The set up is shown in Fig. 2.43.

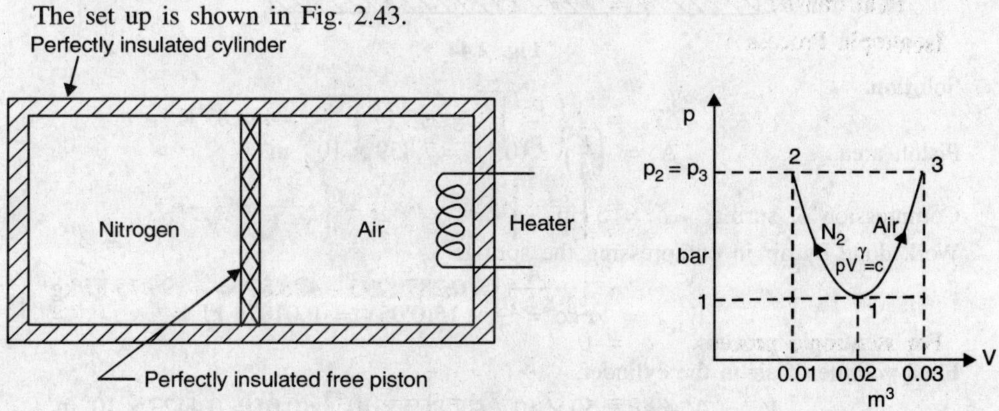

Fig. 2.43

The compression of nitrogen is adiabatic.

Process 1–2:

$$p_2 = p_1\left(\frac{V_1}{V_2}\right)^{\gamma} = 1\times\left(\frac{0.02}{0.01}\right)^{1.4} = 2.639 \text{ bar}$$

$$W_{1-2} = \frac{p_1V_1 - p_2V_2}{\gamma-1}$$

$$= 10^2\left(\frac{0.02-2.639 \times 0.01}{1.4-1}\right) = -1.5975 \text{ kJ}$$

Process 1–3:

$$V_3 = (0.02 + 0.02) - 0.01 = 0.03 \text{ m}^3$$

$$T_3 = \frac{p_3V_3T_1}{p_1V_1} = \frac{2.639\times0.03\times298}{0.02} = 1179.63 \text{ K}$$

$$m = \frac{p_1V_1}{RT_1} = \frac{1\times10^5\times0.02}{287\times298} = 0.0234 \text{ kg}$$

$$dU = U_3 - U_1 = mc_v(T_3 - T_1)$$

$$= 0.0234 \times 0.718\,(1179.63 - 298) = 14.812 \text{ kJ}$$

$$Q_{1-3} = W_{1-3} + dU = W_{1-2} + dU = 1.5975 + 14.812 = 16.41 \text{ kJ}$$

Example 2.32 *A cylinder of 10 cm diameter shown in Fig. 2.44 has a piston loaded with a spring of stiffness 15 kN/m. The initial state of air in the cylinder is 3 bar, 50 × 10^{-6} m^3, and 298 K. Determine the amount of heat added to the system so that piston moves by 5 cm. Assume c_v = 0.718 kJ/kg. K, R = 0.287 kJ/kg. K and γ =1.4.*

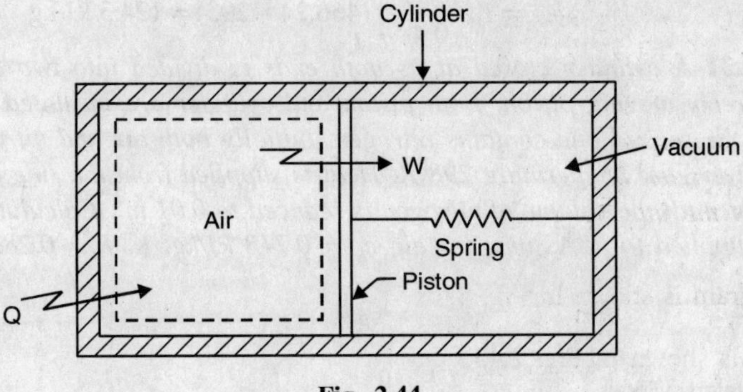

Fig. 2.44

Solution.

Piston area, $A = \left(\frac{\pi}{4}\right) \times (0.1)^2 = 7.859 \times 10^{-3} \text{ m}^2$

Compression of spring, $\delta = 5$ cm

Work done by air in compressing the spring,

$$W_{1-2} = \tfrac{1}{2}k\delta^2 = \frac{1}{2} \times 15(0.05)^2 = 0.01875 \text{ kJ}$$

Final volume of air in the cylinder,

$$V_2 = V_1 + A\delta = 50 \times 10^{-6} + 7.859 \times 10^{-3} \times 0.05 = 0.4427 \times 10^3 \text{ m}^3$$

Final Pressure of air in the cylinder,

$$p_2 = \frac{P_1 + kS}{A} = \frac{3+15 \times 0.05}{(7.859\times10^{-3}\times10^2)} = 3.955 \text{ bar}$$

Now, $$W_{1-2} = \frac{p_1V_1 - p_2V_2}{n-1}$$

$$0.01875 = \frac{[3\times10^2\times50\times10^{-6} - 3.955\times10^2\times 0.4427\times10^{-3}]}{(n-1)}$$

$$n = -7.538$$

Mass of air, $$m = P_1\frac{V_1}{RT_1} = \frac{3\times10^5\times50\times10^{-6}}{(287\times298)} = 5.838 \times 10^5 \text{ kg}$$

$$T_2 = T_1\left(\frac{p_2}{p_1}\right)^{\frac{n-1}{n}} = 298\left(\frac{3.955}{3}\right)^{\frac{-7.538-1}{-7.538}}$$

$$= 298(1.318)^{1.1327} = 407.5 \text{ K}$$

$$Q_{1-2} = m\left[\frac{\gamma-1}{n-1}\right] c_v (T_2 - T_1)$$

$$= 5.838 \times 10^{-5}\left[\frac{1.4-1}{-7.538-1}\right] \times 0.718\ (407.5 - 298)$$

$$= 0.0215 \text{ kJ}$$

Example 2.33. *The heat capacity at constant pressure of a certain system is a function of temperature only and may be expressed as :*

$$c_p = \frac{2.093 + 41.87}{(t+100)} \text{ J/°C}$$

Where t is the temperature of the system in °C. The system is heated while it is maintained at a pressure of 1 atmosphere until its volume increases from 2000 cm^3 to 2400 cm^3 and its temperature increases from 0°C to 100°C. Calculate (a) the magnitude of heat interaciion, and (b) how much does the internal energy of the system increase?

Solution.

The *p–V* diagram is shown in Fig. 2.45.

$$p_1 = 1.013 \times 10^5 \ Pa$$
$$V_1 = 2 \times 10^{-3} \text{ m}^3$$
$$V_2 = 2.4 \times 10^{-3} \text{ m}^3$$
$$t_1 = 0^\circ\text{C}, t_2 = 100^\circ\text{C}$$

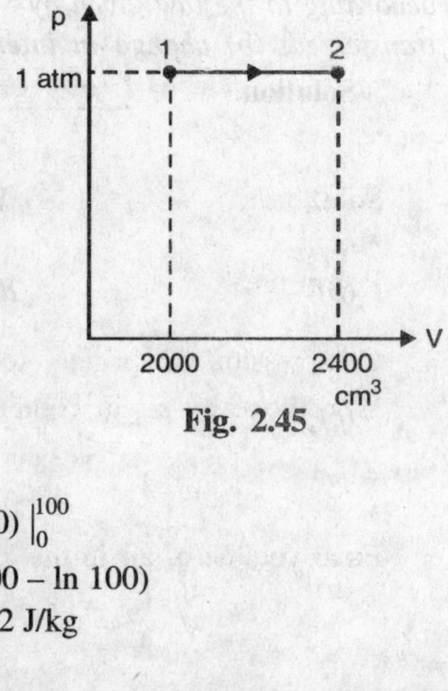

Fig. 2.45

(*a*) $$Q_{1-2} = \int_{t_1}^{t_2} c_p dt$$

$$= \int_0^{100}\left[\frac{2.093 + 41.87}{(t+100)}\right]dt$$

$$= \left| 2.093\,t + 41.87 \text{ In } (t+100) \right|_0^{100}$$

$$= 2.093 \times 100 + 41.87 \ (\text{ln } 200 - \text{ln } 100)$$

$$= 209.3 + 41.87 \text{ ln } 2 = 238.32 \text{ J/kg}$$

(*b*)
$$W_{1-2} = p(V_2 - V_1)$$
$$= 1.013 \times 10^5\,(2.4 - 2)10^{-3} = 40.52 \text{ J}$$
$$Q_{1-2} = U_{1-2} + W_{1-2}$$
$$U_{1-2} = 238.32 - 40.52 = 197.8 \text{ J}$$

Example 2.34 *A fluid system undergoes a non-flow frictionless process, which follows the p–V relationship given by :*

$$p = \frac{5}{V} + 2$$

Where p is in bar and V in m^3. During the process the volume changes from 0.20 m^3 to 0.05 m^3 and the system rejects 50 kJ of heat. Calculate (a) change in internal energy, and (b) change in enthalpy.

Solution.

$$\text{Work done,}\quad W_{1-2} = \int_1^2 p.dV = 10^5 \int_{0.20}^{0.05} \left(\frac{5}{v+2}\right) dV$$

$$= 10^5 \left| 5 \ln V + 2V \right|_{0.20}^{0.05} = 10^5 \left[5 \ln \left(\frac{0.05}{0.20}\right) + 2\,(0.05 - 0.20)\right]$$

$$= -723.1472 \text{ kJ}$$

(*a*)
$$Q = dU + W$$
$$-50 = dU - 723.1472$$
$$dU = 673.1472 \text{ kJ}$$

(*b*)
$$dH = dU + dW = dU + d\,(pV) = dU + (p_2V_2 - p_1V_1)$$

$$= 673.1472 + \left[\left(\frac{5}{0.05} + 2\right) \times 0.05 - \left(\frac{5}{0.20} + 2\right) \times 0.2\right] \times 10^2$$

$$= 673.1472 + [5.1 - 5.4] \times 10^2 = 672.8472 \text{ kJ}$$

Example 2.35 *The specific internal energy of a gas is given by the relation:*

$$u = 3.65\,pv + 100 \text{ kJ/kg}$$

Where p is in kPa and v in m^3/kg.

3 kg and 0.30 m^3 of this gas expands from an initial pressure of 600 kPa to 100 kPa according to the law given by : $pV^{1.25}$ = C. If the expansion is quasi – static, calculate (a) heat transferred, (b) change in internal energy and (c) work transferred.

Solution.

$$V_2 = V_1 \left(\frac{p_1}{p_2}\right)^{\frac{1}{n}} = 0.3 \left(\frac{600}{100}\right)^{\frac{1}{1.25}} = 1.2579 \text{ m}^3$$

(*b*)
$$dU = U_2 - U_1 = 3.65\,(p_2V_2 - p_1V_1)$$
$$= 3.65\,(100 \times 1.2579 - 600 \times 0.3) = -197.87 \text{ kJ}$$

(*c*)
$$W = \int p.dV = \frac{p_1V_1 - p_2V_2}{n-1}$$

$$= \frac{600 \times 0.3 - 100 \times 1.2579}{1.25 - 1} = 216.84 \text{ kJ}$$

(*a*)
$$Q = dU + W = -197.87 + 216.84 = 18.97 \text{ kJ}$$

Example 2.36 *The properties at state 1of a reversible non-flow isobaric process are: p_1 = 1.5 bar, v_1 = 0.3 m^3/kg, t_1 = 20°C and at state 2 are: t_2 = 250°C, v_2 = 0.60 m^3/kg. The specific heat of the fluid is given by:*

$$c_p = \left[1.5 + \frac{80}{t+50}\right] \text{ kJ/kg.°C}$$

Where t is in °C.

Calculate (a) heat added, (b) work done, (c) change in internal energy, and (d) change in enthalpy per kg of fluid.

Solution.

(a)
$$Q = \int_{t_1}^{t_2} c_p \, dt = \int_{20}^{250} \left(1.5 + \frac{80}{t+50}\right) dt$$

$$= \left| 1.5\,t + 80 \ln (t + 50) \right|_{20}^{250} = 1.5(250 - 20) + 80 \ln \frac{300}{70}$$

$$= 461.423 \text{ kJ/kg}$$

(b)
$$w = \int_{v_1}^{v_2} p.dV = p(v_2 - v_1) = 1.5 \times 10^2 \,(0.60 - 0.30) = 45 \text{ kJ/kg}$$

(c)
$$dU = Q - W = 461.423 - 45 = 416.423 \text{ kJ/kg}$$

Example 2.37 *A fluid is contained in a system behind a spring loaded, frictionless piston so that the pressure in the fluid is given by:*

$$p = a + bV$$

The internal energy of the fluid is given by:

$$U = 45 + 3.5\, pV \text{ kJ}$$

Where p is in kPa and V in m^3. The fluid changes from an initial state of 200 kPa, 0.04 m^3 to a final state of 450 kPa, 0.08 m^3. Find the amount of work done and heat transfer.

Solution.

$$dU = U_2 - U_1$$
$$= (45 + 3.5\, p_2V_2) - (45 + 3.5\, p_1V_1)$$
$$= 3.5\,(450 \times 0.08 - 200 \times 0.04) = 98 \text{ kJ}$$

Now
$$p = a + bV$$
$$200 = a + 0.04\, b$$
$$450 = a + 0.08\, b$$
$$b = \frac{250}{0.04} = 6250, \quad a = 200 - 0.04 \times 6250 = -50$$
$$p = -50 + 6250 \text{ V}$$
$$W_{1-2} = \int_1^2 pdV = \int_{0.04}^{0.08} (-50 + 6250 \text{ V})\, dV$$
$$= \left| -50\text{V} + 3125\, \text{V}^2 \right|_{0.04}^{0.08}$$
$$= -50\,(0.08 - 0.04) + 3125\,(0.08^2 - 0.04^2) = -2 + 15 = 13 \text{ kJ}$$
$$Q_{1-2} = dU + W_{1-2} = 98 + 13 = 111 \text{ kJ}$$

Example 2.38 *A gas of mass 1.5 kg undergoes a quasi-static expansion which follows a relationship p = a + bV, where a and b are constants. The initial and final pressures are*

1000 kPa and 200 kPa respectively and the corresponding volumes are 0.20 m^3 and 1.20 m^3. The specific internal energy of the gas is given by the relation

$$u = 1.5\,pv - 85 \text{ kJ/kg}$$

where p is in kPa and v is in m^3/kg. Calculate the net heat transfer and the maximum internal energy of the gas attained during expansion.

Solution.

$$p = a + bV$$
$$1000 = a + 0.2\,b$$
$$200 = a + 1.2\,b$$
$$b = -800,\; a = 1000 + 0.2 \times 800 = 1160$$
$$p = 1160 - 800 \text{ V kPa}$$
$$W_{1-2} = \int_1^2 p.dV = \int_{0.2}^{1.2} (1160 - 800\text{ V})\,dV$$
$$= \left| -1160\text{V} - 400\text{ V}^2 \right|_{0.2}^{1.2} = 1160\,(1.2 - 0.2) - 400(1.2^2 - 0.2^2)$$
$$= 1160 - 560 = 600 \text{ kJ}$$
$$dU = U_2 - U_1 = 1.5\,(p_2V_2 - p_1V_1)$$
$$= 1.5\,(200 \times 1.2 - 1000 \times 0.2) = 60 \text{ kJ}$$
$$Q_{1-2} = dU + W_{1-2} = 60 + 600 = 660 \text{ kJ}$$
$$Q_{1-2} = m\left(\frac{105\,pV}{m}\right) = 1.5\,pV - 85 \text{ m}$$
$$= 1.5\,(1160 - 800\text{ V})\text{ V} - 85 \times 1.5$$
$$= 1.5\,(1160\text{V} - 800\text{ V}^2) - 127.5$$
$$\frac{dU}{dV} = 1.5\,(1160 - 1600\text{V}) - 0 = 0$$
$$V = \frac{1160}{1600} = 0.725 \text{ m}^3$$
$$U_{max} = 1.5\,(1160 - 800 \times 0.725) \times 0.725 - 127.5$$
$$= 503.25 \text{ kJ}$$

2.22 LIMITATIONS OF FIRST LAW OF THERMODYNAMICS

The first law of thermodynamics states as follows :

- Heat and work are mutually convertible.
- It is the law of conversation of energy, *i.e.*, energy can neither be created not destroyed.

The implications of first law are :

- No restriction is placed on the direction of flow of heat and work.
- Potential energy can be transformed into kinetic energy and vice-versa.
- Gas expands from high pressure to low pressure and vice-versa.

The actual observations are :

- Heat is not completely convertible to work.
- Kinetic energy cannot be transformed into potential energy automatically.
- Heat cannot flow automatically from cold to hot body.
- Gas cannot expand from low pressure to high pressure automatically.

The first law fails to explain these observations. The first law does not limit the extent of any energy conversion nor it gives any indication as to whether any particular process under consideration is possible or not. The first law is necessary but not a sufficient condition for a certain process to take place.

2.23 FUNDAMENTAL CONCEPTS ABOUT SECOND LAW

Thermal Reservoir. It is defined as a sufficiently large system in stable equilibrium to which and from which finite amount of heat can be transferred without any change in its temperature.

Heat Source. It is a high temperature reservoir, *e.g.*, boiler furnace, combustion chamber, nuclear reactor, Sun etc.

Heat Sink. It is a low temperature reservoir, *e.g.*, atmospheric air, river, water, ocean water, etc.

Heat Engine. It is defined as a thermodynamic system operating in a cycle in which the heat is transferred and from which net work is delivered. The principle of heat engine is shown in Fig 2.46.

Let Q_1 = Heat added to a system.

Q_2 = Heat rejected from a system.

W_{net} = net work done by a system.

$$\text{Thermal efficiency},\ \eta_{th} = \frac{\text{output}}{\text{input}}$$

$$= \frac{W_{net}}{Q_1}$$

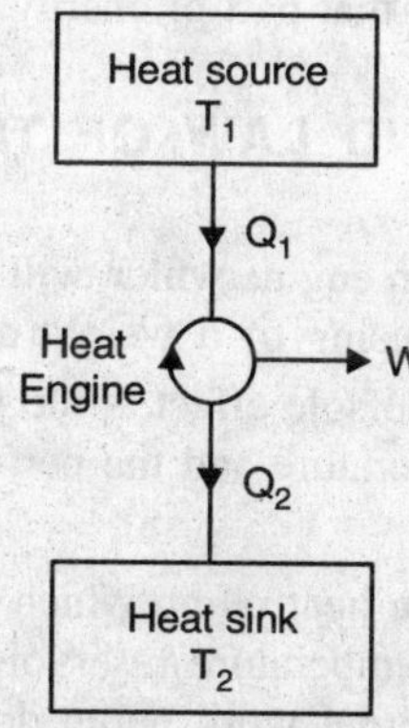

Fig. 2.46 Principle of a heat engine

From first law of thermodynamics.

$$\oint \delta Q = \oint \delta W$$

or $$Q_1 - Q_2 = W_{net}$$

$$\therefore \quad \eta_{th} = \frac{Q_1 - Q_2}{Q_1}$$

$$= 1 - \frac{Q_2}{Q_1}$$

$$= 1 - \frac{T_2}{T_1}$$

Heat Pump. It is defined as a thermodynamic system operating in a cycle which removes heat from a low temperature body taking external energy in the form of work input. The principle of a heat pump is shown in above Fig 2.47.

Let Q_1 = Heat rejected by a system.

Q_2 = Heat supplied to a system.

From first law of thermodynamics,

$$\oint \delta Q = \oint \delta W$$

or $$Q_1 - Q_2 = W_{net}$$

Coefficient of performance (*COP*) of a refrigerator

$$= \frac{\text{Refrigerating effect}}{\text{Work input}}$$

$$= \frac{Q_2}{W_{net}} = \frac{Q_2}{Q_1 - Q_2} = \frac{T_2}{T_1 - T_2}$$

$$COP \text{ of a heat pump} = \frac{\text{Heating effect}}{\text{Work input}} = \frac{Q_1}{W_{net}} = Q_1$$

$$= \frac{T_1}{T_1 - T_2}$$

$$= 1 + (COP)_{refrigerator}$$

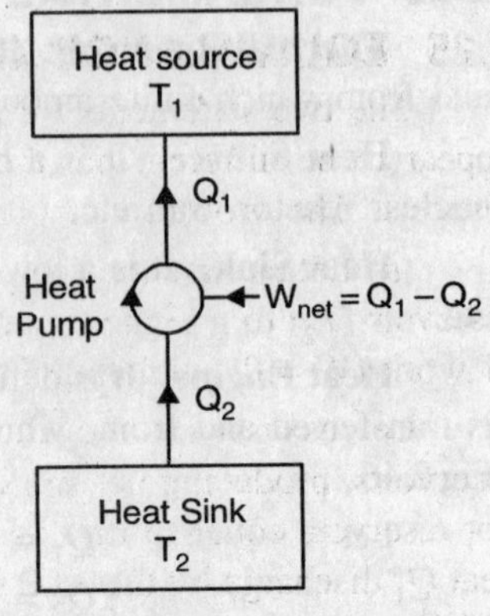

Fig. 2.47 Principle of a heat pump–refrigerator.

Dead State. The dead state of a system is defined as a state of stable equilibrium of the system within the atmosphere, *i.e.*, its capacity to do work ceases and its available energy is zero.

Available Energy. It is defined as that part of energy which can be converted into work by ideal processes.

2.24 STATEMENTS OF SECOND LAW OF THERMODYNAMICS

1. Kelvin – Planck Statement

- It is impossible to construct an engine which will work in a complete cycle and produce no other effect except the raising of a weight and cooling of heat reservoir.
- No process is possible whose sole effect is the removal of heat from a single thermal reservoir at an uniform temperature and the performance of an equal amount of work.

2. Clausius Statement

- It is impossible to construct a heat pump which operating in a cycle will remove heat continuously from a lower temperature reservoir and transfer to a higher temperature without any amount of external work being done on it.
- No process is possible whose sole effect is the removal of heat from a reservoir at lower temperature and the absorption of equal amount of heat by a reservoir at a higher temperature.
- Heat cannot flow from a colder body to a hotter body.

3. Perpetual Motion Machine of Second Kind (PMM-II)

It is impossible to construct a perpetual motion machine of the second kind – which operates in a cycle and delivers an amount of work equal to heat extracted from a single reservoir at a uniform temperature.

4. Principle of Degradation of Energy

- The available energy of an isolated system decreases in all real processes and is conserved in ideal processes.
- As a result of natural processes, energy steadily becomes less available for the performance of work.

5. Principle of Increase of Entropy

- The entropy of an isolated system increases in all real processes and is conserved in ideal processes.
- As a result of natural processes, the entropy of the universe steadily increases. The statement 1 and 2 are widely. used.

2.25 EQUIVALENCE OF KELVIN AND CLAUSIUS STATEMENTS

The two statements for second law of thermodynamics by Kelvin-Planck and Clausius may appear to be different but they are virtually two parallel statements which are equivalent in all respects. The violation of one statement implies the violation of the second and vice-versa.

(*a*) Let us first consider a heat pump (H.P.) which transforms heat Q_1 from a low temperature reservoir (T_2) to a higher temperature reservoir (T_1) with no other effect, *i.e.*, with no expenditure of work ($W = 0$), violating Clausius statement, as shown in Fig 2.48.

Let us assume a cyclic heat engine (H.E.) operating between the same thermal energy reservoirs, producing net work (W_{net}) in one cycle. The H.E. draws an amount of heat Q_1 from the hot reservoir equal to that discharged by H.P. Then the hot reservoir may be eliminated and the heat Q_1 discharge by the H.P is fed to the H.E. so that the H.P. and H.E. acting together constitute a H.E. operating in cycles and producing net work while exchanging heat only with one body at a single fixed temperature. This violates the Kelvin-Planck statement.

(*b*) Now consider a PMM2 which produces net work in a cycle exchanging heat with only one thermal energy reservoir (T_1) and thus violates the Kelvin-Planck statement, as shown in Fig 2.49.

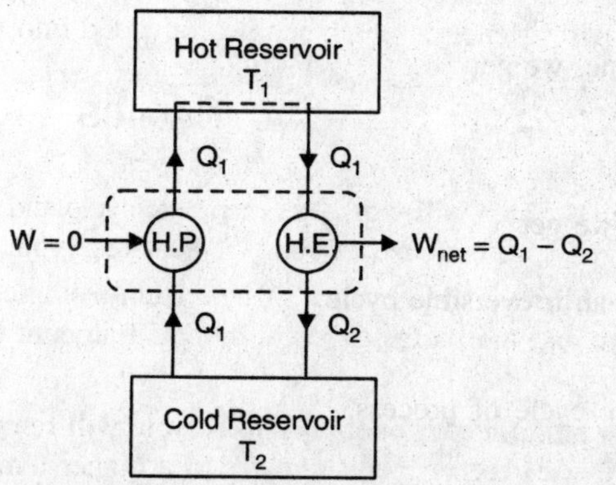

Fig. 2.48 Violation of Clausius statement.

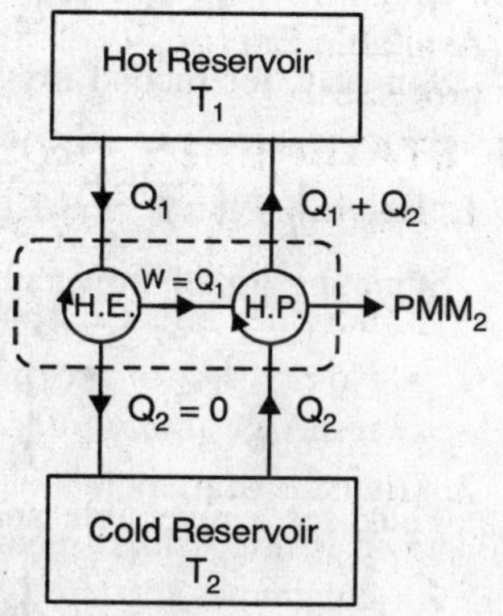

Fig. 2.49 Violation of Kelvin-Planck statement.

Let us assume a cyclic H.P. extracting heat Q_2 from a low temperature reservoir (T_2) and discharging heat to a high temperature reservoir (T_1) with a expenditure of work W equal to that delivered by PMM-2 in a complete cycle. The PMM-2 and H.P. working in cycles and producing the sole effect of transferring heat from a lower to a higher temperature body, thus violating the Clausius statement.

or
$$\frac{\delta Q_1}{T_1} + \frac{\delta Q_2}{T_2} = 0$$

Similarly, for the second cycle,

$$\frac{\delta Q_1'}{T_1'} + \frac{\delta Q_2'}{T_2'} = 0$$

and so on.

Thus, for a reversible cycle *ABCD*, we get

$$\left(\frac{\delta Q_1}{T_1}+\frac{\delta Q_2}{T_2}\right)+\left(\frac{\delta Q_1'}{T_1'}+\frac{\delta Q_2'}{T_2'}\right)+ \ldots\ldots = 0$$

or $$\oint \frac{\delta Q}{T} = 0, \quad \text{for a reversible cycle.}$$

For an irreversible engine *I*, $\eta_I < \eta_R$. Hence

$$\left(1-\frac{\delta Q_{2I}}{\delta Q_{1I}}\right) < \left(1-\frac{\delta Q_{2R}}{\delta Q_{1R}}\right)$$

or $$\left(1-\frac{\delta Q_{2I}}{\delta Q_{1I}}\right) < \left[1-\frac{T_2}{T_1}\right]$$

or $$\left(\frac{\delta Q_{2I}}{\delta Q_{1I}}\right) < \frac{T_2}{T_1}$$

or $$\left(\frac{\delta Q_{1I}}{T_1}-\frac{\delta Q_{2I}}{T_2}\right) < 0$$

Considering the direction of heat addition and rejection, we get

$$\left(\frac{\delta Q_{1I}}{T_1}+\frac{\delta Q_{2I}}{T_2}\right) < 0$$

Similarly, for second irreversible engine, we get

$$\left(\frac{\delta Q_{1I}'}{T_1'}+\frac{\delta Q_{2I}'}{T_2'}\right) < 0$$

Summing up all irreversible engines, we get

$$\oint_I \frac{\delta Q}{T} < 0, \text{ for an irreversible cycle}$$

Thus, for a reversible and irreversible cycle of process,

$$\oint \frac{\delta Q}{T} \leq 0$$

For an irreversible cyclic process, we may write

$$\oint \frac{\delta Q}{T} + I = 0$$

Where *I* indicates the amount by which the given cycle is irreversible. If *I* = 0, the cycle is reversible. **Irreversible process is denoted by dotted lines.**

2.26 CONCEPT OF ENTROPY

The cyclic integral of $\delta q/T$ for a reversible cycle being equal to zero suggests that the quantity $\delta Q/T$ is a point function and is therefore a property of the system. This property is called entropy. It is denoted by 'S'. Mathematically,

$$\int_1^2 \left(\frac{\delta Q}{T}\right)_{\text{rev}} = \int_1^2 dS$$

or
$$\left(\frac{\delta Q}{T}\right)_{rev} = \delta S$$

Change in entropy, δS = Heat supplied or Heat rejected, δQ/Absolute temperature, T

The SI units of entropy is kJ/K. The change in entropy expressed per unit mass of the working substance is called specific entropy (*ds*), and its units are kJ/kg.K.

2.26.1 Entropy as a Property

In order to prove entropy as a property of the system, let us consider arbitrary reversible cycles as shown in Fig. 5.42.

For a reversible cycle,

$$\oint\left(\frac{\delta Q}{T}\right)_{rev} = 0$$

For cycle 1– A – 2 – B – 1, we have

$$\oint\left(\frac{\delta Q}{T}\right)_{rev} = \int_1^2\left(\frac{\delta Q}{T}\right)_A + \int_1^2\left(\frac{\delta Q}{T}\right)_B = 0$$

and for cycle 1 – A – 2 – C – 1, we have

$$\oint\left(\frac{\delta Q}{T}\right)_{rev} = \int_1^2\left(\frac{\delta Q}{T}\right)_A + \int_1^2\left(\frac{\delta Q}{T}\right)_C = 0$$

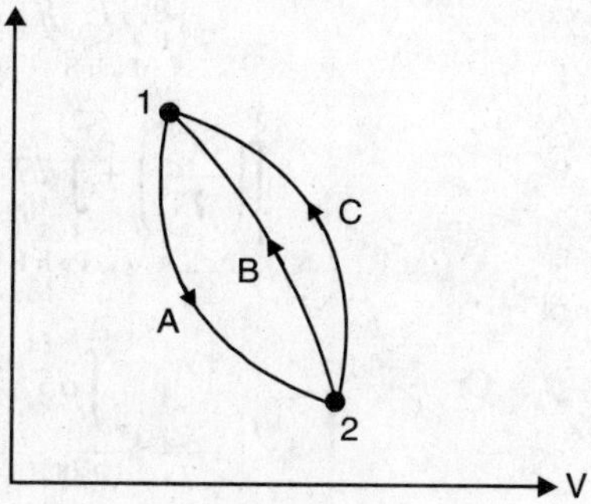

Fig. 2.50 Entropy as a point function.

Comparing the two cycles, we get

$$\int_1^2\left(\frac{\delta Q}{T}\right)_B = \int_1^2\left(\frac{\delta Q}{T}\right)_C = 0$$

Hence $\oint\left(\frac{\delta Q}{T}\right)$ is independent of path. It is a point function (*i.e.* exact differential), and is a property of the system. Thus

$$dS = \left(\frac{\delta Q}{T}\right)_{rev}$$

$$\int_1^2 dS = \oint\left(\frac{\delta Q}{T}\right)_{rev}$$

$$\Delta S = \oint\left(\frac{\delta Q}{T}\right)_{rev}$$

For a reversible adiabatic process, $\delta Q = 0$. Hence

$$\Delta S = S_2 - S_1 = 0.$$

The characteristics of entropy are :

- It increases (or decreases) when heat is supplied (or removed) irrespective of the change in temperature.
- It remains constant during reversible adiabatic process.
- It increases if temperature of heat is lowered without work being done, as in a throttling process.
- It is an extensive property because it depends on the mass of the system.

2.26.2 Entropy Change in an Irreversible Process

Consider an isolated system which changes from state 1 to state 2 by an irreversible process 1-*A*-2, as shown in Fig 2.51. Let the system return to initial state 1 by a reversible process 2-*B*-1. The processes 1-*A*-2 and 2-*B*-1 together from an irreversible cycle. From the Clausius inequality,

Or $$\underbrace{\int_1^2 \left(\frac{\delta Q}{T}\right)_I}_{\text{Path A}} + \underbrace{\int_1^2 \left(\frac{\delta Q}{T}\right)_R}_{\text{Path B}} < 0$$

But $$\underbrace{\int_1^2 \left(\frac{\delta Q}{T}\right)_R}_{\text{Path B}} = \underbrace{\int_1^2 dS}_{\text{Path B}} = -\underbrace{\int_1^2 dS}_{\text{Path B}}$$

$\therefore$ $$\underbrace{\int_1^2 \left(\frac{\delta Q}{T}\right)_I}_{\text{Path A}} + \underbrace{\int_1^2 dS}_{\text{Path B}} < 0$$

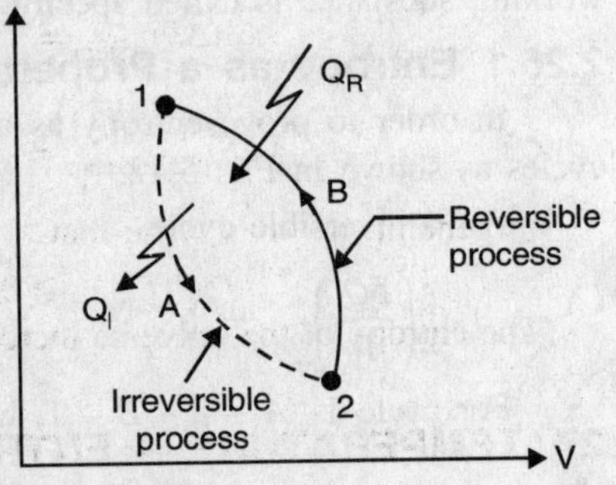

Fig. 2.51 Reversible and irreversible process.

Or $$\underbrace{\int_1^2 dS}_{\text{Path B}} > \underbrace{\int_1^2 \left(\frac{\delta Q}{T}\right)_I}_{\text{Path A}}$$

Or $$dS > \left(\frac{\delta Q}{T}\right)_I$$

But $$dS > \left(\frac{\delta Q}{T}\right)_R$$

Hence, in general, we may write

$$dS \geq \left(\frac{\delta Q}{T}\right)$$

$$\Delta S = (S_2 - S_1) \geq \int_1^2 \left(\frac{\delta Q}{T}\right)$$

Where inequality sign holds good for irreversible process and equality sign for reversible process.

2.26.3 Entropy Change for Adiabatic Process

For adiabatic reversible or irreversible process, $\delta Q = 0$. Hence

$$\Delta S \geq 0$$

i.e., entropy change is zero for reversible adiabatic process and more than zero for irreversible adiabatic process.

2.26.4 Entropy Change for Isolated System

For an isolated system , there is no heat and work interaction with the surrounding. Thus $\delta Q = 0$. Hence

$$(dS)_{\text{isolated}} \geq 0$$

For a reversible process, $(dS)_{\text{iso}} = 0$ or $S = \text{const.}$

For an irreversible process, $(dS)_{\text{iso}} > 0$

Thus, the entropy of an isolated system can never decrease.

2.26.5 Entropy Change of the Universe

The entropy of the universe increases owing to the natural process (irreversible), such that

$$(\Delta S)_{universe} = [(\Delta S)_{system} + (\Delta S)_{environment}] > 0$$

For reversible process,

$$(\Delta S)_{system} = \left(\frac{Q}{T}\right)_{rev}$$

$$(\Delta S)_{environment} = -\left(\frac{Q}{T}\right)_{rev}$$

In general, we may write that

$$(\Delta S)_{universe} \geq 0$$

The entropy of the universe increase in all irreversible process and is conserved in reversible process.

2.27 TEMPERATURE – ENTROPY DIAGRAM (T-S Diagram)

Temperature-entropy diagram is a plot of absolute temperature on vertical axis v's entropy on horizontal axis. It is a plot of isotherms v's reversible adiabatics. During an isothermal process, temperature remains constant but entropy changes, while for reversible adiabatic process entropy remains constant but temperature changes. During a reversible process,

$$(\delta Q)_{rev} = T.dS$$

The amount of heat $(\delta Q)_{rev}$ is represented by the shaded area of height T and width dS under the process path 1-2, as shown in Fig. 2.52. Total amount of heat interaction during the reversible process 1-2 is :

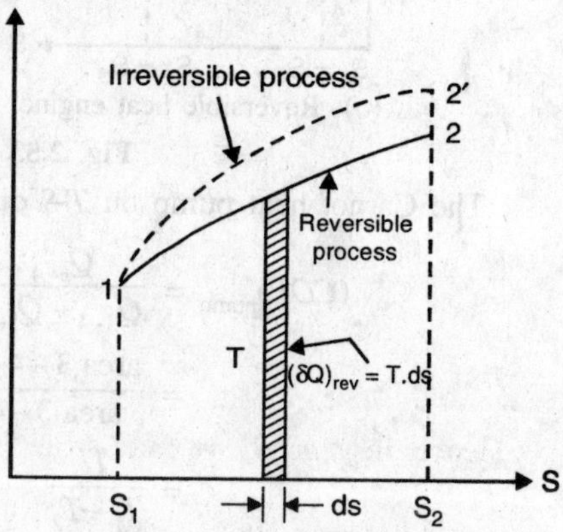

Fig. 2.52 *T-s* diagram.

$$(Q_{1-2})_{rev} = \int_1^2 TdS$$

$$= \text{area } (1 - 2 - S_2 - S_1 - 1)$$

For an irreversible process, the area under the path is greater than the amount of heat interaction.

- Constant entropy process is called isentropic process.
- For a reversible adiabatic process, $(\delta Q)_{rev} = 0$ and since $dS = \left(\frac{\delta Q}{T}\right)_{rev}$, $dS = 0$. Hence, a reversible adiabatic process is an isentropic process. The opposite is not always true.
- An irreversible- adiabatic process requires heat flow from the system.

$$s_2 - s_1 = c_v \cdot \frac{\gamma - n}{\gamma - 1} \cdot \ln \frac{T_2}{T_1}$$

$$= \left[\frac{\gamma - n}{\gamma - 1}\right] \cdot R \ln \frac{v_2}{v_1}$$

2.28 CARNOT CYCLE ON T-S DIAGRAM

The Carnot heat engine on *T-S* diagram is as shown in Fig. 2.53(*a*).

$Q_{1-2} = T_1(S_2 - S_1)$, isothermal heat addition

$Q_{2-3} = 0$, adiabatic expansion

$Q_{3-4} = T_2\,(S_1 - S_2)$, isothermal heat rejection

$Q_{4-1} = 0$, adiabatic compression

Efficiency,
$$\eta_{th} = \frac{Q_{1-2} - Q_{3-4}}{Q_{1-2}}$$
$$= \frac{\text{area } 1-2-3-4-1}{\text{area } 1-2-S_2-S_1-1}$$
$$= \frac{T_1 - T_2}{T_1} = 1 - \frac{T_2}{T_1}$$

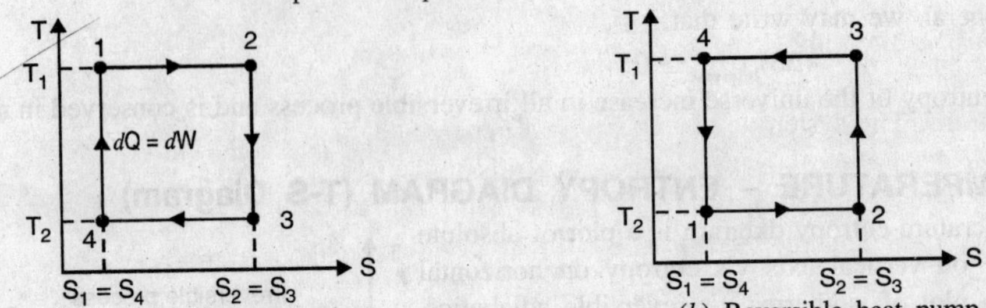

Fig. 2.53 Carnot cycle on *T–S* diagram.

The Carnot heat pump on *T-S* diagram is shown in Fig. 2.53(*b*)

$$(COP)_{pump} = \frac{Q_{3-4}}{Q_{3-4} - Q_{1-2}}$$
$$= \frac{\text{area } 3-4-S_1-S_2-3}{\text{area } 3-4-1-2-3}$$
$$= \frac{T_1}{T_1 - T_2}$$

For the refrigerator,

$$(COP)_{ref} = \frac{Q_{1-2}}{Q_{3-4} - Q_{1-2}}$$
$$= \frac{\text{area } 1-2-S_2-S_1-1}{\text{area } 1-2-3-4-1}$$
$$= \frac{T_2}{T_1 - T_2}$$

2.29 THIRD LAW OF THERMODYNAMICS

The third law of thermodynamics states that the entropy of any pure substance in thermodynamic equilibrium approaches zero as the absolute temperature approaches zero. Mathematically

$$\lim_{T \to 0} S = 0$$

This is also called the **Nernst** theorem.

The third law enables us to calculate the absolute entropy of a substance from the definition of change of entropy.

$$S - S_0 = \int_0^T \frac{(\delta Q)_{rev}}{T}$$

And since a state exists at the absolute zero, where $S_0 = 0$, hence

$$S = \int_0^T \frac{(\delta Q)_{rev}}{T}$$

The third law of thermodynamics is important in respect of the following:

- In calculating thermodynamic properties
- In analyzing chemical and phase equilibria.
- In explaining the behaviour of solids at very low temperature

Example 2.39 *A cyclic heat engine operates between a heat source temperature of 900° C and a sink temperature of 40° C. Calculate the least rate of heat rejection per kW net output of the engine.*

Solution. The system is shown in Fig 2.54

$$T_1 = 900 + 273 = 1173 \text{ K}$$
$$T_2 = 40 + 273 = 313 \text{ K}$$

For a reversible engine, heat rejection will be minimum.

$$\eta_{max} = \eta_{rev} = 1 - \frac{T_2}{T_1}$$
$$= 1 - \frac{313}{1173} = 0.733 \text{ or } 73.3\%$$

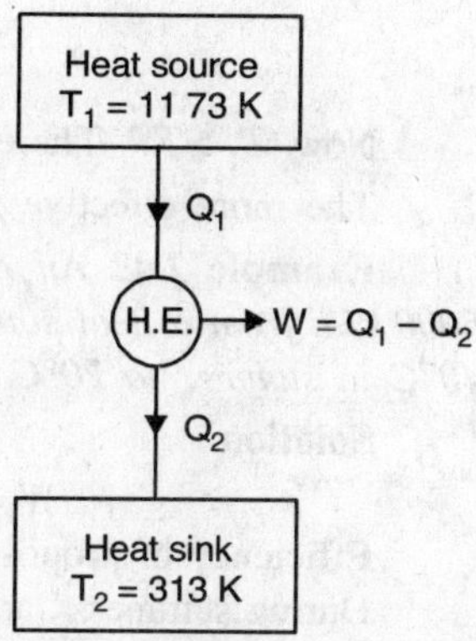

Fig. 2.54

Now
$$\frac{W_{net}}{Q_1} = \eta_{rev} = 0.733$$

$$Q_1 = \frac{1}{0.733} = 1.364 \text{ kW}$$

$$Q_2 = Q_1 - W_{net} = 1.364 - 1 = 0.364 \text{ kW}$$
= Least rate of heat rejection.

Example 2.40 *A freezer maintains a temperature of –20°C. The ambient air temperature is 25°C. If heat leaks into the freezer at a continuous rate of 2 kJ/s, find the least power necessary to pump this heat out continuously.*

Solution. The system is shown in Fig 2.55

$$T_2 = 273 - 20 = 253 \text{ K}$$
$$T_1 = 273 + 25 = 298 \text{ K}$$

Rate of heat removal from freezer = Rate of heat leakage into freezer

For minimum power requirement,

$$\frac{Q_2}{T_2} = \frac{Q_1}{T_1}$$

$$Q_1 = \frac{2 \times 298}{253} = 2.356 \text{ kJ/s}$$

$$W = Q_1 - Q_2 = 2.356 - 2 = 0.356 \text{ kJ/s}$$

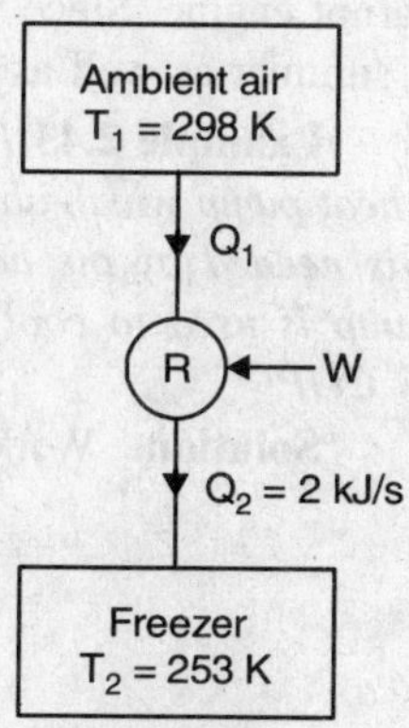

Fig. 2.55

Example 2.41 *The efficiency of a Carnot cycle can be increased either by increasing T_1 or decreasing T_2. Which is more effective way ?*

Solution.

$$\eta = 1 - \frac{T_2}{T_1}$$

(*a*) Let T_1 be increased by ΔT, T_2 remaining the same. Then

$$\eta_1 = 1 - \frac{T_2}{T_1 + \Delta T}$$

(*b*) If T_2 is decreased by ΔT, T_1 remaining the same. Then

$$\eta_2 = 1 - \frac{T_2 - \Delta T}{T_1}$$

$$\eta_2 - \eta_1 = \frac{T_2}{T_1 + \Delta T} - \frac{T_2 - \Delta T}{T_1}$$

$$= \left[\frac{T_1 T_2 - (T_1 + \Delta T)(T_2 - \Delta T)}{T_1(T_1 + \Delta T)}\right]$$

$$= \left[\frac{(T_1 - T_2)\Delta T + (\Delta T)^2}{T_1(T_1 + \Delta T)}\right]$$

Now $T_1 > T_2$. Therefore,$(\eta_2 - \eta_1)$ is positive.

The more effective way to increase the cycle efficiency is to decrease T_2.

Example 2.42 *An engine is to be designed to produce 0.8 kW by taking continuously 6000 kJ/h from a heat source at 500 K. The sink temperature is atmospheric which varies from 40°C in summer to 20°C in winter. Comment on the performance of the engine.*

Solution.

$$W = 0.8 \text{ kW} = 0.8 \times 3600 = 2880 \text{ kJ/h}$$

Efficiency of proposed engine, $\eta = W/Qs$ = 2880/6000 = 48%

During summer, Carnot engine efficiency,

$$\eta_s = 1 - \frac{T_2}{T_1} = 1 - \frac{313}{500} = 37.4\ \%$$

During winter, $\eta_w = 1 - \frac{293}{500} = 41.4\ \%$

According to second law of thermodynamics, no engine can be more efficient than the Carnot engine. Since η_s and η_w are less than η hence the proposed engine will work satisfactorily in summer as well as in winter.

Example 2.43 *For heating a house in winter it requires 200 MJ/h. The heat is supplied by a heat pump which absorbs heat from cold air outside in winter to supply heat to the house. 30 MJ/h is needed by the heat pump for its operation . Calculate the COP of heat pump. If the heat pump is used to cool the house during summer requiring 200 MJ/h of heat rejection, calculate its COP.*

Solution. Work input, $W = Q_1 - Q_2 = Q_r - Q_s$

or $Q_2 = Q_s = Q_r - W = 200 - 30 = 170$ MJ/h

$$(COP)_{HP} = \frac{Q_r}{W} = \frac{200}{30} = 6.67$$

$$(COP)_{Ref} = \frac{Q_s}{W} = \frac{170}{30} = 5.67$$

Example 2.44 *An engine working on Carnot cycle receives 1200 kJ from a heat source at a constant temperature of 1000°C and rejects to a heat sink at a constant temperature of 35°C. Calculate (a) thermal efficiency of the engine, and (b) work done.*

Solution.

$$T_1 = 1000 + 273 = 1273 \text{ K}$$
$$T_2 = 35 + 273 = 308 \text{ K}$$

(*a*) $\eta_{th} = 1-\frac{T_2}{T_1} = 1-\frac{308}{1273} = 0.758$ or 75.8%

(*b*) $\eta_{th} = \frac{W}{Q_1}$

$W = 0.758 \times 1200 = 909.6$ kJ

Example 2.45 *In a Carnot cycle the lowest pressure and temperature are equal to 1 bar and 20°C and isothermal compression pressure is 4 bar. Pressure after isentropic compression is 10 bar and after isothermal heat addition process is 5 bar. Calculate (a) the highest temperature in the cycle, (b) the change in entropy during isothermal expansion, (c) heat addition to the cycle, (d) heat rejected by the cycle,(e) power developed with 150 cycles/min, (f) thermal efficiency of the cycle, and (g) mean effective pressure. Assume γ = 1.4.*

Solution. The cycle is shown in Fig 2.54.

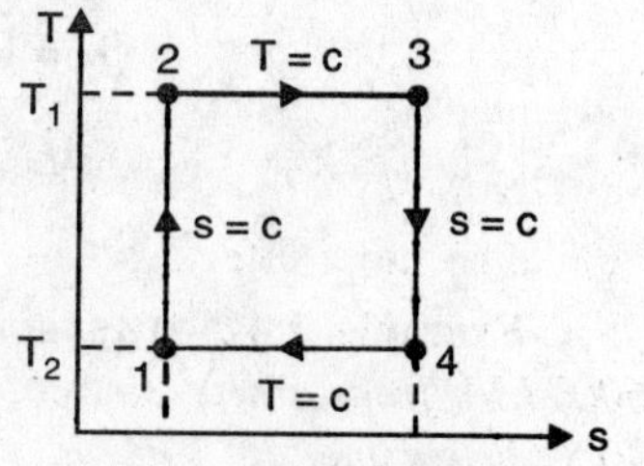

Fig. 2.56

$T_2 = 20 + 273 = 293$ K, $p_4 = 1$ bar

$p_1 = 4$ bar, $p_2 = 10$ bar, $p_3 = 5$ bar

At point 4, $v_4 = \frac{RT_4}{p_4} = \frac{0.287 \times 293}{1 \times 10^2}$

$= 0.8409$ m³/kg

Isothermal compression 4-1 :

$$v_1 = \frac{p_4 v_4}{p_1} = \frac{1 \times 10^5 \times 0.8409}{4 \times 10^5}$$

$= 0.2102$ m³/kg

Isentropic compression 1-2 :

$$v_2 = v_1\left(\frac{p_1}{p_2}\right)^{\frac{1}{\gamma}} = 0.2102\left(\frac{4}{10}\right)^{\frac{1}{1.4}} = 0.1092 \text{ m}^3/\text{kg}$$

Swept volume, $v_s = v_4 - v_2 = 0.8409 - 0.1092 = 0.7317$ m³/kg

(*a*) $$T_1 = T_2\left(\frac{v_1}{v_2}\right)^{\gamma-1} = 293\left(\frac{0.2102}{0.1092}\right)^{0.4} = 380.7 \text{ K}$$

(*b*) Isothermal heat addition 2-3:

$$v_3 = \frac{p_2 v_2}{p_3} = \frac{0.1092 \times 10}{5} = 0.2184 \text{ m}^3/\text{kg}$$

$$s_3 - s_2 = \left(\frac{p_2 v_2}{T_2}\right)\ln\left(\frac{v_3}{v_2}\right)$$

$$= \left[\frac{10 \times 10^2 \times 0.1092}{380.7}\right] \ln\left[\frac{0.2184}{0.1092}\right]$$

$= 0.1988$ kJ/kg K

(*c*) Heat added, $q_s = p_2 v_2 \ln \frac{v_3}{v_2} = T_1(s_3 - s_2) = 380.7 \times 0.1988 = 75.68$ kJ/kg

(*d*) Heat rejected, $q_r = p_4 v_4 \ln \frac{v_1}{v_4} = T_2(s_1 - s_4) = -T_2(s_3 - s_2) = -293 \times 0.1988$

$= -58.25$ kJ/kg

(*e*) Work done per cycle,

$w = q_s - q_r = 75.68 - 58.25 = 17.43$ kJ/kg

$$\text{Power developed} = \frac{1.743 \times 150}{60} = 43.575 \text{ kW}$$

(*f*) Thermal efficiency, $\eta_{th} = 1 - \frac{T_2}{T_1} = 1 - \frac{293}{380.7} = 0.2303$ or 23.03%

(*g*) $p_m = \frac{w}{v_s} = \frac{17.43}{0.7317} = 23.82 \text{ kN/m}^2 = 0.2382$ bar

Example 2.46 *A reversible engine with 45% thermal efficiency discharges 1600 kJ/min at 30°C to a pond. Find the temperature of the source which supplies heat to the engine and power developed by the engine.*

Solution.

$$T_2 = 30 + 273 = 303 \text{ K}$$

$$Q_2 = 1600 \text{ kJ/min}$$

$$\eta = \frac{W}{Q_1} = \frac{W}{W + Q_2} = 1 - \frac{T_2}{T_1}$$

$$0.45 = \frac{W}{W + 1600} = 1 - \left(\frac{303}{T_1}\right)$$

$$T_1 = \frac{303}{0.55} = 550.9 \text{ K or } 277.9°\text{C}$$

Also $0.45\ W + 720 = W$

$$W = \frac{720}{0.55} = 1309.09 \text{ kJ/min}$$

$$\text{Power} = \frac{1309.09}{60} = 21.82 \text{ kW}$$

Example 2.47 *A source S_1 can supply heat energy at the rate of 200 kJ/s at 320° C and source S_2 can supply 2000 kJ/s at 75° C. Which source would you choose to supply energy to an ideal reversible heat engine if the temperature of surroundings is 35° C ?*

Solution. Source S_1 : $Q_1 = 200$ kJ/s

$$T_1 = 320 + 273 = 593 \text{ K}$$

$$T_2 = 35 + 273 = 308 \text{ K}$$

$$\eta_1 = 1 - \frac{T_2}{T_1} = 1 - \frac{308}{593} = 0.4806 \text{ or } 48.06\%$$

Source S_2 : $Q_1 = 200$ kJ/s

$$T_1 = 75 + 273 = 348 \text{ K}$$

$$T_2 = 35 + 273 = 308 \text{ K}$$

$$\eta_2 = 1 - \frac{T_2}{T_1} = 1 - \frac{308}{348} = 0.1149 \text{ or } 11.49\%$$

Work delivered : $W_1 = 200 \times 0.4806 = 96.12$ kJ/s

$$W_2 = 2000 \times 0.1149 = 229.8 \text{ kJ/s}$$

Choose S_2.

Example 2.48 *An engine working on Carnot cycle absorbs Q_1 units of heat from a source at T_1 and rejects Q_2 units of heat to a sink at T_2. The temperature of the working fluid is θ_1 and θ_2, where $\theta_1 < T_1$ and $\theta_2 > T_2$.*

If $\theta_1 = T_1 - kQ_1$ *and* $\theta_2 = T_2 + kQ_2$

where k is constant, then show that efficiency of the engine is given by :

$$\eta = 1 - \frac{T_2}{T_1 - 2kQ_1}$$

Solution.

$$\eta = 1 - \frac{Q_2}{Q_1} = 1 - \frac{\theta_2}{\theta_1}$$

or

$$\frac{Q_2}{Q_1} = \frac{\theta_2}{\theta_1}$$

Also

$$\left(\frac{\theta_2 - T_2}{T_1 - \theta_1}\right) = \frac{Q_2}{Q_1} \quad \text{(given)}$$

$$\left(\frac{\theta_2 - T_2}{T_1 - \theta_1}\right) = \frac{\theta_2}{\theta_1}$$

$$\theta_1\theta_2 - \theta_1 T_2 = \theta_2 T_1 - \theta_1\theta_2$$

$$\theta_2(2\theta_1 - T_1) = \theta_1 T_2$$

$$\theta_2 = \left(\frac{\theta_1 T_2}{2\theta_1 - T_1}\right) = \frac{\theta_1 T_2}{2(T_1 - kQ_1) - T_1}$$

$$= \frac{\theta_1 T_2}{T_1 - 2kQ_1}$$

$\therefore$

$$\eta = 1 - \frac{T_2}{T_1 - 2kQ_1}$$

Hence Proved.

Example 2.49 *2700 kJ/min at 800° C is supplied to an engine which rejects at 125° C. If heat rejected is (a) 1000 kJ/min, (b) 800 kJ/min, and (c) 1200 kJ/min, specify the process.*

Solution.

$$T_1 = 800 + 273 = 1073 \text{ K}$$

$$T_2 = 125 + 273 = 398 \text{ K}$$

(*a*)

$$\oint \frac{\delta Q}{T} = \frac{Q_1}{T_1} - \frac{Q_2}{T_2} = \frac{2700}{1073} - \frac{1000}{398} = 2.516 - 2.512 \approx 0$$

The cycle is reversible.

(*b*)

$$\oint \frac{\delta Q}{T} = \frac{2700}{1073} - \frac{800}{398} = 2.516 - 2.010 = 0.506 > 0$$

The cycle is impossible.

(*c*)

$$\oint \frac{\delta Q}{T} = \frac{2700}{1073} - \frac{1200}{398} = 2.516 - 3.015 = -0.499 < 0$$

The cycle is irreversible.

Example 2.50 *An insulated vessel containing air has two compartments separated by a conducting diaphram. Compartment A is of capacity 0.030 m^3 at 2 bar and 300K. Compartment B is of capacity 0.025 m^3 at 4.5 bar and 450 K. Find (a) equilibrium temperature, and (b) change in entropy of the system. Take C_v = 0.718 kJ/kg.K and R = 0.287 kJ/kg.K.*

Solution.

$$m_A = \frac{p_1 V_1}{RT_1} = \frac{2 \times 10^5 \times 0.030}{287 \times 300} = 0.0697 \text{ kg}$$

$$m_B = \frac{p_2V_2}{RT_2} = \frac{4.5\times10^5\times0.025}{287\times450} = 0.0871 \text{ kg}$$

Let T = equilibrium temperature.

$$m_Ac_v(T-T_1) = m_B\,c_v(T_2-T)$$

$$0.0697(T-300) = 0.0871(450-T)$$

$$0.1568T = 60.105$$

$$T = 383.3 \text{ K}$$

$$\Delta S = m_Ac_v \ln\left(\frac{T}{T_1}\right) + m_Bc_v \ln\left(\frac{T}{T_2}\right)$$

$$= 0.0697\times0.718 \ln\left(\frac{383.3}{300}\right) + 0.0871\times0.718 \ln\left(\frac{383.3}{450}\right)$$

$$= 0.01226 - 0.01003 = 0.00227 \text{ kJ/K}$$

Example 2.51 *A vessel of capacity 3 m^3 contains air at a pressure of 2 bar and temperature 300 K. Additional air is pumped into the system until the pressure rises to 30 bar and temperature to 330 K. Determine (a) the mass of air pumped (b) heat transferred, and (c) the change in entropy if the air is cooled down to 300 K. Take c_v = 0.718 kJ/kg.K.*

Solution. (*a*)

$$m_1 = \frac{p_1V_1}{RT_1} = \frac{2\times10^5\times3}{287\times300} = 6.97 \text{ kg}$$

$$m_2 = \frac{p_2V_2}{RT_2} = \frac{30\times10^5\times3}{287\times330} = 95.027 \text{ kg}$$

Mass of air pumped in, $m = m_2 - m_1 = 88.057$ kg

(*b*) During cooling, $dU = m_2C_v(T_3-T_2) = 95.0\times0.718(300-330) = -2047.5$ kJ

$dQ = dU + dW = -2047.5 + 0 = -2047.5$ kJ

(*c*)

$$\Delta S = S_3 - S_2 = mc_v \ln\frac{T_3}{T_2} = 95.027\times0.718 \ln\frac{300}{330}$$

$$= -6.503 \text{ kJ/K}$$

Example 2.52 *Two identical bodies of constant heat capacity are at the same initial temperature T_i. A refrigerator operates between these two bodies until one body is cooled to temperature T_2. If the bodies remain at constant pressure and undergo no change of phase, show that the minimum amount of work needed to do this is :*

$$W_{min} = c_p\left[\frac{T_i^{\,2}}{T_2} + T_2 - 2T_i\right]$$

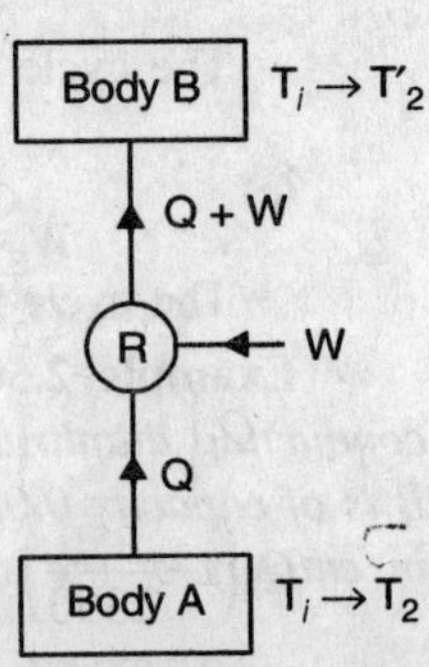

Fig. 2.57

Solution. Let T_i = initial temperature of both the bodies A and B (Fig 2.57)

T_2 = temperature of body A after cooling

T'_2 = final temperature of body B.

Heat removed from body A, $Q = c_p(T_i - T_2)$

Heat discharged to body B, $Q + W = c_p(T_2' - T_i)$

Work input, $W = c_p(T_2' - T_i) - c_p(T_i - T_2)$

$= c_p(T_2' + T_2 - 2T_i)$

Now $$\Delta S_A = c_p \ln\left(\frac{T_2}{T_i}\right)$$

$$\Delta S_B = c_p \ln\left(\frac{T_2'}{T_i}\right)$$

$$\Delta S = c_p \ln\left(\frac{T_2}{T_i}\right) + c_p \ln\left(\frac{T_2'}{T_i}\right) = c_p \ln\left(\frac{T_2 T_2'}{T_i^2}\right)$$

For W to be minimum, T_2' is to be minimum.

$$\therefore \quad c_p \ln\left(\frac{T_2 T_2'}{T_i^2}\right) = 0 = \ln 1$$

$$T_2' = \frac{T_i^2}{T_2}$$

$$W_{min} = c_p \frac{T_i^2}{T_2} + T_2 - 2T_1$$

Example 2.53 *A heat pump working on a reserved carnot cycle takes in energy from a reservoir at 5°C and delivers it to another reservoir where temperature is 77°C. The heat pump derives power for its operation from a reversible engine operating within the higher and lower temperature 1077°C and 77°C. For 100 kJ/kg of energy supplied to reservoir at 77°C, estimate the energy taken from the reservoir at 1077°C.*

Solution. The system is shown in Fig 2.58.

$$T_1 = 273 + 1077 = 1350 \text{ K},$$

$$T_2 = T_4 = 273 + 77 = 350\text{K},$$

$$T_3 = 273 + 5 = 278\text{K}$$

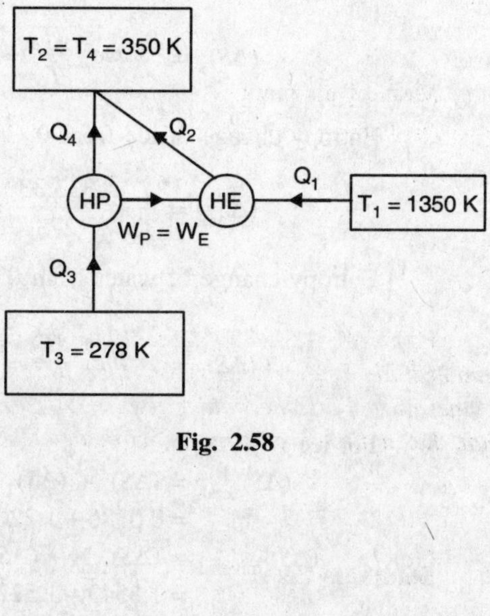

Fig. 2.58

H.E.

$$\eta = 1 - \frac{T_2}{T_1} = 1 - \frac{Q_2}{Q_1}$$

$$\frac{Q_2}{Q_1} = \frac{T_2}{T_1}$$

H.P.

$$COP = \frac{Q_4}{Q_4 - Q_3}$$

$$= \frac{T_4}{T_4 - T_3}$$

$$W_E = W_P$$

$$Q_1 - Q_2 = Q_4 - Q_3$$

$$Q_1\left(1 - \frac{T_2}{T_1}\right) = Q_4\left(1 - \frac{T_3}{T_4}\right)$$

$$Q_1\left(1 - \frac{350}{1350}\right) = Q_4\left(1 - \frac{278}{350}\right)$$

$$0.74074 Q_1 = 0.20571 Q_4$$

$$Q_4 = 3.6 Q_1$$

$$Q_2 = \frac{350Q_1}{1350} = 0.259Q_1$$

$$Q_2 + Q_4 = 100$$

$$(0.259 + 3.6)Q_1 = 100$$

$$Q_1 = 25.9 \text{ kJ}$$

Example 2.54 *1 kg of ice at – 5°C is exposed to the atmosphere which is at 27°C. The ice melts and comes into thermal equilibrium. Determine the entropy increase of the universe. For ice,* c_p *= 2.093 kJ/kg. °C. Latent heat of fusion of ice = 333.33 kJ/kg.*

Solution. The system is shown in Fig 2.59.

$$T_{ice} = -5 + 273 = 268\text{K}$$

$$T_{atm} = 27 + 273 = 300\text{K}$$

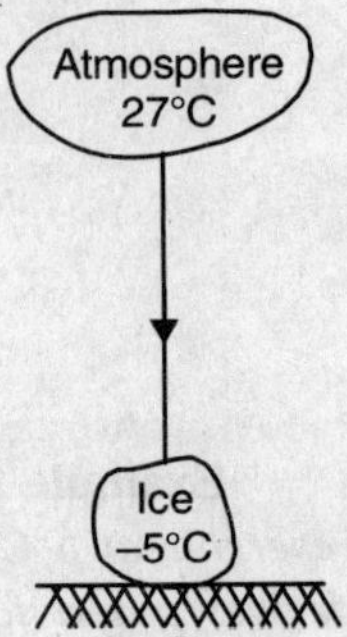

Fig. 2.59

Heat absorbed by ice from atmosphere,

$$Q = mL + mc_p\Delta t + mc_w\Delta t$$

$$= 1 \times 333.33 + 1 \times 2.093 \times [0 - (-5)] + 1 \times 4.187 \times (27 - 0)$$

$$= 333.33 + 10.465 + 113.049 = 456.844 \text{ kJ}$$

(*a*) Entropy change of atmosphere,

$$(\Delta S)_{atm} = \frac{-Q}{T} = \frac{-456.844}{300} = -1.5228 \text{ kJ/K}$$

(*b*) Entropy change of ice from –5°*C* to 0°C,

$$(\Delta S)_1 = \int_{268}^{273} mc_p \frac{dT}{T} = 1 \times 2.093 \ln \frac{273}{268} = 0.0386 \text{ kJ/K}$$

Entropy change of ice from 0°C to water at 0°C,

$$(\Delta S)_2 = \frac{L}{T_0} = \frac{333.33}{273} = 1.2209 \text{ kJ/K}$$

Entropy change of water from 0°C to 27°C,

$$(\Delta S)_3 = \int_{273}^{300} mc_p \frac{dT}{T} = 1 \times 4.187 \ln \frac{300}{273} = 0.3948 \text{ kJ/K}$$

For ice :

$$(\Delta S)_{Total} = (\Delta S)_1 + (\Delta S)_2 + (\Delta S)_3$$

$$= 0.0386 + 1.2209 + 0.3948 = 1.6543 \text{ kJ/K}$$

(*c*) $$(\Delta S)_{Universe} = (\Delta S)_{Total} + (\Delta S)_{atm}$$

$$= 1.6543 - 1.5228 = 0.1315 \text{ kJ/K}$$

FILL IN THE BLANKS

1. If the mass within the boundary of the system remains constant, it is called a .... system.
2. In a closed system, only the .... may transfer across its boundary.
3. The gas confined between a piston and cylinder is a/an .... system.

4. If the mass as well as the energy transfer takes place across the boundary of a system, it is called a/an .... system.
5. Air compressor is a/an .... system.
6. In a/an .... system, neither mass nor energy transfer takes place across its boundary.
7. Thermos flask is an example of .... system.
8. In a ...., mass, momentum and energy may flow.
9. The thermodynamic properties which do not depend on the mass of the system are called .... properties.
10. Temperature is an .... property.
11. The thermodynamic properties which depend on the mass of the system are called .... properties.
12. Potential energy is an .... property.
13. A system consisting of a single phase is called .... system.
14. Water and oil mixture is a .... system.
15. A thermodynamic system passing through a series of states constitutes a ....
16. A transition in which a system changes from one state to another state is called ....
17. If the final state is identical with the initial state, it is called a ....
18. For a reversible process, ... is zero.
19. Throttling is a/an .... process.
20. Heat is energy in ....
21. Internal energy is an .... property.
22. Work is a .... function.
23. Heat is a .... grade energy.
24. work is a .... grade energy.
25. The value of Universal gas constant is ....
26. The first law of thermodynamics is called the law of conservation of ....
27. According to first law of thermodynamics, $\oint \delta Q$ = ...
28. The first law of thermodynamics can be verified by the .... experiment.
29. The .... violates the first law of thermodynamics.
30. Specific internal energy, du = ...
31. For a system undergoing a cycle, $\oint(\delta Q - \delta W)$ ....
32. For a perfect gas, $C_v = R(......................)$.
33. A flow process permit the transfer of mass and .... across its boundary.
34. A nozzle converts .... energy into .... energy.
35. A diffuser converts .... energy into .... energy.
36. For a nozzle, $w = 0$, and q = ...
37. A turbine converts heat energy into ....
38. The non – flow work, W = ...
39. Work done in a steady flow process, W = ...
40. Internal energy of a perfect gas is ....
41. Heat cannot flow automatically from .... to .... body.

42. Gases cannot expand automatically from .... pressure to .... pressure.

43. Heat sink is a .... temperature reservoir.

44. Thermal efficiency of a heat engine = ...

45. A heat pump removes heat from a .... temperature body.

46. Kelvin- Planck and Clausius statements are ....

47. It is .... to construct a PMM2.

48. The Carnot cycle consist of two reversible .... and .... processes.

49. .... heat addition is not possible.

50. Working within the same temperature limits, no engine can be more efficient than the .... engine.

51. The efficiency of any reversible engine does not depend on the nature of the ....

52. Entropy is a .... of the system.

53. For a reversible adiabatic process, entropy change is .... zero.

54. For an irreversible adiabatic process, entropy change is .... zero.

55. Constant volume process is called ____________.

56. Constant pressure process is called ____________.

57. In a constant volume closed system, work done = ____________.

58. In an isobaric process, heat transfer is equal to change in ____________.

59. In an isothermal process, change in internal energy = ____________.

60. A reversible adiabatic process is called ______________.

61. In an isentropic process, change in heat energy = ____________.

62. The work done in a closed system is ______________.

63. In a free expansion process, change in internal energy is ____________.

64. The free expansion process is also called constant ________ process.

65. Throttling process is a/an ______________ process.

66. In a throttling process, there is no change in ______________.

67. The work done in a steady flow process = ____________.

68. The work done in a steady flow isobaric pressure is ____________.

69. The work done in non-flow and flow processes is equal for the ________ case.

ANSWERS

1. closed	**2.** energy	**3.** closed	**4.** open
5. open	**6.** isolated	**7.** isolated	**8.** control volume
9. intensive	**10.** intensive	**11.** extensive	**12.** extensive
13. homogeneous	**14.** heterogeneous	**15.** path	**16.** process
17. cycle	**18.** friction	**19.** irreversible	**20.** in transition
21. intensive	**22.** path	**23.** low	**24.** high
25. 8.314 J/mol.K	**26.** energy	**27.** $\oint \delta W$	**28.** Joule's
29. *PMM–I*	**30.** $\delta q - \delta w$	**31.** $= 0$	**32.** $(\gamma - 1)$
33. energy	**34.** pressure, kinetic	**35.** kinetic, pressure	**36.** $=0$
37. work	**38.** $\int p.dv$	**39.** $-\int v.dp$	**40.** $mC_v dT$

41. cold, hot	**42.** low, high	**43.** low	
44. Net work done / Heat added		**45.** low	**46.** equivalent
47. impossible	**48.** isothermal, adiabatic		**49.** isothermal
50. reversible	**51.** working fluid	**52.** property	**53.** equal to
54. more than	**55.** isochoric	**56.** isobaric	**57.** zero
58. enthalpy	**59.** zero	**60.** isentropic	**61.** zero
62. $\int p.dv$	**63.** zero	**64.** enthalpy	**65.** irreversible
66. enthalpy	**67.** $-\int v.dp$	**68.** zero	**69.** isothermal.

TRUE AND FALSE

State True (T) or False (F) :

1. A system is a finite quantity of matter.
2. For a system to be in thermodynamic equilibrium, velocity gradient may exist.
3. The property of a system depends upon its state.
4. Intensive properties do not depend on the mass of the system.
5. Extensive properties depend on the mass of the system.
6. A process occurs when the system undergoes a change in state.
7. In a reversible process, initial and final states of the system are same.
8. In a closed system, mass does not remain constant.
9. A rubber balloon filled with air represents a closed system.
10. A nozzle represents an open system.
11. A thermos flask is an example of isolated system.
12. In a control volume, mass, momentum and energy may flow.
13. Temperature is an intensive property.
14. Kinetic energy is an extensive property.
15. Work is a path function.
16. The first law of thermodynamics is the law of conservation of energy.
17. According to the first law of thermodynamics, $\oint \delta Q \neq \oint \delta W$.
18. According to the first law, heat and work are mutually convertible.
19. It is impossible to construct a *PMM*-1.
20. For an isolated system, the energy remains constant.
21. In a steady flow process, the conditions in control volume remain unchanged with time.
22. A nozzle converts kinetic energy into pressure energy.
23. A steam turbine converts heat energy into useful work.
24. For a system undergoing a cycle, $\oint (\delta Q - \delta W) = 0$.
25. For a perfect gas, enthalpy, $H = mC_p dT$.
26. Heat can flow automatically from cold to hot body.
27. Heat source is a low temperature reservoir.
28. A heat pump removes heat from a high temperature body.
29. It is impossible to construct a PMM2.
30. A Carnot cycle engine is not practicable.
31. A constant volume cycle is called isobaric.

32. A reversible engine is the most efficient engine working within the same temperature limits.
33. The efficiency of a reversible engine depends on the nature of the working fluid.
34. Entropy is not a property of the system.
35. Entropy is an extensive property of the system.
36. Entropy may e regarded as the degradation of energy.
37. Work done in a closed system isochoric process is zero.
38. An adiabatic process is called isentropic.
39. In an isentropic process, change in heat energy is zero.
40. The work done in a closed system is $-\int v.dp$.
41. There is no change in internal energy in a free expansion process.
42. Throttling process is a reversible process.
43. There is no change in enthalpy in a throttling process.
44. The work done in a steady flow isobaric process is $p\ (v_2 - v_1)$.
45. The work done in a non-flow isochoric process is $v\ (p_1 - p_2)$.
46. The work done in a non-flow process is $\int p.dv$.

ANSWERS

1. T	**2.** F	**3.** T	**4.** T	**5.** T	**6.** T	**7.** T	**8.** F	**9.** T	**10.** T
11. T	**12.** T	**13.** T	**14.** T	**15.** T	**16.** T	**17.** F	**18.** T	**19.** T	**20.** T
21. T	**22.** F	**23.** T	**24.** T	**25.** T	**26.** F	**27.** F	**28.** F	**29.** T	**30.** T
31. F	**32.** T	**33.** F	**34.** F	**35.** T	**36.** T	**37.** T	**38.** F	**39.** T	**40.** F
41. T	**42.** F	**43.** T	**44.** F	**45.** F	**46.** T				

MULTI-CHOICE QUESTIONS

1. In an open system, which crosses the boundary of the system?
(*a*) heat (*b*) work (*c*) mass (*d*) heat and work

2. In an isolated system, what transfers?
(*a*) mass (*b*) energy
(*c*) both mass and energy (*d*) neither mass nor energy

3. Choose the extensive property :
(*a*) pressure (*b*) volume (*c*) temperature (*d*) density

4. Which of the following is an intensive property?
(*a*) volume (*b*) temperature (*c*) mass (*d*) energy

5. A system comprising of a single phase is called :
(*a*) homogeneous (*b*) heterogeneous (*c*) isolated (*d*) closed

6. Consider the following statements :
1. In an open system, mass and energy transfer takes place.
2. In a closed system, energy transfer takes place.
3. The universe can be considered as an isolated system.

Of these statements :
(*a*) only 1 is correct (*b*) only 2 is correct
(*c*) 1 and 2 are correct (*d*) 1,2 and 3 are correct.

7. Consider the following statements :

1. Milk and sugar is a homogeneous system.
2. Free expansion is an irreversible process.
3. Thermos flask is an isolated system.

Of these statements :

(*a*) only 1 is correct (*b*) 1 and 2 are correct
(*c*) 2 and 3 are correct (*d*) 1,2 and 3 are correct.

8. Which of the following is a path function?

(*a*) heat (*b*) temperature (*c*) pressure (*d*) volume.

9. The following are examples of some intensive and extensive properties :

1. Pressure 2. Temperature 3. Volume 4. Velocity
5. Electric charge 6. Magnetisation 7. Viscosity 8. Potential energy.

Which one of the following sets gives the correct combination of intensive and extensive properties?

	Intensive	Extensive
(*a*)	1,2,3,4	5,6,7,8
(*b*)	1,3,5,7	2,4,6,8
(*c*)	1,2,4,7	3,5,6,8
(*d*)	2,3,6,8	1,4,5,7

10. A mixture of gases expands from 0.03 m^3 to 0.06 m^3 at a constant pressure of 1 MPa and absorbs 84 kJ of heat during the process. The change in internal energy of the mixture is :

(*a*) 30 kJ (*b*) 54 kJ (*c*) 84 kJ (*d*) 114 kJ

11. Given that for the path 1 – 2 – 3, a system absorbs 100 kJ as heat and does 60 kJ work. While along the path 1 – 4 – 3, it does 20 kJ work (see figure). The heat absorbed during the cycle 1 – 4 – 3 is :

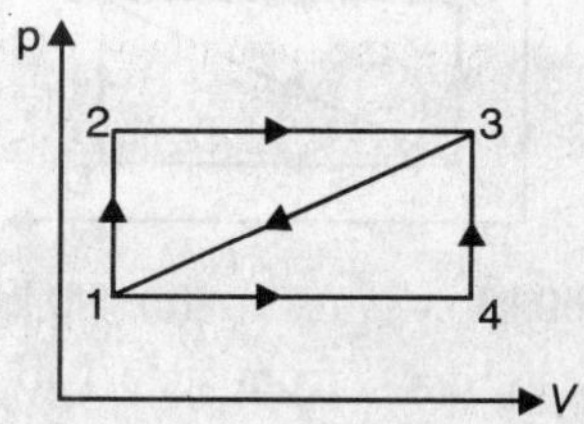

(*a*) – 140 kJ (*b*) – 80 kJ (*c*) – 40 kJ (*d*) + 60 kJ

12. The given figure shows the variation of force in an elementary system which undergoes a process during which the plunger position changes from 0 to 3 m. If the internal energy of the system at the end of the process is 2.5 kJ higher, then the heat absorbed during the process is :

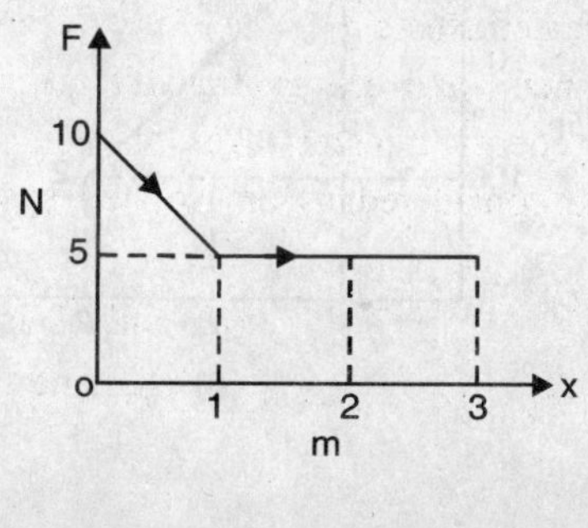

(*a*) 15 J (*b*) 20 J (*c*) 25 J (*d*) 30 J

13. The net work done for the closed cycle shown in the given pressure – volume diagram is :

(*a*) 600 kN.m (*b*) 700 kN.m (*c*) 900 kN.m (*d*) 1000 kN.m

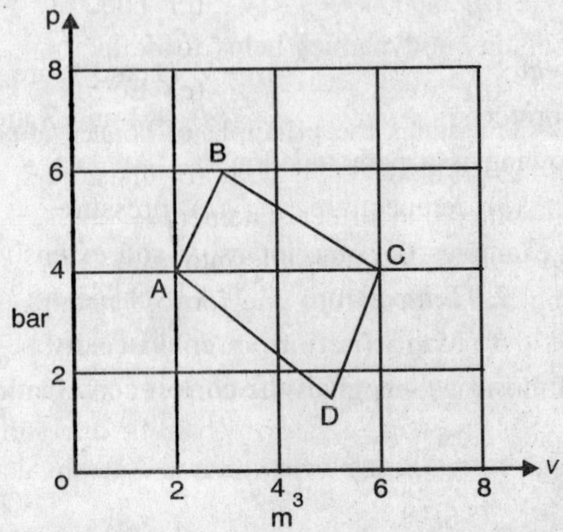

14. The internal energy of a certain system is a function of temperature alone and is given by the formula : $E = 25 + 0.25t$ kJ. If this system executes a process for which the work done by it per degree temperature increase is 0.75 kN.m, the heat interaction per degree temperature increase is :

(*a*) – 1.00 kJ (*b*) – 0.50 kJ (*c*) + 0.05 kJ (*d*) 1.00 kJ

15. When a system is taken from state A to state B along the path $A - C - B$, 180 kJ of heat flows into the system and does 130 kJ of work (see figure). How much heat will flow into the system along the path $A - D - B$ if the work done by it along the path is 40 kJ ?

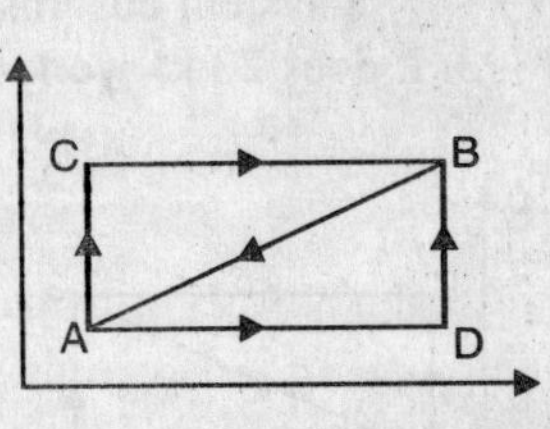

(*a*) 40 kJ (*b*) 60 kJ (*c*) 90 kJ (*d*) 135 kJ

16. A control mass undergoes a process from state 1 to state 2 as shown in the given figure. During this process, if the heat transfers to state 2 from state 1 by another process, then the work interaction during the return process (in kN.m) would be :

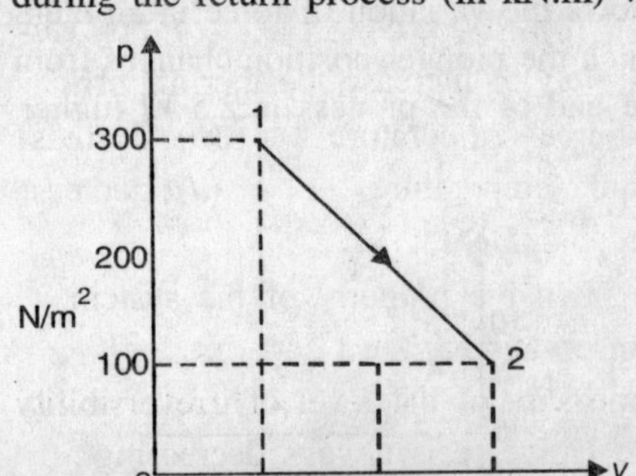

(*a*) – 400 (*b*) – 200 (*c*) 200 (*d*) 400

17. A tank containing air is stirred by a paddle wheel. The work input to the paddle wheel is 9000 kJ and the heat transferred to the surroundings from the tank is 3000 kJ. The external work done by the system is :

(*a*) zero (*b*) 3000 kJ (*c*) 6000 kJ (*d*) 9000 kJ

18. The second law of thermodynamics helps to define

(*a*) Heat (*b*) Work (*c*) Entropy (*d*) Internal energy

19. The Kelvin-Planks law states the principle of conservation of

(*a*) Work (*b*) heat (*c*) mass (*d*) heat into work

20. Which of the following is incorrect statement

(*a*) Heat is not completely convertible into work

(*b*) Energy can be transferred from one form to another automatically

(*c*) Heat cannot flow automatically from cold to hot body

(*d*) Gas cannot expand automatically from low to high pressure

21. It is impossible to construct an engine which while operating in a cycle produces no other effect to extract heat from a single reservoir and do equivalent amount of work. This refers to :

(*a*) Clausius statement (*b*) Kelvin-Planck's statement

(*c*) Carnot theorem (*d*) PMM-2

22. A Carnot cycle comprises of :

(*a*) Two isothermal and two isentropic processes

(*b*) Two constant volume and two isentropic processes

(*c*) Two constant pressure and two isentropic processes

(*d*) One constant volume , one constant pressure and two isentropic processes.

23. A Carnot cycle operates between temperatures of 1000 K and 500 K. Its thermal efficiency is :

(*a*) 50% (*b*) 200% (*c*) 100% (*d*) 150%

24. A Carnot cycle operates between temperatures of 800 K and 400 K. Its efficiency will be 50% if working fluid is :

(*a*) air (*b*) nitrogen (*c*) carbon dioxide (*d*) any fluid

25. A refrigerator and heat pump operate between the same temperature limits. If *COP* of the refrigerator is 3, the *COP* of pump will be :

(*a*) 3 (*b*) 4 (*c*) 5 (*d*) 6

26. A heat engine is supplied with 800 kJ/s of heat at 600 K and rejects at 300 K . For the process to be reversible, heat rejected is :

(*a*) 200 kJ/s (*b*) 400 kJ/s (*c*) 500 kJ/s (*d*) 600kJ/s

27. The more effective way to increase the thermal efficiency of a Carnot cycle is to:

(*a*) increase the source temperature (*b*) decrease the sink temperature

(*c*) increase the sink temperature (*d*) decrease the source temperature

28. Choose the correct statement

(*a*) Entropy is an intensive property of the system

(*b*) Entropy is conservative

(*c*) Entropy is a measure of the level of irreversibility associated with a process

(*d*) Entropy of the Universe is always decreasing

29. Consider the following statements :

The definition of :

1. temperature is due to Zerorth law of Thermodynamics
2. entropy is due to First law of Thermodynamics
3. internal energy is due to Second Law of Thermodynamics
4. reversibility is due to Kelvin- Planck's statement

Of these statements :

(*a*) 1,2 and 3 are correct (*b*) 1,3 and 4 are correct
(*c*) 1 alone is correct (*d*) 2 alone is correct

30. One reversible heat engine operates between 1600 K and T_2 , and another reversible heat engine operates between T_2 and 400 K. If both the engines have the same heat input and output, then the temperature T_2 must be equal to :

(*a*) 1000 K (*b*) 1200 K (*c*) 1400 K (*d*) 800 K

31. A heat engine using lake water at 12°C as source and surrounding atmosphere at 2°C as sink executes 1080 cycles per minute. If the amount of heat drawn per cycle is 57 J, then the output of the engine will be :

(*a*) 66 W (*b*) 56 W (*c*) 46 W (*d*) 36 W

32. In a cyclic heat engine operating between a source temperature of 600°C and a sink temperature of 20°C , the least rate of heat rejection per kW net output of the engine is :

(*a*) 0.460 kW (*b*) 0.505 kW (*c*) 0.588 kW (*d*) 0.650 kW

33. A Carnot engine receiving heat at 400 K has an efficiency of 25% . The *COP* of a Carnot refrigerator working between the same temperature limits is

(*a*) 1 (*b*) 2 (*c*) 3 (*d*) 4

34. The efficiency of a reversible cyclic process undergone by a substance as shown below is :

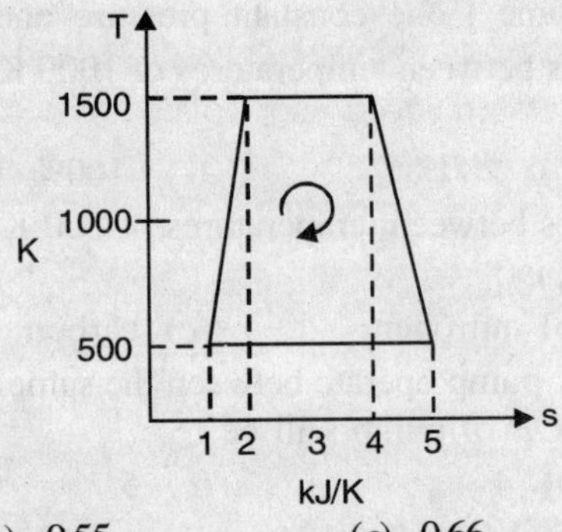

(*a*) 0.40 (*b*) 0.55 (*c*) 0.66 (*d*) 0.80

35. A heat engine is supplied with 250 kJ/s of heat at a constant fixed temperature of 227°C. The heat is rejected at 27°C. The cycle is reversible, if the amount of rejected is :

(*a*) 273 kJ/s (*b*) 200 kJ/ss (*c*) 180 kJ/s (*d*) 150 kJ/s

36. The work done in a steady flow process is:

(*a*) $\int p\,dv$ (*b*) $\int v.dp$ (*c*) $-\int p\,dv$ (*d*) $-\int v\,dp$

37. The following remains constant during a throttling process:

(*a*) internal energy (*b*) enthalpy (*c*) pressure (*d*) temperature

38. In an open system, the following can cross the boundary of the system:

(*a*) heat and work (*b*) mass (*c*) mass, heat and work (d) heat only

39. In an isolated system, the following can cross the boundary of the system:

(*a*) energy (*b*) mass
(*c*) energy and mass (*d*) neither energy nor mass.

40. An isentropic process is
(*a*) irreversible & adiabatic (*b*) reversible & adiabatic
(*c*) irreversible & frictionless (*d*) reversible & isothermal.

41. In the Polytropic Process $pV^n = c$, if n = ¥, the process is called
(*a*) isochoric (*b*) isobaric (*c*) isothermal (*d*) adiabatic

42. During isothermal process, which does not change?
(*a*) pressure (*b*) temperature
(*c*) volume (*d*) both pressure & temperature

43. Choose the incorrect relationship
(*a*) $c_p + c_v = R$ (*b*) $c_p - c_v = R$ (*c*) $\gamma = c_p/c_v$ (*d*) $c_v = R/ (\gamma - 1)$

ANSWERS

1. (*b*) **2.** (*d*) **3.** (*b*) **4.** (*d*) **5.** (*d*) **6.** (*c*) **7.** (*d*) **8.** (*c*) **9.** (*d*)
10. (*d*) **11.** (*b*) **12.** (*b*) **13.** (*a*) **14.** (*d*) **15.** (*c*) **16.** (*a*) **17.** (*c*) **18.** (*c*)
19. (*d*) **20.** (*b*) **21.** (*b*) **22.** (*a*) **23.** (*a*) **24.** (*d*) **25.** (*b*) **26.** (*b*) **27.** (*b*)
28. (*c*) **29.** (*c*) **30.** (*d*) **31.** (*d*) **32.** (*b*) **33.** (*c*) **34.** (*c*) **35.** (*d*) **36.** (*d*)
37. (*b*) **38.** (*c*) **39.** (*d*) **40.** (*b*) **41.** (*a*) **42.** (*b*) **43.** (*a*)

REVIEW QUESTIONS

1. Define the term thermodynamic.
2. Explain the thermodynamic system, surroundings and universe.
3. Distinguish between closed, open and isolated system with examples.
4. Differentiate between homogeneous and heterogeneous systems.
5. What do you mean by phase of a system?
6. What do you mean by thermodynamic equilibrium?
7. What do you mean by property? Distinguish between intensive and extensive properties.
8. Separate the following into intensive and extensive properties :
Mass, weight, volume, pressure, temperature, density, velocity, surface area, internal energy, potential energy.
9. Explain the terms : state, phase, process and cyclic process.
10. What do you understand by reversible and irreversible process? Give examples.
11. Explain cyclic and quasi – static process.
12. Explain the term "Energy". Discuss its various forms.
13. Define work. Show that work is a path function.
14. Define the term heat.
15. Choose the open or closed system from the following :
Water pump, pressure cooker, automobile engine, air compressor, steam turbine, boiler system.
16. State the number of phases in the following :
Pure water, water – alcohol mixture, coffee, water – oil mixture, aqua – ammonia, mild steel, at room temperature.

17. Explain the concept of temperature.
18. State zeroth law of thermodynamics.
19. Write the characteristic gas equation.
20. What is universal gas constant?
21. State first law of thermodynamics.
22. What is PMM-1.
23. Show that energy of an isolated system remains unchanged.
24. Write the steady flow energy equation.
25. Write the relationship between specific heats.
26. What is adiabatic index? Give its usual value.
27. What is a steady flow process?
28. What conditions are fulfilled by a steady flow process?
29. State the limitations of First law of thermodynamics.
30. State Kelvin-Planck statement for second law of thermodynamics.
31. State Clausius statement for the second law of thermodynamic.
32. What is a PMM2?
33. Are Kelvin-Planck and Clausius statements equivalent?
34. Differentiate between a heat engine and a heat sink.
35. What is a thermal reservoir?
36. Differentiate between a heat source and a heat sink.
37. Describe the working of a Carnot cycle.
38. What are the limitations of Carnot cycle?
39. Define thermal efficiency of a heat engine.
40. Differentiate between a heat pump and a refrigerator.
41. Define COP of a heat pump and a refrigerator.
42. Define the term 'Entropy'.
43. Show that the entropy is a property of the system.
44. The entropy of the universe tends to be maximum. Comment.
45. What are the characteristics of entropy?
46. Draw the Carnot cycle on T-S diagram.
47. State Third law of thermodynamics.
48. What is the importance of third law of thermodynamics?
49. Determine an expansion for the heat transfer in a closed system isochoric process.
50. Explain the concept of a closed isobaric process and determine an expansion for its heat transfer.
51. Determine an expansion for the work done in a closed isothermal process.
52. What is an isentropic process? Determine an expansion for the work done in a non-flow isentropic process.
53. Determine an expansion for the heat transfer and work done in a non-flow Polytropic process.
54. What is a free expansion process? What are its characteristics?
55. What is a throttling process? State its characteristics?
56. What is a steady flow process? Determine the work done in such an isochoric process.
57. Determine an expansion for the work done in a steady flow adiabatic process.
58. Compare the work done in a non-flow and flow type polytropic process.

EXERCISES

2.1 5 m^3 of gas at 8 bar and 180°C is heated keeping the pressure same till the volume is doubled. Calculate (*a*) heat added, (*b*) external work done, and (*c*) change in internal energy during the process.

[Ans. 14 MJ, 4 MJ, 10 MJ]

2.2 In a cyclic process, heat transfers are +14.7 kJ, –25.2 kJ, –3.56 kJ and +31.5kJ. What is the net work for this cyclic process?

[Ans. 17.44 kJ]

2.3 A heat engine receives 1000 kW of heat at constant temperature of 285°C and rejects heat at 5°C . The possible heat rejected are: (*a*) 840 kW, (*b*) 442 kW and (*c*) 300 kW. Comment on the results.

[Ans. Not a possible cycle]

2.4 To cool water for drinking during summer, 1 kg of ice at – 2°C is mixed with 5 kg of water at 27°C in an insulated container subjected to 1 bar atmospheric pressure. Calculate the change in temperature of water and the change in entropy of mixture. Take specific heat of ice as 2.093 kJ/kg K and the latent heat of ice as 335 kJ/kg. Specific heat of water is 4.187 kJ/kg K.

[Ans. 9.817°C, 0.0907 kJ/K]

2.5 A heat exchanger handles 40 kg/ min of water which is heated from 20°C to 70°C by a hot gas entering the heat exchanger at 150°C and flow rate 80 kg/min.

Determine the change in entropy. Assume specific heat for water as 4.18 kJ/kg K and for gases as 1.045 kJ/kg K.

[Ans. 0.0632 kW/K]

2.6 A heat pump operates between two identical bodies of specific heat C at T_1. The operation of the pump cools down one of the bodies to T_2. Show that for the operation of pump the minimum work input is given by :

$$W_{min} = C\left[\frac{T_1^2}{T_2} + T_2 - 2T_1\right]$$

2.7 A Carnot engine with an efficiency of 40% receives 400 kJ/h from a high temperature source and rejects heat to a sink at 27°C . What is the power output of the engine and temperature of the source in °C ?

[Ans. 0.0444 kW, 227°C]

2.8 A reverse Carnot engine is used for the heating of a building . The building receives 209 kJ/h of heat at 20°C from a heat pump . The outside air temperature is –5°C. Calculate (*a*) heat taken from outside air per hour, and (*b*) power of motor required

[Ans. 191.17 kJ/h, 4.953 W]

2.9 A piece of red hot iron is suddenly plunged into 10 kg of water at 20°C. The temperature of water rises to 55°C. Calculate the change in entropy.

[Ans. 4.724kJ/K]

2.10 Two reversible heat engines *A* and *B* are arranged in series. A rejects heat directly to *B*. Engine *A* receives 300 kJ at a temperature of 500°C from the heat source, while engine *B* is in communication with a cold sink at a temperature of 5°C. If the work output of *A* is twice that of *B*, calculate

(*a*) the intermediate temperature between *A* and *B*, and

(*b*) thermal efficiency of each engine.

[**Ans.** 170°C, 42.7%, 37.25%]

2.11 A hot iron having specific heat 0.5 kJ/kg K, mass 30 kg and temperature 500°C is dropped into 200 kg of oil of specific heat 2.5 kJ/kg K . Calculate change in entropy of iron, oil and composite system.

[**Ans.** –13.617 kJ/K, 22.690 kJ/K,9.073 kJ/K]

2.12 A heat pump is used to maintain an auditorium hall at 25°C when the atmospheric temperature is – 5°C . The heat load is 2400 kJ/min. Calculate the power required to run the actual heat pump if the *COP* of the actual heat pump is 25% of the Carnot heat pump working between the same temperature limits.

[**Ans.** 16.1 kW]

2.13 An engine manufacturer claims to have developed a heat engine with the following specifications :
Power developed = 75 kW; Fuel burnt = 5kg/h; Heating value of fuel = 75000 kJ/kg
Temperature limits =1000 K and 400 K.
Is the claim of the manufacturer true or false?

[**Ans.** False]

2.14 A heat source S_1 can supply 6000 kJ/min at 300°C and another heat source S_2 can supply 60,000 kJ/min at 100°C. Which source will you choose to supply energy to a Carnot engine to produce a large amount of power if the surroundings are at 27°C ?

[**Ans.** S_2]

2.15 A reversible engine working in a cycle takes 4800 kJ/min of heat from a source at 800 K and develops 45 kW. The engine rejects heat to two sinks at 300 K and 360 K . Determine the heat rejected to each sink.

[**Ans.** 300 kJ/min, 1800 kJ/min]

2.16 A reversible heat engine receives heat from two thermal reservoirs maintained at constant temperatures at 750 K and 500 K. The engine develops 100 kW and rejects 3600 kJ/min of heat to a heat sink at 250 K. Determine thermal efficiency of the engine and heat supplied by each thermal reservoir.

[**Ans.** 62.5%, 7200 kJ/min, 2400 kJ/min]

2.17 A reversible heat engine operates between two reservoirs at temperatures, 700°C and 50°C . The engine drives a reversible refrigerator which operates between reservoirs at temperature of 50°C and –25°C. The heat transfer to the engine is 2500 kJ and the net work output of the combined refrigerator plant is 400 kJ.

(*a*) Determine the net heat transfer to the reservoir at 50°C.

(*b*) Reconsider (*a*) given that the efficiency of the heat engine and the *COP* of the refrigerator are each 45% of their maximum possible values.

[**Ans.** 6300 kJ, 2623 kJ]

2.18 Two Carnot engines work in series between the source and sink temperatures of 500 K and 300 K. If both engines develop equal power determine the intermediate temperature.

[**Ans.** 400 K]

2.19 A reversible heat engine working between two thermal reservoirs at 875 K and 315K drives a reversible refrigerator which operates between the same 315 K reservoir and a reservoir at 260 K. The engine is supplied 2000 kJ of heat and the network output from the composite system is 350 kJ. Calculate the heat transfer to the refrigerator and the net heat interaction with the reservoir at 315 K temperature.

[**Ans.** 4396 kJ, 6046 kJ]

2.20 A heat engine operating between two reservoirs at 1000 K and 300 K is used to drive a heat pump which extracts heat from the reservoir at 300 K, at a rate twice that at which engine rejects heat to it. If efficiency of the engine is 40% of the maximum possible and the co-efficient of performance of the heat pump is 50% of the maximum possible , make calculations for the temperature of the reservoir to which the heat pump rejects heat. Also workout the rate of heat rejection from the heat pump if the rate of supply of heat to the engine is 50 kW.

[**Ans.** 326.6 K,86 kW]

2.21 A domestic food freezer maintains a temperature of –15°C . The ambient air temperature is 30°C. If heat leaks into the freezer at the continuous rate of 1.75 kJ/s, what is the least power necessary to pump this heat out continuously.

[**Ans.** 0.305 kW]

2.22 A reversible heat engine operates between two reservoirs at temperatures of 600°C and 40° C. The engine drives a reversible refrigerator which operates between reservoirs at temperatures of 40°C and – 20°C . The heat transfer to the heat engine is 2000 kJ and the net work output of the combined engine-refrigerator plant is 360 kJ. Calculate the heat transfer to the refrigerant and the net transfer to the reservoir at 40°C.

[**Ans.** 3894.2 kJ, 5534 kJ]

2.23 2 kg of water at 80°C is mixed adiabatically with 3 kg of water at 30°C in a constant pressure process of 1 atmosphere. Find the increase in entropy of the total mass of water due to the mixing process. The c_p of water = 4.187 kJ/kg.K.

[**Ans.** 0.0591 kJ/K]

2.24 Water is heated at a constant pressure of 0.7 mPa . The boiling point is 164.97°C. The initial temperature of the water is 0°C. The latent heat of evaporation is 2066.3 kJ/kg. Find the increase in entropy of water, if the final state is steam.

[**Ans.** 6.697 kJ/kg.K]

2.25 Calculate the entropy change of the universe as a result of the following processes :

(*a*) A copper block of 600 g mass and with c_p of 150 J/kg K at 100°C is placed in a lake at 8°C.

(*b*) The same block at 8°C is dropped from a height of 100 m into the lake.

(*c*) Two such blocks at 100°C and 0°C are joined together.

[**Ans.** 7.2678 J/K, 2.095 J/K, 2.183 J/K]

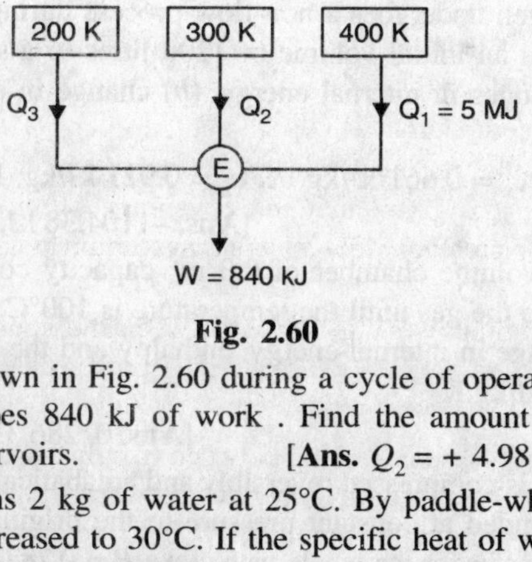

Fig. 2.60

2.26 A reversible engine, as shown in Fig. 2.60 during a cycle of operation draws 5 MJ from the 400K reservoir and does 840 kJ of work . Find the amount and direction of heat interaction with other reservoirs. [**Ans.** Q_2 = + 4.98 MJ, Q_3 = – 0.82 MJ]

2.27 An adiabatic vessel contains 2 kg of water at 25°C. By paddle-wheel work transfer, the temperature of water is increased to 30°C. If the specific heat of water is 4.187 kJ/kg K, find the entropy change of the universe. [**Ans.** 0.139 kJ/K]

2.28 Two blocks of metal, each of mass 10 kg and a specific heat of 0.4 kJ/kg K, are at a temperature of 40°C. A reversible refrigerator receives heat from one block and rejects heat to the other. Calculate the work required to cause a temperature difference of 100°C between the two blocks.

[**Ans.** 32 kJ]

2.29 1 kg of air at 3.5 bar and occupying 0.35 m^3 is heated at constant volume until its temperature has rises to 316^{0}C. Find (*a*) heat added, and (*b*) gain in internal energy per kg of air.Take c_v = 0.718 kJ/kg. K.

[**Ans.** 116.44 kJ/kg, 116.44 kJ/kg]

2.30 1 kg of air having an initial volume of 0.3 m^3 is heated at constant pressure of 3.2 bar until the volume is doubled. Find (*a*) heat added, and (*b*) work done. Take c_p = 1.005 kJ/kg.K, c_v = 0.718 kJ/kg.K.

[**Ans.** 336.17 kJ, 96 kJ]

2.31 5 m^3 of gas at 8 bar and 180°C is heated keeping the pressure same throughout till the volume is doubled. Find the heat added, external work done and the change in internal energy during the process. Take c_p = 1.005 kJ/kg. *K*, c_v = 0.718 kJ/kg. *K*.

[**Ans.** 14006.97 kJ, 4000 kJ, 10008.06 kJ]

2.32 Calculate the difference in non-flow work $\int p\,dv$ and flow work $-\int v.dp$, if 1kg of air having specific volume of 0.4 m^3 at 1.35 bar is compressed to a pressure of 10 bar according to the law $pV^{1.3} = C$. [**Ans.** 31.7 kJ]

2.33 A gas undergoes the following reversible non-flow processes during its initial pressure of 15 bar and volume 1 m^3/kg to final pressure of 6 bar and volume 2 m^3/kg.

(*a*) Constant volume followed by constant pressure.

(*b*) Constant pressure followed by constant volume.

(*c*) Polytropic process, $pV^{1.3} = c$.

Calculate the work done by the gas during each process.

[**Ans.** 600 kJ/kg, 1500 kJ/kg, 1000 kJ/kg]

2.34 A certain gas occupies 0.1 m^3 at 1bar pressure. It is compressed adiabatically to a pressure of 7bar. Find the new volume, and change in internal energy. The density of gas at 0°C and 1bar is 1.5 kg/m^3. Take c_v = 0.728 kJ/kg. *K*. [**Ans.** 0.0249 m^3, 18.847 kJ]

2.35 3 kg of oxygen undergoes a non-flow process during which its pressure remains constant at 7 bar from an initial volume of 1200 litres to a state where the temperature is 520 K. Find (*a*) changes in internal energy, (*b*) change in enthalpy, (*c*) heat transferred, and (*d*) work done.

For oxygen, c_v = 0.661 kJ/kg. *K*, c_p = 0.921 kJ/kg. K.

[**Ans.** –1104.38 kJ, 1538.78 kJ, – 1538.78 kJ, – 434.4 kJ]

2.36 A constant volume chamber of 0.3 m^3 capacity contains 2 kg of a gas at 5°C. Heat is transferred to the gas until the temperature is 100°C. Find the work done, heat transferred, and the change in internal energy, enthalpy and the entropy. Take c_p = 1.968 kJ/kg.K and C_v = 1.507 kJ/kg.K

[**Ans.** 0, 286.33 kJ, 286.33 kJ, 373.92 kJ, 0.886 kJ/K]

2.37 0.5 kg of air is compressed reversibly and adiabatically from 80 kPa, 60°C to 0.4 MPa, and is then expanded at constant pressure to the original volume. Calculate the heat transfer and work transfer for the whole path. Take R = 0.287 kJ/kg. *K*, c_p = 1.005 kJ/kg. *K* and c_v = 0.718 kJ/kg. *K*.

[**Ans.** 571.64 kJ, 93.5 kJ]

2.38 A mass of air is initially at 260^{0}C and 700 kPa, and occupies 0.028 m^3. The air is expanded at constant pressure to 0.084 m^3. A Polytropic process, $pV^{1.5} = C$, is then carried out followed by a constant temperature process which completes the cycle. All the processes are reversible. Find (*a*) the heat received and rejected in the cycle, and (*b*) efficiency of the cycle. Take c_p = 1.005 kJ/kg. *K* and c_v = 0.718 kJ/kg. *K*

[**Ans.** 137.13 kJ, 84.11 kJ, 38.66 %]

2.39 A mass of 8 kg gas expands within a flexible container according to the law, $pV^{1.5}$ = const. The initial pressure is 1 MPa and initial volume is 1 m^3. The final pressure is 5 kPa. If specific internal energy of the gas decreases by 40 kJ/kg, find the heat transfer.

[**Ans.** 2612.5 kJ]

2.40 During a reversible isobaric non-flow process with p = 1.5 bar the properties of the system change from 0.25 m^3/kg, 10°C to 0.45 m^3/kg, 240°C. The specific heat of the fluid is given by c_p = [1.6 + 80 / (t + 40)] kJ/kg.°C. Calculate (*a*) heat added/kg, (*b*) work done /kg, (*c*) change in internal energy/kg, and (*d*)change in enthalpy/kg.

[**Ans.** 505.82 kJ/kg, 30 kJ/kg, 475.82 kJ/kg, 505.82 kJ/kg]

2.41 A closed system passes from state 1 to state 2 while 200 kJ of heat is added and 280 kJ of work is done. As the system is returned to state 1, 160 kJ of work is done on it. Calculate the heat transfer during process 2 to 1. [**Ans.** – 80 kJ]

2.42 A system contains 0.15 $m^3$of a gas at a pressure of 3.8 bar and 423 K. It is expanded adiabatically till the pressure falls to 1 bar. The gas is then heated at a constant pressure till its enthalpy increases by 70 kJ. Calculate the total work done. Assume c_p = 1 kJ/kg. *K* and c_v = 0.714 kJ/kg. *K*.

[**Ans.** 65.27 kJ]

2.43 0.03 m^3 of air at a pressure of 1.2 bar and 298 K is compressed to a volume of 0.0045 m^3 according to the law $pV^{1.3}$ = const. Find the final temperature and work done during compression. [**Ans.** 1050.48 K, – 9.20 kJ]

2.44 1 kg of a perfect gas is compressed from 0.9 bar and 300 K according to the flow $pV^{1.3}$ = const. until the pressure is 5.4 bar. Calculate the heat flow to or from the cylinder well. Take R = 0.277 kJ/kg. *K* and c_p = 1.75 kJ/kg. *K*. [**Ans.** 84.43 kJ]

3 PROPERTIES OF STEAM

3.1 INTRODUCTION

Steam may be defined as water vapours suspended in air. It may be wet, saturated or superheated. Vapours may be considered as a fluid in a gaseous state. It may contain liquid particles in suspension and its temperature is not far from its boiling point. When it is cooled or expanded, it condenses into its liquid state. Vapours in general do not obey gas laws of Boyle and Charles.

Steam has vast industrial applications as a working medium in steam engines and steam turbines. It is also extensively used in processing equipments.

3.2 PURE SUBSTANCE

A pure substance is a substance that has a chemically homogeneous and fixed chemical composition irrespective of the phase or phases in which it exists. The pure substance is sometimes also called one component system. A mixture of ice, water and steam is an example of a pure substance since the chemical composition of the three phases is the same. Stream is regarded as the most important pure substance in engineering practice.

A substance cannot be treated as pure if it undergoes a chemical reaction. A mixture of oxygen and carbon monoxide is a pure substance as long as it remains fixed in composition. If some of the CO combines with some of the O_2 to form CO_2, the system does not involve a pure substance during the process because its chemical composition is changing. The state of a pure substance at rest is specified completely by the values of two independent intensive properties.

3.3 PHASE

The phase of a substance is any homogeneous part of a system that is physically distinct. Each phase in a system is separated from other phases by interfaces called phase boundaries.

In general, a pure substance exists in the following three phases :

(*i*) solid, (*ii*) liquid, and (*iii*) vapour or gaseous phase.

Under certain conditions, all three phases may co-exist. For examples, water has three phases : ice, water, and steam. In a system consisting of an ice cube in liquid water, there are two phases, solid and liquid, and the phase boundary is the surface of the ice cube.

A gas mixture is always a single phase, because the molecules of the constituent gases intermingle completely throughout the entire volume of the gas and make the mixture homogeneous. There could be more than one solid or liquid phase of a substance as in the case of sulphur which exists in three solid phases of monoclinic, rhombic and amorphous forms at room temperature.

A system composed of liquid water with a layer of oil on top of it would clearly involve two phases with a planar phase boundary separating the two.

3.4 CONSTANT PRESSURE FORMATION OF STEAM

Consider a unit mass of ice at –20°C and 1 atm contained in a cylinder-piston system as shown in Fig. 3.1. Let the ice be heated slowly so that its temperature is increased at constant pressure. Let

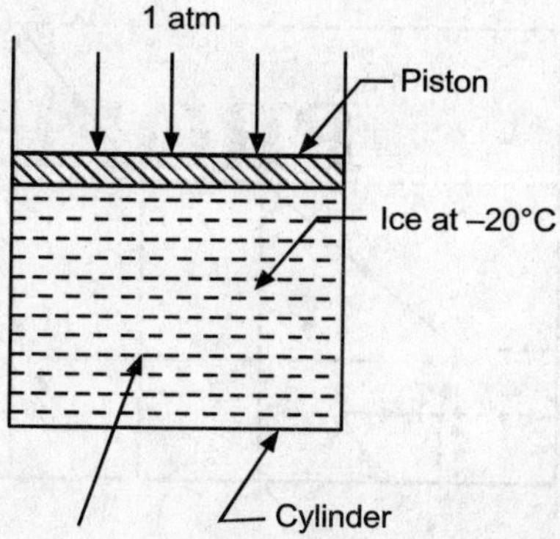

Fig. 3.1

the state changes be plotted on temperature — heat added condiments, as shown in Fig. 3.2. The distinct regimes of heating are :

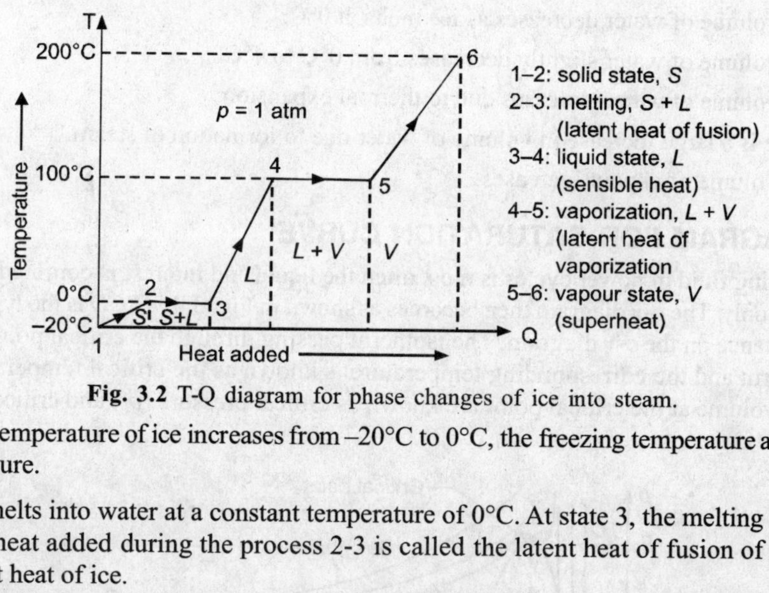

Fig. 3.2 T-Q diagram for phase changes of ice into steam.

1 – 2 : The temperature of ice increases from –20°C to 0°C, the freezing temperature at atmospheric pressure.

2 – 3 : Ice melts into water at a constant temperature of 0°C. At state 3, the melting process ends. The heat added during the process 2-3 is called the latent heat of fusion of ice or simply latent heat of ice.

3 – 4 : The temperature of water increases on gradual addition of more heat from 0°C to 100°C, the boiling point of water.

4 – 5 : On further heating beyond point 4, water starts boiling and ends at state 5. Water will gradually turn into steam and there is no rise in temperature. This steam contains some water particles and is called wet steam. With further addition of heat all the water particles disappear and steam becomes dry saturated at point 5. The heat addition during process 4-5 is called the latent heat of vaporisation of steam or simply latent heat of steam.

5 – 6 : This process is called superheating and the steam obtained is called the superheated steam. Let the temperature of steam corresponding to point 6 be $t°_{sup}$C. Then $(t_{sup} - t_{sat})$°C is called the degree of superheat, where $t°_{sat}$ C is the saturation temperature at the given pressure.

3.5 T-V DIAGRAM FOR FORMATION OF STEAM

The temperature-volume diagram for phase changes of ice into steam is shown in Fig. 3.3.

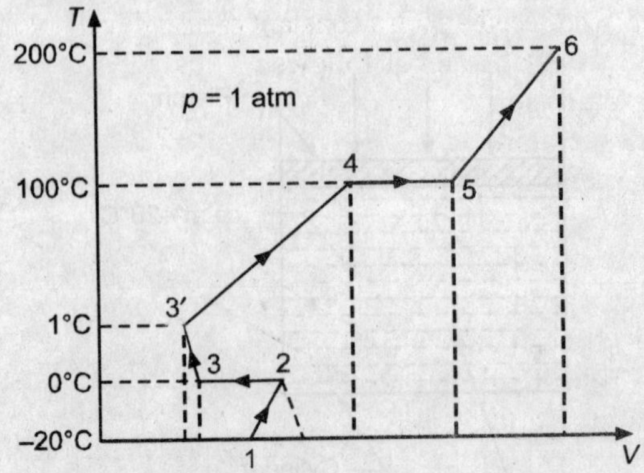

Fig. 3.3 T-V diagram for phase changes of ice into steam.

The volume changes during various process are :

1 – 2 : Volume increases from V_1 to V_2 as ice temperature increases to 0°C.

2 – 3 : The volume of water decreases as ice melts at 0°C.

3 – 3′ : The volume of water slightly decreases from 0°C to 4°C.

3′ – 4 : The volume of water increases due to thermal expansion.

4 – 5 : There is a large increase in volume of water due to formation of steam.

5 – 6 : The volume of vapour increases.

3.6 p-v DIAGRAM FOR SATURATION CURVE

The working fluid in power cycles is most often the liquid and interest is confined to the liquid-vapour region only. The *p-v* diagram then becomes as shown in Fig. 3.4. *ABCD* is the typical isotherm of a pure substance on the *p-v* diagram. The isotherm passing through the critical point is called the critical isotherm and the corresponding temperature is known as the critical temperature (t_c). The pressure and volume at the critical point are known as critical pressure (p_c) and critical volume (v_c) respectively.

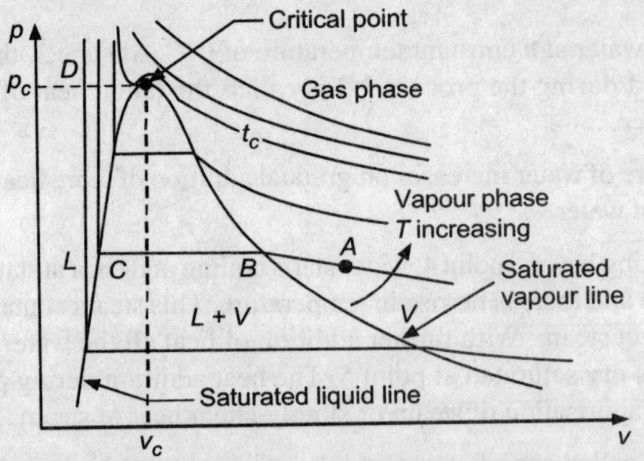

Fig. 3.4 Saturation curve on p-v diagram.

3.7 *T-v* DIAGRAM FOR WATER

Consider 1 kg of water contained in a vessel and heated at constant pressure. Let us plot the constant pressure lines that represents the state through which water passes as it is heated from initial state. The line *ABCD* in Fig. 3.5 represents such a constant pressure line at 1 atm. The various portions of this line represent different states of water as follows :

AB : Heating of liquid to saturation temperature. If the temperature of liquid is less than saturation temperature at a given pressure, the liquid is said to be sub-cooled or compressed.

BC : Charge of phase at constant temperature from liquid to vapour state. The points *B* and *C* are called the saturated liquid and saturated vapour states respectively.

CD : Superheating of vapour and both temperature and volume increase. When the temperature of vapour is higher than saturation temperature at a given pressure, it is called superheated vapour.

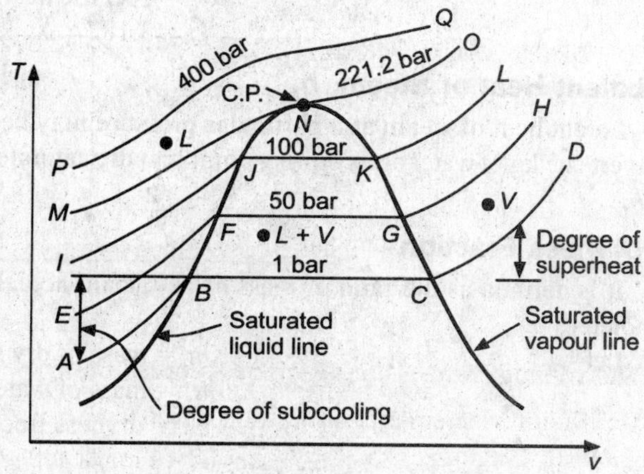

Fig. 3.5 T-v diagram for water.

If the heating process is repeated at different pressures such as 50 bar, 100 bar, curves such as *EFGH* and *IJKL* respectively are obtained. The locus of points like B, F, J etc. is called saturated liquid line and that of *C*, *G*, *K* etc. as saturated vapour line. If the pressure is increased beyond a certain value called the critical pressure (221.2 bar), the constant temperature portion disappears. This corresponds to the critical state.

3.8 T-s DIAGRAM FOR A PURE SUBSTANCE

The *T-s* diagram for a pure substance like water is shown in Fig. 3.6. The *T-s* diagram for liquid-vapour transformation only is shown in Fig. 3.7. At a particular pressure; s_f is the specific entropy of saturated water, and s_g that of saturated vapour. The entropy change of the system during the phase change from liquid to vapour is $s_{fg} = s_g - s_f$.

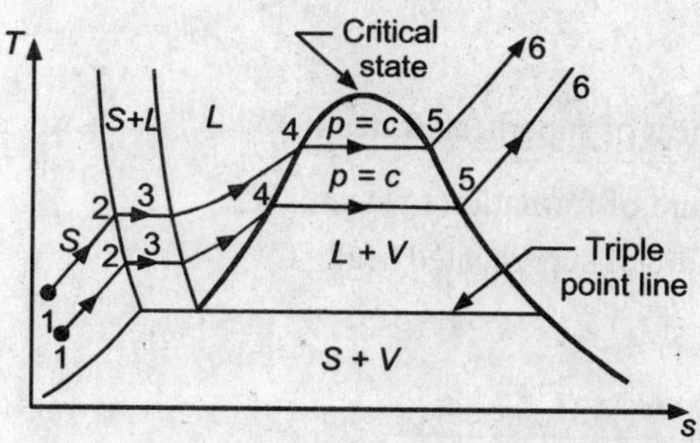

Fig. 3.6 Phase equilibrium diagram on TVs coordinates.

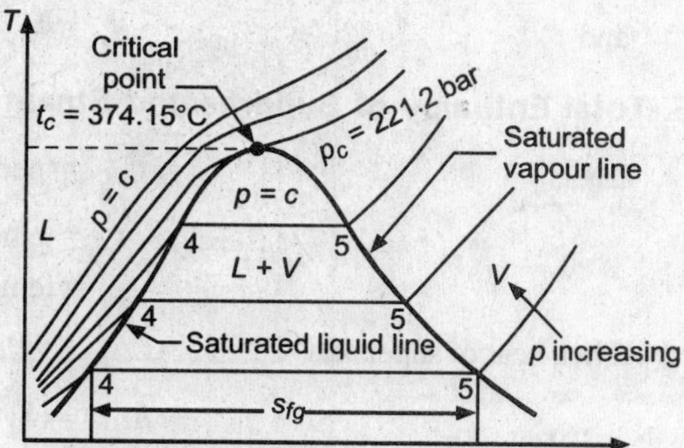

Fig. 3.7 Saturation (or vapour) dome for water.

3.9 PROPERTIES OF STEAM

1. Enthalpy or Total Heat of Water, h_f

It is defined as the quantity of heat required to raise the temperature of one kg of water from 0°C to its boiling point or saturation temperature corresponding to the pressure applied. It is also called enthalpy of saturated water or liquid heat and is represented by h_f.

$$h_f = \text{specific heat of water, } c_{pw} \times \text{rise in temperature}$$
$$= 3.187 \times \Delta t, \text{ kJ/kg}$$

2. Latent Heat of Steam, h_{fg}

Latent heat of steam at a particular pressure may be defined as the quantity of he required to convert one kg of water at its boiling point into dry saturated steam at the same pressure. It is denoted by h_{fg}.

3. Dryness Fraction

It is defined as the ratio of mass of dry steam actually present to the mass of wet steam. It is denoted by x.

Let
m_s = mass of dry steam, kg
m_w = mass of water vapour in steam, kg
x = dryness fraction of the sample

Then
$$x = \frac{m_s}{m_s + m_w} \quad \text{...(3.1)}$$

For dry steam, $m_w = 0$ and $x = 1$

The quality of steam is the dryness fraction expressed as a percentage.

Quality of steam = $100\,x$

Wetness fraction $= 1 - x = m_w / (m_s + m_w)$

The wetness fraction expressed in percentage is called priming.

Priming = $100\,(1 - x)$

3. Enthalpy of wet steam

It may be defined as the quantity of heat required to convert one kg of water at 0°C, at constant pressure, into wet scheme.

$$h = h_f + x h_{fg} \quad \text{...(3.2)}$$

When steam is dry saturated, then $x = 1$

and
$$h_g = h_f + h_{fg} \quad \text{...(3.3)}$$

5. Total Enthalpy of Superheated Steam

Let
c_{ps} = specific heat of superheated steam
t_s = temperature of formation of steam, °C
t_{sup} = temperature of superheated heat, °C

Then heat of superheat $= c_{ps}\,(t_{sup} - t_s)$

$$h_{sup} = (h_f + h_{fg}) + c_{ps}\,(t_{sup} - t_s) \quad \text{...(3.4a)}$$
$$= h_g + c_{ps}\,(t_{sup} - t_s) \quad \text{...(3.4b)}$$

3.10 STEAM TABLES

In actual practice, it is quite cumbersome to calculate the relation between various quantities like pressure, temperature, volume, enthalpy, latent heat and entropy, etc. of steam. Such quantities have been determined experimentally and recorded in the form of tables, known as steam tables. The values of various quantities given in these tables are for 1 kg of steam which is dry saturated. Values for unknown pressure or temperature can be calculated by linear interpolation between two consecutive values. Such tables are given at the end of the book.

3.11 SPECIFIC VOLUME OF WET STEAM

The specific volume of steam is the volume of 1 kg of dry steam, and is represented by v_g in m^3/kg.

$$\text{Density of steam} = \frac{1}{v_g} \text{ kg/m}^3 \quad \text{...(3.5)}$$

Let v_f = volume of 1 kg of water, m^3/kg

Then volume of 1 kg of wet steam $= x\,v_g + (1-x)\,v_f$ ...(3.6*a*)

Now $v_f << v_g$

$\therefore$ Volume of 1 kg of wet steam $= x\,vg$ ...(3.6*b*)

$$\text{Density of wet steam} = \frac{1}{x\,v_g} \text{ kg/m}^3 \quad \text{...(3.7)}$$

3.12 SPECIFIC VOLUME OF SUPERHEATED STEAM

The volume of superheated steam at a given pressure can be found by assuming the superheated steam behaves as a perfect gas obeying gas laws.

Let p = pressure under which the steam is superheated.

v_g = specific volume of dry saturated steam

T_s = temperature of formation of steam in K corresponding to p.

T_{sup} = temperature to which steam is super heated in K.

v_{sup} = superheated volume of steam at pressure p.

Applying Charle's law at the constant pressure p, we have

$$\frac{v_g}{T_s} = \frac{V_{sup}}{T_{sup}}$$

or

$$v_{sup} = \frac{T_{sup}}{T_s} \times v_g \quad \text{...(3.8)}$$

3.13 INTERNAL ENERGY OF STEAM

The internal energy of steam may be defined as the actual heat energy stored in steam above the freezing point of water. It is the difference between the total heat of steam and the external work of evaporation.

Consider 1 kg of steam at a pressure p bar.

Let u = internal energy of steam

h_g = total heat of dry saturated steam

v_g = specific volume of dry saturated steam

External work of evaporation per kg of steam $= pv_g$

Then $h_g = pv_g + u$

or $u = h_g - pv_g$ for dry saturated steam ...(3.9)

For wet steam, $h = h_f + x\,h_{fg}$

and its specific volume $= xv_g$

$\therefore$ $u = h - p\,xv_g$

$= h_f + xh_{fg} - pxv_g$ for wet steam ...(3.10)

For superheated steam,

$$h_{sup} = h_f + h_{fg} + c_{ps}\,(t_{sup} - t_s) = h_g + c_pc\,(t_{sup} - t_s)$$

$\therefore$ $u = h_{sup} - pv_{sup}$

$= h_{sup} - pv_{sup}$ for superheated steam ...(3.11)

3.14 INTERNAL ENERGY OF EVAPORATION

The difference between the internal energy of vapour (u_g) and internal energy of fluid (u_f) is known as internal energy of evaporation or vaporization.

Now $u_g = h_g - pu_g$

$u_f = h_f - pu_f$

$\therefore$ $u_g - u_f = u_{fg} = (h_g - h_f) - p\,(v_g - v_f)$

$= h_{fg} - pu_{fg}$ for dry steam ...(3.12*a*)

$= x(h_{fg} - pu_{fg}) = x(h_{fg} - pu_g)$ for wet steam ...(3.12*b*)

Also $u = u_f + x\,u_{fg} = u_f + x\,(u_g - u_f)$ for wet steam

$= u_g$ for dry saturated steam.

3.15 INTERNAL (LATENT HEAT) HEAT OF EVAPORATION

It may be defined as the sum of external work and internal energy of evaporation.

According to first law of thermodynamics,

$$\delta q = \delta w + du$$

$h_{fg} = pv_{fg} + u_{fg}$ for dry steam

or $h_g - h_f = p(v_g - v_f) + (u_g - u_f)$

Internal heat of evaporation $= h_{fg} - pv_{fg}$ for dry steam ...(3.13)

We observe that internal energy of evaporation and internal heat of evaporation are the same.

For wet steam of dryness fraction x, we have

$$h_{wet} = h_f + x\,h_{fg}$$

$$u_{wet} = u_f + x\,u_{fg}$$

By first law of thermodynamics, we have

$$\delta q = \delta w + du$$

Enthalpy of evaporation = External work + Internal energy of evaporation.

$$h_{wet} - h_f = p\,x\,(v_g - x\,v_{fg}) + \text{internal energy of evaporation}$$

Also $$u_{wet} - u_f = x\,u_{fg}$$

and $$h_{wet} - h_f = x\,h_{fg}$$

Work done $= pxV_{fg} = px\,(v_g - v_f) \simeq pxv_g$ (neglecting v_f being small)

Internal energy of evaporation

$$= \text{Enthalpy of evaporation — external work done}$$

$$= (h_{wet} - h_f) - px\,(v_g - xv_f)$$

$$= x\,h_{fg} - px\,(v_g - xv_f)$$

$$= x\,h_{fg} - px\,v_g \text{ (neglecting } v_f \text{ being small)}$$

$$= x\,(h_{fg} - p\,v_g) \qquad \ldots(3.14)$$

3.16 ENTROPY OF WATER, s_f

Consider 1 kg of water to be heated from a temperature T_s to T at a constant pressure. Then

$$\text{change of entropy} = c_{pw}\ \ln \frac{T_s}{T}$$

where c_{pw} = specific heat of water

The entropy of water is always reckoned above the freezing point of water.

Therefore T_s = saturation temperature of formation of steam

T = absolute temperature of freezing water = 273 K

Entropy of saturated water,

$$s_f = c_{pw}\ \ln \frac{T_s}{273} \qquad \ldots(3.15)$$

3.17 ENTROPY OF EVAPORATION, s_{fg}

$$\text{Change in entropy} = \int \frac{\delta Q}{T}$$

Let Q = heat absorbed during evaporation

Then $$\text{change in entropy} = \frac{Q}{T}$$

During evaporation of water into dry steam, $Q = h_{fg}$

$$\therefore \text{ Change of entropy during evaporation} = \frac{h_{fg}}{T_s}$$

For wet steam change of entropy during evaporation,

$$s_{fg} = \frac{xh_{fg}}{T_s} \qquad \ldots(3.16)$$

3.18 ENTROPY OF DRY STEAM, s_g

Consider 1 kg of wet steam at an absolute pressure p bar.

Let x = dryness fraction of steam.

T_s = temperature of formation of steam, K

h_{fg} = latent heat of steam

Entropy of 1 kg of wet steam reckoned above the freezing point of water,

s_g = Entropy of water + change of entropy during evaporation

$$= s_f + \frac{xh_{fg}}{T_s} = c_{pw} \ln \frac{T_s}{273} + \frac{xh_{fg}}{T_s} \text{ for wet steam} \quad ...(3.17)$$

For dry steam, $x = 1$

$$s_g = s_f + \frac{h_{fg}}{T_s} = c_{pw} \ln \frac{T_s}{273} + \frac{h_{fg}}{T_s} \text{ for dry steam} \quad ...(3.18)$$

3.19 ENTROPY OF SUPERHEATED STEAM

Consider 1 kg of dry saturated steam at T_s K temperature of formation, to be superheated to a temperature T_{sup} K.

Let C_{ps} = specific heat of steam at constant pressure during superheating

Change of entropy per kg of steam during superheating

$$= c_{ps} \ln \frac{T_{sup}}{T_s}$$

Total entropy of superheated steam

s_{sup} = Entropy of dry saturated steam + Change of entropy during superheating

$$= \left(s_f + \frac{h_{fg}}{T_s} \right) + c_{ps} \ln \frac{T_{sup}}{T_s} = s_g + c_{ps} \ln \frac{T_{sup}}{T_s} \quad ...(3.19)$$

3.20 TEMPERATURE-ENTROPY DIAGRAM FOR WATER AND STEAM

It is the plot of the saturation temperature of water and steam corresponding to various absolute pressures v's entropies at those saturation temperatures, as shown in Fig. 3.8. The curve on the left marked water line shows the relationship between entropy and temperature before steam is formed.

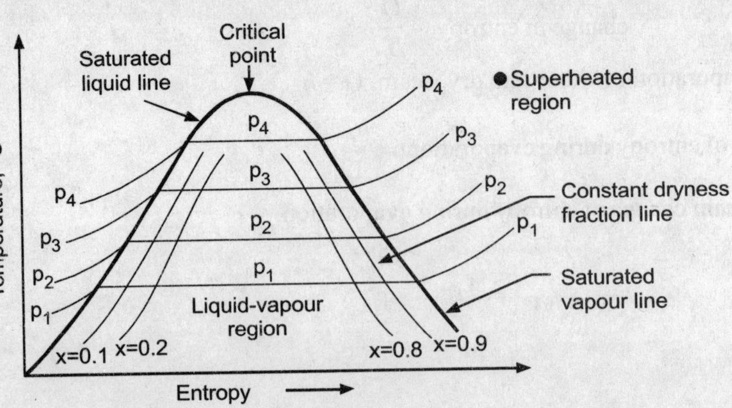

Fig. 3.8 Temperature-entropy diagram for water and steam.

The dry steam line shows the relationship between entropy and temperature when all the water, has been converted into steam. The area in between two regions is called the wet region. The area to the right of the dry steam lines is called the region of superheat. The lines of constant dryness, constant volume and constant total heat are also drawn on the diagram. The temperature at which the distinction between the entropy of water and that of dry steam disappears is called the critical temperature (374.15°C). The pressure corresponding to this temperature is called the critical pressure (221.2 bar)

3.21 MOLLIER DIAGRAM OF STEAM

In this diagram, the vertical ordinate represents the enthalpy and the horizontal ordinate represents the entropy, as shown in Fig. 3.9. The regions above and below the saturation line represent the super

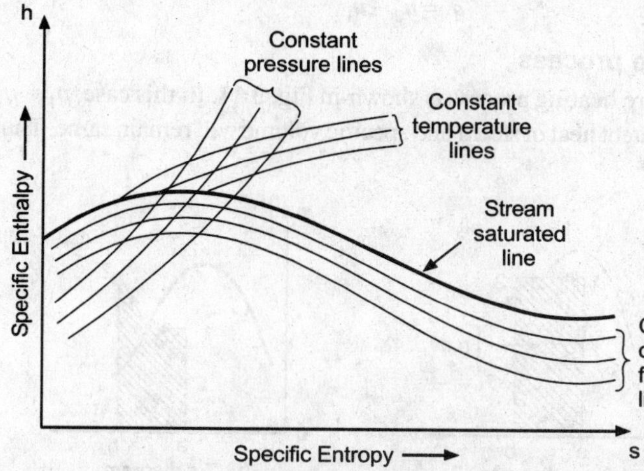

Fig. 3.9 Mollier diagram for steam

heated and wet conditions of steam respectively. The lines of constant dryness fraction are shown in the wet steam, region while the lines of constant temperature in the region of superheat. The lines of constant pressure are straight in the wet region but curved in the superheat region. The drop in enthalpy during an adiabatic process can be directly read from this diagram.

3.22 VARIOUS PROCESSES FOR STEAM

1. Constant Volume process

The constant volume heating process is shown on the p–v and T–s diagrams in Fig. 3.10. Consider 1 kg of wet steam at a certain pressure p_1 and dryness fraction x_1 which is heated to state 2 in the wet region and then to state 3 in the superheated region.

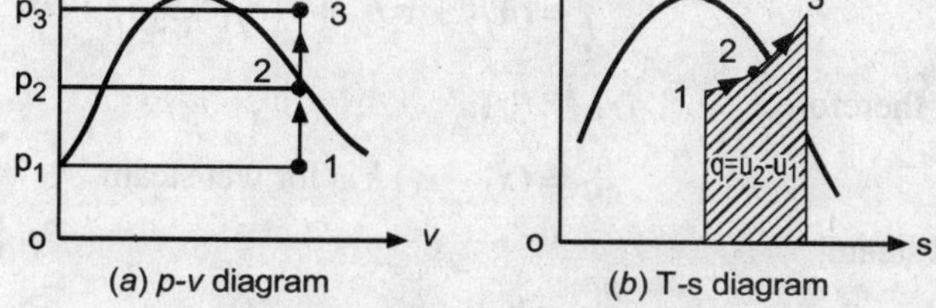

(*a*) p-v diagram (*b*) T-s diagram

Fig. 3.10 Constant volume heating

Initial volume of steam $= x_1 v_{g1}$

Final volume of steam in wet region $= x_2 v_{g2}$

Since the volume remains constant,

$$x_1 v_{g1} = x_2 v_{g2}$$

or

$$x_2 = \frac{v_{g1}}{v_{g2}} . x_1 \quad \text{...(3.20)}$$

Initial internal energy of steam, $u_1 = h_1 - p_1 x_1 v_{g1}$

Final internal energy of steam in the wet region, $u_2 = h_2 - p_2 x_2 v_{g2}$

Since $dv = 0$, therefore $dW = 0$

$$\therefore \quad q = u_2 - u_1 \quad \text{...(3.21)}$$

2. Constant pressure process

The constant pressure heating process is shown in Fig. 3.11. In this case, $p_1 = p_2 = p$ (say). Since p is constant, therefore, latent heat of steam and specific volume will remain same. Thus $v_{g1} = v_{g2} = v_g$ and $h_{fg1} = h_{fg2} = h_{fg}$

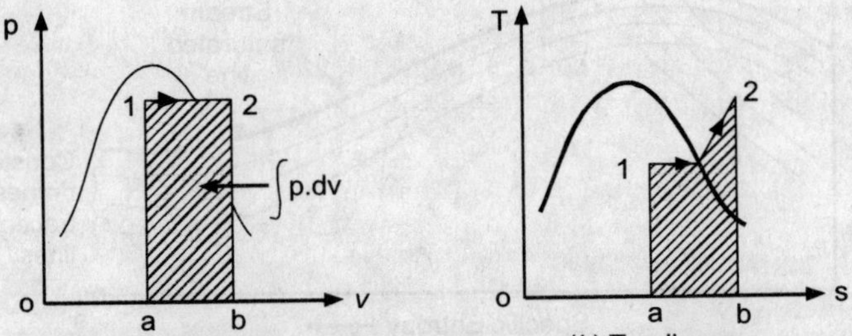

(*a*) p-*v* diagram (*b*) T-s diagram

Fig. 3.11 Constant pressure expansion

Work done per kg of steam, $w = p.dv = p.v_g (x_2 - x_1)$

Initial internal energy of steam, $u_1 = h_1 - p x_1 v_g$

Internal energy after heating, $u_2 = h_2 - p x_2 v_g$

Change in internal energy, $u_2 - u_1 = (h_2 - h_1) - p v_g (x_2 - x_1)$ ...(3.22)

From first law of thermodynamics, $q = w + (u_2 - u_1)$

$$= p v_g (x_2 - x_1) + ([(h_2 - h_1) - p v_g (x_2 - x_1)]$$

$$= h_2 - h_1$$

Now $h_1 = h_{f1} + x_1 h_{fg}$ and $h_2 = h_{f2} + x_2 h_{fg}$

$$q = (h_{f2} + x_2 h_{fg}) - (h_{f1} + x_1 h_{fg})$$

Since p - const., therefore, $h_{f2} = h_{f1}$

$$\therefore \quad q = (x_2 - x_1)\, h_{fg} \text{ for wet steam} \quad \text{...(3.23}a\text{)}$$

For superheated steam,

$$q = h_{\text{sup}} - h_s \quad \text{...(3.23}b\text{)}$$

where h_s = enthalpy of dry saturated steam.

3. Isothermal Process

The isothermal heating process shown in Fig. 3.12. The isothermal process for wet steam will be same as the constant pressure process. But in the superheated region the steam will behave as a gas and shall follow Boyle's law, pv = const. Therefore, the process will be hyperbolic. In the wet region,

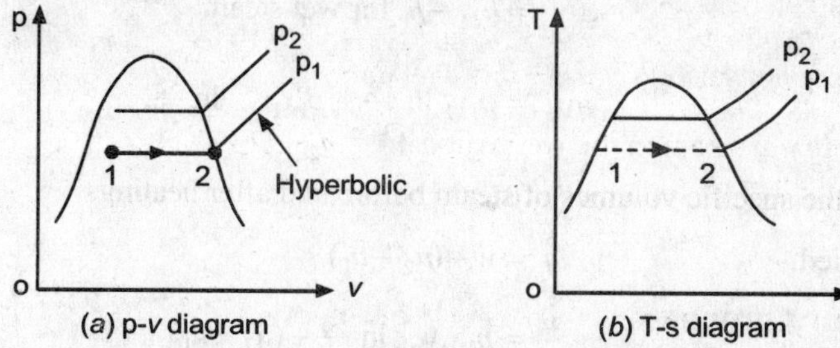

Fig. 3.12 Isothermal process

$$q = h_2 - h_1 \text{ and } w = p_1 v_1 (1 - x_1)$$

4. Hyperbolic Process

During a hyperbolic process, pv = const. The process may be regarded isothermal but only in the superheated region, as shown in Fig. 3.13.

In the wet region,

$$\text{Initial volume of steam} = x_1 v_{g1}$$

$$\text{Final volume of steam} = x_2 v_{g2}$$

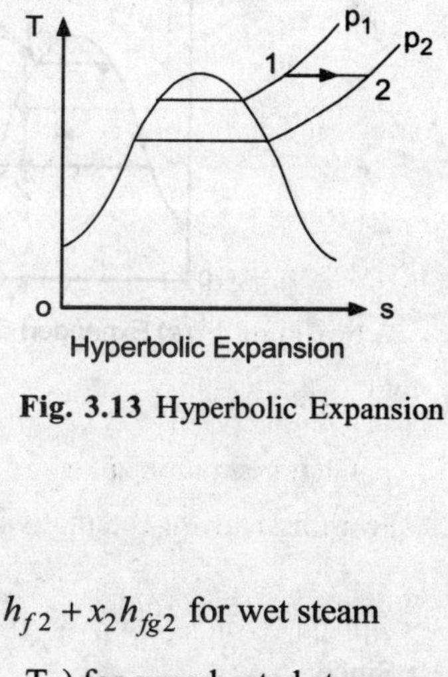

Fig. 3.13 Hyperbolic Expansion

As pv = const., therefore

$$p_1 x_1 v_{g1} = p_2 x_2 v_{g2}$$

or

$$x_2 = \frac{p_1}{p_2} \times \frac{v_{g1}}{v_{g2}} \times x_1$$

For superheated steam,

$$\frac{p_2 v_{g2}}{T_2} = \frac{p_2 v_2}{T_{sup}}$$

or

$$T_{sup} = \frac{v_2}{v_{g2}} \times T_2$$

Now $h_1 = h_{f1} + x_1 h_{fg1}$ and $h_2 = h_{f2} + x_2 h_{fg2}$ for wet steam

and $h_{s2} = h_{f2} + h_{fg2} + c_p (T_{sup} - T_2)$ for superheated steam.

Also $u_1 = h_1 - p_1 x_1 v_{g1}$

$u_2 = h_2 - p_2 x_2 v_{g2}$ for wet steam

$= h_{s2} - p_2 v_2$ for superheated steam.

But $p_1 x_1 v_{g1} = p_2 x_2 v_{g2} = p_2 v_2$

Thus the external work of evaporation is constant during a hyperbolic process.

Change is internal energy,

$$u_2 - u_1 = (h_2 - p_2 x_2 v_{g2}) - (h_1 - p_1 x_1 v g_1) \quad \ldots(3.24a)$$

$$= h_2 - h_1 \text{ for wet steam} \quad \ldots(3.24b)$$

$$= h_{s2} - h_1 \text{ for wet steam} \quad \ldots(3.24c)$$

Work done,

$$W = p_1 v_1 \ln \frac{v_2}{v_1} + p_1 x_1 v_{g1} \ln \frac{v_2}{v_1} \quad \ldots(3.25)$$

where v_1 and v_2 are the specific volumes of steam before and after heating.

Total heat supplied,

$$q = w + (u_2 - u_1)$$

$$= p_1 x_1 v_{g1} \ln \frac{v_2}{v_1} + (u_2 - u_1)$$

$$= (h_2 - h_1) + p_1 x_1 v_{g1} \ln \frac{v_2}{v_1} \quad \ldots(3.26)$$

5. Reversible Adiabatic or Isentropic process

The isentropic process of expansion is shown in Fig. 3.14 on T — s diagram. Let the steam be expanded from state 1 to state 2. Then for wet steam,

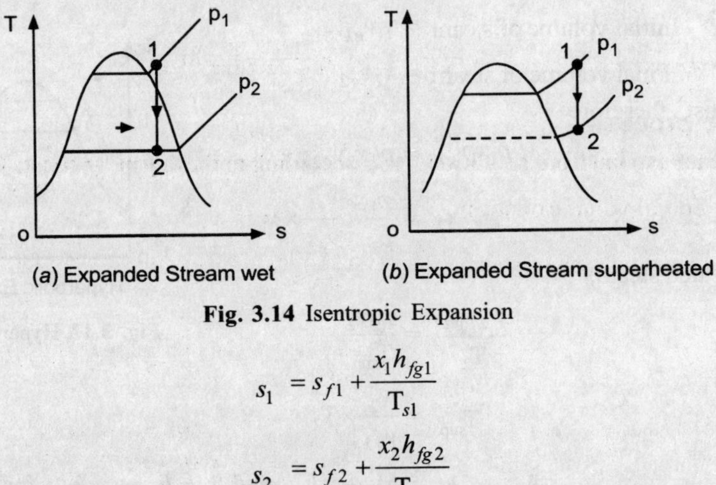

(*a*) Expanded Stream wet (*b*) Expanded Stream superheated

Fig. 3.14 Isentropic Expansion

$$s_1 = s_{f1} + \frac{x_1 h_{fg1}}{T_{s1}}$$

$$s_2 = s_{f2} + \frac{x_2 h_{fg2}}{T_{s2}}$$

Since $s_1 = s_2$, therefore

$$s_{f1} + \frac{x_1 h_{fg1}}{T_{s1}} = s_{f2} + \frac{x_2 h_{fg2}}{T_{s2}}$$

$$x_2 = \left[(s_{f1} - s_{f2}) + \frac{x_1 h_{fg1}}{T_{s1}} \right] \times \frac{T_{s2}}{h_{fg2}} \quad \ldots(3.27)$$

For initially superheated steam at temperature T_{sup},

$$s_1 = s_{f1} + \frac{h_{fg1}}{T_{s1}} + c_{ps} \ln \frac{T_{sup}}{T_{s1}}$$

and $$s_2 = s_{f2} + x_2 \frac{h_{fg2}}{T_{s2}}, \text{ if the steam is wet} \quad \text{...(3.28a)}$$

$\therefore$ $$= s_{f1} + \frac{h_{fg1}}{T_{s1}} + c_{ps} \ln \frac{T_{sup}}{T_{s1}} = s_{f2} + x_2 \frac{h_{fg2}}{T_{s2}} \quad \text{...(3.28b)}$$

$$x_2 = \left[s_{f1} - s_{f2} + \frac{h_{fg1}}{T_{s1}} + c_{ps} \ln \frac{T_{sup}}{T_{s1}} \right] \times \frac{T_{s2}}{h_{fg2}} \quad \text{...(3.29)}$$

Also $$u_1 = h_1 - p_1 x_1 v_{g1}, \text{ for wet steam}$$

and $$u_2 = h_2 - p_2 x_2 v_{g2}$$

Now $$q = w + (u_2 - u_1)$$

But $$q = 0$$

$\therefore$ $$w = u_1 - u_2$$

If the steam is superheated at state 2, then

$$s_2 = s_{f2} + \frac{h_{fg2}}{T_{s2}} + c_{ps} \ln \frac{T_{sup}}{T_{s2}}$$

so that

$$s_{f1} + \frac{h_{fg1}}{T_{s1}} + c_{ps} \ln \frac{T_{sup}}{T_{s1}} = s_{f2} + \frac{h_{fg2}}{T_{s2}} + c_{ps} \ln \frac{T_{sup}}{T_{s2}}$$

6. Polytropic process

Let the steam expand from condition 1 to 2 according to the law pv^n = const. Then for wet steam,

Initial volume of steam, $v_1 = x_1 v_{g1}$

Final volume of steam, $v_2 = x_2 v_{g2}$

Now $$p_1 v_1^n = p_2 v_2^n$$

or $$p_1 (x_1 v_{g1})^n = p_2 (x_2 v_{g2})^n$$

$$x_2 = \left(\frac{p_1}{p_2} \right)^{\frac{1}{n}} \times \frac{v_{g1}}{v_{g2}} \times x_1$$

Also $$h_1 = h_{f1} + x_1 h_{fg1}$$

$$h_2 = h_{f2} + x_2 h_{fg2}$$

and $$u_1 = h_1 - p_1 x_1 v_{g1}$$

$$u_2 = h_2 - p_2 x_2 v_{g2}$$

$$w = \frac{p_1 x_1 v_{g1} - p_2 x_2 v_{g2}}{n-1} \quad \text{...(3.30)}$$

$$q = w + (u_2 - u_1) = \frac{p_1 x_1 v_{g1} - p_2 x_2 v_{g2}}{n-1} + (u_2 - u_1) \quad \text{...(3.31)}$$

n = 1.13 for wet steam and 1.3 for super heated steam.

For superheated steam on heating from wet steam, we have

$$p_1(x_1 v_{g1})^n = p_2 v_2^n$$

where
$$v_2 = v_{s2} \times \frac{T_{sup}}{T_{s2}}$$

7. Throttling Process

The throttling process occurs when steam is expanded through a small aperture as in the case of throat of a nozzle. During this process, no work is done and the enthalpy remains constant. The throttling process is shown on T – s and h – s diagrams in Fig. 3.15.

$$h_1 = h_{f1} + x_1 h_{fg1}$$

$$h_2 = h_{f2} + x_2 h_{fg2} \text{ for wet steam}$$

$$= h_{f2} + h_{fg2} + c_{ps} (T_{sup} - T_s) \text{ for superheated steam}$$

Now
$$h_1 = h_2$$

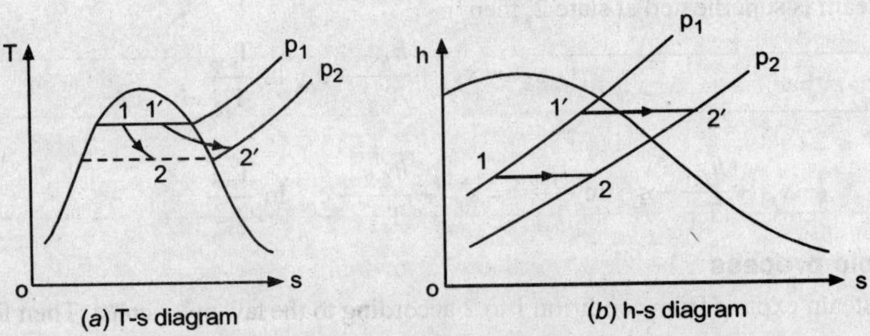

Fig. 3.15 Throttling process

∴
$$h_{f1} + x_1 h_{fg1} = h_{f2} + x_2 h_{fg2}$$

or
$$= h_{f2} + h_{fg2} + c_{ps} (T_{sup} - T_s) \quad \text{...(3.32)}$$

Example 3.1 *Calculate the quantity of the required to generate 1 kg of steam at a pressure of 8 bar from water at 30°C (a) when dryness fraction is 0.9, (b) when steam is just dry, and (c) when it is superheated at constant pressure to 300°C Assume c_{ps}= 2.5 kJ/kg. K.*

Solution.

From steam tables at 8 bar :

$$h_f = 721.1 \text{ kJ/kg}, h_{fg} = 2048.0 \text{ kJ/kg}$$

At 30°C, $h_f = 125.79$ kJ/kg

(*a*) $h = h_f + x\, h_{fg} = 721.1 + 0.9 \times 2048.0 = 2564.3$ kJ/kg

Net heat required $= 2564.3 - 125.79 = 2438.51$ kJ/kg

(*b*) $h = h_f + h_{fg} = 2769.1$ kJ/kg

Net heat required $= 2769.1 - 125.79 = 2643.31$ kJ/kg

(*c*) $t_s = 170.4°C, t_{sup} = 300°C$

$h = h_g + c_{ps}\,(t_{sup} - t_s) = 2769.1 + 2.5\,(300 - 170.4) = 3093.1$ kJ/kg

Net heat required $= 3093.1 - 125.79 = 2967.31$ kJ/kg

Example 3.2 *A pressure cooker contains 2 kg of steam at 5 bar and 0.9 dry. Find the quantity of heat that must be rejected so that the quality of steam becomes 0.5 dry.*

Solution.

The pressure cooker is a constant volume vessel. Thus W = 0

$$\therefore \quad Q = \Delta U = U_2 - U_1$$

$$U_1 = m_s[h_{f1} + x_1 h_{fg1} - p_1 x_1 v g_1 \times 10^2]$$

At $p_1 = 5$ bar, $h_{f1} = 640.2$ kJ/kg, $h_{fg1} = 2108.5$ kJ/kg,

$$v_{g1} = 0.375 \text{ m}^3/\text{kg}$$

$$U_1 = 2\,[640.2 + 0.9 \times 2108.5 - 5 \times 0.9 \times 0.375 \times 10^2]$$

$$= 4738.2 \text{ kJ}$$

For constant volume of steam, we have

$$x_1 v_{g1} = x_2 v_{g2}$$

$$v_{g2} = \frac{0.9 \times 0.375}{0.5} = 0.675 \text{ m}^2/\text{kg}$$

p, bar	v_g, m³/kg	h_f, kJ/kg	h_{fg}, kJ/kg
2.6	0.693	540.9	2177.8
2.7	0.669	546.3	2174.2

For $u_g = 0.675$ m³/kg,

$$p_2 = 2.6 + \left(\frac{0.1}{0.693 - 0.669}\right) \times (0.693 - 0.675) = 2.675 \text{ bar}$$

$$p_{f2} = 540.9 + \left(\frac{5.4}{0.1}\right) \times (2.675 - 2.6) = 541.305 \text{ kJ/kg}$$

$$h_{fg2} = 2177.8 - \left(\frac{3.6}{0.1}\right) \times (2.675 - 2.6) = 2175.1 \text{ kJ/kg}$$

$$U_2 = m\,[h_{f2} + x_2 h_{fg2} - p_2 x_2 v_{g2} \times 10^2]$$

$$= 2\,[541.305 + 0.5 \times 2175.1 - 2.675 \times 0.5 \times 0.675 \times 10^2] = 3077.15 \text{ kJ}$$

$$Q = U_2 - U_1 = 3077.15 - 4738.2 = -\,1661.05 \text{ kJ}$$

Example 3.3 *5400 m³/h of wet steam 0.92 dry and at 10 bar is to be supplied by a boiler for a processing plant. Calculate (a) the mass of steam supplied per hour, and (b) the quantity of coal of calorific value 15 MJ to be burnt in boiler if over all efficiency of boiler is 30%.*

Solution.

(*a*) At 10 bar, $v_g = 0.194$ m³/kg

Mass flow rate of wet steam, $m = \frac{V}{x v_g} = \frac{5100}{0.92 \times 0.194} = 30255.5$ kg/h

(*b*) At $p = 10$ bar, $h_f = 762.8$ kJ/kg, $h_{fg} = 2015.3$ kJ/kg

$$h = h_f + x h_{fg} = 762.8 + 0.92 \times 2015.3 = 2616.876 \text{ kJ/kg}$$

$$H = mh = 30255.5 \times 2616.876 = 79174892 \text{ kJ/h}$$

Heat supplied to boiler, $H_b = \dfrac{H}{\eta_b} = \dfrac{79174892}{0.3} = 263916306$ kJ/h

Quantity of coal burnt $= \dfrac{H_b}{C.V.} = \dfrac{263916306}{15\times10^6} = 17.59$ tonnes/h

Example 3.4 *Find the internal energy of 1 kg of super heated steam at a pressure of 10 bar and 300°C. If this steam is expanded to 1.5 bar and 0.9 dryness, find the change in internal energy.*

Solution.

From steam tables, at $p = 10$ bar, $t_{sup} = 300°C$. For superheated steam, we have

$$h_{sup} = 3051.2 \text{ kJ/kg}$$

For dry saturated steam, at $p = 10$ bar, we have

$$v_g = 0.1944 \text{ m}^3/\text{kg}, t_s = 179.9°C$$

$\therefore$ $T_s = 273 + 179.9 = 452.9$ K

$$T_{sup} = 273 + 300 = 573 \text{ K}$$

$$v_{sup} = \frac{T_{sup}}{T_s} \times v_g = \frac{573}{452.9} \times 0.1944 = 0.24595 \text{ m}^3/\text{kg}$$

$$u_1 = h_{sup} - p_1 v_{sup} = 3051.2 - 10\times10^2 \times 0.24595 = 2805.249 \text{ kJ/kg}$$

At $p_2 = 1.5$ bar, $v_{g2} = 1.159$ m^3/kg, $h_{f2} = 467.1$ kJ/kg,

$$h_{fg2} = 2226.5 \text{ kJ/kg}$$

$$h_2 = h_{f2} + x_2 h_{fg2} = 467.1 + 0.9\times2226.5 = 2470.95 \text{ kJ/kg}$$

$$u_2 = h_2 - p_2 x_2 v_{g2} = 2470.95 - 1.5\times10^2 \times 0.9\times1.159 = 2314.485 \text{ kJ/kg}$$

$$\Delta u = u_2 - u_1 = 2314.485 - 2805.249 = -490.764 \text{ kJ/kg}$$

Example 3.5 *A rigid vessel is initially divided into two parts A and B by a thin partition. Part A contains 1 kg of steam at 4 bar, dry and saturated, and part B 2 kg of steam at 8 bar with dryness fraction of 0.90. The partition is removed and the pressure in the vessel after sometimes is found to be 6 bar. Find (a) the volume of the vessel and the state of steam, and (b) the amount of heat transfer from steam to the vessel and the surroundings.*

Solution.

(*a*) At $p_A = 4$ bar, $v_{gA} = 0.463$ m^3/kg from steam stables.

$$V_A = m_A v_{gA} = 1 \times 0.463 = 0.463 \text{ m}^3$$

$$p_B = 8 \text{ bar}, v_{gB} = 0.240 \text{ m}^3/\text{kg}$$

$$V_B = m_B x_B v_{gB} = 2 \times 0.9 \times 0.240 = 0.432 \text{ m}^3$$

Volume of vessel, $V = V_A + V_B = 0.463 + 0.432 = 0.895$ m^3

At $p = 6$ bar, $v_g = 0.316$ m^3/kg, $m = 1 + 2 = 3$ kg

Dryness fraction at 6 bar, $x = \dfrac{V}{m \times v_g} = \dfrac{0.895}{3\times0.316} = 0.944$

(*b*) Now $U_A = h_{gA} - p_A v_{gA} \times 10^2 = 2553.3$ kJ/kg

$$U_A = m_A u_A = 1\times2553.3 = 2553.3 \text{ kJ}$$

$$u_B = m_B(h_{fB} + x_B h_{fgB} - p_B x_B v_{gB} \times 10^2)$$
$$= 2\ (721.1 + 0.9 \times 2048.0 - 8 \times 0.9 \times 0.240 \times 10^2)$$
$$= 4783.0 \text{ kJ}$$

At 6 bar,
$$U = m[h_f + x\, h_{fg} - pxv_g \times 10^2)$$
$$= 3\,[670.6 + 0.944 \times 2086.3 - 6 \times 0.944 \times 0.316 \times 10^2)$$
$$= 7383.25 \text{ kJ}$$
$$Q = U - (U_A + U_B)$$
$$= 7383.25 - (2553.3 + 4783.0)$$
$$= 46.95 \text{ kJ}$$

Example 3.6 *Calculate enthalpy and entropy for 5 kg of steam at 8 bar under the following conditions : (a) dry and saturated, (b) wet steam having wetness of 40%, and (c) superheated steam at 250°C.*

Solution.

From steam tables, at $p = 8$ bar, we have

$$h_f = 721.1 \text{ kJ/kg},\ h_{fg} = 2048.0 \text{ kJ/kg}$$
$$s_f = 2.046 \text{ kJ/kg.K},\ s_{fg} = 4.617 \text{ kJ/kg.K}$$

At $t_{sup} = 250°$, $p = 8$ bar, we have

$$h = 2950.1 \text{ kJ/kg},\ s = 7.038 \text{ kJ/kg.K}$$

(*a*)
$$H = m\,(h_f + h_{fg}) = 5\ (721.1 + 2048.0) = 13845.5 \text{ kJ}$$
$$S = m\,(s_f + s_{fg}) = 5\ (2.046 + 4.617) = 33.315 \text{ kJ/K}$$

(*b*)
$$x = 1 - 0.4 = 0.6$$
$$H = m\,(h_f + x\,h_{fg}) = 5\ (721.1 + 0.6 \times 2048.0) = 9749.5 \text{ kJ}$$
$$S = m\,(s_f + x\,s_{fg}) = 5\ (2.046 + 0.6 \times 4.617) = 24.081 \text{ kJ/K}$$

(*c*)
$$H = mh = 5 \times 2950.1 = 14750.5 \text{ kJ}$$
$$S = ms = 5 \times 7.038 = 35.19 \text{ kJ/K}$$

Example 3.7 *1 kg of steam 0.85 dry at a pressure of 1 bar is compressed in a cylinder to a pressure of 2 bar according to $pv^{1.25}$ = const. Determine the final condition of steam and the heat transfer through the cylinder walls.*

Solution.

From steam tables at $p_1 = 1$ bar, $v_{g1} = 1.694$ m^3/kg

At $p_2 = 2$ bar, $v_{g2} = 0.886$ m^3/kg

$$x_1 = 0.85$$

Now
$$p_1(x_1 v_{g1})^n = p_2(x_2 v_{g2})^n$$

$$x_2 = \left(\frac{p_1}{p_2}\right)^{\frac{1}{n}} \times \frac{v_{g1}}{v_{g2}} \times x_1 = \left(\frac{1}{2}\right)^{\frac{1}{1.25}} \times \frac{1.694}{0.886} \times 0.85 = 0.934$$

At
$$p_1 = 1 \text{ bar},\ h_{f1} = 417.5 \text{ kJ/kg},\ h_{fg1} = 2258.0 \text{ kJ/kg}$$
$$p_2 = 2 \text{ bar},\ h_{f2} = 504.7 \text{ kJ/kg},\ h_{fg2} = 2201.9 \text{ kJ/kg}$$

$$h_1 = h_{f1} + x_1 h_{fg1} = 417.5 + 0.85 \times 2258.0 = 2336.8 \text{ kJ/kg}$$

$$h_2 = h_{f2} + x_2 h_{fg2} = 504.7 + 0.934 \times 2201.9 = 2561.3 \text{ kJ/kg}$$

$$u_1 = h_1 - p_1\, x_1\, v_{g1} = 2336.8 - 1 \times 10^2 \times 0.85 \times 1.694 = 2192.8 \text{ kJ/kg}$$

$$u_2 = h_2 - p_2 x_2 v_{g2} = 2561.3 - 2 \times 10^2 \times 0.934 \times 0.886 = 2399.5 \text{ kJ/kg}$$

$$\Delta u = u_2 - u_1 = 2399.5 - 2192.8 = 206.7 \text{ kJ}$$

Work done,
$$w = \frac{p_1 x_1 \upsilon_{g1} - p_2 x_2 \upsilon_{g2}}{n-1}$$

$$= \frac{(1 \times 0.85 \times 1.694 - 2 \times 0.934 \times 0.866) \times 10^2}{1.25 - 1} = -71.12 \text{ kJ}$$

Heat transfer through the cylinder walls,

$$q = w + (u_2 - u_1) = -71.12 + 206.7 = 135.58 \text{ kJ/kg}$$

Example 3.8 *Steam at a pressure of 10 bar and 0.9 dry expands to 1 bar hyperbolically. Find (a) work done and heat absorbed, (b) internal energy and change in enthalpy.*

Take specific heat of steam at constant pressure = 2 kJ/kg. K.

Solution. From steam tables :

p bar	h_f kJ/kg	h_{fg} kJ/kg	v_g m^3/kg
10	762.8	2015.3	0.194
1	417.5	2258.0	1.694

$$v_1 = x_1 v_{g1} = 0.9 \times 0.194 = 0.1746 \text{ m}^3/\text{kg}$$

Now
$$p_1 v_1 = p_2 v_2$$

$$\frac{v_2}{v_1} = \frac{p_1}{p_2} = 10$$

Expansion ratio,
$$r = \frac{v_2}{v_1} = 10$$

$$\text{Work done, } w = p_1 v_1 \ln r = 10 \times 10^2 \times 0.1746 \ln 10 = 402 \text{ kJ/kg}$$

$$v_2 = 10\, v_1 = 10 \times 0.1746 = 1.746 \text{ m}^3/\text{kg}$$

Now $v_2 > v_{g2}$. Therefore the steam is superheated at $p_2 = 1$ bar.

$$T_{sup2} = \frac{v_2 T_{s2}}{v_{g2}} = \frac{1.746 \times (273 + 99.6)}{1.694} = 384 \text{ K}$$

$$h_1 = h_{f1} + x_1 h_{fg1} = 762.8 + 0.9 \times 2015.3 = 2576.6 \text{ kJ/kg}$$

$$h_2 = h_{f2} + h_{fg2} + c_{ps}\,(T_{sup2} - T_{s2})$$

$$= 415.5 + 2258.0 + 2\,(384 - 372.6) = 2698.3 \text{ kJ/kg}$$

$$\Delta h = h_2 - h_1 = 2698.3 - 2576.6 = 121.7 \text{ kJ/kg}$$

$$\Delta u = \Delta h \text{ for hyperbolic process}$$

$$q = w + \Delta u = 402 + 121.7 = 523.7 \text{ kJ/kg}$$

Example 3.9 *Steam initially dry saturated, expands isentropically from pressure of 15 bar to 0.15 bar. Find the index of isentropic expansion.*

Solution.

From steam tables, we have

p bar	v_g m^3/kg	s_f kJ/kg.K	s_{fg} kJ/kg.K
15	0.132	2.315	4.130
0.15	10.022	0.7549	7.2536

For isentropic expansion, $s_1 = s_2$

$$s_1 = s_{f1} + s_{fg1} = 2.315 + 4.130 = 6.445 \text{ kJ/kg.K}$$

$$s_2 = s_{f2} + x_2 s_{fg2} = 0.7549 + x_2 \times 7.2536 = 6.445$$

$$x_2 = 0.7844$$

$$v_1 = v_{g1} = 0.132 \text{ m}^3\text{/kg}$$

$$v_2 = x_2 v_{g2} = 0.7844 \times 10.022 = 7.8618 \text{ m}^3\text{/kg}$$

$$p_1 \upsilon_1^n = p_2 \upsilon_2^n$$

$$\left(\frac{\upsilon_1}{\upsilon_2}\right)^n = \frac{p_2}{p_1}$$

$$\left(\frac{0.132}{7.8618}\right)^n = \frac{0.15}{15}$$

$$n \ln\left(\frac{0.132}{7.8618}\right) = \ln\left(\frac{1}{100}\right)$$

$$n = 1.127$$

Example 3.10 *Determine the quality of steam for the following cases :*
(a) p = 10 bar, v = 0.180 m^3/kg *(b) p = 10 bar, t = 200°C*
(c) p = 25 bar, h = 2750 kJ/kg.

Solution. (*a*) Specific volume of dry and saturated steam at 10 bar from steam tables

$$v_g = 0.194 \text{ m}^3\text{/kg and } v_f = 0.001127 \text{ m}^3\text{/kg}$$

As $v < v_g$, therefore, the steam is wet.

Now $v = (1 - x)\, v_f + x v_g$

$$0.180 = (1 - x) \times 0.001127 + x \times 0.194$$

$$x = 0.92741$$

(*b*) Saturation temperature corresponding 10 bar, t_s = 179. 9°C

As $t > t_s$, therefore, steam is superheated

Degree of superheat $= t_{sup} - t_s = 200 - 179.9 = 20.\ 1°C$

(*c*) At 25 bar, $h_f = 962.1$ kJ/kg, $h_g = 2803.1$ kJ/kg

As $h < h_g$, therefore, steam is wet.

$$h = h_f + x h_g$$

$$2750 = 962.\ 1 + x \times 2803.1$$

$$x = 0.63783$$

Example 3.11 *3 kg of steam at a pressure of 30 bar exists in the following conditions :*
(a) wet steam with a dryness fraction of 0. 85.
(b) superheated steam with temperature of 300°C. Assume c_p =2302 kJ/kg.K
Calculate the (i) enthalpy, (ii) volume, (iii) entropy, and (iv) internal energy.

Solution. From steam tables at 30 bar, we have

$$v_f = 0.001217 \text{ m}^3/\text{kg}, v_g = 0.0667 \text{ m}^3/\text{kg}$$
$$h_f = 1008.4 \text{ kJ/kg.K}$$
$$h_{fg} = 1795.7 \text{ kJ/kg}, h_g = 2804.2 \text{ kJ/kg}$$
$$s_f = 2.646 \text{ kJ/kg.K}, s_{fg} = 3.541 \text{ kJ/kg.K}, s_g = 6.187 \text{ kJ/kg.K}$$

(*i*) $$h = h_f + xh_g = 1008.4 + 0.85 \times 2804.2 = 3391.97 \text{ kJ/kg}$$
$$H = mh = 3 \times 3391.97 = 10175.91 \text{ kJ}$$

(*ii*) $$v = (1-x)\, v_f + xv_g = (1-0.85)\times 0.001217 + 0.85\times 0.0667$$
$$- 0.05688 \text{ m}^3/\text{kg}$$
$$V = mv = 3 \times 0.05688 = 0.17063 \text{ m}^3$$

(*iii*) $$s = s_f + xs_g = 2.646 + 0.85 \times 6.187 = 7.90495 \text{ kJ/kg.K}$$
$$S = ms = 3 \times 7.90495 = 23.71485 \text{ kJ/K}$$

(*iv*) $$u = h - pv = 3391.97 - 30 \times 10^2 \times 0.05688 = 3221.33 \text{ kJ/kg}$$
$$U = mu = 3 \times 3221.33 = 9663.99 \text{ kJ}$$

(*b*) At 30 bar, 300°C, for superheated steam, we have

$$v_g = 0.0667 \text{ m}^3/\text{kg}, h_g = 2804.2 \text{ kJ/kg}, s_g = 6.187 \text{ kJ/kg.K}$$
$$t_s = 233.8°\text{C}$$

(*i*) $$h_{\text{sup}} = h_g + C_p\,(t_{\text{sup}} - t_s) = 2804.2 + 2\text{-}8595\,(300 - 233.8) = 2993.5 \text{ kJ/kg}$$
$$H = mh_{\text{sup}} = 3 \times 2993.5 = 8980.5 \text{ kJ}$$

(*ii*) $$v_{\text{sup}}\ v_g \frac{T_{\text{sup}}}{T_s} = 0.0667 \times \frac{573}{506.8} = 0.07541 \text{ m}^3/\text{kg}$$

$$\text{V}_{\text{sup}} = mv_{\text{sup}} = 3\times 0.07541 = 0.22624 \text{ m}^3$$

(*iii*) $$S_{\text{sup}} = s_g + C_p \ln \frac{T_{\text{sup}}}{T_s} = 6.187 + 2.8595 \ln \frac{573}{506.8} = 6.638 \text{ kJ/kg.K}$$

$$S_{\text{sup}} = ms_{\text{sup}} = 3\times 6.538 = 19.614 \text{ kJ/K}$$

(*iv*) $$u_{\text{sup}} = h_{\text{sup}} - pv_{\text{sup}} = 2993.5 - 30\times 10^2 \times 0.07541 = 2767.27 \text{ kJ/kg}$$

$$U_{\text{sup}} = mu_{\text{sup}} = 3\times 2767.27 = 8301.81 \text{ kJ}$$

Example 3.12 *5 kg of steam at 200 kPa occupies a volume of (a) 2.50 m^3, and (b) 5.00 m^3. Determine the temperature in each case and the quality of steam.*

Solution. Specific volume of steam,

$$v = \frac{V}{m} = \frac{2.5}{5} = 0.5 \text{ m}^3/\text{kg}$$

At 200 kPa from steam tables, we have

$$v_f = 0.0010605 \text{ m}^3/\text{kg}, v_g = 0.8858 \text{ m}^3/\text{kg}$$

We see that $v_f < v < v_g$, therefore, the steam is wet, *i.e.*, $x < 1$ and its temperature must be the saturation temperature for 200 kPa, *i.e.*,

$$t_s = 120.24°C.$$

Now

$$v = (1 - x)\, v_f + x v_g$$

$$0.5 = (1 - x) \times 0.0010605 + x \times 0.8858$$

$$x = \frac{0.50 - 0.0010605}{0.8858 - 0.0010605} = 0.5639$$

(*b*) Specific volume, $$v = \frac{V}{m} = \frac{5}{5} = 1.00 \text{ m}^3/\text{kg}$$

$\therefore v > v_g$, so the steam is superheated.

For $p = 200$ kPa, from steam tables, we have,

t, °C	*v*, m³/kg
150	0.9597
175	1.0204

For $v = 1$ m³/kg, by linear interpolation, we get

$$t°C = 175 - \left(\frac{175 - 150}{10204 - 0.9597}\right) \times 0.0204 = 175 - 8.402 = 166.598°C$$

Example 3.13 *A vessel of volume 0.05 m³ contains a mixture of saturated water and saturated steam at a temperature of 300°C. The mass of liquid present is 10 kg. Find the pressure, the mass, the specific volume, the enthalpy, the entropy, and the internal energy.*

Solution. At 300°C, $p_{sat} = 85.812$ bar

$$v_f = 0.00140 \text{ m}^3/\text{kg},\ v_g = 0.02967 \text{ m}^3/\text{kg}$$

$$h_f = 1344.0 \text{ kJ/kg},\ h_{fg} = 1404.9 \text{ kJ/kg},\ h_g = 2749.0 \text{ kJ/kg}$$

$$s_f = 3.2534 \text{ kJ/kg.K},\ s_{fg} = 2.4511 \text{ kJ/kg.K},\ s_g = 5.7045 \text{ kJ/kg.K}$$

$$V_f = m_f v_f = 10 \times 0.00140 = 0.0140 \text{ m}^3$$

$$V_g = 0.05 - 0.0140 = 0.036 \text{ m}^3$$

Mass of vapour, $$m_g = \frac{V_g}{v_g} = \frac{0.036}{0.02167} = 1.66128 \text{ kg}$$

Total mass of mixture, $$m = m_f + m_g = 10 + 1.66128 = 11.66128 \text{ kg}$$

Dryness fraction, $$x = \frac{m_g}{m} = \frac{1.66128}{11.66128} = 0.14246$$

$$v = v_f + x v_{fg} = 0.00140 + 0.14246 \times (0.2167 - 0.00140)$$

$$= 0.0042877 \text{ m}^3/\text{kg}$$

$$h = h_f + x h_{fg} = 1344.0 + 0.14246 \times 1404.9 = 1544.\ 142 \text{ kJ/kg}$$

$$s = s_f + x s_{fg} = 3.2534 + 0.14246 \times 2.4511 = 3.60258 \text{ kJ/kg.K}$$

$$u = u_f + x u_{fg} = h - pv = 1544.142 - 85.812 \times 10^2 \times 0.0042877$$

$$= 1507.348 \text{ kJ/kg}$$

Example 3.14 *The pressure of steam in a condenser is 0.15 bar and the dryness fraction is 0.87. Calculate the amount of heat that must be removed from steam in order to condense (a) 1 kg of steam, and (b) 1 m^3 of steam.*

Solution. From steam tables, at 0.15 bar, h_{fg} = 2375.1 kg/kg

(*a*) Heat to be removed per kg of steam,

$$h = xh_{fg} = 0.87 \times 2373.1 = 2064.6 \text{ kJ/kg}$$

(*b*)

$$v_g = 10.022 \text{ m}^3\text{/kg at } p = 0.15 \text{ bar}$$

$$v = xv_g = 0.87 \times 10.022 = 8.71914 \text{ m}^3\text{/kg}$$

Heat to be removed per m^3 of steam

$$= \frac{h}{v} = \frac{2064.6}{8.71914} = 236.79 \text{ kJ}$$

Example 3.15 *5000 kg/h of steam enters a surface condenser at 0.08 bar. 250 kilolitres/h of cooling water is passed through the tubes and its temperature rises from 25°C to 35°C. The condensed steam leaves the condenser with a temperature of 30°C. Calculate the condition of steam entering the condenser.*

Solution. Volume flow rate of cooling water $= 250 \times 10^3 \times 10^{-3} = 250 \text{ m}^3\text{/h}$

Mass flow rate cooling water $= 250 \times 10^3$ kg/h

Mass flow of cooling water per kg of steam,

$$m = \frac{250 \times 10^3}{5000} = 50 \text{ kg}$$

Flow steam tables at $p = 0.08$ bar

$$h_{f1} = 173.90 \text{ kJ/kg}, h_{fg1} = 2403.1 \text{ kJ/kg}$$

$$h_1 = h_{f1} + x_1 h_{fg1} = 173.90 + x_1 \times 2403.1$$

For condensate at 30°C, $h_{f2} = 12.5.79$ kJ/kg

Heat lost by steam = Heat gained by cooling water

$$h_1 - h_{f2} = m\,C_{pw}\,(t_0 - t_i)$$

$$173.90 + 2403.1 \times x_1 - 125.79 = 50 \times 4.187\,(35 - 25)$$

$$x_1 = 0.851$$

Example 3.16 *2 kg of steam at 15 bar, 300°C undergoes a constant pressure process to give wet steam of 65% of quality. Calculate (a) the work done, (b) change in enthalpy, and (c) heat transferred assuming non-flow process.*

Solution. (*a*) Work done, $w_{1-2} = p_1(v_2 - v_1)$

From steam tables, at 15 bar, 300°C

$$v_1 = u_{\text{sup1}} = 0.169 \text{ m}^3\text{/kg}$$

$$v_2 = (1 - x_2)\,v_{f2} + xv_{g2} = xv_{g2} = 0.65 \times 0.132 = 0.0858 \text{ m}^3\text{/kg}$$

$$w_{1-2} = 15 \times 10^2 (0.0858 - 0.169) = -249.6 \text{ kJ}$$

$$W_{1-2} = mw_{1-2} = -2 \times 124.8 = -249.6 \text{ kJ}$$

(*b*) Heat transferred, $q_{1-2} = h_2 - h_1$

$$h_1 = h_{\text{sup1}} = 3037.6 \text{ kJ/kg}$$

$$h_2 = h_{f2} + xh_{fg2} = 844.9 + 0.65 \times 1947.3$$

$$= 2110.645 \text{ kJ/kg}$$

$$q_{1-2} = 2110.645 - 3037.6 = -926.955 \text{ kJ/kg}$$

$$Q_{1-2} = mq_{1-2} = -2 \times 928.955 = -1853.91 \text{ kJ}$$

Example 3.17 *A cylinder piston assembly contains 2.5 kg of water inside it at a temperature of 25°C. The cylinder is well insulated and its area of cross-section is 0.45 m. A pressure equivalent of 6 bar is applied on water by the piston. Heat is supplied to cylinder by an electric heater at the rate of 1.2 kW. Calculate the time elapsed if the piston rises through a distance of 1.6 m. Also calculate the quality of steam.*

Solution. From steam tables at 25°C, $v_{f1} = 0.001003 \text{ m}^3\text{/kg}$

Total volume of water at state 1, (see Fig. 3.16) $V_1 = 2.5 \times 0.001003 = 0.0025075 \text{ m}^3$

Change in volume due to rise of piston on heating
= distance moved × cross-sectional area
$= 1.6 \times 0.45 - 0.72 \text{ m}^3$

Final volume of steam
$= 0.0025075 + 0.72 = 0.7225075 \text{ m}^3$

Specific volume of steam,

$$v = \frac{0.7225075}{2.5} = 0.289003 \text{ m}^3\text{/kg}$$

At 6 bar, $v_g = 0.3157 \text{ m}^3\text{/kg}$

∴ $v = xv_g$

$$x_2 = \frac{0.289003}{0.3157} = 0.9154$$

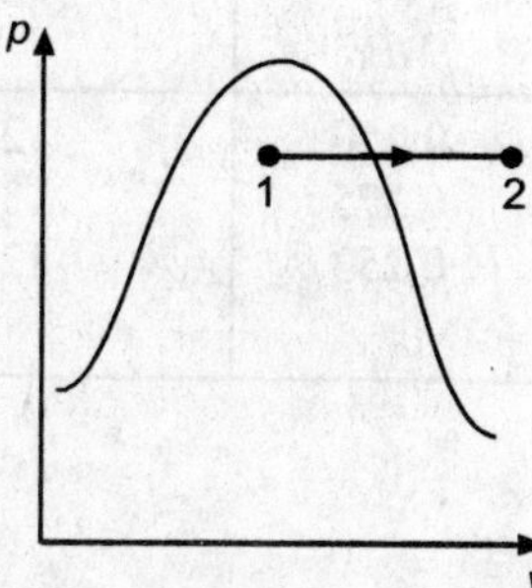

Fig.3.16 *p-v* diagram for constant pressure process.

During heating process the pressure remains constant at 6 bar.

At 25°C, $h_{f1} = 104.87 \text{ kJ/kg}$

At 6 bar, $h_2 = h_{f2} + x_2 h_{fg2} = 670.54 + 0.9154 \times 2086.3 = 2580.34 \text{ kJ/kg}$

Heat added, $q_{1-2} = h_2 - h_{f1} = 2580.34 - 104.87 = 2475.47 \text{ kJ/kg}$

Total heat added by the heater = 2.5 × 2475.47 = 6188.67 kJ

If τ = time elapsed during heating

Then $\tau \times 1.2 \times 3600 = 6188.67$

$$\tau = 1.4326 \text{ hours}$$

Example 3.18 *A house hold pressure cooker contains 2 kg of dry and saturated steam at 4 bar while in operation. Calculate the quantity of heat that must be rejected so that steam becomes wet with 65 present dryness fraction. Also calculate the pressure and temperature of steam that exists in the cooker after the heat rejection.*

Solution. The volume of steam in the pressure cooker before and after the heat rejection remains constant (Fig. 3.17).

Heat transfer during constant volume process,

$$q_{1-2} = (h_2 - p_2v_2) - (h_1 - p_1v_1)$$

From steam tables at 4 bar,

$$h_{f1} = 604.73 \text{ kJ/kg}$$

$$h_{fg1} = 2133.8 \text{ kJ/kg}$$

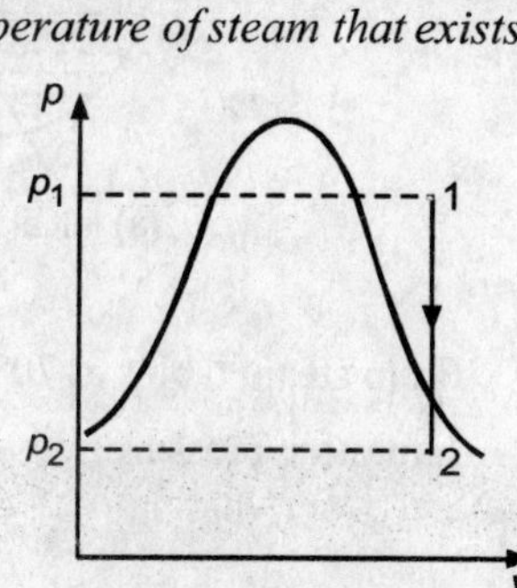

Fig. 3.17 Constant volume process

$$h_{g1} = 2738.5 \text{ kJ/kg}$$
$$v_{f1} = 0.001084 \text{ m}^3\text{/kg}$$
$$v_{g1} = 0.4625 \text{ m}^3\text{/kg}$$

Volume of pressure cooker = Volume of steam

$$V_1 = mv_{g1} = 2 \times 0.4625 = 0.9250 \text{ m}^3$$

Now $\qquad v_1 = v_{g1} = v_2 = v_{f2}(1-x_2) + x_2 v_{g2} \simeq x_2 v_{g2}$

$$0.4625 = 0.65 \times v_{g2}$$
$$v_{g2} = 0.7115 \text{ m}^3\text{/kg}$$

From steam tables, we have

p_2 MPa	t_{s2} °C	h_{f2} kJ/kg	h_{fg2} kJ/kg	v_{g2} m³/kg
0.250	127.43	535.34	2181.5	0.7187
0.275	130.60	548.87	2172.4	0.6573
0.253	127.80	536.92	2180.4	0.7115
				By interpolation

$$h_2 = h_{f2} + x_2 h_{fg2} = 536.92 + 0.65 \times 2180.4 = 1954.18 \text{ kJ/kg}$$
$$v_2 = v_1 = v_{g1} = 0.4625 \text{ m}^3\text{/kg}$$
$$h_1 = h_{g1} = 2738.5 \text{ kJ/kg}$$
$$q_{1-2} = (1954.18 - 0.253 \times 10^3 \times 0.4625) - (2738.5 - 4 \times 10^2 \times 0.4625)$$
$$= 1837.1675 - 2553.5 = -716.3325 \text{ kJ/kg}$$

Total heat rejected, $\qquad Q_{1-2} = mq_{1-2} = -2 \times 716.3325 = -1432.665 \text{ kJ}$

Example 3.19 *A sealed and evacuated drum of volume 0.25 m³ contains saturated water at 70°C. It is heated till the pressure and temperature of steam becomes 14 bar and 250°C respectively. Calculate (a) the mass of water contained in the drum, and (b) the heat required.*

Solution. Refer to Fig. 3.18.

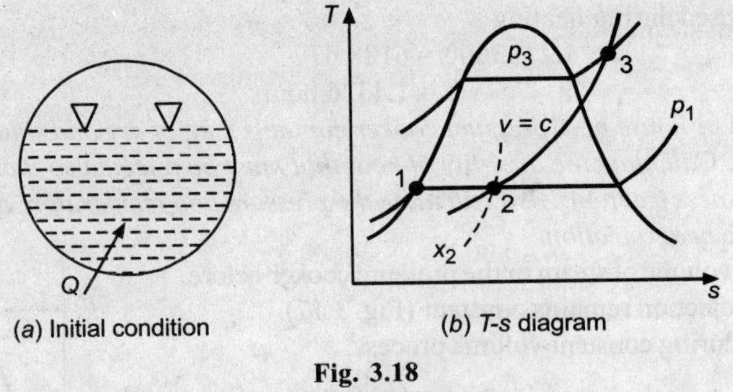

Fig. 3.18

From steam tables at 70°C,

saturation pressure, $\qquad p_1 = 0.31192 \text{ bar}, h_{f1} = 293 \text{ kJ/kg}, h_{fg1} = 2333.8 \text{ kJ/kg},$

$$h_{g1} = 2626.8 \text{ kJ/kg}, v_{g1} = v_{g2} = 5.042 \text{ m}^3\text{/kg}$$

At 14 bar, $250°C, v_3 = v_{sup3} = 0.1635\ m^3/kg$

$$h_{f2} = h_{f1},\ h_{fg1} = h_{fg2},\ h_{sup3} = 2927.2\ kJ/kg$$

(a) Mass of water = mass of steam

$$m = \frac{V}{v_{sup3}} = \frac{0.25}{0.9635} = 1.529\ kg$$

(*b*) Let us assume that first constant pressure heating takes place till whole volume of 0.25 m^3 is filled by wet steam at 0.31192 bar. Thus

$$V = 0.25 = m\, x_2\, v_{g2} = 1.529 \times x_2 \times 5.042$$

$$x_2 = 0.03243$$

Heat added during constant pressure process,

$$q_{1-2} = h_{f2} + x_2 h_{fg2} - h_{f1} = x_2 h_{fg2} = 0.03243 \times 2333.8 = 75.682\ kJ/kg$$

Further addition of heat takes place at constant volume.

Heat added at constant volume,

$$q_{2-3} = (h_3 - p_3 v_3) - (h_2 - p_2 v_2)$$
$$= (2927.2 - 14 \times 10^2 \times 0.1635)$$
$$- (293 + 0.03243 \times 2333.8 - 0.31192 \times 10^2 \times 5.042 \times 0.03243)$$
$$= 2698.3 - 363.585 = 2334.715\ kJ/kg$$

Total heat added per kg during steam generation,

$$q_{1-3} = q_{1-2} + q_{2-3} = 75.682 + 2334.715 = 2410.397\ kJ/kg$$

$$\text{Total heat} = m \times q_{1-3} = 1.529 \times 2410.397 = 3685.497\ kJ$$

Example 3.20 *A spherical vessel of 1 m^3 capacity contains steam at 10 bar and 0.95 dry. Steam is blown off until the pressure drops to 5 bar. The valve is then closed and the steam is allowed to cool until the pressure falls to 2.5 bar. Assuming that the enthalpy of steam in the vessel remains constant during blowing off period, Calculate (a) the mass of steam blown off, (b) the dryness fraction of steam in the vessel after cooling, and (c) the heat lost by steam per kg during cooling.*

Solution. $V = 1\ m^3, p_1 = 10\ bar, x_1 = 0.95, p_2 = 5\ bar, p_3 = 2.5\ bar$

(a) Mass of steam in the vessel (Fig. 3.19),

$$m_1 = \frac{V}{x_1 v_{g1}} = \frac{1}{0.95 \times 0.194} = 7.93\ kg$$

At 10 bar, from steam tables

$$h_1 = h_{f1} + x_1 h_{fg1} = 762.8 + 0.95 \times 2015.3 = 2677.335\ kJ/kg$$

$$h_1 = h_2$$

$$2677.335 = h_{f2} + x_2 h_{fg2} \text{ at 5 bar}$$

$$= 640.2 + x_2 \times 2108.5$$

$$x_2 = 0.966$$

$$m_2 = \frac{1}{0.966 \times 0.375} = 2.76\ kg$$

Mass of steam blown off,

$$m = m_1 - m_2 = 7.93 - 2.76 = 5.17\ kg$$

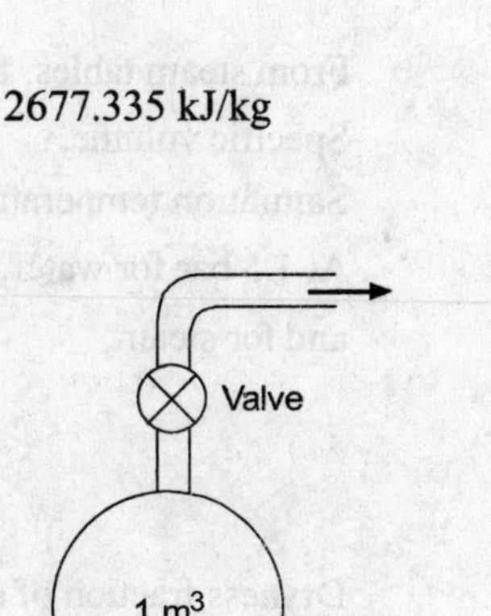

Fig. 3.19

(*b*) $v_2 = v_3$

$$0.966 \times 0.375 = x_3 \times 0.719$$

$$x_3 = 0.5038$$

(*c*) Heat lost during cooling $= m(u_3 - u_2)$

$$u_2 = h_2 - p_2 x_2 v_{g2} = 2677.335 - 5 \times 10^2 \times 0.966 \times 0.375$$

$$= 2496.225 \text{ kJ/kg}$$

$$u_3 = h_3 - p_3 x_3 v_{g3} = (h_{f3} + x_3 h_{fg3}) - p_3 x_3 v_{g3}$$

$$= (535.4 + 0.5038 \times 2181.5) - 2.5 \times 10^2 \times 0.5038 \times 0.719$$

$$= 1543.882 \text{ kJ/kg}$$

Heat lost $= 5.17\,(2496.225 - 1543.882) = 4923.615$ kJ

Example 3.21 *The steam, initially at a pressure of 15 bar and a temperature of 250°C, expand reversibly and polytropically to 1.5 bar. Find the final temperature, work done, and change of entropy, if the index of expansion is 1.25. (Assume 1 bar = 1.013 × 10² kPa). State the assumptions made.*

(IES, 1991)

Solution. $p_1 = 15$ bar,

$t_1 = 250°\text{C}, p_2 = 1.5 \text{ bar}, n = 1.25$

Saturation temperature for steam at 15 bar = 198.3°C

Therefore, the steam at state 1 is superheated (Fig. 3.20).

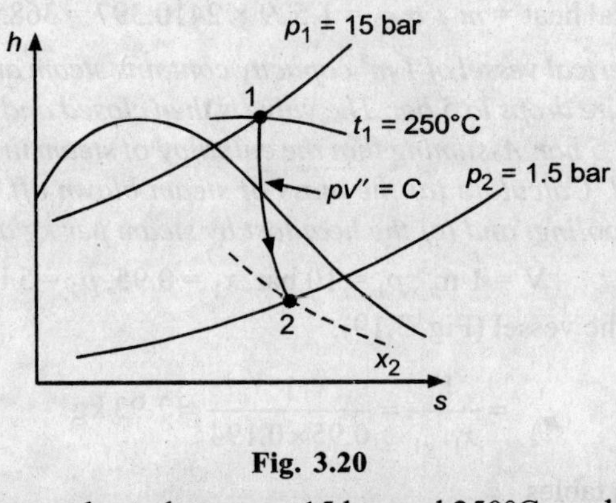

Fig. 3.20

From steam tables, for superheated steam at 15 bar and 250°C, we have

Specific volume, $v_{s1} = 0.152 \text{ m}^3/\text{kg}$

Saturation temperature of steam at 1.5 bar = 111.3°C

At 1.5 bar for water, $v_{f2} = 0.001053 \text{ m}^3/\text{kg}$

and for steam, $v_{g2} = 1.159 \text{ m}^3/\text{kg}$

$$v_{s2} = v_{s1}\left(\frac{p_1}{p_2}\right)^{\frac{1}{n}} = 0.152\left(\frac{15}{1.5}\right)^{\frac{1}{125}} = 0.959 \text{ m}^3/\text{kg}$$

Dryness fraction of steam at state 2,

$$x_2 = \frac{v_{s2}}{v_{g2}} = \frac{0.959}{1.159} = 0.8275$$

$$\text{Work done} = \frac{p_2 v_{s2} - p_1 v_{s1}}{n-1} = \frac{(1.5\times 0.959 - 15\times 0.152)\times 1.013\times 10^2}{1.25-1}$$

$$= -340.976 \text{ kJ/kg}$$

Now $s_1 = 6.709$ kJ/kg.K

$$s_2 = s_{f2} + x_2 s_{fg2} = 1.4336 + 0.8275\times 5.7898 = 6.2246 \text{ kJ/kg.K}$$

Change in entropy, $\Delta s = s_1 - s_2 = 6.709 - 6.2246 = 0.4844$ kJ/kg.K

3.23 DETERMINATION OF DRYNESS FRACTION OF STEAM

There are four methods of measuring the dryness fraction of steam.

1. Barrel Calorimeter
2. Separating Calorimeter
3. Throttling calorimeter
4. Combined separating and throttling calorimeter.

1. Barrel Calorimeter

The barrel calorimeter shown in Fig. 3.21 consists of a copper calorimeter placed on wooden block and covered by a wooden cover. The barrel contains a known quantity of cold water and is

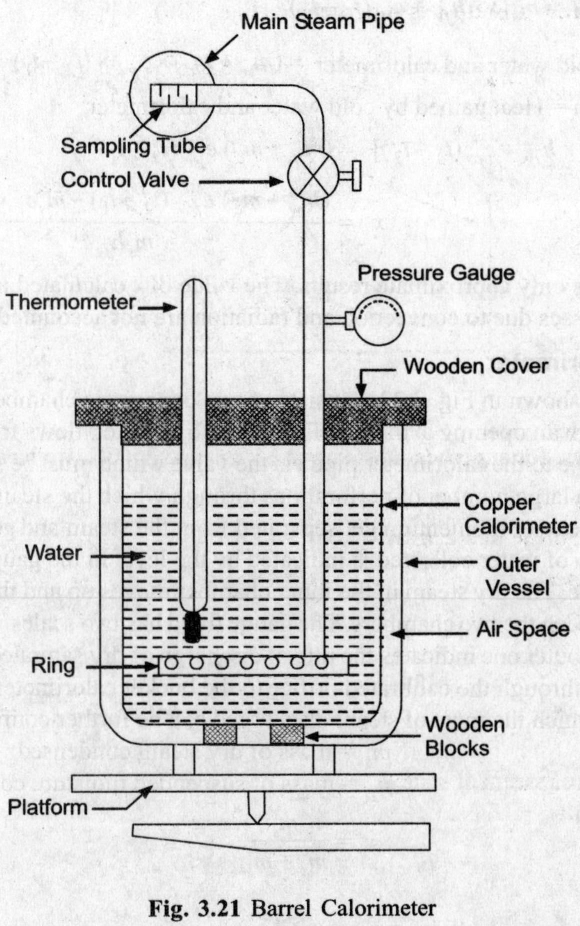

Fig. 3.21 Barrel Calorimeter

surrounded by an outer vessel with air space in between which acts as an insulator. The temperature is measured by a thermometer which passes through one of the holes in the wooden cover. The whole assembly is placed on the platform of a weighing bridge. The steam from the main steam pipe enters through the sampling tube via the control valve, flows into the cold water through fine exit holes of the ring provided at the end of the pipe. The condensation of steam takes place as it comes in contact with cold water, with the result that the temperature of water rises. The quantity of steam condensed can be known from the difference of readings of the weighing bridge before and after the condensation of steam. The pressure of entering steam is indicated by the pressure gauge.

Let m_s = mass of steam condensed

x = dryness fraction of steam

h_{fg} = latent heat of steam

t_s = temperature of formation of steam (corresponding to pressure indicated by the pressure gauge)

m_w = mass of cold water in the calorimeter in the beginning

m_c = water equivalent of calorimeter

t_1 = initial temperature of cold water in the calorimeter

t_2 = final temperature of the mixture after steam condensation.

Heat lost by steam $= m_s[x \times h_{fg} + c_{ps}(t_s - t_2)$

Heat gained by cold water and calorimeter $= (m_w + m_c) \times c_{pw} \times (t_2 - t_1)$

Heat lost by steam = Heat gained by cold water and calorimeter

$$m_s[x\, h_{fg} + c_{ps}(t_s - t_2)] = (m_w + m_c)\, c_{pw}(t_2 - t_1)$$

$$x = \frac{(m_w + m_c)\, c_{pw}\, (t_2 - t_1) - m_s c_{ps}\, (t_s - t_2)}{m_s h_{fg}} \qquad \ldots(3.33)$$

This method gives only approximate results. The value of x calculated is always lower than the actual value as heat losses due to convection and radiation are not accounted for.

2. Separating Calorimeter

This calorimeter, shown in Fig. 3.22, consists of two concentric chambers which communicate with each other through an opening at the top. The steam to be tested flows from the main steam pipe through a sampling tube to the calorimeter pipe via the valve which must be fully open when testing. The metal basket has a large number of perforations through which the steam discharges. The water particles due to their heavier momentum get separated from the steam and get collected in the inner chamber. The quantity of water collected is indicated by the level in the gauge glass and the pointer which moves on a scale. The dry steam in the inner chamber moves up and then down again through the annular space between the two chambers. The gauge fitted has two scales, the inner one indicating the preśsure while the outer one indicates the rate of discharge of dry saturated steam per 10 minutes. Then the steam flows through the calibrated orifice to the bucket calorimeter which is placed on the platform balance by which the most of steam discharged can be further confirmed.

Let m_s = mass of dry steam condensed

m_w = mass of suspended moisture collected

Then
$$x = \frac{m_s}{m_s + m_w} \qquad \ldots(3.34)$$

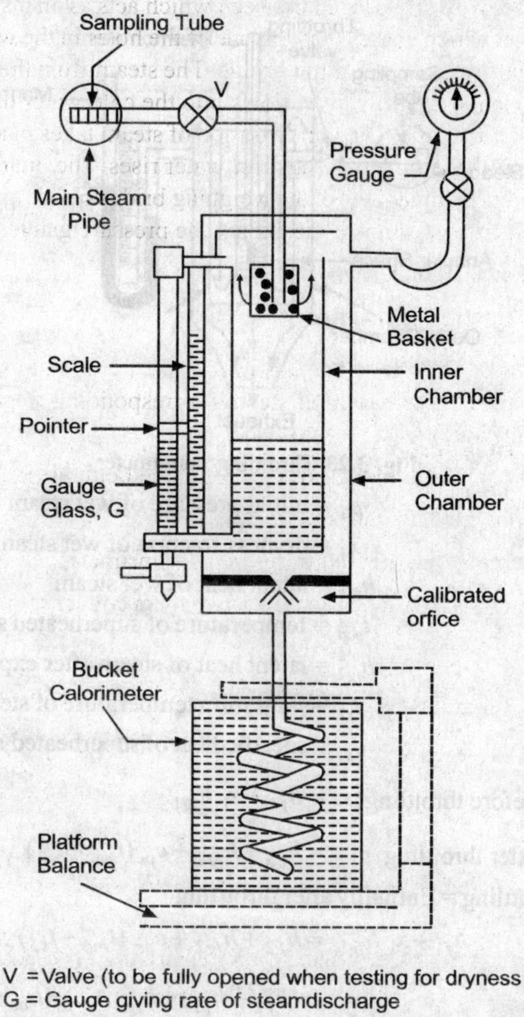

Fig. 3.22 Separating Calorimeter

It is not possible to remove all the water particles from steam and the dryness fraction calculated by this method is always greater than the actual. However it is a quick method and is used for very wet steam.

3. Throttling Calorimeter

The throttling calorimeter, shown in Fig. 3.23, consists of the sampling tube through which the steam from the main steam pipe flows through the throttling valve and becomes superheated. The steam then flows into the inner chamber, flows down and rises up again to enter the annular space. The loss of heat by radiation from the inner chamber is minimum by the hot steam around the outside of the inner chamber. The temperature of throttled steam is measured by the thermometer placed in the pocket filled with cylinder oil. To obtain good results, the steam issuing from the throttle valve must be super heated. To ensure this a manometer is attached to measure the pressure of steam. Corresponding to this pressure the saturation temperature of steam must be lower than the temperature indicated by the thermometer. The steam finally escapes through the exhaust. This is suitable for high dryness steam.

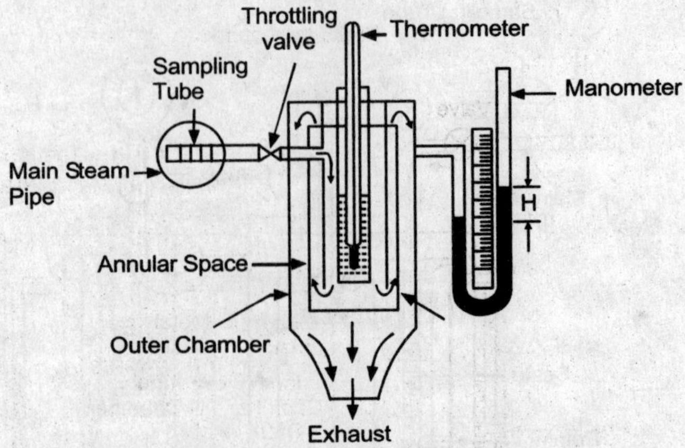

Fig. 3.23 Throttling calorimeter

Let p_1 = initial pressure of wet steam before expansion

x_1 = dryness fraction of wet steam

h_{fg1} = latent heat of wet steam

t_{sup} = temperature of superheated steam after expansion

h_{fg2} = latent heat of steam after expansion

t_{s2} = saturation temperature of steam after expansion

c_{ps} = specific heat of superheated steam at constant pressure

Enthalpy of steam before throttling $= h_{f1} + x_1\, h_{fg1}$

Enthalpy of steam after throttling $= h_{f2} + h_{fg2} + c_{ps}(t_{sup} - t_{s2})$

Enthalpy before throttling = Enthalpy after throttling

$$h_{f1} + x_1\, h_{fg1} = h_{f2} + h_{fg2} + c_{ps}\,(t_{sup} - t_{s2})$$

$$x_1 = \frac{(h_{f2} - h_{f1}) + h_{fg2} + c_{ps}(t_{sup} - t_{s2})}{h_{fg1}} \qquad \ldots(3.35)$$

4. Combined Separating and Throttling Calorimeter

Such a calorimeter is shown in Fig. 3.24. The steam is first passed through the separating calorimeter where it losses most of its moisture and becomes competitively drier. It is then passed through the throttling calorimeter where superheating takes place without change of enthalpy. The temperature and pressure of steam after throttling are measured by using a thermometer and pressure gauge respectively.

Let p_1 = pressure of wet steam entering the throttling calorimeter

x_1 = dryness of steam

h_{fg1} = latent heat of entering steam

m_w = mass of suspended moisture collected

m_s = mass of steam leaving the separating calorimeter and entering the throttling calorimeter

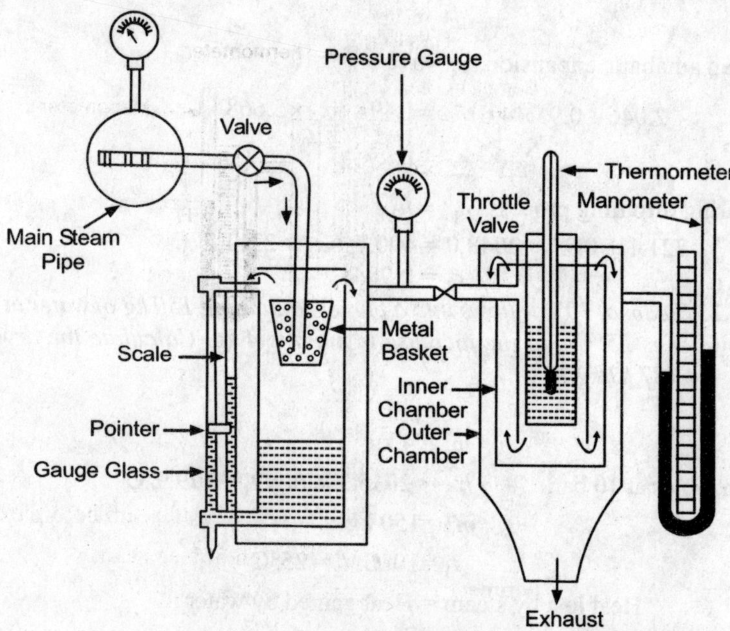

Fig. 3.24 Combined Separating and Throttling Calorimeter

$$p_2 = \text{pressure of steam after throttling}$$

$$h_{fg2} = \text{latent heat of steam at pressure } p_2$$

Enthalpy of steam entering throttling calorimeter = Enthalpy of steam leaving throttling calorimeter

$$h_{f1} + x_1 h_{fg1} = (h_{f2} + h_{fg2}) + c_{ps}(t_{sup} - t_{s2})$$

$$x_1 = \frac{(h_2 - h_{f1}) + c_{ps}(t_{sup} - t_{s2})}{h_{fg1}} \quad \ldots(3.36)$$

Let $x_2 = \dfrac{m_s}{m_s + m_w}$ = apparent dryness fraction steam ...(3.37)

x = dryness fraction of steam entering the separating calorimeter

Mass of dry steam entering the whole apparatus $= m_s x_1$

$$x = \frac{\text{Total mass of dry steam}}{\text{Total mass of steam entering the throttling calorimeter}}$$

$$= \frac{m_s x_1}{m_s + m_w} + x_1\left(\frac{m_s}{m_s + m_w}\right) = x_1 x_2$$

$$\therefore \quad x = x_1 x_2 \quad \ldots(3.38)$$

This calorimeter gives quite satisfactory results when the steam is considerably wet.

Example 3.22 *Steam at a pressure of 8 bar and 0.95 dry is expanded to a pressure of 1.8 bar. Find the final condition of steam and the heat drop for each of the following methods expansion:*

(a) Adiabatic expansion *(b) Throttling expansion*

Solution.

(*a*) During an adiabatic expansion, $s_1 = s_2$

$$2.046 + 0.95 \times 4.617 = 1.494 + x_2 \times 5.668$$

$$x_2 = 0.8712$$

(*b*) During throttling process, $h_1 = h_2$

$$721.1 + 0.95 \times 2048.0 = 490.7 + x_2 \times 2211.2$$

$$x_2 = 0.9841$$

Example 3.23 *Steam at 10 bar flows into a barrel containing 150 kg of water at 10°C. The final temperature of water is 25°C and the increase in mass is 4 kg. Calculate the dryness fraction of steam. Take* $c_{pw} = 4.187$ *kJ/kg.K*

Solution.

$$m_s = 4 \text{ kg}$$

From steam tables at 10 bar, $h_{fg} = 2015.3$ kJ/kg, $t_s = 179.9°C$

$$m_w = 150 \text{ kg}$$

$$t_1 = 10°C, t_2 = 25°C$$

Heat lost by steam = Heat gained by water

$$m_s [x\, h_{fg} + c_{pw} (t_s - t_2)] = m_w (t_2 - t_1)\, c_{pw}$$

$$4\,[x \times 2015.3 + 4.187\,(179.9 - 25)] = 150\,(25 - 10) \times 4.187$$

$$x = 0.847$$

Example 3.24 *In a separating and throttling calorimeter, the steam main pressure is 8 bar absolute and the temperature after throttling is 130°C. The pressure in the throttling calorimeter is 0.01 bar gauge, the barometer reading being 75 cm. 0.15 kg of water is trapped in the separator and 1.8 kg of steam passed through the throttling calorimeter. Determine the dryness fraction of steam in the steam main.*

Solution.

After throttling, $p_2 = 0.01 + 0.75 \times 9.81 \times 13600 \times 10^{-5} = 1.01$ bar

for $p_2 = 1.01$ bar, $t_{s2} = 99.77°C$

Degree of superheat = $t_{sup} - t_{s2} = 130 - 99.77 = 30.23°C$

For superheated steam, from steam tables, for $t_{sup} = 130°C$ and

$p = 1.01$ bar, we have

$$h_{sup2} = 2737 \text{ kJ/kg}$$

$$h_1 = h_{f1} + x_1 h_{fg1}$$

At 8 bar $h_1 = 721.1 + x_1 \times 2048.0$

Now $h_1 = h_{sup2}$

$$721.1 + x_1 + 2048.0 = 2737$$

$$x_1 = 0.984$$

$$x_2 = \frac{m_s}{m_s + m_w} = \frac{1.8}{1.8 + 0.15} = 0.923$$

Dryness fraction, $x = x_1 x_2$

$= 0.984 \times 0.923 = 0.908$

Example 3.25 *A throttling calorimeter is used to measure the dryness fraction of steam in the steam main where the steam is flowing at a pressure of 6 bar. The steam after passing through the calorimeter comes out at 100 kPa, 120°C. Calculate the dryness fraction of steam. Specific heat of superheated steam is 2.09 kJ/kg.K.*

Solution. $(h_{f1} + x_1 h_{fg1})_{p1=6\text{ bar}} = [h_{f2} + h_{fg2} + c_{ps}\ (t_{\text{sup}2} - t_{s2})_{p2=1\text{ bar}}$

$$670.56 + x_1 \times 2086.3 = 417.46 + 2258 + 2.09\ (120 - 99.63)$$

$$x_1 = 0.985$$

Example 3.26 *Steam at 2 MPa is sampled at a flow rate of 3.8 gram/s. Electric power supplied is 580 watts and outlet conditions are 150 kPa, 180°C. Calculate the quality of steam at entry.*

Solution. $m(h_{f1} + x_1 h_{fg1}) + Q = m[h_{f2} + h_{fg2} + C_{ps}\ (t_{\text{sup}2} - t_{g2})]$

At 150 kPa, 180°C is more than $t_{g2} = 111.4$°C

$\therefore$ $$3.8 \times 10^{-3}\ [908.6 + x_1 \times 1818.6] + 580 \times 10^{-3}$$

$$= 3.8 \times 10^{-3}\ [467 + 2226.2 + 2.09\ (180 - 111.4)]$$

$$908.6 + 1818.6x_1 + 152.6 = 467 + 2226.2 + 143.374$$

$$x_1 = 0.976.$$

Example 3.27 *Steam flows in a pipe line at 2 MPa. After expanding to 0.1 MPa in a throttling calorimeter, the temperature is found to be 150°C. Find the quality of steam in the pipe line.*

Solution. At 2 MPa, $h_{f1} = 908.77$ kJ/kg, $h_{fg}'' = 1890.7$ kJ/kg

At 0.1 MPa, 150°C, $h_2 = 2776.4$ kJ/kg

For the throttling process from 1– 2 :

$$h_1 = h_2$$

$$h_1 = h_{f1} + x_1 h_{fg}'' = h_2$$

$$907.77 + x_1 \times 1890.7 = 2776.4$$

$$x_1 = \frac{1868.63}{1890.7} = 0.9883$$

Example 3.28 *The following observations were made with a separating and throttling calorimeter :*

Pressure in pipe line = 1.5 MPa
Condition after throttling = 0. 1 MPa, 110°C
Moisture collected in the separating calorimeter in 5 minutes = 0.150 litre, 70°C
Steam condensed after throttling during 5 minutes = 3.24 kg
Determine the dryness fraction of steam.

Solution. At 1.5 MPa, $h_{f1} = 844.89$ kJ/kg, $h_{fg1} = 1947.3$ kJ/kg

At 0.1 MPa, 110°C, $h_3 = h_2 = 2696.2$ kJ/kg

At 70°C, $v_{f1} = 0.001023\ \text{m}^3/\text{kg}$

Now $h_3 = h_2 = h_{f1} + x_2 h_{fg1}$

$$2696.2 = 844.89 + x_2 \times 1947.3$$

$$x_2 = 0.955$$

Volume of moisture collected in separating calorimeter in 5 min, $V_1 = 0.150$ litre or $150 \times 10^{-6}\ \text{m}^3$

At 70°C, $v_{f1} = 0.001023 \text{ m}^3/\text{hg}$

Mass of moisture collected,

$$m_1 = \frac{V_1}{v_{f1}} = \frac{150\times10^{-6}}{0.001023} = 0.1462 \text{ kg}$$

$$x_1 = \frac{m_2}{m_1+m_2} = \frac{3.24}{0.1462+3.24} = 0.9568$$

$$x = x_1 x_2 = 0.9568\times0.955 = 0.914$$

Example 3.29 *The following data refer to a barrel calorimeter.*
Pressure of steam = 6.5 bar
Mass of water = 50 kg at 20°C
Water equivalent of bucket calorimeter = 3 kg
Temperature of mixture = 38°C
Mass of steam = 1.5 kg
Calculate the dryness fraction of steam.

Solution. From steam tables at 6.5 bar,

$$t_s = 162.0°\text{C},\ h_{fg} = 2076 \text{ kJ/kg}$$

$$m_w = 50 \text{ kg},\ m_c\, c_{pc} = 3 \text{ kg},\ t_1 = 20°\text{C},\ t_2 = 38°\text{C},\ m_s = 4 \text{ kg}$$

Dryness fraction,

$$x = \frac{(c_{pw}m_w + c_{pc}m_c)(t_2-t_1) - m_s(t_s-t_2)\,c_{pw}}{m_s h_{fg}}$$

$$= \frac{(4.18\times50+3)(38-20) - 1.5(162-38)\times4.18}{1.5\times2076} = 0.976$$

Example 3.30 *Steam at a pressure of 6 bar passes into a tank containing water where it gets condensed. The mass and temperature in the tank before the admission of steam are 60 kg and 20°C respectively. Calculate the dryness fraction of steam as it enters the tank if 4 kg of steam gets condensed and resulting temperature of the mixture becomes 45°C. Take water equivalent of tank as 2 kg.*

Solution. $p = 6$ bar, $t_s = 158.9°c$, $m_w = 60$ kg, $t_1 = 20°\text{C}$, $m_s = 4$ kg, $t_2 = 45°\text{C}$, $c_{pc}m_c = 2$ kg

$$x = \frac{(m_w c_{pw} + m_c c_{pc})(t_2-t_1) - m_s c_{pw}(t_s-t_2)}{m_s h_{fg}}$$

$$= \frac{(60\times4.18+2)(45-20) - 4\times4.18\,(158.9-45)}{4\times2086.3} = \mathbf{0.529}$$

Example 3.31 *In a separating and throttling calorimeter for determining the dryness fraction, steam is throttled from 1.5 MPa to 0.15 MPa. The temperature of steam after throttling is 150°C. The discharge from the separator, which is placed ahead of the throttling device, is 0.2 kg/minute of water, while the discharge from the calorimeter is 12 kg/minute. Determine the dryness fraction of the steam entering the separator.* (IES, 1984)

Solution.

$$m = m_s + m_w = 12 \text{ kg/min}$$

$$m_w = 0.2 \text{ kg/min}$$

$$m_s = 12 - 0.2 = 11.8 \text{ kg/min}$$

Dryness fraction for separating calorimeter,

$$x_1 = \frac{m_s}{m} = \frac{11.8}{12} = 0.9833$$

From steam tables at $p_1 = 1.5$ MPa, we have

$h_{f1} = 844.66$ kJ/kg, $h_{fg1} = 1945.2$ kJ/kg

At $p_2 = 0.15$ MPa and 150°C, $h_{g2} = 2772.5$ kJ/kg, $t_s = 111.37$°C

$$h_{f1} + x_2 h_{fg1} = h_{g2} + C_p(t_{sup} - t_s)$$

Let for steam, $c_p = 2.093$ kJ/kg.K

$$844.66 + x_2 \times 1945.2 = 2772.5 + 2.093\,(150 - 111.37)$$

$$x_2 = 1.0326$$

Dryness fraction of steam,

$$x = x_1 x_2 = 0.9833 \times 1.0326 = 1.0153$$

Example 3.32 *1.5 kg of steam at a pressure of 10 bar and temperature of 250°C is expanded until the pressure becomes 2.8 bar. The dryness fraction of the system is then 0.9. Calculate the change of internal energy. The extract from the steam tables is as under. Assume $c_{ps} = 2.1$ kJ/kg.K.*

Pressure	*Saturated temperature*	*Specific volume* m^3/kg		*Specific enthalpy kJ/kg*	
bar	*°C*	v_f	v_g	h_f	h_g
2.8	*131.2*	*0.001071*	*0.64604*	*551.44*	*2721.5*
10	*179.88*	*0.0011274*	*0.19429*	—	*2776.2*

Solution.

Given : $m = 1.5$ kg, $p_1 = 10$ bar, $T_1 = 273 + 250 = 523$ K, $p_2 = 2.8$ bar,

$x_2 = 0.9$, $c_{ps} = 2.1$ kJ/kg.K

At $p_1 = 10$ bar, $T_{sat} = 273 + 179.88 = 452.88 \text{ K} < T_1 = 523$ K.

Hence steam is superheated.

Enthalpy at state 1,

$$h_1 = h_{g1} + c_{ps}(T_1 - T_{sat})$$

$$= 2776.2 + 2.1\,(523 - 452.88)$$

$$= 2923.452 \text{ kJ/kg}$$

Specific volume of superheated steam,

$$v_{sup1} = v_{g1} \times \frac{T_1}{T_{sat\,1}} = 0.19429 \times \frac{523}{452.88} = 0.22437 \text{ m}^3/\text{kg}$$

Internal energy,

$$u_1 = h_1 - \frac{p_1 v_{sup1}}{10^3} = 2923.452 - \frac{10 \times 10^5 \times 0.22437}{10^3}$$

$$= 2699.082 \text{ kJ/kg}$$

Enthalpy of steam at state 2,

$$h_2 = h_{f2} + x_2\, h_{fg2} = h_{f2} + x_2(h_{g2} - h_{f2})$$

$$= 551.44 + 0.9 \times (2721.5 - 551.44)$$

$$= 2504.494 \text{ kJ/kg}$$

Specific volume of steam at state 2 (neglecting volume of water),

$$v_2 = x_2 v_{g2} = 0.9 \times 0.64604 = 0.581436 \text{ m}^3/\text{kg}$$

Internal energy of wet steam,

$$u_2 = h_2 - \frac{p_2 v_2}{10^3}$$

$$= 2504.494 - \frac{2.8 \times 10^5 \times 0.581436}{10^3}$$

$$= 2341.692 \text{ kJ/kg}$$

Change in internal energy, $\Delta U = m(u_2 - u_1)$

$$= 1.5 (2341.692 - 2699.082)$$

$$= -536.085 \text{ kJ}$$

Example 3.33 *Calculate the enthalpy and internal energy of 1 kg of steam at a pressure of 10 bar and dryness fraction as 0.95. Use the following table :*

Pressure	*Saturated temperature*	*Specific volume m^3/kg*		*Specific enthalpy kJ/kg*	
bar	*°C*	v_f	v_g	h_f	h_g
10	*179.88*	*0.0011274*	*0.19429*	*762.61*	*2776.2*

Solution.

Given : $m = 1 \text{ kg}, p = 10 \text{ bar}, x = 0.95$

Enthalpy of steam, $h = h_f + xh_{fg} = h_f + x(h_g - h_f)$

$$= 762.61 + 0.95 (2776.2 - 762.61)$$

$$= 2675.52 \text{ kJ/kg}$$

Internal energy, $u = h - p\,x\,v_g$ (neglecting volume of water)

$$= 2675.52 - 10 \times 10^2 \times 0.95 \times 0.19427$$

$$= 2490.96 \text{ kJ/kg}$$

Example 3.34 *For the data of Example 4.33 with x = 0.8, calculate (a) external work done during evaporation, (b) internal energy of steam, and (c) internal heat of evaporation.*

Solution.

Enthalpy of steam, $h = h_f + x(h_g - h_f)$

$$= 762.61 + 0.8 (2776.2 - 762.61)$$

$$= 2373.482 \text{ kJ/kg}$$

(*a*) External work done during evaporation $= p(v_g - v_f)\,x \simeq p\,x\,v_g$

$$= 10 \times 10^2 \times 0.8 \times 0.19427$$

$$= 155.416 \text{ kJ/kg}$$

(*b*) Internal energy, $u = h - p\,x\,v_g$

$= 2373.482 - 155.416$

$= 2218.066$ kJ/kg

(*c*) Internal heat of evaporation $= x\,(h_{fg} - p\,v_g)$

$= 0.8\,[2776.2 - 762.62) - 10 \times 10^2 \times 0.19427]$

$= 1455.456$ kJ/kg

Example 3.35 *A closed vessel of 0.6 m³ capacity contains dry saturated steam at 3.6 bar pressure. The vessel is cooled until the pressure is reduced to 2 bar. Calculate (a) the mass of steam in the vessel, (b) the final dryness fraction, and (c) the amount of heat transferred during the process in kJ. Use the following table :*

Pressure	*Saturated temperature*	*Specific volume m³/kg*		*Specific enthalpy kJ/kg*		
bar	°C	v_f	v_g	h_f	h_{fg}	h_g
2.0	*120.23*	*0.0010608*	*0.88544*	*504.70*	*2201.6*	*2706.3*
3.6	*139.86*	*0.0010799*	*0.51032*	*588.53*	*2144.4*	*2732.9*

Solution.

$$V_1 = 0.6 \text{ m}^3,\ p_1 = 3.6 \text{ bar},\ x_1 = 1,\ p_2 = 2 \text{ bar}$$

(*a*) Mass of steam in vessel, $m = \dfrac{V_1}{v_{g1}} = \dfrac{0.6}{0.51032} = 1.176$ kg

(*b*) For the closed vessel, $V_2 = V_1$

$$x_1 v_{g1} = x_2\,v_{g2}$$

$$x_2 = \frac{1 \times 0.51032}{0.88544} = 0.576$$

(*c*) Initial internal energy, $u_1 = h_1 - p_1 x_1 v_{g1}$

$= 2732.9 - 3.6 \times 10^2 \times 1 \times 0.51032$

$= 2549.185$ kJ/kg

Final internal energy, $u_2 = h_2 - p_2 x_2 v_{g2}$

$= (h_{f2} + x_2 h_{fg2}) - p_2 x_2 v_{g2}$

$= (504.7 + 0.576 \times 2201.6) - 2 \times 10^2 \times 0.576 \times 0.88544$

$= 1670.82$ kJ/kg

Change in internal energy during the process $= m\,(u_2 - u_1)$

$= 1.176\,(1670.82 - 2549.185) = -\ 1032.96$ hJ

–ve sign indicates decrease in internal energy.

Heat transferred during the process $= m\,(h_2 - h_1)$

$= 1.176\,(1772.82 - 2732.9)$

$= -1129.052$ kJ

–ve sign indicates heat rejected from the vessel.

Example 3.36 *Calculate the internal energy of wet steam with the following data : Pressure = 12 bar, dryness fraction = 0.8. Use the following table :*

Pressure	*Saturated temperature*	*Specific volume m^3/kg*		*Specific enthalpy kJ/kg*		
bar	°C	v_f	v_g	h_f	h_{fg}	h_g
12	*187.96*	*0.0011386*	*0.16320*	*798.43*	*1984.3*	*2782.7*

Solution. $p = 12$ bar, $x = 0.8$

Enthalpy of steam, $h = h_f + x\,h_{fg}$

$= 798.43 + 0.8 \times 1984.3 = 2385.87$ kJ/kg

Flow work, $w_f = px\,v_g = 12 \times 10^2 \times 0.8 \times 0.16320 = 156.672$ kJ/kg

Internal energy, $u = h - w_f$

$= 2385.87 - 156.672 = 2229.198$ kJ/kg

Example 3.37 *Unit mass of steam at a pressure of 10 bar and x = 0.2 is expanded at constant pressure until it becomes 90 dry. Calculate the work done, heat transfer and change in internal energy during the process.*

Solution.

Given : $m = 1$ kg, $p_1 = 10$ bar, $x_1 = 0.2$, $p_2 = p_1$, $x_2 = 0.9$

From steam tables, at $p_1 = 10$ bar, we have

$t_s = 179.88°C$, $v_g = 0.19427$ m³/kg, $h_f = 762.61$ kJ/kg,

$h_{fg} = 2013.6$ kJ/kg, $h_g = 2776.2$ kJ/kg

Work done, $w_{1-2} = p\,(x_2 v_{g2} - x_1 v_{g1})$

$= 10 \times 10^2 (0.9 - 0.2) \times 0.19427$

$= 136$ kJ/kg

$h_1 = h_{f1} + x_1 h_{fg1} = 762.61 + 0.2 \times 2013.6 = 1165.33$ kJ/kg

$h_2 = h_{f2} + x_2 h_{fg2} = 762.61 + 0.9 \times 2013.6 = 2574.85$ kJ/kg

Heat transfer $= h_2 - h_1 = 2574.85 - 1165.33 = 1409.52$ kJ/kg

Change in internal energy $= u_2 - u_1$

$= (h_2 - p_2 x_2 v_{g2}) - (h_1 - p_1 x_1 v_{g1})$

$= (h_2 - h_1) - pv_g\,(x_2 - x_1)$

$$= 1409.52 - 10 \times 10^2 \times 0.19427\ (0.9 - 0.2)$$

$$= 1273.531 \text{ kJ/kg}$$

Example 3.38 *Steam at 8 bar and dryness of 0.9 expands at constant pressure until the dryness fraction is 0.6. Calculate the work done and heat removed per kg of steam during the process.*

Solution.

Given : $p_1 = p_2 = 8$ bar, $x_1 = 0.9$, $x_2 = 0.6$

From steam tables at $p = 8$ bar, we have

$$h_f = 721.1 \text{ kJ/kg},\ h_{fg} = 2048.0 \text{ kJ/kg},\ h_g = 2769.1 \text{ kJ/kg}$$

$$v_g = 0.240 \text{ m}^3\text{/kg}$$

$$h_1 = h_{f1} + x_1\ h_{fg2} = 721.1 + 0.6 \times 2048.0 = 1949.9 \text{ kJ/kg}$$

$$h_2 = h_{f2} + x_2\ h_{fg2} = 721.1 + 0.6 \times 2048.0 = 1949.9 \text{ kJ/kg}$$

Heat removed $= h_1 - h_2 = 2564.3 - 1949.9 = 614.4$ kJ/kg

Work done $= p\ v_g (x_1 - x_2) = 8 \times 10^2 \times 0.240\ (0.9 - 0.6) = 57.6$ kJ/kg

Example 3.39 *Find the external work done during evaporation, internal latent heat (enthalpy) of evaporation and internal energy per kg of steam at a pressure of 15 bar when the steam is (a) 0.9 dry, and (b) dry and saturated.*

Solution.

From steam tables at 15 bar, we have

$$v_f = 0.001154 \text{ m}^3\text{/kg},\ v_g = 0.132 \text{ m}^3\text{/kg},\ h_f = 844.9 \text{ kJ/kg},$$

$$h_{fg} = 1947.3 \text{ kJ/kg},\ h_g = 2792.2 \text{ kJ/kg}$$

(*a*) $x = 0.9$

Work done during evaporation

$$w_{1-2} = p(x\ v_g - v_f)$$

$$= 15 \times 10^2\ (0.9 \times 0.132 - 0.001154)$$

$$= 176.469 \text{ kJ/kg}$$

Internal energy of steam,

$$u = h - px\ v_g$$

$$= (h_f + x\ h_{fg}) - px\ v_g$$

$$= (844.9 + 0.9 \times 1947.3) - 15 \times 10^2 \times 0.9 \times 0.132$$

$$= 2419.27 \text{ kJ/kg}$$

Internal latent heat of evaporation $= x\ (h_{fg} - pv_g)$

$$= 0.9\ (1947.3 - 15 \times 10^2 \times 0.132)$$

$$= 1574.37 \text{ kJ/kg}$$

(*b*) $x = 1.0$

Work done during evaporation, $w_{1-2} = p(v_g - v_f)$

$$= 15 \times 10^2 (0.132 - 0.001154) = 196.269 \text{ kJ/kg}$$

Internal energy of steam, $u = h_g - pv_g = 2792.2 - 15 \times 10^2 \times 0.132 = 2594.2 \text{ kJ/kg}$

Internal latent heat of evaporation $= h_{fg} - p(v_g - v_f)$

$$= 1947.3 - 196.269 = 1751.031 \text{ kJ/kg}$$

Example 3.40 *How much heat is required to convert 4 kg of water at 20°C into steam at 8 bar and 200°C. Take $c_{ps} = 2.1$ kJ/kg.K and $c_{pw} = 4.187$ kJ/kg.K.*

Solution.

Given : $m_w = 4$ kg, $t_w = 20°C$, $p = 8$ bar, $t_{sup} = 200°C$,

$c_{ps} = 2.1$ kJ/kg.K, $c_{pw} = 4.187$ kJ/kg.K

From steam tables at $p = 8$ bar, we have

$t_s = 170.1°C < 200°C$. Hence steam is superheated.

$h_f = 721.1$ kJ/kg, $h_{fg} = 2048.0$ kJ/kg, $h_g = 2769.1$ kJ/kg

$$h_{sup} = h_g + c_{ps}(t_{sup} - t_s)$$

$$= 2769.1 + 2.1(200 - 170.1)$$

$$= 2831.89 \text{ kJ/kg}$$

Heat in 1 kg of water, $h_w = m_w c_{pw}(t_w - 0)$

$$= 1 \times 4.187 \times 20 = 83.74 \text{ kJ/kg}$$

Heat required to be supplied, $h = h_{sup} - h_w$

$$= 2831.89 - 83.74$$

$$= 2748.15 \text{ kJ/kg}$$

Total heat required $= m_h = 4 \times 2748.15 = 10992.6$ kJ

Example 3.41 *The following readings were taken with a separating and throttling calorimeter:*

Initial pressure = 12 bar
Final pressure = 1.2 bar
Water separated = 2 kg
Barometer = 760 mm of Hg

Steam discharged from throttling calorimeter = 20.5 kg

Estimate the quality of steam supplied.

Solution.

Given : $m_s = 20.5$ kg, $m_w = 2$ kg, $p_1 = 12$ bar, $p_2 = 1.2$ bar

Now $m = m_s + m_w$

$\therefore$ $m = 20.5 + 2 = 22.5$ kg

Dryness fraction for separating calorimeter,

$$x_1 = \frac{m_s}{m} = \frac{20.5}{22.5} = 0.91$$

From steam tables at $p_1 = 12$ bar, we have

$$h_{f1} = 798.6 \text{ kJ/kg}, h_{fg1} = 1986.2 \text{ kJ/kg}$$

Now 760 mm Hg = 1.0132 bar and $t_s = 100°C$

At $p_2 = 1.2$ bar, $t_s = 101.8°C$. Therefore, steam is nearly dry saturated.

$$h_{g2} = 2683.5 \text{ kJ/kg}$$

$$\therefore \quad h_{f1} + x_2\, h_{fg1} = h_{g2}$$

$$798.6 + x_2 \times 1986.2 = 2683.5$$

$$x_2 = 0.949$$

$$\therefore \quad x = x_1 x_2 = 0.91 \times 0.949 = 0.863$$

Example 3.42 *Determine the condition of steam in the following cases :*
(a) At a pressure of 10 bar and 200°C.
(b) At a pressure of 8 bar and specific volume of 0.22 m³/kg.

Solution.

(*a*) From steam tables at $p = 10$ bar, $t_s = 179.9° < 200°C$. Therefore steam is superheated.

(*b*) From steam tables at $p = 8$ bar, $v_f = 0.001115 \text{ m}^3/\text{kg}$, $v_g = 0.240 \text{ m}^3/\text{kg}$

Now
$$v = v_f + x\, v_{fg} = v_f + x\,(v_g - v_f)$$

$$0.22 = 0.001115 + x\,(0.240 - 0.001115)$$

$$x = 0.91628$$

Example 3.43 *Determine the state of steam at :*
(a) p = 14 bar and t = 210°C, and (b) p = 8 bar and 2700 kJ/kg of heat supplied to water from 0°C.

Solution.

(*a*) From steam tables at $p = 14$ bar, $t_s = 195°C < 210°C$. Hence steam is superheated.

(*b*) From steam tables at $p = 8$ bar, $t_s = 170.1°C$ and $h_g = 2769.1$ kJ/kg,

$h_f = 721.1$ kJ/kg, $h_{fg} = 2048.0$ kJ/kg. Since $h < h_g$, therefore steam is wet.

$$h = h_f + x\, h_{fg}$$

$$2700 = 721.1 + x \times 2048.0$$

$$x = 0.966$$

Example 3.44 *Steam at 8 bar pressure and dryness fraction of 0.9 expands at constant pressure until the temperature is 200°C. Calculate the work input and heat supplied per kg of steam during the process.*

Solution.

Given : $p_1 = 8 \text{ bar} = p_2$, $x_1 = 0.9$, $t_2 = 200°C$

From steam tables at $p_1 = 8$ bar, we have

$t_s = 170.4°C < 200°C$. Hence steam is superheated at state 2.

$v_{g1} = 0.240 \text{ m}^3/\text{kg}$, $h_{f1} = 721.1$ kJ/kg, $h_{fg1} = 2048.0$ kJ/kg, $h_{g1} = 2769.1$ kJ/kg

For superheated steam tables, for $p = 8$ bar and $t_{sup} = 200°C$, we have

$v_{sup} = 0.2608 \text{ m}^3/\text{kg}$, $h_{sup} = 2839.3$ kJ/kg

Work done, $w_{1-2} = p(v_{sup} - x_1 v_{g1})$

$= 8 \times 10^2 (0.2608 - 0.9 \times 0.240)$

$= 35.84$ kJ/kg

$h_1 = h_{f1} + x_1\, h_{fg1} = 721.1 + 0.9 \times 2048.0 = 2564.3$ kJ/kg

Heat supplied, $q_{1-2} = h_{sup} - h_1 = 2839.3 - 2564.3 = 275$ kJ/kg

Example 3.45 *The steam is generated at 8 bar from water at 32°C. Find the heat required to produce 1 kg of steam.*

(a) When the steam at dryness = 0.9.

(b) When the steam is superheated with a temperature = 300°C.

Take $c_{ps} = 2.3$ kJ/kg.K for superheated steam.

Solution.

Given : $p_1 = 8$ bar, $t_w = 32$°C

(*a*) From steam tables at $p = 8$ bar, we have

$t_s = 170.4$°C, $h_{f1} = 721.1$ kJ/kg, $h_{fg1} = 2048.0$ kJ/kg, $h_{g1} = 2769.1$ kJ/kg

$x = 0.9$

Heat supplied $= m_s (h_{f1} + x_1\, h_{fg1}) - m_w c_{pw} (t_w - 0)$

$= 1 \times (721.1 + 0.9 \times 2048.0) - 1 \times 4.187\,(32 - 0)$

$= 2564.3 - 133.984$

$= 2430.316$ kJ/kg

(*b*) $t_s < 300$°C. Hence steam is superheated.

Heat supplied $= 1 \times [h_g + c_{ps}\,(t_{sup} - t_{sat})] - m_w c_{pw} (t_w - 0)$

$= 1 \times [2769.1 + 2.3(300 - 170.4)] - 1 \times 4.187\,(32 - 0)$

$= 3067.18 - 133.984$

$= 2933.196$ kJ/kg

Example 3.46 *Find the enthalpy and internal energy of steam at a pressure of 10 bar and temperature 200°C. Assume* $c_{ps} = 2.1$ *kJ/kg.K for superheated steam.*

Solution.

Given : $p = 10$ bar, $t = 200$°C, $c_{ps} = 2.1$ kJ/kg.K

From steam tables at $p = 10$ bar, we have

$t_s = 179.9$°C < 200°C. Hence steam is superheated.

$h_g = 2778.1$ kJ/kg, $v_g = v_2 = 0.1944$ m^3/kg

$h = hg + c_{ps}\,(t - t_s)$

$= 2778.1 + 2.1\,(200 - 179.9)$

$= 2820.31$ kJ/kg

$T_s = 273 + 179.9 = 452.9$ K

$$T_{sup} = 273 + 200 = 473 \text{ K}$$

$$v_{sup} = \frac{v_s T_{sup}}{T_s} = \frac{0.1944 \times 473}{452.9} = 0.2030 \text{ m}^3/\text{kg}$$

Flow work = $p\, v_{sup} = 10 \times 10^2 \times 0.2030 = 203$ kJ/kg

Internal energy, $u_{sup} = h_{sup} - pv_{sup}$

$$= 2820.31 - 203$$

$$= 2617.31 \text{ kJ/kg}$$

MULTI-CHOICE QUESTIONS

1. At triple point of a pure substance
(*a*) liquid and vapour exist together
(*b*) solid and liquid exist together
(*c*) solid and vapour exist together
(*d*) solid, liquid and vapour phases exist together

2. A pure substance above critical temperature exists as
(*a*) solid (*b*) liquid (*c*) vapour (*d*) superheated liquid

3. Triple point temperature and pressure for water are
(*a*) 1°C and 1 ata (*b*) 0°C and 1 ata
(*c*) 0°C and 0.006028 ata (*d*) 0.01°C and 0.006028 ata.

4. Partial pressure of a component in a mixture is
(*a*) the average of pressure of all the components
(*b*) the total pressure minus its own pressure
(*c*) the pressure exerted by the component
(*d*) the pressure of the component itself if it was present in the entire volume at the same temperature.

5. The slope of the sublimation curve on *p-T* diagram for all substances is
(*a*) zero (*b*) positive (*c*) negative (*d*) variable.

6. Sublimation is the process of
(*a*) changing from solid phase to vapour phase directly
(*b*) changing from liquid phase to vapour phase
(*c*) supersaturation of vapour
(*d*) changing from solid phase to liquid phase.

7. The process of sublimation is found to occur in
(*a*) liquid N_2 (*b*) solid CO_2 (*c*) solid O_2 (*d*) air

8. When a vapour condenses into a liquid
(*a*) its temperature rises (*b*) its temperature falls
(*c*) it evolves heat (*d*) it absorbs heat

9. 1 kg of steam sample contains 0.8 kg steam. Its dryness fraction is
(*a*) 0.2 (*b*) 0.8 (*c*) 0.9 (*d*) 0.5

10. If x_1 and x_2 are the dryness fractions obtained in separating calorimeter and throttling calorimeter respectively, then the actual dryness fraction of steam is

(*a*) $x_1 x_2$ (*b*) $\frac{x_1 + x_2}{2}$ (*c*) $x_2 - x_1$ (*d*) $\frac{1}{\frac{1}{x_1} + \frac{1}{x_2}}$.

11. The properties of water and steam become identical at
(*a*) 0.1 bar (*b*) 1 bar (*c*) 100 bar (*d*) 212.2 bar.

12. The mass of water and steam collected in 15 minutes in a separating and throttling calorimeter are 1.2 kg and 4.8 kg respectively. The dryness fraction is
(*a*) 50% (*b*) 75% (*c*) 80% (*d*) 90%

13. Consider the following statements about critical point of water :
1. The latent heat is zero.
2. The liquid is denser than its vapour.
3. Steam generators can operate above this point.
Of these statements
(*a*) 1, 2 and 3 are correct (*b*) 1 and 2 are correct
(*c*) 2 and 3 are correct (*d*) 1 and 3 are correct

14. A liquid is boiling in a airtight vessel. Using an exhaust tube the vapour is pumped out at a faster rate. What will happen to the liquid ?
(*a*) Temperature will go down but boiling will continue
(*b*) Temperature will rise and boiling will continue
(*c*) Boiling will stop
(*d*) Whatever vapour inside will condense

15. Steam coming out of the whistle of pressure cooker is
(*a*) Dry saturated vapour (*b*) Wet vapour
(*c*) Superheated vapour (*d*) Ideal gas

16. At its critical point, any substance will
(*a*) exist in all the three phases simultaneously
(*b*) change directly from solid to vapour
(*c*) lose phase distinction between liquid and vapour
(*d*) behave as an ideal gas

17. In thermodynamic analysis, a pure substance is that which
(*a*) Consists of only a single chemical species
(*b*) has no dissolved impurities
(*c*) may have a number of chemical species but the composition remains constant
(*d*) behaves as a perfect gas.

18. A system is said to be consisting of a pure substance when
(*a*) it is homogeneous in composition
(*b*) it is homogeneous and invariable in chemical aggregation
(*c*) it has only one phase
(*d*) it has more than one phase.

19. For a mixture of solid, liquid and vapour phases of a pure substance, in equilibrium, the number of independent intrinsic properties needed will be
(*a*) 0 (*b*) 1 (*c*) 2 (*d*) 3

20. In steam tables, the entropy is shown as zero for
(*a*) saturated vapour at atmospheric pressure
(*b*) saturated liquid at atmospheric pressure
(*c*) saturated vapour at 0°C
(*d*) saturated liquid at 0°C

21. For a pure substance at its triple point, the number of degrees of freedom is
(*a*) 0 (*b*) 1 (*c*) 2 (*d*) none of the above

22. Superheating of steam is done at constant
(*a*) volume (*b*) pressure (*c*) temperature (*d*) enthalpy

23. With the increase of pressure the enthalpy of evaporation of water
(*a*) decreases (*b*) increases (*c*) remains same (*d*) changes randomly

24. Only throttling calorimeter is used for measuring
(*a*) very low dryness fraction upto 0.7
(*b*) very high dryness fraction upto 0.98
(*c*) dryness fraction of only low pressure steam
(*d*) dryness fraction of only high pressure steam

25. Constant pressure lines in the superheated region of the Mollier diagram have
(*a*) a positive sloped (*b*) a negative slope
(*c*) zero slope (*d*) both positive and negative slope

26. Specific volume of wet steam with dryness fraction *x* is
(*a*) $x\,v_f$ (*b*) $x\,v_g$ (*c*) $x^2 v_g$ (*d*) $x^3 v_g$

27. Enthalpy of superheated steam is given by
(*a*) $h_g + c_{ps} \ln \frac{T_{sup}}{T_s}$ (*b*) $h_f + h_{fg}$ (*c*) $s_g + x\,s_{fg}$ (*d*) $x\,s_f$

28. Entropy of wet steam is given by
(*a*) $s_f + x\,s_{fg}$ (*b*) $x\,s_g$ (*c*) $s_g + s_{fg}$ (*d*) $x\,s_f$

29. Entropy of 1 kg of water at TK is given by
(*a*) $c_{pw} \ln \frac{T}{273}$ (*b*) $c_{pw} \log \frac{T}{273}$ (*c*) $c_{pw} \ln \frac{T}{273}$ (*d*) $c_{pw} \log \frac{T}{273}$

30. Water equivalent of calorimeter is given by
(*a*) $\frac{m_{cal} \times c_{pc}}{c_{pw}}$ (*b*) $\frac{m_{cal}}{c_{pw}}$ (*c*) $\frac{m_{cal}}{c_{pc}}$ (*d*) $\frac{m_{cal} \times c_{pw}}{c_{pc}}$

Answers:

1. (*d*) 2. (*d*) 3. (*d*) 4. (*d*) 5. (*b*) 6. (*a*) 7. (*b*) 8. (*c*) 9. (*b*) 10. (*a*)
11. (*d*) 12. (*c*) 13. (*d*) 14. (*a*) 15. (*b*) 16. (*c*) 17. (*c*) 18. (*b*) 19. (*a*) 20. (*d*)
21. (*a*) 22. (*b*) 23. (*a*) 24. (*b*) 25. (*a*) 26. (*b*) 27. (*a*) 28. (*a*) 29. (*a*) 30. (*a*)

REVIEW QUESTIONS

1. Define a pure substance.
2. What are saturation states ?
3. What is the triple point?
4. What is the critical state ?
5. Define normal boiling point.
6. Explain degree of subcooling and degree of superheat.
7. Define dryness fraction of steam.
8. What do you mean by phase of a substance?
9. Explain the term sublimation.
10. Define latent heat of fusion and latent heat of vaporisation.

11. Define dryness fraction and wetness of steam.

12. What is quality of steam ?

13. Write expressions for internal energy of steam when it is wet and when it is superheated.

14. Write expressions for entropy of steam for wet and superheated.

15. What is a Mollier diagram ?

16. When steam is heated at constant volume and its end condition is still wet, what is heat transfer?

17. In the wet region, constant temperature process is also a constant pressure process. Say true or false.

18. Hyperbolic process is also an isothermal process in the super heat region. Say true or false.

19. What remains constant during isentropic process ?

20. What is throttling of steam ?

21. What are the drawbacks of barrel calorimeter ?

22. Which calorimeter is used for a very wet steam ?

23. Distinguish between gas and vapour.

24. Differentiate between wet, dry saturated and superheated steam.

25. Explain the term internal energy of steam.

26. Why steam is superheated ?

27. Describe the process of formation of steam on *T-Q* diagram.

28. Explain the volume changes during formation of steam with the help of *T-V* diagram.

29. What do you understand by the term priming ?

30. State limitations of throttling calorimeter.

EXERCISES

3.1 A rigid vessel of volume 0.86 m^3 contains 1 kg of steam at a pressure of 2 bar. Calculate the specific volume, temperature, dryness fraction, internal energy, enthalpy, and entropy of steam.
[**Ans.** 0.86 m^3/kg, 120.23°C, 0.97, 2468.54 kJ/kg, 2640.54 kJ/kg, 6.9592 kJ/kg.K]

4.2 Water at 40°C is continuously sprayed into a pipe line carrying 5 tonnes per hour of steam at 5 bar, 300°C. At a section downstream where the pressure is 3 bar, the dryness fraction is to be 0.95. Calculate the rate of water supply in kg/h. [**Ans.** 912.67 kg/K]

3.3 A rigid closed tank of volume 3m^3 contains 5 kg of wet steam at a pressure of 200 kPa. The tank is heated until the steam becomes dry saturated. Find the final pressure and the heat transfer to the tank. [**Ans.** 304 kPa, 3346 kJ]

3.4 A sample of steam from a boiler drum at 3 MPa is put through a throttling calorimeter in which the pressure and temperature are found to be 0.1 MPa, 120°C. Calculate the quality of the sample taken from the boiler. [**Ans.** 0.951]

3.5 Two streams of steam, one at 2MPa, 300°C and the other at 2MPa, 400°C, mix in a steady flow adiabatic process. The rates of flow of the two streams are 3 kg/min and 2 kg/min respectively. Find the final temperature of the emerging stream, if there is no pressure drop due to the mixing process. [**Ans.** 340°C]

3.6 Two insulated tanks, *A* and *B*, are connected by a valve. Tank *A* has a volume of 0.70 m^3 and contains steam at 1.5 bar, 200°C. Tank *B* has a volume of 0.35 m^3 and contains steam at 6 bar with a dryness fraction of 0.90. The valve is then opened and the two tanks come to a uniform State. If there is no heat transfer during the process find the final pressure and the entropy change of universe. [**Ans.** 322.6 kPa, 0.1985 kJ/K]

3.7 A mass of wet steam at 165°C is expanded at constant quality 0.8 to 3 bar. It is then heated at constant pressure to a degree of superheat of 66.5°C. Find the enthalpy and entropy changes during expansion and heating. [**Ans.** – 59 kJ/kg, 0.163 kJ/kg.K; 676 kJ/kg, 1.588 kJ/kg.K]

3.8 A cylinder fitted with a piston contains 2 kg steam at 500 kPa, 400°C. Find the entropy change and the work done when the steam expands to a final pressure of 200 kPa (*a*) adiabatically and reversibly, and (*b*) adiabatically and irreversibly to an equilibrium temperature of 300°C. [**Ans.** 0, 386.7 kJ; 0.1976 kJ/K, 309.4 kJ]

3.9 Steam at a pressure of 7 bar and 0.95 dry is supplied to a heater through a pipe of 25 mm internal diameter. The velocity of steam in the pipe is 12 m/s. Water enters the heater at 20°C. The steam is blown into it and the mixture leaves the heater at 90°C. Calculate (*a*) the mass of steam entering the heater per hour, and (*b*) the mass of water entering the heater per hour. [**Ans.** 81.7 kg/h, 636 kg/h]

3.10 A combined separating and throttling calorimeter recorded the following observations :
Pressure in the steam main = 15 bar
Mass of water drained from the separate = 0.55 kg
Mass of steam condensed after passing through the throttle valve = 4.20 kg
Pressure and temperature after throttling = 1 bar, 120°C
Calculate the dryness fraction of steam in the main. [**Ans.** 0.85]

3.11 Steam from an engine exhaust at 1.25 bar flows steadily through an electric calorimeter and comes out at 1 bar, 130°C. The calorimeter has two kW heaters and the flow is measured to be 3.4 kg in 5 min. Find the quality of steam in the engine exhaust. [**Ans.** 0.944]

3.12 In a separating and throttling calorimeter the total quantity of steam passed was 40 kg and 2.2 kg of water was collected from the separator. Steam pressure before throttling was 1.47 MPa and temperature and pressure after throttling were 120°C and 107.88 kPa. Determine the dryness fraction of steam before entering the calorimeter. Take specific heat of superheated steam as 2.09 kJ/kg.K. [**Ans.** 0 9064]

3.13 A vessel containing 5 kg of steam at 8 bar, 250°C is cooled by pouring water over the outer surface till the inside pressure falls to 5 bar. Calculate (*a*) the final state of steam, (*b*) heat loss, and (*c*) loss of internal energy. [**Ans.** wet, 0.7795; 3338.5 kJ; 2900.38 kJ]

3.14 A cylinder fitted with a movable piston contains 1 kg of steam at an initial pressure of 6 bar and of volume 0.26 m^3. Determine the dryness fraction of steam.
If steam expands according to the law $pv^n = c$, to a final volume of 0.52 m^3 and the pressure is 2.5 bar, calculate the value of index *n* and the work done during expansion.
[**Ans.** 0.807, 1.263, 5333.7 kJ]

3.15 A rigid metallic vessel is divided into two parts by a partition. Part *A* contains 1 kg of steam at 3.4 bar dry and saturated. Part *B* contains 2 kg of steam at 6.8 bar having 0.8 dryness fraction. The partition is removed and the pressure in the vessel after steady state is found to be 4.8 bar Calculate (*a*) volume of vessel, (*b*) state of steam at 4.8 bar, and (*c*) heat interchange between steam and the surrounding walls of vessel. [**Ans.** 1.0055 m^3, 0.7649, – 11709.6 kJ]

3.16 In a barrel calorimeter, 5 kg of steam is sampled at 7 bar. The water equivalent of calorimeter at 20°C is 10 kg. The barrel contains 140 kg of water at 20°C. The temperature of water when steam condensed into it was found to be 40°C. Calculate the dryness fraction of steam.
[**Ans.** 0.963]

3.17 A closed vessel having a volume of 2.84 m^3 contains 1.365 kg of mixture of water and steam at a pressure of 7 bar. Find how much heat must be added to the vessel to make the steam 0.9 dry and find the resulting steam pressure. [**Ans.** 3226.5 kJ, 7.47 bar]

3.18 5 kg of steam at 14 bar, 280°C expands to 1.4 bar according to the law $pv^{1.25}$ = constant. Calculate (*a*) the final temperature, (*b*) the work done, (*c*) heat transferred, and (*d*) change in entropy. [**Ans.** 108.75°C, 1807.4 kJ, 466.29 kJ, 1.28 kJ/K]

3.19 Steam enters a turbine at 700 kPa, 250°C and exhausts at 15 kPa. The efficiency of the turbine is 0.8. Calculate (*a*) the power output per kg of steam, and (*b*) the quality of steam leaving the turbine. [**Ans.** 520.12 kJ/kg, 0.93]

3.20 Steam at 10 bar, 240°C is cooled under constant pressure until the steam becomes 0.8 dry. Find (*a*) the work done, change in enthalpy and heat transfer if the process is non-flow one, and (*b*) the quality of steam if 850 kJ/kg is lost during the flow of steam through a pipe under constant pressure. [**Ans.** – 64.56 kJ, – 276.7 kJ/kg; 0.64]

3.21 A spherical vessel of 0.4 m^3 capacity contains 2 kg of wet steam at a pressure of 600 kPa. Calculate (*a*) the volume and mass of liquid, and (*b*) volume and mass of vapour.

[**Ans.** 8×10^{-4} m^3 , 0.736 kg; 0.3992 m^3, 1.264 kg]

3.22 Steam enters a turbine at 10 bar, 300°C with a velocity of 50 m/s. The steam leaves the turbine at 1.5 bar with a velocity of 200 m/s. Determine the work done per kg of steam. Assume the process to be reversible adiabatic and neglect the change in potential energy.

[**Ans.** 375.55 kJ/kg]

3.23 A vessel of 1.35 m^3 capacity is filled with steam at 13 bar absolute and 95% dry. The vessel and its contents cool until the pressure is 2 bar absolute. Calculate the mass of the contents in the vessel and the dryness fraction of steam after cooling. Neglect the volume of water present.

[**Ans.** 9.2 kg, 0.1625]

3.24 Steam at a pressure of 5 bar and temperature 200°C is expanded adiabatically to a pressure of 0.7 bar absolute. Determine the final condition of steam. For superheated steam, c_{ps} = 2.2 kJ/kg. K. [**Ans.** 0.93]

3.25 Steam at a pressure of 20 bar absolute and dryness fraction 0.8 is throttled to a pressure of 0.5 bar. Determine the final condition of steam. [**Ans.** 0.905]

3.26 2 kg of wet steam at 10 bar and 90% dry is expanded according to the law *pv* = const. to a pressure of 1 bar. Determine the final condition of steam and the change in internal energy.

[**Ans.** Superheated to 112°C, 257 kJ]

3.27 While conducting the dryness fraction test with a throttling calorimeter, it was found that the entering steam was at a pressure of 12 bar and a sample after being reduced to 1 bar in the calorimeter was at a temperature of 120°C. Estimate the dryness fraction of steam assuming c_{ps} = 2.1 kJ/kg.K. [**Ans.** 0.965]

3.28 In a test with a separating and throttling calorimeter, the following observations were made :

Water separated = 2.04 kg

Steam discharged from throttling calorimeter = 20.6 kg at 150°C

Initial pressure of steam = 12 bar abs.

Final pressure of steam = 12.3 cm of mercury above atmospheric pressure

Barometer = 76 cm of mercury

Calculate the dryness fraction of steam entering the calorimeter. [**Ans.** 0.905]

3.29 Determine the change in internal energy when one kg of steam expands from 10 bar 300°C to 0.5 bar and 0.9 dry. Take c_{ps} = 2.1 kJ/kg. K. [**Ans.** – 669 kJ/kg]

3.30 One kg of steam at 10 bar exists at 200°C Calculate the enthalpy, specific volume, density internal energy and entropy. Take c_{ps} = 2.1 kJ/kg, K.

[**Ans.** 2818 kJ/kg, 0.203 m^3/kg, 4.93 kg/m^3, 2615 kJ/kg, 6.673 kJ/K]

3.31 A vessel contains one kg of steam which contains $\frac{1}{3}$ rd liquid and $\frac{2}{3}$ rd vapour by volume. The temperature of steam is 151.86°C Calculate the quality, specific volume and specific enthalpy of the mixture. [**Ans.** 0.0576,0.003245 m^3/kg, 650.5 kJ/kg]

3.32 A cylinder fitted with a piston contains 0.5 kg of steam at 4 bar. The initial volume of steam is 0.1 m^3. Heat is transferred to steam at constant pressure until the temperature becomes 300°C. Determine the heat transferred and work done during the process. [**Ans.** 771 kJ, 91 K]

3.33 Steam at 10 bar and 0.9 dry initially occupies 0.35 m^3. It is expanded according to the law $pv^{1.25}$ = const until the pressure falls to 2 bar. Determine (*a*) the mass of steam used in the process, (*b*) the work, done, (*c*) the change in internal energy, and (*d*) the heat exchange between steam and surroundings. [**Ans.** 2 kg, 192 kJ, – 896 kJ, – 704 kJ]

3.34 One kg of steam at 8.5 bar and 0.95 dry expands adiabatically to a pressure of 1.5 bar. The law of expansion is $pv^{1.2}$ = const. Determine (*a*) the final dryness fraction of steam, and (*b*) the change in internal energy during the expansion. [**Ans.** 0.792, – 230 kJ/kg]

3.35 A throttling calorimeter is used to measure the dryness fraction of steam in the steam mass where the steam is flowing at a pressure of 6 bar. The steam after passing through the calorimeter comes out at 100 kPa pressure and 120°C temperature. Calculate the dryness fraction of steam in the main. Assume c_{ps} = 2.09 kJ/kg.K. [**Ans.** 0.98]

3.36 A separating and throttling calorimeter was used to determine the dryness fraction of steam flowing through a steam main at 900 kPa. The pressure and temperature after throttling was 105 kPa and 115°C respectively. The mass of steam condensed after throttling was 1.8 kg and mass of water collected in the separating calorimeter was 0.16 kg. Determine the dryness fraction of steam flowing through the steam main. Take c_{ps} = 2.09 kJ/kg.K. [**Ans.** 0.8]

3.37 (*a*) Explain the terms : melting point, boiling point and critical point.

(*b*) State the difference between

(*i*) Superheated steam and wet steam.

(*ii*) Sensible heat of water and latent heat of evaporation.

(*c*) Define dryness fraction. Explain with a neat sketch the working of throttling calorimeter.

3.38 (*a*) Explain with a neat sketch working of a separating and throttling calorimeter.

(*b*) Explain the following terms : dryness fraction, priming, superheated steam.

4 STEAM BOILERS

4.1 INTRODUCTION

- A boiler may be defined as a closed vessel in which steam is produced from water by combustion of fuel at the desired temperature and pressure.
- According to the Indian Boiler Regulations — 1950 (I.B.R.), a boiler is a closed pressure vessel with capacity exceeding 22.75 litres used for generating steam under pressure. It includes all the mountings fitted to such vessels which remain wholly or partly under pressure when steam is shut off.
- According to American Society of Mechanical Engineers (A.S.M.E.), a steam generating unit is defined as :

  "A combination of apparatus for producing, furnishing or recovering heat together with the apparatus for transferring the heat so made available to the fluid being heated and vaporised".
- The difference between a steam boiler and a steam generator is that a steam boiler consists only of the containing vessel and convection heating surfaces, whereas a steam generator covers the whole unit including water wall tubes, superheater, airheat and economiser.
- The steam generated is employed for the following purposes :
  1. Generating power in steam engines and steam turbines.
  2. Process heating in textile, sugar mills, chemical industries, etc.
  3. Heating of buildings in cold weather and producing hot water for hot water supply.
- The primary requirements of boilers or steam generators are :
  1. The water must be contained safely.
  2. The steam must be safely delivered in desired condition, *i.e.*, pressure, temperature, quality and mass flow rate.

4.2 CLASSIFICATION OF STEAM GENERATORS

Steam generators may be classified on the following basis :

1. Water tube and fire tube,
2. Natural/forced circulation,
3. Horizontal, vertical or inclined,
4. Single/ multiple tube(s),
5. Stationary/mobile
6. Internally/externally fired
7. Source of heat

1. In a water tube boiler, water or steam flows through the tubes and heat is supplied to outside surface. Babcock and Wilcox and Stirling boilers are of this category.

   In case of fire tube boiler, hot gases flow through the tubes and water circulates outside the tubes. There may be one large tube surrounded by water as in Cornish boiler, or two large tubes surrounded by water in a big tube of water as in the case of Lancashire boiler. There may be many smaller tubes through which hot gases pass and all surrounded by water as in the case of vertical Cochran or Locomotive boiler.

2. Natural circulation of water takes place by natural convection currents produced by application of heat as in the case of Lancashire boiler and Babcock and Wilcox boiler.

   In forced circulation, the fluid is forced 'once through' or controlled with partial recirculation as in the case of Lamont boiler, Velox boiler and Benson boiler.

3. The disposition on the principal axis of the boiler decides whether the boiler is horizontal, vertical or inclined.

4. A boiler may have only one fire/water tube or multiple tubes.

5. The boilers are called either stationary (land) or mobile (marine and locomotive). Stationary boilers are used for power plant steam generation. Mobile boilers are portable boilers.

6. The boiler is said to be external combustion takes place outside the region of boiling water. The boiler is said to be internal combustion boiler, if the furnace region is completely surrounded by water-cooled surface as in the case of Lancashire boiler.

7. Boilers may be classified according to the source from which heat is supplied to water for evaporation. For example, coal fired, oil fired, gas fired, electrical energy or nuclear energy.

4.3 COMPARISON OF FIRE TUBE AND WATER TUBE BOILERS

	Criteria	Fire tube Boilers	Water tube Boilers
1.	Raising of steam	Less rapid	More rapid
2.	Suitability for large power plants	Unsuitable because of lower rate of steam raised.	Suitable because of higher rate of steam raised
3.	Operating pressure	$\leq$ 25 bar	> 125 bar
4.	Floor area required per kg of steam	Less	More
5.	Reliability and cost	Greater reliability and low cost	Require careful attention and maintenance costs are higher
6.	Suitability for rapid changes in load	Suitable	Unsuitable
7.	Damage due to failure in feed water supply	No damage to boiler as it contains large quantity of water	The boiler shall be overloaded
8.	Risk of damage to property in case of bursting.	Serious because of its large water capacity	Less serious
9.	Transportation	Very inconvenience because of large size	Can be easily transported and erected as various parts can be separated.
10.	Chances of bursting	Less	More
11.	Maintenance	Difficult for maintenance	All parts are easily accessible for cleaning, inspection and repairing.
12.	Treatment of feed water	Not essential	Essential

4.4 REQUIREMENTS OF A GOOD BOILER

A good boiler should possess the following qualities :

1. It should be able to generate steam at the required pressure and quantity with minimum fuel consumption.
2. Its initial cost, installation, maintenance and operational cost should be as low as possible.
3. It should occupy small floor area and should be light in weight.
4. It should be able to meet the fluctuating demands of steam without fluctuations in steam pressure.
5. All parts of the boiler should be easily accessible for cleaning, inspection and maintenance.
6. It should have minimum of joints to avoid leakage of steam.
7. Its erection time should be reasonable with minimum labour.
8. The water and flue gases velocities should be high for better heat transfer with minimum pressure drop through the system.
9. There should be no deposition of mud and foreign materials on the inside surface and soot deposits on the outside surface of heat transferring parts.
10. The boiler should conform to safety precautions as laid down in the Boiler Regulations.

4.5 FACTORS AFFECTING BOILER SELECTION

1. Power to be generated, *i.e.* steam to be raised per hour.
2. Working pressure.
3. Initial capital investment.
4. Facilities available for erection.
5. Floor air available.
6. Location of power house.
7. Availability of fuel and water.
8. Probable load factor.
9. Operating and maintenance cost.
10. Accessibility for inspection, cleaning and maintenance.

4.6 BOILER TERMINOLOGY

Air-preheater — A device to heat the supply air to boiler by flue gases or exhaust steam.

Ash — The incombustible inorganic matter in the fuel.

Accessories — The additional components used with the boiler to increase its efficiency, *e.g.* superheater, economiser, feed pump, etc.

Blowing off — The removal of mud and other impurities from water at the lowest part of the boiler by means of a blow down cock.

Cinder — Lumps of partially burned coal/coke.

Damper — A device for regulating the mass flow rate of supply air.

Economiser — A device to preheat the feed water by the heat of escaping flue gases.

Fire tube — A tube carrying the products of combustion on the inside and water on the outside.

Flue gas — The gaseous products of combustion.

Fly ash — The suspended ash particles carried in the flue gas.

Furnace (or fire box) — A chamber formed by the space above the grate and below the boiler shell in which combustion takes place.

Grate — It is a platform in the furnace upon which fuel is burnt and air is passed for combustion.

Injector — A device for feeding water into the boiler.

Lagging — Blocks of asbestos or magnesia insulation wrapped on the outside of a boiler shell or steam piping.

Manhole — An access opening to the interior of a boiler for inspection and cleaning purposes.

Mountings — Components mounted on the boiler to work it safely.

Setting — The primary function of setting is to confine heat to the boiler and form a passage for gases. It is made of brick work and may form the wall of the furnace and the combustion chamber.

Refractory — A heat insulating material used for lining combustion chamber.

Scrubber — An apparatus used for the removal of solids from gases by entrainment in water.

Shell — One or more steel plates bent into a cylindrical form and riveted or welded together.

Scale — A deposit of medium to extreme hardness occurring on water heating surface of a boiler.

Slag — A residue deposited by ash particles that have attained their softening temperature.

Soot blower — A tube from which jets of steam or compressed air are blown for cleaning the fire side of tubes or other parts of a boiler.

Superheater — A device for superheating the steam by high temperature gases or radiant heat.

Water tube — A tube in the boiler having water on the inside and products of combustion on the outside.

Water level — The level at which water stands in the boiler.

4.7 DESCRIPTION OF BOILERS

4.7.1 Fire Tube Boilers

1. Simple Vertical Boiler. A simple vertical boiler is shown in Fig. 4.1. It consists of a cylindrical shell the greater portion of which is full of water and remaining is the steam space. At the bottom of the fire box is grate on which fuel is burnt and the ash from it falls in the ash pit. The fire box is provided with two cross tubes. This increases the heating surface and the circulation of water. The cross tubes are fitted inclined for efficient circulation of water. At the ends of each cross tube are provided hand holes to give access for cleaning these tubes. The combustion gases after heating water and converting it into steam escape to the atmosphere through the chimney. Manhole is provided to clean the interior of the boiler and exterior of the combustion chamber and chimney. The various mountings are : pressure gauge, water level indicator, safety valve, steam stop valve, feed check valve, and blow-off cock.

The rate of steam generation of such a boiler does not exceed 2500 kg/h and pressure is generally limited to 7.5 to 10 bar. The efficiency of this boiler is also low. However, it is very compact, occupies less space and easy to transport.

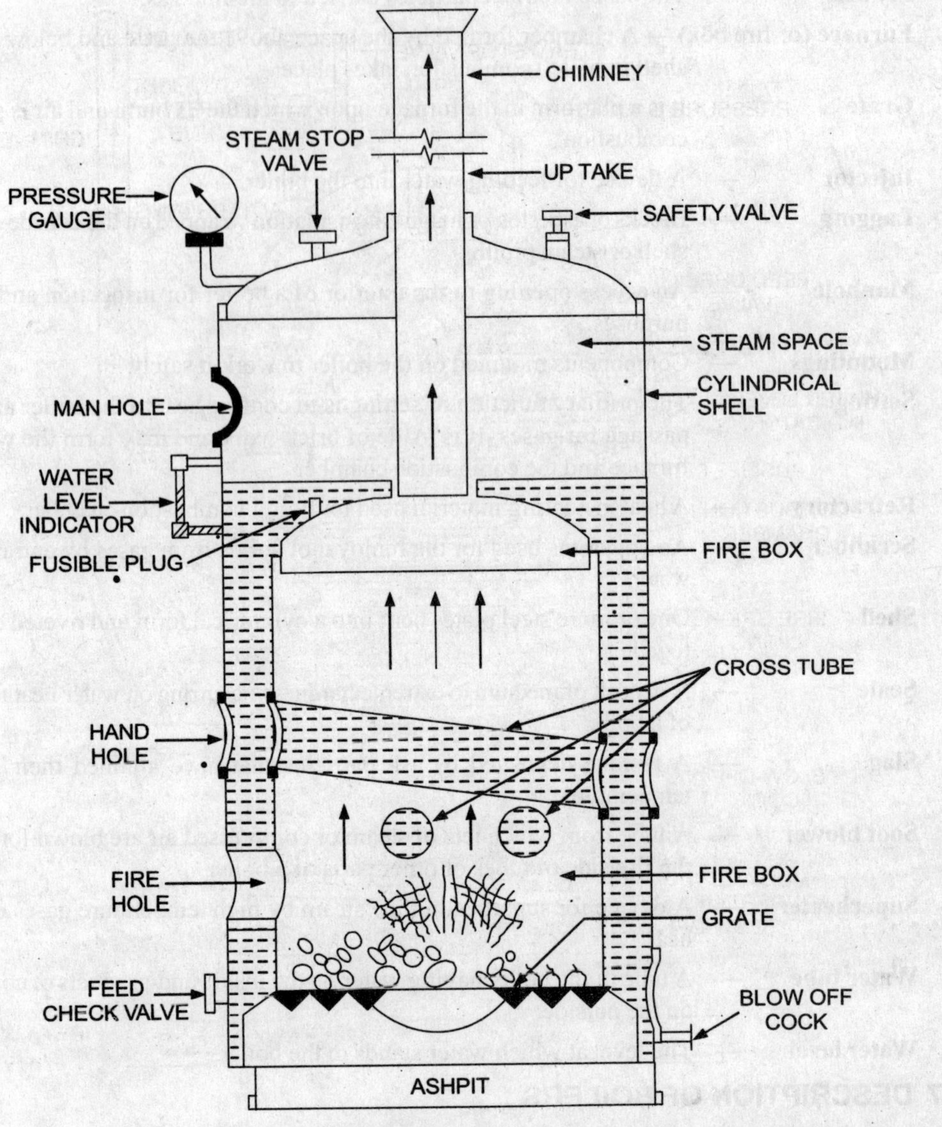

Fig. 4.1 Simple Vertical Boiler

2. Cochran Boiler. A cochran boiler is shown in Fig. 4.2. It consists of a cylindrical shell with a dome shaped top where the space is provided for steam. The furnace is one piece construction and is seamless. Its crown has a hemispherical shape and thus provides maximum volume for steam. The fuel is burnt on the grate and ash is collected and disposed off from the ash pit. The gases of combustion produced by burning of fuel enter the combustion chamber through the flue tube and strike against fire brick lining which directs them to pass through number of horizontal tubes, being surrounded by water. The gases then escape to the atmosphere through smoke box and chimney. A number of hand holes are provided around the outer shell for cleaning purposes. The various boiler mountings are : water level indicator, safety valve, steam stop valve, blow-off cock, manhole, and pressure gauge. It is used to generate steam from 3500 - 4000 kg/h at a pressure of 6.5 to 15 bar. Its efficiency is 70 - 75%.

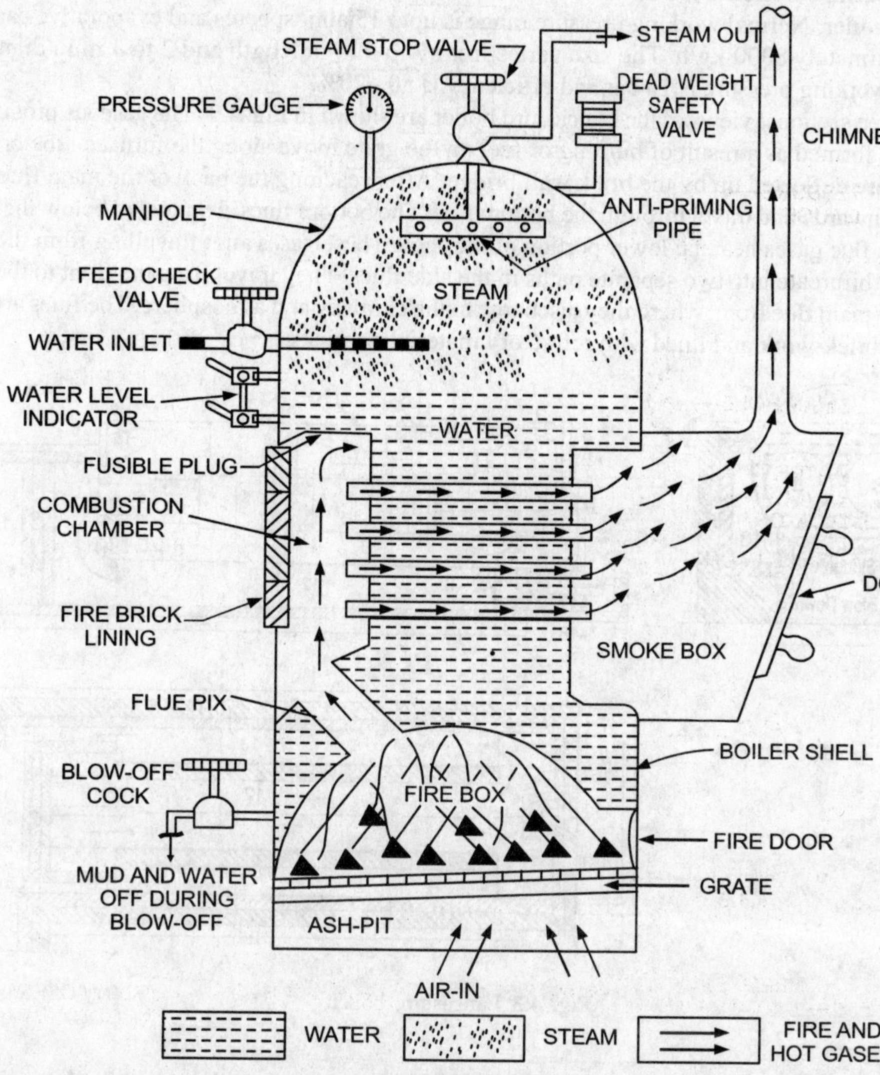

Fig. 4.2 Cochran Boiler

The advantages of this boiler are :

1. It has compact design and requires minimum floor area.
2. Low cost of construction.
3. Easy to transport and install.
4. It has self-contained furnace.
5. Any type of fuel can be used.

Disadvantages :

1. Low steam raising capacity.
2. Difficult to inspect and clean due to vertical design.
3. Limited capacity and pressure.
4. Requires high head room.

3. Lancashire Boiler. It is a stationary, fire tube, horizontal straight tube, internally fired, natural circulation boiler. Normal working pressure range is upto 15 atmospheres and evaporative capacity upto approximately 8000 kg/h. The size varies from 7 to 9 m in length and 2 to 3 m in diameter. Maximum working pressure is 16 bar and efficiency is 50 - 70%.

The three sectional views of the Lancashire boiler are shown in Fig. 4.3. The gaseous products of combustion formed as a result of burning of fuel on the grate move along the furnace tube or main flue. They are deflected up by the brick wall bridge. After reaching the back of the main flue, they deflect downwards and travel through the bottom flue. The bottom flues are situated below the water shell. These flue gases heat the lower portion of the shell. These gases after travelling from the back to the front, bifurcate into two separate paths in the side flues. They travel from the front to the back again in the main flue from, where they discharge to the chimney and atmosphere. The flues are built of ordinary brick-work and lined with refractory material or bricks.

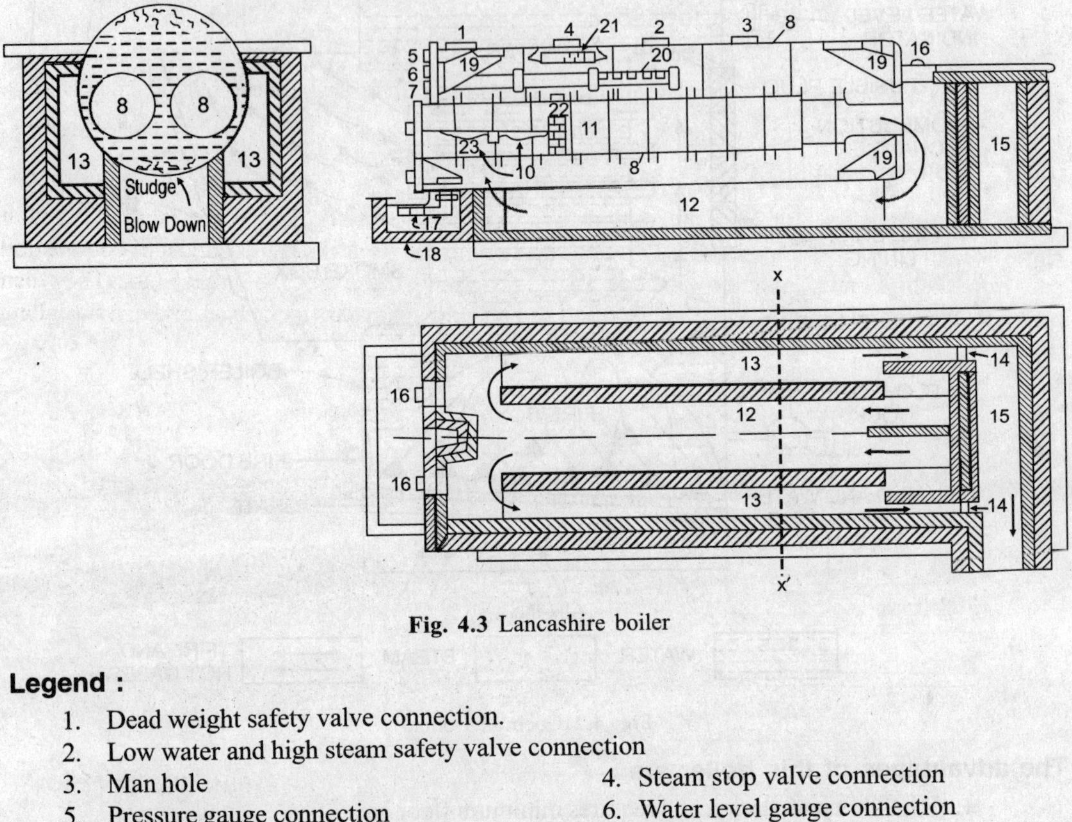

Fig. 4.3 Lancashire boiler

Legend :

1. Dead weight safety valve connection.
2. Low water and high steam safety valve connection
3. Man hole
4. Steam stop valve connection
5. Pressure gauge connection
6. Water level gauge connection
7. Feed check valve connection
8. Flue tubes for first pass
9. Boiler shell
10. Grate
11. Brick-wall bridge
12. Bottom flue
13. Side flue
14. Dampers
15. Main flue
16. Cleaning doors
17. Blow off cock connection
18. Blow off pit
19. Gusset stays
20. Perforated feed pipe
21. Anti-priming device
22. Fusible plug
23. Ash pit.

This type of boilers are used in sugar mills and textile industry where steam is required for power generation as well as for process heating. It is also used in chemical industries.

Advantages

1. Large reservoir capacity to meet load fluctuations.
2. Easy to inspect and clean.
3. Large heating surface area per unit volume of boiler.
4. High steam generation capacity.
5. Low maintenance cost.
6. Easy to operate and reliable.

Disadvantages

1. Low maximum steam pressure.
2. Requires more floor area due to brickwork settings.
3. Possibility of leakage due to large temperature difference between outside and inside.
4. Takes longer time to generate steam.
5. Restricted grate area due to furnace inside the tubes.

4. Cornish Boiler. The Cornish boiler shown in Fig. 4.4 consists of a cylindrical shell with flat ends through which passes a smaller flue tube containing the furnace. The products of combustion pass from the fire grate forward over the brick work bridge to the end of the furnace tube. They then return by the two side flues to the front end of the boiler, and again pass to the back end of a flue along the bottom of the boiler to the chimney.

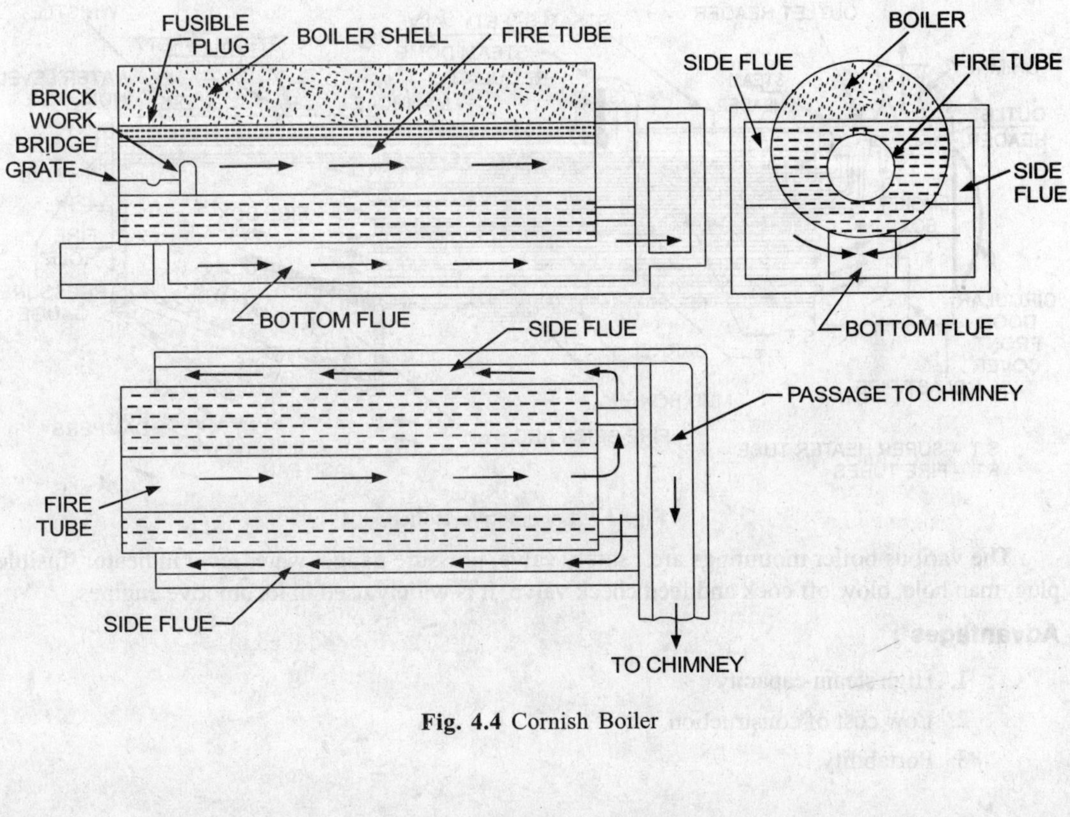

Fig. 4.4 Cornish Boiler

The various boiler mountings are : steam stop valve, pressure gauge, water level indicator, fusible plug, blow off cock, high pressure low water safety valve, feed check valve, and manhole.

Its steam generation capacity is 6500 kg/h at 10.5 bar pressure. The advantage possessed by this type of boiler is that the sediment contained in water falls to the bottom, where the plates are not brought into contact with the hottest portion of the furnace gases.

5. Locomotive Boiler. The locomotive boiler shown in Fig. 4.5 consists of a cylindrical barrel with a rectangular fire box at one end and a smoke box at the other end. The coal is added through the fire hole into the grate placed at the bottom of fire box. The hot gases generated due to burning of coal are deflected by an arch of fire bricks to heat the walls of the fire box properly. The fire box is entirely surrounded by water except for the fire hole and the ash pit. The fire hole and ash pit are situated below the fire box and are fitted with dampers at its front and back ends. The dampers control the flow of air to the grate. The hot gases pass from the fire box to the smoke box through a series of fire tubes and then they are discharged into the atmosphere through the chimney. The fire tubes are placed inside the barrel and are of different sizes. Superheater tubes are placed inside the fire tubes of larger diameter. The heat of hot gases is transmitted to water through the heating surface of the fire tubes. The steam generated is collected over the water surface.

The steam dome is fitted on the upper part of the barrel from where the steam flows through a steam pipe into the chamber. The steam flow is controlled by a regulator. From the chamber steam passes through the super heater tubes and return to the superheated steam chamber from which it is lead to the engine cylinders through separate pipes.

The draught is produced by exhaust steam from the engine cylinders which is discharged through the blast pipes to the chimney. While stationary, fresh steam from the boiler is used for this purpose. Its steam generation capacity is 9000 kg/h at 14 bar, consumes about 1600 kg/h of coal and efficiency is about 70%.

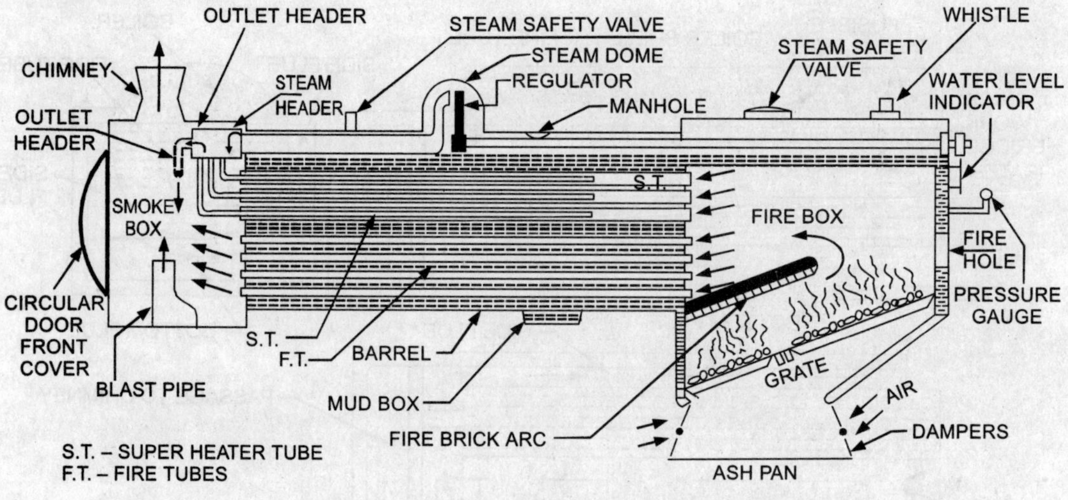

Fig. 4.5 Locomotive Boiler

The various boiler mountings are : safety valve, pressure gauge, water level indicator, fusible plug, man hole, blow off cock and feed check valve. It is widely used in locomotive engines.

Advantages :

1. High steam capacity
2. Low cost of construction.
3. Portability.

4. Low installation cost.
5. Compact size.

Disadvantages :

1. There are chances of corrosion and scale formation in the water spaces due to accumulation of sediments and mud particles.
2. It is difficult to clean some water spaces.
3. Large flat surfaces require bracing.
4. It can not carry high overloads without being damaged by overheating.

6. Scotch Boiler. The Scotch boiler is of the marine type as shown in Fig. 4.6. It consists of a cylindrical shell in which are incorporated one to four cylindrical, corrugated steel furnaces. The furnaces are internally fired and surrounded by water. A combustion chamber is located at the back end of the furnace and is also surrounded by water. A nest of fire tubes run from the front tube plate to the back tube plate. The hot gases produced due to burning of fuel move to the combustion chambers by means of draught. Then they travel to the smoke box through the fire tubes and finally leave the boiler via uptake and chimney. It is used to generate about 1000 kg/h of steam at 17 bar. It is used in marine applications.

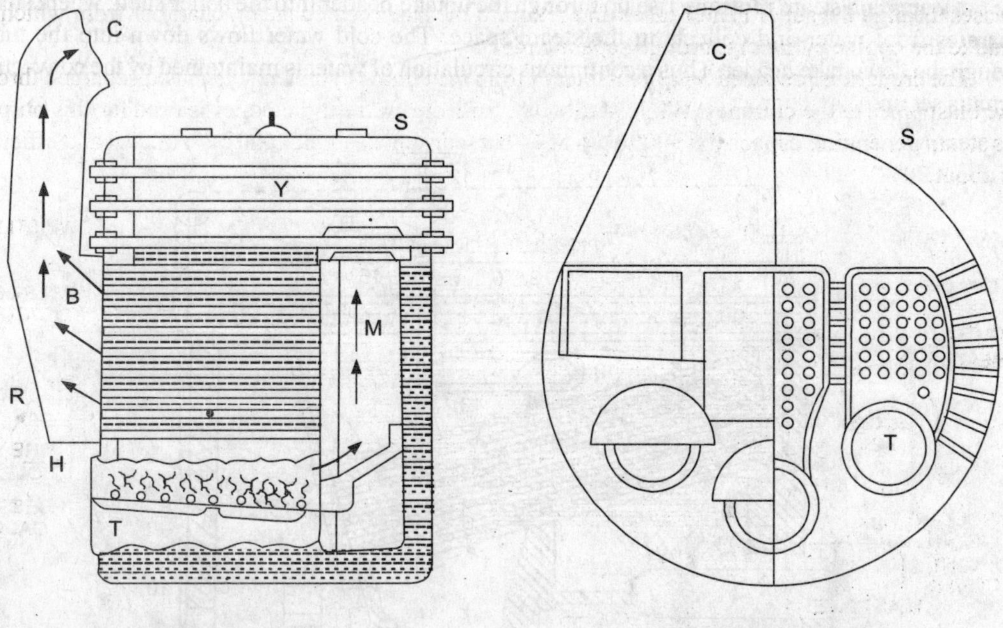

T = Furnace tube
H = Fire hole
M = Combustion chamber
B = Smoke box
R = Smoke box door
C = Chimney
S = Shell
Y = Bar stays

Fig. 4.6 Scotch boiler

Advantages :

1. Compact size and occupies less space.
2. Requires minimum amount of brick work.
3. Requires low head room.
4. All tubes being of same size, only one spare tube is necessary.

Disadvantages :

1. Circulation difficulties are experienced during starting.
2. Low capacity and steam pressure.
3. Low efficiency.
4. It is difficult to inspect and clean.

4.7.2 Water Tube Boilers

For pressures above 10 bar and capacities in excess of 7000 kg of steam per hour, the water tube boiler is used almost exclusively.

1. Babcock and Wilcox Boiler. Fig 4.7 shows a stationary type Babcock and Wilcox boiler. It consists of a large number of parallel tubes inclined at an angle which varies from 5° to 15° to the horizontal which connect the uptake header with the down take header. Both of these are connected to the shell having a substantial quantity of water in it. The uptake header is connected to the shell, through a short tube while a long tube is employed to connect the down take header with the shell. The coal is fed through the fire hole on to the chain grate stoker. The velocity of the chain is so adjusted as to ensure complete combustion of coal by the time it reaches the other end of the grate. The flue gases first rise up then move down and once again rise up due to the presence of the baffles. The hot water and steam moisture rise up through the uptake header into the boiler shell, where steam separates from water and collects in the steam space. The cold water flows down into the tubes through the down take header. Thus a continuous circulation of water is maintained by the convection currents set up.

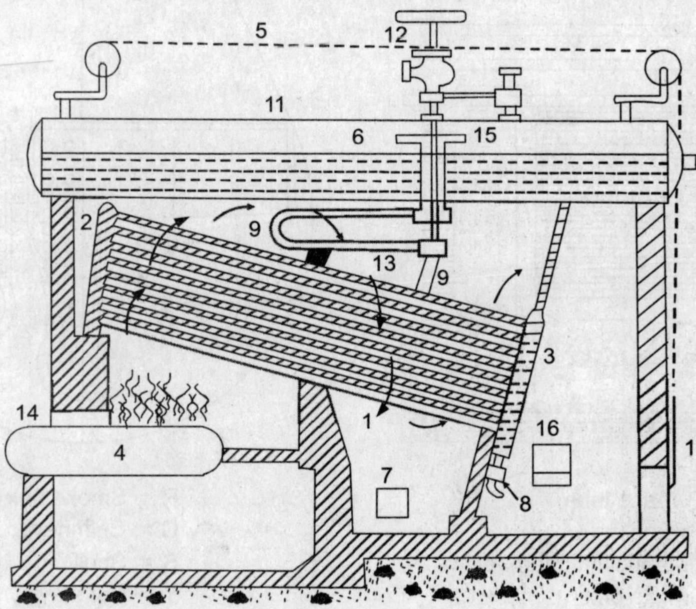

Fig. 4.7 Babcock and Wilcox boiler

Legend

1. Water tubes	5. Damper chain	9. Baffles	13. Super heater tubes
2. Uptake header	6. Steam pipe	10. Damper	14. Fire hole
3. Downtake header	7. Soot doors	11. Shell	15. Tube
4. Grate	8. Mud box	12. Stop valve	16. Side flue

Super heater — A set of super heater tubes is provided to superheat steam which enters these tubes from the steam space in the boiler shell. The superheated steam can be taken out to the steam stop valve, through the steam pipe. The lower part of the down take header has a mud box attached to it to collect the sediments. The damper chain controls the quantity of air flowing through the boiler. Two soot doors are provided for internal cleaning of boiler.

Its steaming capacity is 40,000 kg/h at a maximum pressure of 40 bar. It is used in large power stations due to greater operational safety and lesser space area required as compared to fire tube boilers to generate steam at the same pressure.

Advantages :

1. Both solid and liquid fuels can be burnt.
2. Minimum draught losses.
3. High evaporation capacity.
4. Natural water circulation.
5. Defective tubes can be easily replaced.

2. Stirling Boiler. Fig. 4.8 shows the Stirling boiler which is of water tube type in which three steam drums are connected together by the banks of bent water tubes. These tubes stand nearly in a vertical position near the rear end while they slope steeply over the fire at the front end. The baffles deflect the products of combustion from the furnace over the banks of tubes into the mud drum. The steam generated collects in the steam spaces of the three drums. The steam can be raised at a working pressure of 60 bar with the temperature as high as 450°C. An evaporation capacity of 50,000 kg/h has been achieved with a stirling boiler.

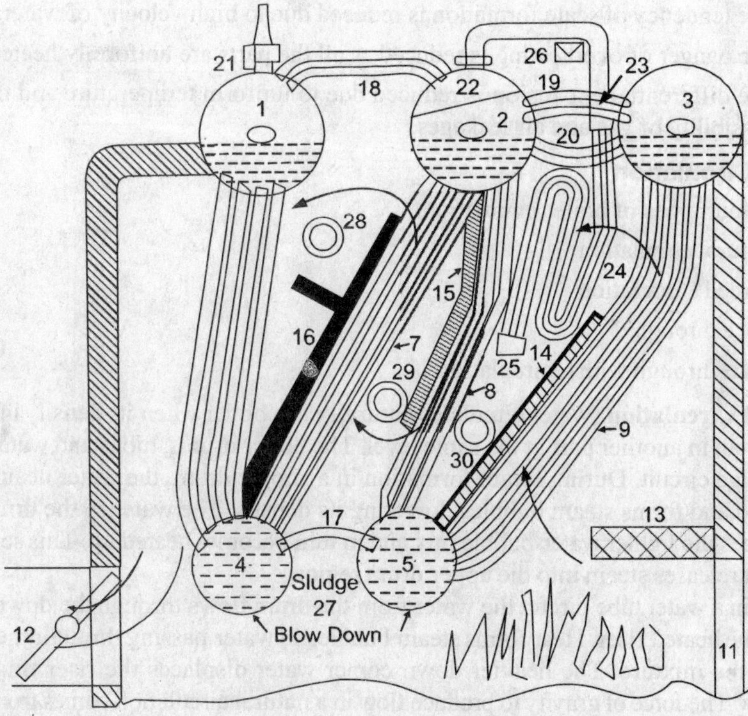

Fig. 4.8 Stirling boiler

Legend

1, 2,3 Steam drums
4, 5 Mud drums
6, 7, 8, 9 Bent tubes banks
10. Brick work
11. Fire door
12. Damper
13. Fire brick arch
14. 15, 16 Baffles
17. Water tubes to mud drums
18, 19 Steam tubes
20. Water tubes connecting front and middle drums
21. Safety valve connection
22. Steam chest
23. Header
24. Superheater
25. Bottom box header for superheater
26. Steam stop valve
27. Blow down valve
28. 29, 30. Inspection and cleaning doors.

This boiler is lighter and more flexible than the straight tube boilers. But it is comparatively more difficult to clean and inspect the bent tubes.

4.8 HIGH PRESSURE BOILERS

In all modern power plants, high pressure boilers raising steam at pressures greater than 100 bar are universally used. They are called high pressure boilers. These boilers offer the following advantages :

1. The efficiency and the capacity of the plant can be increased as reduced quantity of steam is required for the same power generation if high pressure steam is used.
2. The forced circulation of water through the boiler tubes provides freedom in the arrangement of furnace and water walls in addition to the reduction in the heat exchange area.
3. The tendency of scale formation is reduced due to high velocity of water.
4. The danger of overheating is reduced as all the parts are uniformly heated.
5. The differential expansion is reduced due to uniform temperature and this reduces the possibility of gas and air leakages.

4.8.1 Boiler Circulation

There are four types of boiler circulation.

1. Natural circulation
2. Forced Circulation
3. Once through
4. Once through with recirculation

1. Natural Circulation. Water circulates naturally in a boiler when its density in one part of the circuit is less than in another part at the same level. The boiler drums, tubes and water wall make up the multi-passage circuit. During steam formation in a simple drum, the water near the drum sides becomes heated and forms steam bubbles, lowering its density. The water in the drum centre being heavier, displaces the lighter water-bubble mix and in turn becomes heated up. This sets up a constant circulation that releases steam into the upper drum region.

Similarly in a water tube circuit the water from the drum flows through the down-comer tube to the bottom of the heated riser. Heat forms steam bubbles in water passing through the riser lowering the density of the mixture. The heavier down comer water displaces the riser mixer to establish continuous flow. The force of gravity to produce flow in a natural circulation comes from the difference between the densities of the fluids in the down-comer and riser portions of the circuit. The natural circulation is depicted in Fig. 4.9(*a*).

2. Forced Circulation. The forced circulation in various circuits of boiler units is produced by mechanical pumps as depicted in Fig. 4.9(*b*).

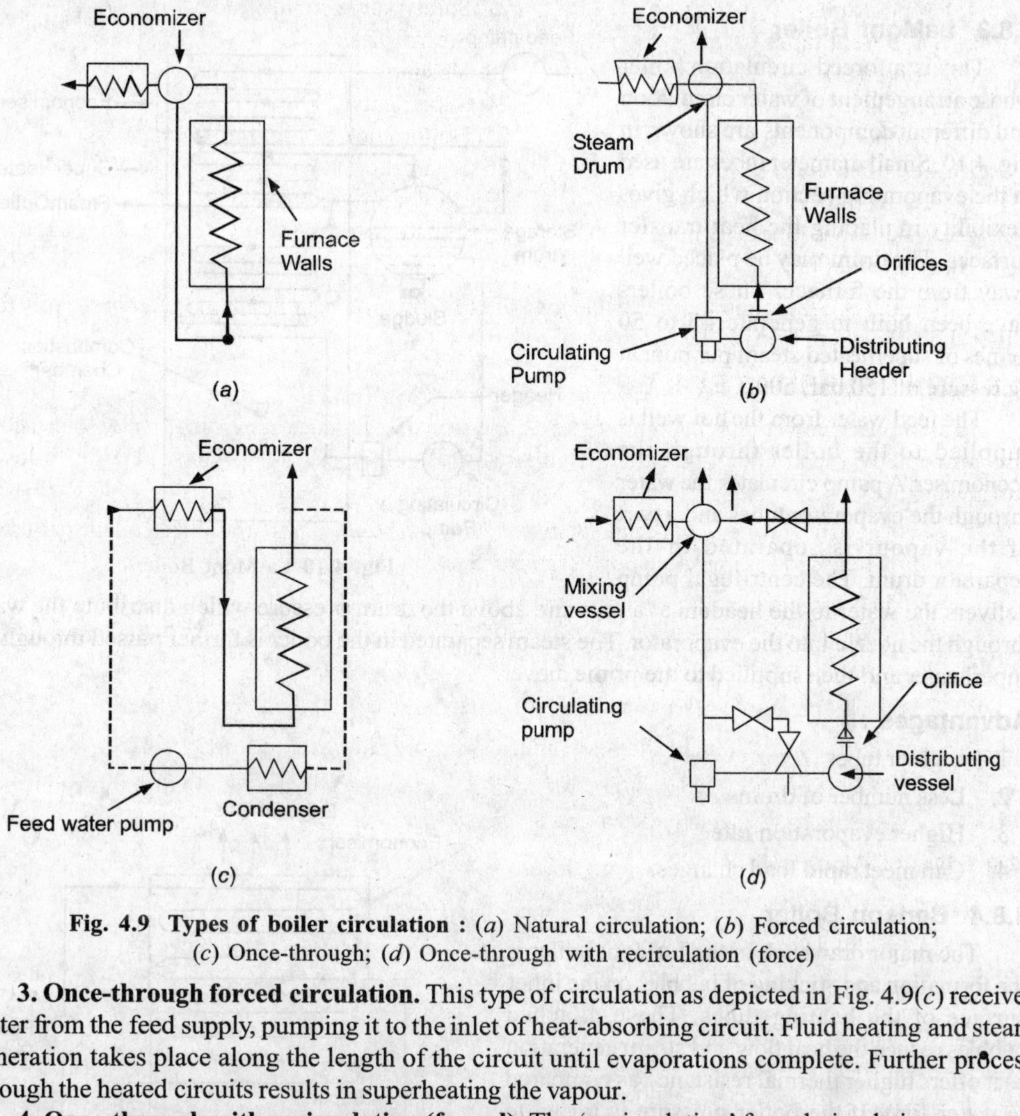

Fig. 4.9 Types of boiler circulation : (*a*) Natural circulation; (*b*) Forced circulation; (*c*) Once-through; (*d*) Once-through with recirculation (force)

3. Once-through forced circulation. This type of circulation as depicted in Fig. 4.9(*c*) receives water from the feed supply, pumping it to the inlet of heat-absorbing circuit. Fluid heating and steam generation takes place along the length of the circuit until evaporations complete. Further process through the heated circuits results in superheating the vapour.

4. Once-through with recirculation (forced). The 'recirculating' forced-circulating type unit has water supplied to the heat absorbing circuits through a separate circulating pump. The water pumped is considerably in excess of steam produced and requires a steam-and-water drum for steam generation. This type of system is depicted in Fig. 4.9(*d*).

4.8.2 Advantages of Forced Circulation Boilers

1. Smaller bore and therefore lighter tubes.
2. Absence from scaling troubles due to high circulation velocity.
3. Lighter for a given output.
4. Steam can be raised quickly and load fluctuations met rapidly.
5. Uniform heating of all parts eliminates the danger of over heating.
6. Greater flexibility in layout of boiler parts.

7. Boiler can be operated at desired conditions.
8. Boiler can be started rapidly.

4.8.3 LaMont Boiler

This is a forced circulation boiler whose arrangement of water circulation and different components are shown in Fig. 4.10. Small diameter tubes are used in the evaporating section which gives flexibility in placing the heat transfer surfaces. The drum may be placed well away from the furnace. These boilers have been built to generate 45 to 50 tonnes of superheated steam per hour at a pressure of 150 bar, 500°C.

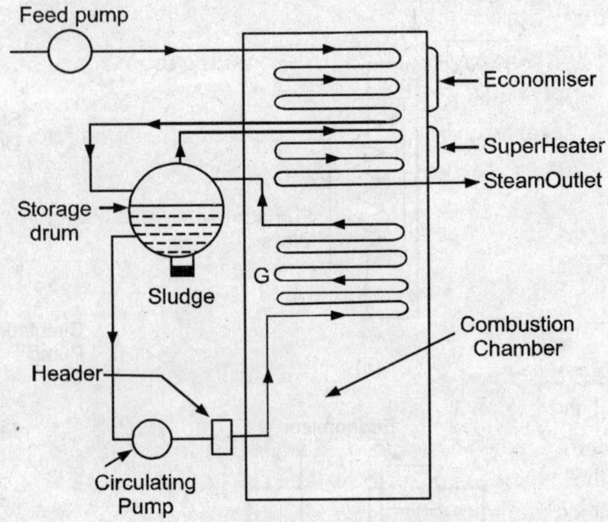

Fig. 4.10 La Mont Boiler

The feed water from the hot well is supplied to the boiler through the economiser. A pump circulates the water through the evaporator tubes and a part of the vapour is separated in the separator drum. The centrifugal pump delivers the water to the headers at a pressure above the drum pressure which distribute the water through the nozzle into the evaporator. The steam separated in the boiler is further passed through the superheater and then supplied to the prime mover.

Advantages :

1. Lighter tubes
2. Less number of drums
3. Higher evaporation rate
4. Can meet rapid load changes.

4.8.4 Benson Boiler

The major drawback in the LaMont boiler is the formation and sticking of bubbles on the inner surface of the heating tubes. These attached bubbles reduce the heat flow and steam generation as it offers higher thermal resistance as compared to water film. If the boiler pressure is raised to critical pressure (221.2 bar), the steam and water would have same density and therefore the danger of bubble formation can be completely eliminated.

The arrangement of the boiler components is shown in Fig. 4.11. The feed water is passed through the economiser and the radiant evaporator where major part of water is converted into steam. The remaining water is evaporated by hot gases in the convective evaporator. The super critical pressure steam is then passed through the super - heater after which it is supplied to the prime mover.

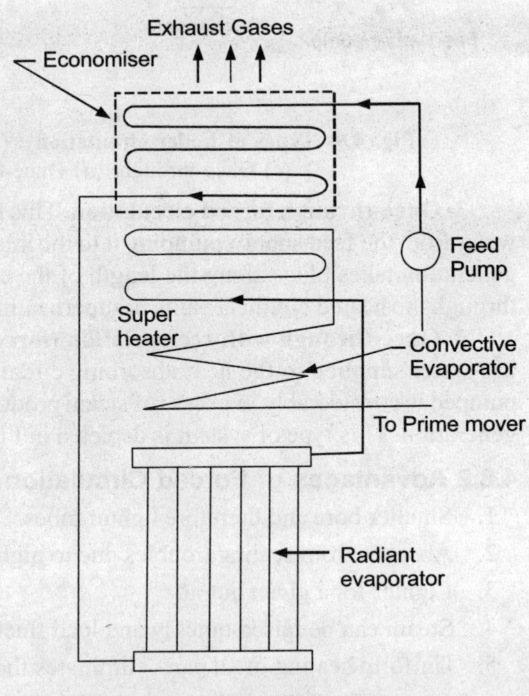

Fig. 4.11 Benson Boiler

The maximum pressure of steam obtained from this boiler is 500 bar, 650°C and generating capacity of 150 tonnes/h. The major difficulty in the operation of this boiler is the deposition of salt and sediments on the inner surface of water tubes.

Advantages :

1. It does not require any evaporating drum since the steam generation is at higher than critical pressure.
2. Lighter in weight with high steam generation rate.
3. Parts can be easily assembled at site.
4. Cheaper since the cost of evaporating drum is eliminated.
5. Occupies less space.
6. Very small starting period of 10-15 minutes.
7. Suitable for power generation.

4.8.5 Loeffler Boiler

The major difficulty of Benson boiler is avoided in the Loeffler boiler by preventing the flow of water into the boiler tubes. Most of the steam is generated outside from the feed water by using part of the superheated steam coming out from the boiler. The arrangement of different components and steam circulations are shown in Fig. 4.12.

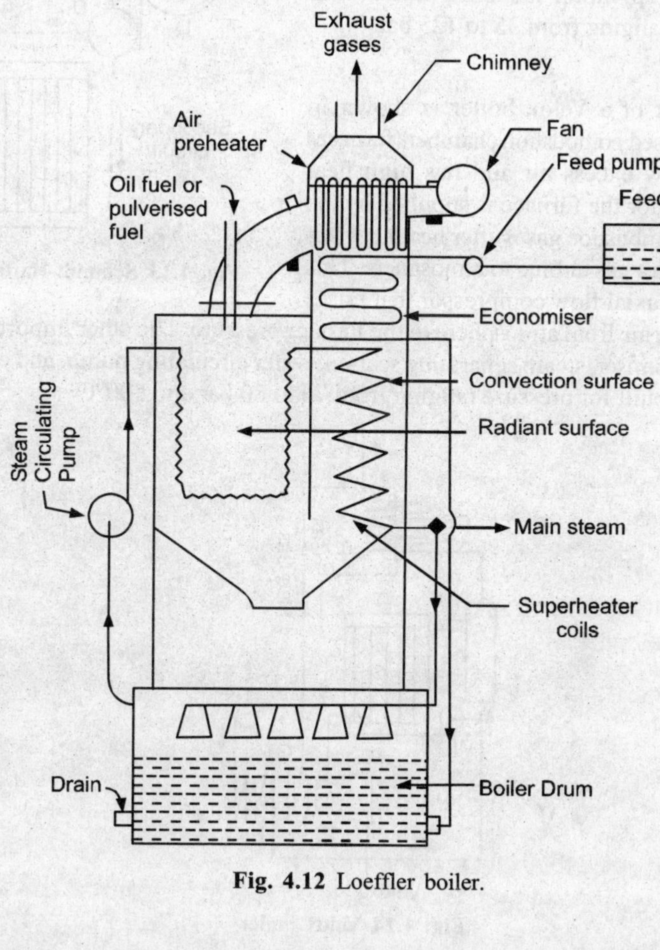

Fig. 4.12 Loeffler boiler.

The pressure feed pump draws the water through the economiser and delivers to the evaporator drum. About 65% of the steam coming out of superheater is passed through the evaporator drum to evaporate the feed water coming from the economiser. This steam is drawn by the steam circulating pump and passed through the radiant superheater and then convective super heater.

Loeffler boilers are available with generating capacity of 94.5 tonnes/h and operating at 140 bar. This boiler can handle high drum salt concentration without trouble.

As compared to LaMont boiler, its advantages are :

1. The evaporating tubes carry only superheated steam, hence there is no salt deposits on the tubes.
2. This boiler is suitable for marine applications.

4.8.6 Schmidt-Hartmann Boiler

The arrangement of this boiler is shown in Fig. 4.13. In the primary circuit, a feed pump supplies water to drum through economiser, and that in turn discharges saturated steam to a convective superheater and then to load. A closed secondary circuit heated by furnace resembles a sectional header boiler. This section supplies the heated vapour in a coil to evaporate the main feed water. It has a tube evaporating section. These boilers are built for pressure ranging from 35 to 125 bar.

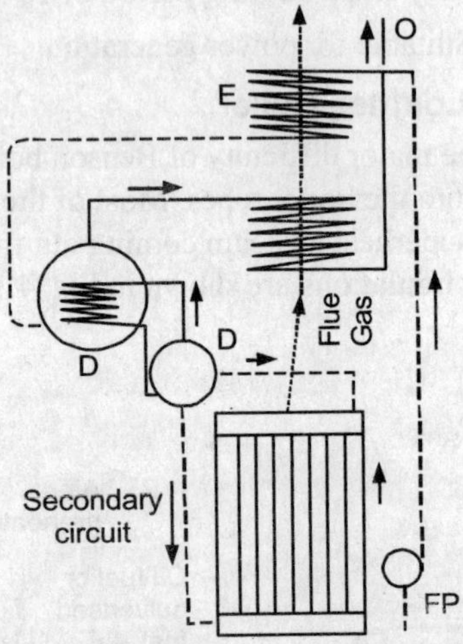

Fig. 4.13 Schmidt-Hartmann boiler

4.8.7 Velox Boiler

The arrangement of a Velox boiler is shown in Fig. 4.14. The pressurised combustion chamber (furnace) of this boiler uses low excess air and has high heat transfer rates. The air for the furnace is supplied by the air compressor. The combustion gases after heating water and steam pass through a gas turbine to atmosphere. This gas turbine drives the axial flow compressor that raises in-coming combustion air from atmosphere to the furnace pressure. The other important, components are : feed pump, economiser, steam separating section, water circulating pump, and convective super-heater. This boiler is built for pressure ranging from 70 to 80 bar and 500°C.

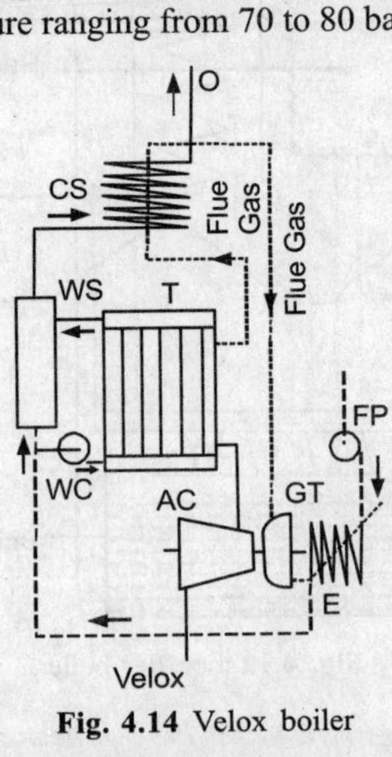

Fig. 4.14 Velox boiler

4.8.8 Once Through Boiler

The arrangement of this boiler is shown in Fig. 4.15. It has inclined coils arranged in a spiral. Forty coils are paralleled around the furnace. Steam generated in the headers flows into headers and then to the convective superheater. Other essential components are the feed pump and economiser.

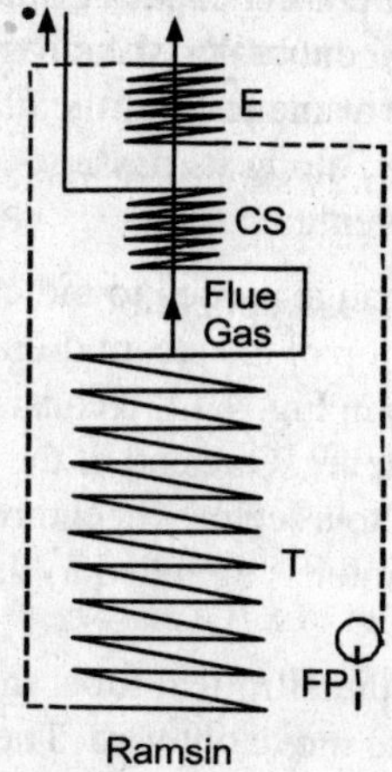

Fig. 4.15 Once-through boiler

4.9 BOILER MOUNTINGS

Different fittings and devices necessary for the operation and safety of a boiler are called boiler mountings. The various boiler mountings are :

1. Water level indicator
2. Pressure gauge
3. Steam stop valve
4. Feed check valve
5. Blow-down cock
6. Fusible plug
7. Safety valve : spring loaded, dead weight, lever type.

1. Water Level Indicator. The function of the water level indicator is to indicate the level of water in the boiler constantly. Every boiler is normally fitted with two water level indicators at its

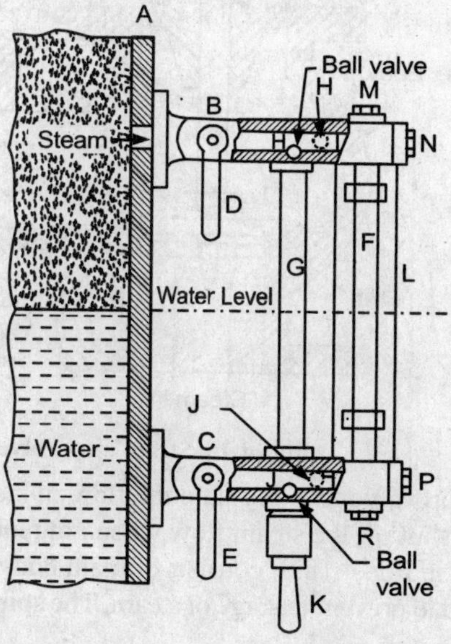

A = End plate of boiler
B, C = Gun metal casting
D, E = Cocks
F = Gauge glass
G = Hollow metal column
H, J = Balls
K = Drain cock
L = Guard glass
M, N, P, R = Screwed caps

Fig. 4.16 Water Level indicator

front end. Fig. 4.16 shows a water level indicator used in low pressure boilers. It consists of three cocks and a glass tube. The steam cock 1 keeps the glass tube in connection with the steam space and cock 2 puts the glass tube in convection with the water space in the boiler. The drain cock 3 is used to drain out the water from the glass tube at intervals to ascertain that the steam and water cocks are clear in operation. The glass tube is generally protected with a shield.

For the observation of water level in the boiler, the steam and water cocks are opened and drain cock is closed. The rectangular passage at the ends of the glass tube contains two balls. In case, the glass tube is broken, the balls are carried along its passage to the ends of the glass tube and flow of water and steam out of the boiler is prevented.

2. Pressure Gauge. The pressure gauge is used to indicate the steam pressure of the boiler. The gauge is normally mounted in the front top of the steam drum. The commonly used pressure gauge is the Bourdon type pressure gauge shown in Fig. 4.17. It consists of an elastic metallic tube of elliptical cross-section bent in the form of circular arc. One end of the tube is fixed and connected to the steam of the boiler and other end is connected to a sector wheel through a link. The sector remains in mesh with a pinion fixed on a spindle. A pointer is attached to the spindle to read the pressure on a dial gauge.

When high pressure steam enters the elliptical tube, the tube section tends to become circular which causes the other end of the tube to move outward. The movement of the closed end of the tube is transmitted and magnified by the link and sector. The sector is hinged at a point on the link. The magnitude of the movement is indicated by the pointer on the dial.

A = Spring tube;
B = block
D = Toothed quadrant;
E = Hinge for quadrant;
F = Pinion
G = Pointer

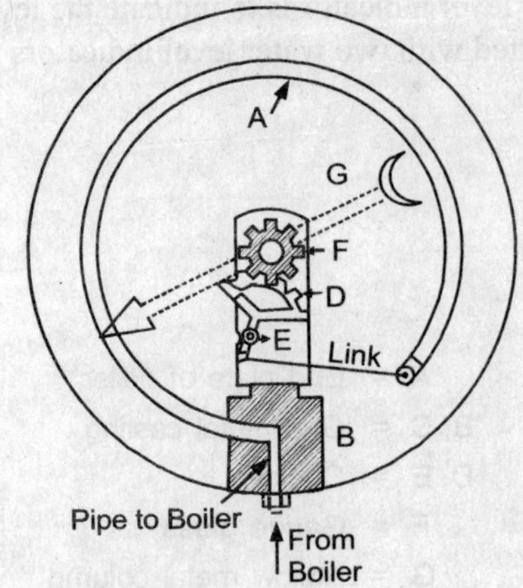

Fig. 4.17 Pressure gauge.

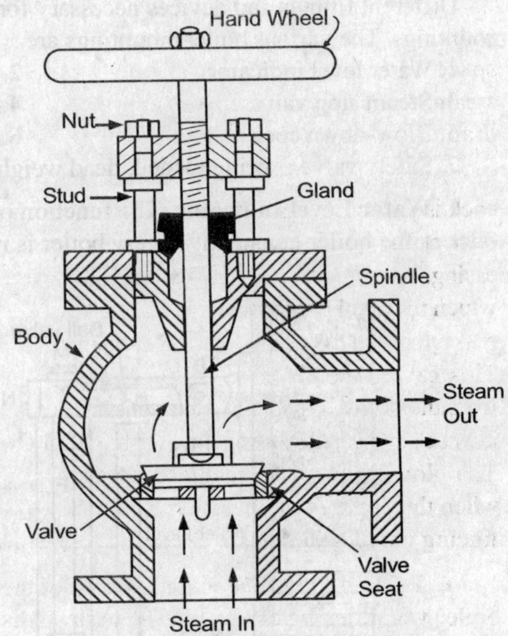

Fig. 4.18 Steam stop valve.

3. Steam Stop Valve. The function of the stop valve is to regulate the flow of steam from the boiler to the prime mover as per requirement and shut off the steam flow when not required.

A commonly used steam stop valve is shown in Fig. 4.18. It consists of main body, valve, valve seat, nut and spindle, which passes through a gland to prevent leakage of steam. The spindle is rotated by means of a hand wheel to close or open.

4. Feed Check Valve. The function of the feed check valve is to allow the supply of water to the boiler at high pressure continuously and to prevent the back flow of water from the boiler when the pump pressure is less than boiler pressure or when pump fails.

A commonly used feed check valve is shown in Fig. 4.19. It is fitted to the shell slightly below the normal water level of the boiler. The lift of the non-return valve is regulated by the end position of the spindle which is attached with the hand wheel. The spindle can be moved up or down with the help of hand wheel which is screwed to the spindle by a nut. Under normal conditions, the non-return valve is lifted due to the water pressure from the pump and water is fed to the boiler. In case pump pressure falls below boiler pressure, valve is closed automatically.

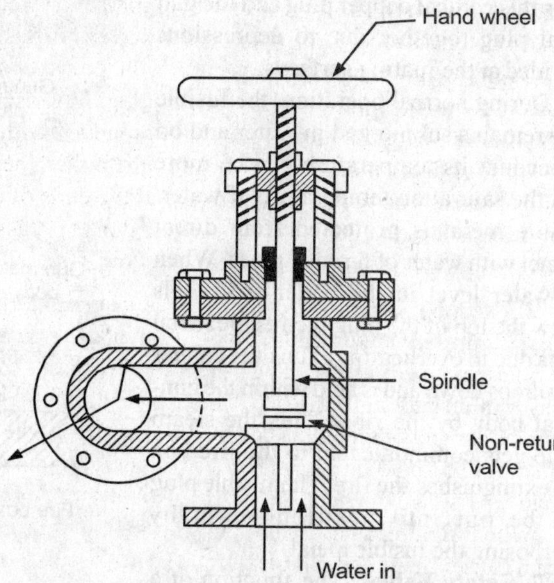

Fig. 4.19 Feed check valve.

5. Blow-Down Cock. The function of blow-down cock is to remove sludge or sediments collected at the bottom-most point in the water space in a boiler, while the boiler is steaming. It is also used for complete draining of the boiler for maintenance.

A commonly used type of blow-down cock is shown in Fig. 4.20. It consists of a conical plug fitted accurately into a similar casing. The plug has a rectangular opening which may be brought with the line of the passage of the casing by rotating the plug. This causes the water to be discharged from the boiler. The discharging of water may be stopped by rotating the plug again. The blow-down cock should be opened when the boiler is in operation for quick forcing out of sediments.

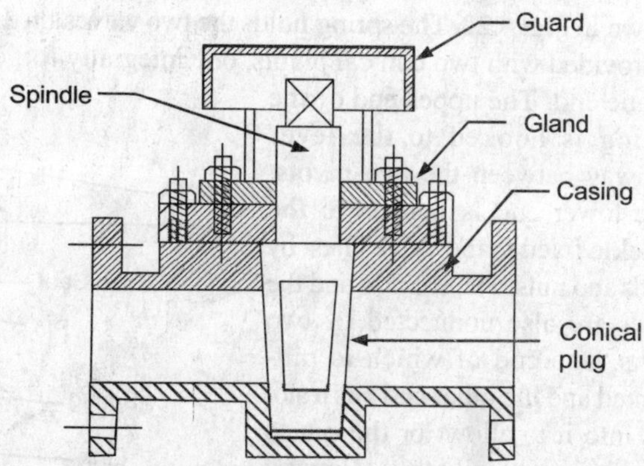

Fig. 4.20 Blow down clock

6. Fusible Plug. The main function of the fusible plug is to put off the fire in the furnace of the boiler when the water level in the boiler falls below an unsafe level and thus avoid the explosion which may occur due to overheating of the tubes and shell. The plug is generally fitted over the crown of the furnace or over the combustion chamber.

A commonly used fusible plug is shown in Fig. 4.21. It consists of a hollow gun metal body screwed into the fire box crown. The body has hexagonal flange to tighten it into the shell. A gun metal plug is screwed into the gun metal body by tightening the hexagonal flange formed into it. There is yet another solid plug made of copper with conical top and rounded bottom. Fusible metal holds this conical copper plug and the gun metal plug together due to depressions provided at the mating surfaces.

During normal operation, the fusible plug remains submerged in water and on no account its temperature can be more than the saturation temperature of water. Fusible metal is protected from direct contact with water of furnace gases. When the water level in the boiler shell falls below the top of the plug the fusible metal melts due to overheating. Thus the copper plug drops down and is held within the gun-metal body by the ribs. Thus the steam space gets communicated to the fire box and extinguishes the fire. The fusible plug can be put into position again by interposing the fusible metal.

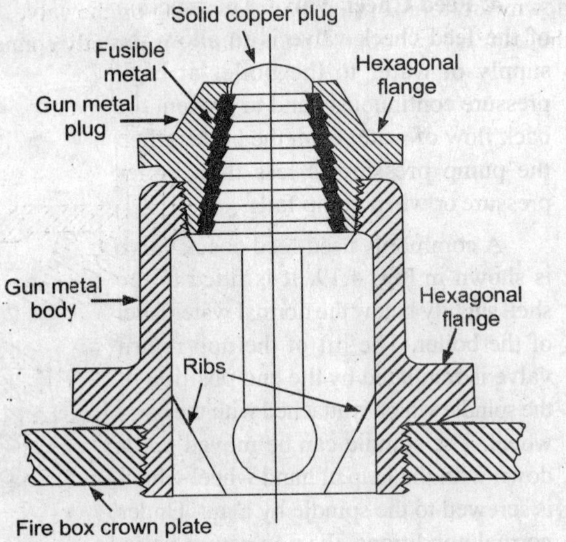

Fig. 4.21 Fusible plug

7. Safety Valves. The function of a safety valve is to prevent the steam pressure in the boiler exceeding the desired rated pressure by automatically opening and discharging steam to atmosphere till the pressure falls back to normal rated value. There are three types of safety valves : spring loaded (Ramsbottom) type, dead weight type, and lever type.

(a) Spring Loaded Safety Valve. The spring loaded safety valve of the Ramsbottom type is shown in Fig. 4.22. The spring holds the two valves on their seats by pulling the lever down. The lever is provided with two conical pivots, one integrally forged with the level and the other pin connected to one end. The upper end of the spring is hooked to the lever midway between the two pivots. The lower end is hooked to the shackle fixed to the valve chest by studs and nuts. The shackle and the lever are also connected by two links, one end of which is pin-jointed and the other end has a slot cut into it to allow for the pin to slide in it vertically thus allowing the lever to be lifted and retaining the link connection. These links are provided as a safety measure in case the spring breaks, the lever should not fly off. The lever projects on one side for manual operation to check the satisfactory working of the device.

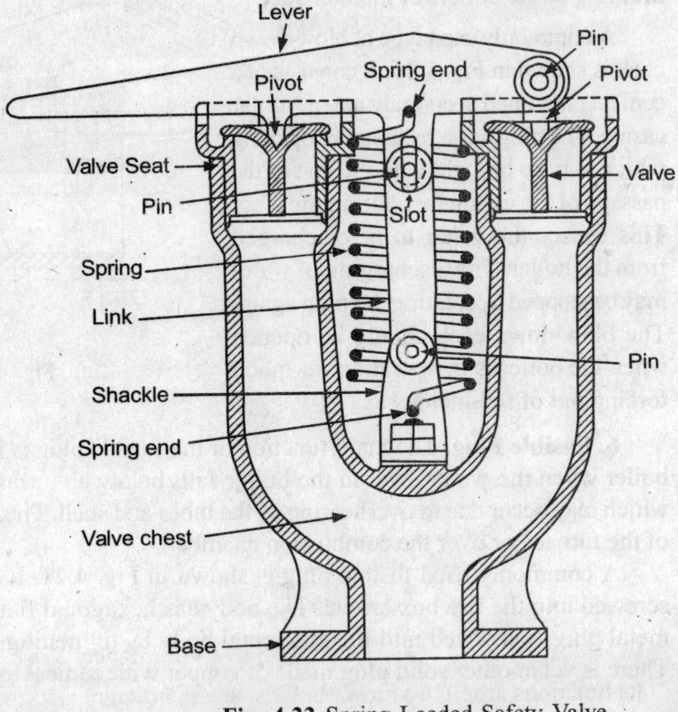

Fig. 4.22 Spring Loaded Safety Valve

(b) Dead Weight Safety Valve. In this type of device, shown in Fig. 4.23, the steam pressure in the upward direction is balanced by the downward force of dead weights acting on the valve. It is generally used on low capacity boilers like the Lancashire boiler. Bottom flange is directly connected to seating block on the boiler shell.

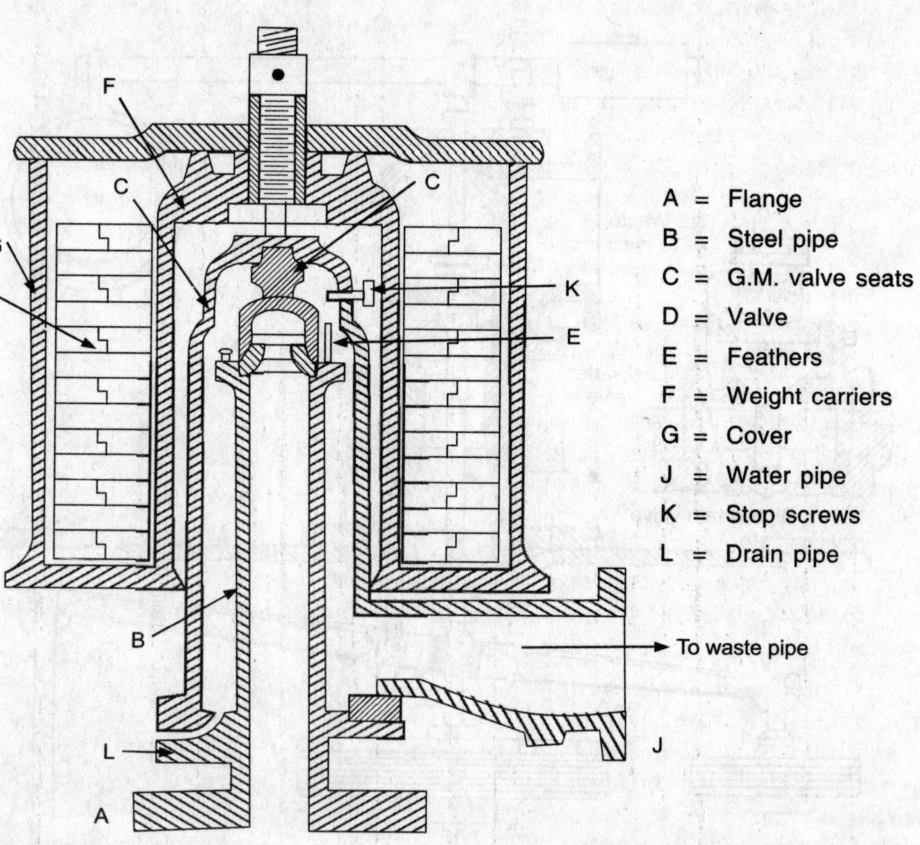

Fig. 4.23 Dead Weight Safety Valve

(c) Lever Safety Valve. A typical lever safety valve is shown in Fig. 4.24. The lever is the second kind with effort in the middle of fulcrum and load. It is suitable for stationary boilers.

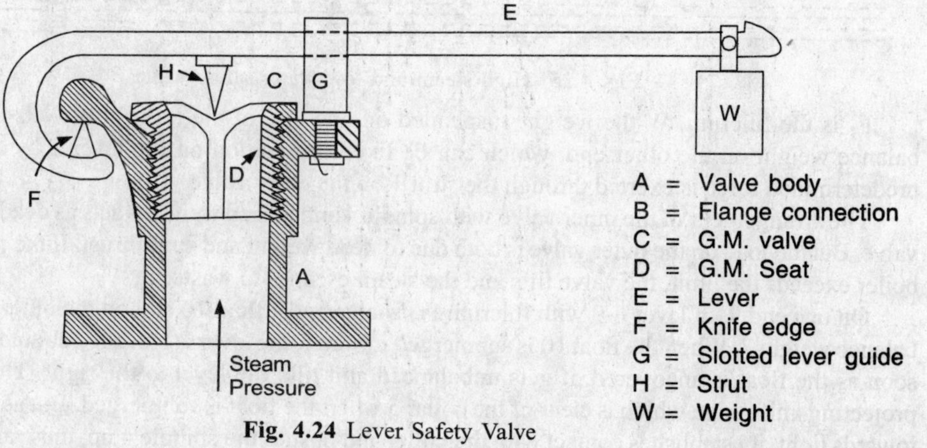

Fig. 4.24 Lever Safety Valve

Its functions are : 1. To blow out if the steam pressure is higher than the working pressure, and 2.

To blow out steam when water level in the boiler is low. It is perhaps the most important mounting on the boiler. Fig. 4.25 shows the details of Hopkinson's high steam and low water safety valve.

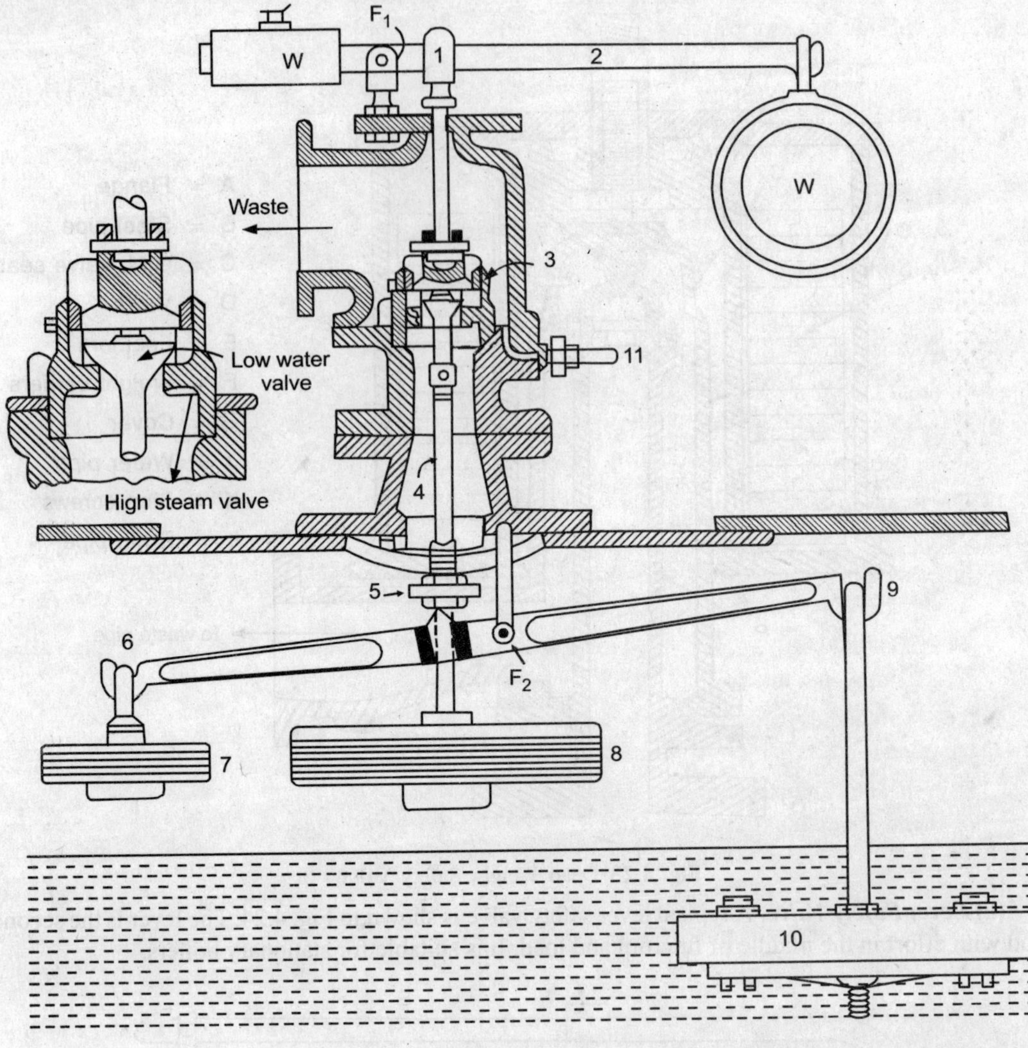

Fig. 4.25 High steam and low water safety valve

F_1 is the fulcrum, W the weight suspended on one end of the lever, and W is the adjustable balance weight on the other end, which can be fixed in position on the lever 2 by a set screw. A predetermined force is exerted through the strut 1 on the outer valve.

The arrangement of the inner valve with spindle 4 and a dead weight 8 acts as dead weight safety valve. But the load on the outer valve is both due to dead weight and strut thrust. If the pressure in the boiler exceeds the limit, the valve lifts and the steam escapes to waste.

On one end 9, of lever 6-9 with fulcrum F_2, is attached a float 10 and on the other end is fixed a balance weight 7. When the float 10 is submerged in water, the lever is balanced about fulcrum F_2. As soon as the float is uncovered, it gets unbalanced and tilts the level to the right. This lever has a projecting knife edge which is clear of the collar 5 when the float is submerged. But as soon as it tilts towards light, it establishes contact with the collar and pushes the spindle 4 up, thus raising the inner valve from its seat Thus steam starts escaping out through the waste pipe. A drain connection 11 is provided to drain off the condensed steam.

4.10 BOILER ACCESSORIES

These are the appliances which are installed to increase the over all efficiency of the steam power plant. The various accessories are :

1. Pressure reducing value
2. Steam trap
3. Steam separator
4. Economiser
5. Air preheater
6. Super heater
7. Feed pump
8. Injector

The functions of the various accessories are :

1. A pressure reducing valve maintains constant pressure on its delivery side with fluctuating boiler pressure.
2. The function of steam trap is to drain off water resulting from the partial condensation of steam from steam pipes and jackets without allowing steam to escape through it
3. The function of the steam separator is to separate suspended water particles carried by steam on its way from the boiler to the prime mover.
4. The function of an economiser is to recover some of the heat carried away in the flue gases up the chimney and utilize for heating the feed water to the boiler.
5. An air preheater recovers some portion of the waste heat of the flue gases by preheating the air supplied to the combustion chamber.
6. Superheaters are used to increase the temperature of steam above its saturation temperature.
7. The function of a feed pump is to pump water to the water space of the boiler.
8. The function of an injector is to feed water to the boiler with the help of a steam jet.

1. Air Preheater. In the air preheater, the air is heated by the heat carried away by the flue gases and going as a waste through the chimney. It is situated between the economiser and the chimney. Air pre-heater are of three types : tubular, plate type, and regenerative.

In the tubular type of air preheater, shown in Fig. 4.26, the air passes down outside the tubes and the flue gases through the tubes before going to the induced draught fan at the base of the chimney. Baffles are provided across the tubes to make the air follow a zig-zag path a number of times to utilize more heat of flue gases.

In the plate type of air preheater, there are alternate chambers provided by the plate surfaces for the flow of flue gases upwards to the induced draught fan and for

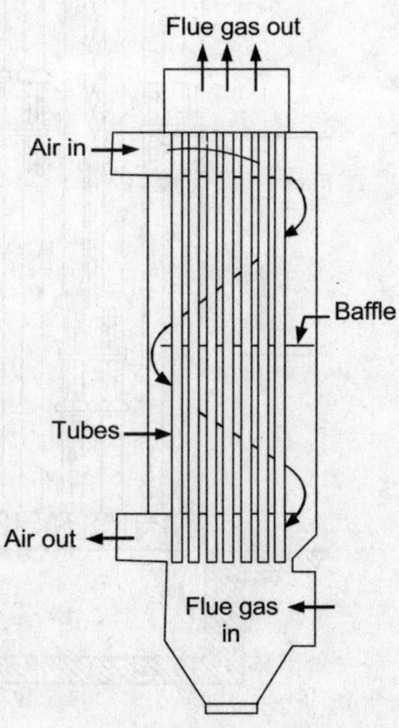

Fig. 4.26 Tubular air pre-heater

the flow of combustion air downwards.

In the regenerative type, the air preheater chamber is divided into two rotating compartments. The heating surfaces are made up of a series of corrugated sheets between which the hot flue gases pass upwards on one side and the combustion air passes down on the other side. Due to the rotation of pre-heater at very low speed, the heater surface comes in contact with combustion air and the cooled surface is again heated by the flue gases. This alternate heating and cooling of heating surface is called regenerative heating and is more effective than other types.

2. Economiser. The Green's economiser shown in Fig. 4.27 is used for boilers of medium pressure range upto about 25 bar. It consists of groups of vertical cast iron pipes A fitted at their two ends to the cast iron boxes, at the top and bottom. All the tubes are enclosed within the brick work of the economiser. There are two pipes C and D outside the brick work E. Rows of top and bottom boxes are connected to these top and bottom pipes. The feed water is pumped to the bottom pipe D. From here it goes to the bottom boxes and rise up in the groups of vertical pipes into the top boxes. From the top boxes it goes to the pipe C from where it passes to the water space of the boiler via the check valve. F is a stop valve for entry and G is a stop valve for exit of the water from the economiser. S is a safety valve. These tubes are continuously scraped by scrapers J to remove soot collected on them due to the passage of flue gases through the economiser. A pair of scrapers is connected by a chain passing over a pulley such that one scraper comes down and the other attached to the same chain goes up. L is the soot chamber where soot collects and is removed through a soot door.

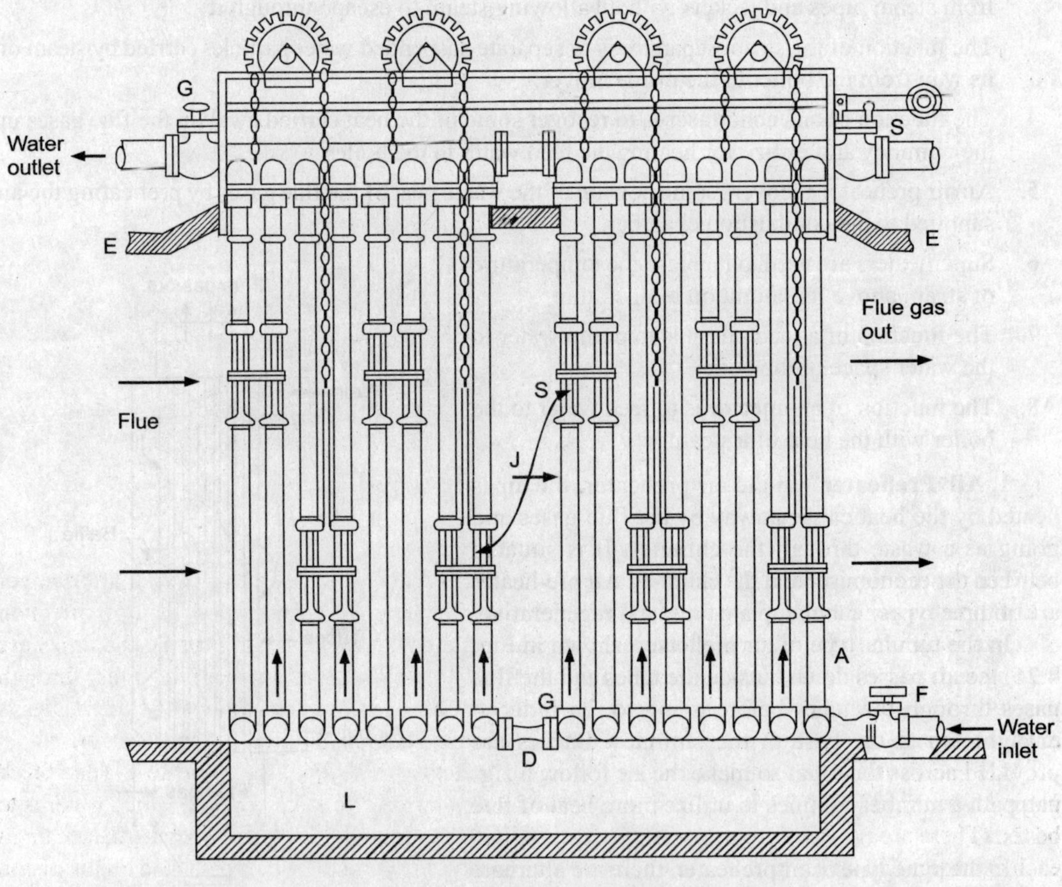

Fig. 4.27 Economiser

3. Super heater. The superheater commonly used in a Lancashire boiler is shown in Fig. 4.28. This superheater consists of two headers and a set of super heater tubes made of high quality steel in the form of U-tube. Super heater is located in the path of furnace gases just before the gases enter the bottom flue. The amount of hot gases passed over the super heater tubes should be in proportion to the amount of super heated steam passing through the tubes. Otherwise, the tubes would be over heated. To avoid this, the amounts of hot gases are diverted as shown in the Figure. The super heater is put out of action by turning tine damper upward to the vertical position. In this position of the damper, the gases coming out from the central flue pass directly into the bottom flue without passing over the super heater tubes.

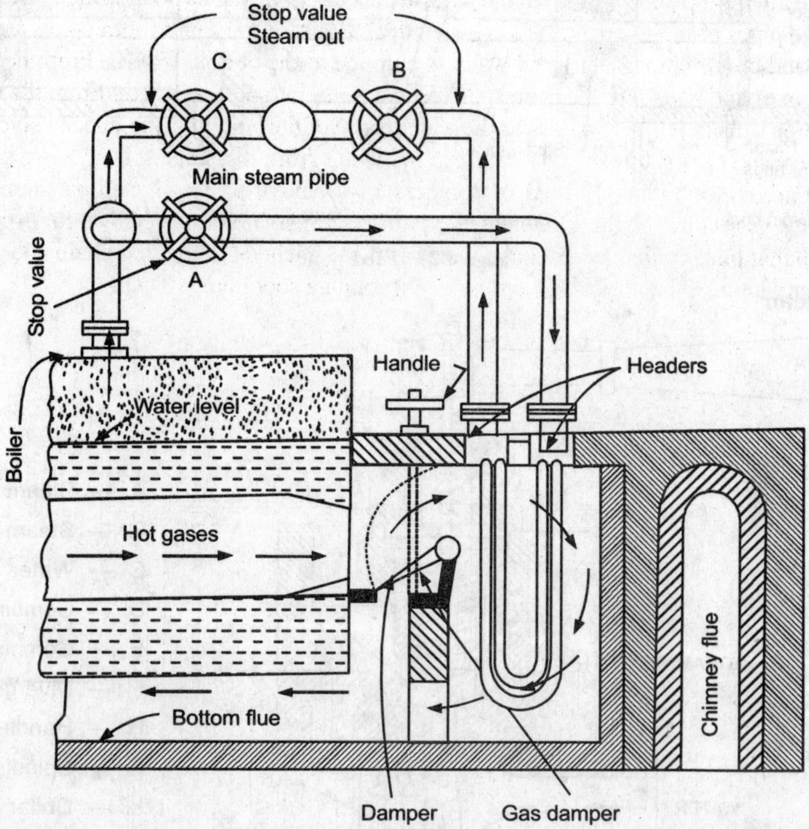

Fig. 4.28 Super heater

For getting superheated steam, the valves A and B are opened and valve C is closed. The damper is kept open as per the quantity of steam flowing through the pipe. For this position, the flow direction of steam is shown in the figure. If wet steam is required, then the valves A and B and gas damper are kept closed, and valve C is kept open. In this case, steam directly comes out from the boiler through the valve C. By adjusting the gas damper, the temperature of steam coming out of superheater is always maintained constant irrespective of amount of steam passing through the superheater.

4. Feed Pump. The feed pump is used to supply water to the boiler. A reciprocating type of feed pump is shown in Fig. 4.29. It is continuously run by steam from the same boiler to which water is to be fed. There are two single steam cylinders placed side by side. Slide values distribute the steam in each cylinder. The slide value in each cylinder steam chest is operated by the crosshead on the piston rod of the opposite cylinder through an arrangement of rods and rocker arms. The feed pump is generally double acting. On each side of the pump plunger there are suction and delivery values.

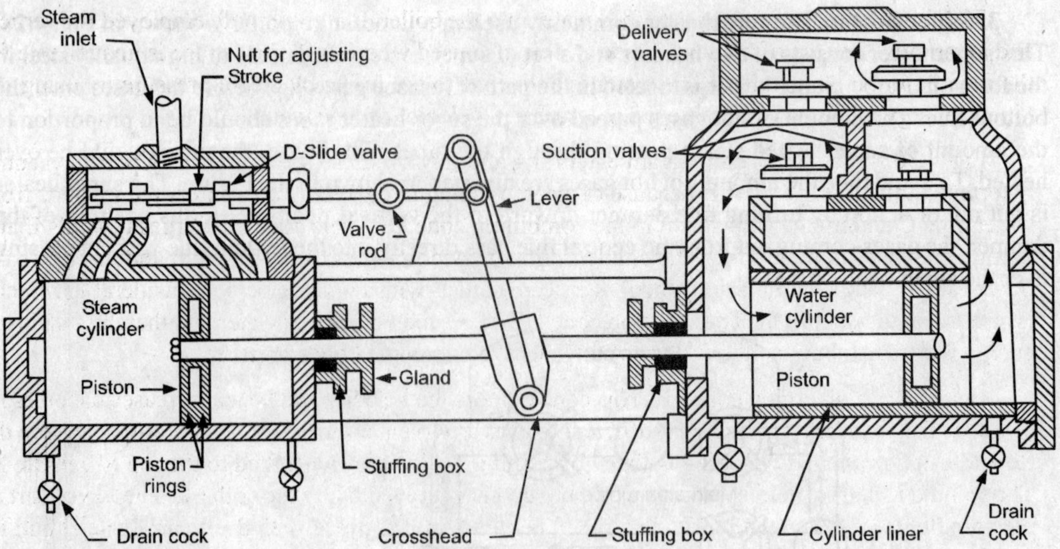

Fig. 4.29 Feed pump

5. Injector

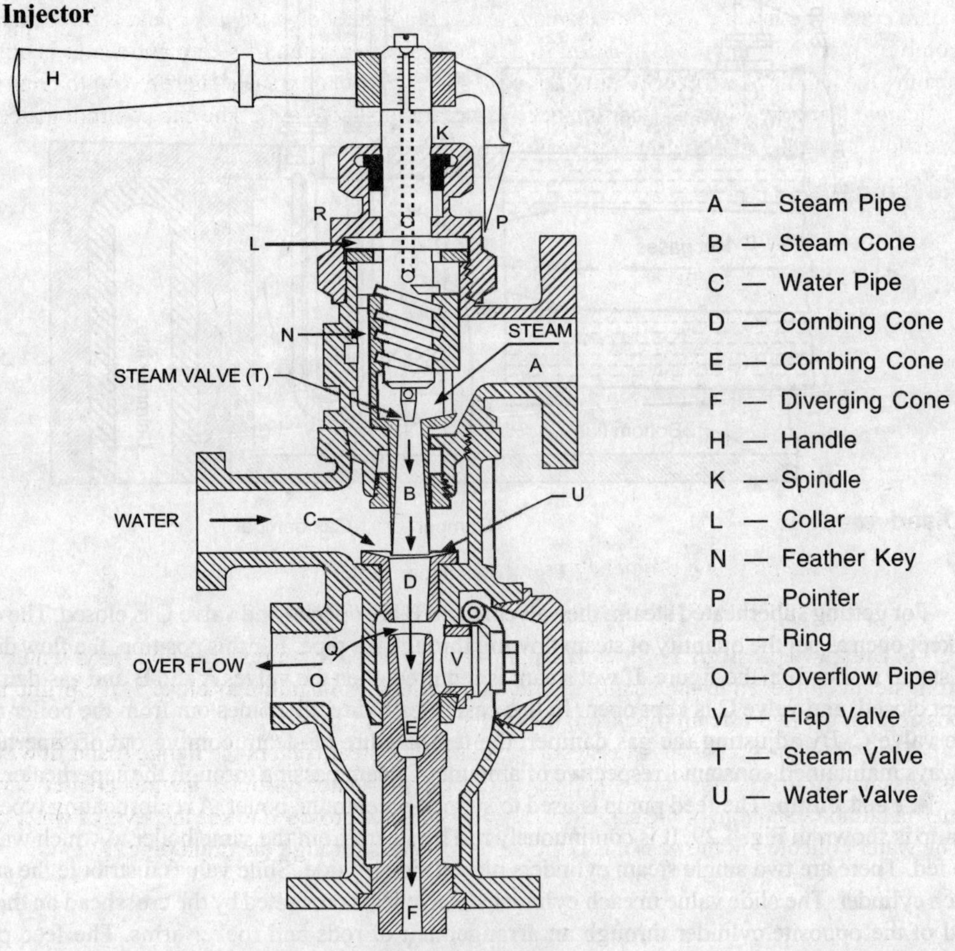

Fig. 4.30 Injector

The function of the injector is to feed water into the boiler. It is commonly employed for vertical and locomotive boilers of medium capacity. It is also used where sufficient space is unavailable for the feed pump to be installed. The water is delivered to the boiler by steam pressure by using the kinetic energy of steam to increase the pressure and velocity of feed water.

Fig. 4.30 shows an injector. Steam enters at A and flows into the steam cone B. A partial vacuum is formed at the lower end of B because of steam rushing in cone B. Water enters the injector from water pipe C and mixes with steam in the combining cone DE. This cone converges towards E and steam condenses in their cone. The condensation of steam increases the vacuum and increases the flow of water. The steam rushes through the steam cone B with a high velocity. Considerable velocity is also given to water in the combination cone DE. At E, the velocity of water and thereby its kinetic energy is greatest, but its pressure is not sufficient to overcome boiler pressure.

Below E is a diverging cone FE. This cone reduces the velocity and hence increases the pressure of water. This increased pressure is sufficient to enable it to enter the boiler. The relation between the flow rates of steam and water is regulated by turning the handle H attached to the top of spindle K. The spindle K can rotate only and its axial movement is prevented by the collar L. The lower part of the spindle has a screw which fits in a nut. The nut is in one piece with the steam cone B and its rotation is prevented by the feather key N. The steam cone B can be raised or lowered by the spindle K. The lower end of the spindle forms a steam valve T. The seat of the valve is the entrance to the steam cone. The lower end of the steam cone forms a water valve U whose seat is the entrance to the combining cone. The proper position to get the desired mixture of steam and water is obtained by turning the handle H until the pointer P is opposite to a point on a scale engraved on the ring R, which indicates steam pressure. The combination cone is interrupted at Q. The gap communicates with the overflow branch O through the flap valve V.

Advantages :

1. Low initial cost.
2. Simple design.
3. Compact size.
4. No moving parts.
5. High thermal efficiency, (about 90%).
6. Easy operation.
7. Low maintenance cost.

Disadvantages :

1. Low pumping efficiency.
2. Irregular operation during extreme pressure variation.

6. Steam Separator. The steam separator is a device to remove the entrained water particles from steam conveyed to the steam engine or turbine. It is installed as close to the prime mover as possible on the main steam pipe from the boiler.

Fig. 4.31 shows a baffle plate steam separator. The steam enters the flange A and flows down. In its passage, it strikes the baffles B, gets deflected and water particles having greater density and inertia fall to the bottom of the separator. A water gauge is provided to see the level of water collected. The water collected in the vessel is removed at intervals through the drain pipe D.

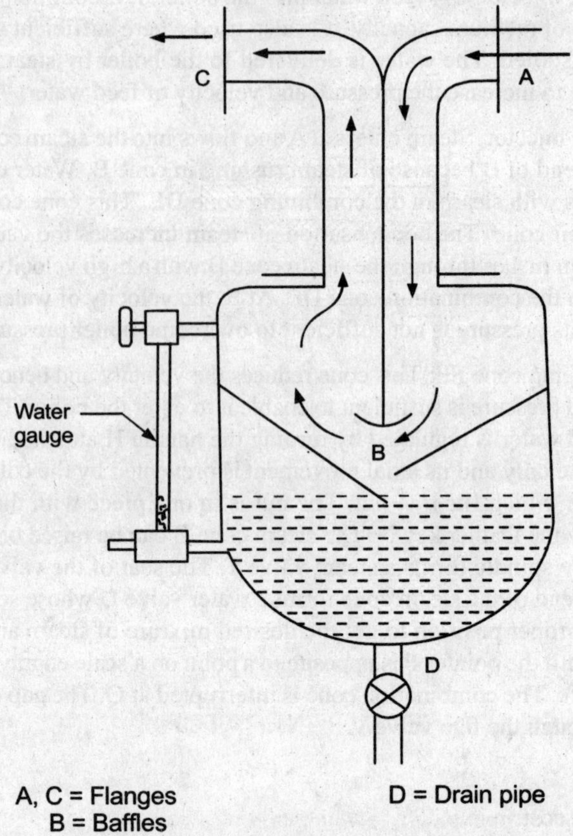

Fig. 4.31 Baffle plate steam separator.

7. Steam Trap. A steam trap is used to drain away automatically the condensed steam from the steam pipes, steam jackets and steam separators without permitting any steam to escape.

Fig. 4.32. shows a steam trap. The water to be drained away enters the casing C at A. The casing always contains sufficient water to float the bucket B. The spindle S is attached to the bucket B.

The valve V is attached to the spindle at the upper edge. D is the seating. The valve V and its seating D are made of pure nickel. The spindle S has spirally formed vanes on it at E and F. G is the guide tube. These vanes can move easily inside tube G. When the water has risen high enough in the casing C, it overflows into the bucket B. The bucket will ultimately sink and opening valve V. The water is then forced up the guide tube G by the steam pressure. The water, after passing through valve V, leaves the trap through the outlet H. By pressing on the knob K, rod L is depressed. This will push the bucket in the downward direction. The valve is open under this condition. The trap may therefore be blown through and tested at any time. The door M is used for cleaning out of any sediment which may accumulate in the casing.

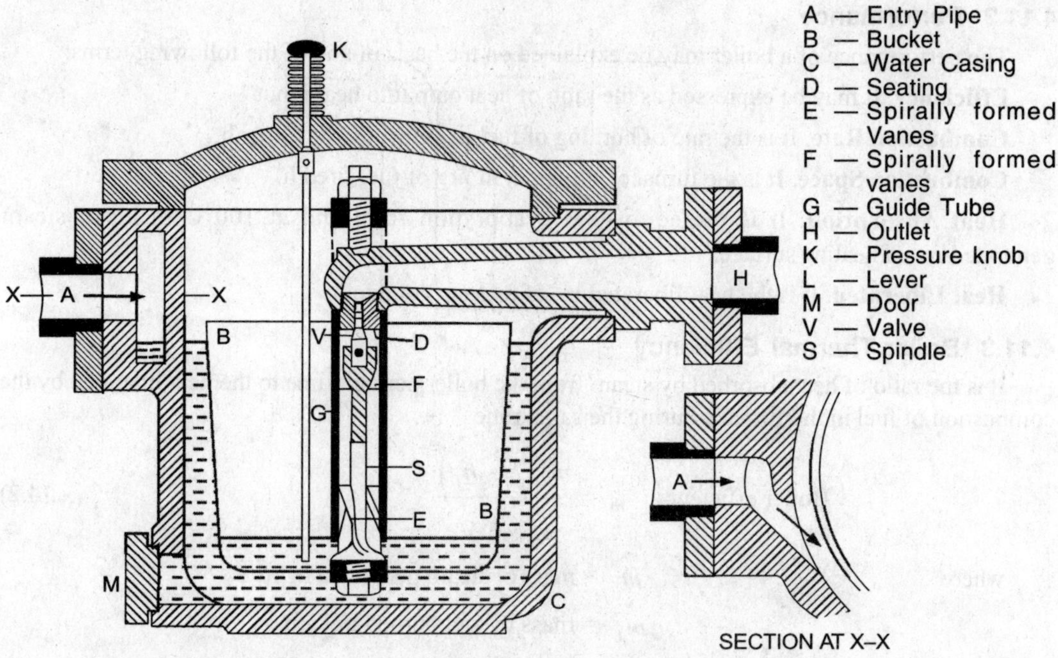

Fig. 4.32 Steam Trap

4.11 PERFORMANCE OF STEAM GENERATOR

4.11.1 Evaporation Rate

The quantity of water evaporated into steam per hour is called the evaporation rate. It is expressed as kg of steam/h, or kg of steam/h/m^2 of heating surface, or kg of steam/h/m^3 of furnace volume, or kg of steam/kg of fuel fired.

Equivalent Evaporation. It is the equivalent of the evaporation of 1 kg of water at 100°C to steam at 100°C. It requires 2256.9 kJ $\simeq$ 2260 kJ.

Factor of Evaporation. The ratio of actual heat absorption above feed water temperature for transformation to steam (wet, dry, or superheated) to the latent heat of steam at atmospheric pressure (1.01325 bar) is known as factor of evaporation. Let

h = specific enthalpy of steam actually produced

hf = specific enthalpy of feed water

$h_{fg(\text{atm})}$ = specific enthalpy of evaporation at standard atmospheric pressure

m_s = actual evaporation expressed in kg/kg of fuel or kg/h of steam

m_e = equivalent evaporation expressed in kg/kg of fuel or kg/h

Then Equivalent evaporation = Actual evaporation × Factor of evaporation

or
$$m_e = \left[\frac{h_1 - h_f}{h_{fg(\text{atm})}}\right] \times m_s = \text{F} \times m_s \qquad \ldots(4.1)$$

where F = factor of evaporation

$h_1 = h_{f1} + xh_{fg1}$ for wet steam actually generated

$= h_{g1}$ for saturated steam actually generated

$= h_{\text{sup}}$ for superheated steam actually generated.

4.11.2 Performance

The performance of a boiler may be explained on thc basis of any of the following terms:

Efficiency. It may be expressed as the ratio of heat output to heat input

Combustion Rate. It is the rate of burning of fuel in kg/m^3 of grate area/h.

Combustion Space. It is the furnace volume in m^3/kg of fuel fired/h.

Heat Absorption. It is the equivalent evaporation from and at 100°C in kg of steam generated/m^2 of heating surface.

Heat Liberated. It is the heat liberated/m^3 of furnace volume/h.

4.11.3 Boiler Thermal Efficiency

It is the ratio of heat absorbed by steam from the boiler per unit time to the heat liberated by the combustion of fuel in the furnace during the same time.

$$\text{Boiler efficiency, } \eta_b = \frac{\dot{m}_2 \left(h_1 - h_f\right)}{\dot{m}_f \times \text{C.V.}} \qquad \ldots(4.2)$$

where $\dot{m}_s$ = mass of steam generated in kg/h

$\dot{m}_f$ = rnass of fuel burned in kg/h

C.V. = calorific value of fuel in kJ/kg

Boiler power = [m_e from and at 100°C/h]/21.296 kW ...(4.3)

4.11.4 Factors Affecting Boiler Efficiency

The factors on which the boiler efficiency mainly depends may be classified as :

1. Fixed factors
2. Variable factors.

1. Fixed Factors

(*a*) **Boiler design**

- Arrangement and effectiveness of the heating surfaces
- Shape and volume of furnace.
- Arrangement of flues
- Arrangement of steam and water circulation.

(*b*) **Built in losses**

- Heat transfer properties of the settings and construction materials.
- Flue gas and ash heat losses.

(*c*) **Heat recovery equipment**

- Economiser
- Superheater
- Air preheaters
- Feed water heater.

(*d*) **Properties and characteristics of fuel burnt.**

(*e*) **Rated firing rate, furnace volume and heating surface.**

2. Variable Factors

(*i*) Fuel condition as it is fired.

(*ii*) Change in draught from the rated value due to atmospheric conditions.

(*iii*) Excess air fluctuations.

(*iv*) Condition of heat absorbing surfaces.

(*v*) Actual firing rate.

(*vi*) Humidity and temperature of combustion air.

(*vii*) Incomplete combustion and combustibles in the refuse.

4.11.5 Heat Losses in a Boiler Plant

1. Heat used to generate steam, $Q = m_s (h - h_{f1})$

2. Heat lost to flue gases.

The flue gases contain dry products of combustion and the steam generated due to the combustion of hydrogen in the fuel.

Heat lost to dry flue gases,

$$Q_1 = m_g\, c_{pg}\, (T_g - T_a) \qquad \ldots(4.4)$$

where m_g = mass of gases formed per kg of fuel

c_{pg} = specific heat of gases

T_g = temperature of gases, °C

T_a = temperature of air entering the combustion chamber of the boiler, °C

3. Heat carried by steam in flue gases,

$$Q_2 = m_{s1} (h_{s1} - h_{f1}) \qquad \ldots(4.5)$$

where m_{s1} = mass of steam formed per kg of fuel due to combustion of H_2 in fuel

h_{f1} = enthalpy of water at boiler house temperature

h_{s1} = enthalpy of steam at the gas temperature and at the partial pressure of the vapour in the flue gas

4. Heat loss due to incomplete combustion

If carbon burns to CO instead of CO_2 then it is known as incomplete combustion. 1 kg of C releases 10,200 kJ/kg of heat if it burns to CO whereas it releases 35,000 kJ/kg if it burns to CO_2. If the percentages of CO and CO_2 in flue gases by volume are known, then

$$\text{Mass of C burnt to CO} = \frac{CO \times C}{CO + CO_2}$$

where CO, CO_2 = % by volume of CO and CO_2 in flue gases

C = fraction of carbon in 1 kg of fuel

Heat lost due to incomplete combustion of carbon per kg of fuel,

$$Q_3 = \frac{CO \times C}{CO + CO_2} \times (35{,}000 - 10{,}200)$$

$$= \frac{CO \times C}{CO + CO_2} \times 24{,}800 \text{ kJ/kg of fuel} \qquad \ldots(4.6)$$

5. Heat lost due to unburnt fuel

$$Q_4 = m_{f1} \times C.V.$$

where m_{f1} = mass of unburnt fuel per kg of fuel burnt.

6. Convection and radiation losses

$$Q_6 = Q - (Q_1 + Q_2 + Q_3 + Q_4 + Q_5) \quad \ldots(4.7)$$

where $Q = m_f \times C.V.$

= heat released per kg of fuel

4.11.6 Boiler Trial and Heat Balance Sheet

There are three purposes of conducting the boiler trial.

1. To determine and check the specified generating capacity of the boiler when working at full load conditions.
2. To determine the thermal efficiency of the plant.
3. To draw up the heat balance sheet so that suitable corrective measures may be taken to improve the efficiency.

The following measurements should be observed during the boiler trial.

1. The fuel supplied and its analysis.
2. Steam generated and its quality or superheat.
3. Flue gases formed from exhaust analysis.
4. Air inlet temperature and gases exhaust temperature.
5. Volumetric analysis of exhaust gases.
6. Mass of fuel left unburnt in ash.
7. Feed water temperature.

The Heat Balance Sheet is a symmetric representation of heat released from burning of fuel and heat distribution on minute, hour or per kg of fuel basis. A proforma for heat balance sheet is given in Table 4.1.

Table 4.1 Heat balance sheet for boiler

Heat supplied	kJ	%	Heat Utilization	kJ	%
Heat supplied by fuel	$Q = m_f \times$ C.V.	100%	1. Heal used to generate steam.	$Q_1 = m_s(h - h_{f1})$	$\frac{Q_1}{Q} \times 100$
			2. Heat carried by dry flue gases.	$Q_2 = m_g c_{pg}(T_g - T_a)$	$\frac{Q_2}{Q} \times 100$
			3. Heat carried by steam in flue gases.	$Q_3 = m_{s1}(h_{s1} - h_{f1})$	$\frac{Q_3}{Q} \times 100$
			4. Heat lost due to incomplete combustion.	$Q_4 = \frac{CO \times C}{CO + CO_2} \times C.V.$	$\frac{Q_4}{Q} \times 100$
			5. Heat lost due to unburnt fuel.	$Q_5 = m_{f1} \times C.V.$	$\frac{Q_5}{Q} \times 100$
			6. Heat unaccounted for	$Q_6 = Q - (Q_1 + Q_2 + Q_3 = Q_4 + Q_5)$	$\frac{Q_6}{Q} \times 100$
	Q	100	—	Q	100%

Example 4.1 *A boiler generates 4000 kg/h of steam at 20 bar, 400°C. The feed water temperature is 50°C. The efficiency of the boiler is 80%. The calorific value of fuel used is 44,500 kJ/kg. The steam generated is supplied to a turbine developing 450 kW and exhausting at 2 bar with dryness fraction 0.96. Calculate the fuel burnt per hour and turbine efficiency. Also find the energy available in exhaust steam above 50°C.*

Solution.

From steam tables at 20 bar, 400°C $h_1 = 3245.5$ kJ/kg

Enthalpy of feed water of 50°C $h_f = 209.3$ kJ/kg

$$\text{Fuel used per hour} = \frac{\dot{m}_s\left(h_1 - h_f\right)}{\eta_s \times \text{C.V.}} = \frac{4000\times(3245.5-209.3)}{0.8\times 44{,}500}$$

$$= 341.146 \text{ kg/h}$$

Enthalpy of exhaust steam at 2 bar, 0.96 dry,

$$h_2 = h_{f2} + x_2 h_{fg2} = 504.7 + 0.96\times 2201.9 = 2618.5 \text{ kJ/kg}$$

$$\text{Enthalpy drop in turbine, } \Delta h = h_1 - h_2 = 3245.5 - 2618.5 = 627.0 \text{ kJ/kg}$$

$$\text{Turbine efficiency, } \eta_t = \frac{k\text{W} \times 3600}{\Delta h \times \dot{m}_s} = \frac{450\times 3600}{627\times 4000} = 0.646 \text{ or } 64.6\%$$

Energy available in the exhaust steam above 50°C

$$= h_2 - h_f = 2618.5 - 209.3 = 2409.2 \text{ kJ/kg}$$

Example 4.2 *The following readings are taken during trial on a boiler for one hour :*

Steam generated	*= 5000kg*
Coal burnt	*= 650 kg*
C.V. of coal	*= 31,500 kJ/kg*
Dryness of steam entering the super heater	*= 0.90*
Rated boiler pressure	*= 10 bar*
Temperature of steam leaving the superheater	*= 250°C*
Temperature of hot well	*= 40°C*

Calculate (a) equivalent evaporation per kg of fuel without and with superheater, (b) thermal efficiency of the boiler without and with superheater, and (c) amount of heat supplied by the super heater per hour.

Solution.

$$\text{Equivalent evaporation, } m_e = \frac{(h_1 - h_f)m_s}{2260}$$

where m_s = mass of steam generated per kg of fuel

$$= \frac{5000}{650} = 7.692 \text{ kg/kg of fuel}$$

(*a*) Without superheater,

$$m_e = \frac{m_s(h_{f1} + x_1 h_{fg1} - h_f)}{2260}$$

At 10 bar, $h_{f1} = 762.8$ kJ/kg, $h_{fg1} = 2015.3$ kJ/kg

At 40°C, $h_f = 167.6$ kJ/kg

$$m_e = \frac{7.692\,[762.8 + 0.9\times 2015.3 - 167.6]}{2260} = 8.2 \text{ kg}$$

$$\text{With super heater, } m_e = \frac{m_s\left[h_{sup} - h_{f1}\right]}{2260}$$

At 10 bar, 250°C, $h_{sup} = 2942.6$ kJ/kg

$$m_e = \frac{7.692[2942.6 - 167.6]}{2,260} = 9.445 \text{ kg}$$

(*b*) $$\text{Thermal efficiency, } \eta_{th} = \frac{m_s\left(h_1 - h_f\right)}{\text{C.V.}}$$

$$\text{Without superheater, } \eta_{th} = \frac{7.692\,[762.8 + 0.9 \times 2015.3 - 167.6]}{31,500}$$

$$= 0.5882 \text{ or } 58.82\%$$

$$\text{With superheater, } \eta_{th} = \frac{7.692\,(2942.6 - 167.6)}{31,500} = 0.6776 \text{ or } 67.76\%$$

(*c*) Heat supplied by the super heater per hour

$$= m_s\,[(1 - x)\,h_{fg1} + h_{sup}]$$

$$= 5000\,[(1 - 0.9) \times 2015.3 + 2942.6]$$

$$= 15720650 \text{ kJ/h}$$

Example 4.3

The following data refer to a boiler trial:

Feed water	*= 700 kg/h*
Feed water temperature	*= 25°C*
Steam pressure	*= 15 bar*
Steam temperature	*= 300°C*
Coal burnt	*= 90 kg*
C.V. of coal	*= 30,500 kJ/kg*
Ash and unburnt coal in ash pit = 4 kg/h with C.V.	*= 2200 kJ/kg*
Flue gas formed	*= 20 kg/kg of coal burnt*
Flue gas temperature at chimney	*= 300°C*
Ambient temperature	*= 30°C*
Mean specific heat of flue gases	*= 1.025 kJ/kg.K*

Calculate (a) boiler efficiency, (b) equivalent evaporation, and (c) percentage heat unaccounted for.

Solution.

From steam tables, at 15 bar, 300°C, $h_{sup} = 3038.9$ kJ/kg

At 25°C, $h_f = 108.77$ kJ/kg

Heat absorbed by 1 kg of feed water to transform into steam at 300°C and 15 bar

$$= h_{sup} - h_f = 3038.9 - 108.77 = 2930.13 \text{ kJ/kg}$$

(*a*) $$\text{Boiler efficiency, } \eta_b = \frac{m_s\left(h_{sup} - h_f\right)}{m_f \times \text{C.V.}} = \frac{700 \times 2930.13}{90 \times 30,500} = 0.7472 \text{ or } 74.72\%$$

(*b*) $$\text{Equivalent evaporation, } m_e = \frac{h_{sup} - h_f}{h_{fg(atm)}} \times \frac{m_s}{m_f} = \frac{2930.13}{2260} \times \frac{700}{90} = 10.084 \text{ kg/kg of fuel}$$

Heat carried away by flue gases $= m_g c_{pg}\,(T_g - T_a)$

$= 20 \times 1.025 \times (300 - 30) = 5535$ kJ/kg of coal

(*c*) Heat unaccounted for per kg of fuel

= C. V. of fuel – heat in steam – heat in flue gases – heat in ash pit

$$= 30{,}500 - 2930.13 \times \frac{700}{90} - 5535 - \frac{4 \times 2200}{90}$$

$$= 30.500 - 22790 - 5535 - 97.8$$

$$= 2077.2 \text{ kJ/kg}$$

Example 4.4 *The following observations were made during a boiler trial:*

Mass of feed water and its temperature = 640 kg/h, 50°C
Steam pressure = 10 bar
Fuel burnt = 55 kg/h
H.C.V. of fuel = 44,100 kJ/kg
Temperature of flue gases = 300°C
Boiler house temperature = 30°C
Dryness fraction of steam = 0.97
Heating surface of boiler = 18.6 m²
Specific heat for flue gases = 1.1 kJ/kg.K
Specific heat for superheated steam = 2.1 kJ/kg.K

Composition of liquid fuel used by mass :

$C = 85\%,\ H_2 = 13\%,\ \text{ash} = 2\%$

The flue gas analysis by volume done by using or set apparatus:

$CO_2 = 12.5\%,\ O_2 = 4.5\%,\ N_2 = 83\%$

Calculate (a) equivalent evaporation per kg of fuel per m² of heating surface area per hour, (b) boiler efficiency, and (c) draw the heat balance sheet on 1 kg of fuel basis.

Take the partial pressure of steam in exhaust gases = 0.07 bar.

Solution.

(*a*) Equivalent evaporation, $m_e = \dfrac{m_s(h_1 - h_f)}{m_f \times 2260}$

$h_1 = h_{f1} + x_1 h_{fg1}$

At 10 bar, $h_{f1} = 762.6$ kJ/kg, $h_{fg1} = 2013.6$ kJ/kg

For feed water at 50°C, $h_f = 209.3$ kJ/kg

$$m_e = \frac{640}{55}\left[\frac{762.6 + 0.97 \times 2013.6 - 209.3}{2260}\right] = 12.9 \text{ kg}$$

(*b*) Boiler efficiency, $\eta_b = \dfrac{m_s(h_1 - h_f)}{m_f \times \text{C.V.}} = \dfrac{640 \times 2506.5}{55 \times 44100} = 0.6613$ or 66.13%

(*c*) Heat used to generate steam per kg of fuel

$$= \frac{m_s}{m_f} \times 2506.5 = \frac{640}{55} \times 2506.5 = 29166 \text{ kJ}$$

$$\text{Air supplied per kg of fuel burnt} = \frac{N \times C}{33(C_1 + C_2)}$$

$$= \frac{0.83 \times 0.85 \times 100^2}{33(0 + 12.5)} = 17.1 \text{ kg}$$

Dry flue gases formed per kg of fuel burnt = 17.1 + 0.85 = 17.95 kg

Heat carried by dry flue gases = $m_g \times c_{pg}\,(T_g - T_a)$

= 17.95 × 1.1 (300 – 30) = 5331 kJ/kg of fuel

Moisture formed per kg of fuel = 0.13 × 9 = 1.17 kg

Heat carried by moisture in the flue gases per. kg of fuel

$= 1 = 1.17\,[h_g + c_{ps}\,(T_{sup} - T_s) - h_f]$ at 0.07 bar

= 1.17 [2572.6 + 2.1 (300 – 39) – 125.8]

= 3504 kJ

Heat unaccounted for = 44,100 – (29166 + 5331 + 3504)

= 6099 kJ

Heat Balance Sheet

Heat supplied	**kJ**	**%**	**Heat utilised**	**kJ**	**%**
Heat supplied by fuel	44,100	100	1. To generate steam	2916666	14
			2. Carried by flue gases	5331	12.09
			3. Carried by moisture in flue gases	3504	7.94
			4. Unaccounted for	6099	13.83
	44,100	100		44,100	100

Example 4.5 *A steam generator delivers steam at 140 bar and 500°C. The feed water inlet temperature = 170°C. The steam generation rate is 110,000 kg/h, and thermal efficiency of steam generator is 90%.*

Calculate (a) the rate of fuel burnt if the calorific value of fuel is 22 MJ/kg, and (b) the percentages of total heat absorbed in economiser, evaporator and superheater.

Assume that only latent heat is absorbed in the evaporator.

Solution.

From steam tables at 140 bar, 500°C, we have

$$h_{sup} = 3322.0 \text{ kJ/kg}$$

For feed water at 170°C, h_f = 719.21 kJ/kg

Heat required for generating steam = 3322 – 719.21 = 2602.79 kJ/kg

$$\eta_{th} = \frac{\text{Heat required for generating steam}}{\text{Heat supplied by fuel}}$$

$$0.9 = \frac{2602.79 \times 110{,}000}{\dot{m}_f \times 22000}$$

$$\dot{m}_f = 14460 \text{ kg/h}$$

Enthalpy of saturated liquid at 140 bar, $h_f = 1571.1$ kJ/kg

$$\text{Heat absorbed in economiser} = \frac{1571.1 - 719.21}{2602.79} = 0.3273 \text{ or } 32.73\%$$

$$\text{Heat absorbed in evaporator} = h_{fg} \text{ at 140 bar}$$

$$= 1066.5 \text{ kJ/kg}$$

$$\text{Heat absorbed in evaporator} = \frac{1066.5}{2602.79} = 0.4097 \text{ or } 40.97\%$$

$$\text{Heat absorbed in superheater} = \frac{3322 - 2637.6}{2602.79} = 0.2629 \text{ or } 26.29\%$$

Example 4.6 *In a boiler test 1200 kg of coal is burnt in 24 hours. The mass of water evaporated is 12,500 kg. Steam pressure is 7.6 bar. The feed water temperature was 37°C. Calorific value of coal is 32 MJ/kg. The steam is dry and saturated. Calculate (a) equivalent evaporation from and at 100°C, (b) boiler efficiency, and (c) boiler power.*

Solution.

Given : $\dot{m}_f = \frac{1200}{24} = 50$ kg/h, $\dot{m}_w = 12500/24 = 520.83$ kg/h,

$$p_1 = 7.5 \text{ bar}, \quad t_f = 37°\text{C}, \quad \text{CV} = 32 \text{ mJ/kg}$$

From steam tables at 7.6 bar for dry and saturated steam,

$$h_{g1} = 2767.0 \text{ kJ/kg}$$

For feed water at 37°C, $h_f = 155.04$ kJ/kg

Heat required for generating steam = 2767.0 – 155.03 = 2611.97 kJ/kg

Mass of water evaporated per kg of fuel $= m_s = \dot{m}_w / \dot{m}_f = 520.83/50 = 10.417$

(*a*) Equivalent evaporation,

$$m_e = \frac{m_s(h_{g1} - h_f)}{2257} = \frac{10.417 \times 2611.97}{2257} = 12.05 \text{ kg/kg of coal}$$

$$\eta_{\text{boiler}} = \frac{m_s(h_{g1} - h_f)}{\text{C.V.}}$$

$$= \frac{10.417 \times 2611.97}{32{,}000} = 0.8503 \text{ or } 85.03\%$$

(*c*) $$\text{Boiler power} = \frac{m_e \times \dot{m}_f}{21.296} = \frac{12.05 \times 50}{21.296} = 28.29 \text{ kW}$$

Example 4.7 *A boiler produces 10 kg of steam/kg of coal burnt at 50°C. The steam pressure is 10 bar. Find the equivalent evaporation from and at 100°C if the dryness fraction of steam is 0.98.*

Solution.

Given : $m_s = 10$ kg/kg of coal, $x = 0.98$

From steam tables at 10 bar, we have

$$h_f = 762.8 \text{ kJ/kg}, \quad h_{fg} = 2015.3 \text{ kJ/kg}$$

For water at 50°C, $h_{fw} = 209.31$ kJ/kg

$$h_1 = h_f + x\,h_{fg} = 762.8 + 0.98 \times 2015.3 = 2737.794 \text{ kJ/kg}$$

$$\text{Equivalent evaporation} = \frac{m_s\,(h_1 - h_{fw})}{2257}$$

$$= \frac{10(2737.794 - 209.31)}{2257} = 11.2 \text{ kg/kg of coal}$$

Example 4.8 *A boiler produces 900 kg of wet steam per hour at a pressure of 7 bar. The dryness fraction of steam is 0.96. Feed water temperature is 20°C. If boiler efficiency is 75%, calculate the amount of coal burnt per hour. Calorific value of coal is 32,500 kJ/kg.*

Solution.

Given : $\dot{m}_s = 900$ kg/h, $p = 7$ bar, $x = 0.96$, $t_w = 20°C$, $\eta_{boiler} = 75\%$, CV = 32,500 kJ/kg

From steam tables at $p = 7$ bar, we have

$$h_f = 697.2 \text{ kJ/kg},\; h_{fg} = 2066.3 \text{ kJ/kg}$$

$$h_1 = h_f + x\,h_{fg} = 697.2 + 0.96 \times 2066.3 = 2680.848 \text{ kJ/kg}$$

For water at 20°C, $h_{fw} = 83.96$ kJ/kg

$$\eta_{boiler} = \frac{\dot{m}_s\,(h_1 - h_{fw})}{\dot{m}_f \times \text{C.V.}}$$

$$0.75 = \frac{900\,(2680.848 - 83.96)}{\dot{m}_f \times 32{,}500}$$

$$\dot{m}_f = 95.885 \text{ kg/h}$$

Example 4.9 *A boiler produces 15000 kg/h of steam at 7 bar, 300°C. The feed water temperature is 70°C. The calorific value of fuel = 30,000 kJ/kg. The grate is to be designed to burn 450 kg of coal/m^2.h. Find the grate area required assuming combustion efficiency = 90% and combined boiler and superheater efficiency = 75%. Take C_{ps} = 2.19 kJ/kg.K.*

Solution.

Given : $\dot{m}_s = 15000$ kg/h, $p = 7$ bar, $t_{sup} = 300°C$, $t_w = 70°C$,

CV = 30,000 kJ/kg, $\dot{m}_f = 450$ kg/m^2.h, $\eta_{comb} = 90\%$,

$\eta_{boiler} = 75\%$, $c_{ps} = 2.1$ kJ/kg.K

From steam tables at $p = 7$ bar, $t_{sup} = 300°C$, we have

$$h_{sup} = 3059.1 \text{ kJ/kg}$$

Otherwise,

$$h_{sup} = h_g + c_{ps}\,(t_{sup} - t_s)$$

$$= 2763.5 + 2.19\,(300 - 165)$$

$$= 3059.15 \text{ kJ/kg}$$

For water at 70°C, $h_{fw} = 293$ kJ/kg

Let A = grate area, m^2

Coal burnt per hour, $\dot{m}_f = 400$ A kg/h

Actual heat evolved $= \dot{m}_f \times \text{C.V.} \times \eta_{comb}$

$= 400 \text{ A} \times 30{,}000 \times 0.9 = 108 \times 10^5 \times \text{A kJ/h}$

Heat utilized for steam generation $=$ Heat evolved $\times\ \eta_{boiler}$

$= 108 \times 10^5 \times 0.75 \times \text{A} = 81 \times 10^5 \times \text{A kJ/h}$

Heat supplied in the boiler and superheater $= \dot{m}_s\ (h_{sup} - h_{fw})$

$= 15000\ (3059.15 - 293)$

$= 41492250$ kJ/h

$\therefore \quad 81 \times 10^5 \times \text{A} = 51492250$

$\text{A} = 5.1225 \text{ m}^2$

Example 4.10 *A boiler generates 8 kg/kg of coal burnt at a pressure of 10 bar. The feed water temperature is 70°C,* $\eta_{boiler} = 75\%$ *and factor of evaporation = 1.15. Take* $c_{ps} = 2.1$ *kj/kg.K. Calculate (a) degree of superheat and temperature of steam generated, (b) Calorific value of coal kJ/kg, and (c) equivalent evaporation in kg of steam/kg of coal.*

Solution.

Given : $m_s = 8$ kg/kg of coal, $p = 10$ bar, $t_w = 70°C$, $\eta_{boiler} = 75\%$,

Evaporation factor, F = 1.15, $c_{ps} = 2.1$ kJ/kg.K

From steam tables at $p = 10$ bar, $h_g = 2778.1$ kJ/kg $t_s = 179.9°C$

For feed water at 70°C, $h_{fw} = 293$ kJ/kg

(*a*) Evaporation factor, $$\text{F} = \frac{\left[h_g + c_{ps}\ (t_{sup} - t_s) - h_{fw}\right]}{2257}$$

$$1.15 = \frac{2778.1 + 2.1\ (t_{sup} - 179.9) - 293}{2257}$$

$t_{sup} = 232.5°C$

Degree of superheat = 232.5 – 179.9 = 52.6°C

(*b*) Boiler efficiency, $$\eta_{boiler} = \frac{m_s\ (h_{sup} - h_{fw})}{\text{C.V.}}$$

$$\text{C.V.} = \frac{8\left[2778.1 + 2.1 \times 52.6 - 293\right]}{0.75}$$

$= 27686$ kJ/kg

(*c*) Equivalent evaporation $= F \times m_s$

$= 1.15 \times 8 = 9.2$ kg/kg of coal burnt.

Example 4.11 *The following data was obtained during trial of a water tube boiler.*

Steam pressure	*= 14 bar*
Degree of superheat	*= 80°C*
Feed water temperature	*= 70°C*
Water evaporated	*= 3000 kg/h*

Coal burnt *400 kg/h*
Ash *= 35 kg/h*
Combustible in ash *= 10%*
Moisture in coal *= 4%*
Calorific value of dry coal *= 30500 kJ/kg*
Calorific value of carbon *= 34000 kJ/kg*
$c_{ps} = 2.1$ *kJ/kg.K*

Calculate (a) the efficiency of boiler plant including superheat, and (b) the efficiency of boiler and furnace combined.

Solution.

Given : $p = 14$ bar, $t_{sup} - t_s = 80°C$, $t_w = 70°C$, $\dot{m}_s = 300$ kg/h,

$\dot{m}_{fw} = 400$ kg/h, ash = 35 kg/h, $(C.V.)_f = 30500$ kJ/kg, $(C.V.)_c = 34000$ kJ/kg,

$c_{ps} = 1$ kJ/kg.K

From steam tables at $p = 14$ bar, $h_g = 2790$ kJ/kg, $t_s = 195.1°C$

$$\therefore \quad h = h_g + c_{ps}\,(t_{sup} - t_s)$$

$$= 2790 + 2.1 \times 80 = 2958 \text{ kJ/kg}$$

For feed water at 70°C, $h_{fw} = 293$ kJ/kg

Mass of dry coal, $\dot{m}_f = 400 - 400 \times \dfrac{4}{100} = 384$ kg/h

$$\text{Boiler efficiency including superheater} = \frac{\dot{m}_s\,(h - h_{fw})}{\dot{m}_f \times (CV)_f}$$

$$= \frac{3000\,(2958 - 293)}{384 \times 30500} = 0.6826 \text{ or } 68.26\%$$

Combustible in ash/h = Ash kg/h × combustible in ash kg/h

$$= 35 \times \frac{10}{100} = 3.5 \text{ kg}$$

Actual heat supplied per hour = heat of dry coal – heat of combustible in ash
= 384 × 30500 – 3.5 × 3400°
= 11593 MJ/h

Heat utilized in the boiler per hour $= \dot{m}_s\,(h_{sup} - h_{fw})$
$= 3000\,(2958 - 293)$
$= 7995$ MJ/h

Efficiency of boiler including furnace

$$= \frac{\text{Heat utilized in boiler per hour}}{\text{Actual heat supplied per hour}}$$

$$= \frac{7995}{11593} = 0.6896 \text{ or } 68.96\%$$

Example 4.12 *A boiler plant supplies 5400 kg of steam per hour at 7.6 bar pressure. The dryness of steam is 0.98. The feed water temperature is 42°C when using 670 kg of coal/hr. The calorific valve of coal is 31000 kJ/kg.*

Calculate (a) The efficiency of the boiler (b) The equivalent evaporation from and at 100°C. The extract from steam table is as under :

Absolute pressure	*Temperature*	*Specific volume, m³/kg*		*Specific Enthalpy, kJ/kg*	
bar	*°C*	*vf*	v_g	*hf*	h_g
0.08198	*42*	*0.0010086*	*17.69*	*175.81*	*2577.9*
7.6	*168.30*	*0.0011123*	*0.25224*	*711.67*	*2765.4*

Solution.

Given : $\dot{m}_s = 5400$ kg/h, $p = 7.6$ bar, $x = 0.98$, $t_w = 42°C$, $\dot{m}_f = 670$ kg/h,

CV = 31000 kJ/kg

Enthalpy of steam at $p = 7.6$ bar,

$$h = h_f + x(h_g - h_f)$$

$$= 711.67 + 0.98\,(2765.4 - 711.67)$$

$$= 2724.32 \text{ kJ/kg}$$

$$h_{fw} = 175.81 \text{ kJ/kg}$$

(*a*) Boiler efficiency,
$$\eta_{\text{boiler}} = \frac{\dot{m}_s(h - h_{fw})}{\dot{m}_f \times \text{CV}}$$

$$= \frac{5400\,(2724.32 - 175.81)}{670 \times 31000} = 0.6626 \text{ or } 66.26\%$$

(*b*) Equivalent evaporation,
$$m_e = \frac{\dot{m}_s(h - h_{fw})}{\dot{m}_f \times 2257} = \frac{5400\,(2724.32 - 175.81)}{670 \times 2257}$$

$$= 9.1 \text{ kg/kg of coal}$$

Example 4.13 *In a boiler test, 1300 kg of coal are consumed in 24 hours. The mass of water evaporated is 13000 kg. Steam pressure is 7 bar. The feed water temperature entering the boiler is 40°C. Calorific value of fuel is 30000 kJ/kg. Dryness fraction of steam = 0.95. Calculate (a) Equivalent evaporation from and at 100°C. (b) Boiler efficiency. Take* c_p *= 4.187 kJ/kg.K for water. Use the following table :*

Pressure bar	*Enthalpy, kJ/kg*			*Specific volume, m³/kg*	*Saturation temperature*
	h_f	h_{fg}	h_g	v_g	°C
7	*697.06*	*2064.94*	*2762.0*	*0.27268*	*164.84*

Solution.

Given : $\dot{m}_f = 1300$ kg/24 h, $\dot{m}_s = 13000$ kg/24 h, $p = 7$ bar, $t_w = 40°C$,

CV = 30,000 kJ/kg, $x = 0.95$, $c_{pw} = 4.187$ kJ/kg.K

Enthalpy of steam at $p = 7$ bar,

$$h = h_f + x\,h_{fg} = 697.06 + 0.95 \times 2064.94 = 2658.753 \text{ kJ/kg}$$

At $t_w = 40°C$, $h_{fw} = c_{pw}\,(t_w - 0) = 4.187 \times 40 = 167.48$ kJ/kg

(*a*) Equivalent evaporation

$$= \frac{\dot{m}_s (h - h_{fw})}{\dot{m}_f \times 2257}$$

$$= \frac{13000\,(2658.753 - 167.48)}{1300 \times 2257}$$

$$= 11.038 \text{ kg/kg of coal}$$

(*b*) Boiler efficiency

$$= \frac{\dot{m}_s (h - h_{fw})}{\dot{m}_f \times \text{CV}}$$

$$= \frac{13000\,(2658.753 - 167.48)}{1300 \times 30,000}$$

$$= 0.8304 \text{ or } 83.04\%$$

Example 4.14 *The following data was obtained in a boiler trial :*

Feed water used	*680 kg/h*
Feed water temperature	*= 20°C*
Steam pressure	*= 15 bar*
Steam temperature	*= 300°C*
Coal used	*= 95 kg/h*
Calorific value of coal	*= 26,500 kJ/kg.*

$c_{ps} = 2.1$ *kJ/kg.K for superheated steam*

$c_{pw} = 4.187$ *kJ/kg.K for water.*

Calculate : (a) Boiler efficiency (b) Equivalent evaporation from and at 100°C. Use the following table :

Pressure	*Saturation temperature*	*Specific volume m^3/kg*		*Specific enthalpy kJ/kg*		
bar	*°C*	v_f	v_g	h_f	h_{fg}	h_g
15	*198.29*	*0.0011539*	*0.13166*	*844.66*	*1945.2*	*2798.9*

Solution.

Given : $\dot{m}_s = \dot{m}_w = 680$ kg/h, $t_w = 20°C$, $p = 15$ bar, $t_{sup} = 300°C$, $\dot{m}_f = 95$ kg/h,

CV = 26500 kJ/kg, $c_{ps} = 2.1$ kJ/kg.K, $c_{pw} = 4.187$ kJ/kg.K

Enthalpy of steam, $h_{sup} = h_g + c_{ps}(t_{sup} - t_s)$

$= 2798.9 + 2.1(300 - 198.29)$

$= 3012.491$ kJ/kg

Enthalpy of feed water, $h_{fw} = c_{pw}(t_w - 0) = 4.187 \times 20 = 83.74$ kJ/kg

(*a*) $$\text{Boiler efficiency} = \frac{\dot{m}_s(h_{sup} - h_{fw})}{\dot{m}_f \times \text{CV}}$$

$$= \frac{680\,(3012.491 - 83.74)}{95 \times 26500} = 0.791 \text{ or } 79.1\%$$

(*b*) $$\text{Equivalent evaporation} = \frac{\dot{m}_s(h_{sup} - h_{fw})}{\dot{m}_f \times 2257}$$

$$= \frac{680\,(3012.491 - 83.74)}{95 \times 2257}$$

$$= 9.29 \text{ kJ/kg of coal}$$

Example 4.15 *The superheated steam is generated from the boiler operation at 12 bar pressure. Steam generated/kg of coal burnt is 10 kg. Coal burnt = 300 kg/h. If the enthalpies of superheated steam and water entering the boiler are 2935.4 kJ/kg and 150.74 kJ/kg respectively, then calculate (a) Equivalent evaporation (b) Boiler power.*

Solution.

Given : $p = 12$ bar, $\dot{m}_s / \dot{m}_f = 10$, $\dot{m}_f = 300$ kg/h, $h_{sup} = 2935.4$ kJ/kg,
$h_{fw} = 150.74$ kJ/kg

$\dot{m}_s = 10 \times 300 = 3000$ kg/h

(*a*) $$\text{Equivalent evaporation from and at } 100°\text{C} = \frac{\dot{m}_s(h_{sup} - h_{fw})}{\dot{m}_f \times 2257}$$

$$= \frac{10\,(2935.4 - 150.74)}{2257}$$

$$= 12.34 \text{ kJ/kg of coal}$$

(*b*) $$\text{Boiler power} = \frac{\text{Equivalent evaporation/h from and at } 100°\text{C}}{21.296} \text{ kW}$$

$$= \frac{12.34 \times 300}{21.296} = 173.835 \text{ kW}$$

Example 4.16 *A boiler produces 1200 kg of steam/h at a pressure of 10 bar and temperature of 200°C. Coal consumed during same period is 150 kg. Feed water temperature is 46°C. C.V. of coal is 45000 kJ/kg. Calculate equivalent evaporation from and at 100°C and boiler efficiency.*

Take $c_{ps} = 2.1$ kJ/kg.K and $c_{pw} = 4.187$ kJ/kg.K.

Solution.

Given : $\dot{m}_s = 1200$ kg/h, $p = 10$ bar, $t_{sup} = 200°$C, $\dot{m}_f = 150$ kg/h,
$t_w = 46°$C, C.V. $= 45000$ kJ/kg, $c_{ps} = 2.1$ kJ/kg.K,
$c_{pw} = 4.187$ kJ/kg.K.

From steam tables at $p = 10$ bar, $h_g = 2778.1$ kJ/kg, $t_s = 179.9$°C

Enthalpy of steam, $h = h_g + c_{ps}\,(t_{\text{sup}} - t_s)$

$= 2778.1 + 2.1\,(200 - 179.9) = 2820.31$ kJ/kg

Enthalpy of feed water, $h_{fw} = c_{pw}\,(t_w - 0) = 4.187 \times 46 = 192.6$ kJ/kg

$$\text{Equivalent evaporation} = \frac{\dot{m}_s(h - h_{fw})}{\dot{m}_f \times 2257} = \frac{1200\,(2820.31 - 192.6)}{150 \times 2257} = 9.314 \text{ kg/kg of coal}$$

$$\text{Boiler efficiency} = \frac{\dot{m}_s(h - h_{fw})}{\dot{m}_f \times \text{CV}}$$

$$= \frac{1200\,(2820.31 - 192.6)}{150 \times 45000} = 0.4671 \text{ or } 46.71\%$$

Example 4.17 *The following observations were taken during a test on a boiler trial.*

Amount of coal consumed in 24 hours = *1200 kg*
Mass of water evaporated = *12000 kg*
Steam pressure = *7 bar*
Feed water temperature = *40°C*
Calorific value of coal = *30,000 kJ/kg*
Dryness fraction of steam produced = *1.0*

Calculate (a) Equivalent evaporation from and at (F & A) 100°C.
(b) Boiler efficiency.

Solution.

Given : $\dot{m}_f = 1200$ kg/24 h, $p = 7$ bar, $t_w = 40$°C, CV = 30,000 kJ/kg,

$\dot{m}_s = 12000$ kg/24 h, $x = 1.0$.

From steam tables at $p = 7$ bar for dry and saturated steam,

Enthalpy of steam, $h = h_g = 2763.5$ kJ/kg

Enthalpy of water at 40°C, $h_{fw} = 167.57$ kJ/kg

$$(a)\ \text{Equivalent evaporation} = \frac{\dot{m}_s(h - h_{fw})}{\dot{m}_f \times 2257}$$

$$= \frac{12000\,(2763.5 - 167.57)}{1200 \times 2257} = 11.5 \text{ kg/kg of coal}$$

$$(b)\ \text{Boiler efficiency} = \frac{\dot{m}_s(h - h_{fw})}{\dot{m}_f \times \text{CV}}$$

$$= \frac{12000\,(2763.5 - 167.57)}{1200 \times 30{,}000}$$

$$= 0.8653 \text{ or } 86.53\%$$

Example 4.18 *In a thermal power plant, the boiler produces 50,000 kg of wet steam per hour with dryness fraction of 0.95. The feed water temperature is 40°C. Calculate the equivalent evaporation if boiler pressure is 10 bar.*

Solution.

Given : $\dot{m}_s = 50,000$ kg/h, $x = 0.95$, $t_w = 40$°C, $p = 10$ bar

From steam tables at $p = 10$ bar, we have

$$h_f = 762.8 \text{ kJ/kg}, h_{fg} = 2015.3 \text{ kJ/kg}$$

Enthalpy of steam, $h = h_f + x\, h_{fg} = 762.8 + 0.95 \times 2015.3 = 2677.335$ kJ/kg

Enthalpy of feed water at 40°C, $h_{fw} = 167.57$ kJ/kg

$$\text{Equivalent evaporation} = \frac{\dot{m}_s (h - h_{fw})}{2257}$$

$$= \frac{50,000\ (2677.335 - 167.57)}{2257}$$

$$= 55599.6 \text{ kg/h}$$

Example 4.19 *A steam generator evaporates 18000 kg/h of steam at 12 bar pressure and steam is 97% dry. Feed water temperature = 105°C. Coal is fired at the rate of 2050 kg/h. C.V. of coal is 27400 kJ/kg.*

Calculate (a) Heat supplied/h (b) Thermal efficiency
(c) Equivalent evaporation.

Solution.

Given : $\dot{m}_s = 18000$ kg/h, $p = 12$ bar, $x = 0.97$, $t_w = 105$°C,

$\dot{m}_f = 2050$ kg/h, CV = 27400 kJ/kg

From steam tables at $p = 12$ bar, we have

$$h_f = 798.6 \text{ kJ/kg}, \quad h_{fg} = 1986.2 \text{ kJ/kg}$$

Enthalpy of steam, $h = h_f + x\, h_{fg} = 798.6 + 0.97 \times 1986.2 = 2725.214$ kJ/kg

Enthalpy of water at 105°C, $h_{fw} = 4.187 \times (105 - 0) = 439.635$ kJ/kg

(*a*) Net heat supplied $= h - h_{fw} = 2725.214 - 439.635 = 2285.579$ kJ/kg

(*b*) Thermal efficiency $= \dfrac{\dot{m}_s (h - h_{fw})}{\dot{m}_f \times \text{CV}} = \dfrac{18000 \times 2285.579}{2050 \times 27400} = 0.7324$ or 73.24%

(*c*) Equivalent evaporation $= \dfrac{\dot{m}_s (h - h_{fw})}{\dot{m}_f \times 2257} = \dfrac{18000 \times 2285.579}{2050 \times 2257} = 8.89$ kg/kg of coal.

Example 4.20 *A boiler produces 1000 kg of steam/h at a pressure of 10 bar abs. and 0.95 dry from water at 40°C. It is ascertained that 10% of coal is unburnt at grate. If the efficiency of boiler alone is 75%, find the efficiency of the boiler and grate combined. C.V. of coal is 30980 kJ/kg. Assume specific heat of water as 4.1868 kJ/kg.K.*

Solution.

Given : $\dot{m}_s = 1000$ kg/h, $p = 10$ bar, $x = 0.95$, $t_w = 40°C$,

Unburnt coal = 10%, $\eta_{boiler} = 75\%$, CV = 30980 kJ/kg,

$c_{pw} = 4.1868$ kJ/kg.K

From steam tables at $p = 10$ bar, we have

$$h_f = 762.8 \text{ kJ/kg}, \; h_{fg} = 2015.3 \text{ kJ/kg}$$

Enthalpy of steam, $h = h_f + x\,h_{fg} = 762.8 + 0.95 \times 2015.3 = 2677.335$ kJ/kg

Enthalpy of water at 40°C, $h_{fw} = 4.1868 \times 40 = 167.472$ kJ/kg

Boiler efficiency, $$\eta_{boiler} = \frac{\dot{m}_s (h - h_{fw})}{\dot{m}_{fa} \times \text{CV}}$$

$$0.75 = \frac{1000\,(2677.335 - 167.472)}{\dot{m}_{fa} \times 30980}$$

Coal actually burnt, $\dot{m}_{fa} = 108.02$ kg/h

Mass of coal fired, $\dot{m}_f = 1.1 \times 108.02 = 118.822$ kg/h

Combined thermal efficiency of boiler and grate

$$= \frac{\dot{m}_s (h - h_{fw})}{\dot{m}_f \times \text{CV}}$$

$$= \frac{1000\,(2677.335 - 167.472)}{118.822 \times 30980}$$

$$= 0.6818 \text{ or } 68.18\%$$

Example 4.21 *A boiler produces 2000 kg of dry saturated steam per hour at 10 bar. The feed water temperature = 110°C. 225 kg of coal having calorific value of 30,000 kJ/kg are fired per hour. If 10% of coal remains unburnt,*

Calculate (a) The thermal efficiency of boiler.

(b) The thermal efficiency of boiler and grate combined.

Solution.

Given : $\dot{m}_s = 200$ kg/h, $p = 10$ bar, $t_w = 110°C$, $\dot{m}_f = 225$ kg/h,

CV = 30,000 kJ/kg, unburnt coal = 10%

From steam tables at $p = 10$ bar for dry and saturated steam,

Enthalpy, $h = h_g = 277.8.1$ kJ/kg

Enthalpy of feed water at 110°C, $h_{fw} = 461.30$ kJ/kg

(*a*) Thermal efficiency of boiler $$= \frac{\dot{m}_s (h - h_{fw})}{\dot{m}_{fa} \times \text{CV}}$$

$$= \frac{2000\,(2778.1 - 461.30)}{0.9 \times 225 \times 30{,}000} = 0.7627 \text{ or } 76.27\%$$

(*b*) Coal fired $= 225$ kg/h

Combined efficiency of boiler and grate

$$= \frac{\dot{m}_s (h - h_{fw})}{\dot{m}_f \times \text{CV}}$$

$$= \frac{2000\,(2778.1 - 461.30)}{225 \times 30{,}000}$$

$$= 0.6864 \text{ or } 68.64\%$$

MULTI-CHOICE QUESTIONS

1. Which of the following are boiler accessories ?
(*a*) Safety valve (*b*) Stop valve (*c*) Economiser (*d*) Blow-off cock

2. The performance of a boiler is measured by
(*a*) Amount of water evaporated/hr (*b*) Steam produced in kg/hr
(*c*) All of the above (*d*) None of above

3. The main objective of a boiler tail is
(*a*) to determine the thermal efficiency of the boiler when working at a definite pressure
(*b*) to draw up heat balance sheet for the boiler
(*c*) both (*a*) and (*b*)
(*d*) none of the above

4. Economiser is generally placed between
(*a*) last superheater/preheater and air preheater
(*b*) air preheater and chimney
(*c*) electrostatic precipitators
(*d*) induced draft fan and forced draft fan

5. At tempering is
(*a*) reduction of steam pressure
(*b*) reduction of steam temperature
(*c*) discharge of flue gases at certain height
(*d*) conditioning of disposable fluids for minimum pollution

6. The air preheater of a boiler plant is located
(*a*) between forced draught fan and furnace
(*b*) between furnace and economiser
(*c*) between economiser and chimney
(*d*) near superheater

7. The following type of safety valve is most suitable for locomotive boiler
(*a*) Dead weight safety valve (*b*) Spring loaded safety valve
(*c*) Lever loaded safety valve (*d*) None of the above

8. The manhole provided in a fire tube boiler is used for
(*a*) Inspecting the drum when the boiler is in operation
(*b*) Inspecting the drum when the boiler is shutdown
(*c*) Inspecting the drum and clean the inside portions of the boiler when the boiler is shutdown
(*d*) none of the above

9. Consider the following statements:
Blow down is necessary on boilers, because

1. the boiler water level is lowered rapidly in case it accidentally rises too high.
2. the precipitated sediment of sludge is removed while the boiler is in service.
3. the concentration of suspended solids in the boiler is controlled.

Of these statements
(*a*) 1,2 and 3 are correct (*b*) 1 and 2 are correct
(*c*) 3 alone is correct (*d*) 1 and 3 are correct

10. A device whose function is to heat feed water by utilizing the heat in the exhaust flue gases before leaving through the chimney is called
(*a*) Superheater (*b*) Economiser (*c*) Air preheater (*d*) Feed pump

11. A device which is used for pumping water in to the boiler is called
(*a*) Economiser (*b*) Feed pump (*c*) Injector (*d*) Air preheater

12. Which of the following are boiler mountings ?
(*a*) Economiser (*b*) Fusible plug (*c*) Super heater (*d*) Injector

13. Which out of the following boiler is capable of generating superheated steam without additional accessories ?
(*a*) Cochran (*b*) Lancashire (*c*) Locomotive (*d*) Cornish

14. The economiser of the boiler plant is used mainly to
(*a*) increase steam capacity (*b*) reduce fuel consumption
(*c*) increase steam pressure (*d*) increase life of the boiler

15. Injector is used in small boilers
(*a*) to inject steam to increase its capacity
(*b*) to inject water like a feed pump
(*c*) to inject air to the furnace
(*d*) none of the above

16. Consider the following
1. Safety valve 2. Steam trap 3. Steam separator 4. Economiser
Among these the boiler accessories would include
(*a*) 1, 2 and 3 (*b*) 2, 3 and 4 (*c*) 1 and 4 (*d*) 1, 2, 3 and 4

17. Match List I and List II and select the correct answer using the codes given below the lists

List I (Type of boiler)	**List II (Classification of boiler)**
A. Babcock and Wilcox	1. Forced circulation
B. Lancashire	2. Fire tube
C. La-mont	3. Water tube
D. Cochran	4. Vertical

Codes:

	A	B	C	D
(*a*)	1	2	3	4
(*b*)	2	3	4	1
(*c*)	3	2	1	4
(*d*)	2	4	1	3

18. The blow down cock is fitted in a boiler because of
(*a*) allowing steam to flow to turbine
(*b*) allowing exhaust gases to go to chimney

(c) removing sludge from the boiler
(d) blowing air in the furnace

19. Duplex feed pumps used in the steam boilers operate on the principle of
(a) centrifugal pump (b) reciprocating pump
(c) ejector pump (d) gear pump

20. Which boilers from the following utilizer the exhaust of the engine for increasing the draught?
(a) Babcock and wilcox (b) Cochran
(c) Locomotive (d) Lancashire

21. Once through boiler is named as such because
(a) flue gas passes only in one direction (b) there is no recirculation of water
(c) air is sent through the same direction (d) steam is sent out only in one direction

22. The gusset stays is provided in a boiler because
(a) of preventing failure due to compression
(b) of preventing leakage of steam
(c) of preventing the bulging of flat surfaces
(d) of providing support to the boiler drum

23. The function of the fusible plug in a boiler is
(a) to prevent excessive pressure in the boiler
(b) to extinguish fire in case when water level is low
(c) to prevent excessive airflow
(d) to prevent excessive steam flow to turbine

24. Glass tube water level indicator of boilers are connected in the following manner
(a) Both ends are connected to water space
(b) One end is connected to water space and the other end is connected to steam space
(c) One end is connected to water space and the other end is connected to the superheater
(d) both ends are connected to steam space

25. The high steam and low water safety valve is not suitable for the following type of boiler
(a) Cochran (b) Lankashire (c) Locomotive (d) Cornish

26. The function of a steam trap in a boiler plant is
(a) to prevent steam from leaking out from steam pipe
(b) to prevent steam from leaking out from stop valve
(c) to prevent steam from leaking out from boiler
(d) to drain off water resulting from partial condensation of steam from steam pipes without allowing steam to escape from it.

27. Match List I (Components) with list II (Functions) and select the correct answer using the codes given below the Lists:

List I	List II
A. Steam trap B. Fusible plug C. Blow-off cock D. Feed check valve	1. Controls steam flow rate 2. Controls rate of water flow to boiler 3. Puts off furnace fire when water level reaches unsafe limit 4. Removes mud and dirt collected at the bottom of boiler 5. Drains off water collected by partial condensation of steam in pipes

Codes:

	A	B	C	D
(*a*)	5	1	4	2
(*b*)	1	3	5	4
(*c*)	5	3	4	2
(*d*)	1	2	5	4

28. The cross section of the water passage in an injector adjacent to feed check valve is
(*a*) divergent towards boiler (*b*) convergent towards boiler
(*c*) parallel passage (*d*) none of the above

29. In a Babcock and Wilcox boiler, baffles are provided in between the water tubes to
(*a*) regulate the water flow
(*b*) regulate the steam flow
(*c*) allow the flue gas to have better contact with water tubes
(*d*) have all the objectives fulfilled

30. The size of the steam drum in a Babcock and Wilcox boiler can be made small because
(*a*) it contains only few water tubes
(*b*) it contains only superheater tube
(*c*) it does not contain any water tube
(*d*) it contains a small number of water tubes and superheater tubes

31. Consider the following statements :
1. Forced circulation is always used in high pressure power boilers.
2. Soot blowers are used for cleaning tube surfaces at regular intervals.
3. Electrostatic precipitator is used to remove fly ash from flue gases.
Which of these statements are correct ?
(*a*) 1, 2 and 3 (*b*) 2 and 3 (*c*) 1 and 3 (*d*) 1 and 2

32. Once-through boilers operate at
(*a*) subcritical pressure
(*b*) super critical pressure
(*c*) subcritical as well as super critical pressures
(*d*) critical pressure only

33. Water after being pumped into a boiler can not come out because of
(*a*) steam stop valve (*b*) feed check valve
(*c*) safety valve (*d*) blow off valve

34. Which of the following is an accessory.
(*a*) Safety valve (*b*) Feed check valve
(*c*) Feed water pump (*d*) Stop valve

35. In Green's Economiser the tubes are scraped
(*a*) for the removal of scales formed due to water passage
(*b*) for the removal of soot collected on the tubes
(*c*) for increasing the circulation of water through tubes
(*d*) none of the above

36. The correct location of an economiser is
(*a*) between the furnace and preheater
(*b*) between the air preheater and chimney
(*c*) between force draught fan and furnace
(*d*) near the superheater

37. The normal working pressure range for Cornish boiler as compared to that of Lancashire boiler is
(*a*) same (*b*) high (*c*) low (*d*) none of this

38. The following mountings are fitted on the drum of a Babcock and Wilcox boiler
(*a*) Safety valve, Pressure gauge, Feed check valve but not the Water valve
(*b*) Safety valve, Pressure gauge, Water gauge but not the feed check valve
(*c*) Safety valve, Feed water check valve, Blow down valve and Water gauge
(*d*) Safety valve, Pressure gauge, Feed check valve Water gauge but not the Blow valve

39. Which of the following is a fire tube boiler
(*a*) Lancashire (*b*) Locomotive (*c*) Cochran (*d*) All of these

40. Which of the following form part(s) of the boiler mountings?
1. Economiser 2. Feed check valve 3. Steam trap 4. Superheater
Select the correct answer using the codes given below :
Codes :
(*a*) 2 alone (*b*) 1 and 3 (*c*) 2, 3 and 4 (*d*) 1, 2, 3 and 4

41. The fitting mounted on the boiler whose function is to put off the fire in the furnace, when level of water falls to an unsafe limit is called
(*a*) Safety valve (*b*) Stop valve (*c*) Fusible plug (*d*) Blowoff cock

42. The fitting mounted on the boiler whose function is to prevent the steam pressure in a boiler exceeding a fixed maximum pressure is called
(*a*) Safety valve (*b*) Stop valve (*c*) Fusible plug (*d*) Blowoff cock

43. The exact level of water in the boiler is indicated by
(*a*) Water level indicator (*b*) Feed check valve
(*c*) Blow off cock (*d*) Stop valve

Answers:

1. (*c*) **2.** (*d*) **3.** (*c*) **4.** (*a*) **5.** (*b*) **6.** (*c*) **7.** (*b*) **8.** (*c*) **9.** (*b*) **10.** (*c*)
11. (*c*) **12.** (*b*) **13.** (*c*) **14.** (*a*) **15.** (*b*) **16.** (*a*) **17.** (*c*) **18.** (*c*) **19.** (*b*) **20.** (*c*)
21. (*b*) **22.** (*c*) **23.** (*b*) **24.** (*b*) **25.** (*c*) **26.** (*d*) **27.** (*c*) **28.** (*a*) **29.** (*c*) **30.** (*c*)
31. (*b*) **32.** (*d*) **33.** (*b*) **34.** (*c*) **35.** (*b*) **36.** (*a*) **37.** (*c*) **38.** (*d*) **39.** (*d*) **40.** (*a*)
41. (*c*) **42.** (*a*) **43.** (*a*)

REVIEW QUESTIONS

1. What is a steam generator ?
2. How do you classify steam generators ?
3. What are the merits of water tube boilers ?
4. What are the demerits of fire tube boilers ?
5. List the requirements of a good boiler.
6. Enumerate the factors affecting boiler selection.
7. What are the advantages of high pressure boilers ?
8. List the advantages of forced circulation boilers.
9. What is a boiler mounting ? List the various boiler mountings ?
10. What are the functions of boiler accessories ?
11. Name the various accessories used on a boiler.

12. Define equivalent evaporation and factor of evaporation.
13. Define boiler efficiency.
14. What are the purposes of carrying out boiler trial ?
15. Differentiate between a boiler and a steam generator.
16. State the primary requirements of a steam generator.
17. Distinguish between water tube and fire tube boilers.
18. Explain the difference between mountings and accessories.
19. What is the function of air preheater ?
20. What is an economiser ? What are its advantages ?

EXERCISES

4.1 Describe the construction features of a Lancashire boiler. How superheated steam can be produced with this boiler ?

4.2 Describe the construction features of a Cochran boiler. Discuss its working briefly.

4.3 Where a water tube boiler is exclusively used ? Discuss the construction and working of Babcock and Wilcox boiler.

4.4 With the help of a neat sketch describe a Loeffler boiler. What is usually the working pressure of such a boiler ?

4.5 Draw a schematic flow diagram of a modem steam generator. Discuss the installation of connection and radiation superheaters and their exit temperature response with varying steam flow.

4.6 With the help of a neat sketch discuss the working principle of Velox high pressure boiler.

4.7 In a boiler trial, 1360 kg of coal were consumed in 24 hours. The mass of water evaporated was 13600 kg and steam pressure was 7.5 bar. The feed water temperature was 35°C Calorific value of 1 kg of coal is 30,000 kJ/kg. Calculate the boiler efficiency and equivalent evaporation. **[Ans.** 87.31%, 11.6 kg**]**

4.8 A boiler generates 8.5 Kg of steam per kg of coal burned at a pressure of 13.5 bar from feed water having absolute temperature of 350 K. The boiler efficiency is 70% and factor of evaporation 1.17. Taking specific heat of steam at constant pressure to be 2.1 kJ/kg.K, calculate

(*a*) Degree of super heat and temperature of steam generated,

(*b*) Calorific value of coal, and

(*c*) Equivalent evaporation **[Ans.** 84°C, 550.4°C, 32,070 kJ/kg, 9.94 kg**]**

4.9 The data obtained during a boiler trial is given below:

Duration	= 1 hour
Steam generated	= 35,500 kg
Steam pressure	= 12 bar
Steam temperature	= 250°C
Temperature of water entering economiser	= 17°C
Temperature of water leaving economiser	= 77°C
Oil burnt	= 3460 kg
C.V. of oil	= 39,500 kJ/kg

Calculate (*a*) equivalent evaporation, (*b*) thermal efficiency of plant, and (*c*) percentage of heat energy of fuel energy utilised by the economiser. **[Ans.** 13.01 kg, 74.4%, 6.52%**]**

4.10 The following data relate to a boiler trial :

Coal burnt per hour	= 6750 kg
Moisture content in coal	= 2%
H.C.V. of coal	= 35490 kJ/kg

Ultimate mass analysis of dry coal : C = 84%, O_2 = 4%, ash = 8%

Dry flue gas analysis by volume : CO_2 = 9.5%, O_2 = 10.48%, N_2 = 80.02%
Temperature of flue gases = 305°C
Room temperature =32°C
Specific heat of flue gases = 1 kJ/kg.K
Specific heat of steam in flue gases = 2 kJ/kg.K
Mass of steam generated at 16.5 bar from feed water at 47°C = 60,500 kg/h
Draw up the heat balance sheet per kg of dry coal. Assume that air contains 23.1% by mass of Oxygen.

Ans.

Heat supplied	**kJ**	**%**	**Heat utilised**	**kJ**	**%**
Heat of dry coal	35,490	98.40	Steam formation	24,365	67.61
Heat in feed water	573.6	1.60	Carried away by dry flue gases	3136	8.70
			Carried by excess air	2877	7.98
			Carried by moisture	1123	3.12
			Unaccounted for	4535.6	12.59
	36,063.6	100		36,063.6	100

4.11 The following data is available from a boiler trial :
Duration = 1 hour, steam generated = 36000 kg, steam pressure = 12.5 bar, steam temperature = 250°C, temperature of water entering the economiser = 16°C, temperature of water leaving the economiser = 75°C, oil burned = 3500 kg, C.V. of oil = 39500 kJ/kg, specific heat of steam = 2.1 kJ/kg.K. Calculate (*a*) equivalent evaporation per kg of fuel, (*b*) thermal efficiency, and (*c*) percentage of heat energy of flue gases utilized by the economiser.

4.12 A boiler generates 850 kg of steam per hour at a pressure of 10 bar and 50°C superheat. It burns 100 kg/h of coal of calorific value 30 MJ/kg. The feed water temperature is 30°C. Calculate (*a*) equivalent evaporation from and at 100°C per kg of coal, and (*b*) boiler efficiency. Take c_{ps} = 2.1 kJ/kg.K and c_{pw} = 4.187 kJ/kg.K.

4.13. A boiler supplies 5000 kg of steam at 7 bar pressure and 0.98 dry. The feed water temperature is 40°C when using 650 kg/h of coal. The calorific value of coal is 31 mJ/kg. Calculate (*a*) the efficiency of boiler, and (*b*) equivalent evaporation from and at 100°C.

4.14 In a boiler trial, the following data is available :
Coal consumed in 24 hours = 1300 kg
Mass of water evaporated = 13000 kg
Steam pressure = 5 bar
Temperature of feed water = 40°C
Calorific value of fuel = 30 MJ/kg
Dryness fraction of steam = 0.95
Calculate (*a*) equivalent evaporation from and at 100°C, and
(*b*) boiler efficiency.
Take c_{pw} = 4.187 kJ/kg.K.

4.15 Steam is generated at 12 bar and 250°C. The feed water temperature is 30°C. The calorific value of coal is 34 MJ/kg. Steam generated per kg of coal burnt is 10 kg. Find (*a*) equivalent evaporation, and (*b*) boiler power if 250 kg/h of coal is burnt.

5 REFRIGERATION AND AIR-CONDITIONING

5.1 REFRIGERATION

Refrigeration may be defined as the artificial withdrawal of heat, producing in a substance or within a space, a temperature lower than that which would exist under the natural influence of surroundings.

Unit of Refrigeration–The capacity of a refrigeration unit is given in tons of refrigeration. A ton of refrigeration is defined as the quantity of heat to be removed in order to form one ton of ice in 24 hours when the initial temperature of water is 0°C. One ton of refrigeration, $TR = 211$ kJ/min or 3.5167 kW

$$1 \text{ tonne of refrigeration} = \frac{1000 \times 336}{24} = 14{,}000 \text{ kJ/h or } 3.8889 \text{ kW}$$

Coefficient of Performance, COP The COP of a refrigeration system is a ratio of the heat removed (cooling effect produced) from a system to the work supplied to achieve the heat removal.

$$\text{COP} = \frac{\text{Heat removal per unit time, Q (kJ/s)}}{\text{Work supplied per unit time, W(kJ/s)}}$$

$$= \frac{211 T_n \ (kJ/\min)}{(60P)\ (kJ/\min)}$$

$$= \frac{3.5167}{\left(\dfrac{P}{T_n}\right)}$$

where P = energy required in kW

T_n = capacity of plant in tons

$$\text{Relative COP} = \frac{\text{Actual COP}}{\text{Theoretical COP}}$$

5.2 VAPOUR COMPRESSION REFRIGERATION SYSTEM

A schematic diagram of a vapour compression refrigeration system is shown in Fig. 5.1. It consists of the following components.

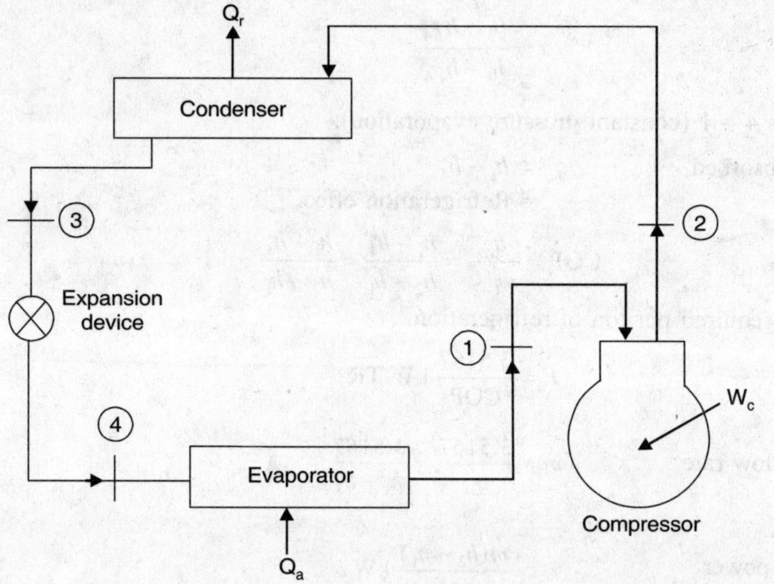

Fig. 5.1 Vapour compression system

1. Compressor
2. Condenser
3. Expansion device for throttling
4. Evaporation.

The working of vapour compression refrigeration system may be explained in the following four processes:

1. Process 1 – 2 (Compression): The vapourized refrigerant enters the compressor from the evaporator where its pressure and temperature increases.

Process 2 – 3 (Condensation): The high pressure and temperature refrigerant enters into the condenser where it is cooled at saturation temperature corresponding to the condenser.

Process 3 – 4 (Throttling): High pressure saturated liquid refrigerant enters the expansion device where its pressure and temperature are reduced.

A portion of liquid refrigerant is converted into vapour.

Process 4 – 1 (Evaporation): The refrigerant at low temperature enters the evaporator where it absorbs heat from the space to be cooled. In absorbing heat the refrigerant is vapourized. The vapourized refrigerant then enters the compressor and the cycle is completed.

The *T* – *s* diagram is shown in Fig. 5.2.

Process 1 – 2 (isentropic compression):

$$s_2 = s_1,\ Q = 0$$

$$w_{1-2} = -\int v.dp = -\int dh = h_1 - h_2$$

Process 2 – 3 (constant pressure condensation):

Heat rejected, $q_r = h_2 - h_3$

Process 3 – 4 (throttling):

$$h_3 = h_4 = h_{f4} + x(h_1 - h_{f4})$$

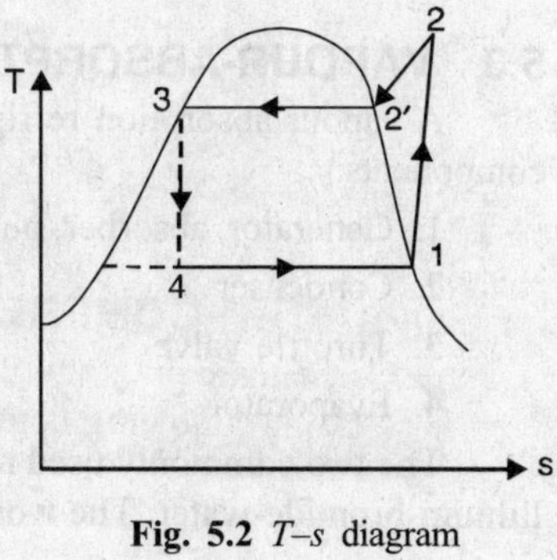

Fig. 5.2 *T–s* diagram

$$x = \frac{h_3 - h_{f4}}{h_1 - h_{f4}}$$

Process 4 – 1 (constant pressure evaporation):

Heat absorbed, $q_a = h_1 - h_4$

= Refrigeration effect

$$\text{COP} = \frac{q_a}{w_{1-2}} = \frac{h_1 - h_4}{h_2 - h_1} = \frac{h_1 - h_3}{h_2 - h_1}$$

Power required per ton of refrigeration,

$$P = \frac{3.5167}{\text{COP}} \text{ kW/TR}$$

Mass flow rate, $\dot{m} = \frac{3.5167}{q_a} = \frac{3.5167}{h_1 - h_3}$

Motor power $= \frac{\dot{m}(h_2 - h_1)}{10^3}$ kW

Volume flow rate $= \frac{3.5167 \times v_1}{h_1 - h_3}$

where v_1 = specific volume of refrigerant at inlet to compressor

$$= \frac{\pi}{4} D^2 L \frac{N}{60} \times n \times \eta_v$$

D, L = diameter and stroke of compressor respectively

N = speed of compressor, rpm

n = number of cylinders in compressor

η_v = volumetric efficiency of compressor

$$= 1 + C - C\left(\frac{p_2}{p_1}\right)^{\frac{1}{n}}$$

C = clearance

Mass flow rate of cooling medium in condenser,

$$\dot{m}_c . c_p . \Delta T = \dot{m}(h_2 - h_3)$$

5.3 VAPOUR-ABSORPTION REFRIGERATION SYSTEM

A vapour-absorption refrigeration system is shown in Fig. 5.3. It consists of the following components:

1. Generator, absorber, pump, heat exchanger, pressure reducing valve, analyser and rectifier.
2. Condenser
3. Throttle valve
4. Evaporator

The two commonly used refrigerant absorbent pairs are ammonia-water (aqua ammonia) and lithium bromide-water. The working of the aqua ammonia absorption system is described below:

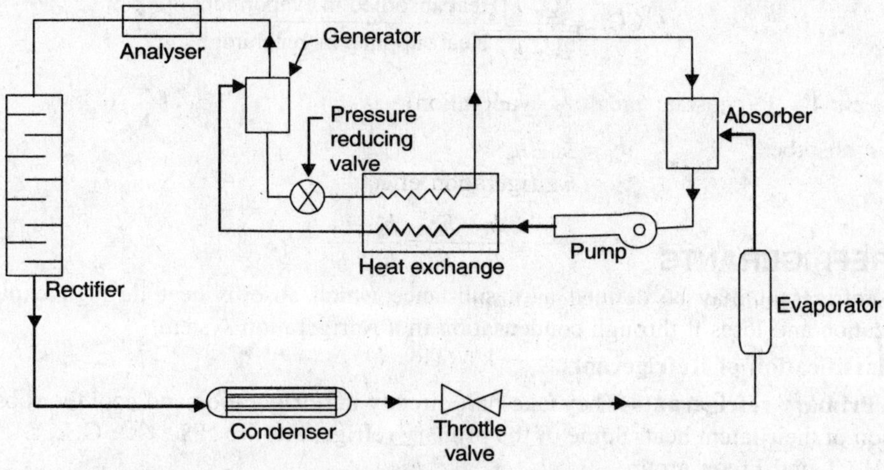

Fig. 5.3 Ammonia absorption refrigeration system

Ammonia vapours are produced in the generator at high pressure from aqua ammonia by external heating. The water vapours are removed by passing through an analyser and rectifier. The dehydrated ammonia vapours enters the condenser where vapours are condensed. The sub-cooled high pressure liquid ammonia is passed through a throttle valve to further lower the temperature. Now they enter the evaporator and leave as saturated vapour. The saturated vapour goes to absorber to become strong after absorbing ammonia vapour. The strong solution is pumped to the generator through the heat exchanger. The solution in the generator becomes weak as ammonia vapour comes out of it. The weak ammonia solution from the generator is passed to the heat exchange through the pressure reducing valve. Thus the cycle is completed.

Let Q_g = heat absorbed by generator

Q_{rc} = heat rejected by condenser

Q_e = heat absorbed by evaporator

Q_p = pump work

Q_{ra} = heat rejected by absorber

Neglecting Q_p, heat rejected to atmosphere,

$$Q_r = Q_{rc} + Q_{ra} = Q_g + Q_e$$

$$\frac{Q_g}{T_g} + \frac{Q_e}{T_e} = \frac{Q_r}{T_r} = \frac{Q_g + Q_e}{T_r}$$

$$\frac{Q_g}{T_g} - \frac{Q_g}{T_r} = Q_e\left(\frac{1}{T_r} - \frac{1}{T_e}\right)$$

$$Q_g\left(\frac{1}{T_g} - \frac{1}{T_r}\right) = Q_e\left(\frac{1}{T_r} - \frac{1}{T_e}\right)$$

$$Q_g\left(\frac{T_r - T_g}{T_r\,T_g}\right) = Q_e\left(\frac{T_e - T_r}{T_e\,T_r}\right)$$

$$Q_g = Q_e\left(\frac{T_g\,(T_e - T_r)}{T_e(T_r - T_g)}\right)$$

$$\text{COP} = \frac{Q_e}{Q_g} = \frac{\text{Heat absorbed in evaporator}}{\text{Heat supplied in generator}}$$

$$= \frac{\frac{T_r}{T_g} - 1}{1 - \frac{T_r}{T_e}}$$

5.4 REFRIGERANTS

A refrigerant may be defined as a substance which absorbs heat through expansion or vapourization and loses it through condensation in a refrigeration system.

Classification of Refrigerants

1. **Primary refrigerants**–They take part directly in refrigeration and cool the substance on absorption of their latent heat. Some of the primary refrigerants are NH_3, CO_2 CH_3, Cl_2, CH_2Cl_2, SO_2, C_2H_5Cl, and Freon group.

2. **Secondary refrigerants**–They are first cooled with the help of primary refrigerant and then used for cooling other substances by absorption of their sensible heat only, *e.g.* ice, solid CO_2.

Primary refrigerants may be grouped as follows:

(*i*) **Halo-Carbon compounds**

They contain one or more of three halogens, fluorine, chlorine and bromine. Commercially they are sold under the names as: Freon, Genetron, Irontron and Areton. They are used in domestic, commercial and industrial refrigerating devices. Some of their important examples are:

Carbon tetrachloride – R10 : CCl_4
Trichloromono-fluro methane – R11 : CCl_3F
Dichloro-difluro methane – R12 : CCl_2F_2
Monobromotri-fluro methane – R13 : $CBrF_3$

(*ii*) **Azeotropes**–They are the mixtures of different refrigerants. They do not undergo any separation with changes in temperature and pressure, e.g.

R — 500 (73.8% R12 + 26.2% R 152) : $CCl_2F_2 + CH_3CHF_2$
R — 502 (48.8% R22 + 51.2% R 115) : $CHClF_2 + CClF_2CF_3$

(*iii*) **Hydrocarbons**–They are hydrocarbons, e.g.

Methane — R 50 : CH_4
Ethane — R 170 : C_2H_6
Propane — R 200 : C_2H_8

(*iv*) **Inorganic compounds**

Ammonia — R 717 : NH_3
Water — R 718 : H_2O
Air — R 720 : —
Carbon dioxide — R 744 : CO_2

(*v*) **Unsaturated organic compounds** – They contain ethylene or propylene as one of the main constituents, e.g.

Ethylene — R 1150 : C_2H_4
Propylene — R 1720 : C_3H_6
Trichloroethylene — R1120 : $C_2H_4Cl_3$

The general chemical formula for refrigerants is:

$$C_mH_nCl_pFq$$

where $n + p + q = 2m + 2$

The number of the refrigerant is : R$(m - 1)(n + 1)$ (q)

5.4.1 Properties of Refrigerants

A good refrigerant should have the following properties:

(*a*) Thermodynamic properties

1. Low boiling point
2. Low freezing point
3. High saturation temperature
4. High latent heat of vaporization
5. High critical pressure and temperature.

(*b*) Chemical properties

1. Non-toxic
2. Non-flammable
3. Non-explosive
4. Non-corrosive
5. Chemically stable
6. Odourless

(*c*) Physical properties

1. Low specific volume
2. High thermal conductivity
3. High electrical insulation
4. Low specific heat
5. Miscibility with lubricating oil.

(*d*) Other properties

1. Ease of leakage detection
2. Ease of handling
3. High COP
4. Ease of availability and of low cost.
5. Low power consumption per ton of refrigeration.

5.4.2 Ozone Layer Depletion

Stratosphere is about 25–30 km away from the earth's surface. It contains ozone (O_3). Ozone is formed in the stratosphere by the reaction that occurs when sunlight interacts with oxygen. This ozone shelters the earth from ultraviolet (UV) radiation. This protection is essential for human health. This ozone is easily destroyed by the UV radiation. The sun's radiation will break down the O_3 molecule into a standard O_2 molecule and an elemental free oxygen (O) atom. More O_3 is produced through photosynthesis and the bonding of O_2 with free oxygen O. Thus ozone is constantly being formed and destroyed in the stratosphere.

In the stratosphere, UV radiation will break off a chlorine atom (Cl) from the CFC and HCFC molecule. The chlorine atom destroys many O_3 molecules The depleted O_3 molecules in the stratosphere let more UV radiation reach the earth. This causes an increase in skin cancers and frequency of cataracts in humans; weakens human immune system and decreases plant and marine life. Ozone depletion potential (ODP) has been used as an index to show the effect of a refrigerant on O_3. Lower the ODP, it will be better. The ODP of some of the refrigerants is given in Table 5.1.

Table 5.1 Ozone Depletion Potential (ODP) of Refrigerants.

Sub Group	Refrigerant Designation	Chemical Formula	Refrigerant Name	ODP
CFCs	R—11	CCl_3F	Trichloro-fluoro methane	1.0
	R—12	CCl_2F_2	Dichlorodi-fluoro methane	1.0
	R—113	$CF_2Cl—CFCl_2$	Trifluoro-trichloro ethane	0.8
	R—114	$CF_2Cl—CF_2Cl$	Tetrafluoro–dichloro ethane	1.0
HCFCs	R—22	$CHClF_2$	Chlorodi-fluoro methane	0.05
	R—123	$CHCl_2F_3$	Dichlorotri-fluoro methane	0.02
HFC	R—134*a*	CH_2FCF_3	Tetra-fluoro ethane	0.0
Azeotropes	R—500	R—12/R—152	—	0.74
	R—717	NH_3	Ammonia	0.0

5.4.3 Eco-Friendly Refrigerants

With the increasing awareness of environmental degradation, the production, use and disposal of Chloro Fluoro Carbons (CFCs) and Hydro Chloro Fluoro Carbons (HCFCs) as refrigerants in mechanical refrigeration systems has become a subject of great concern. However, such systems are being developed using more eco-friendly refrigerants, such as

1. Hydrocarbons (HC–290, HC–600 a).
2. Ammonia (NH_3)
3. Carbon Dioxide (CO_2)
4. Water (R–718)
5. Air.

Eco-friendly refrigerants are also known as natural refrigerants. The important features and applications of eco-friendly refrigerants are given in Table 5.6.

Table 5.2 Important features and applications of eco-friendly refrigerants.

Refrigerant	Important Features	Applications
1. Hydrocarbons	1. Zero Global Warming Potential (GWP) 2. Zero Ozone Depletion Potential (ODP) 3. COP comparable to CFC-12. 4. Flammable 5. Compatible with mineral oils	1. Domestic refrigerator (HC–270, HC-600a) 2. Commercial refrigeration (HC–290 and HC–600a) 3. Transport refrigeration (HC 290/HC–600*a* or HC–290) 4. Cold storage and food processing (HC–290) 5. Industrial refrigeration (HC–290) 6. Air-conditioning (HC-290)
2. Ammonia	1. Excellent thermodynamic and thermophysical properties 2. Zero ODP 3. Zero GWP 4. High efficiency 5. Low cost 6. Flammable	1. In small and large plants. 2. Commercial and cold storage. 3. Super markets 4. Water chillers 5. Centralised air-conditioning

3. Water	1. Thermodynamically attractive 2. Low volumetric refrigeration capacity 3. High COP (7–10) 4. Can be used both as primary and secondary refrigerant 5. Less corrosive	1. Water cooled chillers 2. Eco-chiller
4. Carbon dioxide	1. Zero ODP 2. Low critical temperature 3. High operating pressure 4. Non-flammable, non-toxic and cheap 5. Excellent performance 6. Low COP	1. Commercial air-conditioning system. 2. Mobile air-conditioning system
5. Air	1. Simple and compact 2. Non-toxic and non-flammable 3. Easy repair and maintenance 4. Low COP 5. Requires less space and lighter in weight	1. Used in aircrafts. 2. Used in railways and bus air-conditioning

5.5 DOMESTIC REFRIGERATOR

The primary function of a domestic refrigerator is to provide food storage space maintained at low temperature for the preservation of food. Its essential secondary function is the formation of ice cubes for domestic consumption. They are usually specified by the internal gross volume. A storage

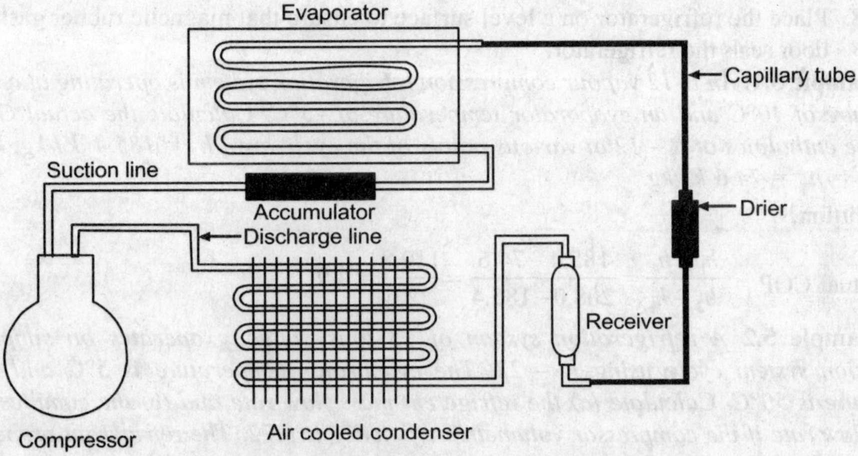

Fig. 5.4 Refrigeration system for a domestic refrigerator

temperature of 0°C to 4°C is satisfactory for the preservation of most of the fresh foods. The freezers are generally provided at the top portion of the refrigerators space.

The refrigerators may be single door, double-door top freezer, double-door bottom freezer, and side-by-side door freezer. The double-door refrigerators are very common because of the need for larger storage space and better preservation of frozen foods. These refrigerators are divided into two separate compartments, one for the fresh food and the other for the storage of frozen foods.

The mechanical vapour compression system shown in Fig. 5.4 is used for the domestic refrigerator. The refrigerants used are generally R-12 or R-22. The compressor is mounted at the bottom of the refrigerator frame. The power of compressor can vary according to the size of the refrigerator (75W to 370W). The condenser is put at the back about 40 to 60 mm away from the cabinet. The condenser may be either chassis type or tube and wire type. The capillary tube is kept in contact with the evaporator inlet pipe. A drier is connected between the receiver and the evaporator to eliminate traces of moisture if any.

The evaporator coil is wrapped around the freezer in a suitable manner to give efficient heat transfer. The cooling of lower space is accomplished by free convection. The thermostatic sensing element is provided to the evaporator coil which can control temperature in the freezer upto –15°C in steps or continuously depending upon the type of controlling switch employed.

The refrigerator body is provided with good quality insulation in order to prevent heat transfer into the system. Usually 60 to 100 mm thick glass wool or thermocole is used.

5.5.1 Precautions for Use of Refrigerator

1. Refrigerator should be placed at least at a distance of 15 to 20 cm from the wall to allow adequate circulation of atmospheric air over the condenser tubes.
2. Keep the condenser tubes clean for adequate heat transfer.
3. Refrigerator should not be kept in a hot room like kitchen.
4. Never store hot food in the refrigerator.
5. Frequent opening of refrigerator should be avoided.
6. All dishes/containers should be stored covered to minimize moisture lose. This would help in decreasing frost formation on evaporator.
7. Use stabilised electric supply to avoid overloading of the compressor motor during low voltage.
8. Place the refrigerator on a level surface to ensure that magnetic rubber gaskets on the door seals the refrigerator.

Example 5.1. *An R-12 vapour compression refrigeration system is operating at a condenser temperature of 40°C and an evaporator temperature of –5°C. Calculate the actual COP of the cycle. The enthalpies of R – 12 at various points of the cycle are :* $h_1 = 185.4$ *kJ/kg,* $h_2 = 208.0$ *kJ/kg,* $h_3 = h_4 = 74.6$ *kJ/kg.*

Solution.

$$\text{Actual COP} = \frac{h_1 - h_4}{h_2 - h_1} = \frac{185.4 - 74.6}{208.0 - 185.4} = \frac{110.8}{22.6} = 4.9$$

Example 5.2 *A refrigeration system of 15 tons capacity operates on simple vapour compression system cycle using R – 22. The evaporator temperature is 5°C and condenser temperature is 50°C. Calculate (a) the refrigerent mass flow rate and (b) the compressor intake volume flow rate if the compressor volumetric efficiency is 0.72. The refrigerant properties data is given in the following table :*

Temperature C°	Pressure bar	Enthalpy kJ/kg		Volume	
		h_f	h_g	v_t l/kg	v_g m^3/hg
5	5.836	205.9	407.1	0.791	0.0404
50	19.423	263.3	417.7	0.922	0.0117

Solution.

(*a*) Capacity of plant $= \dfrac{\dot{m}(h_1 - h_4)}{3,5167}$

$h_1 = 407.1$ kJ/kg, $h_4 = h_3 = 263.3$ kJ/kg

$$\frac{\dot{m}(407.1-263.3)}{3.5167}=15$$

$$\dot{m} = 0.367 \text{ kg/s}$$

(*b*) Volume flow rate, $\dot{v} = \dfrac{\dot{m}v_g}{\eta_c} = \dfrac{0.367\times0.0117}{0.72} = 5.96\times10^{-3}\ \text{m}^3/s$

Example 5.3 *A vapour absorption system has a source temperature of 110°C, the refrigeration temperature of –5°C, and ambient temperature of 35°C. Calculate its COP.*

Solution. Given : $T_r = 273+35 = 308\,\text{K},\ T_g = 273+110 = 383\,\text{K},\ T_e = 273-5 = 268\,\text{K}$

$$COP = \frac{\left(\dfrac{T_r}{T_g}-1\right)}{\left(1-\dfrac{T_a}{T_e}\right)} = \frac{\left(\dfrac{308}{383}-1\right)}{\left(1-\dfrac{308}{268}\right)} = 1.315$$

Example 5.4 *Draw the schematic diagram of simple vapour compression refrigeration system and explain the function of various components.*

Solution.

VAPOUR COMPRESSION REFRIGERATION SYSTEM

The schematic diagram of simple vapour compression refrigeration system is shown in Fig. 5.5. It consists of the following five essential parts :

1. *Compressor.* The low pressure and temperature vapour refrigerant from evaporator is drawn into the compressor through the inlet or suction valve *A*, where it is compressed to a high pressure and temperature. This high pressure and temperature vapour refrigerant is discharged into the condenser through the delivery or discharge valve *B*.

2. *Condenser.* The condenser or cooler consists of coils of pipe in which the high pressure and temperature vapour refrigerant is cooled and condensed. The refrigerant, while passing through the condenser, gives up its latent heat to the surrounding condensing medium which is normally air or water.

3. *Receiver.* The condensed liquid refrigerant from the condenser is stored in a vessel known as receiver from where it is supplied to the evaporator through the expansion valve or refrigerant control valve.

4. *Expansion valve.* It is also called throttle valve or refrigerant control valve. The function of the expansion valve is to allow the liquid refrigerant under high pressure and temperature to pass at a controlled rate after reducing its pressure and temperature. Some of the liquid refrigerant evaporates as it passes through the expansion valve, but the greater portion is vaporised in the evaporator at the low pressure and temperature.

5. *Evaporator.* An evaporator consists of coils of pipe in which the liquid-vapour, refrigerant at low pressure and temperature is evaporated and changed into vapour refrigerant at low pressure and temperature. In evaporating, the liquid vapour refrigerant absorbs its latent heat of vaporisation from the medium (air, water or brine) which is to be cooled.

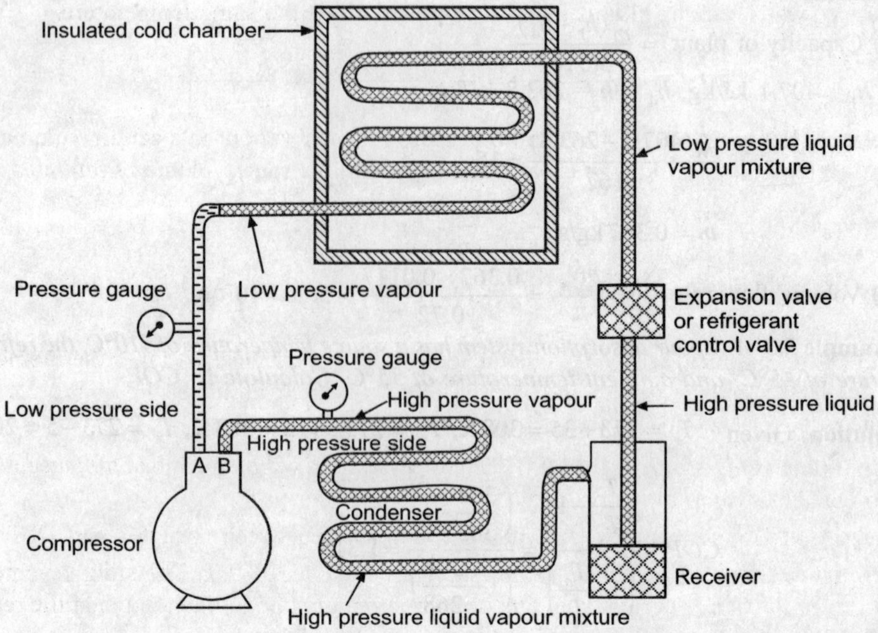

Fig. 5.5 Simple vapour compression refrigeration system.

Note : In any compression refrigeration system, there are two different pressure conditions. One is called the *high pressure side* and other is known as *low pressure side.* The high pressure side includes the discharge line (*i.e.* piping from delivery valve *B* to the condenser), condenser, receiver and expansion valve. The low pressure side includes the evaporator, piping from the expansion valve to the evaporator and the suction line (*i.e.* piping from the evaporator to the suction valve *A*).

5.6 AIR-CONDITIONING FUNDAMENTALS

Air-Conditioning. Is defined as the simultaneous control of temperature, humidity, cleanliness and air motion.

Psychrometry. Is a science which deals with the study of the behaviour of air and water vapour mixture.

Dry air. Is the mixture of nitrogen and oxygen neglecting the water vapour and other gases.

Moist air. Is a mixture of dry air and water vapour.

Moisture. Is the water vapour present in the air.

Saturated air. It is a minute of dry air and water vapour, when the air has diffused maximum amount of water vapour into it.

Humidity. It is mass of water vapour present in 1 kg of dry air. It is also called specific humidity or humidity ratio.

Specific Humidity. Is the mass of water vapour present in one kg of dry air.

$$w = \frac{0.622\, p_v}{p - p_v}$$

Absolute Humidity. Is the mass of water vapour present in unit volume of dry air.

$$p_v = \frac{m_v}{V}$$

Degree of Saturation. Is defined as the ratio of mass of water vapour in unit mass of dry air to mass of water vapour in unit mass of saturated air at the same temperature.

$$\mu = \left[\frac{p_v(p-p_s)}{p_s(p-p_v)}\right] = \phi\left[\frac{p-p_s}{p-p_v}\right]$$

Relative Humidity (RH). Is the ratio of mass of water vapour in a certain volume of moist air at a given temperature to the mass of water vapour in the same volume of saturated air at the same temperature. $\phi = \dfrac{v_s}{v_v} = \dfrac{p_v}{p_s} = \left(\dfrac{w}{0.622}\right)\left(\dfrac{p_a}{p_s}\right) \cong \mu$

Dry Bulb Temperature (DBT). Is the temperature of air measured by an ordinary thermometer.

Wet Bulb Temperature (WBT). Is the temperature of air measured by a thermometer when its bulb is covered with wet cloth and is exposed to a current of air.

Wet Bulb Depression (WBD). Is the difference between DBT and WBT.

Dew Point Temperature (DPT). Is the temperature at which the moisture present in the air begins to condense when the air is cooled.

Dew Point Depression (DPD) – It is the difference between the DBT and DPT air.

Psychrometer – It is a device to measure the DBT and WBT. The sling psychrometer is the most common type. It is rotated in air for approximately one minute and then the readings of DBT and WBT are taken.

Humid Specific Heat. Is the specific heat of moist air per kg of dry air.

Adiabatic Saturation Temperature (AST). Is the temperature at which the air is brought to saturated condition adiabatically by evaporating the water into flowing air.

Enthalpy of dry air, $h = 1.005\, DBT + w\,(2500 + 1.88\, DBT)$ kJ/(kg of d.a.)

$$c_p = 1.005 + 1.88\, w \cong 1.0216 \text{ kJ/(kg of d.a. K)}.$$

Enthalpy of moist air, $h = 1.0216\, DBT + w\,(h_{fg})$ 0°C.

5.6.1 Dalton's law of partial pressures

This law states that the total pressure exerted by the mixture of air and water vapour is equal to the sum of the pressures, which each constituent would exert, if it occupied the same space by itselfs

In other words, the toal pressure exerted by air (p_a) and water vapour (p_v) mixture is equal to the barometric pressure (p_b).

5.6.2 Psychrometric Chart

It is a chart which presents all data essential for the complete thermodynamic and psychrometric analysis of air-conditioning processes. The most common type chart represent specific humidity v's DBT at 760 mm of mercury or 1.01325 bar. The basic psychrometric processes are shown in Fig. 5.6, which represents as follows :

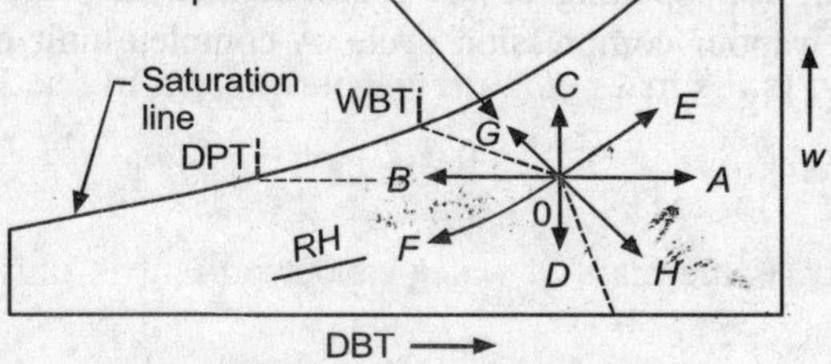

Fig. 5.6 Psychrometric chart.

QA - sensible heating; *OB*-sensible cooling; *OC* - humidifying; *OD* - dehumidifying; *OE* - heating and humidifying; *OF* - cooling and dehumidifying; *OG* - cooling and humidifying ; *OH* - heating and dehumidifying.

Standard Air : $DBT = 20°C$, $RH = 50\%$, $\rho = 1.2$ kg/(m^3 d.a.)

Sensible Heat Factor (SHF). Is the ratio of sensible heat load to the total load (*i.e.*, sensible plus latent).

$$SHF = \frac{Q}{Q_s + Q_L},$$

where

$$Q_s = m_a\,(1.005 + 1.88\,w)\,(t_B - t_A)$$

$$= 0.0204\,(\text{cmm})\,\text{D}t\ \text{kW};\ m_a = (\text{cmm})\,\frac{\rho}{60\ \text{kg d.a.}}\,\text{s}$$

$$Q_L = 50\,(\text{cmm})\,\Delta w\ \text{kW}$$

By Pass Factor (BPF). Represents the fraction of the uncontacted air. $X = \frac{(h_2 - h_s)}{(h_1 - h_s)}$

Contact Factor. Represents the fraction of contacted air, *i.e.*, (1 – *X*).

5.6.3 Comfort Air-Conditioning

Effective Temperature. It is defined as that temperature of saturated air at which subject would experience the same feeling of comfort as experienced in the actual unsaturated environment.

Comfort air-conditioning.

During summer :

Effective temperature, ET = 21.7°C

DBT = 25° ± 1°C

RH = 50 ± 5%

Air velocity = 0.5 m/s

During winter :

DBT = 21°C

RH = 50%

Air velocity = 0.15 to 0.20 m/s.

Metabolic rate – It is the rate of which body produces heat.

Basal metabolic rate — It is the heat produced by a normal healthy person while sleeping. It is approximately 60 W.

Air change — One air change is equivalent to the amount of air supplied per hour equal to the volume of the room.

Infiltration — It is the leapage of outside air through door openings, cracks and interstices around windows and doors into the conditioned space.

5.6.4 Room Air-conditioner

A room air-conditioner is a compact, self contained air-conditioning unit which is normally installed in a window or wall opening of the room. It is widely known as window type air-conditioner. It works on vapour compression cycle. A complete unit of a room air conditioner consists of the following (Fig. 5.7) :

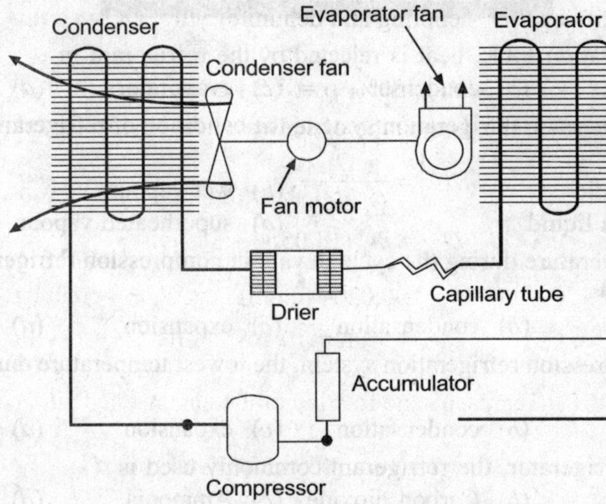

Fig. 5.7 Layout of a room air conditioner

1. Refrigeration system
2. Control system (thermostat and selector switch)
3. Electrical protection system (motor overload switches and thermostat)
4. Air circulation system (fan motor, centrifugal evaporator blower)
5. Ventilation (fresh air damper)
6. Exhaust system.

1. Refrigeration System. It consists of a hermetically sealed compressor, forced air-cooled finned condenser coil, finned cooling coil, capillary tube as the throttling device and a refrigerant drier. The refrigerant used is R-12 or R-22. In hermetically sealed compressor, a winding thermostat is embedded in the compressor motor windings. It puts off the compressor if the winding temperature exceeds the safe limit, thus protecting the winding against high temperature.

The condenser is a continuous coil made of copper tubing with aluminium fins attached to it to increase the heat transfer rate. A propeller type fan provides the necessary air to cool the refrigerant in the condenser and also exhausts air from the air-conditioned space when the exhaust damper is opened.

The evaporator is a cooling coil also made of copper with aluminium fins to increase the heat transfer rate. A centrifugal blower is installed behind the cooling coil which sends cool air in the room. The quality of fresh air may be varied by adjusting the dampers.

2. Control System. A thermostat element is located in the return air passage of the unit. It controls the operation of the compressor based on the return air temperature. When the required temperature is obtained, the compressor is stopped. A selector switch controls the compressor motor, condenser fan motor, and evaporator fan motor. When the control switch is in "Ventilate" position, only evaporator blower motor operates and outside fresh air is supplied in the room. In the "Exhaust" position, the condenser fan motor operates and all the room air is exhausted to the atmosphere. In the "Cool" position, are the motors are in working state and cool air is supplied in the room.

MULTI-CHOICE QUESTIONS

1. During a refrigeration cycle, heat is rejected by the refrigerant in
(*a*) compressor (*b*) condenser (*c*) evaporator (*d*) expansion valve

2. In a vapour compression refrigeration system, the condition of refrigerant before entering the compressor is
(*a*) saturated liquid (*b*) wet vapour
(*c*) dry saturated liquid (*d*) superheated vapour

3. The highest temperature during the cycle in vapour compression refrigeration system occurs after
(*a*) compression (*b*) condensation (*c*) expansion (*d*) evaporation

4. In a vapour compression refrigeration system, the lowest temperature during the cycle occurs after
(*a*) compression (*b*) condensation (*c*) expansion (*d*) evaporation

5. In a domestic refrigerator, the refrigerant commonly used is
(*a*) Air (*b*) Carbon dioxide (*c*) Ammonia (*d*) Freon-12

6. R-12 refrigerant belongs to which group of refrigerants?
(*a*) halo-carbon (*b*) azeotropes (*c*) inorganic (*d*) hydro-carbons

7. Which of the following is an azeotrope refrigerant?
(*a*) R-11 (*b*) R-40 (*c*) R-114 (*d*) R-502

8. The colour of the flame of halide torch in case of leakage of freon refrigerant, will be
(*a*) bright green (*b*) yellow (*c*) red (*d*) orange

9. The use of chlorofluoro carbon refrigerants is banned because these react with
(*a*) water vapours and cause acid rain (*b*) plants and cause green house effect
(*c*) oxygen and cause its depletion (*d*) ozone layer and cause its depletion

10. Expansion device used in a domestic refrigerator of vapour compression type is
(*a*) automatic expansion valve (*b*) thermostatic expansion valve
(*c*) hand operated valve (*d*) capillary tube

11. The bank of tube at the back of a domestic refrigerator of vapour compression type is
(*a*) evaporator (*b*) condenser (*c*) capillary (*d*) electric wire

12. Frosting of evaporator tubes results in
(*a*) increased heat transfer rate (*b*) decreased heat transfer rate
(*c*) increased compressor pressure (*d*) decreased compressor pressure

13. The chemical formula of R-12 is
(*a*) CCl_2F (*b*) $CClF_3$ (*c*) CCl_2F_2 (*d*) $CHClF_2$

14. In a simple vapour compression cycle, the refrigerant is in the form of dry saturated vapour before entering
(*a*) condenser (*b*) evaporator (*c*) throttle valve (*d*) compressor

15. A one ton refrigeration system means that its refrigerating effect is
(*a*) 50 kJ/min (*b*) 210 kJ/min (*c*) 300 kJ/min (*d*) 350 kJ/min

16. Power consumption for the same refrigeration effect of a refrigerating system is
(*a*) directly proportional to COP (*b*) inversely proportional to COP
(*c*) directly proportional to $\sqrt{COP}$ (*d*) inversely proportional to $\sqrt{COP}$

17. The usual expansion device used in a window air conditioner is
(*a*) capillary tube (*b*) automatic expansion valve
(*c*) float valve (*d*) hand expansion valve

18. In a window air-conditioner, air filter is placed
(*a*) in the path of return air (*b*) in the path of supply air
(*c*) after evaporator coil (*d*) before evaporator coil

19. Thermostat in a room air-conditioner controls
(*a*) supply air temperature (*b*) room air temperature
(*c*) outside air temperature (*d*) all of the above

20. The room air-conditioner controls the
(*a*) temperature of air (*b*) humidity of air
(*c*) both temperature and humidity of air (*d*) velocity of air

Answers:

1. (*b*) **2.** (*d*) **3.** (*a*) **4.** (*d*) **5.** (*d*) **6.** (*a*) **7.** (*d*) **8.** (*a*) **9.** (*d*) **10.** (*d*)
11. (*b*) **12.** (*b*) **13.** (*c*) **14.** (*d*) **15.** (*b*) **16.** (*b*) **17.** (*a*) **18.** (*b*) **19.** (*b*) **20.** (*c*)

EXERCISES

5.1 Define refrigeration and refrigeration effect.
5.2 What is the unit of refrigeration?
5.3 With a neat sketch describe vapour compression refrigeration system.
5.4 What is a refrigerant? Give its desirable properties.
5.5 What are the various types of refrigerant?
5.6 Define COP of a refrigerator and a heat pump
5.7 What are the various refrigeration systems?
5.8 Determine the COP for a simple vapour compression refrigeration system.
5.9 Describe the working of a domestic refrigerator with a neat sketch.
5.10 What are the precautions to be taken during use of domestic refrigerator?
5.11 What is air-conditioning? What is comfort air-conditioning?
5.12 Describe the working of a window type air-conditioner with a neat sketch.

6 HYDRAULIC TURBINES, PUMPS AND DEVICES

6.1 HYDRAULIC TURBINES

6.1.1 Classification of Hydraulic Turbines

The hydraulic turbines may be classified as follows:

1. Based upon the head and quantity of water available
   - (*a*) Impulse turbine-requires high head and small quantity of flow.
   - (*b*) Reaction turbine-requires low head and high rate of flow.
2. Based after the name of originator.
   - (*a*) Pelton turbine–named in honour of Lester Allen Pelton, is an impulse turbine used for high head and low discharge.
   - (*b*) Francis turbine–named after James Bichens Francis, is a reaction type of turbine used for medium head and medium discharge.
   - (*c*) Kaplan turbine–named in honour of Dr. Victor Kaplan, is a reaction type of turbine for low head and large quantity of flow.
3. According to action of water on moving blades.

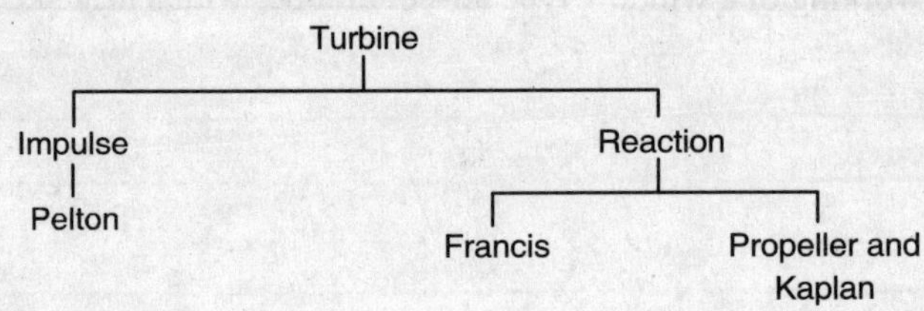

4. According to direction of flow of water in the runner.

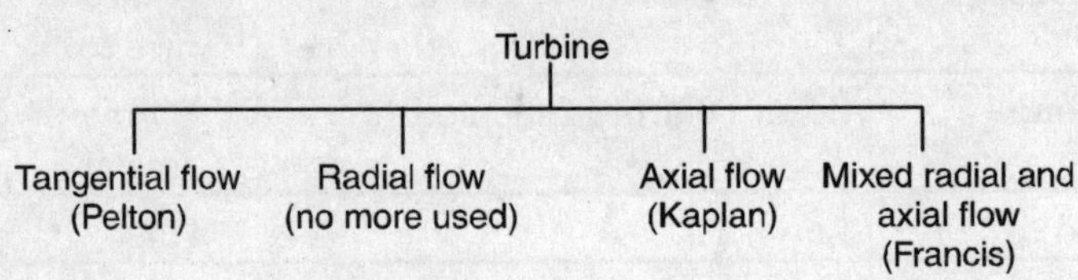

5. According to disposition of shaft.
   Horizontal shaft – Pelton turbine
   Vertical shaft – Francis and Kaplan turbines.
6. According to specific speed.

The specific speed of a turbine is the speed of a geometrically similar turbine that would develop one brake power under a head of one metre.

Specific speed, $N_s = \frac{N\sqrt{P_t}}{H^{5/4}}$

where N = speed of turbine, rpm.

P_t = turbine output in brake power, kW

H = effective head in metres.

Type of turbine	*Specific speed in S.I. Units*
1. Pelton:	
(*a*) Single jet	8.5—30
(*b*) Multiple jets	30—50
2. Francis	50—340
3. Kaplan	250—1025

6.1.2 Comparison Between Impulse and Reaction Turbines

Table 6.1 gives the comparison between impulse and reaction turbines with regard to their operation and application.

Table 6.1. Comparison between impulse and reaction turbines

Aspects	*Impulse turbine*	*Reaction turbine*
1. Conversion of hydraulic energy	Potential energy of water is converted into kinetic energy by a nozzle.	Water potential energy is partly transformed into kinetic energy before water enters the turbine runner.
2. Change in pressure and velocity	No change in pressure but velocity increases.	Both velocity and pressure changes while passing through turbine runner.
3. Necessity of water tight casing	Required	Not necessary.
4. Water head required	High	Medium to low
5. Water discharge required	Low	High
6. Draft tube required	No	Yes
7. Installation of Unit	Above the tail race level	Above or below the tail race level
8. Flow regulation	Spear control in the nozzle	By means of guide vane assembly.
9. Specific speed	Low	High
10. Direction of water flow	Tangential	Mixed flow : radial + axial.

6.2 PELTON TURBINE

The Pelton turbine is an impulse turbine which operates under a high head of running water and requires comparatively less quantity of water. The water is conveyed from a reservoir in the mountains to the turbine in the power house through a long pipe called penstock. A nozzle is fitted at the end of the penstock through which water comes out in the form of a free and compact jet. The water jet having a high velocity is made to impinge on the buckets fixed round the circumference of a wheel mounted on a shaft. The impact of water on the surface of the bucket produces a force which causes the wheel to rotate, thus, supplying a torque or mechanical power on the shaft.

6.2.1 Turbine Components

A pelton turbine consists of the following main components :

1. Guide mechanism. This mechanism controls the quantity of water passing through the nozzle and striking the buckets. It maintains the speed of the turbine constant even when the head varies. It mainly consists of a defector and a spear operated by a governor.

2. Buckets and Runner. Each bucket is divided vertically into two parts by a splitter giving the shape of a double hemispherical cup. The buckets are bolted to the circumference of a sound disc forming the runner of the pelton turbine.

3. Casing. It is necessary only to prevent splashing of water and to lead the water to the tail race, and also as a safeguard against accidents. It has no hydraulic function to perform.

4. Hydraulic brake. It consists of a small nozzle fitted in such a way that on being opened, it directs a jet on the back of the buckets to bring the revolving runner quickly to rest.

The single jet and double jet Pelton turbines are shown in Fig. 6.1.

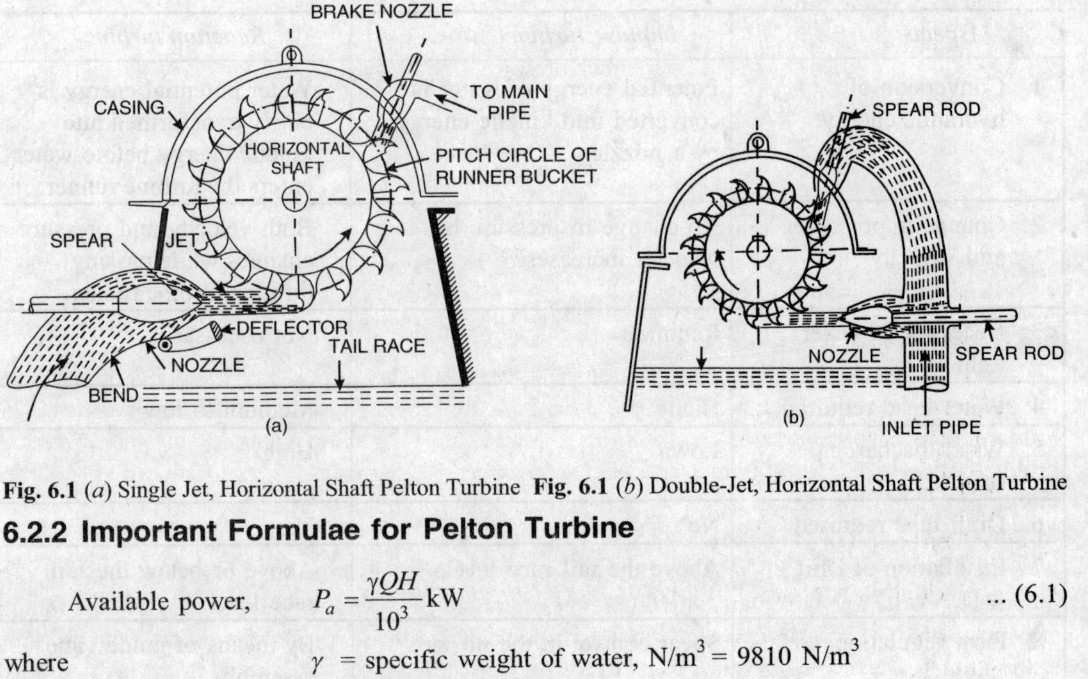

Fig. 6.1 (*a*) Single Jet, Horizontal Shaft Pelton Turbine **Fig. 6.1** (*b*) Double-Jet, Horizontal Shaft Pelton Turbine

6.2.2 Important Formulae for Pelton Turbine

Available power, $P_a = \frac{\gamma QH}{10^3}$ kW ... (6.1)

where γ = specific weight of water, N/m^3 = 9810 N/m^3

Q = rate of flow, m^3/s

H = net head, m

Power supplied by turbine, $P_t = P_a \times \eta_t$ ... (6.2)

where η_t = overall efficiency of turbine

= 0.85 – 0.90

$$Q = \frac{\pi}{4} d_1^2 v_1$$

where d_1 = least diameter of water jet

v_1 = velocity of jet, m/s

$$= k_{v1}\sqrt{2gH} \quad \text{... (6.3)}$$

k_{v1} = coefficient of velocity

= 0.98 – 0.99

Specific speed for multiple jets,

$$N_s = \frac{N\sqrt{nP_t}}{H^{3/4}} \quad \text{... (6.4)}$$

where n = number of jets.

Circumferential velocity of bucket,

$$u_1 = K_{u1}\sqrt{2gH} = \frac{\pi D_1 N}{60} \text{ m/s} \quad \text{... (6.5)}$$

where K_{u1} = speed ratio ≃ 0.45

Frequency of alternator, $$f = \frac{pN}{60} \text{ Hz} \quad \text{... (6.6)}$$

where p = number of pair of poles.

For f = 50 Hz, $N = \dfrac{3000}{p}$

$$\text{Jet ratio, } m = \frac{\text{mean diameter of runner}}{\text{least diameter of jet}} = \frac{D_1}{d_1} \quad \text{... (6.7)}$$

Maximum number of buckets, z = 0.5 m + 15

6.3 FRANCIS TURBINE

It is an inward mixed flow reaction turbine. The water under pressure enters the runner from the guide vanes towards the centre in radial direction and discharges out of the runner axially. The Francis turbine operates under medium heads and requires medium quantity of water. Water is brought down to the turbine and directed to a number of stationary orifices fixed all around the circumference of the runner as shown in Fig. 6.2. These stationary orifices are commonly termed as guide vanes or wicket gates.

A part of the head acting on the turbine is transformed into kinetic energy and the rest remains as pressure head. There is a difference of pressure between the guide vanes and the runner which is called the reaction pressure, and is responsible for the motion of the runner. The pressure at the inlet to the turbine is more than that at the outlet. Therefore, the water in the turbine must flow in a closed conduit. After doing its work the water is discharged to the tail race through a closed tube of gradually enlarging section known as the draft tube. The draft tube converts kinetic head to pressure head. By recovering pressure head in a draft tube the pressure at the runner exit is reduced below atmosphere. This makes it possible to install the turbine above tail race level without any loss in available head.

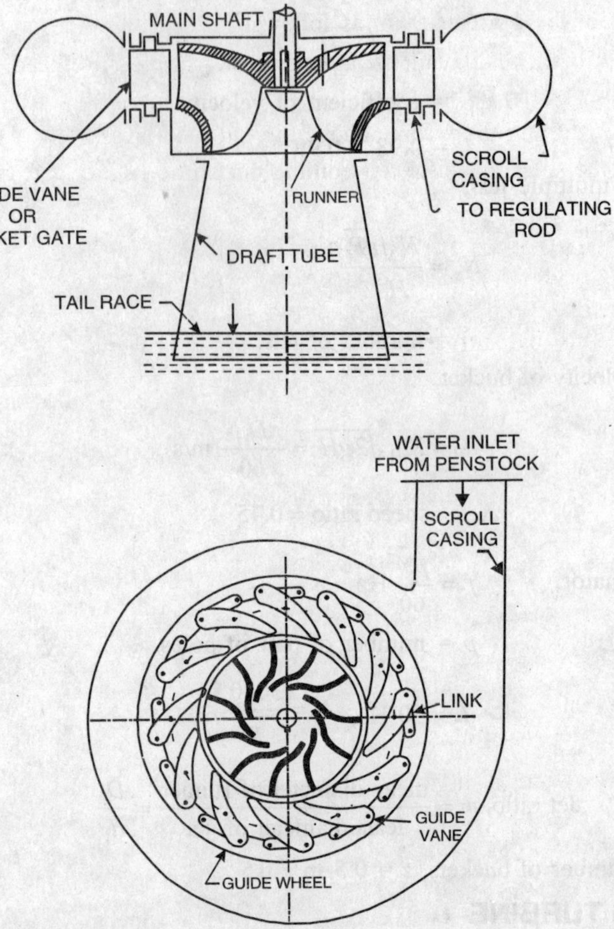

Fig. 6.2 Outlines of a Francis Turbine, Vertical Closed Type

6.3.1 Main Components of Franics Turbine

The main components of a Francis turbine are:

1. **Penstock**– It is a water way to carry water from the reservoir to the turbine casing. Trashracks are provided at the inlet of penstock in order to obstruct the debris entering into it.

2. **Spiral or Scroll Casing**– It is a gradually and uniformly reducing cross-sectional area pipe around the circumference of the turbine which leads water from the penstock to the guide blades. The cross-sectional area is maximum at the entrance and nearly zero at the tip. This gives a spiral shape and hence the casing is named as spiral casing. It distributes the water around the guide ring evenly and presents the formation of eddies to avoid loss of efficiency.

3. **Guide mechanism**– The guide vanes are fixed between two rings in the form of a wheel, known as guide wheel. The guide mechanism controls the quantity of water striking the runner.

4. **Runner and turbine main shaft**– The flow of water through the runner is of the mixed type which is a combination of radial and axial flow. The runner may be made of cast iron, cast steel or stainless steel. It is keyed to the shaft, which may be vertical or horizontal.

5. **Draft tube**– It is a pipe whose cross-section gradually increases towards the outlet. It connects the runner exit to the tail race. The tube should be drowned below the lowest tail race level. The functions of the draft tube are as follows:

(*i*) To convert kinetic head available at turbine exit to pressure head. This increases output and efficiency of the turbine.

(*ii*) It helps in installing the turbine above the tail race level.

(*iii*) It prevents the splashing of water coming out of the runner and guides the water to the tail race.

6.3.2 Important Formulae for Francis Turbine

(*a*) Scroll casing.

Quantity of water flowing per second,

$$Q_0 = \frac{\pi}{4} d_i^2 v_i$$

where d_i = inlet diameter of spiral casing

v_i = inlet velocity of water ... (6.8)

$$= k_{vi}\sqrt{2gH}$$

$$= K_{vi} = 0.15 - 0.20$$

(*b*) Guide vanes.

Tangential velocity, $u_0 = \dfrac{\pi D_0 N}{60}$ m/s $= \mathrm{K}_{u_0}\sqrt{2gH}$ ... (6.9)

where D_0 = diameter of guide vane

$\mathrm{K}_{u_0} = 0.7 - 1.34$

Height of guide vane, $B_0 = \dfrac{Q}{\pi D_0 K_{v_{m0}}\sqrt{2gH}}$ ... (6.10)

$$K_{v_{m0}} = 0.15 - 0.35$$

(*c*) Runner

$$u_1 = \frac{\pi D_1 N}{60} = K_{u1}\sqrt{2gH} \quad \text{... (6.11)}$$

$$K_{u1} = 0.62 - 0.82$$

6.4 KAPLAN TURBINE

The Kaplan turbine, as shown in Fig. 6.3, is an axial flow reaction turbine. It operates entirely in a closed conduit from inlet to tail race. The runner blades are made adjustable. The turbine operates where comparatively low head and large quantity of water are available. The main components of a Kaplan turbine are similar to those of Francis turbine.

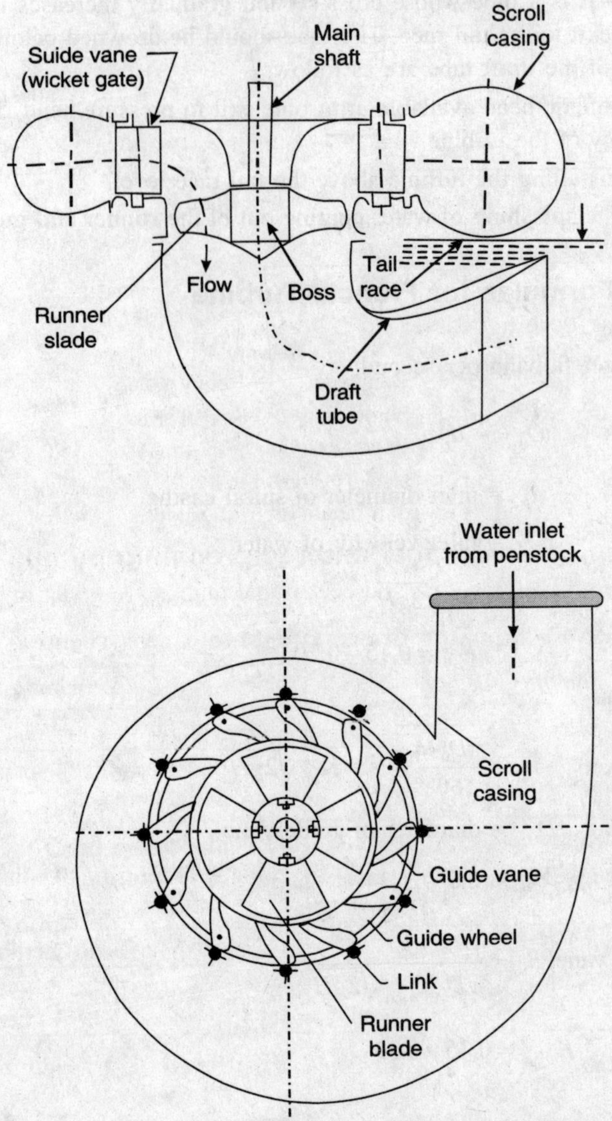

Fig. 6.3 Outlines of a Kaplan Turbine

6.5 PUMPS

Pump is a mechanical device to increase the pressure energy of a liquid. In most of the cases, a pump is used for raising fluids from a lower to a higher level. This is achieved by creating a low pressure at inlet or suction end and high pressure at the outlet or delivery end of the pump. Due to the low inlet pressure the fluid rises from a depth where it is available and the high outlet pressure forces it up to a height where it is required. Work has to be done by a prime-mover on the pump to enable it to impart energy to the fluid.

6.5.1 Classification of Pumps

The pumps may be classified as shown in Fig. 6.4.

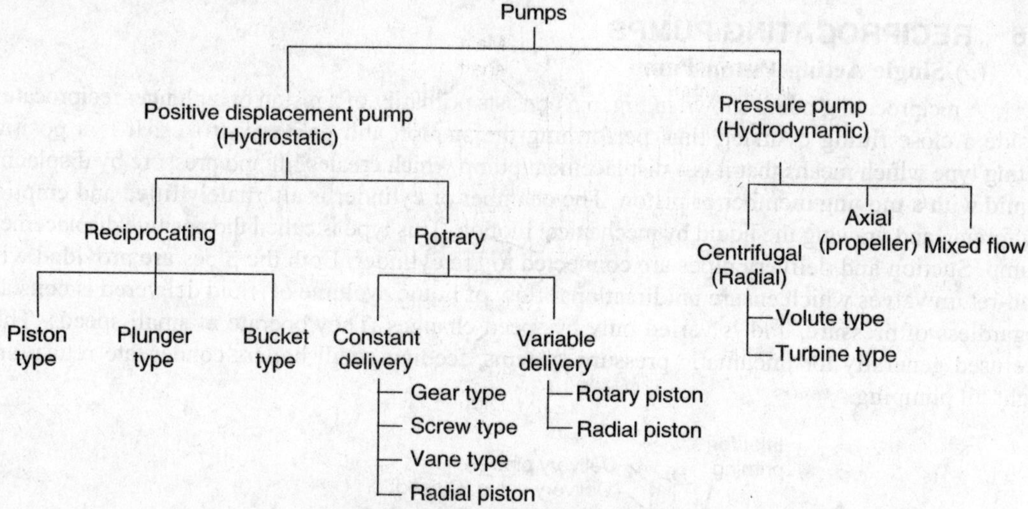

Fig. 6.4 Classification of pumps.

6.5.2 Comparison of Centrifugal and reciprocating pumps

The comparison of reciprocating and centrifugal pumps are given in Table 6.2.

Table 6.2 Comparison of centrifugal and reciprocating pumps

Centrifugal Pump	*Reciprocating Pump*
(*i*) Smooth and even flow.	Pulsating flow.
(*ii*) Low initial cost.	Initial cost as high as four times that of centrifugal pump.
(*iii*) Compact, occupies less floor space. Vertical type requires even lesser space.	Occupies 6 to 8 times the space required for horizontal centrifugal pump.
(*iv*) Gross weight is small.	Gross weight is considerable.
(*v*) Installation is easy.	Installation is more difficult than centrifugal pump.
(*vi*) Efficiency of low head pumps is high.	Efficiency of low head pumps may be as low as 40% chiefly because of relatively higher energy losses *viz.* frictional, valve losses etc.
(*vii*) Construction is simplified by elimination of many parts such as non-return valves, glands, air vessel etc. Therefore less number of spare parts are required.	Complicated construction. Therefore a number of spare parts are necessary.
(*viii*) Low maintenance cost. Periodical check up is sufficient.	Maintenance charges are high because parts like valves require constant attention.
(*ix*) High speed. Can be coupled directly through flanged coupling to electric motors or steam turbines.	Low speed due to separation difficulties. Belt drive indispensable.
(*x*) Uniform torque.	Torque not uniform.
(*xi*) Easy handling of highly viscous fluids such as oils, muddy and sewage water, paper pulp, sugar molasses, chemicals, etc.	Valves and glands cause trouble when required to transmit viscous fluids.

6.6 RECIPROCATING PUMPS

(a) Single Acting Piston Pump

A reciprocating pump shown in Fig. 6.5 consists primarily of a piston or a plunger reciprocating inside a close fitting cylinder, thus performing the suction and delivery strokes. It is a positive acting type which means that it is a displacement pump which creates lift and pressure by displacing liquid with a moving member or piston. The chamber or cylinder is alternately filled and emptied by forcing and drawing the liquid by mechanical motion. This type is called the positive displacement pump. Suction and delivery pipes are connected to the cylinder. Both the pipes are provided with non-return valves which ensure unidirectional flow of liquid. Volume of fluid delivered is constant regardless of pressure, and is varied only by speed changes. They operate at small speeds. They are used generally for pneumatic pressure systems, feeding small boilers condensate return and light oil pumping.

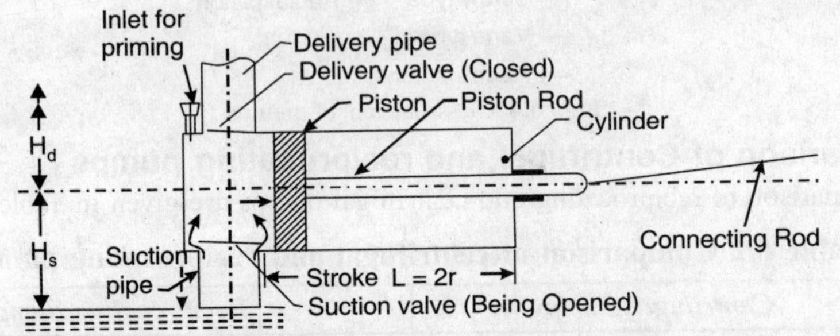

Fig. 6.5 Reciprocating Single Acting Piston Pump.

Working Principle-Movement of the piston or plunger creates a vacuum and atmospheric pressure forces the water up through the suction pipe into the cylinder. Suction pipe and clearance volume of the cylinder are first filled with water to replace air. This is called priming of pump. Once the pump has been primed, water follows closely the piston or plunger on its forward stroke. In the return or backward stroke water is pushed upwards into the delivery pipe. Delivery pipe valve must be opened before starting the pump.

(b) Double Acting Piston Pump

In this pump, shown in Fig. 6.6, in every stroke, there is suction on one side and delivery on the other side. It has two suction and two delivery pipes.

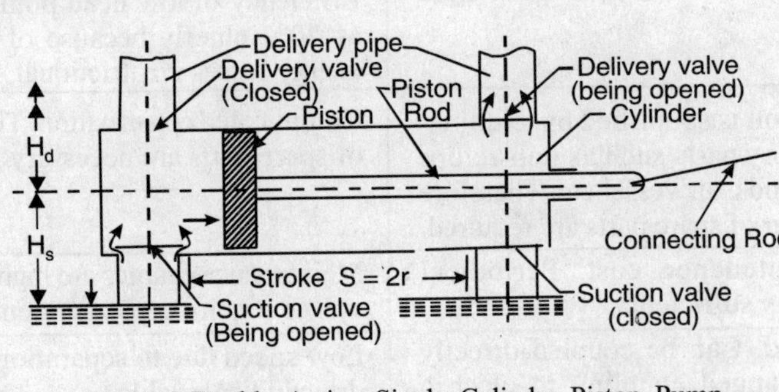

Fig. 6.6 Double Acting Single Cylinder Piston Pump

(c) Plunger pump

It is just a piston pump with a plunger replacing the piston and its rod. It is generally used for rough work and can build up a very high pressure. It is generally preferred for handling water containing sand. Single acting and double acting plunger pumps are illustrated in Fig. 6.7 (*a*) and (*b*) respectively.

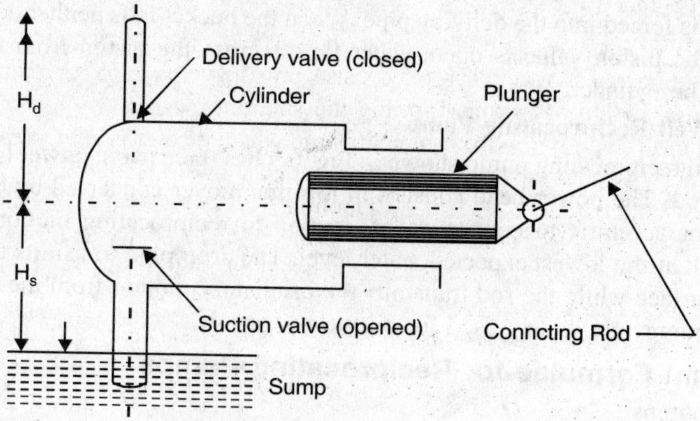

(*a*) Single Acting Plunger Pump

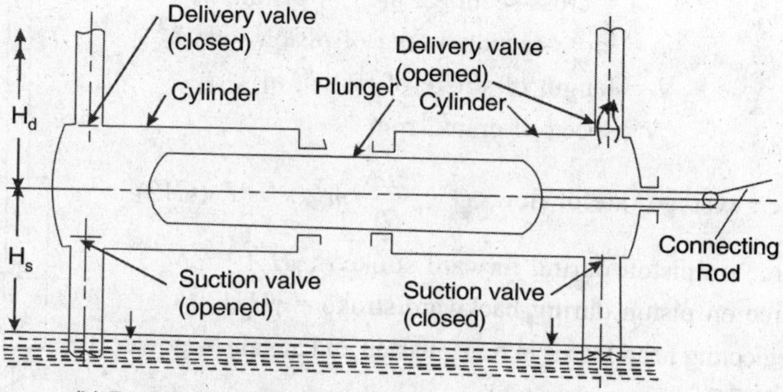

(b) Double Acting Single Cylinder Plunger Pump

Fig. 6.7

(*d*) Bucket (or Hand) Pump

This is a vertical piston pump as shown in Fig. 6.8. The piston which is provided with a value is called a bucket. It is essentially a low speed pump.

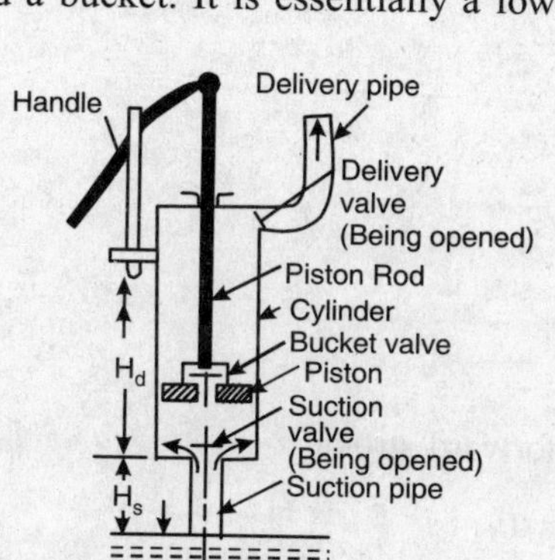

(a) Bucket Pump or Hand Pump

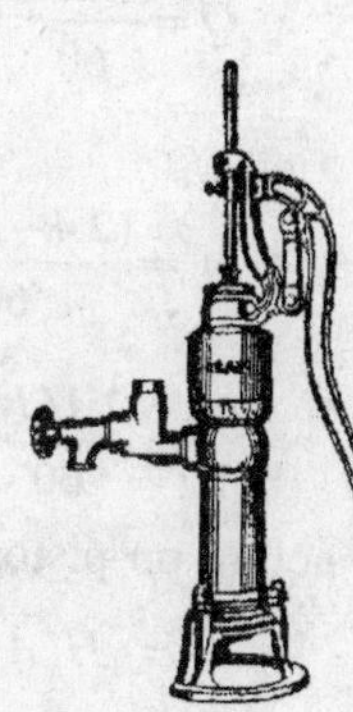

(b) Outside view of Bucket or Hand Pump

Fig. 6.8

When driven by hand it is popularly shown as a hand pump.

When the bucket rises, piston value remains closed. It is the suction-cum-delivery stroke and during this stroke water is drawn into the space under the bucket and simultaneously the water above the bucket is forced into the delivery pipe. When the bucket falls neither delivery not suction takes place. As the piston value is open, water flows across the piston from the suction to the delivery side of the cylinder.

(*e*) **Deep-Well Reciprocating Pump**

A deep-well reciprocating pump shown in Fig. 6.9 has three parts: power head, cylinder drop pipe and pump rod. The power head consists of a prime mover connected by belts and gear box to a crankshaft or eccentric to convert rotary motion to reciprocating motion. The cylinder is placed in the well at the lowest expected water level. The drop pipe transmits the water from the cylinder to the surface while the rod transmits the oscillatory motion from the power head to the plunger. These pumps are widely used for oil wells.

6.6.1 Important Formulae for Reciprocating Pumps

(*a*) Piston pumps

(*i*) Single acting

Let A = cross-sectional area of piston, m^2

a = cross-sectional area of piston rod, m^2

L = length of stroke of piston, m

N = speed of crank, rpm

Then average rate of flow, $Q = \frac{ALN}{60} \ m^3/s$ ... (6.12)

Force on piston during forward stroke = $\gamma H_s A$ N

Force on piston during backward stroke = $\gamma H_d A$ N

Neglecting head losses in transmission and at values, power of pump,

$$P = \frac{\gamma QH}{10^3} = \frac{\gamma \ ALN(H_s + H_d)}{60 \times 10^3} \text{ kW} \quad ... (6.13)$$

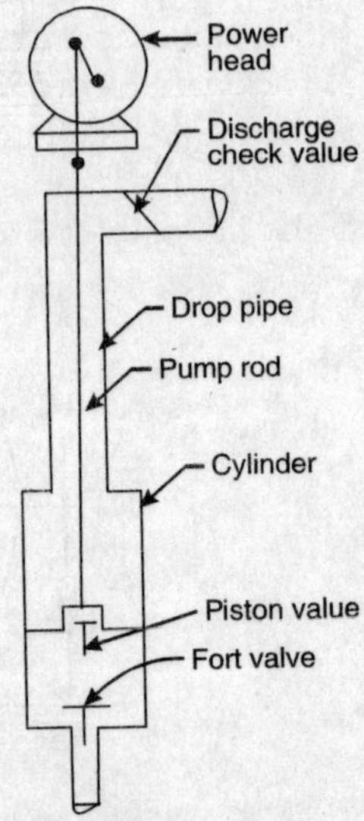

Fig.6.9. Deep-well reciprocating pump.

(*ii*) Double acting single cylinder pump

$$Q = \frac{ALN}{60} + \frac{(A-a)LN}{60} \quad ... (6.13)$$

$$= \frac{(2A-a)LN}{60} \text{ m}^3/\text{s} \quad ... (6.14a)$$

$$\simeq \frac{2ALN}{60} \text{ m}^3/\text{s} \quad ... (6.14b)$$

Force acting on piston during forward stroke

$$= \gamma H_s A + \gamma H_d (A-a) \quad \text{N} \quad ... (6.15a)$$

Force acting on piston during backward stroke

$$= \gamma H_s (A-a) + \gamma H_d A \quad \text{N} \quad ... (6.15b)$$

(*b*) **Slip**– It is the difference of volume swept through the piston and the actual discharge per stroke.

Slip = theoretical discharge – actual discharge = $Q_{th} - Q_{act}$ ... (6.16)

$$\text{Percentage slip} = \frac{\text{Slip}}{\text{Theoretical discharge}} \times 100 \quad \text{... (6.17)}$$

(*c*) **Coefficient of discharge**–It is the ratio of actual discharge per stroke to the volume swept per stroke.

$$C_d = \frac{Q_{act}}{Q_{th}} \quad \text{... (6.18)}$$

6.7 CENTRIFUGAL PUMP

A centrifugal pump is a device to raise liquids from a lower to a higher level by creating the required pressure with the help of centrifugal action. In general, it can be defined as a machine which increases the pressure energy of a fluid. Whirling motion is impacted to the liquid by means of backward curved blades mounted on a wheel known as the impeller. Liquid enters the impeller at the eye of the pump and discharges into the casing surrounding the impeller, as shown in Fig. 6.10. The pressure head developed by centrifugal action is entirely due to the velocity imparted to the liquid by the rotating impeller.

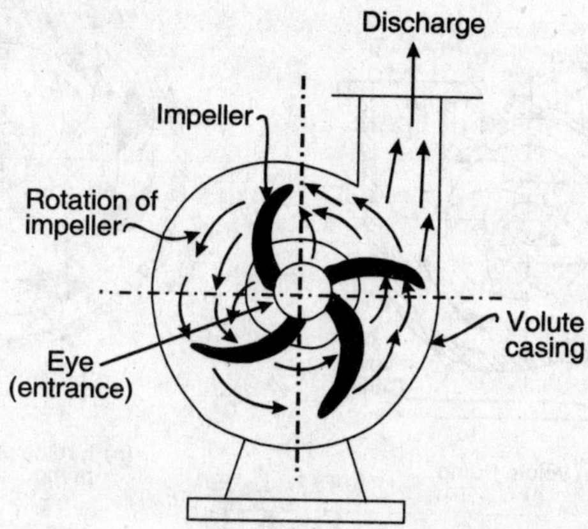

Fig. 6.10 Principle of centrifugal pump

6.7.1 Principle of Operation

First of all priming of the pump is carried out. The power source is switched on and the revolution of the pump impeller inside the casing full of water produces a forced vortex which is responsible for imparting a centrifugal head to water. Rotation of impeller effects a reduction of pressure at the centre. This causes the water in the suction pipe to rush into the eye.

6.7.2 Classification of Centrifugal Pumps

The centrifugal pumps may be classified based on the following characteristic features:

1. Working head.
2. Type of casing.
3. Number of impellers per shaft.
4. Relative direction of flow through impeller.
5. Number of entrances to impeller.
6. Disposition of shaft.
7. Liquid handled.
8. Specific speed

1. **Working head.**

   Low lift : upto 15 m.
   Medium lift : upto 40 m.
   High lift : above 40 m.

2. **Type of casing**

The casings are of two types: Volute and turbine or diffusion type.

(*a*) **Volute Pump-** A volute casing type of pump is shown in Fig. 6.11 (*a*). It has a volute casing into which the impeller discharges water at a high velocity. Volute is of a spiral form and the cross-sectional area of the moving stream gradually increases from the tongue towards the delivery pipe. The cross-sectional area at any point is therefore proportional to the quantity of water flowing across that section and therefore the mean velocity remains constant. This eliminates the possibility of loss of kinetic head. The functions of volute casing are:

(*i*) To collect water from the periphery of impeller and to transmit it to the delivery pipe at a constant velocity.

(*ii*) To elimate the loss of velocity head.

(*iii*) To increase the efficiency of the pump by eliminating loss of velocity head.

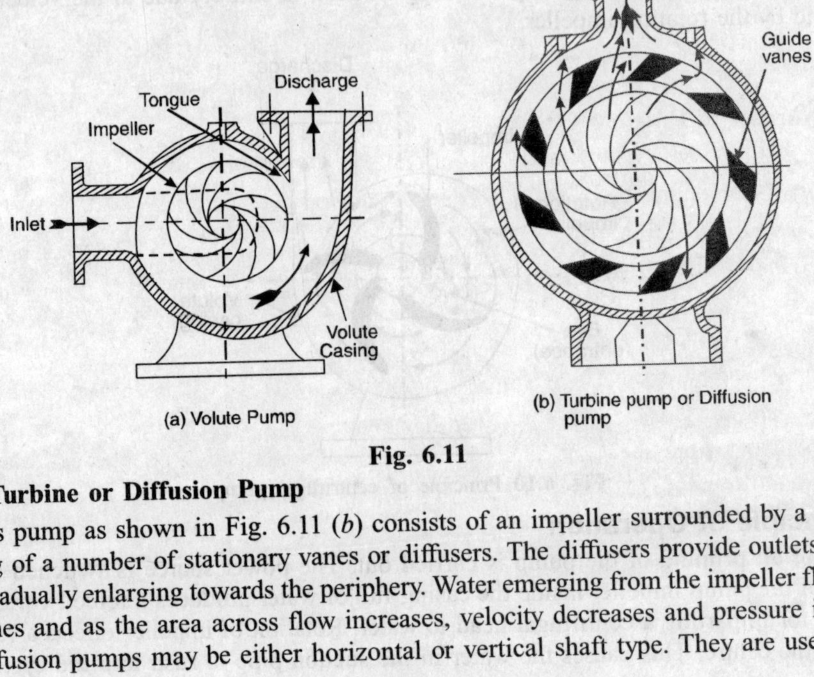

Fig. 6.11

(*b*) **Turbine or Diffusion Pump**

This pump as shown in Fig. 6.11 (*b*) consists of an impeller surrounded by a guide wheel consisting of a number of stationary vanes or diffusers. The diffusers provide outlets with cross-section gradually enlarging towards the periphery. Water emerging from the impeller flows past the guide vanes and as the area across flow increases, velocity decreases and pressure increases.

Diffusion pumps may be either horizontal or vertical shaft type. They are used in narrow wells and mines etc.

(*c*) **Vortex Casing**

The vortex or whirlpool chamber type casing is shown in Fig 6.11 (*c*).

This is an improved type of volute chamber where an annular space is provided between the volute and impeller forming a combination of spiral and circular chamber. This arrangement reduces eddies to a considerable extent and improves the performance of the pump.

3. **Number of Impellers**

(*a*) Single stage – for low lift.

(*b*) Multi-stage – for high lift.

4. **Relative Direction of Flow through Impeller**

(*a*) Radial flow – small discharge at medium head.

(*b*) Axial flow – for large discharge at low head.

(*c*) Mixed flow : Radial + axial flow, large discharge and low head, used for irrigation purposes.

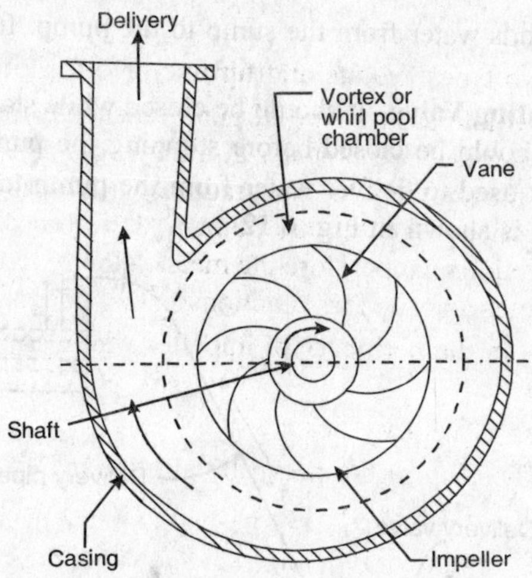

(c) Pump with vortex chamber

Fig.6.11

5. **Number of Entrances**
   (*a*) Single entrance – suction from one side of impeller.
   (*b*) Double entrance – suction from both sides of impeller.
6. Disposition of Shaft
   (*a*) Horizontal – generally used.
   (*b*) Vertical – rarely used.
7. **Liquid Handled**
   (*a*) Closed impeller – for non-viscous liquids such as ordinary water, hot water, hot oil and acids etc.
   (*b*) Open impeller – for handling mixtures of water, sand, pebbles and clay.
8. **Specific speed**

   It is defined as the speed of a geometrically similar pump when delivering one m^3/s against a head of one metre.

   Specific speed, $N_s = \dfrac{N\sqrt{Q}}{H^{3/4}}$

   where N = actual speed of pump in rpm
   Q = quantity of liquid flowing, m^3/s
   H = delivery head, m.

6.7.3 Accessories and Layout of Centrifugal Pump

The main accessories are:

1. Strainer and foot valves.
2. Suction pipe with fittings.
3. Delivery or regulating valve.
4. Delivery pipe with fittings.

1. **Strainer** – It keeps the floating bodies in water, such as leaves, wooden pieces and other kind of rubbish, away from the pump and avoids choking.

   Foot valve– It is a non-return or one way type valve, opening upwards. It does not allow the water to move downwards.

2. **Suction Pipe**– It leads water from the sump to the pump. It should be laid first vertical and then horizontal.
3. **Delivery or Regulating Valve**– It should be closed while starting the pump to allow built up of pressure. It should be closed before stopping the pump.
4. **Delivery pipe**–It is used to deliver water from the pump to the reservoir. The layout of a centrifugal pump is shown in Fig. 6.12.

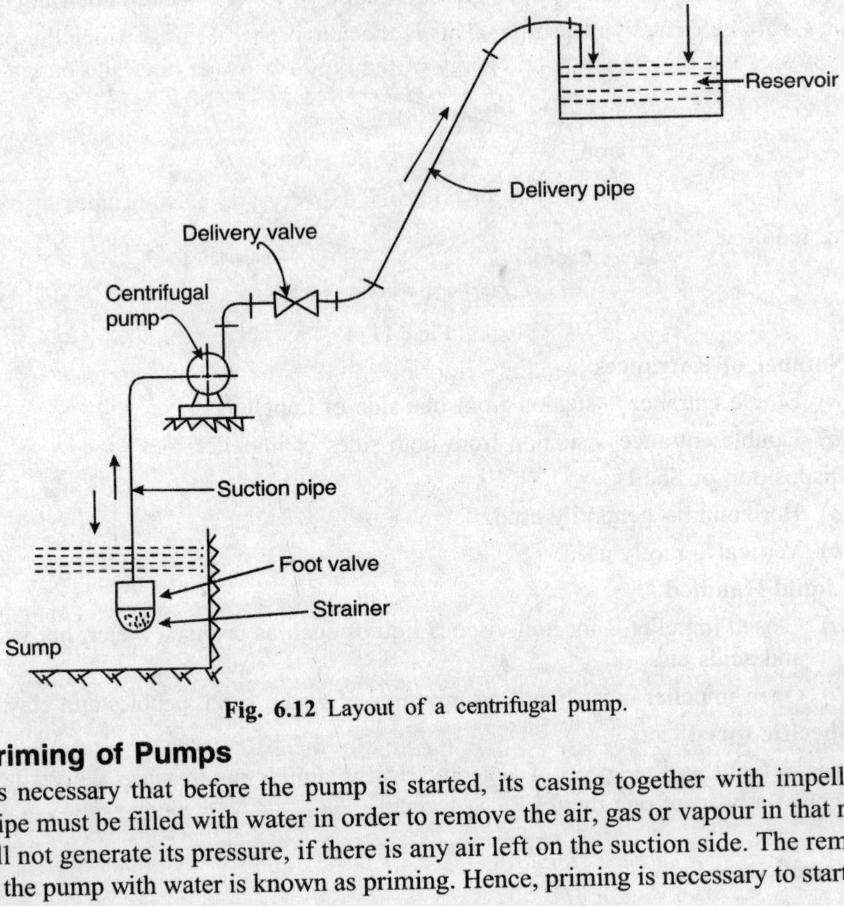

Fig. 6.12 Layout of a centrifugal pump.

6.7.4 Priming of Pumps

It is necessary that before the pump is started, its casing together with impeller and the suction pipe must be filled with water in order to remove the air, gas or vapour in that region. The pump will not generate its pressure, if there is any air left on the suction side. The removal of air by filling the pump with water is known as priming. Hence, priming is necessary to start the pump.

6.7.5 Submersible Pump

In a submersible pump, the pump and the electric motor are placed below the water surface of a well. Delivery of water to the surface is through a vertical pipe on which the assembly is suspended. An advantage of the submersible pump is the elimination of the long drive shaft from the ground surface to the pump. This helps in reducing bearing friction and provides an unobstructed pipe for delivery of water to the surface.

6.7.6 Important Terms and Formulae in Centrifugal Pump

Suction head, H_s – It is the vertical height of the centre line of pump shaft above the surface of liquid from which it is being raised.

Delivery head, H_d – It is the vertical height measured from the centre line of main pump shaft to where the liquid is delivered.

Static head, H_{stat} – It is the sum of static and delivery heads. (6.20)

$$H_{stat} = H_s + H_d$$

Manometric head, H_{mano} – It is the head measured across the pump inlet and outlet flanges. It expresses the increase in pressure energy per unit weight of liquid handled by the impeller.

$$H_{mano} = \frac{p_d - p_s}{\gamma} + h_g$$

$$= (H_{mano})_d - (H_{mano})_s + h_g \tag{6.21}$$

where h_g = vertical distance between the pressure tappings for the suction and delivery gauges.

Total, Gross or Effective head, H– This is the actual head against which the pump has to work. It expresses the increase in total energy of the liquid between inlet and outlet.

$$H = H_{mano} + \frac{v_d^2 - v_s^2}{2g} \tag{6.22}$$

where v_d, v_s = velocity of liquid in delivery and suction pipes respectively.

Power required to drive the pump,

$$P = \frac{\gamma Q H_{mano}}{10^3 \times \eta_{overall}} \text{ kW} \tag{6.23}$$

where γ = specific weight of liquid, N/m^3

Q = discharge, m^3/s

H_{mano} = manometric head, m

$\eta_{overall}$ = overall efficiency of pump.

$$= \frac{\gamma Q H_{mano}}{10^3 (BP - \text{Power lost in coupling mechanism})} \tag{6.24}$$

BP = Brake power of prime mover driving the pump.

Example 6.1 *A double acting reciprocating pump delivering 80 litres per second has the following specifications :*

Diameter of piston = 300 mm; Stroke = 600 mm; Speed = 80 rpm; Suction head = 5 m; Delivery head = 10 m; and Efficiency = 80%

Determine (a) the force required to operate the pump

(b) percentage slip

(c) power input.

Solution.

Given : $D = 0.3$ m, $L = 0.6$ m, $N = 80$ rpm, $H_s = 5$ m, $H_d = 10$ m,

$\eta_p = 0.8$, $Q_a = 80$ litres/s

Cross-sectional area of piston, $A = \frac{\pi}{4} D^2 = \frac{\pi}{4}(0.3)^2 = 0.0707 \text{ m}^2$

(*a*) Suction force, $F_s = \gamma H_s A + \gamma H_d A = \gamma A(H_s + H_d)$

$$= 9.81 \times 1000 \times 0.0707(5 + 10)$$

$$= 10403.5 \text{ N}$$

(*b*) Theoretical discharge, $Q_{th} = \frac{2\, LAN}{60}$

$$= \frac{2\times 0.6\times 0.0707\times 80}{60} = 0.11312 \text{ m}^2/\text{s}$$

or 113.12 litres/s

Percentage slip $= \dfrac{Q_{th} - Q_a}{Q_{th}} \times 100$

$$= \frac{113.12 - 80}{113.12} \times 100 = 29.28\%$$

(c) Power input, $P = \dfrac{\gamma Q_a (H_s + H_d)}{\eta_p \times 10^3}$

$$= \frac{9.81\times 1000\times 80\times 10^{-3}\times 15}{0.8\times 10^3}$$

$= 14.715$ kW

Example 6.2 *A single cylinder, single acting pump lifts 30 litres of water per second against a static head of 25 m. The cylinder bore is 350 mm and stroke is 450 mm. Determine the speed of pump if slip is 2 percent.*

Solution.

Given : $H_s + H_d = 25$ m, $Q_a = 30$ litres/s or $0.03 \text{ m}^3/\text{s}$, $D = 0.35$ m, $L = 0.45$ m, $s = 2\%$

Theoretical discharge, $Q_{th} = \dfrac{LAN}{60} = \dfrac{0.45\times\frac{\pi}{4}(0.35)^2\times N}{60} = 7.216\times 10^{-4}\times N \text{ m}^3/\text{s}$

Slip, $s = \dfrac{Q_{th} - Q_a}{Q_{th}}$

$$0.02 = \frac{7.216\times 10^{-4}\times N - 0.030}{7.216\times 10^{-4}\times N}$$

$$0.1443\times 10^{-4} N = 7.216\times 10^{-4}\times N - 0.030$$

$$N = \frac{0.030}{7.0717\times 10^{-4}} = 42.4 \text{ rpm}$$

Example 6.3 *A single acting reciprocating pump has to deliver water against a total head of 50 m. The plunger diameter is 120 mm. What is the force exerted by the plunger during delivery stroke? Assume pump efficiency to be 80%.*

Solution.

Given : $H = 50$ m, $D = 0.12$ m, $\eta_p = 0.08$

Force on plunger during delivery stroke $= \dfrac{\gamma HA}{\eta_p}$

$$= \frac{9.81\times 10^3\times 50\times\frac{\pi}{4}(0.12)^2}{0.8}$$

$= 6934.3$ N

Example 6.4 *Water is raised to a height of 40 m by means of a centrifugal pump. If the quantity of water raised is 350 m³/h, determine the power of electric motor required to drive the pump. Take efficiency of pump to be 85 percent and that of electric motor 95%.*

Solution.

Given : $H = 40$ m, $Q = 350$ m^3/h, $\eta_p = 0.85$, $\eta_m = 0.95$

Power of motor required,
$$P = \frac{\gamma Q H}{10^3 \times \eta_p \times \eta_m}$$
$$= \frac{9.81 \times 10^3 \times 350 \times 40}{3600 \times 10^3 \times 0.85 \times 0.95}$$
$$= 47.245 \text{ kW}$$

Example 6.5 *A centrifugal pump lifts water against a static head of 30 m of which suction lift is 3 m. The loss of head in suction pipe is 1 m and 5 m in delivery pipe. Speed of pump is 1440 rpm. The combined efficiency of motor and transmission system is 0.85. The flow rate is 60 litres/s. Calculate the power consumed by pump.*

Solution.

Given : $H_{\text{stat}} = H_s + H_d = 30$ m, $H_s = 3$ m, $N = 1440$ rpm, $\eta = 0.85$,
$Q = 60$ litres/s or 0.06 m^3/s, $h_{fs} = 1$ m, $h_{fd} = 5$ m

$$H_{\text{mano}} = H_{\text{stat}} + h_{fs} + h_{fd}$$
$$= 30 + 1 + 5 = 36 \text{ m}$$

Power required to drive the pump
$$= \frac{\gamma Q\, H_{\text{mano}}}{10^3 \times \eta}$$
$$= \frac{9.81 \times 10^3 \times 0.06 \times 36}{10^3 \times 0.85}$$
$$= 24.929 \text{ kW}$$

6.8 HYDRAULIC PRESS/JACK

In Fig. 6.13 is shown a hydraulic device which can operate either as a press or a jack. When working as a press, the object to be pressed is placed between the top rigid support attached to

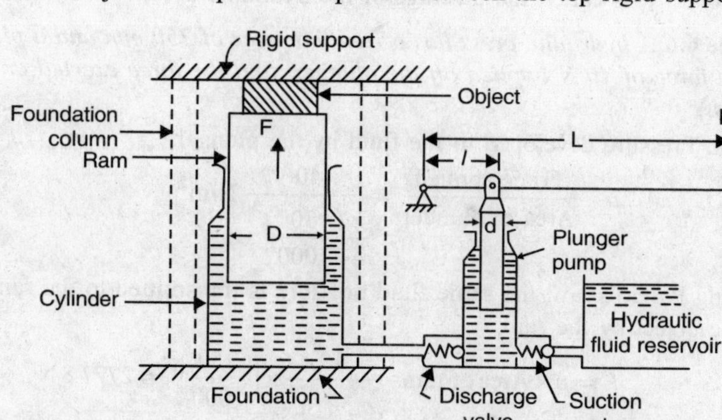

Fig 6.13 Hydraulic Press-Jack-a shcematic diagram.

the foundation by vertical columns and the ram of diameter D. In case of its working as a jack, the load to be lifted will be placed on the movable ram. The hand-driven pump, with a plunger of diameter d having suction and discharge valves as shown, is used to supply pressure into the cylinder, and helps in developing the desired thrust F, by application of effort P on the operating lever at the other end.

Various parts of a hydraulic press are shown in Fig. 6.13. The rigid support is attached to the foundation by means of vertical columns shown by broken lines. The object to be pressed is placed in between the support and the ram as shown. The hand-driven plunger pump is operated by a lever arrangement through which an effort P is applied to move the plunger. When the plunger moves up, the suction valve is opened while the delivery valve remains shut. The hydraulic fluid from the reservoir is sucked into the pump. During the downward movement of the plunger, the suction valve is closed due to the pressure developed and the delivery valve gets opened. The hydraulic fluid is thus pumped into the cylinder. The hydraulic pressure so developed causes an upward thrust F and the ram.

When an effort of magnitude p is applied on the operating lever, the force transmitted on the plunger is

$$P = \left(\frac{L}{l}\right)$$

Here L = total length of the operating lever as shown, and

l = distance between the hinge and the plunger as shown.

The pressure developed in the hydraulic fluid of the pump

$$p = \frac{4PL}{\pi d^2 l} \qquad \text{... (6.25)}$$

is transmitted into the cylinder due to which the ram experience an upward force F

$$F = p.\frac{\pi D^2}{4} = P\left(\frac{L}{l}\right)\left(\frac{D}{d}\right)^2 \qquad \text{... (6.26)}$$

6.8.1 Mechanical Advantage

It is defined as the ratio of the weight lifted to the force applied on the lever.

$$\text{Mechanical Advantage} = \frac{F}{P} \qquad \text{... (6.27)}$$

The ratio $\left(\frac{L}{l}\right)$ is called the leverage of the hydraulic press.

Example 6.6 *A hydraulic press has a ram diameter of 250 mm and a plunger diameter of 30 mm. For a force of 40 N applied on the plunger, find the force exerted on the object by the hydraulic press.*

Solution. Pressure developed in the fluid by the plunger,

$$p = \frac{\text{Force applied}}{\text{Area of plunger}} = \frac{40}{\frac{\pi}{4}\left(\frac{30}{1000}\right)^2} \text{ N/m}^2$$

Following Pascal's law, the same fluid pressure is transmitted to the ram. Hence the force exerted on the object by the ram.

$$F = p \times \text{Area of ram} = \frac{40}{\left(\frac{30}{1000}\right)^2} \times \left(\frac{250}{1000}\right)^2 = 2777.8\,\text{N}$$

Example 6.7 *A hydraulic press has a plunger diameter 20 mm and the ram diameter 160 mm. The stroke of the plunger is 200 mm, and the weight lifted by the press is 10 kN. The* *distance travelled by the weight is 600 mm in 10 minutes. Determine :*

(i) Force applied to the plunger

(ii) power required to move the plunger, and

(iii) the number of strokes of the plunger

Solution. Referring to Fig. 6.13, the pressure developed in the fluid in the ram cylinder

$$p = \frac{10 \times 1000}{\frac{\pi}{4} \times \left(\frac{160}{10 \times 100}\right)^2} \text{N/m}^2 = 497.265 \text{ kN/m}^2$$

On account of continuity of fluid, the same pressure is transmitted in the plunger pump, hence

(i) the force applied to the plunger = p × area of plunger

$$= 497.265 \times \frac{\pi}{4}\left(\frac{20}{10 \times 100}\right)^2 = 0.1562 \text{ kN} = \mathbf{156.2\ N}$$

(ii) the power required to move the plunger = $\frac{\text{Work done}}{\text{Time in sec.}}$

$$= \frac{\text{Force} \times \text{stroke of plunger}}{\text{Time (sec.)}} = \frac{156.2 \times 200/(10 \times 100)}{10 \times 60} \text{ Nm/s (W)}$$

= **0.0521 Nm/s** (Watt)

(iii) the number of strokes of the plunger :

In order to raise the weight, the hydraulic fluid is to be pumped by the plunger into the ram cylinder. Let the number of strokes be n, then

Volume of fluid transferred from the plunger into the ram × n

= Volume of liquid required to raise the weight by 600 mm

or,
$$\frac{\pi}{4}(20)^2 \times 200 \times n = \frac{\pi}{n}(160)^2 \times 600$$

∴
$$n = \left(\frac{160}{20}\right)^2 \times \frac{600}{200} = \mathbf{192}$$

No. of strokes of the plunger = 192

Example 6.8 *In a hydraulic jack, a force of 200 N is applied on the lever having a leverage of 10. If the diameter of ram is 300 mm and that of the plunger 30 mm, determine the weight of the object that can be lifted.*

Solution. Given P = 200 N, Leverage, $\frac{L}{l} = 10, \frac{D}{d} = \frac{300}{30} = 10$

From Eq. (6.26), the weight which can be lifted is :

$$F = P\left(\frac{L}{l}\right)\left(\frac{D}{d}\right)^2 = 200 \times 10 \times (10)^2 = 200 \times 10^3 \text{ N} = \mathbf{200 kN}$$

Example 6.9 *A hydraulic jack with a plunger of 30 mm diameter has a stroke of 100 mm, and a ram of 300 mm diameter. A lever arrangement with a leverage of 10 is used to drive the plunger. If a weight of 5 kN is to be lifted, determine the force to be applied on the lever. Calculate the height to which the weight will be raised in 50 strokes, and the work done involved.*

Solution. The weight to be lifted and the effort to be applied on the operating lever are related by Eq. (6.26). hence

$$F = P\left(\frac{L}{l}\right)\left(\frac{D}{d}\right)^2$$

Making substitutions

$$5 \times 1000 = P(10)\left(\frac{300}{30}\right)^2 = 100P$$

$$P = 5 \text{ N}$$

The force to be applied to the lever = 5 N

Volume of hydraulic flud transferred to the cylinder is 50 strokes of the plunger,

$$V = 50 \times \frac{\pi}{4}(0.03)^2 \times 0.1 = 3.5343 \times 10^{-3} \text{ m}^3$$

$$\text{Area of ram, } A = \frac{\pi}{4}(0.03)^2 = 0.070686 \text{ m}^2$$

Height to which weight will be lifted by the ram,

$$h = \frac{V}{A} = \frac{3.5343 \times 10^{-3}}{0.070686} = 0.05 \text{ m or } 50 \text{ mm}$$

Workdone involved = weight × height lifted

= 5000 × 0.05 = 250 N.m

6.9 HYDRAULIC LIFT

A hydraulic lift is a device used for carrying goods or passengers from one floor to another in a multi-storeyed building. It can be either a direct acting type or a suspended type as described below.

1. **Direct Acting Hydraulic Lift.** It consists of a platform or cage supported by a vertical ram which slides in a fixed vertical cylinder as shown in Fig 6.14. Water under pressure is pumped into the fixed cylinder which exerts a force on the sliding ram causing it to move vertically up. The cage is thus raised to the desired height. Its downward movement is ensured by releasing water from the fixed cylinder.

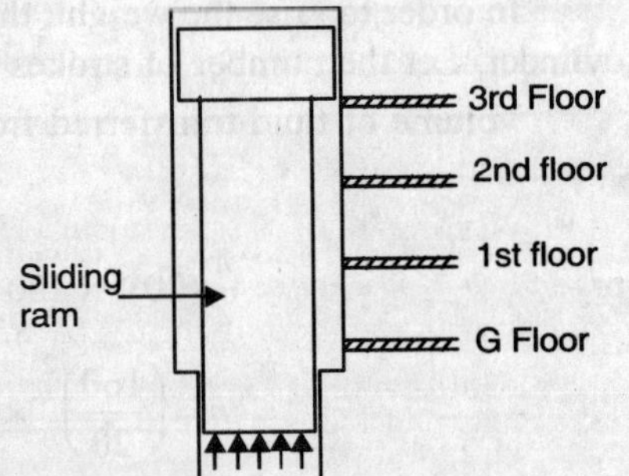

Fig. 6.14 A schematic view of a Direct Acting Hydraulic Lift.

2. **Suspended Hydraulic Lift.** In a suspended type of lift, a jigger arrangement is used to multiply the storke of the sliding ram. The hydraulic jigger consists of a fixed cylinder, a sliding and two sets of pulley blocks as shown in Fig. 6.15. One set of pulley block is movable while the other one is fixed. The movable pulley block is mounted on the sliding ram whereas the fixed pulley block is installed at the head of the fixed cylinder. Each set of pulley block has the same number of pulleys.

A wire rope, one end of which is fixed to the body of the cylinder and the other end is taken around all pulleys of the movable and fixed blocks and finally passes over the guide pulleys to support the cage as shown in the figure.

The raising and lowering of the cage of the lift is done by the hydraulic jigger. When a fluid under pressure is admitted into the cylinder, the sliding ram moves out increasing the distance

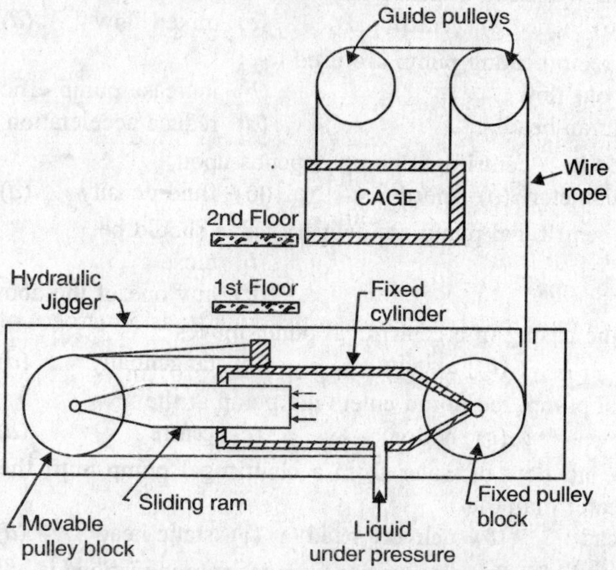

Fig 6.15 Suspended Hydraulic Lift-a line diagram

between the fixed and movable pulley blocks. The wire rope connected to the cage is pulled and the cage is lifted. For lowering the cage, the fluid inside the cylinder is taken out gradually causing the sliding ram to move inwards shortening the distance between the movable and fixed pulleys. This increases the free length of the rope enabling the cage to come downwards. If 'n' is the number of pulleys used in each block, the distance moved by the free end of the rope is 'n' times the stroke of the ram. The hydraulic jigger may be placed horizontally or vertically.

The hydraulic lifts have now become obsolete and have been superseded by electric lifts for use in multistoreyed buildings.

Example 6.10 *A hydraulic lift is required to raise a load of 15 kN through a height of 12 m once in every 90 seconds. The lift moves with a velocity of 0.75 m/s. Determine the power required to drive the lift and its idle period.*

Solution. $W = 15$ kN $= 15 \times 1000$ N, $H = 12$ m, Time = 90 seconds

Power required to drive the lift = Work done per second in lifting the load

$$= \frac{(15 \times 1000) \times 12}{90} = 2000 \text{ Nm/s} = 2000 \text{ W} = 2\text{kW}$$

$$\text{Working period of lift} = \frac{\text{Total lift}}{\text{Velocity}} = \frac{12.0}{0.75} = 16 \text{ seconds}$$

$\therefore$ Idle period of lift = Total time – Working period = 90 – 16 = 74 seconds.

MULTI-CHOICE QUESTIONS

1. Reciprocating pump is started with the delivery valve
   (*a*) closed (*b*) open (*c*) 50% open (*d*) partially open
2. Slip for a reciprocating pump is defined as
   (*a*) $\frac{Q_a - Q_{th}}{Q_{th}}$ (*b*) $\frac{Q_{th} - Q_a}{Q_a}$ (*c*) $\frac{Q_{th} - Q_a}{Q_{th}}$ (*d*) $\frac{Q_a - Q_{th}}{Q_a}$

3. The pump suitable for small discharge at high pressure head is
   (*a*) axial flow (*b*) centrifugal (*c*) mixed flow (*d*) reciprocating
4. Air vessels in reciprocating pump are used to
   (*a*) smoothen out flow (*b*) increase pump efficiency
   (*c*) increase pump head (*d*) reduce acceleration head
5. Head developed by a centrifugal pump depends upon
   (*a*) impeller diameter (*b*) speed (*c*) fluid density (*d*) type of casing
6. For starting a centrifugal pump, its delivery valve should be
   (*a*) open (*b*) close
   (*c*) partially open (*d*) any one of the above
7. The water in the casing of a centrifugal pump moves
   (*a*) radially (*b*) axially (*c*) tangentially (*d*) in a spiral
8. In a centrifugal pump, the liquid enters the pump at the
   (*a*) top (*b*) bottom (*c*) centre (*d*) sides
9. There shall be no flow of water from a centrifugal pump until the pressure rise in the impeller is greater than the
   (*a*) suction head (*b*) delivery head (*c*) static head (*d*) manometric head
10. The suitable pump for flood control and irrigation applications is
    (*a*) reciprocating (*b*) centrifugal (*c*) axial flow (*d*) mixed flow
11. A suitable pump for handling viscous fluid is
    (*a*) reciprocating pump (*b*) centrifugal pump
    (*c*) turbine pump (*d*) screw pump
12. Which one of the following is not a rotary pump?
    (*a*) gear pump (*b*) vane pump (*c*) screw pump (*d*) axial pump
13. Priming is required in a
    (*a*) gear pump (*b*) screw pump
    (*c*) reciprocating pump (*d*) centrifugal pump
14. For pumping molasses, it is preferable to employ
    (*a*) reciprocating pump (*b*) centrifugal pump
    (*c*) open impeller pump (*d*) rotary pump
15. For pumping mustard oil from a drum, a suitable pump is
    (*a*) reciprocating pump (*b*) vane pump
    (*c*) centrifugal pump (*d*) gear pump

Answers:

1. (*b*) **2.** (*c*) **3.** (*d*) **4.** (*d*) **5.** (*a*) **6.** (*b*) **7.** (*d*) **8.** (*c*) **9.** (*d*) **10.** (*b*)
11. (*d*) **12.** (*d*) **13.** (*d*) **14.** (*c*) **15.** (*d*)

REVIEW QUESTIONS

1. What is a pump? What are its functions?
2. How do you classify the pumps?
3. Explain with a neat sketch, the working of a single acting piston pump.
4. Explain with a neat sketch, the working of a double acting piston pump.
5. Explain the working of a single acting plunger pump.
6. Define slip in a reciprocating pump.
7. What is the main difference between a plunger and bucket type pumps?

8. What is the function of air chambers in a reciprocating pump?

9. What is priming? Why priming is required in a centrifugal pump?

10. Describe self priming devices.

11. Explain with a neat sketch the working of a centrifugal pump.

12. How do you classify pumps?

13. How do you classify centrifugal pumps?

14. Explain the main features of volute type and vortex type centrifugal pumps.

15. Give the general classification of rotary pumps.

16. Describe the working of a gear pump.

17. What are the advantages of centrifugal pump over reciprocating pump?

18. Describe the working of (*a*) screw pump (*b*) vane pump (*c*) hydraulic jacts (*d*) hydraulic lift.

EXERCISES

6.1 A double acting piston pump runs at 90 rpm. Its bore and stroke are 150 mm and 450 mm respectively. Determine the theoretical discharge in m^3/h. If the total head to be lifted by the pump is 60 m, determine the power required to drive the pump if its efficiency is 80%.
[**Ans.** 85.88 m^3/h, 17.552 kW]

6.2 A double acting reciprocating pump has a piston diameter of 160 mm and stroke length 350 mm. The suction and delivery heads are 5 m and 18 m respectively. The pump speed is 70 rpm and percentage slip is 3%. Calculate (*a*) the actual discharge per minute, (*b*) coefficient of discharge, and (*c*) power required to drive the pump. Assume efficiency of pump to be 85%. [**Ans.** 0.955 m^3/mm, 0.97, 4.224 kW]

6.3 A single acting reciprocating pump delivers 45 m^3/h of water against a total head of 25 m. The pump speed is 90 rpm. If slip is 5%, calculate (*a*) volume of cylinder in litres, and (*b*) power consumption if overall efficiency is 80%. [**Ans.** 8.77 litres, 4.034 kW]

6.4 A single acting reciprocating pump has its piston diameter as 150 mm and stroke 250 mm. The piston makes 50 double strokes per minute. The suction and delivery heads are 5 m and 15 m respectively. Find the force required to work the piston during the suction and delivery strokes. Assume the efficiency of suction and delivery strokes as 60% and 75% respectively. Determine the power required by the pump. [**Ans.** 1442 N, 3669 N, 1.065 kW]

6.5 A double acting reciprocating pump has a piston of 250 mm diameter and piston rod of 50 mm diameter, which is on one side of the piston only. Length of piston stroke is 380 mm and the speed of the crank moving the piston is 60 rpm. The suction and delivery heads are 5 m and 20 m respectively. Find the force required to work the piston during the inwards and outwards strokes. Determine the quantity of water in litres per minute raised by the pump and power required. [**Ans.** 11632.7 N, 11921 N, 2195 litres/min, 8.97 kW]

7 INTERNAL COMBUSTION ENGINES

7.1 INTRODUCTION

An engine is a device which transforms one form of energy into another form. When the products of combustion generated by the combustion of fuel and air within the cylinder form the working fluid, it is called an internal combustion engine (I.C.E.).

7.2 CLASSIFICATION OF INTERNAL COMBUSTION ENGINES

The internal combustion engines can be classified on the following basis :

1. Basic engine design :
   - (*i*) Reciprocating engines
   - (*ii*) Rotary engines (Wankel)
2. Working cycle :
   - (*i*) Otto cycle — Spark Ignition engines (S.I.)
   - (*ii*) Diesel cycle — Compression Ignition engines (C.I.)
3. Number of strokes :
   - (*i*) Four — stroke engines
   - (*ii*) Two — stroke engines
4. Fuel used :
   - (*i*) Gasoline or petrol engines
   - (*ii*) Compressed natural gas (CNG)
   - (*iii*) Liquefied petroleum gas (LPG)
   - (*iv*) Diesel oil — Light diesel oil : LDO and high speed diesel oil : HSD
   - (*v*) Fuel oil
   - (*vi*) Alcohols — methanol and ethanol
5. Fuel supply and mixture preparation :
   - (*i*) Carburetted type
   - (*ii*) Injection type
   - (*iii*) Multi-point fuel injection (MPFI)
6. Ignition method :
   - (*i*) Battery ignition

(*ii*) Magneto ignition

(*iii*) Compression ignition

7. Cooling method :

(*i*) Water cooled

(*ii*) Air cooled

(*iii*) Evaporative cooled

8. Cylinder arrangement

(*i*) Inline

(*ii*) V

(*iii*) Radial

(*iv*) Opposed

9. Valve or Port design and location :

(*i*) Overhead (I-head)

(*ii*) Side valve (L-head)

10. Engine speed :

(*i*) Low speed - upto 500 rpm

(*ii*) Medium speed - 500 to 1000 rpm

(*iii*) High speed - above 1000 rpm

11. Number of cylinders :

(*i*) Single cylinder

(*ii*) Multi-cylinder

12. Application :

(*i*) Land transport

(*ii*) Marine propulsion

(*iii*) Aircraft propulsion

(*iv*) Industrial purposes and power plants

13. Governing method :

(*i*) Quality governing

(*ii*) Quantity governing

14. Lubrication system

(*i*) Dry sump

(*ii*) Wet sump

(*iii*) Mist

(*iv*) Pressure

7.3 CONSTRUCTIONAL FEATURES OF I.C. ENGINES

(*a*) Four stroke spark Ignition Engine

The cross section of a four-stroke spark ignition engine is shown in Fig. 7.1. The major components of the engine are :

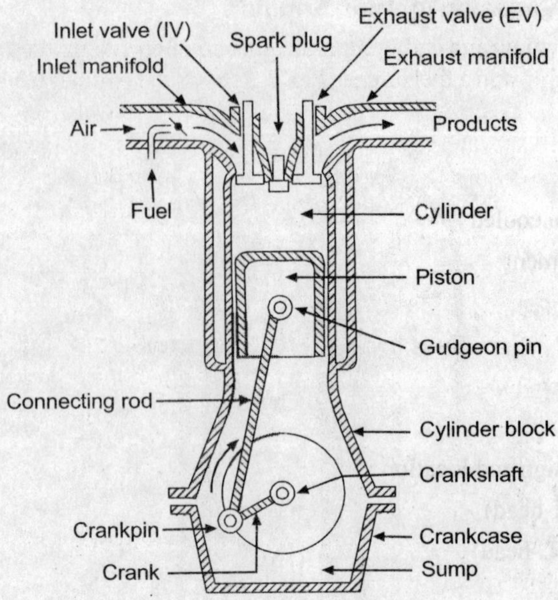

Fig. 7.1 Cross-section of a Spark-Ignition Engine.

Cylinder Block : It is the main supporting structure for the various components. The cylinder head is mounted on the cylinder block. The bottom of the cylinder block is called the crank case. The lubricating oil is kept in the crank case sump.

Cylinder : It is a cylindrical vessel in which the piston reciprocates.

Piston : It is a cylindrical component fitted into the cylinder forming the moving boundary of the combustion system.

Combustion Chamber : It is the space between the cylinder head and the piston top where combustion takes place.

Inlet Manifold : It is the pipe through which air or air-fuel mixture is drawn in the cylinder.

Exhaust Manifold : It is the pipe through which the products of combustion escape into the atmosphere.

Inlet and Exhaust Valves : The valves are used for regulating the incoming charge into the cylinder (inlet valve) or discharging the products of combustion (exhaust valve) from the cylinder.

Spark Plug : It is a component to initiate the combustion process in SI engine.

Carburettor : It is used for mixing fuel and air in correct proportion.

Connecting Rod : It is used to interconnect the piston and the crank to transmit force from the piston to the crankshaft.

Crankshaft : It is used to convert reciprocating motion of the piston into rotary motion of the output shaft.

Piston Rings : They provide a tight seal between the piston and the cylinder wall.

Gudgeon Pin : It connects the small end of die connecting rod to die piston.

Cam Shaft : It is a shaft on which cams are mounted to operate the valves. It is driven from the cam shaft by gears.

Flywheel : It absorbs surplus energy during working stroke and delivers during idle stroke.

(*b*) Four-stroke Compression Ignition Engine

Except for the spark plug and carburettor, all other components are the same as for the S.I. engine. It also requires a fuel pump and a fuel nozzle. Fig. 7.2 shows the outlines of a four-stroke diesel engine.

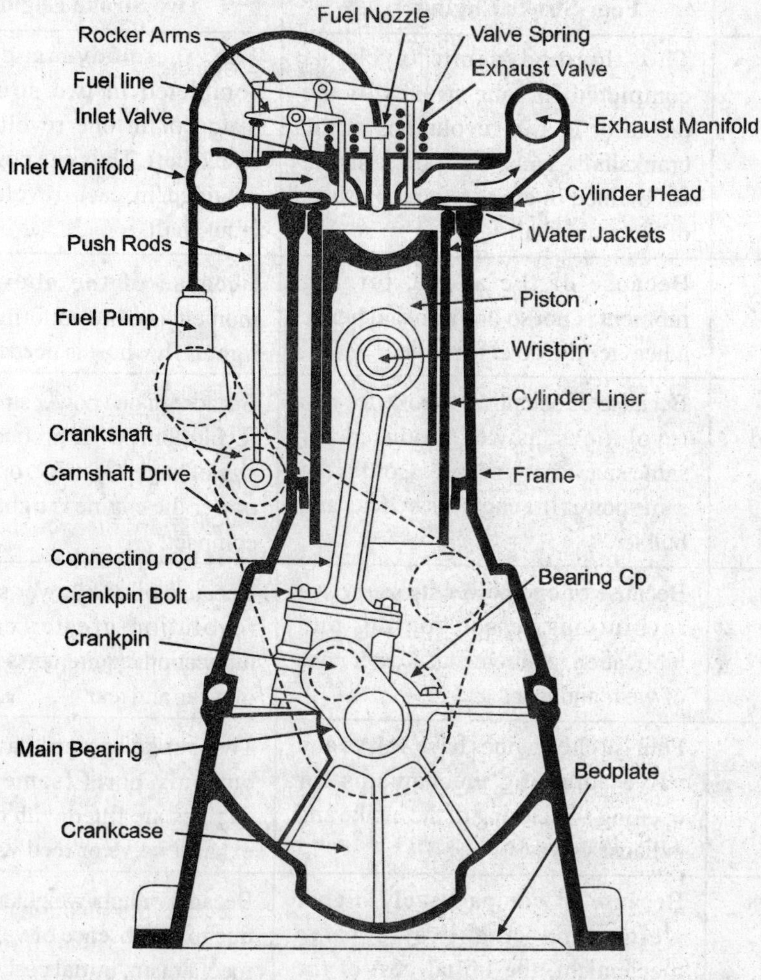

Fig. 7.2 Four-stroke diesel engine outlines.

(*c*) Two-Stroke SI Engine

In the case of two-stroke engine, the valves are replaced by the ports : exhaust port and transfer port. The charging of the cylinder with air-fuel mixture is through the carburettor and ignition by a spark plug. Other components are same as for a four stroke engine.

(*d*) Two-Stroke C.I. Engine

Here the carburettor and spark plug are replaced by a fuel pump and fuel nozzle. All other components are same as for two-stroke S.I. engine.

7.4 COMPARISON BETWEEN FOUR-STROKE AND TWO-STROKE ENGINES

The comparison between four-stroke and two-stroke engines is given in Table 7.1.

Table 7.1 Comparison of two-stroke and four-stroke engines.

Feature	Four-Stroke Engine	Two Stroke Engine
1. Power Strokes	The thermodynamic cycle is completed in four strokes of the piston or in two revolutions of the crankshaft. Thus, one power stroke is obtained in every two revolutions of the crankshaft.	The thermodynamic cycle is completed in two strokes of the piston or in one revolution of the crankshaft. Thus one power stroke is obtained in each revolution of the crankshaft.
2. Size of flywheel	Because of the above, turning moment is not so uniform and hence a heavier flywheel is needed.	Because of the above, turning moment is more uniform and hence a lighter flywheel is needed
3. Power developed	Because of one power stroke for two revolutions, power produced for same size of engine is less, or for the same power the engine is heavier and bulkier	Because of one power stroke for every revolution, power produced for same size of engine is twice, or for the same power the engine is lighter and more compact.
4. Wear and Tear	Because of one power stroke in two revolutions lesser cooling and lubrication requirements. Lower rates of wear and tear	Because of one power stroke in one revolution greater cooling and lubrication requirements. Higher rates of wear and tear.
5. Valves	Four-stroke engines have valves and valve actuating mechanisms for opening and closing of the intake and exhaust valves.	Two-stroke engines have no valves but only ports (some two-stroke engines are fitted with conventional exhaust valve or reed valve)
6. Initial cost	Because of comparatively higher weight and complicated valve mechanism, the initial cost of the engine is more.	Because of light weight and simplicity due to the absence of valve actuating mechanism, initial cost of the engine is less.
7. Volumetric efficiency	Volumetric efficiency is more due to more time for induction.	Volumetric efficiency is low due to lesser time for induction.
8. Thermal efficiency	Thermal efficiency is higher; part load efficiency is better.	Thermal efficiency is lower; part load efficiency is poor.
9. Applications	Used where efficiency is important, viz., in cars, buses, trucks, tractors, industrial engines, aeroplanes, power generation etc.	Used where low cost, compactness and light weight are important, viz., in mopeds, scooters, motorcycles, hand sprayers etc.

7.5 COMPARISON BETWEEN PETROL AND DIESEL ENGINES

The comparison between petrol and diesel engines is given in Table 7.2.

Table 7.2 Comparison between petrol and diesel engines

Parameter	Petrol Engine	Diesel Engine
1. Suction stroke	Mixture of air and petrol	Only air
2. Ignition device/method	Spark plug	Compression ignition
3. Combustion process	Constant volume	Constant pressure
4. Thermal efficiency	Around 25%	Around 40%
5. Maintenance cost	Low	High
6. Speed	High	Low
7. Space occupied	Less	More
8. Compression ratio	6 to 10	14 to 24
9. Cycle of operation	Otto cycle	Diesel cycle
10. Applications	Light vehicles, e.g. scooters, motor cycles, cars, etc.	Heavy vehicles, e.g., buses, trucks, etc.
11. Use of carburettor	Yes	No
12. Running cost	High	Low

7.6 WORKING PRINCIPLE AND PRESSURE-VOLUME DIAGRAMS

7.6.1 Working of Four-Stroke Spark-Ignition Engine

The details of various processes of a four stroke spark ignition engine with over head valves are shown in Fig. 7.3. During the four strokes, there are five events to be completed, viz., suction, compression, combustion, expansion and exhaust.

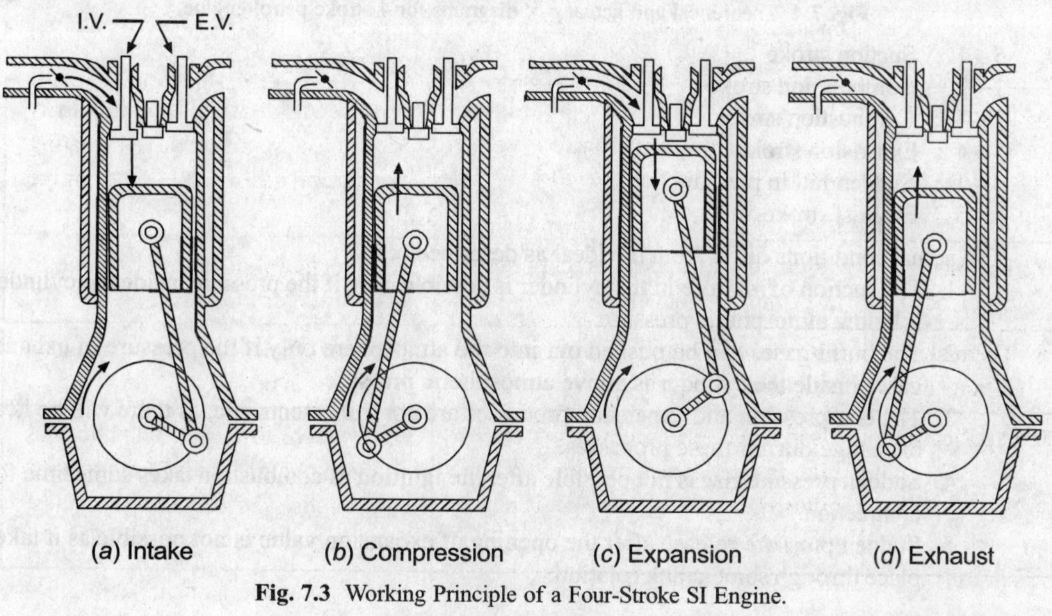

Fig. 7.3 Working Principle of a Four-Stroke SI Engine.

1. Suction Stroke : It starts when the piston is at T.D.C. and about to move downwards. The inlet valve is open and exhaust valve closed at this time. The charge consisting of fuel-air-mixture is drawn into the cylinder. When the piston reaches the B.D.C. the suction stroke ends and the inlet valve closes.

2. Compression Stroke : The charge is compressed to the clearance volume by the return stroke of the piston with both inlet and exhaust valves closed. At the end of the compression stroke the mixture is ignited with the help of a spark plug to convert chemical energy of fuel to heat energy.

3. Expansion Stroke : The high pressure of burnt gases forces the piston towards B.D.C. with both the valves in closed position to produce power.

4. Exhaust Stroke : At the end of the expansion stroke the exhaust valve opens with inlet valve closed. The piston starts moving towards the T.D.C. and sweeps the burnt gases out from the cylinder. The exhaust valve closes when piston has reached T.D.C.

The theoretical and actual *p-V* diagrams for spark ignition engine are shown in Fig. 7.4. The various processes are :

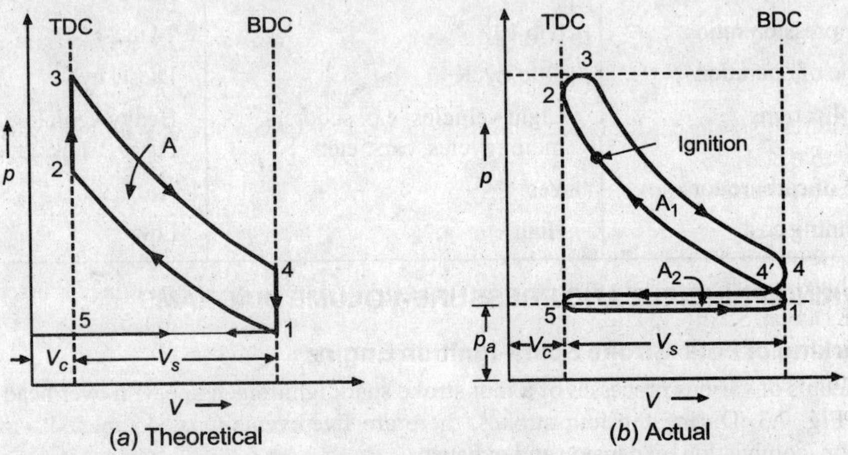

Fig. 7.4 Theoretical and actual *p-V* diagrams for 4-stroke petrol engine.

5 – 1 : Suction stroke
1 – 2 : Compression stroke
2 – 3 : Combustion stroke
3 – 4 : Expansion stroke
4 – 1 : Sudden fall in pressure
1 – 5 : Exhaust stroke.

The actual conditions differ from the ideal as described below :

1. The suction of mixture in the cylinder is possible only if the pressure inside the cylinder is below atmospheric pressure.
2. The burnt gases can be pushed out into the atmosphere only if the pressure of exhaust gases inside the cylinder is above atmospheric pressure.
3. The compression and expansion processes are not truly isentropic, as there will be heat exchange during these processes.
4. Sudden pressure rise is not possible after the ignition as combustion takes some time for completion.
5. Sudden pressure release after the opening of expansion value is not possible as it takes place through some crank rotation.

7.6.2 Working of Four-Stroke Compression-Ignitions Engine

The four cycles of operation of CI engine are shown in Fig. 7.5.

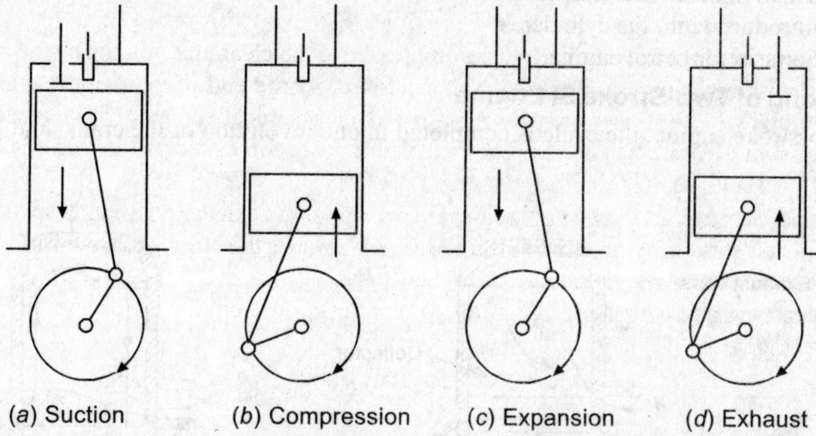

Fig. 7.5 Cycle of Operation of a four stroke C.I. Engine.

***(a)* Suction stroke :** Air alone is induced during the suction stroke with inlet valve open and exhaust-valve closed.

***(b)* Compression Stroke :** The sucked air is compressed into the clearance volume with both valves closed.

***(c)* Expansion Stroke :** Fuel injection starts nearly at the end of the compression stroke resulting into combustion. The products of combustion expand and both the valves remain closed.

***(d)* Exhaust Stroke :** The piston moves from BDC to TDC and pushes out the products of combustion with the exhaust valve open and intake valve closed.

The theoretical and actual *p-V* diagrams for a four stroke diesel engine are shown in Fig. 7.6. The various processes are :

5 – 1 : Suction stroke
1 – 2 : Compression stroke
2 – 3 : Combustion stroke
3 – 4 : Expansion stroke
4 – 1 : Sudden fall in pressure
1 – 5 : Exhaust stroke.

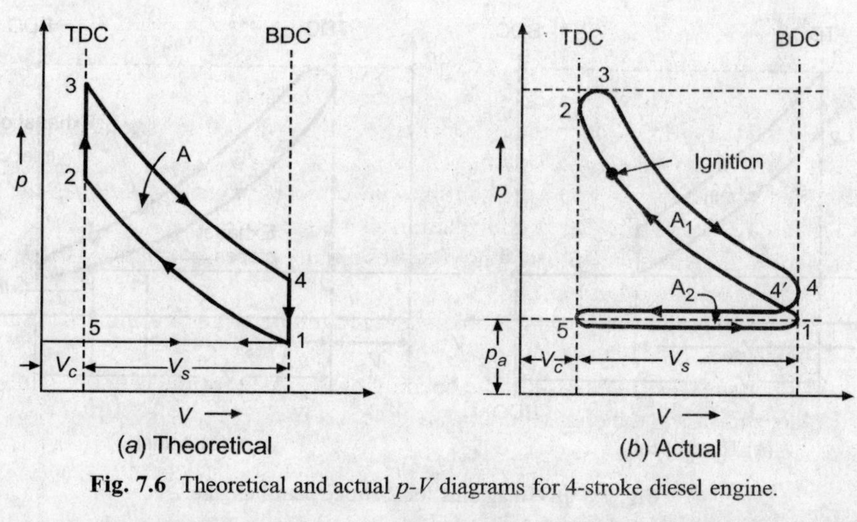

Fig. 7.6 Theoretical and actual *p-V* diagrams for 4-stroke diesel engine.

The actual *p-V* diagram differ from the ideal as described below :

1. Reasons 1, 2 and 3 are the same as for petrol engine.
4. The combustion at constant pressure is not possible as the fuel will not burn as it is introduced into the cylinder.
5. Same as for petrol engine.

7.6.3 Working of Two-Stroke SI Engine

In a two-stroke engine, the cycle is completed in one revolution of the crankshaft. The filling

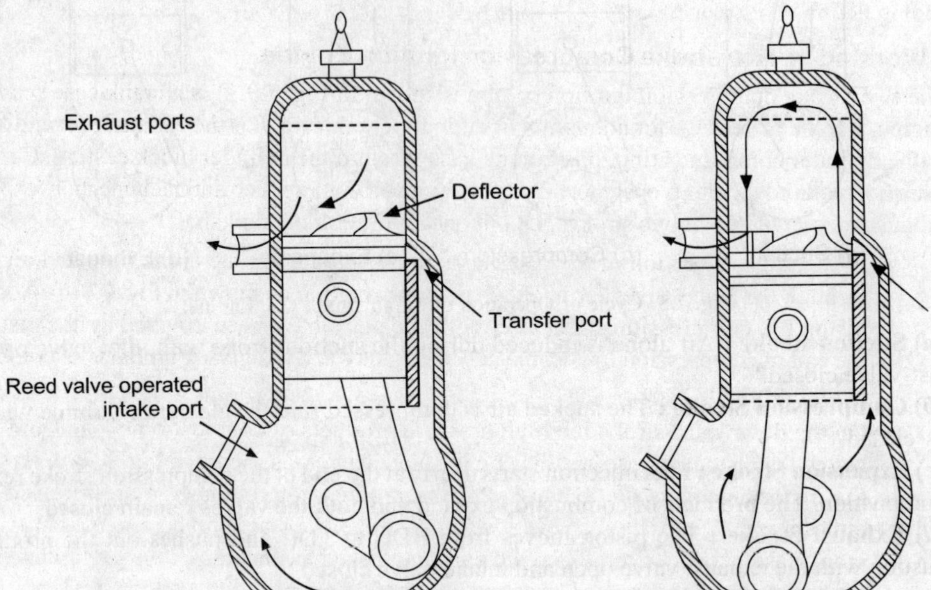

Fig. 7.7 Crankcase-Scavenged Two-Stroke SI Engine.

process is accomplished by the charge compressed in crankcase or by a blower. The induction of the compressed charge moves out the products of combustion through exhaust ports. Therefore, two piston strokes are required for these two operations. Two strokes are sufficient to complete the cycle, one for compressing the fresh charge and the other for expansion or power stroke.

Fig. 7.7 shows the crankcase scavenged two-stroke engine.

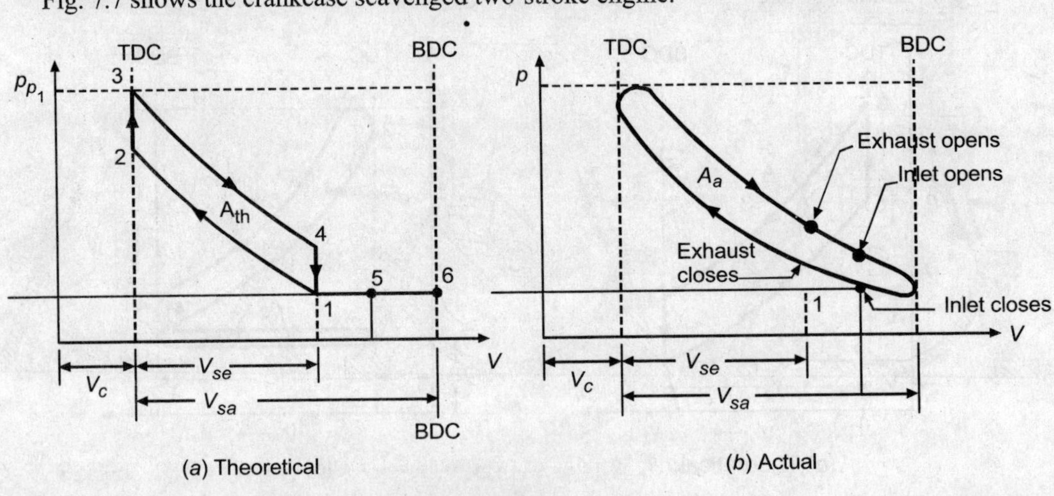

Fig. 7.8 *p-V* diagrams for 2-stroke petrol engine.

The various process are :

Process 1 – 2 : Isentropic compression

Process 2 – 3 : Instantaneous combustion at constant volume

Process 3 – 4 : Isentropic expansion

Process 4 – 1 : Release to atmospheric pressure as the exhaust parts open

Process 1 – 6 & 6 – 1 : Sweeping out the exhaust gases to atmosphere

Process 5 : Opening of the inlet or transfer ports

Process 5 – 6 & 6 – 5 : Charging the cylinder with fresh charge through the transfer ports and scavenging action takes place.

7.6.4 Working of Two stroke Compression Ignition Engine

The two-stroke compression ignition engine is shown in Fig. 7.9. It is a crank case scavenged type engine. S is the plate valve for admission of air in the crank care, *E* is the exhaust port and *A* is the port in the cylinder communicating to the crank case through the cylinder block casting. Cams and valves are not required. The four operations : air induction, air compression and fuel injection, expansion and exhaust are completed in two strokes, *i.e.*, in one revolution of crankshaft.

- In the upward motion of the piston, suction is created in the crank case and air enters through the plate valves for full 180° of crank rotation as shown in Fig. 7.9(*a*). Above the piston the compression starts after both the ports have been covered by the piston. At the end of compression fuel is injected and ignites at TDC giving products of combustion at high pressure.
- On the downward stroke the high pressure products of combustion expand and the air below the piston compresses, closing the plate valves, as shown in Fig. 7.9(*b*). As soon as the piston uncovers the exhaust port the products of combustion are released to the atmosphere. A little latter, on the downward stroke, the other port communicating with the crank case gets uncovered and the air compressed in the crank case gets transferred to the space above the pistons as shown in Fig. 7.9(*c*).

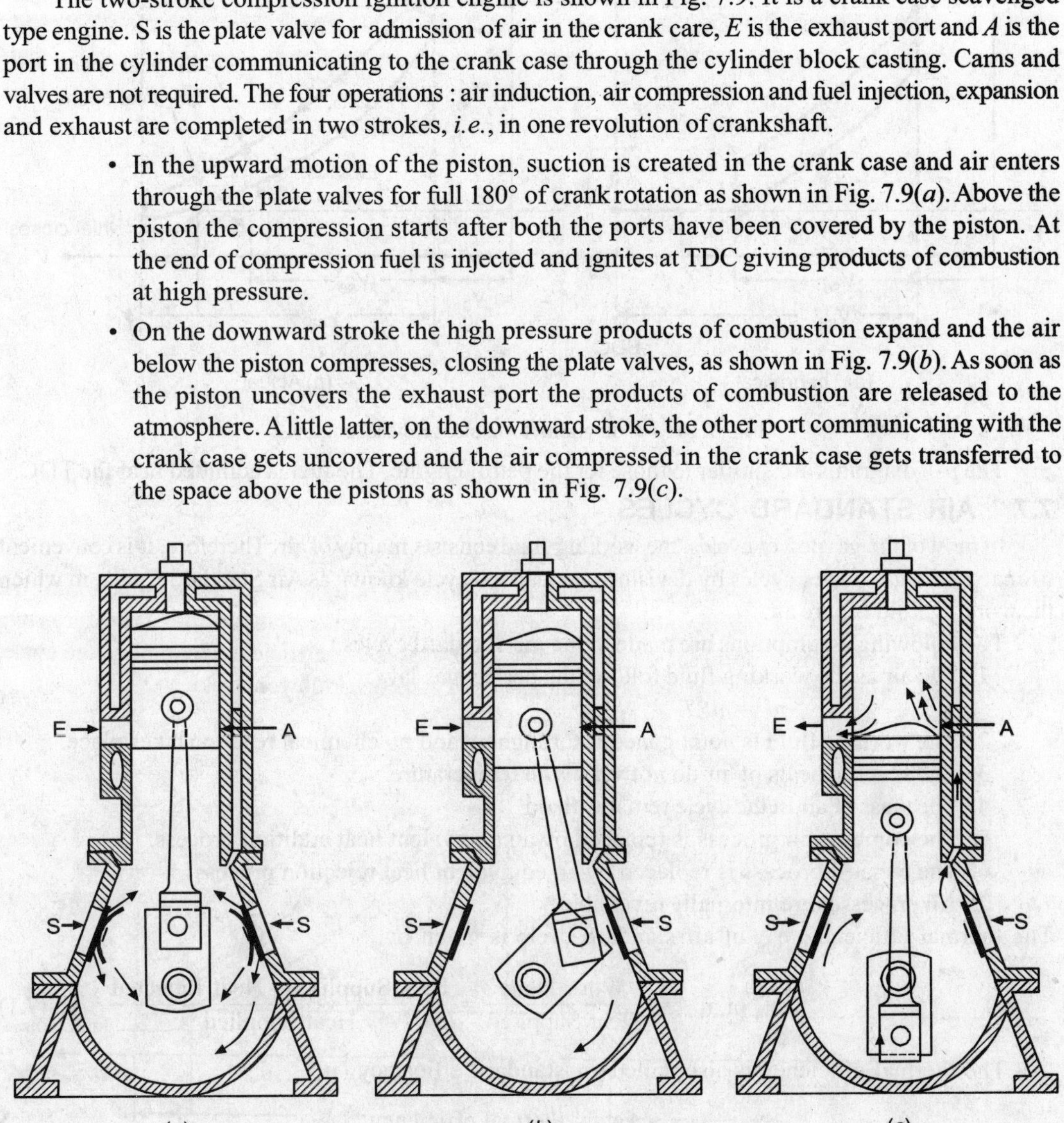

Fig. 7.9 Working of two-stroke C.I. engine.

The *p-V* diagrams for two-stroke diesel engine are shown in Fig. 7.10.

The various processes are :

1 – 2 : Isentropic compression

2 – 3 : Instantaneous combustion at constant pressure

3 – 4 : Isentropic expansion

4 – 1 : Release to atmospheric pressure

1 – 6 & 6 – 1 : Sweeping out of exhaust gases to atmosphere

5 : Inlet or transfer ports open

5 – 6 & 6 – 5 : Charging the cylinder with fresh charge.

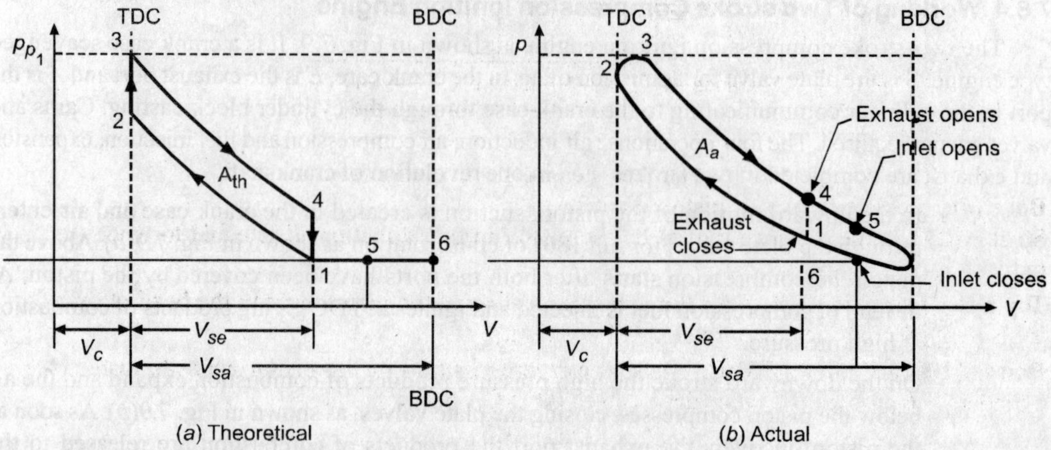

Fig. 7.10 *p-V* diagrams for 2-stroke diesel engine.

The *p-V* diagrams are similar to those for the petrol engine. The fuel is admitted near the TDC.

7.7 AIR STANDARD CYCLES

In most of the gas power cycles, the working fluid consists mainly of air. Therefore, it is convenient to analyse the gas power cycles by devising an idealized cycle known as Air Standard Cycle, in which the working fluid is pure air.

The following assumptions are made in the air-standard cycles :

1. The air as the working fluid follows the perfect gas law :

$$pV = mRT$$

2. The working fluid is homogeneous throughout and no chemical reaction takes place.

3. The specific heats of air do not vary with temperature.

4. The mass of air in the cycle remains fixed.

5. The combustion process is replaced by an equivalent heat addition process.

6. The exhaust process is replaced by an equivalent heat rejection process.

7. All processes are internally reversible.

The thermal efficiency, η_{th}, of air standard cycle is given by

$$\eta_{th} \text{ or } \eta_a = \frac{\text{Work Done}}{\text{Heat Supplied}} = \frac{\text{Heat Supplied} - \text{Heat Rejected}}{\text{Heat Supplied}} \quad \text{... (7.1)}$$

The thermal efficiency is also called air-standard efficiency (η_a)

Relative efficiency, $$\eta_r = \frac{\text{Actual thermal efficiency}}{\text{Air standard efficiency}} \quad \text{... (7.2)}$$

7.7.1 Piston-Cylinder Arrangement

The piston-cylinder arrangement is shown in Fig. 7.11. The various definitions are :

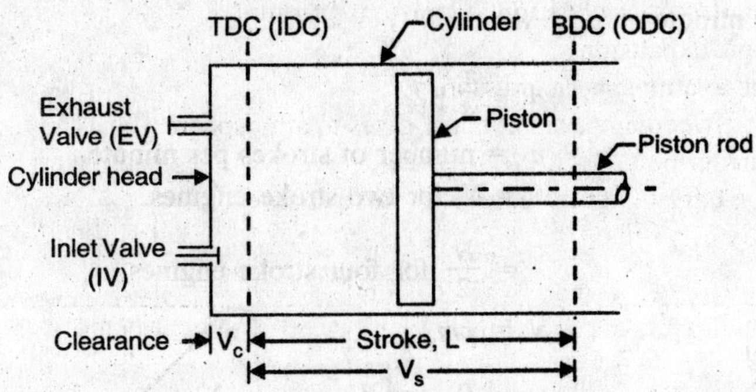

Fig. 7.11 Piston-cylinder arrangement.

Bore (*d*) : is the cylinder or piston diameter.

Stroke (L) : is the distance moved by the piston in one direction. It is equal to twice the crank radius.

Top Dead Centre (T.D.C.) : is the extreme position of the piston near to the head of cylinder. It is also called inner dead centre (I.D.C.) position.

Bottom Dead Centre (B.D.C.) : is the extreme position of the piston opposite to the head of cylinder. It is also called outer dead center (O.D.C.) position.

$$L = \text{T.D.C.} - \text{B.D.C}$$

Clearance Volume (V_c) : is the minimum volume of clearance between the cylinder head and the piston at *T.D.C.* position.

Swept Volume (V_s) : is the maximum volume swept by the piston in moving from T.D.C. to B.D.C. position or vice versa.

$$V_s = \frac{\pi}{4} \cdot d^2 L$$

Cylinder Volume (V) : is the sum of clearance volume and swept volume.

$$V = V_c + V_s$$

Compression ratio (r) : is the ratio of cylinder volume to the clearance volume.

Thus
$$r = \frac{V}{V_c} = \frac{V_c + V_s}{V_c} = 1 + \frac{V_s}{V_c} \quad \text{... (7.3)}$$

Clearance ratio (c) : is the ratio of clearance volume to stroke volume.

Thus
$$c = \frac{V_c}{V_s} \quad \text{... (7.4)}$$

and
$$r = 1 + \frac{1}{c} = \frac{1 + c}{c} \quad \text{... (7.5)}$$

Mean effective pressure (p_m) : is the average pressure which, if acted on the piston during the entire power or outward stroke, would produce the same work output as the net work output for the actual cyclic process. Thus

Work done per cycle,
$$W = \frac{\pi}{4} d^2 L \cdot p_m$$

Or $$p_m = \frac{4W}{\pi\, d^2\, L}$$

Work done per minute = $p_m\, L\, A\, n$

Where $$A = \frac{\pi}{4} d^2$$

n = number of strokes per minute.

= N, for two stroke engines.

= $\frac{N}{2}$ for four stroke engines.

$N = rpm$

Indicated power, $$IP = \frac{p_m\, L\, A\, n}{60} \text{ kW} \qquad \text{... (7.6)}$$

Where, p_m is in kPa.

7.7.2 Otto Cycle (Constant Volume Cycle)

This cycle is composed of four internally reversible processes, two adiabatic and two constant volume processes. The *p-v* and *T-s* diagrams are shown in Fig. 8.6. The various processes are :

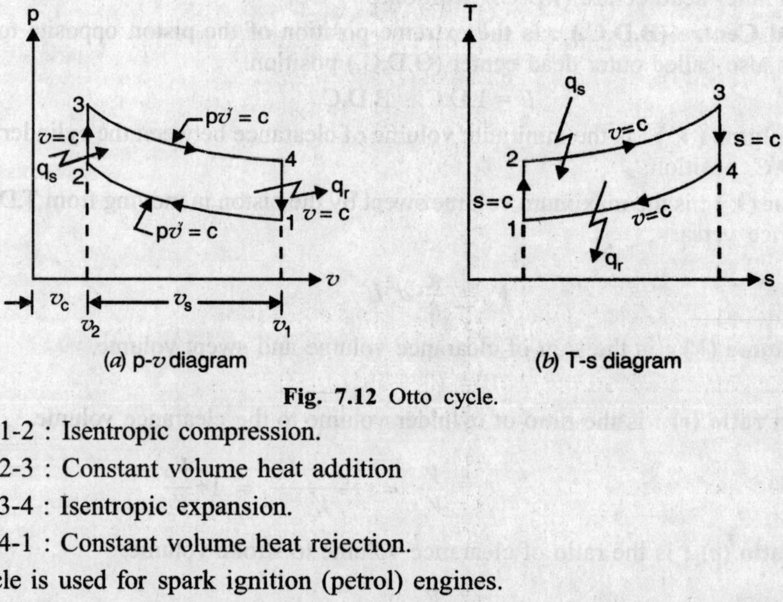

Fig. 7.12 Otto cycle.

Process 1-2 : Isentropic compression.

Process 2-3 : Constant volume heat addition

Process 3-4 : Isentropic expansion.

Process 4-1 : Constant volume heat rejection.

This cycle is used for spark ignition (petrol) engines.

Consider 1 kg of air flowing through the cycle. Since the air in the cylinder act as a closed system, from first law of thermodynamics for isentropic compression and expansion, we have

$$q - w = \Delta u$$

For constant volume heat supplied and rejection processes, since $w = 0$, therefore

$$q = \Delta u = c_v\, \Delta T$$

Heat supplied, $$q_s = q_{2-3} = c_v\, (T_3 - T_2)$$

Heat Rejected, $$q_r = q_{4-1} = c_v\, (T_4 - T_1)$$

Work done per cycle, $$w = q_s - q_r = c_v\, [(T_3 - T_2) - (T_4 - T_1)]$$

Air Standard (or thermal) efficiency = $\frac{\text{Work done}}{\text{Heat supplied}}$

$$\eta_a \text{ or } \eta_{th} = \frac{c_v(T_3 - T_2) - c_v(T_4 - T_1)}{c_v(T_3 - T_2)}$$

$$= 1 - \frac{T_4 - T_1}{T_3 - T_2}$$

Now $$T_2 = \frac{T_1 V_1}{V_2} = T_1 r^{\gamma-1}$$

Where $$r = \frac{V_1}{V_2} = \frac{V_1}{V_c} = \text{compression ratio}$$

$$T_3 = T_2 \frac{p_3}{p_2} = \alpha_p T_2 = \alpha_p T_1 r^{\gamma-1}$$

Where $$\alpha_p = \frac{p_3}{p_2} = \frac{T_3}{T_2} = \text{pressure ratio}$$

$$T_4 = T_3 \left(\frac{1}{r}\right)^{\gamma-1} = \alpha_p T_1$$

∴ Efficiency, $$\eta_a = \eta_{th} = 1 - \frac{\alpha_p T_1 - T_1}{\alpha_p T_1 r^{\gamma-1} - T_1 r^{\gamma-1}}$$

$$= 1 - \frac{\alpha_p - 1}{(\alpha_p - 1) r^{\gamma-1}}$$

$$= 1 - \frac{1}{r^{\gamma-1}} \quad \text{... (7.7)}$$

The air standard efficiency of Otto cycle depends on compression ratio only and increases as compression ratio increases (Fig. 7.13). In actual engine working on Otto cycle, the compression ratio varies from 5 to 8. This engine is used for spark ignition engines working on petrol.

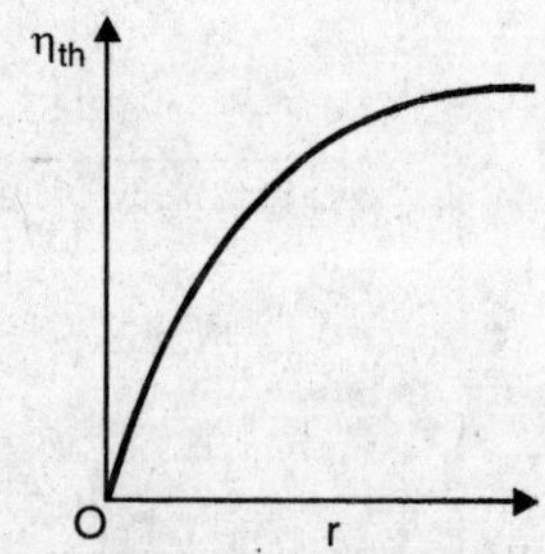

Fig. 7.13 Thermal efficiency v's compression ratio for Otto cycle.

Mean Effective Pressure (M.E.P.) : It may be defined as the ratio of work done to the displacement volume of piston.

M.E.P., $$p_m = \frac{W}{v_s}$$

$$= \frac{c_v[(T_3 - T_2) - (T_4 - T_1)]}{v_1 - v_2}$$

7.7.3 Diesel Cycle

This cycle is used for compression ignition internal combustion engines working on diesel oil. The p–v and T–s diagrams are shown in Fig. 7.14. It consists of four internally reversible processes; two adiabatic, one constant pressure and one constant volume. The various processes are :

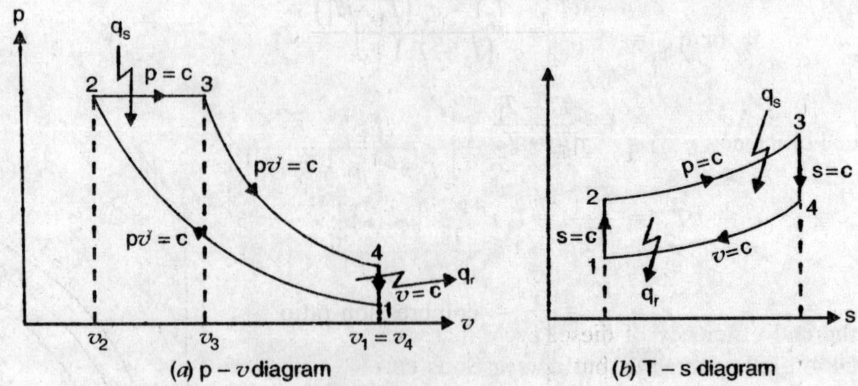

Fig. 7.14 Diesel cycle.

Process 1-2 : Isentropic compression of air.

Process 2-3 : Heat addition at constant pressure.

Process 3-4 : Isentropic expansion of air.

Process 4-1 : Heat rejection at constant volume.

Considering 1 kg of air.

Heat supplied, $q_s = q_{2-3} = c_p (T_3 - T_2)$

Heat rejected, $q_r = q_{4-1} = c_v (T_4 - T_1)$

Work done per cycle, $w = q_s - q_r$

$$= c_p (T_3 - T_2) - c_v (T_4 - T_1)$$

Thermal efficiency, $$\eta_{th} = \frac{\text{Work done}}{\text{Heat supplied}}$$

$$= \frac{w}{q_s}$$

$$= \frac{c_p (T_3 - T_2) - c_v (T_4 - T_1)}{c_p (T_3 - T_2)}$$

$$= 1 - \frac{1}{\gamma}\left[\frac{T_4 - T_1}{T_3 - T_2}\right]$$

Let r $$= \frac{v_1}{v_2} = \frac{v_4}{v_2} = \text{compression ratio}$$

Now $$T_2 = T_1 \left(\frac{v_1}{v_2}\right)^{\gamma-1} = T_1\, r^{\gamma-1}$$

Let $$\rho = \frac{v_3}{v_2} = \text{cut-off ratio}$$

$$\rho = \frac{v_3}{v_2} = \frac{T_3}{T_2}$$

or $$T_3 = T_2\, \rho = T_1\, r^{\gamma-1}\, \rho$$

$$T_4 = T_3 \left(\frac{v_3}{v_4}\right)^{\gamma-1} = T_3 \left(\frac{v_3}{v_2} \times \frac{v_2}{v_4}\right)^{\gamma-1} = T_3 \left(\frac{\rho}{r}\right)^{\gamma-1}$$

$$= T_1 \, r^{\gamma-1} \, \rho \left(\frac{\rho}{r}\right)^{\gamma-1} = T_1 \, \rho^{\gamma}$$

Thermal efficiency,
$$\eta_{th} = 1 - \frac{1}{\gamma}\left[\frac{T_1 \rho^{\gamma} - T_1}{T_1 r^{\gamma-1} \rho - T_1 r^{\gamma-1}}\right]$$

$$= 1 - \frac{1}{r^{\gamma-1}} \frac{\left[\rho^{\gamma} - 1\right]}{\gamma\left(\rho - 1\right)} \quad \text{... (7.9)}$$

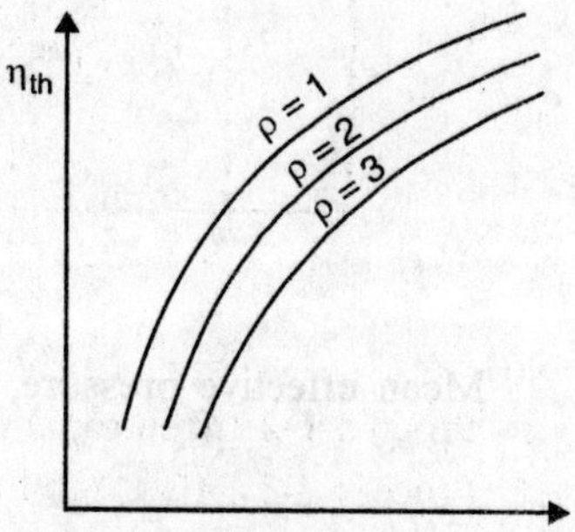

Fig. 7.15 Variation of diesel cycle efficiency.

The thermal efficiency of diesel cycle increases as compression ratio increases but decreases as cut-off ratio increases (Fig 7.15). The thermal efficiency of diesel cycle is less than that of Otto cycle. The compression ratio for diesel cycle varies from 14 to 18.

Mean effective pressure,

$$p_m = \frac{\text{Work done}}{\text{Swept volume}}$$

$$= \frac{c_p\,(T_3 - T_2) - c_v\,(T_4 - T_1)}{v_1 - v_2} \quad \text{...(7.10)}$$

7.7.4 Dual Cycle

In the dual cycle, part of the heat is supplied at constant volume and the rest at constant pressure. This is also called mixed or limited pressure cycle. The p–v and T–s diagrams are shown in Fig. 7.16. The various processes are :

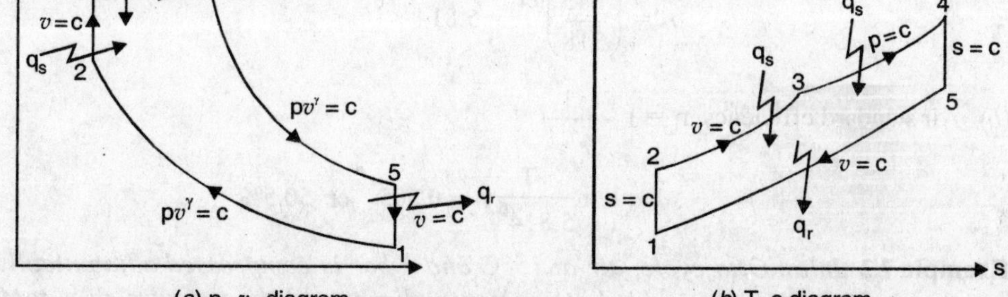

(a) p –v diagram (b) T–s diagram

Fig. 7.16 Dual cycle.

Process 1-2 : Isentropic compression of air.
Process 2-3 : Constant volume heat addition.
Process 3-4 : Constant pressure heat addition.
Process 4-5 : Isentropic expansion of air.
Process 5-1 : Constant volume heat rejection

$$\eta_{th} = 1 - \frac{\left(\alpha_p \, \rho^{\gamma-1}\right)}{\left(\alpha_p \, r^{\gamma-1} - r^{\gamma-1}\right) + \left(\alpha_p \rho \, r^{\gamma-1} - \alpha_p \, r^{\gamma-1}\right)} \quad \text{... (7.11)}$$

The variation of thermal efficiency of dual cycle is shown in Fig. 7.17

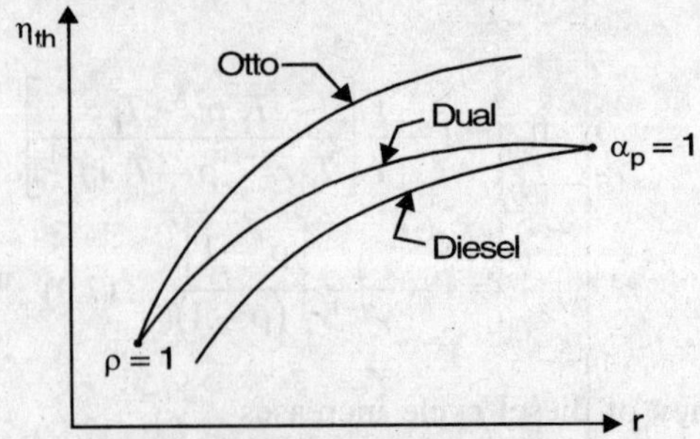

Fig. 7.17 Dual cycle efficiency variation.

Mean effective pressure,

$$p_m = \frac{\text{Work done}}{\text{Swept volume}} = \frac{w}{v_1 - v_2} \qquad \text{... (7.12)}$$

Example 7.1 *In an engine working on ideal Otto cycle, the temperature at the beginning and end of compression are 45° C and 370° C respectively. Find the compression ratio and air-standard efficiency of the engine. Assume* $\gamma = 1.4$.

Solution.

Compression ratio, $$r = \frac{v_1}{v_2} = \left(\frac{T_2}{T_1}\right)^{\frac{1}{\gamma-1}}$$

$$T_1 = 45 + 273 = 318\ K$$
$$T_2 = 370 + 273 = 643\ K$$

$$r = \left(\frac{643}{318}\right)^{\frac{1}{0.4}} = 5.814$$

(*b*) Air standard efficiency, $\eta_a = 1 - \frac{1}{r^{\gamma-1}}$

$$= 1 - \frac{1}{5.814^{0.4}} = 0.505 \quad \text{or } 50.5\%$$

Example 7.2 *In an Otto cycle, air at 15°C and 1 bar is compressed adiabatically until the pressure is 15 bar. Heat is added at constant volume until the pressure rises to 40 bar. Calculate (a) the air-standard efficiency, (b) the compression ratio, and (c) the mean effective pressure for the cycle. Assume* $c_V = 0.718$ *kJ/kg.K,* $\gamma = 1.4$ *and* $R = 8.314$ *kJ/kmol.K.*

Solution.

The Otto cycle is shown in Fig. 7.18.

For isentropic process 1-2,

$$p_1 V_1^{\gamma} = p_2 V_2^{\gamma}$$

Compression ratio, $$r = \frac{V_1}{V_2} = \left(\frac{p_2}{p_1}\right)^{\frac{1}{\gamma}} = 15^{\frac{1}{1.4}} = 6.92$$

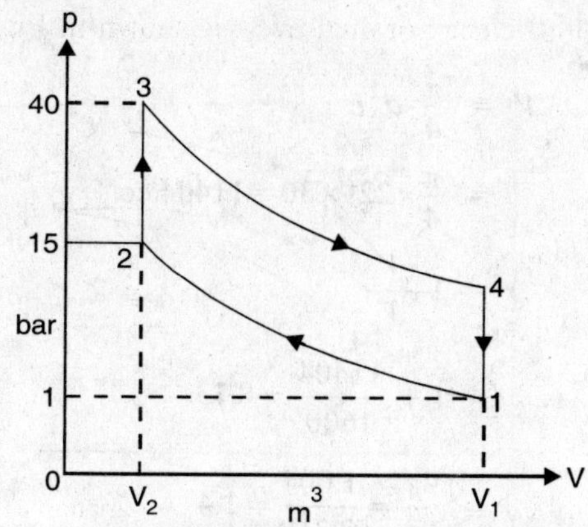

Fig. 7.18

Air standard efficiency, $\eta_a = 1 - \frac{1}{r^{\gamma-1}}$

$$= 1 - \frac{1}{(6.92)^{0.4}} = 0.539 \quad \text{or} \quad 53.9\%$$

$$T_1 = 15 + 273 = 288 \text{ K}$$

$$T_2 = \left(\frac{p_2 V_2}{p_1 V_1}\right) \cdot T_1 = \frac{15 \times 288}{6.92} = 624.3 \text{ K}$$

For constant volume process 2-3 :

$$T_3 = \frac{p_3}{p_2} T_2 = \frac{40}{6.92} \times 624.3 = 1664.7 \text{ K}$$

Heat supplied $q_s = c_v (T_3 - T_2)$

$$= 0.718 (1664.7 - 624.3) = 747.03 \text{ kJ/kg}$$

Work done, $w = \eta_a q_s$

$$= 0.539 \times 747.03 = 402.65 \text{ kJ/kg}$$

Mean effective pressure, $p_m = \frac{\text{Work done}}{\text{Swept volume}} = \frac{w}{v_1 - v_2}$

Specific Volume, $v_1 = \frac{m\,RT_1}{p_1} = \frac{8314 \times 288}{29 \times 1 \times 10^5} = 0.8257 \text{ m}^3/\text{kg}$

$$v_2 = \frac{v_1}{r} = \frac{0.8257}{6.92} = 0.1193 \text{ m}^3/\text{kg}$$

$$v_1 - v_2 = 0.7064 \text{ m}^3/\text{kg}$$

$$p_m = \frac{402.65}{0.7064} \times 10^3 = 5.7 \times 10^5 \text{ N/m}^2 \quad \text{or} \quad 5.7 \text{ bar}$$

Example 7.3 *A gas engine working on Otto cycle has a cylinder of diameter 220 mm and stroke 300 mm. The clearance volume is 1600 cc. Find the air-standard efficiency. Assume c_p = 1.004 kJ/kg.K and c_v = 0.718 kJ/kg.K for air.*

Solution.

Swept volume, $V_s = \frac{\pi}{4} \cdot d^2 L$

$= \frac{\pi}{4} \times 22^2 \times 30 = 11404$ cc

Compression ratio, $r = 1 + \frac{V_s}{V_c}$

$= 1 + \frac{11404}{1600} = 8.13$

Adiabatic index, $\gamma = \frac{c_p}{c_v} = \frac{1.004}{0.718} = 1.4$

Air standard efficiency, $\eta_a = 1 - \frac{1}{r^{\gamma-1}} = 1 - \frac{1}{(8.13)^{0.4}} = 0.5675$ or 56.75%

Example 7.4 *A gas engine working on ideal Otto cycle has a compression ratio of 6.1. The pressure and temperature at the commencement of compression are 1 bar and 27°C. The heat supplied during the constant volume combustion process is 1200 kJ/kg. Determine the peak pressure and temperature, work output per kg of air and air-standard efficiency. Assume* c_v = *0.718 kJ/kg.K and* $\gamma = 1.4$ *for air.*

Solution.

Isentropic compression process 1-2 :

$$\frac{p_2}{p_1} = \left(\frac{v_1}{v_2}\right)^{\gamma} = r^{\gamma} = 6^{1.4} = 12.28$$

$$p_2 = 1 \times 12.28 = 12.28 \text{ bar} = 12.28 \times 10^5 \ Pa$$

$$\frac{T_2}{T_1} = \left(\frac{v_1}{v_2}\right)^{\gamma-1} = r^{\gamma-1} = 6^{0.4} = 2.048$$

$$T_2 = 2.048 \times 300 = 614.3 \text{ K} = 341.3°\text{C}$$

Constant Volume heat addition process 2-3 :

$$q_s = q_{2\text{-}3} = c_v (T_3 - T_2)$$

$$1200 = 0.718 (T_3 - 614.3)$$

$$T_3 = 2285.6 \text{ K}$$

$$\frac{p_3}{p_2} = \frac{T_3}{T_2} = \frac{2285.6}{614.3} = 3.72$$

$$p_3 = 12.28 \times 10^5 \times 3.72 = 45.67 \times 10^5 \ Pa$$

$$= 45.67 \text{ bar}$$

Work output, w = area 1-2-3-4

$$= \frac{1}{\gamma - 1} [(p_3 v_3 - p_4 v_4) - (p_2 v_2 - p_1 v_1)]$$

$$= \frac{R}{\gamma - 1} [(T_3 - T_4) - (T_2 - T_1)]$$

$$\frac{T_3}{T_4} = \left(\frac{v_3}{v_4}\right)^{\gamma-1} = r^{\gamma-1} = 6^{0.4} = 2.048$$

$$R = c_p - c_v = 1.004 - 0.718 = 0.286 \text{ kJ/kg.K}$$

$$w = \frac{0.286}{0.4} [(2285.6 - 1116.0) - (614.3 - 300)]$$

$$w = 611.54 \text{ kJ/kg}$$

Air standard efficiency, $\eta_a = 1 - \frac{1}{r^{\gamma-1}}$

$$= 1 - \frac{1}{6^{0.4}} = 0.5116 \quad \text{or} \quad 51.16\%$$

Example 7.5 *The pressure limit in a Otto cycle are 1 bar and 20 bar. The compression ratio is 5. Calculate (a) thermal efficiency, and (b) mean effective pressure. Assume γ = 1.4 for air.*

Solution.

(*a*) $\eta_{th} = 1 - \frac{1}{r^{\gamma-1}}$

$$= 1 - \frac{1}{5^{0.4}} = 0.4747 \quad \text{or} \quad 47.47\%$$

(*b*) $\frac{p_2}{p_1} = \left(\frac{v_1}{v_2}\right)^{\gamma} = r^{\gamma} = 5^{1.4} = 9.5182$

$$p_2 = 1 \times 9.5182 = 9.5182 \text{ bar}$$

$$p_3 = 20 \text{ bar}$$

Pressure Ratio, $\alpha_p = \frac{p_3}{p_2} = \frac{20}{9.5182} = 2.1$

$$p_m = p_1 \, r \left[\frac{(\alpha_p - 1)(r^{\gamma-1} - 1)}{(r - 1)(\gamma - 1)}\right]$$

$$= 1 \times 5 \left[\frac{(2.1 - 1)(5^{0.4} - 1)}{(5 - 1)(1.4 - 1)}\right]$$

$$= 3.1 \text{ bar}$$

Example 7.6 *An air-standard Otto cycle has a compression ratio of 6. The temperature at the start of compression process is 25°C and the pressure is 1 bar. If the maximum temperature of the cycle is 1150°C, calculate (a) the heat supplied per kg of air, (b) network done per kg* *of air, and (c) thermal efficiency of cycle. Assume γ =1.4, c_v = 0.778 kJ/kg.K for air.*

Solution.

Process 1-2 :

$$\frac{T_2}{T_1} = \left(\frac{v_1}{v_2}\right)^{\gamma-1} = 6^{0.4} = 2.0476$$

$$T_1 = 25 + 273 = 298 \text{ K}$$

$$T_2 = 298 \times 2.0476 = 610.2 \text{ K}$$

Process 3-4 :

$$\frac{T_4}{T_3} = \left(\frac{v_3}{v_4}\right)^{\gamma-1} = \left(\frac{1}{6}\right)^{0.4} = 0.4884$$

$$T_3 = 1150 + 273 = 1423 \text{ K}$$

$$T_4 = 1423 \times 0.4884 = 694.9 \text{ K}$$

(*a*) Heat supplied, $q_s = c_v (T_3 - T_2) = 0.718\ (1423 - 610.2) = 583.59$ kJ/kg

(*b*) Heat Rejected, $q_r = c_v (T_4 - T_1) = 0.718\ (694.9 - 298) = 284.97$ kJ/kg

Work done per cycle, $w = q_s - q_r = 583.59 - 284.97 = 298.62$ kJ/kg.

Thermal Efficiency, $\eta_{th} = \frac{w}{q_s} = \frac{298.62}{583.59} = 0.5117$ or 51.17 %

Example 7.7 *A petrol engine with compression ratio of 5 develops 20 kW indicated power and consumes 8 litres of fuel per hour. The specific gravity of fuel is 0.78 and its calorific value is 44 MJ/kg. Calculate the indicated thermal efficiency and relative efficiency. Take γ = 1.4.*

Solution.

Air standard efficiency, $\eta_a = 1 - \frac{1}{r^{\gamma-1}}$

$$= 1 - \frac{1}{5^{0.4}} = 0.4747 \quad \text{or} \quad 47.47\%$$

Fuel consumption $= 8 \times 0.78 \times 1 = 6.24$ kg/h

$$= \frac{6.24}{3600} = 1.733 \times 10^{-3} \text{ kg/s}$$

Indicated thermal efficiency,

$$(\eta_{th})_i = \frac{\text{Indicated Work}}{\text{Heat Supplied}}$$

$$= \frac{20}{44 \times 10^3 \times 1.733 \times 10^{-3}} = 0.2622$$

$$= 26.22\ \%$$

$$\text{Relative efficiency} = \frac{(\eta_{th})_i}{\eta_a} = \frac{0.2622}{0.4747} = 0.5524 \quad \text{or} \quad 55.24\%$$

Example 7.8 *A diesel engine has a compression ratio of 20 and cut-off takes place at 5% of the stroke. Find the air-standard efficiency. Assume γ = 1.4.*

Solution.

$$\frac{V_1}{V_2} = r = 20$$
$$V_1 = 20\,V_2$$
$$V_s = V_1 - V_2 = 20\,V_2 - V_2 = 19\,V_2$$
$$V_3 = 0.05\,V_s + V_2 = 0.05 \times 19\,V_2 + V_2$$
$$= 1.95\,V_2$$

Cut-off ratio,

$$\rho = \frac{V_3}{V_2} = 1.95$$

$$\eta_a = 1 - \frac{1}{r^{\gamma-1}}\left[\frac{(\rho^\gamma - 1)}{\gamma\,(\rho - 1)}\right]$$

$$\eta_a = 1 - \frac{1}{20^{0.4}}\left[\frac{(1.95^{1.4} - 1)}{1.4\,(1.95 - 1)}\right]$$

$$= 0.649 \quad \text{or} \quad 64.9\ \%$$

Example 7.9 *In an engine working on diesel cycle, inlet pressure and temperature are 1 bar and 20°C. Pressure at the end of adiabatic compression is 40 bar. The ratio of expansion after constant pressure heat addition is 5. Calculate the heat supplied, heat rejected and the efficiency of the cycle. Assume* c_p *= 1.004 kJ/kg.K and* c_v *= 0.717 kJ/kg.K.*

Solution.

Adiabatic index,
$$\gamma = \frac{c_p}{c_v} = \frac{1.004}{0.717} = 1.4$$

Process 1-2 :
$$r = \frac{V_1}{V_2} = \left(\frac{p_2}{p_1}\right)^{\frac{1}{\gamma}} = (40)^{\frac{1}{1.4}} = 13.942$$

Cut-off ratio,
$$\rho = \frac{V_3}{V_2} = \frac{V_3}{V_1} \times \frac{V_1}{V_2} = \frac{13.942}{5} = 2.788$$

$$\frac{T_2}{T_1} = \left(\frac{p_2}{p_1}\right)^{\frac{\gamma-1}{\gamma}} = (40)^{\frac{1.4-1}{1.4}} = 40^{0.286} = 2.872$$

$$T_2 = 2.872 \times 293 = 841.5\ \text{K}$$

Process 2-3 :
$$T_3 = \frac{T_2 V_3}{V_2} = 841.5 \times 2.788 = 2346.1\ \text{K}$$

Process 3-4 :
$$T_4 = T_3\left(\frac{V_3}{V_2}\right)^{\gamma-1} = T_3\left(\frac{V_3}{V_1}\right)^{\gamma-1}$$

$$= 2346.1\left(\frac{1}{5}\right)^{0.4} = 1232.4 \text{ K}$$

Heat supplied, $q_s = c_p(T_3 - T_2) = 1.004\,(2346.1 - 841.5)$

$= 1510.62$ kJ/kg

Heat rejected, $q_r = c_v(T_4 - T_1) = 0.717\,(1232.4 - 293)$

$= 673.55$ kJ/kg

$$\eta_a = \frac{q_s - q_r}{q_s} = 1 - \frac{673.55}{1510.62} = 0.5541 \quad \text{or} \quad 55.41\%$$

Example 7.10 *A diesel cycle operates at a pressure of 1 bar at the beginning of compression and the volume is compressed to 1/15th of the initial volume. Heat is supplied until the volume is twice that of the clearance volume. Calculate the mean effective pressure of the cycle. Assume $\gamma = 1.4$.*

Solution.

$r = 15$

$V_1 = 15\,V_2$

$V_3 = 2\,V_2$

Swept Volume, $V_s = V_1 - V_2$

$= (r - 1)\,V_2$

$= (15 - 1)\,V_2 = 14\,V_2$

or $V_2 = \dfrac{V_s}{14}$

Process 1-2 :

$$p_3 = p_1\left(\frac{V_1}{V_2}\right)^{\gamma} = 1 \times (15)^{1.4} = 44.31 \text{ bar}$$

$$p_2 = p_3 = 44.31 \text{ bar}$$

$$p_4 = p_3\left(\frac{V_3}{V_4}\right)^{\gamma} = p_3\left(\frac{V_3}{V_1}\right)^{\gamma} = p_3\left(\frac{V_3}{V_2} \times \frac{V_2}{V_1}\right)^{\gamma}$$

$$= 44.31\left(\frac{2}{15}\right)^{1.4} = 2.64 \text{ bar}$$

$$p_m = \frac{1}{V_s}\left[p_2\,(V_3 - V_2) + \frac{p_3 V_3 - p_4 V_4}{\gamma - 1} - \frac{p_2 V_2 - p_1 V_1}{\gamma - 1}\right]$$

$$= \frac{V_2}{V_s}\left[p_2\left(\frac{V_3}{V_2} - 1\right) + \frac{p_3 \dfrac{V_3}{V_2} - p_4 \dfrac{V_4}{V_2}}{\gamma - 1} - \frac{p_2 - p_1 \dfrac{V_1}{V_2}}{\gamma - 1}\right]$$

$$= \frac{1}{14}\left[44.31\,(2 - 1) + \frac{44.31 \times 2 - 2.64 \times 15}{0.4} - \frac{44.31 - 1 \times 15}{0.4}\right]$$

$$= 0.071\ [44.31 + 122.55 - 73.27]$$
$$= 6.685 \text{ bar}$$

Example 7.11 *The mean effective pressure of an ideal diesel cycle is 10 bar. If the initial pressure is 1 bar and the compression ratio is 14, determine the cut off ratio and the air – standard efficiency. Assume γ = 1.4.*

Solution.

$$\text{Work output} = p_m\, v_s$$
$$= p_2\,(v_3 - v_2) + \frac{p_3\, v_3 - p_4\, v_4}{\gamma - 1} - \frac{p_2\, v_2 - p_1\, v_1}{\gamma - 1}$$
$$r = \frac{v_1}{v_2} = 1 + \frac{v_s}{v_c} = 14$$
$$v_s = 13 v_c$$

Now $v_2 = v_c$

$\therefore$ $v_s = 13 v_2$

and $v_4 = v_1 = 14 v_2 = 14 v_c$

$$v_3 = \rho v_2 = \rho v_c$$
$$\frac{p_2}{p_1} = r^{\gamma} = 14^{1.4} = 40.23$$
$$p_2 = 1 \times 40.23 = 40.23 \text{ bar}$$
$$\frac{p_3}{p_4} = \left(\frac{r}{\rho}\right)^{\gamma} = \left(\frac{14}{\rho}\right)^{1.4} = \frac{40.23}{\rho^{1.4}}$$
$$p_4 = 40.23 \times \frac{\rho^{1.4}}{40.23} = \rho^{1.4}$$

Work done per cycle $= \dfrac{p_2\,(v_3 - v_2) + (p_3 v_3 - p_4 v_4)}{\gamma - 1} - \dfrac{p_2 v_2 - p_1 v_1}{\gamma - 1}$

$$= 40.23\,(\rho v_c - v_c) + \frac{40.23 \times \rho v_c - \rho \times 14\, v_c}{0.4} - \frac{40.23 v_c - 1 \times 14\, v_c}{0.4}$$
$$p_m v_s = v_c\left[40.23(\rho - 1) + \frac{1}{0.4}(40.23\,\rho - 14\,\rho^{1.4}) - \frac{1}{0.4}(40.23 - 14)\right]$$
$$10 \times 13\ v_c = v_c\,[40.23\,\rho - 40.23 + 100.6\rho - 35\,\rho^{1.4} - 100.6 + 35]$$

$130 = 140.83\,\rho - 35\,\rho^{1.4} - 105.83$

or $f(\rho) = \rho^{1.4} - 4.024\,\rho + 6.738 = 0$

ρ	2	2.5	2.6	2.64
$f(\rho)$	1.329	0.285	0.086	0.007

We take $\rho = 2.64$

$$\eta = 1 - \frac{1}{r^{\rho-1}}\left[\frac{\rho^{\gamma}-1}{\gamma(\rho-1)}\right]$$

$$= 1 - \frac{1}{14^{0.4}}\left[\frac{2.64^{1.4}}{1.4\times 1.64}\right]$$

$$= 0.5616 \quad \text{or} \quad 56.16\%$$

Example 7.12 *In an engine working on the diesel cycle, the air-fuel by weight is 50:1. The temperature of air at the beginning of combustion is 40C and the compression ratio is 15. What is the ideal efficiency of the engine. Calorific value of fuel is 42 MJ/kg.*

Assume. $c_v = 0.717$ kJ/kgK, $c_p = 1.004$ kJ/kgK.

Solution.

$$\gamma = \frac{c_p}{c_v} = \frac{1.004}{0.717} = 1.4$$

$$\rho = \frac{V_3}{V_2} = \frac{T_3}{T_2}$$

Process 1-2 :

$$\frac{T_2}{T_1} = \left(\frac{V_1}{V_2}\right)^{\gamma-1} = r^{\gamma-1} = 15^{0.4} = 2.954$$

$$T_2 = 924.66 \text{ K}$$

Process 2-3 :

$$q_s = c_p(T_3 - T_2) = \frac{F}{A} \times \text{C.V.}$$

$$T_3 - T_2 = \frac{42000}{1.004 \times 50} = 836.66 \text{ K}$$

$$T_3 = 1761.32 \text{ K}$$

$$\rho = \frac{1761.32}{924.66} = 1.905$$

$$\eta = 1 - \frac{1}{r^{\gamma-1}}\left[\frac{\rho^{\gamma-1}}{\gamma(\rho-1)}\right]$$

$$= 1 - \frac{1}{15^{0.4}}\left[\frac{1.905^{1.4}-1}{1.4 \times 0.905}\right] = 0.6085 \quad \text{or} \quad 60.85\%$$

Example 7.13 *An air standard diesel cycle has a compression ratio of 16. The pressure at the beginning of compression stroke is 1 bar and the temperature is 20C. The maximum temperature is 1430C. Determine the thermal efficiency and the mean effective pressure for this cycle. Take $\gamma = 1.4$.*

Solution.

$$r = \frac{V_1}{V_2} = 16, \; p_1 = 1 \text{ bar}, \; T_1 = 20 + 273 = 293 \text{ K}$$

$$T_3 = 1430 + 273 = 1703 \text{ K}$$

$$\frac{T_2}{T_1} = \left(\frac{V_1}{V_2}\right)^{\gamma-1} = r^{\gamma-1} = 16^{0.4} = 3.03$$

$$T_2 = 293 \times 3.03 = 887.79 \text{ K}$$

Process 2-3 :

$$\rho = \frac{V_3}{V_2} = \frac{T_3}{T_2} = \frac{1703}{887.79} = 1.918$$

$$\eta_{th} = 1 - \frac{1}{\gamma r^{\gamma-1}}\left[\frac{\rho^{\gamma} - 1}{\rho - 1}\right]$$

$$= 1 - \frac{1}{1.4 \times 16^{0.4}}\left[\frac{1.918^{0.4} - 1}{1.918 - 1}\right] = 0.618 \quad \text{or} \quad 61.8\%$$

$$p_m = p_1 r\left[\frac{\gamma r^{\gamma-1}(\rho - 1) - (\rho^{\gamma} - 1)}{(r - 1)(\gamma - 1)}\right]$$

$$= 1 \times 16\left[\frac{1.4 \times 16^{0.4}(1.918 - 1) - (1.918^{1.4} - 1)}{(16 - 1)(1.4 - 1)}\right]$$

$$= 6.414 \text{ bar}$$

Example 7.14 *A diesel engine has a compression ratio of 16. The fuel is cut-off at 0.08 stroke. The relative efficiency is 52%. Find the mass of the fuel of calorific value 41800kJ/kg, which would be required per kW per hour.*

Solution.

$$r = \frac{V_1}{V_2} = 16, \quad V_s = V_1 - V_2$$

$$V_3 - V_2 = 0.08\, V_s = 0.08\,(V_1 - V_2)$$

$$\frac{V_3}{V_2} - 1 = 0.08\left(\frac{V_1}{V_2} - 1\right) = 0.08\,(16 - 1)$$

$$\rho = \frac{V_3}{V_2} = 1 + 0.08 \times 15 = 2.2$$

$$\eta_a = 1 - \frac{1}{\gamma r^{\gamma-1}}\left[\frac{\rho^{\gamma} - 1}{\rho - 1}\right]$$

$$= 1 - \frac{1}{1.9 \times 16^{0.9}}\left[\frac{2.2^{1.4} - 1}{2.2 - 1}\right] = 0.5682 \quad \text{or} \quad 56.82\ \%$$

Actual thermal efficiency

$$= \frac{\text{Actual Work Done}}{\text{Heat Added}}$$

$$= \eta_r \times \eta_a = 0.52 \times 0.5682 = 0.2955$$

$$= \frac{\text{Work Done}}{C.V. \times m_f} = \frac{\text{Work Done}}{41800 \times m_f}$$

$$m_f = \frac{3600}{0.295 \times 41800} = 0.2915 \text{ kg/kWh}$$

Example 7.15 *An engine works on dual combustion cycle, the compression ratio being 11. The pressure at the commencement of combustion is 1 bar and the temperature is 90C. The maximum pressure in the cycle is 50 bar and the constant pressure heat addition continues for 1/20 of the stroke. Calculate the work done per kg. of air and the ideal thermal efficiency. Assume c_v = 0.718 kJ/kg K, c_p = 1.005 kJ/kgK.*

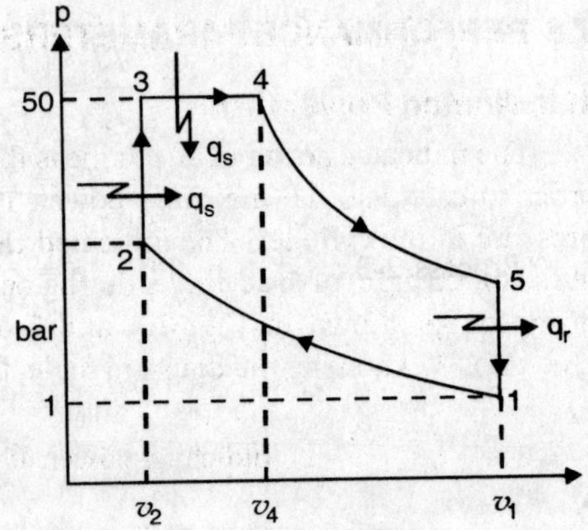

Fig. 7.19

Solution.

$$\gamma = \frac{c_p}{c_v} = \frac{1.005}{0.718} = 1.4$$

$$T_1 = 90 + 273 = 363 \text{ K}$$

Refer to Fig. 7.19.

$$r = \frac{v_1}{v_2} = 11$$

$$v_s = v_1 - v_2 = 10v_2$$

$$v_4 - v_3 = \frac{1}{20}(v_1 - v_2)$$

$$= \frac{v_2}{20}\left(\frac{v_1}{v_2} - 1\right) = \frac{v_2}{20}(11-1) = \frac{v_2}{2} = \frac{v_3}{2}$$

or $$\frac{v_4}{v_3} - 1 = \frac{1}{2} \quad \text{or} \quad \frac{v_4}{v_3} = \frac{3}{2} = 1.5$$

Process 1-2 :

$$T_2 = T_1 r^{0.4} = 363 \times 11^{0.4} = 947.24 \text{ K}$$

$$p_2 = p_1 r^{\gamma} = 1 \times 11^{1.4} = 28.704 \text{ bar}$$

Process 2-3 :

$$T_3 = T_2 \frac{p_3}{p_2} = 947.24 \times \frac{50}{28.704} = 1650 \text{ K } T_2$$

Process 3-4 :

$$T_4 = T_3 \frac{v_4}{v_3} = 1650 \times 1.5 = 2475 \text{ K}$$

Process 4-5 :

$$T_5 = T_3 \left(\frac{v_4}{v_5}\right)^{\gamma-1} = 2475 \left(\frac{1.5}{11}\right)^{0.4} = 1115.46 \text{ K}$$

$$w = q_s - q_r = [c_v (T_3 - T_2) + c_p (T_4 - T_3)] - c_v (T_5 - T_1)$$

$$= [0.718 (1650 - 947.24) + 1.005 (2475 - 1650)]$$
$$- 0.718 (1115.46 - 363)$$

$$= 1333.7 - 540.3$$

$$= 973.49 \text{ kJ/kg}$$

Thermal efficiency, $$\eta_{th} = \frac{w}{q_s} = \frac{973.4}{1333.7} = 0.7298 \quad \text{or} \quad 72.98\%$$

7.8 PERFORMANCE PARAMETERS

1. Indicated Power (IP)

The indicated power of an engine is the power developed within the cylinder of an I.C. engine. In order to calculate the indicated power, it is necessary to determine the indicated mean effective pressure in the cylinder. The indicated mean effective pressure can be determined by plotting the indicator diagram of the cylinder during one cycle of working. An engine indicator can be used for this purpose. As shown in Fig. 7.4 and 7.6, the net area of indicator diagram (p-V diagram) is $(A_1 - A_2)$. By knowing the diagram scale, the indicated mean effective pressure can be determined.

$$\text{Indicated power, IP} = \frac{p_{mi}\,LAN}{n \times 60 \times 10^3}\ \text{kW} \qquad \text{...(7.12)}$$

where

p_{mi} = indicated mean effective pressure, N/m^2
= [area of indicator diagram/length of indicator diagram] × spring constant
L = stroke length, m
A = piston area, $m^2 = \frac{\pi}{4} D^2$
D = Bore, m
n = number of strokes per revolution
= 2 for four-stroke cycle engine
= 1 for two-stroke cycle engine

2. Brake Power (BP)

Brake power of an engine is the net power available at the crankshaft for doing useful work. Generally it is measured with the help of a rope brake friction dynamometer, as shown in Fig. 7.20.

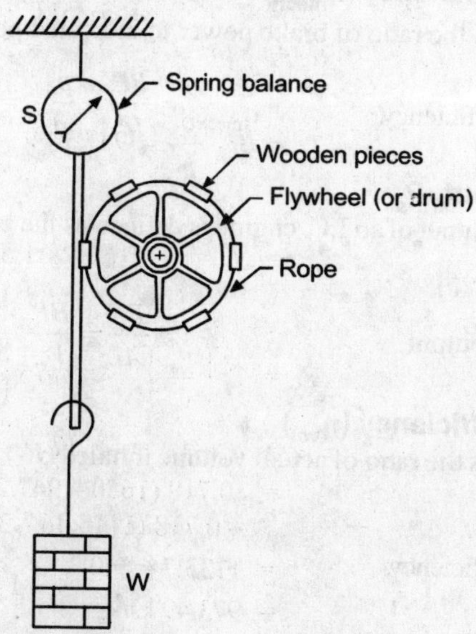

Fig. 7.20 Rope brake friction dynamometer

Let

W = dead weight, N
S = spring balance reading, N
D_b = diameter of brake drum or flywheel, m

d_r = diameter of rope, m

R = effective radius of brake drum, m

$$= \frac{1}{2}(D_b + d_r)$$

N = speed, rpm

Brake or net load = $W - S$

Braking torque, $T = (W - S)\,R$

$$\text{Brake power, } BP = \frac{2\pi NT}{60 \times 10^3} \text{ kW}$$

Also

$$BP = \frac{p_{mb} LAN}{n \times 60 \times 10^3} \text{ kW} \quad \text{...(7.13)}$$

where p_{mb} = brake mean effective pressure.

3. Frictional Power (FP)

The power lost due to friction is called frictional power. Frictional power includes the following losses :

(*i*) Pumping losses due to induction and exhaust.

(*ii*) Friction losses in bearings and other moving parts.

(*iii*) Air friction of flywheel.

(*iv*) Power required to drive the governor, lubricating oil pump, cooling water circulation pump, etc.

$$\text{Frictional power, FP} = \text{IP} - \text{BP} \quad \text{... (7.14)}$$

4. Mechanical Efficiency (η_{mech})

It is defined as the ratio of brake power to indicated power.

$$\text{Mechanical efficiency, } \eta_{\text{mech}} = \frac{BP}{IP} = \frac{p_{mb}}{p_{mi}} = \frac{BP}{BP + FP} \quad \text{... (7.15)}$$

5. Specific Output, P_s

The specific output of an I.C. engine is defined as the brake power per unit of piston displacement or swept volume.

$$\text{The specific output, } P_s = \frac{BP}{AL} = \frac{BP}{V_s} \quad \text{... (7.16)}$$

6. Volumetric Efficiency (η_{vol})

It is defined as the ratio of actual volume inhaled (V_a) during suction at NTP to the swept volume (V_s).

$$\text{Volumetric efficiency, } \eta_{\text{vol}} = \frac{V_a}{V_s} \quad \text{... (7.17)}$$

7. Thermal Efficiency

It is the ratio of work done to the heat energy supplied to the engine. It is either based on IP or BP. Accordingly, we have two types of thermal efficiencies.

$$(a)\text{ Indicated thermal efficiency, } (\eta_{\text{th}})_i = \frac{IP \times 3600}{\dot{m}_f \times LCV} \quad \text{...(7.18)}$$

where IP = indicated power, kW

$\dot{m}_f$ = fuel consumption, kg/h

LCV = lower calorific value of fuel used, kJ/kg

(*b*) Brake thermal efficiency, $$(\eta_{\text{th}})_b = \frac{BP \times 3600}{\dot{m}_f \times LCV} \quad \text{...(7.19)}$$

8. Air Standard Efficiency (η_a)

$$\eta_a = 1 - \frac{1}{r^{\gamma-1}} \text{ for petrol engine} \quad \text{... (7.20a)}$$

$$= 1 - \frac{1}{r^{\gamma-1}} \left[\frac{\rho^\gamma - 1}{\gamma(\rho - 1)} \right] \text{ for diesel engine} \quad \text{... (7.20b)}$$

where r = compression ratio

ρ = cut-off ratio

γ = adiabatic constant = 1.4 for air

9. Relative Efficiency (η_{rel})

It is the ratio of thermal efficiency of an engine to its corresponding air standard efficiency. It is either based on $(\eta_{\text{th}})_i$ or $(\eta_{\text{th}})_b$.

$$(\eta_{\text{rel}})_i = \frac{(\eta_{\text{th}})_i}{\eta_a} \quad \text{... (7.21a)}$$

$$(\eta_{\text{rel}})_b = \frac{(\eta_{\text{th}})_b}{\eta_a} \quad \text{... (7.21b)}$$

10. Specific Fuel Consumption (SFC)

It is defined as the amount of fuel required to be supplied to an engine to develop 1 kW power per hour.

Brake specific fuel consumption, $$\text{BSFC} = \frac{\dot{m}_f}{BP} \text{ kg/kW.}h \text{ or (kg/BP.}h) \quad \text{... (7.22a)}$$

Indicated specific fuel consumption, $$\text{ISFC} = \frac{\dot{m}_f}{IP} \text{ kg/kW.}h \text{ or (kg/IP.}h) \quad \text{... (7.22b)}$$

where $\dot{m}_f$ = fuel consumed, kg/h

11. Air/Fuel Ratio (A/F)

It is defined as the ratio of air supplied to the engine to the fuel consumed.

$$\text{A/F} = \frac{\dot{m}_a}{\dot{m}_f} \quad \text{... (7.23)}$$

12. Mean Piston Speed (SP)

$$\overline{SP} = 2\,LN \quad \text{... (7.24)}$$

Example 7.16 *A six cylinder, four-stroke, spark ignition engine of bore 10 cm and stroke 12 cm with a compression ratio of 6 is tested at 4800 rpm on a dynamometer of arm 55 cm. During a 10 minute test, the dynamometer reads 450 N and the engine consumes 5 kg of petrol of calorific* *value 45 MJ/kg. The carburettor receives the air at the rate of 10 kg/min. Calculate : (a) Brake power, (b) BSFC, (c) BMEP, (d) BSAC, (e) brake thermal efficiency, and (f) air-fuel action.*

Solution.

Given : $d = 10$ cm, $L = 12$ cm, $r = 6$, $N = 4800$ *rpm*, $i = 6$, $n = 2$, $l = 55$ cm, $t = 10$ *min*, $W = 450$ N, $m_f = 5$ kg, $CV = 45$ MJ/kg, $\dot{m}_a = 10$ kg/min.

(*a*) Torque,
$$T = Wl = 450 \times 0.55 = 247.5 \text{ N.m}$$

Brake power,
$$BP = \frac{2\pi NT}{60 \times 10^3} = \frac{2\pi \times 4800 \times 247.5}{60 \times 10^3} = 124.41 \text{ kW}$$

$$BP = \frac{p_{mb} ALNi}{60 \times 10^3 \times n}$$

$$124.41 = p_{mb} \times \frac{\pi}{4}(0.1)^2 \times 0.12 \times \frac{4800 \times 6}{60 \times 10^3 \times 2}$$

(*b*)
$$p_{mb} = 5.5 \text{ bar}$$

(*c*)
$$\dot{m}_f = \frac{m_f}{t} = \frac{5 \times 60}{10} = 30 \text{ kg/h}$$

$$\text{BSFC} = \frac{\dot{m}_f}{BP} = \frac{30}{124.41} = 0.241 \text{ kg/kWh}$$

(*d*)
$$\dot{m}_a = 10 \times 60 = 600 \text{ kg/h}$$

$$\text{BSAC} = \frac{600}{124.41} = 4.823 \text{ kg/kWh}$$

(*e*)
$$\eta_{bt} = \frac{BP}{\dot{m}_f \times CV} = \frac{124.41 \times 3600 \times 10^3}{30 \times 45 \times 10^6} = 0.3317 \text{ or } 33.17\%$$

(*f*)
$$\frac{A}{F} = \frac{\dot{m}_a}{\dot{m}_f} = \frac{600}{30} = 20:1$$

Example 7.17 *A six-cylinder, four-stroke, 10 cm bore and 12.5 cm stroke diesel engine develops 50 kW at 1000 rpm. The various efficiencies are : mechanical = 76%, volumetric = 80% under room conditions, indicated relative = 88%, theoretical thermal = 52%. The calorific value of liquid fuel is 450 kJ/kg. Calculate : (a) BMEP, (b) A : F ratio, (c) SFC.*

Assume air density as 0.12 kg/m³ under room conditions.

Solution.

Given : $i = 6, n = 2, \propto = 0.1$ m, $L = 0.125$ m, $BP = 50$ kW, $N = 1000$ rpm,

$$\rho_a = 0.12 \text{ kg/m}^3, \eta_{\text{mech}} = 0.76, \eta_v = 0.8, \eta_{ri} = 0.88, \eta_{\text{th}} = 0.52, CV = 450 \text{ kJ/kg}$$

$$BP = \frac{p_{mb} ALNi}{60 \times 10^3 \times n}$$

$$50 = \frac{p_{mb} \times \frac{\pi}{4}(0.1)^2 \times 0.125 \times 1000 \times 6}{60 \times 10^3 \times 2}$$

$$p_{mb} = 10.186 \text{ bar}$$

$$IP = \frac{BP}{\eta_{mech}} = \frac{50}{0.76} = 65.79 \text{ kW}$$

$$(\eta_i)_{th} = \eta_{ri} \times \eta_{th} = 0.88 \times 0.52 = 0.4576$$

$$= \frac{IP}{\dot{m}_f \times CV} = \frac{65.79}{\dot{m}_f \times 450 \times 10^3}$$

$$\dot{m}_f = \frac{65.79}{450 \times 10^3 \times 0.4576} = 3.195 \times 10^{-3} \text{ kg/s}$$

Swept volume, $V_s = \frac{\pi}{4}(0.1)^2 \times 0.125 = 9.8175 \times 10^{-4} \text{ m}^3$

Actual air intake per cylinder, $V_a = \eta_v V_s = 0.8 \times 9.8175 \times 10^{-4} = 7.854 \times 10^{-4} \text{ m}^3$

Actual air taken in per cylinder per second, $\dot{V}_a = \frac{V_a \times N}{60 \times n}$

$$= \frac{7.854 \times 10^{-4} \times 1000}{60 \times 2} = 6.545 \times 10^{-3} \text{ m}^3/\text{s}$$

Mass of air taken in per cylinder per second,

$$\dot{m}_a = \dot{V}_a \times P_a = 6.545 \times 10^{-3} \times 0.12 = 7.854 \times 10^{-4} \text{ kg/s}$$

$$\text{A/F} = \frac{\dot{m}_a \times i}{\dot{m}_f} = \frac{7.854 \times 10^{-4} \times 6}{3.195 \times 10^{-3}} = 1.475$$

$$\text{SFC} = \dot{m}_f \times \frac{3600}{BP} = \frac{3.195 \times 10^{-3} \times 3600}{50} = 0.23 \text{ kg/(kWh)}$$

Example 7.18 *A six-cylinder petrol engine operates on the four-stroke cycle. The bore of each cylinder is 70 mm and stroke 100 mm. The clearance volume per cylinder is 67 cm³. At a speed of 3960 rpm the full consumption is 19.5 kg/h and the torque developed is 140 N.m. Calculate (a) BP, (b) BMEP, (c) brake thermal efficiency, if LCV of fuel is 44 MJ/kg, and (d) relative efficiency on brake power basis. The engine works on the constant volume cycle basis. Assume $\gamma = 1.4$ for air.*

Solution.

Given : $i = 6, n = 2, d = 0.07$ m, $L = 0.1$ m, $Vc = 67 \times 10^{-6}$ m³, $N = 3960$ rpm,

LCV = 44 MJ/kg, $\gamma = 1.4$, $\dot{m}_f = 19.5$ kg/h, $T = 140$ N.m

(a) $$\text{BP} = \frac{2\pi NT}{60 \times 10^3} = \frac{2\pi \times 3960 \times 140}{60 \times 10^3} = 58 \text{ kW}$$

(b) $$\text{BP} = \frac{P_{mb} ALNi}{60 \times 10^3 \times n}$$

$$P_{mb} = \frac{58 \times 60 \times 10^3 \times 2}{\frac{\pi}{4}(0.07)^2 \times 0.1 \times 3960 \times 6} = 7.61 \text{ bar}$$

(*c*) $$(\eta_{th})_b = \frac{BP \times 3600}{\dot{m}_f \times LCV} = \frac{58 \times 3600}{19.5 \times 44 \times 10^3} = 0.244 \text{ or } 24.4\%$$

(*d*) Swept volume per cylinder, $$V_s = \frac{\pi}{4} d^2 \times L = \frac{\pi}{4}(0.07)^2 \times 0.1 = 3.848 \times 10^{-4} \text{ m}^3$$

Compression ratio, $$r = \frac{V_s + V_c}{V_c} = \frac{384.8 + 67}{67} = 6.743$$

Air standard efficiency, $$\eta_a = 1 - \frac{1}{r^{\gamma-1}} = 1 - \frac{1}{(6.743)^{0.4}} = 0.534$$

Relative efficiency, $$\eta_r = \frac{(\eta_{th})_b}{\eta_a} = \frac{0.244}{0.534} = 0.457 \text{ or } 45.7\%$$

Example 7.19 *A four-cylinder, two-stroke cycle petrol engine develops 40 kW at 1500 rpm. The mean effective pressure in each cylinder is 7.5 bar. The mechanical efficiency is 85%. Determine the cylinder bore and piston stroke, if stroke to bore ratio is 1.5. Also determine the fuel consumption if brake thermal efficiency is 30%. Take calorific value of petrol as 44 MJ/kg.*

Solution.

Given: $i = 4, n = 1$, BP = 40 kW, $N = 1500$ rpm, $p_{mi} = 7.5$ bar,

$$\eta_{mech} = 0.85,\ L/D = 1.5,\ CV = 44 \text{ MJ/kg},\ (\eta_{th})_b = 0.3$$

$$IP = \frac{BP}{\eta_{mech}} = \frac{40}{0.85} = 47.06 \text{ kW}$$

$$IP = \frac{p_{mi}\ ALNi}{60 \times 10^3 \times n}$$

$$47.06 = \frac{7.5 \times 10^5 \times \frac{\pi}{4} d^2 \times 1.5d \times 1500 \times 4}{60 \times 10^3 \times 1}$$

$$d^3 = 5.3261 \times 10^{-4}$$

$$d = 0.081 \text{ m or } 81 \text{ mm}$$

$$L = 1.5\ d$$

$$= 1.5 \times 81$$

$$= 121.5 \text{ mm}$$

Brake thermal efficiency, $$(\eta_{th})_b = \frac{BP \times 60}{\dot{m}_f \times CV}$$

$$\dot{m}_f = \frac{40 \times 60}{0.3 \times 44 \times 10^3} = 0.1818 \text{ kg/min}$$

Example 7.20 *A four-stroke, four-cylinder spark ignition engine has the following data : Piston diameter = 100 mm; Engine runs at a speed of 800 rpm; Length of stroke = 200 mm. Find the indicated power developed in kW. What should be the value of mean effective pressure?*

Solution.

Given: $d = 0.1$ m, $L = 0.2$ m, $N = 800$ rpm, $i = 4, n = 2$

$$IP = \frac{p_{mi}\ LAN\ i}{60\times10^3\times n}$$

Assuming $IP = 10\ \text{kW}$

$$10 = p_{mi}\times0.2\times\frac{\pi}{4}(0.1)^2\times\frac{800\times4}{60\times10^3\times2}$$

$$p_{mi} = 2.387\ \text{bar}$$

Example 7.21 *A 4-cylinder, two-stroke cycle petrol engine develops 30 kW at 2500 rpm. The mean effective pressure on each piston is 8 bar and mechanical efficiency is 80%. Calculate the dia and stroke of each cylinder if stroke to bore ratio is 1.5. Also calculate the fuel consumption of the engine, if brake thermal efficiency is 28% and calorific value of fuel is 43900 kJ/kg.*

Solution.

Given: $i = 4, n = 1, BP = 30\ \text{kW}, N = 2500\ \text{rpm}, p_{mi} = 8\ \text{bar},$

$\eta_{mech} = 0.8, \text{L/d} = 1.5, (\eta_{th})_b = 0.28, CV = 43900\ \text{kJ/kg}$

$$\eta_{mech} = \frac{BP}{IP}$$

$$IP = \frac{30}{0.8} = 37.5\ \text{kW}$$

$$IP = \frac{p_{mi}\ LAN\ i}{60\times10^3\times n}$$

$$37.5 = \frac{8\times10^5\times1.5d\times\frac{\pi}{4}d^2\times2500\times4}{60\times10^3\times1}$$

$$d^3 = 2.3873\times10^{-4}$$

$$d = 0.062\ \text{m or } 62\ \text{mm}$$

$$L = 1.5d = 1.5\times62 = 93\ \text{mm}$$

Brake thermal efficiency, $$(\eta_{th})_b = \frac{BP\times3600}{\dot{m}_f\times CV}$$

$$\dot{m}_f = \frac{30\times3600}{0.28\times43900} = 8.7862\ \text{kg/h}$$

$$BSFC = \frac{\dot{m}_f}{BP} = \frac{8.7862}{30} = 0.2929\ \text{kg/kWh}$$

Example 7.22 *A six-cylinder, 4-stroke SI engine delivers 400 kW at 2200 rpm. Determine the bore and stroke from the following data :*

Compression ratio = 7.6; stroke to bore ratio = 1.25, i.m.c.p. = 100 N/cm², mechanical efficiency = 0.75, calorific value of fuel = 105000 cal/kg.

What is the brake specific fuel consumption?

Solution.

Given: $i = 6, n = 2, BP = 400\ \text{kW}, N = 2200\ \text{rpm}, r = 7.6, L/d = 1.25,$

$imep = 100\ N/cm^2$ or $10\ Pa$, $\eta_{mech} = 0.75$, $CV = 105000 \times 4.184 = 439320$ kJ/kg,

$$IP = \frac{BP}{\eta_{mech}} = \frac{400}{0.75} = 533.33 \text{ kW}$$

$$IP = \frac{p_{mi} \times ALNi}{60 \times 10^3 \times n}$$

$$533.33 = \frac{10 \times 10^5 \times \frac{\pi}{4} d^2 \times 1.25d \times 2200 \times 6}{60 \times 10^3 \times 2}$$

$$d^3 = 4.9386 \times 10^{-3}$$

$$d = 0.1703 \text{ m or } 170.3 \text{ mm}$$

$$L = 1.25\,d = 1.25 \times 170.3 = 212.8 \text{ mm}$$

Indicated thermal efficiency, $(\eta_{th})_i = 1 - \frac{1}{r^{\gamma-1}}$

$$= 1 - \frac{1}{(7.6)^{0.4}} = 0.5557$$

$$(\eta_{th})_i = \frac{IP}{\dot{m}_f \times CV}$$

$$\dot{m}_f = \frac{533.33}{0.5557 \times 439320} = 0.00218 \text{ kg/s}$$

$$BSFC = \frac{\dot{m}_f \times 3600}{BP} = \frac{0.00218 \times 3600}{400} = 0.0197 \text{ kg/kWh}$$

Example 7.23 *A four stroke petrol engine has a stroke volume of 5 litres. Its indicated mean effective pressure is 0.5 MPa and speed is 750 rpm. Calculate the indicated power of the engine.*

Solution.

Given: $V_s = 5$ litres or 5×10^{-3} m^3, $p_{mi} = 0.5$ MPa, $N = 750$ rpm

Number of working strokes, $n = \frac{N}{2} = \frac{750}{2} = 375$ strokes/min

Indicated power, $IP = \frac{p_{mi}(LA)n}{60 \times 10^3}$

$$= \frac{0.5 \times 10^6 \times 5 \times 10^{-3} \times 375}{60 \times 10^3} = 15.625 \text{ kW}$$

Example 7.24 *A four-cylinder, four stroke cycle petrol engine has 100 mm bore and stroke is 1.25 times the bore. It consumes 4 kg of fuel per hour having a calorific value of 41,000 kJ/kg. The engine speed is 800 rpm. Calculate indicated thermal efficiency if indicated mean effective pressure is 0.75 MPa.*

Solution.

Given: $D = 100$ mm, $L = 1.25\ D = 125$ mm, $\dot{m}_f = 4$ kg/h, $CV = 41{,}000$ kJ/kg,

$N = 800$ rpm, $p_{mi} = 0.75$ MPa, $i = 4$

Indicated power, $IP = \dfrac{ip_{mi}LAN}{2\times 60\times 10^3} = \dfrac{4\times 0.75\times 0.125\times \frac{\pi}{4}\times (100)^2\times 800}{2\times 60\times 10^3} = 19.635$ kW

Indicated thermal efficiency, $(\eta_{th})_i = \dfrac{IP\times 3600}{\dot{m}_f\times CV} = \dfrac{19.635\times 3600}{4\times 41,000} = 0.431$ or 43.1%

Example 7.25 *A 6 cylinder, 4 stroke I.C. engine is to develop 89.5 kW at 800 rpm. The stroke to bore ratio is 1.25:1. Assuming mechanical efficiency of 80% and brake mean effective pressure of 5 bar, determine the diameter and stroke of the engine.*

Solution.

Given: $i = 6$, $IP = 89.5$ kW, $N = 800$ rpm, $\dfrac{L}{D} = 1.25$, $\eta_{mech} = 0.8$,

$p_{mb} = 5$ bar

Brake power, $BP = \dfrac{i\ p_{mb}\ LAN}{2\times 60\times 10^3} = IP\times \eta_{mech}$

$$= \frac{6\times 5\times 10^5\times 1.25D\times \frac{\pi}{4}D^2\times 800}{2\times 60\times 10^3} = 89.5\times 0.8$$

$$D^3 = 3.64656\times 10^{-3}$$

$$D = 0.154 \text{ m or } 154 \text{ mm}$$

$$L = 1.25\times 154 = 192.5 \text{ mm}$$

Example 7.26 *A petrol engine with bore 150 mm and stroke 200 mm has a clearance volume of 700 litres. If the indicated thermal efficiency is 0.30, find the relative efficiency. Take $\gamma = 1.4$.*

Solution.

Given: $D = 150$ mm, $L = 200$ mm, $V_c = 700$ litres or 7×10^{-4} m^3

$(\eta_{th})_i = 0.30$

Swept volume, $V_s = \dfrac{\pi}{4}D^2L = \dfrac{\pi}{4}\times (0.15)^2\times 0.2 = 3.534\times 10^{-3}$ m^3

$$V_c + V_s = 7\times 10^{-4} + 35.34\times 10^{-4} = 42.34\times 10^{-4} \text{ m}^3$$

Compression ratio, $r = \dfrac{V_c + V_s}{V_c} = \dfrac{42.34}{7} = 6.05$

Air standard efficiency, $\eta_a = 1 - \dfrac{1}{r^{\gamma-1}}$

$$= 1 - \frac{1}{(6.05)^{0.4}} = 0.513$$

Relative indicated efficiency, $(\eta_{rel})_i = \dfrac{(\eta_{th})_i}{\eta_a} = \dfrac{0.30}{0.513} = 0.585$ or 58.5%

Example 7.27 *In a trial of a single cylinder oil engine, the following observations were made :*

Oil consumption = 10 kg/h

Calorific value of fuel = *43,800 kJ/kg*
Speed = *1800 rpm*
Torque on brake drum = *180 N.m*
Calculate (a) Brake power
(b) BSFC
(c) Brake thermal efficiency

Solution.

Given: $\dot{m}_f = 10$ kg/h, $CV = 43{,}800$ kJ/kg, $N = 1800$ rpm, $T = 180$ N.m

(*a*) Brake power, $$BP = \frac{2\pi NT}{60\times10^3} = \frac{2\pi\times1800\times180}{60\times10^3} = 33.929 \text{ kW}$$

(*b*) $$BSFC = \frac{\dot{m}_f}{BP} = \frac{10}{33.929} = 0.2947 \text{ kg/kW.h}$$

(*c*) Brake thermal efficiency, $$(\eta_{th})_b = \frac{BP\times3600}{\dot{m}_f\times C.V.}$$

$$= \frac{33.929\times3600}{10\times43{,}800} = 0.3282 \text{ or } 32.82\%$$

Example 7.28 *During testing of single cylinder two stroke oil engine, following data was obtained :*

Brake torque = 640 N.m, Cylinder diameter = 21 cm, Speed = 350 rpm,
Stroke = 28 cm, mean effective pressure = 5.6 bar, oil consumption = 8.16 kg/h,
Calorific value of oil = 42705 kJ/kg

Determine (a) mechanical efficiency, (b) indicated thermal efficiency, (c) brake thermal efficiency, and (d) brake specific fuel consumption.

Solution.

Given: $T = 640$ N.m, $D = 0.21$ m, $L = 0.28$ m, $N = 350$ rpm,
$p_m = 5.6$ bar, $\dot{m}_f = 8.16$ kg/h, $CV = 42705$ kJ/kg

(*a*) Brake power, $$BP = \frac{2\pi NT}{60\times10^3} = \frac{2\pi\times350\times640}{60\times10^3} = 23.457 \text{ kW}$$

Indicated power, $$IP = \frac{p_m LAN}{60\times10^3} = \frac{5.6\times10^5\times0.28\times\frac{\pi}{4}(0.21)^2\times350}{60\times10^3} = 31.680 \text{ kW}$$

Mechanical efficiency, $$\eta_{mech} = \frac{BP}{IP} = \frac{23.457}{31.680} = 0.7404 \text{ or } 74.04\%$$

(*b*) Indicated thermal efficiency, $$(\eta_{th})_i = \frac{IP\times3600}{\dot{m}_f\times CV} = \frac{31.68\times3600}{8.16\times42705} = 0.3273 \text{ or } 32.73\%$$

(*c*) Brake thermal efficiency, $$(\eta_{th})_b = \frac{BP\times3600}{\dot{m}_f\times CV} = \frac{23.457\times3600}{8.16\times42705} = 0.2423 \text{ or } 24.23\%$$

(*d*) $$BSFC = \frac{\dot{m}_f}{BP} = \frac{8.16}{23.457} = 0.3478 \text{ kg/kW.h}$$

Example 7.29 *The following readings were taken during the test on a single cylinder four stroke oil engine :*

Cylinder diameter = 270 mm, stroke length = 380 mm, mean effective pressure = 6 bar, engine speed = 250 rpm, net load on brake = 1000 N, effective diameter of brake of brake = 1.5m, fuel used = 10 kg/h, CV of fuel = 44,400 kJ/kg.

Calculate (a) BP, (b) IP, (c) mechanical efficiency, and (d) thermal efficiency.

Solution.

Given : $D = 0.27$ m, $L = 0.38$ m, $p_m = 6$ bar, $N = 250$ rpm, $W = 1000N$,

$d_b = 1.5$ m, $\dot{m}_f = 10$ kg/h, $CV = 44,400$ kJ/kg

(*a*) Brake torque, $$T = W \times \frac{d_b}{2} = 1000 \times \frac{1.5}{2} = 750 \text{ N.m}$$

$$BP = \frac{2\pi NT}{60 \times 10^3} = \frac{2\pi \times 250 \times 750}{60 \times 10^3} = 19.635 \text{ kW}$$

(*b*) $$IP = \frac{p_m LAN}{2 \times 60 \times 10^3} = \frac{6 \times 10^5 \times 0.38 \times \frac{\pi}{4}(0.27)^2 \times 250}{2 \times 60 \times 10^3} = 27.196 \text{ kW}$$

(*c*) $$\eta_{mech} = \frac{BP}{IP} = \frac{19.635}{27.196} = 0.722 \text{ or } 72.2\%$$

(*d*) Thermal efficiency, $$(\eta_{th})_i = \frac{IP \times 3600}{\dot{m}_f \times CV} = \frac{27.196 \times 3600}{10 \times 44400} = 0.2205 \text{ or } 22.05\%$$

Example 7.30 *During a test on a single cylinder, four stroke engine having a compression ratio of 6, following data were recorded :*

Bore = 10 cm; Stroke = 12.5 cm; Indicated mean effective pressure = 2.6 bar;

Dead load on dynamometer, W = 60 N; Spring balance reading, S = 19 N;

Effective radius of flywheel, R = 40 cm; Fuel consumption, $\dot{m}_f$ = 1.0 kg/h;

Calorific value of fuel used; CV = 42000 kJ/kg; Speed N = 2000 rpm.

Determine its (a) indicated power, (b) brake power, (c) mechanical efficiency, (d) overall efficiency, (e) air standard efficiency, and (f) relative efficiency.

Solution.

Given : $r = 6$, $D = 0.1$ m, $L = 0.125$ m, $p_{mi} = 2.6$ bar, $W = 60N$, $S = 19N$,

$R = 0.4$ m, $\dot{m}_f = 1$ kg/h, $CV = 42000$ kJ/kg, $N = 2000$ rpm,

(*a*) Indicated power, $$IP = \frac{p_{mi} LAN}{2 \times 60 \times 10^3} = \frac{2.6 \times 10^5 \times 0.125 \times \frac{\pi}{4}(0.1)^2 \times 2000}{2 \times 60 \times 10^3} = 4.254 \text{ kW}$$

(*b*) Brake power, $$BP = \frac{(W - S)R \times 2\pi N}{60 \times 10^3} = \frac{(60 - 19) \times 0.4 \times 2\pi \times 2000}{60 \times 10^3} = 3.435 \text{ kW}$$

(*c*) Mechanical efficiency, $$\eta_{mech} = \frac{BP}{IP} = \frac{3.435}{4.254} = 0.8008 \text{ or } 80.08\%$$

(*d*) Overall efficiency, $$\eta_{overall} = \frac{BP}{\dot{m}_f \times CV} = \frac{3.435 \times 3600}{1 \times 42000} = 0.2944 \text{ or } 29.44\%$$

(*e*) Air standard efficiency, $\eta_a = 1-\frac{1}{r^{\gamma-1}} = 1-\frac{1}{6^{0.4}} = 0.5116$ or 51.16%

(*f*) Relative efficiency, $\eta_{\text{relative}} = \frac{\eta_{\text{overall}}}{\eta_a} = \frac{0.2944}{0.5116} = 0.5754$ or 57.54%

Example 7.31 *A four cylinder, four stroke spark ignition engine has bore of 80 mm, stroke of 80 mm, and compression ratio of 8. Calculate cubic capacity of engine and clearance volume of each cylinder.*

Solution.

Given: $i = 4, D = 0.08$ m, $L = 0.08$ m, $r = 8$

Swept volume for 4 cylinders, $V = i \times \frac{\pi}{4} D^2 L = 4 \times \frac{\pi}{4}(0.08)^2 \times 0.08 = 1.6085 \times 10^{-3}\,\text{m}^3$

Cubic capacity of engine $= 1.6085 \times 10^{-3}\,\text{m}^3$ or 1608.5 litres

Compression ratio, $r = \frac{V_c + V_s}{V_c}$

$$8 = \frac{V_c + V_s}{V_c}$$

$$8V_c = V_c + V_s$$

$$V_c = V_s/7 = \frac{V}{4\times 7} = \frac{1.6085\times 10^{-3}}{28} = 5.744\times 10^{-5}\,\text{m}^3 \text{ or } 57.44 \text{ litres}$$

Example 7.32 *A single cylinder, four stroke cycle oil engine is fitted with a rope brake. The diameter of the brake wheel is 600 mm and rope diameter is 26 mm. The dead load on the brake is 200 N and the spring balance reads 30 N. If the engine runs at 450 rpm, what will be the brake power of the engine?*

Solution.

Given: $D_b = 0.6$ m, $d_r = 0.026$ m, $W = 200$ N, $S = 30$ N, $N = 450$ rpm

Brake power, $BP = \frac{(W-S)\,\pi\,(D_b + d_r)N}{60\times 10^3}$

$$= \frac{(200-30)\times \pi\,(0.6+0.026)\times 450}{60\times 10^3}$$

$$= 2.507\,\text{kW}$$

Example 7.33 *A four cylinder, two-stroke cycle petrol engine develops 30 kW at 2500 rpm. The indicated mean effective pressure of each is 800 kPa and η_{mech} = 80%. Calculate the diameter and stroke of each of the cylinder if $\frac{L}{D} = 1.5$. Also calculate the brake specific fuel consumption of the engine if brake thermal efficiency is 28%. The calorific value of petrol is 44,000 kJ/kg.*

Solution.

Given: $i = 4$, $IP = 30$ kW, $N = 2500$ rpm, $p_{mi} = 800$ kPa, $\eta_{\text{mech}} = 0.8$,

$\frac{L}{D} = 1.5$, $(\eta_{\text{th}})_b = 0.28$, $CV = 44{,}000$ kJ/kg

$$IP = \frac{i\,p_{mi}\,LAN}{60\times 10^3}$$

$$30 = \frac{4\times800\times10^3\times1.5D\times\frac{\pi}{4}D^2\times2500}{60\times10^3}$$

$$D^3 = 1.90986\times10^{-4}$$

$$D = 0.0576 \text{ m or } 57.6 \simeq 58 \text{ mm}$$

$$L = 1.5D = 87 \text{ mm}$$

$$\eta_{mech} = \frac{BP}{IP}$$

$$BP = 0.8\times30 = 24 \text{ kW}$$

$$(\eta_{th})_b = \frac{BP\times3600}{\dot{m}_f\times CV}$$

$$0.28 = \frac{24\times3600}{\dot{m}_f\times44{,}000}$$

$$\dot{m}_f = 7.013\ kg/h$$

$$BSFC = \frac{\dot{m}_f}{BP} = \frac{7.013}{24} = 0.292 \text{ kg/(kW.h)}$$

MULTI-CHOICE QUESTIONS

1. The compression ratio of an I.C. engine is defined as
(*a*) Stroke volume (V_s)/Clearance volume (V_c)
(*b*) V_c/V_s (*c*) $\frac{V_c+V_s}{V_c}$ (*d*) $\frac{V_c+V_s}{V_s}$

2. A four-stroke petrol engine sucks during the suction stroke
(*a*) only air (*b*) mixture of fuel and air
(*c*) only fuel (*d*) any one of the above

3. A four-stroke diesel engine sucks during the suction stroke
(*a*) only air (*b*) mixture of fuel and air
(*c*) only fuel (*d*) any one of the above

4. The value(s) closed during compression stroke of an I.C. engine is/are
(*a*) Inlet (*b*) Exhaust
(*c*) Both inlet and exhaust (*d*) none of the above

5. A two-stroke I.C. engine has
(*a*) inlet and exhaust valves (*b*) inlet and exhaust ports
(*c*) inlet, exhaust and transfer ports (*d*) only transfer ports

6. The ratio of BP to IP is called
(*a*) mechanical efficiency (*b*) thermal efficiency
(*c*) volumetric efficiency (*d*) relative efficiency

7. A 100 c.c. I.C. engine means that its
(*a*) swept volume is 100 c.c. (*b*) clearance volume is 100 c.c.
(*c*) clearance + swept volume is 100 c.c. (*d*) swept – clearance volume is 100 c.c.

8. The ratio of brake power to rate of energy input in fuel is called
(*a*) mechanical efficiency (*b*) thermal efficiency
(*c*) relative efficiency (*d*) volumetric efficiency

9. The ratio of indicated thermal efficiency to air standard efficiency is called
(*a*) volumetric efficiency (*b*) thermal efficiency
(*c*) mechanical efficiency (*d*) relative efficiency

10. The widely used engines for earth-moving machines are :
(*a*) petrol engines (*b*) steam engines (*c*) diesel engines (*d*) gasoline engines

11. Brake specific fuel consumption of an I.C. engine is
(*a*) $\frac{\dot{m}_f}{BP}$ (*b*) $\frac{\dot{m}_f}{IP}$ (*c*) $\frac{BP}{\dot{m}_f}$ (*d*) $\dot{m}_f \times BP$

12. The brake thermal efficiency of modern I.C. engines is
(*a*) 10 — 20% (*b*) 25 — 35% (*c*) 50 — 60% (*d*) 60 — 75%

13. Which of the following does not relate to spark ignition engine?
(*a*) ignition coil (*b*) spark plug (*c*) distributor (*d*) fuel injector

14. In a diesel engine, the fuel is ignited by
(*a*) spark (*b*) injected fuel (*c*) ignitor (*d*) compressed air heat

15. Carburettor is used for
(*a*) S.I. engine (*b*) C.I. engine (*c*) Gas engine (*d*) none of the above

16. In which engine values are used ?
(*a*) Two-stroke engine (*b*) Four-stroke engine
(*c*) In both engines (*d*) None of the above

17. With the same revolutions/minute, the ratio of power developed by 4-stroke engine to that developed by 2-stroke engine of same bore and stroke is
(*a*) 2 (*b*) 1 (*c*) $\frac{1}{2}$ (*d*) 4

18. The air/fuel ratio in a petrol engine is controlled by
(*a*) governor (*b*) injector (*c*) carburettor (*d*) flywheel

19. The ratio of stroke to crank radius of an I.C. engine is
(*a*) $\frac{1}{2}$ (*b*) 1 (*c*) 2 (*d*) 4

20. The number of working strokes per minute in case of 4-stroke I.C. engine are
(*a*) N/2 (*b*) N (*c*) 2N (*d*) 4N
where N = rpm.

21. Fuel injector is used in
(*a*) S.I. engine (*b*) C.I. engine (*c*) Gas engine (*d*) Petrol engine

22. In the case of diesel engine, the fuel and air are mixed in
(*a*) injector (*b*) inlet manifold (*c*) carburettor (*d*) engine cylinder

23. Which one of the following does not relate to diesel engines ?
(*a*) fuel pump (*b*) fuel injector (*c*) governor (*d*) carburettor

24. Which one of the following does not relate to petrol engines ?
(*a*) ignition coil (*b*) distributor (*c*) spark plug (*d*) fuel injector

25. The automobiles generally utilise batteries having voltage of
(*a*) 6V (*b*) 12V (*c*) 24V (*d*) 28V

26. The ratio of indicated thermal efficiency to the corresponding air standard cycle efficiency is called
(*a*) net efficiency (*b*) efficiency ratio
(*c*) relative efficiency (*d*) overall efficiency

27. An engine indicator is used to determine
(*a*) speed (*b*) temperature (*c*) pressure (*d*) m.e.p.

28. Air-fuel ratio in petrol engine is controlled by
(*a*) fuel pump (*b*) injector (*c*) controller (*d*) carburettor

Answers:

1. (*c*)	**2.** (*b*)	**3.** (*a*)	**4.** (*c*)	**5.** (*c*)	**6.** (*a*)	**7.** (*a*)	**8.** (*b*)	**9.** (*d*)	**10.** (*c*)
11. (*a*)	**12.** (*b*)	**13.** (*d*)	**14.** (*d*)	**15.** (*a*)	**16.** (*b*)	**17.** (*c*)	**18.** (*c*)	**19.** (*c*)	**20.** (*a*)
21. (*b*)	**22.** (*d*)	**23.** (*d*)	**24.** (*d*)	**25.** (*b*)	**26.** (*c*)	**27.** (*d*)	**28.** (*d*)		

REVIEW QUESTIONS

1. State and describe three major classification of I.C. engines.
2. Give the advantages of I.C. engines over steam engines.
3. How do you classify I.C. engines ?
4. Describe the function of various parts of an I.C. engine.
5. What are the functions of the following in an I.C. engine ?
(*a*) Piston (*b*) Crankshaft (*c*) valve gear mechanism.
6. Define the following :
(*a*) Clearance volume (*b*) Swept volume (*c*) compression ratio
7. Explain with the help of neat sketches, the working of two stroke petrol engine.
8. Explain with the help of neat sketches, the working of four stroke diesel engine.
9. Give the advantages and disadvantages of petrol and diesel engines.
10. State the advantages of two stroke engines over four stroke engines.
11. Define the following :
(*a*) Indicated power (*b*) Brake power (*c*) Mechanical efficiency
(*d*) Indicated and brake thermal efficiency (*e*) Relative efficiency
12. Give the merits and demerits of four-stroke and two-stroke cycle engines.
13. Explain with the help of neat sketches, the working principle of four stroke petrol engine.
14. Draw the theoretical and actual *p-V* diagrams for the following :
(*a*) Four-stroke S.I. engine (*b*) Four-stroke C.I. engine
(*c*) Two-stroke S.I. engine (*d*) Two-stroke C.I. engine
15. Define and explain specific fuel consumption.
16. List applications of I.C. engines.
17. Explain the following terms related to I.C. engines :
(*a*) Piston speed (*b*) Compression ratio
(*c*) Brake power (*d*) Brake thermal efficiency
18. Why petrol engines are called S.I. engines and why diesel engines are called C.I. engines.
19. Explain with neat sketches, the working of two-stroke diesel engine.
20. Name the various components of an I.C. engine.

EXERCISES

7.1 The following results refer to a test on a petrol engine,

indicated power = 42 kW
Brake power = 35 kW
Engine speed = 1800 rpm

Fuel consumed per brake power hour = 0.30 kg
Calorific value of fuel used = 44800 kJ/kg
Calculate (*i*) Indicated thermal efficiency
(*ii*) Brake thermal efficiency
(*iii*) Mechanical efficiency [**Ans.** 32.14%, 26.79%, 83.33%]

7.2 The following data available for four stroke petrol engine : bore = 20 cm stroke to bore ratio = 1.5, speed = 350 rpm, p_m = 2.75 bar, net brake load = 610 N, effective diameter of brake drum = 1 m, petrol consumption = 4.25 kg/hr, C.V. of petrol = 44 MJ/kg.
Determine : (*a*) IP (*b*) BP (*c*) η_{mech} (4) $(\eta_{th})_i$, (5) $(\eta_{th})_b$
[**Ans.** 15.2 kW, 11.2 kW, 73.93%, 29.1%, 21.5%]

7.3 A four stroke six cylinder engine with stroke volume of 1.75 litres develops 26.25 kW. The mean effective pressure is 6 bar. Find speed of the engine. [**Ans.** 500 rpm]

7.4 A four cylinder, four stroke diesel engine runs at 4500 rpm. The dead load on rope brake dynamometer is 200 N and spring balance reading is 22 N. The mean effective radius of the dynamometer is 0.6 m. Find the brake power of the engine and its thermal efficiency if the b.s.f.c. of the engine is 0.26 k/kWh. Assume, C.V. of fuel used 44000 kJ/kg
[**Ans.** 50.33 kW; 31.11%]

7.5 Following data relate to an oil engine working on Otto cycle :
Compression ratio = 6; B.P. = 14.8 kW; Suction pressure = 1 bar and temp. = 27°C; Relative efficiency = 68%; Mechanical efficiency = 78%
Find its air standard and thermal efficiencies.
[**Ans.** η_a = 51.16%, $(\eta_{th})_i$ = 34.79%, $(\eta_{th})_b$ = 27.13%]

7.6 A gasoline engine consumes 8 litres of fuel per hr. having specific gravity of 0.75 while it develops 25 kW power. If the C.V. of fuel used is 44200 kJ/kg, determine its indicated thermal efficiency. [**Ans.** $(\eta_{th})_i$ = 33.94%]

7.7 A four cylinder, four stroke cycle petrol engine develops 40 kW at 2500 rpm. The mean effective pressure is 0.8 MPa and mechanical efficiency is 80%. Calculate diameter and stroke of each cylinder if stroke to bore ratio is 1.5. Also calculate the brake specific fuel consumption of the engine if the brake thermal efficiency is 25%. The calorific value of fuel is 45000 kJ/kg.
[**Ans.** bore, D = 86.025 mm, stroke, L = 129.03 mm, BSFC = 0.32 kg/*BP.h*]

7.8 A six cylinder, four stroke petrol engine has the following data :
bore = 8 cm, stroke = 10 cm, clearance volume = 70 c.c., speed = 4000 rpm, consumption = 20 kg/hr, torque developed = 150 Nm,
C.V. of fuel used = 43 MJ/kg.
Calculate B.P., b.m.e.p., brake thermal efficiency, compression ratio, air standard efficiency and relative efficiency based on brake thermal efficiency. Assume,
γ = 1.4. [**Ans.** B.P. = 62.8 kW; b.m.e.p. = 6.25 bar; $(\eta_{th})_b$ = 26.3%, r = 8.18; η_a = 56.8%; η_{rel} = 46.3%]

7.9 A single cylinder, four stroke C.I. engine has the following data :
bore = 10 cm; stroke = 11 cm, brake torque = 28 Nm, speed = 1600 rpm; i.m.e.p. = 6.5 bar.
Find the B.P., IP., F.P. and its mechanical efficiency.
[**Ans.** B.P. = 4.691 kW; IP. = 7.487; F.P. = 2.796 kW; η_{mech} = 62.66%]

7.10 Following results refer to a test on petrol engine :
I.P. = 32 kW; B.P. = 27 kW; speed = 1200 rpm b.s.f.c. = 0.25 kg/kWh;
C.V. of fuel used = 44 mJ/kg.
Calculate indicated, brake and mechanical efficiencies.
[**Ans.** $(\eta_{th})_b$ = 32.72%, η_{mech} = 84.38%, $(\eta_{th})_i$ = 38.3%]

7.11 A six cylinder four stroke I.C. engine is to develop 100 kW IP at 800 rpm. The stroke to bore ratio is 1.25. Assuming mechanical efficiency of 80% and brake mean effective pressure of 5 bar, find out the diameter and stroke of the engine. **[Ans.** D = 16 cm, L = 20 cm.]

7.12 A six cylinder four stroke petrol engine develops 300 kW (BP) at 2500 rpm. The stroke to bore ratio is 1.25. The mean effective pressure on each piston is 9 bar and mechanical efficiency of the engine is 80%. Indicated thermal efficiency is 30%. Calculate (*a*) stroke and bore of the engine, (*b*) fuel consumption. Take C.V. of petrol equals to 41900 kJ/kg.

[Ans. L = 189 mm, D = 151 mm, $\dot{m}_f$ = 107.4 kg/h]

7.13 During a trial on single cylinder four stroke IC engine, the following observations were taken : Mean effective pressure = 4 bar, speed = 200 RPM, Brake power = 7.5 kW, L/D ratio of piston = 1.5. If mechanical efficiency is 70% find the dimensions of the engine.

[Ans. L = 240 mm, D = 360 mm]

7.14 Following data refers to a test on a petrol engine. Indicated power = 30 kW, Brake power = 26 kW, Engine speed = 1800 RPM, Brake specific fuel consumption = 0.35 kg/kWh, C.V. of fuel = 44,000 kJ/kg

Calculate : (1) η_{mech} (2) $(\eta_{th})_i$ (3) $(\eta_{th})_b$. **[Ans.** 86.67%, 26.97%, 23.37%]

7.15 Give reasons for the following :

(*i*) I.C. Engine cylinders are required to be cooled.

(*ii*) Carburettor and spark plug are provided on petrol engines.

The indicated power of a four stroke cycle petrol engine is 37 kW. The stroke is 280 mm and the bore is 190 mm. If the engine makes 460 explosions per minute, find the mean effective pressure of the engine. **(Ans.** 607.91 1 kN/m^2)

7.16 A 6 cylinder four stroke cycle petrol engine develops 300 kW bp at 2500 r.p.m. The stroke to bore ratio is 1.25. Assuming the mechanical efficiency of 80% and mean effective pressure of 9 bar, determine the bore and stroke of engine.

Also find the fuel consumption in kg/hr if indicated thermal efficiency is 30% and C.V. of fuel used is 41900 kJ/kg. **[Ans.** bore = 151 mm, stroke = 189 mm fuel consumption, $\dot{m}_f$ = 107.4 kg/h]

7.17 An engine is used on a job requiring 110 kW BP. The mechanical efficiency of the engine is 80% and engine consumes 50 kg/hr of fuel. A design improvement is made which reduces the engine friction by 5 kW. If indicated thermal efficiency remains the same, how many kg of fuel per hour will be saved ? **[Ans.** 1.8 kg/h]

7.18 A four cylinder, four stroke petrol engine has the swept volume of 1750 c.c. It develops 27 kW at 1000 rpm. Mean effective pressure is 5 bar. Find the average number of times the cylinder misfires every minute. **[Ans.** : 30.4]

A single cylinder, four stroke, air cooled engine was tested and following observations were made :

Fuel consumption	= 50 c.c. in 1 min. 42 seconds;
Air consumption	= 0.2 m^3 in 32 seconds;
Net brake load	= 8 N;
Speed	= 3000 rpm;
Density of air	= 1.18 kg/m^3,
Specific gravity of fuel	= 0.74;
Heating value of fuel	= 44 MJ/kg;
Dynamometer constant, K	= 5000;

B.P. is given as $\frac{WN}{5000}$ (kW)

Find the B.P., air-fuel ratio, b.s.f.c. and brake thermal efficiency.

[**Ans.** B.P. = 4.8 kW; A.F. ratio = 20.33:1; b.s.f.c. = 0.272 kg/kWh; $(\eta_{th})_b = 30.08\%$]

7.19 A four cylinder two stroke petrol engine with stroke to bore ratio of 1.2. develops 32 kW bp at 2500 r.p.m. The mean effective pressure on each piston is 8 bar and the mechanical efficiency is 85 percent. Determine—

(*i*) Diameter and stroke of each cylinder.

(*ii*) Brake thermal efficiency if the fuel consumption is 9 kg/hr having C.V. of 44000 kJ/kg.

[**Ans.** (*i*) D = 68 mm; L = 82 mm (*ii*) $(\eta_{th})_b = 0.29$]

7.20 A four stroke petrol engine has a stroke volume of 5.7 litres. Its mean effective pressure is 600 kN/m^2 and r.p.m. is 800. Find the indicated power of the engine. [**Ans.** IP= 22.8 kW]

7.21 A four cylinder, four stroke petrol engine is to be design to give brake power of 185 kW at 35 RPS. The stroke bore ratio is to be 1.5. Assuming mechanical efficiency of 75% and indicated mean effective pressure of 830 kPa. Determine the required bore and stroke.

7.22 An oil engine work on diesel cycle with temperature of 27°C at the beginning of compression. If the ratio of adiabatic expansion is 8, find the thermal efficiency of cycle. Assume $\gamma = 1.41$.

A four stroke petrol engine has a stroke - volume of 5.7 litres. Its mean effective pressure is 600 kPa and RPM is 800, find the indicated power of the engine.

7.23 Calculate the ideal efficiency of Otto cycle engine from the following data : Cylinder diameter = 100 mm, Length of stroke = 150 mm clearance volume = 250 cm^3. Assume $\gamma = 1.4$.

7.24 In an ideal diesel cycle the temperature at the beginning and end of compression are 57°C and 603°C respectively. The temperature at the beginning and end of expansion are 1950°C and 870°C respectively. Determine the ideal efficiency of cycle take $\gamma = 1.4$. If the compression ratio is 14 and pressure at the beginning of compression is 1 bar, calculate the maximum pressure in the cycle.

7.25 A petrol engine with a stroke length of 200 mm and diameter of 150 mm has a clearance volume of 5×10^3 mm^3. The indicated thermal efficiency of the engine is 30%. Find the relative efficiency of this engine. If the mean effective pressure is 5 bar and the engine speed is 1000 RPM. Find indicated power of the engine. Take $\gamma = 1.4$ for air. Assume four stroke single cylinder engine.

7.26 (*i*) Describe the construction and working of a spark plug. Where it is used ? Why ?

(*ii*) An engine working on Otto cycle, air has a pressure of 1 bar and temperature of 27°C at the entry. Air is compressed adiabatically with a compression ratio of 7. The heat is added at constant volume till the temperature rises to 2000 K. Find (1) Air standard efficiency (2) Heat supplied (3) Pressure and temperature at the end of compression (4) Mean effective pressure in kPa. Take R = 287 J/kg.K and $\gamma = 1.4$

7.27 The engine is working on ideal Otto cycle. The temperature at the beginning and at the end of compression is 60°C and 400°C. Calculate the air standard efficiency.

7.28 The following data is available for an engine working on ideal Otto cycle Temperature at the beginning of compression = 50°C. Temperature at the end of compressor = 373°C. Calculate (1) Compression ratio (2) Air standard efficiency.

7.29 The following data is available for a four stroke petrol engine stroke volume = 6 Litres. Mean effective pressure = 6 bar. Speed of engine = 750 RPM. Calculate i.p. of the engine.

7.30 A six cylinder four stroke I.C. Engine is to develop 100 kW i.p. at 800 RPM. The stroke to bare ratio is 1.25 Assuming Mechanical efficiency of 80% and break mean effective pressure of 5 bar find out the diameter and stroke of the engine.

7.31 A 4-stroke, 4-cylinder, single acting spark ignition petrol engine develops 20 kW brake power at 3000 rpm. The following data are given :

Bore = 65 mm, Stroke = 95 mm, Clearance volume = 65 cm^3,

Relative efficiency on BP basis = 50%, C.V. of fuel used = 46 mJ/kg
Mechanical efficiency = 80%
Find specific fuel consumption on IP basis.

7.32 A six-cylinder, four-stroke petrol engine develops 40 kW when running at 3000 rpm. The volumetric efficiency at NTP is 85%. Indicated thermal efficiency is 25%. The A/F ratio of the mixture supplied is 15:1. CV of fuel used = 41 MJ/kg. Assume stroke/bore ratio = 1.25. Calculate the bore and stroke of the engine.

7.33 A single cylinder, four-stroke cycle oil engine works on diesel cycle. The following readings were taken when the engine was running at full load :

Area of indicator diagram	= 3 cm^2
Length of diagram	= 4 cm
Spring Constant	= 10 N/cm^2. Cm
Speed of engine	= 400 rpm
Load on brake drum	= 380 N
Spring reading	= 50 N
Diameter of brake drum	= 1.2 m
Fuel consumption	= 2.8 kg/h
CV of fuel	= 42 MJ/kg
Bore	= 16 cm
Stroke	= 20 cm

Calculate (*a*) FP of engine, (*b*) mechanical efficiency, (*c*) brake thermal efficiency, and (*d*) BMEP.

7.34 The following readings were taken during a test on a single-cylinder, four-stroke gas engine :

Speed of engine	16000 rpm
Duration of test	= 1 hour
Missed cycles	= 600
Net brake load	= 1600 N
Brake drum circumference	= 4 m
MEP	= 8 bar
Gas consumption	= 2200 litres
C.V. of gas	= 20 kJ/litre
Diameter of cylinder	= 25 cm
Stroke	= 40 cm
Compression ratio	= 6.5

Calculate (*a*) IP, (*b*) BP, (*c*) BSFC, (*d*) brake thermal efficiency, and (*e*) relative efficiency.

7.35 A 6-cylinder, 4-stroke, single acting petrol engine develops 50 kW brake power at 4200 rpm and brake thermal efficiency is 22%. The temperature and pressure at the end of suction stroke are 60°C and 0.85 bar respectively. The following data are available :

Volumetric efficiency	= 80%
Mechanical efficiency	= 82%
Theoretical A : F ratio	= 14.5:1
Air supplied	= 25% excess than theoretical
C.V. of fuel	= 46 MJ/kg
Density of petrol vapour	= 2 × density of air
Stroke/bore ratio	= 1.25

Calculate bore and stroke of engine.

8 STEAM TURBINES

8.1 PRINCIPLE OF OPERATION

A steam turbine is a device which converts part of the steam energy due to high pressure and temperature into mechanical energy. A steam turbine works upon the dynamic action of steam. Steam is caused to fall in pressure in a passage or nozzle. Due to this fall in pressure certain amount of heat energy is converted into kinetic energy to give it high velocity. The high velocity particles of steam impinge on the moving blades of the turbine and have change in direction of motion. This give rise to change in momentum, therefore, a force. Actually the nozzles direct the steam so that it flows in well-formed high velocity jet. Moving buckets convert this high velocity jet to mechanical work in a rotating shaft.

In the impulse turbines, the steam pressure drop and consequent development of kinetic energy takes place solely in the stationary nozzles. The mechanical work is obtained by the conversion of this kinetic energy on moving blades. In the reaction turbine, only a part of the kinetic energy conversion occurs in the stationary nozzles, the remainder of the kinetic energy conversion being accomplished by a pressure drop in steam as it passes through the moving blades.

8.2 CLASSIFICATION OF STEAM TURBINES

On the basis of principle of operation, steam turbines are classified as:

1. Impulse turbine
2. Reaction turbine
3. Impulse-reaction turbine

1. Impulse Turbine. This type of turbine is called a simple impulse turbine because the expansion of steam takes place in one set of nozzles only, as shown in Fig. 8.1. The pressure of steam falls from that in the steam chest to that existing in the condenser while the steam flows through it. Hence the pressure in the wheel chamber is practically equal to the condenser pressure.

The high velocity jet coming out from the Nozzle impinges on the blades, so that the whole velocity energy is converted into mechanical energy. In practice, this type of turbine is used for small power ratings. The rotor diameter is kept small and consequently the rotational speed becomes very high. This is known as De-Laval turbine.

2. Reaction Turbine. In a reaction turbine, there are guide blades instead of nozzles which convert the pressure energy into kinetic energy. The steam passing over the moving blades have the difference of pressure at the inlet tip and exit tip and hence there is a drop of pressure in stream while passing over the moving blades. The fixed blades serve the purpose of nozzles which change the direction of steam and at the same time allow it to expand to a higher velocity.

The diameter of each stage of a reaction turbine must increase after each group of blade rings in order to accommodate the increased volume of steam at lower pressure.

3. Impulse-Reaction Turbine. If the pressure of steam at outlet from the moving blades of the turbine is less than that at the inlet side of the blades, the drop in pressure suffered by steam during its flow through the moving blades causes a further generation of kinetic energy within the blades. This adds to the generating force which is applied to the turbine shaft. Such a turbine is called impulse-reaction turbine.

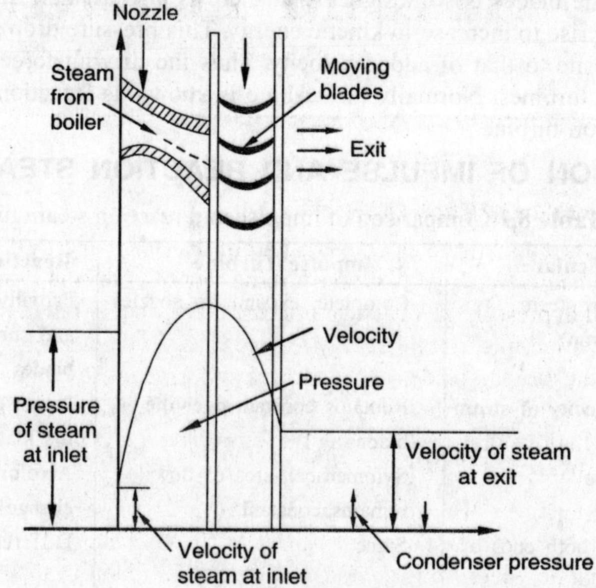

Fig. 8.1 Simple impulse turbine

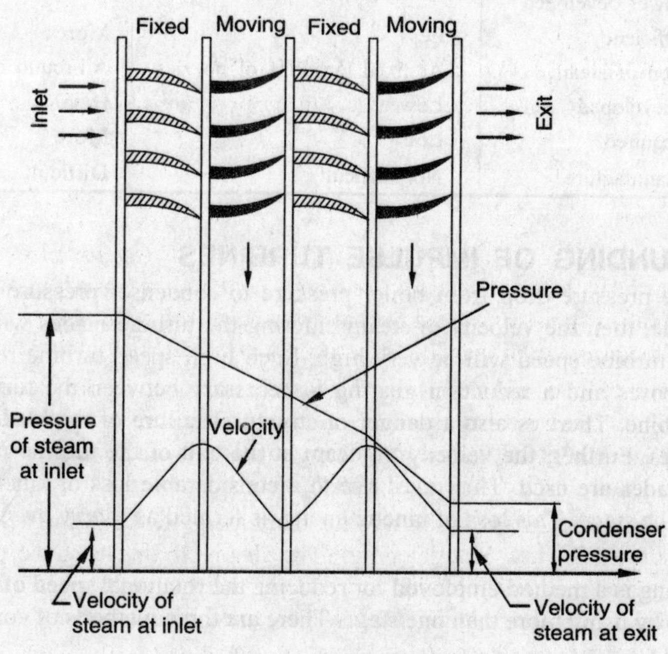

Fig. 8.2 Impulse-reaction turbine

This type of turbine is shown in Fig. 8.2. In this type of turbine, there is application of both principles, namely, impulse and reaction. There are number of rows of moving blades fixed to the shaft and equal number of fixed blades attached to the casing. The fixed blades correspond to the nozzles of an impulse turbine. Steam is admitted for whole of the circumference and therefore there is all round admission. In passing through the first set of fixed blades the steam undergoes small pressure drop and velocity is increased. It then enters the first set of moving blades and suffers the change of direction and therefore momentum. This gives an impulse to the blades. But here the passage to the blades is so designed that there is also a small drop in pressure in the moving blades giving rise to increase in kinetic energy. This pressure drop gives rise to reactions in the direction opposite to that of added velocity. Thus the driving force is the vector sum of impulse and reaction turbines. Normally the turbine is known as Reaction turbine. This is also called Parson's reaction turbine.

8.3 COMPARISON OF IMPULSE AND REACTION STEAM TURBINES

Table 8.1 Comparison of impulse and reaction steam turbines.

S. No.	Particular	Impulse Turbine	Reaction Turbine
1.	Expansion of steam (Pressure drop)	Complete, expands in nozzles	Partially in the fixed blades and subsequently in the rotor blades
2.	Relative velocity of steam	Remains constant over the blades	Increases as steam passes over the blades
3.	Blade profile	Symmetrical, area of flow remains constant	Aerofoil section, area of flow changes
4.	Pressure on both ends of moving blades	Same	Different
5.	Number of stages for same power developed	More	Less
6.	Blade efficiency	Less	More
7.	Admission of steam	At fixed locations of nozzles	All-round or complete
8.	Power developed	Low	High
9.	Space required	Less	More
10.	Blade manufacture	Not difficult	Difficult.

8.4 COMPOUNDING OF IMPULSE TURBINES

If the entire pressure drop from boiler pressure to condenser pressure is carried out in a single stage nozzle, then the velocity of steam entering the turbine blades will be very high and consequently the turbine speed will be very high. Such high speed turbine rotors are not useful for practical purposes and a reduction gearing is necessary between the turbine and generator driven by the turbine. There is also a danger of structural failure of the blades due to excessive centrifugal stresses. Further, the velocity of steam at the exit of the turbine is also high when a single stage of blades are used. This gives rise to a considerable loss of kinetic energy reducing the efficiency of the stage. This loss of kinetic energy is termed as '*carry over losses*' or '*Leaving losses*'.

Compounding is a method employed for reducing the rotational speed of the impulse turbine to practical limits by using more than one stage. There are three methods of compounding impulse turbines.

1. Velocity Compounding
2. Pressure Compounding
3. Pressure and Velocity Compounding.

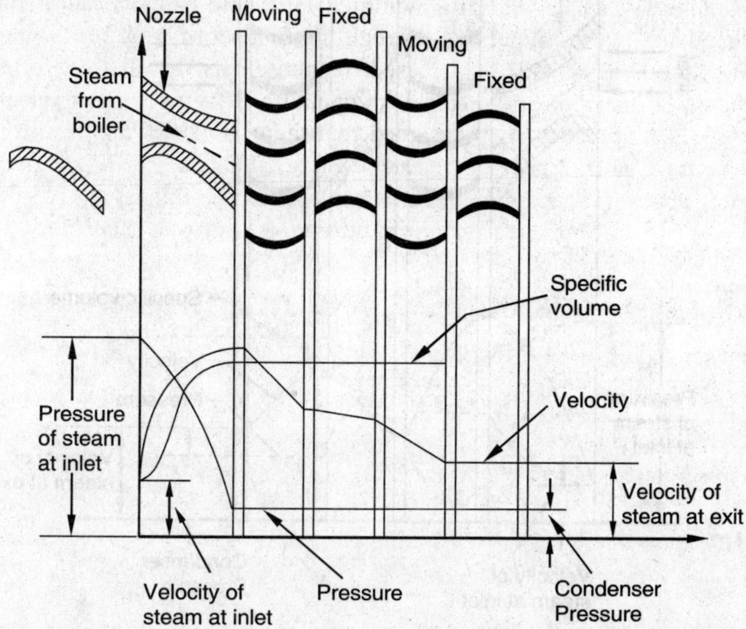

Fig. 8.3 Velocity compounded impulse turbine

1. Velocity Compounding. This system consists of a nozzle or a set of nozzles and a wheel fitted with two or more rows of blades as shown in Fig. 8.3 It has two rings of moving blades and two rows of fixed or guide blades placed as shown. Steam entering the nozzle expands from the initial pressure to the exhaust pressure. On passing through the first row of moving blades, the steam gives only a part of kinetic energy and issues from this row of blades with a fairly high velocity. It then enters first the guide blade and then on moving blades. There is slight drop in velocity in the guide blades due to friction. In passing through the second row of moving blades, the steam suffers a change in momentum and gives a part of kinetic energy to rotor. Again the leaving steam from second row of moving blades is redirected to second row of guide blades. Thus after doing work in third row of moving blades the steam finally leaves the wheel in more or less axial direction with a certain residual velocity, about 2% of the initial velocity of steam at nozzle exit. Thus the whole of velocity is not utilised on one row of moving blades but in a number of stages, as in a Curtis turbine.

2. Pressure Compounding. This arrangement consists of allowing the expansion of steam in a number of steps by having a number of simple impulse turbines in series on the same shaft, as shown in Fig. 8.4. The exhaust steam from one turbine enters the nozzles of succeeding turbines. Each of the simple impulse turbine is called as '*stage*' of the turbine comprising of its set of nozzles and blades. This is equivalent to splitting up the whole pressure drop into a series of smaller pressure drops, hence the term pressure compounding.

The pressure compounding causes a smaller transformation of heat energy into kinetic energy to take place in each stage than in simple impulse turbine. Hence, the steam velocity is much lower with the result that the blade velocity and rotational speed may be lowered. The leaving loss is only one to two percent of the total available energy. Reteau turbine is a pressure compounded turbine.

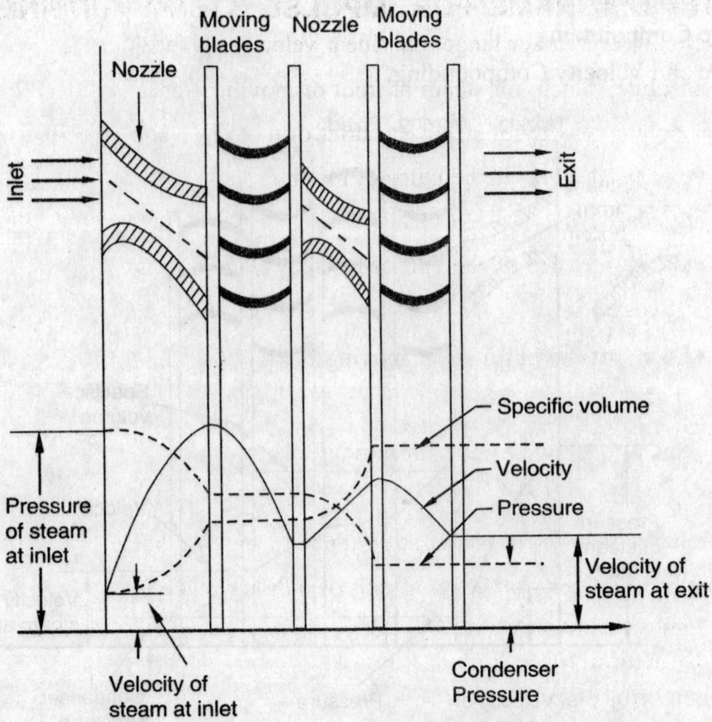

Fig. 8.4 Pressure compounded impulse turbine

3. Pressure and Velocity Compounding. It is a combination of pressure compounding and velocity compounding. The total pressure drop of steam is divided in stages and then velocity of each stage is compounded. This allows a bigger pressure drop in each stage and hence less stages are necessary which require a shorter turbine for a given pressure drop. Such a turbine is shown in Fig. 8.5. The diameter of such a turbine increases in each stage in order to accommodate for a larger volume of steam at a lower pressure. The pressure is constant in each stage.

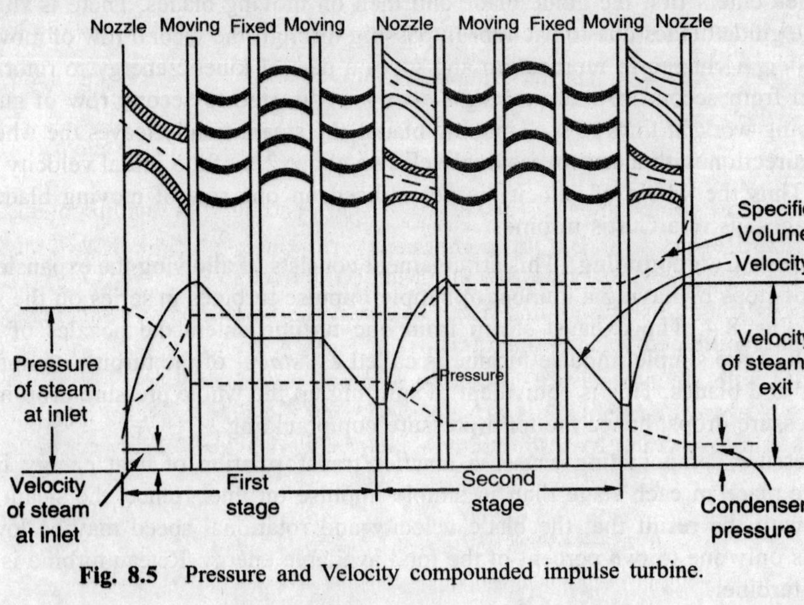

Fig. 8.5 Pressure and Velocity compounded impulse turbine.

8.5 VELOCITY DIAGRAMS FOR IMPULSE STEAM TURBINE

Let u = circumferential or tangential linear velocity of blades

V_{a1} = absolute velocity of steam at inlet of moving blade

V_{a2} = absolute velocity of steam at outlet of moving blade

V_{w1} = velocity of whirl at the entry of moving blade

$= V_{a1} \cos \alpha_1$

V_{w2} = velocity of whirl at the exit of moving blade

$= V_{a2} \cos \alpha_2$

V_{f1} = velocity of flow at inlet of moving blade

$= V_{a1} \sin \alpha_1$

V_{f2} = velocity of flow at outlet of moving blade

$= V_{a2} \sin \alpha_2$

V_{r1} = relative velocity of steam at entrance to moving blade

V_{r2} = relative velocity of steam at exit of moving blade

m = mass rate of flow of steam, kg/s

d = diameter of blade drum

h = height of blade

α_1 = angle which the absolute velocity of steam at inlet makes with the plane of moving blades, or nozzle angle or outlet angle of fixed blades.

α_2 = angle which the absolute velocity at outlet makes with the plane of moving blades or inlet angle of fixed blade.

β_1 = inlet angle of moving blade.

β_2 = exit angle of moving blade.

The jet of steam impinges on the moving blades at angle α_1 to the tangent of wheel with a velocity V_{a1}. This velocity V_{a1} has two components.

1. Tangential or whirl component V_{w1} and since it is in the same direction as the motion of blades, it is the actual component which does work on the blade.

2. Axial or flow component V_{f1} and since it is perpendicular to the direction of motion of blade, therefore it does no work. But this component is responsible for the flow of steam through the turbine. This component also causes axial thrust on the rotor.

The velocity diagrams at inlet and outlet of a moving blade are shown in Fig. 8.6 (*a*) Fig. 8.6 (*b*) shows the combined velocity diagrams.

For blades with smooth surface, it can be assumed that the friction loss is very small or zero. But there is always a certain loss of velocity during the flow of steam over the blade and this loss is taken into account by introducing a factor called blade velocity coefficient, *K*. It is given by

$$K = \frac{V_{r2}}{V_{r1}} \quad \text{where } V_{r2} < V_{r1}$$

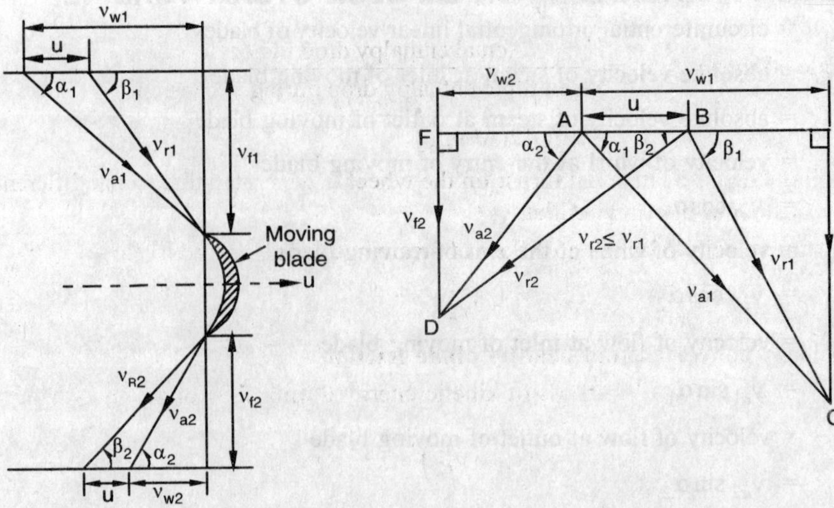

(*a*) Velocity diagrams at inlet and exit (*b*) Combined velocity diagrams

Fig. 8.6 Velocity diagrams for impulse turbine.

(*a*) Power Developed by the Turbine

Work done per kg of steam,

$$w = \text{Force in the direction of blade motion} \times \text{Distance travelled in the direction of force}$$

$$= \text{Rate of change of momentum} \times \text{Distance travelled}$$

$$= [\nu_{w1} - (-\nu_{w2})] \times u$$

$$= (\nu_{w1} + \nu_{w2})u = \nu_w u$$

Where $\nu_w = \nu_{w1} + \nu_{w2}$

Power developed by the turbine for a mass rate of flow of $\dot{m}$ kg/s of steam.

$$P = \frac{\dot{m}(\nu_{w1} + \nu_{w2})u}{10^3} \text{ kW} \qquad \ldots(8.1)$$

(*b*)Diagram or Blade Efficiency, η_b

For a single blade stage,

$$\eta_b = \frac{\text{Work done on blades per kg of steam}}{\text{Energy supplied per keg of steam}}$$

$$= \frac{(\nu'_{w1} + \nu'_{w2})u}{\nu_{a1}^2/2} = \frac{2(\nu_{w1} + \nu_{w2})u}{\nu_{a1}^2} \qquad \ldots(8.2)$$

(*c*) Gross or Stage Efficiency, η_s

$$\eta_s = \frac{\text{Work done per kg of steam}}{\text{Theoretical enthalpy drop in the nozzle per kg of steam}}$$

$$= \frac{(\nu_{w1} + \nu_{w2})u}{(\Delta h)_{isen}} \qquad \ldots(8.3)$$

This is efficiency takes into account the losses in the nozzle also

$$\therefore \qquad \eta_s = \eta_b \eta_n \qquad \text{...(8.4)}$$

where Nozzle efficiency, $\eta_n = \dfrac{\text{Actual enthalpy drop in nozzle}}{\text{Isentropic enthalpy drop during expansion}}$

$$= v_{a1}^2 / [2(\Delta h)_{\text{isen}}]$$

***(d)* Axial Thrust :** The axial thrust on the wheel is generated due to the difference between the velocity of flow at inlet and outlet.

Axial thrust, F_a = Mass rate of flow of steam × Change in axial velocity

$$= \dot{m}(v_{f1} - v_{f2}) \qquad \text{...(8.5)}$$

***(e)* Energy converted into heat by blade friction**

= Loss of kinetic energy during flow of steam over the blades

$$= \frac{1}{2}\dot{m}\left(v_{r1}^2 - v_{r2}^2\right) \qquad \text{...(8.6)}$$

(f) Energy lost at exist $= \dfrac{1}{2}\dot{m}\, v_{r2}^2$

8.5.1 Condition for Maximum Blade Efficiency

Blade efficiency, $\eta_b = \dfrac{2(v_{w1} + v_{w2})u}{v_{a1}^2}$

From velocity diagrams of Fig. 8.6 (*b*), we have

$$v_{w1} + v_{w2} = AE + AF = (AB + BE) + (BF - AB)$$

$$= BE + BF$$

$$= v_{r1}\cos\beta_1 + v_{r2}\cos\beta_2$$

$$= v_{r1}\cos\beta_1\left[1 + \frac{v_{r2}}{v_{r1}} \cdot \frac{\cos\beta_2}{\cos\beta_1}\right]$$

$$= v_{r1}\cos\beta_1[1 + KC]$$

where $K = \dfrac{v_{r2}}{v_{r1}}$ and $C = \dfrac{\cos\beta_2}{\cos\beta_1}$

Now $BE = AE - AB = v_{a1}\cos\alpha_1 - u$

$$\therefore \qquad \eta_b = \frac{2u(v_{a1}\cos\alpha_1 - u)(1 + KC)}{v_{a1}^2} = 2\frac{u}{v_{a1}}\left[\left(\cos\alpha_1 - \frac{u}{v_{a1}}\right)(1 + KC)\right]$$

Let speed ratio, $\rho = \dfrac{u}{v_{a1}}$

$$\therefore \qquad \eta_b = 2\rho[(\cos\alpha_1 - \rho)(1 + KC)] \qquad \text{...(8.8)}$$

For η_b to be maximum,

$$\frac{d}{d\rho}(\eta_b) = 0$$

$$\cos\alpha_1 - 2\rho = 0$$

or
$$\rho = \frac{\cos\alpha_1}{2} \qquad \text{...(8.9)}$$

$$(\eta_b)_{max} = \cos\alpha_1 - \left[\left(\cos\alpha_1 - \frac{\cos\alpha_1}{2}\right)(1+KC)\right]$$

$$= \frac{\cos^2\alpha_1}{2}(1+KC) \quad \ldots(8.10)$$

For symmetrical blades, $\beta_1 = \beta_2$ so that $C = 1$ and with no friction over the blade, $K = 1$

$$(\eta_b)_{max} = \cos^2\alpha_1 \quad \ldots(8.11)$$

8.5.2 Maximum Work Done

Now $\quad w = (v_{w1} + v_{w2})u$

$$= (v_{a1}\cos\alpha_1 - u)(1+KC)u$$

$$= 2u(v_{a1}\cos\alpha_1 - u) \text{ for } K = 1, C = 1$$

$$w_{max} = 2u[v_{a1} \times 2\rho - u]$$

$$= 2u\left[v_{a1} \times \frac{2u}{v_{a1}} - u\right]$$

$$= 2u^2 \quad \ldots(8.12)$$

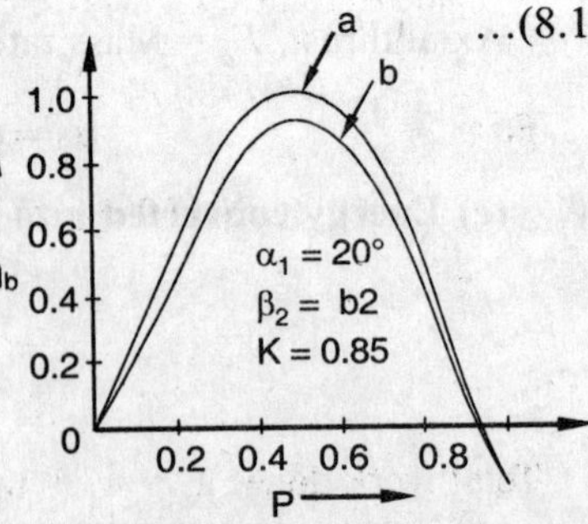

Fig. 8.7 Blade efficiency v's speed ratio (a) without losses (b) with losses

The Eqs. (8.11) and (8.12) represent a parabola. The blade efficiency has been plotted against speed ratio in Fig. 8.7, without losses and with losses for $\alpha_1 = 20°$, $\beta_1 = \beta_2$ and $K = 0.85$.

8.5.3 Velocity Diagrams for Velocity Compounded Impulse Turbine

The velocity diagrams for the first and second stage moving blades of a velocity compounded impulse turbine are shown in Fig. 8.8 (*a*) and Fig. 8.8(*b*) respectively. The final absolute velocity of steam leaving the second row is axial. The velocity *u* of the blades for both the rows is same as they are mounted on the same shaft and are of equal height.

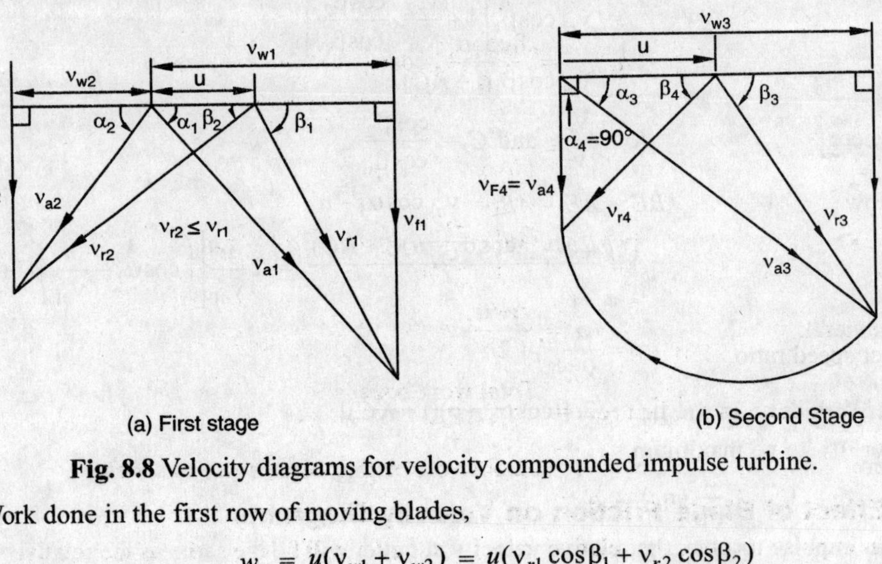

(a) First stage (b) Second Stage

Fig. 8.8 Velocity diagrams for velocity compounded impulse turbine.

Work done in the first row of moving blades,

$$w_1 = u(v_{w1} + v_{w2}) = u(v_{r1}\cos\beta_1 + v_{r2}\cos\beta_2)$$

If there is no friction loss, $v_{r1} = v_{r2}$, and for symmetrical blades, $\beta_1 = \beta_2$.

$$w_1 = 2u\ v_{r1}\cos\beta_1 = 2u(v_{a1}\cos\alpha_1 - u) \quad \ldots(1)$$

Now $\quad v_{a3} = v_{a2}$

Work done in second row of moving blades,

$$w_2 = u v_{w3} \text{ as } v_{w2} = 0 \text{ and } \alpha_4 = 90°$$
$$= u(v_{r3}\cos\beta_3 + v_{r4}\cos\beta_4)$$

For no friction, $v_{r3} = v_{r4}$, and for symmetrical blades, $\beta_3 = \beta_4$.

$$w_2 = 2u v_{r3}\cos\beta_3 = 2u(v_{a3}\cos\alpha_3 - u) \quad \ldots(2)$$

For $\quad \alpha_3 = \alpha_2$

$$v_{a3}\cos\alpha_3 = v_{a2}\cos\alpha_2$$
$$= v_{r2}\cos\beta_2 - u = v_{r1}\cos\beta_1 - u$$
$$= (v_{a1}\cos\alpha_1 - u) - u = v_{a1}\cos\alpha_1 - 2u$$
$$w_2 = 2u[(v_{a1}\cos\alpha_1 - 2u) - u] = 2u(v_{a1}\cos\alpha_1 - 3u) \quad \ldots(3)$$

Total work done, $w_t = w_1 + w_2$

$$= 2u(v_{a1}\cos\alpha_1 - u) + 2u(v_{a1}\cos\alpha_1 - 3u)$$
$$= 2u(2v_{a1}\cos\alpha_1 - 4u) = 4u(v_{a1}\cos\alpha_1 - 2u)$$

$$\text{Blade efficiency, } \eta_b = \frac{w_t}{v_{a1}^2/2} = \frac{8u}{v_{a1}^2}(v_{a1}\cos\alpha - 2u) = 8\rho(\cos\alpha_1 - 2\rho) \quad \ldots(8.13)$$

For η_b to be maximum, $\dfrac{d\eta_b}{d\rho} = 0$

$$\cos\alpha_1 - 4\rho = 0$$

$$\rho = \frac{\cos\alpha_1}{4} \quad \ldots(8.14)$$

$$(\eta_b)_{max} = \frac{8\cos\alpha_1}{4}\left(\cos\alpha_1 - \frac{\cos\alpha_1}{2}\right) = \cos^2\alpha_1 \quad \ldots(8.15)$$

For $\quad v_{a1} = \dfrac{4u}{\cos\alpha_1}$

$$(w_t)_{max} = 4u\left[\frac{4u}{\cos\alpha_1}\cdot\cos\alpha_1 - 2u\right] = 8u^2 \quad \ldots(8.16)$$

In general, $\quad \rho = \dfrac{\cos\alpha_1}{2n} \quad \ldots(4.17)$

and Work done in the last row $= \dfrac{\text{Total work done}}{2^n} \quad \ldots(4.18)$

where $\quad n$ = number of rotating blade rows in series.

8.5.4 Effect of Blade Friction on Velocity Diagrams

In an impulse turbine, the relative velocity at outlet will be the same as the relative velocity at inlet, if friction is neglected. In practice, there is a frictional resistance to the flow of steam jet over the blade, the effect of which is to cause a slowing down of the relative velocity. Usually, there is a loss of 10 to 15 percent in the relative velocity due to friction. Owing to the frictional resistance of the blades, it will be found that $v_{r2} = K v_{r1}$, where K is a blade velocity coefficient which takes into account the blade velocity loss due to friction.

The velocity diagrams considering blade friction are shown in Fig. 8.9. In this diagram, inlet diagram is first drawn and the line BC, of an unknown length, is drawn at the correct angle β_2. Mark, off on line $BD = v_{r1}$, the friction loss of relative velocity DD', then $BD' = K v_{r1}$. With B as centre, draw an arc of radius BD' to cut BC at C. Then $BC = v_{r2} = K v_{r1}$.

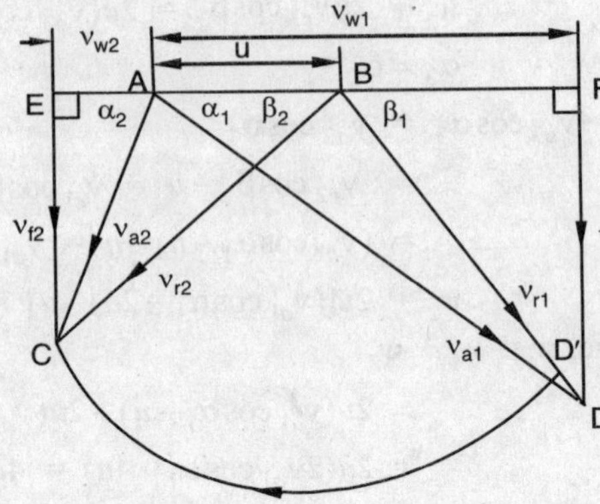

Fig 8.9 Velocity diagrams for impulse turbine considering blade friction

By joining A and C, the line AC representing v_{a2} is obtained. This completes the outlet velocity diagram.

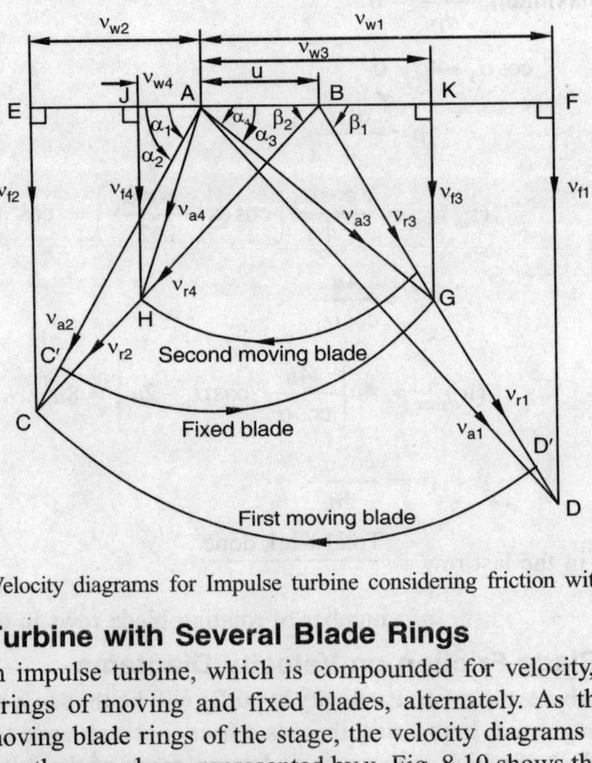

Fig. 8.10 Velocity diagrams for Impulse turbine considering friction with several blade rings.

8.5.5 Impulse Turbine with Several Blade Rings

The stage of an impulse turbine, which is compounded for velocity, consists of a ring of nozzles followed by rings of moving and fixed blades, alternately. As the blade velocity u is constant for all the moving blade rings of the stage, the velocity diagrams for all the blade rings can be superimposed on the same base, represented by u. Fig. 8.10 shows the velocity diagram for a stage consisting of two moving and one fixed blade rings. Let us assume that following data is known :

1. Blade velocity, u
2. Nozzle angle, α
3. The moving blade angles, β_1 and β_2, which are assumed to be same for both blade rings.
4. The velocity of steam v_{a1} discharged from nozzle.
5. Blade friction loss 10% i.e., $K = 0.9$.

The following steps may be followed for drawing the velocity diagrams :

1. Draw $AB = u$ to any convenient scale. Draw AD inclined at angle α_1 to AB so that $AD = v_{a1}$ to the same scale. Join BD. Then $BD = v_{r1}$ for the first moving blade ring.
2. Mark off $DD' = 0.1 \times BD$ so that $v_{r2} = 0.9\, v_{r1} = BD'$. With B as centre, draw an arc of radius BD' to cut BC at C where the line BC is drawn at angle of β_2 to BA. Joint AC. Then $AC = v_{a2}$ for the first moving blade ring.
3. The steam now flows over the fixed blade ring and will lose 10% of its velocity during the passage. Hence, mark off $CC' = 0.1 \times AC$. With A as centre, draw an arc of radius AC' to cut BD at G. Then $AG = v_{a3}$ representing the steam velocity when entering the second moving blade ring. The velocity diagram for the second blade ring is triangle AGB. $BG = v_{r3}$, the relative velocity.
4. The steam now flow over the second moving blade and loses one-tenth of its relative velocity due to friction. Hence, mark off $GG' = 0.1 \times BG$ so that $BG' = 0.9\, BG$. With B as centre and radius BG', draw an arc to cut BC at H, then $AH = v_{a4}$ and $BH = v_{r4}$.

α_2 = angle of discharge from first moving blade

α_4 = angle of discharge from second moving blade

α_3 = outlet angle of fixed blade.

Work done per kg of steam for first moving blade ring.

$$w_1 = (v_{w1} + v_{w2})u$$

Work done per kg of steam for second moving blade ring,

$$w_2 = (v_{w3} + v_{w4})u$$

Total work done, $w = w_1 + w_2$

$$= [(v_{w1} + v_{w2}) + (v_{w3} + v_{w4})]u$$

$$\text{Power developed per stage} = \frac{\dot{m}u\Sigma(v_{w1} + v_{w2})}{10^3} \text{ kW}$$

$$\text{Blade efficiency, } \eta_b = \frac{2u\Sigma(v_{w1} + v_{w2})}{v_{a1}^2}$$

$$\text{Stage efficiency, } \eta_s = \frac{u\Sigma(v_{w1} + v_{w2})}{(\Delta h)_{\text{isen}}}$$

$$\text{Total axial thrust} = \dot{m}\Sigma(v_{f1} - v_{f2})$$

8.6 ADVANTAGES AND LIMITATIONS OF VELOCITY COMPOUNDING

Advantages :

1. Fewer number of stages are required and therefore initial cost is less.
2. The space required is less.

3. The pressure in the housing is considerably less which requires cheaper and thin casing for the turbine.
4. All the stages are not used with equal economy.

Limitations.

1. The friction losses are larger due to the high velocity of steam.
2. The maximum blade efficiency and efficiency range decrease with the increase in number of stages.
3. The power developed in each successive blade row decrease with increase in number of rows
4. All the stages are not used with equal economy.

8.7 VELOCITY DIAGRAMS FOR IMPULSE-REACTION TURBINE

Steam passes to the fixed blades with velocity v_{a1}. Fixed blades also act as nozzles with pressure drop occurring while steam passes through them. So there is gain in kinetic energy and steam leaves the fixed blades with velocity v_{a2}. The expansion of steam and the enthalpy drop occur both in fixed and moving blades. The velocity diagrams for an impulse-reaction turbine are shown in Fig. 8.11.

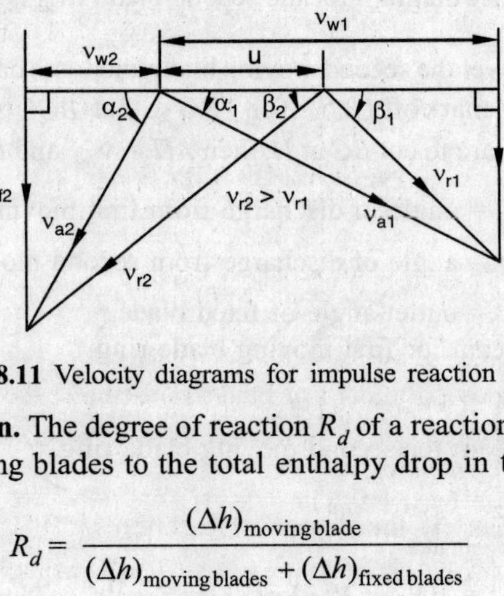

Fig. 8.11 Velocity diagrams for impulse reaction turbine

***(a)* Degree of Reaction.** The degree of reaction R_d of a reaction turbine is defined as the ratio of enthalpy drop over moving blades to the total enthalpy drop in the stage. Thus,

$$R_d = \frac{(\Delta h)_{\text{moving blade}}}{(\Delta h)_{\text{moving blades}} + (\Delta h)_{\text{fixed blades}}}$$

$$= \frac{\Delta h_m}{\Delta h_m + \Delta h_f} \quad \text{...(8.19)}$$

Enthalpy drop through fixed blades, $\Delta h_f = \dfrac{v_{a1}^2 - v_{a2}^2}{2}$

Enthalpy drop along the moving blades, $\Delta h_m = \dfrac{v_{r2}^2 - v_{r1}^2}{2}$

Kinetic energy supplied to moving blades, K.E. $= \dfrac{1}{2}[v_{a1}^2 + (v_{r2}^2 - v_{r1}^2)]$

Neglecting friction of blade surface, $\dfrac{v_{r2}^2}{2} = \dfrac{v_{r1}^2}{2} + \Delta h_m$.

Axial thrust on rotor = $(\nu_{f1} - \nu_{f2}) + (\Delta p)_m \times$ area of blade disc.

Now total enthalpy drop in the stage = Work done by steam in the stage

$$\Delta h_f + \Delta h_m = u(\nu_{w1} + \nu_{w2})$$

$$R_d = \frac{\nu_{r2}^2 - \nu_{r1}^2}{2u(\nu_{w1} + \nu_{w2})} \quad ...(8.20)$$

From Fig. 8.9, we have

$$\nu_{r2} = \nu_{f2} \operatorname{cosec} \beta_2 \text{ and } \nu_{r1} = \nu_{f1} \operatorname{cosec} \beta_1$$

and $\quad \nu_{w1} + \nu_{w2} = \nu_{f1} \cot \beta_1 + \nu_{f2} \cot \beta_2$

The velocity of flow generally remains constant through the blades.

$$\therefore \quad \nu_{f1} = \nu_{f2} = \nu_f$$

$$R_d = \frac{\nu_f^2 (\operatorname{cosec}^2 \beta_2 - \operatorname{cosec}^2 \beta_1)}{2u\nu_f (\cot \beta_1 + \cot \beta_2)} = \frac{\nu_f}{2u}\left[\frac{(\cot^2 \beta_2 + 1) - (\cot^2 \beta_1 + 1)}{\cot \beta_1 + \cot \beta_2}\right]$$

$$= \frac{\nu_f}{2u}\left[\frac{\cot^2 \beta_2 - \cot^2 \beta_1}{\cot \beta_1 + \cot \beta_2}\right] = \frac{\nu_f}{2u}(\cot \beta_2 - \cot \beta_1) \quad ...(8.21)$$

If $\quad R_d = 0.5$, then $\frac{1}{2} = \frac{\nu_f}{2u}(\cot \beta_2 - \cot \beta_1)$

or $\quad u = \nu_f (\cot \beta_2 - \cot \beta_1) \quad ...(8.22)$

From Fig. 8.9, we have

$$u = \nu_f (\cot \beta_2 - \cot \alpha_2) = \nu_f (\cot \alpha_1 - \cot \beta_1)$$

Hence $\quad \beta_1 = \alpha_2$ and $\beta_2 = \alpha_1 \quad ...(8.23)$

Which means that the moving and fixed blades must have the same shape if the degree of reaction is 50%. This condition gives symmetrical velocity diagrams. This type of turbine is known as '*Pearson's Reaction Turbine*'.

(*b*) Efficiency. We assume that (*i*) the degree of reaction is 50%, *i.e.*, $\Delta h_f = \Delta h_m$, (*ii*) the moving and fixed blades are of the same shape, and (*iii*) the velocity of steam at exit from the preceding stage is same as the velocity of steam at the entrance to the succeeding stage.

The work done per kg of steam, $w = u(\nu_{w1} + \nu_{w2}) = u[\nu_{a1} \cos \alpha_1 + (\nu_{r2} \cos \beta_2 - u)]$

Now $\beta_2 = \alpha_1$ and $\quad \nu_{r2} = \nu_{a1}$

$$w = u(2\nu_{a1} \cos \alpha_1 - u) = \nu_{a1}^2 \left[\frac{2u\,\nu_{a1} \cos \alpha_1}{\nu_{a1}^2} - \frac{u^2}{\nu_{a1}^2}\right]$$

$$= \nu_{a1}^2 (2\rho \cos \alpha_1 - \rho^2) \quad ...(8.24)$$

where $\quad$ speed ratio, $\rho = \frac{u}{\nu_{a1}}$

K.E. supplied to fixed blade $= \frac{\nu_{a1}^2}{2}$

K.E. supplied to moving blade $= \frac{\nu_{r2}^2 - \nu_{r1}^2}{2}$

Total energy supplied to stage, $\Delta h = \dfrac{v_{a1}^2}{2} + \dfrac{v_{r2}^2 - v_{r1}^2}{2}$

For symmetrical blades, $v_{r2} = v_{a1}$

$$\Delta h = \frac{v_{a1}^2}{2} + \frac{v_{a1}^2 - v_{r1}^2}{2} = v_{a1}^2 - \frac{v_{r1}^2}{2} \qquad \ldots(8.25)$$

From Fig. 8.9, we have

$$v_{r1}^2 = v_{a1}^2 + u^2 - 2v_{a1}u\cos\alpha_1$$

$$\therefore \qquad \Delta h = v_{a1}^2 - \frac{1}{2}\left[v_{a1}^2 + u^2 - 2v_{a1}u\cos\alpha_1\right]$$

$$= \frac{1}{2}[v_{a1}^2 + 2v_{a1}u\cos\alpha_1 - u^2]$$

$$= \frac{v_{a1}^2}{2}\left[1 + \frac{2u}{v_{a1}}\cos\alpha_1 \frac{u^2}{v_{a1}^2}\right]$$

$$= \frac{v_{a1}^2}{2}[1 + 2\rho\cos\alpha_1 - \rho^2] \qquad \ldots(8.26)$$

Blade efficiency,

$$\eta_b = \frac{w}{\Delta h} = \frac{2\rho\cos\alpha_1 - \rho^2}{\frac{1}{2}(1 + 2\rho\alpha_1 - \rho^2)} = \frac{2\rho(\cos\alpha_1 - \rho)}{1 + 2\rho\cos\alpha_1 - \rho^2} \qquad \ldots(8.27)$$

For η_b to be maximum, $\dfrac{d\eta_b}{d\rho} = 0$

$$(1 + 2\rho\cos\alpha_1 - \rho^2)(4\cos\alpha_1 - 4\rho) - 2\rho(2\cos\alpha_1 - \rho)(2\cos\alpha_1 - \rho) = 0$$

$$4(\cos\alpha_1 - \rho)(1 + 2\rho\cos\alpha_1 - \rho^2) - 4\rho(\cos\alpha_1 - \rho)(2\cos\alpha_1 - \rho) = 0$$

$$(\cos\alpha_1 - \rho)[(1 + 2\rho\cos\alpha_1 - \rho^2) - \rho(2\cos\alpha_1 - \rho)] = 0$$

$$\cos\alpha_1 - \rho = 0$$

$$\rho = \cos\alpha_1 \qquad \ldots(8.28)$$

$$(\eta_b)_{max} = \frac{2\cos\alpha_1(2\cos\alpha_1 - \cos\alpha_1)}{1 + 2\cos^2\alpha_1 - \cos^2\alpha_1} = \frac{2\cos^2\alpha_1}{1 + \cos^2\alpha_1} \qquad \ldots(8.29)$$

(c) Height of Turbine Blading

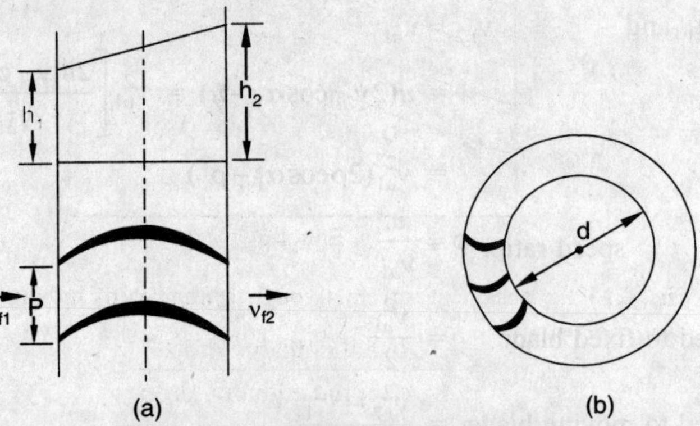

Fig. 8.12 Height of Turbine Blading.

The height and thickness of blading are shown in Fig. 8.12.

Volume flow = Area × flow velocity

$$\therefore \qquad \dot{m}\, x v_s = (\pi d - nt_1) h_1 v_{f1} = (\pi d - nt_2) h_2 v_{f2} \qquad \ldots(8.30)$$

where d = mean diameter of turbine wheel

h_1, h_2 = blade heights at inlet and outlet respectively

t_1, t_2 = thickness of blades at inlet and outlet respectively.

u = number of blades

For most of the turbines,

$$v_{f1} = v_{f2} = v_f \text{ and } h_1 = h_2 = h$$

and

$$t_1 = t_2 = t$$

$$\therefore \qquad \dot{m}\, x v_s = (\pi d - nt) h v_f \qquad \ldots(8.31)$$

Eq. (8.31) gives height of blades.

The pitch of blades, p is given by :

$$\dot{m}\, x v_s = [n(p-t)h] v_f$$

Generally $t << p$,

$$\dot{m}\, x v_8 = n\, p h v_f = n d h v_f$$

$$p = \frac{\pi d}{n} \qquad \ldots(8.32)$$

8.8 REHEAT FACTOR

The expansion of steam through a number of stages of turbine is shown in Fig. 8.13. A_1B_1 represents isentropic expansion in the first stage. The actual state of steam with frictional reheating is shown by point A_2. Therefore the actual heat drop in first stage is A_1C_1. Similarly the isentropic and actual heat drops in the succeeding stages are shown by A_2B_2, A_3B_3, ..., and A_2C_2, A_3C_3, ..., and so on respectively. A_1D represents the isentropic heat drop between supply and condenser stage of steam.

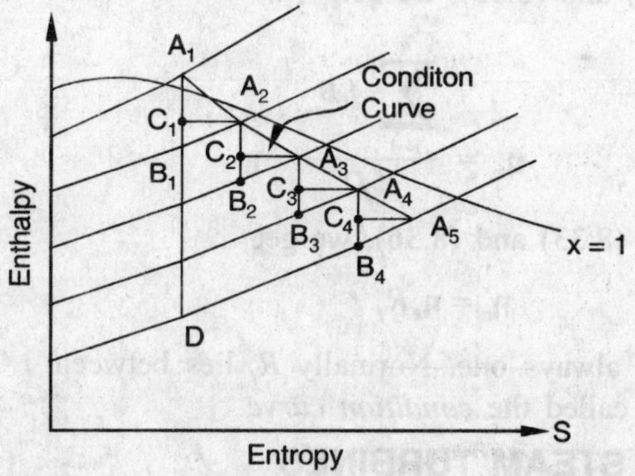

Fig. 8.13 Expansion of stream through a number of turbine stages

$$\text{Stage efficiency,} \qquad \eta_s = \frac{\text{Actual enthalpy drop}}{\text{Isentrophic enthalpy drop}}$$

For first stage, $$\eta_{s1} = \frac{A_1C_1}{A_1B_1}$$

For second stage, $$\eta_{s1} = \frac{A_2C_2}{A_2B_2} \text{ and so on.}$$

Reheat factor $$= \frac{\text{Cumulative enthalpy drop}}{\text{Isentropic heat drop}}$$

$$R_f = \frac{A_1B_1 + A_2B_2 + \ldots}{A_1D} = \frac{\sum_{i=1}^{n} A_iB_i}{A_1D} \quad \ldots(8.33)$$

Turbine internal efficiency $$= \frac{\text{Total actual turbine work}}{\text{Isentropic work}}$$

$$\eta_i = \frac{A_1C_1 + A_2C_2 + \ldots}{A_1D} = \frac{\sum_{i=1}^{n} AiCi}{A_1D} \quad \ldots(8.34)$$

If $$\eta_{s1} = \eta_{s2} = \ldots = \eta_s, \text{ then}$$

$$\eta_s = \frac{A_1C_1}{A_1B_1} = \frac{A_2C_2}{A_2B_2} = \ldots = \frac{A_1C_1 + A_2C_2 + \ldots}{A_1B_1 + A_2B_2 + \ldots}$$

or $$\eta_s = \frac{\sum_{i=1}^{n} A_iC_i}{\sum_{i=1}^{n} A_iB_i} \quad \ldots(8.35)$$

From Eqs. (8.34) and (8.35), we get

$$\eta_i = \frac{\sum_{i=1}^{n} A_iB_i}{A_1D} \times \eta_s \quad \ldots(8.36)$$

Combining Eqs. (8.33) and (8.36), we get

$$\eta_i = \eta_s R_f$$

Reheat factor is always one. Normally R_f lies between 1.02 to 1.05. The line passing through A_1, A_2, ... is called the *condition curve.*

8.9 LOSSES IN STEAM TURBINES

The various losses in steam turbines and their causes are as listed below :

1. Residual velocity loss. This loss is due to the absolute exit velocity of steam and is equivalent to $v_{a2}^2/2$, where v_{a2} is the absolute exit velocity of steam. In a single stage turbine it may be about 10 to 12% and can be reduced by using multistages.

2. Frictional and turbulence loss. Friction loss mainly occurs in nozzles and turbine blades. The friction loss is accounted for by nozzle efficiency. The loss due to friction and turbulence is about 10%.

3. Leakage loss. It occurs at the following points:

(*i*) Between the turbine shaft and bearings.

(*ii*) Between the shaft and stationary diaphragms carrying nozzles, and blade tips.

(*iii*) Through the labyrinth glands.

4. Mechanical Friction Loss. Friction between shaft and bearing and regulating valves account for this loss. It can be reduced by proper lubrication.

5. Wet steam loss. The velocity of water particles is less than that of steam and therefore the water particles have to be dragged with the steam causing loss of kinetic energy.

6. Radiation loss. This is due to the much higher temperature of turbine as compared to the surroundings. This loss can be reduced by proper insulation.

8.10 TURBINE EFFICIENCIES

1. Blade or diagram efficiency, $\eta_b = \dfrac{\text{Work done on the blade}}{\text{Energy entering the blade}}$

2. Stage efficiency, $\eta_s = \dfrac{\text{Net workdone on shaft per stage}}{\text{Adiabatic heat drop per stage}}$

$$= \frac{\text{Network done on blades – Disc friction and windage}}{\text{Adiabatic heat drop per stage}}$$

3. Internal efficiency, $\eta_i = \dfrac{\text{Heat converted into useful work}}{\text{Total isentropic heat drop}}$

4. Overall or turbine efficiency, $\eta_{\text{overall}} = \dfrac{\text{Work delivered at turbine coupling}}{\text{Total adiabatic heat drop}}$

5. Net efficiency or efficiency ratio, $\eta_{\text{net}} = \dfrac{\text{Brake thermal efficiency}}{\text{Rankine thermal efficiency}}$

8.11 GOVERNING OF STEAM TURBINES

The purpose of governing is to maintain the speed of a turbine sensibly constant irrespective of the load. The various methods of governing are :

1. Throttle governing
2. Nozzle control governing
3. By-pass governing
4. Combination of 1 and 2
5. Combination of 1 and 3.

1. Throttle Governing. The line diagram of throttle governing is shown in Fig. 8.14. The steam flow to the turbine is throttled by a balanced throttle valve actuated by a centrifugal governor. An oil differential relay is incorporated to magnify the small force produced by the governor for a small change of speed to actuate the throttle valve. The throttle valve is moved by a relay piston. A floating differential lever is fixed to its one end and at some intermediate point a piston valve is fixed. The pilot piston valve consists of two piston valves covering ports without any overlap. The piston valves are also operated by lubricating oil supplied by a pump at about 3 to 4 bar. The oil from this chamber is returned by the oil drain.

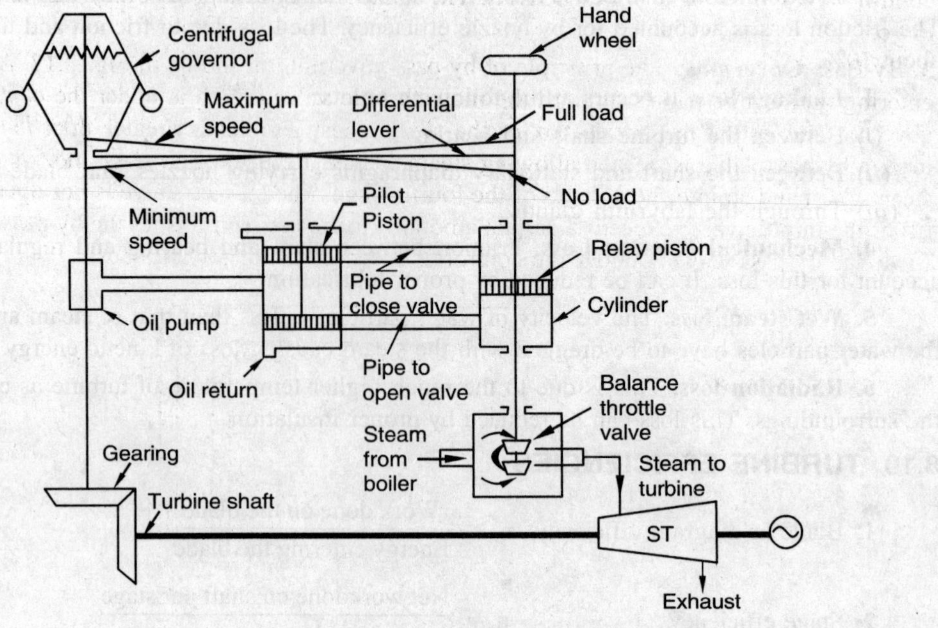

Fig. 8.14 Throttle governing

Operation. Let the turbine work at full rated load at the rated constant speed. If the load is reduced the energy supplied to the turbine will be in excess and the turbine rotor will accelerate. Thus the governor sleeve will lift. Since the throttle valve position is assumed same momentarily, the pilot piston valve spindle will get lifted opening the upper port to oil pressure and lower port to oil return. The relay piston will thus close the throttle valve partially. The lowering of throttle valve spindle will lower the pilot piston spindle and closes the ports. As soon as the ports are closed, the relay piston gets stabilized in one position corresponding to the reduced load. The change in available enthalpy for throttle governing is shown in Fig. 8.15.

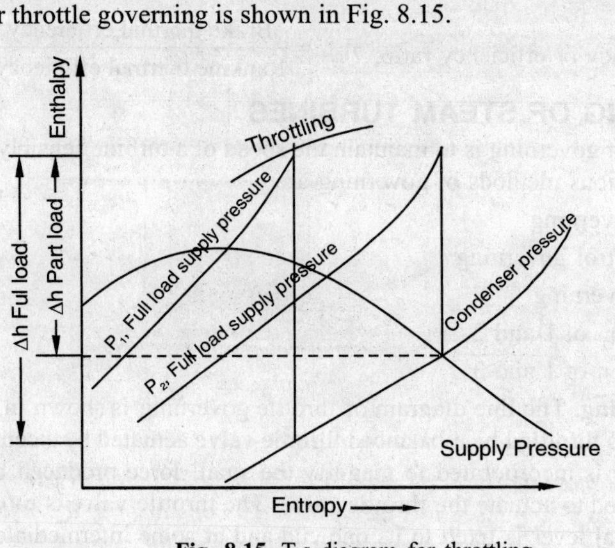

Fig. 8.15. T-s diagram for throttling

2. Nozzles Control Governing. The principle of nozzle control governing is accomplished by uncovering as many steam passages as are necessary to meet the load by poppet valves. An arrangement is shown in Fig. 8.16. The nozzles are divided into groups N_1, N_2 and N_3 under the control valves V_1, V_2 and V_3 respectively. The number of nozzle groups may vary from 3 to 5 or more.

3. By-Pass Governing. The principle of by-pass governing is shown in Fig. 8.17. All the steam entering the turbine passes through the main throttle valve which is under the control of speed governor and enters the nozzle box or the steam chest. For loads greater than the economical load, a by-pass valve is opened allowing steam to pass from the first stage, nozzle box into the steam chest and so into the nozzles of the fourth stage. The by-pass valve is not opened until the lift of the throttle valve exceeds a certain amount. As the load diminishes the by-pass valve closes first. The by-pass valve is under the control of speed governor for all loads.

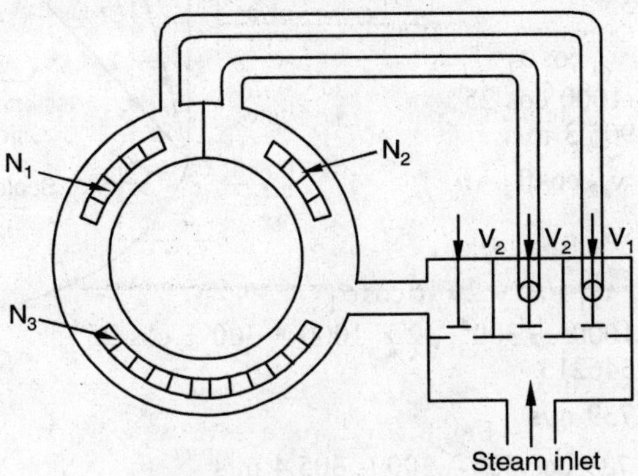

Fig. 8.16 Nozzle governing

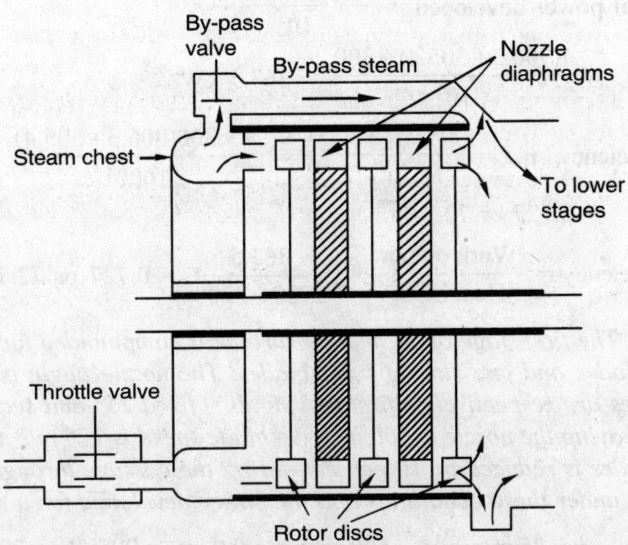

Fig. 8.17 By-pass governing.

Example 8.1 *Steam at 49 N/cm² and 160° C is supplied to a single stage impulse turbine at the rate of 60 kg/min. From there it is exhausted to a condenser at a pressure of 1.96 N/cm². The blade speed is 300 m/s. The nozzles are inclined at 25° to the plane of the wheel and outlet blade angle is 35°. Neglect friction losses and estimate (a) the theoretical power developed by the turbine, (b) the diagram efficiency, and (c) the stage efficiency.*

Solution: From Mollier diagram,

At 49 N/cm^2 and 160°C, h_1 = 2700 kJ/kg

Assuming isentropic expansion, at 1.96 N/cm^2, h_2 = 2200 kJ/kg

Inlet velocity of steam to turbine = outlet velocity of steam from nozzle.

$$\therefore \quad v_{a1} = v_2 = \sqrt{2(h_1 - h_2)} = \sqrt{2(2700 - 2200) \times 10^3} = 1000 \text{ m/s}$$

The velocity diagrams are shown in Fig. 8.18.

Blade velocity, u = 300 m/s

$$\alpha_1 = 25°$$

$$\beta_1 = 35°$$

$$v_{w1} = v_{a1} \cos \alpha_1$$
$$= 1000 \cos 25°$$
$$= 905.3 \text{ m/s}$$

$$v_{w2} = v_{r2} \cos \beta_2 - u$$

$$v_{r2} = v_{r1}$$

$$v_{r1}^2 = v_{a1}^2 + u^2 - 2v_{a1}u \cos \alpha_1$$
$$= 1000^2 + 300^2 - 2 \times 1000 \times 300 + \cos 25°$$
$$= 546215$$

$$v_{r1} = 739 \text{ m/s}$$

$$v_{w2} = 739 \cos 35° - 300 = 305.4 \text{ m/s}$$

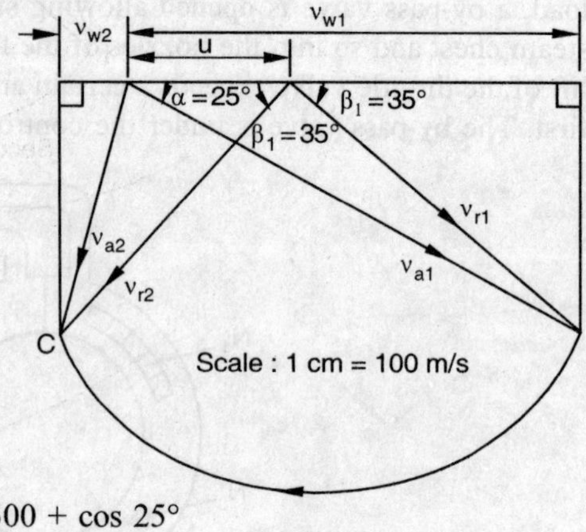

Fig. 8.18.

(*a*) Theoretical power developed $= \dfrac{\dot{m}(v_{w1} + v_{w2})u}{10^3}$

$$= \frac{1(906.3 + 305.4) \times 300}{10^3} = 363.5 \text{ kW}$$

(*b*) Blade efficiency, $\eta_b = \dfrac{2u(v_{w1} + v_{w2})}{v_{a1}^2} = \dfrac{2 \times 300(906.3 + 305.4)}{1000^2}$

$$= 0.727 \text{ or } 72.7 \text{ \%}.$$

(*c*) Stage efficiency $= \dfrac{\text{Work output}}{\text{Heat input}} = \dfrac{363.5}{2700 - 2200} = 0.727$ or 72.7%.

Example 8.2 *The first stage of an impulse turbine is compounded for velocity and has two rings of moving blades and one ring of fixed blades. The nozzle angle is 20° and the leaving angles of the blades are respectively : first moving 20°, fixed 25° and second moving 30°. The velocity of steam leaving the nozzle is 600 m/s, the blade speed is 125 m/s and the steam velocity relative to the blades is reduced by 10 percent during the passage through each ring. Find the diagram efficiency under these conditions and the power developed for a steam flow of 4 kg/s.*

Solution. $u = 25$ m/s, $v_{a1} = 600$ m/s, $K = 0.9$, $\alpha_1 = 20°$, $\beta_2 = 20°$, $\beta_3 = 25°$, $\beta_4 = 30°$

Choose a scale of 1 cm = 50 m/s. The velocity diagrams are drawn in Fig. 8.19

$ab = u$ = 125 m/s = 2.5 cm

$\angle bad$ = 20°, $ad = v_{a1}$ = 600 m/s = 12 cm

Join bd, bd = 9.7 cm, bd' = 9.7 × 0.9 = 8.7 cm

$\angle abc$ = 20°, $bc = bd'$

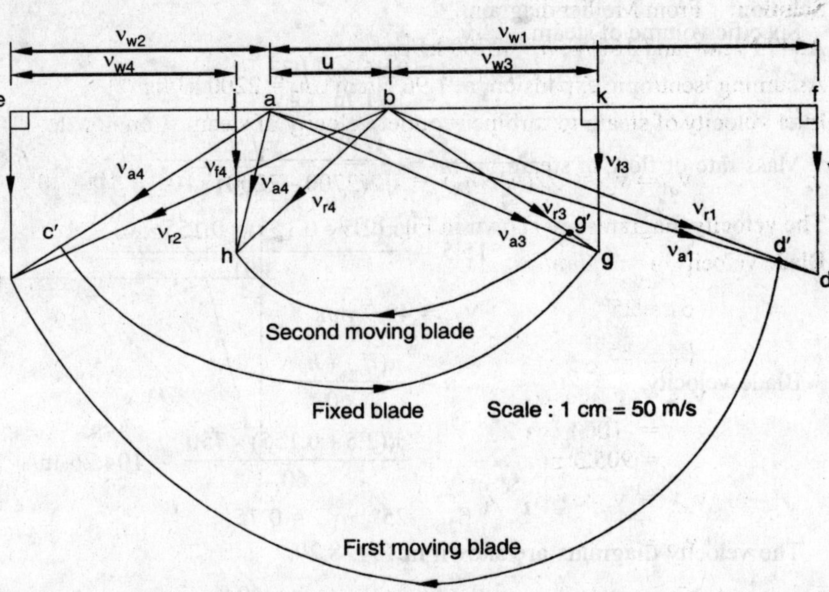

Fig. 8.19

Join ac, ac = 5.4 cm, $ac' = 0.9 \times 5.4 = 5.7$ cm

$$\angle bag = 25°$$

$$bg = ac'$$

$$bg = 3.6 \text{ cm}, \ bg' = 0.9 \times 3.6 = 3.2 \text{ cm}$$

$$\angle abh = 30°, \ bh = bg'$$

Join ah

$$\nu_{w1} = af = 11.3 \text{ cm}, \ \nu_{w2} = ca = 5.8 \text{ cm},$$

$$\nu_{w3} = ah = 5.1 \text{ cm}, \ \nu_{w4} = af = 0.4 \text{ cm},$$

Blade efficiency, $\eta_b = \dfrac{2u\Sigma(\nu_{wl} + \nu_{w2})}{\nu_{al}^2}$

$$= \frac{2 \times 125 \times 50 \times (11.3 + 5.8 + 5.1 \times 0.4)}{600^2} = 0.785 \text{ or } 78.5\%$$

$$\text{Power developed} = \frac{\dot{m}u\Sigma(\nu_{wl} + \nu_{w2})}{10^3} \text{ kW} = \frac{4 \times 125 \times 50 \times 22.6}{10^3} = 565 \text{ kW}$$

Example 8.3 *The drum diameter in a reaction turbine is 2.5 m. The speed is 750 rpm and the steam consumption is 15.5 kg/s. At a particular ring the blade height is 15.5 cm and the discharging angle is 25°. The pressure at this place in the turbine is 0.4 bar and the dryness is 0.95. Estimate the power developed in this particular ring. Assuming a turbine efficiency of 75 percent find the heat drop through the turbine pair.*

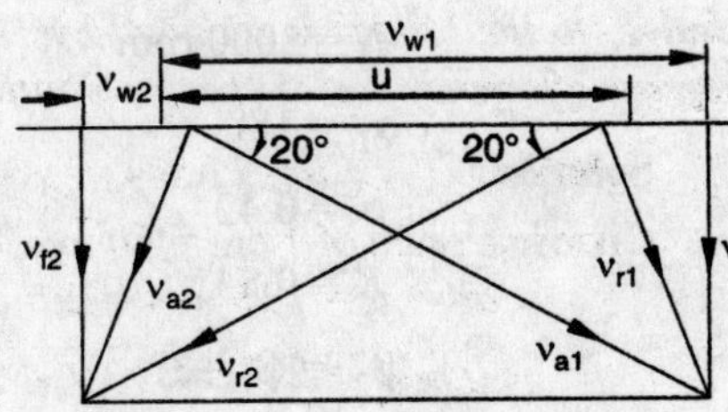

Fig. 8.20

Solution:

From steam tables at 4 bar, $\nu_g = 4.073 \text{ m}^3/\text{kg}$

Specific volume of steam, $\quad v_s = xv_g$

$$= 0.96 \times 4.073$$
$$= 3.91 \text{ m}^3/\text{kg}$$

Mass rate of flow of steam, $\quad \dot{m} = \dfrac{A_f v_f}{v_s} = \dfrac{\pi d_m h v_f}{3.91}$

$$15.5 = \frac{\pi(2.5+0.155)\times 0.155 \times v_f}{3.91}$$

$$v_f = 49.9 \text{ m/s}$$

Blade velocity, $\quad u = \dfrac{\pi(d_m + h)N}{60}$

$$= \frac{\pi(2.5+0.155)\times 750}{60} = 104.26 \text{ m/s}$$

$$\beta_2 = 25°, \; \eta_b = 0.75$$

The velocity diagrams are shown in Fig. 8.20.

$$v_{w1} = \frac{v_f}{\tan 25°} = \frac{49.9}{\tan 25°} = 107.01 \text{ m/s}$$

$$v_{w2} = \frac{v_f}{\tan 25°} - u = 107 - 104.26 = 2.75 \text{ m/s}$$

Power per pair $= \dot{m}\, u\, (v_{w1} + v_{w2})$

$$= \frac{16.5 \times 104.26(107.01+2.75)}{10^3} = 188.82 \text{ kW}$$

Turbine efficiency $\quad = \dfrac{u(v_{w1} + v_{w2})}{(\Delta h)_{isen}}$

$$(\Delta h)_{isen} = \frac{104.26 \times 109.76 \times 10^{-3}}{0.75} = 15.26 \text{ kJ.}$$

Example 8.4 *In an impulse turbine, the mean diameter of the blades is 1.05 m and the speed is 3000 rpm. The nozzle angle is 18°, speed ratio is 0.42 and the friction factor is 0.84. The outlet blade angle is to be made 3° less than the inlet angle. The steam flow as 10 kg/s. Draw the velocity diagrams for the blades and calculate (a) tangential thrust, (b) axial thrust, (c) resultant thrust, (d) power developed, and (e) blading efficiency.*

Solution. $\quad d_m = 1.05$ m

$N = 3000$ rpm

$\alpha_1 = 18°$

$\rho = 0.42$

$K = 0.84$

$\beta_2 = \beta_1 - 3°$

$\dot{m} = 10$ kg/s

Blade speed, $\quad u = \dfrac{\pi d_m N}{60}$

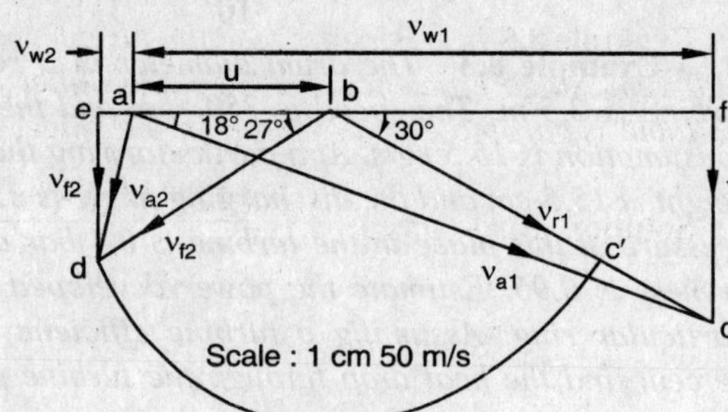

Fig. 8.21

$$= \frac{\pi \times 1.05 \times 300}{60} = 164.93 \text{ m/s}$$

$$\rho = \frac{u}{v_{a1}}, \quad v_{a1} = \frac{164.93}{0.42} = 392.7 \text{ m/s}$$

Draw the velocity diagrams as shown in Fig. 8.21.

$$\begin{aligned} v_{r1}^2 &= u^2 + v_{a1}^2 - 2uv_{a1}\cos\alpha_1 \\ &= (164.93)^2 + (392.7)^2 - 2 \times 164.93 \times 392.7 \times \cos 18° \\ &= 58219 \end{aligned}$$

$$v_{r1} = 241.28 \text{ m/s}$$

$$v_{f1} = v_{a1}\sin 18° = 392.7 \sin 18° = 121.35 \text{ m/s}$$

$$\sin\beta_1 = \frac{v_{f1}}{v_{r1}} = \frac{121.35}{241.28} = 0.50295$$

$$\beta_1 = 30.2°$$

$$\beta_2 = 30.2° - 3 = 27.2°$$

$$v_{r2} = 0.84 v_{r1} = 0.84 \times 241.28 = 202.67 \text{ m/s}$$

$$v_{f2} = v_{r2}\sin\beta_2 = 202.67 \sin 27.2° = 92.64 \text{ m/s}$$

$$v_{w1} = v_{a1}\cos\alpha_1 = 392.7 \cos 18° = 373.48 \text{ m/s}$$

$$v_{w2} = v_{r2}\cos\beta_2 - u = 202.67 \cos 27.2° - 164.93 = 15.33 \text{ m/s}$$

(*a*) Tangential thrust $= \dot{m}(v_{w1} + v_{w2}) = 10(373.48 + 15.33) = 3888$ N

(*b*) Axial thrust $= \dot{m}(v_{f1} - v_{f2}) = 10(121.35 - 92.64) = 287.1$ N

(*c*) Resultant thrust $= [(3888)^2 + (287.1)^2]^{1/2} = 3898.6$ N

(*d*) Power developed $= \dfrac{\dot{m}u(v_{w1} + v_{w2})}{10^3} = \dfrac{10 \times 164.93(373.48 + 15.33)}{10^3}$

$$= 641.26 \text{ kW}$$

(*e*) Blade efficiency, $\eta_b = \dfrac{2u(v_{w1} + v_{w2})}{v_{a1}^2} = \dfrac{2 \times 164.93(373.48 + 15.33)}{(392.7)^2}$

$$= 0.8316 \text{ or } 83.16\%$$

Example 8.5 *In a Parson's reaction turbine, mean diameter of the rotor is 1m, speed of turbine 1500 rpm, stage efficiency = 0.8, blade outlet angle = 20°, speed ratio = 0.7 Calculate the available isentropic enthalpy drop in the stage.*

Solution. Blade speed, $u = \dfrac{\pi d_m N}{60} = \dfrac{\pi \times 1 \times 1500}{60} = 78.54$ m/s

Speed ratio, $\rho = \dfrac{u}{v_{a1}} = 0.7$

$$v_{a1} = \frac{78.54}{0.7} = 112.2 \text{ m/s}$$

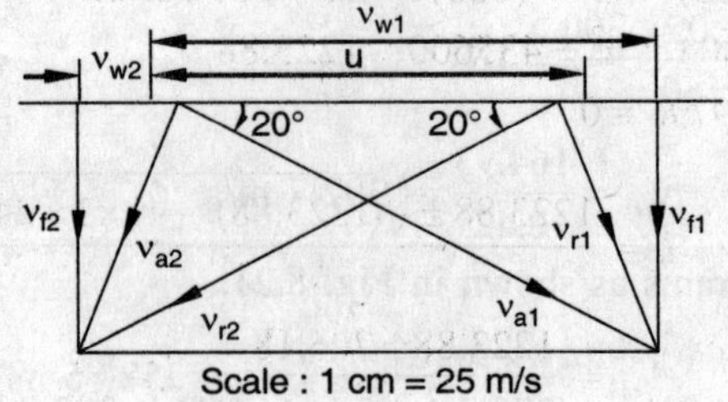

Fig. 8.22

In a Parson's turbine, $\alpha_1 = \beta_2 = 20°$ (see Fig. 8.22)

$$V_{r1}^2 = u^2 + V_{a1}^2 - 2uV_{a1}\cos\alpha_1$$
$$= (78.54)^2 + (112.2)^2 - 2 \times 78.54 \times 112.2 \cos 20°$$
$$= 2195.87$$

$$V_{r1} = 45.86 \text{ m/s}, \ V_{a1} = V_{f1}\sin\alpha_1 = 112.2 \sin 20° = 38.37 \text{ m/s}$$

$$V_{f2} = V_{f1} = 38.37 \text{ m/s}, \ V_{r2} = V_{f2}/\sin\beta_2 = 38.37/\sin 20° = 112.2 \text{ m/s}$$

$$V_{wl} = V_{al}\cos\alpha_1 = 112.2 \cos 20° - 78.54 = 25.89 \text{ m/s}$$

$$V_w = V_{wl} + V_{w2} = 105.4 + 25.89 = 132.29 \text{ m/s}$$

Stage efficiency, $\eta_s = \dfrac{uV_m}{\Delta h}$

$$0.8 = \frac{78.54 \times 132.29}{\Delta h \times 10^3}$$

Enthalpy drop in the stage, $\Delta h = 12.987$ kJ

Example 8.6 *One stage of impulse turbine consists of a covering nozzle ring of moving blades. The nozzles are inclined at 22° to the blades whose tip angles are both 35°. If the velocity of steam at exit from nozzle is 660 m/s, find the blade speed so that the steam passes on without shock. Find the diagram efficiency neglecting losses if the blades are run at this speed.*

Solution. $\alpha_1 = 22°$,

$$\beta_1 = \beta_2 = 35°$$

$$V_{a1} = 660 \text{ m/s}$$

The velocity diagrams are shown in Fig. 8.23.

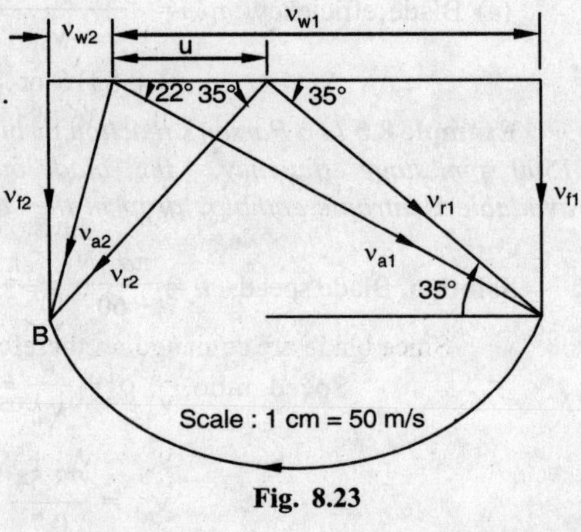

Fig. 8.23

$$V_{f1} = V_{a1}\sin 22° = V_{r1}\sin 35°$$

$$V_{r1} = \frac{660 \sin 22°}{\sin 35°}$$
$$= 431.08 \text{ m/s}$$

$$V_{w1} = V_{a1}\cos\alpha_1$$
$$= 660 \cos 22° = 611.94 \text{ m/s}$$

$$V_{r1}^2 = u^2 + V_{a1}^2 - 2u\,V_{a1}\cos\alpha_1$$

$$(431.05)^2 = u^2 + (660)^2 - 2u \times 660 \cos 22°$$

$$185804 = u^2 + 435600 - 1223.88\,u$$

$$u^2 - 1223.88\,u + 249796 = 0$$

$$u = \frac{1223.88 \pm \sqrt{(1223.88)^2 - 4 \times 1 \times 249796}}{2}$$

$$= \frac{1223.88 \pm 706.18}{2} = 258.85 \text{ m/s (taking –ve sign)}$$

As $K = 1$, therefore $\quad v_{r2} = v_{r1} = 431.05$ m/s

$$v_{w2} = v_{r2} \cos\beta_2 - u = 431.05 \cos 35° - 258.85 = 94.24 \text{ m/s}$$

Diagram efficiency, $\quad \eta_d = \dfrac{2u(v_{w1} + v_{w2})}{v_{a1}^2} = \dfrac{2 \times 258.85(611.94 + 9.24)}{660^2}$

$$= 0.8392 \text{ or } 83.92\%.$$

Example 8.7 *The velocity of steam exiting the nozzle of the impulse stage of a turbine is 400 m/s. The blades operate close to the maximum blading efficiency. The nozzle angle is 20°. Considering equiangular blades and neglecting blade friction, calculate for a steam flow of 0.6 kg/s, the diagram power and the diagram efficiency.*

Solution. $\quad v_{a1} = 400$ m/s, $\alpha_1 = 20°$, $\dot{m} = 0.6$ kg/s

$$\beta_1 = \beta_2, \; K = 1$$

For maximum blade efficiency, speed ratio,

$$\rho = \frac{u}{v_{a1}} = \frac{\cos\alpha_1}{2}$$

$$\frac{u}{400} = \frac{\cos 20°}{2}$$

$$u = 187.94 \text{ m/s}$$

$$v_{w1} = v_{a1}\cos\alpha_1 = 400 \cos 20° = 375.88 \text{ m/s}$$

$$v_{f1} = v_{a1}\sin\alpha_1 = 400 \sin 20° = 136.8 \text{ m/s}$$

$$\tan\beta_1 = \frac{v_{f1}}{v_{w1} - u} = \frac{136.8}{375.88 - 187.94} = \frac{136.8}{187.94} = 0.72789$$

$$\beta_1 = 36.05°$$

$$v_{r1} = \frac{v_{f1}}{\sin\beta_1} = \frac{136.8}{\sin 36.05°} = 232.46 \text{ m/s}$$

$$v_{r2} = v_{r1} = 232.46 \text{ m/s}$$

Since blade are equiangular, therefore $\beta_2 = \beta_1 = 35.05°$

$$v_{w2} = v_{r2}\cos\beta_2 - u = 232.46 \cos 35.05° - 187.94 \sim 0$$

$$v_w = v_{w1} + v_{w2} = 375.85 \text{ m/s}$$

Diagram power, $P = \frac{\dot{m} u v_w}{10^3} = \frac{0.6 \times 187.94 \times 375.85}{10^3} = 42.38$ kW

Diagram efficiency, $\eta_d = \frac{2u v_w}{v_{a1}^2} = \frac{2 \times 187.94 \times 375.85}{400^2} = 0.883$ or 88.3%

SUMMARY

1. Steam turbines are of three types: impulse turbines, impulse reaction turbines and reaction turbines.
2. Compounding of steam turbines is a method employed for reducing the rotational speed of the impulse turbine to practical limits by using more than one stage.
3. Impulse steam turbines can be compounded by : velocity compounding, pressure compounding, and mixed velocity and pressure compounding.
4. Impulse turbine :

Power developed, $P = \frac{\dot{m}(v_{w1} + v_{w2})u}{10^3}$ kW

Diagram of blade efficiency, $\eta_b = \frac{2(v_{w1} + v_{w2})u}{v_{a1}^2}$

Gross or stage efficiency, $\eta_s = \frac{(v_{w1} + v_{w2})u}{(\Delta h)_{\text{isen}}}$

Nozzle efficiency, $\eta_n = \frac{v_{a1}^2}{2(\Delta h)_{\text{isen}}}$

Axial thrust, $F_a = \dot{m}(v_{f1} - v_{f2})$

Energy converted into heat by blade friction

$$= \frac{1}{2}\dot{m}(v_{r1}^2 - v_{r2}^2)$$

For maximum blade efficiency, $\rho = \frac{u}{v_{a1}} = \frac{\cos \alpha_1}{2}$

$$(\eta_b)_{\max} = \frac{\cos^2 \alpha_1}{2}(1 + KC)$$

where $K = \frac{v_{r2}}{v_{r1}}$, $C = \frac{\cos \beta_2}{\cos \beta_1}$

For symmetrical blades, $\beta_1 = \beta_2$ and $C = 1$

$$(\eta_b)_{\max} = \cos^2 \alpha_1$$

Work done, $w = 2u(v_{w1} - u)(1 + KC)$

$$w_{\max} = 2u^2$$

5. Reaction turbine:

Degree of reaction, $R_d = \dfrac{(\Delta h)_{\text{moving blade}}}{(\Delta h)_{\text{moving blade}} + (\Delta h)_{\text{fixed blade}}}$

$$\frac{v_{r2}^2 - v_{r1}^2}{2u(v_{w1} + v_{w2})} = \frac{v_f}{2u}(\cot\beta_2 - \cot\beta_1)$$

For $R_d = 0.5$

$$u = v_f(\cot\beta_2 - \cot\alpha_2) = v_f(\cot\alpha_1 - \cot\beta_1)$$

$\therefore$ $\beta_1 = \alpha_2$ and $\beta_2 = \alpha_1$

Work done per kg of steam,

$$w = v_{a1}^2(2\rho\cos\alpha_1 - \rho^2)$$

where $\rho = \dfrac{u}{v_{a1}}$ = speed ratio

Blade efficiency, $n_b = \dfrac{2\rho(\cos\alpha_1 - \rho)}{1 + 2\rho\cos\alpha_1 - \rho^2}$

For η_b to be maximum,

$$\rho = \frac{\cos\alpha_1}{4}$$

and $(\eta_b)_{\max} = \cos^2\alpha_1$

where n = number of blade rows.

Work done in the last row $= \dfrac{\text{Total work done}}{2^n}$

6. Reheat factor $= \dfrac{\text{Cumulative heat drop}}{\text{Isentropic heat drop}}$

7. Steam turbines can be governed by: Throttle governing, nozzle control governing, and by-pass governing.

FILL IN THE BLANKS

1. The expansion of steam in an impulse turbine occurs in _____.
2. A steam turbine converts pressure/kinetic energy into _____.
3. _____ takes into account the loss in blade velocity due to friction.
4. Blade or diagram efficiency is defined as the ratio of _____ to _____
5. Maximum blading efficiency of a single stage impulse turbine having symmetrical blades and nozzle angle α is _____
6. In a velocity compounded steam turbine. _____ blade rings are separated by _____ blade rings.
7. In a reaction turbine, steam is expanded partially in the _____ and remaining expansion takes place in _____ blades.
8. Degree of reaction = _____/_____.
9. For symmetrical blading, the degree of reaction is _____.
10. DeLaval turbine is a/an _____.

ANSWERS

1. nozzle **2.** mechanical energy **3.** blade velocity coefficient
4. work done by blades, energy supplied to blades **5.** $\cos^2 \alpha$ **6.** moving, fixed
7. nozzle, moving **8.** $\Delta h_m / (\Delta h_m + \Delta h_f)$ **9.** 50%
10. impulse turbine

MULTI-CHOICE QUESTIONS

1. De-Laval turbine is a:
(*a*) simple single wheel reaction turbine
(*b*) simple single wheel impulse turbine
(*c*) velocity compounded impulse turbine
(*d*) pressure compounded impulse turbine.

2. Curtis turbine is a:
(*a*) reaction steam turbine
(*b*) pressure compounded impulse steam turbine
(*c*) velocity compounded impulse turbine
(*d*) pressure-velocity compounded steam turbine.

3. Parson's turbine is :
(*a*) impulse-reaction turbine (*b*) impulse turbine
(*c*) reaction turbine (*d*) velocity compounded impulse turbine

4. Blade or diagram efficiency is:
(*a*) $\dfrac{(V_{w1} + V_{w2})u}{V_{a1}^2}$ (*b*) $\dfrac{2(V_{w1} + V_{w2})u}{V_{a1}^2}$
(*c*) $\dfrac{u^2}{V_{a1}^2}$ (*d*) $\dfrac{V_{a1}^2}{u^2}$

5. Axial thrust on rotor of steam turbine is
(*a*) $\dot{m}(V_{f1} - V_{f2})$ (*b*) $\dot{m}(V_{f1} + V_{f2})$
(*c*) $\dot{m}(V_{f1} + V_{f2})$ (*d*) $\dot{m}(V_{f1} - 2V_{f2})$

6. For maximum blade efficiency of single stage impulse turbine
(*a*) $\rho\left(= \dfrac{u}{V_{al}}\right) = \dfrac{\cos^2 \alpha_1}{2}$ (*b*) $\rho = \cos\alpha_1$
(*c*) $\rho = \dfrac{\cos\alpha_1}{2}$ (*d*) $\rho = \cos^2 \alpha_1$

7. Degree of reaction of steam turbine is
(*a*) $\dfrac{(\Delta h)_m}{(\Delta h)_m + (\Delta h)_f}$ (*b*) $\dfrac{(\Delta h)_f}{(\Delta h)_m + (\Delta h)_f}$
(*c*) $\dfrac{(\Delta h)_m}{(\Delta h)_f}$ (*d*) $\dfrac{(\Delta h)_f}{(\Delta h)_m}$

8. The maximum blade efficiency of impulse turbine is

(*a*) $\frac{\cos^2 \alpha_1}{2}(1+KC)$ (*b*) $\cos^2 \alpha_1(1+KC)$

(*c*) $\cos \alpha_1(1+KC)$ (*d*) $\frac{\cos \alpha_1}{2}(1+KC)$

ANSWERS

1. (*b*) **2.** (*c*) **3.** (*a*) **4.** (*b*) **5.** (*a*)
6. (*c*) **7.** (*a*) **8.** (*a*)

EXERCISES

8.1 In a single stage impulse turbine the blade angles are equal and the nozzle angle is 20°. The velocity coefficient for the blade is 0.83. Find the mzximum blade efficiency possible. If the actual blade efficiency is 90% of maximum blade efficiency, find the possible ratio of blade speed to steam speed. [**Ans.** 0.813, 0.3235]

8.2 The following data refer to a compound impulse turbine having two rows of moving blades and one row of fixed blades in between them.

Nozzle angle = 15°, Exit velocity of steam from the nozzle = 450 m/s

Moving blades tip discharge angles = 30°

Fixed blade discharge angle = 20°

Friction loss in each blade rows = 10% of the relative velocity

Calculate the blade velocity, blade efficiency and specific steam consumption for the turbine. [**Ans.** 98 m/s, 78.8%, 45 kg/k Wh]

8.3 In a stage of impulse-reaction turbine, steam enters with a speed of 250 m/s at an angle of 30° in the direction of blade motion. The mean blade speed is 150 m/s when the rotor is running as 3000 rpm. The blade height is 10 cm. The specific volume of steam at nozzle outlet and blade outlet are 3.5 m^3/kg and 4m^3/kg respectively. The turbine develops 250 kW. Assuming the efficiency of nozzle and blades combinedly considered as 90% and carry over coefficient as 0.8. Find (*a*) the enthalpy drop in each stage, (*b*) degree of reaction, and (*c*) stage efficiency. [**Ans.** 29.3 kJ/kg, 0.39 71%]

8.4 At a stage of reaction turbine, the mean diameter of rotor is 1.4 m. the speed ratio is 0.7. Determine the blade inlet angle if the blade outlet angle is 20°. The rotor speed is 3000 rpm. Also find the diagram efficiency the percentage increase in it and rotor speed if the rotor is designed to run at the best theoretical speed, the exit angle being 20°. [**Ans.** 53.8°, 90.5%, 3.65%, 93.8%]

8.5 In a 50% reaction turbine stage running at 50 rps, the exit angles are 30° and the inlet angles are 50°. The mean diameter is 1 m. The steam flow rate is 10^4 kg/min and the stage efficiency is 85%. Calculate (*a*) the power output of the stage, (*b*) the specific enthalpy drop in the stage and (*c*) the percentage increase in the relative velocity of steam when it flows over the moving blades. [**Ans.** 11.6 kW, 82 kJ/kg 52.2%]

8.6 A single row impulse turbine develops 130 kW at a blade speed of 180 m/s using 2 kg/s of steam. The steam leaves the nozzle at 400 m/s. The friction coefficient for blades is 0.9 and steam leaves the blades axially. Determine (*a*) nozzle angle, and (b) blade angles at entry and exit, assuming no shock. [**Ans.** 25.47°, 43.5°, 35.8°]

8.7 A simple impulse turbine has one ring of moving blades running at 150 m/s. The absolute

velocity of steam at exit from the stage is 80 m/s at an angle of 75° from the tangential direction. Blade speed coefficient is 0.85 and the rate of steam flowing through the stage is 3 kg/s. If the blades are equiangular, determine (*a*) blade angles, (*b*) nozzle angle, (*c*) absolute velocity of steam issuing from the nozzle, and (*d*) axial thrust.
[**Ans.** 24.35°, 14.5°, 362.4 m/s]

8.8 In a single stage impulse turbine, nozzle angle is 20° and blade angles are equal. The velocity coefficient for blades is 0.85. calculate the maximum blade efficiency possible. If the actual blade efficiency is 92% of the maximum blade efficiency, find the possible ratio of blade speed to steam speed. [**Ans.** 81.6%, 0.336 or 0.603]

8.9 In a single stage steam turbine saturated steam at 10 bar is supplied through a convergent divergent steam nozzle of 20° angle. The mean blade speed is 400 m/s. The steam pressure leaving the nozzle is 1 bar. Find (*a*) the best blade angles if they are equiangular, and (*b*) the maximum power developed by the turbine if 5 nozzles are used of 0.6 cm^2 area at the throat. Assume nozzle efficiency 90% and blade friction coefficient 0.87.
[**Ans.** 35.46°, 115.8 kW]

8.10 The first stage of an impulse turbine is compounded for velocity and has two rows of moving blades and one ring of fixed blades. The nozzle angle is 15° and leaving angles of blades are respectively, first moving 30°, fixed 20, second moving 30°. The velocity of steam leaving the nozzle is 540 m/s. The friction loss in each blade row is 10% of the relative velocity. Steam leaves the second row of moving blades axially. Calculate (*a*) blade velocity, (*b*) blade efficiency and (*c*) specific steam consumption.
[**Ans.** 117.3 m/s, 78.6%, 31.39 kg/kWh]

9 CONDENSERS

9.1 INTRODUCTION

A condenser is a closed vessel in which steam is condensed by abstracting the heat. In a condenser, the pressure is maintained below atmospheric pressure. It is used in steam power plants to condense the exhaust steam which is used as feed water for the boiler.

9.2 FUNCTIONS OF A CONDENSER

There are two main functions of a condenser, as explained below:

1. To reduce the back pressure considerably upon the prime mover and there by increasing the work output of steam during expansion.

   Consider the ideal $p-v$ diagram of a non-condensing steam engine, as shown in Fig. 9.1 (*a*). Here the steam is supplied at pressure p_1 and the exhaust takes place at atmospheric pressure p_2. The work output is given by the area 1–2–3–4–5–1. If we reduce the back pressure p_2 to p_2', which is below the atmospheric pressure, keeping p_1 same, then the work output will be given by the area $1-2-3'-4'-5'-1$. It is shown in Fig. 9.1 (*b*). This work output is obviously greater, than the previous work output. The additional work output is $4-4'-5'-5-4$.

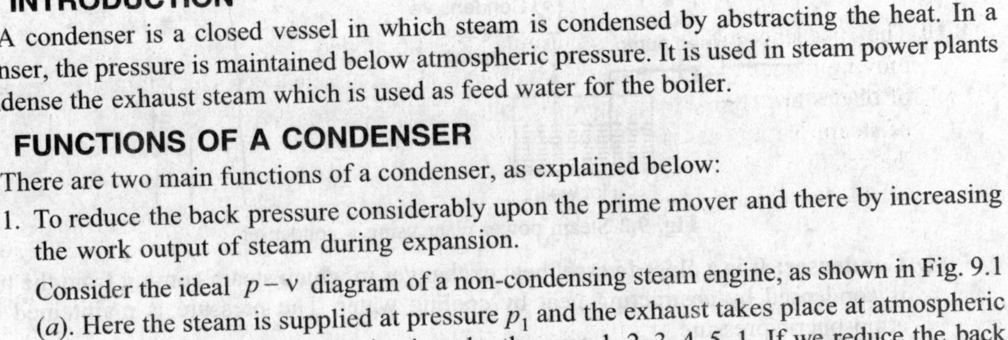

(a) Non-condensing steam engine

(b) Condensing steam engine

Fig. 9.1 Effect of back pressure on work output of a compressor

2. The condensate condensed from exhaust steam in the condenser is free from impurities and can be directly used as feed water for the boiler without any treatment. This recovery of feed water is an incidental advantage.

9.3 ELEMENTS OF STEAM CONDENSING PLANT

A steam power plant using a condenser is shown in Fig. 9.2. The various components of a steam condensing plants are described below :

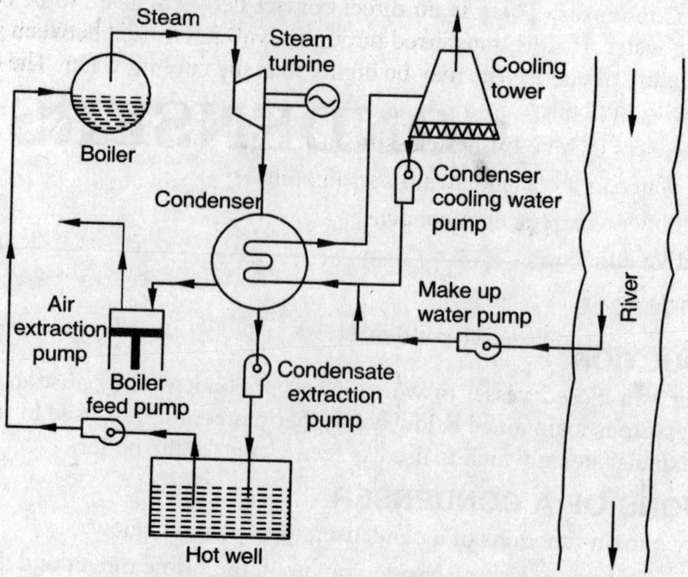

Fig. 9.2 Steam power plant using a condenser

1. **Condenser:** It is a closed vessel heat exchanger in which steam coming from the turbine is condensed by abstracting heat by cooling water. The pressure is maintained below atmospheric pressure.
2. **Condenser cooling water pump:** The cooled water from the cooling tower is circulated through the condenser by this pump. This pump is not required if water from the cooling tower is not re-circuited.
3. **Make up water pump:** This pump circulates the required quantity of water through the condenser.
4. **Condensate extraction pump:** The function of the condensate pump is to extract the condensate from the condenser where the pressure is below atmospheric. This condensate is then fed it to the hot well, where the pressure is atmospheric.
5. **Hot well:** It is a sump between the condenser and the boiler where the condensate coming from the condenser is collected.
6. **Air-extraction pump:** Its function is to extract the air which has leaked into the condenser by various paths and to maintain the required vacuum inside.
7. **Boiler feed pump:** Its function is to pump the condensate from hot well into the boiler by increasing the pressure of condensate above boiler pressure.
8. **Cooling tower:** It is essential when the large amount of cooling water required is not available from the river throughout the years. Its function is to cool the hot water coming out of condenser instead of discharging is to waste. The hot water is cooled by evaporative cooling.

9.4 CLASSIFICATION OF STEAM CONDENSERS

The steam condensers may be broadly classified into two categories.

1. Jet Condensers: It is contact type condensers in which the steam to be condensed mixes with the cooling water. And the temperature of the condensate and the cooling water is the same when leaving the condenser. The condensate cannot be recovered for use as feed water to the boiler. Heat transfer is by direct conduction between steam and water.

These condensers may be further classified as:

(*i*) Low level, counter flow type condenser (*ii*) Low level, parallel flow type condenser

(*iii*) High level jet condenser (*iv*) Ejector condenser

2. Surface Condenser: There is no direct contact between steam to be condensed and the circulating cooling water. Heat is transferred through a wall interposed between steam and cooling water. The temperature of condensate may be higher than the cooling water. The condensate can be used as feed water to the boiler.

These condensers may be further classified as:

(*i*) Cooling water in tubes and steam surrounding it

(*ii*) Central flow or regenerative condenser

(*iii*) Evaporative condenser

9.4.1 Jet Condenser

(*i*) Low level counter-flow jet condenser

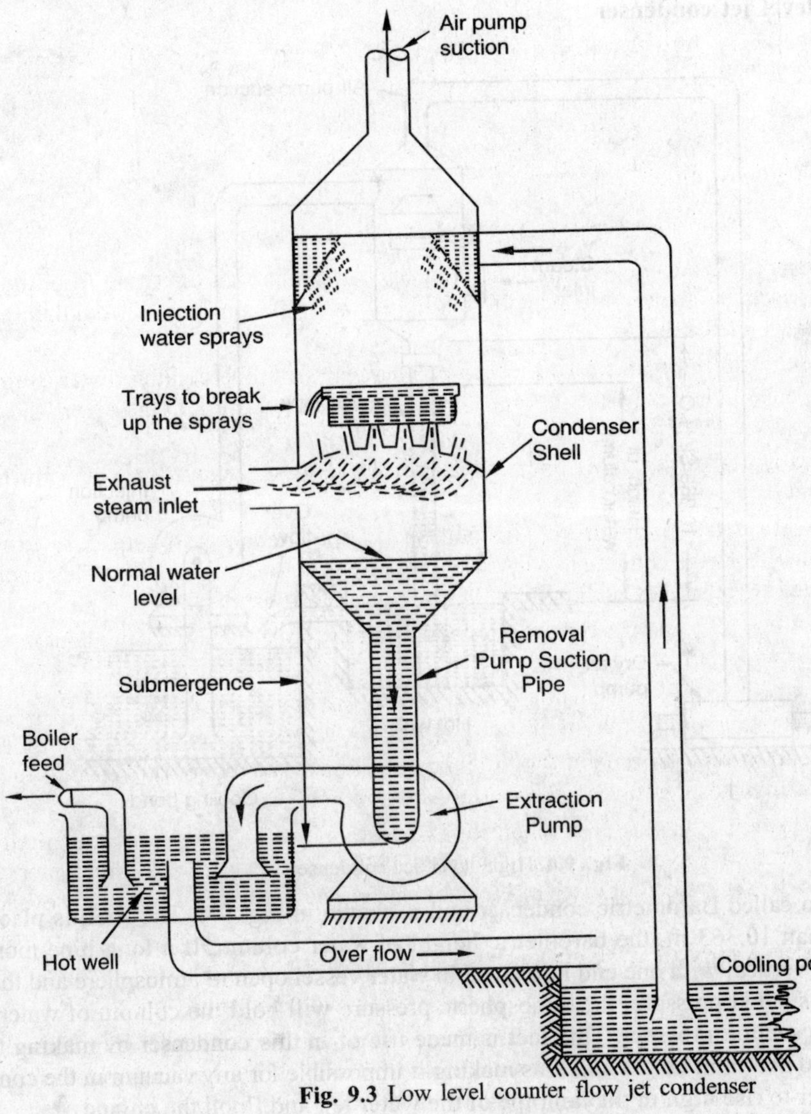

Fig. 9.3 Low level counter flow jet condenser

This type of condenser is shown in Fig. 9.3. The cold water is drawn up in the condenser shell from the cooling pond due to the vacuum head created in the shell. The level of the cooling pound should not be much below the delivery pipe of shell as no pump is used. The shell is arranged with two or three water trays with perforations to break up water into small jets. The exhaust steam and any mixed air enters at lower portion of the shell and tries to ascend up through the falling sprays. Thus steam gets condensed and air ascends up and cools down. The air is removed by a separate suction pump at the top. The mixture of condensate and cooling water descends down through a vertical pipe to the extraction pump and pumped to the hot well. From the hot well, the boiler feed pump delivers water to the boiler and surplus water overflows to the cooling pond. Since the boiler feed and the cool injection water mix, this type of condenser may be used only where cheap pure water is available. The non-condensable air and gas are continuously removed by air pump suction of maintain vacuum.

(*ii*) Low level parallel flow jet condenser

It is similar to the counter flow jet condenser except that steam enters the condenser shell at the top and just below it the cooling water is delivered. It is loss efficient than the counter flow type.

(*iii*) High level jet condenser

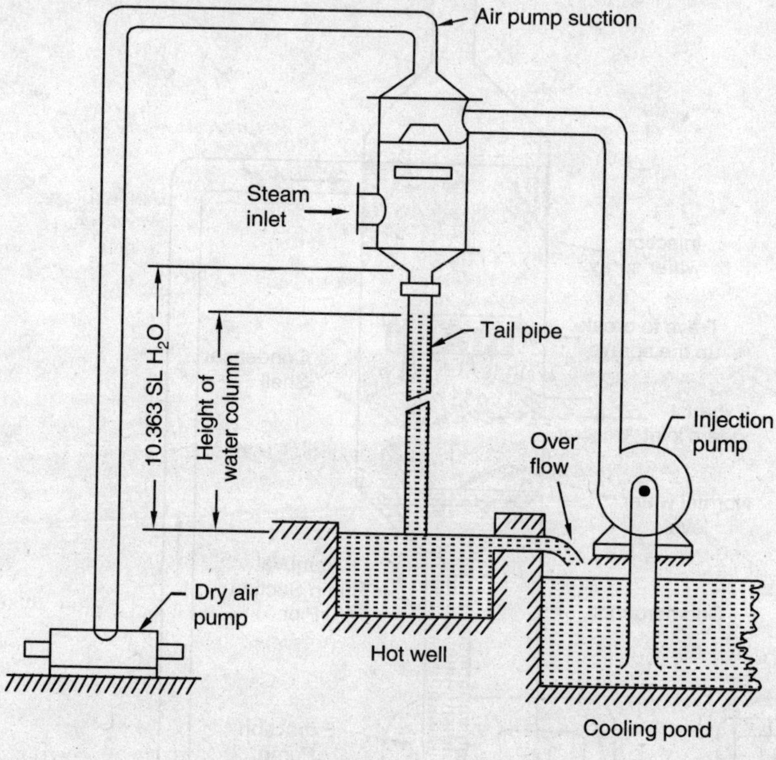

Fig. 9.4 High level jet condenser

This is also called Barometric condenser and is shown in Fig. 9.4. The shell is placed at a height greater than 10.363 m, the barometric height of water column. If a long pipe more than 10.363 m is held vertical with one end immersed in water vessel open to atmosphere and the other end subjected to suction pressure, the atmospheric pressure will hold the column of water in the pipe equal to the suction pressure. This fact is made use of in this condenser by making the tail pipe more than 10.363 m in height and thus making it impossible for any vacuum in the condenser to cause the water to rise high in the tail pipe of the water leg and flood the engine.

The height of shell much above the barometric height requires a separate pump for injecting cool water. The condensate and water will fall under gravity to the hot well and maintain a column of water in the water leg depending on the vacuum in the condenser. A separate air pump to remove air is employed.

(*iv*) Ejector Condenser

It is a low level type of condenser as shown in Fig. 9.5. This condenser works on the principle that by discharging a smooth jet of cold water under a head of about 6m through a series of converging guide cones, the steam and associated air are drawn in through the hollow truncated cones and led to the diverging cones. In the converging cones the pressure energy is partly converted to kinetic energy. In the diverging cones, the kinetic energy is again partly converted to pressure energy so as to obtain pressure greater than the atmospheric to enable the condensate and water mixture to be discharged to the hot well exposed to atmospheric pressure.

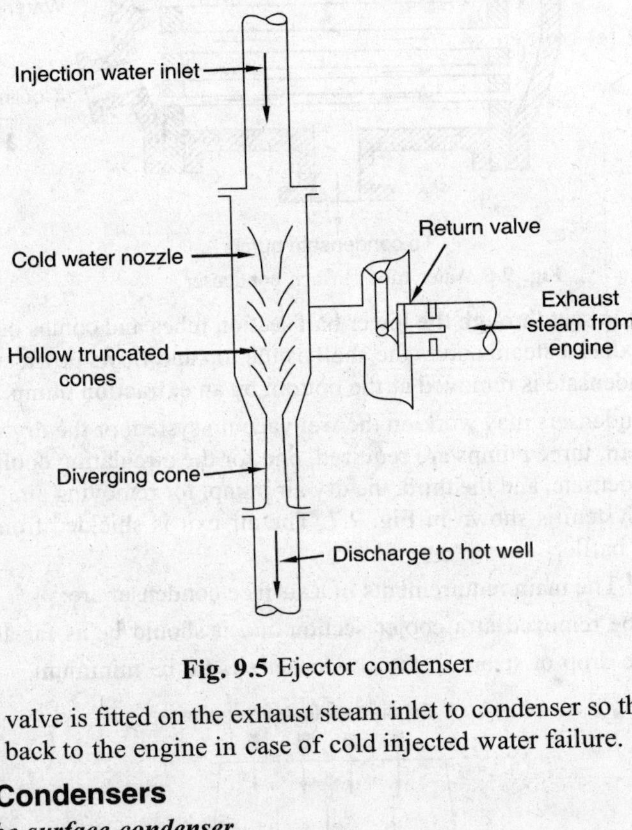

Fig. 9.5 Ejector condenser

A non-return valve is fitted on the exhaust steam inlet to condenser so that water from the hot well does not rush back to the engine in case of cold injected water failure.

9.4.2 Surface Condensers

(i) Water tube surface condenser.

In this type surface condenser, water flows in the tubes and steam surrounds it. Thus the condensate is directly available as an ideal boiler feed. A surface condenser produces more vacuum and any type of cooling water can be used. However, they are bulky and initial capital cost is higher than jet condensers. But this is justified by the saving in the running cost due to high efficiency.

This type of condensers may be further classified as:

(*a*) Single pass, two pass condensers.

(*b*) Down flow : central flow regenerative condensers.

A longitudinal section of a two pass, down flow condenser is shown in Fig. 9.6. It consists of a cast iron shell, cylindrical in shape, with the two ends covered by cover plates. A nest of brass tubes is fixed in the two tube plates at the end. They are fixed by brass ferrules so that they can be replaced easily. The space between the tube plates and the cover plates are known as water boxes. One of the water boxes has baffle partitioning the box into two sections, one upper half and the other lower half.

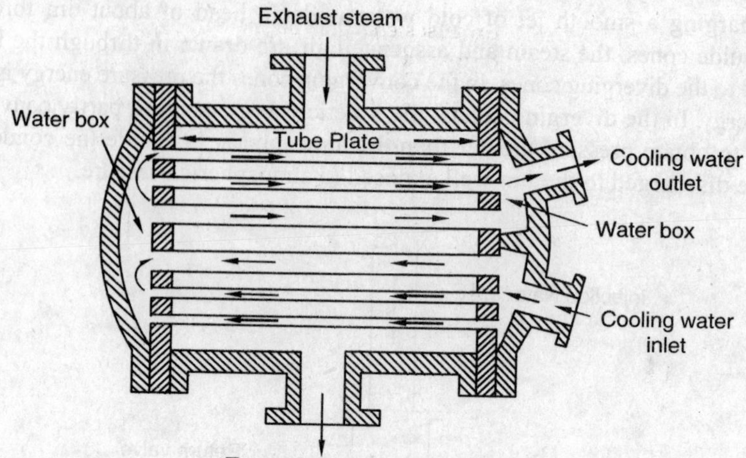

Fig. 9.6 Water tube surface condenser

The cold water is sent through the lower half section tubes and comes out through the upper half section tubes. Exhaust steam enters the shell as the top and flows down surrounding the cold water tubes. The condensate is removed at the bottom by an extraction pump.

The surface condensers may work on the wet vacuum system or the dry vacuum system. For the dry vacuum system, three pumps are required, one for the circulating cooling water, the other for removing the condensate, and the third, the dry air pump, for removing air. The cross-section of such a dry vacuum system is shown in Fig. 9.7. The air exit is shielded from the down flow or steam by means of a baffle.

Requirements: The main requirements of a surface condenser are:

1. Air should be removed at a cooler section and it should be as far as possible dry.

2. The pressure drop of steam in the condenser should be minimum.

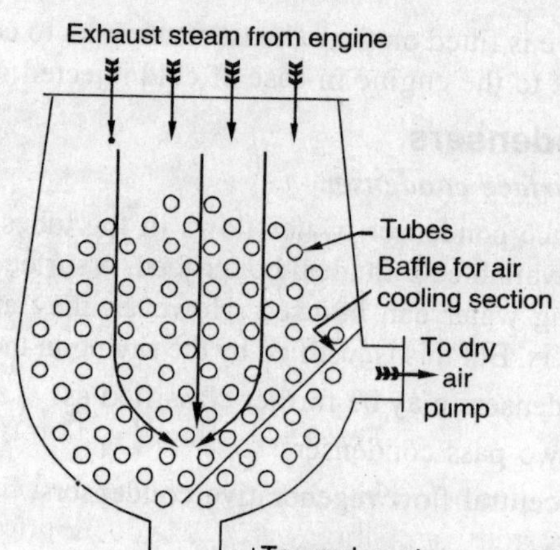

Fig. 9.7 Down flow surface condenser

(ii) **Central flow or regenerative condenser**

In such a condenser shown in Fig. 9.8, the steam passages are all around the periphery of the shell. The exhaust steam converges to the central of the shell where there are no tubes. The air is removed by the air pump suction from the centre. Some of the exhaust steam also passes to the centre and gets condensed. The condensate is reheated to nearly the temperature of the exhaust steam. The pressure drop of steam is also reduced due to large flow passages.

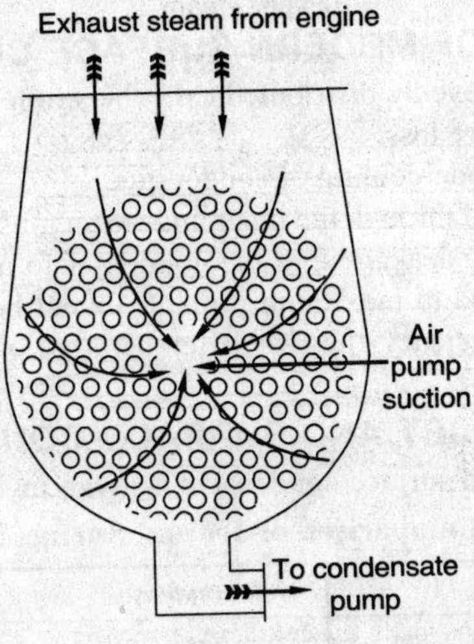

Fig. 9.8 Central flow surface condenser

(iii) **Evaporative condenser**

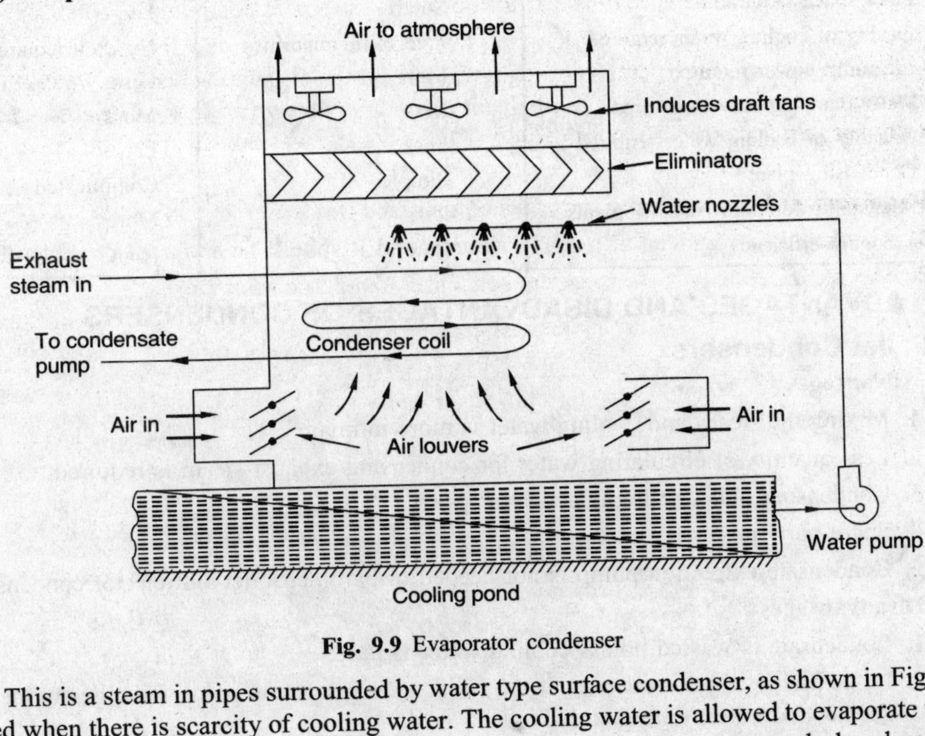

Fig. 9.9 Evaporator condenser

This is a steam in pipes surrounded by water type surface condenser, as shown in Fig. 9.9. It is used when there is scarcity of cooling water. The cooling water is allowed to evaporate under a small partial pressure. It consists of gilled piping bent to form many rows and placed vertically.

Steam passes through these pipes. Pumped water is sprayed from the top and descends down forming a thin film over the pipes. A natural or forced air draught causes the evaporation of water film. Latent heat of evaporation is removed from steam circulating through the inside of the pipes and steam gets condensed. This mode of heat transfer reduces the cooling water requirement of condenser to 0% of the requirement for surface condensers. The water particles carried with air due to high velocity of air are removed with the help of eliminator. This type of condenser works better in dry weather.

9.5 REQUIREMENTS OF MODERN SURFACE CONDENSERS

1. The steam should be evenly distributed over the whole cooling surface of the condenser with minimum pressure loss.
2. There should be no undercooling of condensate.
3. Water should be passed through the tubes and steam must surround the tubes from outside.
4. There should be no air leakage into the condenser to maintain proper vacuum.
5. The air must be cooled to maximum possible extent before extraction.
6. For better thermal efficiency, the rise in temperature of cooling water should be limited to 10° C.

9.6 COMPARISON OF JET AND SURFACE CONDENSERS

The comparison of jet and surface condensers is given in Table 9.1.

Table 9.1 Comparison of Jet and Surface Condensers

Particular's	*Jet Condenser*	*Surface Condenser*
1. Mixing of steam and cooling water	Yes	No
2. Manufacturing cost	Low	High
3. Upkeep	Lower	Higher
4. Floor space required	Small	Large
5. Quality of cooling water required	Free from impurities	No such requirement
6. Air pump power required	More	Less
7. Water pumping power required	Less	More
8. Quantity of cooling water required	Less	Large
9. Condensing plant	Simple	Complicated
10. Suitability for high capacity plants	Less	More
11. Vacuum efficiency	Low	High

9.7 ADVANTAGES AND DISADVANTAGES OF CONDENSERS

9.7.1 Jet Condensers

Advantages

1. Mixing the steam and cooling water is more intimate.
2. Less quantity of circulating water for condensing exhaust steam is required.
3. Condensing plant is simple and less costly.
4. Spare parts required are less.
5. Condensate extraction pump is not needed in barometric jet and ejector condensers.

Disadvantages

1. Condensate is wasted unless cooling water is pure.
2. In the low level jet condenser, if the condensate extraction pump fails the engine would get flooded unless safe guarded.
3. Barometric condenser cost is high due to long piping.

4. There is vacuum loss in barometric condenser due to leakage in long piping.
5. There are limitations of vacuum of upto 650 mm of Hg due to liberation of dissolved gases in cooling water.
6. The air extraction pump require comparatively high power.

9.7.2 Surface Condensers

Advantages

1. Vacuum obtained is comparatively very high resulting in high efficiency.
2. Condensate can be used directly as boiler feed.
3. There is no need of water treatment plant.
4. More suitable for high capacity plants.

Disadvantages

1. Requires more space.
2. Costly maintenance.

9.8 VACUUM AND ITS MEASUREMENT

When the operating pressure of a condenser is lowered, vacuum is created.

A vacuum gauge is used to measure the vacuum in the condenser. Fig. 9.10 shows a vacuum gauge. It consists of a glass tube containing mercury to indicate the barometric pressure and another mercury tube connected to the condenser and vacuum gauge. The vacuum gauge had readings marked in cm of Hg on it. It represents the height at which a column of mercury, the upper surface of which is in communication with the cylinder, will stand when supported by barometric pressure. The vacuum is dependent upon the barometric pressure and the absolute pressure in the condenser. Thus

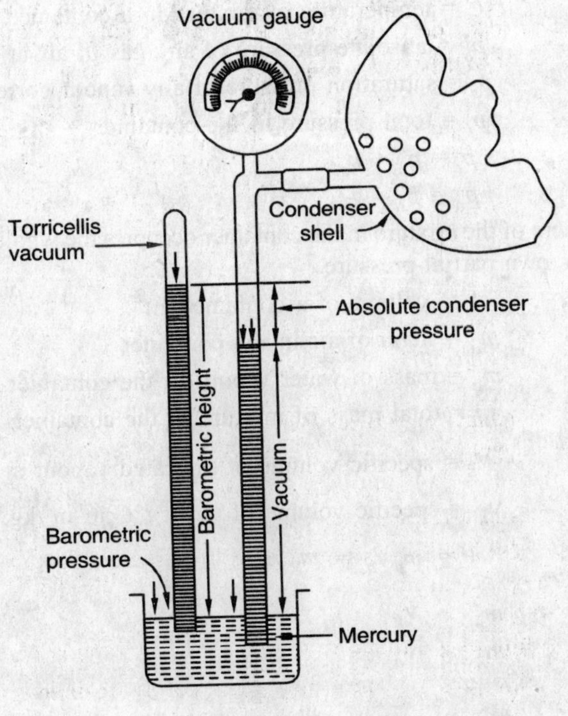

Fig. 9.10 Vacuum gauge

Absolute pressure in the condenser = barometric pressure – vacuum gauge pressure

Usually the vacuum gauge reading are corrected to standard barometer reading of 76 cm of Hg. Thus

Corrected vacuum is cm Hg = 76 – absolute pressure in cm of mercury or Hg
= 76 – (actual barometric height – vacuum)

Let H_b = actual barometric height in cm of Hg

H_g = actual vacuum gauge reading in cm of Hg

$H_b - H_g$ = absolute pressure in condenser in cm of Hg

p = absolute condenser pressure in bar

Now 76 cm of Hg = 1.01325 bar

$$1 \text{ cm of Hg} = \frac{1.01325}{76} = 0.0133322 \text{ bar}$$

$$\therefore \quad p = 0.0133322\,(H_b - H_g) \text{ bar} \quad \ldots(9.1)$$

Corrected or standard vacuum = 76 – $(H_b - H_g)$ cm of Hg ...(9.2)

9.9 DALTON'S LAW OF PARTIAL PRESSURES

Dalton's law states that

1. The total pressure in a container having mixture of gas and vapour is the sum of the partial pressure of the vapour at the common temperature and the partial pressure of the gas.

If the degree of saturation of air at any point inside the condenser is 100%, the partial pressure of vapour is the same as the saturation pressure.

Let $t°C$ = temperature of mixture in a container.

p_a = absolute pressure of any gas of air at t°C.

p_s = saturation pressure of any vapour corresponding to $t°C$.

p_t = total pressure in the container

$$\therefore \quad p_t = p_a + p_s \quad \ldots(9.3)$$

or

$$p_a = p_t - p_s \quad \ldots(9.4)$$

2. Each constituent of the mixture in the container occupies the whole volume of the container and exerts its own partial pressure.

Let V = volume of a container, m^3

m_a = mass of air in the container

m_g = mass of water vapour in the container

m = total mass of mixture in the container

v_g = specific volume of saturated vapour at $t°C$ and p_a in m^3/kg.

v_a = specific volume of air at $t°C$ in m^3/kg.

Then

$$V = m_a v_a = m_g v_g \quad \ldots(9.5a)$$

or

$$\frac{m_a}{m_g} = \frac{v_g}{v_a} \quad \ldots(9.5b)$$

$$\text{Mass of air per m}^3 \text{ of container} = \frac{m_a}{V} = \frac{1}{v_a} \quad \ldots(9.6)$$

Mass of vapour per m^3 of container $= \dfrac{m_g}{V} = \dfrac{1}{v_g}$ ...(9.7)

Mass of mixture, $m = m_a + m_g = m_g\left(1+\dfrac{m_a}{m_g}\right) = m_g\left(1+\dfrac{v_g}{v_a}\right)$ ...(9.8)

or $m = m_a\left(1+\dfrac{m_g}{m_a}\right) = m_a\left(1+\dfrac{v_a}{v_g}\right)$ ...(9.9)

Example 9.1 *A vacuum of 67 cm of Hg was obtained with the barometer reading of 75 cm of Hg. the condensate temperature is 20°C. Correct the vacuum to a standard barometer of 76 cm and hence determine the partial pressure of air and steam. Also find the mass of air present with 1 kg of steam.*

Solution. Corrected vacuum = 76 – (actual barometric reading – actual vacuum)

= 76 – (75 – 76) = 68 cm of Hg

Pressure of dry saturated steam at 20°C from steam tables is 0.02337 bar

or $\dfrac{76 \times 0.02337}{1.01325} = 1.753$ cm of Hg. and specific volume is 57.838 m^3/kg.

Total absolute pressure in the condenser = 75 – 67 = 8 cm of Hg

Partial pressure of steam, P_g = 1.753 cm of Hg

$\therefore$ Partial pressure of air, p_a = 8 – 1.753 = 6.247 cm of Hg

According to Dalton's law the volume of air present per kg of steam

= 57.838 m^3/kg at 20°C.

Also for air $p_a V = m_a RT$

$$m_a = \frac{(0.0133322 \times 6.247 \times 10^5) \times 57.838}{287 \times 293}$$

= 5.728 kg of air per kg of steam.

Example 9.2 *If a barometer stands at 760 mm and condenser vacuum is at 700 mm and temperature is 27°C, calculate the mass of air per kg of uncondensed steam.*

Solution :

Barometric reading = 760 mm of Hg

Condenser vacuum = 700 mm of Hg

Absolute pressure in the condenser

= 760 – 700 = 60 mm of Hg

$= 60 \times \dfrac{1.01325}{760} = 0.08$ bar

Temperature in the condenser = 27° C or T = 27 + 273 = 300 K

Pressure exerted by air, p_a = 0.08 – 0.03567 = 0.04433 bar

Specific volume of steam at 27°C from steam tables,

= 38.774 m^3/kg

Now, $p_a V = m_a RT$

$$m_a = \frac{pV}{RT} = \frac{0.04433 \times 38.774}{287 \times 300} = 1.996 \text{ kg}$$

9.10 MASS OF COOLING WATER REQUIRED IN A CONDENSER

Let

m_a = mass rate of circulating water, kg/h

m_s = mass rate of steam condensed, kg/h

t = temperature of wet exhaust steam, °C

t_c = temperature of condensate, °C

h_f = enthalpy of water at t°C. kJ/kg

h_{fg} = enthalpy of evaporation of steam at t°C, kJ/kg

x = dryness fraction of exhaust steam

t_{iw} = inlet temperature of cooling water, °C

t_{ow} = output temperature of cooling water, °C

h_c = enthalpy of condensate, kJ/kg = $c_{pw}.t_c$

c_{pw} = specific heat of cooling water, 6.1868 kJ/kg. K

Now Heat given by steam = Heat absorbed by water

For jet condenser,

$$t_{ow} = t_c$$

$$m_w = \frac{m_s(h_f + xh_{fg} - h_c)}{c_{pw}(t_c - t_{iw})} \quad ...(9.10)$$

For surface condenser,

$$m_s\,(h_f + xh_{fg} - h_c) = m_w c_{pw}(t_{ow} - t_{iw})$$

$$m_w = \frac{m_s(h_f + xh_{fg} - h_c)}{c_{pw}(t_c - t_{iw})} \quad ...(9.11)$$

Example 9.3 *A 200 kW steam engine has a steam consumption of 10 kg/kWh. The back pressure of the engine is equal of the condenser pressure of 0.15 bar. The condensate temperature is 32°C. The cooling water temperature at inlet and outlet are 20°C and 30°C respectively. Calculate the mass of cooling water required per hour if the exhaust steam is dry saturated.*

Solution. At 0.15 bar, h_f = 226 kJ/kg, h_{fg} = 2373.2 kJ/kg, t_c = 332°C, t_{ow} = 30°C, b_{iw} = 20°C.

$$m_w = \frac{m_s(h_f + h_{fg} - h_c)}{c_{pw}(t_{ow} - t_{iw})} = \frac{1(226 + 1 \times 2373.2 - 4.1868 \times 32)}{4.1868(30 - 20)}$$

$$= 58.88 \text{ kg/h/Kg of steam}$$

Total steam used per hour = 200 × 10 = 2000 kg

Total circulating water used = 58.88 × 2000 = 117760 kg/h

9.11 AIR REMOVAL FROM THE CONDENSER

To maintain in desired vacuum in the condenser air is removed from the condenser.

9.11.1 Sources of Air Infiltration in Condenser

1. Air may pass into the condenser with the boiler steam.
2. Air leakge through the packings.

The amount of air leakage should not be more than 5 to 15 kg/10,000 kg of steam.

9.11.2 Effect of Air Infiltration in Condensers

1. The leaked in air in the condenser increases the back pressure on the prime mover and consequently the thermal efficiency of the plant is lowered.

2. The infiltrated air lowers the partial pressure of steam which means a lower saturation temperature and increased latent heat. So it will require grater amount of cooling water.
3. Because of poor thermal conductivity of air, the rate of heat transfer from the vapour is reduced which requires increased surface area of tubes for a given condenser duty.

9.12 AIR PUMP

The function of the air pump is to maintain a desired vacuum in the condenser by removing both air and condensed steam. Edward's air pump is commonly used for this purpose.

9.12.1 Edward's Air Pump

This pump is shown in Fig. 9.11. It consists of a plunger or piston *D* having a conical head *E* which slides inside a barrel or cylinder liner *C* having a cover *B* which has a number of delivery valves *A*. The passage *G* is connected to the condenser. The barrel has number of parts *F*, *F* pierced in the cylinder liner *C*. These parts are communicating to the air pump suction pipe. *H* is the water weir where the discharged condensate collects and over flows to the hot well. This retention of sufficient water above the cylinder head acts as a water seal against leakage of air. *J* is the relief valve.

On the downward stroke of the piston, partial vacuum is created above the piston because the head valves are closed and sealed by water. As soon as the ports uncovered, the water vapour and air push into the space above the piston. The condensate collected at the piston is displaced by further motion of the piston with conical head and pushed rapidly through the ports to the top of the piston. On the upward stroke, the water, air and water vapour above the piston are compressed to pressure above the atmospheric pressure and discharged outside through the head valves.

$$\text{Capacity of the pump, } V = \frac{\pi}{4} D^2 L N \eta_v$$

where D = piston dia, L = stroke length, N = rpm, η_v = volumetric efficiency.

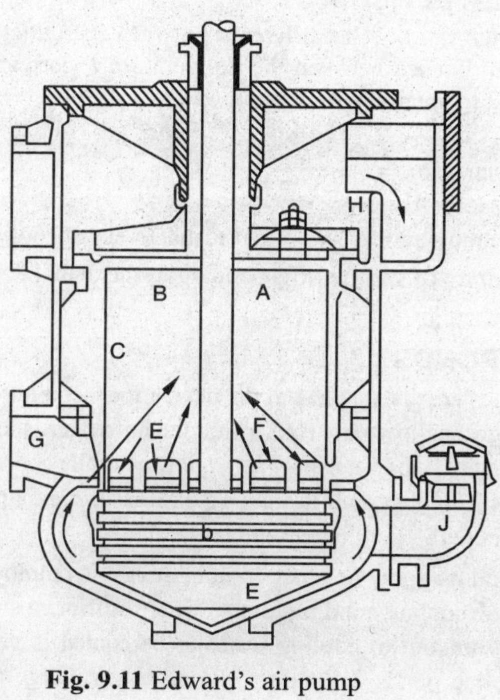

Fig. 9.11 Edward's air pump

9.13 VACUUM EFFICIENCY

The lowest pressure which can exist in a condenser is the situation pressure of steam corresponding to the temperature of water entering the condenser. But the actual pressure is always greater than the ideal pressure by an amount equal to the partial pressure of air present in the condenser.

Let p_s = saturation pressure of steam corresponding to the temperature of water entering the condenser.

p_a = partial pressure of air in the condenser.

p_b = barometric pressure

$p_t = b_a + p_s$

= total pressure of air and steam in the condenser

The ideal vacuum possible without air leakage = $p_b - p_s$

Actual vacuum existing in condenser due to air leakage

$$= p_b - p_t = p_b - (p_a + p_s)$$

The "vacuum efficiency" is defined as the ratio of actual vacuum to ideal vacuum.

$$\therefore \text{Vacuum efficiency, } \eta_{\text{vacuum}} = \frac{p_b - (p_a + p_s)}{p_b - p_s} = \frac{p_b - p_t}{p_b - p_s} \quad \text{...(9.12)}$$

The factors affecting the vacuum efficiency are :

1. Air leakage increases p_a and vacuum efficiency decreases.
2. The vacuum efficiency decrease with increase in barometric pressure keeping p_a and p_s same.
3. With insufficient cooling water the pressure in the condenser increases and subsequently reduces the vacuum efficiency of the condenser.

9.14 CONDENSER EFFICIENCY

It is defined as the ratio of the difference between the outlet and inlet temperatures of the cooling water to the difference between the temperature t_s corresponding to the vacuum in the condenser and the inlet temperature of cooling water.

$$\text{Condenser efficiency, } \eta_c = \frac{t_{ow} - t_{iw}}{t_s - t_{iw}} \quad \text{...(9.13)}$$

where t_{iw}, t_{ow} = temperature at inlet and outlet respectively of cooling water

t_s = temperature of steam corresponding to the actual absolute pressure in the condenser.

9.15 COOLING PONDS

The cooling ponds are the simplest of the device for recooling of the cooling water. A cooling pond consists of a large, shallow pool into which the hot water is allowed to come in contact with the atmospheric air. Fig. 9.12 show a *non-directed flow* naturally cooling pond and Fig. 9.13 illustrates a *directed flow* natural cooling pond. Its cooling effect is more than the non-directed flow pond as water comes in contact with air at lower temperature.

In a cooling pond no spray or other cooling device is employed thus keeping the operating expenses quite low. The cooling pond is suitable where sufficient supply of circulating water is not available. In case the amount of cooling water to be cooled is very large, as in the case of big power plants, the cooling ponds will have to be made very large thus making it prohibitive. For large power plants therefore, spray ponds should be used. The depth of cooling pond should be around one metre.

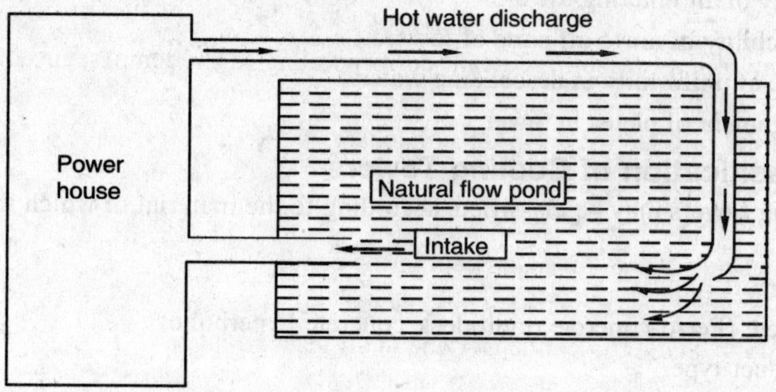

Fig. 9.12. Non-directed flow natural cooling pond.

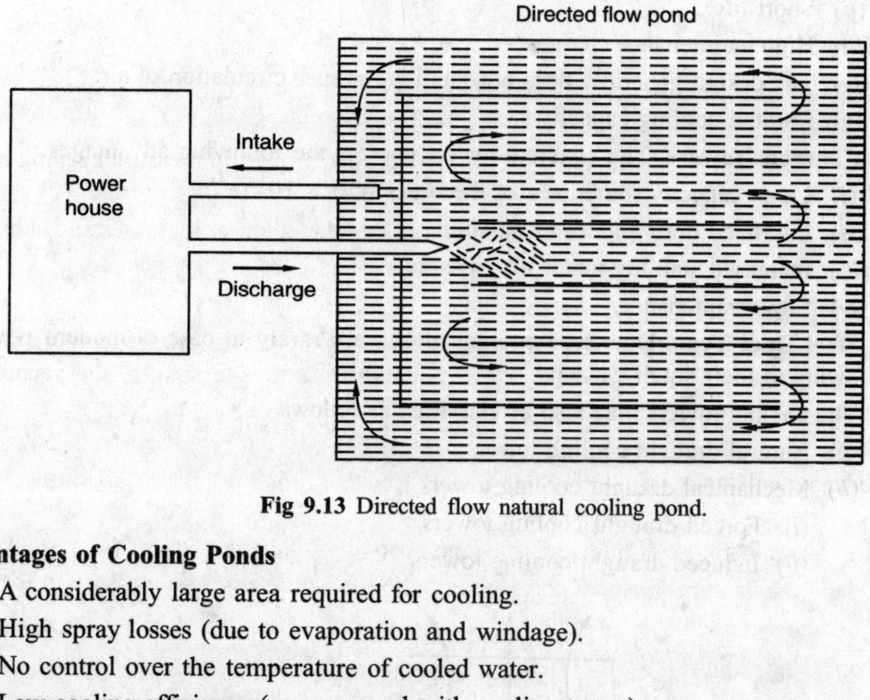

Fig 9.13 Directed flow natural cooling pond.

Disadvantages of Cooling Ponds

1. A considerably large area required for cooling.
2. High spray losses (due to evaporation and windage).
3. No control over the temperature of cooled water.
4. Low cooling efficiency (as compared with cooling tower).

9.16 COOLING TOWERS

In power plants the hot water from condenser is cooled in cooling tower, so that it can be reused in condenser for condensation of steam. In a cooling tower water is made to trickle down drop by drop so that it comes in contact with the air moving in the opposite direction. As a result of this some water is evaporated and is taken away with air. In evaporation the heat is taken away from the bulk of water, which is thus cooled.

Factor affecting cooling of water in a cooling tower are :

1. Temperature of air.
2. Humidity of air.
3. Temperature of hot air.
4. Size and height of tower.

5. Velocity of air entering tower.
6. Accessibility of air to all parts of tower.
7. Degree of uniformity in descending water.
8. Arrangement of plates in tower.

9.16.1 Classification of Cooling Towers

1. Cooling towers may be classified, according to the material of which these are made, as follows:

(*a*) Timber

(*b*) Concrete (Ferro-concrete, multideck concrete hyperbolic)

(*c*) Steel duct type.

(*a*) **Timber Towers.** These towers are rarely used due to the following disadvantages:

(*i*) Due to exposure to sun, wind, water etc., timber rots easily.

(*ii*) Short life.

(*iii*) High maintenance charges.

(*iv*) The design generally does not facilitate proper circulation of air.

(*v*) Limited cooling capacity.

(*b*) **Concrete Towers.** The concrete towers possess the following advantages:

(*i*) Large capacity sometimes of the order of 5×10^3 m^3/h.

(*ii*) Improved draft and air circulation.

(*iii*) Increased stability under air pressure.

(*iv*) Low maintenance.

(*c*) **Steel Duct Type.** Duct type cooling towers are rarely in case of modern power plants owing to their small capacity.

2. The cooling towers may also be classified as follows :

(*a*) Natural draught cooling towers

(*b*) Mechanical draught cooling towers :

(*i*) Forced draught cooling towers

(*ii*) Induced draught cooling towers.

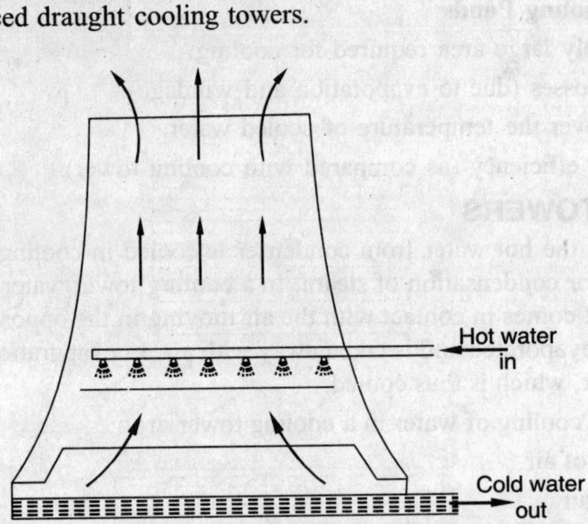

Fig. 9.14 Hyperbolic cooling tower

(*a*) **Natural Draught Cooling Tower.** In this type of tower, the hot water from the condenser is pumped to the troughs and nozzles situated near the bottom. Troughs spray the water falls in the form of droplets into a pond situated at the bottom of the tower. The air enters the cooling tower from air opening provided near the base, rises upward and takes up the heat of falling water. A concrete hyperbolic cooling tower is shown in Fig. 9.14. This tower has the following advantages over mechanical towers:

(*i*) Low operating and maintenance cost.

(*ii*) It gives more or less trouble free operation.

(*iii*) Considerable less ground area required.

(*iv*) The towers may be as high as 125 m and 100 m in diameter at the base with the capability of withstanding winds of very high speed. These structures are more or less self-supported structures.

(*v*) The enlarged top of the tower allows water to fall out of suspension.

The main *drawbacks* or this tower are listed below:

(*i*) High initial cost.

(*ii*) Its performance varies with the seasonal changes in DBT (dry bulb temperature) and R.H. (relative humidity) of air.

While initial cost may be higher, the saving in fan power, longer life and less maintenance always favour for this type of tower. It is also more favourable over mechanical draught cooling towers as central station size increases.

(*b*) **Mechanical Draught Cooling Towers.** In these towers the draught of air for cooling the tower is produced mechanically by means of *propeller fans*. These towers are usually built in cells or units, the capacity depending upon the number of cells used. Fig. 9.15 shows a *forced draught cooling tower*. It is similar naturally draught tower as far as interior construction is concerned, but the sides of the tower are closed and form an air and water tight structure, except for fan openings at the base for the inlet of fresh air, and the outlet at the top for the except of air and vapours. There are hoods at the base projecting from the main portion of the tower where the fans are placed for forcing the air, into the tower.

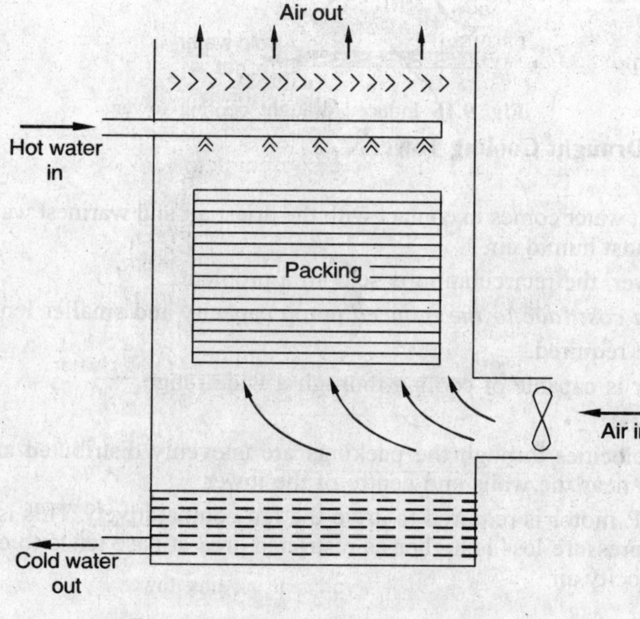

Fig. 9.15 Forced draught cooling tower

Fig. 9.16 shows an *induced draught cooling tower*. In these towers, the fans are placed at the top of the tower and they draw the air in through louvers extending all around the tower at its base.

9.16.2 Comparison of Forced and Induced Draught Towers

(a) Forced Draught Cooling Towers

Advantages:

1. More efficient (than induced draught).
2. No problem of fan blade erosion (as it handles dry air only).
3. More safe.
4. The vibration and noise are minimum.

Disadvantages:

1. The fan size is limited to 4 metres.
2. Power requirement high (approximately double that of induced draught system for the same capacity).
3. In the cold weather, ice is formed on nearby equipments and buildings or in the fan housing itself. The frost in the fan outlet can break the fan blades.

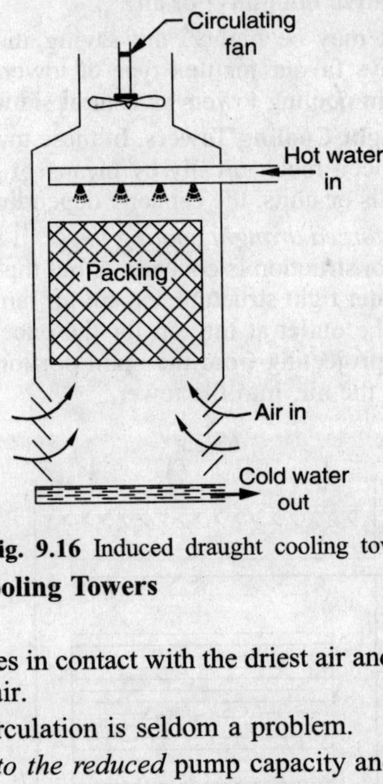

Fig. 9.16 Induced draught cooling tower

(b) Induced Draught Cooling Towers

Advantages

1. The coldest water comes in contact with the driest air and warmest water comes in contact with the most humid air.
2. In this tower, the recirculation is seldom a problem.
3. *Lower first cost* (*due to the reduced* pump capacity and smaller length of water pipes).
4. Less space required.
5. This tower is capable of cooling thorugh a wide range.

Disadvantages:

1. The air velocities through the packings are unevenly distributed and it has very little movement near the walls and centre of the tower.
2. Higher H.P. motor is required to drive the fan comparatively. This is due to the fact that the static pressure loss is higher as restricted area at base tends to choke off the flow of higher velocity air.

9.16.3 Comparison between Natural and Mechanical Draught Towers

(a) Mechanical Draught Towers

Advantages

1. These towers require a small land area and can be built at most locations.
2. The fans give a good control over the air flow and thus the water temperature.
3. Less costly to install than natural draught towers.

Disadvantages

1. Fan power requirements and maintenance costs make them more expensive to operate.
2. Local fogging and icing may occur in winter season.

Example 9.4 *The following readings were taken during a test on a surface condenser:*

Vacuum in condenser = 71 cm of Hg
Barometer reading = 76 cm of Hg
Temperature in condenser = 33°C
Hot well temperature = 30°C
Cooling water circulated = 48,000 kg/h
Inlet temperature of cooling water = 15°C
Outlet temperature of cooling water = 28°C
Condensate = 1500 kg/h

Calculate: (a) the mass of air in kg/m³ of condensate volume.
(b) dryness fraction of steam entering the condenser, and
(c) the vacuum efficiency.

Solution: (*a*) Absolute pressure in condenser,

$$p_t = (76 - 71) \times 0.0133322 = 0.0667 \text{ bar}$$

At $t = 33°C$, $p_s = 0.05029$ bar from steam tables = 3.772 cm of Hg

Partial pressure of air, $p_a = p_t - p_s = 0.0667 - 0.05029 = 0.01654$ bar

Mass of air, $$m_a = \frac{p_a V}{RT} = \frac{0.0164 \times 10^5 \times 1}{287 \times (273 + 33)} = 0.0187 \text{ kg/m}^3$$

(*b*) $$\frac{m_w}{m_s} = \frac{h_f + xh_{fg} - h_c}{c_{pw}(t_{ow} - t_{iw})}$$

At $p = 0.667$ bar, $h_f = 159.47$ kJ/kg, $h_{fg} = 2411.4$ kJ/kg

$$\frac{48,000}{1500} = \frac{159.47 + x \times 2411.4 - 4.1868 \times 30}{4.1868(28 - 15)}$$

$$x = 0.708$$

(*c*) Vacuum efficiency $$= \frac{p_b - p_t}{p_b - p_s} = \frac{76 - 5}{76 - 3.772} = 0.983 \text{ or } 98.3\%$$

Example 9.5 *A steam turbine discharges 6000 kg of steam per hour at 45°C and 0.82 dryness. The estimated air leakage is 15 kg/h. The temperature at the suction of air pump is 30°C and temperature of condensate is 33°C. Find (a) the vacuum gauge reading, (b) capacity of air pump, (c) loss of condensate per hour, and (d) the quality of cooling water required by limiting its temperature rise by 10°C.*

Solution:

(*a*) Total pressure in the condenser, $p_t = p_a + p_s$

At 45°C, $p_s = 0.09582$ bar from steam tables

$$v_g = 15.276 \text{ m}^3\text{/kg}$$

Volume of 6000 kg steam, $V = 6000 \times xv_g = 6000 \times 0.82 \times 15.276$

$$= 75158 \text{ m}^3\text{/h}$$
$$= \text{Volume of air}$$

Partial pressure of air, $p_a = \dfrac{m_a RT}{V} = \dfrac{15 \times 287 \times (45+273)}{75158 \times 10^5}$

$$= 18.215 \times 10^{-5} \text{ bar}$$

$$p_t = 18.215 \times 10^{-5} + 0.09582 = 0.096 \text{ bar}$$

Vacuum in condenser $= \dfrac{1.01325 - 0.096}{0.0133322} = 68.8$ cm of Hg

(*b*) Partial pressure of steam at air pump suction at 30°C = 0.04242 bar

$$p_a = p_t - p_s = 0.096 - 0.04242 = 0.0536 \text{ bar}$$

Volume of air at 30°C and 0.0536 bar,

$$V = \frac{m_a RT}{p_a} = \frac{15 \times 287 \times (30+273)}{0.0536 \times 10^5} = 243.36 \text{ m}^3$$

Air pump capacity = 243.36 m³/h

(*c*) Specific volume of steam at 30°C, $v_g = 32.929$ m³/kg

Loss in condensate $= \dfrac{V}{v_g} = \dfrac{243.36}{32.929} = 7.39$ kg/h

(*d*) At 45°C, $h_f = 188.4$ kJ/kg, $h_{fg} = 2396.9$ kJ/kg

Mass of cooling water, $m_w = \dfrac{m_s(h_f + xh_{fg} - h_c)}{c_{pw}(t_{ow} - t_{iw})}$

$$= \frac{6000(188.4 + 0.82 \times 2394.9 - 4.1868 \times 33)}{4.1868 \times 10 \times 10^3}$$

$$= 288.63 \text{ tonnes/h}$$

Example 9.6 *The air leakage into a surface condenser operating with a steam turbine is estimated as 80 kg/h. The vacuum near the inlet of air pump is 70 cm of Hg when barometer reads 76 cm of Hg. The temperature at the inlet of vacuum pump is 20°C. Calculate (a) the minimum capacity of the air pump, (b) dimensions of the reciprocating air pump to remove air if it runs at 200 rpm. Take L : D : : 3 : 2 and volume efficiency = 98%, (c) mass of vapours extracted per hour.*

Solution: (*a*) *Total pressure at inlet of air pump at 1 – 1, (Fig. 9.17)*

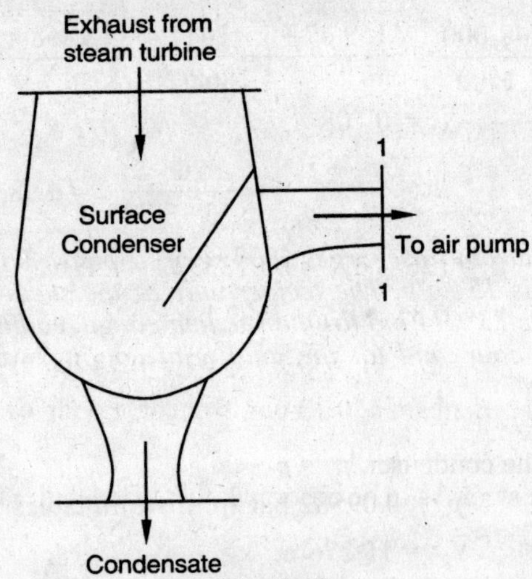

Fig. 9.17

$$p_t = \left(\frac{76-70}{76}\right) \times 1.01325 = 0.08 \text{ bar}$$

Partial pressure of water vapour at 1–1 corresponding to saturation temperature of 20°C, from steam tables,

$$p_s = 0.0234 \text{ bar}$$

Partial pressure of air at 1–1,

$$p_a = p_t - p_s = 0.08 - 0.0234$$

Volume of 80 kg/h of air at 20°C and 0.0566 bar,

$$V = \frac{m_a RT}{p_a} = \frac{80 \times 287 \times 293}{0.0566 \times 10^5} = 1188.6 \text{ m}^3\text{/h}$$

Capacity of air pump = 1188.6 m³/h

(*b*) Now $V = \frac{\pi}{4} D^2 L N \eta_v$

$$\frac{1188.6}{60} = \frac{\pi}{4} \times D^2 \times 1.5D \times 200 \times 0.98$$

$$D^3 = 0.0858$$

$$D = 0.441 \text{ m or } 46.1 \text{ cm}$$

$$L = 66.2 \text{ m}$$

(*c*) Mass of water vapour going with air in the air pump,

$$m_s = \frac{V}{v_g}$$

At 20°C, $v_g = 57.84 \text{ m}^3\text{/kg}$

$$m_g = \frac{1188.6}{57.84} = 20.55 \text{ kg/h}$$

Example 9.7 *The equality of steam entering a jet condenser is 0.92. It is condensed by using cooling water at 15°C. The mass of air in the condenser is 35% of the mixture. Assuming that only latent heat of steam is absorbed by the cooling water, calculate (a) the temperature of mixture of condensate and water leaving the condenser, and (b) the mass of water required per kg of steam condensed.*

Take for air, R = 287 J/kg. K and for steam, R = 463 J/kg.K. Condenser vacuum = 61 cm of Hg, Barometer reading = 76 cm to Hg.

Solution: (*a*) Given $p_a = 0.35\, p_t$

$$p_t = (76 - 61) \times 0.0133322 = 0.2 \text{ bar}$$

$$p_a = 0.35 \times 0.2 = 0.07 \text{ bar}$$

$$p_s = p_t - p_a = 0.2 - 0.07 = 0.13 \text{ bar}$$

Saturation temperature of steam at 0.13 bar, from steam tables.

$$t_s = 51.06°\text{C}$$

As only latent heat of steam is to be absorbed, therefore, outlet temperature of cooling water,

$$t_{ow} = t_s = 51.06°\text{C}$$

Since steam and cooling water mix together in the jet condenser,

Mixture temperature coming out of condenser = 51.06°C

(*b*) Now
$$\frac{m_w}{m_s} = \frac{h_f + xh_{fg} - c_{pw} \times t_{ow}}{c_{pw}(t_{ow} - t_{iw})}$$

At $p_s = 0.13$ bar, $h_f = 213.7$ kJ/kg, $h_{fg} = 2380.3$ kJ/kg from steam tables

$$\frac{m_w}{m_s} = \frac{213.7 + 0.92 \times 2380.3 - 4.1868 \times 51.06}{4.1868(51.06 - 15)} = 14.5$$

Minimum quantity of cooling water required per kg of steam condensed = 14.5 kg.

Example 9.8. *In trial on a surface condenser, the vacuum gauge reading was 660 mm of Hg when barometer read 760 mm of Hg. The temperature in the condenser was 45°C.*

(*a*) *Determine the alteration in vacuum if the quantity of air entering the condenser is reduced by 100 kg/h.*

(*b*) *If mass of condensate is 960 kg/h, calculate the quantities of air and vapour which the pump has to handle.*

Solution.

Absolute condenser pressure = 760 – 660 = 100 mm of Hg

$$p_t = \frac{100}{760} \times 1.01325 = 0.1333 \text{ bar}$$

At 45°C, from steam tables,

Partial pressure of steam, $p_s = 0.0958$ bar

Specific volume of steam, $v_{sg} = 15.276$ m^3/kg

Partial pressure of air, $p_a = 0.1333 - 0.0958 = 0.0375$ bar

Volume of steam per hour, $V_s = 960 \times 15.276 = 14665$ m^3

= Volume of air, V_a

Mass of air, $$m_a = \frac{p_a V_a}{RT} = \frac{0.0375 \times 14665 \times 10^5}{287 \times (273 + 45)} = 602.6 \text{ kg/h}$$

If the mass of air reduced to 100 kg/h, then partial pressure of air,

$$p'_a = \frac{100}{602.6} \times 0.0375 = 0.00622 \text{ bar}$$

Mew Absolute pressure of condenser, $p'_2 = 0.0958 + 0.00622 = 0.102$ bar

$$= \frac{0.102 \times 760}{1.01325} = 76.52 \text{ mm of hg}$$

New vacuum reading = 760 – 76.52 = 683.48 mm Hg

Alteration in vacuum = 683.48 – 660 = 23.48 mm Hg

Capacity of pump = 960 + 100 = 1060 kg/h

Example 9.9 *Following data are recorded from a surface condensing plant:*

Steam condensed = 50,000 kg/h

Temperature of steam in condenser = 40°C

Quality of steam entering into condenser = 0.86

Air leakage in the condenser = 150 kg/g

Condensate temperature = 35°C

Temperature at the suction of the air pump = 32°C

Calculate (i) Vacuum gauge reading in the condenser

(ii) Capacity of dry-air pump

(iii) Loss of steam in kg/hr.

Amount of cooling water circulated through the condenser per hour, if the rise in cooling water temperature is 10°C.

Solution. *The pressure in the condenser corresponds to saturation temperature at 40°C,*

$$p_c = 0.07375 \text{ bar} \qquad \text{(from steam tables)}$$

The pressure near suction of air pump,

$$p_t = p_s + p_a$$

where p_s = steam pressure at 32°C

$= 0.04753$ bar [from steam tables]

$\therefore$ Air pressure, $p_a = 0.07375 - 0.04753 = 0.02622$ bar

(*a*) Vacuum gauge reading $= p_{atm} - p_c$

$$= 1.01325 - 0.07375 = 0.9395 \text{ bar}$$

$$= \frac{0.9395}{0.0136} = 69 \text{ cm of Hg} \qquad [\because \text{ 1cm of Hg} = 0.0136 \text{ bar}]$$

(*b*) Capacity of air pump = Volume of air removed by air pump

$$= \frac{m_a R_a T_a}{p_a} = \frac{150 \times 287 \times (32 + 273)}{0.02622 \times 10^5}$$

$$= 5007.7 \text{ m}^3\text{/hr}$$

(*c*) Volume of steam carried away with air

$= 5007.7$ m^3/hr at p_s corresponding to 32°C

Specific volume of steam at 32°C = 29.57 m^3/kg [from steam tables]

$$\text{Loss of steam mass carried away with air} = \frac{5007.7}{29.57} = 169.35 \text{ kg/hr}$$

$$\text{Cooling water required} = \frac{(m_s + m_c)[(h_{f1} + x_1 h_{fg1}) - c_{pw} T_c]}{c_{pw} \times (T_{wo} - T_{wi})}$$

$$= \frac{(5{,}000 - 169.35)[(167.45 + 0.86 \times 2406.9) - 4.2 \times 35]}{4.2 \times 10}$$

$$= 248 \times 10^4 \text{ kg/hr.}$$

Example 9.10 *The steam condensed in a surface condenser is 12500 kg/hour and the amount of air leakage is 5 kg/hr. The vacuum near the suction pump is 70 cm of Hg and temperature is 34°C. The air and condensate is removed by a wet air pump. Find the capacity of wet air pump if the volumetric efficiency of the pump is 80% take the barometric pressure = 76 cm of Hg.*

Solution. *Steam condensed,* m_s *= 12500 kg/hr, Air leakage,* $\dot{m}_a$ *= 5kg/hr,*

Vacuum near suction pump = 70 cm of Hg

Temperature near suction pump = 34°C

Volumetric efficiency, $\eta_{vol} = 0.80$

Barometric pressure, $p_{atm} = 76$ cm of Hg

$p_{atm} = 76 \times 0.01326 = 1.0336$ bar

70 cm of Hg $= 70 \times 0.0136 = 0.952$ bar

Vacuum gauge reading $= p_{atm} - p_t$

$0.952 = 1.0336 - p_t$

Total pressure near the suction of air pump,

$$p_t = 1.0336 - 0.952 = 0.0816 \text{ bar}$$

Saturation pressure of steam at 34°C, $p_s = 0.05324$ bar (from steam tables)

Now $p_t = p_s + p_a$

Partial pressure of air, $p_a = 0.0816 - 0.05324 = 0.02836$ bar.

Capacity of air pump, $$V_a = \frac{\dot{m}_s R_a T_a}{p_a}$$

$$= \frac{5 \times 287 \times (273 + 34)}{0.02836 \times 10^6} = 155.34 \text{ m}^3\text{/hr}$$

Actual pump capacity $$= \frac{V_a}{\eta_{vol}} = \frac{155.34}{0.80} = 194.17 \text{ m}^3\text{/hr.}$$

Example 9.11 *The air leakage into the condenser operating in conjunction with a steam turbine is estimated at 0.681 kg per minute. The vacuum near the outlet to the air pump is 710 mm when the barometer reads 760 mm and temperature at this point is 18°C find:*

(i) The minimum capacity of air-pump in m^3/min.

(ii) The mass of vapour extracted with the air/min.

Solution.

Pressure in the condenser $= 760 - 710 = 50$ mm of Hg

$= 50 \times 0.001333 = 0.0666$ bar.

Partial pressure of steam 18°C = 0.0206 bar.

Hence partial pressure of air, $p_a = 0.0666 - 0.0206 = 0.046$ bar

Mass of air leakage, $m_a = 0.681$ kg/min.

Volume of air, $$V_a = \frac{m_a RT}{P_a} = \frac{0.681 \times 287 \times (273 + 18)}{0.046 \times 10^5} = 12.36 \text{ m}^3\text{/min}$$

As per Dalton's law of partial pressures, $v_s = v_a$

$\therefore$ Mass of steam condensed, $$m_s = \frac{\text{Volume of steam } V_s}{\text{Specific volume of steam at 0.0206 bar}}$$

$$= \frac{12.36}{65.04} = 0.19 \text{ kg/min}$$

Example 9.12 *In a small capacity of thermal power plant the condenser is provided with a separate air cooling section. The temperature of steam entering the condenser in 56°C and temperature at the air pump suction is 46°C. The barometer read 76 cm of Hg.*

Find:

(a) The vacuum in the condenser.

(b) If the discharge of dry air pump is 75 m^3/min find the air leakage in condenser in kg/hr.

(c) Loss of condensate due to water vapour ended with air through air pump.

Solution. Saturation pressure corresponding to temperature of steam entering the condenser at 46°C = pressure in the condenser, p_t

$$p_t = 0.16530 \text{ bar (from steam tables)}$$

Total pressure near the suction of air pump,

$$p_t = p_s + p_a$$

Where p_s = partial pressure of steam corresponding to temperature of air at air pump suction of 46°C

= 0.10101 bar (from steam tables)

Partial pressure of air, $p_a = 0.16530 - 0.10101 = 0.06429$ bar

(*a*) Vacuum gauge reading $= p_{atm} - p_t = 1.01325 - 0.16530 = 0.84795$ bar

$$\therefore \text{ Vacuum in condenser} = \frac{0.84795}{0.0136} = 62.35 \text{ cm of hg}$$

Because 1 cm Hg = 0.0136 bar.

(*b*) Discharge of dry air pump, $V_a = 75$ m^3/min

Let m_a = mass of air leakage in condenser, kg/hr

Then $p_a V_a = m_a R_a T_a$

$$m_a = \frac{0.06439 \times 10^5 \times 75 \times 60}{287 \times (273 + 46)} = 316 \text{ kg/hr}$$

(*c*) Specific volume of steam at 46°C, v_s = 14.540 m^3/kg (from steam tables)

Volume of steam carried with air =Volume of air removed by pump = 75 m^3/min

$$\text{Loss of condensate mass carried air} = \frac{V_a}{v_s} = \frac{75 \times 60}{14.540} = 309.5 \text{ kg/hr.}$$

MULTI-CHOICE QUESTIONS

1. In a steam condenser, the partial pressure of steam and air are 0.060 bar and 0.007 bar respectively. The condenser pressure is

(*a*) 0.067 bar (*b*) 0.060 bar

(*c*) 0.053 bar (*d*) 0.007 bar

2. Cooling tower in a steam power plant is a device for

(*a*) condensing steam into water

(*b*) cooling the exhaust gases coming out the boiler

(*c*) reducing the temperature of superheated steam

(*d*) reducing the temperature of cooling water used in condenser.

3. The function of the surface condenser is to

(*a*) lower the engine thermal efficiency

(*b*) increase the engine thermal efficiency

(*c*) increase the black pressure of the engine

(*d*) cool the exhaust gases

4. In a jet condenser

(*a*) steam and cooling water mix together

(*b*) steam and cooling water do not mix together

(*c*) steam passes through tubes and cooling water surrounds them

(*d*) cooling water passes through tubes and steam surrounds them

5. A surface condenser is a

(*a*) water tube device

(*b*) steam tube device

(*c*) steam and cooling water mix to give the condensate

(*d*) all of the above

6. The air removal from the surface condenser leads to

(*a*) fall in condenser pressure (*b*) rise in condenser pressure

(*c*) no change in condenser pressure (*d*) rise in condenser temperature

7. Edward's air pump

(*a*) removes only air from the condenser

(*b*) removes air and vapour from the condenser

(*c*) removes only uncondensed vapour from condenser

(*d*) removes air along with vapour and condensed water from the condenser.

8. Vacuum efficiency of a condenser is defined as:

(*a*) $\dfrac{\text{Actual vacuum in condenser with air present} = p_b - p_t}{\text{Theoretical vacuum in condenser without air present} = p_b - p_s}$

(*b*) $\dfrac{\text{Theoretical vacuum in condenser without air present}}{\text{Actual vacuum in condenser with air present}}$

(*c*) $\dfrac{p_s + p_a}{p_s}$ (*d*) $\dfrac{p_s}{p_s + p_a}$

9. Condenser efficiency is defined as :

(*a*) $\dfrac{t_{ow} - t_{iw}}{t_s - t_{iw}}$ (*b*) $\dfrac{t_s - t_{iw}}{t_{ow} - t_{iw}}$

(*c*) $\dfrac{t_{ow} - t_{iw}}{t_s}$ (*d*) $\dfrac{t_s}{t_{ow} - t_{iw}}$

10. The function of a condenser is to

(*a*) condense steam

(*b*) maintain pressure below atmospheric pressure

(*c*) increase work output of prime mover

(*d*) all of the above

11. Evaporative condenser has

(*a*) steam in pipes (*b*) meter in pipes

(*c*) steam mixed with meter in pipes (*d*) all of the above

ANSWERS

1. (*a*)	**2.** (*d*)	**3.** (*b*)	**4.** (*a*)	**5.** (*a*)	**6.** (*a*)	**7.** (*d*)
8. (*a*)	**9.** (*a*)	**10.** (*d*)	**11.** (*a*)			

REVIEW QUESTIONS

1. Define a condenser.
2. What are the functions of a condenser ?
3. Name the various elements of a steam condensing plant.
4. What are the various types of steam condenser ?
5. Differentiate between jet and surface condensers.
6. What is an evaporative condenser ?
7. What are the requirements of a modern surface condenser ?
8. List the advantages and disadvantages of a jet condenser.
9. What are the advantages and disadvantages of surface condensers ?
10. State Dalton's law of partial pressures.
11. How the mass of cooling water can be estimated in a surface condenser?
12. What are the sources of air infiltration in condenser ?
13. What are the effects of air infiltration in condensers ?
14. Define vacuum efficiency and condenser efficiency.
15. What is the role of cooling towers in surface condenser ?
16. Describe various types of cooling towers.

EXERCISES

9.1 The vacuum in a condenser handling 9000 kg/h of steam is found to be 72 cm of Hg when Barometer reading is 76 cm of Hg and the temperature is 25°C. The air leakage amounts to 2kg for every 5000 kg of steam. Determine the capacity of a suitable dry air pump in m^3/min required for the condenser. Take volumetric efficiency of the pump as 0.85.
[**Ans.** 2.83 m^3/min]

9.2 A steam condenser has separate air and condensate pumps. The entry to air pump suction is screened. Steam enters the condenser at 38°C and the condensate is removed at 37°C. The air removed has a temperature of 36°C. If the quantity of air infiltration from various sources is 5 kg/h, determine the volume of air handled by the air pump per hour. Compare this with the quantity that would have to be dealt with by using a combined air and condensate pump. Neglect the pressure due to air at the entry of steam and assume uniform pressure in the condenser. [**Ans.** 647.3 m^3/h, 1267.38 m^3/h]

9.3 A surface condenser deals with 13625 kg of steam per hour at a pressure of 0.09 bar. The steam enters 0.85 dry and the temperature at the condensate and air extraction pipes is 36°C. The air leakage amounts to 7.26 kg/hour. Determine:

(*a*) The surface are required if the average heat transmission rate in 4 kJ/cm per second.

(*b*) The cylinder diameter for the dry air pump, if it is to be single acting and running at 60 rpm with a stroke-bore ratio of 1.25 and volumetric efficiency of 0.85.

[**Ans.** 1956.57 cm^2, 41 cm]

9.4 A surface condenser having an absolute pressure of 0.10 bar is supplied with cooling water at the rate of 40 kg of steam condensed. The rise in the temperature of cooling water is 14°C. Find the dryness fraction of steam entering the condenser. The condensate leaves at 44°C.
[**Ans.** 0.98]

9.5 A steam turbine discharges 5000 kg/h of steam at 40°C and 0.85 dry. The air leakage in the condenser is 15 kg/h. The temperature at the suction of air pump is 32°C and the temperature of condensate is 35°C. Find (*a*) vacuum gauge reading, (*b*) capacity of air pump, (*c*) loss of condensate in kg/h, and (*d*) quantity of cooling water if its temperature rise is limited to 10°C.

[**Ans.** 70.45 cm, 502.5 m3/h, 17 kg/h, 254 × 10^3 kg/h]

9.6 A steam turbine uses 50,000 kg/h of steam. The exhaust steam with dryness fraction 0.9 enters the condenser fitted with water extraction and air pumps. When the barometer reads 76 cm of Hg, vacuum of air pump suction is 72 cm of Hg and temperature is 32°C. The air leakage is estimated at 50 kg/h. Calculate

(*a*) Net capacity of air pump.

(*b*) Quantity of water circulated per minute if temperature rise is limited to 15°C.

[**Ans.** 1291.83 m^3/min, 27160 kg/min]

9.7 The temperature in a surface condenser is 40°C and the vacuum is 69 cm of Hg while the barometer reads 75 cm of Hg. Determine the partial pressure of steam and air and the mass of air present per kg of steam.

[**Ans.** 0.07375 bar, 0.00625 bar, 0.136 kg/kg of vapour]

9.8 In a condenser, to check the leakage of air, the following procedure is adopted:

After running the plant to reach the steady conditions, the steam supply to the condenser and also the air and condensate pumps are shut down, thus completely isolating the condenser. The temperature and vacuum readings are noted at shut down and also after a period of 5 minutes. They are 39°C and 68.5 cm Hg; and 28°C and 27 cm Hg respectively. The barometer reads 75 cm Hg. The effective volume of condenser is 1 m^3. Calculate (*a*) the quantity of air leakage into the condenser during the period of observation, and (*b*) the quantity of water vapour condensed during the period.

[**Ans.** 0.3523 kg, 0.0214 kg]

9.9 A surface condenser is required to deal with 20,000 kg of steam per hour. The air leakage is estimated at 0.3 kg per 1000 kg of steam. The steam enters the condenser dry saturated of 38°C. The condensate is extracted at the lowest point of the condenser at a temperature of 36°C. The condensate loss is made up with water at 7°C. It is required to find the saving in condensate and the saving in heat supplied in the boiler, by fitting a separate air extracting pump which draws the air over an air cooler. Assume that the air leaves the cooler at 27°C. The pressure in the condenser can be assumed to remain constant.

[**Ans.** 28.1 kg/h, 3411 kJ/h]

9.10 A surface condenser deals with 13000 kg/h of steam. The leakage air in the system amounts to 1 kg per 2700 kg of steam. The vacuum in the air pump suction is 705 mm of mercury (barometer 760 mm of Hg), and temperature is 34.6°C. Determine the discharging capacity of the wet air pump which removes both air and condensate, taking the volumetric efficiency of the pump as 90 percent. [**Ans.** 4.539 m^3]

10 GAS TURBINES

10.1 INTRODUCTION

A gas turbine is the most satisfactory way of producing very large power in a self-contained and compact unit. The gas turbine obtains its power by utilizing the energy of a jet or burnt gases and air, the velocity of the jet being absorbed as it flows over several rings of moving blades which are fixed to a common shaft. The gas turbine requires an air compressor which is driven off its own shafting. This absorbs a considerable proportion of the power produced and thus lowers the overall efficiency, which is of the order of 20 to 30%.

10.2 FIELDS OF APPLICATIONS OF GAS TURBINE

The following are the major fields of application of gas turbines :

1. **Aviation** — A gas turbine is used in aviation because it is self-contained, light weight not requiring cooling water and generally fit into the overall shape of the aircraft structure. It is also suitable for turbo-jet and turbo-propeller engines.

2. **Power generation** — It is used for electric power generation because of its simplicity, lack of cooling water, quick installation and starting. Also suitable for supercharging, road transport and railways.

3. **Oil and gas industry** — It is used in oil and gas industry because of cheaper supply of fuel and low installation cost.

4. **Marine propulsion** — It is used in marine propulsion because it is of light weight and not requiring cooling water.

10.3 LIMITATIONS OF GAS TURBINES

The limitations of gas turbines are :

1. They are not self starting.
2. Low efficiency at part loads.
3. Non-reversible.
4. Higher rotor speeds, and
5. Low overall plant efficiency.

10.4 COMPARISON OF GAS TURBINES WITH I.C. ENGINES

(a) Advantages

1. Perfect balancing of the rotating parts is possible.
2. Mechanical efficiency of very high (95%) compared with 85% of I.C. engine because of its large number of sliding parts.

3. Torque on the turbine shaft is continuous and uniform, therefore flywheel is not required.
4. Work developed is more because expansion of gases upto atmospheric pressure is possible.
5. The weight of gas turbine per kW of power developed is less and are therefore superior for use in aircraft.
6. The compression, combustion and expansion take place in different units and therefore these units can be designed, tested and developed individually.
7. The gas turbine can be driven at much higher speeds.
8. The maximum pressure used in gas turbines is very low, therefore its components can be made lighter.
9. The lubrication and ignition systems are much more simple.
10. Cheaper fuels can be used in a gas turbine.
11. The installation and maintenance costs are less.
12. The exhaust is free from smoke and is less polluting.

(*b*) Disadvantages

1. The thermal efficiency of a simple gas turbine is low.
2. Very poor thermal efficiency at part loads.
3. The temperature of gases supplied to the turbine is limited to 100 K as there are no metals which can be used for blades to sustain this temperature at high centrifugal stresses.
4. The manufacture of blades is difficult and costly as Ni-Cr alloy is used.
5. The control of fuel is difficult with wide operating speeds and output.
6. A special cooling system is required for the blades.
7. The starting is more difficult.
8. It produces five times exhaust gases than I.C. engine for same output
9. Speed reduction device is always required due to higher operating speeds.
10. The life of blades and combustion chamber is low due to high temperature.

10.5 ADVANTAGES OF GAS TURBINES OVER STEAM TURBINES

1. Boiler along with accessories are not required.
2. The capital and running cost is less.
3. The space required is less for the plant.
4. Starting of the plant is more easy and quick.
5. The weight of the plant per kW generated is less.
6. The control of plant is much easier.
7. The plant can be installed anywhere as cooling water is not required
8. Gas turbine plant is used as a best type of peak load plant.

10.6 CLASSIFICATION OF GAS TURBINES

The gas turbines may be classified as:

1. Constant pressure combustion (or continuous combustion) gas turbine.
   (*a*) open cycle constant pressure gas turbine.
   (*b*) closed cycle constant pressure gas turbine.
2. Constant volume combustion (or explosion) gas turbine.

1. Constant Pressure Combustion Gas Turbine

In this type, the fuel is burned at constant pressure and the cycle used is the Joule cycle. The turbine is of the reaction type using oil fuel and is fitted with an axial flow rotary air compressor which is coupled to the rotor shaft. The heat is supplied to the working fluid at constant pressure.

The open and closed cycle gas turbines are shown in Fig. 10.1 (*a*) and Fig. 10.1 (*b*) respectively.

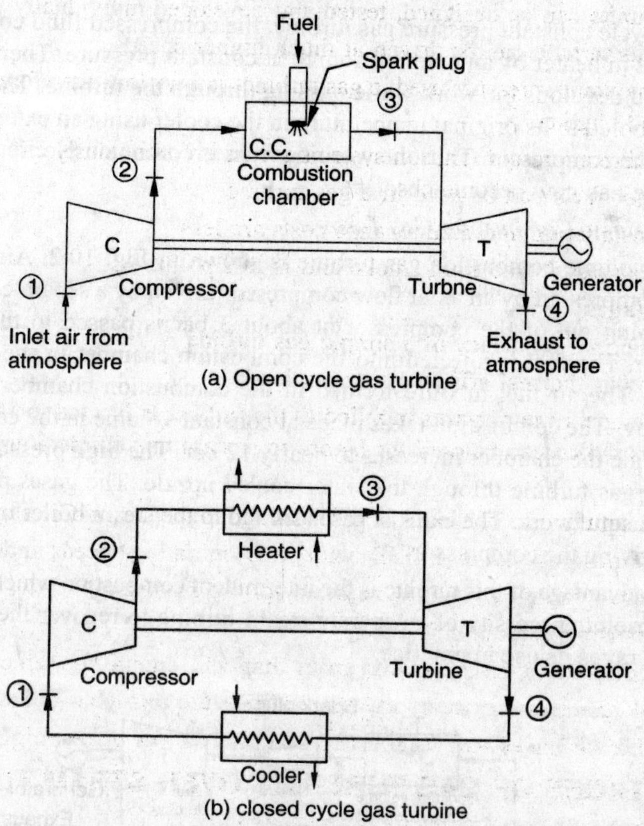

Fig. 10.1 Constant pressure combustion gas turbines.

A simple open cycle constant pressure gas turbine consists of

* a compressor,
* combustion chamber and
* turbine.

The air is taken from the atmosphere by the compressor and is compressed to the combustion pressure of 1 to 4 atom. It is then forced into the combustion chamber. Part of this air is used as combustion air for the oil which enters the burner, the remainder is forced through the annular space between the wall of the combustion chamber and the burner jacket. The air receives heat from the burner jacket and also mixes with the products of combustion chamber and the burner jacket. This raises the temperature and volume of the air. The use of a very large quantity of air in excess of the combustion air prevents the temperature of the mixture from reaching values which are too high for the metal of the rotor blades. It also prevents the burner from becoming too hot.

The high pressure and high temperature gases now enter the turbine and flow through the blade rings. It continuously expands there and its pressure energy is converted into kinetic energy, which in turn is absorbed by the rotor. On leaving turbine, the spent gases pass away to exhaust. The speed of the turbine is regulated by controlling the fuel oil supply and by the safety valve which by-passes some of the mixture if the pressure becomes too high. The net power developed by the turbine is utilized in driving the generator but a considerable amount of the total power produced by the rotor is absorbed in driving the compressor. The turbine is started by the electric starter motor.

In the closed cycle constant pressure gas turbine, the compressed fluid coming out from the compressor is heated in heater by an external source at constant pressure. Then the high pressure and temperature fluid develops the work while passing through the turbine. The fluid coming out from the turbine is cooled to its original temperature in the cooler using an external cooling source before passing into the compressor. Thus the working fluid is continuously circulated in the closed cycle. Such a turbine has now become obsolete.

2. Constant Volume Combustion Gas Turbine

The constant volume combustion gas turbine is shown in Fig. 10.2. Air is taken from the atmosphere and is compressed by an axial flow compressor driven by a separate steam turbine. The compressed air coming out of the compressor at about 3 bar is passed to the constant volume combustion chamber. The fuel is injected into the combustion chamber in the current of air by a separate fuel pump. The air-fuel mixture formed in the combustion chamber is then ignited by means of a spark plug. The combustion takes place at constant volume in the combustion chamber and the pressure inside the chamber increases to nearly 12 bar. The high pressure and temperature gases are fed to the gas turbine through the water cooled nozzle. The gases passing through the turbine develop the useful work. The exhaust gases are fed to the steam boiler to generate steam for the steam turbine driving the compressor.

The major disadvantage of this turbine is the intermittent combustion which impairs its smooth functioning. The absolute necessity of a separate steam turbine to recover the heat from exhaust gases further discourages its use in practice.

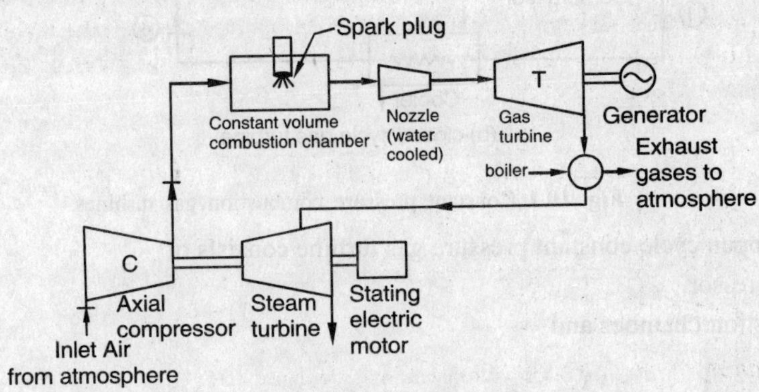

Fig. 10.2 Constant volume combustion gas turbine.

10.7 COMPARISON OF OPEN AND CLOSED CYCLES GAS TURBINES

Particular	*Closed Cycle*	*Open Cycle*
1. Cycle pressure	Higher which results in a reduced size of plant.	Lower which results in a bigger size of plant
2. Corrosion and erosion of blades	Avoided due to indirect heating.	More due to direct heating

Particular	Closed Cycle	Open Cycle
3. Filtration of incoming air	Not required.	Poses a severe problem.
4. Density of working fluid	High density reduces the size of compressor, turbine and heat exchanger	Low, which increases the size of various components
5. Type of fuel used	Solid fuel can be used more economically	Fuel has to be a fluid, which is comparatively costlier.
6. Working fluid	Any thermodynamically better fuel than air can be used	Working fluid is restricted to air only
7. Maintenance cost	Low	High
8. Use	Suitable for stationary plants and marine power as it requires abundance quantities of water	Suitable for aircraft turbines
9. Weight	High, not suitable for moving vehicles	Low, suitable for moving vehicles also
10. heat exchange coefficient	Low	High
11. Installation cost	High	Low
12. Response to changing load	Poor	Good

10.8 POSITION OF GAS TURBINE IN POWER INDUSTRY

Constant pressure gas turbines are now used for generating electricity, driving locomotives and aeroplane propellers producing the air steam for wind tunnels and for the jet propulsion of air craft. In some types of aircraft, the exhaust gases from the main engines are used for driving a gas turbine which in turn drives the super-charger. Gas turbines driven by exhaust gases are also used for supplying a pressure charge to diesel engines.

Overall efficiencies of constant pressure gas turbine plants vary between 18 to 25 percent. This is based on a turbine efficiency of 78%. The efficiency of this type of turbine can be improved by using heat from exhaust gases to preheat the compressed air before it enters the combustion chamber. The hot mixture of air and burnt gases is then expanded through the turbine from which they are exhausted into the heat exchanger. Their available heat is now given to the high pressure air from the compressor. Such a system is shown in Fig. 10.3.

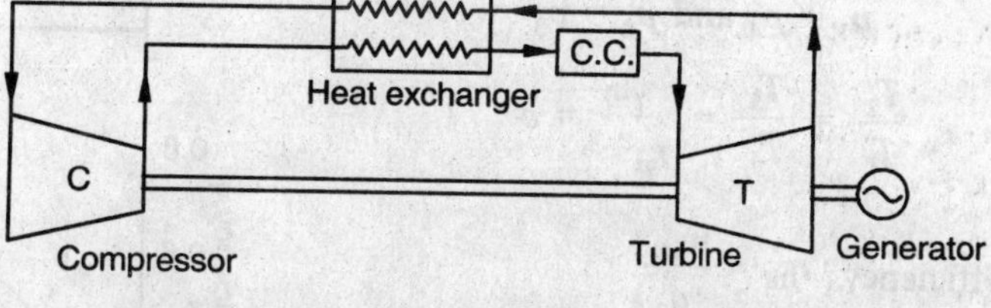

Fig. 10.3 Gas turbine utilizing exhaust heat.

10.9 THERMODYNAMICS OF CONSTANT PRESSURE GAS TURBINE-BRAYTON CYCLE

(*a*) Efficiency

The Brayton cycle is shown on $p-v$ and T-s diagrams in Fig. 10.4. During this cycle, air is drawn into the compressor at point 1 and then compressed isentropically along the process 1-2. Process 2-3 represents the burning of the oil at constant pressure p_2. Process 3-4 is the isentropic expansion of the gases through the turbine. Process 4-1 represents the exhausting and cooling of the gases at constant pressure p_1.

$$\text{Compressor work, } w_c = w_{1\text{-}2} = h_2 - h_1 = c_p\,(T_2 - T_1)$$

$$\text{Heat supplied, } q = q_{2\text{-}3} = h_3 - h_2 = c_p(T_3 - T_2)$$

$$\text{Turbine work, } w_t = w_{3\text{-}4} = h_3 - h_4 = c_p(T_3 - T_4)$$

$$\text{Net work done, } w_{\text{net}} = w_t - w_c = \text{cp}\;[(T_3 - T_4) - (T_2 - T_1)]$$

$$= c_p T_1\left[\frac{T_3}{T_1} - \frac{T_4}{T_1} - \frac{T_2}{T_1} + 1\right]$$

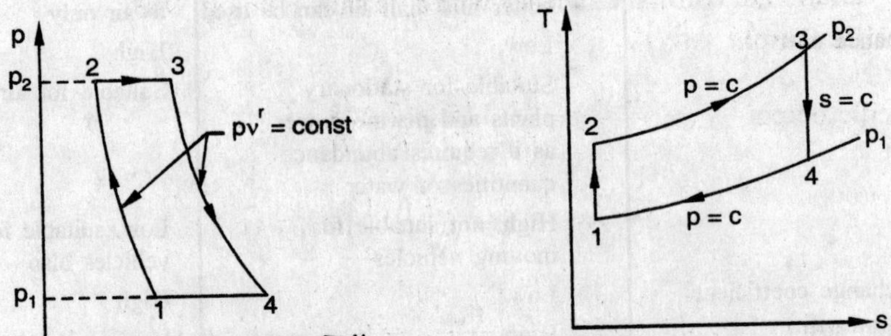

(a) p-u diagram (b) T-u diagram

Fig. 10.4 Brayton cycle for constant pressure gas turbine.

Let $r_p = \dfrac{p_2}{p_1}$ be the pressure ratio and $\alpha = \dfrac{T_3}{T_1}$

For the adiabatic compression process 1–2, we have

$$\frac{T_1}{T_2} = \frac{v_2}{v_1} = \left(\frac{p_1}{p_2}\right)^{\frac{\gamma-1}{\gamma}} \qquad \ldots(1)$$

and for the adiabatic expansion process 3–4, we have

$$\frac{T_4}{T_3} = \frac{v_3}{v_4} = \left(\frac{p_4}{p_3}\right)^{\frac{\gamma-1}{\gamma}} \qquad \ldots(2)$$

But $p_1 = p_4$ and $p_2 = p_3$

$$\therefore \quad \frac{T_2}{T_1} = \frac{T_3}{T_4} = r_p^{\frac{\gamma-1}{\gamma}} = c$$

$$\text{Thermal efficiency, } \eta_{th} = \frac{w_{\text{net}}}{q}$$

$$= \frac{c_p T_1\left[\alpha\left(1-\dfrac{1}{c}\right)-(c-1)\right]}{c_p(T_3 - T_2)}$$

$$= \frac{\alpha\left(1-\dfrac{1}{c}\right)-(c-1)}{\dfrac{T_3}{T_1} - \dfrac{T_2}{T_1}}$$

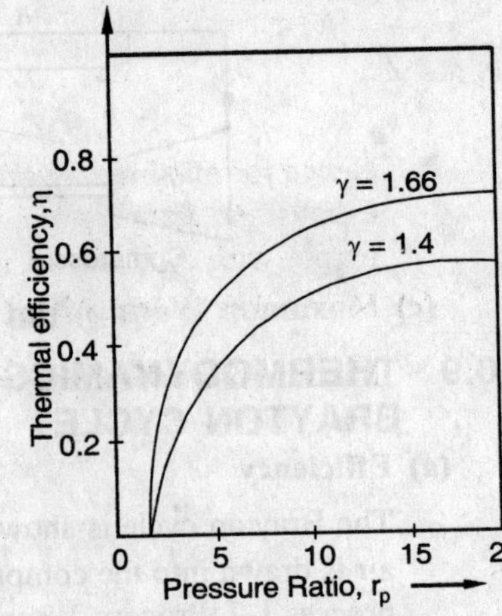

Fig. 10.5 Thermal efficiency v's pressure ratio

$$= \frac{\alpha\left(1-\frac{1}{c}\right)-(c-1)}{\alpha-c} = \frac{(\alpha-c)-\left(\frac{\alpha-c}{c}\right)}{\alpha-c}$$

$$= 1-\frac{1}{c} = 1-\frac{1}{r_p^{\frac{\gamma-1}{\gamma}}} \qquad \text{...(10.1)}$$

The variation of thermal efficiency with r_p is shown in Fig. 10.5.

(b) Specific Output

$$\text{Specific output} = \frac{w_{net}}{c_p T_1} = \frac{T_3}{T_1} - \frac{T_4}{T_3} \times \frac{T_3}{T_1} - \frac{T_2}{T_1} + 1$$

$$= \alpha - \frac{\alpha}{c} - c + 1 = \alpha\left(1-\frac{1}{c}\right)-(c-1) \qquad \text{...(10.2)}$$

For maximum specific output, $\frac{d}{dc}\left[\frac{w_{net}}{c_p T_1}\right] = 0$

$$\therefore \quad \alpha\left(\frac{1}{c^2}\right) - 1 = 0$$

$$\alpha = c^2 = \left[r_p^{\frac{\gamma-1}{\gamma}}\right]^2 \qquad \text{...(10.3)}$$

or

$$\left(r_p^{\frac{\gamma-1}{\gamma}}\right)_{opt} = \sqrt{\alpha}$$

Also

$$r_p^{\frac{\gamma-1}{\gamma}} = \frac{T_2}{T_1} = \frac{T_3}{T_4}$$

$$\alpha = \frac{T_2}{T_1} \times \frac{T_3}{T_4} = \frac{T_2}{T_4}\alpha$$

$$T_2 = T_4 \qquad \text{...(10.4)}$$

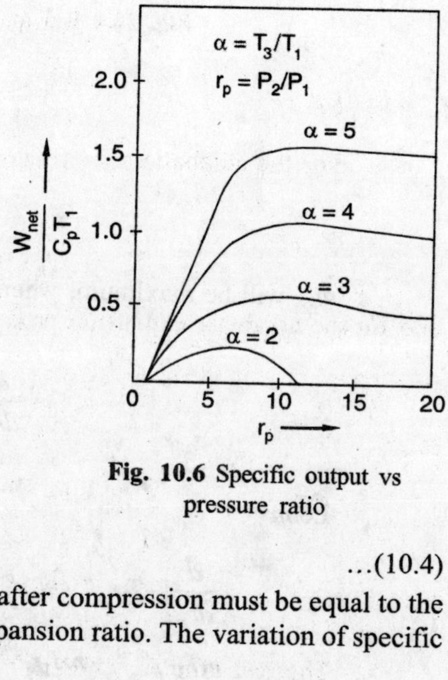

Fig. 10.6 Specific output vs pressure ratio

Hence for maximum work output the temperature after compression must be equal to the exhaust gas temperature, which depends on the expansion ratio. The variation of specific output with r_p is shown in Fig. 10.6.

(c) Maximum Work output

$$\text{Maximum work output, } (w_{net})_{max} = c_p[(T_3 - T_4) - (T_2 - T_1)]$$

$$= c_p[T_3 + T_1 - T_4 - T_2]$$

$$= c_p[T_3 + T_1 - 2T_2] \qquad (\because T_4 = T_2)$$

$$= c_p\left[\sqrt{T_3^2} + \sqrt{T_1^2} - 2\sqrt{T_2 T_4}\right] \qquad \left(\because \frac{T_3}{T_4} = \frac{T_2}{T_1}\right)$$

$$= c_p[\sqrt{T_3} - \sqrt{T_1}]^2 = c_p[\sqrt{T_{max}} - \sqrt{T_{min}}]^2 \qquad \text{...(10.5)}$$

(d) Work Ratio

Work ratio, $$R_w = \frac{w_{net}}{w_t} = \frac{w_t - w_c}{w_t} = 1 - \frac{w_c}{w_t}$$

$$= 1 - \frac{T_2 - T_1}{T_3 - T_4} = 1 - \frac{T_1 (r_p)^{\frac{\gamma-1}{\gamma}} - T_1}{T_3 - \frac{T_3}{r_p^{\frac{\gamma-3}{\gamma}}}}$$

$$= 1 - \left[\frac{T_1}{T_3}\right]\left[\frac{r_p^{\frac{\gamma-1}{\gamma}}}{1 - \frac{1}{r_p^{\frac{\gamma-1}{\gamma}}}}\right] = 1 - \frac{T_1}{T_3}(r_p)^{\frac{\gamma-1}{\gamma}} = 1 - \frac{c}{\alpha} \quad \text{...(10.6)}$$

(e) Optimum Pressure Ratio for Maximum specific Work Output

$$w_{net} = w_t - w_c = c_p(T_3 - T_4) - c_p(T_2 - T_1)$$

$$= cpT_3\left(1 - \frac{T_4}{T_3}\right) - c_p T_1 \left(\frac{T_2}{T_1} - 1\right)$$

$$= c_p T_3 \left[1 - \frac{1}{r_p^{\frac{\gamma-1}{\gamma}}}\right] - c_p T_1 \left[r_p^{\frac{\gamma-1}{\gamma}} - 1\right]$$

w_{net} will be maximum, when $\frac{dw_{net}}{dr_p} = 0$

$$\therefore \quad \frac{d}{dr_p}\left[c_p T_3 - c_p T_3 r_p^{-\left(\frac{\gamma-1}{\gamma}\right)} - c_p T_1 r_p^{\frac{\gamma-1}{\gamma}} + c_p T_1\right] = 0$$

Letm $$= \frac{\gamma - 1}{\gamma}$$

$$\therefore \quad \frac{d}{dr_p}[-T_3 r_p^{-m} - T_1 r_p^{m}] = 0$$

$$m T_3 r_p^{-m-1} - m T_1 r_p^{m-1} = 0$$

$$\frac{T_3}{T_1} = r_p^{2m} = r_p^{\frac{2(\gamma-1)}{\gamma}} \quad \text{...(10.7)}$$

$$r_p = \left[\frac{T_3}{T_1}\right]^{\frac{\gamma}{2(y-1)}}$$

From Eq. (10.3), we have

$$(\boldsymbol{r_p})_{opt} = \left(\frac{T_3}{T_1}\right)^{\frac{\gamma}{\gamma-1}}$$

Combining with Eq. (10.7), we get

$$r_p = \sqrt{(r_p)_{\text{opt}}} \qquad \ldots(10.8)$$

10.10 CYCLE OPERATION WITH MACHINE EFFICIENCY

(a) Maximum Pressure ratio for maximum specific work

It is not possible to achieve the isentropic compression and expansion in the actual Brayton cycle due to inevitable losses due to friction and turbulence in the compressor and turbine. Therefore, the temperature at the end of compression and expansion are higher than that in an ideal cycle for the same pressure ratio. The actual Brayton cycle is shown in Fig. 10.7.

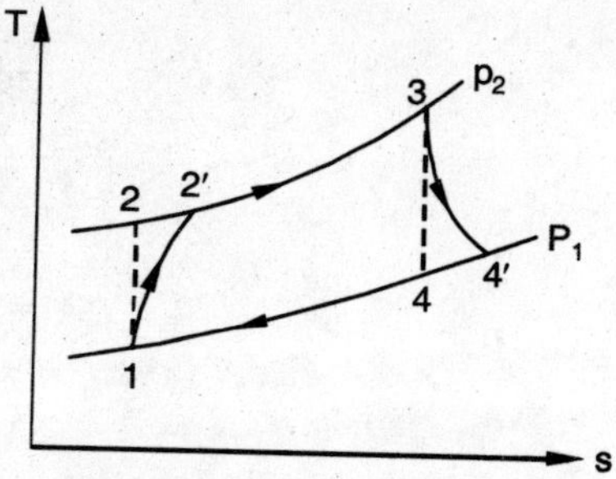

Fig. 10.7 Actual Brayton cycle with machine efficiencies

Isentropic efficiency of compressor, $\eta_c = \dfrac{T_2 - T_1}{T_{2'} - T_1}$

Isentropic efficiency of turbine, $\eta_t = \dfrac{T_3 - T_4'}{T_3 - T_4}$

Net specific work available, $w_{\text{net}} = w_t - w_c = c_p(T_3 - T_4') - c_p(T_2' - T_1)$

$$= c_p\eta_t(T_3 - T_4) - c_p\frac{(T_2 - T_1)}{\eta_c}$$

$$= c_p\eta_t T_3\left(1 - \frac{T_4}{T_3}\right) - \frac{c_pT_1}{\eta_c}\left(\frac{T_2}{T_1} - 1\right)$$

$$= c_p\eta_t T_3\left[1 - \frac{1}{r_p^{\frac{\gamma-1}{\gamma}}}\right] - \frac{c_pT_1}{\eta_c}\left[r_p^{\frac{\gamma-1}{\gamma}} - 1\right] \qquad \ldots(10.9)$$

For specific work output to be maximum for given temperature limits,

$$\frac{d}{dr_p}(w_t) = 0$$

$$\eta_t T_3\left[\left(\frac{\gamma-1}{\gamma}\right) r_p^{\frac{\gamma-1}{\gamma}-1}\right] - \frac{T_1}{\eta_c}\left(\frac{\gamma-1}{\gamma}\right) r_p^{\frac{\gamma-1}{\gamma}-1} = 0$$

$$n_t T_3 r_p^{\frac{\gamma-1}{\gamma}-1} - \frac{T_1}{\eta_c} r_p^{-\frac{1}{\gamma}} = 0$$

$$r_p^{2\left(\frac{1-\gamma}{\gamma}\right)} = \frac{T_1}{\eta_c\eta_t T_3}$$

$$r_p = \left[\eta_c\eta_t\left(\frac{T_3}{T_1}\right)\right]^{\frac{\gamma}{2(\gamma-1)}} \qquad \ldots(10.10)$$

(b) Optimum pressure ratio for maximum cycle thermal efficiency

Heat supplied, $q = c_p(T_3 - T_2') = c_p(T_3 - T_2)$

$$\text{Thermal efficiency, } \eta_{th} = \frac{w_{net}}{q} = \frac{c_p(T_3 - T_4') - c_p(T_2' - T_1)}{c_p(T_3 - T_2)}$$

$$= \frac{\eta_t(T_3 - T_4) - \dfrac{1}{\eta_c}(T_2 - T_1)}{T_3 - T_2}$$

$$= \frac{\eta_t T_3 \left[1 - \dfrac{T_4}{T_3}\right] - \dfrac{T_1}{\eta_c}\left[\dfrac{T_2}{T_1} - 1\right]}{T_1\left[\dfrac{T_3}{T_1} - \dfrac{T_2}{T_1}\right]}$$

Now $\dfrac{T_2}{T_1} = \dfrac{T_3}{T_4} = r_p^{\frac{\gamma-1}{\gamma}}$

$$\eta_{th} = \frac{\eta_t T_3\left[1 - \dfrac{1}{r_p^{\frac{\gamma-1}{\gamma}}}\right] - \dfrac{T_1}{\eta_c}[r_p^{\frac{\gamma-1}{\gamma}} - 1]}{T_1\left[\dfrac{T_3}{T_1} - r_p^{\frac{\gamma-1}{\gamma}}\right]} \quad \ldots(10.11)$$

The thermal efficiency of the cycle is dependent only on r_p for the fixed values of T_1 and T_3. The condition for the maximum value of thermal efficiency for given temperature limits is:

$$\frac{d\eta_{th}}{dr_p} = 0$$

Let $c = r_p^{\frac{\gamma-1}{\gamma}}$ and $\alpha = \dfrac{T_3}{T_1}$, then $\eta_{th} = \dfrac{\alpha\eta_t\left(1 - \dfrac{1}{c}\right) - \dfrac{1}{\eta_c}(c-1)}{\alpha - c}$

$$\frac{d\eta_{th}}{dc} = (\alpha - c)\left[\alpha\eta_t \times \frac{1}{c^2} - \frac{1}{\eta_c}\right] + 1\left[\alpha\eta_t\left(1 - \frac{1}{c}\right) - \frac{1}{\eta_c}(c-1)\right] = 0$$

$$\frac{\alpha^2\eta_t}{c^2} - \frac{\alpha}{\eta_c} - \frac{\alpha\eta_t}{c} + \frac{c}{\eta_c} + \alpha\eta_t - \frac{\alpha\eta_t}{c} - \frac{c}{\eta_c} + \frac{1}{\eta_c} = 0$$

$$\frac{\alpha^2\eta_t}{c^2} - \frac{\alpha}{\eta_c} - \frac{2\alpha\eta_t}{c} + \alpha\eta_t + \frac{1}{\eta_c} = 0$$

$$\alpha^2\eta_c\eta_t - \alpha c^2 - 2\alpha\eta_c\eta_t c + \alpha\eta_c\eta_t c^2 + c^2 = 0$$

$$c^2(1 + \alpha\eta_c\eta_t - \alpha) - 2\alpha\eta_c\eta_t c + \alpha^2\eta_c\eta_t = 0$$

$$c = \frac{2\alpha\eta_c\eta_t \pm [4\alpha^2\eta_c^2\eta_t^2 - 4\alpha^2\eta_c\eta_t(1 + \alpha\eta_c\eta_t - \alpha)]^{\frac{1}{2}}}{2(1 + \alpha\eta_c\eta_t - \alpha)}$$

$$= \frac{\alpha\eta_c\eta_t \pm \alpha[\eta_c^2\eta_t^2 - \eta_c\eta_t(1+\alpha\eta_c\eta_t - \alpha)]^{\frac{1}{2}}}{(1+\alpha\eta_c\eta_t - \alpha)} = \frac{\alpha\left[1 \pm \left\{1 - \frac{1+\alpha\eta_c\eta_t - \alpha}{\eta_c\eta_t}\right\}^{\frac{1}{2}}\right]}{\frac{1+\alpha\eta_c\eta_t - \alpha}{\eta_c\eta_t}}$$

Taking –ve sign

$$c = \frac{\alpha\left[1 - \left\{1 - \frac{1}{\eta_c\eta_t} - \alpha + \frac{\alpha}{\eta_c\eta_t}\right\}^{\frac{1}{2}}\right]}{\left[1 + \left\{\frac{1}{\eta_c\eta_t} - \alpha + \frac{\alpha}{\eta_c\eta_t}\right\}^{\frac{1}{2}}\right]\left[1 - \left\{1 - \frac{1}{\eta_c\eta_t} - \alpha + \frac{\alpha}{\eta_c\eta_t}\right\}^{\frac{1}{2}}\right]}$$

$$r_p^{\frac{\gamma-1}{\gamma}} = \frac{\alpha}{1 + \left\{\frac{1}{\eta_c\eta_t}(\alpha-1)(\alpha-1)\right\}^{\frac{1}{2}}} = \frac{\alpha}{1 + \left\{(\alpha-1)\frac{1}{\eta_c\eta_t} - 1\right\}^{\frac{1}{2}}}$$

$$(r_p)_{\text{opt}} = \left[\frac{\frac{T_3}{T_1}}{1 + \left\{\left(\frac{T_3}{T_1} - 1\right)\left(\frac{1}{\eta_c\eta_t} - 1\right)\right\}^{\frac{1}{2}}}\right]^{\frac{\gamma}{\gamma-1}} \quad \text{...(10.12)}$$

10.11 OPEN CYCLE CONSTANT PRESSURE GAS TURBINE

The open constant pressure gas turbine plant is shown in Fig. 10.8. The fuel is burned with air coming out form the compressor into the combustion chamber. The products of combustion from the combustion chamber are passed through the turbine and develop the required power. Part of the power developed is used to run the compressor. The gases leaving the turbine are exhausted to atmosphere.

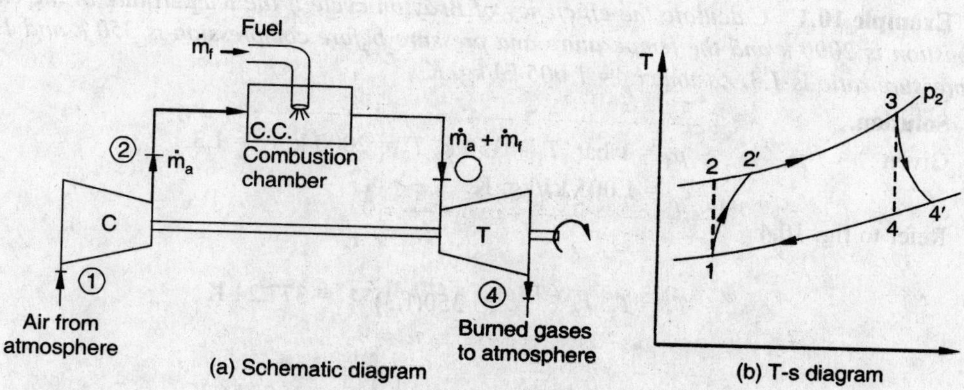

Fig. 10.8 Open cycle constant pressure gas turbine.

Let $\dot{m}_a$ = mass rate of flow of air

$\dot{m}_f$ = mass rate of flow of fuel

Compressor work, $W_c = \dot{m}_a c_{pa}(T_2' - T_1)$

Turbine work, $W_t = (\dot{m}_a + \dot{m}_f)c_{pg}(T_3 - T_4')$

where $c_{pa}c_{pg}$ = specific heat of air and gases respectively

Network, $w_{net} = W_t - W_c = (\dot{m}_a + \dot{m}_f)c_{pg}(T_3 - T_4') - \dot{m}_a c_{pa}(T_2' - T_1)$

Heat supplied, $Q = \dot{m}_f \times \text{C.V.}$

C.V. = calorific value a of fuel

Thermal efficiency, $\eta_{th} = \dfrac{W_{net}}{Q}$

$$= \frac{(\dot{m}_a + \dot{m}_f)c_{pg}(T_3 - T_4') - \dot{m}_a c_{pa}(T_2' - T_1)}{\dot{m} f \times \text{C.V.}} \quad \ldots(10.13)$$

Now $\eta_c = \dfrac{T_2 - T_1}{T_2' - T_1}$

$\therefore \quad T_2' - T_1 = \dfrac{T_2 - T_1}{\eta_c}$

and $\eta_t = \dfrac{T_3 - T_4'}{T_3 - T_4}$

$\therefore \quad T_3 - T_4' = \eta_t (T_3 - T_4)$

Hence
$$\eta_{th} = \frac{(\dot{m}_a + \dot{m}_f)c_{pg}\eta_t(T_3 - T_4) - \dfrac{\dot{m}_a c_{pa}}{\eta_c}(T_2 - T_1)}{\dot{m}_f \times \text{C.V.}} \quad \ldots(10.14)$$

Example 10.1 *Calculate the efficiency of Brayton cycle if the temperature at the end of combustion is 2000 k and the temperature and pressure before compression is 350 k and 1 bar. The pressure ratio is 1.3. Assume c_p = 1.005 kJ/kg. K.*

Solution.

Given : $p_1 = 1$ bar, $T_1 = 350$ k, $T_3 = 2000$ k, $r_p = 1.3$, $c_p = 1.005$ kJ/kg. K.

Refer to fig. 10.4

$$T_2 = T_1,\ r_p^{(\gamma-1)/\gamma} = 350(1.3)^{\frac{1}{3.5}} = 377.24 \text{ K}$$

$$T_4 = T_3\left(\frac{1}{r_p}\right)^{\frac{1}{3.5}} = 200\left(\frac{1}{1.3}\right)^{\frac{1}{3.5}} = 1855.56 \text{ K}$$

Compressor work per kg of air, $w_c = c_p\,(T_2 - T_1)$

$$= 1.005\ (377.24 - 350) = 27.376\ \text{kJ/kg}$$

Turbine work per kg of air, $w_t = c_p(T_3 - T_4)$

$$= 1.005\ (2000 - 1856.56) = 145.162\ \text{kJ/kg}$$

Net work done, $w_{net} = w_t - w_c = 145.162 - 27.376 = 117.786$ kJ/kg

Heat supplied, $q_s = c_p(T_3 - T_2) = 1.005\ (2000 - 377.24) = 1630.874$ kJ/kg

$$\text{Thermal efficiency,}\quad \eta_{th} = \frac{w_{net}}{q_s} = \frac{117.786}{1630.874} \times 100 = 7.22\ \%$$

Example 10.2 *The ratio of network to turbine work of an ideal gas turbine plant is 0.563 and T_1 = 300 K. Calculate the temperature drop across the turbine if thermal efficiency of the unit is 35%. Assume a mass flow rate of 10kg/s, c_p = 1 kJ/kg. K and r = 1.4.*

Solution.

$$\text{Thermal efficiency,}\quad \eta_{th} = 1 - \frac{1}{(rp)^{(\gamma-1)/\gamma}}$$

$$0.35 = 1 - \frac{1}{r_p^{0.286}}$$

$$r_p = 4.51$$

$$T_2 = T_1 r_p^{(\gamma - 1/\gamma)} = 300(4.51)^{0.286} = 461.54\ \text{k}$$

Compressor work, $w_c = c_p(T_2 - T_1) = 1.0\ (461.54 - 300) = 161.54$ kJ/kg

$$\frac{w_{net}}{w_t} = \frac{w_t - w_c}{w_t} = 0.563$$

$$1 - \frac{w_c}{w_t} = 0.563$$

$$w_t = \frac{161.54}{0.437} = 369.66\ \text{kJ/kg}$$

$$= c_p \cdot \Delta t$$

Temperature μ_{op} of across the turbine, $\Delta T = \dfrac{369.66}{1} = 369.66°C$

Examples 10.3 *In a gas turbine plant the air enters to the compressor at 1 bar and 300 K. The pressure ratio is 5. The temperature at the turbine inlet is 1200 K. The mass flow rate is 12 kg/s. Determine :*

(a) Compressor work (b) turbine work (c) net work (d) heat supplied, and (e) thermal efficiency

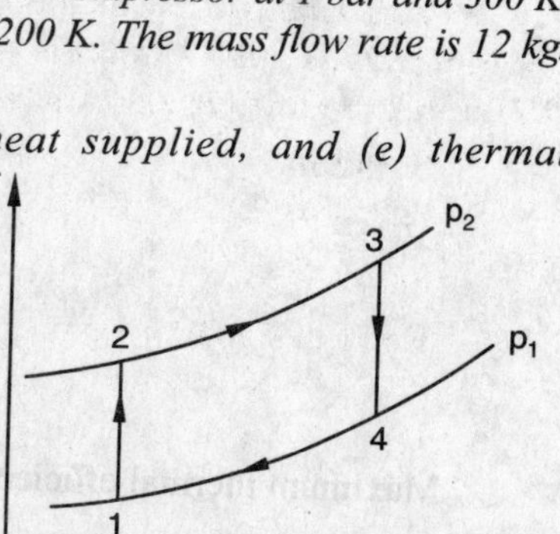

Fig. 10.9

Solution. Refer to Fig. 10.9.

Given: p_1 = 1 bar, T_1 = 300 K,

$r_p = 5$, T_3 = 1200 K, $\dot{m} = 12$ kg/s

$$T_2 = T_1(r_p)^{(r-1)r} = 300\ (5)^{0.286} = 475.15\ \text{K}$$

$$T_4 = T_3\left(\frac{1}{r_p}\right)^{(\gamma-1)/\gamma} = 1200\left(\frac{1}{5}\right)^{0.286} = 757.7\ \text{K}$$

(*a*) Compressor work, $W_c = \dot{m} c_p (T_2 - T_1)$

$$= 12 \times 1.005\ (475.15 - 300) = 2112.3 \text{ kW}$$

(*b*) Turbine work, $w_t = \dot{m} c_p (T_3 - T_4)$

$$= 12 \times 1.005\ (1200 - 757.7) = 5334.14 \text{ kW}$$

(*c*) Net work, $w_n = w_t = w_c = 5334.14 - 2112.3 = 3221.84$ k/W

(*d*) Heat supplied, $Q_s = \dot{m} c_p (T_3 - T_2)$

$$= 12 \times 1.005\ (1200 - 475.1) = 8741.7 \text{ kW}$$

(*e*) Thermal efficiency, $\eta_{th} = \dfrac{W_n}{Q_s} = \dfrac{3221.84}{8741.7} \times 100 = 36.8\%$

Example 10.4 *A gas turbine is operating on Brayton cycle between 27°C and 827°C. Determine the maximum net work per kg and cycle efficiency.*

Solution. Given: $T_1 = 273 + 27 = 300$ K, $T_3 = 273 + 827 = 1100$ K

$$\alpha = \frac{T_3}{T_1} = \frac{1100}{300} = 3.667$$

for maximum network, $r_p = (\alpha)^{\frac{\gamma}{2(\gamma-1)}}$

or $(r_p)^{\frac{\gamma-1}{\gamma}} = \sqrt{\alpha}$

Work done per kg, $w_{net} = c_p[(T_3 - T_4) - (T_2 - T_1)]$

$$= c_p T_1 \left[\left(\frac{T_3}{T_1} - \frac{T_4}{T_1}\right) - \left(\frac{T_2}{T_1} - 1\right)\right]$$

$$= c_p T_1 \left[\frac{T_3}{T_1}\left(1 - \frac{T_4}{T_3}\right) - \left(\frac{T_2}{T_1} - 1\right)\right]$$

$$= c_p T_1 \left[\alpha\left\{1 - \frac{1}{r_p^{(\gamma-1)/\gamma}}\right\} - \left\{r_p^{(\gamma-1)/\gamma} - 1\right\}\right]$$

$$= c_p T_1 \left[\alpha(1 - \frac{1}{\alpha}) - (\sqrt{\alpha} - 1)\right]$$

$$= 1.005 \times 300 \left[3.667\left(1 - \frac{1}{\sqrt{3.667}}\right) - (\sqrt{3.667} - 1)\right]$$

$$= 252.34 \text{ kJ/kg}$$

Maximum thermal efficiency, $(\eta_{th})_{max} = 1 - \dfrac{1}{\sqrt{\alpha}}$

$$= 1 - \frac{1}{\sqrt{3.667}} = 0.4777 \text{ or } 47.77\%$$

Example 10.5 *Air enters the compressor of an open cycle constant pressure gas turbine at a pressure of 1 bar and temperature 20°C. The air is compressed to 4 bar pressure. The isentropic efficiencies of compressor and turbine are 80% and 85% respectively. The air-fuel ratio used is 90 : 1. If flow rate of air is 3 kg/s, calculate :*

(a) power developed, and (b) thermal efficiency of the cycle.

Take c_p = 1.0 kJ/kg.K, r = 1.4 for air and gases, and calorific valve of fuel = 41800 kJ/kg.

Solution. Given : $p_1 = 1$ bar, $T_1 = 273 + 20 = 293$ K, $p_2 = 4$ bar, $\eta_c = 0.80$, $\eta_t = 0.85$,

$\dot{m}_a/\dot{m}_f = 90$, $\dot{m}_a = 3$ kg/ s, $c_{pa} = c_{pg} = 1.0$ kJ/kg.K, $r_a = r_g = 1.4$, $CV = 41800$ kJ/kg

Refer to fig. 10.10

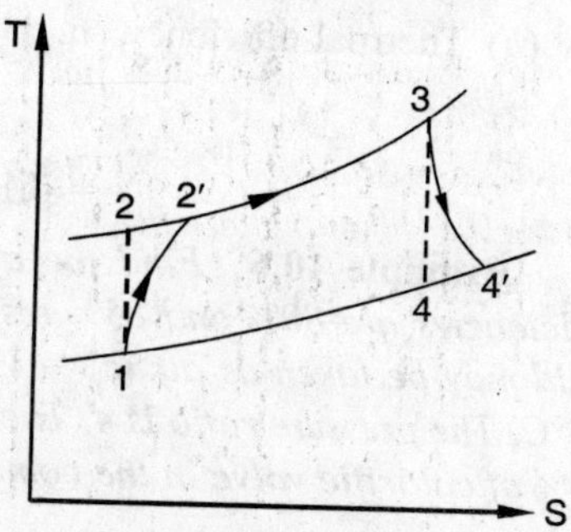

Fig. 10.10

(a)

$$T_2 = T_1\left(\frac{p_2}{p_1}\right)^{(\gamma-1)/\gamma}$$

$$= 293(4)^{(1.4-1)/1.4}$$

$$= 435.4 \text{ k}$$

$$\eta_c = \frac{T_2 - T_1}{T_2' - T_1}$$

$$0.80 = \frac{435.4 - 293}{T_2' - 293}$$

$$T_2' = 471 \text{ K}$$

$$\dot{m} = \frac{\dot{m}_a}{\dot{m}_f} + 1 = 90 + 1 = 91$$

Heat supplied by fuel = Heat absorbed by burning gases

$$\dot{m}_f \times CV = (\dot{m}_a + \dot{m}_f)c_{pg}(T_3 - T_2')$$

$$CV = \left(1 + \frac{\dot{m}_a}{\dot{m}_f}\right)c_{pg}(T_3 - T_2')$$

$$41800 = 91 \times 1.0\,(T_3 - 471)$$

$$T_3 = 930\text{K}$$

Now

$$T_4 = T_3\left(\frac{p_4}{p_3}\right)^{(\gamma-1)/\gamma} = 930\left(\frac{1}{4}\right)^{0.4/1.4} = 624.9 \text{ K}$$

$$\eta_t = \frac{T_3 - T_4'}{T_3 - T_4}$$

$$0.85 = \frac{930 - T_4'}{930 - 624.9}$$

$$T_4' = 670.6 \text{ K}$$

$$W_t = \dot{m}_g\, c_{pg}(T3 - T_4') = \left(\frac{90+1}{90}\right) \times 1 \times (930 - 670.6)$$

$$= 262.28 \text{ kJ/kg of air}$$

$$W_c = \dot{m}_a\, c_{pa}(T_2' - T_1) = 1 \times 1.0\,(471 - 293) = 178 \text{ kJ/kg of air}$$

$$W_{net} = W_t - W_c = 262.28 - 178 = 84.28 \text{ kJ/kg of air}$$

Power developed, $p = W_{net} \times \dot{m}_a = 84.28 \times 3 = 252.84$ kW

(*b*) Thermal efficiency, $\eta_{th} = \dfrac{W_{net}}{Q_s} = \dfrac{84.28}{\left(\dfrac{41800}{90}\right)}$

$$= 0.1814 \text{ or } 18.14\ \%$$

Example 10.6 *Find the air-fuel ratio in a gas turbine whose compressor and turbine efficiencies are 80% and 85% respectively. Maximum cycle temperature is 875°C. The working fluid may be taken as air (c_p = 1.0 kJ/kg.K, r = 1.4) which enters the compressor at 1 bar and 27°C. The pressure ratio is 4. The fuel used has calorific valve of 42000 kJ/kg. There is a loss of 10% of calorific valve in the combustion chamber.*

Solution. Given: $\eta_c = 0.80$, $\eta_t = 0.85$, $T_3 = 875 + 273 = 1148$ K,

$T_1 = 27 + 273 = 300$ K, $p_1 = 1$ bar, $r_p = 4$, $c_p = 1.0$ kJ/kg. K,

$r = 1.4$, $CV = 42000$ kJ/kg, $\eta_{comb} = 0.90$

Refer to Fig. 10.11

$$T_2 = T_1(r_p)^{(\gamma-1)\gamma} = 300(4)^{0.4/1.4} = 445.8 \text{ K}$$

$$\eta_c = \frac{T_2 - T_1}{T_2' - T_1}$$

$$0.80 = \frac{445.8 - 300}{T_2' - 300}$$

$$T_2' = 482.2 \text{ K}$$

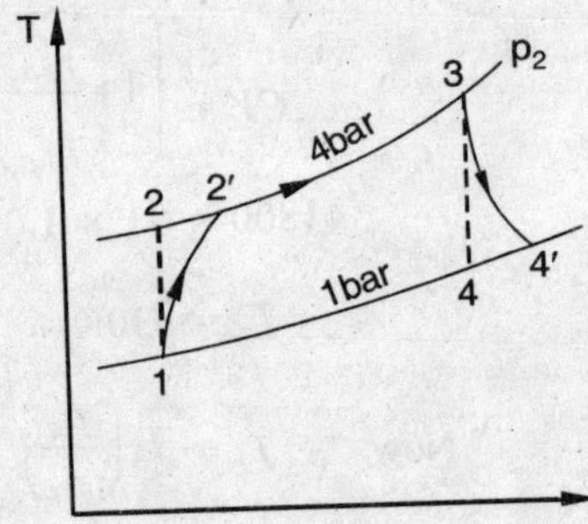

Fig. 10.11

Heat supplied by fuel = Heat absorbed by burning gases.

$$\eta_{comb} \times \dot{m}_f \times CV = (\dot{m}_a + \dot{m}_f)c_p(T_3 - T_2')$$

$$0.9 \times 42000 = \left(\frac{\dot{m}_a}{\dot{m}_f} + 1\right) \times 1.0(1148 - 482.2)$$

$$\frac{\dot{m}_a}{\dot{m}_f} = 55.77 \simeq 56 : 1$$

MULTI-CHOICE QUESTIONS

1. A gas turbine works on
   (*a*) Otto cycle (*b*) Rantine cycle
   (*c*) Brayton cycle (*d*) Stirling cycle
2. In a gas turbine cycle, the turbine output is 600 kJ/kg, compressor work is 400 kJ/kg and heat supplied is 1000 kJ/kg. The thermal efficiency of the cycle is:
   (*a*) 80% (*b*) 60%
   (*c*) 40% (*d*) 20%
3. The thermal efficiency of gas turbine as compared to diesel engine is
   (*a*) higher (*b*) lower
   (*c*) same (*d*) unpredictable.
4. For a gas turbine the pressure ratio may be in the range
   (*a*) 2 to 3 (*b*) 3 to 5
   (*c*) 6 to 10 (*d*) 12 to 16
5. With the increase in pressure ratio thermal efficiency of a simple gas turbine plant with fixed turbine inlet temperature
   (*a*) increases (*b*) decreases
   (*c*) first increases and then decreases (*d*) first decreases and then increases.
6. The thermal efficiency of closed cycle gas turbine is given by
   (*a*) $1-\frac{1}{r_p^{\gamma-1}}$ (*b*) $1-\frac{1}{r_p^{(\gamma-1)/\gamma}}$
   (*c*) $1-r_p$ (*d*) r_p-1

ANSWERS

1. (*c*) **2.** (*d*) **3.** (*b*) **4.** (*c*) **5.** (*c*) **6.** (*b*)

REVIEW QUESTIONS

1. Give the fields of application of gas turbines.
2. What are the limitations of gas turbines ?
3. Compare gas turbines with I.C. engines.
4. How gas turbines are classified ?
5. Explain the working of constant pressure combustion gas turbine with (*a*) open cycle, and (*b*) closed cycle.
6. Explain the working of constant volume combustion gas turbine.
7. Compare open cycle and closed cycle gas turbines.
8. Define an expression for thermal efficiency of Brayton cycle.

EXERCISES

10.1 In a gas turbine, air at a temperature of 15°C and 1 bar pressure enters the compressor, where it is compressed to 5 bar. Air enters the turbine at a temperature of 815°C and expands to original pressure of 1 bar. Calculate the ratio of turbine work to compressor work and thermal efficiency. **[Ans.** 2.39, 37%]

10.2. A gas turbine works on Brayton cycle between 27°C and 827°C. Determine the maximum not work per kg and cycle efficiency. Take c_p = 1.005 kJ/kg. K.

[**Ans.** 252.34 kJ/kg, 47.77%]

10.3. A gas turbine set takes in air at 27°C and 1 atm. The pressure ratio is 4 and the maximum temperature is 560°C. The compressor and turbine efficiencies are 0.83 and 0.85 respectively. Determine the overall efficiency of the set.

10.4. The maximum and minimum temperature occurring in a closed cycle gas turbine plant are 927°C and 37°C. The pressure at the outlet and the inlet of the compressor are 5 bar and 1 bar respectively per kg of air flow in the cycle. Determine the thermal efficiency of the cycle and the optimum pressure ratio for maximum power output for the given temperatures.

11 SIMPLE LIFTING MACHINES

11.1 FUNDAMENTAL CONCEPTS

Machine. A machine may be defined as a device which is capable of doing some useful work or overcoming resistance or lifting load at some desirable point by the application of an effort at some other convenient point.

Simple machine. If there is one point of application for the effort and one point for the load to be lifted, it is called a simple machine, *e.g.* a lever, screw jack, inclined plane, bicycle, etc.

Resistance or Load (W). It is the amount of load to be lifted by the machine or the frictional resistance to be overcome.

Effort (P). It is the force applied to a machine to do some useful work.

Input. It is the amount of work done on the machine. It is the product of effort P and distance *D* through which it has moved in the direction of the effort.

Output. It is the amount of workdone or available from a machine. It is the product of load lifted *W* and the distance *d* through which it is lifted.

Mechanical Advantage (MA). It is the ratio of load lifted *W* to effort applied *P*. Thus

$$\text{M.A.} = \frac{W}{P} \qquad \text{... (11.1)}$$

Velocity Ratio (R). It is the ratio of the distance *D* moved by the effort to the distance *d* moved by the load lifted or resistance overcome. Thus

$$\text{V.R.} = \frac{D}{d} \qquad \text{... (11.2)}$$

Efficiency (η). It is the ratio of mechanical advantage to velocity ratio. Thus,

$$\eta = \frac{\text{M.A.}}{\text{V.R.}} = \frac{\text{W}}{\text{P}} \cdot \frac{d}{D} = \frac{\text{Output}}{\text{Input}} \qquad \text{... (11.3)}$$

Therefore, efficiency may also be defined as the ratio of output to input.

Lifting Machine. A machine used for lifting of loads only is called a lifting machine.

Ideal Machine. A machine whose efficiency is 100 percent is called an ideal machine. Practically speaking, no machine is ideal.

Reversible Machine. A machine which is capable of doing some work in the reverse direction is called a reversible machine. The efficiency of such a machine is more than 50 percent.

Irreversible (or self-locking) Machine. A machine which is not capable of doing any work in the reverse direction, when the effort is removed, is called irreversible or self-locking machine. The efficiency of such a machine is less than 50 percent.

11.2 LAW OF A MACHINE

The law of a machine is the relationship between the effort applied and the load lifted. If an experiment is performed on the given machine where a set of readings for P and W are taken then an equation of the following form is obtained:

$$P = mW + c \quad \text{... (11.4)}$$

where m is the slope of the line and c is a constant representing effort required to overcome resistance due to friction in the machine when $W = 0$.

This equation represents a straight line as shown in Fig.11.1, and is called the law of a machine.

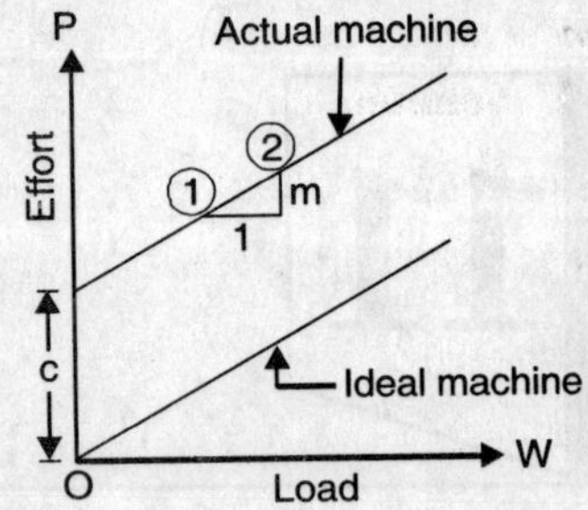

Fig. 11.1 Law of a machine.

The slope m can be expressed as :

$$m = \frac{P_2 - P_1}{W_2 - W_1}$$

11.3 RELATION BETWEEN EFFICIENCY AND LOAD LIFTED

The relation between efficiency and load lifted is shown in Fig. 11.2. As W increase, efficiency increases and then attains a constant value.

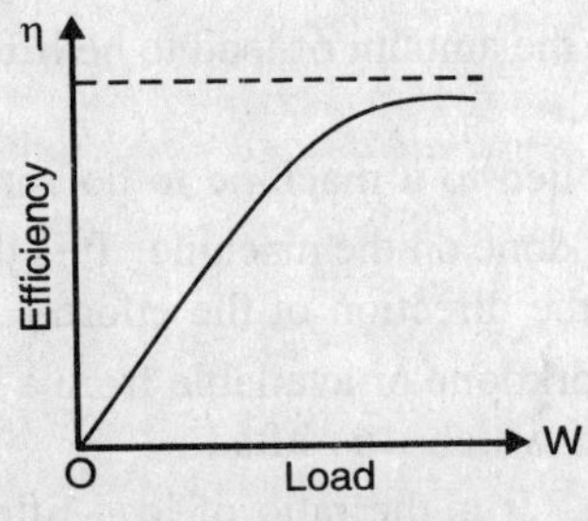

Fig. 11.2 Efficiency v's load.

11.4 MAXIMUM MECHANICAL ADVANTAGE AND VELOCITY RATIO

The law of the machine is :

$$P = mW + c$$

$$\text{M.A.} = \frac{W}{P} = \frac{W}{mW + c}$$

For M.A. to be maximum, $\dfrac{d(\text{M.A.})}{dW} = 0$

$$\frac{(mW + c) \cdot 1 - W \cdot m}{(mW + c)^2} = 0$$

or $c = 0$

Maximum mechanical advantage, $(\text{M.A})_{max} = \dfrac{1}{m}$ ... (11.5)

Efficiency, $\eta = \dfrac{\text{M.A.}}{\text{V.R.}}$

Velocity ratio of a machine is fixed. Therefore, efficiency will be maximum when mechanical advantage is maximum.

$$\eta_{max} = \frac{1}{m \times V.R.}$$

Example 11.1 *In a lifting machine, an effort of 200 N raises a load of 800 N. What is the mechanical advantage? If the efficiency of the machine is 50 percent, what is the velocity ratio ?*

Solution.

Here $P = 200$ N, $W = 8000$ N, $\eta = 0.5$

$$\text{M.A.} = \frac{W}{P} = \frac{8000}{200} = 40$$

$$\eta = \frac{\text{M.A.}}{\text{V.R.}}$$

$$\text{V.R.} = \frac{40}{0.5} = 80\,.$$

Example 11.2 *In a lifting machine, the following readings were taken :*

Effort, P N	*Load lifted, W* N
200	*8000*
250	*15000*

Calculate the maximum mechanical advantage. If velocity ratio is 80, what is the maximum efficiency ?

Solution.

$$m = \frac{P_2 - P_1}{W_2 - W_1} = \frac{250-200}{15000-8000} = \frac{50}{7000} = \frac{1}{140}$$

$$\text{Maximum mechanical advantage} = \frac{1}{m} = 140$$

$$\text{Maximum efficiency} = \frac{1}{m \times VR} = \frac{100}{140 \times 80} = 8.93\%$$

Example 11.3 *In a lifting machine, the following readings were taken :*

Effort, P *N*	*Load lifted, W* *N*
350	*10,000*
650	*20,000*

Determine the law of the machine.

Solution. The law of the machine is :

$$P = mW + c$$

$$m = \frac{P_2 - P_1}{W_2 - W_1} = \frac{650-350}{20,000-10,000} = \frac{300}{10,000} = 0.03$$

$$350 = 0.03 \times 10,000 + c = 3000 + c$$

$$c = 50$$

$$P = 0.03\text{ W} + 50$$

11.5 LOSS OF EFFORT IN FRICTION

Let P_a = actual effort applied

W_a = actual load lifted

P_i = ideal effort required

W_i = ideal load lifted

Effort lost due to friction, P_f = Pa – Pi

Loss in load lifted due to friction, $W_f = W_i - W_a$

Efficiency,
$$\eta = \frac{\left(\frac{W_a}{P_i}\right)}{\text{V.R.}}$$

For an ideal machine, $\eta = 1$

$$\therefore \quad \frac{W_a}{P_i} = 1 \times \text{V.R.} = \text{V.R.}$$

$$P_i = \frac{W_a}{\text{V.R.}}$$

$$P_f = P_a - \frac{W_a}{\text{V.R.}} = \frac{P_a \times \text{V.R.} - W_a}{\text{V.R.}} \quad ...(1)$$

Also
$$\frac{W_a}{P_i} = \eta \times \text{V.R.} = 1 \times \text{V.R.} = \text{V.R.}$$

$$W_i = P_a \times \text{V.R.}$$

$$W_f = W_i - W_a = P_a \times \text{V.R.} - W_a$$

From Eqns. (1) and (2), we get

$$W_f = P_f \times \text{V.R.}$$

Now
$$\eta = \frac{P_i}{P_a} = \frac{W_a}{W_i}.$$

Example 11.4 *In a simple machine, an effort of 200 N is just sufficient to lift a load of 1200 N. The velocity ratio is 20. Calculate the efficiency and the loss of effort and load due to friction.*

Solution.

Here, P = 200 N, W = 1200 N, V.R. = 20

$$\text{Efficiency} = \frac{\text{M.A.}}{\text{V.R.}} = \frac{W}{P \times \text{V.R.}}$$

$$= \frac{1200}{200 \times 20} = 30\%$$

Loss of effort due to friction,

$$P_f = P - \frac{W}{\text{V.R.}} = 200 - \frac{1200}{20} = 140\,\text{N}$$

Loss of load due to friction,

$$W_f = P \times \text{V.R.} - W = 200 \times 20 - 1200 = 2800 \text{ N}$$

11.6 CONDITION FOR IRREVERSIBILITY OF A MACHINE

If a machine is to be irreversible (or self-locking) then the frictional resistance of the machine will have to be such that even when the effort P is removed there is enough frictional resistance in the machine to prevent the load W from slipping down.

Effort lost due to friction, $P_f = P_a - \frac{W_a}{\text{V.R.}}$

On removal of P_a, if W_a is not to slip down, the friction of the machine must supply a resistance of atleast $\frac{W_a}{\text{V.R.}}$, *i.e.*

$$P_f = \left(P_a - \frac{W_a}{\text{V.R.}}\right) \geq \frac{W_a}{\text{V.R.}}$$

i.e. $$\frac{2W_a}{\text{V.R.}} \leq \text{Pa}$$

or $$\frac{W_a}{P_a \times \text{V.R.}} \leq \frac{1}{2}$$

or $$\eta \leq 50\%$$

Therefore, for a machine to be irreversible or self-locking, its efficiency should be less than 50 percent. Otherwise, for a reversible machine efficiency should be more than 50 percent.

11.7 PULLEYS

11.7.1 Single Pulley

Pulley can be used as single unit for lifting of load. Consider a single pulley system shown in Fig. 11.3. Let T be the tension in the rope. Considering the equilibrium of load W being lifted, we have

$$T = W$$

Also $$T = P$$

$\therefore$ $$W = P$$

Mechanical advantage, $\text{M.A.} = \frac{W}{P} = 1$

Distance moved by the load is same as that moved by the effort. Therefore,

$$\text{V.R} = 1$$

Efficiency, $$\eta = \frac{\text{M.A.}}{\text{V.R.}} = 100\% \quad \text{... (11.9)}$$

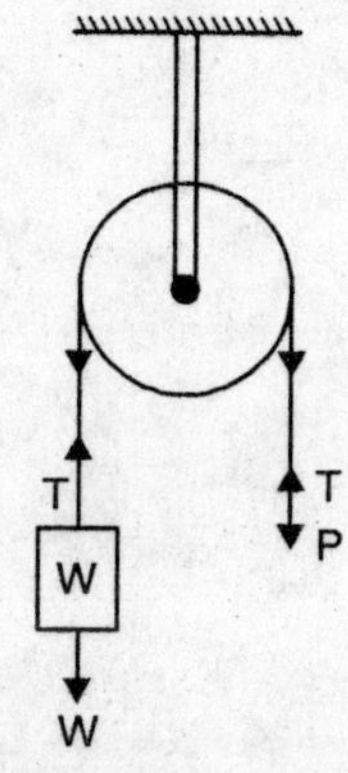

Fig. 11.3 Single pulley.

11.7.2 System of Pulleys

To obtain a higher mechanical advantage a system consisting of several pulley is often used. When two or more pulleys are combined together in such a manner so as to obtain higher mechanical advantage, it is called a system of pulleys. There are three systems of pulleys commonly used, as given below :

1. First system of pulleys.
2. Second system of pulleys.
3. Third system of pulleys.

The following assumptions are made while dealing with the pulleys :

1. The weight of the pulley is small as compared to the load lifted.
2. The friction between the pulley and the rope is negligible and is ignored. Therefore, tension in the rope is same throughout.
3. The loss due to friction in the pulley bearing is negligible.

1. First System of Pulleys. In the first system of pulleys, there are as many fixed ropes as the number of movable pulleys. As shown in Fig. 11.4, there are three fixed ropes and three movable pulleys 1 to 3. The tensions in the various parts of the ropes are as shown in the figure. Finally at the fixed pulley 4, we have

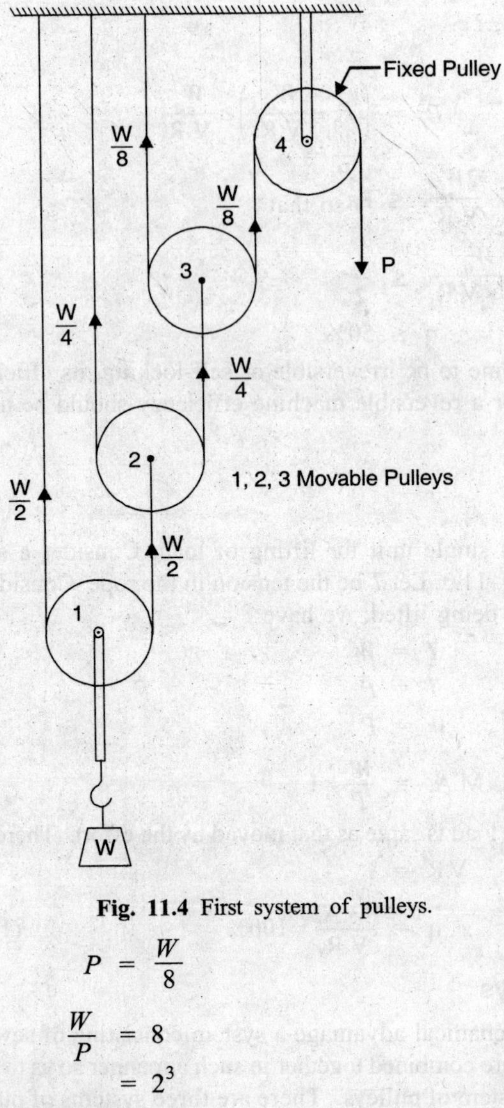

Fig. 11.4 First system of pulleys.

$$P = \frac{W}{8}$$

or
$$\frac{W}{P} = 8$$
$$= 2^3$$

In general, for n movable pulleys,

$$\frac{W}{P} = 2^n$$

For an ideal system, efficiency is 100 percent.

Thus,
$$\text{M.A.} = \text{V.R.}$$
$$= 2^n \quad \text{... (11.10)}$$

2. Second System of Pulleys. In the second system of pulleys, there are two pulley blocks, one fixed and the other movable. The number of pulleys in the fixed block may be either more by one than the number of pulleys in the movable block as shown in Fig. 11.5(*a*) or equal to the number of pulleys in the movable block as shown in Fig. 11.5(*b*). In the first case, the rope is tied to the movable block whereas in the second case, the rope is tied to the fixed block. Only one rope passes over all the pulleys.

If n is the total number of pulleys in the fixed and movable block considering the equilibrium of weight W,

$$nT - W = 0$$

$$T = P$$

or
$$nP - W = 0$$

$$\text{M.A.} = \frac{W}{P} = n$$

For an ideal system, $\eta = 1$ so that

$$\text{V.R.} = n \qquad \text{... (11.11)}$$

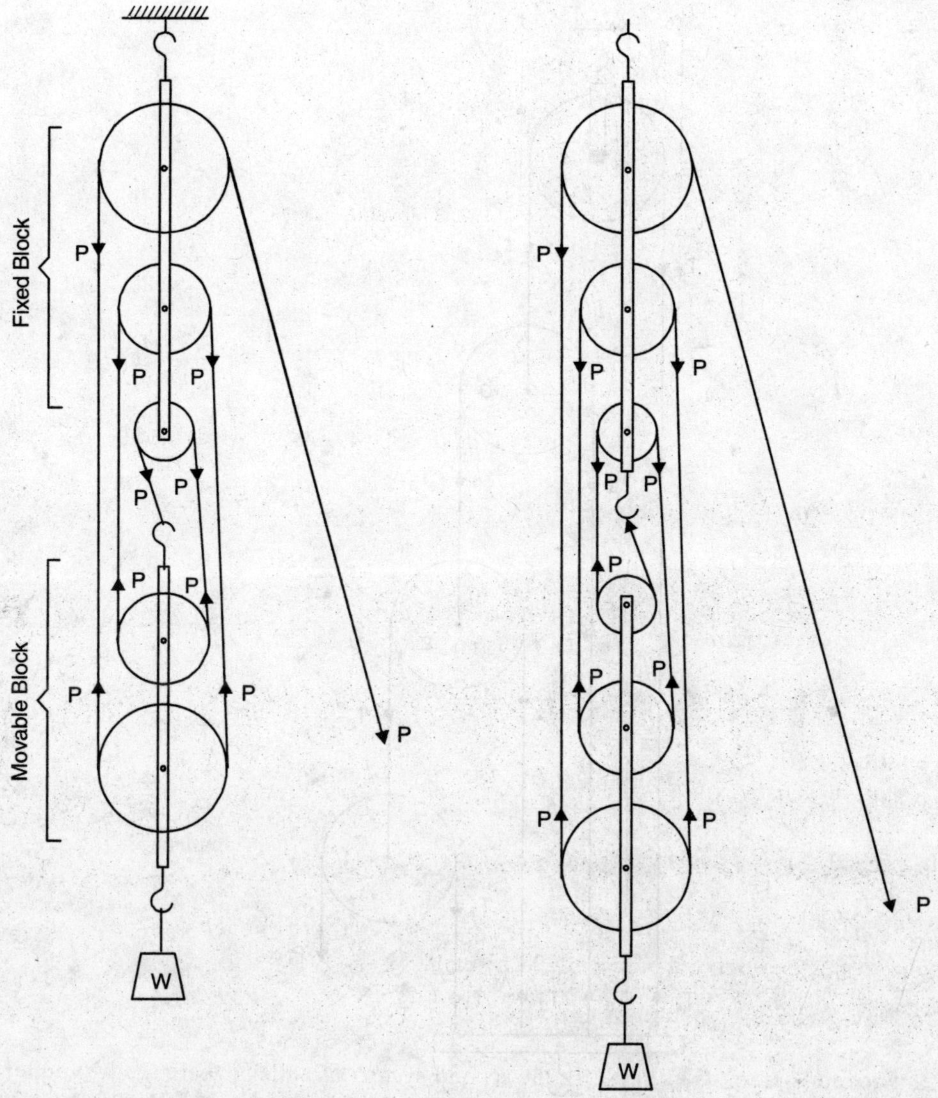

Fig. 11.5 Second system of pulleys.

3. Third System of Pulleys. In the third system of pulleys, the number of rope parts fixed to the load is equal to the number of pulleys, as shown in Fig. 11.6. Considering an ideal system, for the equilibrium of the system,

$$W = T_1 + T_1 + T_3 + \ldots + T_n$$

But

$$T_1 = P$$

$$T_2 = 2\,T_1 = 2\,P = (2)^1\,P$$

$$T_3 = 2\,T_2 = 4\,P = (2)^2\,P \quad \text{and so on.}$$

$$T_n = (2)^{n-1}\,P$$

$$W = P + (2)^1\,P + (2)^2\,P + \ldots + (2)^{n-1}\,P$$

$$\text{M.A.} = \frac{W}{P} = 1 + (2)^1 + (2)^2 + \ldots + 2^{n-1} \qquad \ldots (11.12)$$

$$= 2^n - 1$$

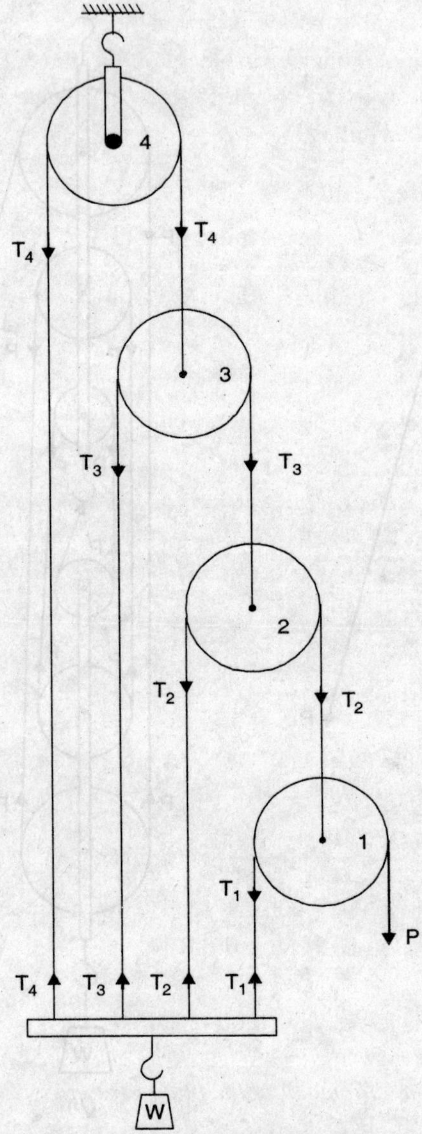

Fig. 11.6 Third system of pulleys.

For an ideal, $\eta = 1 = \dfrac{\text{M.A.}}{\text{V.R.}}$

$$\text{V.R.} = 2^n - 1$$

11.7.3 Differential Pulley Block

It consists of two pulley blocks, upper and lower as shown in Fig. 11.7. The upper block carries two pulleys and remains fixed, whereas the lower block has only one pulley and is movable. These pulleys have teeth on their groove periphery over which an endless chain can fit.

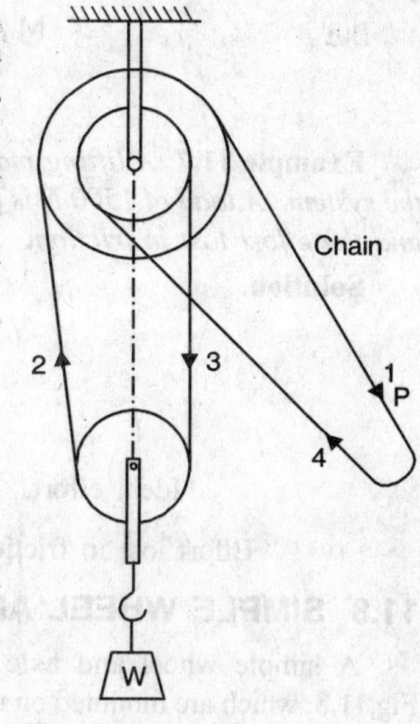

Fig. 11.7 Differential pulley block

When effort P is applied to the chain as shown, the chain moves by arrows 1 to 4 to lift the load W.

Let d_1 = diameter of bigger pulley in upper block

d_2 = diameter of smaller pulley in upper block

In one revolution of bigger pulley in upper block,

$$\text{Distance moved by effort} = \pi d_1$$

$$\text{Distance moved by smaller pulley} = \pi d_2$$

$$\text{Distance moved by load upwards} = \frac{\pi}{2}(d_1 - d_2)$$

$$\text{Velocity ratio} = \frac{\pi d_1}{\frac{\pi}{2}(d_1 - d_2)} = \frac{2d_1}{d_1 - d_2}$$

$$\text{Mechanical advantage} = \frac{W}{P}$$

$$\text{Efficiency} = \frac{W}{P} \times \left(\frac{d_1 - d_2}{2d_1}\right) \quad \text{... (11.14)}$$

Example 11.5 *In a first system of pulleys, there are 4 movable pulleys. If an effort of 80 N lifts a load of 1000 N, find (a) the effort lost in friction, and (b) the load lost in friction.*

Solution.

Actual load, W_a = 1000 N

Actual effort, P_a = 80 N

Here $n = 4$

Velocity ratio, V.R. $= 2^n = 2^4 = 16$

For an ideal system, $\eta = 1$

$\therefore$ M.A. = V.R. = 16

$$\text{Ideal effort, } P_i = \frac{W_a}{\text{M.A.}} = \frac{1000}{16} = 62.5 \text{ N}$$

Effort lost in friction = 80 – 62.5 = 17.5 N

Ideal load lifted, W_i = 16 × 80 = 1280 N

Load lost in friction = $W_i - W_a$

= 1280 – 1000 = 280 N

Example 11.6 *In a second system of pulleys, there are 3 pulleys in the upper block and 2 pulleys in the lower block. If the efficiency of the pulley system is 70%, find the effort required to lift a load of 800 N.*

Solution.

For the second system of pulley, V.R. = $n = 3 + 2 = 5$

$$\text{Efficiency, } \eta = \frac{\text{M.A.}}{\text{V.R.}}$$

$$\text{M.A.} = 0.7 \times 5 = 3.5$$

$$P = \frac{\text{W.}}{\text{M.A.}} = \frac{800}{3.5} = 228.6 \text{ N}$$

Example 11.7 *A lifting machine consists of third system of pulleys. There are 3 pulleys in the system. A load of 1500 N is lifted by an effort of 250 N. Find the efficiency of the pulley system and the effort lost in friction.*

Solution.

$$\text{V.R.} = 2^n - 1 = 2^3 - 1 = 7$$

$$\eta = \frac{W}{P \times \text{V.R.}} = \frac{1500}{250 \times 7} = 0.8571 \text{ or } 85.71\%$$

$$\text{Ideal effort, } P_i = \frac{W}{\text{V.R.}} = \frac{1500}{7} = 214.3 \text{ N}$$

$$\text{Effort lost in friction} = 250 - 214.3 = 35.7 \text{ N}$$

11.8 SIMPLE WHEEL AND AXLE

A simple wheel and axle is shown in Fig.11.8, which are mounted on the same shaft supported on two bearings. A rope is wound round the wheel whose one end is fixed to the wheel and to the other end an effort P is applied. Another rope is wound on the wheel so that when the rope is on the wheel unwinds the rope on the axle winds. To the other end of the rope on the axle a load W is attached. The wheel and the axle will have same revolutions, being on the same shaft.

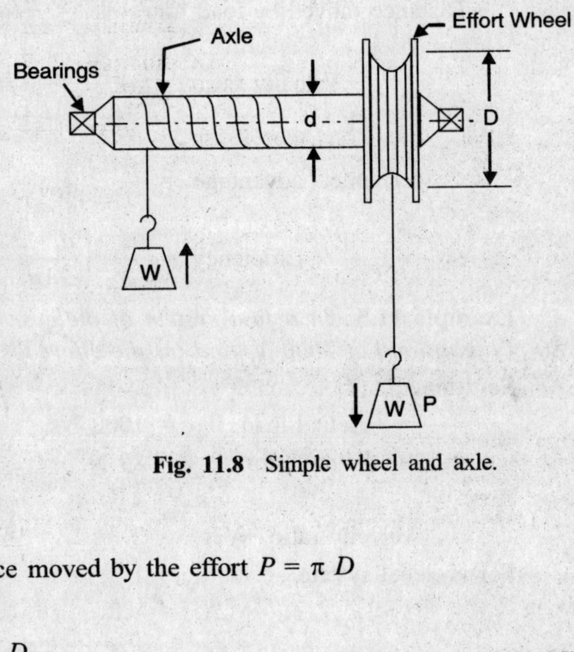

Fig. 11.8 Simple wheel and axle.

Let D = diameter of the wheel

d = diameter of the axle

Distance moved by the rope in one revolution of the wheel

= distance moved by the effort $P = \pi D$

$$\text{Distance moved by } W = \pi d$$

$$\text{Velocity ratio} = \frac{\pi D}{\pi d} = \frac{D}{d} \quad \text{... (11.15)}$$

$$\text{Mechanical advantage} = \frac{W}{P}$$

$$\text{Efficiency, } \eta = \frac{W d}{P D} \quad \text{... (11.16)}$$

Example 11.8 *A weight of 600 N is to be lifted by means of a simple wheel and axle. The diameter of the wheel is 500 mm and that of the axle 150 mm. An effort of 200 N is required to lift this load.*

Determine the mechanical advantage, velocity ratio and efficiency of the machine.

Solution.

Here $W = 600$ N, $P = 200$ N, $D = 500$ mm, $d = 150$ mm

$$\text{M.A.} = \frac{W}{P} = \frac{600}{200} = 3$$

$$\text{V.R.} = \frac{D}{d} = \frac{500}{150} = 3.33$$

$$\eta = \frac{\text{M.A.}}{\text{V.R.}} = \frac{Wd}{PD} = \frac{600 \times 150}{200 \times 500} = 0.9 \text{ or } 90\%$$

11.9 WHEEL AND DIFFERENTIAL AXLE

The wheel and differential axle is shown in Fig. 11.9.

A rope is wound on the wheel of diameter D whose one end is tied to the wheel and to the other end an effort P is applied. Another rope is wound on the axle with diameter d_2 which after passing over the pulley is wound round the axle with diameter d_1 in the opposite direction to that if on d_2. The direction of winding of rope on D and d_1 should be same.

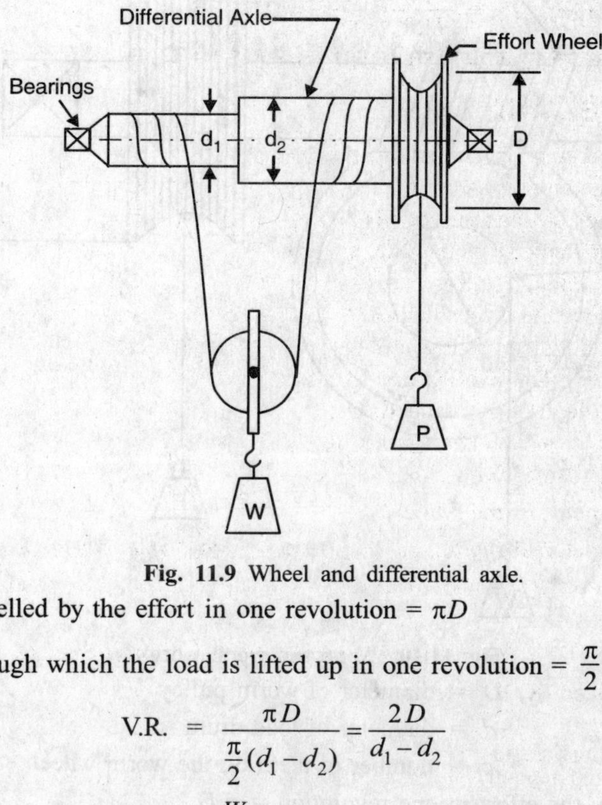

Fig. 11.9 Wheel and differential axle.

Distance travelled by the effort in one revolution = πD

Distance through which the load is lifted up in one revolution = $\frac{\pi}{2}(d_2 - d_1)$

$$\text{V.R.} = \frac{\pi D}{\frac{\pi}{2}(d_1 - d_2)} = \frac{2D}{d_1 - d_2} \qquad \text{... (11.17)}$$

$$\text{M.A.} = \frac{W}{P}$$

$$\eta = \frac{W(d_2 - d_1)}{2PD} \qquad \text{... (11.18)}$$

Example 11.9 *In a differential wheel and axle, the diameter of the wheel is 500 mm. The diameters of the differential axles are 200 mm and 150 mm. Find the load which can be lifted by an effort of 200 N, assuming the efficiency of machine to be 80%.*

Solution.

$$\eta = \frac{W(d_2 - d_1)}{2PD}$$

$$W = \frac{2 \times 200 \times 500 \times 0.8}{200 - 150} = 3200 \text{ N}$$

11.10 WORM AND WORM WHEEL

The worm and worm wheel is shown in Fig. 11.10. It consists of a horizontal shaft on which a worm is provided. The shaft is supported on two bearings and is overhanging from one bearing, at the end of which a pulley is provided. A cord is wound over the pulley, one end of which is tied to the pulley and to the other end an effort P is applied. The worm meshes with the worm wheel which is provided with a load drum. A cord is attached to the load drum to the other end of which load W is attached. The worm can be either of the single or multiple start type.

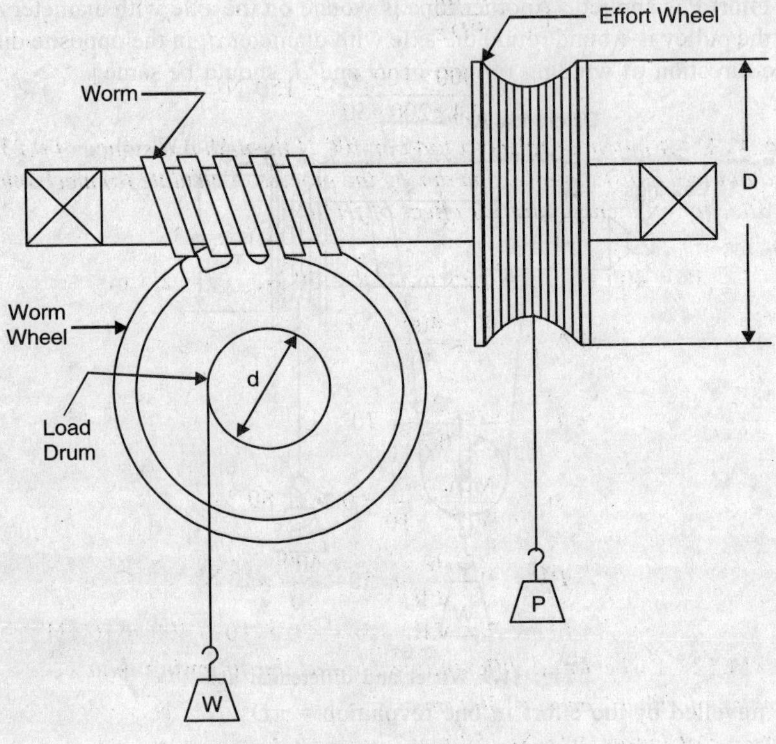

Fig. 11.10 Worm and worm wheel.

Let D = diameter of worm pulley

d = diameter of load drum

z = number of teeth on the worm wheel.

Distance moved by the effort in one revolution = πD

$$\text{Distance moved by the load} = \frac{\pi d}{z}$$

$$\text{V.R.} = \frac{W}{P}$$

$$\eta = \frac{Wd}{PDz}$$

For an n start worm, $\text{V.R.} = \dfrac{Dz}{dn}$

$$\eta = \frac{Wdn}{PDz}$$

Example. 11.10 *In a double start worm and worm wheel, the number of teeth on the worm wheel is 50. The diameter of the worm pulley is 200 mm and that of the load drum is 100 mm. Calculate the velocity ratio. If the efficiency of the machine is 40 percent, calculate the effort required to lift a load of 3 kN.*

Solution.

$$\text{V.R.} = \frac{Dz}{dn} = \frac{200 \times 50}{100 \times 2} = 50$$

$$\eta = \frac{W\,dn}{PDz}$$

$$P = \frac{3000 \times 100 \times 2}{0.4 \times 200 \times 50} = 150 \text{ N}$$

Example. 11.11 *A machine raised a load of 400 N through a distance of 0.25 m. The effort of 50 N moved through a distance of 2.5 m during the process. Calculate (a) mechanical advantage, (b) velocity ratio, (c) efficiency, and (d) effect of friction.*

Solution.

$$W = 400 \text{ N}, \quad x = 0.25 \text{ m}, \quad P = 50 \text{ N}, \quad y = 2.5 \text{ m}$$

(*a*) $$\text{M.A.} = \frac{W}{P} = \frac{400}{50} = 8$$

(*b*) $$\text{V.R.} = \frac{y}{x} = \frac{2.5}{0.25} = 10$$

(*c*) $$\eta = \frac{\text{M.A.}}{\text{V.R.}} = \frac{8}{10} = 0.8 \text{ or } 80\ \%$$

(*d*) $$P_f = P - \frac{W}{\text{V.R.}} = 50 - \frac{400}{10} = 10 \text{ N}$$

$$W_f = P \times \text{V.R.} - W = 50 \times 10 - 400 = 100 \text{ N}$$

Example. 11.12 *A wheel and differential axle has the following data:*

Diameter of effort wheel = 300 mm
Diameter of larger axle = 150 mm
Diameter of smaller axle = 100 mm

The effort to lift a load of 2 kN is 600 N while an effort of 800 N lifts a load of 4 kN. Calculate:

(a) The law of the machine (b) Effort lost in friction at the load of 4 kN (c) Limiting maximum efficiency.

Solution.

$$D = 300 \text{ N}, \quad d_1 = 150 \text{ mm}, \quad d_2 = 100 \text{ mm}$$

(*a*) Law of machine

$$P = mW + c$$

$$600 = 2000\,m + c$$

$$800 = 4000\,m + c$$

Solving we get, $c = 400$, $m = 0.1$

$$\therefore \quad P = 0.1\ W + 400$$

(*b*)
$$P_f = P - \frac{W}{\text{V.R.}}$$

$$\text{V.R.} = \frac{2D}{d_1 - d_2} = \frac{2 \times 300}{150 - 100} = 12$$

$$P_f = 800 - \frac{4000}{12} = 466.67 \text{ N}$$

(*c*)
$$\eta_{max} = \frac{1}{m \times \text{V.R.}} = \frac{1}{0.1 \times 12} = 0.833 \text{ or } 83.3\%$$

Example. 11.13 *What load will be lifted by an effort of 12 N if the velocity ratio is 18 and efficiency of the machine at this load is 60%?*

If the machine has a constant frictional resistance, determine the law of the machine and the effort required to run this machine at (a) no load, and (b) a load of 900 N.

Solution.

$$P = 12 \text{ N}, \quad \text{V.R.} = 18, \quad \eta = 0.6$$

Now
$$\eta = \frac{\text{M.A.}}{\text{V.R.}} = \frac{W}{P \times \text{V.R.}}$$

$$W = 0.6 \times 12 \times 18 = 129.6 \text{ N}$$

Effort lost in friction,

$$P_f = P - \frac{W}{\text{V.R.}} = 12 - \frac{129.6}{18} = 4.8 \text{ N}$$

Law of machine,

$$P = mW + c$$

For a constant frictional resistance, $c = 4.8$ N

$$\therefore \quad 12 = m \times 129.6 + 4.8$$

$$m = \frac{7.2}{129.6} = \frac{1}{18}$$

$$\therefore \quad P = \frac{W}{18} + 4.8$$

(*a*) At no load, W = 0

$$\therefore \quad P = 4.8 \text{ N}$$

(*b*) At W = 900 N

$$P = \frac{900}{18} + 4.8 = 54.8 \text{ N}$$

Example. 11.14 *The law of a machine is given by:*

$$P = 0.05\ W + 7.5$$

where P and W are in N.

(a) What is the mechanical advantage and efficiency of the machine, when the load is 2 kN

and velocity ratio is 40? What is the maximum efficiency of the machine?

(b) Find the relation between effort lost in friction and W. Also find the effort lost in friction when W is 2 kN.

Solution.

(*a*) For $W = 2$ kN

$$P = 0.05 \times 2000 + 7.5 = 107.5 \text{ N}$$

$$\text{M.A.} = \frac{W}{P} = \frac{2000}{107.5} =$$

$$\eta = \frac{\text{M.A.}}{\text{V.R.}} = \frac{18.6}{40} = 0.465 \text{ or } 46.5\ \%$$

$$\eta_{max} = \frac{1}{m \times \text{V.R.}} = \frac{1}{0.05 \times 40} = 0.5 \text{ or } 50\%$$

(*b*)

$$P_f = P - \frac{W}{\text{V.R.}} = (0.05\,W + 7.5) - \frac{W}{\text{V.R.}}$$

$$= W\left(0.05 - \frac{1}{40}\right) + 7.5$$

$$= 0.025\ W + 7.5$$

For $W = 2$ kN

$$P_f = 0.025 \times 2000 + 7.5 = 57.5 \text{ N}$$

Example. 11.15 *The larger and smaller diameter of a differential axle are 100 mm and 90 mm respectively. The effort is applied to the wheel at the end of a lever which is 125 mm long. What is the velocity ratio? Find also the efficiency and effort lost in friction, when the load lifted is 800 N with an effort of 35 N.*

Solution.

$d_2 = 100$ mm, $d_1 = 90$ mm, $l = 125$ mm or $D = 2\,l = 250$ mm,
W = 800 N, P = 35 N

$$\text{V.R.} = \frac{2D}{d_2 - d_1} = \frac{2 \times 250}{100 - 90} = 50$$

$$\text{M.A.} = \frac{W}{P} = \frac{800}{35} = 22.86$$

$$\eta = \frac{\text{M.A.}}{\text{V.R.}} = \frac{22.86}{50} = 0.457 \text{ or } 45.7\ \%$$

$$P_f = P - \frac{W}{\text{V.R.}} = 35 - \frac{800}{50} = 19 \text{ N}$$

Example. 11.16 *In a differential wheel and axle, the diameter of the effort wheel is 450 mm. The radii of axles are 150 mm and 100 mm respectively. The rope diameter is 10 mm. Find the load which can be lifted by an effort of 100 N, assuming the efficiency of the machine to be 75%.*

Solution.

Effective diameter of effort wheel, $D = 450$ + rope diameter of 10 mm

$= 460$ mm

Similarly, $d_1 = 2 \times 150 + 10 = 310$ mm

$d_2 = 2 \times 100 + 10 = 210$ mm

$$\text{V.R.} = \frac{2D}{d_1 - d_2} = \frac{2 \times 460}{310 - 210} = 9.2$$

$$\text{M.A.} = \frac{W}{P} = \frac{W}{100}$$

$$\eta = \frac{\text{M.A.}}{\text{V.R.}}$$

$$0.75 = \frac{W}{100 \times 9.2}$$

$$W = 690 \text{ N}$$

Example. 11.17 *A Weston differential pulley consists of a lower block and an upper block. The upper block has two cogged grooves, one of which has a radius of 125 mm and the other has a radius of 115 mm. If the efficiency of the machine is 45%, calculate the effort required to raise a load of 200 N.*

Solution.

$d_1 = 250$ mm, $d_2 = 230$ mm, $\eta = 0.45$, $W = 200$ N

$$\text{V.R.} = \frac{2d_1}{d_1 - d_2} = \frac{2 \times 250}{250 - 230} = 25$$

$$\text{M.A.} = \frac{W}{P} = \frac{200}{P}$$

$$\eta = \frac{\text{M.A.}}{\text{V.R.}}$$

$$0.45 = \frac{200}{P \times 25}$$

$$P = 17.78 \text{ N}$$

Example. 11.18 *In a double threaded worm and worm wheel, the number of teeth on the worm wheel is 50. The diamter of the effort wheel is 200 mm and that of the load drum is 100 mm. Calculate the velocity ratio. If the efficiency of the machine is 20%, determine the effort required to lift a load of 30 N.*

Solution.

$z = 50$, $n = 2$, $D = 200$ mm, $d = 100$ mm, $\eta = 0.20$, $W = 30$ N

$$\eta = \frac{Wdn}{PDz}$$

$$P = \frac{30 \times 100 \times 2}{0.20 \times 200 \times 50} = 3 \text{ N}$$

Example. 11.19 *In a system of pulleys of the first type, there are three pulleys, and a weight of 300 N can just be supported by an effort of 50 N. Calculate the efficiency of the machine and* the amount of friction.

Solution.

$$n = 3, \quad W = 300 \text{ N}, \quad P = 50 \text{ N}$$

$$\text{M.A.} = \frac{W}{P} = \frac{300}{50} = 6$$

$$\text{V.R.} = 2^n = 2^3 = 8$$

$$\eta = \frac{\text{M.A.}}{\text{V.R.}} = \frac{6}{8} = 0.75 \text{ or } 75\%$$

$$W_f = P \times \text{V.R} - W$$

$$= 50 \times 8 - 300 = 100 \text{ N}$$

$$P_f = P - \frac{W}{\text{V.R.}} = 50 - \frac{300}{8} = 12.5 \text{ N}$$

Example. 11.20 *A weight of 1.5 kN is lifted by an effort of 160 N by second system of pulleys having 5 pulleys in each block. Calculate the amount of effort wasted in friction and the frictional load.*

Solution.

$$W = 1.5 \text{ kN}, \quad P = 160 \text{ N}, \quad n = 5 + 5 = 10$$

$$\text{V.R.} = n = 10$$

$$P_f = P - \frac{W}{\text{V.R.}} = 160 - \frac{1500}{10} = 10 \text{ N}$$

$$W_f = P \times \text{V.R} - W = 160 \times 10 - 1500 = 100 \text{ N}$$

Example. 11.21 *In a third system of pulleys, there are 4 pulleys. Find the effort required to lift a load of 2 kN, if the efficiency of the machine is 75%. Also calculate the effort wasted in friction.*

Solution.

$$n = 4, \quad W = 2000 \text{ N}, \quad \eta = 0.75$$

$$\text{V.R.} = 2^n - 1 = 2^4 - 1 = 15$$

$$\text{M.A.} = \frac{W}{P} = \frac{2000}{P}$$

$$\eta = \frac{\text{M.A.}}{\text{V.R.}} = \frac{2000}{P \times 15} = 0.75$$

$$P = 177.78 \text{ N}$$

$$P_f = P - \frac{W}{\text{V.R.}} = 177.78 - \frac{2000}{15} = 44.44 \text{ N}$$

11.11 SINGLE PURCHASE CRAB

A single purchase crab is shown in Fig. 11.11. It consists of an effort axle on which a pinion having z_1 number of teeth and a pulley having mean diameter D are keyed. The pinion meshes with a spur gear having z_2 number of teeth. The spur gear is keyed to the load axle having diameter d. A cord is tied to the load axle, wound up a few turns and to the free end, the load W to be lifted is attached. The effort and load axles are supported on bearings mounted in the housing frame. The effort can be applied either by using a pulley, as shown in figure, or by a handle.

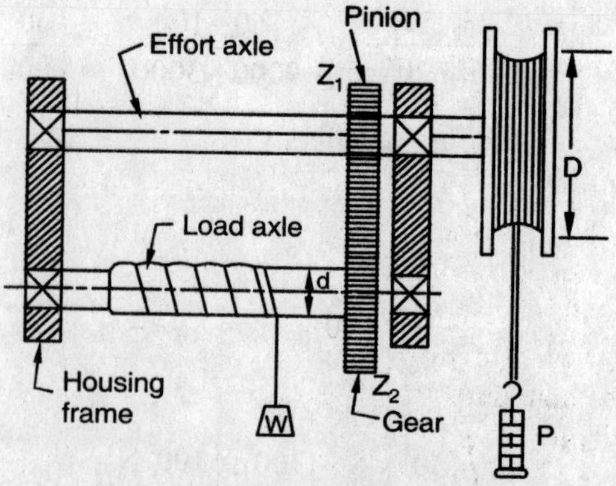

Fig. 11.11 Single purchase crab.

For one revolution of the pulley, distance moved by the effort $P = \pi D$.

Now
$$z_1 n_1 = z_2 n_2$$

∴
$$n_2 = \frac{z_1}{z_2} \text{ as } n_1 = 1$$

Hence distance moved by the load W

$$= \pi d \times \frac{z_1}{z_2}$$

$$\text{Velocity ratio} = \frac{\pi D}{\pi d} \times \frac{z_2}{z_1} = \frac{D}{d} \times \frac{z_2}{z_1} \qquad ...(11.21)$$

$$\text{Mechanical advantage} = \frac{W}{P}$$

$$\eta = \frac{W}{P} \times \frac{d}{D} \times \frac{z_1}{z_2} \qquad ...(11.22)$$

Example 11.22 *Following are the specifications for a single purchase crab :*

Diameter of load drum, $d = 20$ cm

Length of lever, $l = 120$ cm

Number of teeth on pinion, $z_1 = 10$

Number of teeth on spur wheel, $z_2 = 100$

Find the velocity ratio of this machine. On this machine, efforts of 100 N and 160 N are required to lift the loads of 3kN and 9kN respectively. Find the law of machine and the efficiencies at the above loads.

Solution.
$$d = 20 \text{ cm}, D = 2l = 2 \times 120 = 240 \text{ cm}$$
$$z_1 = 10, z_2 = 100$$

$$\text{Velocity ratio} = \frac{D}{d} \times \frac{z_2}{z_1} = \frac{240}{20} \times \frac{100}{10} = 120$$

The law of the machine is

$$P = mW + c$$

where $$m = \frac{P_2 - P_1}{W_2 - W_1} = \frac{160-100}{9000-3000} = \frac{60}{6000} = 0.01$$

$$100 = 0.01 \times 3000 + c$$

$$c = 70$$

$$P = 0.01W + 70$$

First Case : $$\text{M.A.} = \frac{3000}{100} = 30$$

$$\text{V.R.} = 120$$

$$\eta = \frac{130}{120} = 25 \text{ per cent}$$

Second Case : $$\text{M.A.} = \frac{9000}{160} = 56.25$$

$$\text{V.R.} = 120$$

$$\eta = \frac{56.25}{120} = 46.87 \text{ per cent}$$

11.12 DOUBLE PURCHASE CRAB

A double purchase crab is shown in Fig. 11.12. Here two sets of spur gears are used. Thus, for one revolution of the pulley, distance moved by effort $P = \pi D$,

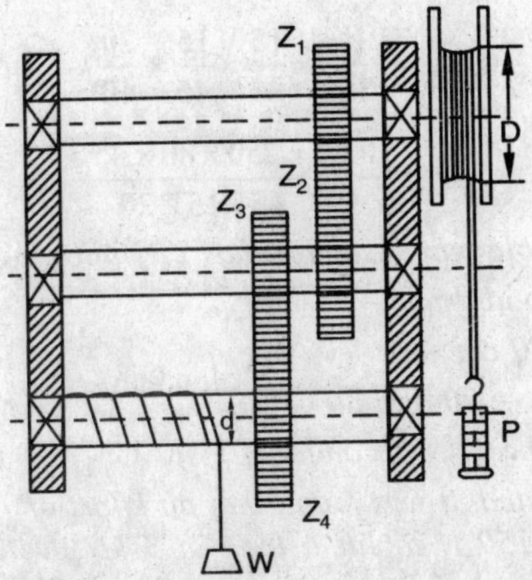

Fig. 11.12 Double purchase crab.

Now $$n_1 z_1 = n_2 z_2$$

∴ $$n_2 = \frac{z_1}{z_2} \quad \text{as } n_1 = 1$$

Also $$n_3 = n_2$$

and $$n_3 z_3 = n_4 z_4$$

$$n_4 = \frac{n_2 z_3}{z_4} = \frac{z_1}{z_2} \times \frac{z_3}{z_4}$$

Hence, distance moved by load W

$$= \pi d \times \frac{z_1}{z_2} \times \frac{z_3}{z_4}$$

$$\text{Velocity ratio} = \frac{\pi D}{\pi d} \times \frac{z_2 z_4}{z_1 z_3} = \frac{D}{d} \times \frac{z_2}{z_1} \times \frac{z_4}{z_3} \quad \text{...(11.23)}$$

$$\text{Mechanical advantage} = \frac{W}{P}$$

$$\eta = \frac{W}{P} \times \frac{d}{D} \times \frac{z_1}{z_2} \times \frac{z_3}{z_4} \quad \text{...(11.24)}$$

Example 11.23 *In a double purchase crab, the pinions have 15 and 200 teeth while the spur wheels have 45 and 40 teeth. The effort handle is 40 cm while the effective diameter of the drum is 15 cm. If the efficiency of the winch is 40 per cent, what load will be lifted by an effort of 250 N applied at the end of the handle ?*

Solution. Here $z_1 = 15$, $z_3 = 20$, $z_2 = 45$, and $z_4 = 40$

$$D = 2l = 2 \times 40 = 80 \text{ cm}$$

$$d = 15 \text{ cm}$$

$$\eta = 40 \text{ per cent}$$

$$P = 250 \text{ N}$$

$$\eta = \frac{W}{P} \times \frac{d}{D} \times \frac{z_1}{z_2} \times \frac{z_3}{z_4}$$

$$0.40 = \frac{W}{250} \times \frac{15}{80} \times \frac{15}{45} \times \frac{20}{40}$$

$$W = \frac{0.40 \times 250 \times 80 \times 45 \times 40}{15 \times 15 \times 20} = 3200 \text{ N.}$$

Example 11.24 *A double purchase crab has the following specifications :*

Effective diameter of load drum = 15 cm

Diameter of pulley = 40 cm

No. of teeth on pinions = 20 and 30

No. of teeth on spur wheels = 80 and 90

When the winch was tested, it was found that an effort of 100N was required to lift a load of 1900 N and an effort of 150 N to lift a load of 3 kN. Determine (i) law of the machine, (ii) probable effort to lift a load of 5 kN, (iii) efficiency in the above case, and (iv) maximum efficiency.

Solution.

$$\text{Velocity ratio} = \frac{D}{d} \times \frac{z_2}{z_1} \times \frac{z_4}{z_3} = \frac{40}{15} \times \frac{80}{20} \times \frac{90}{30} = 32$$

(*i*) Law of the machine is

$$P = mW + c$$

$$m = \frac{P_2 - P_1}{W_2 - W_1} = \frac{150 - 100}{3000 - 1900} = \frac{50}{1100} = 0.04545$$

$$100 = \frac{5}{110} \times 1900 + c$$

$$c = 13.64$$

$$\therefore \quad P = 0.04545W + 13.64$$

(*ii*) $$P = 0.04545 \times 5000 + 13.64 = 240.9 \text{ N}$$

(*iii*) $$\text{M.A.} = \frac{W}{P} = \frac{5000}{240.9} = 20.7555$$

$$\eta = \frac{\text{M.A.}}{\text{V.R.}} = \frac{20.7555}{32} = 64.86 \text{ per cent}$$

(*iv*) $$\eta_{max} = \frac{1}{m \times \text{V.R.}}$$

$$= \frac{110}{5 \times 32} = 68.75 \text{ per cent}$$

11.13 SIMPLE SCREW JACK

A simple screw jack is shown in Fig 11.13. It consists of a screw having square threads. The screw can be rotated by a handle of length l, at the end of which effort P can be applied. The screw rotates in a nut seated in the body of the screw.

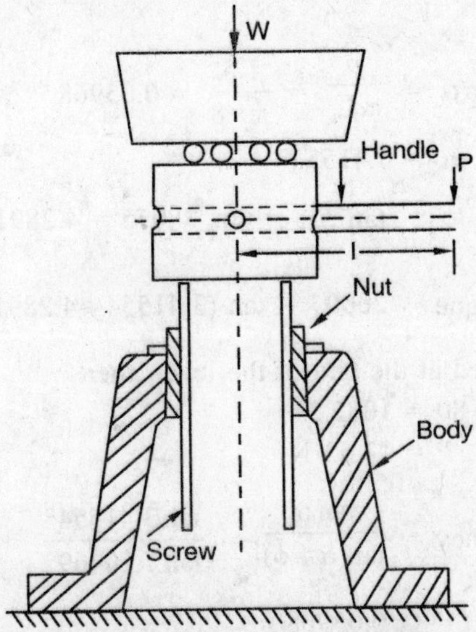

Fig. 11.13 Simple screw jack

Let p = pitch of the screw threads. In one revolution of the screw, effort P moves through a distance $= 2\pi l$ and the load W moves be a distance $= p$. Here

$$\text{Velocity ratio} = \frac{2\pi l}{p} \qquad \text{...(11.25)}$$

$$\text{Mechanical advantage} = \frac{W}{P}$$

Efficiency, $$\eta = \frac{W}{P} \cdot \frac{p}{2\pi l} \quad \ldots(11.26)$$

Now $$P = W \tan (\phi \pm \alpha) \quad \ldots(11.27)$$

where $$\phi = \tan^{-1} \mu$$

$$\alpha = \tan^{-1}\left[\frac{p}{\pi d_m}\right]$$

μ = coefficient of friction between the screw and the nut

d_m = mean diameter of screw.

Example 11.25 *A screw jack has mean diameter = 8 cm and pitch = 1.5 cm. The coefficient of friction between its screw and nut is 0.075. Find the effort required to be normally applied at the end of its operating lever 80 cm long to (i) raise a load of 2 kN, and (ii) lower the same load. Find the efficiency under these loads.*

Solutions. $d_m = 8$ cm, $p = 1.5$ cm

$\mu = 0.075$, $l = 80$ cm, $W = 2$ kN

(*i*) Torque required to raise the load

$$= W\frac{d_m}{2}\tan(\alpha + \phi)$$

Now $$\tan \alpha = \frac{p}{\pi d_m} = \frac{1.5}{\pi \times 8} = 0.05968$$

$$\alpha = 3.41554°$$

$$\phi = \tan^{-1} \mu = \tan^{-1} 0.075 = 4.28915°$$

$$\text{Tarque} = 2000 \times \frac{8}{2} \tan (3.41554 + 4.28915)° = 1082.31 \text{ N.cm}$$

If P is the effort applied at the end of the lever, then

$$P \times 80 = 1082.31$$

$$P = 13.53 \text{ N}$$

$$\text{Efficiency} = \frac{\tan \alpha}{\tan(\alpha + \phi)} = \frac{\tan 3.41554°}{\tan 7.70469°}$$

$$= \frac{0.05968}{0.13529} = 44.11 \text{ per cent}$$

(*ii*) Torque required for lowering of the load

$$= W \times \frac{d_m}{2}\tan(\phi - \alpha)$$

$$= 2000 \times \frac{8}{2} \tan (4.28915 - 3.41554)° = 121.99 \text{ N cm}$$

$$= P \times 80$$

$$\therefore \quad P = 1.525 \text{ N}$$

$$\text{Efficiency} = \frac{\tan(\phi - \alpha)}{\tan \alpha} = \frac{0.01525}{0.05968} = 25.55 \text{ per cent}$$

Example 11.26 *A screw jack with square threads has a mean diameter of 6.25 cm. The pitch of the threads is 12.5 mm. The load on the jack revolves with screw. Coefficient of friction at the screw thread is 0.05.*

(a) Find the tangential force which is applied to it at 30 cm radius required to lift a load of 5440 N.

(b) State whether the jack is self-locking. If it is, find the torque necessary to lower the load. If it is not, find the torque which must be applied to keep the load from descending.

Solution. $d_m = 6.25$ cm, $p = 1.25$ cm

$\mu = 0.05$, $l = 30$ cm, $W = 5440$ N, $P = ?$

$$\tan\alpha = \frac{p}{\pi d_m} = \frac{1.25}{\pi \times 6.25} = 0.06366$$

$$\alpha = 3.64265°$$

$$\phi = \tan^{-1}\mu = \tan^{-1}0.05 = 2.86240°$$

(*a*) Now

$$P \times l = W \times \frac{d_m}{2}\tan(\alpha+\phi)$$

$$P = \frac{5440 \times 6.25}{30 \times 2}\tan(3.64265 + 2.86240)° = 64.6 \text{ N}$$

(*b*)

$$\text{Efficiency} = \frac{\tan\alpha}{\tan(\alpha+\phi)} = \frac{0.06366}{0.11402} = 55.83 \text{ per cent.}$$

As the efficiency is more than 50 per cent the screw jack is not self-locking.

Torque required to be applied at the circumference to prevent the load from descending

$$= W \times \frac{d_m}{2}\tan(\alpha-\phi)$$

$$= 5440 \times \frac{6.25}{2}\tan(3.64265 - 2.86240)° = 231.52 \text{ Ncm}$$

11.14 DIFFERENTIAL SCREW JACK

A differential screw jack has been shown in Fig. 11.14. It consists of two screws, one placed inside the other. When the lower screw is rotated by means of a handle the upper screw moves up and down in the reverse direction, thus providing differential action.

In one revolution of the effort P, the distance moved by effort $= 2\pi l$

Distance moved by load $= p_1 - p_2$

where p_1 and p_2 are the pitches of the threads of the lower and upper screws respectively.

$$\text{Velocity ratio} = \frac{2\pi l}{p_1 - p_2} \quad \text{...(11.28)}$$

$$\text{Mechanical advantage} = \frac{W}{P}$$

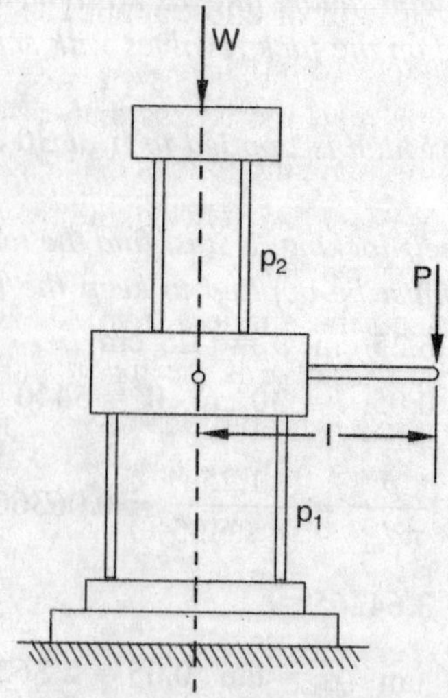

Fig. 11.14 Differential screw jack.

$$\text{Efficiency} = \frac{W}{P} \times \frac{(p_1 - p_2)}{2\pi l} \qquad \text{...(11.29)}$$

FILL IN THE BLANKS

1. A screw jack is a _____ machine.
2. A centre lathe is a _____ machine.
3. Mechanical advantage is the ratio of _____ to _____.
4. Velocity ratio is the ratio of distance moved by _____ to distance moved by _____.
5. Efficiency is the ratio of _____ to _____.
6. The efficiency of an ideal machine is _____
7. The efficiency of a reversible machine is _____ 50 percent.
8. The efficiency of a self-locking machine is _____ 50 percent.
9. Maximum mechanical advantage is _____.
10. The velocity ratio of first system of pulleys is _____.
11. The velocity ratio of second system of pulley is _____.
12. The velocity ratio of third systemof pulleys is _____.

ANSWER

1. Simple	**2.** Compound	**3.** load, effort	**4.** effort, load
5. M.A., V.R.	**6.** 100%	**7.** more than	**8.** less than
9. $\frac{1}{m}$	**10.** 2^n	**11.** n	**12.** $2^n - 1$.

TRUE AND FALSE

State True (T) or False (F) :

1. The efficiency of an ideal machine is 50%.
2. The bicycle is a simple machine.
3. The milling machine is a compound machine.

4. Mechanical advantage is the ratio of effort applied to load lifted.
5. Efficiency is the ratio of output to input.
6. The efficiency of a reversible machine is less than 50%.
7. The efficiency of self-locking machine is less than 50%.
8. An irreversible machine is also called self-locking machine.
9. The law of a machine is $W = mP + c$.
10. The efficiency of an actual machine is less than 100%.
11. The maximum mechanical advantage is the reciprocal of PV's W slope.
12. Velocity ratio of a machine is a variable quantity.
13. The velocity ratio of first system of pulleys is 2^n.
14. The velocity ratio of third system of pulleys is $2^n - 1$.

ANSWER

1. F	**2.** T	**3.** T	**4.** F	**5.** T	**6.** F	**7.** T
8. T	**9.** F	**10.** T	**11.** T	**12.** F	**13.** T	**14.** T

MULTI-CHOICE QUESTIONS

1. In actual machines, mechanical advantage is
   (*a*) greater than velocity ratio (*b*) equal to velocity ratio
   (*c*) less than velocity ratio (*d*) unity
2. In ideal machines, mechanical advantage is
   (*a*) greater than velocity ratio (*b*) equal to velocity ratio
   (*c*) less than velocity ratio (*d*) unity
3. The velocity ratio of an inclined plane inclined at angle α to the horizontal and the weight being pulled up the inclined plane by vertical effort is
   (*a*) $\sin\alpha$ (*b*) $\dfrac{1}{\sin\alpha}$
   (*c*) $\cos\alpha$ (*d*) $\dfrac{1}{\cos\alpha}$
4. The law of a machine is given by
   (*a*) $W = mP - c$ (*b*) $W = mP + c$
   (*c*) $P = mW - c$ (*d*) $P = mW + c$
5. For a machine to be of the self-locking type, its efficiency should be
   (*a*) $W = mP - c$ (*b*) $W = mP + c$
   (*c*) $P = mW - c$ (*d*) $P = mW + c$
6. A machine is said to be reversible, if its efficiency is
   (*a*) equal to 100% (*b*) less than 50%
   (*c*) more than 50% (*d*) equal to 50%
7. The velocity ratio of first system of pulleys is
   (*a*) 2^n (b) n
   (*c*) $2^n - 1$ (*d*) $2^n + 1$

8. The velocity ratio of second system of pulleys is

(*a*) 2^n (b) n

(*c*) $2^n - 1$ (*d*) $2^n + 1$

9. The velocity ratio of thirds system of pulleys is

(*a*) 2^n (b) n

(*c*) $2^n - 1$ (*d*) $2^n + 1$

10. If a machine is lubricated, then its

(*a*) velocity ratio increases (*b*) velocity ratio decreases

(*c*) mechanical advantage increases (*d*) efficiency decreases

11. If D is the diameter of wheel and d_1 and d_2 diameters of two axles, then velocity ratio is

(*a*) $\dfrac{D}{d_2 - d_1}$ (*b*) $\dfrac{D}{2(d_2 - d_1)}$ (*c*) $\dfrac{2D}{d_2 - d_1}$ (*d*) $\dfrac{2D}{d_2 + d_1}$

ANSWER

1. (*c*) **2.** (*b*) **3.** (*b*) **4.** (*d*) **5.** (*b*) **6.** (*c*) **7.** (*a*)
8. (*b*) **9.** (*c*) 10. (*c*)

REVIEW QUESTIONS

1. What is a machine ?
2. Define a simple machine.
3. Define a compound machine.
4. Differentiate between load and effort.
5. Define mechanical advantage, velocity ratio and efficiency.
6. What is a lifting machine? Give exaples.
7. What is a reversible machine ?
8. What is a self-locking machine ?
9. Define law of the machine.
10. What is the maximum mechanical advantage of a machine ?
11. Why a system of pulleys is used for lifting load ?
12. What is the velocity ratio of the first system of pulleys ?
13. What is the velocity ratio of the second system of pulleys ?
14. What is the velocity ratio of the third system of pulleys ?
15. What is the velocity ratio of the differential pulley block ?
16. What is the condition on efficiency for a machine to be reversible ?
17. What is the condition on efficiency for a machine to be self-locking ?

EXERCISE

11.1. In a lifting machine, efforts required to lift loads of 250 N and 360 N were 65 N and 80N respectively. The velocity ratio of the machine is 25. Determine the law of the machine and the efficiency to lift 250 N. [**Ans.** $P = 0.136\ W + 31$, 15.38 %]

11.2 The following table gives the experimental values for W and P of a lifting machine :

W, N	50	100	150	200	250
P, N	6.5	11	14	18	22

Determine the law of the machine. [**Ans.** P = 0.0767 W + 2.56]

11.3 In a wheel and axle, the diameter of wheel is 500 mm and that of the axle 150 mm. Find the velocity ratio. If the load to be lifted is 1500 N, find the effort required for machine efficiency of 60%. [**Ans.** 3.33, 750 N]

11.4 In a differential wheel and axle, the diameter of the larger axle is 250 mm and that of the smaller 225 mm. The diameter of effort wheel is 500 mm. Find the velocity ratio. If an effort of 150 N lifts a load of 3.5 kN, what is the efficiency and effort lost in friction. [**Ans.** 40, 16.67, 41.67%, 87.5 N]

11.5 In a first system of pulleys, there are 4 movable pulleys. If the efficiency of the system is 80%, find the effort required to lift a load of 5 kN. [**Ans.** 390.625 N]

11.6 In a second system of pulleys, there are 5 pulleys in each block. If an effort of 150 N can just lift a load of 1400 N, find (*a*) the effort wasted in friction, and (*b*) load wasted in friction. [**Ans.** 10 N, 100 N]

11.7 In the third system of pulleys, 4 pulleys are arranged. Find the effort required to lift a load of 1 kN if the efficiency of the pulley system is 80%. [**Ans.** 83.33 N]

11.8 In a differential pulley block, the diameter of the smaller pulley is $\frac{9}{10}$*th* of the larger pulley. Find the load lifted by the block with an effort of 200 N if the efficiency is 45%. [**Ans.** 1620 N]

11.9 In a worm geared pulley block, the effective diameter of the effort wheel is 250 mm. The worm is single threaded and the worm wheel has 100 teeth. The effective diameter of load drum fixed to it is 100 mm. If an effort of 200 N is required to lift a load of 12.5 kN, what is the efficiency of the block ? [**Ans.** 25%]

12 POWER TRANSMISSION METHODS AND DEVICES

12.1 INTRODUCTION

The mechanical power is transmitted in machines from one component to another depending upon the requirement. Various types of power and motion transmission systems are available.

12.2 POWER TRANSMISSION ELEMENTS

"In order to transmit power from one machine member to another, either flexible or non-flexible type of elements may be used. By a flexible machine element we mean, in which the centre distance between the driving and the driven member can be changed conveniently, and in the case of non-flexible machine elements the centre distance cannot be changed. These machine members can be classified as shown in Fig. 12.1. The flexible machine elements are used for the transmission of power over comparatively long distances. Since these elements are elastic and usually long, they play an important part in absorbing shock loads and damping out the effects of vibrating forces. In this chapter we shall study the design of flat and V-belts.

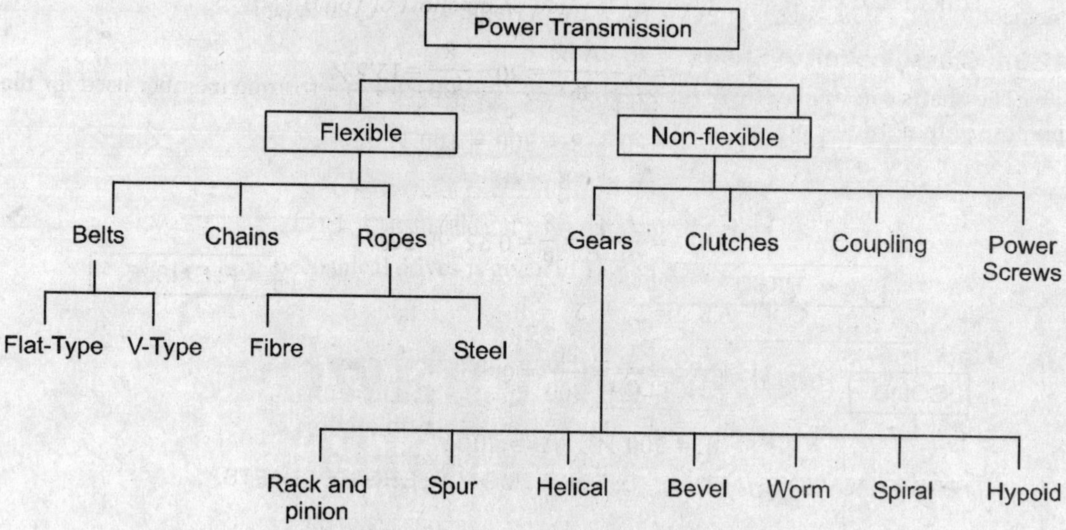

Fig. 12.1 Classification of power transmission elements.

A comparison of the mechanical drives is given in Table 12.1.

Table 12.1 Comparison of Mechanical Drives

Type of drive	*Centre distance*	*Speed ratio*	*Shifting*	*Maintenance cost*	*Initial cost*	*Power rating kW*	*Peripheral speed m/s*	*Efficiency percent*
Flat belts	Long	4-10	Yes	Low	1	upto 100	5-30	92-98
V-belts	Short	7-15	No	Low	3	50	5-30	87-97
Roller chains	Short	6-10	No	Medium	5	3500	Upto 25	94-98
Silent chains	Short	5-15	No	Medium	7	3000	Upto 25	94-98
Ropes	Very long	5-10	No	Medium	10	2500	6-15	92-96
Spur gears	Short	upto 10	Yes	Medium	15	10000	25	92-99
Helical gears	Short	upto 10	Yes	Medium	25	35000	25	94-99
Bevel gears	Short	upto 20	No	Medium	30	30000	30	90-98
Worm gears	Short	8-100	No	High	35	100	Upto 35	75-90

12.3 SHAFT AND AXLE

The shaft is one of the most common and basic machine elements. Failure of a shaft usually necessitates a costly and time consuming major overhaul. It is essential, therefore, that the designer be able to provide a suitable shaft design to meet the operating conditions to which the shaft will be subjected. The shaft must not only be strong enough to provide an adequate factor of safety from the static and dynamic stress view points but it must also be sufficiently rigid to prevent harmful torsional and lateral deflections. Provisions must usually be made for the proper positioning of the shaft and the components mounted on it. Finally, the shaft must have such dimensions that its natural frequency of vibration is sufficiently remote from the operating frequency to avoid resonance with its consequent destructive stresses and displacements.

The shaft may be attached to other components either by keys, pins, splines or force-fitted and shrink-fitted. The length of the shaft may be extended by a coupling.

A shaft may be defined as a machine member which is supported on bearings, supports radial and thrust loads, rotates and transmits power. It is subjected to both bending moment and twisting moment.

12.3.1 Classification of Shafts

The shafts can be classified as shown in Fig. 12.2. A shaft is a rotating member used for the purpose of transmitting power.

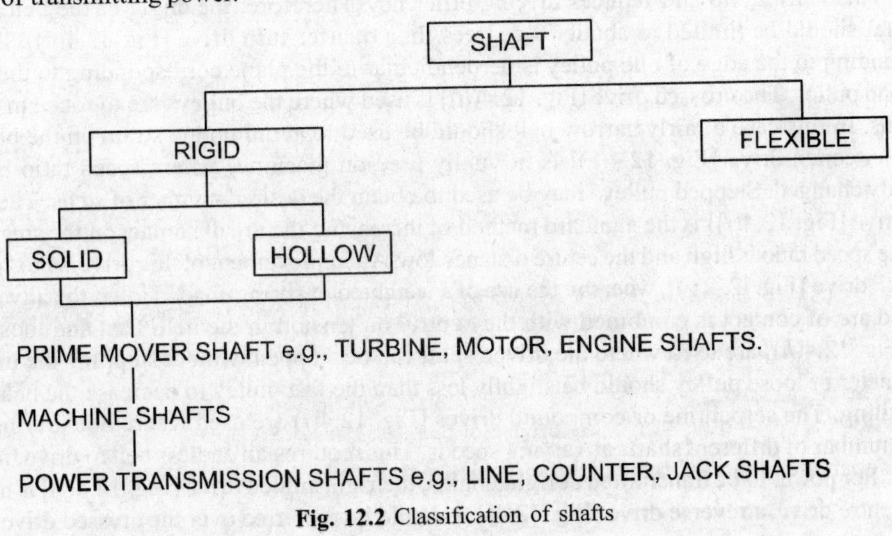

Fig. 12.2 Classification of shafts

A line shaft is a long and continuous shaft used in workshops and factories for the distribution of power. A counter shaft is used in between a line shaft and a machine. A shaft directly coupled to the power source is called a jack shaft. Flexible shafts are used where the axis of rotation of the driving and driven members are not coaxial.

12.3.2 Axle

An axle is a stationary member used as a support for rotating elements such as wheels, idler gears, and so on. A spindle is a short shaft. For example, the wheels of a cycle are supported on axles. In some cases, the axle may also rotate, as in the case of railway bogies. An axle does not transmit power. It only supports transverse loads.

A spindle normally refers to a machine tool. It is generally hollow and is subjected to both bending moment and twisting moment.

12.4 BELTS DRIVE

12.4.1 Flat Belts Drive

A flat belt drive is shown in Fig. 12.3. Belts are used for power transmission from one pulley to another.

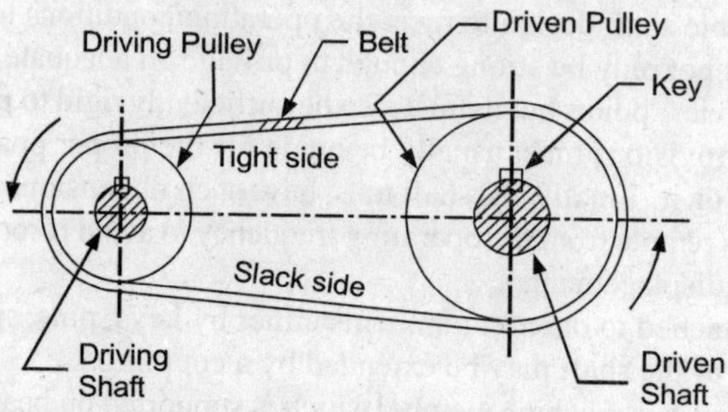

Fig. 12.3 Belt Drive

The layout of a flat belt drive may be of many types. The simplest drive is where the centres are horizontal as shown in Fig. 12.4(*a*). This drive can be laid with the slack side either at the top or below the centres. The vertical drive shown in Fig. 12.4(*b*) should be avoided wherever operating conditions permit. Due to the force of gravity the natural tendency for the belt is to fall away from the lower pulley surface. This causes slip and reduces driving efficiency. Therefore, the angle of the belt with the horizontal should be limited to about 70 degrees. In a quarter-turn drive [Fig. 12.4(*c*)], the plane corresponding to the edge of one pulley is perpendicular to the plane corresponding to the edge of the second pulley. The crossed drive [Fig. 12.4(*d*)] is used where the pulleys are to rotate in opposite directions. In this case a fairly narrow belt should be used to avoid undue strain on the belt while turning. A coned drive [Fig. 12.4(*e*)] is normally used on machines where speed ratio has to be constantly changed. Stepped pulleys may be used to obtain the desired number of steps. The jockey-pulley drive [Fig. 12.4(*f*)] is the standard method of increasing the arc of contact on the small pulley where the speed ratio is high and the centre distance low. An improvement of this drive is the "pulmax" or "lenix" drive [Fig. 12.4(*g*)], whereby the use of a weighted or spring-loaded idler, the advantage of increased arc of contact is combined with the control on tension in die belt. Fast and loose pulley drives [Fig. 12.4(*h*)] are used where die driven shaft can be stopped without stopping the line shaft. The diameter of loose pulley should be slightly less than the fast pulley to decrease the belt tension during idling. The serpentine or compound drives [Fig. 12.4(*i*) are used where one driving pulley drives a number of different shafts at varying speeds. This requires an endless belt to drive from both the sides. For power to be transmitted at right angles, the right angled drive [Fig. 12.4(*j*)] is used. For a short centre drive, a reverse drive [Fig. 12.4(*k*)] should be preferred over the crossed drive.

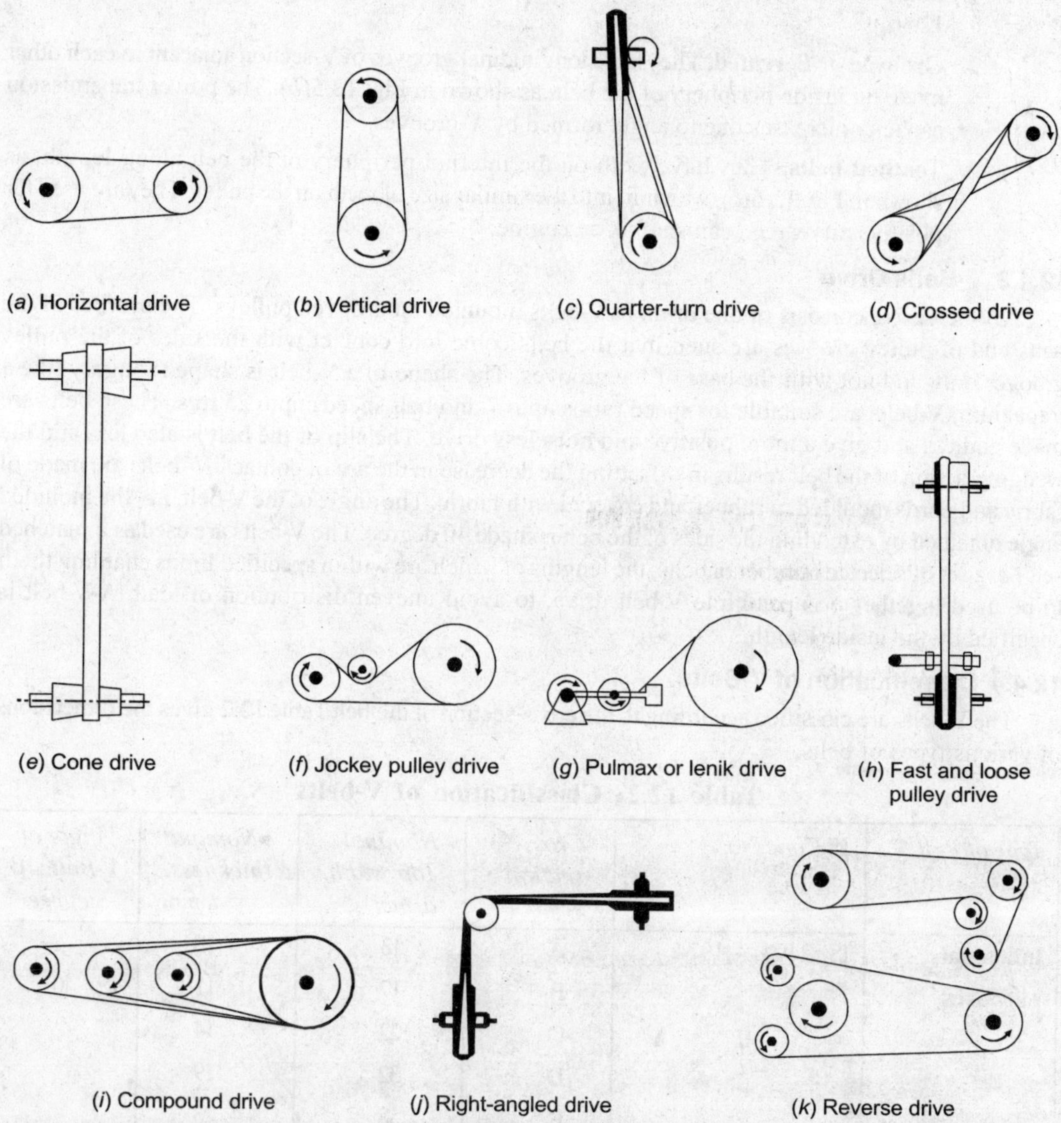

Fig. 12.4 Types of flat belt drive

12.4.2 Types of Flat Belts

Flat belts are made of leather, fabric, rubberized fabric, non-reinforced rubber and reinforced leather. Flat belts are of three types as shown in Fig. 12.5.

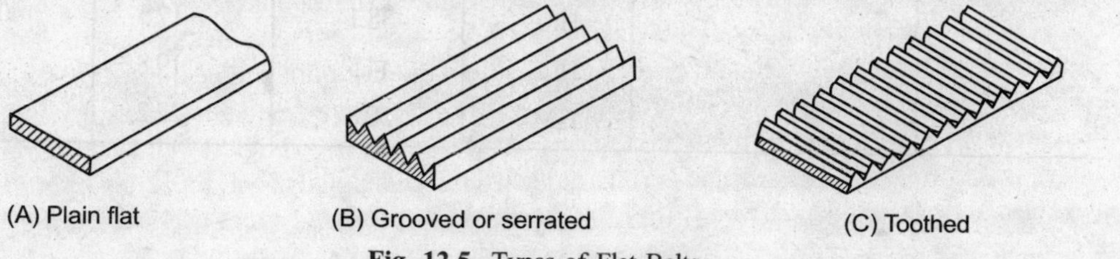

Fig. 12.5 Types of Flat Belts

1. **Plain flat.** They are of rectangular cross-section with no teeth or groove, as shown in Fig. 12.5(*a*).
2. **Grooved or Serrated.** They have longitudinal grooves of V-section adjacent to each other made on inside periphery of the belt, as shown in Fig. 12.5(*b*). The power transmission capacity increases due to angle formed by V-grooves.
3. **Toothed belts.** They have teeth on the internal periphery of the belt along length, as shown in Fig. 12.5(*c*), which fit into the similar size of teeth on the pulley. They are used for positive drive, *e.g.*, camshaft of an engine.

12.4.3 V-Belts Drive

A V-belt drive consists of one or more V-belts mounted on grooved pulleys. The profiles of the belts end of pulley grooves are such that the belts come into contact with the sides of the pulley grooves only and not with the base of the grooves. The shape of a V-belt is shaped roughly like a trapezium. V-belts are suitable for speed ratios upto 7 and belt speeds upto 25 m/s. These belts are made endless and give a more positive and noiseless drive. The slip of the belt is also less and the wedging action of the belt results in offsetting the decrease in the arc of contact. V- belts are made of fabric and cords moulded in rubber and covered with fabric. The angle of the V-belt, *i.e.* the included angle obtained by extending the sides of the belt is made 40 degree. The V-belts are used as a matched set, *i.e.* a set of selected number of belts, the lengths of which are within specified limits enabling them to be used together on a multiple V-belt drive, to avoid uneven distribution of load. A V-belt is specified by the inside length.

12.4.4 Classification of V-Belts

The V-belts are classified according to the cross-section of the belt Table 12.2 gives the dimensions of various types of belts.

Table 12.2 Classification of V-belts

Type of Belt	*IS Code No.*	*Cross-section symbol*	*Nominal Top width, a, mm*	*Nominal thickness, t, mm*	*Angle of V-Belt, 2β degree*
Industrial purposes	IS : 2494 — 1974	A	13	8	
		B	17	11	40
		C	22	14	
		D	32	19	
		E	38	23	
Automotive fan	IS : 5635—1970	FB 10	10	8	
		FB 13	13	11	40
Agricultural machinery	IS : 8777—1978	HI	25.4	12.7	
		HJ	31.8	15.1	
		HK	38.1	17.5	40
		HL	44.5	19.8	
		HM	50.8	22.2	

The cross-section of V-belts is shown in Fig. 12.6.

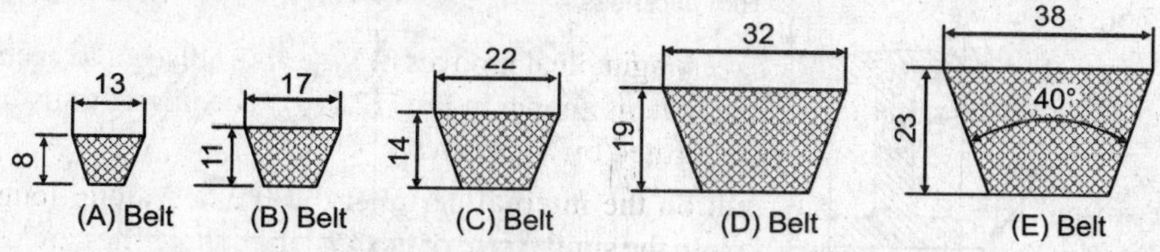

Fig. 12.6 V Belt Cross-Sections

12.5 ROPES DRIVE

For transmission of power in cases where centre distance between driver and driven shafts is quite large, rope drive is used. Ropes are made from cotton, manila, sisal, and steel wire. Steel wire ropes are made either with fibre core or steel core. The rope drive is used in cranes, elevators, aerial ropeways, drilling and mines.

12.6 PULLEYS FOR BELTS

They are used to transmit power with the help of belts. A pulley is a circular machine element having a hole in the centre with a keyway that fits on the shaft. The outer rim is plain or arced (crowned) for flat belts. For a grooved pulley, the number of grooves can be one or more than one. Toothed belts use a pulley that have teeth on its outer periphery to fit in the teeth of the belt for positive power transmission. Various types of pulleys are shown in Fig. 12.7.

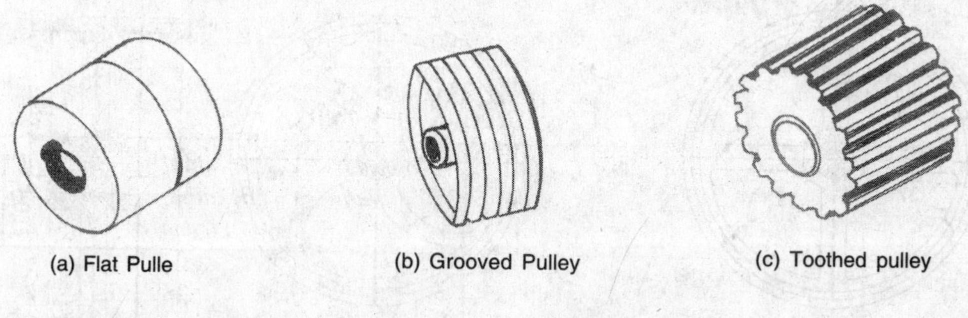

Fig. 12.7 Various Types of Pulleys for belts.

Small pulley can be made of forged steel, but large pulleys are made of cast iron.

12.6.1 Flat Belt Pulleys

Flat belt pulley has its outer periphery almost flat. A slight camber is given to make the outer periphery slightly convex (Fig. 12.8). This helps in keeping the belt positioned centrally on the pulley. Flat pulleys are of many types, e.g. solid, webbed, straight armed, curved armed, built up, stepped, fast and loose. Each type is described in the following sections.

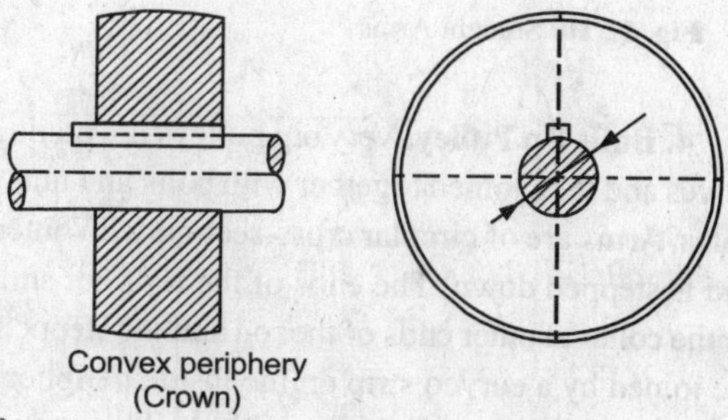

Fig. 12.8 Solid Pulley

1. Solid Pulley. Small pulleys are made solid as shown in Fig. 12.8. The hole is of the shaft size according to the fit required and has a keyway or a collar to fit by a screw.

2. Webbed Pulley. Medium sized pulleys are made webbed to reduce weight and save material (Fig. 12.9). Sometimes the web is provided with holes to decrease weight and save material.

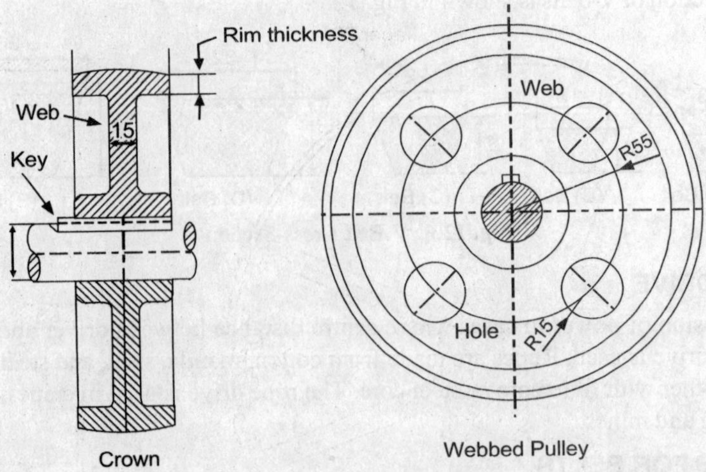

Fig. 12.9 Webbed Pulley

3. Armed Pulley. Large sized pulleys have a rim and arms joining the rim and central portion called the hub. The arms can be straight (Fig. 12.10) or curved (Fig. 12.11). Number of arms can be 4, 6 or more depending upon the size of pulley. Cross-section of the arm is generally elliptical but can be circular. The shaft hole and keyway is cut to suit the shaft size.

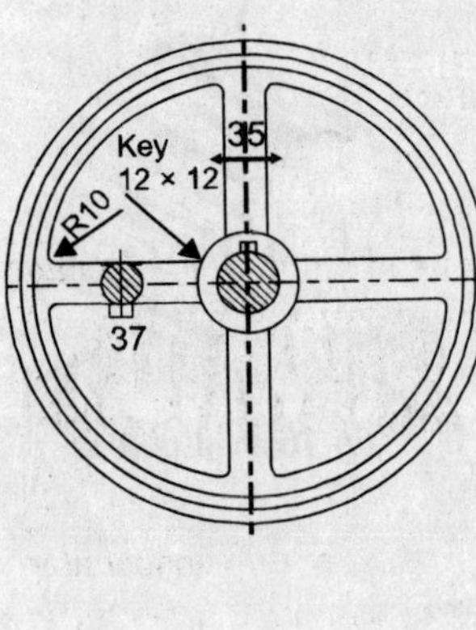

Fig. 12.10 Straight Arms

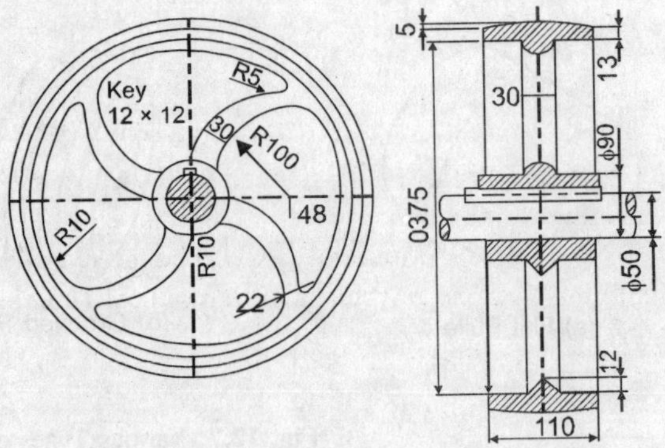

Fig. 12.11 Curved Arms

4. Built Up Pulley. Very big pulleys are made as built up type. Both hub and rim are made in two halves and then joined together with bolts and nuts (Fig. 12.12). Hub and rim are provided with radial holes. Arms are of circular cross-section with outer end having a collar and step at one end and inner end is stepped down. The ends of the rods are shrunk fit into holes of the hub. The rim is supported by the collar at outer ends of the rod and the stepped portion of rod is riveted to the rim. The rim halves are joined by a curved strip on the inside periphery of the rim. The strap is riveted to one side of the rim using counter such rivets and bolted by a counter sunk bolt at the other side. The two halves of hub are bolted on shaft with spacer in between.

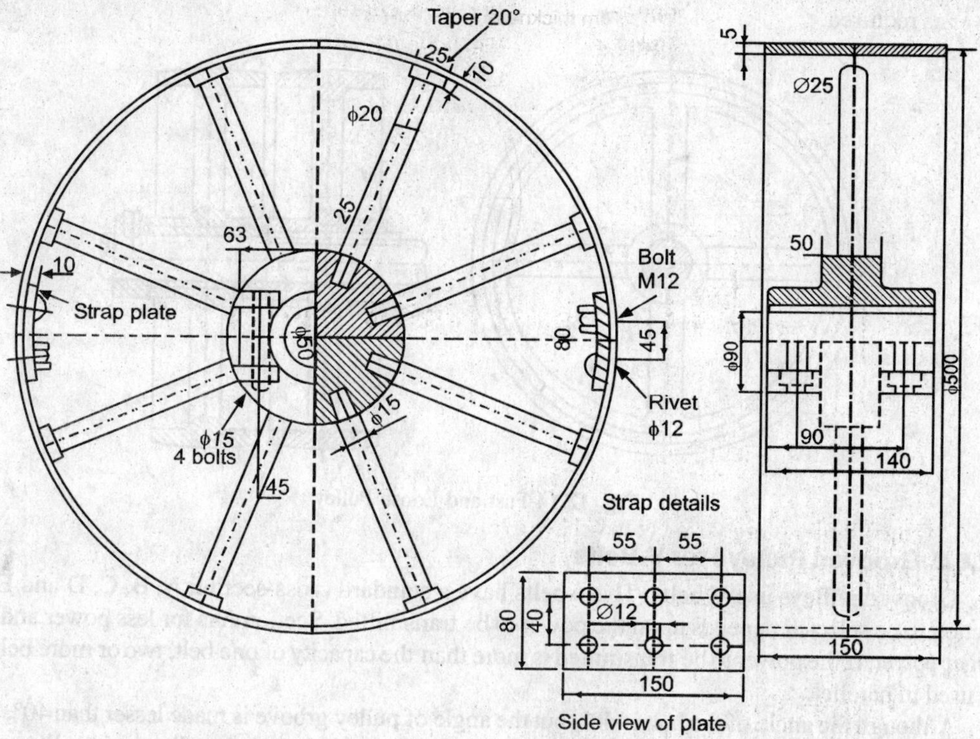

Fig. 12.12 Built up Pulley

5. Stepped Pulley. With stepped pulleys, speed of the driven pulley can be changed by changing the position of belt from one step to another step. The steps are generally three or four (Fig. 12.13). More than four steps commonly used. The pulleys are so arranged that the belt at smallest step of the driver aligns with biggest step of driven pulley. Their sizes are so adjusted that the belt length required remains same.

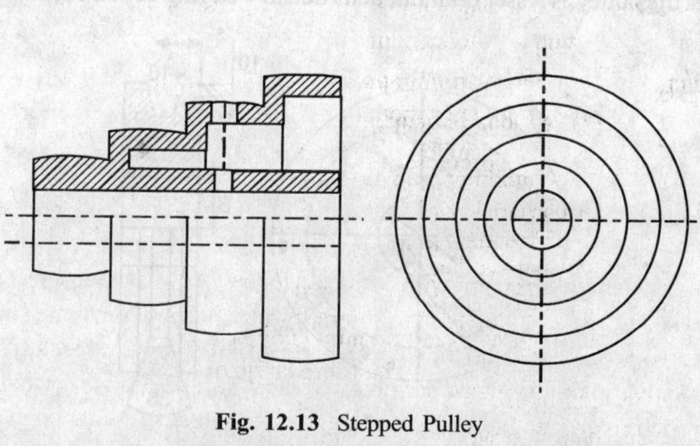

Fig. 12.13 Stepped Pulley

6. Fast and Loose Pulley. Sometimes two pulleys are adjacent to each other. One of the pulleys is keyed to the shaft called the fast **pulley,** while the other is without key called **loose pulley.** Loose pulley runs freely on a bush without transmitting power to shaft. This arrangement is known as Fast and Loose pulley (Fig. 12.14). Whenever the power transmission is not required, belt is made to slip from fast to loose pulley. Some machine tools use this type of power transmission to connect and disconnect power as required.

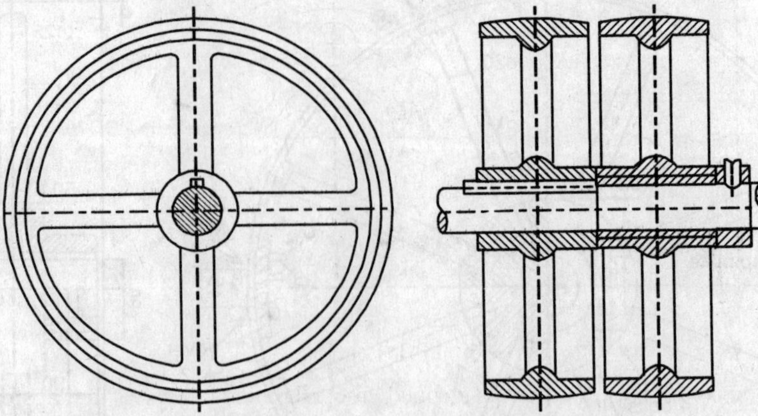

Fig. 12.14 Fast and Loose Pulley

12.6.2 Grooved Pulleys for V-Belts

Grooved pulleys use V belts. These belts have a standard cross-section A, B, C, D and E. The section to be selected depends upon the power to be transmitted. Section A is for less power and E for larger power. If the power to be transmitted is more than the capacity of one belt, two or more belts can be used in parallel.

Although the angle of the belt is 40°, but the angle of pulley groove is made lesser than 40°, as the wedging action causes more gripping force between pulley and belt. Driver pulleys generally smaller) have included angle between 32° and 34° and large pulley 36° and 38°. Depth of the groove is also kept about 10% more than the actual depth of belt so that when the belt gets compressed due to continuous use, it may not touch the bottom surface of the groove, else the wedging action will not take place. Grooved pulleys can have one groove or more than one if more power is to be transmitted.

1. Single Belt Grooved Pulley. Only one V belt is used (Fig. 12.15). It may be solid or webbed. The slip of the belt over the pulley is lesser than flat belts due to wedging action of the belt into the groove.

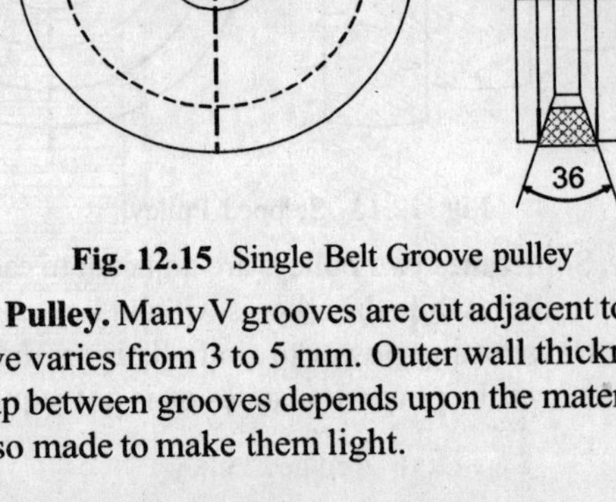

Fig. 12.15 Single Belt Groove pulley

2. Multi Belt Grooved Pulley. Many V grooves are cut adjacent to each other (Fig. 12.16). Wall thickness between the groove varies from 3 to 5 mm. Outer wall thickness is kept slightly more than inner wall thickness. The gap between grooves depends upon the material also. Larger size of pulleys are webbed and holes are also made to make them light.

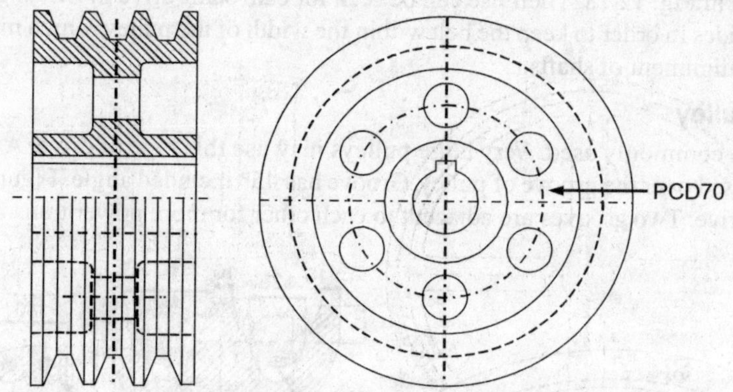

Fig. 12.16 Multi Belt Groove Pulley

3. Stepped Grooved Pulley. Purpose of stepped grooved pulley is same as that of stepped pulley for flat belt. These pulleys are also used in pair, aligned such that belt of smallest pulley aligns with biggest pulley. Angle of groove lies between 34° and 38° (Fig. 12.17). The groove is kept deeper than the height of belt. Size of groove depends upon cross-section of belt A, B or C.

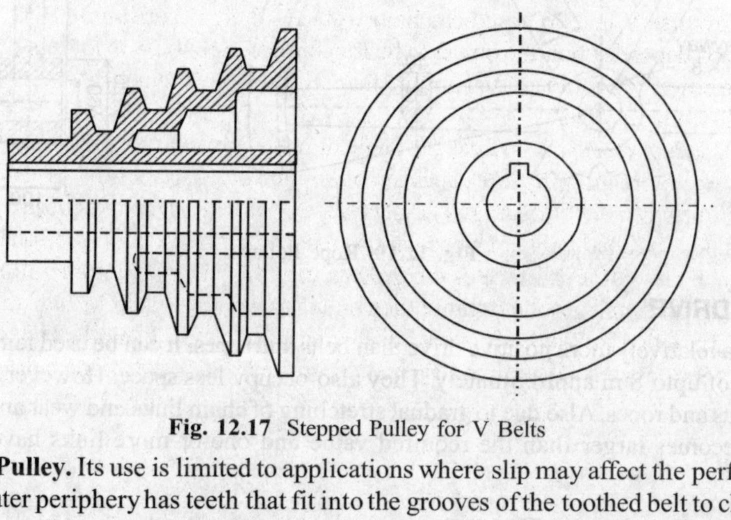

Fig. 12.17 Stepped Pulley for V Belts

4. Toothed Pulley. Its use is limited to applications where slip may affect the performance of the machine. The outer periphery has teeth that fit into the grooves of the toothed belt to check slip. Such

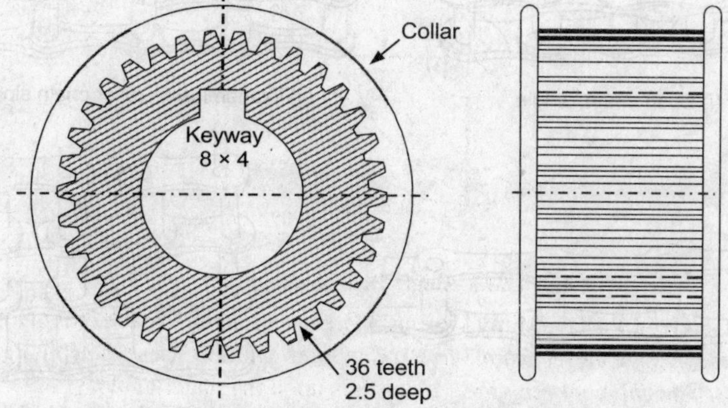

Fig. 12.18 Toothed Pulley

a pulley is shown in Fig. 12.18. Their use can be seen for cam shaft drive in I.C. engines. A collar is provided on the sides in order to keep the belt within the width of the pulley, which may shift on sides due to slight misalignment of shafts.

12.6.3 Rope Pulley

They are not commonly used. Very large pulleys may use this type of drive. A rope of circular cross-section is used over the groove of pulley. Groove has 45° included angle. Figure 12.19 shows a pulley for rope drive. Two grooves are adjacent to each other for more power transmission.

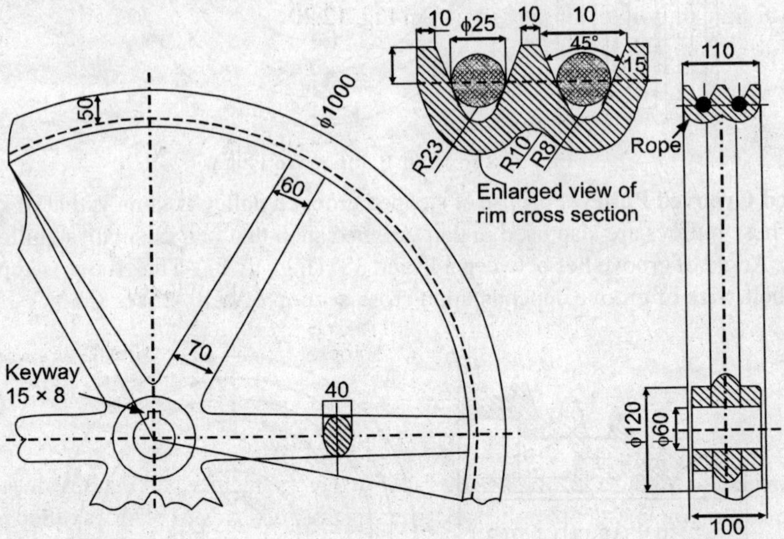

Fig. 12.19 Rope Pulley

12.7 CHAIN DRIVE

A chain is a relatively more positive drive than belts and ropes. It can be used fairly over short and long distances of upto 8 m approximately. They also occupy less space. However, they have more weight than belts and ropes. Also due to gradual stretching of chain links and wear and tear, the length of the chain becomes larger than the required value and one or more links have to be removed occasionally.

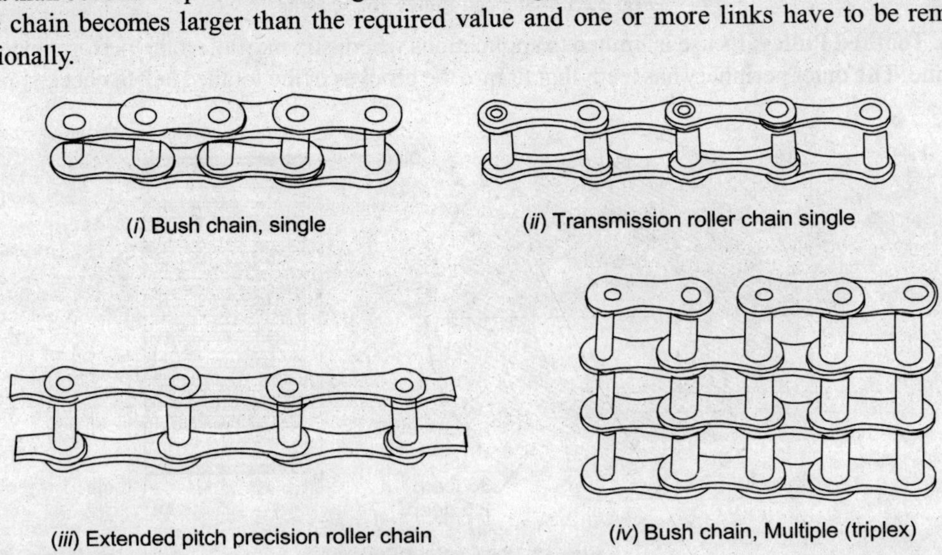

Fig. 12.20 Power transmitting chains

Chain drives find applications in agricultural machinery, bicycles, motor cycles, conveying machinery rolling mills etc.

Power transmitting chains are of the types as shown below :

Power Transmitting Chains

- Block Chain
- Roller Chain
- Silent Chain

The power transmitting chains are shown in Fig. 12.20.

A chain drive is shown in Fig. 12.21.

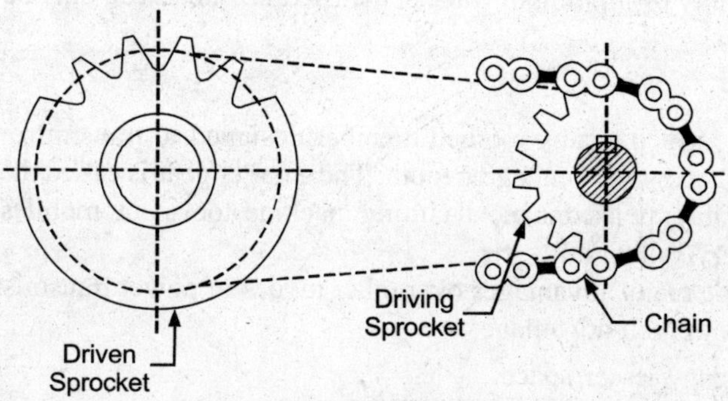

Fig. 12.21 Chain Drive

The transmission steel roller chains consist of two rows of outer and inner plates. The outer row of plates is known as pin link or coupling link whereas the inner row of plates is called roller link. The pins are press-fitted in the pin links and pass through the bushings which are a press-fit in the roller and join them. The chain rollers are mounted on the bushings and roll over the sprocket teeth during motion. They are made in simple, duplex and triplex construction.

12.8 FRICTION DRIVE

Consider two plain circular wheels *A* and *B* mounted on parallel shafts as shown in Fig. 12.22. The wheels have sufficient rough surfaces and press against each other.

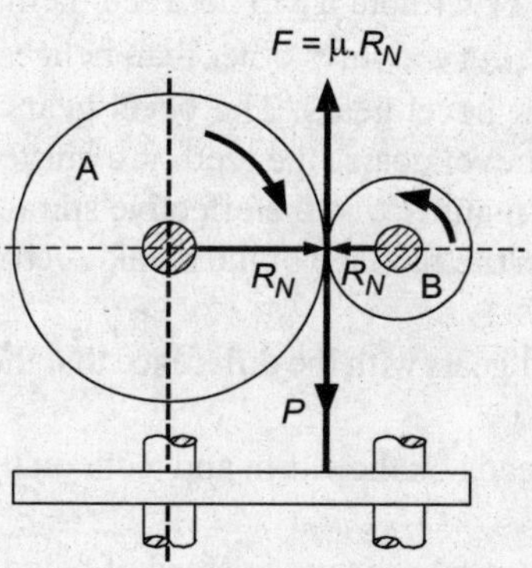

Fig. 12.22 Friction wheels.

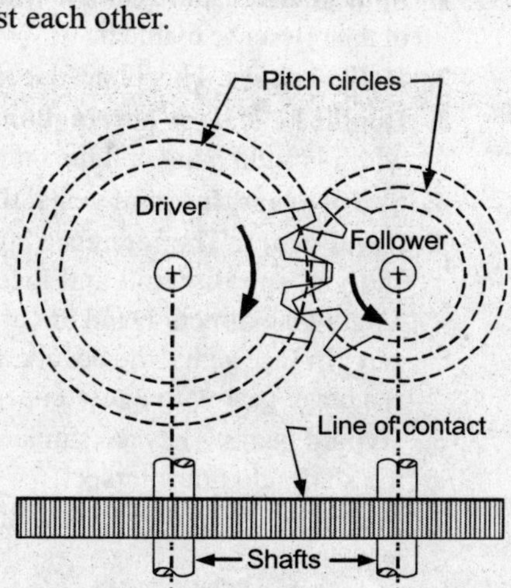

Fig. 12.23 Gear or toothed wheels

Let the wheel A is keyed to the rotating shaft and the wheel B to the shaft to be rotated. When the wheel A is rotated by a rotating shaft, it will rotate the wheel B in the opposite direction as shown in Fig. 12.23. The wheel B will be rotated by the wheel A so long as the tangential force exerted by the wheel A does not exceed the maximum frictional resistance between the two wheels. But when the tangential force (P) exceeds the *frictional resistance (F), slipping will take place between the two wheels.

In order to avoid the slipping, a number of projections (called teeth) as shown in Fig. 12.23 are provided on the periphery of the wheel A which will fit into the corresponding recesses on the periphery of the wheel B. A friction wheel with the teeth cut on it is known as ***gear*** or ***toothed wheel.*** The usual connection to show the toothed wheels is by their pitch circles.

Note : Kinematically, the friction wheels running without slip and toothed gearing are identical. But due to the possibility of slipping of wheels, the friction wheels can only be used for transmission of small powers.

12.9 GEAR DRIVE

A gear may be defined as any toothed member designed to transmit or receive motion from another member by successively engaging tooth. The smaller gear is called the pinion and the bigger one the gear wheel. They are used in metal cutting machine |tools, automobiles, tractors, hoisting and transporting machinery, rolling mills, etc.

The gears provide many advantages over other modes of power transmission likes belts, ropes, and chains etc. Some of their advantages are :

1. They occupy lesser space.
2. There is no slip between the gears in mesh and provide exact speed ratio.
3. They can transmit higher power.
4. Their efficiency is higher.

However, the error in tooth meshing may cause undesirable vibration and noise during operation.

12.9.1 Classification of Gears

The various types of gears are shown in Fig. 12.24.

The gears may be classified as follows :

1. **Spur gears.** A spur gear is a cylindrical gear whose tooth traces are straight line generators of the reference cylinder.
2. **Helical gears.** This is similar to the spur gear in which the tooth traces are helices.
3. **Double helical (or herringbone) gears.** It is a cylindrical gear in which a part of the face width is right hand and the other left hand/ with or without a gap between them.
4. **Spiral gears.** In spiral gears, the tooth traces are curved lines other than helices.
5. **Bevel gears.** The reference surface is a cone in bevel gears. The bevel gears may be straight, spiral, zerol, and face gears. In zerol bevel gears, the teeth are curved in the lengthwise direction and are arranged in such a manner that the effective spiral angle is zero. In face gears, the bevel gear teeth are cut on the flat face of the blank. A crown gear is a bevel gear with a reference cone angle of 90°.
6. **Hypoid gears.** They are similar to the spiral bevel gears with the difference that the axes of the shafts do not intersect.
7. **Worm gears.** In these gears, there are screw threads on the worm and teeth on the worm wheel.
8. **Planetary gears.** A gear pair or a gear train one of whose axes, instead of being fixed in position in the mechanism of which the gear pair is a part, moves around the other is called planetary gear train. These gears are discussed in chapter 5.

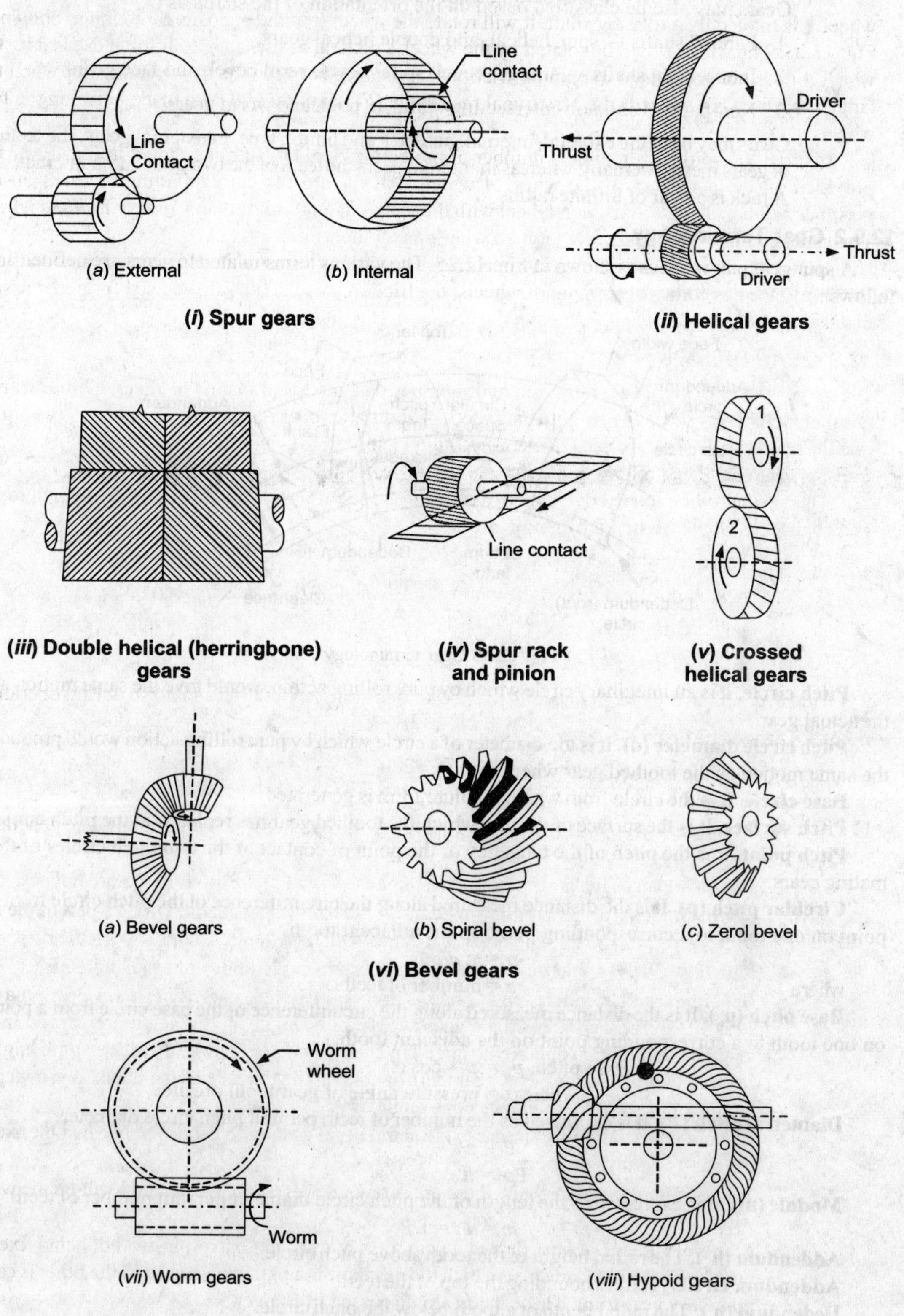

Fig. 12.24 Types of gears

Gears may also be classified based on the orientation of the shafts as :

1. Parallel shafts — spur, helical, and double helical gears.
2. Intersecting shafts — straight bevel, spiral bevel, zerol bevel, and face gears.
3. Non-parallel and non-intersecting-spiral, hypoid, and worm gears.

Gears may be of the external, internal, and rack and pinion type. In external gears, the teeth of gears mesh externally, whereas in internal gears the teeth of the two gears mesh internally. A rack is a gear of infinite radius.

12.9.2 Gear Terminology

A spur gear pair in mesh is shown in Fig. 12.25. The various terms related to gears are defined as follows :

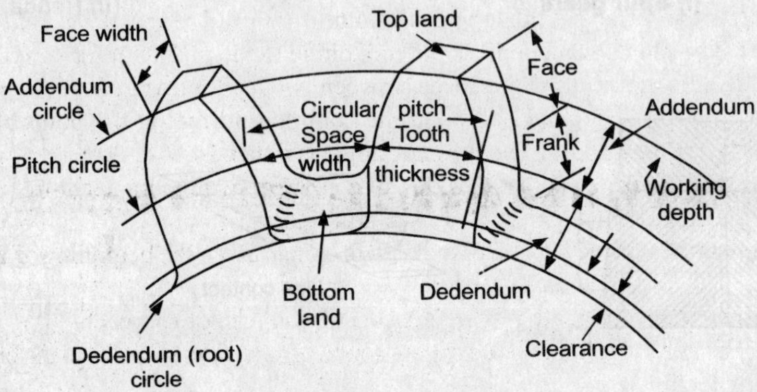

Fig. 12.25 Gear terminology

Pitch circle. It is an imaginary circle which by pure rolling action, would give the same motion as the actual gear.

Pitch circle diameter (d). It is the diameter of a circle which by pure rolling action would produce the same motion as the toothed gear wheel.

Base circle. It is the circle from which involute form is generated.

Pitch surface. It is the surface of the disc which the toothed gear has replaced at the pitch circle.

Pitch point. It is the pitch of the tangency or the point of contact of the two pitch circles of the mating gears.

Circular pitch (p). It is the distance measured along the circumference of the pitch circle from a point on one tooth to a corresponding point on the adjacent tooth.

$$p = \pi\, d/z$$

where z = number of teeth.

Base pitch (p_b). It is the distance measured along the circumference of the base circle from a point on one tooth to a corresponding point on the adjacent tooth.

$$\text{Base pitch, } p_b = p \times \cos \alpha$$

where α = pressure angle of gear tooth profile.

Diametral pitch (P). It is expressed as the number of teeth per unit pitch circle diameter.

$$\text{P} = z/d$$

$$\text{P}p = \pi$$

Module (m). It is expressed as the length of the pitch circle diameter per unit number of teeth.

$$m = d/z = 1/\text{P}$$

Addendum (h_a). The radial height of the tooth above pitch circle.

Addendum circle. A circle bounding the top of the teeth.

Dedendum (h_f). The radial depth of a tooth below the pitch circle.

Dedendum circle. A circle passing through the roots of all the teeth.

Clearance (c). The radial height difference between addendum and dedendum of a teeth.

Working depth. It is the radial distance of tooth from addendum circle to clearance circle.

Total depth. It is the sum of addendum and dedendum or the radial distance from dedendum circle to addendum circle.

Face. The part of the tooth surface lying below the pitch surface.

Backlash. The minimum distance between the non driving side of a tooth and adjacent side of the mating tooth at the pitch circle.

Profile. The curve forming face and flank.

Tooth thickness (t). This is the arc distance measured along the pitch circle from its intercept with one flank to its intercept with the other flank of the same tooth.

Face width (b). The width of the gear tooth measured axially along the pitch surface.

Top land. It is the surface of the top of the tooth.

Tooth fillet. The radius that connects the root circle to the profile of the tooth.

Tooth Space. It is the width of the space between two teeth measured on the pitch circle.

Pressure angle (α). The angle between the common normal at the point of contact and the common tangent at the pitch point The pressure angle is either 14.5° or 20°.

Path of contact. It is the locus of the point of contact of two mating teeth from the beginning of engagement to the end of engagement It is a straight line.

Path of approach. It is the portion of the path of contact from the beginning of engagement to the pitch point.

Angle of approach. The angle turned by gears during path of approach.

Path of recess. It is the portion of the path of contact from the pitch point to the end of engagement of the two mating teeth.

Angle of recess. The angle turned through during path of recess.

Arc of contact. It is the locus of a point on the pitch circle, from the beginning of engagement to the end of engagement of pair of teeth in mesh.

Involute. The involute of a circle is the curve traced by the end of a thread as it is unwound from a stationary cylinder.

Base Circle Diameter (d_b). It is the diameter of the base circle.

$$d_b = d \cos \alpha$$

Cycloid. It is the locus of a point on the circumference of a circle which rolls without slipping on a fixed straight line.

Centre distance (C). It is the distance between the centres of rotation of the two gears in mesh.

$$C = (d_1 + d_2)/2 = m\,(z_1 + z_2)/2$$

12.10 GEAR TRAINS

A gear train is composed of two or more gears in mesh for the purpose of transmitting motion from one shaft to another. A gear train enables to have larger centre distance between the driving and driven shafts, provides control on the direction of rotation of the driven gear, and facilitates increased transmission ratio.

12.10.1 Types of Gear Trains

There are four types of gear trains :

1. Simple gear train
2. Compound gear train
3. Reverted gear train
4. Planetary (or epicyclic) gear train.

1. Simple Gear Train. A simple gear train is one in which there is only one gear on each shaft. A simple gear train is often used to change the direction of rotation of a gear without changing its angular velocity. This can be done by placing an idler gear between the driving and driven gears.

Consider the simple gear train shown in Fig. 12.26.

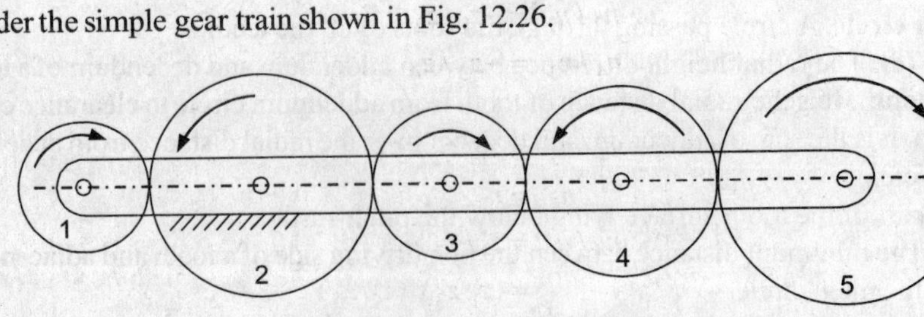

Fig. 12.26 Simple gear train

$$n_1 / n_2 = z_2 / z_1$$

$$n_2 / n_3 = z_3 / z_2$$

— — — — — —

— — — — — —

$$n_{m-1} / n_m = z_m / z_{m-1}$$

Multiplying, we get

$$n_1 / n_m = z_m / z_1$$

Therefore, the velocity ratio of a simple gear train is the ratio of the angular velocity of the first gear in the train to the angular velocity of the last gear. We find that the intermediate gears do not in any way affect the velocity ratio. These gears are called **idler gears.** If the number of gears in the train are even then the direction of rotation of the last gear is reversed and if the number of gears in the train are odd then the direction of rotation of the last gear remains the same. Idler gears are used for two purposes : to connect gears where a large centre distance is required, and to control the directional relationship between gears.

2. Compound Gear Train. A pair of gears is compound if they have a common axis and are integral. A compound gear train is a gear train containing compound gears. Consider the compound gear train shown in Fig. 12.27. Gears 2 and 3 are on the same shaft.

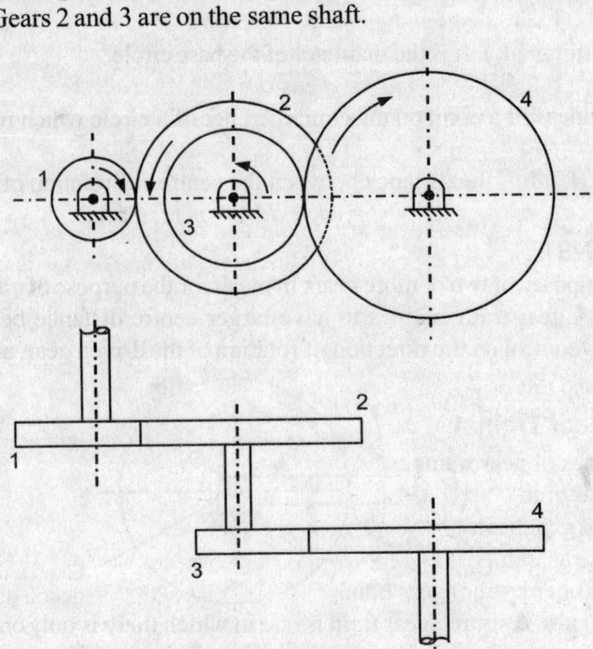

Fig. 12.27 Compound gear train

$$n_2 / n_1 = -z_1 / z_2$$
$$n_4 / n_3 = -z_3 / z_4$$

Multiplying, we get

$$n_2 / n_1 \times n_4 / n_3 = z_1 / z_2 \times z_3 / z_4$$

But $n_2 = n_3$

Thus
$$n_4 / n_1 = z_1 / z_2 \times z_3 / z_4 = (z_1 \ z_3)/(z_2 \ z_4)$$

Speed of driven gear/Speed of driving gear
= Product of teeth of driving gears/Product of teeth of driven gears

The ratio given by above equation is called the train value.

3. Reverted Gear Train. In a reverted gear train, the first and the last gear are coaxial. Reverted gear trains are used in automotive transmission, lathe back gears, industrial speed reducers, and in clocks where the minute and hour hand shafts are coaxial. In the reverted gear train shown in Fig. 12.28, we have

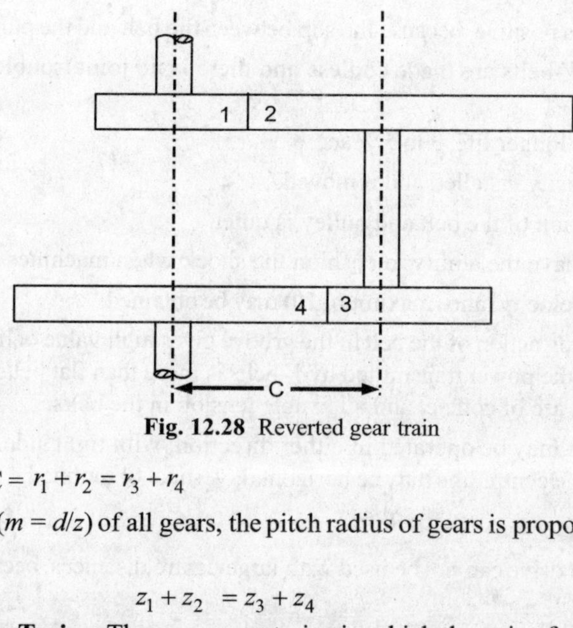

Fig. 12.28 Reverted gear train

Centre distance, $C = r_1 + r_2 = r_3 + r_4$

For same module ($m = d/z$) of all gears, the pitch radius of gears is proportional to the number of teeth. Hence

$$z_1 + z_2 = z_3 + z_4$$

4. Planetary Gear Trains. These are gear trains in which the axis of one or more gears move relative to the frame. The gear at the centre is called the **Sun,** and the gears whose axes move are called the **planets.**

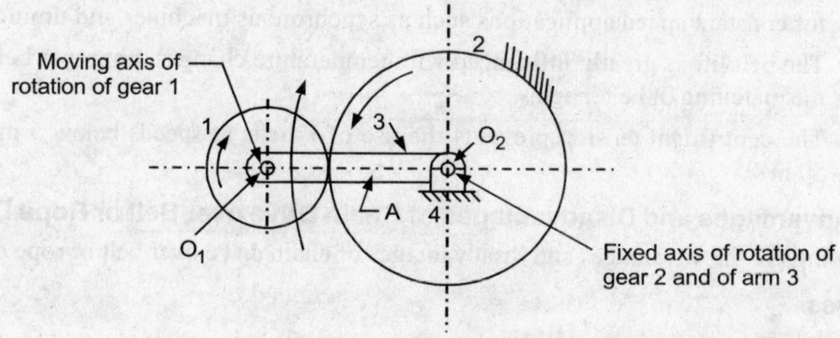

Fig. 12.29 Planetary gear train

In Fig. 12.29, arm 3 drives gear 1 about gear 2, which is a fixed external gear. Gear 1 rotates about its centre O_1 while this centre rotates about centre O_2 of the fixed gear. As gear 1 rolls on the outside of gear 2, a point on its surface will generate an epicycloid. If gear 2 happens to be an internal gear and gear 1 rolls on the inside of gear 2, then a point on the surface of gear 1 will generate a hypocycloid. Because of the curves generates a planetary gear train is often called as an epicyclic, or cyclic, gear train.

Large speed reductions are possible with epicyclic gears and if the fixed wheel is annular, a more compact unit could be obtained. Important applications of epicyclic gears are in transmission, computing devices, etc.

12.11 COMPARISON OF DRIVES

12.11.1 Advantages and Disadvantages of V-belt Drive over Flat Belt Drive

Following are the advantages and disadvantages of the V-belt drive over flat belt drive :

Advantages

1. The V-belt drive gives compactness due to the small distance between centres of pulleys.
2. The drive is positive, because the slip between the belt and the pulley groove is negligible.
3. Since the *V*-belts are made endless and there is no joint trouble, therefore the drive is smooth.
4. It provides longer life, 3 to 5 years.
5. It can be easily installed and removed.
6. The operation of the belt and pulley is quiet.
7. The belts have the ability to cushion the shock when machines are started.
8. The high velocity ratio (maximum 10) may be obtained.
9. The wedging action of the belt in the groove gives high value of limiting ratio of tensions. Therefore the power transmitted by *V*-belts is more than flat belts for the same coefficient of friction, arc of contact and allowable tension in the belts.
10. The *V*-belt may be operated in either direction, with tight side of the belt at the top or bottom. The centre line may be horizontal, vertical or inclined.

Disadvantages

1. The *V*-belt drive can not be used with large centre distances, because of larger weight per unit length.
2. The *V*-belts are not so durable as flat belts.
3. The construction of pulleys for *V*-belts is more complicated than pulleys of flat belts.
4. Since the *V*-belts are subjected to certain amount of creep, therefore these are not suitable for constant speed applications such as synchronous machines and timing devices.
5. The belt life is greatly influenced with temperature changes, improper belt tension and mismatching of belt lengths.
6. The centrifugal tension prevents the use of *V*-belts at speeds below 5 m/s and above 50 m/s.

12.11.2 Advantages and Disadvantages of Chain Drive over Belt or Rope Drive

Following are the advantages and disadvantages of chain drive over belt or rope drive :

Advantages

1. As no slip takes place during chain drive, hence perfect velocity ratio is obtained.

2. Since the chains are made of metal, therefore they occupy less space in width than a belt or rope drive.
3. It may be used for both long as well as short distances.
4. It gives a high transmission efficiency (upto 98 percent).
5. It gives less load on the shafts.
6. It has the ability to transmit motion to several shafts by one chain only.
7. It transmits more power than belts.
8. It permits high speed ratio of 8 to 10 in one step.
9. It can be operated under adverse temperature and atmospheric conditions.

Disadvantages

1. The production cost of chains is relatively high.
2. The chain drive needs accurate mounting and careful maintenance, particularly lubrication and slack adjustment.
3. The chain drive has velocity fluctuations especially when unduly stretched.

12.11.3 Advantages and Disadvantages of Gear Drives

The following are the advantages and disadvantages of the gear drive as compared to other drives, *i.e.* belt, rope and chain drives :

Advantages

1. It transmits exact velocity ratio.
2. It may be used to transmit large power.
3. It may be used for small centre distances of shafts.
4. It has high efficiency.
5. It has reliable service.
6. It has compact layout.

Disadvantages

1. Since the manufacture of gears require special tools and equipment, therefore it is costlier than other drives.
2. The error in cutting teeth may cause vibrations and noise during operation.
3. It requires suitable lubrication and reliable method of applying it, for the proper operation of gear drive.

12.12 CLUTCHES

The main function of the clutch is to control the flow of power from the driving member to the driven member as and when desired. The driven member can be stopped or started without stopping the driving member (prime mover).

The clutches may be classified as given below :

1. **Mechanical Clutches**
   (*a*) Positive clutches
      (*i*) Jaw clutch
      (*ii*) Toothed clutch

(*b*) Friction clutches

(*i*) Single disc clutch

(*ii*) Multi-disc clutch

(*iii*) Cone clutch

(*iv*) Centrifugal clutch

2. **Pneumatic clutch**
3. **Hydraulic clutch**
4. **Electromagnetic clutch**

The mechanical clutches are commonly used and we shall discuss them only.

12.12.1 Positive Clutches

These type of clutches ensure a positive engagement without any slip.

1. Jaw Clutch. The jaw clutch shown in Fig. 12.30 consists of segmental projections or dogs on one side flange and corresponding recesses on other side flange on the driven shaft. Driving flange is rigidly attached to the end of driving shaft while the other flange is keyed to the other shaft by a feather key so that it can slide on that shaft. The shifting device is accommodated on the hub of the sliding flange which makes it possible to engage with the driven flange to transmit the motion and power to the driven shaft. It is easy to operate, do not slip and cheaper. They are of the positive type.

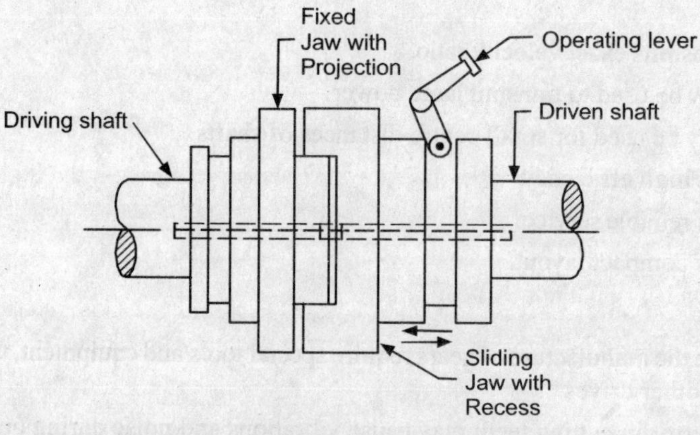

Fig. 12.30 Jaw clutch

2. Toothed Clutch. A gear-tooth shaped clutch is shown in Fig. 12.31. They have better engagement and disengagement as compared to square shape jaws.

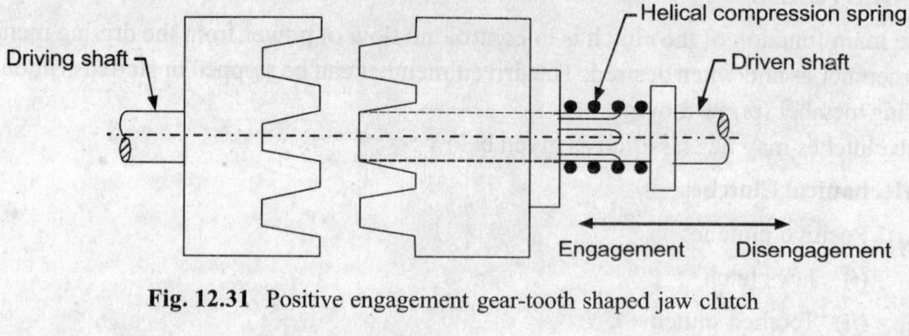

Fig. 12.31 Positive engagement gear-tooth shaped jaw clutch

12.12.2 Friction Clutches

Friction clutches transmit motion and power by virtue of friction force developed between contacting surfaces. They are used when sudden and complete disconnection of two rotating shafts is necessary.

1. Single Disc Clutch. A single disc clutch is shown in Fig. 12.32. There are two flanges, one keyed rigidly to the driving shaft by feather key or splines so that it may be moved along the shaft. Driven flange may be pressed against the driving flange by the operating lever. The torque is transmitted by friction between the flanges with friction disc. They are used in trucks and buses.

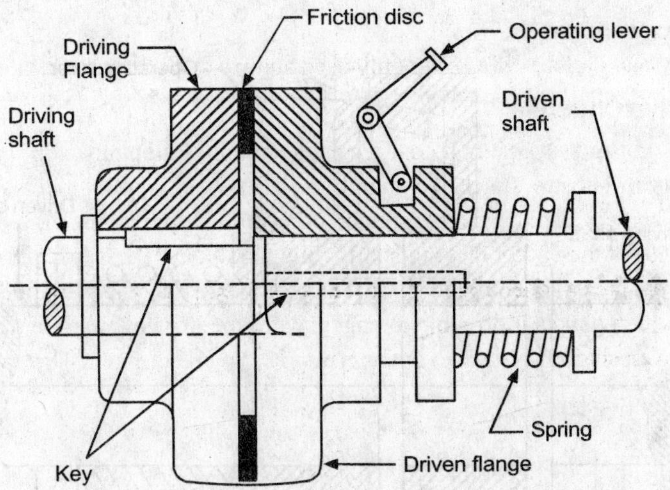

Fig. 12.32 Single disc clutch

2. Multi-plate Clutch. In a multi-plate clutch, the number of frictional linings and the metal plates are increased which increases the capacity of the clutch to transmit torque. Fig. 12.33 shows a simplified diagram of a multi-plate clutch.

The friction rings are splined on their outer circumference and engage with corresponding splines on the flywheel. They are free to slide axially. The friction material thus, rotates with the flywheel and the engine shaft. The number of friction rings depends upon the torque to be transmitted.

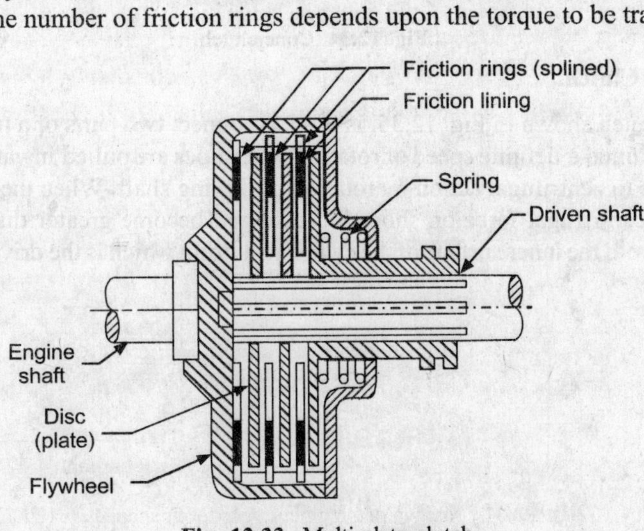

Fig. 12.33 Multi-plate clutch

The driven shaft also supports discs on the splines which rotate with the driven shaft and can slide axially. If the actuating force on the pedal is removed, a spring presses the discs into contact with the friction rings and the torque is transmitted between the engine shaft and the driven shaft. They are used in small automobiles like scooters, motor cycles, etc.

3. Cone Clutch. In cone clutch shown in Fig. 12.34 torque is transmitted using friction between internal surface of one cone mounted on driving shaft and external surface of another cone mounted on driven shaft. The cone on driven shaft can move axially by a suitable mechanism. Cone clutch can couple two shafts without excessive axial pressure but it is difficult to disengage. Therefore, it has become obsolete.

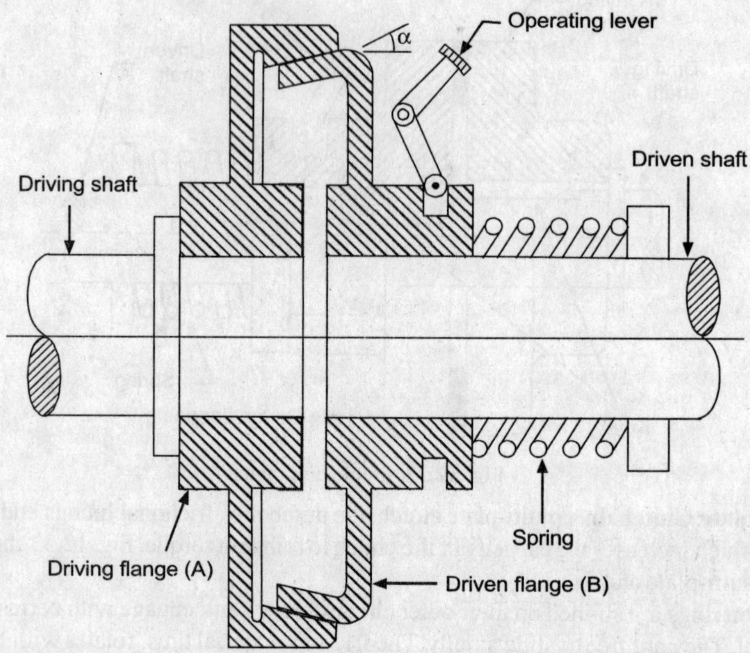

Fig. 12.34 Cone clutch

4. Centrifugal Clutch.

This type of clutch shown in Fig. 12.35, is used to connect two parts of a transmission after the driving part has attained a definite speed of rotation. The shoes are pulled inward by the springs and forced outward due to centrifugal action by rotation of driving shaft. When the driving shaft attains certain speed, the centrifugal force on shoes increase and become greater than spring force. The shoes are engaged with the inner surface of the cylindrical drum which is the driven member. It is used in mopeds.

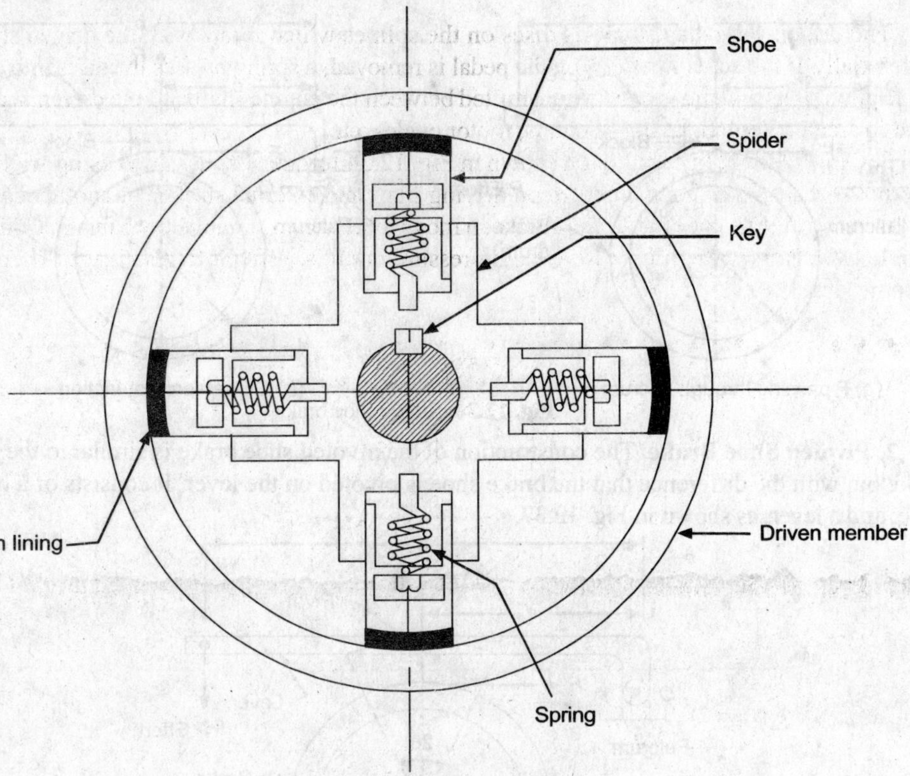

Fig. 12.35 Centrifugal Clutch

12.13 BRAKES

Brakes are the devices either to reduce or stop the speed of a moving machine component. It does so by absorbing kinetic energy or the potential energy of objects being lowered by hoists, elevators, etc.

The brakes may be classified as :

1. **Mechanical brakes**
   (*i*) Block or shoe brake
   (*ii*) Band brake
   (*iii*) Band and block brake
   (*iv*) Internal and external expanding shoe brake
2. **Hydraulic brake**
3. **Vacuum brake**
4. **Air brake**

We shall discuss only mechanical brakes.

12.13.1 Block or Shoe Brake

These brakes may be classified as :

1. **Single shoe brake**
2. **Pivoted shoe brake**
3. **Double shoe brake**

1. Single Shoe Brake. A single shoe brake is shown in Fig. 12.36. It consists of a brake drum, shoe or block, and a lever. One end of the lever is hinged at fulcrum *O* and effort *P* is applied at the other end. On the application of effort *P*, the drum motion can be slowed down or completely brought to stand still.

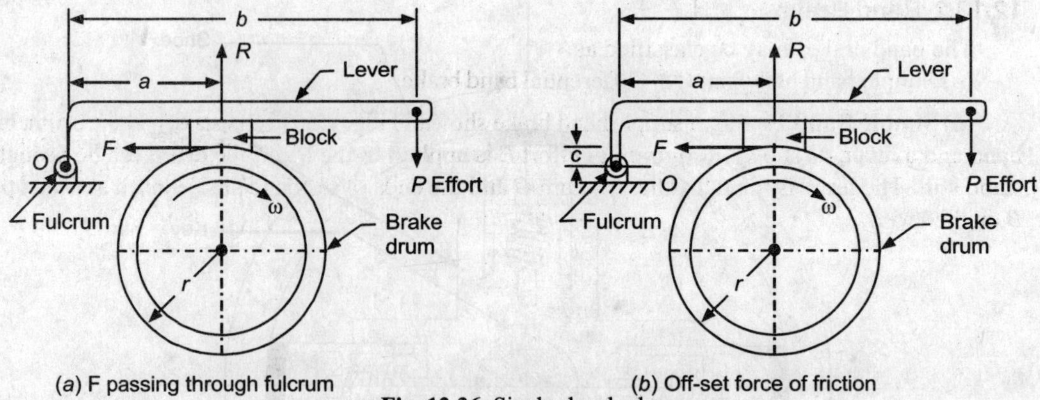

Fig. 12.36 Single shoe brake

2. Pivoted Shoe Brake. The construction of the pivoted shoe brake is similar to the single shoe brake but with the difference that the brake shoe is pivoted on the lever. It consists of a brake drum, shoe, and a lever, as shown in Fig. 12.37.

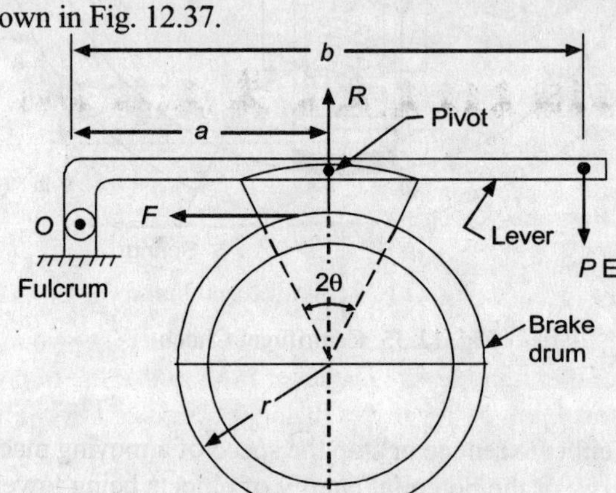

Fig. 12.37 Pivoted shoe brake

3. Double Shoe Brake. A double shoe brake is shown in Fig. 12.38. It consists of two brake shoes applied at the opposite sides of the brake drum which more or less eliminate the unbalance force on the shaft due to normal reaction.

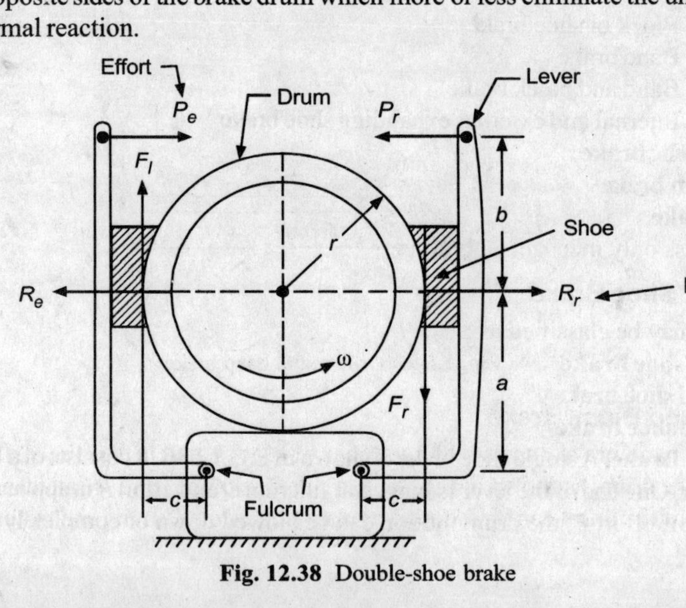

Fig. 12.38 Double-shoe brake

12.13.2 Band Brake

The band brakes may be classified as:

(*a*) Simple band brake, and (*b*) Differential band brake.

***(a)* Simple Band Brake.** A simple band brake shown in Fig. 12.39 consists of a brake drum, brake band and a lever, as shown in figure. As effort *P* is applied to the lever, the drum can be brought to stand still. The lever is hinged at the fulcrum *O* and the ends of the band are fastened at *O* and point *B* on the lever.

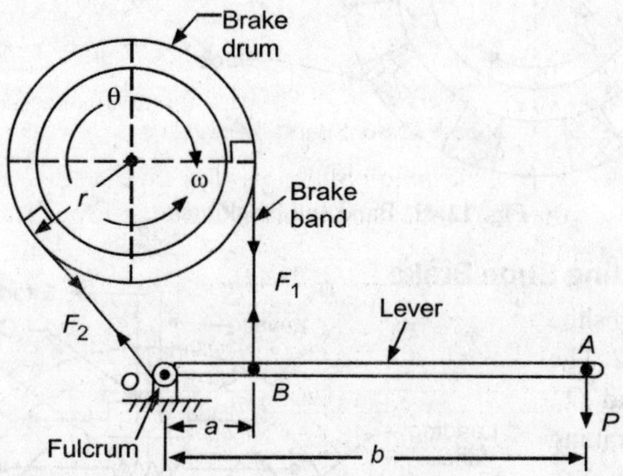

Fig. 12.39 Simple band brake

***(b)* Differential Band Brake.** The differential band brake is shown in Fig. 12.40. It consists of a brake drum, brake band a differential lever. The lever is hinged at the fulcrum *O* and ends of the band are fastened at points *B* and *C* on the lever. On the application of effort *P* to the lever to point *A,* the drum speed can be reduced.

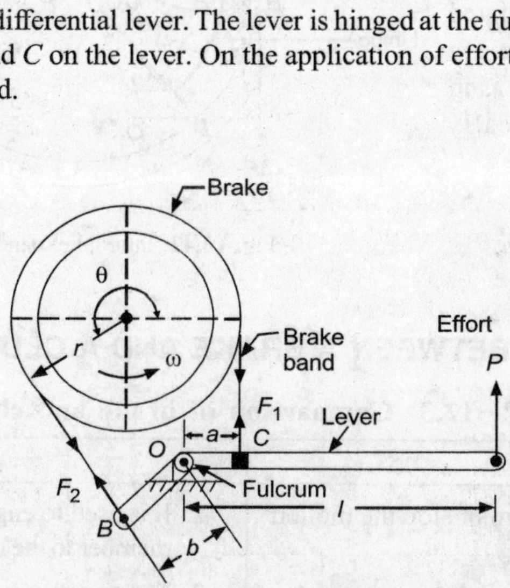

Fig. 12.40 Differential band brake

12.13.3 Band and Block Brake

A block and band brake consists of a number of blocks attached to the inside of the band and resting on the brake drum, as shown in Fig. 12.41. The ends of the band are attached to the differential lever. The speed of the drum can be reduced by applying effort *P* to the end of the lever.

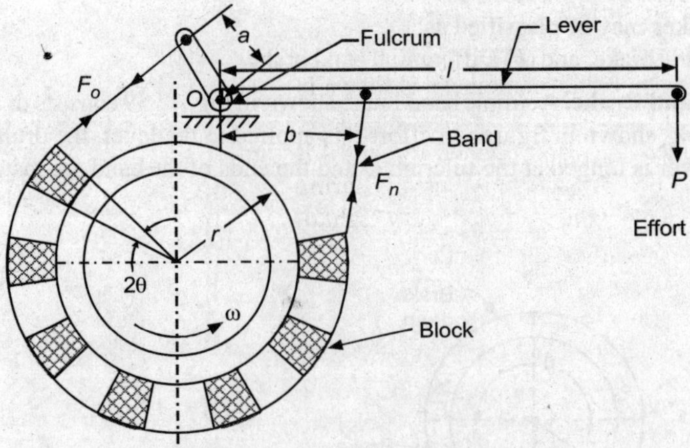

Fig. 12.41 Band and block brake

12.13.4 Internal Expanding Shoe Brake

The internal expanding shoe brake is shown in Fig. 12.42. The shoes are pinned at O_1 and O_2. The shoes are kept in non-braking position by the spring.

The brakes are applied when the cam is pressed down, in the case of a mechanically operated brake, or when the shoes are pressed on the brake drum, as in the case of a hydraulically operated brake.

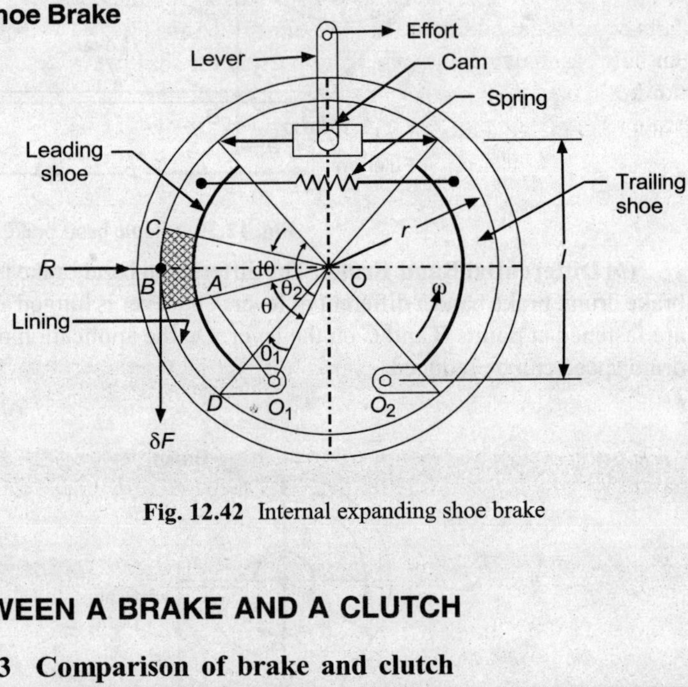

Fig. 12.42 Internal expanding shoe brake

12.14 DIFFERENCE BETWEEN A BRAKE AND A CLUTCH

Table 12.3 Comparison of brake and clutch

Brake	*Clutch*
1. It is used to slow down or stop the motion of a moving member.	1. It is used to engage or disengage the driven member to the driving member.
2. The energy of the machine member is dissipated as heat energy due to braking action.	2. Most of the energy of driving member is transmitted to the driven member. However, a small part is lost in friction.
3. The brake generally remains in disengaged position in automobiles.	3. The clutch generally remains in engaged position in automobiles.
4. In hoist, the brake normally remains in engaged position.	4. In machine tools, presses, etc. the clutch normally remains in disengaged position.

12.15 DYNAMOMETERS

1. **Belt Dynamometer**

A transmission type belt dynamometer is shown in Fig. 2.69. It consists of driving pulley 1, 2 and 3 are guide pulleys mounted on a lever pivoted at P and 4 is the driven pulley. If the effective belt tensions during running conditions are F_1 and F_2 and W the weight of the mass required for balance, then

$$Wl = (F_1 - F_2) \times 2x$$

Output torque, $T = (F_1 - F_2) \times R_4 = \dfrac{WlR_4}{2x}$

Power, $P = Tw_4/10^3$ kW

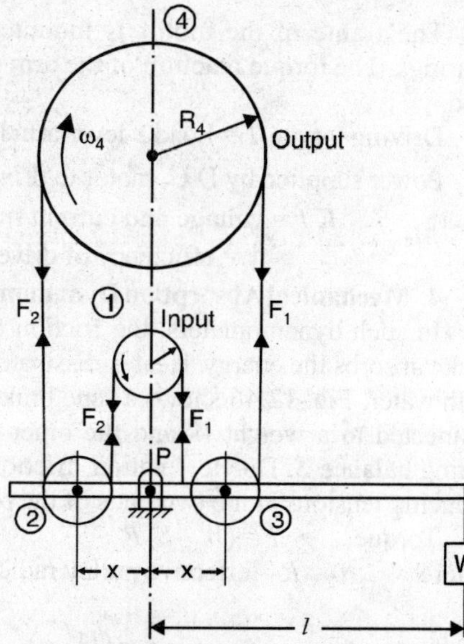

Fig. 12.43 Belt dynamometer

2. **Gear Dynamometer**

This type of arrangement, shown in Fig. 12.44, is used for small power measurement. The input and output shafts are coaxial and rotate in the same direction. The gear box casing is mounted on trunnions on the shafts. In order to prevent the casing from rotating, an external torque is applied by a weight W at distance l.

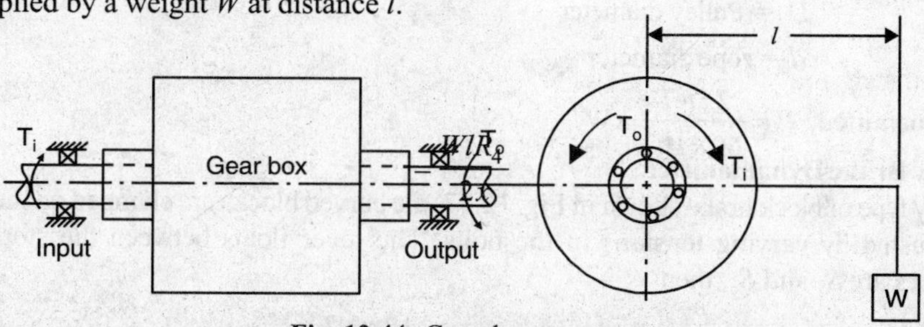

Fig. 12.44 Gear dynamometer

If T_i and T_0 are torques on the input and output shafts respectively and corresponding speeds are N_1 and N_2 with η as the overall mechanical efficiency, then

$$T_0 N_0 = \eta T_i N_i$$

$$T_0 - T_i = Wl$$

$$\therefore \quad T_i = \frac{Wl}{\eta \dfrac{N_i}{N_0} - 1}$$

Thus input torque can be obtained.

3. **Driving Type Dynamometer**

When a motor drives a driven member like a pump or a compressor, the motor can be used as both a driver and a dynamometer, as shown in Fig. 12.45.

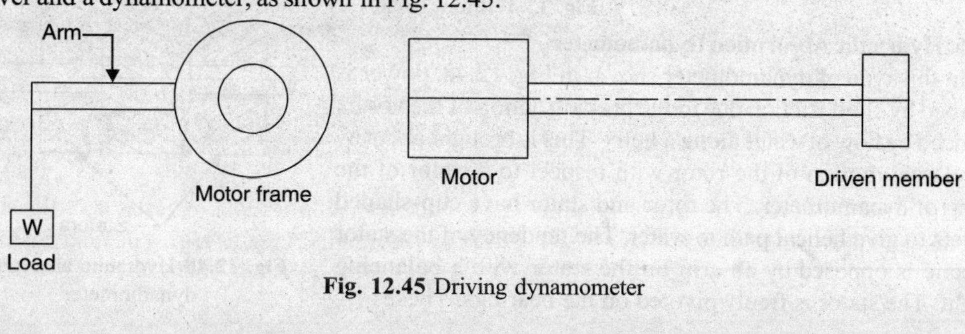

Fig. 12.45 Driving dynamometer

The frame of the motor is mounted freely on trunnion bearings. The torque reaction of the arm is balance by external load.

Driving torque, T = Load × lever arm length

Power supplied by D.C. motor to driven member = ηVI

where V, I = voltage and current supplied to motor.

η = efficiency of drive.

4. Mechanical Absorption Dynamometer

In such dynamometers, the friction of rope, band or block brake absorbs the energy. Heat is dissipated by cooling the brake with water. Fig. 12.46 shows a rope brake, one end of which is connected to a weight W and the other end is connected to a spring balance S. Due to rotation, frictional force is generated, inducing tensions at the two ends of the rope.

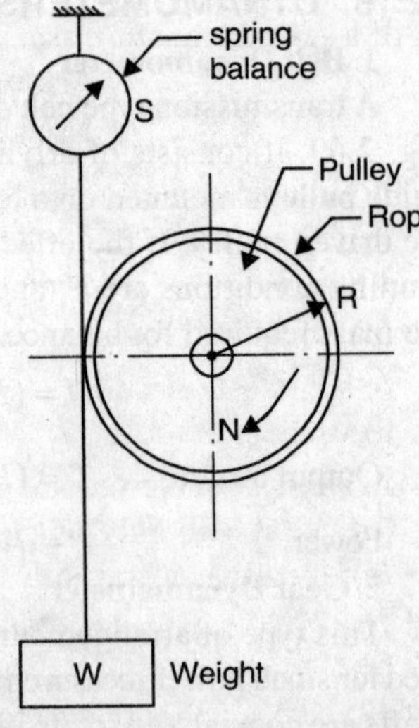

Fig. 2.72 Rope brake dynamometer

Torque, $T = (W - S)R$

where R = effective pulley radius.

$$= \frac{1}{2}(D_p + d_r)$$

D_p = Pulley diameter

d_r = rope diameter

Power transmitted, $P = \dfrac{2\pi NT}{60 \times 10^3}$ kW

5. **Prony Brake Dynamometer**

In prony type of block brake shown in Fig. 12.47, the curved blocks are clamped on the pulley rim. Load is adjusted by varying tensions in the bolts. The lever floats between the stops. If spring balance forces are S_1 and S_2, then

Torque, $T = S_1x_1 + S_2x_2$ Power, $P = \dfrac{2\pi \text{NT}}{60 \times 10^3}$ kW

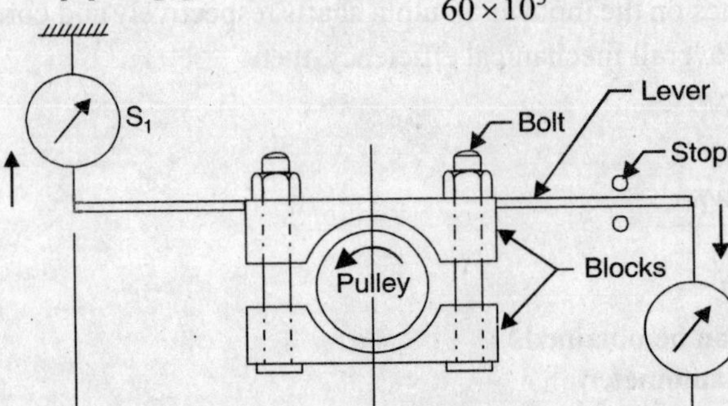

Fig. 12.47 Prony brake

6. **Hydraulic Absorption Dynamometer**

In this type of dynamometer shown in Fig. 12.48, power is absorbed by fluid friction due to the braking action of the vortex produced by flow of water along a helix. This is brought about by the relative motion of the rotor with respect to a stator of the casing of dynamometer. The rotor and stator have cup-shaped pockets to give helical path to water. The tendency of the stator to rotate is opposed by an arm on the stator with a balancing weight. The stator is freely pivoted on the bearings. These

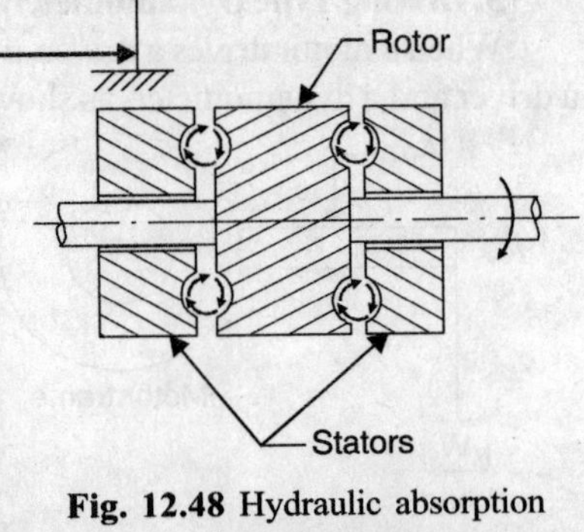

Fig. 12.48 Hydraulic absorption dynamometer

dynamometers are used in engine testing and high speed-high power applications. There is no need of a separate cooling arrangement as the control fluid acts as the coolant.

MULTIPLE CHOICE QUESTIONS

1. A suitable drive for comparatively larger centre distance is
(*a*) chain drive (*b*) gear drive (*c*) belt drive (*d*) rope drive
2. An axle is subjected to
(*a*) axial load (*b*) bending load (*c*) torsional load (*d*) bending and torsion
3. A spindle is used on
(*a*) machine tools (*b*) I.C. engines (*c*) cranes (*d*) automobiles
4. A shaft is subjected to
(*a*) axial load only (*b*) bending loads only
(*c*) torsional loads only (*d*) all of the above
5. The convex periphery of a flat belt pulley is called
(*a*) crown (*b*) bulge (*c*) web (*d*) depression
6. The standard angle of a *V*-belt is
(*a*) 30° (*b*) 35° (*c*) 40° (*d*) 45°
7. The pulley groove angle for a *V*-belt is
(*a*) equal to *V*-belt angle (*b*) less than *V*-belt angle
(*c*) more than *V*-belt angle (*d*) any one of the above
8. A *V*-belt is specified by the
(*a*) outside length (*b*) pitch length (*c*) inside length (*d*) average length
9. The drive used in a bicycle is
(*a*) belt drive (*b*) rope drive (*c*) gear drive (*d*) chain drive
10. The vertical feed mechanism in a drilling machine uses
(*a*) rack and pinion (*b*) spur gears (*c*) helical gears (*d*) bevel gears
11. The type of clutch used in trucks and buses is
(a) single plate clutch (b) multi-plate clutch
(c) cone clutch (d) centrifugal clutch
12. The clutch used in mopeds is
(*a*) single plate clutch (*b*) multi-plate clutch
(c) cone clutch (*d*) centrifigal clutch
13. The clutch normally remains in engaged position in
(*a*) automobiles (*b*) machine tools
(*c*) elevators (*d*) cranes
14. The clutch normally remains in disenged position in
(*a*) buses (*b*) machine tools
(*c*) scooters (*d*) tractors

Answers:

1. (*c*) **2.** (*b*) **3.** (*a*) **4.** (*d*) **5.** (*a*) **6.** (*c*) **7.** (*b*) **8.** (*c*) **9.** (*d*) **10.** (*a*) **11.** (*a*) **12.** (*d*) **13.** (*a*) **14.** (*b*)

EXERCISES

12.1 Differentiate between a shaft and an axle?

12.2 What are the various types of shafts?

12.3 What are the power transmission elements?

12.4 What are the various types of flat belt drives?

12.5 What are the various types of flat belts?

12.6 What are the various types of pulley used for flat belt?

12.7 Differentiate between a fast and loose pulley.

12.8 Describe various types of grooved pulleys for *V*-belt.

12.9 What are the advantages of gear drive over other drives?

12.10 List various types of gears.

12.11 Describe various types of gear trains.

12.12 State the (*i*) applications (*ii*) Advantages (*iii*) disadvantages of (*a*) Rope drive (*b*) Chain drive (*c*) Gear drive.

12.13 State the advantages of chain drive over belt drive. Draw neat sketches of simple and compound gear train.

12.14 Explain with neat sketch the working of various types of belt drives used for motion and power transmission.

12.15 State the different types of pulleys used in power transmission and define the velocity ratio of pulleys.

12.16 What are the main elements of power transmission ? Mention them.

12.17 What is belt drive ? Describe briefly types of belts.

12.18 With a neat sketch describe open belt and cross belt drive.

12.19 Explain with a neat sketch the working of fast and loose pulley.

12.20 Write a short note on : Types of belt drive.

12.20 Write a short note on : Types of belt drive.11

12.21 Explain with a neat sketch fast and loose pulley.

12.22 How do you classify mechanical clutches ? Describe multi plate type friction clutch.

12.23 Describe the working of a cone clutch. Why a cone clutch has become obsolete ?

12.24 What is the function of clutch in automabile vehicle ? Which type of clutch is used in two - wheelers and why ?

12.25 With a simple stretch explain the working of disc clutch.

12.26 List the various types of dynamometers.

12.27 Explain the working of an absorption type dynamometer.

12.28 Explain the working of a transmission type dynamometer.

13 STRESSES AND STRAINS

13.1 INTRODUCTION

The subject of ***Mechanics of Solids*** involves analytical methods for the determination of strength, stiffness (deformation characteristics), and stability of the various load carrying members. The subject has wide applications in all branches of engineering.

The behaviour of a member subjected to the action of force depends not only on the fundamental laws of a Newtonian mechanics that govern the equilibrium of forces, but also on the mechanical characteristics of the materials of which the member is fabricated. The latter part comes from laboratory experiments. Mechanics of solids is a blended science of experimental and Newtonian postulates of analytical mechanics.

13.2 ENGINEERING MATERIALS

Materials useful for making engineering products are generally solids. They may be broadly classified as:

- Metals and their alloys
- Plastics or polymers
- Ceramics, and
- Composites

The metallic materials may be further classified as:

- Ferrous materials
- Non-ferrous materials

Ferrous materials are mainly cast irons and steel. Steel is mainly used for making engineering products. It is ductile in nature. Cast irons have applications in special situations where energy is to be absorbed or vibrations are to be decreased. Cast irons are brittle in nature.

Non-ferrous metals having engineering application are : Copper, aluminum, lead, tin, nickel, antimony, magnesium, manganese, zinc and titanium. They are basically used as alloying elements in ferrous materials.

Plastics are synthetic organic materials, resins, which are solid in finished from but, at some stage in their processing, are fluid enough to be shaped by application of heat and pressure. There are two basic types of plastics.

Thermoplastics: which can be resoftened repeatedly without undergoing a change in chemical composition, *e.g*, nylon, polyethylene, PVC, Plexiglass, Teflon, etc.

Thermosetting Plastics : which undergo a chemical change with application of heat and pressure and cannot be resoftened, *e.g*, araldite, bakelite, duraplex, beckosol, etc.

Ceramic: is a solid formed by the combination of metallic and non-metallic elements. They are well known for their thermal resistance, insulative properties, high hardness, and light weight and brittleness. Typical examples of ceramics are : alumina, silica, chromium oxide, tungsten carbide, boron carbide, boron nitride, etc.

Composite Materials : are multifunction systems that provide characteristics not obtainable from any discrete material. They are cohesive structures made by physically combining two or more compatible materials, differing in composition and characteristics and sometimes in form. They are ideal materials where high strength – to – weight and stiffness ratios are required. The most important class of composite materials is fibre reinforced plastics (FRP). The commonly used fibres are : glass, silica, boron, carbon, nylon and metals, etc.

13.3 PROPERTIES OF ENGINEERING MATERIALS

The major properties of engineering materials are:

- **Chemical :** Composition, microstructure, crystal structure and corrosion resistance.
- **Physical :** Melting point, density, specific gravity, curie point, refractive index, thermal conductivity, thermal expansion, heat dissipation, water absorption, dielectric strength, electrical resistivity, specific heat, Poisson's ratio and colour.
- **Mechanical :** Tensile (ultimate) strength, compressive strength, yield strength, Young's modulus, flexural strength, shear strength, percentage elongation, percentage reduction in area, hardness, impact strength, endurance limit, creep strength,
- **Dimensional :** Roughness, waviness, lay, camber and microtopography .
- **Technological :** Machinability, weldability, castability, ductility and hardenability.

13.4 CONCEPT OF MECHANICAL PROPERTIES

Tension: When the external forces acting on a body are trying to pull or elongate it, the forces are said to be tensile, as shown in Fig 13.1 A tensile force produces tension (or tensile stress) in the body. Tensile (or ultimate) strength is the ratio of maximum tensile load, which a member can withstand, to its original area of cross-section.

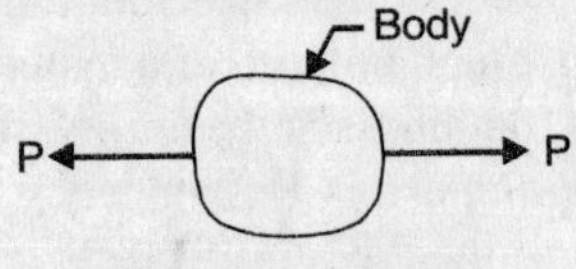

Fig. 13.1 Tensile force.

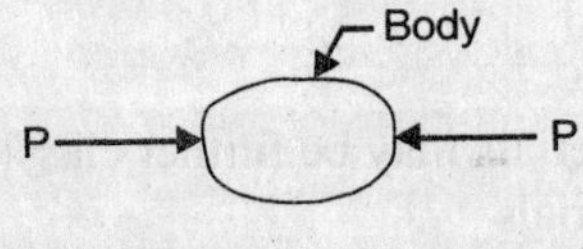

Fig. 13.2 Compressive force.

Compression : When the external forces acting on a body are trying to push or shorten the body, the forces are said to be compressive, as shown in Fig. 13.2. A compressive force produces compression (or compressive stress) in the body. Compressive strength is the ratio of the maximum compressive load, which a member can withstand, to its original area of cross-section.

Torsion : Two equal and opposite forces (torque) acting on a body produces twist in a body, as shown in Fig 13.3. Torsion produces torsional shear stress in a body.

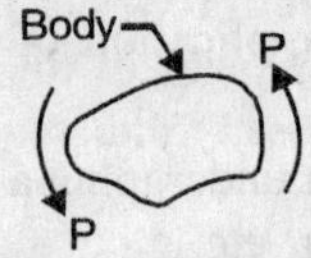

Fig. 13.3 Torsion.

Fig. 13.4 Bending loads.

Bending : When a body is subjected to such a system of loads so that the top and bottom fibres experience opposite nature of stresses, the body is said to be subjected to bending. Consider a rectangular beam simply supported at the ends and carrying two concentrated loads, as shown in Fig. 13.4. The bent shape of the beam is shown by dotted lines. The top fibres are under compression and bottom fibres under tension.

Hardness : It is the resistance of a material to plastic deformation, abrasion, scratching or indentation. Hardness scales used are: Rockwell, Brinell, Vickers's and Moh's scale.

Fatigue : It is a phenomenon leading to fracture (or reduced strength) under repeated or fluctuating cyclic stresses below the tensile strength of the material. The number of cycles of stress that can be sustained prior to failure of a specified nature for a stated stress condition is called fatigue life. The maximum stress below which a material can presumably endure an infinite number of stress cycles, is called endurance limit. The plot of log of stress v's log of number of cycles is called the Wohler curve, as shown in Fig. 13.5.

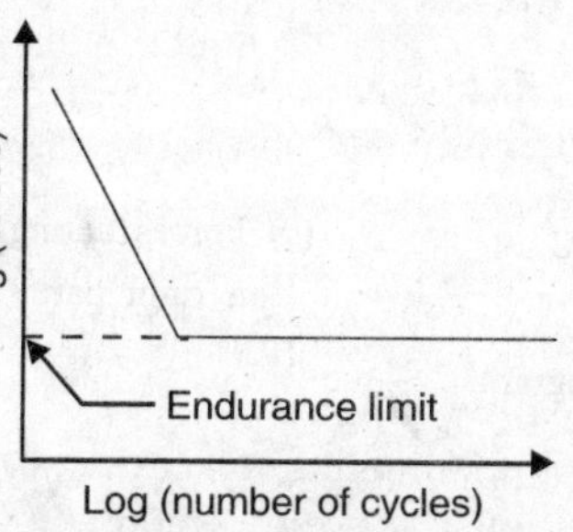

Fig. 13.5 Wohler curve.

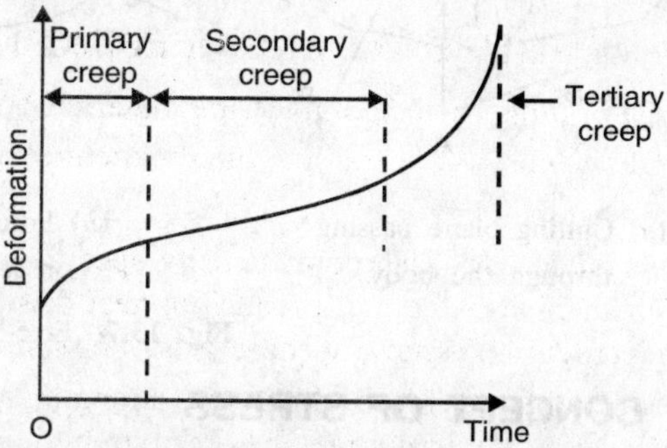

Fig. 13.6 The creep curve.

Creep : It is the continuous flow of a material under constant temperature and stress. Creep strength is the constant nominal stress that will cause a specified quantity of creep in a given time at constant temperature. Creep is of three types. Primary (or transient or cold) creep is decreasing creep rate, because of the work hardening resulting from deformation. Secondary creep is a steady state creep. It is also called hot or viscous creep. Tertiary or accelerated creep is the final stage at an accelerating rate, preceding fracture. The various types of creep are shown in Fig. 13.6.

Impact : When a member is subjected to sudden loading (load falling on to the member from a height), usually in bending, tension or torsion, the member is said to be subjected to impact load. Impact strength is the amount of energy required to fracture a given volume of a material. Impact tests are performed to determine the toughness of a material. Toughness is the ability of a material to absorb energy and deform plastically before fracture. Brittleness is the tendency to fracture without appreciable deformation. The tests performed to determine the impact strength are : Izod, Charpy, tension and torsion impact tests.

3.5 FREE-BODY DIAGRAM

Mechanics of solids is concerned with the investigation of the internal resistance of a ody; that is, the nature of forces set up within a body to balance the effect of the externally pplied forces. For this purpose, a complete diagrammatic sketch of the member to be investigated s prepared on which all the external forces acting on it are shown at their respective points f application. Such a sketch is called a free-body diagram. All forces acting on the body, ncluding the reactive forces caused by the supports and the self weight of the body due to its ass are considered external forces. The forces acting on the body satisfy the equations of static quilibrium.

Consider a body shown in Fig. 13.7 (*a*) subjected to forces P_1 to P_4. Pass an arbitrary section *BCD* through the body, completely separating it into two parts as shown in Fig. 13.7(*b*) and (*c*). the body as a whole is in equilibrium, any part of it must also be in equilibrium. The external rces are balanced by the internal forces F_1 to F_3. The free-body diagram is very helpful in etermining the stresses in machine members and structural components.

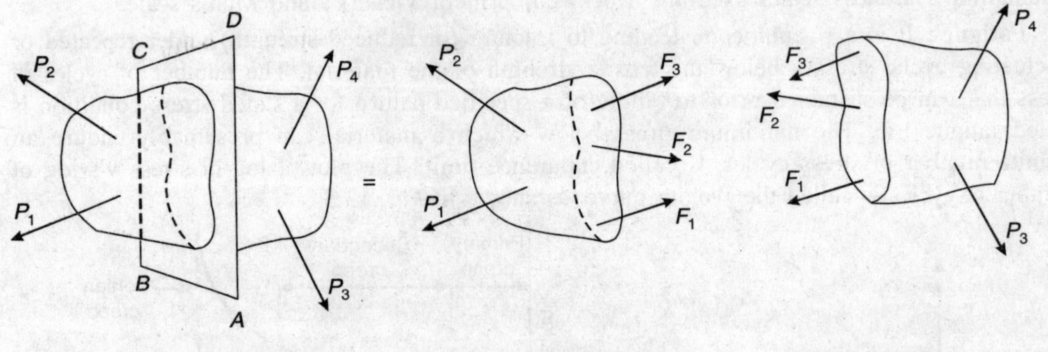

(*a*) Cutting plane passing through the body (*b*) Forces acting on left part (*c*) Forces acting on right part

Fig. 13.7 Free-body diagram.

13.6 CONCEPT OF STRESS

Consider a body subjected to the action of external forces P_1 to P_5 as shown in Fig. 13.8(*a*). If a section 1-1 is passed to divide the body into two parts *I* and *II*, then the free body diagrams of both the parts are shown in Fig. 13.8 (*b*) and (*c*) respectively. The internal forces F_1 to F_3 developed at the cut must balance the external forces applied on the body.

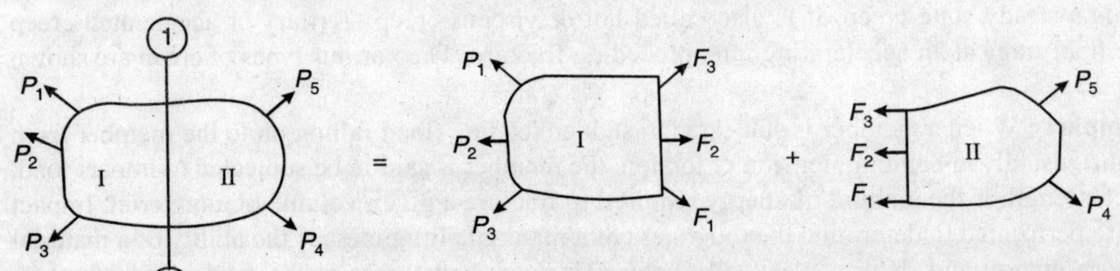

Fig. 13.8 Free body diagrams.

In general, the internal forces acting on infinitesimal areas of a cut are of varying magnitude and directions. They vary from point to point and are inclined with respect to the plane of th section. In is advantageous to resolve these forces perpendicular and parallel to the sectio considered. Now again consider part *I* of the body with the force ΔP acting on an infinitesima area ΔA as shown in Fig. 13.9(*a*). The components of this force ΔP are shown in Fig. 13.9 (*b* along the *x*-axis perpendicular to the cutting plane and the other two perpendicular axes *y* an *z*, lying in the plane of the section. By definition, normal stress along the *x*-axis is:

$$\tau_{xx} \text{ or } \sigma_x = \lim_{\Delta A \to 0} \frac{\Delta P_x}{\Delta A}$$

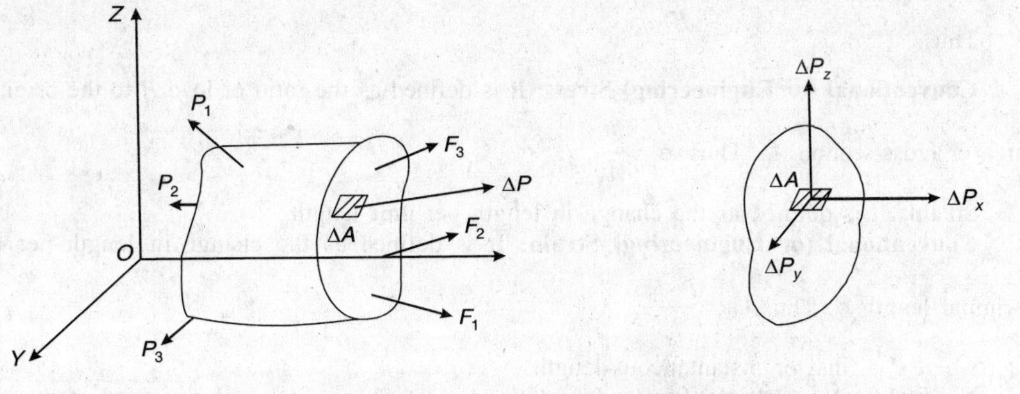

(*a*) Force ΔP acting on infinitesimal area ΔA.

(*b*) Components of force ΔP along rectangular axes

Fig. 13.9 Defining stress at a point

and is called the normal stress in the *x*-direction at a point. It is customary to refer to normal stresses that cause tension on the surface of a section as **tensile stresses.** On the other hand, those that are pushing against it are called **compressive stresses.** By definition, stress is considered to act at a point.

The stresses lying in the plane of the section are called **shear stresses.** By definition, shear stress in the plane *xy* is:

$$\tau_{xy} = \lim_{\Delta A \to 0} \frac{\Delta P_y}{\Delta A}$$

and in the plane *xz* is:

$$\tau_{xz} = \lim_{\Delta A \to 0} \frac{\Delta P_z}{\Delta A}$$

Stress is a tensor quantity i.e. its magnitude, plane as well as direction have to be stated to define the stress completely. The first subscript indicates the plane parpendicular to the axis and the second subscript designates the direction of the stress component. Thus we find that stress is the internal resistance offered by the body and is the force per unit area. The units of stress are N/m^2 or MPa.

13.7 DEFINITIONS

Stress: All bodies offer an equal internal resistance to the externally applied forces. The magnitude of the resisting force per unit area is called stress. Its SI unit are N/mm^2 or N/m^2.

- **Tensile (or ultimate) strength:** It is the ratio of maximum load to original area of cross-section.
- **Yield Strength:** It is the stress at which a material exhibits a specifed deviation from proportionality of stress and strain.
- **Compressive strength:** It is the maximum compressive stress based upon the original area that a material is capable of withstanding.

Normal Stress: The stress developed on a plane normal to it is called normal stress. It is equal to the forces acting on the body per unit normal area. Thus, $\sigma_n = \frac{P}{A}$. The normal stress may be tensile or compressive depending upon the force to be either of the pull or push type. Tensile and compressive stresses together area called ***direct stresses***.

Shear Stress: It is defined as the ratio of shear force to area parallel to the force.

Thus, $\tau = \frac{F}{A}.$

Conventional (or Engineering) Stress: It is defined as the ratio of load P to the origianal area or cross-section A_0. Thus $\sigma = \frac{P}{A_0}$.

Strain: It is defined as the change in length per unit length.

Conventional (or Engineering) Strain: It is defined as the change in length per unit original length l_0. Thus, $\varepsilon = \frac{l - l_0}{l_0}$.

where l = final or instantaneous length

Natural (or logarithmic) Strain: It is defined as the change in length per unit instantaneous length. Thus, $\bar{\varepsilon} = \ln\left(\frac{l}{l_0}\right) = \ln(1+\varepsilon)$.

True Stress: It is defined as the ratio of load P to the instantaneous area of cross-section. Thus, $\bar{\sigma} = \frac{P}{A} = \sigma(1+\varepsilon)$.

Normal Strain: It is the strain produced under the action of direct stresses.

Shear Strain: It is the strain produced under the action of shear stresses. Shear strain, $\gamma = \tan\phi = \phi = \frac{\delta l}{l}$.

Percentage Elongation: It is the change in length per unit original length expressed as percentage, *i.e.*, $\frac{l - l_0}{l_0} \times 100$, where l = final length and l_0 = original length.

Percentage Reduction of Area: It is defined as the change in area per unit original area of cross-section A_0 expressed as a percentage, *i.e.*, $\frac{A - A_0}{A_0} \times 100$, where A = final area of cross-section.

- **Gauge length:** It is the specified length of the test piece on which elongation is measured during the test. The gauge length is generally $5.65\ \sqrt{A_0}$, where A_0 is the original area of cross-section of the specimen.
- **Ductility:** It is the ability of a meterial to deform plastically before fracture. This enables the materials to be drawn into wires.
- **Melleability:** It is the property of metal and alloys to deform plastically under compression without rupture. This enables the material to be rolled into thin sheets.
- **Yield point:** The first stress in a material at which an increase in strain results without an increase in stress. Mild steel has two yield points : Upper and lower yield points. This term refers to ductile materials only.
- **Necking:** A form of shape of the localized reduction in cross-section occurring in a ductile material untenal under tension before fracture is called necking.
- **Permanent set:** The plastic deformation which persists after the removal of the applied load is called the permanent set.

- **Modulus of Elasticity, E:** It is defined as the ratio of normal stress to normal strain within the elastic limits. $E = \frac{\sigma}{\varepsilon} = \frac{Pl_0}{A_0 \delta l}$.

Modulus or Rigidity, G: It is defined as the ratio of shearing stress τ to shearing strain γ.

$$G = \frac{\tau}{\gamma}$$

13.8 BAR OF VARYING CROSS-SECTION

Consider a bar of varying cross-section shown in Fig. 13.11 (*a*). The free body diagrams for the various parts are shown in Fig. 13.10 (*b*). The stresses in the various parts are:

$$\sigma_{AB} = \frac{P_1}{A_1},\ \sigma_{BC} = \frac{P_1 - P_2}{A_2},\ \sigma_{CD} = \frac{P_4}{A_3}$$

The elongation of various parts are:

$$(\delta l)_{AB} = \frac{P_1 l_1}{A_1 E'} (\delta l)_{BC} = \frac{(P_1 - P_2) l_2}{A_2 E},\ (\delta l)_{CD} = \frac{P_4 l_3}{A_3 E}$$

In general, total elongation,

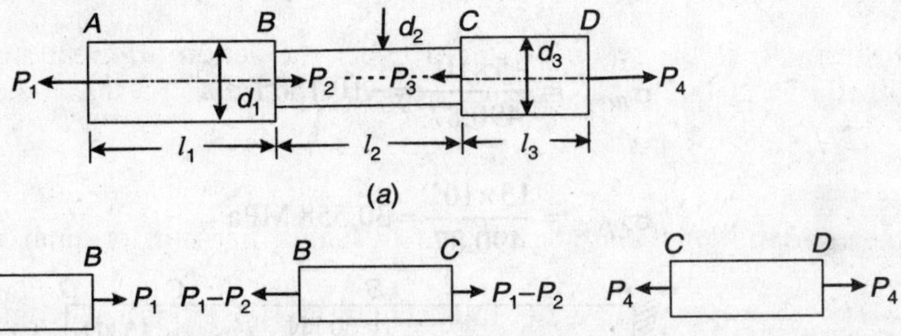

(*a*)

(*b*) Face body diagrams

Fig 13.10 Bar of varying cross-section.

$$\left[\delta l = \sum_{i=1}^{n} \frac{P_i l_i}{A_i E_i} \right] \quad ...\ (13.1)$$

Example 13.1 *A steel bar 20 mm diameter is loaded as shown in Fig. 13.11 (a). Determine the stresses in each part.*

Solution. The free body diagrams of the three portions are shown in Fig. 13.11 (*b*).

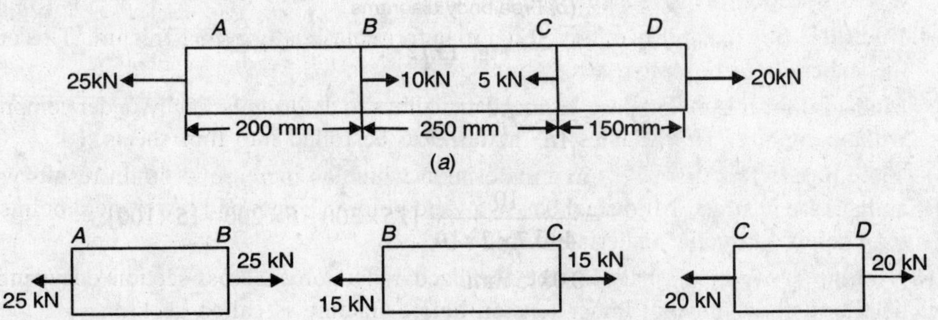

(*a*)

(*b*) Free body diagrams

Fig. 13.11

Area of cross-section, $A = \frac{\pi}{4} \times 400 = 314.16 \text{ mm}^2$

Stresses in various parts are :

$$\sigma_{AB} = \frac{25 \times 10^3}{314.16} = 79.59 \text{ MPa}$$

$$\sigma_{BC} = \frac{15 \times 10^3}{314.16} = 47.75 \text{ MPa}$$

$$\sigma_{CD} = \frac{20 \times 10^3}{314.16} = 63.66 \text{ MPa}$$

Example 13.2 *A steel bar of 25 mm diameter is loaded as shown in Fig. 13.12 (a). Calculate the stress in each portion and the total elongation. Take E = 200 GPa.*

Solution. The free body diagram of various portions is shown in Fig. 13.12 (*b*).

$$A = \frac{\pi}{4} \times 25^2 = 490.87 \text{ mm}^2$$

$$\sigma_{AB} = \frac{25 \times 10^3}{490.87} = 50.93 \text{ MPa}$$

$$\sigma_{BC} = \frac{-5 \times 10^3}{490.87} = -10.186 \text{ MPa}$$

$$\sigma_{CD} = \frac{15 \times 10^3}{490.87} = 30.558 \text{ MPa}$$

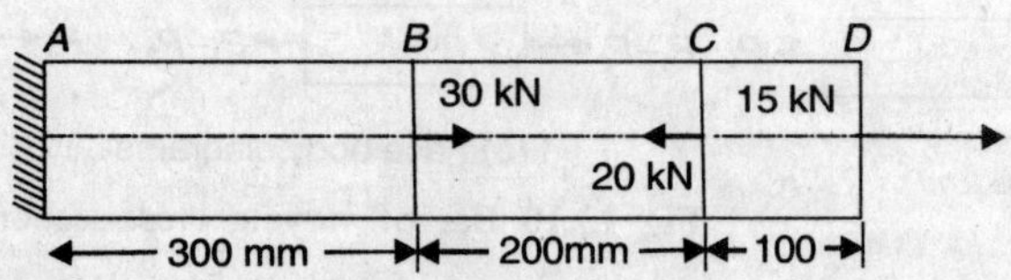

(*a*)

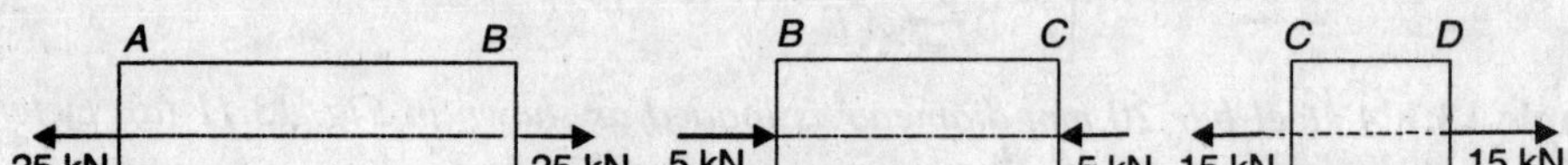

(*b*) Free body diagrams

Fig. 13.12

$$\delta l = \frac{1}{AE} \Sigma P_i l_i$$

$$= \frac{10^3}{490.7 \times 2 \times 10^5} [25 \times 300 - 5 \times 200 + 15 \times 100]$$

$$= 0.0815 \text{ mm.}$$

Example 13.3 *A stepped bar is loaded as shown in Fig. 13.13 (a). Calculate the stress in each part and total elongation. E = 200 GPa.*

Solution. The free body diagram is shown in Fig. 13.13 (*b*)

$$A_1 = \frac{\pi}{4} \times 25^2 = 490.87 \text{ mm}^2$$

$$A_2 = \frac{\pi}{4} \times 20^2 = 314.16 \text{ mm}^2$$

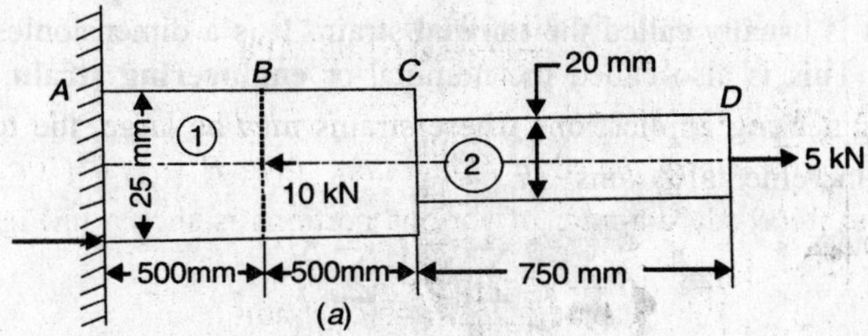

(*a*)

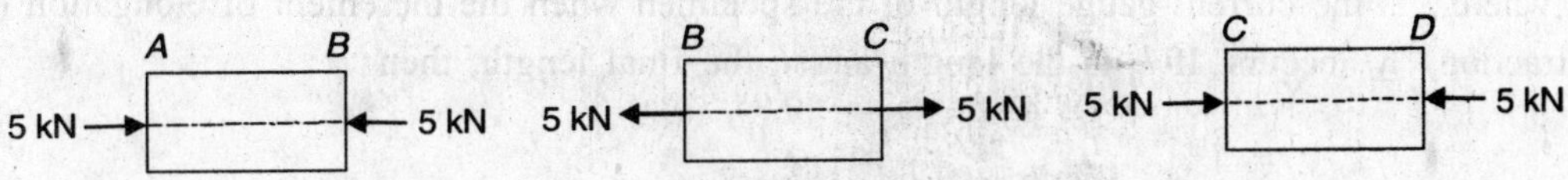

(b) Free body diagrams

Fig. 13.13

$$\sigma_{AB} = -\frac{5000}{490.97} = -10.186 \text{ MPa}$$

$$\sigma_{BC} = +\ 10.186 \text{ MPa}$$

$$\sigma_{CD} = +\frac{5000}{314.16} = +15.915 \text{ MPa}$$

$$\delta l = \frac{1}{E}\sum \sigma_i l_i = \frac{1}{2\times 10^5}[-10.186\times 500 + 10.186\times 500 + 15.915\times 750]$$

$$= 0.0597 \text{ mm}$$

Example 13.4 *A pillar is shown in Fig. 13.14. The total contraction of the pillar is 0.25 mm. If E_s = 200 GPa and E_{CI} = 120 GPa, find the value of load P.*

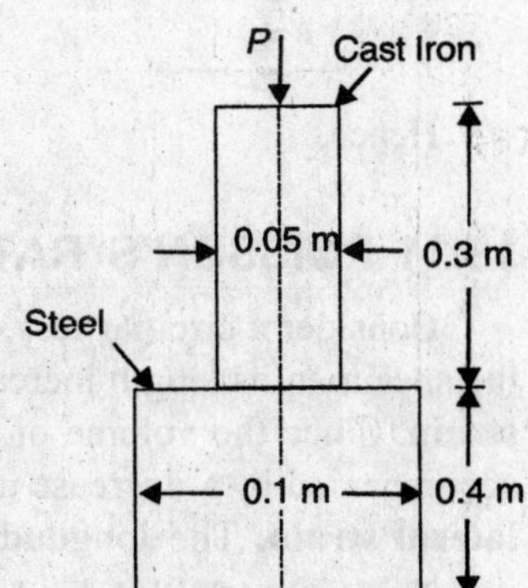

Fig. 13.14

Solution. Total contraction, $\delta = \frac{Pl_1}{A_1E_1} + \frac{Pl_2}{A_2E_2}$

$$0.25 \times 10^{-3} = \frac{4P}{\pi}\left[\frac{0.3}{(0.05)^2\times 120\times 10^9} + \frac{0.4}{(0.1)^2\times 200\times 10^9}\right]$$

$$= \frac{4P}{\pi}[1 + 0.2]\times 10^{-9}$$

$$P = 0.25\times 10^{-3}\times 10^9 \times \frac{\pi}{4}\times\frac{1}{1.2}$$

$$= 163.6246 \text{ kN}$$

13.9 CONCEPT OF STRAIN

If l_0 is the initial gauge length and l is the observed length under a given load, the gauge elongation, $\Delta l = l - l_0$. The elongation (or contraction) per unit of the initial gauge length is given as:

$$\varepsilon = \frac{\Delta l}{l_0} \quad \text{... (13.2)}$$

This expression defines the tension (or compression) strain. Since this is associated with the normal stress, it is usually called the **normal** strain. It is a dimensionless quantity (mm/mm or µm/m or m/m). This is also called the nominal or **engineering strain**.

In some engineering applications where strains may be large, the total strain is defined as the sum of the incremental strains $\Delta\bar{\varepsilon}$.

Thus,
$$\bar{\varepsilon} = \sum \Delta\bar{\varepsilon} = \sum \frac{\Delta l}{l} \quad \text{... (13.3)}$$

where l is the current gauge length of the specimen when the increment of elongation (or contraction) Δl occurs. If l_0 is the length and l_f the final length, then

$$\bar{\varepsilon} = \int_{l_0}^{l_f} \frac{dl}{l} = \ln\left[\frac{l_f}{l_0}\right] = \ln(1+\varepsilon) \quad \text{... (13.4)}$$

This strain, obtained by adding up the increments of strains which are based on the current dimension of a specimen is called a **natural** or **true** or **Logarithmic strain.**

13.10 TRUE STRESS

The true stress $\bar{\sigma}$ is related to the instantaneous cross-sectional area A. For the applied force P,

$$\bar{\sigma} = \frac{P}{A} \quad \text{... (13.5)}$$

For volume constancy of specimen, $A_0 l_0 = Al$ and $l = l_0(1+\varepsilon)$.

$$\therefore \quad \frac{A_0}{A} = \frac{l}{l_0} = 1+\varepsilon$$

Hence,
$$\bar{\varepsilon} = \frac{P}{A} = \frac{P}{A_0}(1+\varepsilon) = \sigma(1+\varepsilon) \quad \text{... (13.6)}$$

13.11 POISSON'S RATIO

Consider a circular test specimen of diameter d and length l. If a tensile load P is applied to the specimen its length increases. The increase in length per unit length is called the **longitudinal strain**. Since the volume of the specimen remains constant, therefore, the increase in length is accompanied by a decrease in diameter. This decrease in diameter per unit diameter is called the **lateral strain.** The longitudinal and lateral strains are of opposite nature.

The ratio of the lateral strain to the longitudinal strain is called the Poisson's ratio. It is denoted by ν(Greek Nue). Thus

$$\text{Poission's ratio, } \nu = \frac{\text{Lateral strain}}{\text{Longitudinal strain}} = \frac{-(\delta d / d)}{\delta l / l} \quad \text{... (13.7)}$$

Poisson's ratio for most of the materials varies from 0.25 to 0.40.

13.12 VOLUMETRIC STRAIN

It is the ratio of the change in volume ΔV of the body to its original volume V_0, when subjected to hydrostatic stress.

$$\text{Volumeter strain, } \varepsilon_v = \frac{\Delta V}{V_0} \quad \text{... (13.8)}$$

13.13 HOOKE'S LAW AND ELASTIC MODULI

According to this Law, normal stress is directly proportional to normal strain, within the elastic (strictly speaking proportional) limits.

$$\sigma \propto \varepsilon$$

or

$$\sigma = E\varepsilon \quad \text{... (13.9)}$$

The constant of porportionality, E, is called the **modulus of elasticity** or elastic modulus or Young's modulus.

$$E = \frac{\text{normal stress}}{\text{normal strain}} = \frac{\sigma}{\varepsilon} = \frac{Pl_0}{A_0 \Delta l} \quad \text{... (13.10)}$$

The ratio of shearing stress to shearing strain is called the **modulus of rigidity**, G.

$$G = \frac{\text{shear stress}}{\text{shear strain}} = \frac{\tau}{\gamma} \quad \text{... (13.11)}$$

The ratio of the uniform hydrostatic stress of volumetric strain is called the **bulk modulus**, K.

$$K = \frac{\text{hydrostatic stress}}{\text{volumetric strain}} = \frac{\sigma_h}{\varepsilon_v} \quad \text{... (13.12)}$$

13.14 EXTENSION OF A TAPERED BAR

Consider a bar of length L tapering from d_2 to d_1 and subjected to axial tensile load P, as shown in Fig. 13.15. At a distance x from diameter d_2, the diameter of the bar is:

$$d = d_2 - \left(\frac{d_2 - d_1}{L}\right)x$$

Area of cross-section of bar at section $x - x$, $A = \frac{\pi}{4}d^2 = \frac{\pi}{4}\left[d_2 - \left(\frac{d_2 - d_1}{L}\right)x\right]^2$

Stress in the bar, $\sigma = \frac{P}{A}$

Extension of elementary length dx is,

$$\delta u = \frac{Pdx}{AE} = \frac{\sigma dx}{E}$$

Now $$P = \sigma A = \sigma_1 A_1 = \sigma_2 A_2$$

and $$\frac{A_1}{A} = \left(\frac{d_1}{d}\right)^2$$

Also $$\frac{d_1}{d} = \frac{l}{L+l-x}$$

$\therefore$ $$\delta u = \frac{\sigma_1}{E}\left[\frac{l}{L+l-x}\right]^2 dx$$

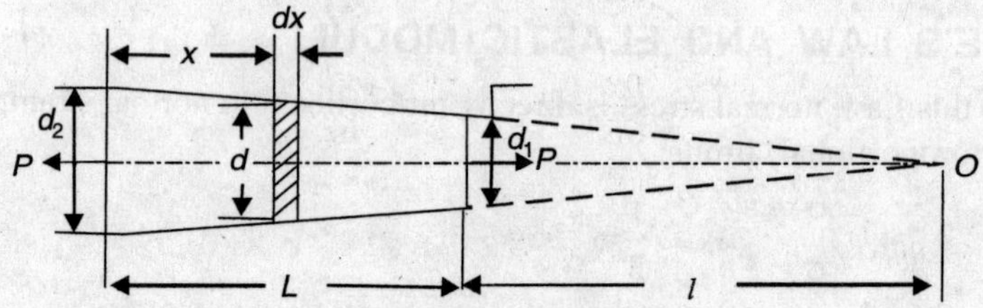

Fig. 13.15 Tapered bar

Total elongation, $$u = \int \delta u = \int_0^L \frac{\sigma_1}{E}\left[\frac{l^2}{(L+l-x)^2}\right] dx$$

$$= \frac{\sigma_1 l^2}{E}\left|\frac{1}{L+l-x}\right|_0^L$$

$$= \frac{\sigma_1 l^2}{E}\left[\frac{1}{l} - \frac{1}{L+l}\right] = \frac{\sigma_1 L l}{E(L+l)}$$

$$= \frac{PL}{A_1 E} \times \frac{d_1}{d_2} = \frac{4PL}{\pi d_1 d_2 E} \qquad \text{... (13.13)}$$

For a square tapering bar,

$$u = \frac{PL}{d_1 d_2 E} \qquad \text{...(13.14)}$$

13.15 EXTENSION OF A CONICAL BAR

Consider a conical bar of base radius r, height h and of mass density ρ, as shown in Fig. 13.16, hanging under its own weight. Consider an elementary strip at a distance x of width dx from the base.

$$\text{Radius of strip} = \left(\frac{h-x}{h}\right) r$$

Load acting on the elementary strip, $$\Delta P = \frac{1}{3}\pi\left[\left(\frac{h-x}{h}\right) r\right]^2 (h-x)\,\rho g$$

Area of strip, $\Delta A = \pi\left[\frac{(h-x)r}{h}\right]^2$

Stress in the strip, $\sigma = \frac{\Delta P}{\Delta A} = \frac{1}{3}(h-x)\rho g$

Extension of strip, $dx = \frac{\sigma}{E}dx = \frac{(h-x)\rho g}{3E}dx$

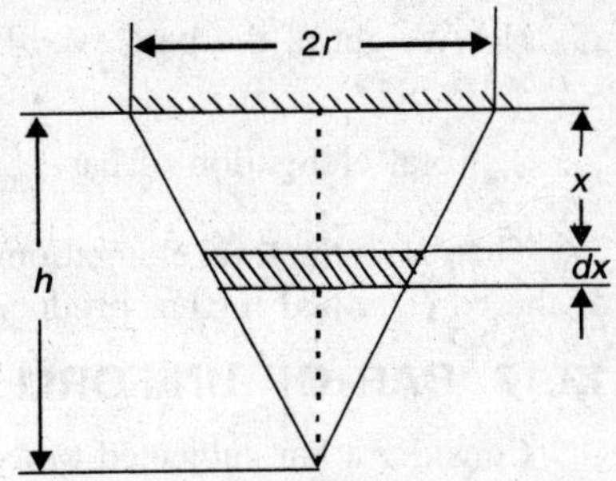

Fig. 13.16 Conical bar

Extension of conical bar, $\delta = \frac{\rho g}{3E}\int_0^h (h-x)\,dx$

$$= \frac{\rho g}{6E}\left|-(h-x)^2\right|_0^h = \frac{\rho g h^2}{6E} \qquad \text{... (13.15)}$$

Now weight of bar, $W = \frac{1}{3}\pi r^2 h\rho g$

Area of base, $A = \pi r^2$

$\therefore \quad W = \frac{1}{3}Ah\rho g$

or $\quad h\rho g = \frac{3W}{A}$

$$\therefore \quad \delta = \frac{Wh}{2AE} \qquad \text{... (13.16)}$$

13.16 EXTENSION OF UNIFORM BAR HANGING UNDER ITS OWN WEIGHT

Consider a bar of uniform area of cross-section A and length L hanging under its own weight as shown in Fig. 13.17. Consider a strip of the bar of thickness dx at a distance x from the bottom. The downward force acting on this strip is due to the weight of the bar that lies below this strip and is equal to $Ax\rho g$, where ρ = density of its material, and g = acceleration due to gravity.

Stress of the strip, $\sigma = \frac{\rho g\, Ax}{A} = \rho g x$

Strain in the strip, $\varepsilon = \frac{\sigma}{E} = \frac{\rho g x}{E}$

Extension of the strip, $\varepsilon\, dx = \frac{\rho g x}{E}dx$

Total elongation of the bar $= \int_0^L \frac{\rho g x}{E}dx = \frac{\rho g}{E}\int_0^L x\,dx$

$$= \frac{\rho g}{E}\left|x^2\right|_0^L = \frac{\rho g L^2}{2E} \qquad \text{... (13.5)}$$

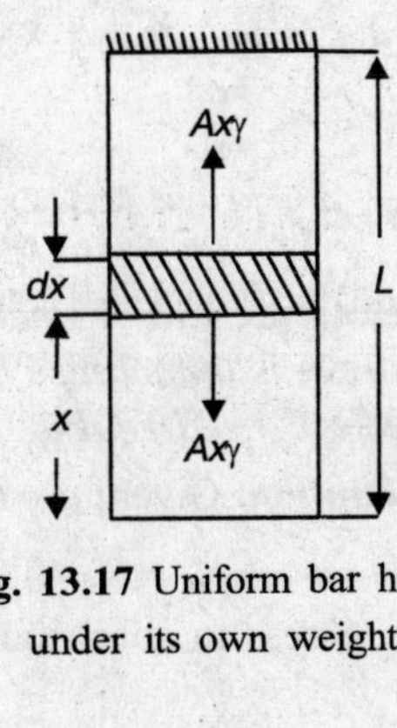

Fig. 13.17 Uniform bar hanging under its own weight

Weight of the bar, $W = \rho g A L$

$\therefore$ Total elongation of bar $= \dfrac{1}{2} \times \dfrac{WL}{AE}$

Therefore, the total elongation of the bar of uniform cross section produced by the self weight of the bar is equal to that produced by a load of half of its weight applied at the lower end.

13.17 BAR OF UNIFORM STRENGTH

Consider a bar subjected to axial load P shown in Fig. 13.18 that stress throughout the bar is σ. Consider an elementary strip of the bar between x and $x + \delta x$ from the lower end. Let the cross-sectional area change from A to A $+ \delta A$ in going from x to $x + \delta x$.

Upward force on strip $= \sigma(A + \delta A)$

Downward on strip $= \sigma A + \gamma \delta x \times \delta A$

where γ = specific weight of bar material

For equilibrium of strip,

$$\sigma(A + \delta A) = \sigma A + \gamma \delta x \times \delta A$$

$$\frac{\delta A}{A} = \frac{\gamma}{\sigma} \times \delta x$$

or

$$\frac{dA}{A} = \frac{\gamma}{\sigma} \times dx$$

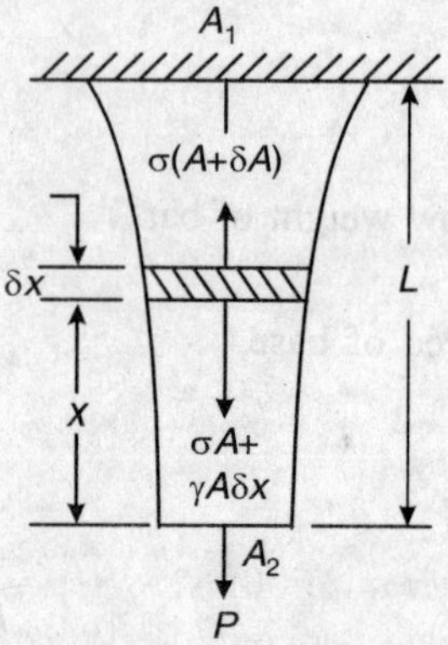

Fig. 13.18 Bar of Uniform strength

Intergrating, we get

$$\int_{A_2}^{A} \frac{dA}{A} = \frac{\gamma}{\sigma} \int_0^x dx$$

$$\ln\left(\frac{A}{A_2}\right) = \frac{\gamma}{\sigma} x$$

$$A = A_2 \exp\left(\frac{\gamma}{\sigma} x\right)$$

or

$$A = A_2 \exp\left(\frac{\gamma L}{\sigma}\right) \quad \text{... (13.18)}$$

Example 13.5 *A steel rod 60 mm in diameter and 2 m long is subjected to a pull of 120kN. It the strain is to increase by 10 percent, find the length of 20mm diameter bore under the same pull. Take E = 210 GPa*

Solution: Given: $d = 60$ mm, $l = 2$m, $P = 120$ kN $d_0 = 20$ mm, 210 GPa, $\Delta\varepsilon = 10\%$.

Stress in solid rod, $\sigma_s = \dfrac{4P}{\pi d^2} = \dfrac{4 \times 120 \times 10^3}{\pi \times 60^2} = 42.441 \text{N/mm}^2$

Extension of solid rod $\delta l = \frac{\sigma_s l}{E} = \frac{42.441 \times 2 \times 10^3}{210 \times 10^3} = 0.4042\text{mm}$

Strain in solid rod $= \varepsilon_s = \frac{\delta l}{l} = \frac{0.4042}{2 \times 10^3} = 2.021 \times 10^{-4}$

Strain in bored rod $\varepsilon_0 = 1.1\ \varepsilon_s = 2.223 \times 10^{-4}$

Let x =length of bore, *m*

Stress in the bored protion, $\sigma_b = \frac{4P}{\pi(d^2 - d_0^2)} = \frac{4 \times 120 \times 10^3}{\pi(60^2 - 20^2)} = 47.764\ \text{N/mm}^2$

$$\text{Strain in bored rod} = \frac{1}{lE}[\sigma_s(2-x) + \sigma_b x] \times 10^3$$

$$= \frac{1}{2 \times 10^3 \times 210 \times 10^3}[42.441(2-x) + 47.746] \times 10^3$$

$$= 2.223 \times 10^{-4}$$

or
$$42.441(2-x) + 47.764 = 93.366$$

$$x = 0 \cdot 925\text{m}$$

Example 13.6 *A stepped bar as shown in Fig. 13.19 is held between two rigid supports such that there is a gap of 1.0 mm between the lower support and the bar. Calculate the stresses in the bar. Take E = 200 GPa.*

Solution: d_1 = mm, d_2 = 20 mm l_1 = 1 m, l_2 = 2m, Δ = 1.0 mm,

E = 200 GPa

Let R_A = reaction of support at A

R_C = reaction of support at C

$A_1 = \frac{\pi}{4} \times 10^2 = 78.54\ \text{mm}^2$

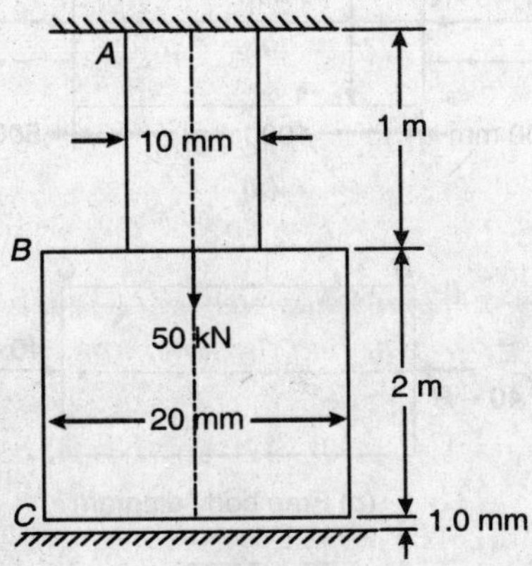

Fig 13.19

$A_2 = \frac{\pi}{4} \times 20^2 = 314.16\ \text{mm}^2$

$R_A + R_C = 50$ kN

$R_A = 50 - R_C$ (tensile) and R_C is compressive.

$$\delta l = \frac{(50 - R_c) \times 10^3 \times 1}{78.54 \times 10^{-6} \times 200 \times 10^9} \text{ (tensile)}$$

$$\delta l_2 = \frac{R_c \times 10^3 \times 2}{314.16 \times 10^{-6} \times 200 \times 10^9} \text{ (compressive)}$$

$$\delta l_1 - \delta l_2 = \Delta$$

$$\frac{10^3}{200 \times 10^3}\left[\frac{50 - R_c}{78.54} - \frac{2R_c}{314.16}\right] = \frac{1.0}{10^3}$$

$$314.16\,(50 - R_C) - 2 \times 78.54 \times R_C = 0.2 \times 78.54 \times 314.16$$

$$15708 - 314.16\,R_c - 157.08\,R_c = 4934.825$$

$$R_c = 22.861\ kN$$

$$R_A = 27.139\ kN$$

$$\text{Stress in } AB, \sigma_1 = \frac{R_A}{A_1} = \frac{27.139 \times 10^3}{78.54} = 345.54 \text{ N/mm}^2$$

$$\text{Stress in } AB, \sigma_2 = \frac{R_c}{A_2} = \frac{22.861}{314.16} = 72.77 \text{ N/mm}^2$$

Example 13.7 *Calculate the stresses in AB and BC of the bar ABC as loaded in Fig. 13.20(a).*

Solution: The free-body diagrams are shown in Fig. 13.20(*b*)

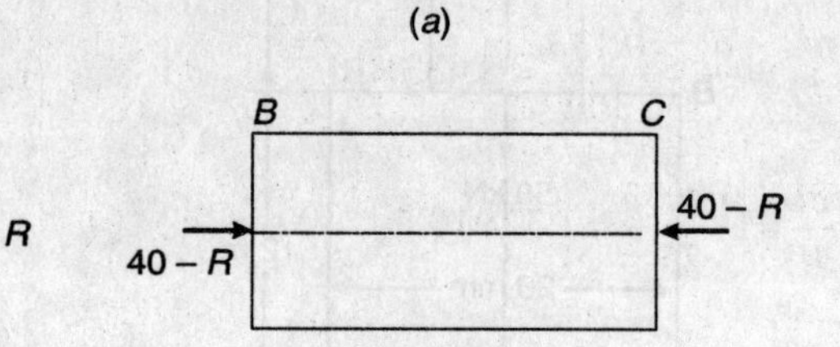

(*a*)

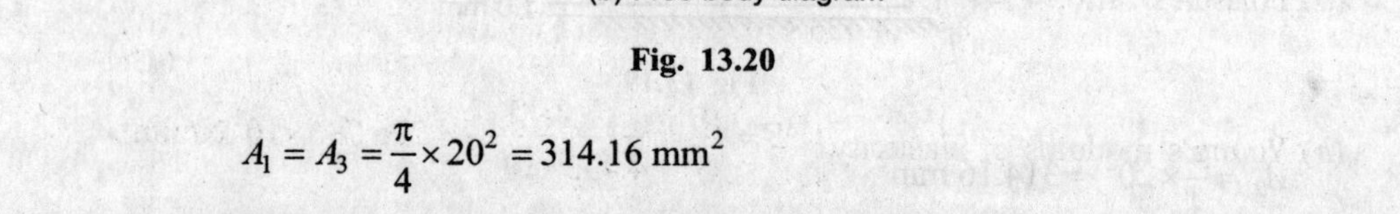

(*b*) Free body diagram

Fig. 13.20

$$A_1 = A_3 = \frac{\pi}{4} \times 20^2 = 314.16 \text{ mm}^2$$

$$A_2 = \frac{\pi}{4} \times 40^2 = 1256.64 \text{ mm}^2$$

Let R = reaction of supports, kN

$$\delta l_1 = \delta l_3 = \frac{R \times 10^3 \times l_1}{A_1 E}$$

$$\delta l_2 = \frac{(40 \times R) \times 10^3 \times l_2}{A_2 E}$$

$$2\delta l_1 - \delta l_2 = 0$$

$$\frac{2 \times 10^3 \times R \times 200}{314.16} - \frac{(40 - R) \times 10^3 \times 1000}{1256.64} = 0$$

$$3183.1R - (40 - R) \times 795.773 = 0$$

$$R = 8 \text{ kN}$$

Stress in AB, $\sigma_1 = \frac{8 \times 10^3}{314.16} = 25.465 \text{ N/mm}^2$

Stress in BC, $\sigma_2 = \frac{32 \times 10^3}{1256.64} = 25.465 \text{ N/mm}^2$

Example 13.8 *The following data refer to a circular bar subjected to a tensile test:*

Diameter of bar = 25 mm
Tensile load = 45 kN
Gauge length = 300 mm
Extension = 0.133 mm
Decrease in diameter = 0.0033 mm

Determine (a) Posisson's ratio, and (b)Young's modulus of elasticity.

Solution: Given: d = 25 mm, P = 45 kN, l = 300 mm, δl = 0.133 mm, δd = 0.0033 mm

Tensile stress, $\sigma = \frac{P}{A} = \frac{4 \times 45 \times 10^3}{\pi \times 25^2} = 91.673 \text{ N/mm}^2$

Longitudinal strain, $\varepsilon_l = \frac{\delta l}{l} = \frac{0.133}{300} = 4.433 \times 10^{-4}$

Lateral strain, $\varepsilon_d = \frac{\delta l}{d} = \frac{0.0033}{25} = 1.32 \times 10^{-4}$

(*a*) Poisson's ratio, $v = \frac{\delta d}{\varepsilon_l} = \frac{1.32 \times 10^{-4}}{4.433 \times 10^{-4}} = 0.298$

(*b*) Young's modulus of elasticity, $E = \frac{\sigma}{\varepsilon_l} = \frac{91.673}{4.433 \times 10^{-4}} = 206.796 \times 10^3 \text{ N/mm}^2$

13.18 STRESS-STRAIN DIAGRAM FOR MILD STEEL

The stress-strain diagram for a ductile material like mild steel is shown in Fig 13.21. The curve starts from the origin *O,* showing thereby that there is no initial stress or strain in the specemen.Upto point *A*, Hooke's law is obeyed and stress is proportional to strain. Therefore, *OA* is a straight line.Point *A* is called the **limit of proportionality.** Upto point *B*,the material remains elastic, *i.e.* on removal of the load, no permanent set is formed. *AB* is not a straight line. Point *B* is called the elastic **Limit point.** Beyond point *B*, the material goes to the plastic stage until the **upper yield point** *C* is reached. At this point the cross-sectional area of the material starts decreasing and the stress decreases to a lower value to point *D*, called the **lower yield** point. Between *DE*, the specimen elongates by a considerable amount, without and increase in stress. From point *E* onwards, the strain hardening phenomena becomes predominant and the strength of the material increases thereby requiring more stress for deformation, until point *F* is reached. Point *F* is called the ultimate point and the corresponding stress is called the **ultimate strength.** At point *F*, necking of the material begins and the cross-sectional area decreases at a rapid rate. The apparent stress decreases but the actual or true stress goes on increasing until the specimen breaks at point *G*, called the point of fracture. The fracture of ductile material is of the cup and cone type.

The phenomena of yielding and necking is not exhibited by brittle materials. The ultimate strength is calculated at 0.2 percent of maximum strain.

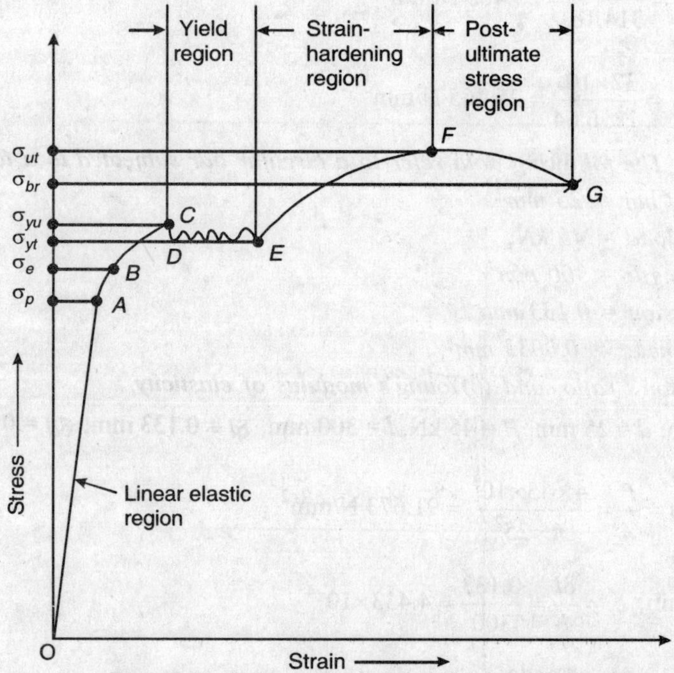

Fig. 13.21 Stress-strain diagram for ductile materials.

The plot of the normal stress *v's* engineering strain can be divided into four well-defined regions, as shown in Fig. 13.21.

1. Linear elastic region.
2. Yield region.
3. Strain-hardening region.
4. Post-ultimate stress region.

13.18.1 True Stress-Strain Diagrams

During plastic strain of a uniform specimen subjected to an axial tension (compression) , the cross-sectional area gets smaller (larger) as the specimen elongates (shortens).The two stress-strain diagrams for ductile materials (such as mild steel), the compression and the tension true stress-strain diagrams practically coincide, where as the two engineering stress strain diagrams drift apart as shown in Fig.13.22

Comparison of the engineering and the true stress-strain diagrams shown that in tensinon, since the cross-sectional area dedcreases as the specimen elongates, the true stress is greater than engineering stress, whereas in compression as the specimen shortens, the cross-sectional area increases and thus the true stress is less than the corresponding engineering stress.

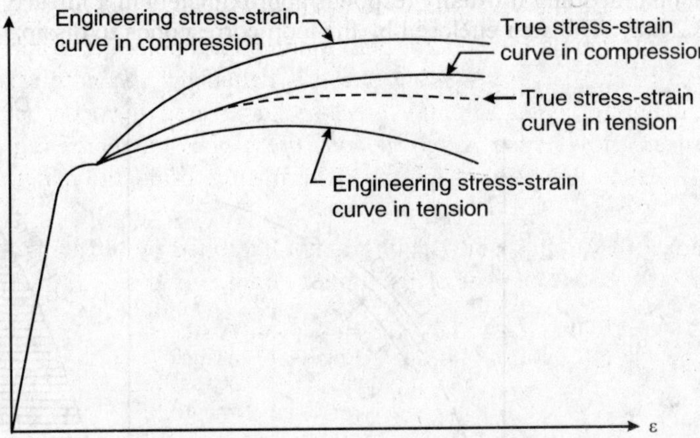

Fig. 13.22 Comparison of tension and compression stress-strain diagrams for ductile material

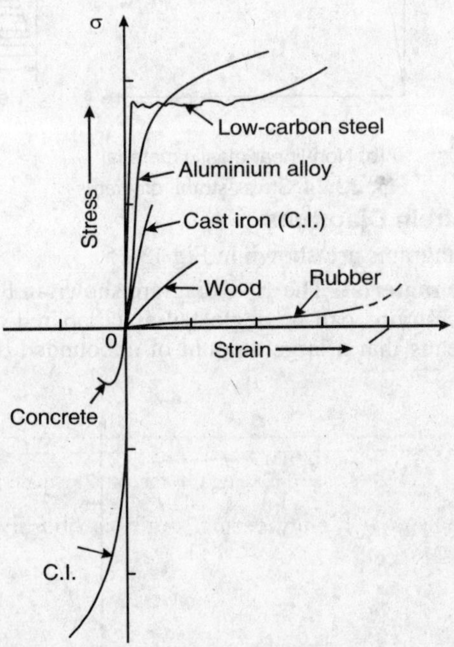

Fig. 13.23 Typical stress-strain diagrams for different materials.

13.18.2 Stress-Strain Diagrams For other Materials

The engineering stress-strain diagrams for a few representative materials are shown in Fig 13.23. For most engineering applications, to limit the deformation, the lower range of strains is particularly important. The large depormations of materials is of great importance in the analysis of metal forming processes like forging, drawing and rolling etc.

13.18.3 Linear And Non-Linear Materials

An elastic material usually implies that stress is directly proportional to strain. Such materials are called ***linearly elastic*** or Hookean. a material responding in a nonlinear manner and yet,when unloaded, returning back along the loading path to its initial stress-free state of deformation is also an elastic material. Such materials are called ***nonlinear elastic.*** The difference between the two types of elastic materials is shown in Fig. 13.24. (*a*) and (*b*). If in stressing a material its elastic limit is exceeded, on unloading it usually responds approximately in a linearly elastic manner, as shown in Fig. 13.24.(*c*). The area enclosed by the loop corresponds to dissipated energy released through heat.

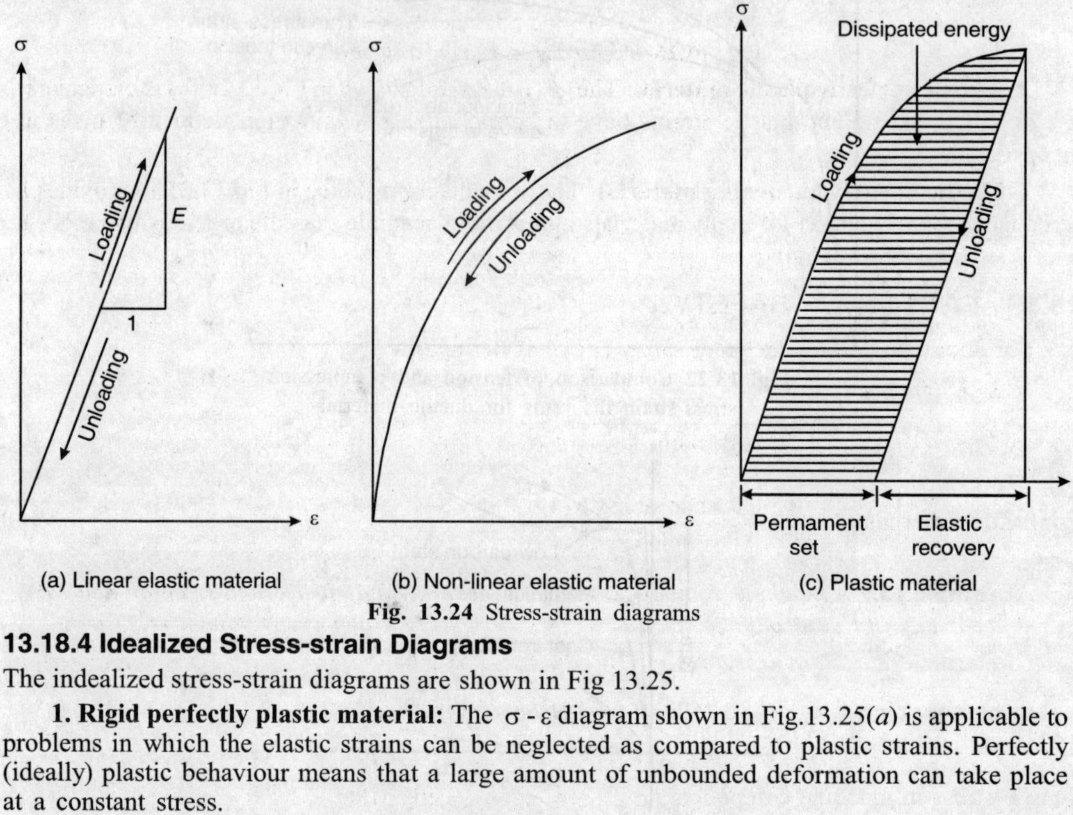

Fig. 13.24 Stress-strain diagrams

13.18.4 Idealized Stress-strain Diagrams

The indealized stress-strain diagrams are shown in Fig 13.25.

1. Rigid perfectly plastic material: The σ - ε diagram shown in Fig.13.25(*a*) is applicable to problems in which the elastic strains can be neglected as compared to plastic strains. Perfectly (ideally) plastic behaviour means that a large amount of unbounded deformation can take place at a constant stress.

(a) Rigid perfectly plastic material (b) Elastic perfectly plastic material

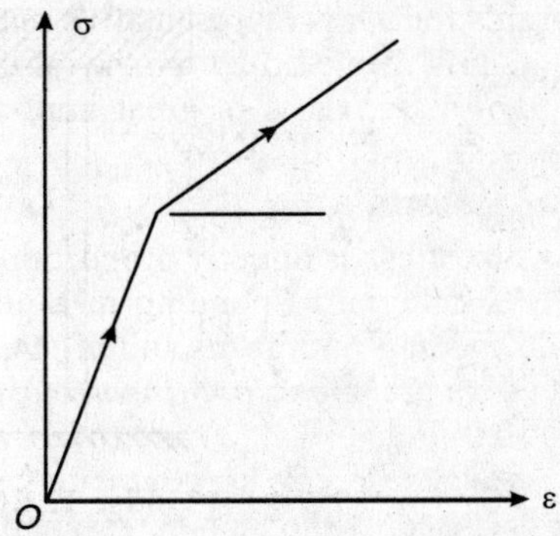

(c) Elastic linearly hardening material

Fig. 13.25 Idealized stress-strain diagrams

2. Elastic perfectly plastic material: The σ - ε diagrams shown in Fig 13.25 (*b*) is particularly useful if both elastic and plastic strains have to be included. This situation frequently arises in analysis.

3. Elastic linearly hardening material: The σ - ε diagram shown in Fig.13.25(*c*) provides a reasonable approximation for many materials and is more accurate than the previous two models over a wider range of strain.

13.19 FACTOR OF SAFETY

For ductile materials, factor of safety (*FOS*) is defined as:

$$FOS = \frac{\text{Yield strength}}{\text{Allowable stress}} = \frac{\sigma_y}{\sigma_a}$$

For brittle materials, $FOS = \dfrac{\text{Ultimate strength}}{\text{Allowable stress.}} = \dfrac{\sigma_u}{\sigma_a}$

Example 13.9 *A machine component made of steel having tensile yield strength 380 MPa is subjected to axial load of 5 kN. The area of cross- section of the component is 25 mm².*

Determine the factor of safety.

Solution: Given: $\sigma_y = 380$ MPa, $P = 5$ kN, $A = 25$ mm^2

Tensile stress in machine component, $\sigma_t = \dfrac{P}{A} = \dfrac{5\times10^3}{25} = 200$ MPa

Factor of safety $\dfrac{\sigma_y}{\sigma_t} = \dfrac{380}{200} = 1.9$

13.20 COMPOSITE SYSTEM OF EQUAL LENGTHS SUBJECTED TO LOAD

Consider a composite system shown in Fig.13.26 consisting of different materials and different area of cross-section, subjected to load *P*.

Now $\quad P = P_1 + P_2 + P_3$

Let $\quad \Delta l$ = change in length l

Strain, $\varepsilon = \dfrac{\Delta l}{l}$

Then $P_i = \varepsilon A_i E_i$

$\therefore$ $\varepsilon \Sigma A_i E_i = P$

or $\varepsilon = \dfrac{P}{\Sigma A_i Ei}$

$$P_i = \frac{P A_i E_i}{\Sigma A_i E_i}$$

and $\sigma_i = \dfrac{P E_i}{\Sigma A_i E_i}$

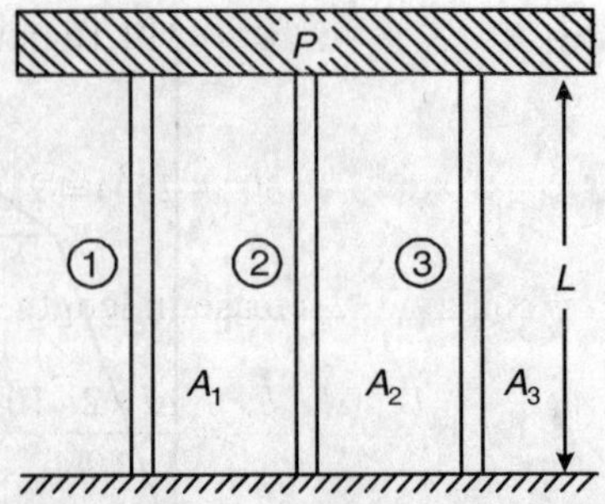

Fig. 13.26 Composite system of equal lengths under load

13.21 COMPOSITE SYSTEM OF UNEQUAL LENGTHS SUBJECTED TO LOAD

Consider a composite system of unequal lengths shown in Fig. 13.27 subjected to load P. Before both the members start sharing the load,

Strain in inner tube, $\varepsilon = \dfrac{\Delta}{l}$

Initial load taken by inner tube, $P_i = \varepsilon A_2 E_2$ (–*ve*)

Remaining load, $P_0 = P - P_i$

The load P_0 is then shared as per art. 13.20.

$$P_1 = \frac{P_0 A_1 E_1}{A_1 E_1 + A_2 E_2}. \quad \text{... (13.20}a\text{)}$$

$$P_2 = \frac{P_0 A_2 E_2}{A_1 E_1 + A_2 E_2} + p_i \quad \text{... (13.20}b\text{)}$$

$$\sigma_1 = \frac{P_1}{A_1}, \quad \sigma_2 = \frac{P_2}{A_2}$$

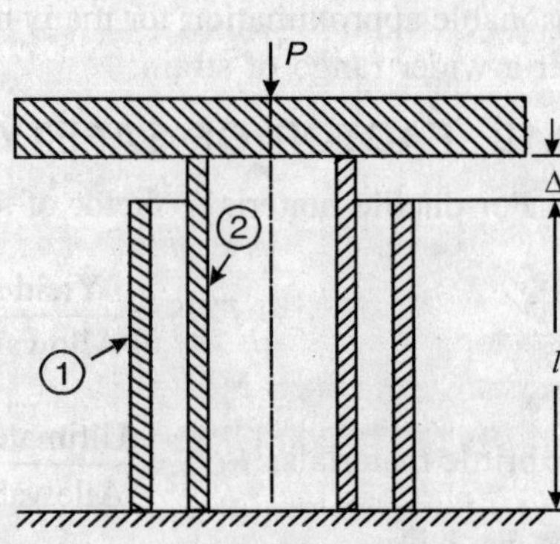

Fig. 13.27 Composite system of unequal lengths under load

Example 13.10 *A steel rod 20 mm diameter is passed through a brass tube 25 mm internal diameter and 30 mm external diameter. The tube is 1 m long and is closed by thin rigid washers and fastened by nuts, screwed to the rod, as shown in Fig. 13.28. the nuts are tightened until the compressive force in the tube is 5 kN. Calcutate the stresses in the rod and the tuble.* E_s *= 200 GPa,* E_b *= 80 GPa.*

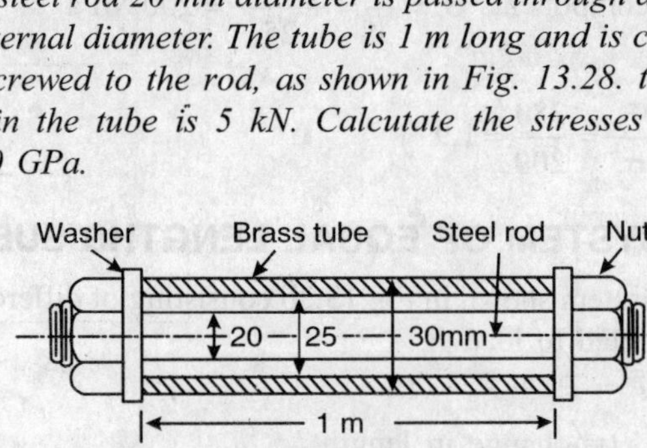

Fig. 13.28

Solution: $A_s = \frac{\pi}{4} \times 20^2 = 314.16 \text{ mm}^2$

$$A_b = \frac{\pi}{4} \times (30^2 - 25^2) = 216 \text{ mm}^2$$

Let Δl = common decrease in length, mm

$$\sigma_s = \frac{\Delta l \times E_s}{l} = \frac{\Delta l \times 2 \times 10^5}{1000} = 200\,\Delta l\,\text{MP}a$$

$$\sigma_b = \frac{\Delta l \times E_b}{l} = \frac{\Delta l \times 0.8 \times 10^5}{1000} = 80\,\Delta \text{l}\,\text{MP}a$$

Total force, $P_s + P_b = P$

$$\sigma_s A_s + \sigma_b A_b = P$$

$$\Delta l\,(200 \times 314.16 + 80 \times 216) = 5000$$

$$\Delta l = 0.0642 \text{ mm}$$

$$\sigma_s = 200 \times 0.0624 = 12.48 \text{ MP}a\,(+\text{ve})$$

$$\sigma_b = 80 \times 0.0624$$

$$= 4.993 \text{ MPa } (-\text{ ve})$$

Example 13.11 *A rigid beam BD is suspended from two rods AB and CD as shown in Fig. 13.29. Rod AB is of steel of 20 mm diameter and rod B is of copper of 25 mm diameter. At what distance from rod AB the load P be applied if the beam is to remain horizontal ? Calculate the stresses in the rods if P = 25 kN.*

Take E_s =200 *GPa* and E_c = 100 GPa

Solution: Let P_s, P_c = forces in rods *AB* and *CD* respectively. Taking moments about rod *AB*, we have

$$Px = 2P_c$$

or $$P_c = \frac{Px}{2}$$

and $$P_s = P - \frac{Px}{2}$$

For the beam *BD* to remain horizontal,

$$\delta_s = \delta_c$$

Now $A_s = \frac{\pi}{4} \times 20^2 = 314.16 \text{ mm}^2$

$$A_c = \frac{\pi}{4} \times 25^2 = 490.87 \text{ mm}^2$$

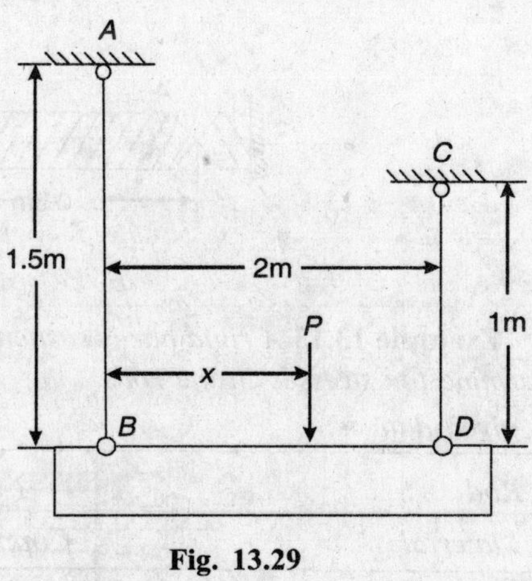

Fig. 13.29

$$\frac{\left(P-\frac{Px}{2}\right)\times 1.5\times 10^3}{314.16\times 2\times 10^5}=\frac{\frac{Px}{2}\times 1\times 10^3}{490.87\times 1\times 10^5}$$

$$1-\frac{x}{2}=0.42667x$$

$$x = 1.079 \text{ m}$$

$$P_s = 25\left(1-\frac{1.079}{2}\right)=11.51\,\text{kN}, P_c = 13.49\,\text{kN}$$

$$\sigma_s = \frac{P_s}{A_s} = 36.64\,\text{MP}a, \sigma_c = \frac{P_c}{A_c} = 27.48\,\text{MP}a$$

Example 13.12 *Determine the stresses in the rods loaded as shown in Fig .13.30 and having the same area of cross-section of 5 cm².*

Solution: $\Sigma M_A = 0$ gives

$$0.5\,P_1 + P_2 = 150 \times 1$$

or $\quad P_1 + 2P_2 = 300$

Also $\quad P_1 + P_2 = 150$

$\therefore \quad P_2 = 150$ kN and $P_1 = 0$

$$\sigma_1 = 0, \sigma_2 = \frac{150\times 10^3}{5\times 10^2} = 300 \text{ MPa}$$

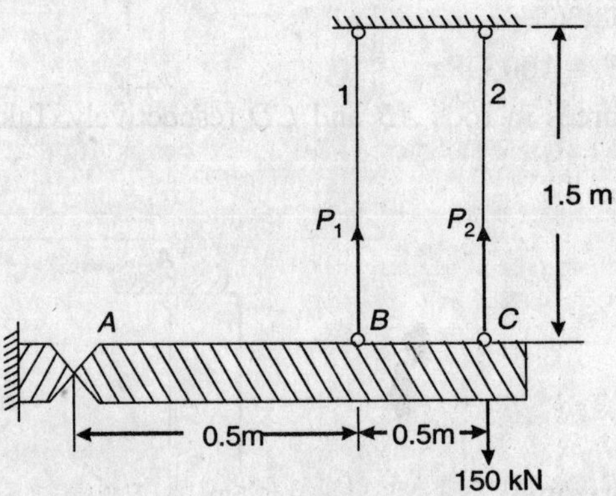

Fig. 13.30

Example 13.13 *A rigid bar is suspended from three rods and loaded as shown in Fig.13.31. Determine the stresses in the rods.*

The given data is:

Rod	*1*	*2*	*3*
Material	*Copper*	*Steel*	*Aluminium*
Area of cross-section, cm²	*1*	*1.5*	*2*
E, GPa	*100*	*200*	*70*

Solution: $\Sigma M_3 = 0$ gives

$$2P_1 + P_2 = 80 \times 0.5 = 120 \quad ...(1)$$

$\Sigma M_1 = 0$ gives

$$P_2 + 2P_3 = 80 \times 1.5 = 120 \quad ...(2)$$

$$\delta_1 = \frac{P_1 l_1}{A_1 E_1} = \frac{P_1 \times 2 \times 10^6}{1 \times 10^2 \times 1 \times 10^5}$$

$$= \frac{P_1}{5} \text{ mm}$$

$$\delta_2 = \frac{P_2 \times 1.5 \times 10^6}{1.5 \times 10^2 \times 2 \times 10^5}$$

$$= \frac{P_2}{20} \text{ mm}$$

$$\delta_3 = \frac{P_3 \times 2 \times 10^6}{2 \times 10^2 \times 0.7 \times 10^5}$$

$$= \frac{P_3}{7} \text{ mm}$$

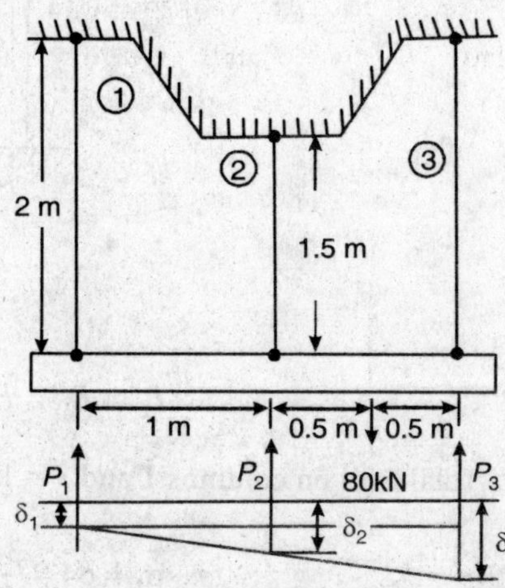

Fig. 13.31

From Fig. 13.31(*b*)we have

$$\frac{\delta_2 - \delta_1}{1} = \frac{\delta_3 - \delta_2}{1}$$

or $$\delta_2 - \delta_1 = \delta_3 - \delta_2$$

$$2\delta_2 - \delta_1 - \delta_3 = 0$$

$$\frac{P_2}{10} - \frac{P_1}{5} - \frac{P_3}{7} = 0$$

or $$7P_2 - 14P_1 - 10P_3 = 0 \quad ...(3)$$

From Eqn. (2), $P_3 = \frac{120 - P_2}{2} = 60 - 0.5P_2$

Eqn. (3) becomes,

$$7P_2 - 14P_1 - 600 + 5P_2 = 0$$

or $$12P_2 - 14P_1 = 600$$

$$6P_2 - 7P_1 = 300$$

From Eqn. (1) $6P_2 + 12P_1 = 720$

$$P_1 = \frac{420}{19} = 22.105 \text{ kN}$$

$$P_2 = 75.79 \text{ kN}, P_3 = 22.105 \text{ kN}$$

$$\sigma_1 = 221.05 \text{ MPa}, \sigma_2 = 505.93 \text{ BPa}, \sigma_3 = 110.52 \text{ MPa}$$

Example 13.14 *A rigid beam is placed on there columns of identical cross- sectional areas of 200cm² each as shown in Fig. 13.32. Calculate the stresses in the columns if there was a gap of 2 mm between the beam and the middle column before the load wass applied. Take E = 18.GPa.*

Solution: Initial compression of columns 1 and 3 are:

$$\delta_1 = \delta_3 = 2\,\text{mm}$$

Initial load, $$P_1 = P_3 = \frac{\delta AE}{l} = \frac{2\times 200\times 10^2 \times 10^3}{3\times 10^3}$$

$$= 240 \text{ kN}$$

Remaining load = 800 – 2 × 240 = 320 kN

$$\therefore \quad P_1' + P_2' + P_3 = 320\,\text{kN}$$

Now $\delta_1' = \delta_2' = \delta_3 =$ as the further compression of columns.

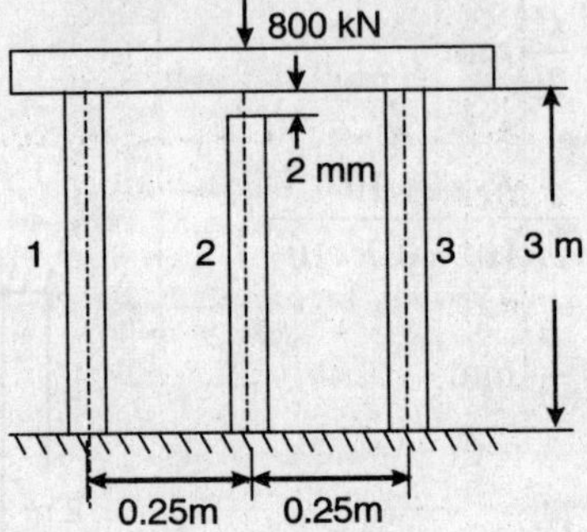

Fig. 13.32

$$\therefore \quad P_1' = P_2' = P'_3 = \frac{320}{3} = 106.67\text{kN}$$

Total load on columns 1 and 3 = 106.67 kN + 240 = 346.67 kN

$$\sigma_1 = \sigma_3 = \frac{346.67\times 10^3}{200\times 10^2} = 17.67\,\text{MPa}\,(-ve)$$

$$\sigma_2 = \frac{106.67\times 10^3}{200\times 10^2} = 5.33\,\text{MPa}\,(-ve)$$

Example 13.15 *A solid steel bar 0.5m long and 50 mm diameter is placed inside and aluminium tube having 60 mm inside and 80 mm outside diameter. the aluminium tube is 0.2 mm longer than the steel bar. An axial load of 500 kN is applied to the bar and tube through rigid cover plates as shown in Fig 13.33. Find the stresses developed in steel bar and aluminium tube.* $E_s = 220$ GPa, $E_a = 70$ GPa

Solution: $$A_s = \frac{\pi}{4}\times 50^2 = 1963.5\,\text{mm}^2$$

$$A_a = \frac{\pi}{4}\times(80^2 - 60^2) = 2199.1\,\text{mm}^2$$

Initial stress in aluminium tube $= \dfrac{\delta E}{l} = \dfrac{0.2 \times 70 \times 10^3}{0.5 \times 10^3} = 28\,\text{MPa}$

Initial load taken by aluminium tube $= 28 \times 2199.1 = 61575$ N

Remaining load $= 500 - 61.575 = 438.425$ kN

$$\therefore \quad P_a + P_s = 438.425 \qquad \ldots(1)$$

Now
$$\frac{P_s}{A_s E_s} = \frac{P_a}{A_a E_a}$$

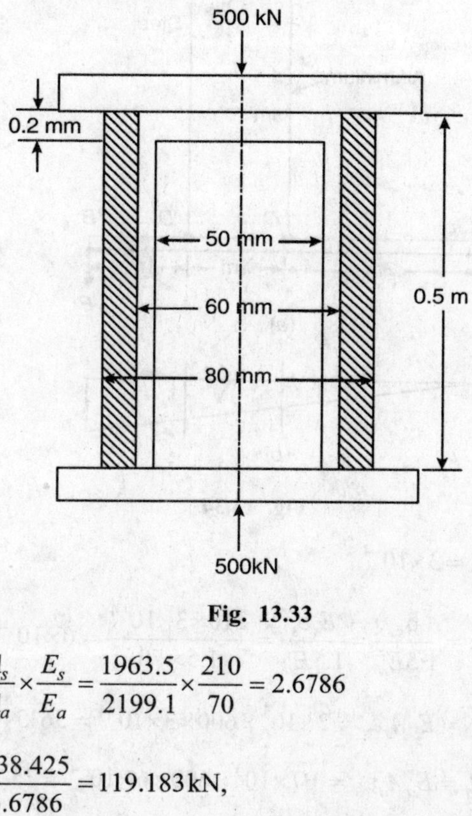

Fig. 13.33

$$\frac{P_s}{P_a} = \frac{A_s}{A_a} \times \frac{E_s}{E_a} = \frac{1963.5}{2199.1} \times \frac{210}{70} = 2.6786 \qquad \ldots(2)$$

$$p_a = \frac{438.425}{3.6786} = 119.183\,\text{kN},$$

$$P_s = 319.243 \text{ kN}$$

$$\sigma_s = \frac{319.243 \times 10^3}{1963.5} = 162.589\,\text{kNa}$$

$$\sigma_s = 28 + \frac{119.183 \times 10^3}{2199.1} = 82.196\,\text{MPa}$$

Example 13.16 *A horizontal beam AB is supported on two cables CE and DF as shown in Fig 13.34(a) Calculate the values of P when the strain in DF is 3×10^{-4} Assume that beam AB does not bend. $A_a = A_s = 600$ mm^2 ,$E_a = 70$ GPa, $E_s = 210$ GPa.*

Solution: From Fig. 13.34(*b*), we have

$$\frac{\delta_a}{2} = \frac{\delta_s}{3}$$

$$\delta_s = 1.5\delta_a$$

$$\varepsilon_s = \frac{\delta_s}{E_s}, \varepsilon_a = \frac{\delta_a}{E_a}$$

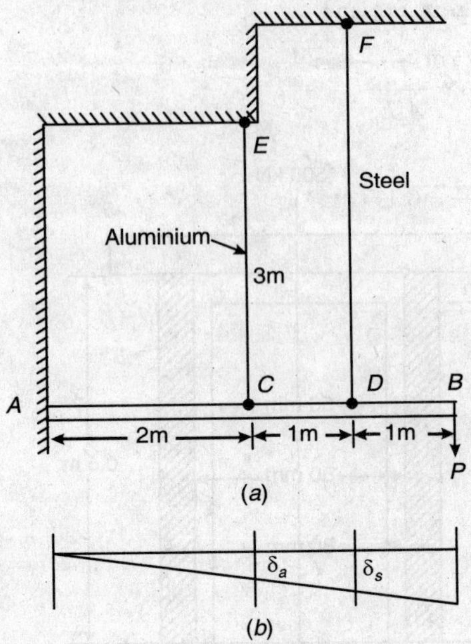

Fig. 13.34

Now $\varepsilon_s = 3\times10^{-4}$

$$\varepsilon_s = \frac{\delta_s}{1.5E_a} = \frac{E_s\varepsilon_s}{1.5E_a} = \frac{210\times3\times10^{-4}}{1.5\times70} = 6\times10^{-4}$$

$$P_s = E_s A_s \varepsilon_s = 2\times10^5 \times 600\times3\times10^{-4} = 36\,\text{kN}$$

$$P_a = E_a A_a \varepsilon_a = .07\times10^5 \times 600\times6\times10^{-4} = 25.2\,\text{kN}$$

$\Sigma M_A = 0$ given

$P \times 4 = 25.2 \times 2 + 36 \times 3$

$P = 39.6$ kN

Example 13.17 *A compound bar loaded as shown in Fig.13.35 has a gap of 1.0 mm. Calculate The stresses in the two bars. E_s = 200 $GP_{a'}$ E_{ci} = 105 GPa.*

Solution: The steel bar is subjected to a tensile load of 100kN. Its free extension is:

$$\delta l_s = \frac{Pl_s}{A_s E_s} = \frac{100\times10^3\times0.75}{\frac{\pi}{4}\times(0.02)^2\times200\times10^9} = 1.194\,\text{mm}$$

Restricted extension = 1.94 – 1.0 = 0.194 mm

The support *B* will exert compressive force in the bars. Let *F* be the compressive force introduced in the bars, then

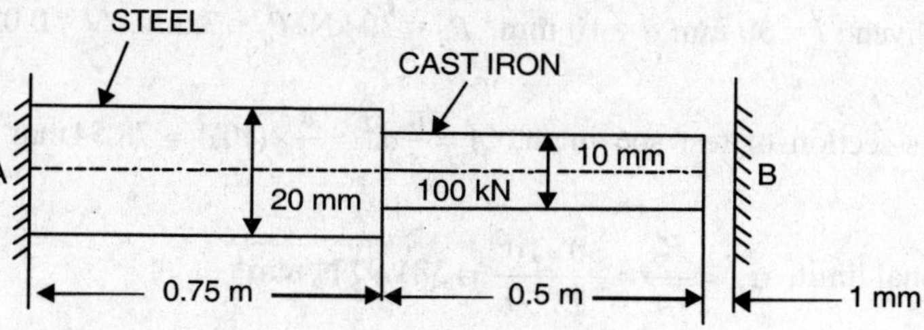

Fig. 13.35

$$\frac{Fl_s}{A_s E_s} + \frac{Fl_{CI}}{A_{CI} E_{CI}} = 0.194\,\text{mm}$$

or
$$F\left[\frac{0.75}{(0.02)^2 \times 200} + \frac{0.5}{(0.01)^2 \times 105}\right] \times \frac{\pi}{4} \times 10^{-9} = 0.194 \times 10^{-3}$$

$$F[9.376 + 47.619] \times \frac{\pi}{4} \times 10^{-9} = 0.194 \times 10^{-3}$$

$$72.567 \times 10^{-9}\, F = 0.194 \times 10^{-3}$$

$$F = 2673.4\,\text{N}$$

Net force in steel bar = 100 – 2.6734 = 97.3266 kN

Stress in steel bar $= \dfrac{97.3266 \times 10^3}{\frac{\pi}{4} \times (0.02)^2} = 309.8\,\text{MPa}$ (tensile)

Stress in cast iron bar $= \dfrac{2673.4}{\frac{\pi}{4} \times (0.01)^2} = 34.04\,\text{MPa}$ (compressive)

Example 13.18 *A rectangular steel bar 25 mm × 12 mm and 0.5 m long is subjected to an axial tensile load of 10 kN. Calculate the change in dimensions if E = 200 GPa and poisson's ratio = 0.30*

Solution: Longitudinal strain, $\varepsilon = \dfrac{PL}{AE} = \dfrac{10 \times 10^3 \times 0.5}{25 \times 12 \times 10^{-6} \times 200 \times 10^9} = 83.33 \times 10^{-6}$

Leteral strain, $\varepsilon_l = -\upsilon\varepsilon = -0.3 \times 83.33 \times 10^{-6} = -25 \times 10^{-6}$

Change in length, $\delta l = \varepsilon l = 83.33 \times 10^{-6} \times 0.5 = 0.0417\,\text{mm}$

Change in width, $\delta b = \varepsilon_l b = -25 \times 10^{-6} \times 25 = -0.625\,\text{mm}$

Change in thickness, $\delta l = \varepsilon_l t = -25 \times 10^{-6} \times 12 = -0.3\,\text{mm}$

– ve sign indicates decrease in dimension.

Example 13.19 *In a tensile test on mild test on mild steel, the gauge length was 50 mm and diameter 10 mm. The load at the proportional limit was 30 kN and at the upper yield point 32 kN. The extension on 50 mm gauge length was 0.028 mm under a load of 8.8 kN. Calculate the (a)Proportional limit, (b) upper yield point stress, and (c)modulus of elasticity.*

Solution: Given : $l = 50$ mm $d = 10$ mm, $P_p = 30$ kN $P_u = 32$kN, $\Delta l = 0.028$ mm, $P = 8.8$ kN

Area of cross-section of test specimen, $A = \frac{\pi}{4} d^2 = \frac{\pi}{4} \times (10)^2 = 78.54\,\text{mm}^2$

(*a*) Proportional limit, $\sigma_p = \frac{P_p}{A} = \frac{30 \times 10^3}{78.59} = 381.97\,\text{N/mm}^2$

(*b*) Upper yield point stress, $\sigma_{yu} = \frac{P_u}{A} = \frac{32 \times 10^3}{78.54} = 407.43\,\text{N/mm}^2$

(*c*) Strain, $\varepsilon = \frac{\Delta l}{l} = \frac{0.028}{50} = 5.6 \times 10^{-4}$

Stress, $\sigma = \frac{P}{A} = \frac{8.8 \times 10^3}{78.59} = 112.045\text{N/mm}^2$

Modulus of elasticity, $E = \frac{\sigma}{\varepsilon} = \frac{112.045}{5.6 \times 10^{-6}} = 2 \times 10^5\,\text{N/mm}^2$ or 200 GPa.

Example 13.20 *A steel column 4 m hight 250 mm exetrnal diameter and 200 mm inside diameter is subjected to an axial compressive load of 40 kN at the top. The density of steel is 7470 kg.m^3.*

Calculate the stress at the top and bottom of the column.

Solution: Given: $l = 4$ m, $d_0 = 250$ mm, $d_i = 200$ mm, $P = 40$ kN, $\rho = 7470$ kg/m^3

Area of coross-section of column, $A = \frac{\pi}{4}[(250)^2 - (200)^2] = 17671\text{mm}^2$

Weight of column, W = $Al\rho g = 17671 \times 10^{-6} \times 4 \times 7470 \times 9.81 = 5180$ N

Compressive stress at the top column $= \frac{P}{A} = \frac{40 \times 10^3}{17671} = 2.26\text{N/mm}^2$

Total load at the bootom $P + W = 40{,}000 + 5180 = 45{,}180$ N

Compressive stress at the bootom of column $= \frac{45{,}180}{17671} = 2.557\,\text{N/mm}^2$

Example 13.21 *A short reinforced concrete column has 600cm^2 area of cross-section. The column is reinforced with 4 steel rods arranged symmetrically, eafch having 10 cm^2 cross-sectional area. The column is subjected to an axial load of 800 kN. E_s = 200 GPa and E_c =20 GPa. Calculate the load shared by concrete and steel.*

Solution: $A_s = 4 \times 10 = 40$ cm^2. $A_c = 600 - 40 = 560$ cm^2

Load shared by steel rods, $P_s = \frac{PE_s A_s}{A_s E_s + A_c E_c}$

$$= \frac{50 \times 10^3 \times 200 \times 10^9}{176.71 \times 200 \times 10^3 + 113.1 \times 105 \times 10^3} = 211.785\,\text{MPa}$$

$$\sigma_{cu} = \frac{PE_{cu}}{A_s E_s + A_{cu} E_{cu}}$$

$$= \frac{50 \times 10^3 \times 105 \times 10^9}{176.71 \times 200 \times 10^3 + 113.1 \times 105 \times 10^3} = 111.187 \text{ MPa}$$

Resultant stresses are :

Steel bolt = 100.6 + 211.785 = 312.385 MPa

Copper tuble = – 157.18 + 111.187

= – 45.993 MPa

Example 13.22 *A load of 2 kN is suspended from two rods as shown is Fig. 13.36. Rod AB is of steel of 15 mm diameter and rod CD of copper having 10 mm diameter.Calculate the distance x so that the the bar BD remains horizontal. Also calculate the stresses production in the rods.* E_s *= 200 GPa,* E_{cu} *= 110 GPa.*

Solution: For the bar *BD* to remain horizontal,

$$\delta l_s = \delta l_{cu}$$

Let P_s and P_{cu} be the loads shared by steel and copper rods.

$$\frac{P_s l_s}{A_s E_s} = \frac{P_{cu} l_{cu}}{A_{cu} E_{cu}}$$

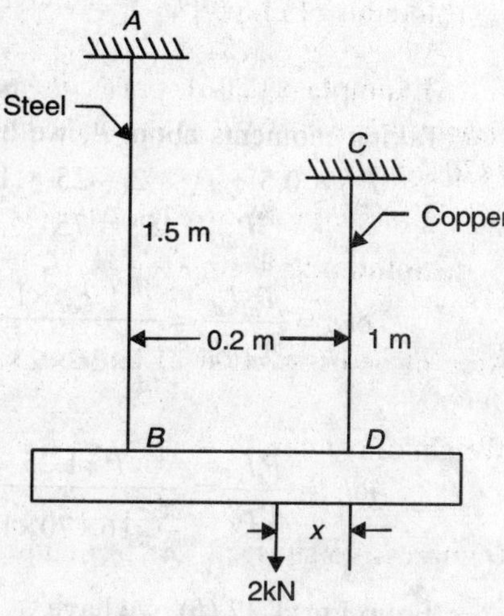

Fig. 13.36

$$\frac{P_s \times 1.5}{\frac{\pi}{4} \times 225 \times 2 \times 10^5} = \frac{P_{cu} \times 1}{\frac{\pi}{4} \times 100 \times 1.1 \times 10^5}$$

$$\frac{P_s}{P_{cu}} = \frac{225 \times 2}{100 \times 1.1 \times 1.5} = 2.7273$$

$$P_s + P_{cu} = 2000$$

$$P_{cu} = \frac{2000}{3.7273} = 563.6 \text{ N}$$

$$P_s = 1463.4 \text{ N}$$

Taking moments about *D*, we have

$$P_s \times 0.2 = 2000 \times x$$

$$x = \frac{1463.4 \times 4 \times 0.2}{2000} = 0.146 \text{ m}$$

$$\sigma_s = \frac{P_s}{A_s} = \frac{146.4}{\frac{\pi}{4} \times 225} = 8.28 \text{N/mm}^2$$

$$\sigma_{eu} = \frac{P_{cu}}{A_{cu}} = \frac{536.6}{\frac{\pi}{4} \times 100} = 6.83 \text{N/mm}^2$$

Example 13.23 *A rigid bar ABC is hinged at A and suspended at two points B and C by two bars BD and CE made of aluminium and steel respectively as shown in Fig. 13.37. The aluminium bar is of 6 mm diameter and steel bar of 4 mm diameter. Calculate the stress developed in each bar.* E_{al} = *70 GPa*, E_s = *200 GPa..*

Solution: Let P_{al} and P_s be the loads shared by *BD* and *CE* respectively.

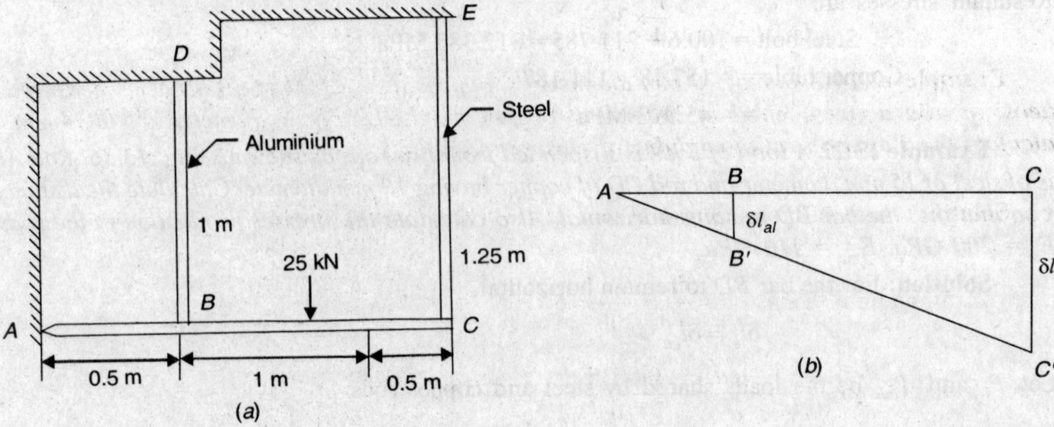

Fig. 13.37

Taking moments about *A*, we have

$$P_{al} \times 0.5 + P_s \times 2 = 25 \times 1.5$$

$$P_{al} + 4P_s = 75$$

$$\delta l_{al} = \frac{P_{al} l_{al}}{A_{al} E_{al}} = \frac{P_{al} \times 1}{\frac{\pi}{4} \times 36 \times 2 \times 10^5} = 0.01768 \times 10^{-5} \times P_{al}$$

$$\delta l_s = \frac{P_s l_s}{A_s E_s} = \frac{P \times 1.25}{\frac{\pi}{4} \times 16 \times 70 \times 10^3} = 0.14210 \times 10^{-5} \times P_s$$

From Fig. 1.37 (*b*), we have

$$\frac{\delta l_{al}}{\delta l_s} = \frac{0.5}{2} = 0.25$$

$$\frac{0.01768 P_{al}}{0.14210 P_s} = 0.25$$

$$\frac{P_{al}}{P_s} = 2.0088$$

From Eqs. (1) and (2), we get

$$P_s = \frac{75}{6.0088} = 12.4817 \text{ kN}$$

$$P_{al} = 25.0732 \text{ kN}$$

$$\sigma_s = \frac{12.4817 \times 10^3}{\frac{\pi}{4} \times 16} = 993.26 \text{ N/mm}^2,$$

$$\sigma_{al} = \frac{25.0732 \times 10^3}{\frac{\pi}{4} \times 36} = 886.78 \text{ N/mm}^2$$

Example 13.24 *A bar of 20 mm diameter is subjected to a pull of 30 kN. The measured extensions over a guage length of 200 mm is 0.1 mm and the change in diameter is 0.0034 mm. Calculate the Poisson's ratio, modulus of elasticity and bulk modulus.*

Solution: Area of cross-section of bar, $A = \frac{\pi}{4} \times (20)^2 = 314.16 \text{ mm}^2$

Longitudinal strain, $\varepsilon_d = \frac{\delta l}{l} = \frac{0.1}{200} = 5 \times 10^{-4}$

Lateral strain, $\varepsilon_d = \frac{-\delta d}{d} = -\frac{0.0035}{20} = -1.75 \times 10^{-4}$

Poisson's ratio, $v = \frac{\varepsilon_d}{\varepsilon_l} = \frac{-1.75 \times 10^{-4}}{5 \times 10^{-4}} = -0.35$

Modulus of elasticity, $E = \frac{P}{A\varepsilon_l} = \frac{30 \times 10^3}{314.16 \times 5 \times 10^{-4}} = 1.91 \times 10^5 \text{ N/mm}^2$

Bulk modulus, $K = \frac{E}{3(1-2v)} = \frac{1.91 \times 10^5}{3(1-2 \times 0.35)} = 2.122 \times 10^5 \text{ N/mm}$

13.22 TEMPERATURE STRESSES AND STRAINS

When the temperature of a material is changed its dimensions change. A stress is setup in the material if this change in dimension due to temperature change is prevented. This is called the temperature stress.

Let L = length of the member at temperature t

Δt = change in temperature

α = Coeffecient of linear expansion for the material,

Change in length $\Delta L = \alpha L.\Delta t$

Expanded length = $L(1 + \alpha.\Delta t)$

$$\text{Temperature strain, } \varepsilon_l = \frac{\alpha l.\Delta t}{L(1+\alpha.\Delta t)} = \frac{\alpha \Delta t}{1+\alpha.\Delta t} = \alpha.\Delta t(1 - \alpha \Delta t)$$

$$= \alpha \Delta t, \text{ neglecting } (\alpha.\Delta t)^2, \text{ being small.} \quad \text{... (13.21)}$$

Temperature stress, $\sigma_t = E\alpha\Delta t$ ... (13.22)

When the temperature increases, length increases. Since this increase in length is prevented, compressive stress is developed in the material. The reverse phenomena occurs when the temperature is decreased and tensile stress is developed.

13.23 COMPOSITE SYSTEM OF EQUAL LENGTHS SUBJECTED TO VARIATION OF TEMPERATURE

For the composite system shown in Fig. 13.38 Let $\alpha_1 > \alpha_2$, then on temperature rise by Δt member 2 will be under tension and member 1 under compression. If P is the common force developed then,

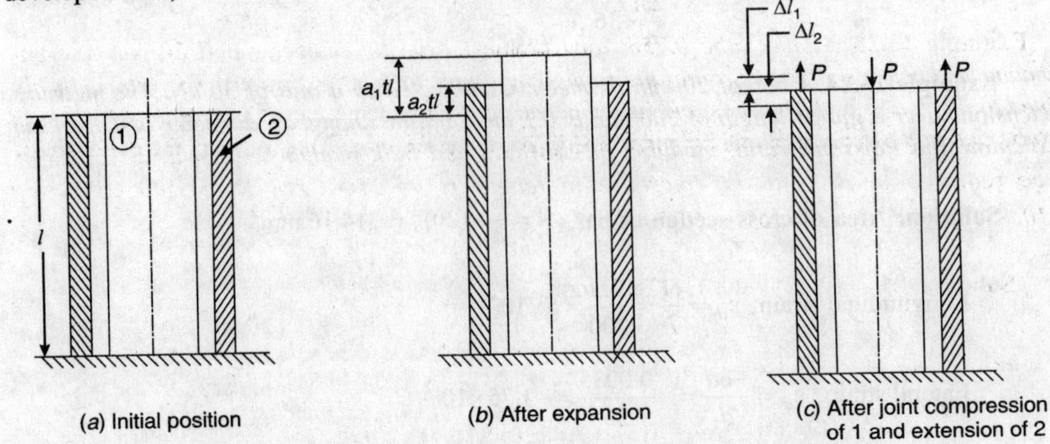

Fig 13.38 Composite system of equal lengths subjected to temperature rise

$$\Delta l_1 = \frac{Pl}{A_1E_1}(-ve), \Delta l_2 = \frac{Pl}{A_2E_2}(+\upsilon e)$$

$$\therefore \quad (\alpha_1 - \alpha_2)l.\Delta t = Pl\left(\frac{1}{A_1E_1} + \frac{1}{A_2E_2}\right)$$

$$\text{or} \quad P = \frac{(\alpha_1 - \alpha_2).\Delta t}{\left(\frac{1}{A_1E_1} + \frac{1}{A_2E_2}\right)} \qquad \text{... (13.23)}$$

$$\sigma_1 = \frac{P}{A_1} \text{ and } \sigma_2 = \frac{P}{A_2}$$

13.23.1 More Than Two Members

Consider three members shown in Fig 13.39 subjected to temperature rise.

$$\text{Increase in length, } \Delta l = \frac{l.\Delta t \Sigma \alpha_i A_i E_i}{\Sigma A_i E_i}$$

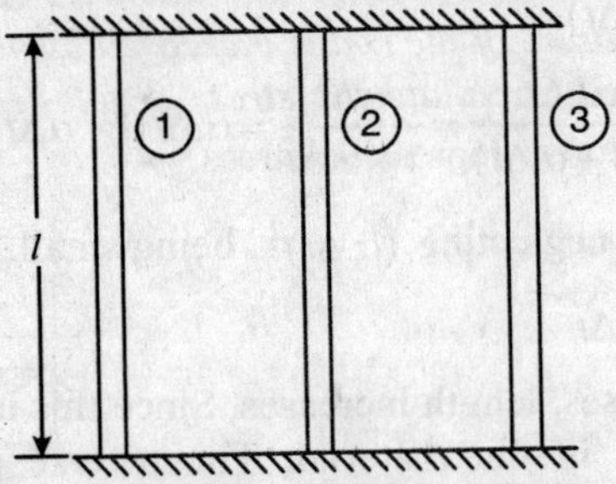

Fig. 13.39 Three members of equal lengths subjected to temperature rise

Stress in each member, $\sigma_i = \left(\frac{\Delta l - \alpha_i l.\Delta t}{l}\right)E_i$ ...(13.24)

If a load P is also applied in addition to temperature rise, then

$$\Delta l = \frac{l[P + \Delta t \Sigma \alpha_i A_i E_i]}{\Sigma A_i E_i} \quad ...(13.25)$$

Example 13.25 *A steel tube 0.75m long, 25 mm external diameter and 20 mm internal diamameter encloses a copper rod of the same length and 15 mm in diameter. The tube is firmly joined to the rod at both the ends and its temperature is raised by 120° C. Calculate the stresses in tube and rod. Also calculate the increase in length of the composite system and the external force required to prevent this increase in length* E_s *= 210 GPa,* E_c *= 100 GPa,* $\alpha_s = 10 \times 10^{-6}$ *per* ^{0}*C,* $\alpha_c = 15 \times 10^{-6}$ *per °C*

Solution: $A_s = \frac{\pi}{4}(25^2 - 20^2) = 176.71 \text{mm}^2$

$$A_c = \frac{\pi}{4} \times 15^2 = 176.71 \text{mm}^2$$

$$P = \frac{(\alpha_c - \alpha_s)\Delta t}{\frac{1}{A_c E_c} + \frac{1}{A_s E_s}} = \frac{(15-10) \times 120 \times 176.71 \times 10^5}{\frac{1}{1} + \frac{1}{2.1}} = 7182.36 \text{N}$$

$$\sigma_c = \sigma_s = \frac{7182.36}{176.71} = 40.645 \text{MPa}$$

$$\Delta l = \frac{l.\Delta t \Sigma \alpha_i A_i E_i}{\Sigma A_i E_i}$$

$$= \frac{750 \times 120(15 \times 1 + 10 \times 2.1) \times 10^{-6} \times 10^5 \times 176.71}{176.71 \times 10^5 (1 + 21)}$$

$$= 1.045 \text{ mm}$$

$$\Delta l = \frac{l[-P + \Delta t \Sigma \alpha_i A_i E_i]}{\Sigma A_i E_i}$$

$$= \frac{750[-P + 120(15 \times 1 + 10 \times 2.1) \times 10^5 \times 10^{-6} \times 176.71}{176.71 \times 10^5 (1 + 2.1)}$$

$$P = 76.34 \text{ kN}$$

Example 13.26 *A steel rod 1 m long and 15 mm diameter is held between rigid supports as shown in Fig. 13.40 The temperature of the rod is increased by 40° C. The coefficient of linear expansion for steel is* 12×10^{-6}*°C. Caclulate the stress developed in the rod.* E_s *= 200 GPa.*

Solution: The rod will develop compressive stress.

Thermal strain, $\varepsilon_t = \alpha.\Delta_t = 12 \times 10^{-6} \times 40$

$= 480 \times 10^{-6}$

Thermal stress $= E_s \varepsilon_t$

$= 200 \times 12^9 \times 480 \times 10^{-6}$

$= 96$ MPa

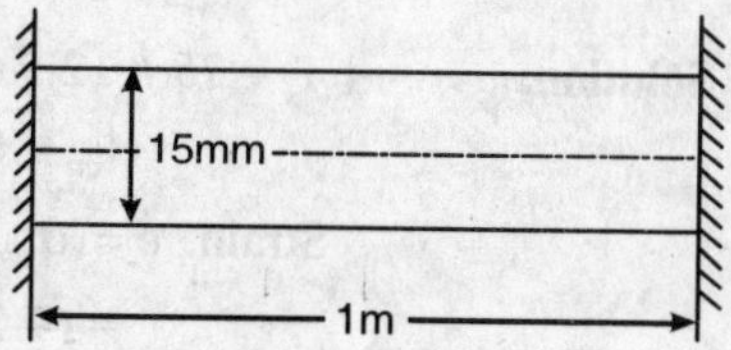

Fig. 13.40

Example 13.27 *Three identical vertical wires of 1 m length each and 4mm diameter are suspended from a horizontal support as shown in Fig.13.41. a load if 5 kN is applied at their lower ends by means of a rigid cross bar E_s = 210 GPa, E_{cu} = 105GPa. If the cross bar remains horizontal, find the stresses developed in the wries.*

Solution:

$$P = 2P_s + P_{cu} \quad ...(1)$$

$$\delta l_s = \delta l_{cu}$$

$$\frac{P_s l_s}{A_s E_s} = \frac{P_{cu} l_{cu}}{A_{cu} E_{cu}}$$

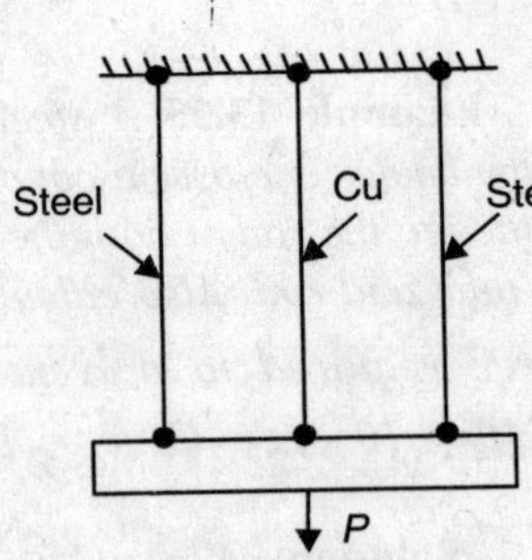

Fig. 13.41

Now

$$l_s = l_{cu}, A_s = A_{cu}$$

$$\frac{P_s}{E_s} = \frac{P_{cu}}{E_{cu}}$$

$$\frac{P_s}{p_{cu}} = \frac{E_s}{E_{cu}} = \frac{210}{105} = 2 \quad ...(2)$$

From Eqs. (1) and (2), we have

$$5P_{cu} = 5$$

$$P_{cu} = 1 \text{ kN}$$

$$P_s = 5 - 1 = 4 \text{ kN}$$

$$\sigma_s = \frac{4\times10^3}{\frac{\pi}{4}\times16} = 318.3\,\text{N/mm}^2$$

$$\sigma_{eu} = \frac{1\times10^3}{\frac{\pi}{4}\times16} = 79.58\,\text{N/mm}^2$$

Example 13.28 *A compound bar is shown in Fig. 13.42. Its temperature is raised by 120°F. Calculate the stresses in each metal and the change in length.E_s = 210 GPa E_{cu} 105 GPa. $\alpha_s = 6.5\times10^{-6}$ per ^0F, $\alpha_{cu} = 9.5\times10^{-6}$ per ^0F.*

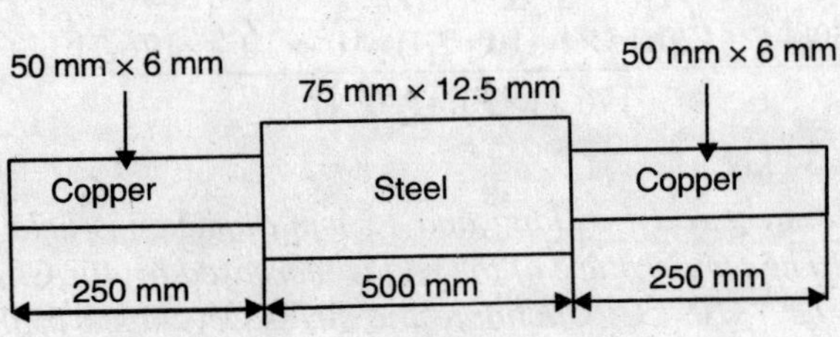

Fig. 13.42

Solution:

$$A_s = 75 \times 12.5 = 973.5 \text{ mm}^2$$

$$A_{cu} = 50 \times 6 = 300 \text{ mm}^2$$

$$\text{Strain, } \varepsilon = (\alpha_{cu} - \alpha_s).\Delta t$$

$$= (9.5 - 6.5) \times 10^{-6} \times 120$$

$$= 360 \times 10^{-6}$$

Now $$\frac{\sigma_s}{E_s}+\frac{\sigma_{cu}}{E_{cu}}=\varepsilon$$

$$\frac{\sigma_s}{2.1}+\frac{\sigma_{cu}}{1.05}=360\times10^{-6}\times10^5$$

or $$\sigma_s+2\sigma_{cu}=75.6 \qquad ...(1)$$

Also $$\sigma_s A_s=\sigma_{cu}A_{cu}$$

$$\sigma_{cu}=\frac{937.5}{300}\times\sigma_s=3.125\sigma_s \qquad ...(2)$$

From Eqs. (1) and (2), we get

$$\sigma_s=10.427\,\text{MPa}\,(+ve),\sigma_{cu}=32.586\text{MPa}\,(-ve)$$

Change in length, $\Delta l=\varepsilon l=360\times10^{-6}\times10^3=0.36\,\text{mm}$

Example 13.29 *A composite rod of steel and copper of lengths 1.5 m and 1 m respectively is clamped at the ends, as shown in Fig. 13.43. Its temperature is then raised by 50° C. Calculate the stresses in the two metels of rod E_s = 210 GPa, E_c = 105 GPa, $\alpha_s=10\times10^{-6}$ per °C and $\alpha_c=15\times10^{-6}$ per °C.*

Solution: Free extensions are:

$$\delta_{cu}=\alpha_{cu}l_{cu}.\Delta t=15\times10^{-6}\times1\times10^3\times50=0.75\text{mm}$$

$$\delta_s=\alpha_s l_s.\Delta t=10\times10^{-6}\times1.5\times10^3\times50=0.75\text{mm}$$

Total extension, $\delta=\delta_{cu}+\delta_s=1.5\,\text{mm}$

Let P = compressive force developed in bar.

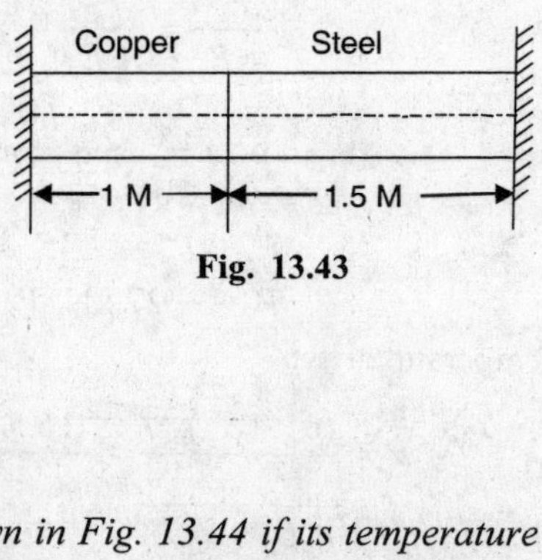

Fig. 13.43

$$\delta=\frac{p}{A}\left[\frac{l_{cu}}{E_{cu}}+\frac{l_s}{E_s}\right]$$

$$1.5=\sigma\left[\frac{1\times10^3}{1.05\times10^5}+\frac{1.5-10^3}{2.1\times10^5}\right],$$

$\sigma=90\text{MPa}$

Example 13.30 *Determine the stresses in the bar shown in Fig. 13.44 if its temperature is increased by 30°C. $E_s=2E_c=200\,\text{GPa}$ $\alpha_s=12.5\times10^{-6}$ per °C, $\alpha_c=16.5\times10^{-6}$ per °C.*

Solution: $$A_c=\frac{\pi}{4}\times150^2=17671\text{mm}^2$$

$$A_s=\frac{\pi}{4}\times100^2=7854\,\text{mm}^2$$

Free extensions:

$$\delta_c=\alpha_c l_c.\Delta t=16.5\times10^{-6}\times1.5\times10^3\times30$$
$$=0.7425\text{ mm}$$

$$\delta_s = \alpha_s l_s .\Delta t = 12.5\times10^{-6}\times1\times10^3\times30$$
$$= 0.375 \text{ mm}$$

Total extension, $\delta_s = \delta_c + \delta_s = 1.1175\,\text{mm}$

Constrained compression = 1.1175 – 0.03 = 1.0875 mm

Let P = compressive force developed in the bars.

$$P\Sigma\frac{l_i}{A_iE_i} = 1.0875$$

$$\frac{P}{10^2}\left[\frac{1.5}{17671\times1}+\frac{1}{7854\times2}\right] = 1.0875$$

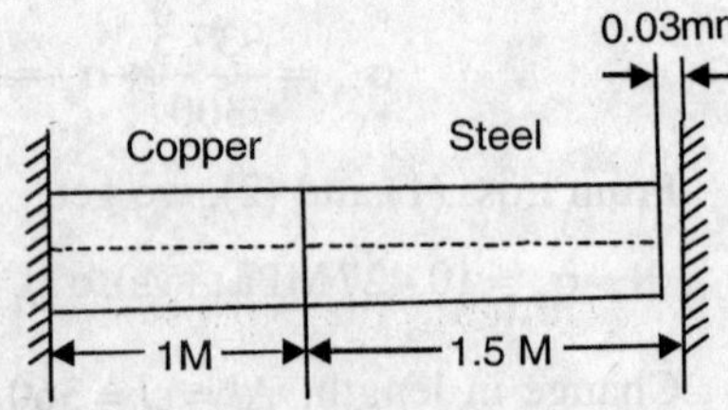

Fig. 13.44

$$P = 732093 \text{ N}$$

$$\sigma_c = \frac{P}{A_c} = \frac{732093}{17671} = 41.429\text{MMa (–ve)}$$

$$\sigma_s = \frac{P}{A_s} = \frac{732093}{7854} = 93.213\text{MPa (–ve)}$$

Example 13.31 *A load of 45 kN is transmitted through a slab to a composite solid steel cylinder (A_s = 15cm²) and hollow copper cylinder (A_c = 20 cm²). After the application of the load the temperature was increased by 30° C. Determine the distribution of load before and after the change in temperature. $E_s = 2E_c$ = 200 GPa, α_s = 12.5 × 10^{-6} per °C, α_c = 16.5 × 10^{-6} per °C.*

Solution: Load $E_s = E_c$

$$\frac{P_s}{A_sE_s} = \frac{P_c}{A_cE_c}$$

$$P_s = \frac{15}{20}\times2P_c = 1.5P_c \qquad \ldots(1)$$

$$P_s + P_c = -45 \qquad \ldots(2)$$

$$\therefore \quad P_s = -27 \text{ kN}, P_c = -18 \text{ kN}$$

Temperature rise:

$$P = \frac{(\alpha_c - \alpha_s).\Delta t}{\frac{1}{A_sE_s}+\frac{1}{A_cE_c}} = \frac{(16.5-12.5)\times10^{-6}\times30\times10^{-7}}{\frac{1}{15\times2}+\frac{1}{20\times2}} = 14.4\text{kN}$$

$$P_c = -14.4 \text{ kN}$$
$$P_s = +14.4 \text{ kN}$$

Resultant load, P_c = – 18 – 14.4 = – 32.4 kN

P_s = + 27 – 14.4 = + 12.6 kN

Example 13.32 *A tapered circular bar is rigidly fixed at both the ends as shown in Fig. 13.45. If the temperature is raised by 50° C, calculate the stress in the bar. Take E = 202 GPa and $\alpha = 12\times10^{-6}$ per °C.*

$$d_l = 80 \text{ mm}, d_2 = 120 \text{ mm}, l = 1.0 \text{ m}.$$

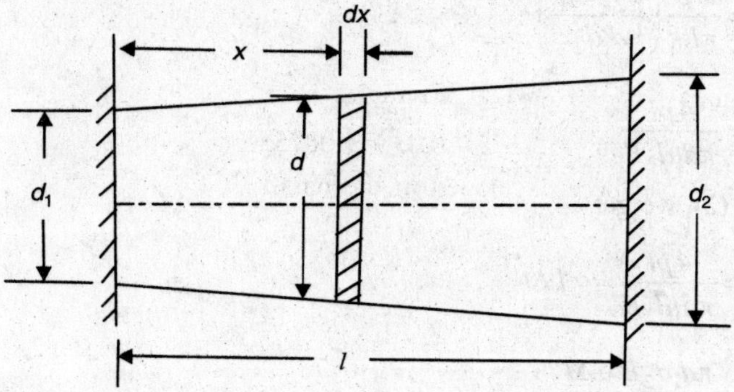

Fig. 13.45

Solution: Free expansion of bar $\delta l = \alpha l.\Delta t$

$$= 12 \times 10^{-6} \times 1.0 \times 50 = 6 \times 10^{-4} \text{ m} \quad \ldots(1)$$

Let P = force developed in the bar in preventing its free expansion.

Consider an element of thickness dx at a distance x from the smaller end of the bar. Expansion of element dx due to P,

$$\delta u = \frac{P.dx}{A_x E}$$

Diameter of element, $d = d_1 + \left(\dfrac{d_2 - d_1}{l}\right)x$

Area of cross-section of element,

$$A_x = \frac{\pi}{4}d^2 = \frac{\pi}{4}\left[d_1 + \left(\frac{d_2 - d_1}{l}\right)x\right]^2$$

$$= \frac{\pi}{4}[d_1 + kx]^2$$

where $\quad k = \dfrac{d_2 - d_1}{l}$

Total extension, $\delta l = \displaystyle\int_0^l \delta u = \frac{4p}{\pi E}\int_0^1 \frac{dx}{(d_1 + kx)^2}$

$$= \frac{-4p}{\pi Ek}\left|\frac{1}{d_1 + kx}\right|_0^1$$

$$= \frac{-4p}{\pi Ek}\left|\frac{1}{d_1 + kl} - \frac{1}{d_1}\right|$$

$$= \frac{-4p}{\pi Ek}\left|\frac{1}{d_1 + d_2 - d_1} - \frac{1}{d_1}\right|$$

$$= \frac{4p}{\pi Ek}\left(\frac{d_2 - d_1}{d_1 d_2}\right)$$

$$= \frac{4pl}{\pi d_1 d_2 E} \qquad \ldots(2)$$

From Eqs. (1) and (2), we get

$$= \frac{4pl}{\pi d_1 d_2 E} = \infty l.\Delta t$$

$$P = \frac{\pi d_1 d_2 E \infty.\Delta t}{4}$$

Maximum stress, $\sigma_{max} = \frac{P}{A_1} = \frac{4P}{\pi d_1^2} = E \propto .\Delta t \times \frac{d_2}{d_1}$

$$= 202 \times 10^9 \times 12 \times 10^{-6} \times 50 \times \frac{120}{80}$$

$$= 180.8 \times 106 \text{ N/mm}^2$$

Example 13.33 *A trapezoidal flat steel plate of thickness 10 mm tapers uniformly from a width 150 mm to 100 mm in a length of 500 mm. Calculate the extension of the plate when subjected to axial tensile force of 50 kN. Take E = 200 GPa*

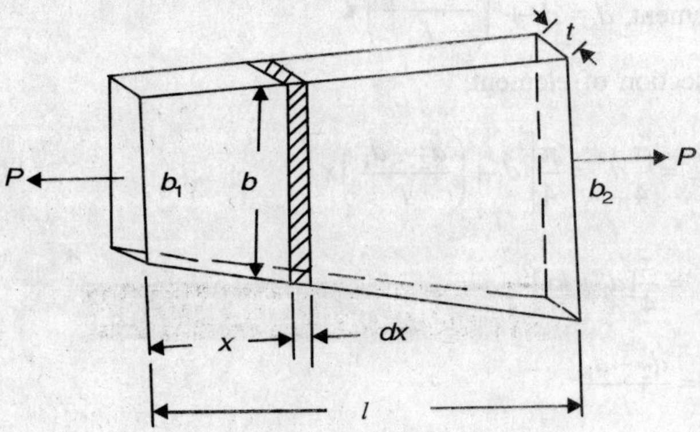

Fig. 13.46

Solution: $b_1 = 100$ mm $b_2 = 150$ mm, $l = 500$ mm, $t = 10$ mm

$P = 50$ kN, $E = 200$ GPa

Consider a small element of the plate at a distance x and of thickness dx from the small end as shown in Fig. 13.46.

Width of element, $b = b_1 + \left(\frac{b_2 - b_1}{l}\right)x$

$$= b_1 + kx$$

where $k = \frac{b_2 - b_1}{l}$

Area of element, $A_x = (b_1 + kx)t$

Extension of dx, $\delta u = \dfrac{p.dx}{(b_1 + kx)tE}$

Total extention of plate, $\delta l = \int_o^l \delta u = \int_o^l \dfrac{P.dx}{(b_1 + kx)tE}$

$$= \frac{P}{tE}\int_o^l \frac{dx}{b_1 + kx}$$

$$= \frac{P}{kte}\left|\text{In}\,(b_1 + kx)\right|_o^l$$

$$= \frac{P}{ktE}[\text{In}\,(b_1 + kl) - l_n b_1]$$

$$= \frac{P}{ktE}[l_n b_2 - l_n b_1]$$

$$= \frac{P}{ktE} l_n\left(\frac{b_2}{b_1}\right)$$

$$= \frac{Pl}{(b_2 - b_1)tE} l_n\left(\frac{b_2}{b_1}\right)$$

$$= \frac{50\times10^3\times500}{(150-100)\times10\times2\times10^5}\times l_n\left(\frac{150}{100}\right)$$

$$= 0.1014 \text{ mm}$$

MULTI – CHOICE QUESTIONS

1. Choose the brittle marterial.
   (*a*) Mild steel (*b*) Cast iron (*c*) Aluminium (*d*) Copper
2. Which is not a physical property of a material?
   (*a*) Density (*b*) refractive index
   (*c*) Roughness (*d*) Thermal expansion
3. Which is not a mechanical property of a material?
   (*a*) Hardness (*b*) Yield strength
   (*c*) Poisson's ratio (*d*) Endurance limit
4. Toughness of a material means
   (*a*) High strength (*b*) High hardness
   (*c*) Ability to absorb strain energy (*d*) Limit of proportionality
5. Hooke's law is obeyed up to the
   (*a*) Yield point (*b*) Ultimate point

(*c*) Breaking point (*d*) Limit of proportionality

6. If equal and opposite forces are applied to a body to elongate it, they produce
(*a*) Zero stress (*b*) Tensile stress
(*c*) Compressive stress (*d*) Shear stress

7. If the radius of a wire stretched by a load is doubled. then Young's modulus of elasticity will be
(*a*) Doubled (*b*) Halved (*c*) Four times (*d*) Unaffected

8. If a material is restricted to deform and heated, it will develop
(*a*) Tensile stress (*b*) Compressive stress
(*d*) Shear stress (*d*) No stress

9. If a material expands freely due to heating, it will develop
(*a*) Tensile stress (*b*) Compressive stress
(*c*) Shear stress (*d*) No stress

10. Longitudinal and lateral strains are
(*a*) Same in nature (*b*) Opposite in nature
(*c*) Same in magnitude (*d*) None of the above

11. Thermal strain is expressed by.
(*a*) $\alpha.\Delta t$ (*b*) $\alpha l.\Delta t$ (*c*) $E\alpha.\Delta t$ (*d*) $E.\Delta t$

12. True strain is given by.
(*a*) $\bar{\varepsilon} = 1+\varepsilon$ (*b*) $\bar{\varepsilon} = ln+(1+\varepsilon)$ (*c*) $\bar{\varepsilon} = \frac{1}{1+\varepsilon}$ (*d*) $\bar{\varepsilon} = \ln \varepsilon$

13. True stress is given by
(*a*) $\bar{\sigma} = \sigma(1+\varepsilon)$ (*b*) $\bar{\sigma} = \frac{\sigma}{1+\varepsilon}$ (*c*) $\bar{\sigma} = \sigma(1-\varepsilon)$ (*d*) $\bar{\sigma} = \frac{\sigma}{1-\varepsilon}$

14. The SI units of stress are:
(*a*) kg/ mm^2 (*b*) N/mm (*c*) N/mm^2 (*d*) N.mm

15. The ratio of sudden load to equivalent static load is.
(*a*) 3 (*b*) 2 (*c*) 1.5 (*d*) 1/2

ANSWERS:

1. (*b*) **2.** (*c*) **3.** (*c*) **4.** (*c*) **5.** (*d*) **6.** (*b*) **7.** (*d*) **8.** (*b*) **9.** (*d*) **10.** (*b*)
11. (*a*) **12.** (*b*) **13.** (*a*) **14.** (*c*) **15.** (*d*)

REVIEW QUESTIONS

1. Define true and engineering stresses.
2. State Hooke's law.
3. Define engineering and natural strains.
4. Define Poisson's ratio.
5. What is a thermal stress?
6. Define hardness.
7. Define toughness and brittleness.
8. Define fatigue and endurance limit.
9. What is a permanent set ?
10. What is sudden load?
11. Define creep. What are its stages?

12. What is a Wohler curve?

13. Define Young's modulus of elasticity.

14. Define strain energy and resilience.

EXERCISES

13.1 A steel bar of diameter 15 mm is loaded as shown in Fig. 13.47. Calculate the stresses in each part and total change in length. Take E = 200 GPa.

[**Ans.** 226.35/N/mm^2, 141.47/N/mm^2/0.438mm]

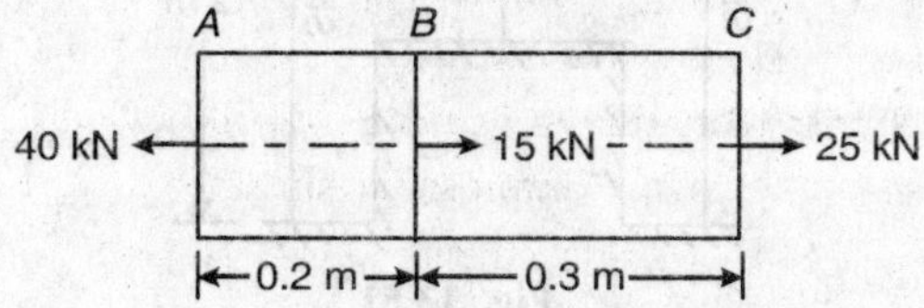

Fig. 13.47

13.2 Find the change in length of the bar loaded as shown in Fig. 13.48. Take E = 200 GPa.

[**Ans.** 0.924mm]

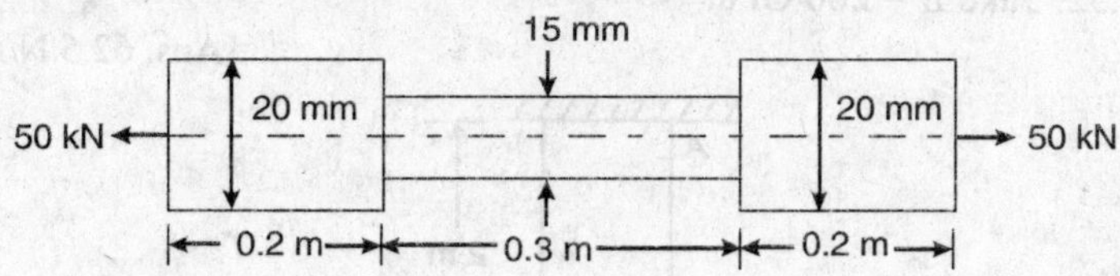

Fig. 13.48

13.3 Calculate the change in length of the bar shown in Fig. 13.49. Take E = 200 GPa.

[**Ans.** No change]

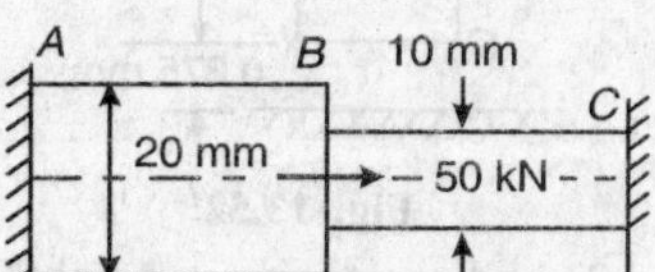

Fig. 13.49

13.4 A block weighing 42 kN is supported by three wires as shown in Fig. 13.50. The two outer steel wires have an area of 100 mm^2 each, whereas the middle aluminium wire has an area of 200 mm^2. Calculate the stresses in the wires. Take E_s = 200 GPa and E_a = 80 GPa.

[**Ans.** 150 MPa, 60 MPa]

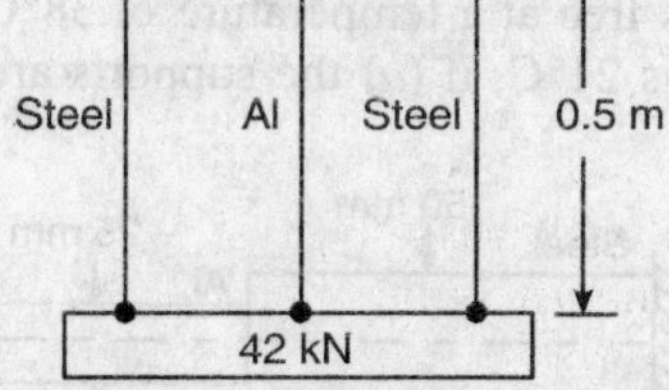

Fig. 13.50

13.5 Two steel wires and a copper wire support a load of 50 kN as shown in Fig. 13.51. The diameter of steel wire is 20 mm and that of copper wire 15 mm. If $E_s = 200$ GPa and $E_c = 100$ GPa, calculate the stresses induced in the wire.

[**Ans.** 67.01 N/mm^2, 44.67 N/mm^2]

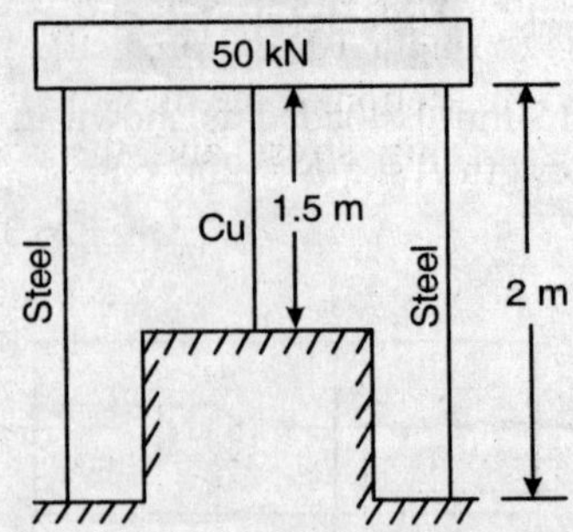

Fig. 13.51

13.6 A steel rod clamped as its upper end, has a cross-sectional area of 20 cm^2. Between its lower end and a rigid plane, there was a clearance of 0.375 mm before the load was applied. Determine the stresses in the upper and lower parts of the rod due to the force $P = 225$ kN, as shown in Fig. 13.52. Take $E = 200$ GPa.

[**Ans.** 62.5 N/mm^2, –50 N/mm^2]

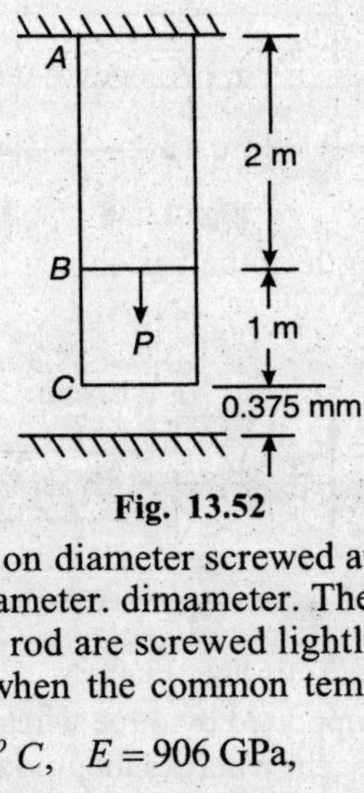

Fig. 13.52

13.7 A gun metal rod 2 on diameter screwed at the ends, passes through a steel tube 2.5 cm internal and 3 cm external diameter. dimameter. The temperature of the whole assembly is raised to 130° C and nuts on the rod are screwed lightly home on the ends of the tube. Find. the stress in the rod and the tube when the common temperature has fallen to 15°C.

For gun meter : $\infty = 18 \times 10^{-6}\ /^{o}\,C,\quad E = 906$ GPa,

For steel : $\propto = 12 \times 10^{-6}\ /^{o}\,C, E = 210$ GPa.

[**Ans.** 38.28 MPa. – 55.69 MPa]

13.8 A composite bar made up of aluminium and steel is held between two supports as shown in Fig. 13.53.The bars are stress free at a temperature of 38°C. What will be the stresses in the two bars when the temperature is 21°C, if (*a*) the supports are rigid, and (*b*) the supports come nearer to each other by 0.1 mm.

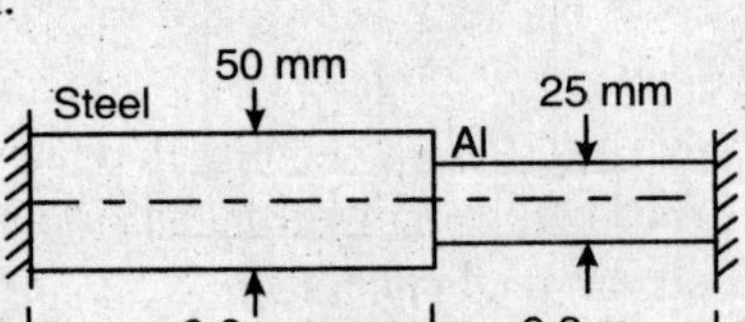

Fig. 13.53

For steel : $\propto = 11.7 \times 10^{-6} /^{o} C, E = 210$ GPa.

For aluminium: $\propto = 23.4 \times 10^{-6} /^{o} C, E = 74$ GPa.

[**Ans.** (*a*) 12.51 MPa, 50.04 MPa (*b*) 7.27 MPa, 29.08 MPa]

13.9 An unknown weight falls throught 10 mm on a collar rigidly attached to the lower end of a vertical bar, 3m long and 6 cm^2 in section. If the maximum instantaneous extension is known to be 2 mm, what is the corresponding stress and the value of unknown weight ? take $E = 200$ GPa.

[**Ans.** 133.33 MPa, 6.667 kN]

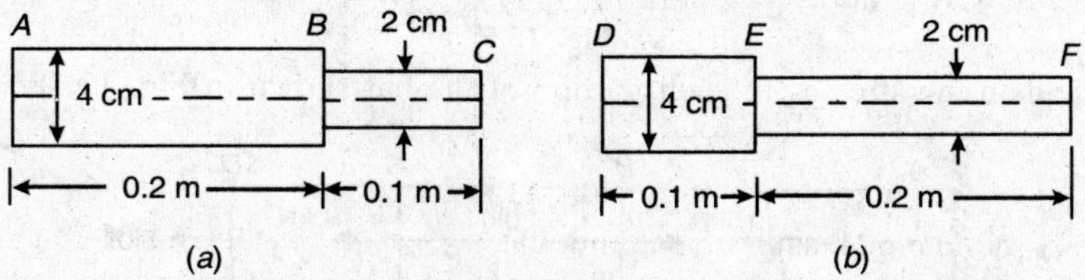

Fig. 13.54

13.10 Two similar round bars "*a*" and "*b*" are each 0.3 m long a shown in Fig 13.54 The bar "*a*" receives an axial blow which produces a maximum stress of 200 MPa. Find the maximum stress produced in bar ''b'' by the same blow. [**Ans.** 163.3 Mpa]

13.11 Calculate the total elongation of the bar loaded or shown in Fig.13.55. E = 200 GPa, area of cross-section of bar is 500 mm^2 [**Ans.** – 0.13 mm]

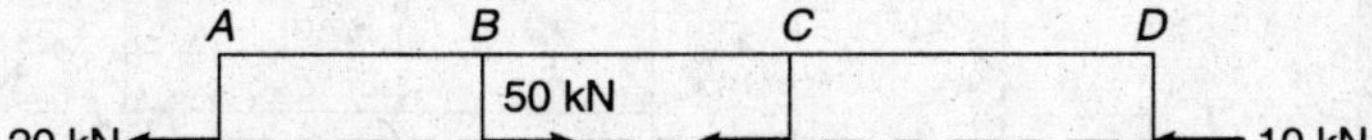

Fig. 13.55

13.12 Calculate the stresses in the bars loaded as shown in Fig.13.56. E = 200 GPa [**Ans.** 385.71 MPa, –57.143 MPa]

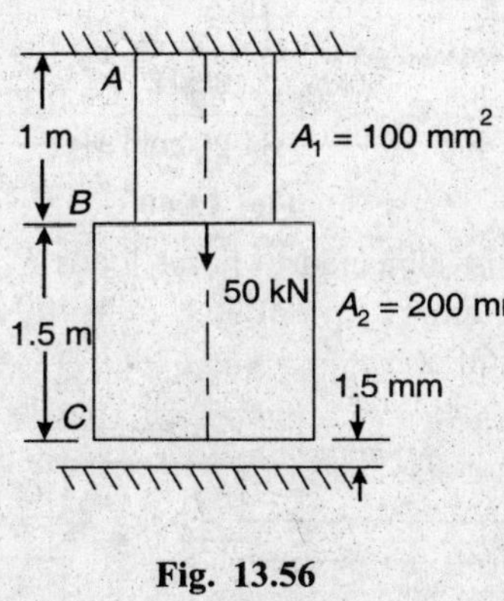

Fig. 13.56

13.13 A steel specimen 2 cm^2 are of cross-section elongates by 0.1 mm over a gauge length of 50mm under an axial load of 40 kN. Calculate the strain energy in the specimen. If its limit of proportionality is 45 MPa, find the resilience. [**Ans.** 2 N.m 10125 N.mm/mm^3]

13.14 A bar 0.5 m long has 2 cm^2 area of cross-section for 0.4 m of its length and 1 cm^2 for the remaining length. A load of 40 kN falls on the collar provided at the lower end of the bar, the other end being fixed, from a height of 5cm. Find the maximum stress induced in the bar. E = 200 GPa. **[Ans.** 163.7 MPa]

13.15. Calculate the total elongation of the steel bar shown in Fig. 13.57. E = 200GPa.

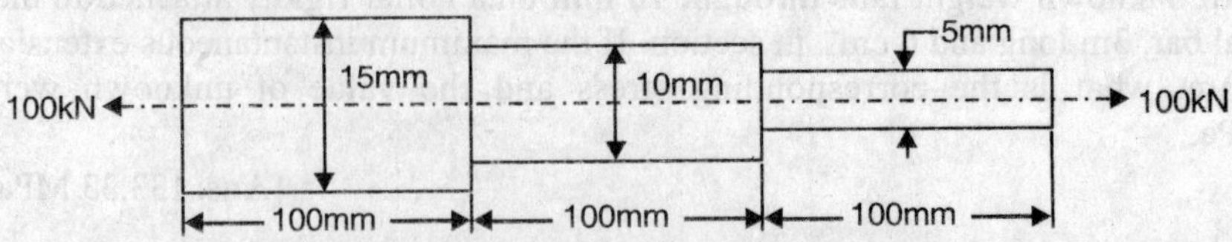

Fig. 13.57

13.16 Calculate the stresses in each portion of the bar shown in Fig. 13.58

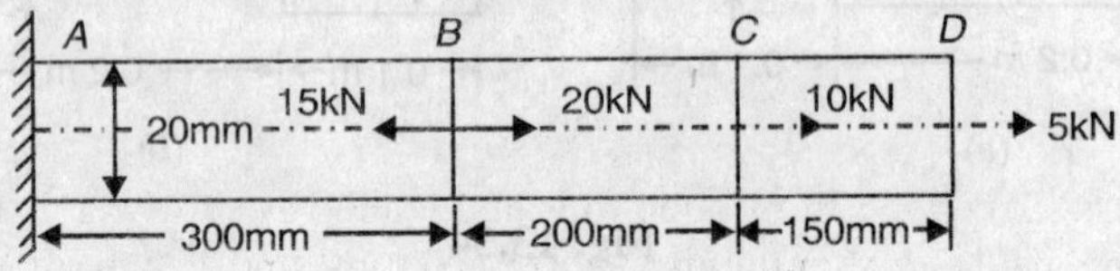

Fig. 13.58

13.17 A stepped bar is loaded as shown in Fig.13.59. Calculate the stresses in each portion and total elongation. E = 100 GPa.

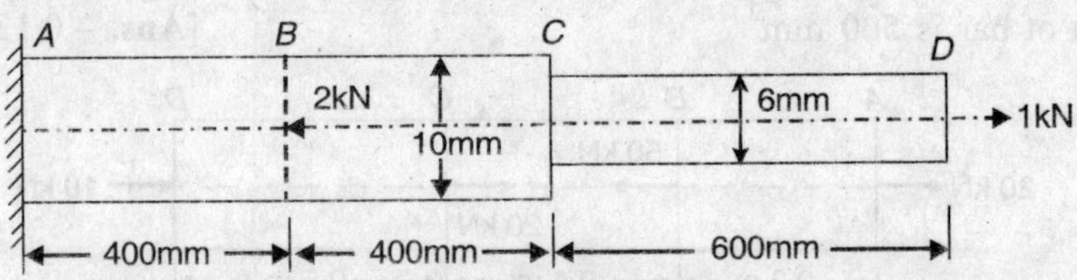

Fig. 13.59

13.18 Determine the stresses in all parts of the bar shown in Fig. 13.60 and rigidly clamped at the ends.

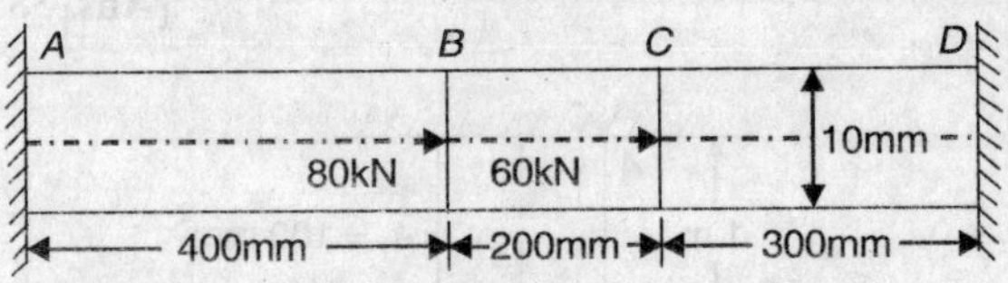

Fig. 13.60

13.19 A member $A\ B\ C\ D$ is subjected to point loads P_1 to P_4 as shown in Fig. 13.61 Find the value of force P_2 for equilibrium, if P_1 = 40 kN, P_2 = 400 kN and P_4 = 120 kN. Also calculate the total elongation, if E = 200 GPa.

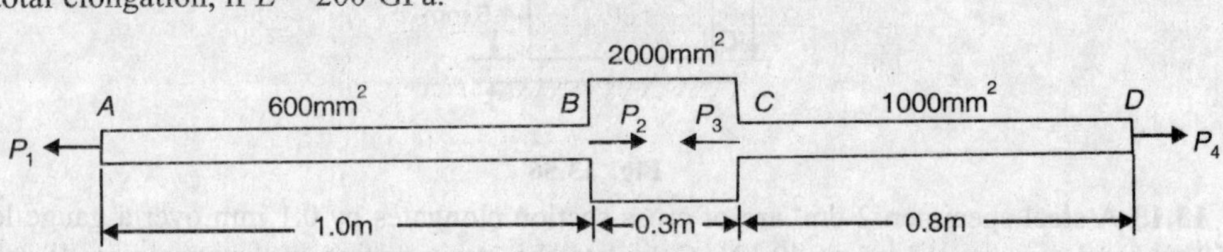

Fig. 13.61

13.20 A rigid beam *AB* is suspended horizontally from rods 1and 2 as shown in Fig.13.62 Rod 1 is of steel and has 15 mm diameter, while rod 2 is of copper and has 20 mm diameter. At what distance from rod 1 must the load *P* be applied if the beam *AB* is to remain horizontal after the deformation or rods.If P = 30 kN, calculate the stresses in the rods.

Take E_s = 200 GPa and E_c = 100GPa.

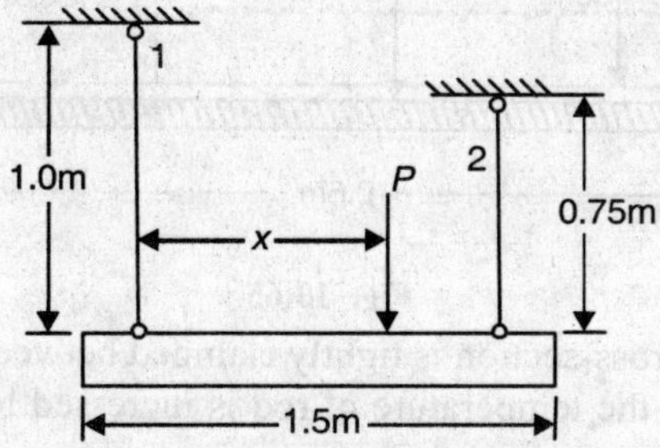

Fig. 13.62

13.21 A rigid bar is suspended from three rods and loaded as shown in Fig.13.63. The crosss-sectional area of copper ,steel and aluminium rods are 100 mm^{2,} and 150 mm^2 respectively Calculate the stresses in the rods. Take E_s = 200 GPa, E_c = 100 GPa, and E_a = 70 GPa.

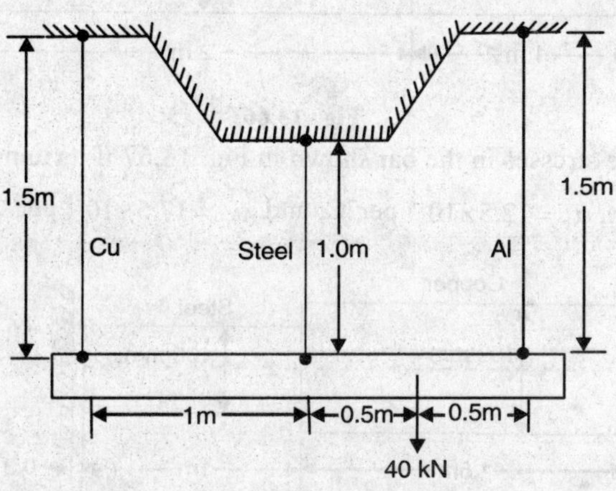

Fig. 13.63

13.22 A steel bolt is put through a copper tube as shown in Fig. 13.64. The pitch of screw threads of bolt is 5 mm. Calculate the stresses in the bolt and tube when the nut is turned through one-quarter of a revolution, E_s = 2 E_c =200 GPa.

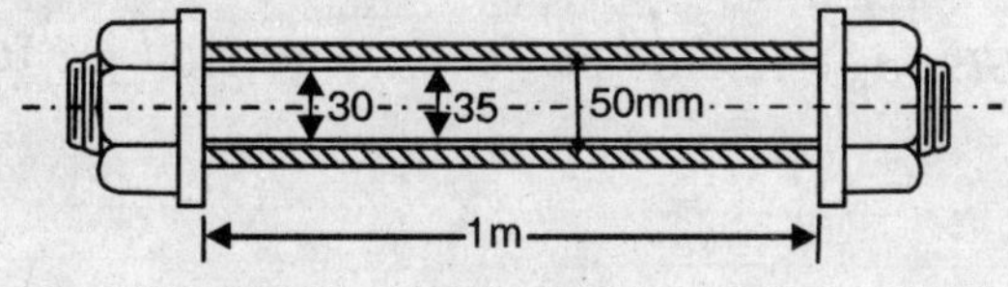

Fig. 13.64

13.23 A solid steel bar 0.5 m long and 50 mm diameter is placed inside and aluminium tube having inside and outside diameters 60 mm and 80 mm respectively. The aluminium tube is 0.2 mm longer than steel bar. An axial load of 750 kN is applied to the composite system through rigid cover plates as shown in Fig. 13.65. Calculate the stresses developed in steel bar and aluminium tube. $E_s = 200$ GPa, $E_a = 70$ GPa.

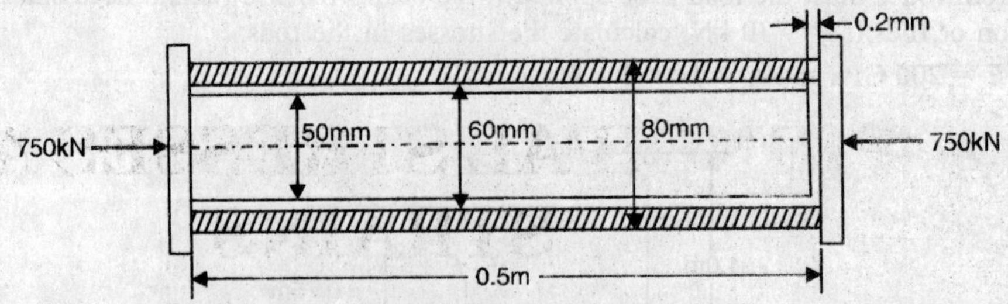

Fig. 13.65

13.24 A rod of constant cross-section is tightly clamped between two rigid supports as shown in Fig. 13.66. After clamping the temperature of rod is incressed by 60°C. Calculate the stresses in two parts of the rod. $E_s = 200$ GPa. $E_c = 100$ GPa, $\alpha_s = 12 \times 10^{-6}$ per °C, and $\alpha_c = 15 \times 10^{-6}$ per °C.

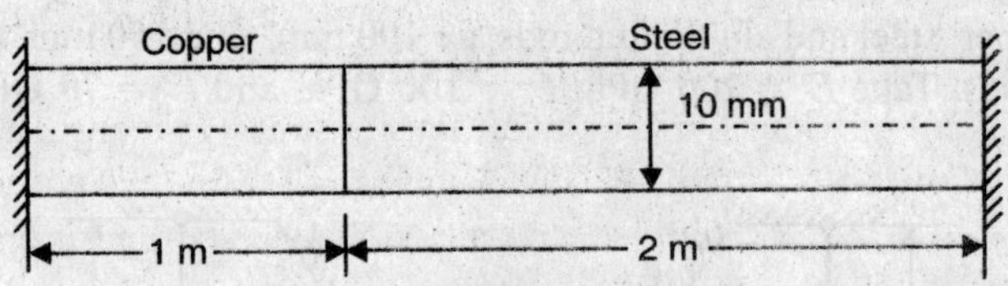

Fig. 13.66

13.25 Calculate the stresses in the bar shown in Fig. 13.67 if its temperature is raised by 40° C. $E_s = 2E_c = 200\ GPa$, $\alpha_s = 12.5 \times 10^{-6}$ per°C, and $\alpha_c = 17.5 \times 10^{-6}$ per °C.

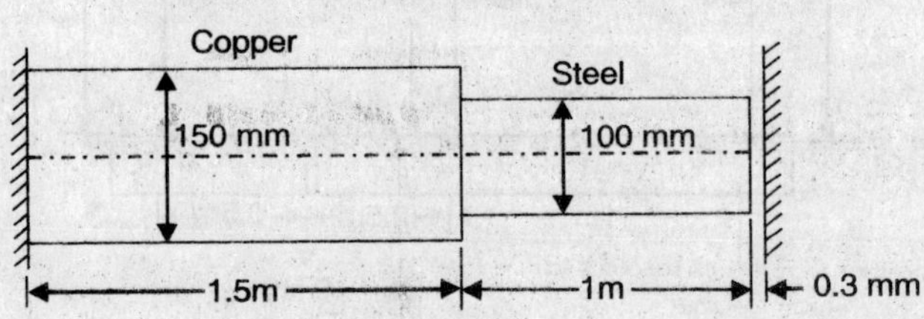

Fig. 13.67

13.26 Two rods are joined together one behind the other and the extreme ends are built in the fixed supports. One rod is of steel 3 m long and 600 mm^2 cross-sectional area. The other rod is of bronze 0.8 m long and 1200 mm^2 cross-sectional area. There is no stress initially in the rods. Calculate the stresses set up in the rods if the temperature is dropped by 50° C.

$E_s = 200\ GPa$, $E_b = 70\ GPa$, $\alpha_a = 12 \times 10^{-6}$ per °C, $\alpha_b = 18 \times 10^{-6}$ per °C.

14 PRINCIPAL STRESSES AND STRAINS

14.1 INTRODUCTION

In this chapter, we shall study the stresses and strains on an oblique plane under uniaxial, biaxial and complex stresses and strains. The determination of principal stresses and elastic constants shall also be discussed.

14.2 STRESSES ON AN OBLIQUE PLANE

(*a*) **Uniaxial Loading:** Consider a uniaxial tensile stress σ acting on plane *BC* of a bar of uniform area of cross-section, as shown in Fig. 14.1 (*a*). We are interested to determine the normal and shear stresses on plane *AC*, whose normal is inclined at an angle θ with the line of action of σ. The forces acting at a point on plane *AC* are shown in Fig. 14.1 (*b*)

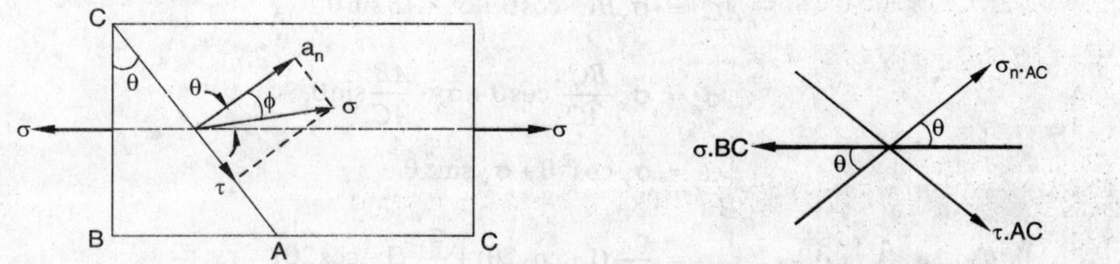

(a) Uniaxial loading on a bar (b) Forces on oblique plane

Fig. 14.1 Stresses on an oblique plane due to uniaxial loading.

Resolving the forces perpendicular and parallel to the plane *AC*, we have

$$\sigma_n \times AC = \sigma \times BC \cdot \cos\theta$$

$$\sigma_n = \sigma \cdot \frac{BC}{AC} \cdot \cos\theta = \sigma\cos^2\theta = \frac{1}{2}\sigma(1+\cos 2\theta) \qquad ...(14.1)$$

$$\tau AC = \sigma BC \cdot \sin\theta$$

$$\tau = \sigma\frac{BC}{AC}\sin\theta = \sigma\cos\theta \cdot \sin\theta = \frac{1}{2}\sigma\sin 2\theta \qquad ...(14.2)$$

At $\theta = 45°$ and $135°$, $\tau_{max} = \frac{1}{2}\sigma$

Resultant stress, $\sigma_r = \sqrt{\sigma_n^2 + \tau^2} = \frac{1}{2}\sigma\sqrt{(1+\cos 2\theta)^2 + \sin^2 2\theta}$

$$= \frac{1}{2}\sigma\sqrt{1+\cos^2 2\theta + 2\cos 2\theta + \sin^2 2\theta}$$

$$= \frac{1}{2}\sigma\sqrt{2+2\cos 2\theta} = \frac{1}{2}\sigma\sqrt{2\times 2\cos^2\theta} = \sigma\cos\theta \quad \text{...(14.3)}$$

$$\tan\phi = \frac{\tau}{\sigma_n} = \tan\theta$$

or $$\phi = \theta \quad \text{...(14.4)}$$

(*b*) **Biaxial Loading:** Now consider the case of biaxial loading, as shown in Fig. 14.2. Resolving the forces perpendicular and parallel to the plane *AC*, we have

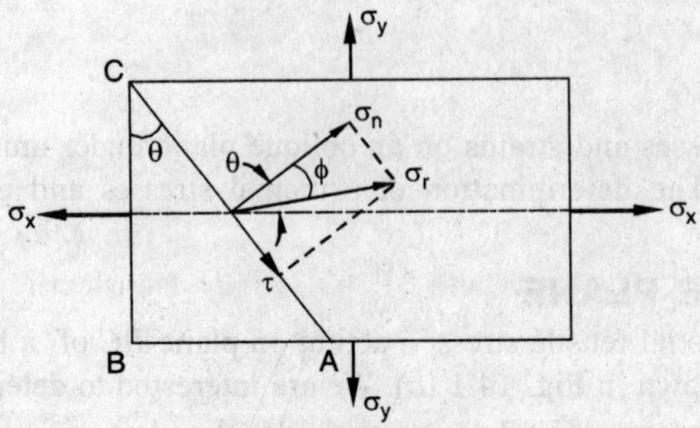

(a) Biaxal loading on a bar

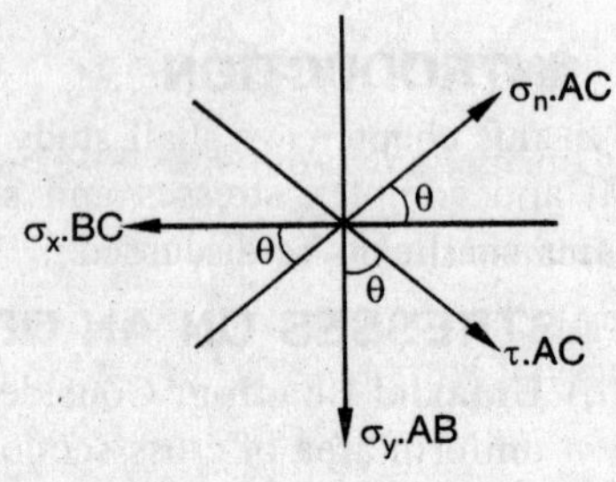

(b) Forces on oblique plane

Fig. 14.2 Stresses on an oblique plane due to biaxial loading.

$$\sigma_n AC = \sigma_x BC\cdot\cos\theta + \sigma_y\cdot AB\sin\theta$$

$$\sigma_n = \sigma_x\frac{BC}{AC}\cdot\cos\theta + \sigma y\cdot\frac{AB}{AC}\sin\theta$$

$$= \sigma_x\cos^2\theta + \sigma_y\sin^2\theta$$

$$= \frac{\sigma_x}{2}(1+\cos 2\theta) + \frac{\sigma_y}{2}(1-\cos 2\theta)$$

$$= \frac{1}{2}(\sigma_x+\sigma_y) + \frac{1}{2}(\sigma_x-\sigma_y)\cos 2\theta \quad \text{...(14.5)}$$

$$\tau\cdot AC = \sigma_x BC\sin\theta - \sigma_y AB\cdot\cos\theta$$

$$\tau = \sigma_x\cos\theta\sin\theta - \sigma_y\sin\theta\cos\theta = \frac{1}{2}(\sigma_x-\sigma_y)\sin 2\theta \quad \text{...(14.6)}$$

Maximum shear stress, $\tau_{max} = \frac{1}{2}(\sigma_x - \sigma_y)$ at $\theta = 45°$ ...(14.7)

Resultant stress, $\sigma_r = \sqrt{\sigma_n^2+\tau^2} = \sqrt{\sigma_x^2\cos^2\theta + \sigma_y^2\sin^2\theta}$ ...(14.8)

$$\tan\phi = \frac{\tau}{\sigma_n} = \frac{(\sigma_x-\sigma_y)\sin\theta\cos\theta}{\sigma_x\cos^2\theta+\sigma_y\sin^2\theta} = \frac{\sigma_x-\sigma_y}{\sigma_x\cot\theta+\sigma_y\tan\theta} \quad \text{...(14.9)}$$

For ϕ to be maximum, $\frac{d(\tan\phi)}{d\theta} = 0$

or $$\frac{d}{d\theta}(\sigma_x \cot\theta + \sigma_y \tan\theta) = 0$$

$$-\sigma_x \operatorname{cosec}^2\theta + \sigma_y \sec^2\theta = 0$$

$$\tan\theta = \sqrt{\frac{\sigma_x}{\sigma_y}} \qquad ...(14.10)$$

$$\tan\phi_{max} = \frac{\sigma_x - \sigma_y}{2\sqrt{\sigma_x\sigma_y}} \qquad ...(14.11)$$

14.3 SHEAR STRESSES

Consider a two-dimensional lamina shown in Fig. 14.3 (*a*) subjected to shear stresses τ_{xy} and τ_{yx}. Because shear stresses are complementary in nature, therefore, $\tau_{xy} = \tau_{yx}$. To prove the shear stresses to be complementary in nature, let $AB = DC = dx$ and $BC = AD = dy$. Considering unit thickness of lamina *ABCD* and taking moments about *A*, we have

$$\tau_{xy} \times BC \times AB = \tau_{yx} \times DC \times AD$$

or $$\tau_{xy} \times dy \times dx = \tau_{yx} \times dx \times dy$$

$\therefore$ $$\tau_{xy} = \tau_{yx} \qquad ...(14.12)$$

Considering the force normal to the diagonal AC shown in Fig. 14.3 (*b*), we have

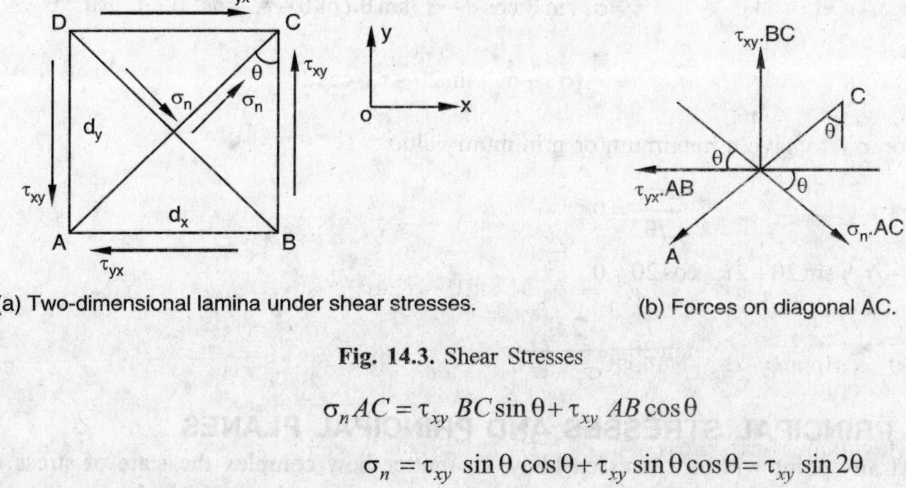

(a) Two-dimensional lamina under shear stresses. (b) Forces on diagonal AC.

Fig. 14.3. Shear Stresses

$$\sigma_n AC = \tau_{xy}\, BC \sin\theta + \tau_{xy}\, AB \cos\theta$$

$$\sigma_n = \tau_{xy}\, \sin\theta\, \cos\theta + \tau_{xy}\, \sin\theta \cos\theta = \tau_{xy}\, \sin 2\theta$$

For $$\theta = 45°, \sigma_n = \tau_{xy}$$

The normal stress σ_n is compressive on diagonal *AC* and is equal to the shear stress. Similarly on diagonal *BD*, $\sigma_n = \tau_{xy}$ and is tensile in nature.

14.4 COMPLEX STRESSES

Consider the complex stresses acting on a body, as shown in Fig. 14.4 (a). The forces on AC are shown in Fig. 14.4 (b). Resolving the forces perpendicular and parallel to the plane AC, we have

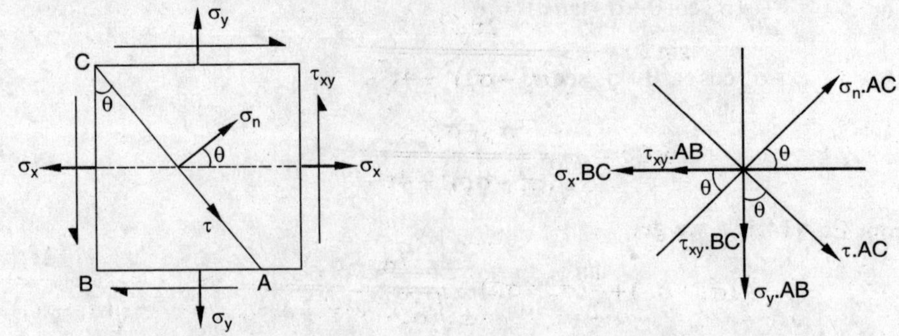

(a) Complex stresses on a body (b) Forces on oblique plane

Fig. 14.4 Complex stresses

$$\sigma_n AC = \sigma_x BC.\cos\theta + \sigma_y.AB \sin\theta + \tau_{xy}.BC.\sin\theta + \tau_{xy}.AB\cos\theta$$

$$\sigma_n = \sigma_x \cos^2\theta + \sigma_y \sin^2\theta + 2\tau_{xy}\sin\theta\cos\theta$$

$$= \frac{1}{2}\sigma_x(1+\cos 2\theta) + \frac{1}{2}\sigma_y(1-\cos 2\theta) + \tau_{xy}\sin 2\theta$$

$$= \frac{1}{2}(\sigma_x+\sigma_y) + \frac{1}{2}(\sigma_x-\sigma_y)\cos 2\theta + \tau_{xy}\sin 2\theta \quad ....(14.13)$$

$$\tau.AC = \sigma_x.BC\sin\theta - \sigma_y.AB\cos\theta - \tau_{xy}.BC\cos\theta + \tau_{xy}.AB\sin\theta$$

$$\tau = \sigma_x \sin\theta\cos\theta - \sigma_y\sin\theta\cos\theta - \tau_{xy}\cos^2\theta + \tau_{xy}\sin^2\theta$$

$$= \frac{1}{2}(\sigma_x-\sigma_y)\sin 2\theta - \tau_{xy}\cos 2\theta \quad ....(14.14)$$

For σ_n to have a maximum or minimum value,

$$\frac{d\sigma_n}{d\theta} = 0$$

$$-(\sigma_x-\sigma_y)\sin 2\theta + 2\tau_{xy}\cos 2\theta = 0$$

$$\tan 2\theta = \frac{2\tau_{xy}}{\sigma_x-\sigma_y} \quad ....(14.15)$$

14.5 PRINCIPAL STRESSES AND PRINCIPAL PLANES

At any point within a stressed body, no matter how complex the state of stress may be, there always exist three mutually perpendicular planes on each of which the resultant stress is normal stress. These mutually perpendicular planes are called principal planes, and the resultant normal stresses acting on them are called principal stresses.

For σ_n to be either maximum or minimum from Eq. (14.15), we have

$$\tan 2\theta = \frac{2\tau_{xy}}{\sigma_x - \sigma_y}$$

$$\therefore \quad \sin 2\theta = \frac{2\tau_{xy}}{\pm\sqrt{(\sigma_x - \sigma_y)^2 + 4\tau^2_{xy}}}$$

$$\cos 2\theta = \frac{\sigma_x - \sigma_y}{\pm\sqrt{(\sigma_x - \sigma_y)^2 + 4\tau^2_{xy}}}$$

From Eq. (14.13), we get

$$\sigma_n = \frac{1}{2}(\sigma_x + \sigma_y) + \frac{1}{2}(\sigma_x - \sigma_y)\frac{(\sigma_x - \sigma_y)}{\pm\sqrt{(\sigma_x - \sigma_y)^2 + 4\tau^2_{xy}}} + \tau_{xy} \times \frac{2\tau_{xy}}{\pm\sqrt{(\sigma_x - \sigma_y)^2 + 4\tau^2_{xy}}}$$

$$= \frac{1}{2}(\sigma_x + \sigma_y) - \frac{1}{2}\left[\frac{(\sigma_x - \sigma_y)^2 + 4\tau^2_{xy}}{\pm\sqrt{(\sigma_x - \sigma_y)^2 + 4\tau^2_{xy}}}\right]$$

$$\sigma_{1,2} = \frac{1}{2}(\sigma_x + \sigma_y) \pm \frac{1}{2}[(\sigma_x - \sigma_y)^2 + 4\tau^2_{xy}]^{1/2} \qquad \text{....(14.16)}$$

Eq. (14.16) gives two principal stresses and Eq. (14.15) gives their orientation. For shear stress τ to be maximum or minimum,

$$\frac{d\tau}{d\theta} = 0$$

Using Eq. (14.14), we have

$$(\sigma_x - \sigma_y)\cos 2\theta + 2\tau_{xy}\sin 2\theta = 0$$

$$\tan 2\theta = \frac{-(\sigma_x - \sigma_y)}{2\tau_{xy}} \qquad \text{....(14.17)}$$

The planes of maximum shear stress are θ_3 and $\theta_4 = \theta_3 + 90°$. These planes are inclined at 45° to the principal planes

$$\tau_{max} = \frac{1}{2}(\sigma_x - \sigma_y)\frac{(\sigma_x - \sigma_y)}{\pm\sqrt{(\sigma_x - \sigma_y)^2 + 4\tau^2_{xy}}} + \tau_{xy} \times \frac{2\tau_{xy}}{\pm\sqrt{(\sigma_x - \sigma_y)^2 + 4\tau^{2xy}}}$$

$$= \pm\frac{1}{2}\sqrt{(\sigma_x - \sigma_y)^2 + 4\tau^2_{xy}} \qquad \text{....(14.18)}$$

$$\tau_{max} = \frac{1}{2}(\sigma_1 - \sigma_2) \qquad \text{.....(14. 19)}$$

14.6 MOHR'S CIRCLE

Mohr's circle is a circular diagram to determine the normal, shear and resultant stresses on an oblique plane when a body is subjected to biaxial or complex state of stress. The values of principal stresses and principal planes can also be determined from this diagram. The radius of the circle gives the maximum shear stress.

(a) Biaxial stresses (Fig. 14.5)

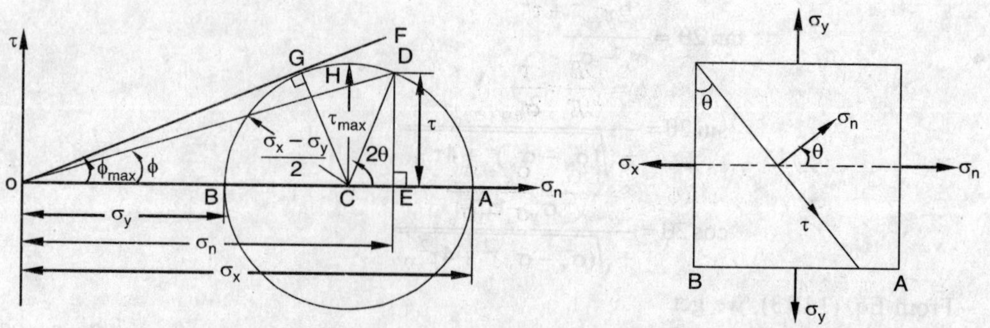

Fig 14.5 Mohr's stress circle for biaxial stresses when both stresses are tensile.

Form Eqs. (14.5) and (14. 6), we have

$$\sigma_n = \frac{1}{2}(\sigma_x + \sigma_y) + \frac{1}{2}(\sigma_x - \sigma_y)\cos 2\theta$$

$$\tau = \frac{1}{2}(\sigma_x - \sigma_y)\sin 2\theta$$

$$\left(\sigma_n - \frac{\sigma_x + \sigma_y}{2}\right) + \tau^2 = \left(\frac{\sigma_x - \sigma_y}{2}\right)^2 \quad .....(14.20)$$

Eq. (14.20) represents a circle in the $\sigma_n - \tau$ plane. The centre of the circle is $\left(\frac{\sigma_x + \sigma_y}{2}, 0\right)$ and radius $\frac{\sigma_x - \sigma_y}{2}$

Construction

(*i*) $\sigma_x > \sigma_y$

1. On a convenient scale, take $OA = \sigma_x, OB = \sigma_y, AC = CB = \frac{\sigma_x - \sigma_y}{2}$
2. With *C* as centre and *CA* as radius, draw the circle.
3. Make $\angle ACD = 2\theta$ anticlockwise.
4. Drop $DE \perp OA$ and join *OD*. Then $OE = \sigma_n, DE = \tau$ and $OD = \sigma_r$, $\angle AOD = \phi$.
5. Draw *OF* tangential to the circle, touching the circle at *G*. Join *CG*, then $\angle AOF = \phi_{max}$, the maximum obliquity of the resultant stress.
6. Draw $HC \perp OA$. Then $HC = \tau_{max}$

Proof:

$$DE = CD\sin 2\theta = \frac{(\sigma_x - \sigma_y)}{2}\sin 2\theta = \tau$$

$$OE = OC + CE = \frac{\sigma_x + \sigma_y}{2} + \frac{(\sigma_x - \sigma_y)}{2}\cos 2\theta = \sigma_n$$

$$OD^2 = OE^2 + DE^2$$

$$\sigma_r^2 = \sigma_n^2 + \tau^2$$

$$\tan\phi = \frac{DE}{OE} = \frac{\tau}{\sigma_n}$$

$$\sin\phi_{max} = \frac{CG}{OC} = \frac{\sigma_x - \sigma_y}{\sigma_x + \sigma_y}$$

(*ii*) σ_y is compressive

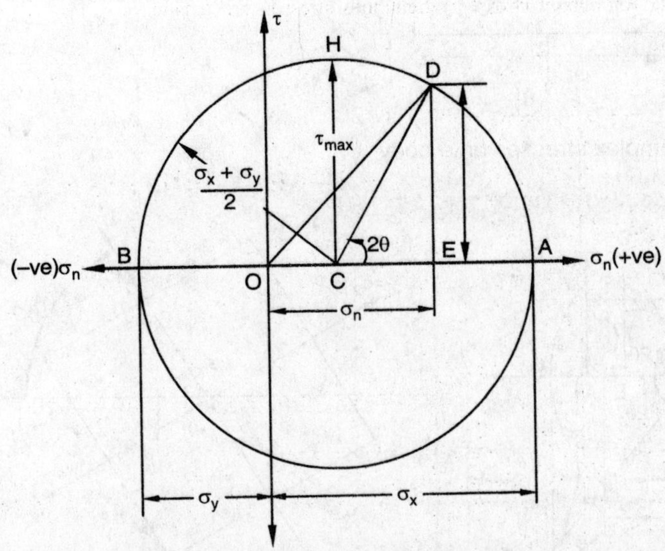

Fig. 14.6 Mohr's stress circle for biaxial stresses when σ_y is compressive

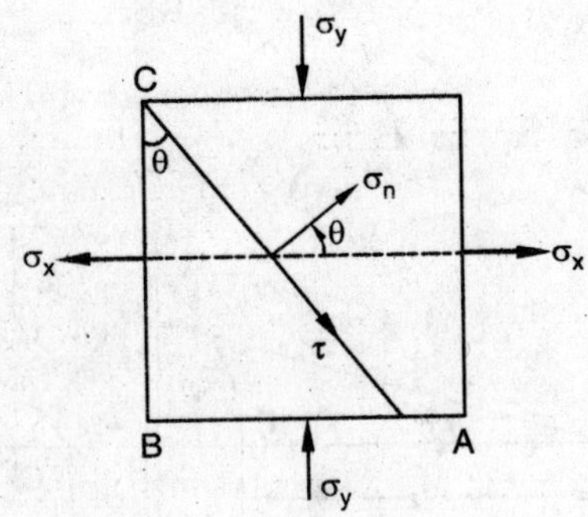

Fig. 14.7 Mohr's stress circle when σ_x is compressive.

In this case, take σ_y along the –ve σ_n axis. Then draw the Mohr's circle as explained in (*i*).

(*iii*) σ_x is compressive.

The Mohr's circle is shown in Fig. 14.7

Complex Stresses

Stress conventions:

1. Normal stresses are plotted as horizontal coordinates, with tensile stresses (positive) plotted to the right of origin and compressive stresses (negative) plotted to the left.
2. Shearing stresses are plotted as vertical coordinates, with those tending to produce a clockwise rotation of the stress element plotted above the σ_n - axis, and those tending to produce a counterclockwise rotation of the stress element plotted below the σ_n - axis. Consider the stress system shown in Fig. 14.8(*a*). The Mohr's circle is drawn in Fig. 14.8(*b*) following the above conventions.

3. σ_x and σ_y are both tensile $(\sigma_x > \sigma_y)$ and τ_{xy} positive.

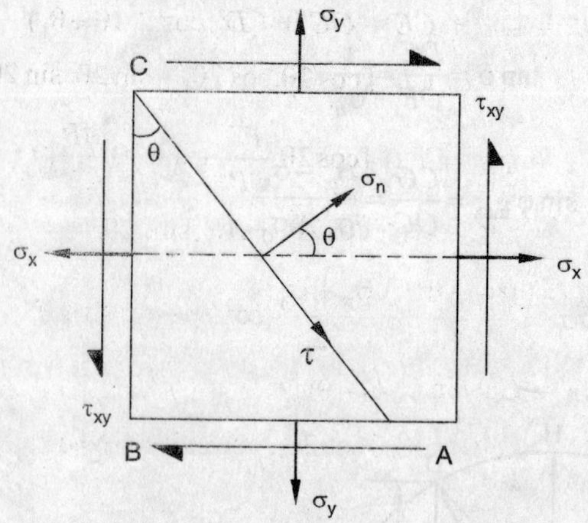

(a) Complex stresses on a body

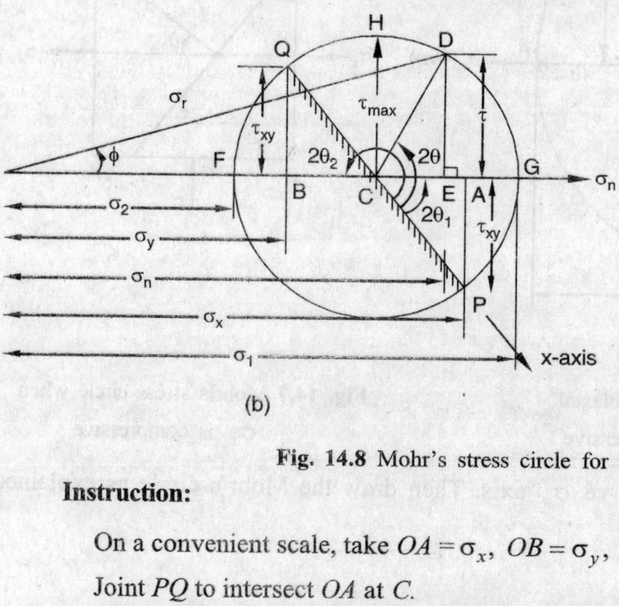

(b)

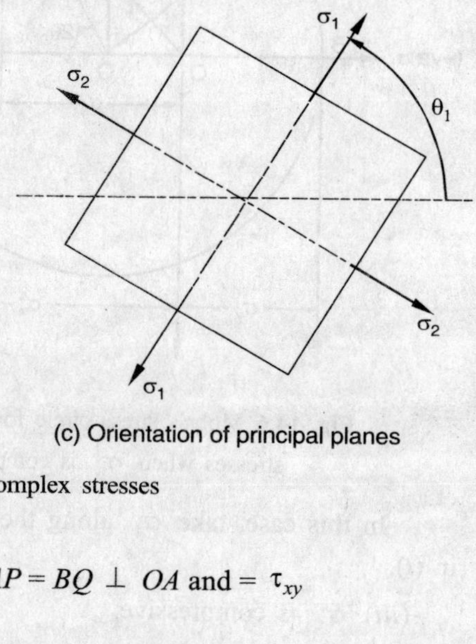

(c) Orientation of principal planes

Fig. 14.8 Mohr's stress circle for complex stresses

Instruction:

On a convenient scale, take $OA = \sigma_x$, $OB = \sigma_y$, $AP = BQ \perp OA$ and $= \tau_{xy}$

Joint PQ to intersect OA at C.

With CP as radius, draw the circle. Then $OG = \sigma_1$, $OF = \sigma_2$

4. Make $\angle PCD = 2\theta$. Drop $DE \perp OA$ and join OD. Then $OE = \sigma_n$, $DE = \tau$ and $OD = \sigma_r$. $\angle AOD = \phi$.

5. $\angle PCG = 2\theta_1$ and $PCF = 2\theta_2$. The orientation of principal planes is shown in Fig. 14.8 (c)

Proof:

$$AB = \sigma_x - \sigma_y$$

$$AC = CB = \frac{\sigma_x - \sigma_y}{2}$$

$$OC = OB + BC = \sigma_y + \frac{\sigma_x - \sigma_y}{2} = \frac{\sigma_x + \sigma_y}{2}$$

$$OE = OC + CE = OC + CD \ \cos 2\ (\theta - \theta_1)$$

$$= OC + CD\ [\cos 2\theta . \cos 2\theta_1 + \sin 2\theta .\ \sin 2\theta_1]$$

$$= OC + CD\ [\cos 2\theta \frac{AC}{CP} + \sin 2\theta \frac{AP}{CP}]$$

$$= OC + AC . \cos 2\theta + AP . \sin\ 2\theta \qquad [\because CD = CP]$$

$$\sigma_n = \frac{\sigma_x + \sigma_y}{2} + \frac{\sigma_x - \sigma_y}{2} \cos 2\theta + \tau_{xy} \sin 2\theta$$

$$DE = CD \sin 2\ (\theta - \theta 1)$$

$$= CD (\sin 2\theta \cos 2\theta_1 - \cos 2\theta . \sin 2\theta_1)$$

$$= CD (\sin 2\theta . \frac{AC}{CP} - \cos 2\theta .\ \frac{AP}{CP})$$

$$= AC \sin 2\theta - AP \cos 2\theta$$

$$\tau = \frac{\sigma_x - \sigma_y}{2} \sin 2\theta - \tau_{xy} \cos 2\theta$$

$$CH = \tau_{max} = CP = \sqrt{AC^2 + AP^2} = \sqrt{\left(\frac{\sigma_x - \sigma_y}{2}\right)^2 + \tau^2{}_{xy}}$$

$$= \frac{1}{2}\sqrt{(\sigma_x - \sigma_y)^2 + 4\tau^2{}_{xy}}$$

$$\sigma_1 = OG = OC + CG = OC + CP$$

$$= \frac{\sigma_x + \sigma_y}{2} + \frac{1}{2}\sqrt{(\sigma_x - \sigma_y)^2 + 4\tau^2{}_{xy}}$$

$$\sigma_2 = OF = OC - FC = OC - CP$$

$$= \frac{\sigma_x + \sigma_y}{2} - \frac{1}{2}\sqrt{(\sigma_x - \sigma_y)^2 + 4\tau^2{}_{xy}}$$

$$\sigma_r = OD = \sqrt{OE^2 + DE^2} = \sqrt{\sigma_n^2 + \tau^2}$$

$$\tan \phi = \frac{DE}{OE} = \frac{\tau}{\sigma_n}$$

$$\tan 2\theta_1 = \frac{AP}{AC} = \frac{2\tau_{xy}}{\sigma_x - \sigma_y}$$

(*ii*) σ_x and σ_y both positive ($\sigma_x > \sigma_y$) and τ_{xy} negative.

Consider the stress system shown in Fig. 14.9(*a*). The Mohr's circle is drawn in Fig. 14.9(*b*).

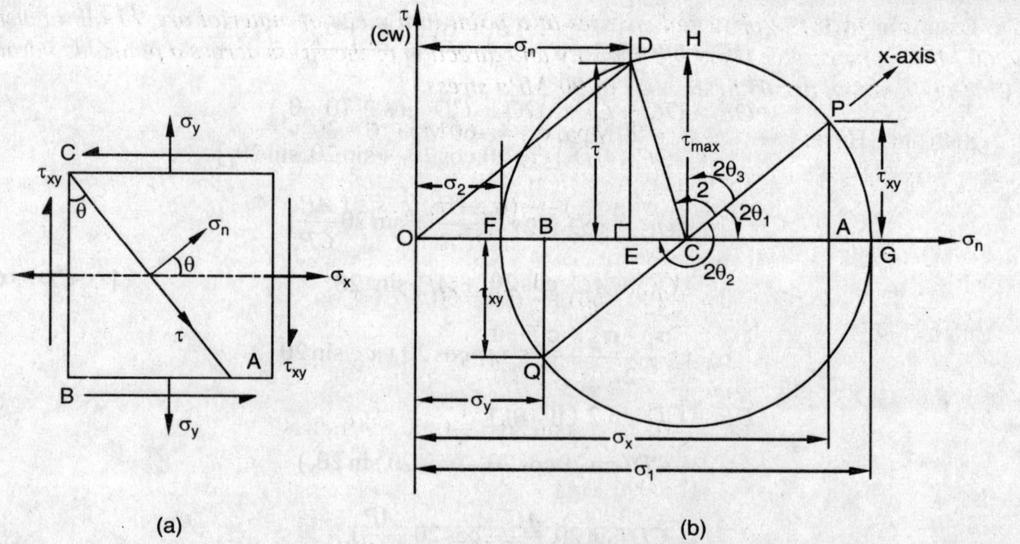

Fig. 14.9 Mohr's stress circle for –ve τ_{xy}.

(*iii*) σ_x is compressive: The Mohr's circle is shown Fig. 14. 10 (*b*).

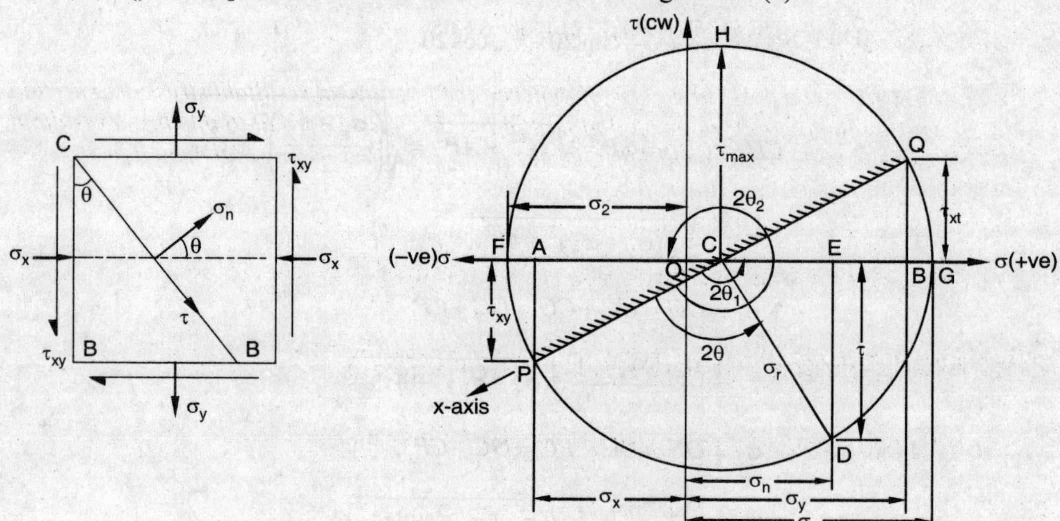

Fig. 14.10 Mohr's circle for σ_x compressive.

The principal planes are shown in Fig. 14.10 (c).

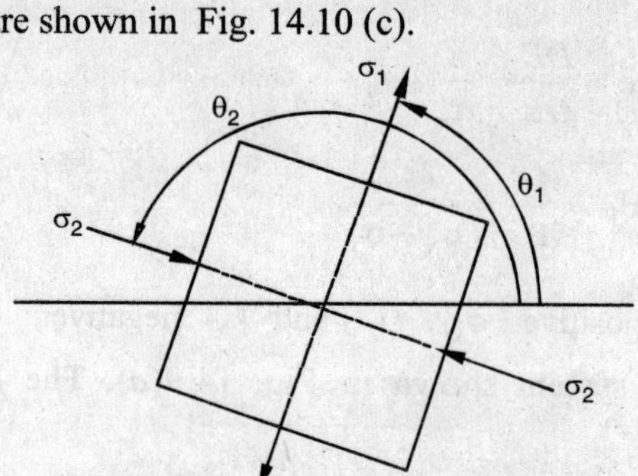

Fig. 14.10 (c) Orientation of principal planes

Example 14.1 *The principal stresses at a point in a piece of material are 90 MPa tensile and 60 MPa compressive. Find the intensity and direction of the stress across a plane the normal of which is inclined at 30° to the axis of 90 MPa stress.*

Solution: Here, $\sigma_x = 90\,\text{Mpa}, \sigma_y = -60\,\text{Mpa}, \theta = 30°$

$$\sigma_n = \frac{1}{2}(\sigma_x + \sigma_y) + \frac{1}{2}(\sigma_x - \sigma_y)\cos 2\theta$$

$$= \frac{1}{2}(90 - 60) + \frac{1}{2}(90 = 60)\cos 60°$$

$$= 15 + 37.5 = 52.5\ \text{MPa}$$

$$\tau = \frac{1}{2}(\sigma_y - \sigma_x)\sin 2\theta = \frac{1}{2}(-60 - 90)\sin 60°$$

$$= -\ 64.95\ \text{Mpa}$$

$$\sigma_r = \sqrt{\sigma_n^2 + \tau^2} = \sqrt{(52.5)^2 + (-64.95)^2}$$

$$= 82.52\ \text{Mpa}$$

$$\tan\phi = \frac{\iota}{\sigma_n} = \frac{64.95}{52.5} = 1.23714,\ \phi = 51.05°$$

Example 14.2 *At a point in a piece of material the intensity of resultant stress on a certain plane is 50 Mpa inclined at 30° to normal to that plane. The plane normal to this has a resultant stress whose normal component is 30 MPa. Find (a) the principal planes and principal stresses, and (b) the plane of maximum shear and its intensity.*

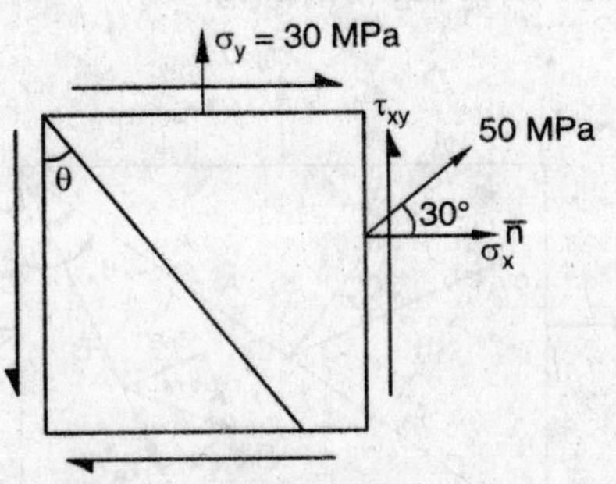

Fig. 14.11

Solution: The stress system is shown in Fig. 14.11

$$\sigma_x = 50\cos 30°$$

$$= 43.3\ \text{Mpa}$$

$$\tau_{xy} = 50\sin 30°$$

$$= 25\ \text{Mpa}$$

$$\sigma_y = 30\,\text{MPa}$$

(a) Principal stresses

$$\sigma_{1,2} = \frac{1}{2}(\sigma_x + \sigma_y) \pm \frac{1}{2}\sqrt{(\sigma_x - \sigma_y)^2 + 4\tau_{xy}^2}$$

$$= \frac{1}{2}(43.3+30) \pm \frac{1}{2}\sqrt{(43.3-30)^2+4\times 25^2}$$

$$= 62.52 \text{ MPa}, 10.76 \text{ Mpa}$$

Principal Planes are given by,

$$\tan 2\theta = \frac{2\tau_{xy}}{\sigma_x - \sigma_y} = \frac{2\times 25}{43.3-30} = 3.7594$$

$$2\theta = 75.1°$$

$$\theta_1 = 37.55°$$

$$\theta_2 = 90° + \theta_1 = 17.55°$$

$$\tau_{max} = \frac{\sigma_1 - \sigma_2}{2} = \frac{62.52-10.76}{2} = 25.88 \text{ MPa}$$

Planes of maximum shear stress are given by

$$\theta_3 = \theta_1 + 45° = 37.55° + 45° = 82.55°$$

$$\theta_4 = \theta_3 + 90° = 172.55°$$

Example 14.3 *At a point in a material subjected to two direct stresses on planes at right angles, the resultant stress on a plane A is 80 MPa inclined at 30° to the normal, and on plane B is 40 MPa inclined at 45° to the normal. Find the principal stresses and show the position of the two planes A and B relative to the principal stresses.*

Solution: (A) Analytical Method

The planes *A* and *B* are shown in Fig. 14.12 (*a*).

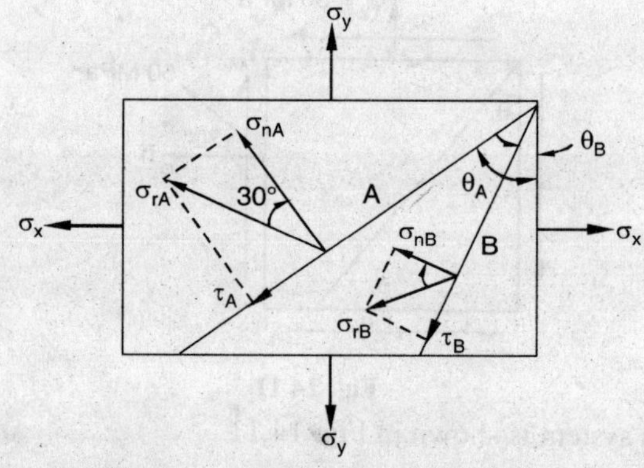

Fig. 14.12.(*a*)

Plane A: $\sigma_{nA} = \sigma_{rA}\cos\phi_A = 80\cos 30°$

$$= 69.282 \text{ MPa}$$

$$\tau_A = \sigma_{rA}\sin\phi_A = 80\sin 30°$$

$$= 40 \text{ MPa}$$

$$\sigma_{nA} = \frac{1}{2}(\sigma_x + \sigma_y) + \frac{1}{2}(\sigma_x - \sigma_y)\cos 2\theta_A$$

$$\therefore \quad 138.564 = (\sigma_x + \sigma_y) + (\sigma_x - \sigma_y)\cos 2\theta_A \quad ....(1)$$

$$\tau_A = \frac{1}{2}(\sigma_x - \sigma_y)\sin 2\theta_A$$

$$\therefore \quad 80 = (\sigma_x - \sigma_y)\sin 2\theta_A \qquad \text{....(2)}$$

Plane B

$$\sigma_{nB} = \sigma_{rB}\cos\phi_B = 40\cos 45° = 28.284 \text{ MPa}$$

$$\tau_B = \sigma_{rB}\sin\phi_B = 40\sin 45° = 28.284 \text{ MPa}$$

$$\therefore \quad 56.568 = (\sigma_x + \sigma_y) + (\sigma_x - \sigma_y)\cos 2\theta_B \qquad \text{....(3)}$$

$$56.568 = (\sigma_x - \sigma_y)\sin 2\theta_B \qquad \text{....(4)}$$

From Eq. (2), we have

$$\sin 2\theta_A = \frac{80}{\sigma_x - \sigma_y}$$

$$\cos 2\theta_A = \sqrt{1 - \sin^2 2\theta_A} = \sqrt{1 - \frac{6400}{(\sigma_x - \sigma_y)^2}}$$

Eq. (1) becomes

$$138.564 = (\sigma_x + \sigma_y) + \sqrt{(\sigma_x - \sigma_y)^2 - 6400}$$

or

$$[138.564 - (\sigma_x + \sigma_y)]^2 = (\sigma_x - \sigma_y)^2 - 6400$$

$$19200 + \sigma_x{}^2 + \sigma_y{}^2 + 2\sigma_x\sigma_y - 277.128(\sigma_x + \sigma_y) = \sigma_x{}^2 + \sigma_y{}^2 - 2\sigma_x\sigma_y - 6400$$

$$25600 + 4\sigma_x\sigma_y = 2177.128(\sigma_x + \sigma_y)$$

$$\sigma_x + \sigma_y - 0.01443\,\sigma_x\sigma_y = 92.376$$

From Eq. (4), we have

$$\sin 2\theta_B = \frac{56.568}{\sigma_x - \sigma_y}$$

$$\cos 2\theta_B = \sqrt{1 - \sin^2 2\theta_B} = \sqrt{1 - \left(\frac{56.568}{\sigma_x - \sigma_y}\right)^2}$$

Eq. (3) becomes

$$56.568 = (\sigma_x + \sigma_y) + (\sigma_x - \sigma_y)\sqrt{1 - \left(\frac{56.568}{\sigma_x - \sigma_y}\right)^2}$$

$$= (\sigma_x + \sigma_y) + \sqrt{(\sigma_x - \sigma_y)^2 - 3200}$$

$$[56.568 - (\sigma_x + \sigma_y)]^2 = (\sigma_x - \sigma_y)^2 - 3200$$

$$3200 + \sigma_x{}^2 + \sigma_y{}^2 + 2\sigma_x\sigma_y - 1113.136(\sigma_z + \sigma_y) = \sigma_x{}^2 + \sigma_y{}^2 - 2\sigma_x\sigma_y - 3200$$

$$6400 + 4\sigma_x\sigma_y = 113.136(\sigma_x + \sigma_y)$$

$$\sigma_x + \sigma_y = -\,0.03535\,\sigma_x\,\sigma_y = 56.569 \qquad \text{....(6)}$$

Subtracting Eq. (6) From (5), we get

$$0.02092\ \sigma_x\sigma_y = 35.807$$

$$\sigma_x\sigma_y = 1711.6 \qquad ......(7)$$

From Eq. (5), we get

$$\sigma_x + \sigma_y = 92.376 + 0.01443 \times 1711.6 = 117.075 \qquad ......(8)$$

Now $$138.564 = 117.075 + \sqrt{(\sigma_x - \sigma_y)^2 - 6400}$$

$$(\sigma_x - \sigma_y)^2 - 6400 = (138.564 - 117.075)^2 = 461.794$$

$$\sigma_x - \sigma_y = 82.836 \qquad ......(9)$$

Adding Eqs. (8) and (9), we get

$$\sigma_x = 99.955 = 100\,\text{MPa}$$

$$\sigma_y = 17\,\text{MPa}$$

$$\sin 2\theta_A = \frac{80}{82.836} = 0.96576$$

$$\theta_A = 37.482°$$

$$\sin 2\theta_B = \frac{56.568}{82.836} = 0.68289$$

$$\theta_B = 21.535°$$

Graphical Method:

1. Draw OA = 8 cm = 80 MPa at 30° to σ_n – axis and OB = 4 cm = 40 MPa at 45°, as shown in Fig. 14.12 (b)

   $\sigma_x = OD = 10\text{cm} = 100$ MPa

   $\sigma_y = OE = 1.7$ cm 17 MPa

2. Join AB and draw the right bisector of AB to intersect σ_n – axis at C.
3. With C as centre an $CA = CB$ as radius draw the circle. Join CA and CB.

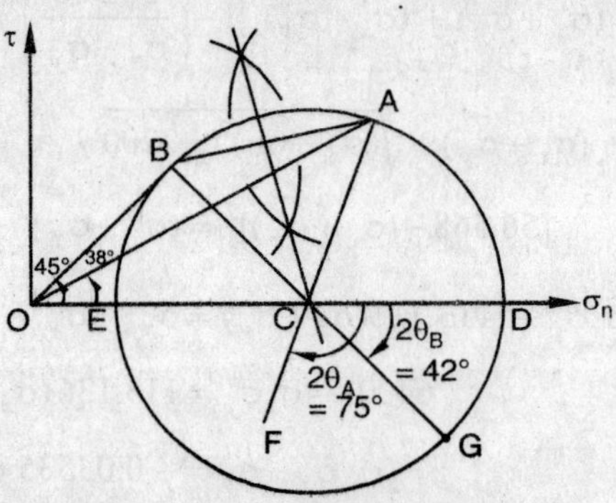

Fig. 14.12(*b*)

4. $OD = \sigma_x = 10\text{cm} = 100$ MPa , $OE = \sigma_y = 1.7$ cm 17 MPa.

$$\angle DCF = 2\theta_A = 75° \text{ so that } \theta_A = 37.5°$$

$$\angle DCG = 2\theta_B = 42° \text{ so that } \theta_B = 21°$$

Example 14.4 *The principal stresses at a point in a material are 120 MPa and 60 Mpa. Find the magnitude and direction of stress on a plane inclined at 30° to the direction of 60 MPa stress. Find also the plane on which the resultant stress is most oblique and its value.*

Solution: $\sigma_x = 120\,\text{MPa}, \sigma_y = 60\,\text{MPa}, \theta = 90° - 30° = 60°$

$$\sigma_n = \frac{1}{2}(\sigma_x + \sigma_y) + \frac{1}{2}(\sigma_x - \sigma_y)\cos 2\theta$$

$$= \frac{1}{2}(120+60) + \frac{1}{2}(120-60)\cos 120°$$

$$= 90 - 15 = 75 \text{ MPa}$$

$$\tau = \frac{1}{2}(\sigma_x - \sigma_y)\sin 2\theta$$

$$= \frac{1}{2}(120-60)\sin 120°$$

$$= 25.98 \text{ MPa}$$

$$\sigma_r = (\sigma_n^2 + \tau^2)^{1/2} = [75^2 + (25.98)^2]^{1/2} = 79.37\,\text{MPa}$$

$$\tan\phi = \frac{\tau}{\sigma_n} = \frac{25.98}{75} = 0.3464$$

$$\phi = 19.1°$$

$$\tan\phi_{max} = \frac{\sigma_x - \sigma_y}{2\sqrt{\sigma_x \sigma_y}} = \frac{120-60}{2\sqrt{120\times 60}} = 0.35355$$

$$\phi_{max} = 19.47°$$

$$\tan\theta = \sqrt{\frac{\sigma_x}{\sigma_y}} = \sqrt{\frac{120}{60}} = \sqrt{2}, \theta = 54.73°$$

$$(\sigma_r)_{max} = [\sigma_x^2 \cos^2\theta + \sigma_y^2 \sin^2\theta]^{1/2}$$

$$= [120^2 \cos^2 54.73° + 62\sin^2 54.73°]^{1/2}$$

$$= \sqrt{4801 + 2400} = 84.85\,\text{MPa}$$

Example 2.5. *The state of stress at a point is given by:*

$$\sigma_x = 60 \text{ MPa}, \sigma_y = -30 \text{ MPa}, \tau_{xy} = 30 \text{ Mpa(ccw)}$$

Calculate (a) principal stresses, (b) principal planes, (c) maximum shear stress, shown the orientation of principal planes and planes of maximum shear stress. Use Mohr's circle.

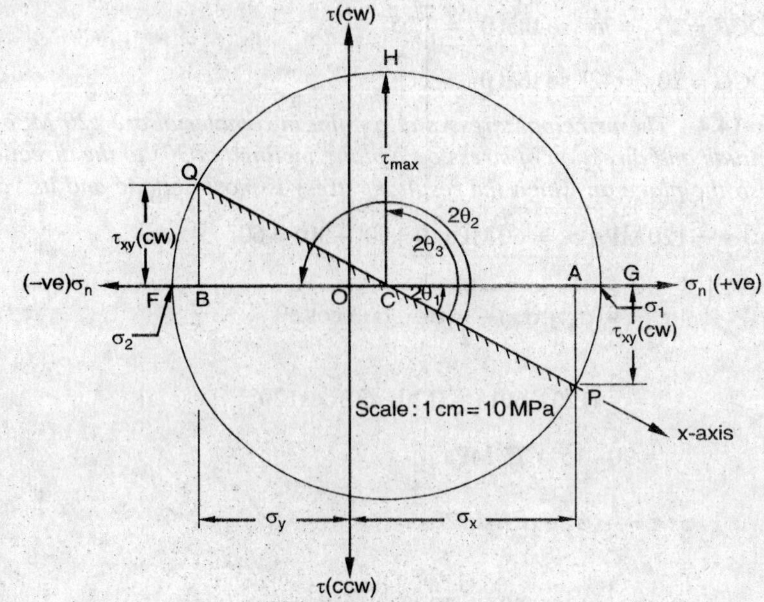

(a) Mohr's stress circle

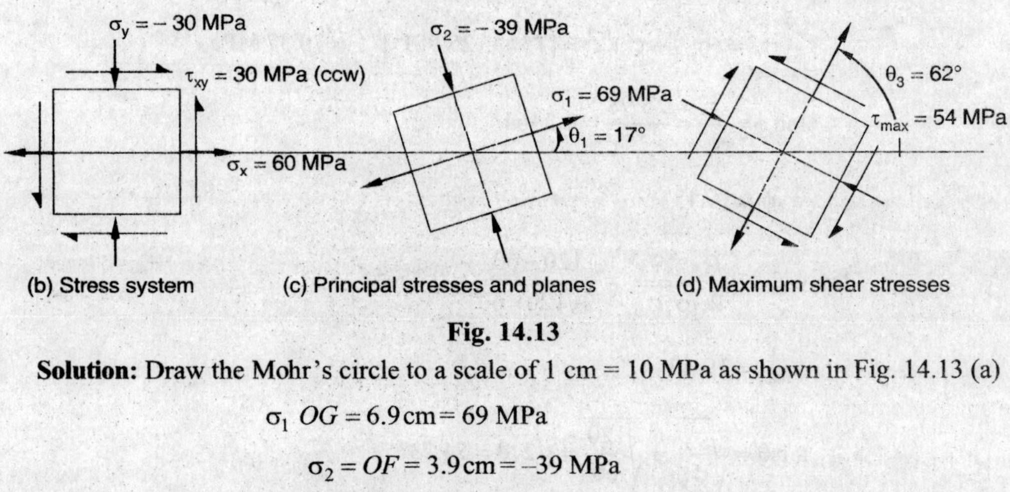

(b) Stress system (c) Principal stresses and planes (d) Maximum shear stresses

Fig. 14.13

Solution: Draw the Mohr's circle to a scale of 1 cm = 10 MPa as shown in Fig. 14.13 (a)

$$\sigma_1\ OG = 6.9\,\text{cm} = 69\ \text{MPa}$$

$$\sigma_2 = OF = 3.9\,\text{cm} = -39\ \text{MPa}$$

$$\theta_1 = 17°, \theta_2 = 107°\ (\text{ccw})$$

$$\tau_{max} = CH = 5.4\,\text{cm} = 54\ \text{MPa}$$

$$\theta_3 = 62°, \theta_4 = 152°\ (\text{ccw})$$

The orientation of principal stresses and principal planes is shown in Fig. 14.13 (b) to (d).

Example 14.6 *At a point in a load carrying member, the state of stress is as given below.*

$$\sigma_x = 400\ MPa,\ \sigma_y = -300\ MPa,\ \tau_{xy} = 200\ MPa\,(ccw)$$

Draw the Mohr's circle and find the principal stresses and principal planes. Also draw the principal and maximum shear stress elements.

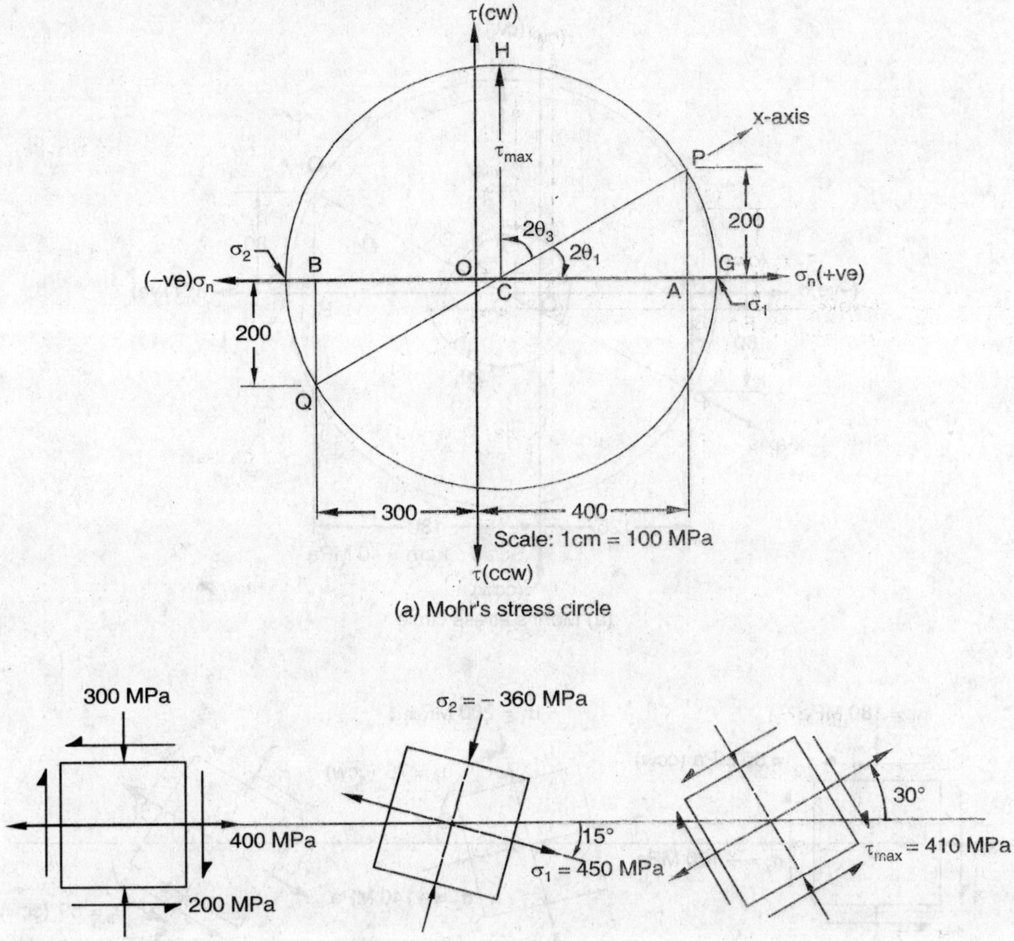

(a) Mohr's stress circle

(b) Initial stress (c) Principle stress element (d) Maximum shear stress element

Fig. 14.14

Solution: The Mohr's circle has been drawn in Fig. 14.14 (a) to a scale of 1 cm = 100 MPa. The stress elements are shown in Figs. 14.14 (b) to (d)

$$\sigma_1 = OG = 4.5\,\text{cm} = 450\text{ MPa}, \sigma_2 = OF = -3.6\,\text{cm} = -360\text{ MPa}$$

$$2\theta_1 = 30°, \theta_1 = 15°, \theta_2 = 105°\text{ (ccw)}$$

$$\tau_{max} = CH = 4.1\,\text{cm} = 410\text{ MPa}, 2\theta_3 = 60°, \theta_3 = 30°, \theta_4 = 120°\text{ (ccw)}$$

Example 14.7. *The state of stress at a point is given by*

$$\sigma_x = -120\text{ MPa}, \sigma_y = 180\text{ MPa}, \sigma_{xy} = 80\text{ MPa (ccw)}$$

Draw the Mohr's circle the find the principal stress, principal planes and maximum shear stress. Show the results on stress element.

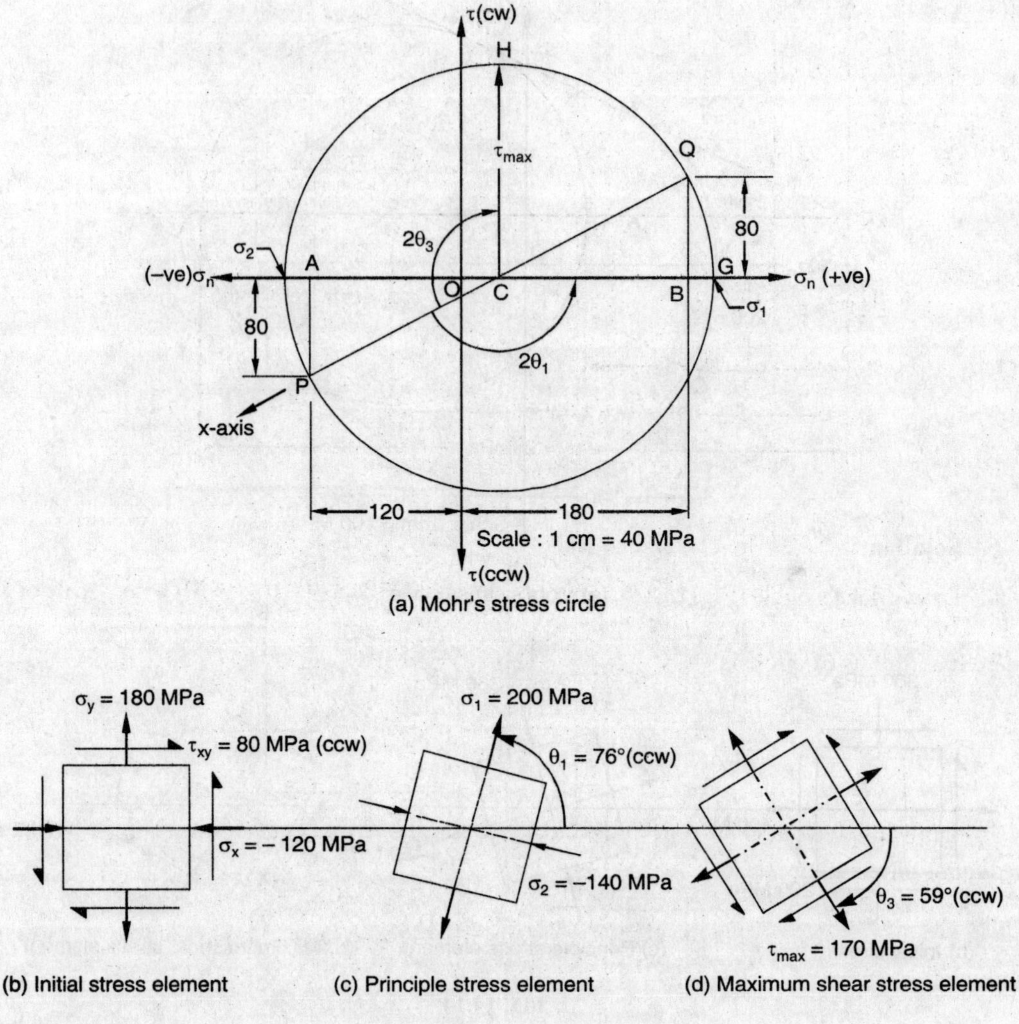

Fig. 14.15

Solution:

Draw the Mohr's circle as shown in Fig. 14.15 (a) to a scale of 1 cm = 40 MPa

The stress elements are shown in Fig. 14.15 (*b*) to (*d*)

$$\sigma_1 = OG = 5\text{cm} = 200\,\text{MPa},\ \sigma_2 = OF = 3.5\text{ cm} = 140\text{ MPa}$$

$$\theta_1 = 76°, \theta_2 = 166°\,(\text{ccw})$$

$$\tau_{max} = CH = 4.3\,\text{cm} = 172\,\text{MPa}$$

$$\theta_3 = 59°, \theta_4 = 149°\,(\text{ccw})$$

Example 14.8 *The principal stresses at a point in a stressed body are 20 MPa and 8 MPa. Determine the maximum obliquity of the resultant stress*

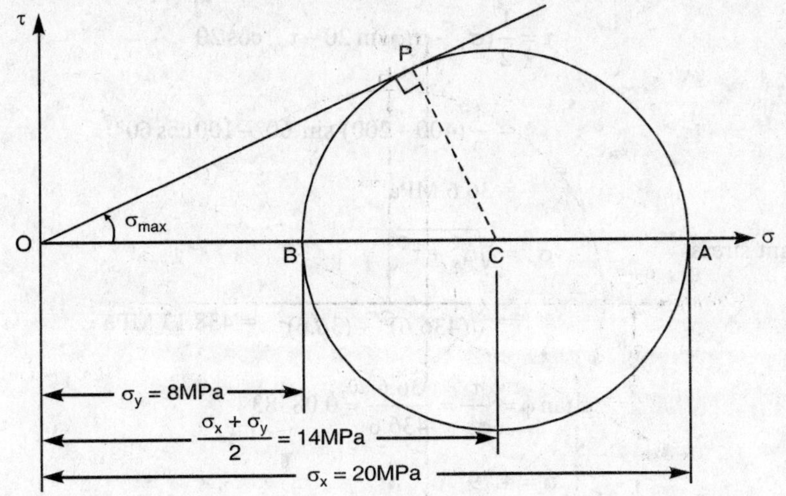

Fig. 14.16

Solution:

Draw Mohr's circle (Fig. 14.16) with $OA = \sigma_x = 20\,\text{MPa}, OB = \sigma_y = 8\,\text{MPa}$ to a scale of 1 cm = 2 MPa. Radius of circle is $\frac{1}{2}(\sigma_x - \sigma_y) = \frac{1}{2}(20-8) = 6\,\text{MPa}$

Centre of circle, C is $\left(\frac{\sigma_x + \sigma_y}{2}, 0\right) = (14, 0)$

Draw a line from O tangent to the circle touching it at P. Then $\angle POC = \phi_{max}$ = maximum obliquity of resultant stress.

$$\sin \phi_{max} = \frac{CP}{OC} = \frac{6}{14} = 0.42857$$

$$\phi_{max} = 25.38°$$

By measurement, $\phi_{max} = 25°$

Example 14.9 *The state of stress at a point is given by:*

$$\sigma_x = 400\,MPa,\ \sigma_y = 200\,MPa,\ \tau_{xy} = 100\,MPa\,(ccw)$$

Calculate the normal, shear and resultant stresses on a plane making an angle of 30° with the x-axis.

Solution: Refer to Fig. 14.17 (a)

$$\sigma_n = \frac{1}{2}(\sigma_x + \sigma_y) + \frac{1}{2}(\sigma_x - \sigma_y)\cos 2\theta + \tau_{xy}\sin 2\theta$$

$$= \frac{1}{2}(400+200) + \frac{1}{2}(400-200)\cos 60° + 100\sin 60°$$

$$= 436.6\,\text{MPa}$$

$$\tau = \frac{1}{2}(\sigma_x - \sigma_y)\sin 2\theta - \tau_{xy}\cos 2\theta$$

$$= \frac{1}{2}(400 - 200)\sin 60° - 100\cos 60°$$

$$= 36.6 \text{ MPa}$$

Resultant stress, $\sigma_r = \sqrt{\sigma_n^2 + \tau^2}$

$$= \sqrt{(436.6)^2 + (36.6)^2} = 438.13 \text{ MPa}$$

$$\tan\phi = \frac{\tau}{\sigma_n} = \frac{36.6}{436.6} = 0.08383$$

$$\phi = 4.79°$$

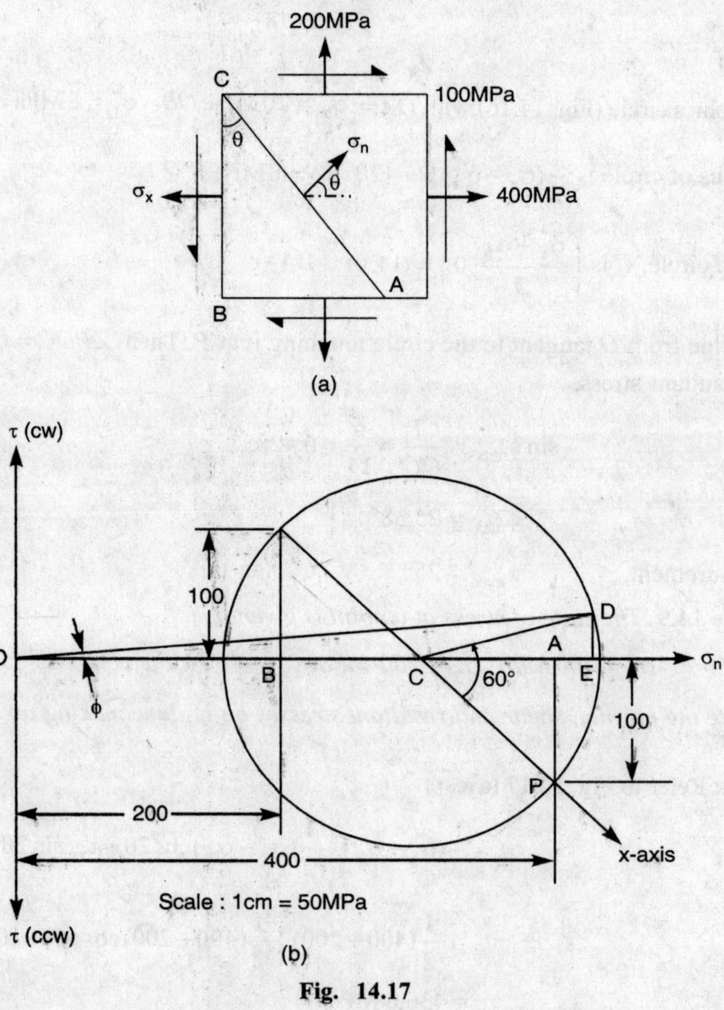

Fig. 14.17

Draw the Mohr's circle as shown in Fig. 14.17 (b) to a scale of 1 cm = 50 MPa

$$\sigma_n = OE = 8.13\,\text{cm} = 436.5 \text{ MPa}$$

$$\tau = DE = 0.7\,\text{cm} = 35\,\text{MPa}$$

$$\phi = 5°$$

Example 14.10 *The normal stresses on two neutrally perpendicular planes are 100 MPa (tensile) and 40 MPa (compressive) together with shear stress of 70 MPa. Determine graphically or otherwise, (a) the magnitude and direction of principal stresses, and (b) the magnitude of normal and shear stresses on a plane on which the shear stress is maximum.*

Solution: Given : $\sigma_x = 100\,\text{MPa}, \sigma_y = -40\,\text{MPa}, \tau_{xy} = 70\,\text{MPa}$

(*i*) Analytical Method:

(*a*)
$$\sigma_{1,2} = \frac{1}{2}(\sigma_x + \sigma_y) \pm \frac{1}{2}\sqrt{(\sigma_x - \sigma_y)^2 + 4\tau^2{}_{xy}}$$

$$= \frac{1}{2}(100 - 40) \pm \frac{1}{2}\sqrt{(100+40)^2 + 4 \times 70^2}$$

$$= 30 \pm 99$$

$$= 129 \text{ MPa}, -69 \text{ MPa}$$

$$\tan 2\theta = \frac{2\tau_{xy}}{\sigma_x - \sigma_y} = \frac{2 \times 70}{100+40} = 1$$

$$2\theta = 45°$$

$$\theta_1 = 22.5°,\ \theta_2 = 112.5°$$

(*b*) Planes of maximum shear stress are : $\theta_3 = 45° + 22.5° = 67.5°, \theta_4 = 157.5°$

$$\sigma_n = \frac{1}{2}(\sigma_x + \sigma_y) + \frac{1}{2}(\sigma_x - \sigma_y)\cos 2\theta + \iota_{xy} \sin 2\theta$$

$$= \frac{1}{2}(100-40) + \frac{1}{2}(100+40)\cos 135° + 70 \sin 135°$$

$$= 30 \text{ MPa}$$

$$\tau = \frac{1}{2}(\sigma_x - \sigma_y)\sin 2\theta - \tau_{xy} \sin 2\theta$$

$$= \frac{1}{2}(100+40) \sin 135° - 70 \cos 135°$$

$$= 99 \text{ MPa}$$

(*ii*) Graphical Method

The Mohr's circle is shown in Fig. 14.18

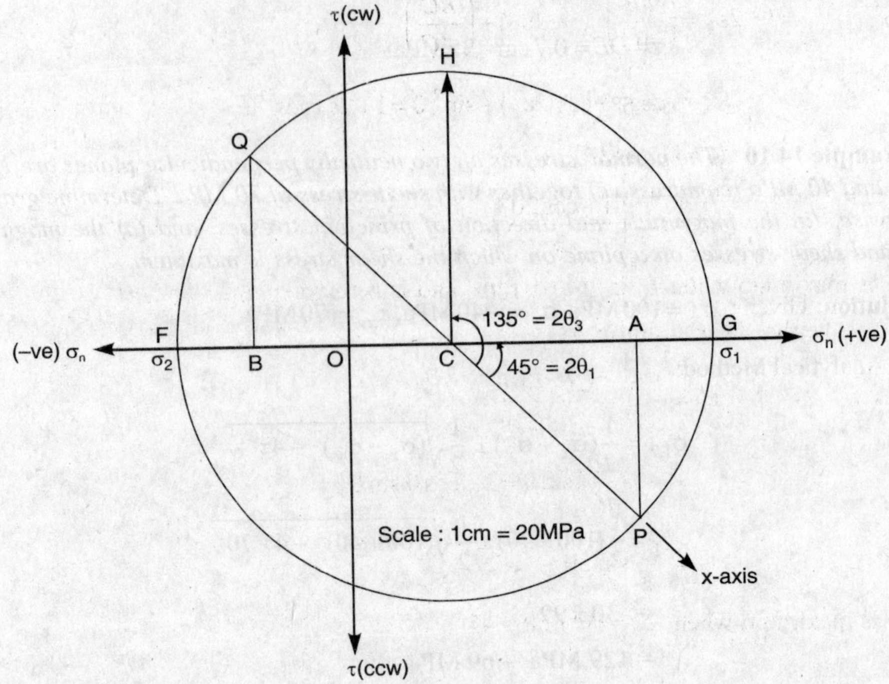

Fig. 14.18

$$\sigma_1 = OG = 6.5\,\text{cm} = 130 \text{ MPa}$$

$$\sigma_2 = OF = -3.5\,\text{cm} = -70 \text{ MPa}$$

$$2\theta_1 = 45°, \theta_1 = 22.5°, \theta_2 = 112.5°$$

$$\tau_{max} = CH = 4.95 = 99 \text{ MPa}$$

$$2\theta_3 = 135°, \theta_3 = 67.5°, \theta_4 = 157.5°$$

On the plane of maximum shear stress,

$$\sigma_n = OC = 1.5 \text{ cm} = 30 \text{ MPa}$$

$$\tau = \tau_{max} = CH = 99\,\text{MPa}$$

14.7 NORMAL AND SHEAR STRAINS ON AN OBLIQUE PLANE

14.7.1 Uniaxial Strains

Let the rectangular body $ABCD$ be extended to $AB_1C_1D_1$ under uniaxial strain ε_x as shown in Fig. 14.19. Let ε_n be the normal strain in the diagonal AC. Hence

$$AC_1 = AC\ (1 + \varepsilon_n)$$

Now $$(AC_1)^2 = (AB_1)^2 + (B_1C_1)^2$$

$$= (AB)^2\,(1+\varepsilon_x)^2 + (BC)^2$$

$$= (AC)^2(1+2\varepsilon_n) = (AB)^2(1+2\varepsilon_x) + (BC)^2$$

(Neglecting $\varepsilon_x{}^2, \varepsilon_n{}^2$, being small)

Fig. 14.19 Strains on an oblique plane under uniaxial strain

$$1+2\varepsilon_n=\left(\frac{AB}{AC}\right)^2(1+2\varepsilon_x)+\left(\frac{BC}{AC}\right)^2$$

$$=\cos^2\theta\,(1+2\varepsilon_x)+\sin^2\theta=1+2\varepsilon_x\cos^2\theta$$

$$\therefore\quad \varepsilon_n=\varepsilon_x\cos 2\theta$$

$$=\varepsilon_x\left(\frac{1+\cos 2\theta}{2}\right) \quad \text{.... (14.21)}$$

ε_n is maximum when θ is 0° or 180° and is zero when $\theta=90°$

Let ε_s be the shear strain in AC.

$$\therefore\quad \varepsilon_s=\frac{CE}{AC}=\frac{CC_1\sin\theta}{AC}$$

$$=\varepsilon_x.\frac{AB}{AC}.\sin\theta=\varepsilon_x.\cos\theta.\sin\theta$$

$$=\frac{\varepsilon_x}{2}\sin 2\theta \quad \text{.... (14.22)}$$

ε_s is maximum when $2\theta=90°$ or $\theta=45°$

$$(\varepsilon_s)_{\max}=\frac{\varepsilon_x}{2} \quad \text{.... (14.23)}$$

14.7.2 Shear Strain

Let the rectangular body $ABCD$ be deformed to ABC_1D_1 under the action of a shearing force so that the shear strain is γ_{xy} as shown in Fig. 14.20.

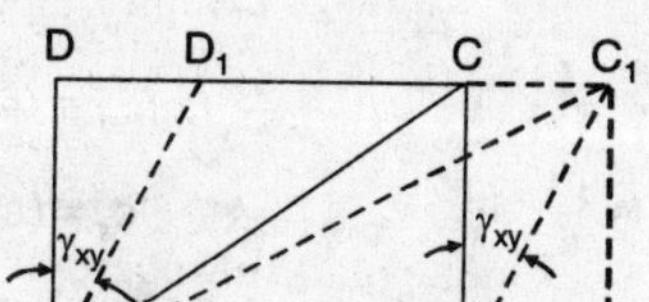

Fig. 14.20 Normal strain under shear strain

Now $\quad CC_1=BC.\gamma_{xy}$

$$AC_1=(1+\varepsilon_n)AC$$

$$(AC_1)^2=(AB_1)^2+(B_1C_1)^2$$

$$=(AB+CC_1)^2+(BC)^2$$

$$(AC)^2(1+\varepsilon_n)^2=(AB+BC.\gamma_{xy})^2+(BC)^2$$

Neglecting small quantities, we get

$$1+2\varepsilon_n=\left(\frac{AB}{AC}\right)^2+2.\frac{AB}{AC}.\frac{BC}{AC}.\gamma_{xy}+\left(\frac{BC}{AC}\right)^2$$

$$=\cos^2\theta+2.\cos\theta.\sin\theta.\gamma_{xy}+\sin^2\theta=1+\gamma_{xy}.\sin 2\theta$$

$$\varepsilon_n=\frac{\gamma_{xy}}{2}\sin 2\theta \quad \text{... (14.24)}$$

ε_n is maximum when $\theta=45°$ and

$$(\varepsilon_n)_{\max}=\frac{\gamma_{xy}}{2} \quad \text{.... (14.25)}$$

14.7.3 Biaxial Strains

Consider a rectangular plate *ABCD* acted upon by strains ε_x and ε_y as shown in Fig. 14.21. After information, the rectangle takes up the shape $A_1B_1C_1D_1$.

Normal strain in OC

$$\varepsilon_n = \frac{HC_1}{OC} = \frac{HE + EC_1}{OC} = \frac{GF + EC_1}{OC} = \frac{CF\cos\theta + C_1F\sin\theta}{1/2\,AC}$$

$$= \frac{1/2\,DC.\varepsilon_x.\cos\theta + 1/2\,BC.\varepsilon_y.\sin\theta}{1/2\,AC} = \varepsilon_x.\frac{DC}{AC}\cos\theta + \varepsilon_y.\frac{BC}{AC}.\sin\theta$$

$$= \varepsilon_x.\cos^2\theta + \varepsilon_y.\sin^2\theta$$

$$= \frac{1}{2}(\varepsilon_x + \varepsilon_y) + \frac{1}{2}\left(\varepsilon_x - \varepsilon_y\right)\cos 2\theta \qquad \text{... (14.26)}$$

Shear strain in OC,

$$\varepsilon_s = \frac{CH}{OC} = \frac{CG - HG}{OC} = \frac{CG - EF}{OC}$$

$$= \frac{CF\sin\theta - C_1F\cos\theta}{OC}$$

$$= \frac{1/2\,DC.\varepsilon_x.\sin\theta - 1/2\,BC.\varepsilon_y.\cos\theta}{1/2\,AC}$$

$$= \varepsilon_x.\frac{DC}{AC}.\sin\theta - \varepsilon_y.\frac{BC}{AC}.\cos\theta$$

$$= \varepsilon_x.\cos\theta.\sin\theta - \varepsilon_y.\sin\theta.\cos\theta$$

$$\varepsilon_s = (\varepsilon_x - \varepsilon_y)\sin\theta\cos\theta$$

$$\varepsilon_s = \frac{1}{2}(\varepsilon_x - \varepsilon_y)\sin 2\theta \qquad \text{......(14.27)}$$

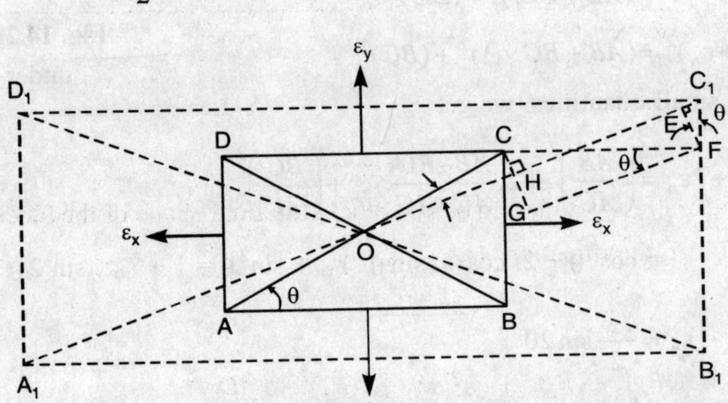

Fig. 14.21 Strains on an oblique plane under biaxial strains.

Now $\varepsilon_s = \frac{\gamma}{2}$, the angular unit shear. For shear strain to be maximum, sin 2θ = 1.0 or 2θ = 90°. Hence θ = 45° and

$$(\varepsilon_s)_{max} = \frac{1}{2}(\varepsilon_x - \varepsilon_y) \quad \text{.... (14.28)}$$

Resultant strain, $\varepsilon_r = \sqrt{\varepsilon_n^2 + \varepsilon_s^2} \; \sqrt{(\varepsilon_x \cos^2\theta + \varepsilon_y \sin^2\theta)^2 + (\varepsilon_x - \varepsilon_y)^2 \sin^2\theta.\cos^2\theta}$

$$= [\varepsilon_x^2 \cos^4\theta + \varepsilon_y^2 \sin^4\theta + 2\varepsilon_x\varepsilon_y \sin^2\theta.\cos^2\theta + (\varepsilon_x{}^2 + \varepsilon_y{}^2 - 2\varepsilon_x\varepsilon_y)\sin^2\theta.\cos^2\varepsilon]^{½}$$

$$\varepsilon_r = \sqrt{\varepsilon_x^2 \cos^2\theta + \varepsilon_y^2 \sin^2\theta} \quad \text{.... (14.29)}$$

14.7.4 Biaxial Normal Strains Combined with Shear Strains

Consider a block *ABCD* subjected to direct stresses σ_x and σ_y and shear stress τ_{xy} as shown in Fig. 14.22 (a). Let σ_x be the total normal strain in *AB*, then form Fig. 14.22 (b), we have

$$\varepsilon_x = \frac{AB_1 - AB}{AB} = \frac{BB_1}{AB} = \frac{\sigma_x}{E} - \upsilon\frac{\sigma_y}{E}$$

and ε_y be the total normal strain in *AD*, i.e.,

$$\varepsilon_y = \frac{AD_1 - AD}{AD} = \frac{DD_1}{AD} = \frac{\sigma_y}{E} - \upsilon\frac{\sigma_x}{E}$$

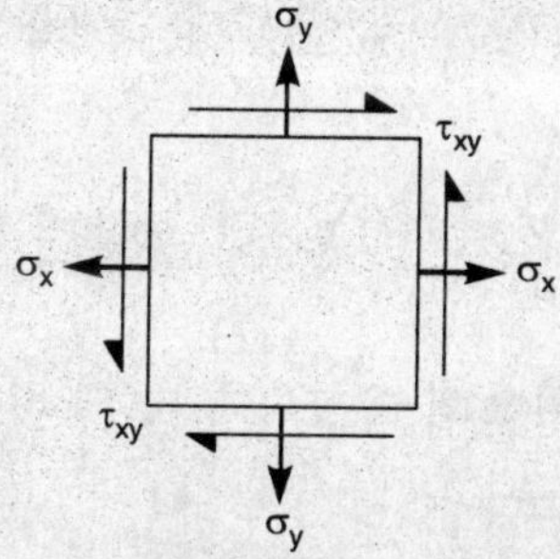

(a) Stress element

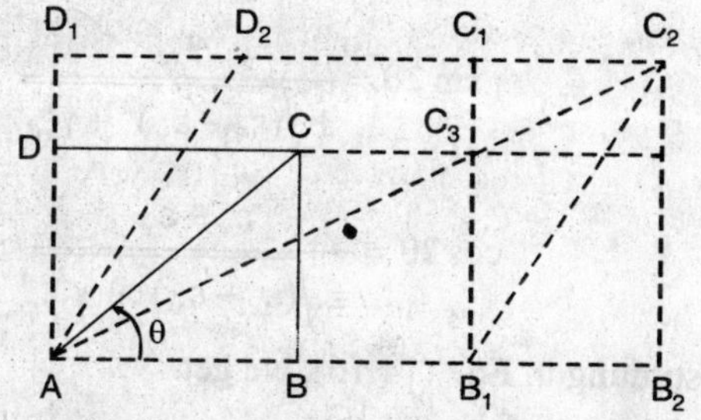

(b) Complex strains

Fig. 2.22.

Due to the shear stress τ_{xy} acting on the block, the final shape of the block will be $AB_1C_2D_2$. The direct strain in the direction *A* to *C* is,

$$\varepsilon_n = \frac{AC_2 - AC}{AC}$$

Now $(AC_2)^2 = (AB_2)^2 + (B_2C_2)^2 = (AB_1 + B_1B_2)^2 + (AD_1)^2$

$$= (AB_1 + C_1C_2)^2 = (AD_1)^2 = [AB.(1+\varepsilon_x) + \gamma_{xy}(1+\varepsilon_y)AD]^2 + (AD)^2(1+\varepsilon_y)^2$$

$$= (AB)^2(1+\varepsilon_x)^2 + \gamma_{xy}^2(1+\varepsilon_y)^2(AD)^2 + 2\gamma_{xy}(1+\varepsilon_x)(1+\varepsilon_y)AB.AD + (AD)^2(1+\varepsilon_y)^2$$

Simplifying and neglecting small quantities, we get

$$(AC_2)^2 = (AB)^2(1+2\varepsilon_x) + 2\gamma_{xy}.AB.AD + (AD)^2(1+2\varepsilon_y)$$

$$= 2(AB)^2.\varepsilon_x + 2\gamma_{xy}.AB.AD + [(AB)^2 + (AD)^2] + 2(AD)^2.\varepsilon_y$$

$$= 2(AB)^2.\varepsilon_x + 2\gamma_{xy}.AB.AD + (AC)^2 + 2(AD)^2.\varepsilon_y$$

Now $(AC_2)^2 = (AC)^2(1+\varepsilon_n)^2 = (AC)^2(1+2\varepsilon_n)$

$$\therefore \quad 1+2\varepsilon_n = 2c_x\left(\frac{AB}{AC}\right)^2 + 2\gamma_{xy}.\frac{AB}{AC}.\frac{AD}{AC} + 1 + 2\varepsilon_y.\left(\frac{AD}{AC}\right)^2$$

$$\varepsilon_n = \varepsilon_x \cdot \cos^2\theta + \gamma_{xy}\cos\theta \cdot \sin\theta + \varepsilon_y \cdot \sin^2\theta$$

or $$\varepsilon_n = \varepsilon_x.\cos^2\theta + \varepsilon_y.\sin^2\theta + \gamma_{xy}\sin\theta.\cos\theta$$

or $$\varepsilon_n = \frac{1}{2}(\varepsilon_x+\varepsilon_y) + \frac{1}{2}(\varepsilon_x-\varepsilon_y)\cos 2\theta + \frac{\gamma_{xy}}{2}\sin 2\theta \qquad (14.30)$$

The maximum and minimum value of ε_n are given when $\frac{d\varepsilon_n}{d\theta} = 0$

$$-\frac{1}{2}(\varepsilon_x-\varepsilon_y)2\sin 2\theta + \frac{1}{2}\gamma_{xy}.2\cos 2\theta = 0$$

$$\tan 2\theta = \frac{\gamma_{xy}}{\varepsilon_x - \varepsilon_y} \qquad ...(14.31)$$

$$\sin 2\theta = \frac{\gamma_{xy}}{\pm\sqrt{(\varepsilon_x-\varepsilon_y)^2+\gamma_{xy}^2}}$$

$$\cos 2\theta = \frac{\varepsilon_x-\varepsilon_y}{\pm\sqrt{(\varepsilon_x-\varepsilon_y)^2+\gamma_{xy}^2}}$$

Substituting in Eq. (14.16), we get

$$\varepsilon_{1,2} = \frac{1}{2}(\varepsilon_x+\varepsilon_y) \pm \sqrt{(\varepsilon_x-\varepsilon_y)^2+\gamma_{xy}^2} \qquad ...(14.32)$$

the two principal strains.

In order to determine the shearing strain, which is defined as the change in angle between to mutually perpendicular planes, the relative displacement of two points C and E will be considered as shown in Fig. 14.23.

Now $\alpha \cong \tan\alpha$ for small α

$$= \frac{CF}{AC} = \frac{CM+MF}{AC}$$

$$= \frac{C_1H - C_1L + CM}{AC}$$

$$= \frac{C_1C_2 \sin\theta - C_1C_3 \cos\theta + CC_3 \sin\theta}{AC}$$

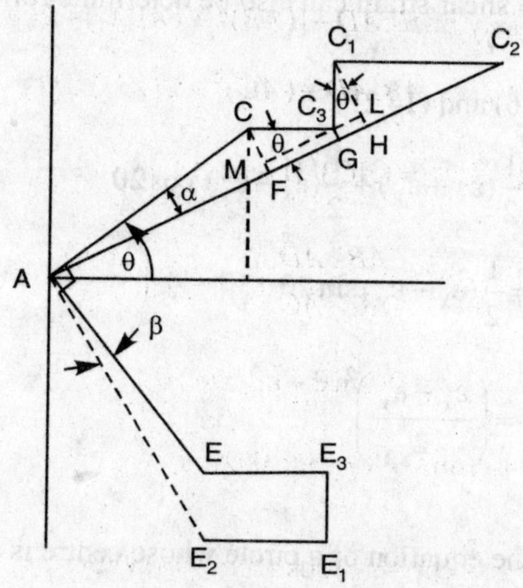

Fig. 14.23

$$= \frac{\gamma_{xy}(1+\varepsilon_y)BC\sin\theta - \varepsilon_y.BC.\cos\theta + \varepsilon_x.AB.\sin\theta}{AC}$$

$$= \gamma_{xy}\sin^2\theta - \varepsilon_y \sin\theta \times \cos\theta + \varepsilon_x \cos\theta \sin\theta$$

$$\alpha = \gamma_{xy}\sin^2\theta + (\varepsilon_x - \varepsilon_y)\sin\theta \cos\theta$$

Similarly by putting $\theta = -(90° - \theta)$, we get

$$\beta = \gamma_{xy}\sin^2\theta - (\varepsilon_x - \varepsilon_y)\sin\theta \cos\theta$$

Shear strain, $\gamma = -\beta + \alpha$

$$2\varepsilon_s = \gamma = 2(\varepsilon_x - \varepsilon_y)\sin\theta \cos\theta - \gamma_{xy}\cos 2\theta$$

$$= (\varepsilon_x - \varepsilon_y)\sin 2\theta - \gamma_{xy}\cos 2\theta$$

$$\therefore \quad \varepsilon_s = \frac{1}{2}(\varepsilon_x - \varepsilon_y)\sin 2\theta - \frac{\gamma_{xy}}{2}\cos 2\theta \qquad ....(14.33)$$

For shearing strain to be maximum or minimum $\frac{d\varepsilon_s}{d\theta} = 0$

$$\therefore \quad (\varepsilon_s - \varepsilon_y)\cos 2\theta + \gamma_{xy}.\sin 2\theta = 0$$

$$\tan 2\theta = -\left(\frac{\varepsilon_x - \varepsilon_y}{\gamma_{xy}}\right) \qquad ...(14.34)$$

Substituting Eq. (14.19) in Equ. (14.18), we get

$$(\varepsilon_s)_{\max} = \left(\frac{\gamma}{2}\right)_{\max} = \pm\,[(\varepsilon_x - \varepsilon_y)^2 + \gamma_{xy}^2\,]^{1/2} \qquad .... (14.35)$$

14.8 MOHR'S CIRCLE OF STRAIN

Mohr's strain circle is a circular diagram to determine the normal, shear and resultant strains it a point in a body when subjected to biaxial or complex state of strain. The principal strains, principal planes and maximum shear strain can also be determine form this circle.

14.8.1 Biaxial Strains

We have from Eqs. (14.26) and (14.27)

$$\varepsilon_n = \frac{1}{2}(\varepsilon_x + \varepsilon_y) + \frac{1}{2}(\varepsilon_x - \varepsilon_y)\cos 2\theta$$

$$\varepsilon_s = \frac{1}{2}(\varepsilon_x - \varepsilon_y)\sin 2\theta$$

$$\therefore \left(\varepsilon_n - \frac{\varepsilon_x - \varepsilon_y}{2}\right)^2 + \varepsilon_s^{\,2} = \left(\frac{\varepsilon_x - \varepsilon_y}{2}\right)^2 \qquad \text{....(14.36)}$$

Eq. (14.21) represents the equation of a circle whose centre is at $\left(\frac{\varepsilon_x + \varepsilon_y}{2}, 0\right)$ in the ε_n and $\varepsilon_s = \frac{\gamma}{2}$ coordinate system and with radius of the circle equal to $\left(\frac{\varepsilon_x - \varepsilon_y}{2}\right)$. This leads us to represent the state of strain at a point by a circular diagram.

Construction:

(*a*) Like strains

Take $OA = \varepsilon_s, OB = \varepsilon_y$. Let C be the centre of AB. With C as centre and CA as radius, describe the circle as shown in Fig. 14.24. Make $\angle ACD = 2\theta$. Drop $DE \perp BA$. Join OD. Then $OE = \varepsilon_n$, $OE = \varepsilon_s$, $OD = \varepsilon_r$ and $CF = (\varepsilon_s)_{\max}$.

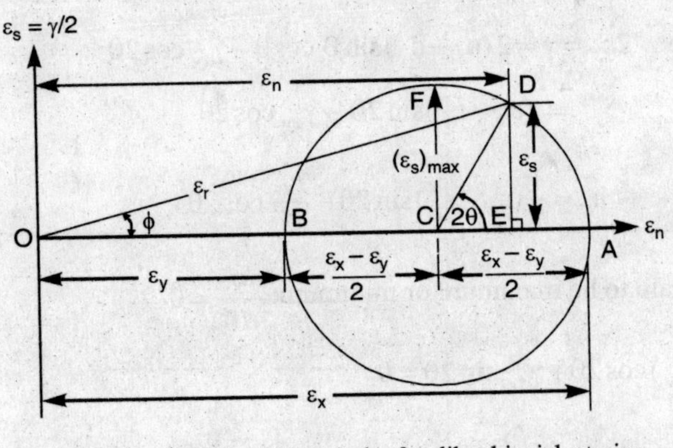

Fig. 14.24 Mohr's strain circle for like biaxial strains.

Proof:

$$OC = OB + BC$$

$$= \varepsilon_y + \frac{\varepsilon_x - \varepsilon_y}{2} = \frac{1}{2}(\varepsilon_x + \varepsilon_y)$$

$$CA = \frac{1}{2}(\varepsilon_x - \varepsilon_y)$$

$$OE = OC + CE = OC + CD\cos 2\theta$$

$$\varepsilon_n = \frac{1}{2}(\varepsilon_x + \varepsilon_y) + \frac{1}{2}(\varepsilon_x - \varepsilon_y)\cos 2\theta$$

$$\varepsilon_s = DE = CD\sin 2\theta$$

$$= \frac{1}{2}(\varepsilon_x - \varepsilon_y)\sin 2\theta$$

Maximum shear strain

$$(\varepsilon_s)_{\max} = CF = \frac{1}{2}(\varepsilon_x - \varepsilon_y)$$

Resultant strain $\varepsilon_r = OD = \sqrt{OE^2 + DE^2}$

$$\varepsilon_r = \sqrt{\varepsilon_n^2 + \varepsilon_s^2}$$

***(b)* Unlike strains**

Let ε_y be compressive and ε_x tensile in nature.

Draw $OA = \varepsilon_x$, $OB = \varepsilon_y$. Let *C* be the centre of *AB*. With *C* as centre and *CA* as radius, describe the circle. Make $\angle ACD = 2\theta$. Drop $DE \perp CA$. Join *OD*. Then $OE = \varepsilon_n$, $DE = \varepsilon_s$, $OD = \varepsilon_r$ as shown in Fig. 14.25.

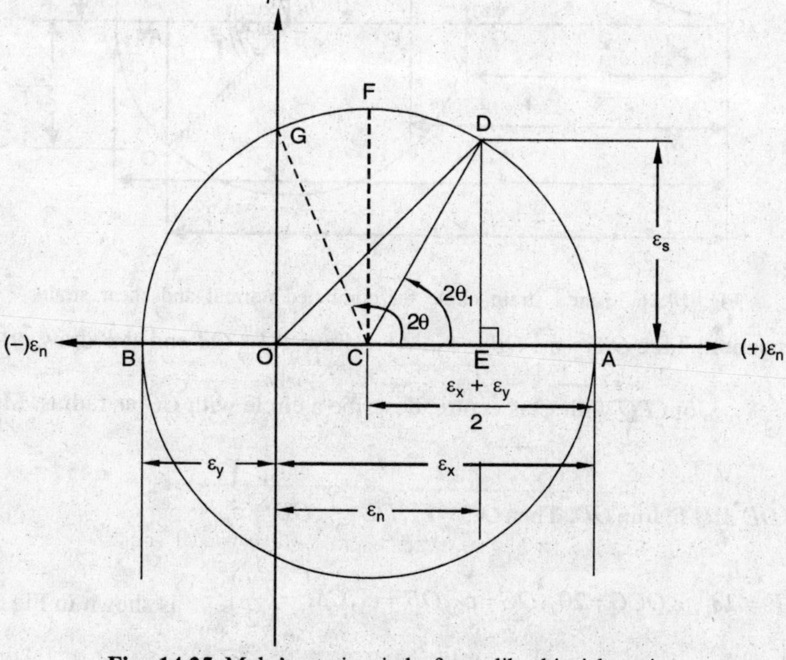

Fig. 14.25 Mohr's strain circle for unlike biaxial strains

Beyond $\angle ACG = 2\theta_1$, the normal strain will become compressive.

14.8.2. Normal Strains Combined with Shear Strains

In this case, we have from Eqs. (14.30) and (14.33)

$$\varepsilon_n = \frac{1}{2}(\varepsilon_x + \varepsilon_y) + \frac{1}{2}(\varepsilon_x - \varepsilon_y)\cos 2\theta + \frac{1}{2}\gamma_{xy}\sin 2\theta$$

$$\varepsilon_s = \frac{1}{2}(\varepsilon_x - \varepsilon_y)\sin 2\theta - \frac{1}{2}\gamma_{xy}\cos 2\theta$$

$$\left(\varepsilon_n - \frac{\varepsilon_x + \varepsilon_y}{2}\right)^2 + \varepsilon_s^2$$

$$= \left[\frac{1}{2}\left((\varepsilon_x - \varepsilon_y)^2 + \gamma_{xy}^2\right)^{1/2}\right]^2 \qquad .....(14.37)$$

Eq. (14.37) represents a circle whose centre is at $\left(\frac{\varepsilon_x + \varepsilon_y}{2}, 0\right)$ and radius $\frac{1}{2}[(\varepsilon_x - \varepsilon_y)^2 + \gamma_{xy}^2]^{1/2}$

in the $\varepsilon_n - \varepsilon_s$ plane.

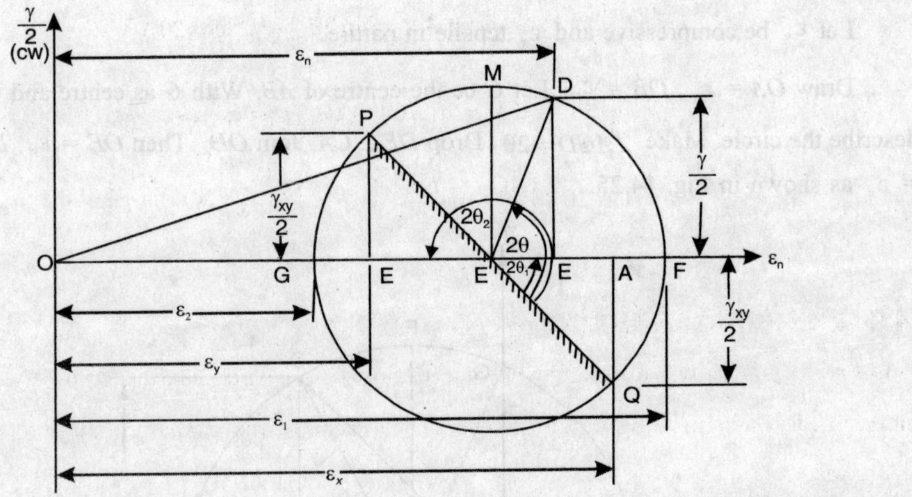

Fig. 14.26 Mohr's strain circle for combined normal and shear strains.

Construction: Take $OA = \varepsilon_x$, $OB = \varepsilon_y$. At A draw $AQ \perp OA$ and at B draw $BP \perp OA$ so that $AQ = BP = \frac{1}{2}\gamma_{xy}$. Join PQ with C as centre, describe a circle with CO as radius. Make $\angle QCD = 2\theta$.

Drop $DE \perp OA$. Join OD. Then $OE = \varepsilon_n, DE \frac{-\gamma}{2}, OD = \varepsilon_r$

$\angle QCF = 2\theta_1$, $\angle QCG = 2\theta_2 . OG = \varepsilon_2, OF = \varepsilon_1, CM = \left(\frac{\gamma}{2}\right)_{max}$ as shown in Fig. 14.26

Proof:

$$OE = OC + CE = OC + CD\cos 2(\theta - \theta_1)$$

$$= OC + CD\,[\cos 2\theta \cdot \cos 2\theta_1 + \sin 2\theta \cdot \sin 2\theta_1]$$

$$= OC + CD\left[\cos 2\theta . \frac{CA}{CQ} + \sin 2\theta . \frac{AQ}{CQ}\right]$$

$$= OC + CD.\cos 2\theta + AQ.\sin 2\theta \qquad [\therefore\ CD = CQ]$$

$$\varepsilon_n = \frac{1}{2}(\varepsilon_x + \varepsilon_y) + \frac{1}{2}(\varepsilon_x - \varepsilon_y)\cos 2\theta + \frac{1}{2}\gamma_{xy}\sin 2\theta$$

$$DE = CD\sin 2\,(\theta - \theta_1)$$

$$= CD[\sin 2\theta . \cos 2\theta_1 - \cos 2\theta . \sin 2\theta_1]$$

$$= CD\left[\sin 2\theta . \frac{CA}{CQ} - \cos 2\theta . \frac{AQ}{CQ}\right]$$

$$= CA.\sin 2\theta - AQ\cos 2\theta$$

$$= \frac{1}{2}(\varepsilon_x - \varepsilon_y)\sin 2\theta - \frac{1}{2}\gamma_{xy}\cos 2\theta$$

$$\left(\frac{\gamma}{2}\right)_{\max} = CM = \sqrt{(AC)^2 + (AQ)^2} = \pm\frac{1}{2}\sqrt{(\varepsilon_x - \varepsilon_y)^2 + \gamma_{xy}^2}$$

$$\tan 2\theta_1 = \frac{AQ}{AC} = \frac{\gamma_{xy}}{\varepsilon_x - \varepsilon_y}$$

$$\theta_2 = \theta_1 + 90°$$

14.9 PRINCIPAL STRAINS

Consider a body subjected to normal strains $\varepsilon_x, \varepsilon_y$ and shearing strain γ_{xy} in the two-dimensional system. Then it can be shown that on a plane inclined at an angle θ with ε_x.

Normal strain $$\varepsilon_n = \frac{1}{2}(\varepsilon_x + \varepsilon_y) + \frac{1}{2}(\varepsilon_x - \varepsilon_y)\cos 2\theta + \frac{\gamma_{xy}}{2}\sin 2\theta$$

The principal strains are given by

$$\varepsilon_{1,2} = \frac{1}{2}(\varepsilon_x + \varepsilon_y) \pm \frac{1}{2}\sqrt{(\varepsilon_x - \varepsilon_y)^2 + \gamma^2{}_{xy}}$$

The principal planes are given by,

$$\tan 2\theta = \frac{\gamma_{xy}}{\varepsilon_x - \varepsilon_y}$$

Maximum shearing strain;

$$\left(\frac{\gamma}{2}\right)_{\max} = \sqrt{(\varepsilon_x - \varepsilon_y)^2 + \gamma_{xy}^2} = \frac{1}{2}(\varepsilon_1 - \varepsilon_2)$$

In a three-dimensional state of stress, the principal strains are:

$$\left.\begin{aligned} \varepsilon_1 &= \frac{1}{E}[\sigma_1 - \upsilon(\sigma_2 + \sigma_3)] \\ \varepsilon_2 &= \frac{1}{E}[\sigma_2 - \upsilon(\sigma_3 + \sigma_1)] \\ \varepsilon_3 &= \frac{1}{E}[\sigma_3 - \upsilon(\sigma_1 + \sigma_2)] \end{aligned}\right\} \quad \text{....(14.38)}$$

14.10 RELATIONSHIP BETWEEN ELASTIC CONSTANTS

Consider a cube of unit size acted upon by simple shear stress τ_{xy}, as shown in Fig. 14.27 on its faces *AB* and *DC*.

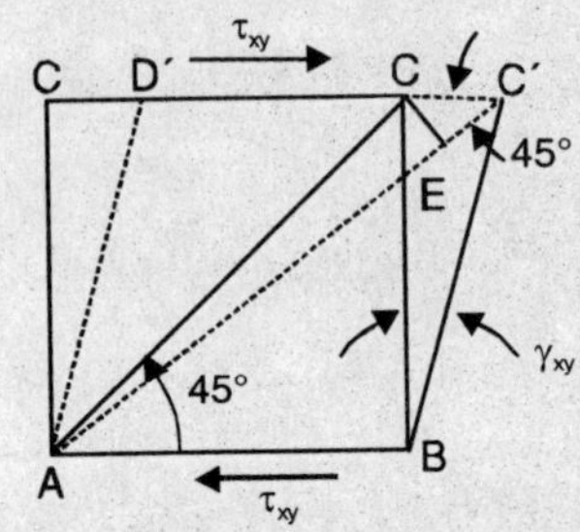

Fig.14.27 Simple shear of a cubic element

Shear strain, $\tau_{xy} = \dfrac{CC'}{BC} = \dfrac{\tau_{xy}}{G}$

From C and $CE \perp AC'$, so that $AC = AE$

Elongation of diagonal $AC = EC'$

$$= CC' \cos 45° = \frac{CC'}{\sqrt{2}}$$

Now $AC = \sqrt{2}\, AB$

Strain in the diagonal $AC = \dfrac{EC}{AC} = \dfrac{CC'}{\sqrt{2}} \times \dfrac{1}{\sqrt{2}\, AB} = \dfrac{1}{2} \cdot \dfrac{CC'}{AB}$

$$= \frac{1}{2} \cdot \frac{CC'}{BC} = \frac{1}{2}\gamma_{xy} = \frac{1}{2} = \frac{\tau_{xy}}{G}$$

Now if the cube is subjected to simple shear, then the normal stress on the diagonal of the cube is equal to the shear stress, i.e.

$$\sigma_n = \tau_{xy}$$

as shown in Fig. 14.28. This can be shown as explained below.

For the ΔACB, resolving forces along the normal, we have

$$\sigma_n\, AC = \tau_{xy} . AB \cos 45° + \tau_{xy} . BC \sin 45°$$

$$\sigma_n = \tau_{xy} \cdot \frac{AB}{AC} \cos 45° + \tau_{xy} . \frac{BC}{AC} \sin 45°$$

$$= \tau_{xy} \cos^2 45° + \tau_{xy} \sin^2 45°$$

$$= \frac{\tau_{xy}}{2} + \frac{\tau_{xy}}{2} = \tau_{xy}$$

$\therefore$ strain in the diagonal $= \dfrac{1}{2} \dfrac{\sigma_n}{G}$

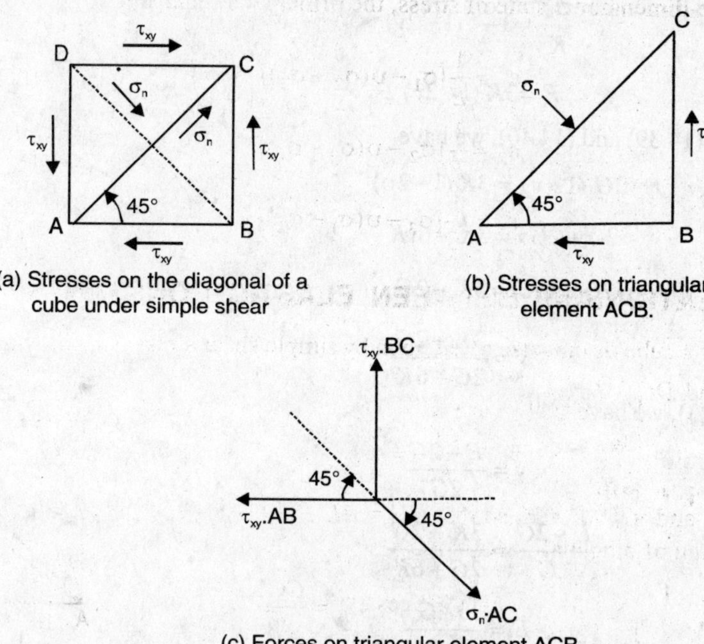

(a) Stresses on the diagonal of a cube under simple shear

(b) Stresses on triangular element ACB.

(c) Forces on triangular element ACB.

Fig. 14.28

If σ_n acted alone in the direction of the diagonal *AC*, then the resultant strain would be $\frac{\sigma_n}{E}$, but there is an equal and opposite normal stress in the direction of the diagonal *BD* also. The strain caused by this in the direction $AC = -\nu\frac{\sigma_n}{E}$.

Total strain in the direction of diagonal $AC = \frac{\sigma_n}{E} - \left(-\frac{\nu\sigma_n}{E}\right) = \frac{\sigma_n}{E}(1+\nu)$

$$\therefore \qquad \frac{1}{2}\frac{\sigma_n}{G} = \frac{\sigma_n}{E}(1+\nu)$$

or

$$G = \frac{E}{2(1+\nu)} \qquad \text{.... (14.39)}$$

Now consider cube under an equal compressive stress acting on all its faces, as shown in Fig. 14.29.

Then, volume strain, $\varepsilon_\upsilon = \frac{\sigma_n}{K}$

when K = bulk modulus

Total strain along the edge $= \frac{\sigma_n}{E}(1-2\nu)$

$$\varepsilon_\upsilon = 3\frac{\sigma_n}{E}(1-2\nu)$$

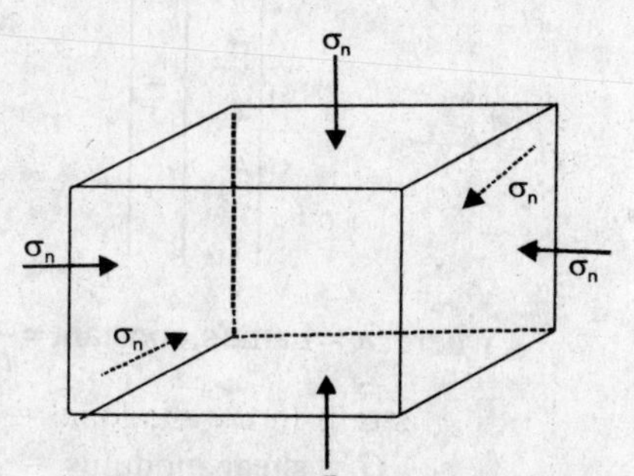

Fig.14.29 Cubic element under hydrostatic stress.

$$\frac{\sigma_n}{K} = 3\frac{\sigma_n}{E}(1-2v)$$

$$E = 3K(1-2v) \qquad \text{.... (14.40)}$$

From Eqs. (14.39) and (14.40), we have

$$2G(1+v) = 3K(1-2\upsilon)$$

$$2G + 2Gv = 3K - 6vK$$

$$\upsilon(2G+6K) = 3K - 2G$$

$$v = \frac{3K-2G}{2G+6K} \qquad \text{... (14.41)}$$

From Eq. (3), we have

$$v = \frac{E-2G}{2G} \qquad \text{....(14.42)}$$

$$\therefore \quad \frac{E-2G}{2G} = \frac{3K-2G}{2G+6K}$$

$$E = \frac{9KG}{G+3K} \qquad \text{....(14.43)}$$

14.11 GENERALISED HOOKE'S LAW

In a three-dimensional state of stress at a point, there are 9 stress components and 9 strain components. Out of these 9 components, 3 are normal components and 6 are shear components. Therefore, to correlate stresses with strains, there has to be 81 elastic constants for an elastic anisotropic media. But the shear components are complementary in nature, therefore, there are 6 independent stress components and 6 independent strain components. This reduces the number of independent elastic constants to 36. In an orthotropic material, symmetry is exhibited about the three diagonal planes. This reduces the member of independent elastic constants to 9. In cubic materials, the mechanical properties are same along three orthogonal directions. This reduces the number of elastic constants to 3.

The isotropic materials possess elastic properties which are independent of the orientation of the axes. Thus we get only two independent elastic constants λ and G.

The stress-strain relationship for three-dimensional state of stress for isotropic material is :

$$\begin{Bmatrix} \sigma_x \\ \sigma_y \\ \sigma_z \\ \tau_{xy} \\ \tau_{yz} \\ \tau_{xz} \end{Bmatrix} = \begin{bmatrix} 2G+\lambda & \lambda & \lambda & 0 & 0 & 0 \\ & 2G+\lambda & \lambda & 0 & 0 & 0 \\ & & 2G+\lambda & 0 & 0 & 0 \\ & \text{symmetric} & & G & 0 & 0 \\ & & & & G & 0 \\ & & & & & 0 \end{bmatrix} \begin{Bmatrix} \varepsilon_x \\ \varepsilon_y \\ \varepsilon_z \\ \gamma_{xy} \\ \gamma_{yz} \\ \gamma_{xz} \end{Bmatrix}$$

where $\lambda = \text{Lame's constant} = \dfrac{Ev}{(1+v)(1-2v)}$

$$G = \text{shear modulus} = \frac{E}{2(1+v)}$$

Example 14.11 : *A bar of 20 mm diameter is subjected to a pull of 20 kN. The measured extension over a gauge length of 200 mm is 0.1 mm and the change in diameter is 0.0035 mm. Calculate the value of modulus of elasticity, bulk modulus, and Poisson's ratio.*

Solution: Given: $d = 20$ mm, $P = 20$ kN, $l = 200$ mm, $\delta l = 0.1$ mm, $\delta d = 0.0035$ mm

Area of cross-section of bar, $A = \frac{\pi}{4}d^2 = \frac{\pi}{4}(20)^2 = 314.16\,\text{mm}^2$

Modulus of elasticity, $E = \frac{Pl}{A\delta l} = \frac{20\times10^3\times200}{314.16\times0.1} = 127\times10^3\ N/mm^2$

Longitudinal strain, $\varepsilon_l = \frac{\delta l}{l} = \frac{0.1}{200} = 5\times10^{-4}$

Lateral strain, $\varepsilon_d = \frac{\delta d}{d} = \frac{0.0035}{20} = 1.75\times10^{-4}$

Poisson's ratio, $\nu = \frac{\varepsilon_d}{\varepsilon_l} = \frac{1.75}{5} = 0.35$

Bulk modulus, $K = \frac{E}{3(1-2\nu)} = \frac{127\times10^3}{3(1-0.7)} = 141.1\times10^3\ \text{N/mm}^2$

Example 14.12 : *A material is subjected to two mutually perpendicular linear strains together with a shear strain. One of the linear strain is 250×10^{-6} tensile. Determine the magnitude of the other linear strain and shear strain if the principal strains are -100×10^{-6} and $+300\times10^{-6}$.*

Solution : Given. $\varepsilon_x = +250\times10^{-6}, \varepsilon_1 = 300\times10^{-6}, \varepsilon_2 = -100\times10^{-6}$

$$\varepsilon_y = ?,\ \gamma_{xy} = ?$$

$$\varepsilon_1 = \frac{1}{2}(\varepsilon_x+\varepsilon_y)+\frac{1}{2}\sqrt{(\varepsilon_x-\varepsilon_y)^2+\gamma_{xy}^2}$$

$$300 = \frac{1}{2}(250+\varepsilon_y)+\frac{1}{2}\sqrt{(250-\varepsilon_y)^2+\gamma^2_{xy}} \quad ....(1)$$

$$\varepsilon_2 = \frac{1}{2}(\varepsilon_x+\varepsilon_y)-\frac{1}{2}\sqrt{(\varepsilon_x-\varepsilon_y)^2+\gamma_{xy}^2}$$

$$-100 = \frac{1}{2}(250+\varepsilon_y)-\frac{1}{2}\sqrt{(250-\varepsilon_y)^2+\gamma_{xy}^2} \quad ....(2)$$

Subtracting Eq. (2) from (1), we have

$$400 = \sqrt{(250-\varepsilon_y)^2+\gamma_{xy}^2} \quad ....(3)$$

Substituting Eq. (3) in (1), we get

$$300 = \frac{1}{2}(250+\varepsilon_y)+200$$

$$600 = 650+\varepsilon_y$$

$$\varepsilon_y = -50\ \mu\ \text{strain or} -0.00005$$

Substituting the value of ε_y in Eq. (3), we have

$$400 = \sqrt{(250+50)^2 + \gamma_{xy}^2}$$

$$= \sqrt{(300)^2 + \gamma_{xy}^2}$$

Squaring both sides, we have

$$400^2 = 300^2 + \gamma_{xy}{}^2$$

$$\gamma_{xy} = 264.6\mu - \text{strain or } 0.0002646$$

Example 14.13 : *A circle of 100 mm diameter is scribed on an aluminium plate before it is loaded as shown in Fig. 14.30. After stressing the circle deforms into an ellipse. Determine the lengths of major and minor axes of the ellipse and there orientation.*

For aluminium $E = 10$ GPa and $\upsilon = 0.32$.

Solution: Given : $\sigma_x = 80\,\text{MPa}$, $\sigma_y = 60\,\text{MPa}$, $\tau_{xy} = 25\,\text{MPa}$, $E = 70$ GPa, $\upsilon = 0.32$

Principal stresses in aluminium plate,

$$\sigma_{1.2} = \frac{1}{2}(\sigma_x + \sigma_y) \pm \frac{1}{2}\sqrt{(\sigma_x - \sigma_y)^2 + 4\tau_{xy}^2}$$

$$= \frac{1}{2}(80+60) \pm \frac{1}{2}\sqrt{(80-60)^2 + 4\times 25^2}$$

$$= 70 \pm 26.926$$

$$= 96.926 \text{ MPa, } 43.074 \text{ MPa}$$

$$\tan 2\theta = \frac{2\tau_{xy}}{\sigma_x - \sigma_y} = \frac{2\times 25}{80-60} = 2.5$$

$$\theta_1 = 34°, \theta_2 = 124°$$

Principal strains are:

$$\varepsilon_1 = \frac{1}{E}[\sigma_1 - \upsilon\sigma_2]$$

$$= \frac{1}{70\times 10^3}[96.926 - 0.32\times 43.074]$$

$$= 1.188 \times 10^{-3}$$

Increase in diameter of circle, $\delta d_1 = \varepsilon_1 d = 1.188\times 10^{-3} \times 100 = 0.1188\,\text{mm}$
Major diameter of ellipse$= d + \delta d_1$, $= 100.1188$ mm

$$\varepsilon_2 = \frac{1}{E}[\sigma_2 - \upsilon\sigma_1]$$

$$= \frac{1}{70\times 10^3}[43.074 - 0.32\times 96.926]$$

$$= 1.7225\times 10^{-4}$$

Increase in diameter of circle perpendicular to previous diameter

$$\delta d_2 = \varepsilon_2 d = 1.7225\times 10^{-4} \times 100 = 0.017225\,\text{mm}$$

Minor diameter of ellipse $= d + \delta d_2 = 100.0172 \text{ mm}$

The deformed shape of hole is shown in Fig. 14.30.

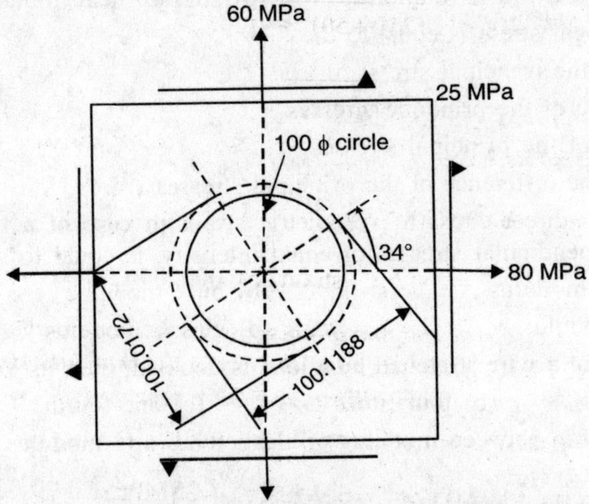

Fig. 14.30

MULTIPLE CHOICE QUESTIONS

1. In a uniaxial state of stress, the maximum normal stress occurs on a plane which is inclined to the load line at

(a) 0° (b) 45° (c) 60° (d) 90°

2. In a uniaxial state of stress, the maximum shear stress occurs on a plane which is inclined to the load line at

(a) 0° (b) 45° (c) 60° (d) 90°

3. In a uniaxial state of stress, the ratio of maximum shear stress to normal stress is :

(a) 1 : 2 (b) 2 : 1 (d) 1 : 4 (d) 4 : 1

4. On the principal planes, the maximum stress is the

(a) normal stress (b) shear stress (c) both normal and shear stress

(d) either normal or shear stress

5. In a uniaxial state of stress, the resultant stress on a plane inclined at an angle θ is :

(a) $\sigma \sin\theta$ (b) $\sigma \cos\theta$ (c) $0.5\sigma \sin 2\theta$ (d) $0.5\sigma \cos 2\theta$

6. In a biaxial state of stress, the shear stress is maximum on a plane making an angle with the stresses, equal to:

(a) 22.5° (b) 30° (c) 45° (d) 67.5°

7. In a biaxial state of stress, the maximum shear stress is equal to:

(a) sum of the stresses (b) difference of the stresses

(c) average of the stresses (d) half of the difference of stresses

8. In a body subjected to a complex state of stress, the planes of maximum shear stress are inclined to the principal planes at:

(a) 22.5° (b) 45° (c) 67.5° (d) 90°

9. For a complex state of stress, the radius of the Mohr's circle is:

(a) $0.5[(\sigma_x - \sigma_y)^2 + 4\tau_{xy}{}^2]^{1/2}$ (b) $[(\sigma_x - \sigma_y)^2 + 4\tau_{xy}{}^2]^{1/2}$

(c) $0.5[(\sigma_x - \sigma_y)^2 + \tau_{xy}{}^2]^{1/2}$ (d) $[(\sigma_x - \sigma_y)^2 + \tau_{xy}{}^2]^{1/2}$

10. Pure shear is equal to
(a) simple shear (b) simple shear plus rotation
(c) simple shear minus rotation (d) simple shear minus twice the rotation

11. Maximum shear stress is equal to the
(a) sum of the principal stress
(b) difference of the principle stresses
(c) average of the principal stresses
(d) half of the difference of the principal stresses

12. The ratio of direct stress to volumetric strain in case of a body subjected to three mutually perpendicular stresses of equal intensity, is equal to
(a) Young's modulus (b) bulk modulus
(c) shear modulus (d) elastic modulus

13. If the radius of a wire stretched by a load is doubled, then its Young's modulus will be
(a) doubled (b) four times (c) one-fourth (d) unaffected

14. The relationship between modulus of elasticity E and modulus of rigidity G is :
(a) $E = G\ (1+ v)$ (b) $G = E\ (2 - v)$
(c) $G = 0.5\ E/\ (1+ v)$ (d) $G = E\ (1+ 2\ v)$
wher v = Poisson's ratio

15. The ratio of shear modulus to the modulus of elasticity, if Poisson's ratio is 0.25, will be
(a) 0.40 (b) 0.25 (c) 4.00 (d) 0.50

16. The value of Poisson's ratio for steel lies in the range
(a) 0.10 to 0.20 (b) 0.21 to 0.25 (c) 0.26 to 0.30 (d) 0.30 to 0.35

17. If the modulus of elasticity of a material is 200 GPa and Poisson's ratio is 0.25, then the modulus of rigidity will be
(a) 80 GPa (b) 125 GPa (c) 133.3 GPa (d) 160 GPa

18. Choose the correct relationship between E, G and K.
(a) E = (G + 3K)/ 9GK
(b) E = (3G + K) / 9GK
(c) E = 9 GK / (G + 3K)
(d) E = 9 GK / (3G + K)

19. Choose the correct relationship between E, K, and v.
(a) $E = 3K\ (1–2v)$ (b) $E = 3K\ (1–v)$
(c) $E = K\ (1–2v)$ (d) $E = K\ (1–v)$

20. Maximum shear stress in Mohr's circle is equal to:
(a) radius of the circle (b) diameter of the circle
(c) centre of circle from y-axis (d) chord of circle

ANSWERS

1. (d)	2. (b)	3. (a)	4. (a)	5. (b)	6. (c)
7. (d)	8. (b)	9. (a)	10. (b)	11. (d)	12. (b)
13. (d)	14. (c)	15. (a)	16. (d)	17. (a)	18. (c)
19. (a)	20. (a)				

REVIEW QUESTIONS

1. Derive an expression for normal and shear stresses in a body subjected to biaxial loading.
2. Defined principal stresses and principal planes.

3. Derive and expression for principal stresses in a body subjected to complex state of stress.
4. Explain the procedure for drawing the Mohr's circle under biaxial loading.
5. Explain the procedure for drawing Mohr's circle under complex stresses.
6. Derive the relationship between elastic constants.

EXERCISES

14.1 The biaxial stresses in a material are +100 MPa and – 50 MPa. Determine the normal shear and resultant stresses on a plane whose normal is inclined at 30° to the axis of first stresses.

14.2 At a certain point in a material the resultant stress on a plane is 80 MPa tensile inclined at 30° to the normal to the plane. The stress on a plane at right angles to this plane has a normal tensile stress 50 MPa. Find (a) the resultant stress on second plane, (b) principal stresses, and (c) principal planes.

14.3 The state of stress at a point is given by:

$$\sigma_x = 90 \text{ MPa}, \sigma_y = -50 \text{ MPa}, \tau_{xy} = 40 \text{ MPa}$$

Determine (a) the planes of maximum shear stress, and (b) resultant stress on these planes.

14.4 The resultant stress at a point on plane A is + 90 MPa having an obliquity of 15°. On another plane B, the resultant stress is + 50 MPa having an obliquity of 35°. Determine (a) the angle between the planes A and B, and (b) the maximum obliquity the resultant stress can have with any plane at the point.

14.5 The state of stress at a point is given by:

$$\sigma_x = 140 \text{ MPa}, \sigma_y = 100 \text{ MPa}, \tau_{xy} = 60 \text{ MPa}$$

Calculate (a) the principal stresses, and (b) the principal planes.

14.6 The state of stress at a point is given by

$$\sigma_x = 300 \text{ MPa}, \sigma_y = -200 \text{ MPa}, \tau_{xy} = 100 \text{ MPa (cw)}$$

Draw the Mohr's circle and determine (a) the principal stresses and principal planes, (b) normal and shear stresses on a plane inclined at 40° with σ_x.

14.7 The state of stress at a point is given by:

$$\sigma_x = -150 \text{ MPa}, \sigma_y\ 250 \text{ MPa}, \tau_{xy} = 100 \text{ MPa (ccw)}$$

Draw the Mohr's circle and determine (a) the principal stresses and principal planes, (b) the maximum shear stress and its planes, and (c) draw the orientation of principal and maximum shear stress planes on the stress element.

14.8 A rectangular block 250 mm × 100 mm × 50 mm is subjected to the following state of stress.

$$\sigma_x = 100 \text{ MPa}, \sigma_y = -60 \text{ MPa}, \sigma_z = 80 \text{ MPa}$$

If v = 0.25 and E = 200 GPa, determine its change in volume.

14.9 A square bar of 20 mm side and 200 mm long is subjected to a compressive load of 200 KN applied in the direction of its length. If all the strains in the other two directions are prevented, determine the contraction in length.

14.10 A bar is subjected to the following stresses

$$\sigma_x = 50\,\text{MPa},\ \sigma_y = -40\,\text{MPa},\ \sigma_z = 70\,\text{MPa}$$

If $v = 0.25$ and $E = 200$ GPa, calculate (a) the principal strains, (b) volumetric strain, (c) shear modulus, and (d) bulk modulus.

14.11 The following strains were measured on pressures vessel

$$\varepsilon_x = 500\times10^{-6},\ \varepsilon_y = -100\times10^{-6}$$

and at 45° with x-axis, $\varepsilon_z = -200\times10^{-6}$ Calculate (a) the principal strains, and (b) principal stresses.

Take $v = 0.3$ and $E = 210$ GPa.

14.12 The state of strain at a point is given by

$$\varepsilon_x = 400\times10^{-6},\ \varepsilon_y = -200\times10^{-6},\gamma_{xy} = 250\times10^{-6}$$

Calculate (a) the principal strains, and (b) principal strains.

14.13 The principal strains at a point are given by

$$\varepsilon_1 = 800\times10^{-6},\ \varepsilon_2 = -600\times10^{-6},\varepsilon_3 = 400\times10^{-6}$$

If $v = 0.25$ and E = 200 GPa, calculate the principal stresses.

14.14 A circle 50 mm in diameter is scribed on a mild steel plate before it is subjected to the following state of stress:

$$\sigma_x = 200\,\text{MPa},\ \sigma_y = 100\,\text{MPa} = \tau xy = 80\,\text{MPa}$$

Calculate the length of the major and minor axes of the ellipse after deformation. Take $v = 0.28$ and $E = 210$ GPa.

15 SHEAR FORCE AND BENDING MOMENT

15.1 CONCEPT OF BEAM

Any member of a machine or structure whose one dimension (length) is very large as compared to the other two dimensions (width and thickness) and which can carry lateral or tranverse loads in the axial plane is called a beam. A beam may be of rectangular, square, triangular, hexagonal and circular, etc. cross-sections. A beam is a very important member in structural mechanics to withstand the tranverse loads. A beam may be made of timber, flitched beam, *i.e.*, timber reinforced with mild steel strips, steel and reinforced concrete. The reinforced concrete beams are mostly used in building construction, bridges and flyovers.

15.2 CONCEPT OF BENDING MOMENT AND SHEAR FORCE

To draw the bending moment and shear force diagrams it is essential to understand the concept of bending moment and shear force. The sign conventions to be followed are also very important for drawing these diagrams. Now we study these concepts.

15.2.1 Bending Moment

The bending moment (B.M.) at any point along a loaded beam is the algebraic sum of the moments of all the vertical forces acting to one side of the point about the point. Consider a simply supported beam *AB* carrying concentrated loads as shown in Fig. 15.1

Let R_A and R_B be the vertical reactions at supports *A* and *B* respectively. Consider a section *x-x* at a distance *x* from end *A*. The clockwise moment at this section due to all the loads acting on the beam to the left of the section is:

$$M_x = R_A x - W_1 (x - a) - W_2 (x - b) \circlearrowright \qquad \text{... (1)}$$

If we consider the forces to the right of the section *x-x*, then anticlockwise moment is:

$$M_x = R_B (l - x) - W_3 (l - x - c) \circlearrowleft \qquad \text{... (2)}$$

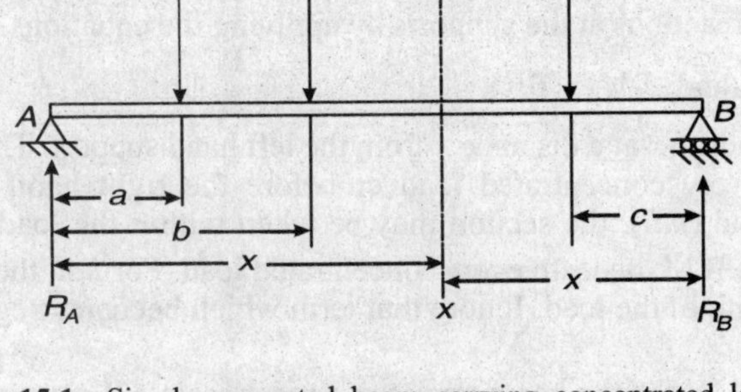

Fig. 15.1 Simply supported beam carrying concentrated loads.

For equilibrium of the beam, the B.M. given by equations (1) and (2) are equal. The SI units of B.M. are N.m or kN.m.

15.2.2 Shear Force

The shear force (S.F.) at any point along a loaded beam is the algebraic sum of all the vertical forces acting to one side of the point. Thus, for the beam *AB* shown in Fig. 15.2, the shear force at cross-section *x-x* as measured from the left hand side is:

$$F_x = R_A - W_1 - W_2 \quad \text{... (1)}$$

The shear force as measured from the right hand side is:

$$F_x = W_3 - R_B \quad \text{... (2)}$$

Since the beam is in equilibrium, the S.F. given by equations (1) and (2) are equal.

15.2.3 Sign conventions

A.B.M. is considered positive if it produces compression on the top fibres of the beam (Fig. 15.2*a*) and negative if it produces tension on the top fibres of the beam (Fig. 15.2*b*). Positive B.M. is called sagging moment and negative B.M. as hogging moment.

Fig. 15.2 Sign convention for B.M.

The S.F. is positive if the net resultant force to the left of the section is upward (Fig. 15.3a) and negative if it is downward (Fig. 15.3b). A positive S.F. produces clockwise rotation of the stress element and –ve S.F. produces counter-clockwise rotation.

Fig. 15.3 Sign convention for S.F.

15.3 B.M. AND S.F. DIAGRAMS

15.3.1 B.M. Diagram

To draw the B.M.D, the following procedure may be followed:

1. Take sheet of graph paper. Draw the beam along with loading to an appropriate scale.
2. Calculate the reactions at the supports by applying the equations of equilibrium, *i.e.*, $\Sigma F_x = 0$, $\Sigma F_y = 0$ and $\Sigma M_0 = 0$.
3. Choose a section *x-x* at a distance *x* from the left hand support. The section may be chosen either after every concentrated load or before the right hand support. For uniformly distributed load (*udl*), the section may be taken within the load.
4. Calculated the B.M. beneath every concentrated load. For *udl*, the B.M. may be calculated along the length of the load. Ignore that term which becomes negative on substituting the value of *x*.
5. Draw the B.M.D. for the beam on a convenient scale. Of course, sign convention has to

be followed for B.M.

Point of Inflextion: It is the point on the beam in the B.M.D. where the bending moment becomes zero.

Point of Contraflexure: It is the point in the B.M.D. where the B.M. changes slope from an increasing one to a decreasing one. Contraflexure means opposite and flexture means bending. Some authors consider point of inflexion and point of contraflexure to be synonymous.

15.3.2 S.F. Diagram

The first three steps for the S.F.D. are the same as for the B.M.D., and need not be repeated.

1. Calculate the S.F. beneath every concentrated load just to the left and to the right. For *u.d.l.*, the S.F. has to be calculated along the length of the laod.
2. Draw the S.F.D. to a convenient scale using the sign convention.

15.4 B.M. AND S.F. DIAGRAMS FOR A SIMPLY SUPPORTED BEAM

A simply supported beam is that which has pinned or hinged ends. These type of ends are obtained by supporting the beam on rollers. There are only normal reactions at such supports and no couple or moment acts at the supports.

15.4.1 Concentrated Load at Midspan

Consider a beam *AB* simply supported at the ends and carrying a concentrated load *W* at midspan *C* as shown in Fig. 15.4 (*a*).

$$\Sigma F_y = 0; \qquad R_A + R_B = W$$

$$\Sigma M_B = 0 \qquad R_A \times l = W \times \frac{l}{2}$$

$$R_A = \frac{W}{2}$$

$$R_B = \frac{W}{2}$$

S.M.D.

Take a section *x-x* at a distance *x* from *A*.

$$M_x = R_A x - W\left(x - \frac{l}{2}\right)$$

$$= \frac{W}{2}x - W\left(x - \frac{l}{2}\right)$$

This represents a straight line.

x	0	$\frac{l}{2}$	l
M_x	0	$\frac{Wl}{4}$	0

The B.M.D is shown in Fig. 15.4 (*b*) and is +ve through out the beam span.

$$M_{max} = \frac{Wl}{4}$$

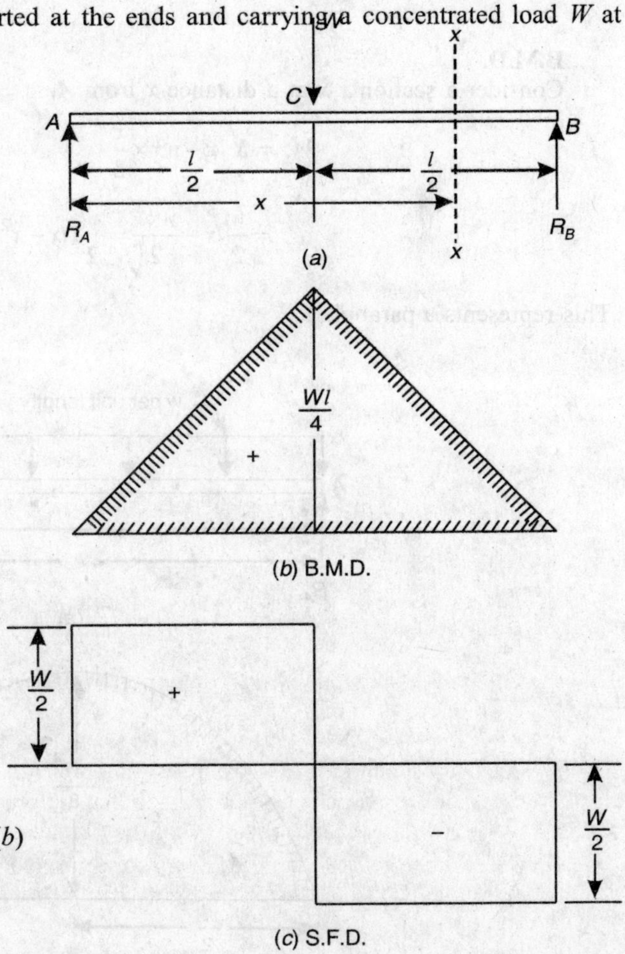

Fig. 15.4

and occurs at the mid-span.

S.F.D.

Between AB, S.F. $= \frac{W}{2} \uparrow$ and positive in nature.

Between BC, S.F. $= \frac{W}{2} - W = -\frac{W}{2} \downarrow$, *i.e.*, S.F. is negative in nature.

The S.F.D. is shown in Fig. 15.4 (*c*).

15.4.2 Uniformly Distributed Load (*udl*)

Consider a beam AB of span l simply supported at the ends and carrying a uniformly distributed load of intensity w per unit length as shown in Fig. 15.5 (*a*).

$$\Sigma F_y = 0; \quad R_A + R_B = wl$$

$$\Sigma M_B = 0; \quad R_A \times l = wl \times \frac{l}{2}$$

$$R_A = \frac{wl}{2}$$

$$R_B = \frac{wl}{2}.$$

B.M.D.

Consider a section x-x at a distance x from A.

$$M_x = R_A x - wx \times \frac{x}{2}$$

$$= \frac{w}{2} lx - \frac{wx^2}{2} = \frac{w}{2}(lx - x^2)$$

This represents a parabola

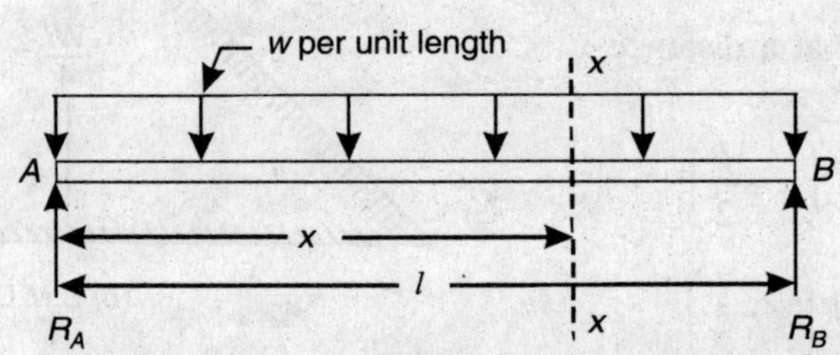

(*a*)

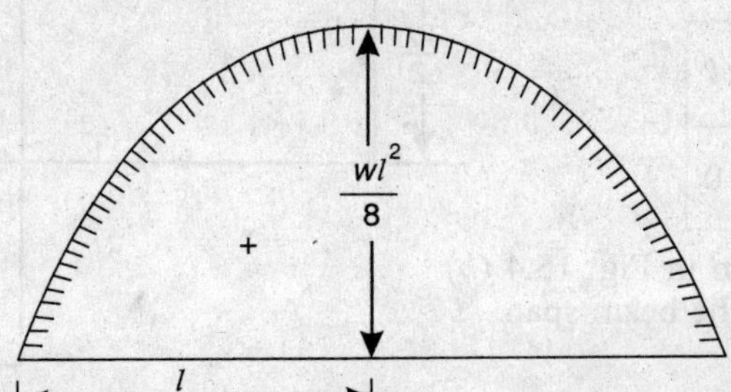

(*b*) B.M.D.

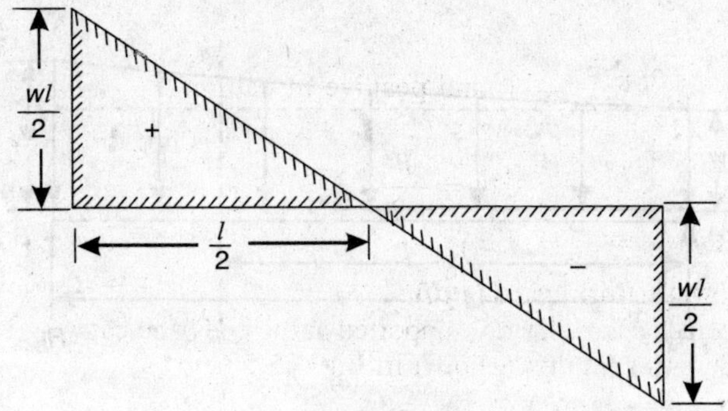

(c) S.F.D.

Fig. 15.5

x	0	$\frac{l}{2}$	l
M_x	0	$\frac{wl^2}{8}$	0

This B.M.D. is shown in Fig. 15.5 (*b*) and is +ve throughout the beam span.

S.F.D.

$$F_x = R_A - wx = \frac{wl}{2} - wx = w\left(\frac{1}{2} - x\right)$$

This represents a straight line.

x	0	$\frac{l}{2}$	l
F_x	$\frac{wl}{2}$	0	$\frac{-wl}{2}$

The S.F.D. is shown in Fig. 3.5 (*c*).

15.4.3 Uniformly Varying Loads

Consider a simply supported beam *AB* of span l carrying *udl* which varies from w_1 per unit length at end *A* to w_2 per unit length at end *B* as shown in Fig. 15.6 (*a*). The varying load can be considered as the sum of two loads, one of uniform intensity w_1 and the other of triangular variation from zero at *A* to $(w_2 - w_1)$ at *B*.

$$\Sigma F_y = 0; \quad R_A + R_B = \frac{1}{2}(w_1 + w_2)\, l$$

$$\Sigma M_B = 0; \quad R_A \times l = w_1 l \times \frac{1}{2} + \frac{1}{2}(w_2 - w_1)\, l \times \frac{1}{3} l$$

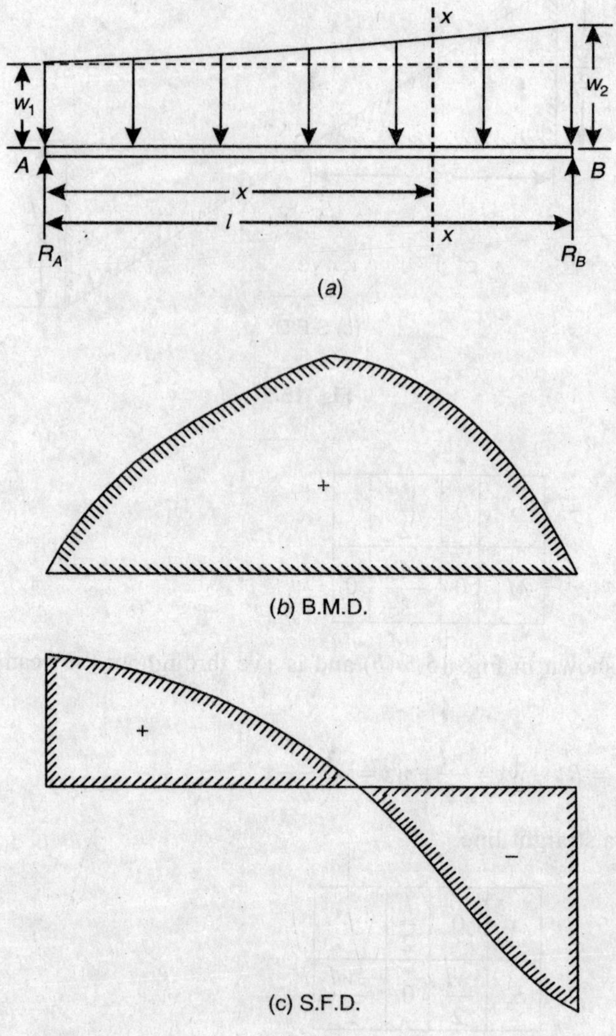

(a)

(b) B.M.D.

(c) S.F.D.

Fig. 15.6

$$= \frac{1}{2} w_1 l^2 + \frac{1}{6} w_2 l^2 - \frac{1}{6} w_1 l^2$$

$$= \frac{1}{3} w_1 l^2 + \frac{1}{6} w_2 l^2$$

$$R_A = \left(\frac{1}{3} w_1 + \frac{1}{6} w_2\right) l = \frac{1}{6}(2w_1 + w_2)\, l$$

$$R_B = \frac{1}{2}(w_1 + w_2) l - \frac{1}{6}(2w_1 + w_2)\, l$$

$$= \left(\frac{1}{6} w_1 + \frac{1}{3} w_2\right) l = \frac{1}{6}(w_1 + 2w_2)\, l$$

B.M.D.

Consider a section x-x at a distance x from end A.

Intensity of load, $w = w_1 + (w_2 - w_1)\frac{x}{l}$

$$M_x = R_A \cdot x - w_1 x \cdot \frac{x}{2} - \frac{1}{2}(w_2 - w_1)\frac{x}{l} \cdot x \times \frac{x}{3}$$

$$= \left(\frac{1}{3}w_1 + \frac{1}{6}w_2\right)lx - w_1\frac{x^2}{2} - \frac{1}{6}(w_2 - w_1)\frac{x^3}{l}$$

This represent the equation of a cubic curve.

x	0	$\frac{l}{2}$	l
M_x	0	$\frac{1}{16}(w_1 + w_2)\, l^2$	0

For M_x to be maximum, $\frac{dM_x}{dx} = 0$

$$\left(\frac{1}{3}w_1 + \frac{1}{6}w_2\right)l - w_1 x - \frac{1}{2}(w_2 - w_1)\frac{x^2}{l} = 0$$

or $$(w_2 - w_1)x^2 + 2w_1 lx - 2\left(\frac{1}{3}w_1 + \frac{1}{6}w_2\right)l^2 = 0$$

or $$(w_2 - w_1)x^2 + 2w_1 lx - \frac{1}{3}(2w_1 + w_2)l^2 = 0$$

$$x = \frac{-2w_1 l \pm \sqrt{4w_1^2 l^2 + \frac{4}{3}(2w_1 + w_2)\,(w_2 - w_1)\, l^2}}{2(w_2 - w_1)}$$

$$= l\left[\frac{-w_1 + \sqrt{(w_1^2 + w_2^2 + w_1 w_2)^3}}{(w_2 - w_1)}\right]$$

The B.M.D. is shown in Fig. 15.6 (*b*) and is positive throughout the beam span.

S.F.D.

$$F_x = R_A - w_1 x - \frac{1}{2}(w_2 - w_1)\frac{x}{l} \times x$$

$$= \left(\frac{1}{3}w_1 + \frac{1}{6}w_2\right)l - w_1 x - \frac{1}{2}(w_2 - w_1)\frac{x^2}{l}$$

It is a parabola.

0	0	$\frac{l}{2}$	l
F_x	$\left(\frac{1}{3}w_1 + \frac{1}{6}w_2\right)l$	$-\frac{1}{24}(5w_1 + w_2)$	$-\frac{1}{6}(w_1 + 2w_2)l$

The S.F.D. is shwon in Fig. 15.6 (*c*).

S.F. shall be equal to zero where B.M. is maximum.

15.4.4 Concentrated Load off Mid-Span

Consider a simply supported beam carrying concentrated load W at a distance 'a' from left hand support, as shown in Fig. 15.7(a). Taking moments about end B, we have

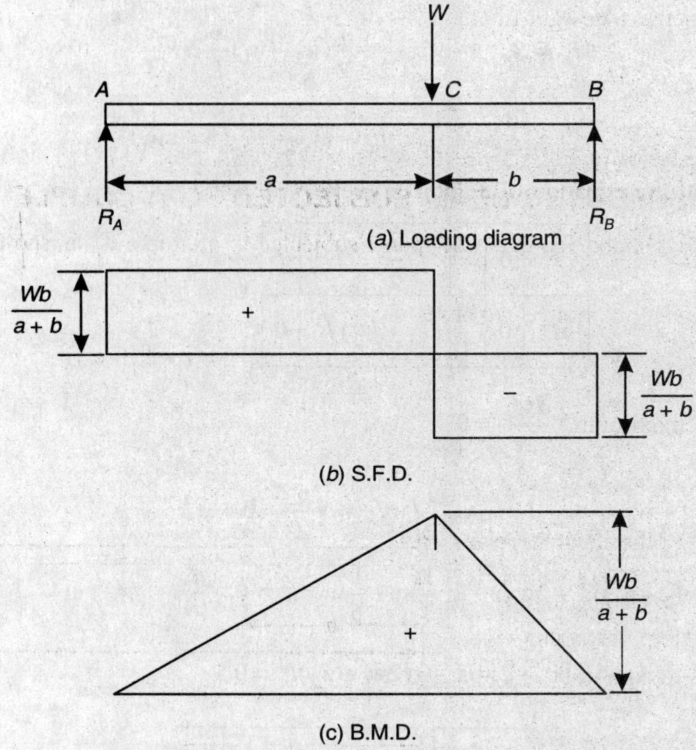

Fig. 15.7

$$RA(a+b) = Wb$$

$$R_A = \frac{Wb}{a+b}$$

and
$$R_B = \frac{Wa}{a+b}$$

S.F.D.

Between span AC, the shear force is R_A upwards and is positive. Between span CB, the shear force is R_B and is negative. The S.F.D. is shown in Fig. 15.7 (b).

B.M.D.

At a distance x from A, the bending moment is,

$$M_x = R_A \cdot x = \left(\frac{Wb}{a+b}\right)x$$

This represents a straight line.

At
$$x = a$$

$$M_C = \frac{Wab}{a+b}$$

This is a positive bending moment.

Between span *CB*,

$$M_x = R_A x - W(x - a)$$

$$= \left(\frac{Wb}{a+b}\right)x - W(x-a) \;= Wa - \left(\frac{Wa}{a+b}\right)x$$

This also represents a straight line.

At $\qquad x = a,\ M_C = \dfrac{Wab}{a+B}$

At $\qquad x = a + b,\ M_B = 0$

The B.M.D. is shown in Fig. 3.7. (*c*).

15.5 SIMPLY SUPPORTED BEAM SUBJECTED TO A COUPLE

Consider a simply supported beam of span l subjected to a couple C_0 as shown in Fig. 15.8 (*a*).

Taking moments about end *B*, we have

$$R_A l = C_0$$

$$R_A = \frac{C_0}{l}$$

$$R_B = -R_A$$

S.F.D.

Between span *AC*, the shear force is equal to R_A and is positive. For span *CB*, the shear force is $-R_B = R_A$, which is again positive. Therefore, throughout the span, the shear force is C_0/l. The S.F.D. is shown in Fig. 15.8. (*b*).

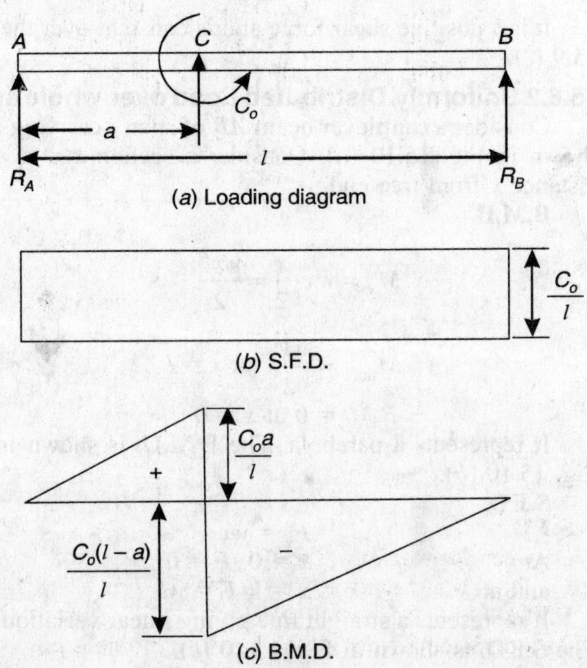

(*a*) Loading diagram

(*b*) S.F.D.

(*c*) B.M.D.

Fig. 15.8

B.M.D.

Between span *AC*,

$$M_x = R_A x = \frac{C_0 x}{l}$$

At $\qquad x = a,\ M_C = \dfrac{C_0 a}{l}$

This is a linearly varying and positive bending moment.

For span *CB*,

$$M_x = R_A x - C_0 = \frac{C_0 x}{l} - C_0 = C_0\left(\frac{x}{l} - 1\right)$$

At $\ x = a,\ M_C = C_0\left(\dfrac{a}{l} - 1\right) = -C_0\left(\dfrac{l-a}{l}\right)$

This is a linearly varying negative bending moment.

At $\qquad x = l,\ M_B = 0$

The B.M.D. is shown in Fig. 15.8 (*c*).

15.6 CANTILEVER BEAM

A cantilever is a beam whose one end is fixed and the other free, *i.e.*, without any support. By a fixed end we mean that the beam at this end is resting on long bearing without rollers. At the end there shall be fixing moment acting on the beam to beep it in position.

15.6.1 Concentrated Load at Free End

Consider a cantilever beam *AB* of span l carrying a concentrated load P at the free end as shown in Fig. 15.9. (*a*). Consider a section x-x at distance x from the free end.

B.M.D.

$$M_x = P_x$$

It is a negative B.M. as it produces tension on the top fibres of cantilever. It represents a straight line.

$$M_{max} = Pl$$

The B.M.D. is shown in Fig. 15.9 (*b*).

S.F.D.

The shear force at section x-x is:

$$F_x = P$$

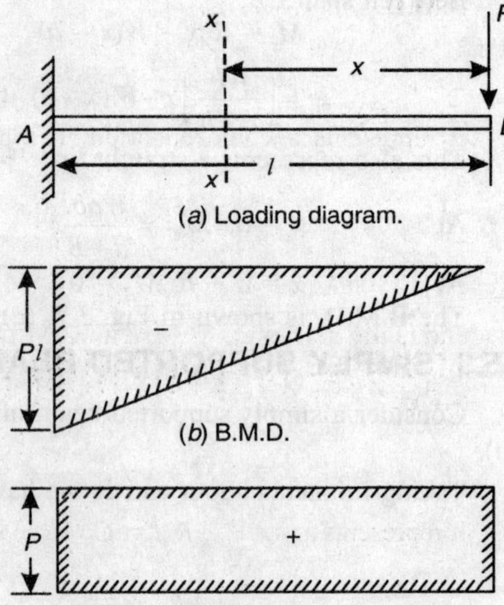

(*a*) Loading diagram.

(*b*) B.M.D.

(*c*) S.F.D.

Fig. 15.9

It is a positive shear force and is constant over the whole length. The S.F.D. is shown in Fig. 15.9 (*c*).

15.6.2 Uniformly Distributed Load over whole span

Consider a cantilever beam *AB* of span l carrying a *udl* of intensity ω over the whole span as shown in Fig. 15.10 (*a*). Consider a section x-x at a distance x from free end.

B.M.D.

$$M_x = wx \cdot \frac{x}{2} = \frac{wx^2}{2}$$

$$M_{max} = \frac{wl^2}{2} \text{ at } x = l$$

$$B.M. = 0 \text{ at } x = 0$$

It represents a parabola. The B.M.D. is shown in Fig. 15.10 (*b*).

S.F.D.

$$F_x = wx$$

At $x = 0,\ F = 0$

and at $x = 1,\ F = wl$

It represents a straight line giving linear variation. The S.F.D. is shown in Fig. 15.10 (*c*).

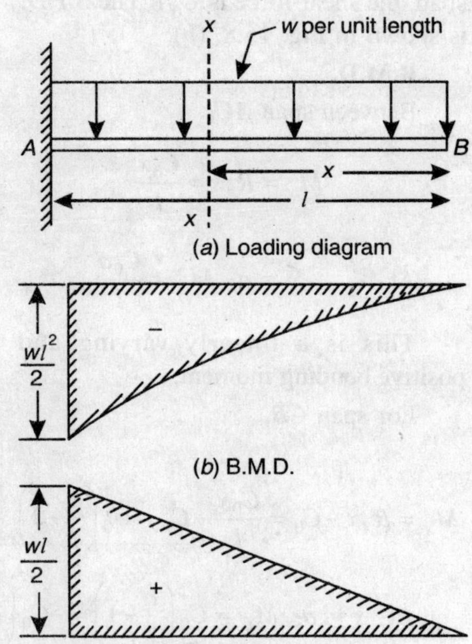

(*a*) Loading diagram

(*b*) B.M.D.

(*c*) S.F.D.

Fig. 15.10

15.6.3 Uniformly Varying Load

Consider a cantilever beam *AB* of span l carrying a uniformly varying load of intensity zero at the free end to w at the fixed end as shown in Fig. 15.11 (*a*). Consider a section x-x at a distance x from the free end. Intensity of load at distance x is $\frac{wx}{l}$.

B.M.D.

$$M_x = \frac{1}{2} \times \frac{wx}{l} \times x \times \frac{x}{3} = \frac{1}{6}\frac{wx^3}{l}$$

It represents a cubic equation. It is a negative B.M

At $x = 0$, $BM = 0$

At $x = l$, $BM = \frac{wl^2}{6}$,

and is the maximum B.M. The B.M.D. is shown in Fig. 15.11 (b).

S.F.D.

$$F_x = \frac{1}{2}\frac{wx}{l} \cdot x = \frac{1}{2}\frac{wx^2}{l}$$

It represents a parabola. It is a positive shear force.

At $x = 0$, $SF = 0$

At $x = l$, $SF = \frac{1}{2} wl$

The S.F.D. is shown in Fig. 15.11 (c).

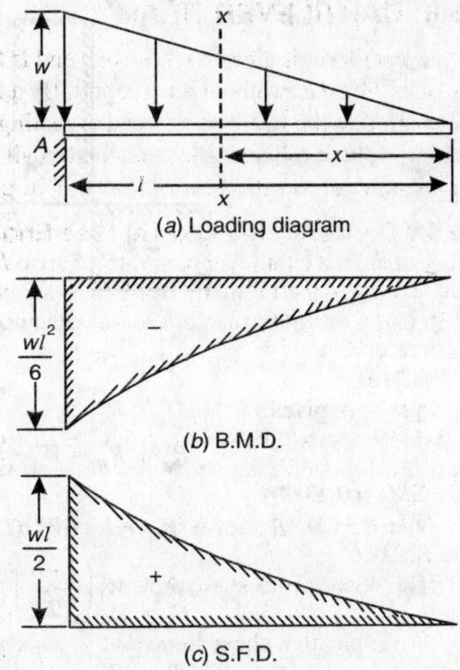

Fig. 15.11

15.7 OVERHANGING BEAMS

An overhanging beam is that which is projecting beyond the supports. Such like beams are encountered is residential buildings in balconies.

15.7.1 Concentrated loads

Consider a beam AB of span l having overhangs CA and BD, each equal to b. It carries concentrated loads W_1 to W_4 as shown in Fig 15.12 (a).

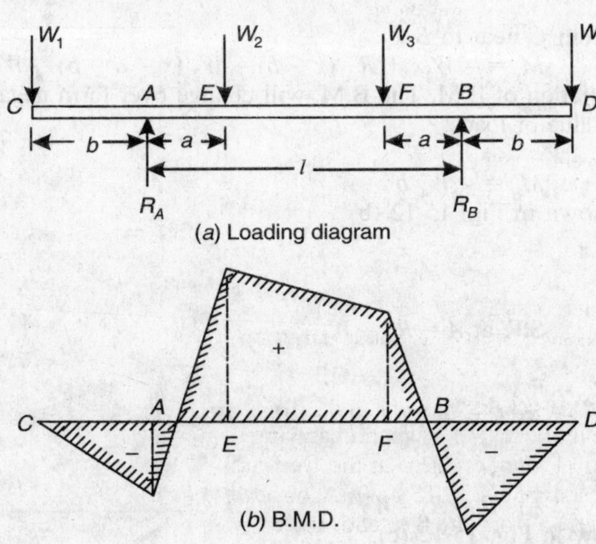

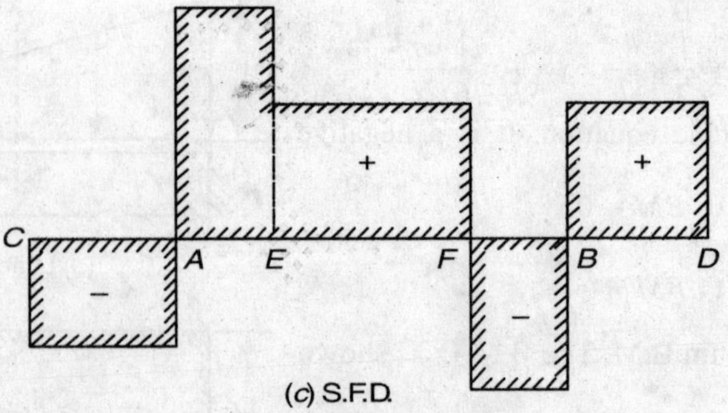

(c) S.F.D.

Fig. 15.12

$\Sigma F_y = 0$ gives,

$$R_A + R_B = W_1 + W_2 + W_3 + W_4$$

$\Sigma M_B = 0$ gives,

$$R_A \times l + W_4 \times b = W_1 (l + b) + W_2 (l - a) + W_3 a$$

$$R_A = W_1\left(\frac{l+b}{2}\right) + W_2\left(\frac{l-a}{l}\right) + W_3\frac{a}{l} - W_4\frac{b}{l}$$

or

$$R_A = W_1\left(1+\frac{b}{l}\right) + W_2\left(1-\frac{a}{l}\right) + W_3\frac{a}{l} - W_4\frac{b}{l}$$

B.M.D.

Span CA:

At a distance x from C,

$$M_x = -W_1 x$$

At $x = 0, M = 0; x = b, M_A = W_1 b$

This is a negative B.M. as it produces tension on the top fibres. It is a linearly verying B.M.

Span AB:

At a distance x from C near to B,

$$M_x = -W_1 x + R_A (x - b) - W_2 (x - a - b) - W_3 (x - l + a - b)$$

It gives linear variation of B.M. The B.M. will change over form negative to positive value depending upon the value of loads.

Span BD:

$$M_B = -W_4 b$$

The B.M.D. is shown in Fig. 15.12 (b).

S.F.D.

Span CA:

S.F. in $CA = -W_1$

S.F. at $A = R_A - W_1$

Span AB:

At $E = -W_1 + R_A - W_2$

At $F = -W_1 + R_A + W_2 - W_3$

Span BD:

S.F. in $DB = +W_4$

At $B = R_B - W_4$

The S.F.D. is show in Fig. 15.12 (c).

15.7.2 Uniformly Distributed Load

Consider a beam *ABCD* (Fig. 15.13a) of span l and having overhangs equal to 'a' on both sides and subjected to a *udl* of intensity w per unit length.

$\Sigma F_y = 0;$ $\qquad R_B + R_C = w\,(l + 2a)$

$\Sigma M_C = 0;$ $\qquad R_B \times l = \dfrac{w(a+l)^2}{2} - \dfrac{wa^2}{2}$

$$R_A = R_B$$

$$= \frac{w(a+l)^2}{2l} - \frac{wa^2}{2l}$$

$$= \frac{w(l+2a)}{2}$$

B.M.D.

Span AB:

At a distance x form A,

$$M_x = wx \times \frac{x}{2} = -\frac{wx^2}{2}$$

At A, $x = 0$, $M_A = 0$

At B, $x = a$, $M_B = -\dfrac{wa^2}{2}$

The variation of B.M. is parabolic, and it is a –ve B.M.

span BC:

$$M_x = -\frac{wx^2}{2} + R_B\,(x-a)$$

$$= -\frac{wx^2}{2} + \frac{w(l+2a)}{2}(x-a)$$

The variation of B.M. is parabolic. The B.M. will change over from –ve to +ve value depending on the magnitude of w and a.

At $x = a + \dfrac{l}{2}$,

$$M = w\left(\frac{l^2}{8} - \frac{a^2}{2}\right)$$

Span CD:

At a distance x from D,

$$M_x = -\frac{wx^2}{2}$$

The variation of B.M. is parabolic.

At D, $x = 0$, $M_D = 0$

At C, $x = a$, $\qquad M_C = -\dfrac{wa^2}{2}$

The B.M.D. is shown in Fig. 15.13 (b).

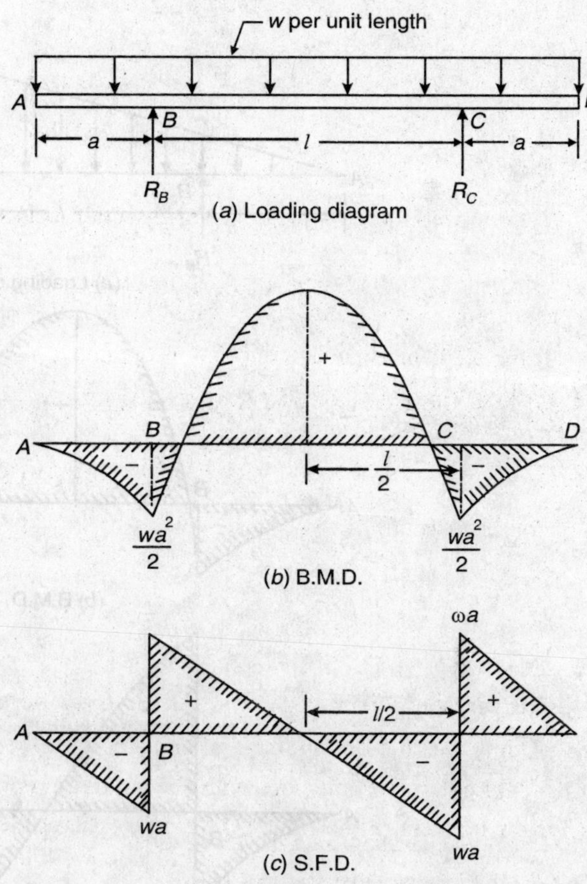

(a) Loading diagram
(b) B.M.D.
(c) S.F.D.

Fig. 15.13

S.F.D.

At a distance x from A.

Span AB:

$F_x = -wx$

At $x = 0$, $F_A = 0$

$x = a$, $F_B = -wa$

Span BC:

$$F_x = -wx + R_B = -wx + \frac{w(l+a)}{2}$$

Span CD:

At a distance x form D,

$F_x = wx$

$F_D = 0$, $F_C = wa$

The variation of shear forces is linear. The S.F.D. is shown in Fig. 15.13 (*c*)

15.7.3 Uniformly Varying Load

Consider a beam *ABCD* with equal overhangs on both sides of the supports and carrying uniformly varying load form zero at and *A* to *w* per unit length at end *D* as shown in Fig 15.14 (*a*).

$$\Sigma F_y = 0; \quad R_B + R_C = \frac{1}{2} w(2a+l)$$

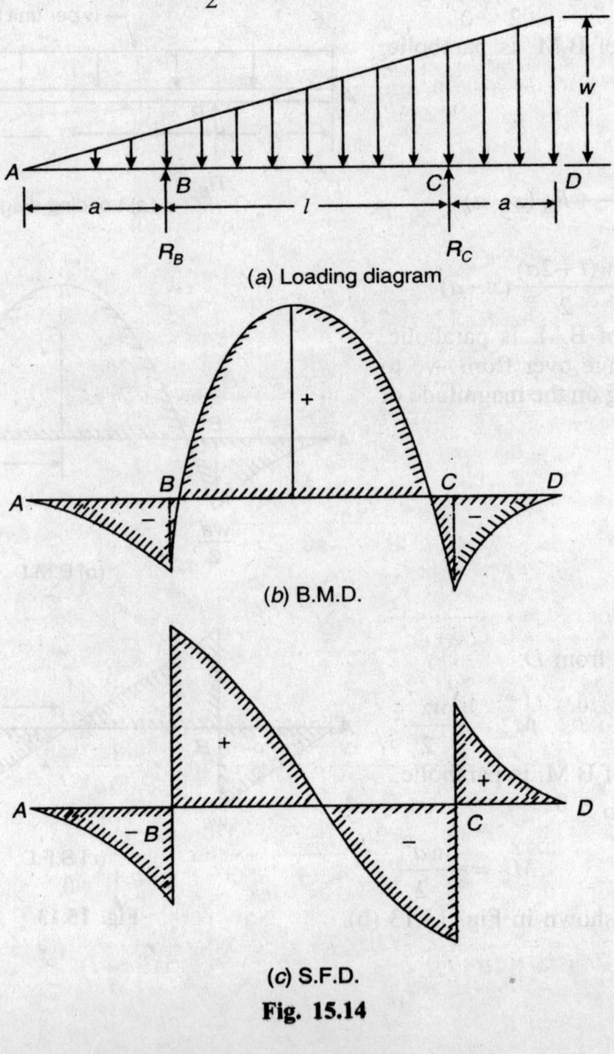

(*a*) Loading diagram

(*b*) B.M.D.

(*c*) S.F.D.

Fig. 15.14

$$\Sigma M_C = 0;\quad R_B \times l = \frac{1}{2} w\left(\frac{a+l}{2a+l}\right)(a+l)\times\frac{1}{3}(a+l)$$

$$-w\left(\frac{a+l}{2a+l}\right)\times a\times\frac{a}{2} - w\left(1-\frac{a+l}{2a+l}\right)\times\frac{1}{2}a\times\frac{2a}{3}$$

$$= \frac{1}{6}w\times\frac{(a+l)^3}{2a+l} - \frac{1}{2}wa^2\left(\frac{a+l}{2a+l}\right) - \frac{1}{3}\frac{wa^3}{(2a+l)}$$

$$R_B = \frac{1}{6}\frac{w}{l(2a+l)}(3al^2 + l^3 - 4a^3 l)$$

B.M.D.

At a distance x from A, we have

Span AB:

$$M_x = \frac{1}{2}\times\frac{wx}{(2a+l)}\times x\times\frac{x}{3} = \frac{1}{6}w\frac{x^2}{(2a+l)}$$

It represents a cubic curve. The B.M. is negative as it produces tension on the top fibres of the beam.

At A, $x = 0$, $M_A = 0$

At B, $x = a$, $M_B = \frac{1}{6}w\left(\frac{a^3}{2a+l}\right)$

Span BC:

$$M_x = -\frac{1}{2}w\frac{x}{(2a+l)}\times x\times\frac{x}{3} + R_B\,(x-a)$$

$$= -\frac{1}{6}w\left(\frac{x^3}{2a+l}\right) + R_B\,(x-a)$$

$$= -\frac{1}{6}w\left(\frac{x^3}{2a+l}\right) + \frac{1}{6}\frac{w(x-a)}{l(2a+l)}(3al^2 + l^3 - 4a^3)$$

At C, $x = a + l$,

$$M_C = \frac{1}{6}w\frac{(a+l)^3}{(2a+l)} + \frac{1}{6}\frac{w}{(2a+l)}(3al^2 + l^3 - 4a^3)$$

$$= \frac{1}{6}\frac{w}{(2a+l)}[-(a+l)^3 + 3al^2 + l^3 - 4a^3] = -\frac{1}{6}\frac{wa^2}{(2a+l)}(5a+3l)$$

Span CD:

At a distance x from D we have

$$M_x = \frac{w(2a+l-x)}{(2a+l)}\times x\times\frac{x}{2} + \frac{1}{2}w\left[1-\frac{2a+l-x}{2a+l}\right]\times x\times\frac{2}{3}x$$

$$M_x = \frac{1}{2}wx^2\left(\frac{2a+l-x)}{2a+l}\right) + \frac{1}{3}wx^3\left(\frac{1}{2a+l}\right)$$

$$= \frac{wx^2}{(2a+l)}\left[\frac{1}{2}(2a+l-x) + \frac{1}{3}x\right] = \frac{1}{6}\frac{wx^2}{(2a+l)}(6a+3l-x)$$

At D, $x = 0$, $M_D = 0$

At C, $x = a$, $M_C = \frac{1}{6}\frac{wa^2}{(2a+l)}(5a+3l)$

The B.M. is a cubic curve

The B.M.D. is shown in Fig. 15.14 (b).

S.F.D.

At a distance x form A, we have

Span AB:

$$F_x = \frac{1}{2}\frac{wx^2}{(2a+l)}$$

At A, $x = 0$, $F_A = 0$

At B, $x = a$ $F_B = \frac{1}{2}\frac{wa^2}{(2a+l)}$

The S.F. is parabolic in nature and is negative.

Span BC:

$$F_x = \frac{1}{2}\frac{wx^2}{(2a+l)} + R_B$$

$$= -\frac{1}{2}\frac{wa^2}{(2a+l)} + \frac{1}{6}\frac{w}{l(2a+l)}(3al^2 + l^3 - 4a^3)$$

At $x = a$, $F_B = -\frac{1}{2}\frac{wa^2}{(2a+l)} + \frac{1}{6}\frac{w}{l(2a+l)}(3al^2 + l^3 - 4a^3)$

$$= \frac{1}{6}\left(\frac{w}{2a+l}\right)\left[-3a^2 + \frac{1}{l}(3al^2 + l^3 - 4a^3)\right]$$

AT $x = a + l$, $F_C = -\frac{1}{2}\frac{w(a+l)^2}{(2a+l)} + \frac{1}{6}\frac{w}{(2a+l)}(3al^2 + l^3 - 4a^3)$

$$F_C = \frac{w}{6(2a+l)}\left[-3(a+l)^2 + \frac{1}{l}(3al^2 + l^3 - 4a^3)\right]$$

The S.F. is parabolic in nature.

Span CD:

At a distance x from D, we have

$$F_x = \frac{w(2a+l-x)x}{(2a+l)} + \frac{1}{2}w\left[1 - \frac{2a+l-x}{2a+l}\right]\times x$$

$$= \frac{w(2a+l-x)x}{(2a+l)} + \frac{1}{2}\frac{wx^2}{(2a+l)} = \frac{w(2a+l-x)}{(2a+l)} + \frac{1}{2}\frac{wx^2}{(2a+l)}$$

$$= \frac{wx}{(2a+l)}\left[2a + l - x + \frac{1}{2}x\right] = \frac{wx}{2(2a+l)}(4a + 2l - x)$$

At $x = 0$, $F_D = 0$

At $x = a$, $F_C = \frac{wa}{2(2a+l)}(3a + 2l)$

The S.F. is parabolic in nature. The S.F.D. is shown in Fig. 15.14 (*c*).

15.8 RELATIONSHIP BETWEEN LOAD, SHEAR FORCE AND BENDING MOMENT

Cosider a simply supported beam carrying a *udl* of intensity δ per unit length as shown as Fig. 15.15. Consider an elementary length of the beam of length δx between cross-sections x and $x+\delta x$.

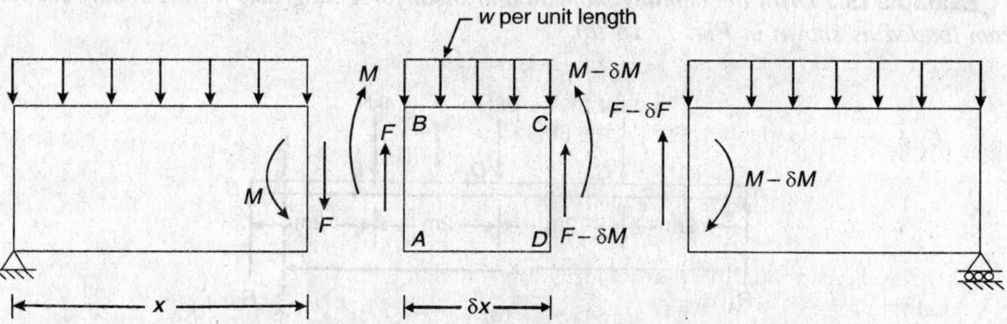

Fig. 15.15 Simply supported beam carrying udl.

Let the bending moment at x be M and at $x+\delta x$ it be $M-\delta M$. The shear force at x is F and at $x+\delta x$ it is $F-\delta F$. For the equilibrium of the beam element *ABCD*, we have

$$\Sigma F_y = 0;\quad F-(F-\delta F)-w\cdot\delta x = 0$$

$$\delta F - w\cdot\delta x = 0 \quad \text{or} \quad \frac{\delta F}{\delta x} = w$$

In the limit as $\delta x \to 0$, we get

$$\frac{dF}{dx} = w$$

Therefore, the first devivative of shear force with respet to x at a point given the intensity of loading at the point.

$$\Sigma M_A = 0;\quad (F-\delta F).\delta x-(M-\delta M)+w.dx.\frac{\delta x}{2}+M=0$$

$$F.\delta x-\delta F.\delta x-M+\delta M+w.\frac{\delta x^2}{2}+M=0$$

Neglecting small quantities, we get

$$F.\delta x+\delta M=0$$

or

$$\frac{\delta M}{\delta x} = -F$$

In the limit as $\delta x \to 0$, we get

$$\frac{dM}{dx} = -F$$

Therefore, the rate of change of bending moment with respect to x is equal to the shear force. Whenever, bending moment is maximum or minimum, the shear force is zero.

Taking the derivative again, we get

$$\frac{d^2 M}{dx^2} = \frac{dF}{dx} = -w$$

Example 15.1 *Draw the bending moment and shear force diagrams for the simply supported beam loaded as shown in Fig. 15.16 (a).*

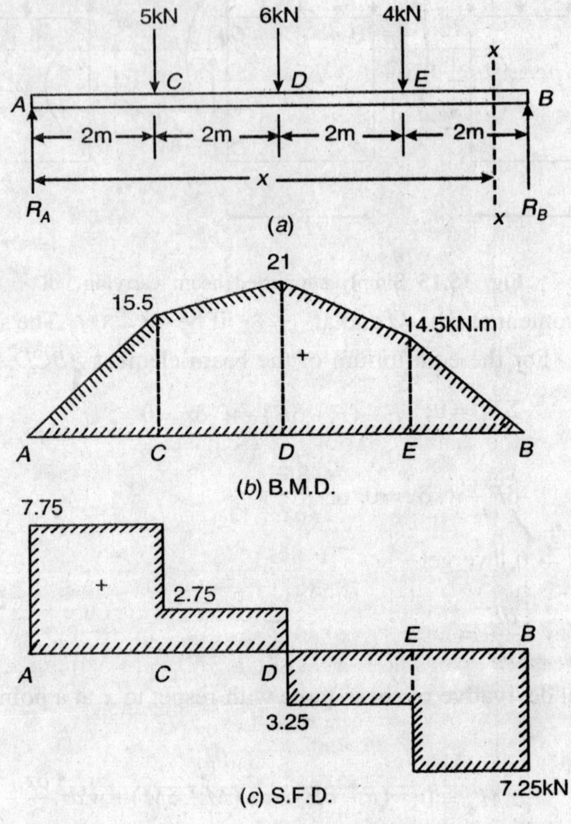

Fig. 15.16

Solution: $\Sigma F_y = 0$:

$$R_A + R_B = 5 + 6 + 4 = 15 \text{ kN}$$

$$\Sigma M_B = 0; \qquad R_A \times 8 = 5 \times 6 + 6 \times 4 + 4 \times 2 = 62$$

$$R_A = 7.75 \text{ kN}$$

$$R_B = 7.25 \text{ kN}$$

Consider a cross-section x-x of the beam near of the support B at a distance x from A.

B.M.D.

$$M_x = R_A \times x - 5\ (x - 2) - 6\ (x - 4) - 4\ (x - 6)$$

$$= 7.75\ x - 5\ (x - 2) - 6\ (x - 4) - 4\ (x - 6)$$

The variation of B.M. is linear throughout. A bracket is ignored, if on substituting the value of x, it becomes negative.

x, m	0	2	4	6	8
M, kNm	0	15.50	21.0	14.5	0

The B.M.D. is shown in Fig. 15.16 (b) to a scale of 1mm = 1kN.m.

S.F.D.

Point	A	–	C	–	D	–	E	–	B
Span	–	AC	–	CD	–	DE	–	EB	–
F, kN	7.75	7.75	2.75	2.75	–3.25	–3.25	–7.25	–7.25	–7.25

The S.F.D. is shown in Fig. 15.16 (*c*) to a scale of 1mm = 0.5 kAN

Example 15.2 *Drawn the bending moment and shear force diagrams for the simply supported beam shown in Fig 15.17 (a).*

Solution: $\Sigma F_y = 0$;

$$R_A + R_B = 5 + 2 + 2 + 10 = 19 \text{ kN}$$

$\Sigma M_B = 0$:

$$R_A \times 6 = 5 \times 5 + 2 \times 2 \times 3 + 10 \times 1 = 47$$

$$R_A = 7.833 \text{ kN}$$

$$R_B = 11.167 \text{ kN}$$

B.M.D.

Taking a section x-x at a distance x from A, we have

$$M_x = R_A \times x - 5(x-1) - 2 \times 2 \times (x-3) - 10(x-5)$$
$$= 7.833\,x - 5(x-1) - 4(x-3) - 10(x-5)$$

x, m	0	1	2	3	4	5	6
M, kN.m	0	7.833	10.666	13.4999	12.332	11.165	0

The B.M.D. is shown in Fig. 15.17 (*b*) to a scale of 1mm = 0.5 kN.m.
The B.M. curve between span *DE* is a parabola. In the remaining span it is linear.

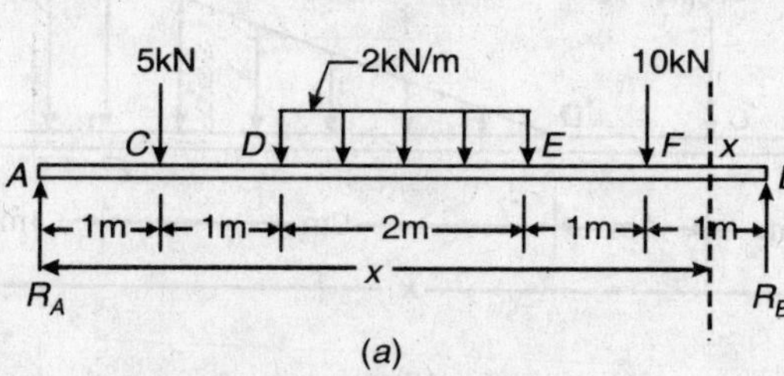

(*a*)

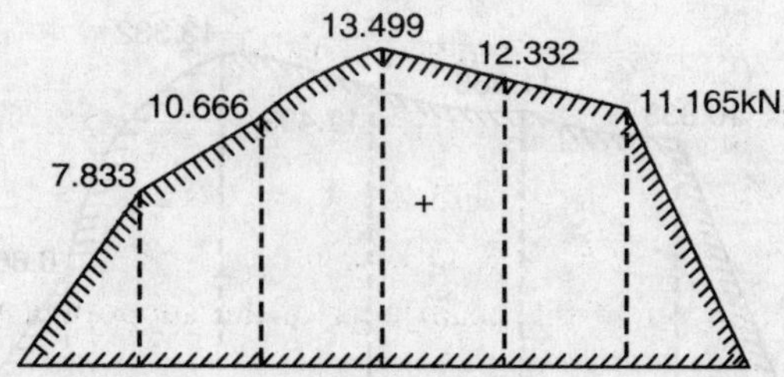

(*b*) B.M.D.

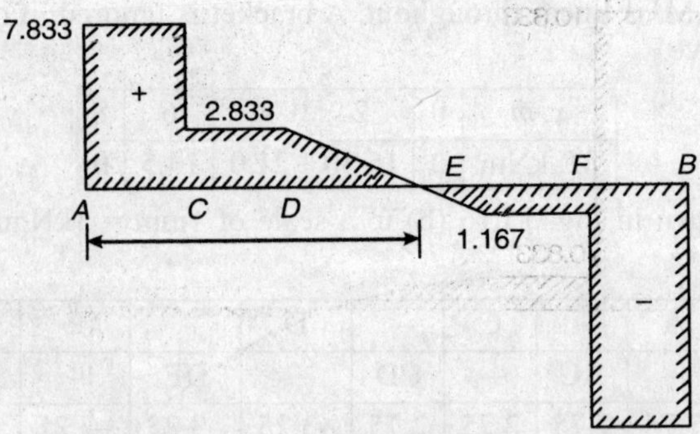

(c) S.F.D.

Fig. 15.17

S.F.D.

S.F. at A, $F_A = 7.833$ kN

S.F. between span $AC = 7.833$ kN

S.F. at C, $F_C = 7.833 - 5 = 2.833$ kN

S.F. between span $CD = 2.833$ kN

$F_D = 2.833$ kN

S.F. between span $DE = 2.833 - 2\ (x - 2)$

At $x = 3$ m, S.F. $= 2.833 - 2 \times 1 = 0.833$ kN

At $x = 4$ m, S.F., $F_E = -\ 1.167$ kN

S.F. between span $E_F = -\ 1.167$ kN

$F_F = -\ 1.167 - 10 = -\ 11.167$ kN

S.F. between span $FB = -\ 11.167$ kN

$F_B = -\ 11.167$ kN

The S.F.D. is shown in Fig 15.17 (c) to a scale of 1 mm = 0.5 kN.

For S.F. to be zero between DE,

$$2.833 - 2\ (x - 2) = 0;\ x = 3.4165 \text{ m.}$$

Example 15.3 *Draw the B.M. and S.F. diagrams for the simply supported beam shown in Fig. 15.18 (a).*

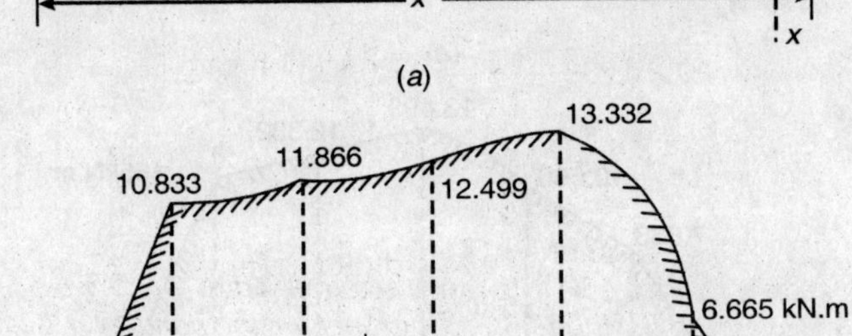

(a)

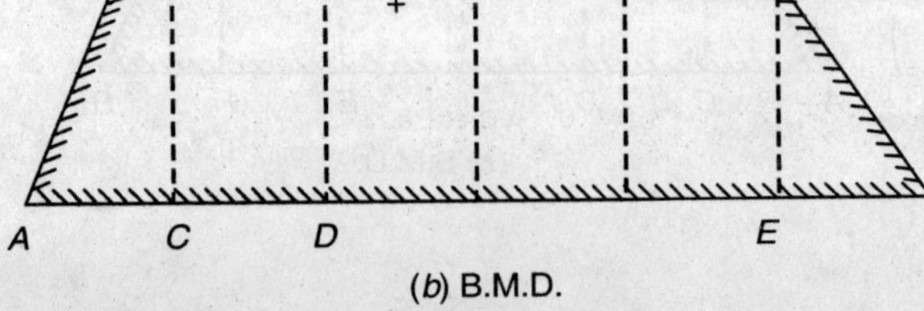

(b) B.M.D.

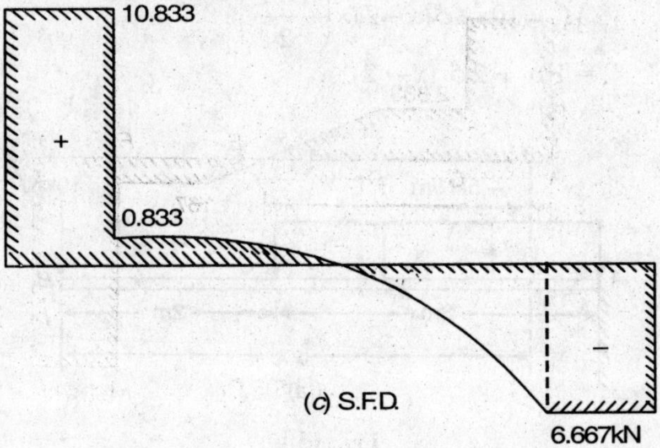

(c) S.F.D.

Fig. 15.18

Solution: $\Sigma F_y = 0;$

$$R_A + R_B = 10 + \frac{1}{2} \times 5 \times 3$$
$$= 17.5 \text{ kN}$$

$$\Sigma M_B = 0; \quad R_A \times 6 = 10 \times 5 + \frac{1}{2} \times 5 \times 3 \times \left(\frac{1}{3} \times 3 + 1\right)$$
$$= 65$$
$$R_A = 10.833 \text{ kN}$$
$$R_B = 6.667 \text{ kN}$$

B.M.D.

Consider a section *x-x* at a distance *x* from *A*.

$$M_x = R_A \times x - 10(x-1) - \frac{1}{2} \times 5 \times 3 \times \left(x - 2 - \frac{2}{3} \times 3\right)$$
$$= 10.833\, x - 10\,(x-1) - 7.5\,(x-4)$$

x, m	0	1	2	3	4	5	6
M, kN.m	0	10.833	11.666	12.499	13.332	6.665	0

The B.M.D. is shown in Fig. 15.18 (*b*) to a scale of 1mm = 0.5 kN.m.

S.F.D.

$$F_A = 10.833 \text{ kN}$$
$$F_C = 10.833 - 10 = 0.833 \text{ kN}$$

S.F. between *DE* at a distance *x* from *D* is,

$$F_x = 0.833 - \frac{1}{2} \times \frac{5}{3} x^2 = 0.833 - 0.833 x^2$$

At $x = 1$m, $\quad F = 0$

$x = 2$m, $\quad F = -2.5$ kN

$x = 3$m, $\quad F_E = -6.667$ kN

$F_B = -6.667$ kN

The S.F.D. is shown in Fig. 15.18 (*c*) to a scale of 1mm = 0.5 kN.

Example 15.4 *Drawn the B.M. and S.F. diagrams for the cantilever beam shown in Fig 15.19 (a).*

Solution. B.M.D.

At a distance *x* from free end *B*, we have

$$M_x = 10x + 5(x-2) \times \left(\frac{x-2}{2}\right)$$

$= 10x + 2.5\ (x - 2)^2$

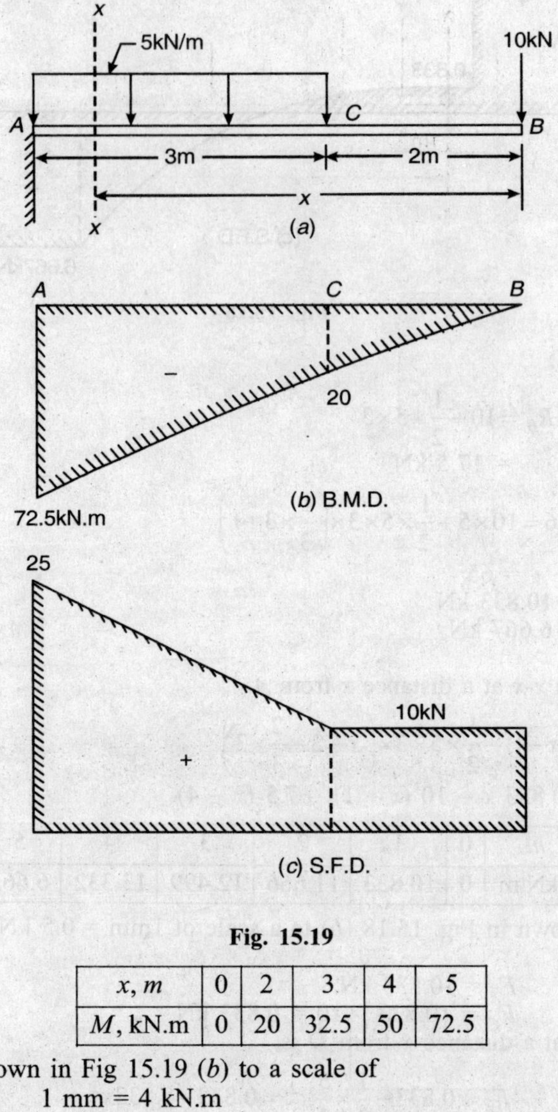

Fig. 15.19

x, m	0	2	3	4	5
M, kN.m	0	20	32.5	50	72.5

The B.M.D. is shown in Fig 15.19 (b) to a scale of

1 mm = 4 kN.m

S.F.D.

$F_B = F_C$
$= 10$ kN

Between C_A, $F = 10 + 5\ (x - 2)$

x, m	3	4	5
F, kN	15	20	25

The S.F.D. is shown in Fig. 15.19 (c) to a scale of 1 mm = 1 kN.

Example 15.5 *Draw the B.M. and S.F. diagrams for the cantilever beam shown in Fig. 15.20 (a)*

Solution: B.M.D.

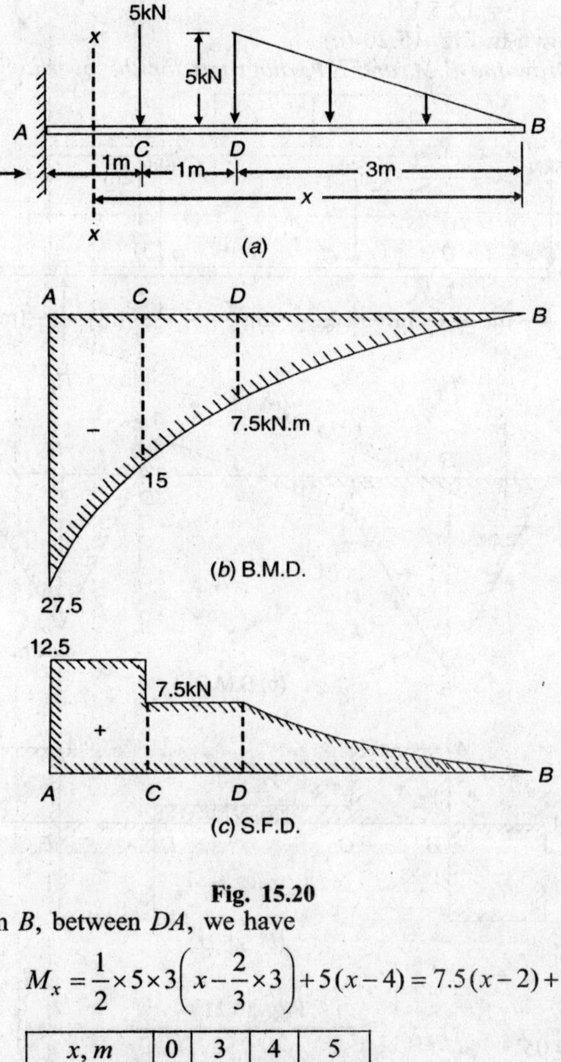

Fig. 15.20

At a distance x from B, between DA, we have

$$M_x = \frac{1}{2}\times 5\times 3\left(x - \frac{2}{3}\times 3\right) + 5(x-4) = 7.5(x-2) + 5(x-4)$$

x, m	0	3	4	5
M, kN.m	0	7.5	15	27.5

When x lies between BD, we have

$$M_x = \frac{1}{2}\times\frac{5}{3}x^2 + \frac{x}{3} = \frac{5}{18}x^3$$

x, m	0	1	2	3
M, kN.m	0	0.278	1.111	7.5

The B.M.D. is shown in Fig. 15.20 (b) to a scale of
1 mm = 1kN.m

S.F.D.

Span BD: $F_x = \frac{1}{2}\times\frac{5}{3}x^2 = \frac{5}{6}x^2$

x, m	0	1	2	3
F, kN	0	0.833	3.333	7.5

$$F_C = 7.5 + 5$$
$$= 12.5 \text{ kN}$$

The S.F.D. is shown in Fig. 15.20 (*c*).

Example 15.6 *Draw the B.M. and S.F. diagrams for the overhanging beam shown in Fig. 15.21 (a)*

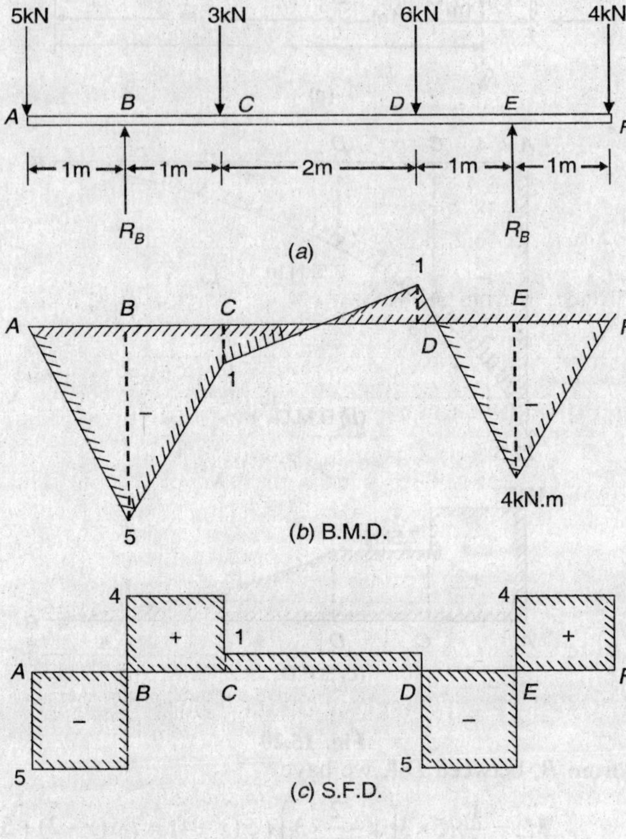

Fig. 15.21

Solution: $\Sigma F_y = 0;$

$$R_A + R_E = 5 + 3 + 6 + 4$$
$$= 18 \text{ kN}$$

$\Sigma M_E = 0;$ $\quad R_B \times 4 = 5 \times 5 + 3 \times 3 + 6 \times 1 - 4 \times 1$

$$= 36$$
$$R_B = 9 \text{ kN}$$
$$R_E = 9 \text{ kN}$$

B.M.D.

Span AB: $\quad M_B = 5 \times 1 = -5$ kN.m

Span BC: $\quad M_C = R_B \times 1 - 5 \times 2 = 9 - 10$

$$= -1\text{kN. m}$$

Span CD: $\quad M_D = R_B \times 3 - 5 \times 4 - 3 \times 2$

$$= 9 \times 3 - 20 - 6 = +1 \text{ kN.m}$$

Span EF: $\quad M_E = -4 \times 1 = 4$ kN.m

The *B.M.D.* is shown in Fig. 15.21 (*b*)

S.F.D.

$F_A = -5$ kN
$F_B = -5 + 9 = +4$ kN
$F_C = 4 - 3 = +1$ kN
$F_D = 1 - 6 = -5$ kN
$F_E = -5 + 9 = +4$ kN
$F_B = +4$ kN

The S.F.D. is shown in Fig. 15.21 (*c*)

Example 15.7 *Draw the B.M and S.F. diagrams for the beam shown in Fig. 15.22 (a)*

Solution: $\Sigma F_y = 0;$

$$R_B + R_C = 10 + 15 \times 4 + 15 = 85 \text{ kN}$$

$\Sigma M_C = 0;$ $R_B \times 4 = 10 \times 5 + 15 \times 4 \times 2 - 15 \times 1 = 155$

$R_B = 38.75$ kN
$R_A = 46.25$ kN

B.M.D.

Span AC: At a distance x from A, we have

$$M_x = -10x + 38.75(x-1) - 7.5(x-1)^2$$

x, m	0	1	2	3	4	5
M, kN.m	0	−10	11.25	17.5	8.75	−15

Span CD: $M_C = -15 \times 1 = -15$ kN.m

The *B.M.D.* is shown in Fig. 15.22 (*b*).

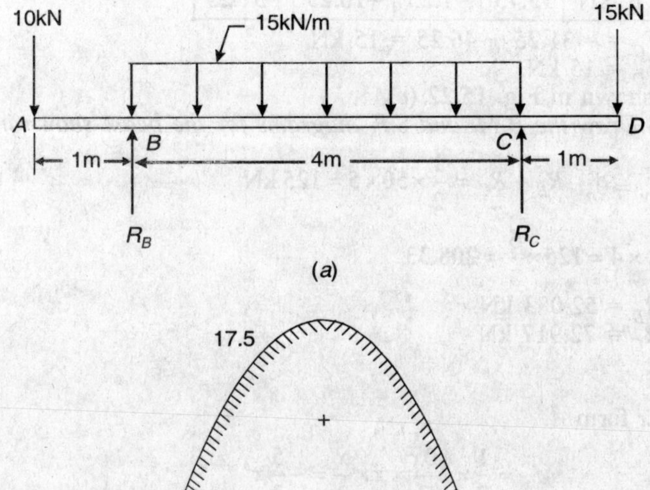

(*a*)

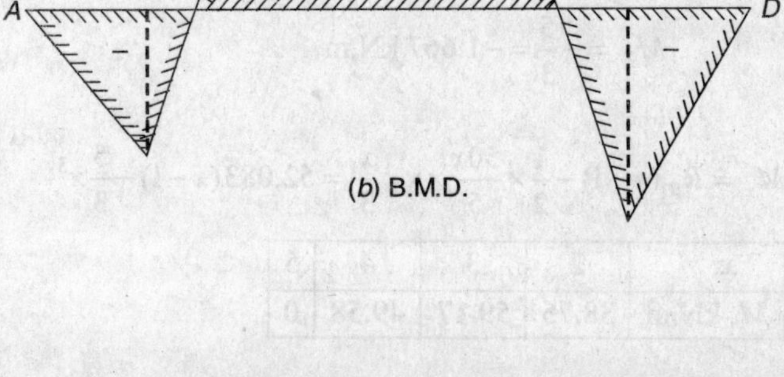

(*b*) B.M.D.

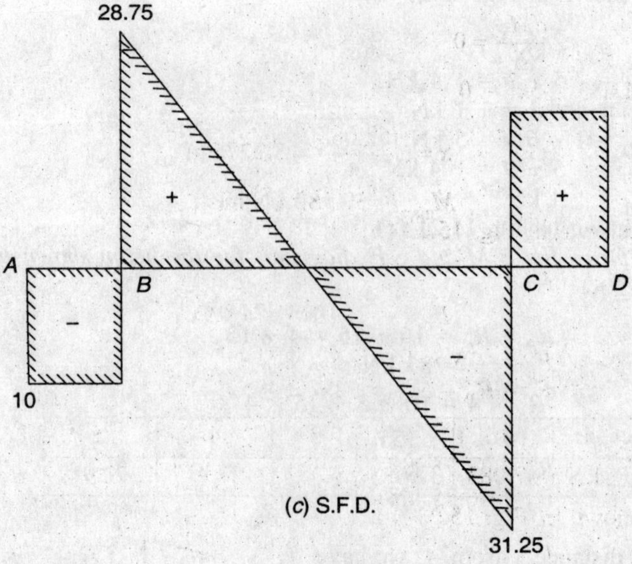

(c) S.F.D.

Fig. 15.22

S.F.D.

$F_A = -10$ kN

$F_B = -10 + 38.75 = 28.75$ kN

At a distance x from B,

x, m	1	2	3	4
F, kN	13.75	−1.25	−16.25	−31.25

$F_C = -31.25 + 46.25 = 15$ kN

$F_D = 15$ kN

The *S.F.D.* is shown in Fig. 15.22 (*c*)

Example 15.8 *Draw the B.M. and S.F. diagrams for the beam shown in Fig 15.23 (a).*

Solutlion: $\Sigma F_y = 0;\ \ R_B + R_C = \frac{1}{2} \times 50 \times 5 = 125$ kN

$\Sigma M_B = 0:\ \ R_B \times 4 = 125 \times \frac{5}{3} = 208.33$

$R_B = 52.083$ kN

$R_C = 72.917$ kN

B.M.D.

Span AB:

At a distance x form A.

$$M_x = \frac{1}{2} \times \frac{50x}{5} \times x \times \frac{x}{3} = -\frac{5}{3} x^3$$

At x = 1m, $$M_B = -\frac{5}{3} = -1.667 \text{ kN,m}$$

Span BC:

$$M_x = R_B\,(x-1) - \frac{1}{2} \times \frac{50x}{5} \times x \times \frac{x}{3} = 52.083\,(x-1) - \frac{5}{3} x^3$$

x, m	2	3	4	5
M, kN.m	38.75	59.17	49.58	0

For *B.M.* to be maximum,

$$\frac{dM_x}{dx} = 0$$

$$52.083 - 5x^2 = 0$$

$$x = \sqrt{\frac{52.083}{5}} = 3.2275 \text{ m}$$

$$M_{max} = 59.982 \text{ kN,m}$$

The *B.M.D.* is shown in Fig. 15.23 (*b*)

S.F.D.

Span AB: $F_x = -\frac{1}{2} \times 10x \times x = -5x^2$

At x = 1m, $F_B = -5 \times 1 = -5$ kN

Span BC: $F_x = R_B - 5x^2 = 52.083 - 5x^2$

x, m	1	2	3	4	5
F, kN	47.083	32.083	7.083	–27.917	–72.917

The *S.F.D.* is shown in Fig. 15.23 (*c*).

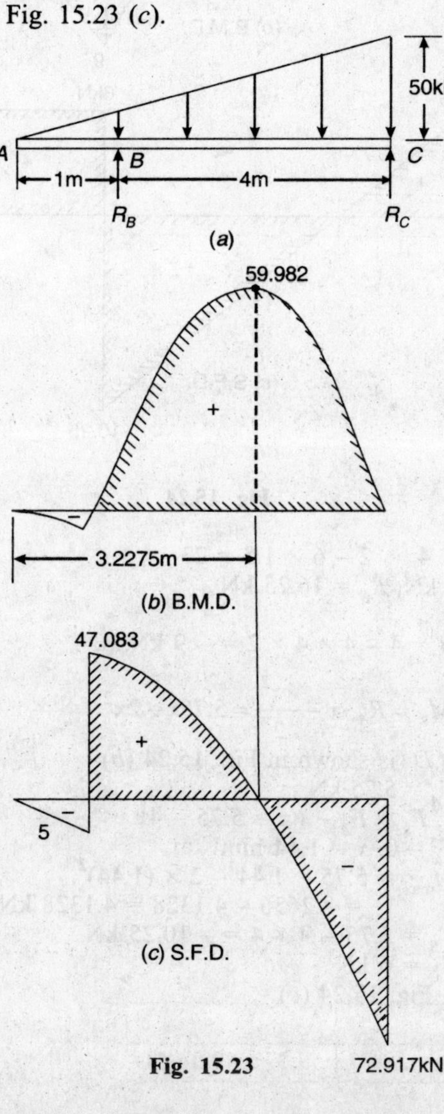

Fig. 15.23

Example 15.9 *Draw the B.M. and S.F. diagrams for the beam shown in Fig 15.24 (a).*
Solution: $R_A + R_B = 4 \times 4 + 6 = 22$

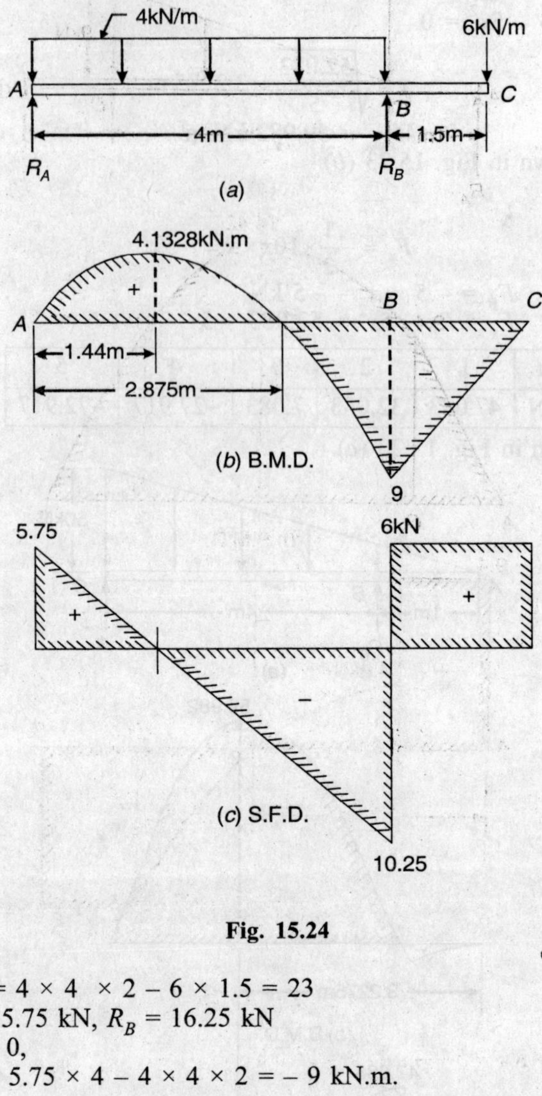

Fig. 15.24

$$4R_A = 4 \times 4 \times 2 - 6 \times 1.5 = 23$$
$$R_A = 5.75 \text{ kN}, R_B = 16.25 \text{ kN}$$

B.M.D. $M_A = 0,$
$$M_B = 5.75 \times 4 - 4 \times 4 \times 2 = -9 \text{ kN.m.}$$

$$M_x = R_A \cdot x - \frac{wx^2}{2} = 5.75x - 2x^2.$$

It is parabola. The *B.M.D.* is shown in Fig. 15.24 (*b*)

S.F.D. $F_A = 5.75$ kN

At a distance x from A, $F_x = F_A - wx = 5.75 - 4x$

For $F_x = 0$, $x = 1.44$ mm

$$M_{max} = 5.75 \times 1.44 - 2 \times (1.44)^2$$
$$= 8.2656 - 4.1328 = 4.1328 \text{ kN.m}$$
$$F_B = 5.75 - 4 \times 4 = -10.25 \text{ kN}$$
$$F_C = 6 \text{ kN}$$

The *S.F.D.* is shown in Fig, 15.24 (*c*)

Example 15.10 *Draw the B.M. and S.F. diagrams for the beam shown in Fig. 15.25 (a)*

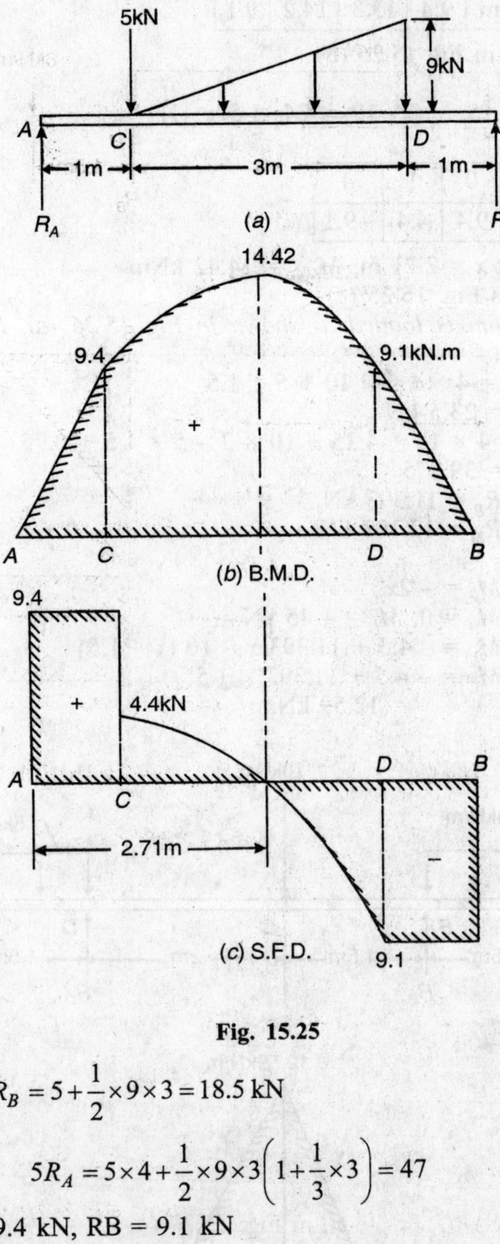

Fig. 15.25

Solution: $R_A + R_B = 5 + \frac{1}{2} \times 9 \times 3 = 18.5 \text{ kN}$

$$5R_A = 5 \times 4 + \frac{1}{2} \times 9 \times 3\left(1 + \frac{1}{3} \times 3\right) = 47$$

R_A = 9.4 kN, RB = 9.1 kN

B.M.D. $M_x = 9.4x - 5(x-1) - \frac{1}{2} \times 3(x-1)^3 \times \frac{1}{3}$

$= 9.4x - 5(x-1) - \frac{1}{2}(x-1)^3$

$M_A = 0$, $M_C = 9.4$ kN,m.

$M_D = 9.4 \times 4 - 5 \times 3 - \frac{1}{2} \times 27$

= 9.1 kN.m

x, m	1	2	3	4
M, kN.m	9.4	13.3	14.2	9.1

The *B.M.D.* is shown in Fig 15.25 (*b*).

S.F.D. $F_x = 9.4 - 5 - \frac{1}{2}(x-1)^2 = 4.4 - 1.5(x-1)^2$

x, m	0	1	4
F, kN	9.4	4.4	–9.1

$F_x = 0$ at $x = 2.71$ m; $M_{max} = 14.42$ kNm

The *S.F.D.* is shown is Fig. 15.25 (*c*)

Example 15.11 *A beam is loaded as shown in Fig 15.26 (a). Draw the B.M. and S.F. diagrams.*

Solultion: $R_B + R_D = 4 \times 1.5 + 10 + 5 \times 1.5$
$= 23.5$ kN

$3.5\ R_B = 4 \times 15 + 4.25 + 10 \times 2 - 5 \times 1.5 + 0.75$
$= 39.875$

$R_B = 11.393$ kN,

$R_D = 12.107$ kN

S.M.D.

Span AB: $M_x = -\ 2x^2$

$M_A = 0$, $M_B = -\ 45$ kN.m

Span BD: $M_x = -\ 4.5 + 11.393\ x - 10\ (x - 1.5)$

$M_C = -\ 4.5 + 11.393 \times 1.5$
$= 12.59$ kN.m

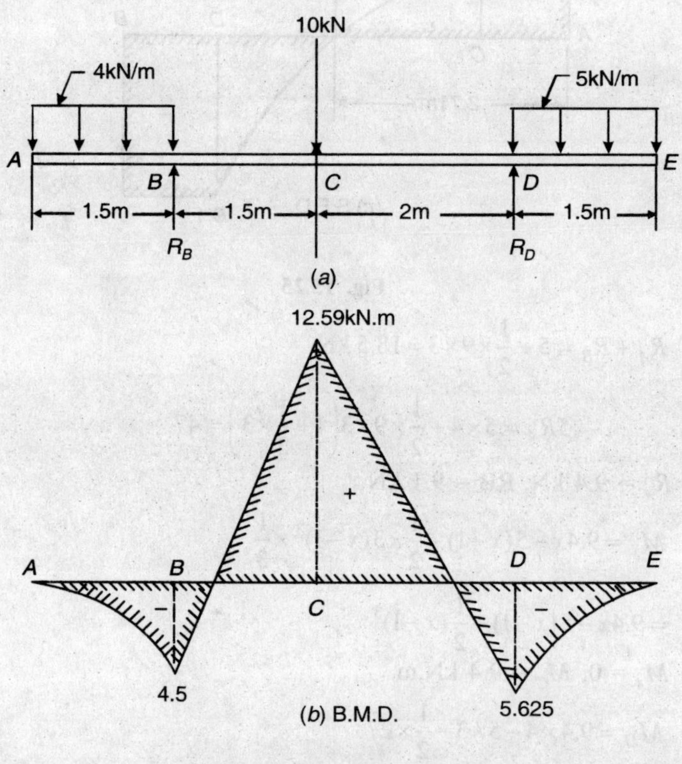

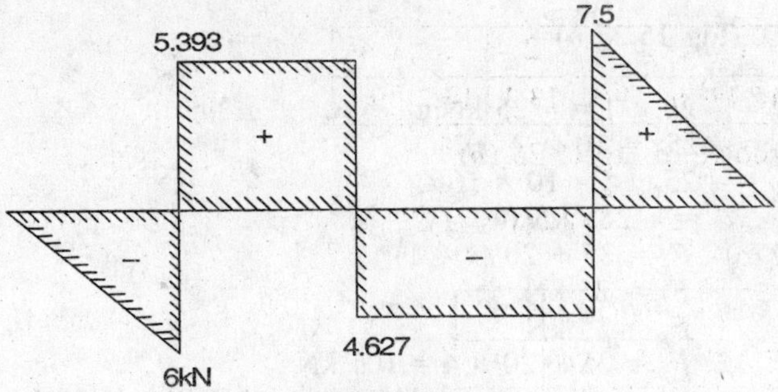

(c) S.F.D.

Fig. 15.26

Span DE: $M_x = -2.5x^2$

$M_D = -5.625$ kN.m

The *B.M.D.* is shown in Fig. 15.26 (*b*)

S.F.D.

Span AB: $F_x = -4x,\ F_A = 0,$

$F_B = -4 \times 1.5 = -6$kN

Span BD: $F_B = -6 + 11.393 = 5.393$ kN

$F_C = 5.393 - 10 = -4.627$ kN

$F_D = -4.627 + 12.107$

$= 7.48$ kN $= 7.5$ kN

Span DE: $F_x = 5x,\ F_C = 0,\ F_D = 5 \times 1.5 = 7.5$ kN

The S.F.D. is shown in Fig. 15.26(*c*).

Example 15.12 *A cantilever beam is loaded as shown in Fig 15.27(a). Draw the B.M. and S.F. diagrams.*

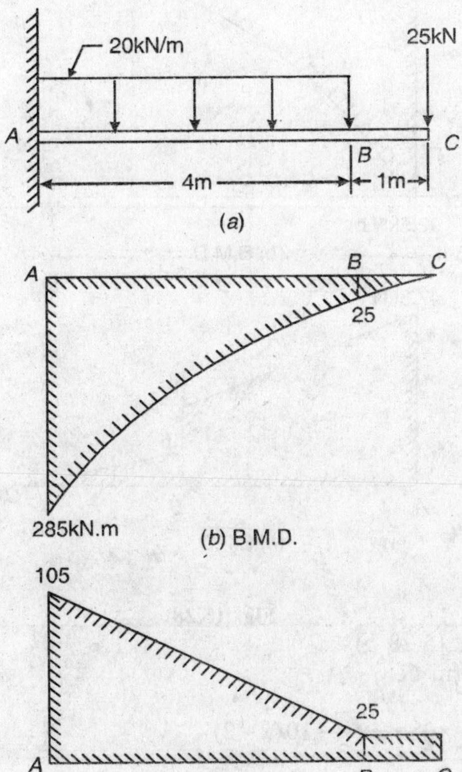

(c) S.F.D.

Fig. 15.27

Solution: *B.M.D.* (Fig 15.27 *b*)

Span BC: $M_C = 0,$

$M_B = -25 \times 1 = -25$ kN.m

$M_x = -25x - 10(x-1)^2$

$M_A = -25 \times 5 - 10 \times 16$

$= -285$ kN.m

S.F.D. (Fig 15.27*c*) $F_x = 25 + 20(x-1)$

$F_C = 25$ kN,

$F_B = 25$ kN

$F_A = 25 + 20 \times 4 = 105$ kN

Example 15.13 *A cantilever beam is loaded as shown in Fig. 15.28 (a). Draw the B.M. and S.F. diagrams.*

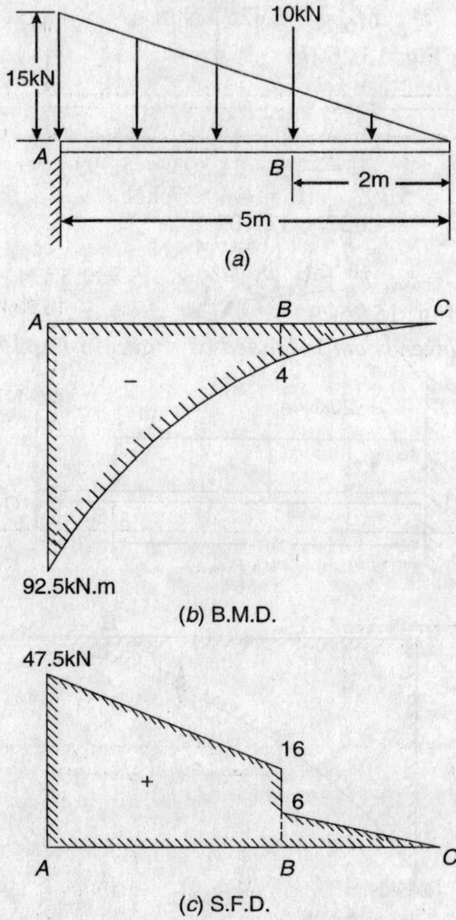

Fig. 15.28

Solution B.M.D. (Fig. 15.28 *b*)

At a distance *x* form *C*,

$$M_x = -\frac{1}{2} \times 3x \times x \times \frac{x}{3} - 10(x-2)$$

$$= -\frac{1}{2}x^3 - 10(x-2)$$

x, m	0	1	2	3	4	5
M, kN.m	0	−0.5	−4	−23.5	−52	−92.5

S.F.D. (Fig 15.28 *c*)

$$F_x = \frac{1}{2} \times 3x \times x + 10(x-2)^0$$
$$= 1.5x^2 + 10\ (x-2)^0$$

x, m	0	1	2	3	4	5
F, kN	0	1.5	6	23.5	34	47.5

Example 15.14 *Draw the B.M and S.F. diarams for the beam loaded as shown in Fig 15.29.*

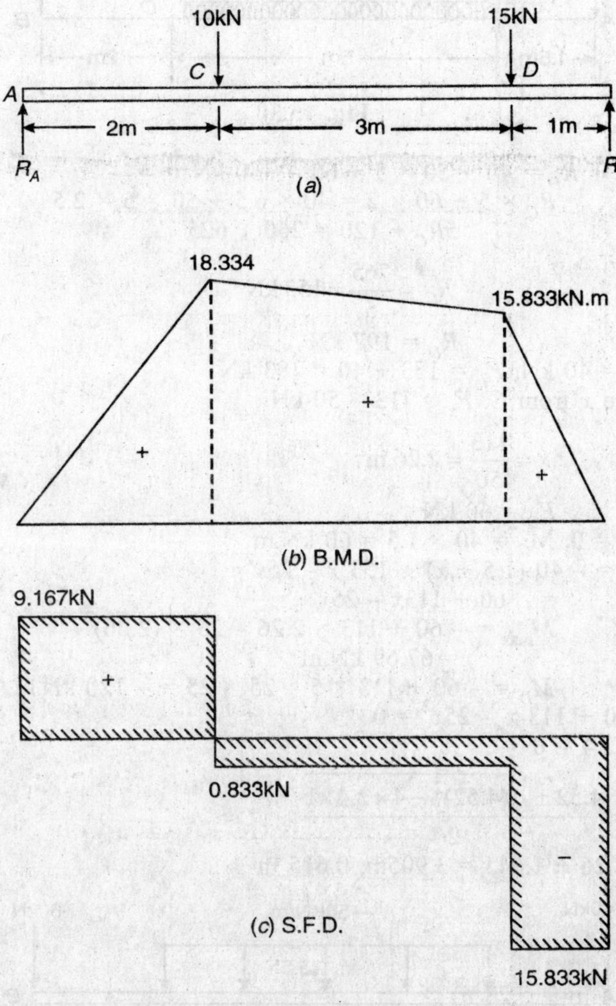

Fig. 15.29

Solution:

$$R_A + R_B = 10 + 15 = 25 \text{ kN}$$
$$6R_A = 10 \times 4 + 15 \times 1 = 55$$
$$R_A = 9.167 \text{ kN},\ R_B = 15.833 \text{ kN}$$

B.M.D. (Fig. 15.29 *b*) $M_A = 0$

S.F.D. (Fig, 15.29 *c*) $M_C = 9.167 \times 2 = 18.33$ kN

$M_D = 9.167 \times 5 - 10 \times 3$

$= 15.833$ kN.m

$M_B = 0$

$F_A = + 9.167$ kN

$F_C = 9.167 - 10 = - 0.833$ kN

$F_D = - 0.833 - 15 = - 15.833$ kN

$F_B = - 15.833$ kN

Example 15.15 *Draw S.F. and B.M. Diagrams for the beam shown in Fig 15.30 Find the position and magnitude of bending moment and the points of contraflexure.*

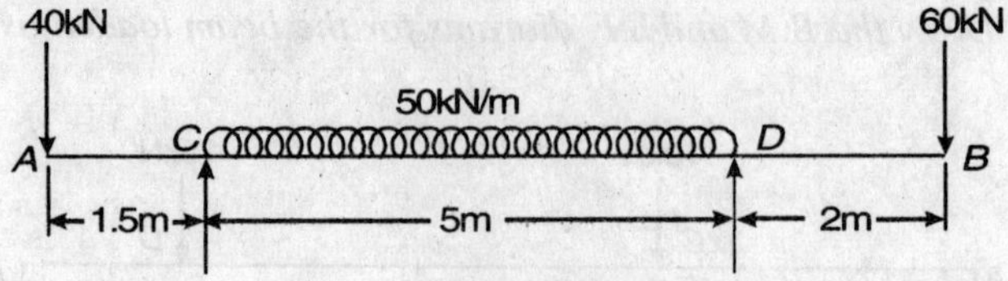

Fig. 15.30

Solution: $R_C + R_D = 40 + 50 \times 5 + 60 = 350$ kN

$R_C \times 5 + 60 \times 2 = 40 \times 6.5 + 50 \times 5 \times 2.5$

$5R_C + 120 = 260 + 625$

$$R_C = \frac{765}{5} = 153 \text{ kN}$$

$R_D = 197$ kN

S.F.D. $F_A = 40$ kN, $F_C = 153 + 40 = 193$ kN

At a distance x from C, $F_x = 113 - 50$ kN

For $F_x = 0$, $$x = \frac{113}{50} = 2.26 \text{ m}$$

$F_B = 60$ kN

B.M.D. $M_A = 0$, $M_C = 40 \times 1.5 = 60$ kN.m

$M_x = - 40\,(1.5 + x) + 153\,x - 25x^2$

$= - 60 + 113x - 25x^2$

At $x = 2.26$ m, $M_{max} = - 60 + 113 \times 2.26 - 25 \times (2.26)^2$

$= 67.69$ kN.m

At $x = 5$ m, $M_D = - 60 + 113 \times 5 - 25 \times 25 = - 120$ kN.m

For $M_x = 0$, $- 60 + 113\,x - 25x^2 = 0$

or $x^2 - 4.52x + 2.4 = 0$

$$x = \frac{4.52 \pm \sqrt{(4.52)^2 - 4 \times 2.4 \times 1}}{2}$$

$= 2.26 \pm 1.645 = 3.905$m, 0.615 m

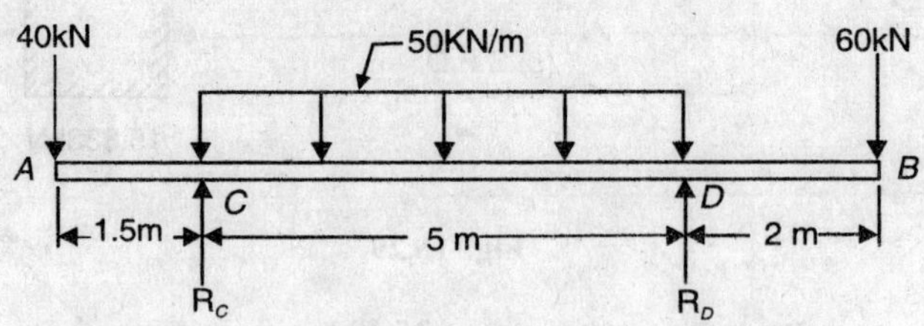

(*a*)

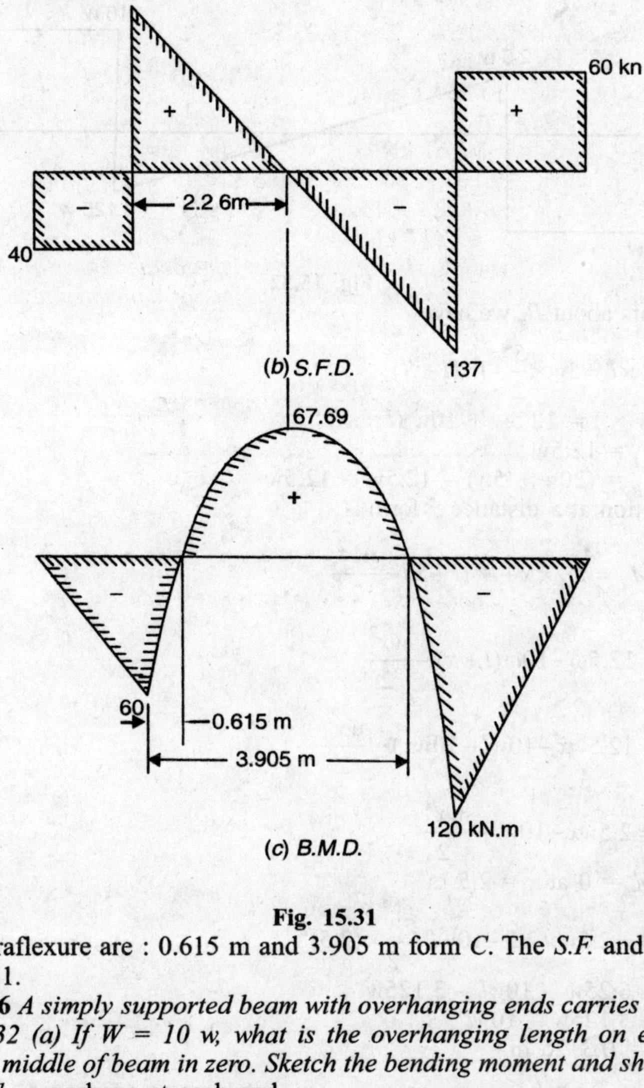

Fig. 15.31

Points of contraflexure are : 0.615 m and 3.905 m form *C*. The *S.F.* and *B.M.* diagrams are shwon in Fig. 15.31.

Example 15.16 *A simply supported beam with overhanging ends carries transverse loads as shown in Fig. 15.32 (a) If W = 10 w, what is the overhanging length on each side such that bending noment at middle of beam in zero. Sketch the bending moment and shear force diagrams.*

Solution. Let l = overhang at each end, *m*

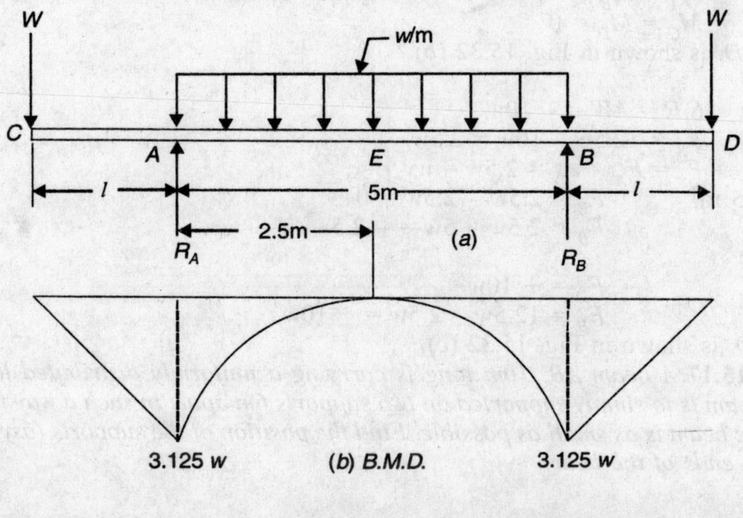

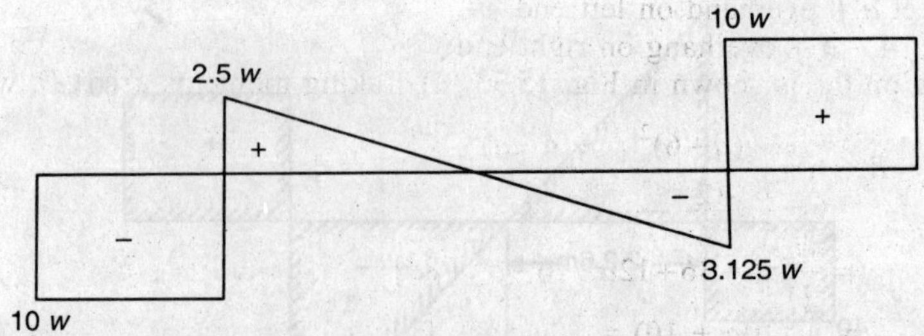

Fig. 15.32

Taking moments about B, we have

$$R_A \times 5 + W \times l = 5w \times \frac{5}{2} + W\,(l+S)$$

$$SR_A + 10w \times 1 = 12.5w + 10w\,(l + S)$$

$$R_A = 12.5w$$

$$R_B = (20w + 5w) - 12.5w = 12.5w$$

Consider a section at a distance x form A,

$$M_x = R_A \times x - W\,(l+x) - \frac{wx^2}{2}$$

$$= 12.5\omega - 10\omega(l+x) - \frac{wx^2}{2}$$

$$= 12.5wx - 10wl - 10wx) - \frac{wx^2}{2}$$

$$= 2.5wx - 10wl - \frac{wx^2}{2}$$

$M_x = 0$ at $x = 2.5$ m

$$0 = 2.5w \times 2.5 - 10wl - \frac{w}{2} \times (2.5)^2$$

$$= 6.25w - 10wl - 3.125w$$

$$= 3.125w - 10wl$$

$$l = 0.3125 \text{ m}$$

B.M.D.

$$M_A = M_B = Wl = -\,10\,w \times 0.3125 = -\,3.125\,w$$

$$M_C = M_D = 0$$

The *B.M.D.* is shown in Fig. 15.32 (*b*)

S.F.D.

Span CA: $S.F. = -W = -\,10w$

Span AB: $F_A = 12.5w - 10w = 2.5w$

$F_x = F_A - wx = 2.5w - wx$

At $x = 2.5$ m, $F_E = 2.5w - 2.5w = 0$

$x = 5$m, $F_B = 2.5w - 5w = -\,2.5w$

Span BD:

$F_D = +\,10w$

$F_B = 12.5w - 2.5w = +\,10w$

The *S.F.D.* is shown in Fig. 15.32 (*c*).

Exmple 15.17 *A beam AB, 10m long is carrying a uniformly distributed load of intensity w N/m. The beam is to simply supported on two supports 6m apart in such a way that the bending moment on the beam is as small as possible. Find the position of the supports (asymmetrical) with respect to the ends of the beam.*

Solution: Let a = overhand on left end, m

$4 - a$ = overhang on right end

The loading on the is shown in Fig. 15.33 (a). Taking moments about D, we have

$$R_c \times 6 = \frac{w(a+6)^2}{2} - \frac{w(4-a)^2}{2}$$

$$= \frac{w}{2}[a^2 + 36 + 12a - 16 - a^2 + 8a]$$

$$= w\,(10a + 10) = 10w\,(a + 1)$$

$$R_c = \frac{5}{2} w(a+1)$$

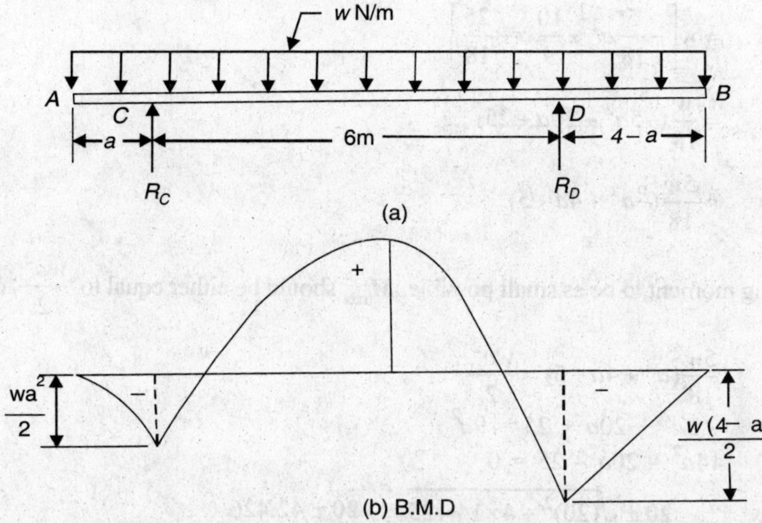

Fig. 15.33

B.M.D. [Fig. 15.33 (b)] $M_c = \dfrac{-wa^2}{2}$ and $M_D = \dfrac{-w(4-a)^2}{2}$

At a section distant x from C,

$$M_x = R_c \times x - \frac{w(4-a)^2}{2}$$

$$= \frac{5}{3} w(a+1)x - \frac{w(a-x)^2}{2}$$

For M_x to be maximum,

$$\frac{dM_x}{dx} = 0$$

$$= \frac{5}{3} w(a+1) - w(a+x) = 0$$

or $5a + 5 - 3a - 3x = 0$

$$2a + 5 = 3x$$

$$x = \frac{2a+5}{3}$$

$$M_{max} = \frac{5}{3}w(a+1)\times\left(\frac{2a+5}{3}\right)-\frac{w}{2}\left[a+\frac{2a+5}{3}\right]^2$$

$$= \frac{5}{9}w(2a^2+7a+5)-\frac{w}{2}\left[\frac{5a+5}{3}\right]^2$$

$$= \frac{5}{9}w(2a^2+7a+5)-\frac{25w}{18}(x^2+1+2a)$$

$$= w\left[a^2\left(\frac{10}{9}-\frac{25}{18}\right)+a\left(\frac{35}{9}-\frac{25}{9}\right)+\left(\frac{25}{9}-\frac{25}{18}\right)\right]$$

$$= w\left[-\frac{5}{18}a^2+\frac{10}{9}a+\frac{25}{18}\right]$$

$$= \frac{w}{18}(-5a^2+20a+25)$$

$$= \frac{5w}{18}(-a^2+4a+5)$$

For bending moment to be as small possible, M_{max} should be either equal to $\frac{wa^2}{2}$ or $\frac{w(4-a)^2}{2}$.

For $\frac{5w}{18}(a^2+4a+5) = \frac{wa^2}{2}$

$$-5a^2 + 20a + 25 = 9a^2$$

$$14a^2 - 20a - 25 = 0$$

$$a = \frac{20 \pm \sqrt{(20)^2+4\times14\times25}}{28} = \frac{20 \pm 42.426}{28}$$

$$= 2.23 \text{ m}$$

For $\frac{5w}{18}(-a^2+4a+5) = \frac{w(4-a)^2}{2}$

$$-5a^2 + 20a + 25 = 9(16 + a^2 - 8a)$$

$$= 144 + 9a^2 - 72a$$

$$14a^2 - 92a + 119 = 0$$

$$a = \frac{92 \pm \sqrt{(92)^2-4\times14\times119}}{28}$$

$$= \frac{92 \pm 42.426}{28}$$

$$= 4.8\text{m}, 1.77\text{m}$$

Permissible value of overhang on loft end, $a = 1.77$ m

REVIEW QUESTIONS

1. Define a beam.
2. Define sagging and hogging bending moment.
3. Define point of inflation.
4. Define point of ontraflexure.
5. Explain the sign convention for shear force.
6. What is the solution ship between load, shear force and bending moment?

EXERCISES

15.1 Draw the shear force and bending moment diagrams for the simply supported beam loaded as show in Fig. 15.34.

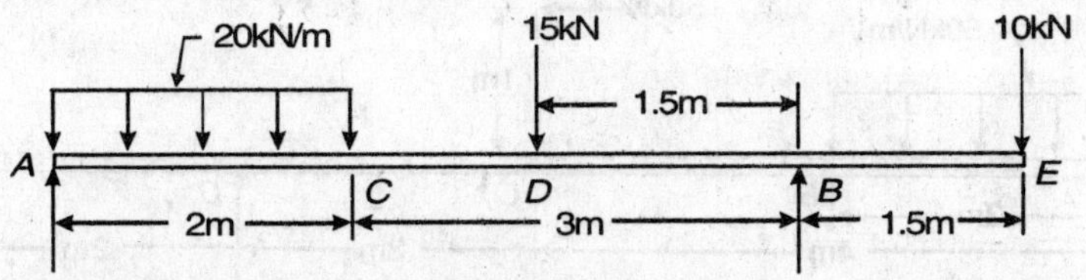

Fig. 15.34

15.2 Draw the B.M. and S.F. diagrams for the simply supported beam loaded as shown in Fig 15.35.

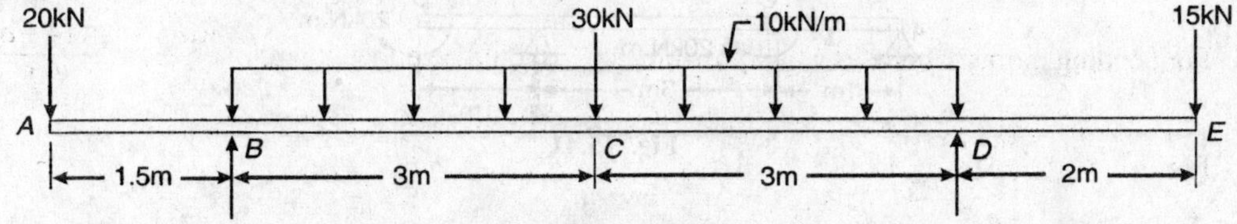

Fig. 15.35

15.3 Draw the B.M. and S.F. diagrams for the simply supported beam loaded as shown in Fig. 15.36.

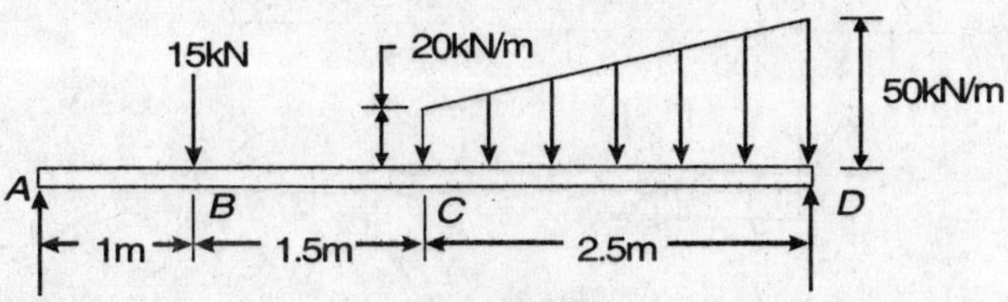

Fig. 15.36

15..4 Draw the B.M. and S.F. diagrams for the beam loaded as shown in Fig. 15.37.

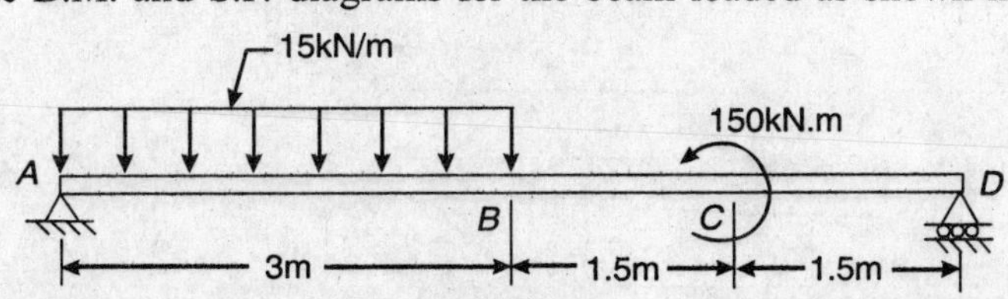

Fig. 15.37

15..5 Draw the B.M. and S.F. diagrams for the beam shown in Fig. 15.38.

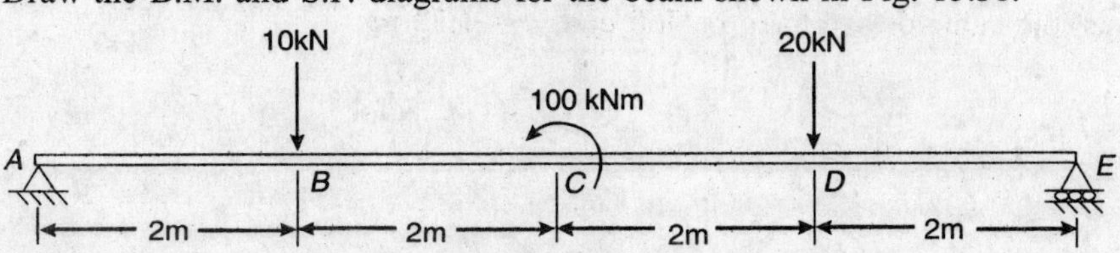

Fig. 15.38

15..6 Calculate the value of 'a' so that maximum positive bending moment is equal to the maximum negative bending moment for the beam shown in Fig. 15.39.

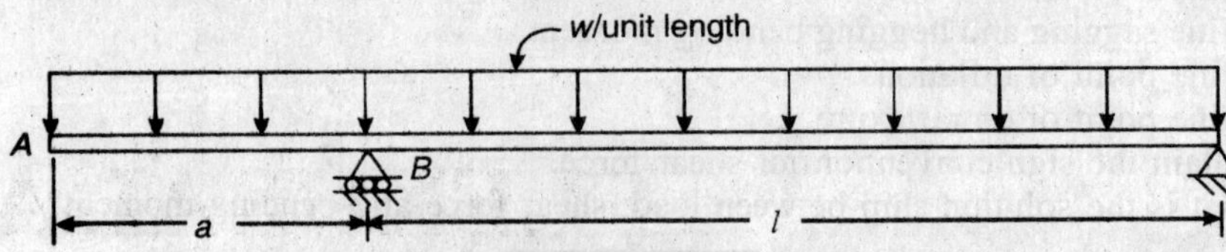

Fig. 15.39

15..7 Draw the *B.M.* and *S.F.* diagrams for the beam shown in Fig. 15.40.

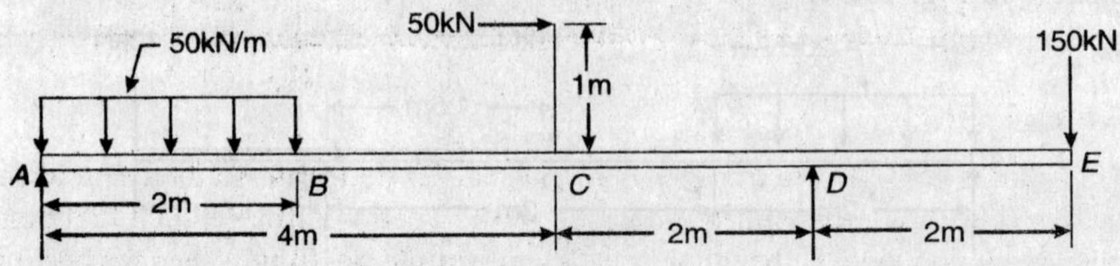

Fig. 15.40

15..8 Draw the *B.M.* and *S.F.* diagrams for the beam shown in Fig 15.41.

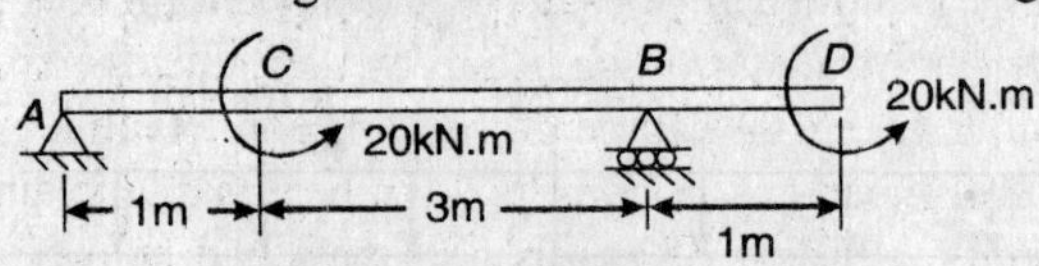

Fig. 15.41

16 FLEXURAL AND TORSIONAL LOADING

16.1 DEFINITIONS

When a beam is subjected to transverse loads it causes deflection in the beam which produces bending. When the loads producing bending lie in the centroidal plane such that bending is not accompanied by torsion then the bending is said to be **simple bending.** When the beam is subjected to such a system of loads so that the shear force in the beam is zero then the beam is said to be subjected to **pure bending.** Bending is quite a common phenomenon. Beams and plates used in civil engineering construction, transission shafting, tall structures like chimneys and distillation columns subjected to wind load, etc. are some of the common examples of bending. We shall study the tensile, compressive and shear stresses produced in beams due to simple bending. Some of the terms associated with simple bending of beams are explained below.

Consider a beam subjected to bending as shown in Fig. 16.1.

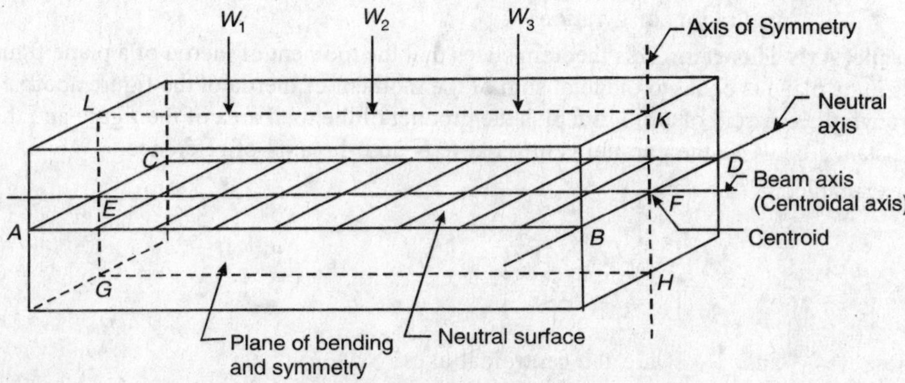

Fig. 16.1 Terms relating to bending of beam

Axis of Beam (or Centroidal axis): It is a line passing through centroids of all cross sections of the beam. Line *EF* is the axis of beam.

Axis of symmetry: It is a line about which the area of cross-section of the beam is symmetrical. For example, circular, rectangular and square sections have two axes of symmetry, where as T-and I-sections have only on axis of symmetry. *AC*, *BD*, *GL* and *HK* are the axes of symmetry.

Plane of Symmetry: It is a plane passing through the axis of symmetry. *GHKL* is the plane of symmetry.

Plane of Bending: It is a plane in which bending takes place. For vertical loading, the plane of symmetry and plane of bending coincide. GHKL is the plane of bending.

Natural Surface: It is a surface on which the bending stress is zero and the beam fibres do not undergo any tension or compression. *ABCD* is the natural surface.

Neutral axis: It is a line where the neutral surface intersects with the area of cross-section of the beam. *AC* and *BD* are the neutral axes.

Centroid: The centroid of a plane area is that point at which the whole of the area of the plane may be assumed to be concentrated. If a plane area is subdivided into small areas A_1, A_2 ..., and the perpendicular distances of the centroids of these area from two chosen coordinate axes x and y are y_1, y_2 ..., and x_1, x_2 ..., respectively, then the coordinates of the centroid $(\bar{x}, \bar{y})$ of the whole area are given by:

$$\bar{x} = \frac{A_1x_1 + A_2x_2 + ...}{A_1 + A_2 + ...} = \frac{\Sigma A_i x_i}{A}, \quad A = \Sigma A_i \quad ... (16.1a)$$

$$\bar{y} = \frac{A_1y_1 + A_2y_2 + ...}{A_1 + A_2 + ...} = \frac{\Sigma A_i y_i}{A} \quad ... (16.1b)$$

Moment of Inertia: The moment of inertia or the second moment of area of a plane figure about a given axis is the sum of the product of the elementary areas into which plane figure may be divided and the square of the distance of centroids of elementary areas from the chosen axis. Thus

$$I_x = \Sigma y^2 \cdot dA \quad ... (16.2a)$$

and

$$I_y = \Sigma x^2 \cdot dA \quad ... (16.2b)$$

Moment of inertia, $I = AK^2$

where A = area of the figure

K = radius of gyration

Parallel Axis Theorem: This theorem states that the moment of inertia of a plane figure about any axis in its plane is equal to the total sum of the moment of inertia of the figure about a parallel axis through the centroid of the figure plus the product of the total area of the figure and the square of the distance between the parallel centroidal axis and the axis of reference.

For example,

$$\left.\begin{aligned} I_{xx} &= I_{\bar{x}\bar{x}} + A\bar{y}^2 \\ I_{yy} &= I_{\bar{y}\bar{y}} + A\bar{x}^2 \end{aligned}\right\} \quad ... (16.3)$$

where $\bar{x} - \bar{x}$ and $\bar{y} - \bar{y}$ are the centroidal axes.

16.1.1 Theory of Simple Bending

The theory of simple bending was developed by Bernoulli. The following assumptions are made in this theory:

1. The material of the beam is assumed to be homogeneous, perfectly elastic and isotropic. Homogeneous material means that the composition is uniform throughout. Perfectly elastic means, Hooke's law is obeyed and no permanent set is formed on unloading. Isotropic means that the elastic constants like E, G, K are same in all directions in the beam.
2. All transverse sections of the beam, which are planes before bending, remain plane after bending.
3. The radius of curvature of the beam before bending is very large in comparison to its transverse dimensions. This implies that the beam is initially straight.

4. The rasultant pull or push across any transverse section of the beam is zero. This implies that total tensile force is equal to the total compressive force in the beam cross-section.
5. The Young's modulus of elasticity is same in tension and compression.
6. The stresses are within the proportional limit.

Consider the simple bending of a beam as shown in Fig. 16.2 (*a*). In the unloaded state, let *GH* be the fibres at a distance *y* from the centroidal axis *KL*, its length being determined by two transverse parallel planes *AD* and *BC*. Fig. 16.2 (*b*) shows the beam after bending, when loaded, and the planes *AD* and *BC* assume the positions A_1D_1 and B_1C_1 respectively. The planes A_1D_1 and B_1C_1 intersect at point *O*, the centre of curvature, and are inclined at an angle θ. Let R be the radius of curvature of the centroidal fibres at E_1F_1. Then

$$G_1H_1 = (R+y)\theta \quad \text{and} \quad E_1F_1 = R\theta$$

$$\therefore \quad \frac{G_1H_1}{E_1F_1} = \frac{(R+y)\theta}{R\theta} = \frac{R+y}{R} = \frac{G_1H_1}{EF}$$

Strain in the fibres at GH,

$$\varepsilon = \frac{G_1H_1 - GH}{GH} = \frac{G_1H_1 - EF}{EF} = \frac{G_1H_1}{EF} - 1 = \frac{R+y}{R} - 1 = \frac{y}{R}$$

Let σ = bending stress in the fibres at *GH*

Then $\sigma = E\varepsilon = \dfrac{Ey}{R}$

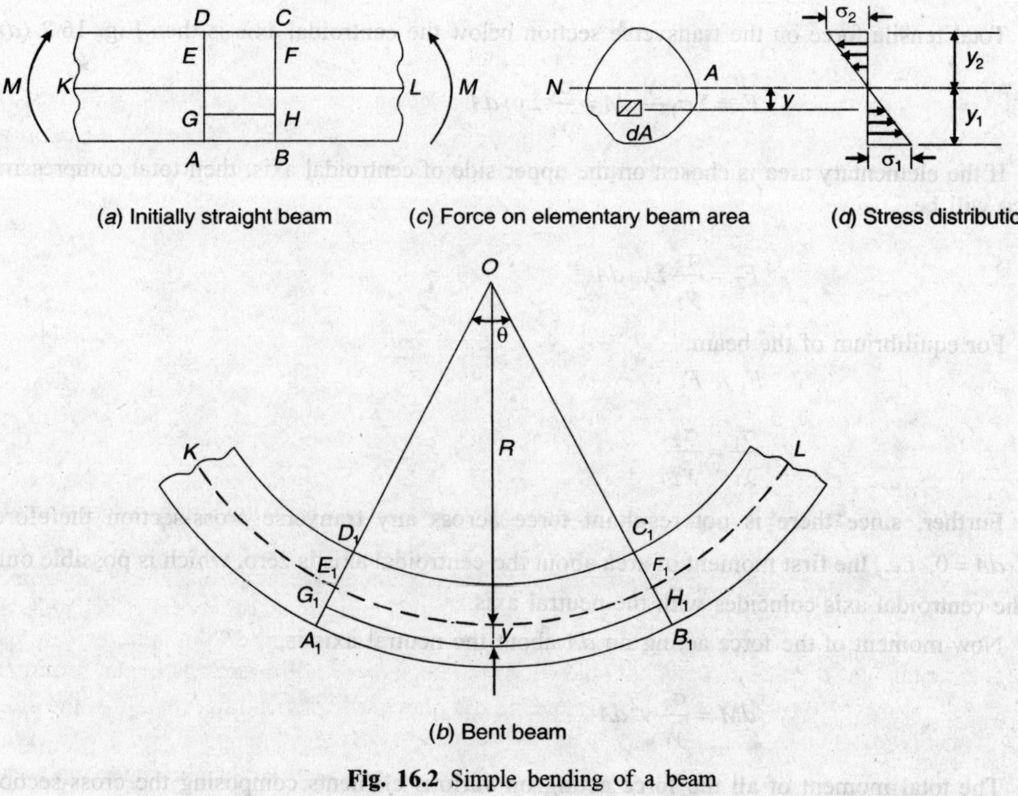

(*a*) Initially straight beam (*c*) Force on elementary beam area (*d*) Stress distribution

(*b*) Bent beam

Fig. 16.2 Simple bending of a beam

or $$\frac{\sigma}{y} = \frac{E}{R} \qquad \text{... (1)}$$

For a given load on a beam,

$$\frac{E}{R} = \text{const.} = k \text{ (say)}$$

$$\therefore \qquad \sigma = ky$$

Therefore, the bending stress at any point in the cross-section is proportional to its distance from the centroidal axis.

Now consider an elementary area *dA* at a distance y from the centroidal axis, as shown in Fig. 16.2 (*c*). Force acting on *dA* is,

$$dF = \sigma \cdot dA$$

But $$\frac{\sigma}{y} = \frac{\sigma_1}{y_1}$$

or $$\sigma = \sigma_1 \frac{y}{y_1}$$

$$\therefore \qquad dF = \sigma_1 \frac{y}{y_1} \cdot dA$$

Total tensile force on the transverse section below the centroidal axis is then Fig. 16.2 (*d*).

$$F_1 = \Sigma \sigma_1 \frac{y}{y_1} dA = \frac{\sigma_1}{y_1} \Sigma y \cdot dA$$

If the elementary area is chosen on the upper side of centroidal axis, then total compressive force will be,

$$F_2 = \frac{\sigma_2}{y_2} \Sigma y \cdot dA$$

For equilibrium of the beam,

$$F_1 = F_2$$

$$\therefore \qquad \frac{\sigma_1}{y_1} = \frac{\sigma_2}{y_2}$$

Further, since there is not resultant force across any tranverse cross-section therefore, $\Sigma y \cdot dA = 0$, *i.e.*, the first moment of area about the centroidal axis is zero, which is possible only if the centroidal axis coincides with the neutral axis.

Now moment of the force acting on *dA* about the neutral axis is,

$$dM = \frac{\sigma_1}{y_1} y^2 dA$$

The total moment of all the force acting on various elements composing the cross-section forms a couple which is equal to the bending moment M. This total moment is called the moment of resistance.

$$\therefore \quad M = \frac{\sigma_1}{y_1} \Sigma y^2 \cdot dA$$

$$= \frac{\sigma_1}{y_1} I$$

or
$$\frac{M}{I} = \frac{\sigma_1}{y_1} = \frac{\sigma}{y} \quad \text{... (2)}$$

where $I = \Sigma y^2 \cdot dA$ = moment of inertial of the cross-section about the neutral axis.

From Eqs. (1) and (2), we get

$$\frac{M}{I} = \frac{\sigma}{y} = \frac{E}{R} \quad \text{... (16.4)}$$

Eq. (16.4) represents the bending formula.

The implications of the bending formula are:

- The variation of bending stress (tensile or compressive) through the beam cross-section is linear.
- The bending stress is directly proportional to the distance of the beam fibres from the neutral axis. Therefore, maximum stress occurs on the outermost fibres of the beam.
- The neutral axis coincides with the centroidal axis.

16.1.2 Section Modulus

To correlate the bending tensile and compressive stresses with the direct tensile and compressive stresses, a quaintly called section modulus is defined. This is analogous to the area of cross-section. As direct stresses, $\sigma_d = \frac{P}{A}$, similarly bending stress, $\sigma_b = \frac{M}{z}$, where z = section modulus.

$$\frac{M}{I} = \frac{\sigma}{y}$$

$$\sigma_{max} = \frac{My_{max}}{I} = \frac{M}{\left(\frac{I}{y_{max}}\right)} = \frac{M}{z} \quad \text{... (16.5a)}$$

where $z = \frac{I}{y_{max}}$ is called the "section modulus".

The SI units of z are mm^3 or m^3.

$$M = \sigma_{max} \times z \quad \text{... (16.5b)}$$

Those sections which have higher value of section modulus will develop lower value of bending stress for the same value of bending moment applied and same area of cross-section.

16.1.3. Calculation of Bending Stresses

(a) Circular Section [Fig. 16.3]

$$\sigma = \frac{M}{z}$$

For a solid circular section beam of diameter d,

$$z = \frac{I}{y_{max}} = \frac{\pi d^4}{64} \times \frac{2}{d} = \frac{\pi d^3}{32}$$

$$\sigma = \frac{32\,M}{\pi\,d^3}$$

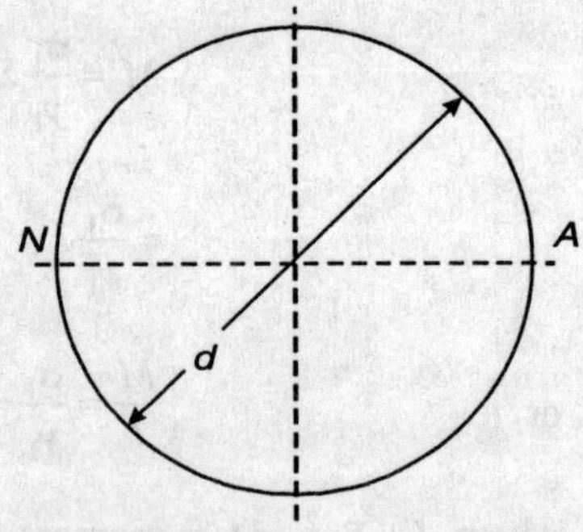

Fig. 16.3 Circular section

For a hollow circular section beam of outer diameter d_0 and inner diameter d_i, we have

$$z = \frac{\pi}{64}(d_0^4 - d_i^4) \times \frac{2}{d_0} = \frac{\pi}{32} d_0^3 \left[1 - \left(\frac{d_i}{d_0}\right)^4\right]$$

If $k = \frac{d_i}{d_0}$, then

$$z = \frac{32M}{\pi(1-k^4)d_0^3}$$

(b) Rectangular Section [Fig. 16.4]

For a rectangular section of width b and depth h, we have

$$z = \frac{I_{NA}}{y_{max}} = \frac{bh^3}{12} \times \frac{2}{h} = \frac{bh^2}{6}$$

$$\sigma = \pm \frac{6M}{bh^2}$$

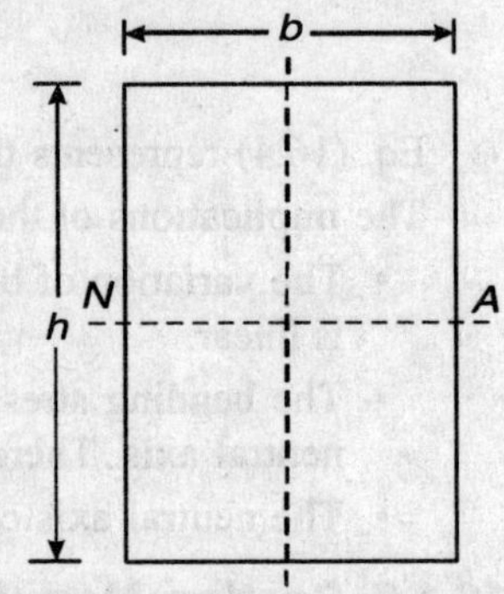

Fig. 16.4 Rectangular section

(c) I-Section

Consider a symmetrical I-section as shown in Fig. 16.5.

For bending about the neutral-axis, we have

$$I_{NA} = \frac{b(h+2t)^3}{12} - \frac{(b-t_1)h^3}{12}$$

$$y_{max} = \frac{h}{2} + t; \quad z = \frac{I_{NA}}{y_{max}}; \quad \sigma = \pm \frac{M}{z}$$

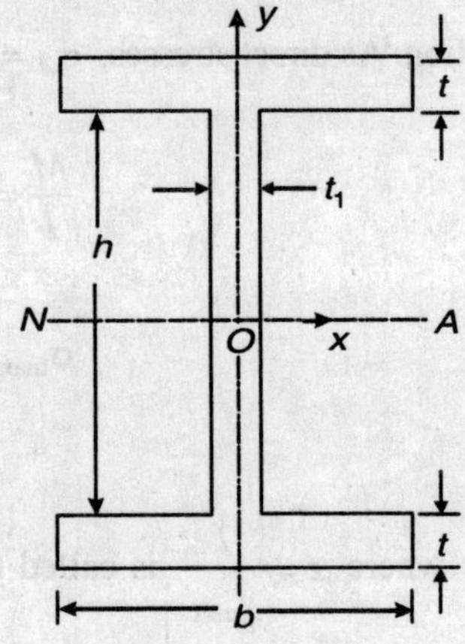

Fig. 16.5 *I*-section

(d) T-Section

The T-section [Fig. 16.6] has y-axis as the axis of symmetry.

$$\bar{y} = \frac{\frac{1}{2}b_1 t_1^2 + b_2 t_2\left(t_1 + \frac{b_2}{2}\right)}{(b_1 t_1 + b_2 t_2)}$$

$$I_{NA} = \frac{b_1 t_1^3}{12} + b_1 t_1 \left(\overline{y} - \frac{t_1}{2}\right)^2 + \frac{t_2 b_2^3}{12} + t_2 b_2 \left(t_1 + \frac{b_2}{2} - \overline{y}\right)^2$$

Compressive bending stress,

$$\sigma_c = \frac{M\overline{y}}{I_{NA}}$$

Tensile bending stress, $\sigma_t = \dfrac{M(b_2 + t_1 - \overline{y})}{I_{NA}}$

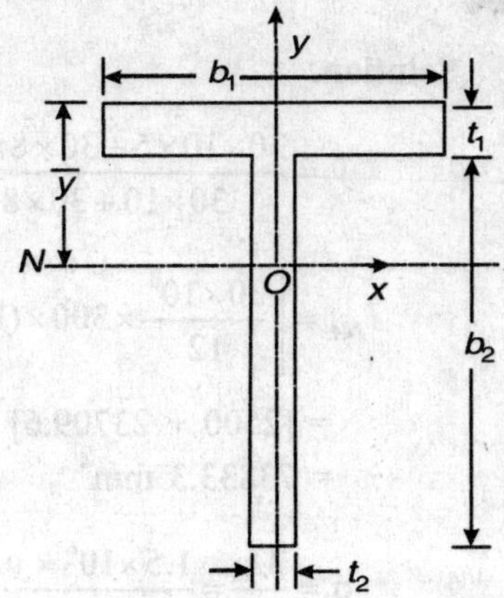

Fig. 16.6 *T*-section

(e) Channel Section [Fig. 16.7

$$A_1 = ht$$

$$A_2 = (b - t)\, t$$

Total area, $A = A_1 + 2A_2 = ht + 2\,(b - t)\, t$

$$\overline{x} = \frac{A_1\overline{x}_1 + 2A_2\overline{x}_2}{A}$$

$$\overline{x}_1 = \frac{t}{2} \qquad \overline{x}_2 = t + \frac{b-t}{2}$$

$$\overline{y} = \frac{h}{2} \qquad I_{NA} = I_{\overline{x}}$$

$$= \frac{th^3}{12} + 2\left[\frac{(b-t)t^3}{12} + (b-t)t\left\{\frac{h}{2} - \frac{t}{2}\right\}^2\right]$$

$$\frac{M}{I} = \frac{\sigma}{\left(\frac{h}{2}\right)} \quad \Rightarrow \quad \sigma = \frac{Mh}{2I}$$

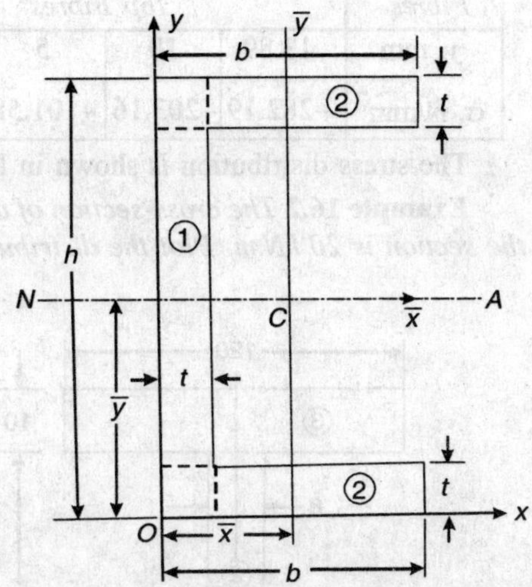

Fig. 16.7 Channel section

Example 16.1 *The cross-section of a T-section beam is shown in Fig. 16.8 (a). It is subjected to a bending moment of 1.5 kNm. Plot the distribution of stress due to bending in the beam.*

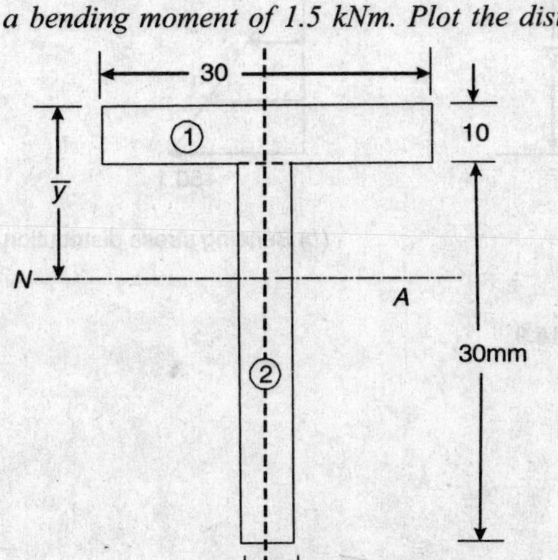

(*a*) T-section

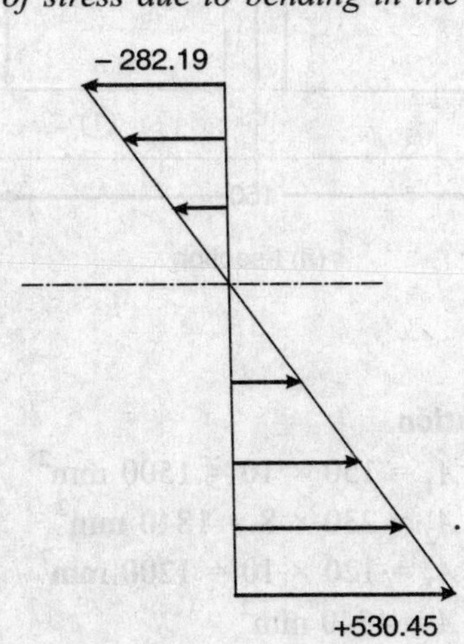

(*b*) Bending stress distribution

Fig 16.8

Solution.

$$\bar{y} = \frac{30\times10\times5+30\times8\times25}{30\times10+30\times8} = \frac{7500}{540} = 13.89 \text{ mm}$$

$$I_{NA} = \left[\frac{30\times10^3}{12}\times300\times(13.89-5)^2\right]+\left[\frac{8\times30^2}{12}+240\times(25-13.89)^2\right]$$

$= [2500 + 23709.6] + [18000 + 29623.7]$
$= 73833.3 \text{ mm}^4$

$$\sigma = \frac{My}{I} = \frac{1.5\times10^6\times y}{73833.3} = 20.316\, y \text{ N/mm}^2$$

Fibres	*Top Fibres*			*Bottom Fibres*				
y, mm	13.89	10	5	5	10	15	20	26.11
σ, N/mm²	–282.19	–203.16	–101.58	101.58	203.16.	304.74	406.32	530.45

The stress distribution is shown in Fig. 16..8 (*b*).

Example 16.2 *The cross-section of an I-beam is shown in Fig. 16.9. The bending moment at the section is 20 kN.m. Plot the distribution of bending stress in the beam.*

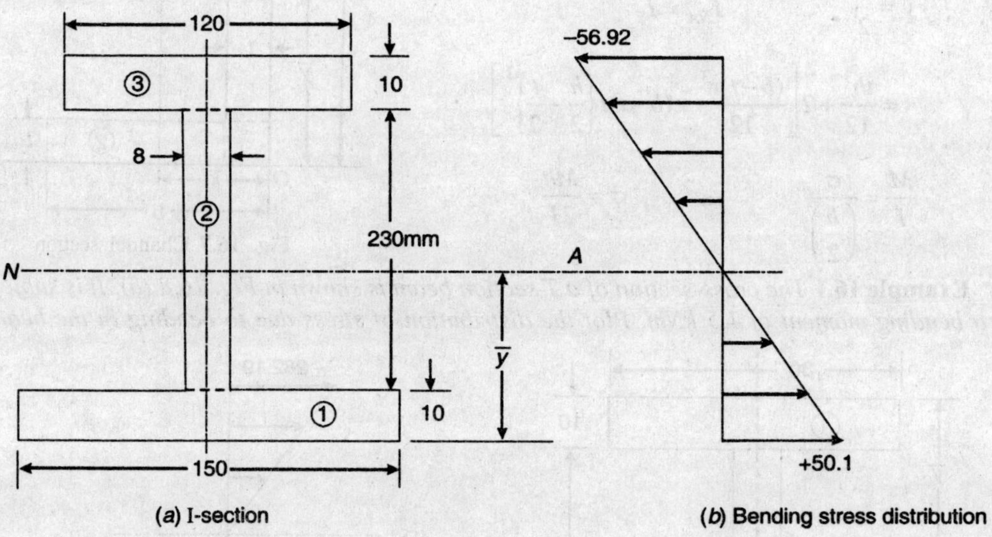

(*a*) I-section

(*b*) Bending stress distribution

Fig. 16.9

Solution.

$A_1 = 150 \times 10 = 1500 \text{ mm}^2$
$A_2 = 230 \times 8 = 1840 \text{ mm}^2$
$A_3 = 120 \times 10 = 1200 \text{ mm}^2$
$A = 4540 \text{ mm}^2$

$$\bar{y} = \frac{1500\times5+1840\times125+1200\times245}{4540} = \frac{531500}{4540} = 117 \text{ mm}$$

$$I_{NA} = \left[\frac{150\times10^3}{12} + 1500\times(117-5)^2\right] + \left[\frac{8\times230^2}{12} + 1840\times(125-117)^2\right]$$

$$+\left[\frac{120\times10^3}{12} + 1200\times(245-117)^2\right]$$

$$= [12500 + 11816 \times 10^3] + [8111333 + 117760] + [10000 + 196608 \times 10^2]$$

$$= 46727393 \text{ mm}^4$$

$$\sigma = \frac{My}{I} = \frac{20\times10^6\times y}{46728393} = 0.428\ y \text{ N/mm}^2$$

Fibres	*Top Fibres*				*Bottom Fibres*		
y, mm	133	100	50	0	50	100	117
σ, N/mm²	–56.92	–42.8	–21.4	0	21.4	42.8	50.1

The stress distribution is shwon in Fig. 16.9 (*b*).

Example 16.3 *A cast iron channel carries water as shown in Fig. 16.10. It is supported at two points 12 m apart. Density of water is 1000 kg/m³ and that of cast iron 7000 kg/m³. Calculate the depth of water in the channel if the tensile and compressive stresses in bending are not to exceed 20 MPa and 50 MPa respectively.*

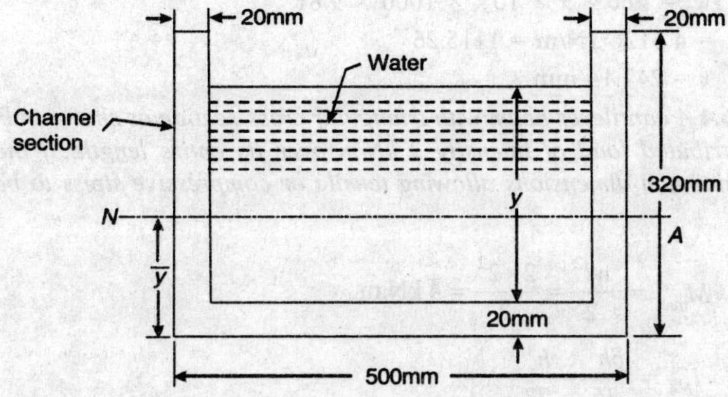

Fig. 16.10 Channel Section containing water.

Solution.

$$\bar{y} = \frac{500\times20\times10 + 2\times300\times20\times170}{500\times20 + 2\times300\times20} = 97.3 \text{ mm}$$

$$I_{NA} = \frac{500\times20^3}{12} + 500\times20\,(97.3-10)^2 + 2\left[\frac{20\times30^3}{12} + 20\times300(170-97.3)^2\right]\times10^{-12}$$

$$= 23 \times 10^{-5} \text{ m}^4$$

Moment of resistance in Compression,

$$M_c = \frac{\sigma_E I_{NA}}{y_c} = \frac{50\times23\times10^{-5}\times10^6}{(320-97.3)\times10^{-3}} = 51.639 \text{ kN.m}$$

Moment of resistance in tension,

$$M_t = \frac{\sigma_t I_{NA}}{y_t} = \frac{20\times10^6\times23\times10^{-5}}{97.3\times10^{-3}} = 47.276 \text{ kN.m}$$

Safe bending moment = 47.276 kN.m

$$\therefore \quad \sigma_t = 20 \text{ MPa}$$

$$\sigma_c = \frac{20\times222.7}{97.3} = 45.776 \text{ MPa}$$

Let w = Total weight of channel and water per metre length, kN/m

$$M = \frac{wl^2}{8} = \frac{w\times12^2}{8} = 18w \text{ kN.m}$$

$$\therefore \quad 18w = 47.276$$

$$w = 2.626 \text{ kN/m}$$

Weight of channel per meter,

$$w_1 = (500 \times 20 + 2 \times 300 \times 20) \times 10^{-6} \times 7000 \times 9.81$$

$$= 1510.74 \text{ N}$$

Weight of water,

$$w_2 = 2626 - 1510.74 = 1115.26 \text{ N/m}$$

Let $\quad x$ = depth of water in the channel, mm

Then $\quad w_2 = 460 \times x \times 10^{-6} \times 1000 \times 9.81$

$$= 4.5126\, x\text{N/m} = 1115.26$$

$$x = 247.14 \text{ mm}$$

Example 16.4 *A cantilever beam with triangular cross-section as shown in Fig. 16.11 carries a uniformly distributed load of intensity 2 kN/m over its entire length. If the depth h = 2b, determine the minimum dimensions allowing tensile or compressive stress to be 25 MPa.*

Solution.

$$M_{max} = \frac{wl^2}{2} = \frac{2\times2^2}{2} = 4 \text{ kN.m}$$

$$I_{NA} = \frac{bh^3}{36} = \frac{h^4}{72}$$

$$\sigma_t = \frac{4\times10^3\times\dfrac{h}{3}}{\dfrac{h^4}{72}} = \frac{96\times10^3}{h^3} \text{ N/m}^2$$

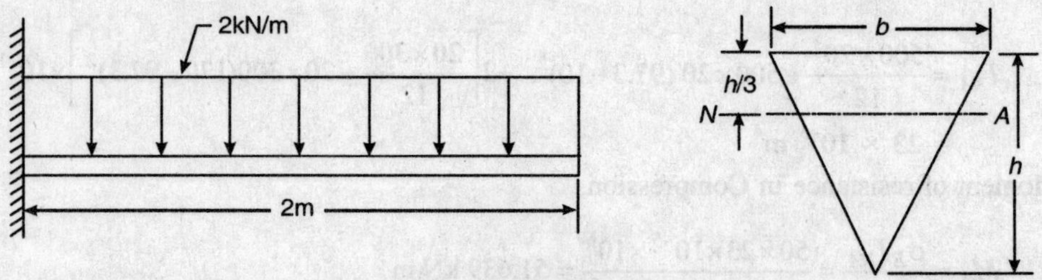

Fig. 16.11

$$\sigma_c = \frac{4\times10^3\times\dfrac{2h}{3}}{\dfrac{h^4}{72}} = \frac{192\times10^3}{h^3}\ \text{N/m}^2$$

For minimum dimension of cross-section,

$$\frac{96\times10^3}{h^3} = 25\times10^6$$

$$h = 3.84 \times 10^{-3}\ \text{m}$$

$$h = 0.1566\ \text{m}$$

$$b = 0.0783\ \text{m}$$

Example 16.5 *A timber beam as shown in Fig. 16.12 is simply supported on a span of 4 m. The weight of the beam is 10 kN/m^3 and the permissible stress in timber is 10 MPa. Calculate the safe point load at mid-span.*

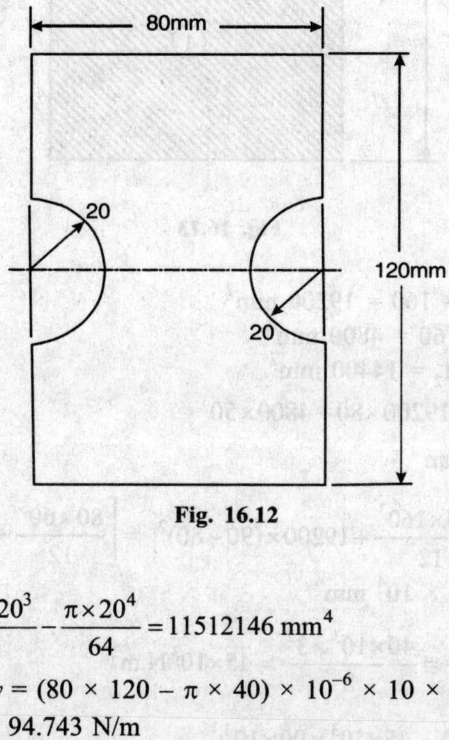

Fig. 16.12

Solution.

$$I = \frac{80\times120^3}{12} - \frac{\pi\times20^4}{64} = 11512146\ \text{mm}^4$$

Weight of beam, $w = (80 \times 120 - \pi \times 40) \times 10^{-6} \times 10 \times 10^3$

$$= 94.743\ \text{N/m}$$

$$M_{\max} = \frac{Wl}{4} + \frac{wl^2}{8}$$

$$= \frac{W\times4}{4} + \frac{94.743\times4^2}{8} = W + 189.487\ \text{N.m}$$

$$z = \frac{I}{y} = \frac{11512146}{60} = 191869\ \text{mm}^3$$

$$M_{\max} = \sigma\times z$$

$$W + 189.487 = 10 \times 10^6 \times 191869 \times 10^{-9}$$
$$W = 1729.2 \text{ N}$$

Example 16.6 *A beam of 3 m length is simply supported at the ends and carries a uniformly distributed load of 40 kN/m over its whole span. The cross-section of the beam is shown in Fig. 16.13. Calculate the maximum tensile and compressive stresses in the beam.*

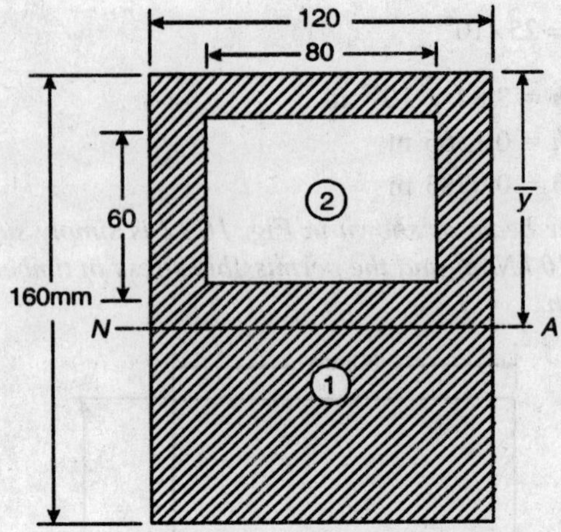

Fig. 16.13

Solution.

$$A_1 = 120 \times 160 = 19200 \text{ mm}^2$$
$$A_2 = 80 \times 60 = 4800 \text{ mm}^2$$
$$A = A_1 - A_2 = 14400 \text{ mm}^2$$
$$14400\,\bar{y} = 19200 \times 80 - 4800 \times 50$$
$$\bar{y} = 90 \text{ mm}$$

$$I_{N\text{-}A} = \frac{120 \times 160^3}{12} + 19200 \times (90-80)^2 = \left[\frac{80 \times 60^3}{12} + 4800 \times (90-50)^2\right]$$
$$= 3376 \times 10^4 \text{ mm}^4$$

$$M_{max} = \frac{wl^2}{8} = \frac{40 \times 10^3 \times 3^2}{8} = 45 \times 10^3 \text{ N.m}$$

$$\text{Compressive stress} = \frac{45 \times 10^3 \times 90 \times 10^3}{3376 \times 10^4} = 119.96 \text{ MPa}$$

$$\text{Tensile stress} = 119.96 \times \frac{70}{90} = 93.3 \text{ MPa}$$

Example 16.7 *A log of timber 300 mm in diameter is available to be used as a beam. Calculate the following:*

(a) Dimensions of the strongest rectangular section that can be cut out of it.

(b) Maximum span to carry a uniformly distributed load of 3 kN/m to limit the bending stress to 10 MPa.

(c) Longest span to use the log as such.

Solution.

(*a*) Let b = width of section (Fig. 16.14)

h = depth of section

Diameter of log, d = 300 m

$$d^2 = b^2 + h^2$$

$$h^2 = 300^2 - b^2$$

$$z = \frac{bh^2}{6} = \frac{b(300^2 - b^2)}{6}$$

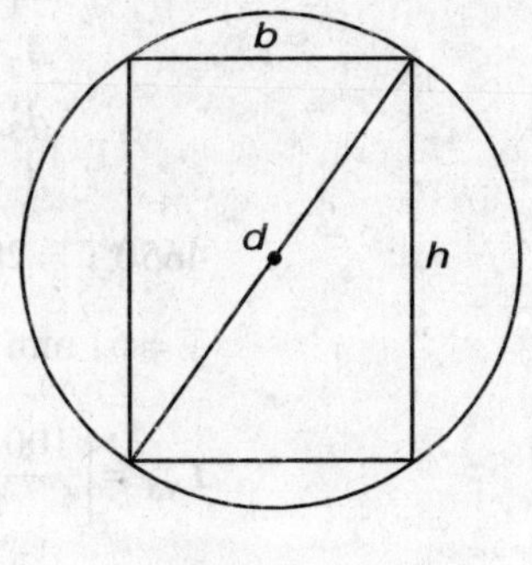

Fig 16.14

For a strongest section, z should be maximum.

$$\frac{dz}{db} = 0$$

$$300^2 - 3b^2 = 0$$

$$b = 173.2 \text{ mm}$$

$$h^2 = 300^2 - 30000 = 60000$$

$$h = 244.95 \simeq 245 \text{ mm}$$

(*b*) $$z = \frac{173.2 \times 60000}{6} = 1732000 \text{ mm}^3$$

$$M = \sigma z = 10 \times 10^6 \times 1732000 \times 10^{-9} = 17320 \text{ N.m} = \frac{\omega l^2}{8} = \frac{3 \times 10^3 \times l^2}{8}$$

$$l = 6.796 \text{ m}$$

(*c*) $$Z = \frac{\pi d^3}{32} = \frac{\pi \times 300^3}{32} = 2650719 \text{ mm}^3$$

$$M = 10 \times 10^6 \times 2650719 \times 10^{-9} = 26507 \text{ N.m} = \frac{3 \times 10^3 \times l^2}{8}$$

$$l = 8.4 \text{ m}$$

Example 16.8 *A beam of I-section shown in Fig. 16.15 is subjected to be bending moment of 12 kN.m. Find the maximum tensile and compressive stresses induced in the beam.*

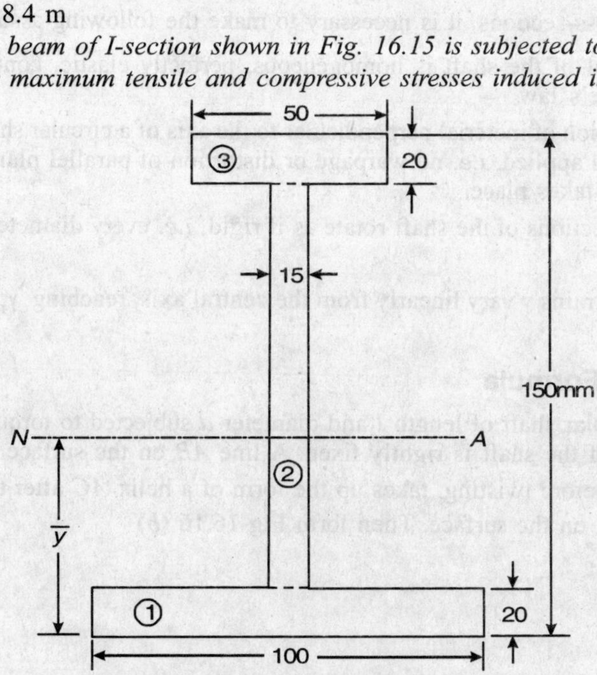

Fig. 16.15 Unequal *I*-section beam.

Solution.

$$A_1 = 100 \times 20 = 2000 \text{ mm}^2$$
$$A_2 = 110 \times 15 = 1650 \text{ mm}^2$$
$$A_3 = 50 \times 20 = 1000 \text{ mm}^2$$
$$A = 2000 + 1650 + 1000 = 4650 \text{ mm}^2$$

$$4650\,\bar{y} = 2000 \times 10 + 1650 \times 75 + 1000 \times 140$$

$$\bar{y} = 61 \text{ mm}$$

$$I_{NA} = \left[\frac{100 \times 20^3}{12} + 2000 \times (61-10)^2\right] + \left[\frac{15 \times 110^3}{12} + 1650 \times (75-61)^2\right] + \left[\frac{50 \times 20^3}{12} + 1000 \times (140-61)^2\right]$$

$$= 13530150 \text{ mm}^4$$

$$\sigma_t = \frac{12 \times 10^6 \times 61}{13530150} = 54.1 \text{ MPa}$$

$$\sigma_c = \frac{54.1 \times 89}{61} = 78.93 \text{ MPa}$$

16.2 TORSION OF CIRCULAR BARS

A Shaft is a rotating machine member which is required to transmit power in the industry from one unit to another. During this process it is subjected to twisting action, *i.e.* shearing of fibres of the shaft. The power is transmitted on account of a couple acting in a plane perpendicular to the longitudinal axis of the shaft. This couple is called the torque. The angular twist caused by the torque is called torsion.

To establish a relation between the torque and the shear stress set up in members with circular solid and tubular cross-sections, it is necessary to make the following assumptions:

1. The material of the shaft is homogeneous, perfectly elastic, continuous, isotropic and obeys Hooke's law.
2. A plane section of meterial perpendicular to the axis of a circular shaft remains plane after the torque is applied, *i.e.* no warpage or distortion of parallel planes normal to the axis of the shaft takes place.
3. All cross- sections of the shaft rotate as if rigid, *i.e.* every diameter rotates thorough the same angle.
4. The shear strains γ vary linearly from the central axis, reaching γ_{max} at the periphery of the shaft.

16.2.1 Torsion Formula

Consider a circular shaft of length l and diameter d subjected to torque T, as shown in Fig. 16.16(*a*). One end of the shaft is rightly fixed. A line AB on the surface of the shaft, which is parallel to the axis before twisting, takes up the form of a helix AC after twisting. Let ϕ be the angle of shear strain on the surface. Then form Fig 16.16 (*b*)

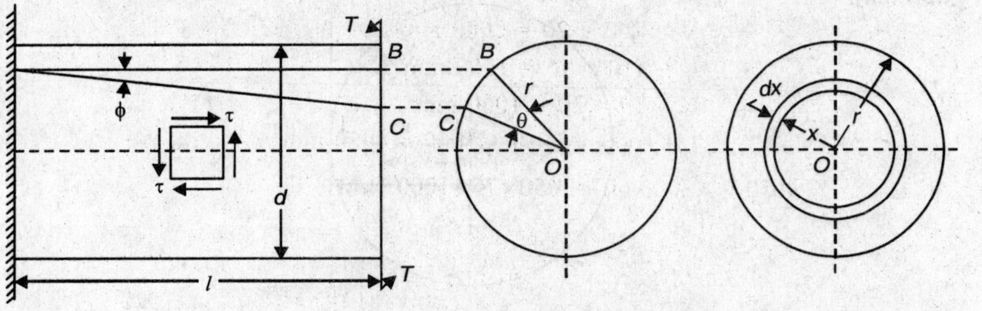

(a) Circular shaft under torsion (b) Twist of shaft (c) Elementary ring of shaft.

Fig. 16.16 Torsion of a circular shaft.

$$BC = l\phi$$

or
$$\phi = \frac{BC}{l}$$

But
$$\phi = \frac{\tau}{G}$$

where τ = shear stress in the shaft

G = modulus of rigidity of shaft material

$\therefore$
$$\tau = G\phi$$

Now
$$BC = \frac{d}{2}\cdot\theta$$

where θ = angle of twist of shaft

$\therefore$
$$\tau = G\times\frac{d}{2}\times\frac{\theta}{l} = \left(\frac{G\theta}{l}\right)\cdot\frac{d}{2} = k.r$$

where $k = \dfrac{G\theta}{l}$ is a constant for a given shaft

$r = \dfrac{d}{2}$ is the shaft radius.

Therefore, we find that shear stress in the shaft is proportional to the radius of the shaft, as shown in Fig. 16.17. Hence

$$\frac{\tau}{r} = \frac{G\theta}{l}$$

At any radius x, we have

$$\frac{\tau_x}{\tau} = \frac{x}{r}$$

or
$$\tau_x = \frac{\tau x}{r}$$

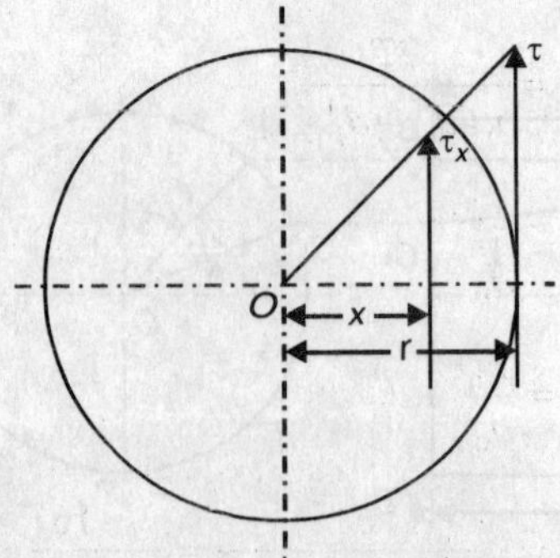

Fig. 16.17 Linear variation of torsional shear stress.

Now consider the elementary ring of the shaft at a radius x and of thichness dx, as shown in Fig. 16.16 (c).

Total force on the ring $F = \tau_x \times$ area of the ring $= \tau_2 \times 2\pi x dx$

Moment of force F about the axis of the shaft,

$$dM = F \times x = \tau_2 \times 2\pi x \, dx \times x = 2\pi x^2 \tau_x .dx$$

$$= 2\pi x^2 \times \frac{\tau x}{r} \times dx = \frac{2\pi\tau}{r}.x^3 dx$$

Total resisting moment of the shaft,

$$M = \frac{2\pi\tau}{r}\int_0^r x^3 dx = \frac{2\pi\tau}{r}\left|\frac{x^4}{4}\right|_0^4 = \frac{\pi r^3}{2}\tau$$

Total resisting moment, M = Applied torque T,

$$\therefore \qquad T = \frac{\pi r^3}{r}.\tau = \frac{\pi d^3}{16}\tau = \frac{\tau}{(d/2)} \times \frac{\pi d^4}{32} = \frac{2\tau}{d}.J = \frac{\tau}{r}.J$$

where $J = \dfrac{\pi d^4}{32}$ is the polar moments of inertia of shaft cross-section.

$$\therefore \qquad \frac{T}{J} = \frac{\tau}{r} = \frac{G\theta}{l} \qquad ...(16.6)$$

This is the well known torsion formula for a circular shaft. The SI units of various quantities are:

T = torque, N.m.
J = polar moment of inertia, m^4
τ = shear stress, MN/m^2 (MPa)
r = radius, m
G = shear modulus, GPa
θ = angle of twist, radians
l = length, m

16.2.2 Hollow Circular Shaft

For a hollow circular shaft of outer diameter d_0 and inner diameter d_i, we have

$$J = \frac{\pi}{32}(d_0^4 - d_i^4)$$

$$\frac{T}{J}=\frac{32T}{\pi(d_0^4-d_i^4)}$$

$$\frac{\tau}{r}=\frac{2\tau}{d_0}$$

$$\therefore \quad \frac{32T}{\pi(d_0^4-d_i^4)}=\frac{2\tau}{d_0}$$

$$\tau=\frac{16T}{\pi d_0^3\left[1-\left(\frac{d_i}{d_0}\right)^4\right]}=\frac{16T}{\pi d_0^3(1-k^4)} \quad ...(16.9)$$

where $k=\frac{d_i}{d_0}$ is the ratio of inner to outer diameters.

16.2.3 Comparision of Solid and Hollow Shafts

Let d_0 = outer diameter of hollow shaft

d_i = inner diameter of hollow shaft

d = diameter of solid shaft

Let the density of the shaft material and length be the same.

(a) For the Same Weight

Weight of shaft is proportional to area of cross-section. For same weight and material of solid and hollow cirular shafts, we have

$$\frac{\pi}{4}d^2=\frac{\pi}{4}(d_0^2-d_i^2)$$

or $$d=\sqrt{d_0^2-d_i^2}$$

For a solid shaft, $$T_s=\frac{\pi d^3}{16}\tau_s$$

For a hollow shaft, $$T_h=\frac{\pi(d_0^4-d_i^4)\tau_h}{16d_0}$$

$$\frac{T_h}{T_S}=\left(\frac{d_0^4-d_i^4}{d_0}\right)\times\frac{1}{d^3}\times\frac{\tau_h}{\tau_S}=\left(\frac{d_0^2-d_i^4}{d_0}\right)\times\frac{1}{(d_0^2-d_i^2)d}\times\frac{\tau_h}{\tau_s}$$

$$=\left(\frac{d_0^2-d_i^2}{d_0 d}\right)\times\frac{\tau_h}{\tau_s}=\frac{d_0}{d}(1+k^2)\times\frac{\tau_h}{\tau_s}$$

where $$k=\frac{d_i}{d_0}$$

Now $$\left(\frac{d_0}{d}\right)^2=\frac{d_0^2}{d_0^2-d_i^2}=\frac{1}{1-k^2}$$

$$\frac{d_0}{d} = \frac{1}{\sqrt{1-k^2}}$$

$$\frac{T_h}{T_s} = \frac{1+k^2}{\sqrt{1-k^2}} \times \frac{\tau_h}{\tau_s} \qquad \text{...(16.8a)}$$

If $\tau_h = \tau_s$, then

$$\frac{T_h}{T_s} = \frac{1+k^2}{\sqrt{1-k^2}} \qquad \text{...(16.8b)}$$

(b) For the Same Strength

Here $\tau_h = \tau_s$

$$\frac{T_h}{T_s} = \left(\frac{d_0^4 - d_i^4}{d_0}\right) \times \frac{1}{d^3}$$

For the same outer diameter, $d = d_0$.

$$\frac{T_h}{T_s} = \frac{d_0^4 - d_i^4}{d_0^4} = 1 - \left(\frac{d_i}{d_0}\right)^4 = 1 - k^4 \qquad \text{...(16.9)}$$

16.2.4. Power TRANsMitted

The power transmitted by a shaft is given by:

$$P = \frac{\omega T_{\text{mean}}}{10^3} - \frac{2\pi N T_{\text{mean}}}{60 \times 10^3} kW \qquad \text{...(16.10)}$$

where N = revolution/min

$$\omega = \frac{2\pi N}{60} \text{ rad/s}$$

T_{mean} = mean torque transmitted, N.m.

T_{max} = maximum torque transmitted, N.m.

= (1+ over load factor) × T_{mean}

16.2.5 Torsional Stiffness

It is the torqure per unit angle of twist.

Form torsional formula, we have

$$\frac{T}{J} = \frac{G\theta}{l}$$

Torsional stiffness,

$$k_t = \frac{T}{\theta} = \frac{GJ}{l} \qquad \text{...(16.11)}$$

16.2.6 Torsional Section Modulus

The torisonal section modulus,

$$z_p = \frac{J}{r}$$

$$= \frac{\pi d^4}{32} \times \frac{2}{d} = \frac{\pi d^3}{16} \text{ for a solid shaft.} \qquad \text{...(16.12a)}$$

$$= \frac{\pi}{16}\left(\frac{d_0^4 - d_i^4}{d_0}\right) = \frac{\pi}{16} d_0^3 (1 - k^4) \text{ for a hollow shaft.} \quad ...(16.12b)$$

where
$$k = \frac{d_i}{d_0}$$

Torque = Torsional section modulus × shear stress

$$T = z_p \tau \quad ...(16.13)$$

Example 16.9 *A hollow shaft of diameter ratio 0.6 is required to transmit 500 kW at 100 rpm. The maximum torque being 15% greater than the mean. The shear stress is not to exceed 60 MPa and the angle of twist in a length of 3 m is limited to 1°. Calculate the minimum external diameter satisfying these conditions G = 80 GPa.*

Solution:
$$P = \frac{2\pi N T_{\text{mean}}}{60 \times 10^3} \text{ kW}$$

$$500 = \frac{2\pi \times 100 \times T_{\text{mean}}}{60 \times 10^3}$$

$$T_{\text{mean}} = \frac{500 \times 60 \times 10^3}{2\pi \times 100} = 47{,}746.5 \text{ N.m}$$

$$T_{\max} = 1.15 \times 47{,}746.5 = 54{,}908.5 \text{ N.m}$$

$$d_i = 0.6 \text{ d}_0$$

$$J = \frac{\pi d_0^4}{32}[1 - (0.6)^4] = 0.08545 \, d_0^4 \text{ mm}^4$$

$$\frac{T_{\max}}{J} = \frac{\tau}{r}$$

$$\frac{54{,}908 \times 10^3}{0.0854 d_0^4} = \frac{60 \times 2}{d_0}$$

$$d_0^3 = \frac{54{,}908.5 \times 10^3}{0.08545 \times 120} = 5354.75 \times 10^3$$

$$d_0 = 175 \text{ mm}$$

$$\frac{T_{\max}}{J} = \frac{G\theta}{l}$$

$$\frac{54{,}908.5 \times 10^3}{0.08545 d_0^4} = \frac{80 \times 10^3 \times \pi}{180 \times 3 \times 10^3}$$

$$d_0^4 = \frac{54{,}908.5 \times 10^3 \times 180 \times 3}{0.08545 \times 80 \times \pi} = 1380643056$$

$$d_0 = 192.8 \text{ mm} \simeq 193 \text{ mm}$$

Hence d_0 = 193 mm will satisfy both the conditions.

Example 16.10 *Compare the weights of two solid shafts of identical length, one of steel and the other of aluminium alloy, both designed for the same angle of twist per unit length when subjected to the same torque.* $G_s = 3G_{al}$ *and* $\gamma_s = 2\gamma_{al}$, *where* γ *specific weight.*

Solution: Weight of steel shaft, $W_s = \frac{\pi}{4} d_s^2 l_s \gamma_s$,

Weight of aluminium shaft, $W_{al} = \frac{\pi}{4} d_{al}^2 l_{al} \gamma_{al}$

Now $T_s = T_{al} = T$,

and $l_s = l_{al} = l$

$$\frac{\theta}{l} = \frac{T}{JG}$$

$$J_s = \frac{\pi d_s^4}{32}, \quad J_{al} = \frac{\pi d_{al}^4}{32}$$

$$\left(\frac{\theta}{l}\right)_s = \frac{32T}{\pi d_s^4 G_s},$$

$$\left(\frac{\theta}{l}\right)_{al} = \frac{32T}{\pi d_{al}^t G_{al}}$$

$$\left(\frac{\theta}{l}\right)_s = \left(\frac{\theta}{l}\right)_{al}$$

$$\frac{1}{d_s^4 G_s} = \frac{1}{d_{al}^4 G_{al}}$$

$$\left(\frac{d_s}{d_{al}}\right)^4 = \frac{G_{al}}{G_s} = \frac{1}{3} \text{ or } \left(\frac{d_s}{d_{al}}\right)^2 = \frac{1}{\sqrt{3}}$$

$$\frac{W_s}{W_{al}} = \left(\frac{d_s}{d_{al}}\right)^2 \times \frac{\gamma_s}{\gamma_{al}} = \frac{1}{\sqrt{3}} \times 3 = \sqrt{3} = 1.732$$

Example 16.11 *A solid aluminium shaft 1m long and 45 mm outside diameter is to be replaced by a tubular steel shaft of the same length and the same outside diameter so that either shaft could carry the same torque and have the same angle of twist over the whole length. What must be the inner radius of the tubular shaft?* $G_s = 3\ G_{al} = 84$ *GPa.*

Solution: Given: $l = 1\text{m}, d_0 = 45\text{mm}, d_i = ?$

$$l_{al} = l_s, \ T_{al} = T_s, \ \theta_{al} = \theta_s$$

$$J_{al} = \frac{\pi d_0^4}{32} = \frac{\pi}{32} \times 45^4 \text{ mm}^4$$

$$J_s = \frac{\pi}{32}(d_0^4 - d_i^4) = \frac{\pi}{32}(45^4 - d_i^4) \text{ mm}^4$$

$$\frac{T}{J} = \frac{G\theta}{l}, \quad \frac{\theta}{l} = \frac{T}{GJ}$$

$$\left(\frac{T}{GJ}\right)_{al} = \left(\frac{G\theta}{GJ}\right)_s$$

$$\frac{32T_{al}}{G_{al} \times \pi \times 45^4} = \frac{32T_s}{G_s \times \pi \times (45^4 - d_i^4)}$$

$$\frac{G_{al}}{G_s} = \frac{45^4 - d_i^4}{45^4} = \frac{1}{3}$$

$$3(45^4 - d_i^4) = 45^4$$

$$3d_i^4 = 45^4\,(3-1) = 8201250$$

$$d_i^4 = 2733750$$

$$d_i = 40.66 \text{ mm}$$

$$r_i = 20.33 \text{ mm}$$

Example 16.12 *A solid shaft is to transmit 300 kW at 80 rpm. The shear stress is not to exceed 60 MPa. Find the shaft diameter. Calculate the percentage saving in weight if this shaft were replaced by a hollow one whose internal diameter equals 60 per cent of external diamter. The length, material and maximum shear stress remain the same.*

Solution: Given P = 300 kW, N = 80 rpm, $\tau = 60$ MPa, k = 0.6

Power transmitted $$P = \frac{2\pi NT}{60\times10^3} \text{ kW}$$

$$T = \frac{300\times60\times10^3}{2\pi\times80} = 35809.86 \text{ N.m}$$

For the hollow shaft, $$J_h = \frac{\pi}{32}\left(d_0^4 - d_i^4\right) = \frac{\pi}{32} d_0^4\,[1-(0.6)^4]$$

$$= 0.0272\times\pi d_0^4$$

For the solid shaft, $$J_s = \frac{\pi}{32} d^4$$

Now $$T_s = T_h \text{ and } \tau_h = \tau_s$$

$$\frac{T}{J} = \frac{\tau}{r}$$

or $$\frac{T}{\tau} = \frac{J}{r}$$

or $$\frac{j_h}{d_0} = \frac{J_s}{d}$$

$$\frac{0.0272\times\pi d_0^4}{d_0} = \frac{\frac{\pi}{32}d^4}{d}$$

$$d^3 = 0.8704\,d_0^3$$

Also $$\tau = \frac{16T}{\pi d^3}$$

$$60\times10^6 = \frac{16\times35809.86}{\pi d^3}$$

$$d^3 = 3039.6\times10^{-6}$$

$$d = 14.486\times10^{-2}\ \text{m}$$

$$\therefore \quad 3039.6\times10^{-6} = 0.8704\, d_0^3$$

$$d_o = 15.17\times10^{-2}\ \text{m}$$

$$d_i = 9.10\times10^{-2}\ \text{m}$$

Area of solid shaft, $$A_s = \frac{\pi}{4}\times(14.486\times10^{-2})^2 = 164.81\times10^{-4}\,\text{m}^2$$

Area of hollow shaft, $$A_h = \frac{\pi}{4}\times[15.17)^2-(9.10)^2]\times10^{-4} = 115.7\times10^{-4}\,\text{m}^2$$

For same length and material of solid and hollow shaft, the weights are proportional to areas. Hence,

$$\text{Percentage saving in weight} = \frac{A_s - A_h}{A_s}\times100$$

$$= \frac{164.81-115.7}{164.81} = 29.8\%$$

Example 16.13 *A hollow shaft of diameter ratio 0.6 is required to transmit 450 kW at 100 rpm, the maximum torque being 20 % greater than the mean value. The shear stress is not to exceed 56 MPa and the angle of twist in a length of 3m is not to exceed 1.5 degrees. Calculate the external diameter satisfying above conditions.*

Take G = 84 GPa

Solution: Given k = 0.6, P = 450 kW, N = 100 rpm, overload = 0.2, $\tau_{max} = 56 MPa$, 1 = 3m, $\theta = 1.5°$.

Power transmitted, $$P = \frac{2\pi N T_{mean}}{60\times10^3}$$

$$T_{mean} = \frac{4580\times60\times10^3}{2\pi\times100} = 42972\,\text{N.m}$$

$$T_{max} = (1+0.2)\times42972 = 51566.2\ \text{N.m}$$

(*i*) For a hollow shaft,

$$\tau_{max} = \frac{16T_{max}}{\pi d_0^3\,(1-k^4)}$$

$$d_0^3 = \frac{16\times51566.2\times10^3}{\pi\times56\times[1-(0.6)^4]} = 5388\times10^3$$

$$d_o = 175.3\ \text{mm}$$

(*ii*) $$\frac{T}{J} = \frac{G\theta}{l}$$

$$\frac{T}{\frac{\pi}{32}d_0^4(1-k^4)} = \frac{G\theta}{l}$$

$$d_0^4 = \frac{51566.2\times3\times32\times180}{\pi[1-(0.6)^4]\times84\times10^9\times\pi\times1.5} = 82322.7\times10^{-8}$$

$$d_0 = 0.1694 \text{ m or } 169.4 \text{ mm}$$

Adopt, $d_0 = 175.3$ m

Example 16.14 *A solid circular shaft is to transmit 250 kW at 120 rpm. (a) If shear stress is not the exceed 63 MPa, find its diameter. (b) If this shaft is replaced by a hollow shaft whose diameter ratio is 0.6, what percentage saving in weight shall be obtained keeping the material and maximum shear stress same.*

Solution: Given: $P = 250$ kW, $N = 120$ rpm, $\tau_{max} = 63$ MPa, $k = 0.6$

(*a*)

$$P = \frac{2\pi NT}{60\times10^3}$$

$$T = \frac{250\times60\times10^3}{2\pi\times120} = 19894 \text{ N.m}$$

Now

$$T = \frac{\pi}{16}d^3\tau$$

$$d = 3\sqrt{\frac{16\times19894}{\pi\times63\times10^6}} = 117 \text{ mm}$$

(*b*) For $\tau_s = \tau_h$ and $T_s = T_h$

$$d^3 = d_0^3(1-k^4)$$

$$d_0^3 = \frac{117^3}{1-(0.6)^4} = 1840088$$

$$d_0 = 122.54 \text{ mm}$$

$$\frac{W_h}{W_s} = \frac{A_h}{A_s} = \frac{d_0^2-d_i^2}{d^2} = \frac{(122.54)^2[1-(0.6)^2]}{117^2} = 0.702$$

Percentage saving in weight $= \left(1-\frac{W_h}{W_s}\right)\times100$

$= (1 - 0.702) \times 100$

$= 29.8$ %

Example 16.15 *A hollow steel shaft is to be replaced by an aluminium shaft of the same external diameter. The material of steel shaft is 25% stronger than aluminium shaft in shear. Calculate the diameter ratios. Also find the percentage saving in weight assuming that steel is 3 times heavier than aluminum.*

Solution:

Let d_0, d_i = outside and inside diameters of shaft respectively

$d = d_0$ diameter of aluminium shaft

For same torpue to be tramsmitted by both the shafts,

$$T_s = T_{al}$$

$$\frac{\pi}{16} d_0^3 (1-k^4)\tau_s = \frac{\pi}{16} d_0^3 \tau_{al}$$

$$(1-k^4)\tau_s = \tau_{al}$$

Now $\tau_s = 1.25\,\tau_{al}$

$$1.25\,(1 - k^4) = 1$$

$$1 - k^4 = 0.8$$

$$k^4 = 0.2$$

$$k = 0.6687$$

Weight $\propto$ (area of cross-section × density)

$$W_s = \frac{\pi}{4} d_0^2 (1-k^2)\rho_s$$

$$W_{al} = \frac{\pi}{4} d_0^2 \rho_{al}$$

$$\frac{W_s}{W_{al}} = (1-k^2)\frac{\rho_s}{\rho_{al}}$$

$$= [1 - (0.6687)^2] \times 3 = 1.65852$$

or $$\frac{W_{al}}{W_s} = 0.603$$

Percentage saving in weight = (1 – 0.603) × 100 = 39.7 %

Example 16.16 *A solid shaft of medium carbon steel 100 mm diameter is to be replaced by a hollow shaft of alloy steel. The shear strength of alloy steel is 30% greater than medium carbon steel. The power to be transmitted is increased by 20 % and the speed of rotation decreased by 5%. Determine the maximum internal diameter of hollow shaft if its external diameter is to be 100 mm.*

Solution:

Given d_s = 100 mm, $d_0 = d_s$ = 100 mm, d_i = ? $\tau_h = 1.3\tau_s$, $P_h = 1.2\,P_s$, $N_h = 0.95\,N_s$

$$P = \frac{2\pi NT}{60 \times 10^3}$$

$$P_s = \frac{2\pi N_s T_s}{60 \times 10^3}$$

$$P_h = \frac{2\pi N_h T_h}{60 \times 10^3}$$

$$ph = 1.2\,P_s$$

$$\therefore \quad 2\pi N_h T_h = 1.2 \times 2\pi N_s T_s$$

$$N_h T_h = 1.2\,Ns\,T_s$$

$$0.95\ T_h = 1.2\ T_s$$

or
$$T_h = 1.263\ T_s$$

$$T_s = \frac{\pi}{16} d_s^3\, \tau_s$$

$$T_h = \frac{\pi}{16} d_s^3\,(1-k^4)\,\tau_h$$

$$\frac{T_h}{T_s} = \frac{(1-k^4)\,\tau_h}{\tau_s} = 1.3(1-k^4)$$

Substituting Eq. (1) in (2), we have

$$1.263 = 1.3\ (1 - k^4)$$
$$0.9715 = 1 - k^4$$
$$k^4 = 0.02846$$
$$k = 0.41$$
$$d_i = 0.41 \times 100 = 41 \text{ mm}$$

MULTI-CHOICE QUESTIONS

1. At the neutral axis of a beam
(*a*) bending stress is zero (*b*) bending stress is maximum
(*c*) shear force is zero (*d*) shear force is maximum

2. In a beam subjected to simple bending, the bending stress is
(*a*) directly proportional to the distance from the neutral axis
(*b*) inversely proportional to the distance from the neutral axis
(*c*) directly proportional to the square of the distance from the neutral axis
(*d*) inversely proportional to the square of the distance from the neutral axis.

3. In the simple bending of beam theory, the plane of leads
(*a*) does not coincide with the centroidal plane
(*b*) coincides with the centroidal plane
(*c*) is inclined at 45° to the centroidal plane.
(*d*) is inclined at 90° to the centroidal plane.

4. In the simple bending of beams, the bending stress is the beam varies
(*a*) linearly (*b*) parabolically
(*c*) hyperbolically (*d*) elliptically

5. the moment of resistance (m_r) of a beam is
(*a*) directly proportional to the section modules (z) of the beam
(*b*) $m_r \times \frac{1}{z}$ (*c*) $m_r \times z^2$ (*d*) $m_r \times \frac{1}{z^2}$

6. The top fibres of a cantilever beam carrying end load develope
(*a*) compressive stress (*b*) tensile stress
(*c*) strear stress (d) bending stress

7. The shear stress in a circular shaft under torsion varies
(*a*) uniformly (*b*) lineraly

(*c*) parabolically (*d*) hyperbolically

8. The maximum shear stress in a shaft under torsion occurs at the

(*a*) centre of shaft (*b*) outermost fibres

(*c*) mid-radius (*d*) square root of radius

9. The torsional rigidity of a shaft is given by

(*a*) $\frac{T}{J}$ (*b*) $\frac{T}{r}$ (*c*) $\frac{T}{\theta}$ (*d*) $\frac{T}{l}$

10. The torsional section modules of a circular shaft is

(*a*) $\frac{J}{\theta}$ (*b*) $\frac{J}{l}$ (*c*) $\frac{J}{r}$ (*d*) $\frac{T}{G}$

Answers

1. (*a*) 2. (*a*) 3. (*b*) 4. (*a*) 5. (*a*) 6. (*b*) 7. (*b*) 8. (*b*) 9. (*c*) 10. (*c*)

REVIEW QUESTIONS

1. State parallel axis theorem.

2. What is neutral axis.

3. State the assumptions made in simple bending theory.

4. Define section modulus.

5. Define bending stiffness

6. What is moment of resistance ?

7. What are the assumptions made in deriving the torsion formula ?

8. Define torsional stiffness.

9. Define torsional section modulus.

10. White the torsion formula and give S.I. units of various terms.

EXERCISES

16.1 An inverted *T*-beam of 4 m span is simply supported at the ends. Its cross-section is 300 mm × 250 mm × 10 mm. If permissible stresses in tension and compression and 60 MPa and 80 MPa respectively, calculate the safe uniformly distributed load on the whole span.

16.2 A rolled steel joist 400 mm × 200 mm × 20 mm is freely supported on a span of 4 m. The flanges are strengthened by two 250 mm × 20 mm plates, one riveted to each flange. Calculate (*a*) the maximum central load the beam can carry if the stress in the plate is not to exceed 125 MPa, (*b*) the minimum length of the strengthening plates required to restrict the maximum stress to 125 MPa.

16.3 A rectangular cross-section beam has a width of 100 mm. Determine the depth of the beam so that maximum bending stress in the beam does not exceed 35 MPa. The maximum bending moment in the beam is 8.4 kN.m.

16.4 A cast iron water main pipe 12 m long, 500 mm inside diameter and 25 mm wall thickness runs full of water. It is supported at its ends. If the density of cast iron and water are 7200 kg/m^3 and 1000 kg/m^3 respectively, calculate the maximum stress induced in cast iron.

16.5 A uniform heavy girder of length 10 m is to be placed over two supports 6 m apart. Calculate the length of overhang for the girder at both ends so that the maximum bending moment, positive or negative, due to self weight shall be as small as possible.

16.6 A solid steel shaft is required to transmit 45 kW at 150 rpm. If the permissible shear stress is 75 MPa, find the diameter of the shaft.

16.7 A shaft is required to transmit 7.5 kW at 600 rpm. The maximum torque exceeds the mean torque by 40%. Calculate the diameter of the shaft if shear stress is not to exceed 70 MPa. Also calculate the angle of twist in a length of 2m. Take G = 84 Gpa.

16.8 A hollow steel shaft of external diameter 400 mm and metal thichness 25 mm is required to transmit power at 240 rpm. If the shear stress is not to exceed 63 MPa and maximum torque exceeds the mean by 35%, find the power it can transmit.

16.9 A solid shaft of 100 mm diameter is to be replaced by a hollow shaft with internal diameter equal to 40% of external diameter. Find the size of hollow shaft and saving in material. Both the shafts are made of same material.

17 INTRODUCTION TO MANUFACTURING SYSTEMS

17.1 FUNDAMENTALS OF NUMERICAL CONTROL

Numerical control, NC, can be defined simply as control by numbers. Electronics Industries Association defines NUMERICAL CONTROL as " a system in which actions are controlled by the direct insertion of numerical data at some point. The system must automatically interpret at least some portion of these data." In numerically controlled machine tools, the input information for controlling the machine tool motion is proved by means of punched paper tapes, plastic tapes (Mylar), floppy disk, hard disk or magnetic tapes in a coded language. Thus, with numercial control, the operation and motions of a machine tool are controlled electronically. NC machine tools are thus automated production machines.

The principles of NC were established in 1950s. However, with the availability of low-cost programmable control, based around the microprocessor in 1970s, this technology diffused widely. This opened up the development of the concept of Computer Numerical Control (CNC), and other functions such as automatic tool change or part manipulation. All this led to highly sophisticated, multipurpose machining centres, complete with a range of support functions such as tool change, head change, transport and manipulation and so on-all under computer control. The next development was the idea of Direct (or distributed) Numerical Control (DNC), in which more than one machine tool (alongwith associated functions) could be grouped into a manufacturing cell under the overall control of a larger, supervisory computer. The early numerically controlled machine tools were milling and profiling machines that provided faster and less expensive means for producing aircrafts parts. Now, NC, has been applied to other machine tools, such as lathes, drilling and boring machines, welding and flame cutting machines, punching machines and inspection devices.

17.2 WORKING OF NC MACHINE TOOL

To understand the working of a *NC* machine tool, let us refer to Fig. 17.1. The first two steps, component drawing and process planning are similar in both operator and *NC* machine tools. In the operator controlled machine tools, the operator controls the cutter position during machining. He also makes the necessary adjustments and corrections to produce the desired component. However, in numerically controlled system, the place of the operator is taken by the data processing part of the system and the control unit. In the data processing unit, the co-ordinate information regarding the component is recorded on a tape by means of a teleprinter. Tape is fed to the control unit which sends the position command signals to slid-way transmission elements of the machine tool. At the same time, the command signal is constantly compared with the actual position achieved, with the help of position feedback signal derived from automatic monitoring of the machine tool side position. The difference in the two signals, if any, is corrected until the desired component is produced.

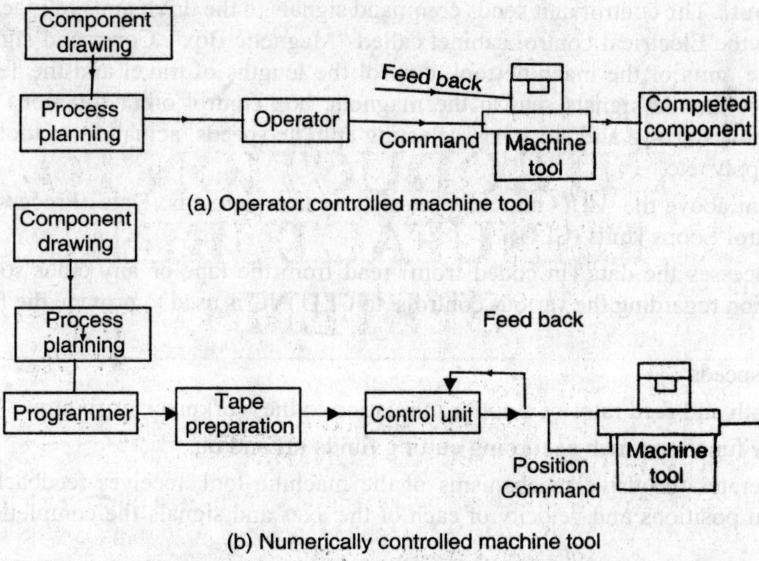

(a) Operator controlled machine tool

(b) Numerically controlled machine tool

Fig. 17.1 Operator and numerically controlled machine tools.

The conventional *NC* system consists of basically two parts: **Machine Tool and Machine Control Unit (MCU).**

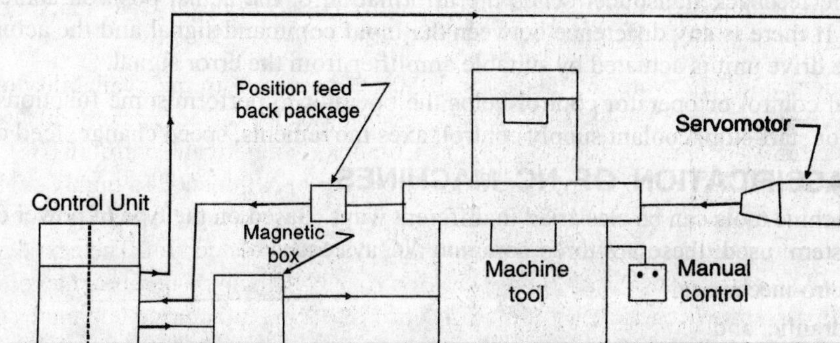

Fig. 17.2 *Main Elements of a NC machine*

The main elements of a NC machine tool are shown in Fig. 17.2. These are:

1. The machine control unit (MCU), also known as N.C. console or Director.
2. The drive units, or actuators.
3. The position feed back package
4. Magnetic Box
5. Manual control

The hardware also includes the associated circuits.

The functional elements involved in *MCU* are:

1. **Data input :** The instructions for manufacturing the component are written in a Coded language (Paper tape is the most commonly used device) are ready by a tape reader.
2. **Data processing :** The instructions undergo electronic processing resulting in information in the form of electrical signals (pulsed commands).

3. **Data Output :** The control unit sends command signals to the drive units of machine tool and also to the Electrical control cabinet called "Magnetic Box". Command signals sent to the drive units of the machine tool, control the lengths of travel and the feed rates, while the command signals sent to the magnetic box control other functions such as: spindle motor starting and stopping, selecting spindle speeds, actuation of tool change, coolant supply etc.

It is clear from above the *MCU* basically consists of two elements: Date Processing Unit (DPU) and the Control Loops Units (CLU).

The DPU processes the data (in coded from) read from the tape or any other source and passes the information regarding the various controls to CLU. NC is used to provide the following controls:

– all cutting speeds
– complete path and feed rates of a cutter in relation to the workpiece or fixture
– all auxiliary functions such as turning cutting fluids off and on.

The *CLU* operates the drive mechanisms of the machine tool, receives feedback signals regarding the actual positions and velocity of each of the axes and signals the completion of the operations.

The *DPU* reads and process the data sequentially. When a line has completed execution as signalled by C*LU*, and *DPU* reads and process the next line of the programme and so on.

To check whether the required lengths of travel have been obtained, a feedback transducer is provided. The feedback transducer sends the information of the actual position achieved to the control unit. If there is any difference between the input command signal and the actual position achieved, the drive unit is actuated by suitable amplifier from the error signal.

Manual control or operator control helps the operator to perform some functions manually such as: motor start-stop, coolant supply control, axes movements, speed change, feed change etc.

17.3 CLASSIFICATION OF NC MACHINES

NC machine tools can be classified in different ways. Based on the type of power drive or the actuation system used, these are three common *NC* systems:

1. Electro-mechanical
2. Hydraulic, and
3. Pneumatic

The range of the machines performance capabilities and its applications will normally depend upon the type of the power source.

Hydraulic power provides the largest power. The movement of the machine slides is more uniform in speed. However, the drawbacks are: Higher cost, noise, hydraulic contamination from leaking fluid and the need for additional equipment such as reservoir, intensifier, valves etc.

Pneumatic power is least expensive, because the shop air can be tapped for this purpose. However, the motion of the worktable is not uniform.

Electro-mechanical drives are the most precise and the most commonly used power sources.

The group of electro-mechanical, hydraulic or 'pneumatic equipment used to control the motion of an NC machine is called the "Servo".

Based on control system features, NC machines are classified as:

1. Point-to-Point system
2. Straight line system
3. Contour system

These three systems are illustrated in Fig. 17.3.

In point-to-point system, the machining is done at specific positions. The working-piece remains unaffected as the tool moves from one positions to the next. This system is the simplest. In Fig. 17.3(a), after drilling the hole at position A, the tool moves to position B, along the dotted line. A drilling machine is the best example of point-to-point system.

Straight line system, Fig.17.3 (b) is an extension of point-to-point system. Here the tool moves at a controlled feed rate in one axis direction at a time. The examples of straight line system are; stepped turning on lathe, pocket milling etc.

In contour or continuous path system, Fig 17.3 (c), there are continuous, simultaneous and co-ordinated motions of the tool and the workpiece along different co-ordinate axes. The various profiles, contours and curved surfaces are machined by this system.

Depending on the feedback, NC machine systems are of two types:

1. Open loop system
2. Closed loop system

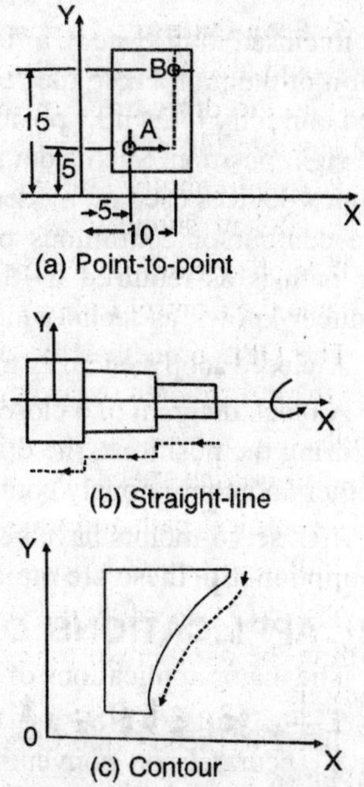

Fig. 17.3 NC Control System.

As discussed above, a command signal is sent to the machine tool to carry out a certain operation. In the open loop system, there is no 'feedback' and no return signal to indicate whether the tool has reached the correct position at the end of the operation, or not, Hence, there are no means of knowing whether there is an error or not between the input command signal and the result achieved. An example of open loop system is a co-ordinate drilling machine.

Fig.17.4 (a) illustrates an "Open loop system". The command signal from the MCU is given to the servo-motor (here a stepper motor). The motor is driven a precise angular rotation for every pulse issued by CLU. So, the response of the motor is in incremental steps. An incremental step of 1.8° is common. This will result in a corresponding linear movement of the lead screw (depending upon its lead) and hence of the machine slide.

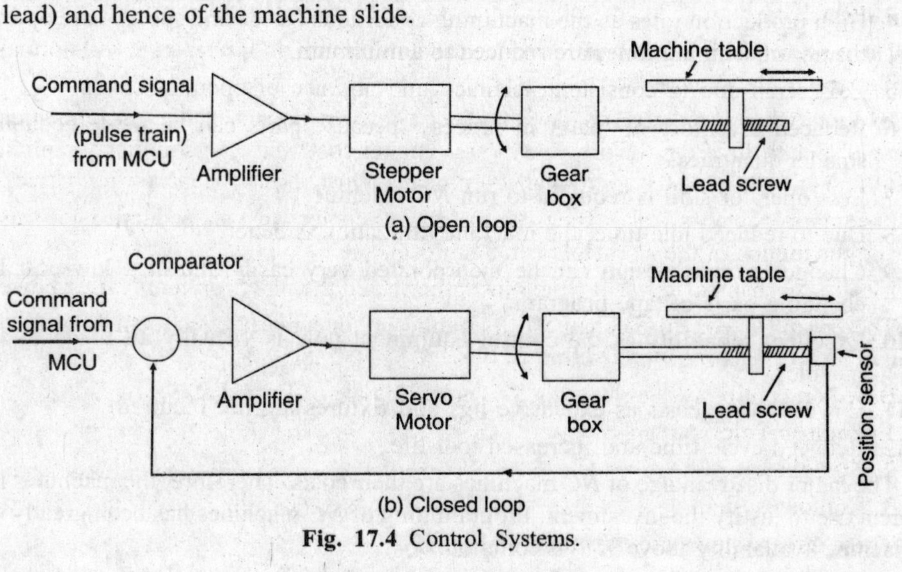

Fig. 17.4 Control Systems.

In closed loop system, a 'feedback' is built into the system, which automatically monitors the position of the tool. From the feedback signal to the control unit, it is known whether the tool has moved correctly to its new position or not. If not, its position is automatically controlled until it is in the right position. Such a device is known as a position feed back transducer. Common position feed back devices used on modern *NC* machines are : shaft encoders, linear scales or Inductosyns. In the contour or continuous path system, velocity control is also essential to ensure that the cutter path is as required by the profile. In this system, a velocity feed back is provided by a transducer known as tacho-generator attached to the feed drive system.

A closed loop system is more expensive than an open loop system.

A block diagram of a closed-loop system is shown in Fig. 17.4(b). Since, there is a continuous monitoring the positions, the drive is a continuous-positions device. These servomotors provide a smoother and continuously controllable movement of the work-table.

A.C. servo-motors have better reliability, better performance-to-weight ratio and lower power consumption. So, these are rapidly replacing the *DC* servomotor on new *NC* machine.

17.4 APPLICATIONS OF NC MACHINES

The major applications of NC machines are :

1. For parts, which are complex and it will not be possible to manufacture them very accurately on conventional machines, due to human error involved.
2. For parts which are frequently subjected to design changes.
3. Repetitive and precision quality parts which are to be produced in low to medium batch quantity.
4. In situations where the investment on tooling and fixture inventory will be high if the parts are made on conventional machine tools.
5. To cut down 'lead time' in manufacture.

17.5 ADVANTAGES AND DISADVANTAGES OF NC MACHINES

Advantages:

1. Greater accuracy.
2. Lesser production cost per piece due to reduction in lead time and also set up time.
3. Improved product quality and provision of high order of repeatability.
4. High production rates as the machining conditions (feeds and speeds) are optimised and the non-machining times are reduced to a minimum.
5. Less scrap due to consistent accuracy and absence of operator errors.
6. Reduced inventory of 'parts in process' because parts can be made economically in smaller quantities.
7. Less operator skill is required to run *NC* machine.
8. Due to reduced idle time, the machine utilization is better.
9. Changes in part design can be incorporated very easily and at a low cost by simply changing parts of tape program.
10. Excellent reliability as the control equipment now is virtually all made of solid state modules.
11. Lower tooling costs as expensive jigs and fixtures are not required.
12. Reduced cycle time and increased tool life.

The major disadvantage of *NC* machines are their costs. Therefore, the machines must have sufficient use to justify the investment. Programming of *NC* machines has been greatly simplified and machine availability above 95% is common.

17.6 COMPUTER NUMERICAL CONTROL (*CNC*)

Computer has been associated with *NC* right from the beginning, and over the years this association has resulted in many new conceptions like : Computer Numercial Control (*CNC*). Direct Numerical Control (*DNC*), Computer Aided Design (*CAD*), Computer Aided Manufacture (*CAM*), Adaptive Control (*AC*), and Integrated Manufacturing System.

It is clear from the above discussion that in conventional *NC* systems, the instructions (program) stored on a paper tape (in a coded language) are read by tape reader. The instructions are processed by *DPU* and the command signals are controlled by *CLU*. There is no provision for storage or memory in *MCU*. For every new component, the tape needs to be loaded again, since the program is not stored in *MCU* and so can not be called back. The entire data input and data handling sequence including control functions are determined only by the fixed circuit interconnections of *DPU* and *CLU*. Any modifications will need a corresponding change in the circuitry of the system. Thus, the conventional NC system, which is also known as "Hard-wired system" is a rigid one. Reloading of tape each time a component is to be manufactured results in loss of time and sometimes errors in reading. Also, modifications and editing in the program is not possible.

To overcome the above mentioned drawbacks of conventional *NC* systems, Computer Numerical Control (*CNC*) has been developed. Here, a bare minimum of electronic hardware is used and the functions of *MCU* of the conventional *NC* system are partly or wholly taken over by a dedicated micro-computer of the machine tool.

The control program (called Executive programme) is stored which performs all the basic *NC* functions. Thus, in this system, the machine control data comes directly from a micro-mini-computer, instead of from a continuously read tape (as in conventional *NC* system). By changing or modifying the executive programme (through paper tape), the system can be used for another machine tool. Thus, this system is flexible or 'soft wired system'. For repetitive production, the memory of the dedicated mini-computer can be utilized for temporary storage of the part programme being run. However, tape or disc is still the medium of transmitting the part programme from the programming computer to the dedicated micro-mini-computer of the machine tool, Fig.17.6.

However, the instructions (program) can also be entered in the memory of the micro-computer with the help of a key-board (known as direct programming). As Computer-Aided Design (*CAD*) systems become more common generating data bases, the use of tape for storing program will decrease with the corresponding increase in the use of floppy disk, hard disk and other computer storage devices for this purpose. Again, computer-aided programming is being replaced by computer-aided manufacturing (*CAM*), where the programs for the *NC* machine tools can be generated on the graphic terminal directly from the data base created by *CAD* (*CAD* file), with the help of *CAD/CAM* software.

17.6.1 General Configuration of *CNC* System

The general configuration of a *CNC* system is shown in Fig.17.5.

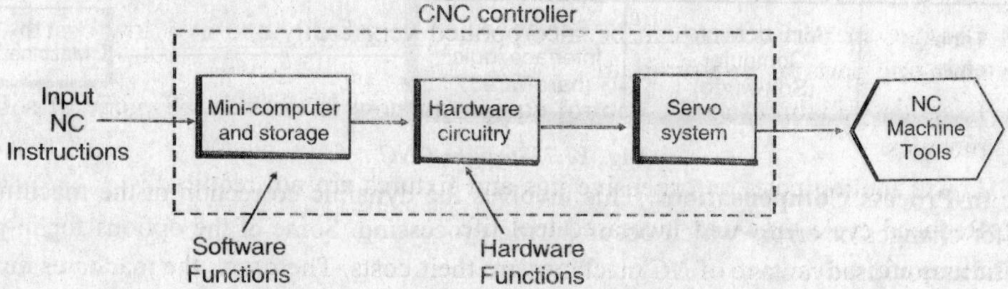

Fig. 17.5 *General Configuration of CNC.*

The principle functions of *CNC* are:

1. Machine tool control
2. In-process compensation
3. Improved programming and operating features
4. Diagnostics

1. Machine Tool Control: This is the primary function of the *CNC* system. This involves conversion of the part program instructions into machine tool motions through the computer interface and servosystem. There are two controller designs in *CNC*: hybrid *CNC* and straight *CNC*. In the hybrid *CNC* system (Fig. 17.6), the controller consists of the soft-wired computer plus hard-wired logic circuits. The hard-wired components perform functions, such as feed rate generation and circular interpolation. The computer performs the remaining control functions plus other duties not normally associated with a conventional hard-wired controller.

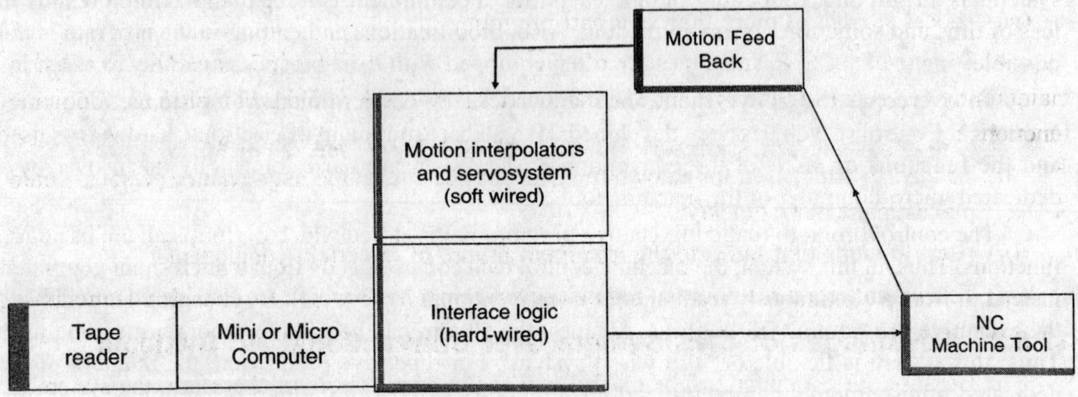

Fig. 17.6 *Hybrid CNC.*

The straight *CNC* system (Fig. 17.7) uses a computer to perform all the *NC* functions. The only hard-wired elements are those required to interface the computer with the machine tool and the operator's console. Interpolation, tool position feedback, and all other functions are performed by computer software.

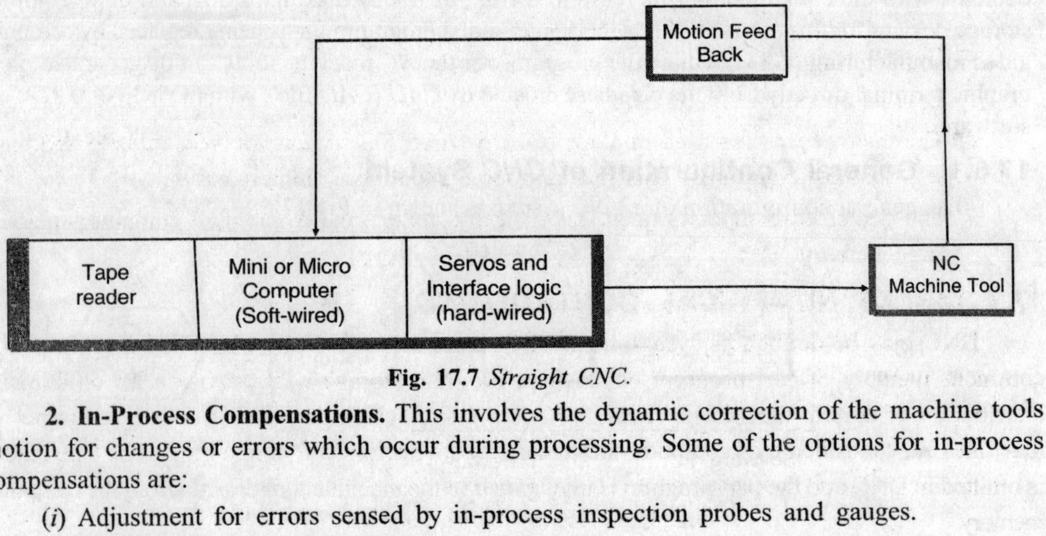

Fig. 17.7 *Straight CNC.*

2. In-Process Compensations. This involves the dynamic correction of the machine tools motion for changes or errors which occur during processing. Some of the options for in-process compensations are:

(*i*) Adjustment for errors sensed by in-process inspection probes and gauges.

(*ii*) Recomputation of axis positions

(*iii*) Offset adjustments for tool radius and lengths.

(*iv*) Adaptive control adjustments to speed and /or feed.

(*v*) Computation of predicted tool life

3. Improved programming and operating features

The programming and operating features of *CNC* are:

(*i*) Editing of part programs at the machine

(*ii*) Graphic display of the tool path of verify the tape

(*iii*) Circular, parabolic, and cubic interpolation.

(*iv*) Support of both metric and SI units.

(*v*) Use of specially written subroutines.

(*vi*) Manual data input.

(*vii*) Local storage of more than one part program.

4. Diagnostics : *CNC* machines are often equipped with a diagnostics capability to assist in maintaining and repairing the system. The diagnostics subsystem would accomplish the following functions :

(*i*) To identify the reason for a down time occurrence so that the maintenance personal could make repairs more quickly.

(*ii*) Alert to signs that indicate the imminent failure of a certain component.

(*iii*) To contain a certain amount of redundancy of components which are considered unreliable.

17.6.2 Advantages of *CNC* System over Conventional *NC* Systems

1. Because the computer can be readily and easily reprogrammed, therefore, the system is very flexible. The machine can manufacture a part followed by other parts of different designs.
2. More versatility. Editing and debugging programs, reprogramming; and plotting and printing part shapes are simpler.
3. Program to manufacture a component can be easily called. This saves time and eliminates errors due to tape reading.
4. Greater accuracy
5. Ease of operation
6. The microprocessors used in place of hard-wired *NC* circuits are very reliable and have self diagnostic features. This makes trouble shooting extremely easy.
7. The dedicated computers need less memory storage and even small computers possess large memory.

17.7 DIRECT NUMERICAL CONTROL-*DNC*

DNC may be defined as system connecting a set of numerically controlled machines to a common memory of part program or machine program storage with provision for on-demand distribution of data to machines. Thus *DNC* is a manufacturing system in which a number of machines are controlled by a computer through direct connection and in real time. The tape reader is omitted in DNC and the part program is transmitted to the machine tool directly from the computer memory.

The configuration of the basic *DNC* system is shown in Fig. 17.8 . A DNC system consists of four basic components.

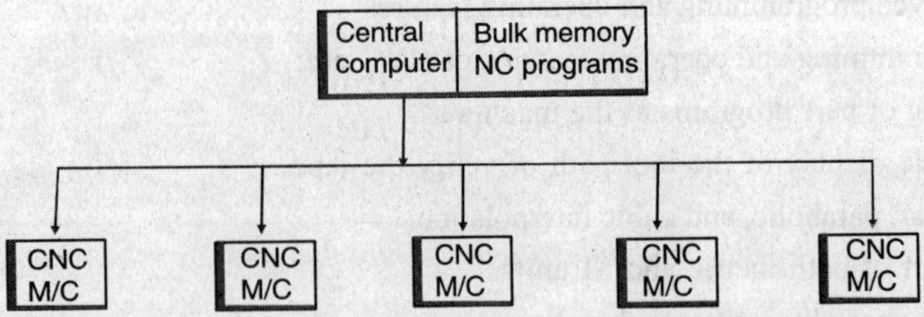

Fig. 17.8 *General configuration of DNC system.*

1. Central computer
2. Bulk memory to store the *NC* part programs
3. Telecommunication lines
4. Machine tools

17.7.1 Advantages of DNC

The major advantages of DNC system are:

1. Elimination of punched tapes and tape readers
2. Greater computational capability and flexibility
3. Convenient storage of *NC* part programs in computer files
4. Programs stored as CLFILE, i.e., cutter path data rather than post processed programs for specific machine tools.
5. Reporting of shop performance
6. Establishes the framework for the evolution of the future computer automated factory.

EXERCISE

17.1 Justify the necessity for *NC* machines.

17.2 Define *HC, CNC* and *DNC.*

17.3 Explain the working of a *NC* machine tool with the help of a neat diagram.

17.4 What are the main components of a *NC* machine tool ? Explain their functions.

17.5 Which are the two control loop systems for a *NC* machine ? Explain briefly.

17.6 How do you classify *NC* machine tools ?

17.7 Give the applications and advantages of *NC* machines.

17.8 List the advantages of *CNC* systems over the *NC* systems.

17.9 List the main advantages of *DNC.*

17.10 Compare the *NC* and *CNC* machine tools.

APPENDIX-I

Sample Question Paper of MDU, Rohtak (Haryana)

B.Tech. Ist Semester (F. Scheme)

Examination – December, 2009

BASICS OF MECHANICAL ENGINEERING

Paper : ME - 101-F

Time : Three hours ***Maximum Marks : 100***

Before answering the question, candidates should ensure that they have been supplied the correct and complete question paper. No complaint in this regard, will be entertained after examination.

Note : Attempt five questions in all. **Q.** No. **1** is **compulsory.** Attempt at least **one** question from each Section. All questions carry equal marks.

1. (i) Why quick return machanism is used in shaper ?
(ii) Define zeroth law of thermodynamics.
(iii) What are open, closed and isolated systems ?
(iv) Why we use Psychometric charts ?
(v) What is air conditioning ?
(vi) Define specific speed of turbine.
(vii) Explain rope and chain drives.
(viii) What is Poisson's ratio ?
(ix) What is MCU in NC machines ?
(x) Define stress and strain. $2 \times 10 = 20$

SECTION – A

2. (a) Explain drilling m/c with diagram. 10
(b) Expain different types of chips in metal cutting. 10

3. (a) Define laws of thermodynamics. 10
(b) Explain throttling calorimeter. 10

SECTION – B

4. (a) Explain simple refrigeration vapour compression cycle. 15
(b) What is coefficient of performanc (COP) ? 05

5. (a) Explain pelton turbine with diagram. 10
(b) Explain rotary water pumps. 10

SECTION – C

6. (a) Explain single plate clutch. 15
(b) Explain belt drive for power transmission. 05

7. (a) What is Hooke's Law ? Derive an expression for relationship between elastic constants. 15
(b) Explain stress & strain. 05

SECTION – D

8. Explain NC machines in detail. 20

9. (a) Write down difference between NC & CNC machines. 10
(b) Expalain manufacturing systems. 10

MDU
December, 2009
Basics of Mechanical Engineering (ME-101-F)

Solutions.

Sol. 1. (*i*) Refer to art. 1.4.5.

(*ii*) Refer to art. 2.13.

(*iii*) Refer to art 2.2.1.

(*iv*) Refer to art. 5.6.2.

(*v*) Refer to art. 5.6.

(*vi*) The specific speed of a turbine is defined as the speed of a geometrically similar turbine of such a size as to generate one unit power under one unit of head at maximum efficiency.

(*vii*) In a rope drive, either fibre or steel wire ropes are used for transmitting power. The ropes run on grooved pulleys.

For chain drive, refer to art. 12.7.

(*viii*) Refer to art. 13.11.

(*ix*) Refer to art. 17.2.

(*x*) Refer to art. 13.7.

Sol. 2. (*a*) Refer to art. 1.7.2.

(*b*) Refer to art. 1.9.5.

Sol.3. (*a*) Refer to art. 2.16, 2.24 and 2.29.

(*b*) Refer to art. 3.23 (3)

Sol. 4. (*a*) Refer to art. 5.2.

(*b*) Refer to art. 5.1.

Sol. 5. (*a*) Refer to art. 6.2.1.

(*b*) Refer to art. 6.7.2.

Sol. 6. (*a*) Refer to art. 2.12.2 (1)

(*b*) Refer to art. 12.4

Sol. 7. (*a*) Refer to art. 14.10.

(*b*) Refer to art. 13.7.

Sol. 8. (*a*) Refer to art. 17.2.

Sol. 9. (*a*) Refer to art. 17.6.2.

(*b*) Refer to art. 17.8.

B. Tech 2nd Semester "F Scheme" Examination – May, 2010

BASICS OF MECHANICAL ENGINEERING

Paper : ME-101-F

Time : Three hours ***Maximum Marks : 100***

Before answering the questions, candidates should ensure that they have been supplied the correct and complete question paper. No complaint in this regard, will be entertained after examination.

Note : Attempt *five* question in all, selecting *one* question from each Section. Question No. 1 is compulsory. All questions carry equal marks.

1. (*i*) Explain working principle of drilling machine.

(*ii*) Define internal energy and enthalpy.

(*iii*) Define refrigeration and air conditioning.

(*iv*) Why Francis turbine is called mixed flow turbine.

(*v*) Explain belt drive.

(*vi*) What is Poission's ratio ?

(*vii*) Explain *four* advantages of NC machine.

(*viii*) What is Zeroth law of thermodynamics.

(*ix*) Why we use Psycheometric charts ?

(*x*) What is the difference between spur and helical gear ? $2 \times 10 = 20$

SECTION – A

2. (*a*) Explain different types of chips in metal cutting? 10

(*b*) Define I, II and III law of Thermodynamics. 10

3. (*a*) What is throttling calorimeter ? 10

(*b*) Explain 5 operations of lathe machine. 10

SECTION – B

4. (*a*) Explain simple refrigeration vapour compression cycle. 15

(*b*) Define human comfort. 5

5. (*a*) Derive an expression for specific speed of turbine. 10

(*b*) Explain reciprocating water pumps. 10

SECTION – C

6. (*a*) Explain with neat diagram, multiplate clutch. 15

(*b*) Define different types of gear. 5

7. Explain elastic constants and their relationship. 20

SECTION – D

8. Explain NC machine in detail. 20

9. Define manufacturing system and compare NC and CNC machines. 20

MDU
May, 2010
Basics of Mechanical Engineering

Solutions.

Sol. 1. (*i*) Refer to art. 1.7.1

(*ii*) Refer to art. 3.13 and 3.9.

(*iii*) Refer to art. 5.1. and 5.6.

(*iv*) refer to art. 6.3.

(*v*) Refer to art. 12.4.

(*vi*) Refer to art. 13.11.

(*vii*) Refer to art. 17.5.

(*viii*) Refer to art. 2.13.

(*ix*) Refer to art. 5.6.2.

(*x*) Refer to art. 12.9.1.

Sol. 2. (*a*) Refer to art. 1.9.5.

(*b*) Refer to art. 2.16, 2.24 and 2.29.

Sol. 3. (*a*) Refer to art. 3.23 (3)

(*b*) Refer to art. 1.3.7.

Sol. 4. (*a*) Refer to art. 5.2.

(*b*) Refer to art. 5.6.3

Sol. 5. (*a*) Refer to art. 6.2.3.

(*b*) Refer to art. 6.6.

Sol. 6. (*a*) Refer to art. 12.12.2.

(*b*) Refer to art. 12.9.1.

Sol. 7. (*a*) Refer to art. 14.10

Sol. 8. Refer to art. 17.2.

Sol. 9. Refer to art. 17.6.2.

APPENDIX – II

SOLVED QUESTIONS FROM UNIVERSITY EXAMINATIONS

Example 1. *A simply supported horizontal beam, 5 m long carries concentrate load of 70, 90, 50 and 80 kN at distances 1, 3, 4 and 4.5 m respectively from the left hand support. Find support reactions and draw S.F. at B.M. diagrams.* **[KU, May 2006]**

Solution. Draw the loading diagram as shown in fig. 1(*a*).

Taking moments about *B*, we have

$$R_A \times 5 = 70 \times 4 + 90 \times 2 + 50 \times 1 + 80 \times 0.5$$

$$= 550$$

$$R_A = \frac{550}{5} = 110 \text{ kN}$$

$$R_B = (70 + 90 + 50 + 80) - 110$$

$$= 290 - 110 = 180 \text{ kN}$$

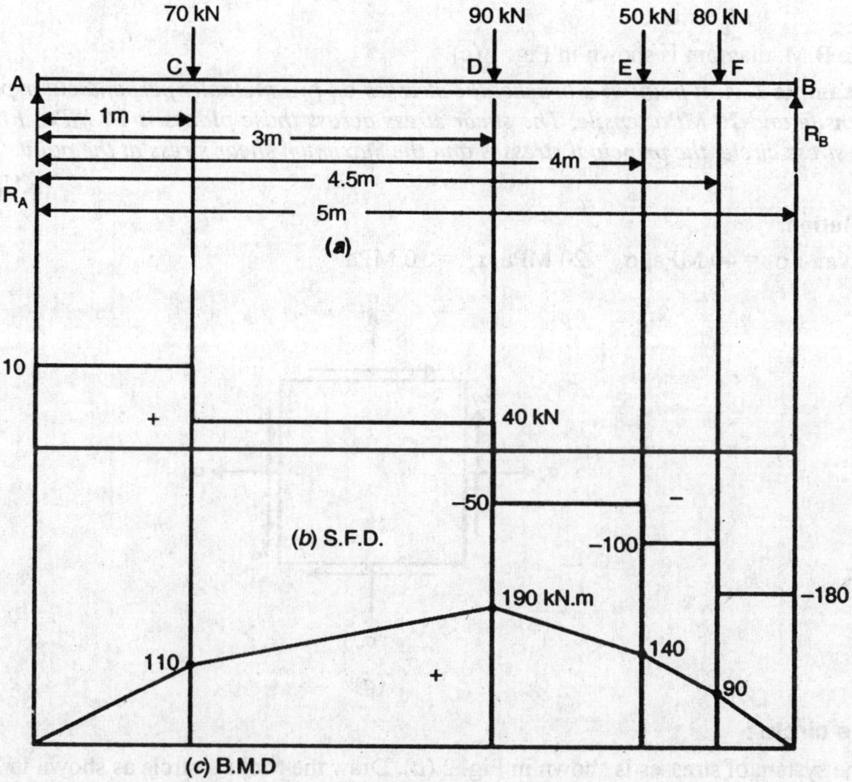

Fig. 1.

S.F.D.

Span AC, S.F. = R_A = 110 kN
Span CD, S.R.= R_A – 70 = 110 – 70 = 40 kN
Span DE, S.F. = 40 – 90 = – 50 kN
Span EF, S.F. = –50 – 50 = – 100 kN
Span FB, S.F. = –100 – 80 = – 180 kN
The S.F. diagram is shown in Fig. 1(*b*).

B.M.D.

Span AC, $M_x = R_A \times x = 110\,x$

where x = distance from A.

$$M_A = 0, M_C = 110 \times 1 = 110 \text{ kN.m}$$

Span CD:

$$M_x = 110\,x - 70\,(x - 1)$$
$$M_D = 110 \times 3 - 70\,(3 - 1) = 330 - 140 = 190 \text{ kN.m}$$

Span DE:

$$M_x = 110 \times x - 70\,(x - 1) - 90\,(x - 3)$$
$$M_E = 110 \times 4 - 70\,(4 - 1) - 90\,(4 - 3) = 440 - 210 - 90 = 140 \text{ kN.m}$$

Span EF:

$$M_x = 110 \times x - 70\,(x - 1) - 90\,(x - 3) - 50\,(x - 4)$$
$$M_f = 110 \times 4.5 - 70 \times 3.5 - 90 \times 1.5 - 50 \times 0.5 = 495 - 245 - 135 - 25 = 90 \text{ kN.m}$$

Span FB:

$$M_B = 0$$

The B.M. diagram is shown in Fig. 1 (c).

Example 2. *At a point in a bracket the stresses on two mutually perpendicular planes are 40 MPa tensile and 20 MPa tensile. The shear stress across those planes is 10 MPa. Find using the Mohr's stress circle, the principal stresses and the maximum shear stress at the point.*

[KU, May 2006]

Solution.

Given : $\sigma_x = 40$ MPa, $\sigma_y = 20$ MPa, $\tau_{xy} = 10$ MPa

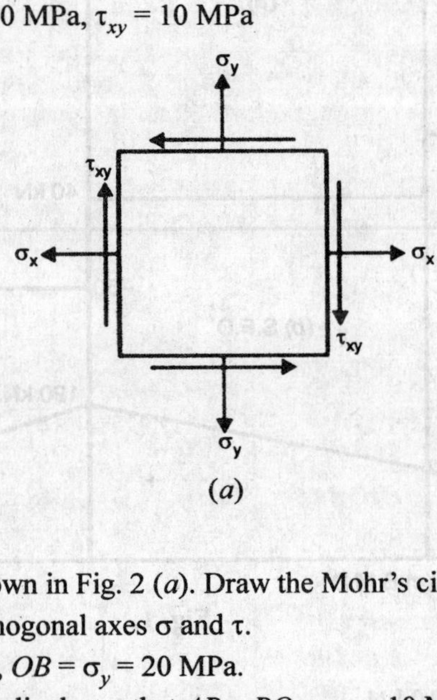

(*a*)

Mohr's circle :

The system of stresses is shown in Fig. 2 (*a*). Draw the Mohr's circle as shown in Fig. 2(*b*).

1. Draw two mutually orthogonal axes σ and τ.
2. Cut $OA = \sigma_x = 40$ MPa, $OB = \sigma_y = 20$ MPa.
3. At A and B draw perpendicular so that $AP = BQ = \tau_{xy} = 10$ MPa.
4. Join PQ cutting σ – axis at C.
5. With C as centre and CP as radius draw the circle. Then

   $OF = \sigma_1 = 9$ cm $= 45$ MPa

   $OG = \sigma_2 = 3.2$ cm $= 16$ MPa

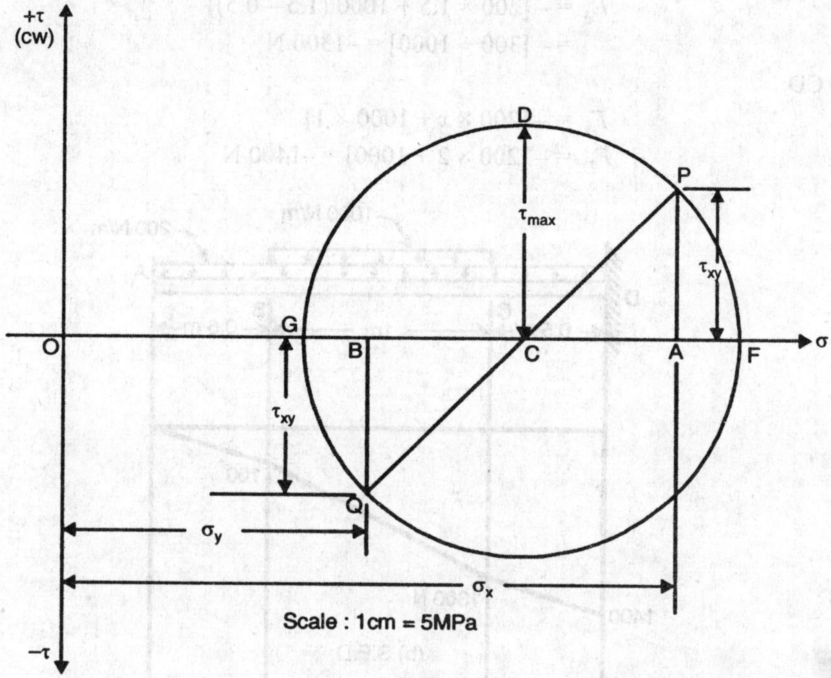

Fig. 2 (b)

6. At C draw a perpendicular CD. Then

$$\tau_{max} = CD = 2.9 \text{ cm} = 14.5 \text{ MPa}$$

Analytically :

$$\sigma_{1,2} = \frac{1}{2}(\sigma_x + \sigma_y) \pm \frac{1}{2}\sqrt{(\sigma_x - \sigma_y)^2 + 4\tau_{xy}^2}$$

$$= \frac{1}{2}(40 + 20) \pm \frac{1}{2}\sqrt{(40 - 20)^2 + 4 \times 10^2}$$

$$= 30 \pm 14.14$$

$$= 44.14 \text{ MPa}, 15.86 \text{ MPa}$$

$$\tau_{max} = \frac{\sigma_1 - \sigma_2}{2} = \frac{44.14 - 15.86}{2} = 14.14 \text{ MPa}$$

Example 3. *A cantilever beam 2 m long carries a uniformly distributed load of 1000 N per meter on a span of 1 m at a distance of 0.5 m from the free end. The beam itself weighs 200 N per meter run. Construct the SF and BM diagrams for the cantilever.* **[KU, Dec. 2006]**

Solution.

The loading diagram is shown in Fig. 1(a).

S.F.D.

Span AB : Distance x is measured from free end A.

$$F_x = -200\,x$$

$$F_A = 0, F_B = -200 \times 0.5 = -100 \text{ N}$$

Span BC :

$$F_x = -[200 \times x + 1000\,(x - 0.5)]$$

$$F_c = -[200 \times 1.5 + 1000\,(1.5 - 0.5)]$$
$$= -[300 + 1000] = -1300 \text{ N}$$

Span CD:

$$F_x = -[200 \times x + 1000 \times 1]$$
$$F_D = -[200 \times 2 + 1000] = -1400 \text{ N}$$

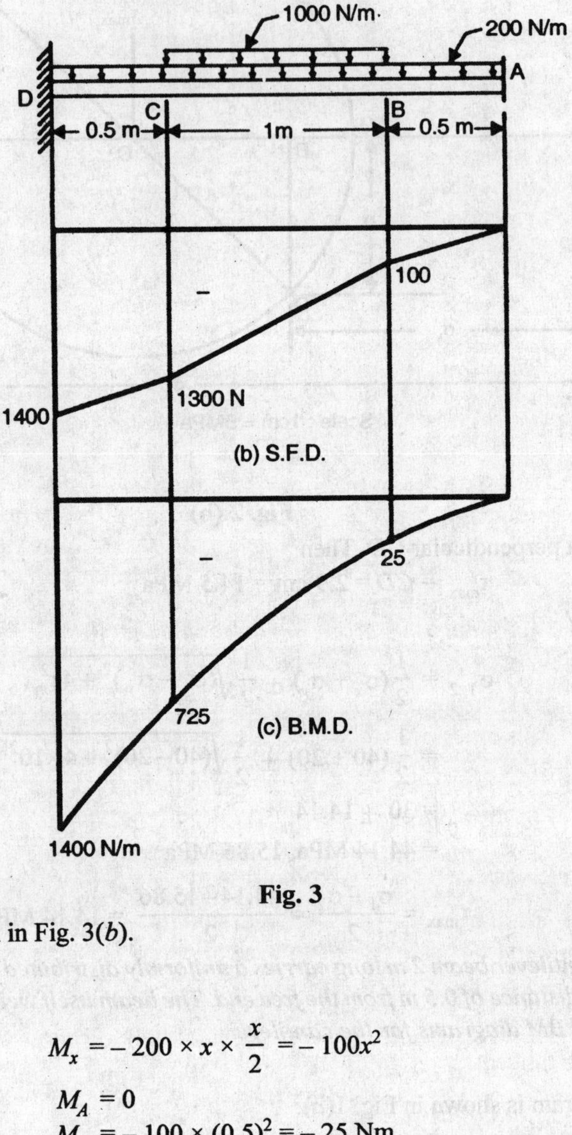

Fig. 3

The S.F.D. is shown in Fig. 3(*b*).

B.M.D.

Span AB :

$$M_x = -200 \times x \times \frac{x}{2} = -100x^2$$
$$M_A = 0$$
$$M_B = -100 \times (0.5)^2 = -25 \text{ Nm}$$

Span BC :

$$M_x = -\left[200 \times x \times \frac{x}{2} + 1000(x - 0.5)\frac{(x-0.5)}{2}\right]$$
$$= -\left[100x^2 + 500(x - 0.5)^2\right]$$

$$M_c = -[100 \times (1.5)^2 + 500\,(1.5 - 0.5)^2]$$
$$= -[225 + 500] = -725 \text{ N.m}$$

Span CD:

$$M_x = -\left[200 \times x \times \frac{x}{2} + 100 \times 1\{x - (0.5 + 0.5)\}\right]$$
$$= -[100\,x^2 + 1000\,(x - 1)]$$
$$M_D = -[100 \times 2^2 + 1000\,(2 - 1)]$$
$$= -[400 + 1000] = -1400 \text{ N.m}$$

The B.M.D. is shown in Fig. 3(*c*).

Example 4. *A member formed by connecting a steel bar to an aluminum bar is shown in Fig. 4. Assuming that the bars are prevented from bucking sidewise, calculate the magnitude of force P, that will cause the total length of the member to decrease 0.25 mm. The values of elastic modulus for steel and aluminium are 210 kN/mm² and 70 kN/mm² respectively.* **[KU, Dec. 2006]**

Solution.

$$A_s = 50 \times 50 = 2500 \text{ mm}^2$$
$$L_s = 380 \text{ mm}$$
$$E_s = 210 \times 10^3 \text{ N/mm}^2$$
$$A_a = 100 \times 100 = 10^4 \text{ mm}^2$$
$$L_a = 380 \text{ mm}$$
$$E_a = 70 \times 10^3 \text{ N/mm}^2$$
$$\delta L = 0.25 \text{ mm}$$
$$\delta L = P\left[\frac{L_s}{A_s E_s} + \frac{L_a}{A_a E_a}\right]$$

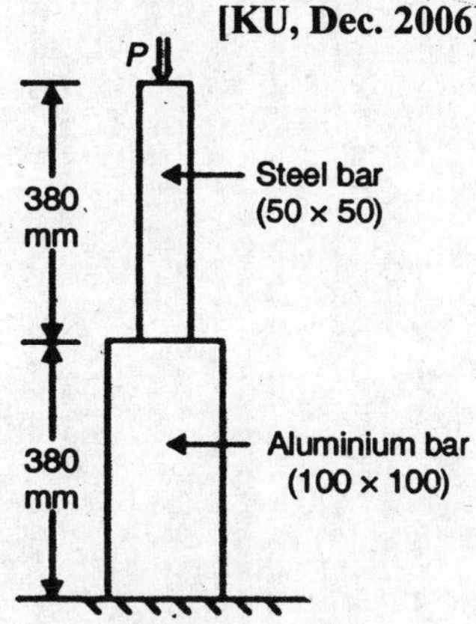

Fig. 4

$$0.25 = P\left[\frac{380}{2500 \times 210 \times 10^3} + \frac{380}{10^4 \times 70 \times 10^3}\right] = \frac{380P}{10^6}\left[\frac{1}{525} + \frac{1}{700}\right]$$
$$= \frac{1.2667P}{10^6}$$
$$P = 197368 \text{ N}$$

Example 5. *A cantilever 1.5 m long loaded with a uniformly distributed load of 2 kN/m run over a length of 1.25 m from the free end. It also carries a point of 3 kN at a distance of 0.25 m from the free end. Draw the SF and BM diagrams for the cantilever.* **[KU, June 2007]**

Solution. The loading diagram is shown in Fig. 5(*a*).

S.F.D.

Span AB :

$$F_x = -2x$$
$$F_A = 0,\; F_B = -2 \times 0.25 = -0.5 \text{ kN}$$

Span BC :

$$F_x = -(2x + 3)$$
$$F_B = -(2 \times 0.25 + 3) = -3.5 \text{ kN}$$
$$F_c = -(2 \times 1.25 + 3) = -(2.5 + 3) = -5.5 \text{ kN}$$

Span CD :

$$F_x = -(2 \times 1.25 + 3) = -5.5 \text{ kN}$$

$$F_C = F_D = -(2 \times 1.25 + 3) = -5.5 \text{ kN}$$

The S.F. diagram is shown in Fig. 5(*b*).

B.M.D.

Span AB:

$$M_x = -2x \times \frac{x}{2} = -x^2$$

$$M_A = 0,\ M_B = -(0.25)^2 = -0.0625 \text{ kN.m}$$

The B.M.D. is shown in Fig. 5(*c*).

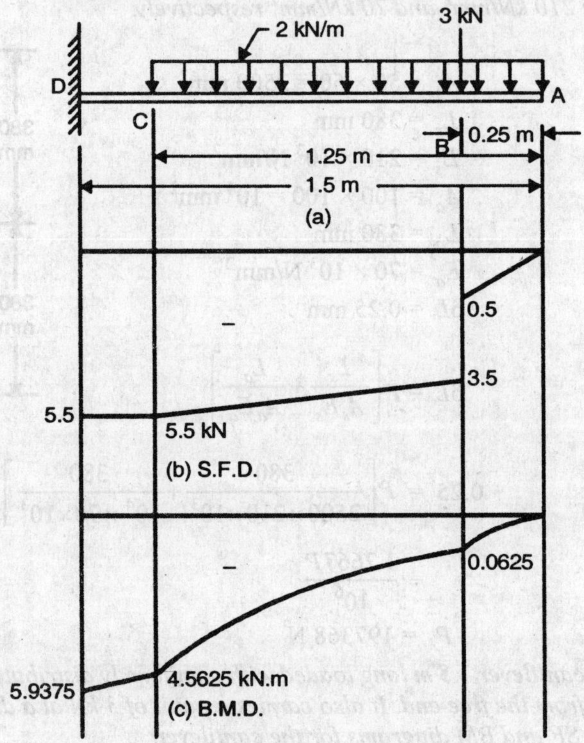

Fig. 5

Span BC :

$$M_x = -(x^2 + 3(x - 0.25))$$

$$M_B = -[(0.25)^2 + 0] = -0.0625 \text{ kN.m}$$

$$M_C = -[(1.25)^2 + 3\,(1.25 - 0.25)] = -4.5625 \text{ kN.m}$$

Span CD :

$$M_x = -[2 \times 1.25\,(x - 0.625) + 3(x - 0.25)]$$

$$M_D = -[2.5\,(1.5 - 0.625) + 3\,(1.5 - 0.25)] = -5.9375 \text{ kN.m}$$

Example 6. *A steel bar 10 cm long and 5 cm square section is subjected to pull of 10000 kg and 12000 kg respectively in axial and transverse directions. Find graphically the resultant stress on a plane inclined at 30° to the axis of the greater stress.* **[KU, June 2007]**

Solution. The loaded bar is shown in Fig. 6(*a*).

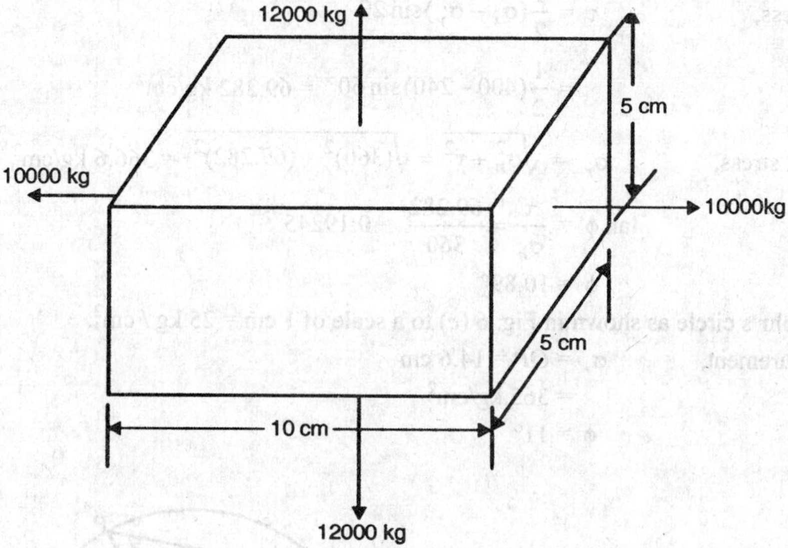

(*a*) Loaded bar

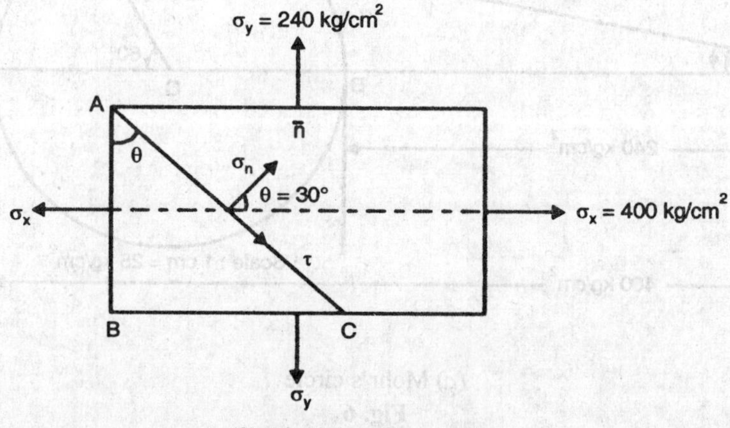

(*b*) Biaxial stresses on bar

Stress in axial direction,

$$\sigma_x = \frac{10,000}{5\times5} = 400 \text{ kg/cm}^2$$

Stress in transverse direction,

$$\sigma_y = \frac{12,000}{10\times5} = 240 \text{ kg/cm}^2$$

Let *AC* be the plane whose normal $\bar{n}$ is inclined at 30° to the axis of greater stress $\sigma_x = 400$ kg cm^2, as shown in Fig. 6(*b*).

Normal stress, $$\sigma_n = \frac{1}{2}(\sigma_x+\sigma_y)+\frac{1}{2}(\sigma_x-\sigma_y)\cos 2\theta$$

$$= \frac{1}{2}(400+240)+\frac{1}{2}(400-240)\cos 60°$$
$$= 320 + 40 = 360 \text{ kg/cm}^2$$

Shear stress, $\tau = \frac{1}{2}(\sigma_x - \sigma_y)\sin 2\theta$

$$= \frac{1}{2}(400-240)\sin 60° = 69.282 \text{ kg/cm}^2$$

Resultant stress, $\sigma_r = \sqrt{\sigma_n^2+\tau^2} = \sqrt{(360)^2+(69.282)^2} = 366.6 \text{ kg/cm}^2$

$$\tan\phi = \frac{\tau}{\sigma_n} = \frac{69.282}{360} = 0.19245$$
$$\phi = 10.89°$$

Draw Mohr's circle as shown in Fig. 6 (*c*) to a scale of 1 cm = 25 kg / cm^2.

By measurement, $\sigma_r = OP = 14.6$ cm
$$= 365 \text{ kg/cm}^2$$
$$\phi = 11°$$

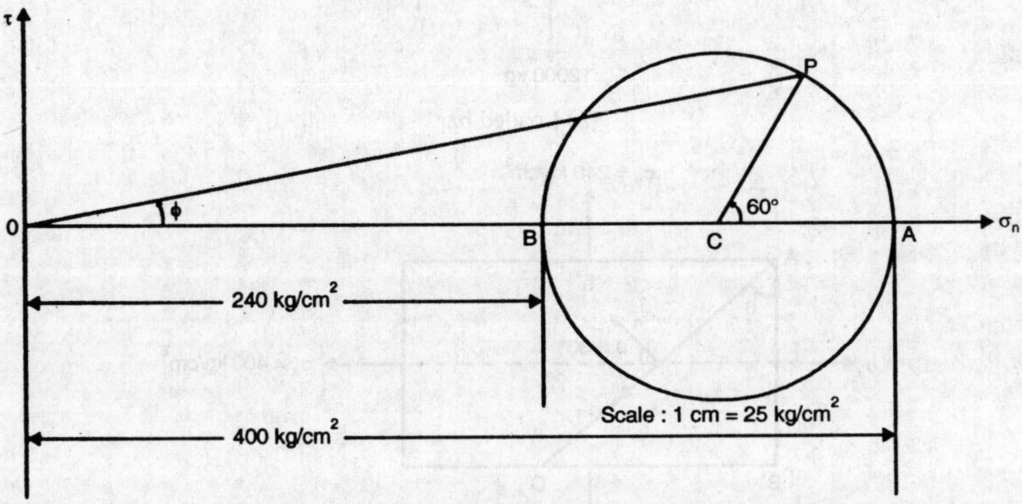

(*c*) Mohr's circle

Fig. 6

Example 7. *A four stroke engine working on a diesel cycle has a piston diameter of 25 cm; a stroke of 40 cm and a clearance volume of 1200 cc. The fuel injection takes place for 5% of the stroke. If the induction pressure corresponds to 1 bar and engine turns 5 rev/sec. Find the air standard efficiency, mean effective pressure and power developed.* **[KU, June 2007]**

Solution. Given : $D = 25$ cm, $L = 40$ cm, $V_c = 1200$ c.c., fuel injection = 5% of stroke, $p_1 = 1$ bar, $N = 5$ rps

Refer to Fig. 7.

Stroke volume, $V_s = V_1 - V_c = \frac{\pi}{4}{}^2 DL$

$$= \frac{\pi}{4}\times 25^2 \times 40 = 19635 \text{ c.c.}$$
$$V_1 = 19635 + 1200 = 20835 \text{ c.c.}$$

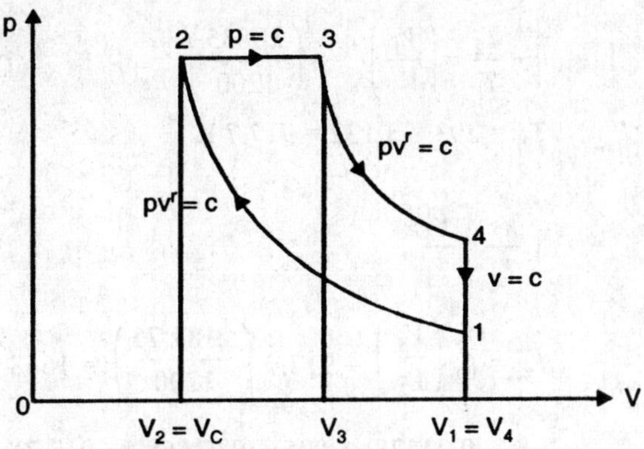

Fig. 7. Diesel cycle

Cut-off volume, $V_3 - V_2 = 0.05\ V_s = 0.05 \times 19635 = 981.75$ c.c.

$$V_3 = 2181.75 \text{ c.c.}$$

Compression ratio, $r = \dfrac{V_1}{V_c} = \dfrac{20835}{1200} = 17.36$

Cut-off ratio, $\rho = \dfrac{V_3}{V_c} = \dfrac{2181.75}{1200} = 1.818$

Air standard efficiency, $\eta_a = 1 - \dfrac{1}{r^{r-1}}\left[\dfrac{\rho^r - 1}{r(\rho - 1)}\right] = \dfrac{\text{Work done}}{\text{Heat supplied}}$

$$= 1 - \frac{1}{(17.36)^{1.4-1}}\left[\frac{(1.818)^{1.4} - 1}{1.4(1.818 - 1)}\right]$$

$$= 0.635 \text{ or } 63.5\ \%$$

Mean effective pressure, $p_m = \dfrac{\text{Work done}}{\text{Swept volume}} = \dfrac{m[C_p\,(T_3 - T_2) - C_v(T_4 - T_1)]}{V_1 - V_2 = V_s}$

Let $T_1 = 293$ K

$$p_1 v_1 = mRT_1$$

$$1 \times 10^5 \times 20835 \times 10^{-6} = m \times 287 \times 293$$

$$m = 0.02478 \text{ kg}$$

$$\text{Work done} = mC_p\,(T_3 - T_2)\eta_a$$

Process 1 → 2 :

$$\frac{T_2}{T_1} = \left(\frac{p_2}{p_1}\right)^{\frac{r-1}{r}}$$

Also $p_1 v_1^r = p_2 v_2^r$

$$\frac{p_2}{p_1} = \left(\frac{V_1}{V_2}\right)^r$$

$$\therefore \quad \frac{T_2}{T_1} = \left(\frac{V_1}{V_2}\right)^{r-1} = \left(\frac{20835}{1200}\right)^{r-1} = r^{r-1} = (17.36)^{0.4} = 3.132$$

$$T_2 = 293 \times 3.132 = 917.7 \text{ K}$$

Process 2 → 3:

$$\frac{V_2}{T_2} = \frac{V_3}{T_3}$$

$$T_3 = T_2\left(\frac{V_3}{V_2}\right) = 917.7\left(\frac{2181.75}{1200}\right) = 1668.5 \text{ K}$$

$$p_m = \frac{0.02478 \times 1.005 \times 10^3 (1668.5 - 917.7) \times 0.635}{19635 \times 10^{-6}}$$

$$= 0.6047 \times 10^6 \text{ N/m}^2 \text{ or } 6.047 \text{ bar}$$

Power developed, p = Work done per cycle × Number of strokes per second

$$= \frac{mC_p(T_3 - T_2) \times \eta_a \times \frac{N}{2}}{10^3}$$

$$= \frac{0.02478 \times 1.005 \times 10^3 (1668.5 - 917.7) \times 0.635 \times \frac{5}{2}}{10^3} = 29.68 \text{ kw}$$

Example 8. *A beam 6 m long is simply supported at the ends, and carries a uniformly distributed load of 1.5 kN/m (including its own weight) and three concentrated loads of 1,2 and 3 kN acting respectively at the left quarter point, centre point and right quarter point. Draw the SF and BM diagrams and determine the maximum bending moment.* **[KU, June 2008]**

Solution.

The loading diagram on the beam is shown in Fig. 8 (*a*).

$$R_A + R_B = 6 \times 1.5 + 1 + 2 + 3 = 15 \text{ kN}$$

Taking moments about B, we have

$$6R_A = 1 \times 4.5 + 2 \times 3 + 3 \times 1.5 + 1.5 \times 6 \times 3 = 42$$

$$R_A = 7 \text{ kN}$$

$$R_B = 15 - 7 = 8 \text{ kN}$$

S.F.D.

Span AC :

$$F_x = R_A - 1.5x = 7 - 1.5x$$

$$F_A = 7 \text{ kN}, F_c = 7 - 1.5 \times 1.5 = 4.75 \text{ kN}$$

Span CD:

$$F_x = 7 - 1 - 1.5x = 6 - 1.5x$$

$$F_c = 6 - 1.5 \times 1.5 = 3.75 \text{ kN}$$

$$F_D = 6 - 1.5 \times 3 = 1.5 \text{ kN}$$

Span DE :

$$F_x = 7 - 1 - 2 - 1.5x = 4 - 1.5x$$

$$F_D = 4 - 1.5 \times 3 = -0.5 \text{ kN}$$

$$F_E = 4 - 1.5 \times 4.5 = -2.75 \text{ kN}$$

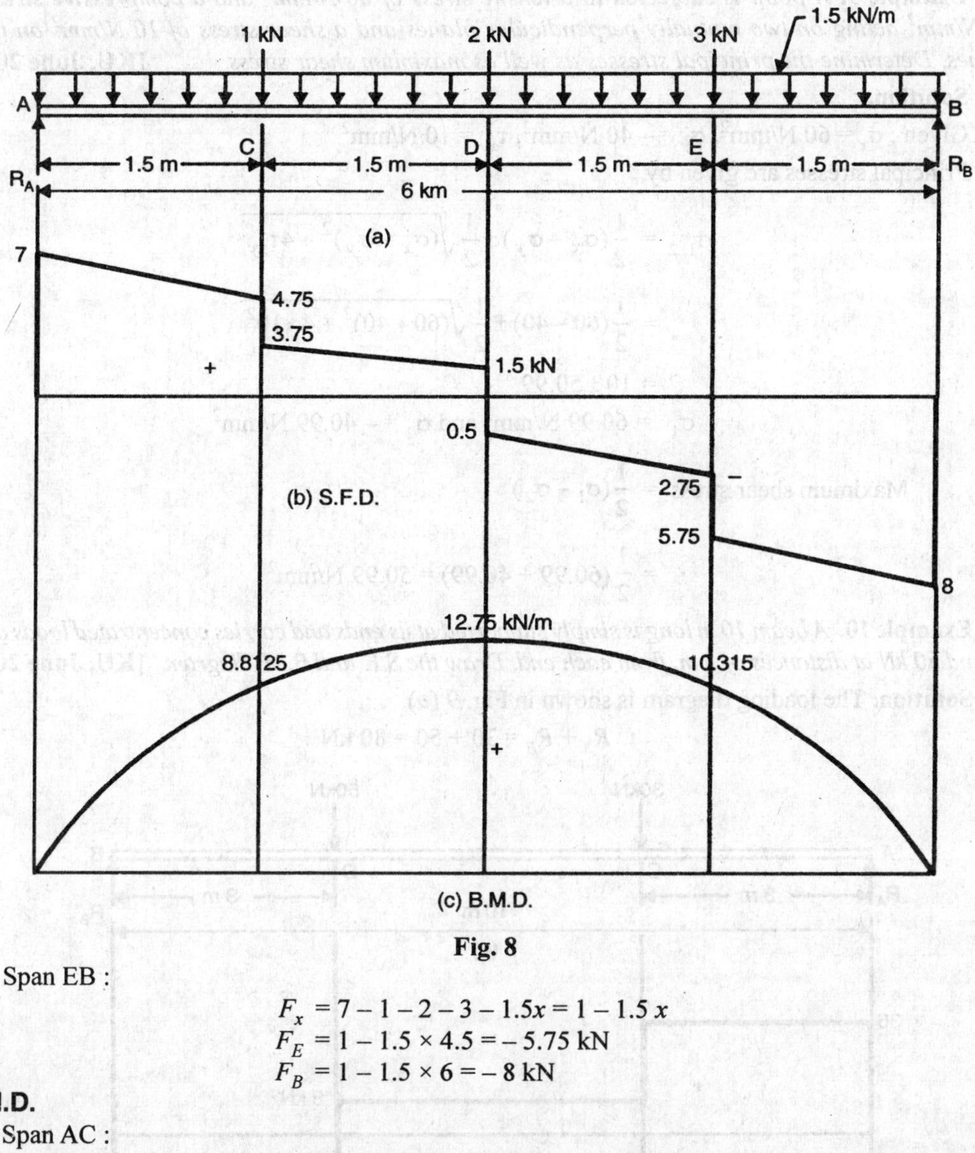

Fig. 8

Span EB :

$$F_x = 7 - 1 - 2 - 3 - 1.5x = 1 - 1.5\,x$$
$$F_E = 1 - 1.5 \times 4.5 = -\,5.75 \text{ kN}$$
$$F_B = 1 - 1.5 \times 6 = -\,8 \text{ kN}$$

B.M.D.

Span AC :

$$M_x = 7\,x - 1.5\,x \times \frac{x}{2} = 7x - 0.75\,x^2$$
$$M_A = 0$$
$$M_C = 7 \times 1.5 - 0.75\,(1.5)^2 = 8.8125 \text{ kN.m}$$

Span CD :

$$M_x = 7x - 1\,(x - 1.5) - 0.75\,x^2$$
$$M_D = 7 \times 3 - 1\,(3 - 1.5) = 0.75 \times 3^2 = 12.75 \text{ kN.m}$$

Span DE :

$$M_x = 7x - 1\,(x - 1.5) - 2\,(x - 3) - 0.75\,x^2$$
$$M_E = 7 \times 4.5 - 1\,(4.5 - 1.5) - 2\,(4.5 - 3) - 0.75 \times (4.5)^2 = 10.315 \text{ kN.m}$$

Span EB :

$$M_B = 0$$

The S.F.D. and B.M.D. are shown in fig. 8 (*b*) and 8 (*c*) respectively.

Example 9. *A point is subjected to a tensile stress of 60 N/mm² and a compressive stress of 40 N/mm² acting on two mutually perpendicular planes and a shear stress of 10 N/mm² on these planes. Determine the principal stresses as well as maximum shear stress.* **[KU, June 2008]**

Solution.

Given : $\sigma_x = 60$ N/mm², $\sigma_y = -40$ N/mm², $\tau_{xy} = 10$ N/mm²

Principal stresses are given by,

$$\sigma_{1,2} = \frac{1}{2}(\sigma_x + \sigma_y) \pm \frac{1}{2}\sqrt{(\sigma_x - \sigma_y)^2 + 4\tau_{xy}^2}$$

$$= \frac{1}{2}(60 - 40) \pm \frac{1}{2}\sqrt{(60 + 40)^2 + 4 \times 10^2}$$

$$= 10 \pm 50.99$$

$$\sigma_1 = 60.99 \text{ N/mm}^2 \text{ and } \sigma_2 = -40.99 \text{ N/mm}^2$$

$$\text{Maximum shear stress} = \frac{1}{2}(\sigma_1 - \sigma_2)$$

$$= \frac{1}{2}(60.99 + 40.99) = 50.99 \text{ N/mm}^2$$

Example 10. *A beam 10 m long is simply supported at its ends and carries concentrated loads of 30 kN and 50 kN at distances of 3 m, from each end. Draw the S.F. and B.M. diagram.* **[KU, June 2009]**

Solution. The loading diagram is shown in Fig. 9 (*a*).

$$R_A + R_B = 30 + 50 = 80 \text{ kN}$$

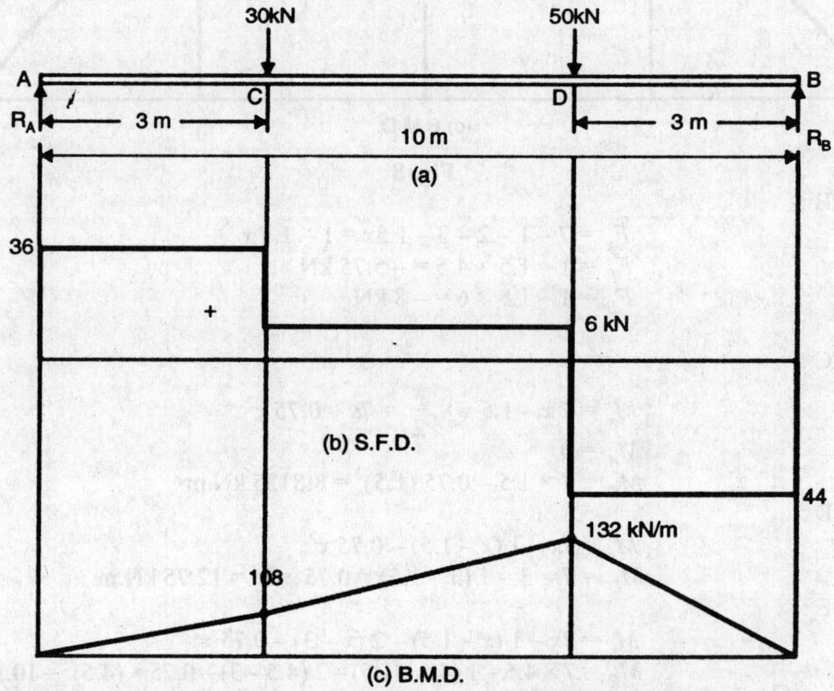

Fig. 9

Taking moments about *B*, we have

$$10\,R_A = 30 \times 7 + 50 \times 3 = 360$$

$R_A = 36$ kN
$R_B = 80 - 36 = 44$ kN

S.F.D.

Span AC :

$F_x = R_A = 36$ kN
$F_A = F_C = 36$ kN

Span CD :

$F_x = 36 - 30 = 6$ kN
$F_C = F_D = 6$ kN

Span DB:

$F_x = 36 - 30 - 50 = -44$ kN
$F_B = -44$ kN

The S.F.D. is shown in Fig. 9 (*b*).

B.M.D.

Span AC :

$M_x = 36\,x$
$M_A = 0, M_C = 36 \times 3 = 108$ kN.m

Span CD :

$M_x = 36\,x - 30\,(x - 3)$
$M_D = 36 \times 7 - 30\,(7 - 3) = 252 - 120 = 132$ kN.m

Span DB :

$M_B = 0$

Example 11. *Calculate the thermal efficiency of an engine working on the Otto cycle. The bore and the stroke of the cylinder are 17 cm and 30 cm respectively. The clearance volume of 0.002025 m*^{3}. *Assume* $\gamma = 1.4$. **[KU, June 2009]**

Solution.

Given : $D = 17$ cm, $L = 30$ cm, $V_C = 0.002025$ m^3 or 2025 cm^3

Stroke volume, $V_s = V_1 - V_2 = \frac{\pi}{4} D^2 L = \frac{\pi}{4} (17)^2 \times 30 = 6809.4$ cm^3

$V_1 = 6809 + 2025 = 8834.4$ cm^3

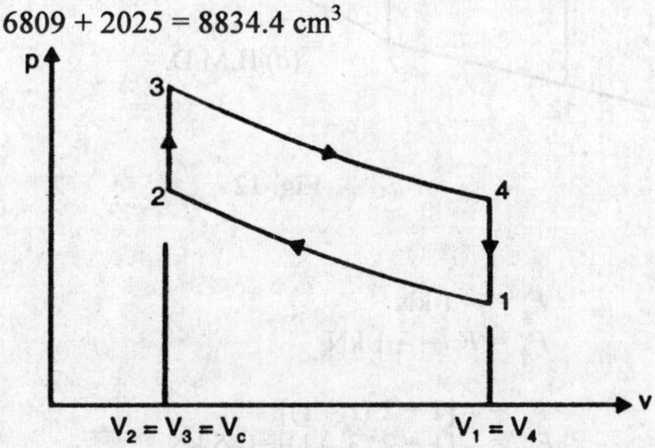

Fig. 10 Otto Cycle

Compression ratio, $r = \frac{V_1}{V_2} = \frac{8834.4}{2025} = 4.36.$

Thermal efficiency of otto cycle, $\eta_{th} = 1 - \frac{1}{r^{r-1}}$

$$= 1 - \frac{1}{(4.36)^{0.4}} = 0.4452 \text{ or } 44.52\%$$

Example 12. *Draw the SF and BM diagrams for a cantilever loaded as shown in Fig. 11.*

[KU, June 2009]

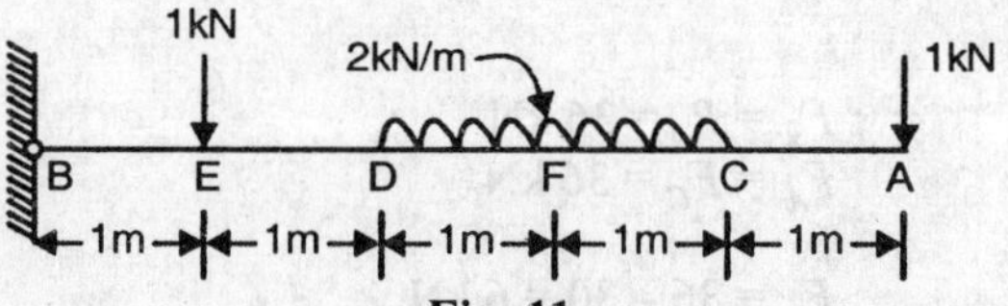

Fig. 11

Solution.

The loading diagram is shown in Fig. 12 (*a*).

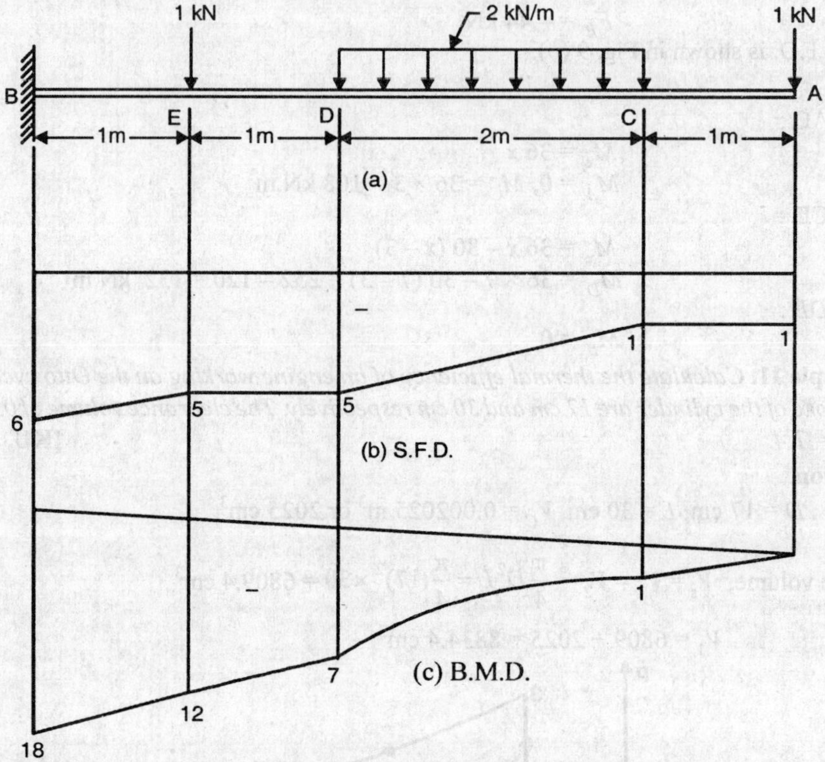

Fig. 12

S.F.D.

Span AC :

$$F_x = -1 \text{ kN}$$
$$F_A = F_C = -1 \text{ kN}$$

Span CD:

$$F_x = -[1 + 2(x-1)]$$
$$F_D = -[1 + 2(3-1)] = -5 \text{ kN}$$

Span DE:

$$F_x = -[1 + 2 \times 2] = -5 \text{ kN}$$
$$D_E = -5 \text{ kN}$$

Span EB :

$$F_x = -[1 + 2 \times 2 + 1] = -6 \text{ kN}$$
$$F_B = -6 \text{ kN}$$

The S.F.D. is shown in Fig. 12 (*b*).

B.M.D.

Span AC:

$$M_x = -1 \times x = -1\,x$$
$$M_A = 0,\ M_C = -1 \times 1 = -1 \text{ kN.m}$$

Span CD:

$$M_x = -\left[1 \times x + 2(x-1)\frac{(x-1)}{2}\right] = -[x + (x-1)^2]$$
$$M_D = -[3 + (3-1)^2] = -7 \text{ kN.m}$$

Span DE:

$$M_x = -[1 \times x + 2 \times 2\,(x-2)] = -[x + 4\,(x-2)]$$
$$M_E = -[4 + 4\,(4-2)] = -12 \text{ kN.m}$$

Span EB :

$$M_x = -[1 \times x + 2 \times 2\,(x-2) + 1\,(x-4)] = -[x + 4\,(x-2) + (x-4)]$$
$$M_B = -[5 + 4\,(5-2) + (5-4)] = -[5 + 12 + 1] = -18 \text{ kN.m}$$

The B.M.D. is shown in Fig. 12 (*c*).

Example 13. *A 6 m long cantilever carries loads of 2 kN at 2 m and 5 m respectively from the fixed end and a UDL of 10 kN over its entire length. Draw shear force and bending moment diagrams.*

[KU, June 2010]

Solution.

The loding diagram is shown in Fig. 13 (*a*).

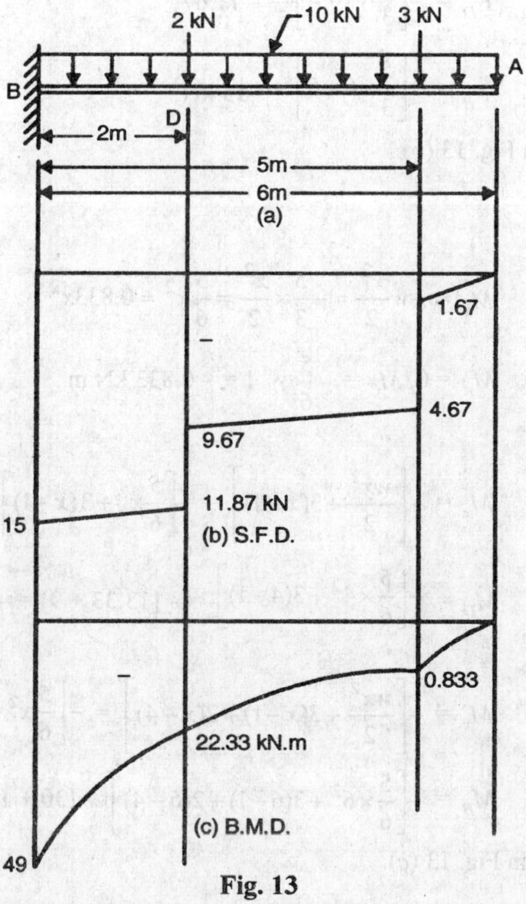

Fig. 13

UDL, $w = \frac{10}{6} = \frac{5}{3}$ kN/m

Span AC :

$$F_x = -wx = -\frac{5}{3}x$$

$$F_A = 0, F_C = -\frac{5}{3}\times 1 = -\frac{5}{3} \text{ kN}$$

Span CD :

$$F_x = -\left[\frac{5}{3}x + 3\right], \; F_c = -\left[\frac{5}{3} + 3\right] = -4.67 \text{ kN}$$

$$F_D = -\left[\frac{5}{3}\times 4 + 3\right] = -9.67 \text{ kN}$$

Span DB :

$$F_x = -\;-\left[\frac{5}{3}x + 3 + 2\right] = -\left[\frac{5}{3}x + 5\right]$$

$$F_D = -\left[\frac{5}{3}\times 4 + 5\right] = -11.67$$

$$F_B = -\left[\frac{5}{3}\times 6 + 5\right] = -15 \text{ kN}$$

The S.F.D. is shown in Fig. 13 (*b*).

B.M.D.

Span AC :

$$M_x = -w\frac{x^2}{2} = -\frac{5}{3}\times\frac{x^2}{2} = \frac{5}{6}x^2 = 0.833x^2$$

$$M_A = 0, M_C = -\frac{5}{6} \times 1 = -0.833 \text{ kN.m}$$

Span CD :

$$M_x = -\left[\frac{wx^2}{2} + 3(x-1)\right] = -\left[\frac{5}{6}x^2 + 3(x-1)\right]$$

$$M_D = -\left[\frac{5}{6}\times 4^2 + 3(4-1)\right] = -[13.33 + 9] = -22.33 \text{ kN.m}$$

Span DB :

$$M_x = -\left[\frac{wx^2}{2} + 3(x-1) + 2(x-4)\right] = -\left[\frac{5}{6}x^2 + 3(x-1) + 2(x-4)\right]$$

$$M_B = -\left[\frac{5}{6}\times 6^2 + 3(6-1) + 2(6-4)\right] = [30 + 15 + 4] = -49 \text{ kN.m}$$

The B.M.D. is shown in Fig. 13 (*c*).

Example 14. *At a point in an elastic material under strain, there are normal stresses of 50 N/mm² and 30 N/mm² respectively at right angles to each other with a shearing stress of 25 N/mm². Find the principle stresses the position of principal planes if*

(*a*) *50 N/mm² and 30 N/mm² are both tensile.'*

(*b*) *50 N/mm² is tensile and 30 N/mm² is compressive.*

Find also the maximum shear stress and its plane in both the cases. **[KU, June 2010]**

Solution.

(*a*) $\sigma_x = 50$ N/mm², $\sigma_y = 30$ N/mm², $\tau_{xy} = 25$ N/mm²

$$\sigma_{1,2} = \frac{1}{2}(\sigma_x + \sigma_y) \pm \frac{1}{2}\sqrt{(\sigma_x - \sigma_y)^2 + 4\tau_{xy}^2}$$

$$= \frac{1}{2}(50+30) \pm \frac{1}{2}\sqrt{(50-30)^2 + 4\times 25^2}$$

$$= 40 \pm 26.926$$

$$\sigma_1 = 66.926 \text{ N/mm}^2, \sigma_2 = 13.074 \text{ N/mm}^2$$

$$\tau_{max} = 26.926 \text{ N/mm}^2$$

Principle planes are given by,

$$\tan 2\theta = \frac{2\tau_{xy}}{\sigma_x - \sigma_y} = \frac{2\times 25}{50-30} = \frac{50}{20} = 2.5$$

$$2\theta = 68.2°$$

$$\theta_1 = 34.1°, \theta_2 = 34.1° + 90° = 124.1°$$

(*b*) $\sigma_x = 50$ N/mm², $\sigma_y = -30$ N/mm², $\tau_{xy} = 25$ N/mm²

$$\sigma_{1;2} = \frac{1}{2}(50-30) \pm \frac{1}{2}\sqrt{(50+30)^2 + 4\times 25^2}$$

$$= 10 \pm 47.17$$

$$\sigma_1 = 57.17 \text{ N/mm}^2, \sigma_2 = -37.17 \text{ N/mm}^2$$

$$\tau_{max} = 47.17 \text{ N/mm}^2$$

$$\tan 2\theta = \frac{2\times 25}{50+30} = 0.625$$

$$2\theta = 32°, \theta_1 = 16°, \theta_2 = 106°$$

Example 15. *A flat belt is required to transmit 35 kW from a pulley of 1.5 m effective diameter running at 300 rpm. The angle of contact is spread over 11/24 of th e circumference and the coefficient of friction between belt and pulley surface is 0.3. Taking centrifugal tension into account, determine the width of the belt. Take belt thickness as 9.5 mm, density 1100 kg/m³ and permissible stress as 2.5 MN/m².* **[KU, June 2010]**

Solution. Given : $P = 35$ kW, $d = 1.5$ m, $n = 300$ rpm, $\theta = \frac{11}{24}\times 360 = 165°$, $\mu = 0.3$, $t = 9.5$ mm, $\rho = 1100$ kg/m³, $\sigma_w = 2.5$ MN/m².

$$\text{Belt speed, } v = \frac{\pi d n}{60} = \frac{\pi\times 1.5\times 300}{60} = 23.56 \text{ m/s}$$

Let b = width of belt, mm

Mass of belt per metre length, $m = bt \times 1 \times \rho = b \times 9.5 \times 10^{-6} \times 1 \times 1100 = 0.0104.5 \times b$ kg/m

Centrifugal tension, $T_c = mv^2 = 0.01045 \times 6 \times (23.56)^2 = 5.8 \times b$ N

Maximum tension in belt, $T = \sigma_w \times bt = 2.5 \times b \times 9.5 = 23.75 \times b$ N

Tension on right side of belt, $T_1 = T - T_C = 23.75 \times b - 5.8 \times b = 17.95 \times b$ N

Now, $$\frac{T_1}{T_2} = \exp(\mu\theta) = \exp\left(0.3 \times \frac{\pi}{180} \times 165\right) = \exp(0.86394) = 2.3725$$

$$T_2 = \frac{T_1}{2.3725} = \frac{17.95 \times b}{2.3725} = 7.566 \times b$$

Power transmitted, $$P = \frac{(T_1 - T_2)v}{10^3}$$

$$35 = \frac{(17.95 - 7.566) \times b \times 23.56}{10^3} \qquad b = 143 \text{ mm}$$

Example 16. *The intensity of loading in a simply supported beam of 8 m span varies gradually from 2 KN/m at one end to 6 k N/m at the other end. Draw the shear force and bending moment diagram.*

[KU, Jan 2010]

Solution. The loading diagram is shown in Fig. 14 (*a*). The intensity of loading at a distance *x* from *A* is

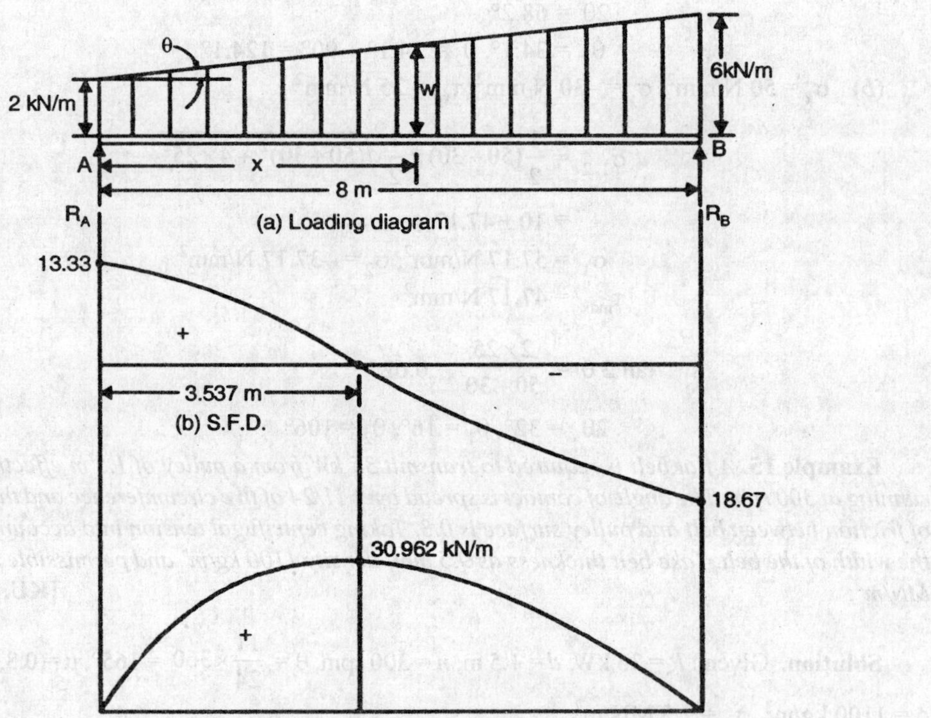

Fig. 14

$$\tan\theta = \frac{6-2}{8} = \frac{4}{8} = \frac{1}{2}$$

$$w = \left(2+\frac{x}{2}\right) \text{ kN/m}$$

Taking moments about *B*, we have

$$8R_A = 2 \times 8 \times 4 + \frac{1}{2} \times 8 \times 4 \times \frac{8}{3}$$

$$= 64 + 42.67 = 106.67$$

$$R_A = 13.33 \text{ kN}$$

$$R_B = \frac{1}{2}(2+6) \times 8 - 13.33 = 18.67 \text{ kN}$$

S.F.D.

$$F_x = R_A - \frac{1}{2}\left(2+2+\frac{x}{2}\right) \times x$$

$$= 13.33 - \left(2+\frac{x}{4}\right)x = 0$$

$$F_A = 13.33 \text{ kN and } F_B = -18.67 \text{ kN}$$

$$F_x = 0 \text{ when } 13.33 - \left(2+\frac{x}{2}\right)x = 0$$

or

$$\frac{x^2}{2} + 2x - 13.33 = 0$$

$$x^2 + 4x - 26.66 = 0$$

$$x = \frac{1}{2}[-4 \pm \sqrt{16+4\times 26.66}]$$

$$x = 3.537 \text{ m}$$

The S.F.D. is shown in Fig. 14 (*b*).

B.M.D.

$$M_x = \frac{40}{3}\times x - 2\times x\times\frac{x}{2} - \frac{1}{2}\times x\times\frac{x}{2}\times\frac{x}{3}$$

$$= \frac{40}{3}x - x^2 - \frac{x^3}{12}$$

$$M_A = 0, M_B = 0$$

At $x = 3.537$ m, $M_{max} = \frac{40}{3}\times 3.537 - (3.537)^2 - \frac{1}{12}(3.537)^3 = 30.962$ kN.m

Example 17. *For the beam shown in Fig. 15, draw SFD and BMD.* **[KU, Dec. 2009]**

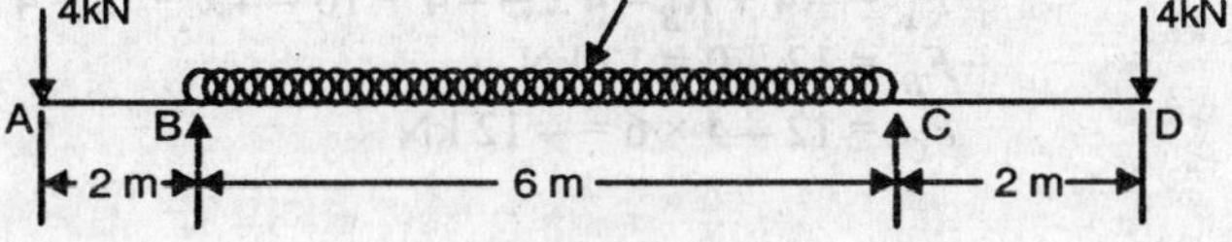

Fig. 15

Solution. The loading diagram is shown in fig. 16 (*a*).

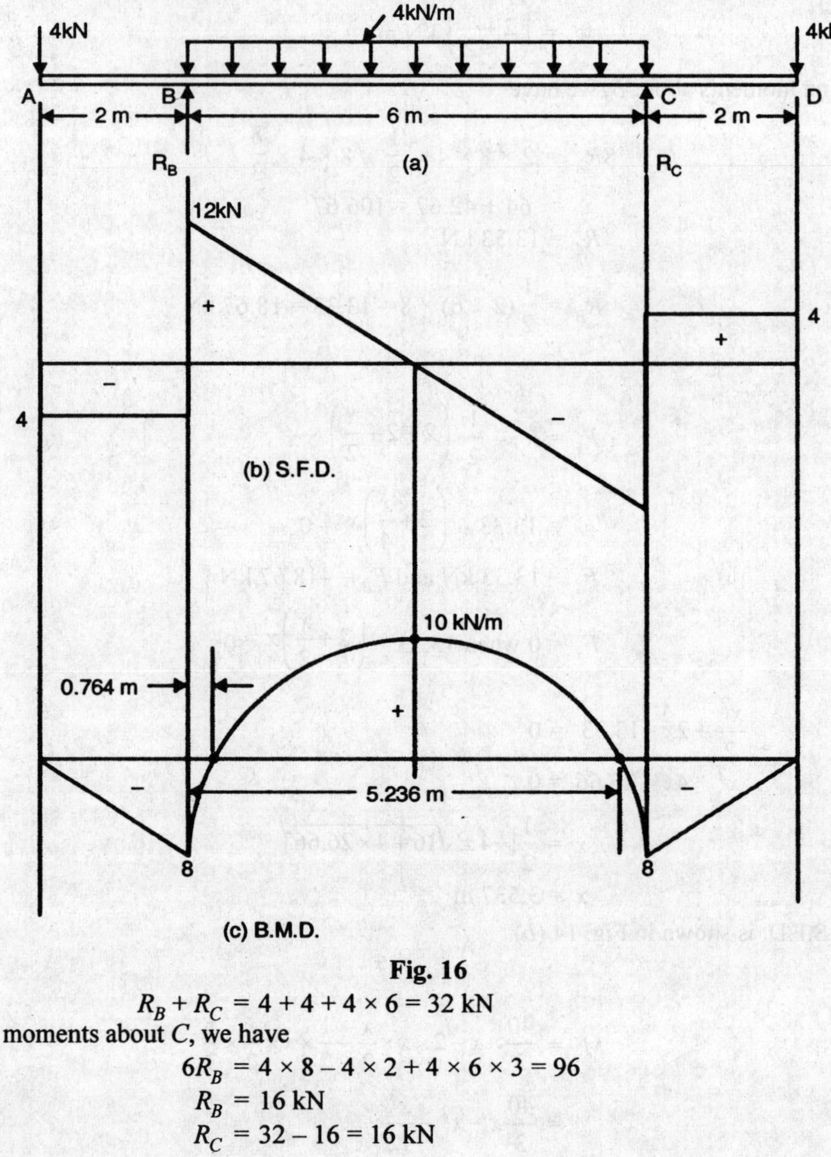

Fig. 16

$$R_B + R_C = 4 + 4 + 4 \times 6 = 32 \text{ kN}$$

Taking moments about C, we have

$$6R_B = 4 \times 8 - 4 \times 2 + 4 \times 6 \times 3 = 96$$

$$R_B = 16 \text{ kN}$$

$$R_C = 32 - 16 = 16 \text{ kN}$$

S.F.D.

Span AB :

$$F_A = F_B = -4 \text{ kN}$$

Span BC:

Let x = distance from B.

$$F_x = -4 + R_B - 4x = -4 + 16 - 4x = 12 - 4x$$

$$F_B = 12 - 0 = 12 \text{ kN}$$

$$F_C = 12 - 4 \times 6 = -12 \text{ kN}$$

Span CD:

$$F_D = 4 \text{ kN}$$

$$F_C = 4 - 16 = -12 \text{ kN}$$

The S.F.D. is shown in Fig. 16 (b).

B.M.D.

Span *AB* :

$$M_A = 0, M_B = -4 \times 2 = -8 \text{ kN.m}$$

Span *BC*:

$$M_x = -4(2+x) + 16 \times 6 - 2 \times 6^2 = -32 + 96 - 72 = -8 \text{ kN.m}$$

At $x = 3$ m from *B*

$$M = -4 \times 5 + 4\,8 - 18 = 10 \text{ kN.m}$$

For

$$M_x = 0, 2x^2 - 12x + 8 = 0 \text{ or } x^2 - 6x + 4 = 0$$

$$x = \frac{1}{2}(6 \pm \sqrt{36-16}) = \frac{1}{2}(6 \pm 4.47) = 0.764 \text{ m and } 5.236 \text{ m}$$

Span *CD*:

$$M_D = 0, M_C = -4 \times 2 = -8 \text{ kN.m}$$

The B.M.D. is shown in Fig. 16 (*c*).

Example 18. *A simply supported beam carries point loads as shown in Fig. 17. Draw SFD & BMD.* **[MDU, Dec. 2009]**

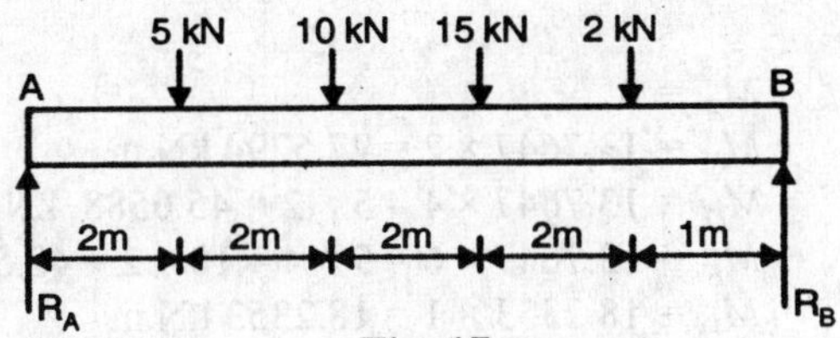

Fig. 17

Solution.

The loading diagram is shown in Fig. 18 (*a*).

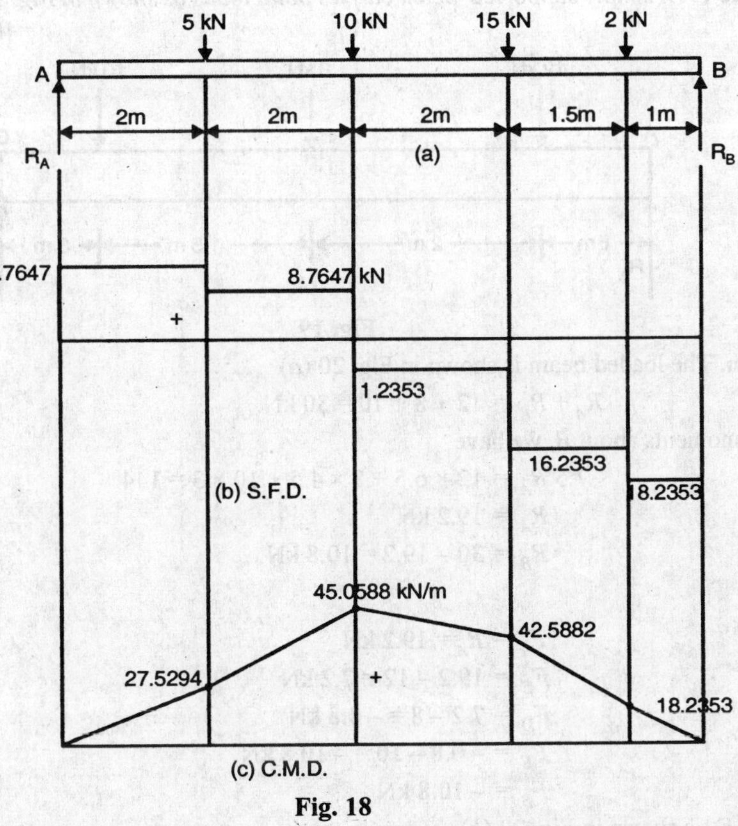

Fig. 18

$$R_A + R_B = 5 + 10 + 15 + 2 = 32 \text{ kN}$$

Taking moments about B, we have

$$8.5 R_A = 5 \times 6.5 + 10 \times 4.5 + 15 \times 2.5 + 2 \times 1 = 117$$
$$R_A = 13.7647 \text{ kN}$$
$$R_B = -32 - 13.7647 = 18.2353 \text{ kN}$$

S.F.D.

$$F_A = 13.7647 \text{ kN}$$
$$F_C = 13.7647 - 5 = 8.7647 \text{ kN}$$
$$F_D = 8.7647 - 10 = -1.2353 \text{ kN}$$
$$F_E = -1.2353 - 15 = -16.2353 \text{ kN}$$
$$F_F = F_B = -16.2353 - 2 = -18.2353 \text{ kN}$$

The S.F.D. is shown in Fig. 18 (*b*).

B.M.D.

$$M_A = 0$$
$$M_C = 13.7647 \times 2 = 27.5294 \text{ kN.m}$$
$$M_D = 13.7647 \times 4 - 5 \times 2 = 45.0588 \text{ kN.m}$$
$$M_E = 13.7647 \times 6 - 5 \times 4 - 10 - 2 = 42.5882 \text{ kN.m}$$
$$M_F = 18.2353 \times 1 = 18.2353 \text{ kN.m}$$
$$M_B = 0$$

The B.M.D. is shown in Fig. 18 (*c*).

Example 19. *A simply supported beam carries point loads as shown in Fig. 19. Draw SFD and BMD.* **[MDU, May 2010]**

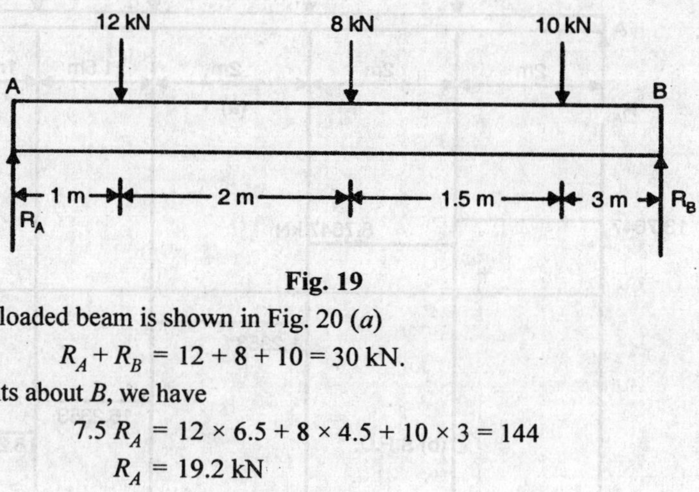

Fig. 19

Solution. The loaded beam is shown in Fig. 20 (*a*)

$$R_A + R_B = 12 + 8 + 10 = 30 \text{ kN.}$$

Taking moments about B, we have

$$7.5 R_A = 12 \times 6.5 + 8 \times 4.5 + 10 \times 3 = 144$$
$$R_A = 19.2 \text{ kN}$$
$$R_B = 30 - 19.2 = 10.8 \text{ kN}$$

S.F.D.

$$F_A = R_A = 19.2 \text{ kN}$$
$$F_C = 19.2 - 12 = 7.2 \text{ kN}$$
$$F_D = 7.2 - 8 = -0.8 \text{ kN}$$
$$F_E = -0.8 - 10 = -10.8 \text{ kN}$$
$$F_B = -10.8 \text{ kN}$$

The S.F.D. is shown in Fig. 20 (*b*).

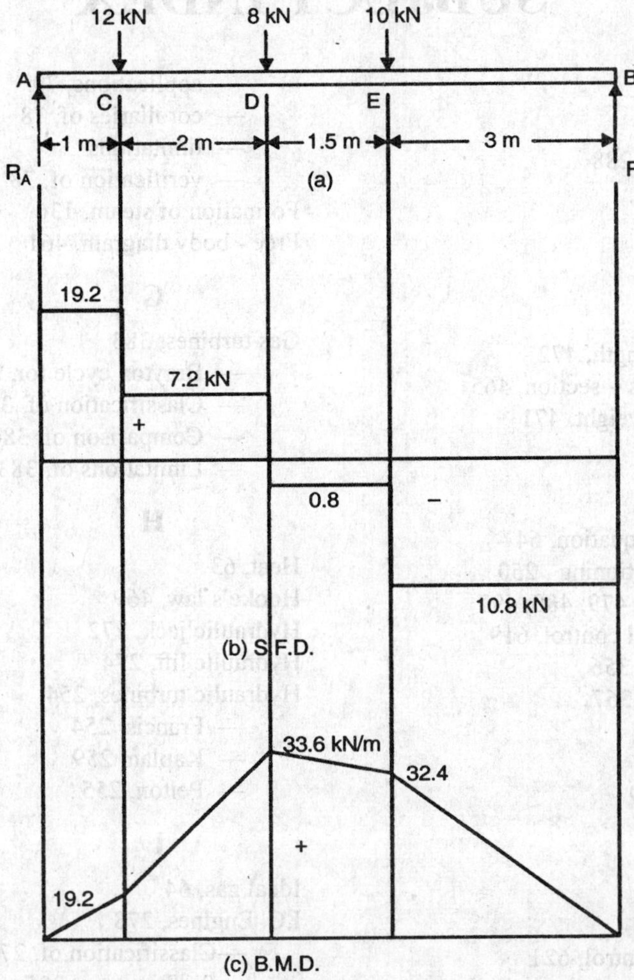

Fig. 20

B.M.D.

$$M_A = 0$$
$$M_C = 19.2 \times 1 = 19.2 \text{ kN.m}$$
$$M_D = 19.2 \times 3 - 12 \times 2 = 57.6 - 24 = 33.6 \text{ kN.m}$$
$$M_E = 19.2 \times 4.5 - 12 \times 3.5 - 8 \times 1.5 = 32.4 \text{ kN.m}$$
$$M_B = 0$$

The B.M.D. is shown in Fig. 20 (*c*).

SUBJECT INDEX

T

V

W

Z